IRON ORE CONFERENCE 2023

Transitioning to a green future

18–20 September 2023
PERTH, AUSTRALIA

The Australasian Institute of Mining and Metallurgy
Publication Series No 5/2023

≋ AusIMM

Published by:
The Australasian Institute of Mining and Metallurgy
Ground Floor, 204 Lygon Street, Carlton Victoria 3053, Australia

ISBN 978-1-922395-24-5

ORGANISING COMMITTEE

Ralph Holmes
HonFAusIMM(CP)
Conference Co-Chair

Erick Ramanaidou
Conference Co-Chair

Keith Vining
Conference Co-Chair

Brian McDonald
MAusIMM

Camilla Stark

Clayton Simpson
MAusIMM

John Clout
FAusIMM

Nishka Piechocka

Peter Fisher

Tom Honeyands
MAusIMM

Will Patton

Alan Ooi
MAusIMM

AUSIMM

Julie Allen
Head of Events

Fiona Geoghegan
Manager, Events

Kathryn Laslett
Conference Program Manager

REVIEWERS

We would like to thank the following people for their contribution towards enhancing the quality of the papers included in this volume:

Heath Arvidson

Cameron Boyle

Joffre Buswell

John Clout

Jim Cribbes

Peter Fisher

Ralph Holmes

Tom Honeyands

James Manuel

Angus McFarlane

Alan Ooi

Will Patton

Wade Perrin

Michael Peterson

Brian Povey

Mark Pownceby

Clayton Simpson

Camilla Stark

Adel Vatandoost

Michael Verrall

Keith Vining

John Visser

FOREWORD

On behalf of the Organising Committee, we are delighted to welcome you to Iron Ore 2023, the tenth in the now well-established and very successful, biennial international Iron Ore Conference series, jointly hosted by the Australasian Institute of Mining and Metallurgy (AusIMM) and Australia's Commonwealth Scientific and Industrial Research Organisation (CSIRO).

Global steel demand remains strong despite a gradual production decline in China and is expected to increase driven by economic expansion in other parts of the world including India, North Africa and ASEAN countries. Despite the healthy outlook for iron ore exports, the iron and steel making industry is under increased scrutiny for its contribution to global CO_2 emissions and the domestic iron ore industry faces the challenge of meeting the needs of steel-makers evolving to meet emissions reduction targets. Consequently, the theme of Iron Ore 2023 is *Transitioning to a green future*.

The Conference Organising Committee has lined up an impressive range of keynote speakers and over 65 technical papers showcasing the latest developments in the key areas of exploration, geology, characterisation, geometallurgy, mining and process optimisation, as well as health and safety practices, automation, logistics and new equipment.

We would like to thank the conference Organising Committee, the AusIMM Events team, authors, paper reviewers, and all the sponsors and exhibitors, without whom, the conference would not be possible.

We welcome you to Iron Ore 2023 and trust that you will find it an enjoyable and rewarding event.

Yours faithfully,

Ralph Holmes HonFAusIMM(CP)   Erick Ramanaidou   Keith Vining
Conference Co-Chair   Conference Co-Chair   Conference Co-Chair
CSIRO Mineral Resources   CSIRO Mineral Resources   CSIRO Mineral Resources

SPONSORS

Major Conference Sponsor

BHP

Gold Sponsor

PLOTLOGIC

Silver Sponsor

Mineral Technologies

Coffee Cart and Breaks

ALS

Conference App Sponsor

AMC consultants
mine smarter

Welcome Reception Sponsor

Fortescue.

Name Badge and Lanyard Sponsor

RioTinto

Destination Partner

BUSINESS EVENTS PERTH

CONTENTS

Exploration and geology – geostatistics and ore reserve estimation

Exploration and geology – new iron ore deposits, including magnetite

Greener downstream processing – decarbonisation

Greener downstream processing – DRI, HBI and pig iron production

Greener downstream processing – green iron and steel production

Greener downstream processing – sintering and pelletising

Health, safety, environment and community – dust mitigation and reduction of water consumption

Health, safety, environment and community – health and safety practices

Health, safety, environment and community – licence to operate

Mining and processing – new equipment

Mining and processing – ore processing and beneficiation

Mining and processing – tailings processing and storage

Ore characterisation and geometallurgy – method development

Ore characterisation and geometallurgy – new applications

Process optimisation – a greener and cleaner iron ore value chain

Process optimisation – automation and machine learning

Process optimisation – data science and predictive analytics

Exploration and geology – geostatistics and ore reserve estimation

Homogeneity assessment of crushed and pulverised iron ore certified reference materials – implications for laboratory quality control

J Carter[1] and B J Armstrong[2]

1. General Manager, Independent Mineral Standards, Bayswater WA 6053.
   Email: john@imstandards.com.au
2. Operations Manager, Independent Mineral Standards, Bayswater WA 6053.
   Email: bruce@imstandards.com.au

ABSTRACT

Best practice for the submission of samples to an internal or commercial laboratory requires quality control materials to be included for the purpose of independently and comprehensively assessing laboratory performance. These quality controls include the use of certified reference materials (CRMs) and may be blind to the laboratory. Iron ore companies purchase these materials from reputable manufacturers from stock of run-of-mine materials at the appropriate grades. The international standard, ISO 17034 (ISO, 2016), specifies the competency requirements for organisations manufacturing reference materials, and includes rigorous requirements for the assessment of homogeneity, stability and establishment of property values and their uncertainties. Homogeneity is one of the most important measures of suitability for all certified reference materials and refers to the degree to which the composition and properties of a sample are uniform throughout the batch, relative to the analytical precision of the method. More specifically, homogeneity studies of candidate reference materials should be assessed for both within sample variance and between sample variance with appropriate statistical controls. The design and application of homogeneity studies are discussed as they apply to both crushed and pulverised reference materials.

Crushed certified reference materials have been utilised by the iron ore industry for a number of years and provide the user with a critical assessment of variances in sample preparation in addition to analysis. Crushed reference materials are similar in presentation to an RC sample and are generally used as discrete parcels of material submitted to the laboratory from the field. Pulverised certified reference materials miss the sample preparation step and are subsampled from the delivered sachet at the point of analysis. This paper compares the relative homogeneity performance of crushed and pulverised certified reference materials by grade across key iron ore chemical constituents. Outcomes are benchmarked to the international standard ISO 9516 (ISO, 2003) for the determination of various elements in iron ores by lithium borate fusion and X-ray fluorescence. Practical implications are discussed with reference to a case study from an iron ore laboratory in the Pilbara.

INTRODUCTION

Iron ore companies conducting exploration, strategic or tactical mine planning, collect samples from the field and submit them to in-house or commercial laboratories for analysis. In order to obtain confidence in the results from the contract laboratory and assess their performance, the mining company needs to implement a quality assurance regime throughout the custody chain from sample collection to analytical stage. Part of this quality assurance (QA) regime typically includes insertion of duplicate samples and reference materials into the sample stream submitted to the laboratory for analysis alongside the regular drill samples (Abzalov, 2016; Sterk, 2015). The analytical results of these quality control samples are regularly monitored and benchmarked. Quality control through the use of reference materials is therefore fundamental to exploration and mine planning activities and are a critical activity for establishing confidence and reliability of the estimations made.

The submission of a certified reference material (CRM) is the primary tool for the assessment of accuracy and bias in analytical results. Purchased from a reputable supplier, CRMs are a stable material, sufficiently homogenised and characterised by a metrologically traceable procedure (ISO 17034; ISO, 2016). The CRM is accompanied by a certificate containing consensus values, their associated uncertainties and a statement of metrological traceability. Uncertainties can be used to establish control limits that are monitored by the exploration companies throughout the drilling program. These CRMs are ideally 'blind' to the laboratory and may have been sourced from the exploration company's ore or manufactured from appropriate material. The CRMs from the field are usually submitted in the calico bags that would normally contain a field sample.

Within the iron ore mining industry, there are two types of CRMs used; pulverised and crushed. Pulverised iron ore CRMs are sourced from natural ores and are usually milled and size screened to P_{95} of 54 µm, followed by homogenisation prior to packaging in typically 10 g units for submission to the laboratory along with the drill samples. The pulverised CRM represent a dilemma for the laboratory as they are separated from the drill samples as they undergo drying, particle size reduction, and sub-sampling before being reintroduced prior to the final sub-sampling to typically 0.5 g for creation of the fused bead from which the XRF analysis is performed. The obvious difference in visual presentation between drill sample and a pulverised CRM allows identification by the laboratory of CRM samples, which may be flagged as client CRM and placed at the end of analytical sequences.

Crushed CRMs are also manufactured from natural ores which have been dried, crushed and screened to typically P_{95} of 3 mm or 5 mm followed by homogenisation and packaging into 2 kg to 4 kg units. Crushed CRMs when submitted are similar to an RC drill sample with similar visual presentation, mass and particle size. The samples undergo identical sample preparation processes as submitted field samples such as sorting, drying, crushing, pulverising with typically two sub-sampling steps in the process prior to presentation for analysis. The laboratory should not be able to identify and separate the crushed CRM from the field samples within the laboratory workflow, nor allocate it to a different analytical sequence to the field samples.

The difference between the processing of a pulverised and crushed CRM within the laboratory is significant as illustrated in Figure 1. While both CRMs assess the quality of the analytical steps, the crushed CRM also includes variance contributions from the sample preparation process. Sample preparation includes various manual or automated handling steps where sampling errors due to particle size reduction and/or mass reduction can and does occur (Gy, 1982). While visibility of the additional contribution to quality of the final result through sample preparation is important with a crushed CRM, one of the key components being assessed arises due to the differences in homogeneity between crushed and pulverised CRMs.

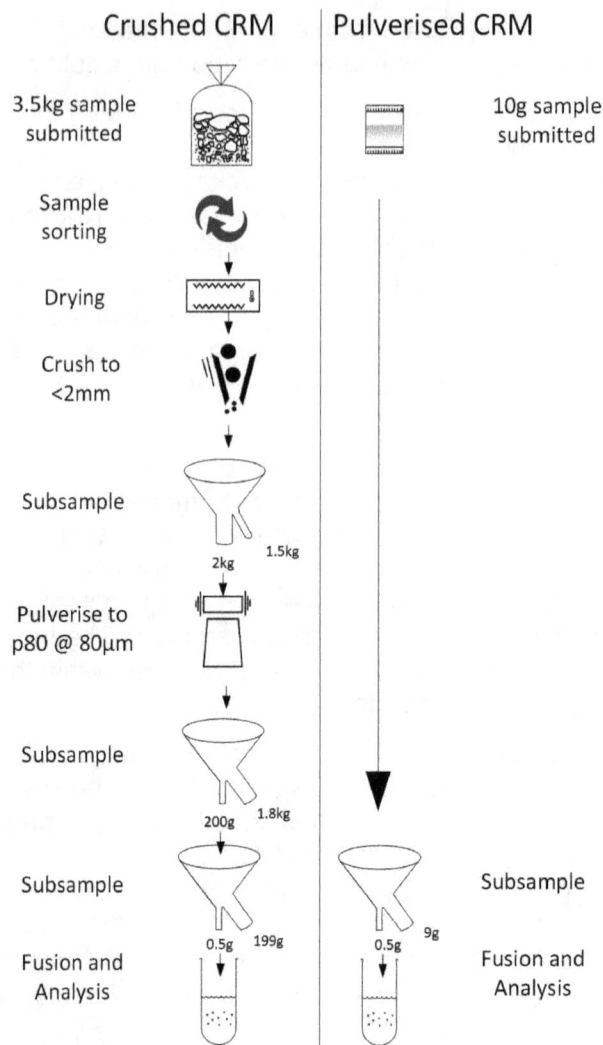

FIG 1 – Process of crushed and pulverised CRMs within the laboratory workflow.

While it is critical CRMs are homogeneous between sample units, the significant difference between crushed and pulverised CRM is the internal heterogeneity within the sample. This comes about because a crushed CRM includes a wide range of particle sizes. Pulverised CRMs are nominally <54 μm and are subsampled during analysis. Crushed CRMs include particles from <5 μm to 5 mm, with batch specific particle size distributions and are analysed in their entirety.

The propensity for particle size segregation during transport, transfer, and storage occurs by a number of mechanisms such as percolation and rolling and is widely described (Levy and Kalman, 2001). The freely segregating material of heterogenous samples requires good laboratory practice in sub-sampling and materials handling to obtain a final pulverised sample for analysis that is representative of the original submitted sample. It is important to point out that the potential for sample heterogeneity to cause erroneous results are the same for field samples as they are for crushed CRMs, hence why this is an important addition to the overall quality control plan.

Good laboratory sub-sampling practice is vital because the mineral concentrations of crushed iron ore are not consistent by particle size. The degree of mineral concentration by particle size varies between iron-ores, typically with relatively lower Fe concentrations in finer products, and corresponding elevated SiO_2 and Al_2O_3 concentrations in the finer products as shown in Figure 2.

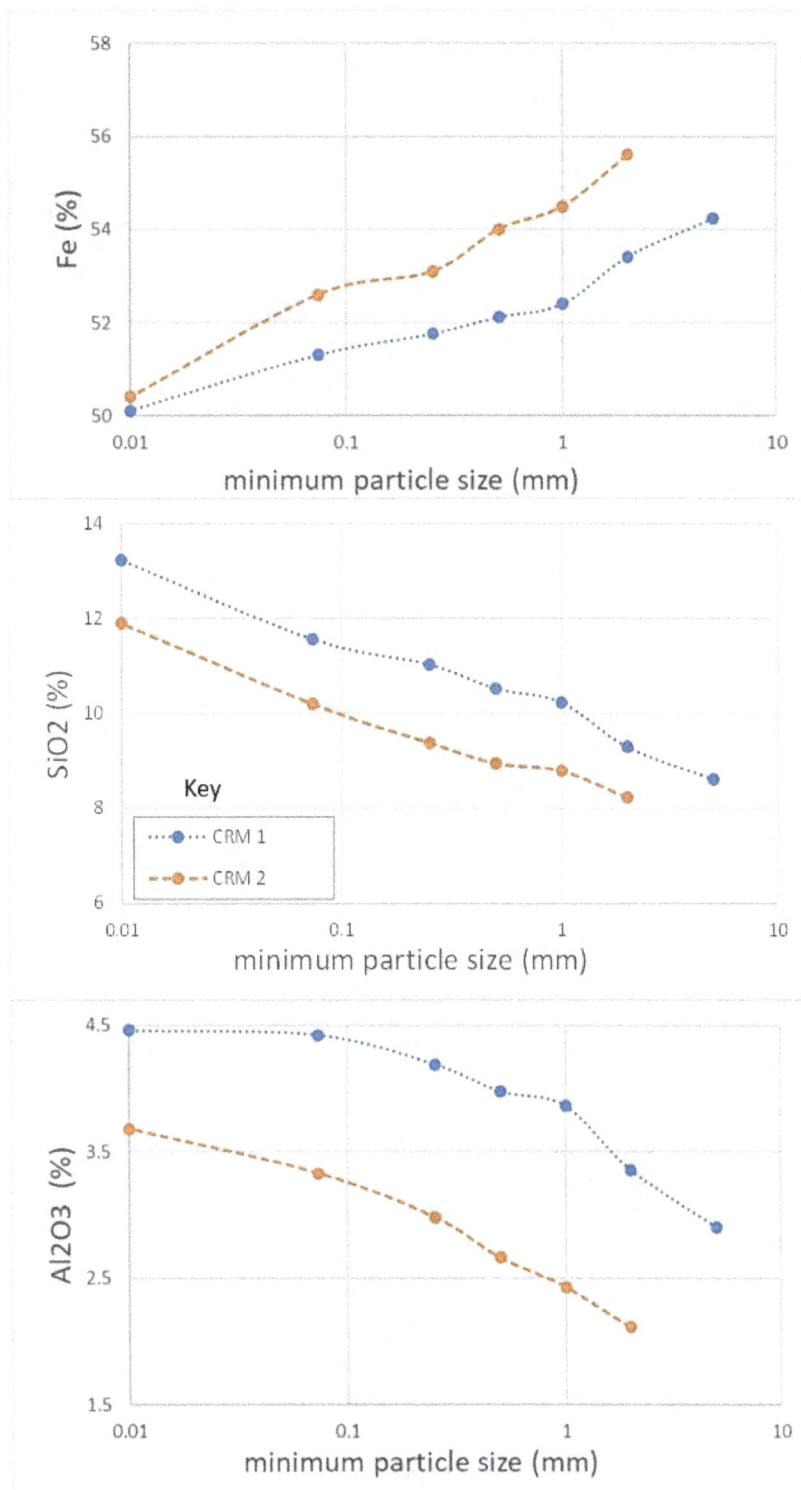

FIG 2 – Fe, SiO$_2$ and Al$_2$O$_3$ concentrate of two crushed CRMs by particle size.

The data in Figure 2 was derived by bench screen tests of stacked sieves of decreasing mesh openings across a 2 kg bag randomly selected from two different crushed CRM batches manufactured from iron ores sourced from the Pilbara. The heterogeneity of the CRM by particle size trends in a consistent way between each CRM analysed.

MANUFACTURE AND CERTIFICATION OF REFERENCE MATERIALS

The manufacture and certification of CRMs are guided by ISO 17034 (ISO, 2016), the general requirements for the competence of reference material producers and Guide 35 (ISO, 2017b) providing guidance to characterisation studies, the assessment of homogeneity and stability, along with the treatment of data. ISO 17034 (ISO, 2016) accredited products provide the user with additional confidence and assurance that the manufacturer has complied with an internationally establish best practice approach. The demonstration of the competence of the reference material producer is a basic requirement for ensuring the quality of the reference materials are fit for purpose for the application.

Within the ISO 17034 (ISO, 2016) framework, the reference material producer is fully responsible for the planning, management and assignment of property values of the CRM. In the minerals sector, results from multiple commercial and mine site laboratory round-robins are often used for the determination of certified values and their associated uncertainties, following a rigorous statistical process. Confidence in the results and metrological traceability are also established through the use of ISO 17025 (ISO, 2017a) accredited laboratories. The certified, consensus or target value for each analyte of interest is established from the mean or median of laboratory means. Each certified value must also include an assessment of its uncertainty or variance. Table 1 shows a typical CRM where a number of statistical parameters are shown.

TABLE 1

Extract from reference material certificate PBS-222, manufactured by IMS.

Analyte	Certified value (y)	Standard deviation		95% Confidence interval (CI)		u_{CRM}	k	U_{CRM}	No. of labs (ISO 17025)	No. samples
		1 SD (s)	1 SD within lab (s_w)	lower	upper					
Fe	52.08	0.168	0.113	51.97	52.18	0.14	2	0.28	10	30
SiO_2	9.33	0.049	0.023	9.30	9.36	0.040	2	0.081	10	29
Al_2O_3	9.60	0.082	0.034	9.54	9.65	0.038	2	0.077	10	30
TiO_2	0.760	0.0098	0.0020	0.752	0.768	0.0056	2	0.011	10	30
Mn	0.126	0.0065	0.0022	0.121	0.131	0.0024	2.26	0.0054	9	27
CaO	0.073	0.0045	0.0027	0.070	0.076	0.010	2.26	0.023	9	27
P	0.039	0.0009	0.0004	0.038	0.040	0.0011	2	0.0023	10	30
S	0.034	0.0025	0.0004	0.031	0.037	0.0018	2.31	0.0042	8	24

Accompanying each analyte certified value (y), a number of statistical parameters related to uncertainty are also listed. Uncertainty is fundamentally a measurement of the doubt in the certified value and each of these parameters represent uncertainty in different ways.

- Standard deviation (s) is the measure of spread of analyte determinations and includes inter-laboratory bias, method uncertainty and material homogeneity uncertainty. Approximately 95 per cent of determinations using the same analytical method are expected to be between two standard deviations either side of the certified value. The standard deviation is calculated from the validated laboratory group data less outlier laboratory and individual determinations.

- Within laboratory standard deviation (s_w) is the average spread of determination values across the reporting laboratories, less outlier laboratory and individual determinations. This is calculated by single factor ANOVA of the participating laboratory groups.

- Confidence Interval (CI) is an estimate of the true (unknowable) analyte concentration in the material at the 95 per cent confidence interval. For example, a 95 per cent CI could be interpreted as there is a 0.95 probability that the true value is between certified value ± CI. The narrower the interval, the more precise the certified value. The 95 per cent CI should not be used for determination of quality control gates.

- Standard Uncertainty (u_{CRM}) is the sum of variance from characterisation, homogeneity and stability studies as shown in the formula below. The coverage Factor (k) is the students t-distribution value for two tailed test at 95 per cent. The expanded Uncertainty (U_{CRM}) is the product of coverage factor and standard uncertainty, and represents the 95 per cent confidence interval of the true unknowable analyte concentration of the batch combined with the bias from individual samples.

$$U_{CRM} = k\sqrt{u_{char}^2 + u_{hom}^2 + u_{stab}^2}$$

where:

- Uncertainty of characterisation (u_{char}) is the standard error of the mean of each laboratory's average analyte value. This is an estimate of the uncertainty of the CRM's true analyte value. A robust characterisation study that includes laboratories that are ISO 17025 (ISO, 2017a) accredited satisfy the metrological traceability requirements of characterisation. The uncertainty of characterisation relates primarily to the uncertainties of the laboratory methods.

- Uncertainty of homogeneity (u_{hom}) is an estimate of the material variance, which incorporates analytical, within-unit and between-unit uncertainty. This parameter is a property largely controlled by the quality of the certified reference material.

- Uncertainty of stability (u_{stab}) which includes uncertainties from both short-term and long-term storage, transport and degradation variances. This is typically insignificant in an iron ore CRM as the material as packaged is inherently stable with respect to the other uncertainties.

HOMOGENEITY STUDY DESIGN

It follows, therefore, during the manufacture and certification of CRMs, u_{hom} is an important contributor to the total uncertainty budget. This parameter is directly related to the inherent quality of the reference material. Materials of natural origin like iron ores, are typically heterogeneous by nature. This heterogeneity may be evident chemically, mineralogically and across particle size divisions as shown in Figure 3. During manufacture, the aim of the reference material producer is to minimise the magnitude of the between-unit differences so that they are insignificant compared to uncertainties that arise from characterisation. In the case of crushed CRMs additional challenges arise because of the inherent particle and mineralogical properties of the natural materials when handled. For a crushed CRM to be useful as a laboratory quality control tool, the homogeneity needs to carefully assessed and include both sample preparation and analytical steps.

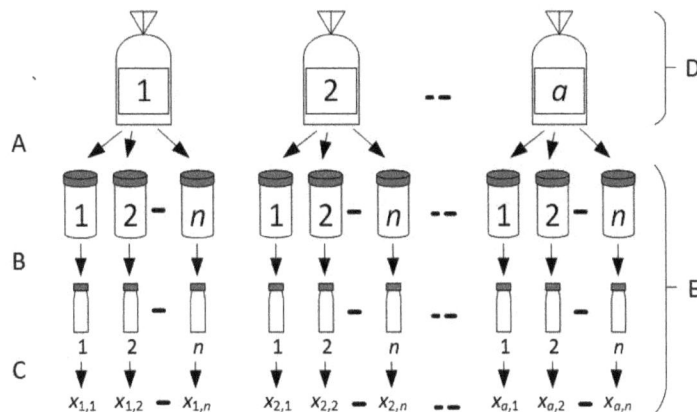

Key
A Subsampling
B Preparation
C Measurement
D Contributes to the observed between-unit variation
E Operations contributing to observed within-unit variation
a $x_{1,j}$ denotes the jth aliquot for unit i

FIG 3 – Schematic of a homogeneity study.

Using a homogeneity study layout, such as in Figure 3, estimation of both *within sample variance* (S_w) and *between sample variance* (S_b) of multiple CRM units using analysis of variance (ANOVA) statistical techniques is possible. The apportioning of variance within and between a CRM unit is facilitated by the duplicate analysis of many samples. The total uncertainty of homogeneity (u_{hom}) is estimated as the square root of the sum of squared S_w and S_b variances (ISO Guide 35, 2017b).

$$U_{hom} = \sqrt{S_b^2 + S_w^2}$$

Now that we have a measurement of the contributions to reference material homogeneity it is possible to use this figure to assess its quality and suitability for use in the laboratory as a QC tool. Uncertainty of homogeneity of a reference material can be assessed via either a statistical inference such as an F-test at the 95 per cent level of confidence, or via reference to external criteria such as characterisation uncertainty or analytical measurement uncertainty by which the CRM will be used as a quality control tool. In the case of iron ore samples routinely measure using a lithium borate fusion and XRF analysis, ISO 9516-1 (ISO, 2003) can be referenced and compared.

ISO 9516-1 (ISO, 2003) is a comprehensive procedure for the determination of the common elements required in the iron ore industry and is applicable regardless of mineralogical type. The procedure documents sources of error in the method and tabulates permissible tolerances related to the uncertainty contributions of various components of the test method, including, but not exclusively, weighing, sample matrix and instrument conditions. The precision of the method is expressed by a series of regression equations unique to each element. The independent duplicate limit (R_d) is particularly relevant as the minimum uncertainty level of the method. This value can be used to compare to the homogeneity uncertainties determined by candidate reference materials as a measure of suitability, or fit for purpose, of the CRM as a laboratory quality control tool.

CRUSHED VERSUS PULVERISED CRM PERFORMANCE

The uncertainty of homogeneity (u_{hom}) for a number of CRMs are shown in Figure 4. Pulverised (n = 10) and crushed (n = 28) iron-ore CRMs are plotted against the analyte concentration and compared to the ISO 9516-1 (ISO, 2003) independent duplicate tolerance (R_d) for a few of the common elements significant to iron ore analysis.

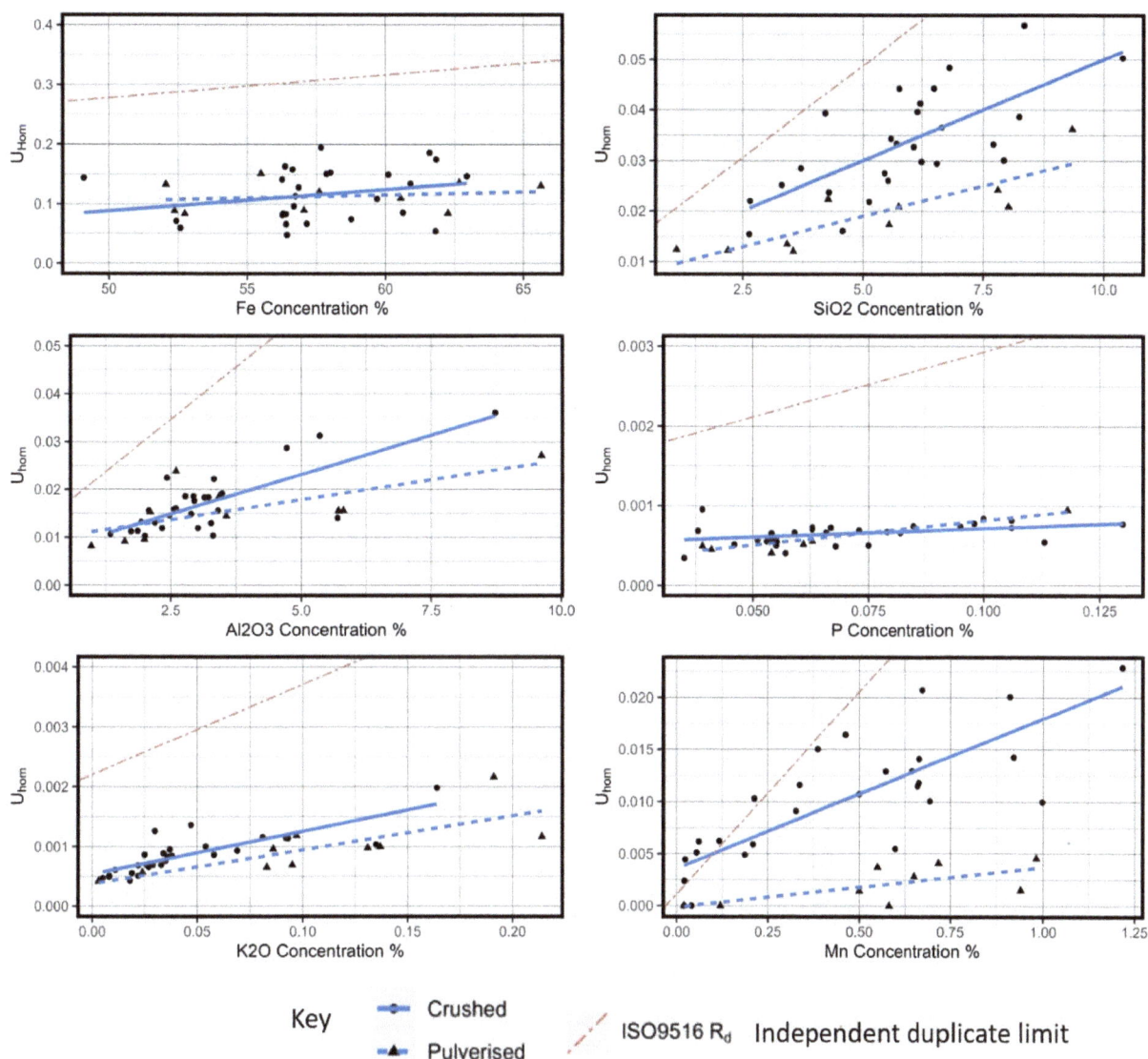

FIG 4 – Comparison of uncertainty of homogeneity (u_{hom}) of pulverised CRMs and crushed CRMs to the independent duplicate limit of the method for some key iron ore elements.

The data presented in Figure 4 has been collated from commercially produced batches for which a detailed homogeneity study was completed. Randomly selected samples from the production of each reference material were submitted to a single laboratory and multiple analyses completed utilising a regime schematically shown in Figure 3. Where a crushed CRM was submitted, the process included sample preparation steps (crushing, splitting and pulverisation) with duplicate analysis occurring at the analytical stage. All participating laboratories in these homogeneity studies have ISO 17025 (ISO, 2017a) accreditation for iron ore by lithium borate fusion – XRF. Linear regression lines are displayed for both pulverised (dashed) and crushed (solid) CRM.

Figure 4 demonstrates that the comparison of pulverised to crushed CRM homogeneity are not consistent across each element. In the case of iron (Fe), the performance of a crushed reference material is very similar to a pulverised reference material, and trends similarly to the method tolerance, in a slight upward trend as iron concentration increases. In both pulverised and crushed CRM types, the material homogeneity is significantly less than the ISO 9516-1 (ISO, 2003) method tolerance.

Overall, pulverised and crushed CRMs have no significant difference in total homogeneity for almost all analytes. This is an important point as it demonstrates a crushed CRM is a valuable quality control tool for determining whole of laboratory performance and can provide confidence and assurance of quality during both sample preparation and analytical processes.

In all cases, for crushed and pulverised CRMs, the homogeneity uncertainty was under the ISO 9516-1 (ISO, 2003) independent duplicate limit for XRF analysis, with the exception of Mn at the low level for crushed CRMs.

The notable exceptions to the u_{hom} parity between pulverised and crushed CRM are SiO_2 and Mn. An investigation into the source of the discrepancy for SiO_2 and Mn is not within the scope of this paper, however, by comparing the trends across multiple analytes we can surmise there may be a combination of mineralogical or textural characteristics that have an impact on adverse sample preparation outcomes.

CASE STUDY – SAMPLE PREPARATION BIAS

To demonstrate the effectiveness of crushed reference materials as a quality control tool, a case study was published for sample preparation and analysis in an iron ore laboratory (Independent Mineral Standards, 2021). In the laboratory performing the analysis, an automated crushing and pulverisation system was employed to prepare the samples, followed by fused bead-XRF analysis. The results for a crushed CRM were plotted over time on a quality control, or Shewhart chart, as shown in Figure 5. An excursion outside the control limits for key elements Fe, SiO_2 and Al_2O_3 was observed over a particular time period.

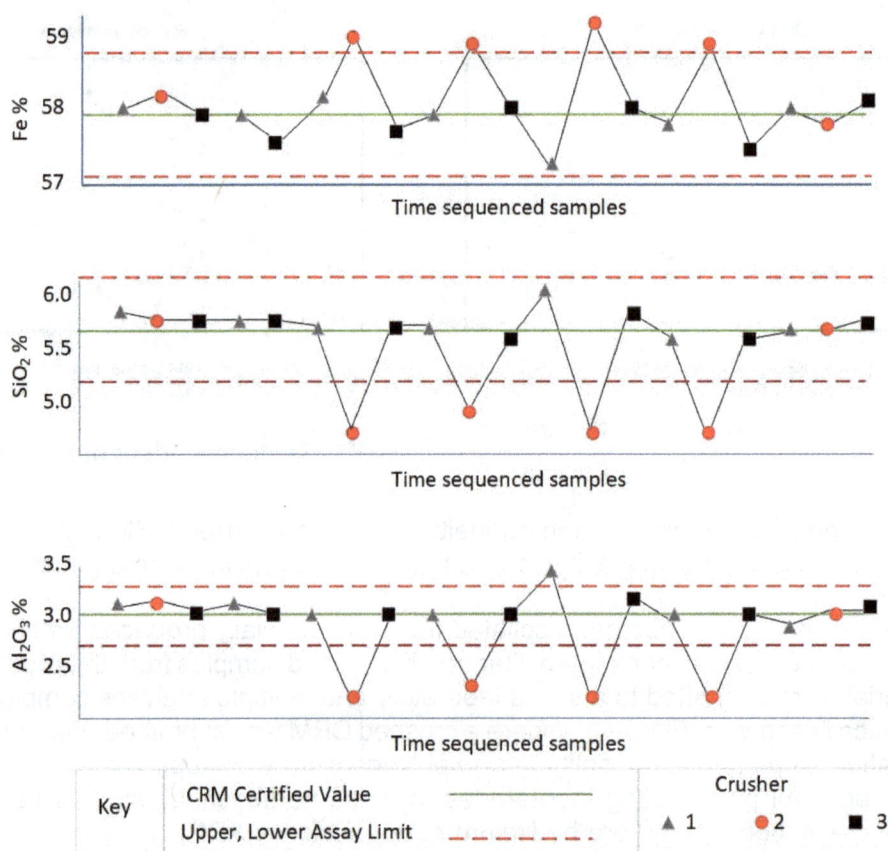

FIG 5 – Shewhart chart for Fe, SiO_2 and Al_2O_3 for 20 samples demonstrating a sample bias due to a crusher fault.

The control charts identify four samples failing with high Fe (>3 × standard deviation from the expected value). The same samples also fail low for both SiO_2 and Al_2O_3. The fines of this CRM are known to contain elevated concentrations of SiO_2 and Al_2O_3 compared to that of the entire sample. Interrogation of the sample preparation system log files showed that all four samples failing QC passed through the same crusher. The laboratory had three crushers installed. Investigation of this crusher showed that a vacuum valve, open during cleaning, was damaged and was not closing fully while processing the sample. This was causing loss of the fines in the sample, leading to depression of SiO_2 and Al_2O_3 and elevation of Fe.

The case study demonstrates that a systematic sample preparation bias was occurring due to the preferential loss of material during crushing. The crushed CRM, when used in conjunction with a pulverised CRM for analysis, will therefore provide visibility of both precision and bias throughout the entire laboratory, and can help identify sample preparation issues that have a bearing on the quality of the analytical results. It would be very difficult to detect this issue in routine samples. If there were sample duplicates or splits taken, they would be collected after the crushing step and as a result, both samples would be biased in this case.

The detection of loss in SiO_2 in the crusher is more interesting considering the u_{hom} data in Figure 4. In this case the between sample homogeneity, and analytical method precision are both sufficiently low for the detection of a bias caused by the crusher. The result demonstrates the quality of the CRM used has a level of homogeneity sufficiently fit for purpose to detect an adverse condition during the preparation of routine samples.

CONCLUSIONS

The use of certified reference materials (CRM) is critical to establishing a robust quality control program for iron ore projects. It is important for reference material manufacturers to assess the homogeneity characteristics of the CRM and demonstrate fit for purpose quality. A comparison of the homogeneity of crushed CRMs to pulverised CRMs demonstrates they are a suitable choice for whole of laboratory quality assessment, and are capable of detecting failures within sample preparation that may contribute to data quality.

ACKNOWLEDGEMENTS

The authors wish to thank Independent Mineral Standards for permission to publish this data.

REFERENCES

Abzalov, M, 2016. Quality control and assurance (QAQC), *Applied Mining Geology*, pp 135–159.

Gy, P, 1982. *Sampling of Particulate Materials Theory and Practice* (Elsevier: Amsterdam).

Independent Mineral Standards, 2021. Case Study: Coarse CRM Analysis Identifies Crusher Malfunction [online]. Available from: <https://imstandards.com.au/content-hub/#brochures> [Accessed: 9 January 2023].

International Organisation for Standardization (ISO), 2003. ISO 9516-1 – Iron ores – Determination of various elements by X-ray fluorescence spectrometry – Part 1: Comprehensive procedure.

International Organisation for Standardization (ISO), 2016. ISO 17034 – General requirements for the competence of reference material producers.

International Organisation for Standardization (ISO), 2017a. ISO 17025 – General requirements for the competence of testing and calibration laboratories.

International Organisation for Standardization (ISO), 2017b. ISO Guide 35 – Reference materials – Guidance for characterisation and assessment of homogeneity and stability.

Levy, A and Kalman, H, 2001. Segregation of powders – mechanisms, processes and counteraction, *Handbook of Conveying and Handling of Particulate Solids*, 9:589 (Elsevier Science BV).

Sterk, R, 2015. Quality control on assays: addressing some issues, paper presented to 2015 AusIMM New Zealand Branch Annual Conference, Dunedin.

Case studies illustrating the role of F-series mine reconciliations in production forecast improvement initiatives

S J Loach[1]

1. Senior Geologist; Reconciliations, BHP, Western Australian Iron Ore, Perth WA 6000.
   Email: steve.loach@bhp.com

ABSTRACT

F-Series reconciliations are formal comparisons of estimates and measurements of ore quantity and quality made at pre-determined points along the mining value chain. F1 is the ratio of 'Depleted Grade Control Ore/Depleted Long-term Model Ore' and validates the models upon which Resource and Reserve statements are based. F2 is the ratio of 'Produced Tonnes and Grade/Depleted Grade Control Ore (-) Changes in Pre-Crusher Stockpiles' and validates that the ore tonnes and grade delineated in the pit are being accurately extracted and correctly delivered to the processing plant. These factors provide important corporate assurance for a mining company by providing quantitative feedback on model-predicted tonnes and block grade estimates.

Since 2009, BHPs Western Australian Iron Ore F-Series reconciliations have been calculated in a single, centralised sequel-server database. This software routinely collates data regarding mineral resource block model depletions, grade control model depletions, pre-crusher stockpile inventories and production tonnes and grade measurements and then calculates F-Series factors for both tonnes and product quality.

This paper will present case studies where reconciliation factors were used to identify improvement opportunities leading to projects that improved our models and subsequent forecasts of production tonnes and quality. In addition, it will show how reconciliation factors are tools that monitor control during periods of rapid change and provide feedback on the effectiveness of improvement projects.

INTRODUCTION

Since 2009, BHPs Western Australian Iron Ore (WAIO) F-Series reconciliations have been calculated in a single, centralised sequel-server database. A variety of variables are reconciled in this environment including tonnes, Fe, P, SiO_2, Al_2O_3, LOI and geometallurgical variables including lump% yield.

The reconciliation system was built around a company-mandated requirement to calculate three basic factors (see Figure 1). These factors are defined as follows:

- F1: The ratio of 'Depleted Grade Control Ore/Depleted Long-term Model Ore'. This factor validates the models upon which Resource and Reserve statements are based and which inform strategic decisions and long-term plans.

- F2: The ratio of 'Produced Tonnes and Grade/Depleted Grade Control Ore (-) Delta Pre-Crusher Stockpiles' This validates that ore delineated in the pit is successfully delivered to the processing plant and that value is not lost through ore-loss or dilution.

- F3: The ratio of 'Shipped Product/Depleted Long-term Model Ore (-) Delta Stockpiles'. This validates that the entire value chain is delivering the full value represented in the Reserve Model. It is a test of the validity of geoscience inputs to the business case.

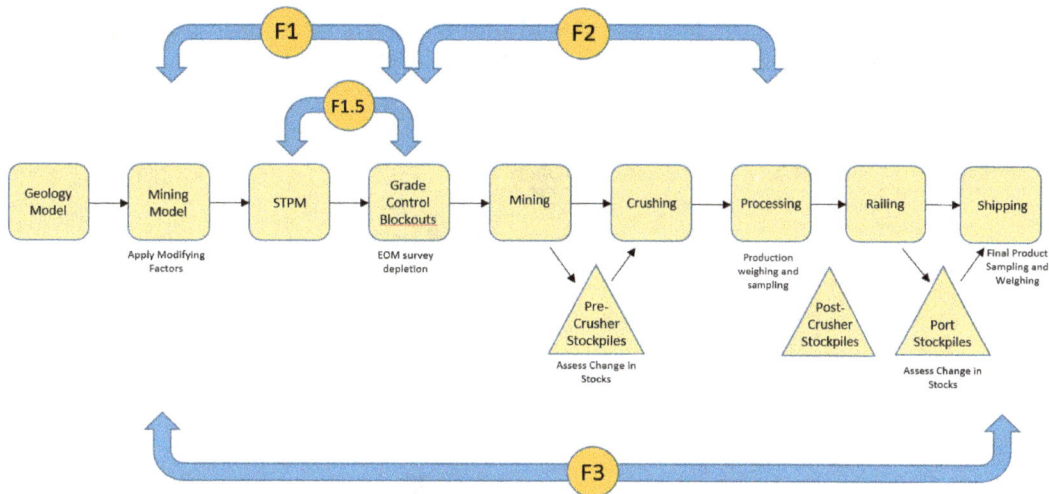

FIG 1 – Conceptual overview of the WAIO value/supply chain from Resource estimation to shipping showing the reconciliation factors referred to in this paper.

In addition to these three factors several bespoke factors are calculated in this system. In 2013 across WAIO, Short-Term Geology Models (STGM) were introduced into the business. The purpose of these models was to bridge the information gap between the long-term Reserve model (locally known as the mining model) and the final grade control mark-out and thereby provide updated information to the six-month planning horizon. This method of providing updated information to the planning team proved valuable and the STGM has subsequently evolved into the Short-Term Planning Model (STPM) which now provides geoscience input to each operation's 12-month tactical plan and annual product-quality guidance to market. To assess the accuracy of the STPM a new factor 'F1.5' has been developed.

- F1.5: The ratio of 'Depleted Grade Control Ore/Depleted Short-Term Planning Model Ore'. This factor validates the current iteration of the STPM upon which tactical plans and decisions are based.

Despite the addition of this factor and several others outside of the scope of this paper, the core of the system remains F1, F2 and F3 calculation, reporting and analysis.

After 14 years of operation this system has now collated a large, continuous and highly detailed data set that allows in-depth analysis of the performance of the various models that provide production forecasts to our business. Analysis of this data set allows us to identify opportunities for improvement of these models so that production forecasts can be improved and risk reduced. In addition to the data contained within the reconciliation system a comprehensive library of quarterly reconciliation presentations and reports is maintained to provide commentary on the results.

It is important to note that the F-Series factors are calculated independently of compliance to plan, product quality compliance and budget. F-Series factors are intended to isolate the model under analysis from these considerations. The isolation of the model from other measures of operational and business performance is the essential characteristic that makes F-Series reconciliations suitable for providing feedback on model performance.

The following four case studies all outline instances where F-Series reconciliation has provided motivation to start important improvement initiatives and has provided feedback on the effectiveness of these initiatives.

In this paper the name of each operation has been obfuscated by replacing the actual name with a number.

CASE STUDY 1 – DENSITY ESTIMATION AT 'MINE SITE ONE'

Mine Site One is a significant iron ore fines producer in the Central and Northern Pilbara. Up until Q3 2010 Mine Site One's F2 tonnes was under strong control, consistently returning values at or near 1.00. However, from Q3 2010 until Q1 FY2014 this site's F2 tonnes suffered a gradual downward trend (see Figure 2). In the last five of these 14 financial quarters Mine Site One's F2 tonnes were either just above or slightly below the pre-determined acceptable lower limit.

'Minesite One' F2 Tonnes

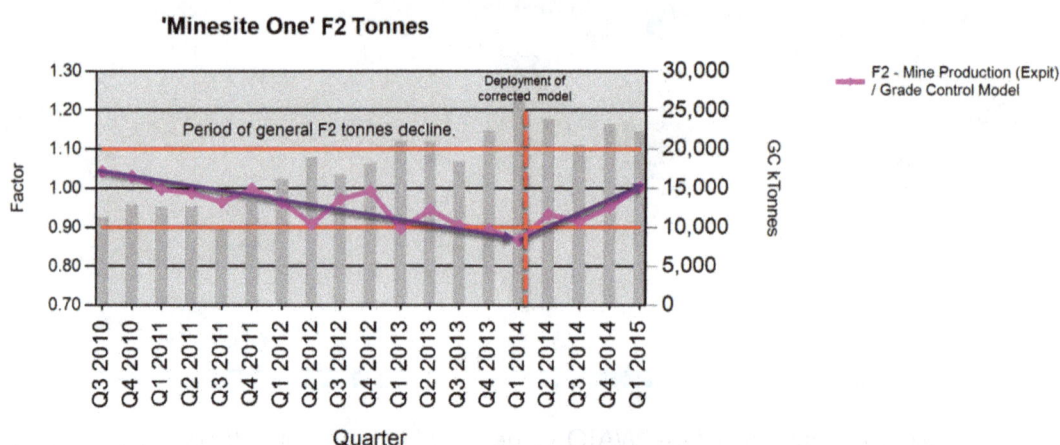

FIG 2 – From Q3 2010 to Q1 2014 Mine Site One mine site suffered a trend of deteriorating F2 tonnes results. Deployment of a new mining model with revised densities based on corrected density data reversed the trend and returned this factor to satisfactory control.

In-depth investigation indicated that the lowest F2 tonnage results coincided with a period of increasing proportion of tonnes drawn from Mine Site One's Western Deposit pits. In the Western deposits, density determination had a complex history. From the years 1987 to 2006 there were several campaigns of density data gathering in and around these deposits. A variety of density determination techniques were used including laboratory core measurements, small trial pits and geophysical wire-line logging. At least two different geophysical surveying companies were employed in this work, each using their own instruments, procedures and personnel. By 2003 a series of default densities for different geological units within the Western Deposit pits had been developed and these were used for Resource estimation and grade control. These default densities were based on dry densities multiplied by 1.086 to account for moisture.

In 2009 there was a perceived risk that the estimated densities across several WAIO deposits could be incorrect. This perception was due to the wide variety of data sources and the perceived questionable quality of some of data sets, especially the older ones, many of which had been gathered before rigorous QAQC protocols had been established. To address this risk, BHP geophysicists from the WAIO Resource Evaluation Group conducted a campaign of very careful density determinations across many WAIO deposits. This involved re-entering selected holes – including calibration holes – with modern, high-quality, wire-line instrumentation. This study provided evidence that one contractor's density readings, for holes surveyed between 2004 and 2008, were on average too high by 0.09 g/cc. With this bias firmly established an appropriate correction was applied to the WAIO density data set. This change represented a significant improvement in the assurance of WAIO's estimated tonnages, however as this correction applied only to data gathered after the Mine Site One Western default densities were set, no change was made to them. Furthermore, few holes were available for re-entry in the Western Deposits so direct testing of the default densities was difficult.

In Q2 2013, Mine Site One's sustained downward F2 tonnage trend was raising questions regarding the density estimates at the site and an in-depth investigation began. This investigation compared recent downhole density data with the historical data set and found evidence that a second – previously undetected – calibration bias may have existed. This bias appeared to affect all pre-2008 data from a second geophysics company. To check the original readings a series of twin-holes were drilled nearby and parallel to the original, now inaccessible, holes and new readings taken. These check measurements confirmed that the old readings were over-representing density by 0.15 to 0.22 g/cc.

BHP WAIO's first response was to issue a revised set of default densities for use in grade control. The impact on F2 tonnes performance was almost immediate with a discernible improvement seen just six weeks later. Meanwhile appropriate corrections were made to the density data set and a new

Resource estimate was generated using corrected measurements rather than default values. The revised mining model was deployed into operation during Q2 2014. F2 tonnage results continued to improve over the next three quarters and a value of 1.00 was achieved for Q1 2015. With rare exceptions, F2 tonnage results at Mine Site One have been satisfactory ever since.

The implications of this correction to the business were significant. The downward revision of tonnes in the Resource estimate led to a reduction of tonnes in Mine Site One's Reserve and – critically – led to shortening of estimated remaining life of that operation. This in-turn shortened the time frame to select and build a replacement for Mine Site One.

To prevent the risk of recurrence of incorrect default densities ever being used in grade control block-outs, in FY18 Q1 WAIO ceased using default densities. Now the density written into the grade control model is derived by an interrogation of the estimated densities in the current STPM. During the block-out process a lava script interrogates the STPM for equivalent material types either within or near the volume blocked out and that density value is passed to the appropriate grade control blocks. This ensures alignment between the densities estimated in the mining model, the STPM and the grade control block-outs so that production forecasts are calculated using the best available density and tonnage estimates. Furthermore, prior to analysis of each quarter's reconciliation results the 'F1 density' (Grade Control density/Mining Model density) for each site is now formally reviewed. Each site's F1 density must fall between 0.98 and 1.02 for other tonnage results to be considered valid.

CASE STUDY 2 – LOI DETERMINATION ALIGNMENT BETWEEN MODELS

Prior to the XRF analysis of any iron ore sample the sample must be carefully prepared. In addition to the usual reduction of sample particle size and mass, each sample must be thoroughly dried so that any free moisture is removed. Once dried the sample is then heated to a sufficient temperature to turn it into a glass bead suitable for XRF analysis. The mass of the sample is measured both prior to and after vitrification and the difference is attributed to 'LOI' (loss on ignition) which is material of any composition that is vapourised by this final heating process. LOI is comprised of the volatile component of the rock mass that is lost when heating to 1000 degrees (ISO, 2015). In iron ore, much – but not necessarily all – of this will be OH ions within hydrous Iron oxide minerals, such as limonite or goethite.

The accurate determination of LOI is critically important to the accurate determination of all other analytes because the LOI value is used in the normalisation of each assay to achieve a total composition equalling 100 per cent (ISO, 2003). Consequently, the systematic under-measurement of LOI can lead to an over-measurement of other important analytes such as Fe, SiO_2 and Al_2O_3. In the iron ore context, the greatest impact of incorrect data feeding into the normalisation falls on the Fe value, simply because Fe is generally the highest percentage analyte and is therefore subject to the greatest adjustment during normalisation.

Without exception, for the period from FY16 Q1 to FY20 Q1 WAIO's F1 LOI results were below 1.00. From 01 July 2015 to 31 March 2020 the average LOI in the mining model's depleted high-grade (HG) was 6.21 per cent while in the grade control model it was 6.03 per cent. This equated to an F1 LOI value of 0.97 for the entire period. The consistency of this bias in the F1 results led to a suspicion that there may be a systematic difference between the performance of the laboratory that assayed our exploration and resource definition samples against that which assayed our grade control samples. A review of routine Certified Reference Material (CRM) QAQC results at the grade control laboratory indicated that there was an ongoing tendency to under-measure LOI. Various round-robin checks between this laboratory and accredited external laboratories, including the one that analysed our exploration and resource definition drilling, confirmed this difference.

The ongoing campaign of F1 reconciliation calculation and reporting increased the visibility of this subtle but important bias and increased the business's motivation to resolve this issue. The root cause was found to be thermo-gravimetric analysis (TGA) ovens set to 900°C rather than 1000°C as required. Investigation revealed that because these units had proven susceptible to breakdown if routinely run at 1000°C, a decision had been made to reset their temperature to 900°C. As the required LOI determination temperature was not met, these ovens caused a bias toward low LOI determination.

In FY19 Q3 (February 2019) capital funds were allocated to upgrade the laboratory's TGA ovens but the capital works could not be completed immediately. Over the following year routine laboratory maintenance was increased and greater focus placed on LOI determination QAQC. Some improvement was noted however the root cause of the problem could only be addressed by replacing the ovens. In the period from June 2020 to December 2020 each of four ovens were individually replaced. Comparison of the grade control laboratory's LOI and Fe determinations against an external laboratory shows that by 2021 Q3 the long-standing difference between the laboratories had been resolved (see Figure 3) and an F1 LOI of 0.99 was achieved (see Figure 4).

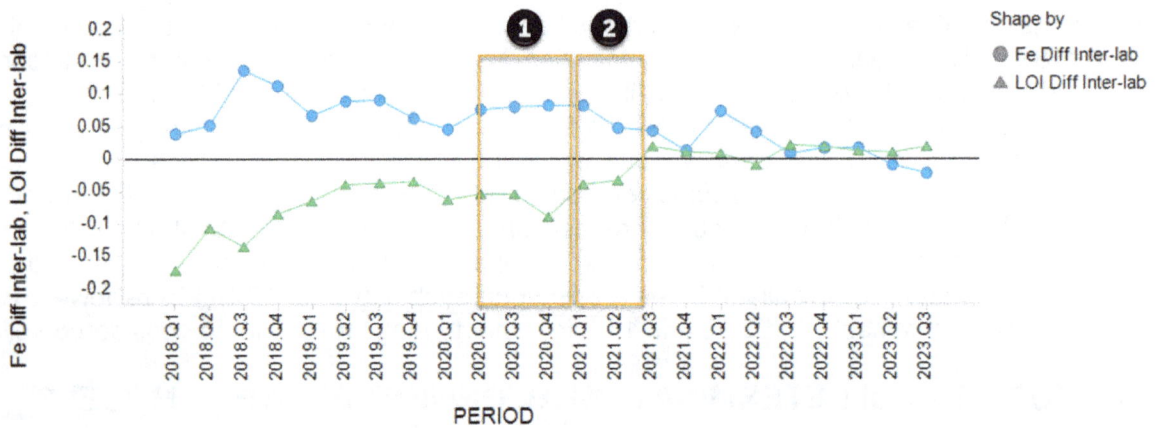

FIG 3 – Difference in LOI and Fe grade determinations between WAIO's grade control laboratory and an external 'umpire' laboratory. These results are based on 81 296 separate analyses of certified reference materials. *Period 1* is the time during which increased attention was paid to QAQC regarding both sample-drying and TGA temperature. *Period 2* is the time during which the TGA ovens were progressively replaced.

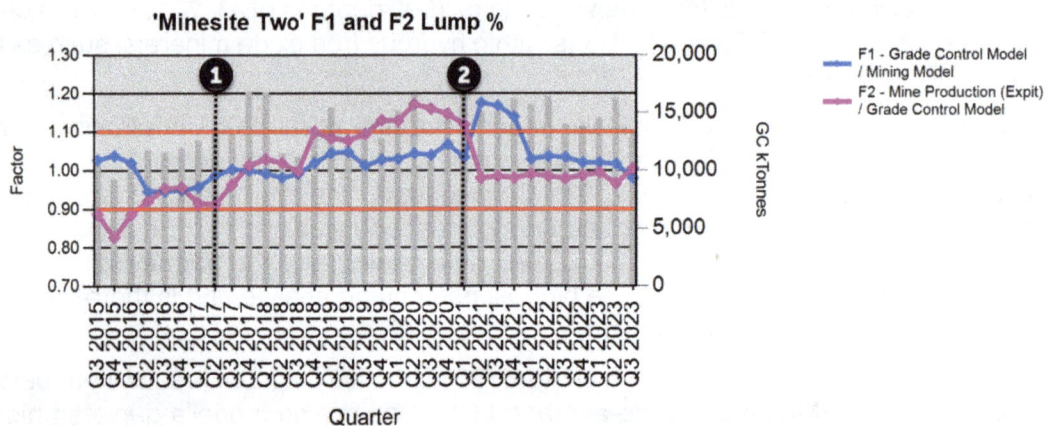

FIG 4 – F1 and F2 lump% reconciliations at 'Mine Site Two'. Event 1 is the adjustment to the deployed geomet algorithm which had a temporary positive impact on F2 lump% performance. Event 2 is the deployment of a completely revised algorithm which placed much greater importance on density as an input.

Although the absolute grade difference in this case was small (0.18 per cent absolute under-determination of LOI) the difference is significant when applied to the total grade control depletion. For the period 01 July 2015 to 31 March 2020, 1 279 110 kt (1.28 Bt) of HG material was depleted from the grade control model. When the established bias in the laboratory results is applied to that mass it indicates a tonnage underestimation of LOI units in the grade control model of 2302 kt (2.3 Mt). This is indicative of the mass of material that had been incorrectly characterised as Fe, SiO_2 or Al_2O_3 units (with the greatest error attributing to Fe) within the grade control model's HG estimates. This error will have led to an over estimation of valued metal units passing into the supply chain. In addition, the erroneous determination of Fe grades will have led to the incorrect characterisation of an unknown, but potentially large, tonnage of marginal LG material as HG. This problem is likely to have been most severe in cases where the HG cut-off is close to the average grade of the deposit or where the ore is 'patchy'. This will have led to dilution of the ore-stream. This dilution could not be detected by F2 analysis because the production samples that inform the F2 grade measurements were analysed by the same laboratory as the grade control samples and were therefore subject to the same analytical bias.

This case study underlines the importance of calculating reconciliation factors over long periods so that small, subtle but consistent signals in the data can be detected and the root cause corrected.

CASE STUDY 3 – LUMP% ESTIMATION AT 'MINE SITE TWO'

'Mine Site Two' is a significant iron ore producer in the Eastern Pilbara. The measured percentage of total production that is lump ore (Lump%) is an important economic parameter for this mine and is a valuable parameter for us to forecast. Lump% is estimated in all our models. The geometallurgical group within the Geoscience Department are responsible for the generation and maintenance of these estimates. The estimates are generated by deposit-specific geometallurgical algorithms; mathematical formulae that review the geological information for each estimation cell within the model and from that information generate a predicted lump% for each cell. These algorithms are developed by exposing large volumes of data, generated by the destructive testing of core in a geometallurgical laboratory, to highly specialised machine-learning software.

Prior to the calculation of F-Series factors for geomet, improvement of these algorithms was extremely difficult as the only feedback available was whether the processing plant was producing the *planned* lump%. Typically, the benchmark plan was the annual budget. As there are many variables outside of the geomet algorithm that influence whether-or-not the plant produces the budgeted lump% it was very difficult to determine the root cause of any non-compliance. Particularly difficult cases occurred when after the budget plan was set new geological information changed the prediction of the geomet algorithm. For example, if lithology, stratigraphy, or grade were different in the grade control model to that which was in the mining model a geomet algorithm when applied to the grade control model would produce a new lump% estimate. This new estimate – although valid – would no longer match the budget plan. Since the 'benchmark test' of geomet was whether the plant's lump% matched the budget lump%, in such cases it *appeared* that the geomet algorithm was in error.

To counter these issues, in Sept 2014 WAIO upgraded its F-Series reconciliation system to calculate F1, F1.5, F2 and F3 for lump%. F-Series reconciliation offers important advantages over the previous straightforward comparison with the budget lump%:

- Geomet F1 compares the lump% estimation made by the algorithm as applied to the grade control model with the lump% estimation made by the same algorithm when applied to the mining model. Geomet F1 therefore gives a strong indication of how much the algorithm's lump% estimation has changed due to changes in geological interpretations and estimations between these two classes of model.

- Geomet F1.5 compares the lump% estimation made by the algorithm as applied to the grade control model with the lump% estimation made by the same algorithm when applied to the STPM. Geomet F1.5 therefore gives a strong indication of how much the algorithm's lump% estimation has changed due to revisions of geological interpretations and estimations between these two classes of model.

- Geomet F2 compares measured production lump% against the lump% estimation made by applying the algorithm to the grade control model.

- Geomet F2 takes the change in pre-crusher stockpiles and the estimated lump% in those stockpiles into account.

- These factors are all calculated with reference to the volume extracted by mining rather than the volume that was planned to be extracted, this removes any complications due to mining outside of the planned volume.

- The final lump% measurement can be influenced by changes to physical aspects of mining and processing. If for example blasting powder factors change or the jaw settings on crushers change or the screen-deck size changes, then F2 lump% will deteriorate but F1 and F1.5 will remain within control. This assists in root cause analysis.

When F1 and F2 lump% were first calculated at WAIO the results were encouraging with WAIO-level F-Series lump% values generally falling within pre-set tolerances. However, it soon became apparent that this satisfactory result was only achieved by high results at some sites countering low results at others and that significant improvement was possible.

In the first quarter of calculations (FY15 Q3) Mine Site Two returned a low F2 lump% which dropped further in the next quarter. To some extent these low F2 results were countered by slightly elevated F1 results which meant that, from the perspective of the business, the overall lump% proportion was acceptable. However, from a technical perspective the reconciliations had revealed a weakness in Mine Site Two's ability to recover the lump% indicated by grade control.

During FY17 Q2 a revision of the existing geomet algorithm was completed and the effect on F2 lump% reconciliations was positive with an F2 of 1.00 returned in Q3 2018. However, in the period from Q4 2018 to Q1 2021 the F1 lump% consistently remained slightly elevated while the F2 was consistently high. Combined, these two factors led to a very high F3 lump%, indicating that a far higher proportion of product was shipped as lump than indicated in any of our forecast models. Superficially this was advantageous since lump product attracts a higher price than fines, but our inability to accurately predict lump yield created difficulties for plant and supply-chain optimisation as well as for technical marketing forecasts.

The elevated F2 lump% result during this period was, in part, caused by a higher than usual proportion of crusher feed derived from old long-term, lower-grade, pre-crusher stockpiles. These stockpiles typically have poorly constrained geological and geometallurgical estimates so feeding large proportions from them often results in poor F2 performance. Nevertheless, within the geometallurgical group suspicions were aroused that there was a deeper cause behind these unsatisfactory results and so began a fundamental review of the inputs and the weighting of the inputs in the geomet algorithm.

Independently, the geophysics team were concluding a long campaign of QAQC on geophysics density data across WAIO. This work included a correction of old, uncompensated, single-detector density wireline data to align with modern dual-head density data. As a result, confidence in this data set had markedly improved and the geomet team decided to give the geomet machine-learning tool access to this recently 'cleaned' density data. This process generated a series of completely revised algorithms which greatly increased the weighting of downhole density relative to other input variables. Following a period of testing and assurance these new algorithms were passed for deployment into production.

In Q1 2021 a completely new algorithm was deployed into the grade control model at Mine Site Two. The impact on F2 lump% was pronounced with the previously high F2 Lump% (1.12) dropping to tight control (0.98) in the following quarter. Following this marked improvement, the same algorithm was deployed into the mining model and the STPM so that both F1 and F1.5 lump% also improved. Since this change these factors have remained well controlled at Mine Site Two (see Figure 4).

CASE STUDY 4 – DUAL-SHELL TO SINGLE-SHELL ESTIMATION AT 'MINE SITE TWO'

Historically across WAIO F2 tonnes control has generally been satisfactory proving that the grade control model and in-pit ore-control procedures are sufficiently well governed to ensure that the total of ore tonnes delineated by grade control are delivered to the supply chain. By comparison, historically WAIO operations have had greater difficulties with F2 Fe, SiO_2 and Al_2O_3 control, with F2 Fe often low and F2 SiO_2 and Al_2O_3 often high. SiO_2 has proven especially difficult to control with F2 SiO_2 at 'Mine Site Three' (a large Central Pilbara iron ore operation) and Mine Site Two elevated for many years. Numerous investigations and projects to address this issue have been attempted, many focusing on grade control sample quality. These improvement initiatives have met with varying degrees of success and none have provided a decisive improvement. Within the last two years however focus has turned to the estimation method used to derive the grade in our grade control block-outs.

The grade control blockout process, the process that creates the grade control model, involves collating blasthole and reverse-circulation sample assays and logs on a computerised bench-plan. On this plan a geological interpretation is created that delineates mineable volumes of different classes of material according to pre-determined cut-off grades. The resulting grade control blocks distinguish ore from waste and delineate zones of different types of ore within each blast pattern.

Historically, the grade of each block was estimated by calculating the average of the assays within its volume. This method was flawed for at least three reasons:

- The polygon delineating the HG material was – with occasional exceptions forced by mineable geometry considerations – specifically drawn to contain only assays that are above the Fe cut-off. This introduces a high likelihood of a selection error; the possibility that the exact shape of the block is influenced by chance groupings of high-grade assays. If the shape is influenced by such non-representative groupings, then the estimated grade will not be recoverable.

- The polygon boundary was determined by the cut-off grade which is an entirely artificial economic boundary imposed upon the data set. The resulting polygons do not delineate any natural geological entity or identifiable domain.

- Geostatistical aspects of the data set were not considered. All assays within each polygon were equally weighted regardless of their position relative to polygonal boundaries, other assays, or any discernible geostatistical trends. Regardless of their proximity, assays that fell outside of each block's polygonal boundary had no weighting whatsoever on the grade within the polygon.

These considerations tend towards an overestimation of the Fe grade within the HG blocks. This leads to a grade control estimation that cannot be delivered to the processing plant regardless of how diligently the mining is executed.

To address this issue, during 2022, WAIO's grade control grade estimation method changed. Now we use a geostatistical grade estimate made by ordinary kriging within a 'local pattern block model' This model generates kriged grade estimates for each grade control block's volume and then transfers those estimates to the grade control model.

The positive impact of this change on F2 grade reconciliations at our sites was clear and provided significant evidence that the revised estimation method was superior to the previous classical estimation (see Figure 5).

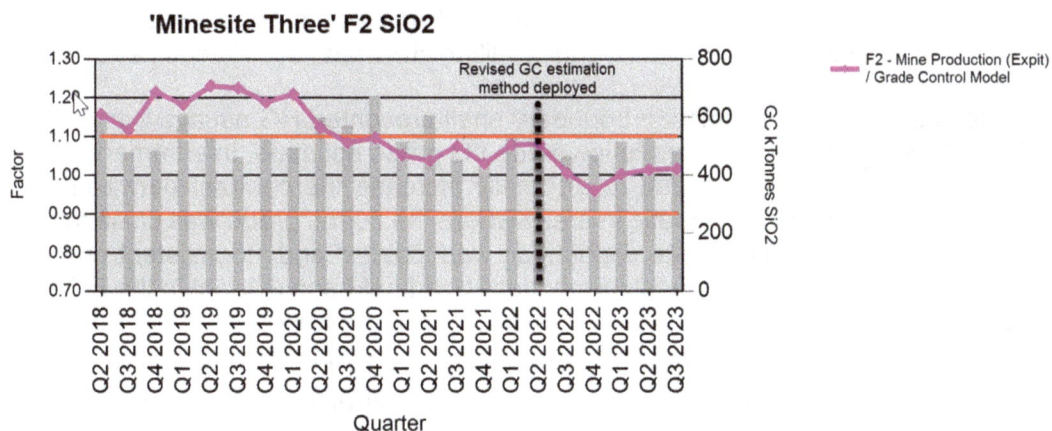

FIG 5 – The vertical marker represents the change from polygonal grade control to grade control informed by local pattern models with geostatistical estimations derived from a single-shell model. The resulting improvement is visible in Mine Site Three's F2 SiO$_2$ results.

The improved F2 results were however matched with a deterioration in F1 and F1.5. This change indicated that our strategic and tactical horizon models were now less capable of predicting the grade control estimates despite all three classes of model now estimating using ordinary kriging. Consequently, attention turned to a detailed review of the estimation methods used in the mining models and STPMs.

It was found that the critical difference was that the LPM is informed by assays within a single LG shell that delineated a volume of grade enrichment above background, while both the STPM and Mining Model estimates were constrained by two shells:

1. A LG shell, very similar to the shell used in the grade control model, that delineates the volume of material that has been subjected to grade enrichment that raised the Fe grade above natural background.

2. An inner HG shell that delineates a volume of material above the economic cut-off grade.

The presence of the HG shell was the critical factor. This hard estimation boundary at the economic cut-off grade was introducing a selection error. Since the hard boundary's wireframe was specifically drawn to contain almost exclusively HG assays, this introduced a similar geometrical selection bias that which had previously been present in the grade control estimate. Furthermore, despite geostatistical estimation methods being used in all these models, assays that fell outside of the HG shell could play no role in the estimation of HG ore within the inner shell. This was because the HG shell was used as a 'hard boundary' across which no external assay could exert any influence. Any samples that fell outside of the inner shell, even samples that were very proximal to the HG ore, had no influence on the grade estimation within the shell. This tended to artificially raise the ore's estimated grade (see Figure 6).

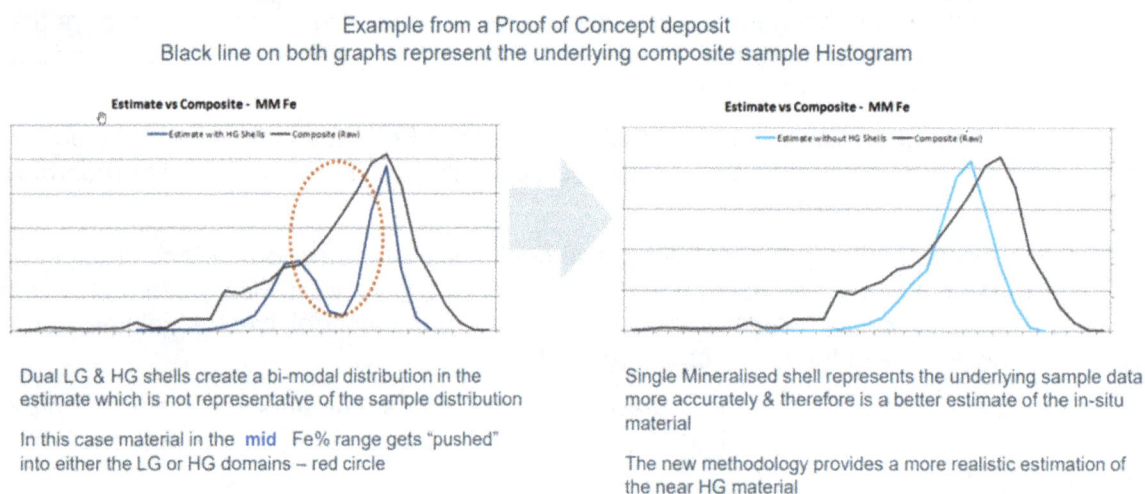

Example from a Proof of Concept deposit
Black line on both graphs represent the underlying composite sample Histogram

Dual LG & HG shells create a bi-modal distribution in the estimate which is not representative of the sample distribution

In this case material in the mid Fe% range gets "pushed" into either the LG or HG domains – red circle

Single Mineralised shell represents the underlying sample data more accurately & therefore is a better estimate of the in-situ material

The new methodology provides a more realistic estimation of the near HG material

FIG 6 – Dual-shell models force an artificial bimodal distribution in the estimate that does not reflect the underlying sample distribution. The resulting HG material's Fe estimation is too high.

In the opposite case, the hard estimation boundary shell prevented HG assays from informing low-grade (LG) block estimates. This tended to lower the LG grade estimate. In summary, the presence of this hard estimation boundary tended to artificially increase the contrast between LG and HG materials. This was especially problematic in areas where the cut-off grade was close to the average grade of assays or where the ore-grade distribution was 'patchy' and inconsistent or the transition from LG to HG was gradational. Previously such biases had not been evident in the F-Series results because a similar error was made in the grade control model and hence F1 and F1.5 indicated very little error although F2 indicated that the estimated grades were not deliverable.

To address these issues a decision was made to replace all dual-shell estimation models (both mining models and STPM) across WAIO with single-shell models. On 14 Feb 2023 Mine Site Two was the first site to deploy these revised models and by May 2023 the positive impact of this change was evident in a markedly improved set of F-Series Reconciliation factors for this site (see Figure 7).

| Location: | Minesite Two | | Start Date: | 01-Jan-23 | End Date: | 31-Jan-23 |

TOTAL	kTonnes	Fe %	P %	SiO2 %	Al2O3 %	LOI %
F1 - Grade Control Model / Mining Model	0.98 🙂	0.990 ☹️	1.03 🙂	1.11 ☹️	1.10 ☹️	1.03 🙂
F1.5 - Grade Control Model (with STM) / STM	0.94 😐	0.991 😐	1.05 😐	1.08 😐	1.14 ☹️	1.03 🙂
F2 - Mine Production (Expit) / Grade Control Model	1.04 🙂	1.004 🙂	0.95 🙂	0.97 🙂	0.98 🙂	0.97 🙂

| Location: | Minesite Two | | Start Date: | 01-Apr-23 | End Date: | 30-Apr-23 |

TOTAL	kTonnes	Fe %	P %	SiO2 %	Al2O3 %	LOI %
F1 - Grade Control Model / Mining Model	1.01 🙂	0.996 🙂	0.98 🙂	1.05 🙂	1.02 🙂	1.01 🙂
F1.5 - Grade Control Model (with STM) / STM	0.95 🙂	1.000 🙂	0.96 🙂	1.00 🙂	1.02 🙂	1.00 🙂
F2 - Mine Production (Expit) / Grade Control Model	1.01 🙂	0.999 🙂	1.04 🙂	0.96 🙂	1.03 🙂	1.04 🙂

FIG 7 – Comparison of reconciliation results at 'Mine Site Two' prior to and after deployment of single-shell estimation models. The later reconciliation factors provide strong evidence of the improved capacity of these models to provide more reliable production forecasts.

These results provide strong evidence that the change to single-shell estimation has had a marked positive impact on the ability of each of our production planning models to predict the grade of our products. This is a valuable improvement for WAIO allowing us to plan and set quality targets with greater confidence and lower risk.

DISCUSSION

F-Series Reconciliations can monitor and report on a very large number of business-critical estimates each of which are difficult – and in some cases perhaps impossible – to improve without these calculations.

The improvements that are shown here are the result of long-term analysis directed at identifying and addressing the root-cause of each problem. Rarely, if ever, are valuable improvements made in the short-term. A common misconception is that upon detection of a problem an 'action plan' should immediately be developed and executed to 'fix' the problem in the shortest possible time. Given that these values are calculated within a complex and frequently changing business and operational environment, the accurate diagnosis of each problem and the development and execution of a plan to address the true cause of a problem can be a drawn-out and complex process. Often the correct response to a problem is highly technical, demanding expertise, time and the commitment of capital. F-Series Reconciliation is therefore best thought of as a strategic tool capable of supporting long-term, enduring improvements rather than tactical, short-term gains.

It is imperative to avoid the temptation of applying an arbitrary 'correction' or 'weighting' or 'bias' to any model in order to 'eliminate' a problem. It is reasonable to apply such factors or 'biases' to *plans* and *forecasts* or *targets* so that historical performance issues are projected into the future, but this is a different proposition to arbitrarily altering the models themselves or applying 'corrections' to measurements. To alter the estimations generated by models or the data generated from the mining process or the processing plant, only serves to obscure any underlying problem rather than solve it. Such actions drive the operation further from any chance of real improvement. In extreme cases such manipulation of data could be considered fraudulent.

It is critical that the definition of these factors and the calculation methods are completely stable so that comparisons between various periods and locations are always valid and meaningful. To this end it is necessary that the factor calculations are carefully and correctly designed and that the software that performs these calculations is stable, well-governed and not subject to any casual manipulation. Certainly, the system must be flexible and robust enough to survive the various changes that inevitably occur throughout the life of a mining operation, but throughout all such changes the aim must be to preserve the integrity and intent of the underlying factors and the quality and completeness of the data that informs them. Data completeness and quality is of critical importance. It is essential that the data in the reconciliation system be cross-referenced against the original 'parent' data sets to ensure that a full and correct set of data is present. Ideally all the data in the reconciliation system will be transferred to it by automated interfaces. Manual data transfer and data entry is to be avoided.

For the reconciliation process to deliver its full value it must be conducted routinely, frequently and on an ongoing basis, typically once per month for the entire life of the operation. The temptation to only calculate these factors at the end of the financial year or 'only when there is a problem' greatly undermines the value of the process.

To maintain the consistent effort required to run a reconciliation system it is essential that reconciliations are valued by the organisation and are established as an inherent part of the modelling and mining cycle. Although central governance is essential to make this process work, reconciliations must be perceived by all geoscientists as a necessary part of the organisation's workflow. To be fully successful reconciliations must become embedded in the organisation's culture so that they become a natural part of the business landscape.

CONCLUSION

In the mining environment, F-Series reconciliations provide a practical and powerful quantitative test of the accuracy of various critically important models. It also provides feedback on the ability of the mining operation to deliver the value represented in those models. The reconciliation process is therefore well placed to assist in the:

- Detection of opportunities for improvement.
- Diagnosis of root causes of problems.
- Development of corrective actions.
- Assessment of the effectiveness of those actions.

F-Series reconciliations can play a role in every step of the continuous improvement of models and mining processes. The four case studies discussed in this paper give concrete, real-world examples of this.

ACKNOWLEDGEMENTS

The author gratefully acknowledges the assistance of Will Patton, Bojan Colac, Henning Reichardt, Jan Nortje, Dorothee Mittrup and Nelson Mora in reviewing this paper. I also wish to recognise the efforts made every day by the Mine Geology team across WAIO without whose efforts our reconciliation system would not function. I would also like to thank BHP WAIO for granting permission for this paper to be published.

REFERENCES

International Organisation for Standardization (ISO), 2003. ISO 9516-1:2003 – Iron ores – Determination of various elements by X-ray fluorescence spectrometry – Part 1: Comprehensive procedure, April 2003.

International Organisation for Standardization (ISO), 2015. ISO 11536: Iron ores – Determination of loss on ignition – Gravimetric method, August 2015.

Exploration and geology – new iron ore deposits, including magnetite

New occurrence of banded iron formations in the Minvoul Region, Ntem Complex (Northern Gabon)

M Iglesias-Martínez[1]

1. Research Scientist, CSIRO Mineral Resources, Kensington WA 6151.
   Email: mario.iglesiasmartinez@csiro.au

ABSTRACT

The potential for iron ore in the Central African Archean belts has received much less attention by comparison to their equivalents in other cratonic regions of West Africa, South Africa, Australia and Brazil. Historically, iron ore is known to occur in South Cameroon and North Gabon with significant iron ore deposits (eg Mbalam in Cameroon, Avima in the Republic of Congo, Minkébé, Bélinga and Boka-boka in Gabon), having been the subject of geological exploration programs carried out by government institutions and exploration companies. Although the presence of strong aeromagnetic anomalies, undeniably related to the presence of greenstone belt formations, has been recognised, this study is the first attempt to assess the potential of banded iron formations (BIF) in the Northern part of the Gabonese Republic in the region between Minvoul and Minkébé (>10.000 km^2). Deep lateritic weathering, inaccessible equatorial rain forest and high operational costs are some of the reasons why only limited exploration has been undertaken in the area. This work aims to improve the geological knowledge and to highlight the potential for iron ore of the eastern margin of the Ntem Complex units dated between 3.3–2.7 Ga. Field work showed a series of mafic rocks (amphibolite, anorthosite and gabbro), schist and BIF (described as itabirites in neighbouring areas). Samples from Minvoul BIF showed an extensive and deep weathering to powdery hematite revealing the presence, in thin sections, of iron oxides including magnetite, hematite and martite, primary and secondary quartz, garnet, hornblende, anthophyllite, cummingtonite and epidote. Mineral assemblages and coarse grain size suggest a metamorphic grade ranging from medium to upper amphibolite facies. The existence of known economic iron ore deposits in neighbouring regions combined with an extensive weathering enrichment process are promising factors for potential iron ore discovery in this remote area.

INTRODUCTION

The Minvoul region is located in northern Gabon, adjacent to the Gabon-Cameroon border. The study area is part of the Ayina unit, which represents the eastern edges of the Congo craton's NW margin, known as the Ntem complex (Maurizot *et al*, 1986; Tchameni *et al*, 2001; Shang *et al*, 2004) (Figure 1 and Table 1). Locally, the geology of the Minvoul area is characterised by Neoarchean to Palaeoproterozoic terrain overlying a Mesoarchean granite-gneissic basement, within which two series of medium-grade supracrustal metamorphic rocks. The main one consists of amphibolites and metagabbros which seem to represent mafic metavolcanic rocks associated with metapelites and banded iron formations. For the first time, it has been registered the existence of banded iron formations (BIF) occurring as lenses or narrow bands of metric to decametric thickness, strongly folded, of complex geometry and spatially associated with metasediments and amphibolites. The local relief can be considered as a peneplain crossed by long continuous NE-SW ridges composed of supracrustal rocks including BIF which stand as much as 100 m above the rolling terrain that is underlain by gneiss. These resistant ridges are the result of etching by chemical weathering and present steep slopes and cliffs where relict iron duricrust deposits are preserved. The lateritic deposits are well developed over the BIF (up to 5 m thick) and most of the times are covered by pisolitic gravels and surficial clays. Residual and transported lateritic covers are extensively developed, which hampers near-surface exploration.

The contribution of airborne magnetic surveys has been especially useful for outlining the greenstone-belt boundaries, particularly, in the case of those containing iron formations. The strong contrast in magnetic response between the BIFs, the rest of the greenstone-belt sequences and the granite-gneissic country rock allows a fairly clear delineation. These structures and geometries revealed by magnetism offer a picture of the Archean basement tectonics (CGG, 1984).

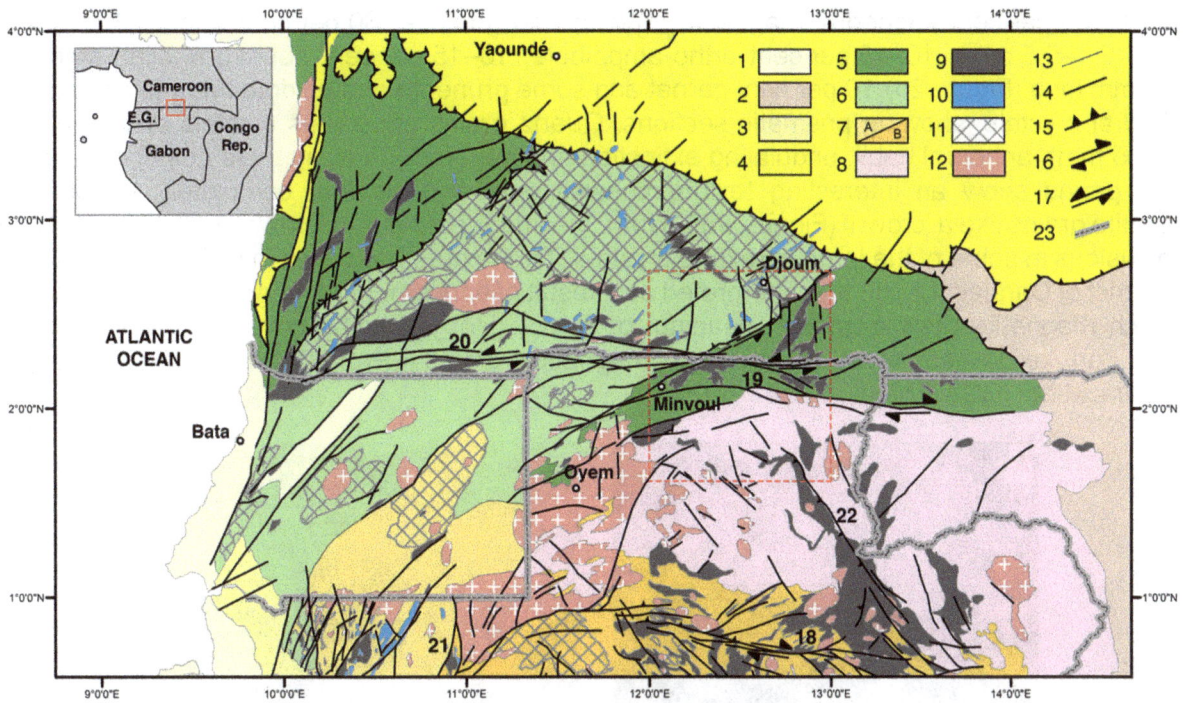

FIG 1 – Geological map of the Ntem complex and neighbouring units. Dark grey entities [9] represent Mesoarchean to Palaeoproterozoic greenstone-belts containing BIFs. See Table 1 for a description of the regions (Iglesias-Martínez and Amandi, 2016).

TABLE 1

Breakdown of Ntem complex geological regions (Iglesias-Martínez and Amandi, 2016).

Region	Represents	Region	Represents
[1]	Phanerozoic cover	[13]	Geological contour
[2]	Yakadouma et Dja Series (Neoproterozoic)	[14]	Major fault
[3]	Noya Basin (Neoproterozoic)	[15]	Major thrust
[4]	Yaoundé Series (Neoproterozoic, Pan-African)	[16]	Dextral transcurrent movement
[5]	Ntem complex, Ayina unit and Nyong unit (Neoarchean to Palaeoproterozoic) Foliated series: amphibolite/garnet gneiss	[17]	Sinistral transcurrent movement
[6]	Ntem Complex, Ntem unit (Mesoarchean) banded series: undifferentiated granulitic gneiss	[18]	Nkol shear zone
[7]	Monts de Cristal migmatitic belt (Mesoarchean?)	[19]	Nord Gabon shear zone
[8]	Nord Gabon plutonic Complex (3.1–2.8 Ga) granitoid, tonalite, granodiorite	[20]	Ntem shear zone
[9]	Greenstone belts (Mesoarchean to Palaeoproterozoic): Amphibolite, BIFs, micaschist	[21]	Abanga fault
[10]	Dolerites and Gabbros [Pre-Eburnean (2.2 Ga) and Pre-Pan-Africain (1.0 Ga)]	[22]	Nouna Fault
[11]	Magmatic charnockitic suite and TTG suite (2.9–2.8 Ga)	[23]	National border
[12]	K-rich granite and porphyritic monzogranite (2.7–2.5 Ga)		

PETROLOGICAL CHARACTERISATION OF BIF

The Minvoul BIF are metamorphosed rocks in which layers (1–3 mm) of quartz alternate with iron oxide-rich laminae (magnetite and its alteration products) (Figure 2a–2b) presenting tectonic structures such as kink bands. The freshest hand samples are weathered but still magnetic. It was identified 30 per cent subhedral magnetite, martite and goethite, 30 per cent anhedral crystals of quartz (0.1–0.5 mm), 10–15 per cent ortho-amphibole, 10–15 per cent subhedral clino-amphibole 0.5–3 mm long (Figure 2d), 10 per cent garnet and some grunerite and epidote. Magnetite occurs as subhedral crystals, showing prismatic sections. Quartz grains appear as crystals of variable size (0.1–0.5 mm), anhedral habit, undulating extinction and generally showing triple point contacts. The quartz grains show an interesting texture consisting of crystals that completely surround the magnetite grains like a 'crown' (Figure 2h). In turn, the 'magnetite-quartz crown' assemblage is found as inclusions in some of the large pyroxene crystals. This texture described in itabirites from the iron ore district of Carajás (Brazil) was interpreted as a result of enrichment or recrystallisation processes, in which magnetite grew from the 'cleaning' of the original jasper by collective recrystallisation, leaving an oxide-free area around the magnetite grains (Prof Carlos Rosiere, personal communication, IGC/UFMG).

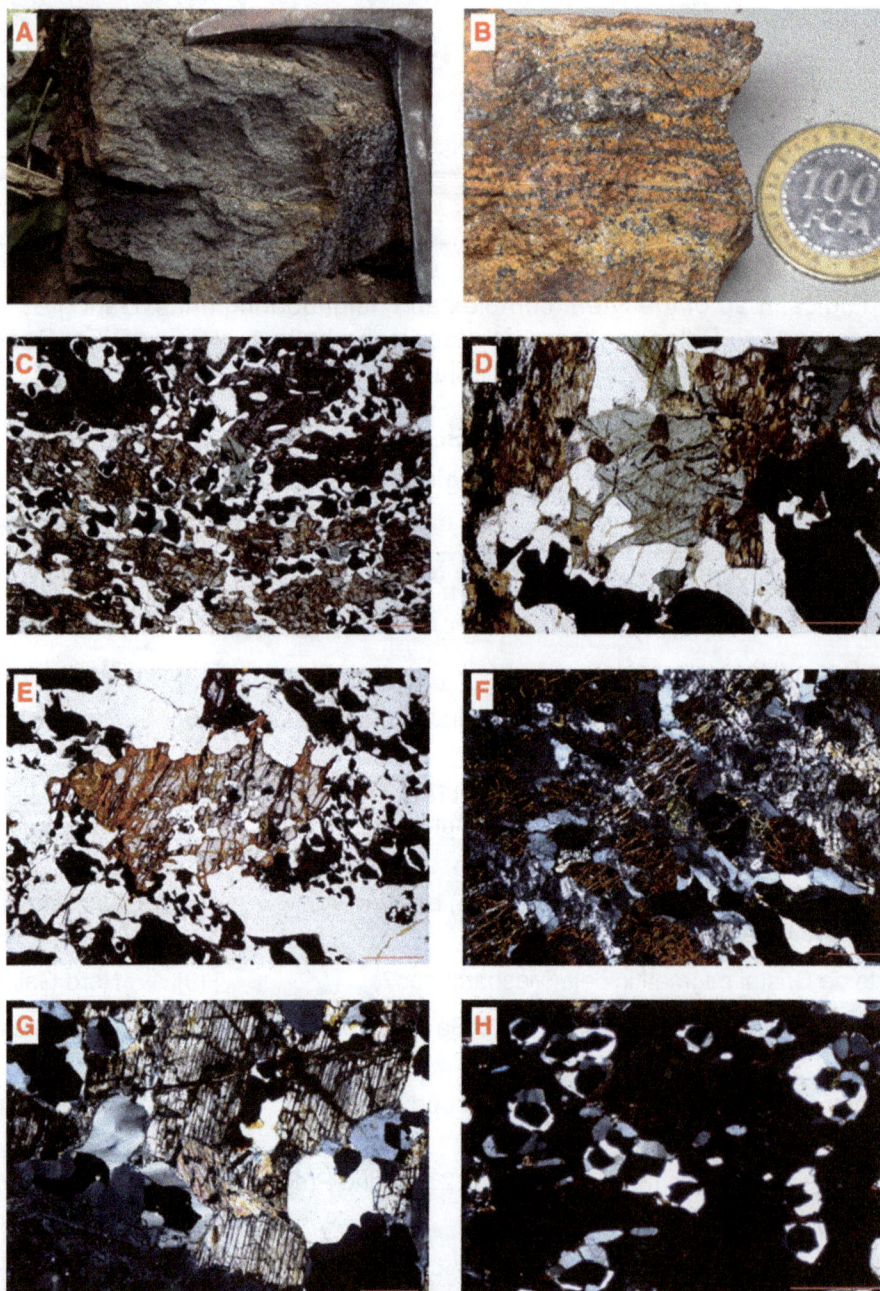

FIG 2 – Outcrop (a) and hand sample (b) of BIFs in Minvoul. (c) Photomicrograph showing mineral lamination; (d) Ferrohornblende crystal; (e) Strongly fissured and altered garnet. The inner part of the garnet shows drop-like quartz inclusions; (f) Deeply weathered iron oxides by martitisation processes; (g) Anthophyllite crystal partially replaced by white mica; (h) Corona-type texture with quartz grains rimming magnetite crystals.

CONCLUSIONS

The mineral association (clinoamphibole + orthoamphibole + grunerite + garnet) and the relatively coarse size of the crystals indicate that BIF have been subjected to amphibolite facies metamorphism. This contrasts with the low degree of metamorphism (greenschist facies) characteristic of nearby deposits such as Belinga. This variability, like the dispersion of radiochronological ages (Bassot et al, 1987), has led some authors to consider the existence of two episodes of deposition of iron formations for high metamorphism rocks and recent (Mesoarchaean) for epi- to mesozonal rocks. These formations, in neighbouring regions, have very high concentrations of hematite and other iron oxides that have been evaluated as potentially economic deposits. All the studied BIF outcrops in Minvoul showed quite extensive alteration and the initial magnetite is almost systematically recrystallised as iron oxides (hematite) or hydroxides (goethite). This alteration is crucial to produce a mineable ore as it is often accompanied by desilicification, which leads to a concomitant iron enrichment. The existence of these known economic iron ore deposits combined with an extensive weathering enrichment process are promising factors for potential iron ore discovery in this remote area.

ACKNOWLEDGEMENTS

This publication was authorised by the management of Craton del Congo Exploraciones SARL. All the data compiled and the results of fieldwork contained in this paper were collected as part of the exploration activities carried out for the aforementioned company during 2015 and 2016. Special thanks are due to the National Directorate of Mines and Geology of the Ministry of Industry and Mines of Gabon and to Javier Riera Táboas, Paul Hugman, Dixon Porter, Marc Vanhoutte, Julio García, Laura Gonzalez Acebrón and Carlos Pérez Garrido for their valuable personal, technical and scientific support. This paper was written during the postdoctoral stay of the author at CSIRO financed by the Spanish Program 'Ayuda para la Recualificación del Sistema Universitario Español (Margarita Salas)'.

REFERENCES

Bassot, J P, Caen-Vachette, M, Vialette, Y and Vidal, P, 1987. Géochronologie du socle gabonais (Geochronology of the Gabonese Archean bedrock), Final Report (unpublished), Department of Mineral Geology, University of Clermont-Ferrand II, 228 p.

CGG, 1984. Interprétation de la prospection aéroportée de magnétisme et spectrométrie gamma du Gabon (Magnetic and gamma spectrometric airborne interpretation of Gabon), Geophysics report (CGG), Ministry of Mines and Petroleum of Gabon, 54 p.

Iglesias-Martínez, M and Amandi, B, 2016. Rapport annuel d´activités 2015–2016 du Permis de Recherche G9–580 Minvoul (Annual report of activities 2015–2016 of the G9–580 Exploration Prospect of Minvoul), Libreville-Madrid.

Maurizot, P, Abessolo, A, Feybesse, J L, Johan, V and Lecomte, P, 1986. Étude et prospection minière du Sud-Ouest Cameroun, Synthèse des travaux de 1978 à 1985 (Study and Mining Prospection of the South-West Cameroon: Synthesis of the works carried out from 1978 to 1985), Geological and Mining Research Department (BRGM), Report 85, 66 p.

Shang, C K, Satir, M, Siebel, W, Nsifa, E N, Taubald, H, Liegeois, J P and Tchoua, F M, 2004. Major and trace element geochemistry, Rb-Sr and Sm-Nd systematics of the TTG magmatism in the Congo craton: case of the Sangmelima region, Ntem complex, southern Cameroon, Journal of African Earth Sciences, 40:61–79.

Tchameni, R, Mezger, K, Nsifa, N E and Pouclet, A, 2001. Crustal origin of early Proterozoic syenites in the Congo craton (Ntem Complex), southern Cameroon, Lithos, 57:23–42.

An iron ore detective story in Papua New Guinea

E R Ramanaidou[1]

1. Commodity Research Leader Iron Ore, CSIRO Mineral Resources, Kensington WA 6151.
Email: erick.ramanaidou@csiro.au

ABSTRACT

Two unknown iron ore samples were found by a geologist from Ok Tedi Mining Limited copper, gold and silver mine at an old wharf site dedicated to a 1970s drilling program in Lake Murray in Papua New Guinea (PNG). These iron ore samples could not come from a PNG mine and it was decided that a study should be undertaken to understand their origin.

Lake Murray is the largest lake in PNG with a total surface area of 647 km^2. It has a high-water 2038 km long convoluted shoreline with seasonal water level changes (Osborne, Kyle and Abramski, 1987) and it is located around 200 km south of the Ok Tedi mine (5°13'25"S, 141°09'37"E). The petrology, mineralogy and chemistry of two Fe samples (Ok Tedi 1 and Ok Tedi 2) included optical microscopy, X-ray fluorescence (XRF) analysis and a comparison with worldwide iron ore deposits. The two rock samples were cut to produce two subsamples: (1) a polished thin section; and (2) a pulp that was used for XRF analysis including major, minor and trace elements. Ok Tedi 1 and Ok Tedi 2 were petrologically and chemically compared to a range of known iron ore samples from Australia, Brazil and South Africa.

The petrological study of Ok Tedi 1 indicates that it is composed of granoblastic recrystallised hematite and the overall texture is uniform without any obvious original 'sedimentary' texture. Ok Tedi 2 is a 'specular' or platy hematite with recrystallised and fused grains. Using both petrological and chemical clues, the two Ok Tedi samples likely originated from the Iron Quadrangle in Brazil.

INTRODUCTION

Two unidentified iron ore samples were discovered by an Ok Tedi Mining Limited geologist at an old wharf site dedicated to a 1970s drilling program in Lake Murray in Papua New Guinea (PNG) located around 200 km South of the Ok Tedi mine. Lake Murray, the largest lake in PNG has a total surface area of 647 km^2 with a 2038 km long high-water complex shoreline. As PNG does not host iron ore deposits, these iron ore samples could not be derived from a PNG mine, it was decided that an investigation should be undertaken to evaluate their petrology, mineralogy and chemistry to uncover their origin.

SAMPLES AND METHODS

The two iron ores samples, Ok Tedi 1 and Ok Tedi 2 (Figures 1a and 1b) were cut to produce two subsamples: (1) a polished thin section (Figures 1c and 1d); and (2) a pulp that was used for XRF analysis including major, minor and trace elements. Mosaicked images of the samples were generated using a Zeiss microscope and an Axio Cam Mrc colour camera. Reflected optical images were generated using a Nikon Eclipse LV100N POL and QCapture Pro7 software. The XRF analysis including major, minor and trace elements, was conducted by Bureau Veritas (Perth, Western Australia).

RESULTS

Ok Tedi 1 is composed of granoblastic recrystallised hematite (Figure 1c). Some alignment and lineation of grains in thin bands indicate compression. The lineation may align with the original bedding. The overall texture is uniform in this case, without any obvious original 'sedimentary' (ie from the original banded iron-formation) texture. The hematite grains are smaller and are less intergrown than Ok Tedi 2 (Figure 1d). Ok Tedi 2 is a 'specular' or platy hematite (Figure 1d). The recrystallised and fused hematite grains are clearest in the cross polar image (Figure 1d). Encapsulated silica inclusions occur in the hematite crystals. The platy hematite grains are aligned in a preferred orientation and this lineation is more strongly developed than in Ok Tedi 1.

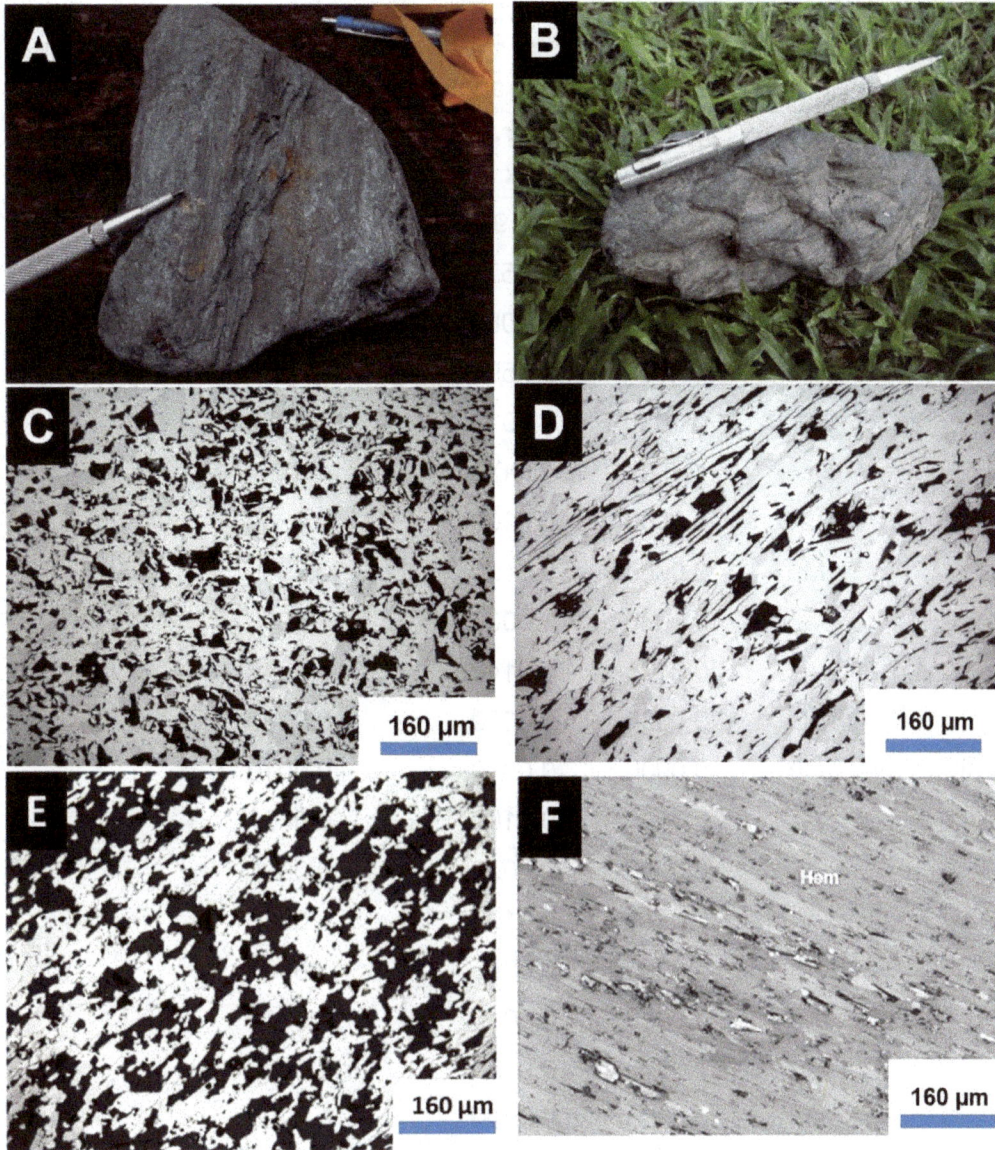

FIG 1 – (a) Ok Tedi 1 hand sample; (b) Ok Tedi 2 hand sample; (c) Ok Tedi 1 photomicrograph reflected light; (d) Ok Tedi 2 photomicrograph reflected light; (e) Aguas Claras mine photomicrograph reflected light; and (f) Cauê mine photomicrograph reflected light.

The chemical composition of both the Ok Tedi 1 and Ok Tedi 2 samples is very similar and is also close to the chemical composition of our standard iron ore samples from the Iron Quadrangle in Brazil (Table 1).

TABLE 1

Chemical composition of Ok Tedi 1 and Ok Tedi 2 and an iron ore sample from the Iron Quadrangle.

ID	Ok Tedi 1	Ok Tedi 2	Iron ore iron quadrangle sample
Fe	69.31	68.91	68.04
SiO_2	0.47	0.57	0.49
P	0.0020	0.0080	0.0402
Al_2O_3	0.36	0.59	1.02
S	0.0010	0.0010	0.0032
MgO	0.025	0.005	0.033
CaO	0.0000	0.0000	0.0280
Mn	0.0000	0.0400	0.2500
K_2O	0.0030	0.0000	0.0060
Na_2O	0.0125	0.0140	0.0040
TiO_2	0.0000	0.0000	0.0290
Cu	0.0000	0.0000	0.0010
Cr	0.0000	0.0100	0.0030
Ni	0.0005	0.0020	0.0001
V	0.0040	0.0100	0.0050
Pb	0.0050	0.0055	0.0001
As	0.0000	0.0000	0.0014
Zn	0.0020	0.0020	0.0020
Co	0.0000	0.0000	0.0005
Ba	0.0045	0.0045	0.0070

DISCUSSION

Ok Tedi 1 and Ok Tedi 2 samples were petrologically compared to a range of known iron ore samples from Australia including Mt Whaleback and Channar (Hamersley Province, Western Australia) Koolyanobbing (Yilgarn, Western Australia) and Koolan Island (Kimberley, Western Australia), from Brazil including Sierra Sul (Carajas), Aguas Claras and Cauê mine (Iron Quadrangle) and South Africa including Sishen. Out of all the iron ore samples studied here for textural and chemical composition analyses only the samples for the Iron Quadrangle in Brazil matched our OK Tedi 1 and OK Tedi 2 samples. Ok Tedi 1 consists of granoblastic hematite indicating that there was some recrystallisation under oxidising conditions of previous martite/magnetite. Ok Tedi 1 is petrologically close to some of our reference samples sourced from the Aguas Claras mine (Figure 1e) in the Western Low Strain Domain (Hensler *et al*, 2014) of the iron Quadrangle where granoblastic hematite ore dominates (Hensler *et al*, 2014). Ok Tedi 2 consists of 'specular' or platy hematite similar to Eastern high Strain Domain of the iron Quadrangle where tabular and elongated hematite dominates (Figure 1f) particularly in the Cauê mine.

Additionally, the chemical composition of the Ok Tedi 1 and Ok Tedi 2 samples are very similar to our standard iron ore sample from the Iron Quadrangle.

Finally, the Ok Tedi 1 and Ok Tedi 2 samples were found at an old wharf site dedicated to a 1970s drilling program. The only Brazilian iron ore mines open at that time are Cauê, Dois Corregos, Conceicão mine, Pico mine, Aguas Claras mine, Mutuca mine and Carajas (Table 2).

TABLE 2

Opening year of iron ore mines from Brazil.

Mines	Region	Opening year
Cauê	Iron Quadrangle	1942
Dois Corregos	Iron Quadrangle	1942
Conceicão	Iron Quadrangle	1951
Pico	Iron Quadrangle	1973
Aguas Claras	Iron Quadrangle	1973
Mutuca	Iron Quadrangle	1973
Carajas	Para	1974

Based on the petrology, the mineralogy, the chemical composition and the operational Brazilian mines in the 1970s, it is likely that the samples are derived from banded iron formation-hosted iron ore from the Itabira District in the Iron Quadrangle of Minas Gerais State in Brazil. Ok Tedi 1 is probably from the Aguas Claras mine and Ok Tedi 2 likely from the Cauê mine.

The Ok Tedi 1 and Ok Tedi 2 iron ore samples were probably left over samples from Brazil that were thrown away in the lake during the drilling program. These samples could have been brought by a geologist and disposed of in Lake Murray or could have been used as ship ballast as historically, ballast was solid such as sand, rocks, cobble and iron (David and Gollasch, 2015).

CONCLUSIONS

By combining location and historical information and, petrological and chemical data, it was demonstrated that two unknown iron ore samples from Lake Murray in Papua New Guinea, a country without iron ore deposits, could be accurately linked to two iron ore mines (Aguas Claras and Cauê) from the iron Quadrangle in Brazil.

ACKNOWLEDGEMENTS

The author thanks Ok Tedi Mining Limited for providing the samples and believing that I could provide the why and where from two out of context samples.

REFERENCES

David, M and Gollasch, S (eds), 2015. *Global Maritime Transport and Ballast Water Management, Invading Nature – Springer Series in Invasion Ecology 8*, 306 p. doi:10.1007/978-94-017-9367-4.

Hensler, A-S, Hagemann, S G, Brown, P E and Rosière, C A, 2014. Using oxygen isotope chemistry to track hydrothermal processes and fluid sources in itabirite-hosted iron ore deposits in the Quadrilátero Ferrífero, Minas Gerais, Brazil, *Mineralium Deposita*, 49:293–311.

Osborne, P L, Kyle, J H and Abramski, M S, 1987. Effects of seasonal water level changes on the chemical and biological limnology of Lake Murray, Papua New Guinea, *Australian Journal of Marine and Freshwater Research,* 38:397–408.

Characterisation of the Iron Bridge magnetite project

E R Ramanaidou[1], B Godel[2] and M Verrall[3]

1. Commodity Research Leader Iron Ore, CSIRO Mineral Resources, Kensington WA 6151. Email: erick.ramanaidou@csiro.au
2. 3D Characterisation Laboratory Manager, CSIRO Mineral Resources, Kensington WA 6151. Email: belinda.godel@csiro.au
3. Manager Electron Beam and X-ray laboratories, CSIRO Mineral Resources, Kensington WA 6151. Email: michael.verrall@csiro.au

ABSTRACT

Fortescue Metals Group Ltd Iron Bridge (IB) operation is a magnetite deposit located in the Pilbara region of Western Australia situated around 145 km south of Port Hedland. This Mesoarchean deposit belongs to the 3230 Ma Pincunah banded-iron formations (BIF), a member of the Kangaroo Caves Formation, Sulfur Springs Group (van Kranendonk, 2000).

A study was undertaken to improve the understanding of the geometallurgical properties of the IB magnetite deposit by ranking and providing recommendations on the best characterisation methods. To achieve this goal, diamond core was selected and analysed with an array of instruments to assess the following:

- The techniques that can be used on a large scale to determine the geochemical, mineralogical, petrophysical and textural characteristics.

- The resolution and timing required to obtain fit-for-purpose analyses.

- How advanced 3D characterisation techniques can be used to obtain additional information from routine analysis.

The non-destructive instruments included:

- HyLogging system for the estimation of the mineralogy.

- Minalyzer Core Scanning for the estimation of chemical composition.

- X-ray computed tomography (XCT) for the 2D/3D visualisation and quantification of texture (minerals + pores) and the density from the entire diamond core volume.

- Geotek Multi-Sensor Core Logger (MSCL): which comprises a range of geophysical sensors including magnetic susceptibility, photo spectrometry, density, resistivity, P-wave velocity, spectral natural gamma and core thickness.

A combination of HyLogger™ 3 and Minalyzer was shown to best measure quickly large volumes of core with limited handling and to provide quantitative mineralogy and chemistry. The XCT is the only technique that provides full 3D textural information and density variation at a 100 μm resolution. Although the XCT scanning speed is fast, the data processing requires extensive skills. Nonetheless, the processing can be automated once a workflow has been developed. Using the Geotek:

- density and magnetic susceptibility measurements are ideal for detecting magnetite

- resistivity is positively correlated with stilpnomelane (strong) and orthoclase (weaker)

- total natural Y (particularly K per cent and Th ppm) provides the exact location of orthoclase.

In addition to the core scanning technologies reviewed, Tescan Integrated Mineral Analyzer (TIMA) chemical/mineralogical images were produced on selected intervals for microscopy analysis as a validation method.

INTRODUCTION

The Fortescue Metals Group Ltd IB magnetite deposit is in the Pilbara region of Western Australia at approximately 145 km south of the port of Port Hedland. This 3230 Ma Mesoarchean deposit is located in the East Pilbara terrane of the Pilbara craton. The region contains two volcano-sedimentary formations, namely the Sulfur Springs group and Soanesville group. The Sulfur Springs group is a steeply dipping sequence of ultramafic to mafic volcanics, its youngest member being the Kangaroo Caves Formation, which includes the uppermost Pincunah banded-iron formation (BIF) member. The Soanesville group consists of terrigenous clastic sediments which conformably overlie the Pincunah BIF which include the Corboy Formation existing within the IB project area (van Kranendonk, 2000).

This project aimed to improve the understanding of the geometallurgical properties of the IB magnetite deposit by ranking the best characterisation methods and providing recommendations on how to best characterise the deposit. To achieve this goal, selected diamond core was analysed with a range of instruments to assess the techniques that can be used on large volumes of core to determine the deposit's geochemical, mineralogical, petrophysical and textural characteristics.

SAMPLES AND METHODS

FMGL provided 11 PQ diamond drill cores for a total of 1.3 km. Non-destructive instruments include:

- The Minalyzer Core Scanning that provides XRF analytical results, specific gravity and 1D and 3D images (Scanning directly from the tray).

- HyLogger™ 3 providing the mineralogy using reflectance spectroscopy visible near, short and thermal infrared (VNIR-SWIR and TIR) (Scanning from the tray).

- X-ray computed tomography (XCT) providing 2D/3D Visualisation and quantification of texture and density from the entire diamond core volume (Need to transfer samples to the XCT).

- The Geotek multi-Sensor Core Logger (MSCL) that comprises a range of geophysical sensors including magnetic susceptibility, density, resistivity, P-wave velocity, spectral natural gamma and core thickness (need to transfer samples to the MSCL).

Analyses requiring destructive testing include some cylindric plugs (14 mm diameter and 30 mm length) drilled in the cores for validation analyses (Figure 1). The selection of the plugs was based on a combination of Minalyzer chemistry and HyLogger™ 3 mineralogy. The top 3 mm of the plugs was cut to be mounted on a polished section for further petrological and mineral analyses; the remaining part of the plugs was powdered (<75 µm) for X-ray diffraction (mineralogy) and XRF analyses and the remaining sub-samples were kept as standards. The polished sections were measured using the TESCAN Integrated Mineral Analyzer or TIMA.

FIG 1 – IB diamond core with drilled plugs.

HyLogger methodology

Spectral mineralogy was acquired using a HyLogger™ 3, scanning the visible and near infrared or VNIR and SWIR (400 to 2500 nm) and thermal infrared or TIR (6000 to 14500 nm) with a spatial resolution of 12 × 8 mm. The use of the TIR was required to detect quartz and orthoclase as they are not detected in the VNIR-SWIR wavelength range. The Spectral Geologist (TSG) has been used to process the HyLogger Data.

Minalyzer methodology

The Minalyzer core scanning conditions were 2 mm/s; measurement acquisition scales: 1 cm, 2 cm, 5 cm, 10 cm and 1 m and acquired with both Ag and Cr X-ray tubes. Na is not detectable and Mg is only detectable in high concentrations with the Cr X-ray tube. Ti, K and Ca analyses are more accurate with the Cr X-ray tube, but Zr, Mn, Cr, V, Rb, Sr and Ag are not detectable. However, the Cr X-ray tube has lower detection limits for some lighter elements, such as Al_2O_3, K_2O and SiO_2 (Figure 2). The Ag X-ray tube is a better overall detector and can detect both Cr and Mn. The data was calibrated using OREAS24B and OREAS624 certified reference materials.

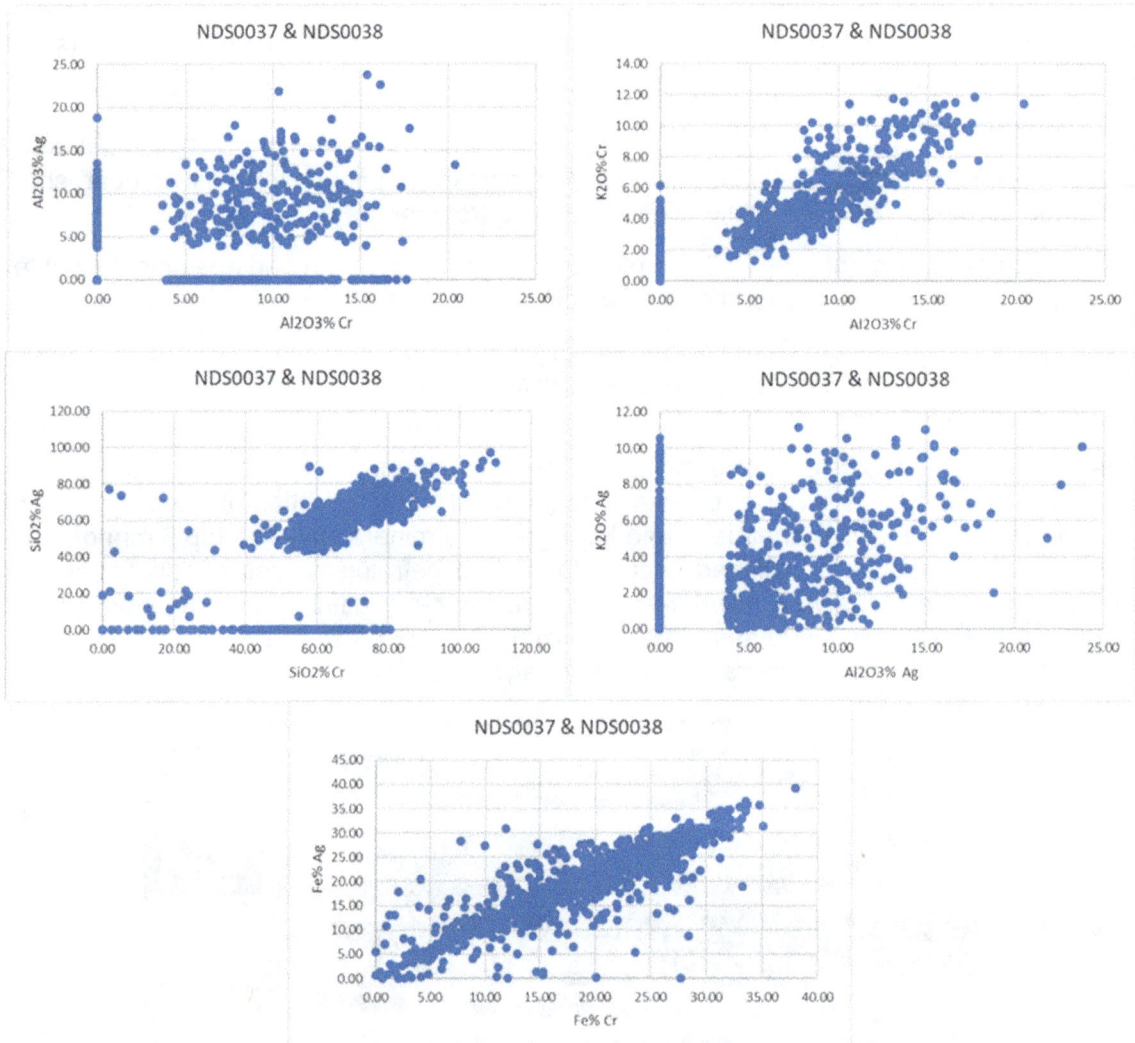

FIG 2 – Comparison of Ag and Cr tubes for Al_2O_3, SiO_2, K_2O and Fe for the diamond cores NDS0037 and NDS0038.

Of note is that the elemental results from this method are surficial only (akin to portable XRF results, but with a higher energy source and precision). Results are still deemed highly useful for domaining purposes.

XCT methodology

XCT was acquired with a voxel size of 200 × 200 × 100 µm, calibrated with 19 reference materials. Density values were reported as a bulk density over the given depth interval. More than 7000 images are generated for 700 mm. Each raw image is in a Siemens format and is then reconstructed in the

Dicom format (standard medical format). The slices have been preprocessed in Avizo to improve the features.

Geotek MSCL methodology

The Geotek MSCL characteristics are shown in Table 1. The Geotek multi-Sensor Core Logger offers a range of measurements (Table 1) including core thickness, density, magnetic susceptibility, P wave, resistivity, spectrophotometer and natural gamma. Geotek measurements are more time consuming than Minalyze and HyLogger analyses. However, Geotek measurements are generally volume analyses whereas HyLogger and Minalyze measurements are surface analyses.

TABLE 1

Geotek MCLS analyses duration (s) and volume or area measured.

Sensor	Analyses (s)	Volume/area
Core thickness	6	Through the sides; pad is about 3 cm
Core thickness	6	Through the sides; pad is about 3 cm
Density	5	Through the sides; Y rays from a 0.5 cm collimator, but the detector is 7.6 cm
Magnetic susceptibility point (P) and loop (L)	1	1 cm^3 volume on top surface (M); 2–4 cm segment through full thickness (L)
P-wave	8	Through the sides contact point is 3 cm
Resistivity	1	2–4 cm, ~2 cm penetration on the bottom surface
Spectrophotometer	15	0.8 cm diameter on the top surface
Natural gamma	50	7.6 cm crystal on each side and on top

Validation (TIMA) methodology

TIMA is a fully automated, high throughput, analytical scanning electron microscope. TIMA is based on a completely integrated Energy Dispersive X Ray (EDX) system that performs full spectrum imaging at very fast scan speeds. TIMA image analysis is performed simultaneously with SEM backscatter electron images and a suite of X-ray images. The resolution of the maps was at 5 μm steps.

RESULTS

The HyLogger™ 3 spectra in both the VNIR-SWIR and TIR show that all the minerals occurring in the IB deposits have been detected including magnetite, hematite, carbonate, quartz, orthoclase and stilpnomelane (Figures 3 and 4). The Spectral Geologist (TSG) images display the relative mineral abundance for different scale acquisitions. TSG images show that as scale increases, detailed mineralogical information is disappearing but that the dominant mineralogy is still showing (Figure 5), providing useful information on optimum sampling intervals. The core spectral mineralogy distribution using the TSG software shows a good correlation between mineralogy (HyLogger™ 3) and chemistry (Minalyzer). Note the increase of K_2O and Al_2O_3 where orthoclase occurs (Figure 6). The combination of HyLogger™ 3 and Minalyze data provides a way of objectively domaining the different zones of the diamond cores. TIR1 (quartz and orthoclase) and TIR2 (carbonate, stilpnomelane and chlorite) correspond to the two dominant mineralogical parageneses (Figure 6).

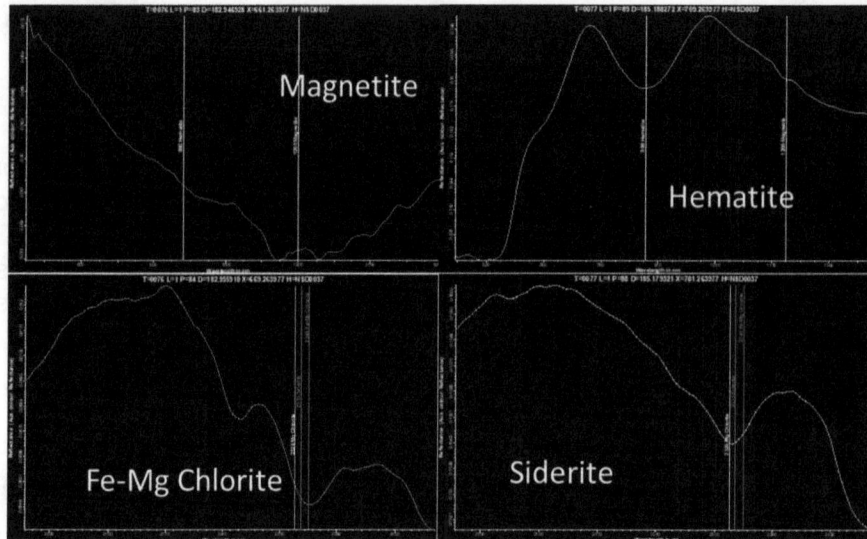

FIG 3 – HyLogger™ 3 spectra of magnetite, hematite, chlorite and siderite in the VNIR-SWIR wavelength range.

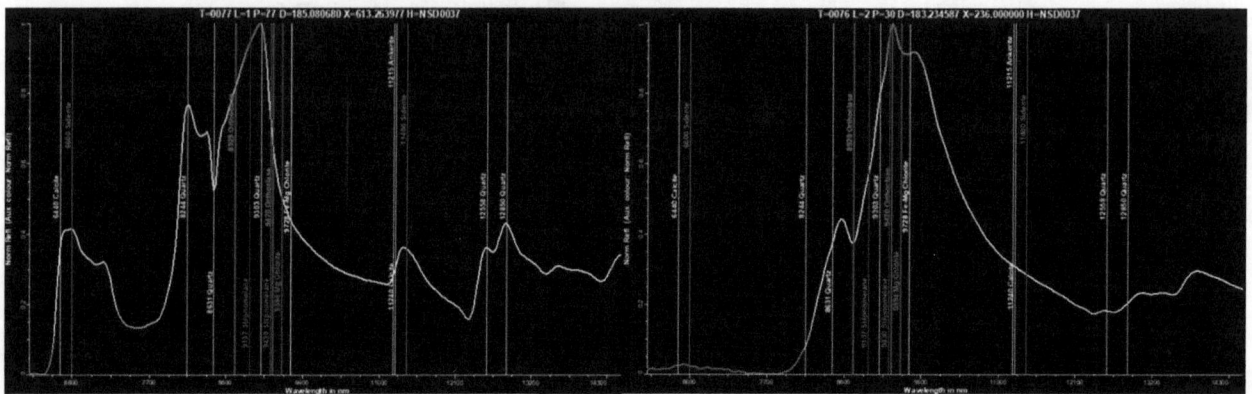

FIG 4 – HyLogger™ 3 spectra of carbonate (calcite and siderite), quartz, orthoclase, stilpnomelane and chlorite in the TIR wavelength range.

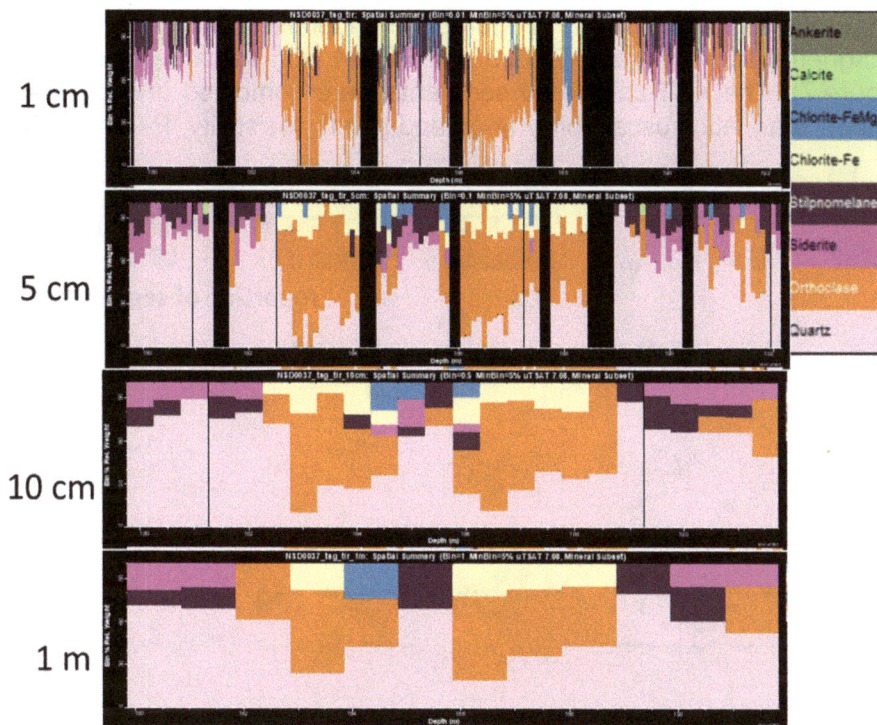

FIG 5 – TSG images displaying the relative mineral abundance for different scale acquisitions for a specific core.

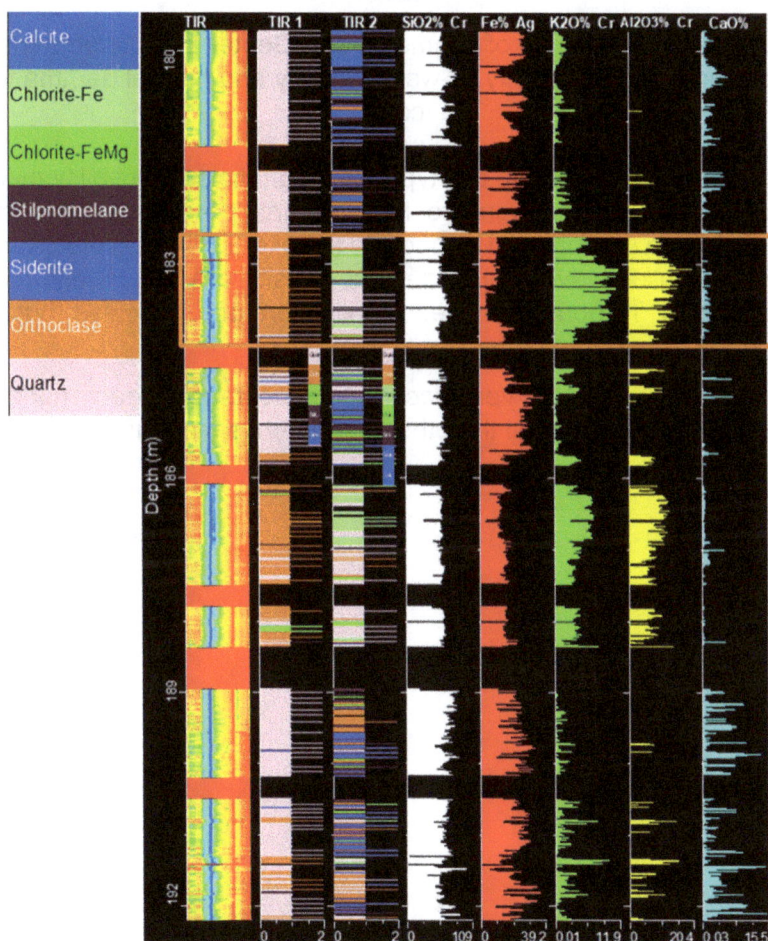

FIG 6 – HyLogger™ 3 TIR spectra and Minalyze chemical composition for a selected area of a diamond core.

XCT shows that four segmented (SEG) textures were identified, including SEG-1 with magnetite, SEG-2 with magnetite and chert and SEG-3 and SEG-4 with various percentages of silicate and carbonate (Figure 7). The XCT-calculated segments were combined and compared with Geotek-calculated density, magnetic susceptibility (loop and point), resistivity, P- wave velocity and natural gamma.

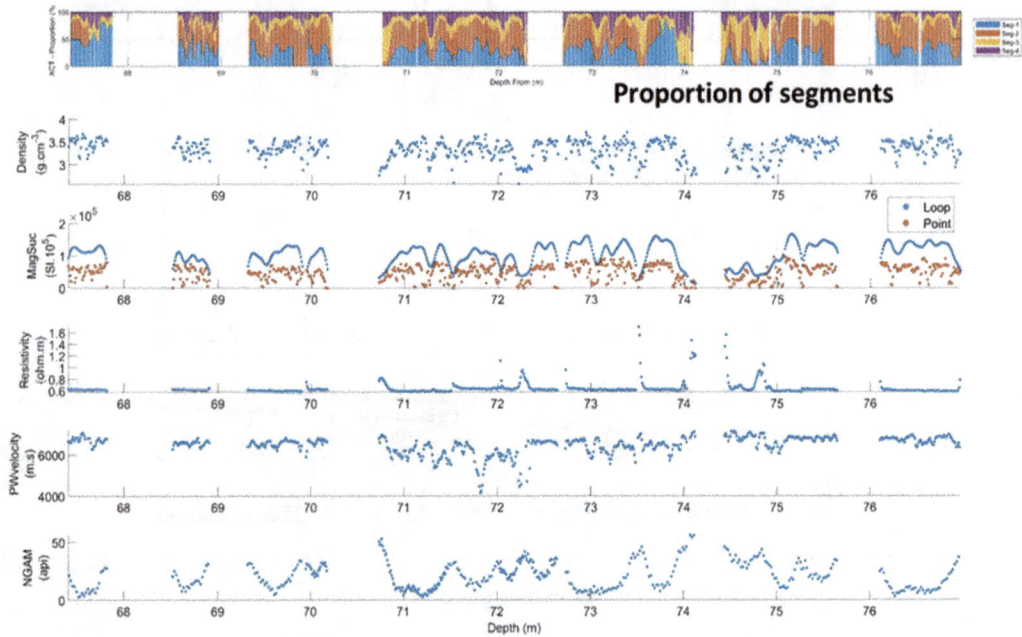

Proportion of segments

FIG 7 – XCT-calculated proportion of segments and Geotek-calculated density, magnetic susceptibility (loop and point), resistivity, P-wave velocity and natural gamma for a IB diamond core.

The Geotek results show that density and two types of magnetic susceptibility measurements (Point and Loop) are ideal for detecting magnetite location and percentage. Resistivity is positively correlated with stilpnomelane (strong) and orthoclase (weaker), P wave velocity in combination with density provides domaining, P wave amplitude and density are negatively correlated and Total Natural Y (K per cent and Th ppm) provides localisation of orthoclase but not so much for stilpnomelane.

The TIMA mineralogical maps offer a very accurate percentage of each mineral (Figure 8) and provide an effective way of calibrating all the other instrument's measurements.

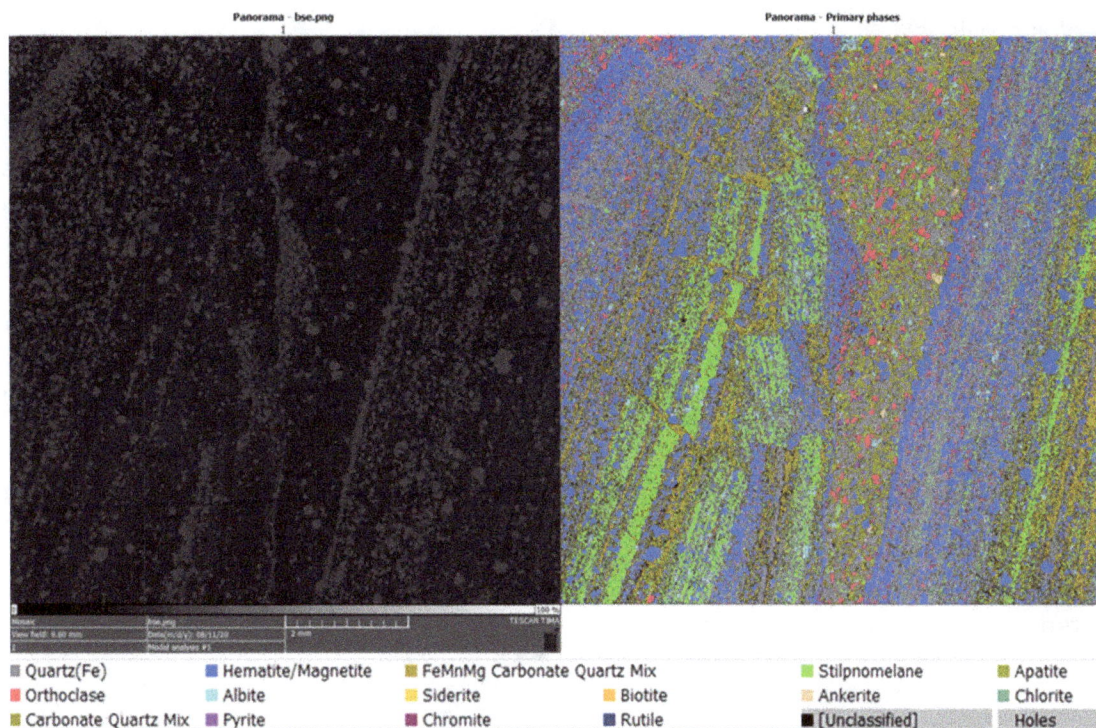

Quartz(Fe)	Hematite/Magnetite	FeMnMg Carbonate Quartz Mix	Stilpnomelane	Apatite
Orthoclase	Albite	Siderite	Biotite	Ankerite	Chlorite
Carbonate Quartz Mix	Pyrite	Chromite	Rutile	[Unclassified]	Holes

FIG 8 – TIMA backscattered image (left) and quantified mineralogy (right) of a polished section.

A summary of each technique's characteristics in Table 2 includes the core tray handling requirement, the speed of each method, the mineralogical, chemical and textural information provided and the measurement type. This offers a fit for purpose practical ranking for the characterisation of magnetite deposits.

TABLE 2

Summary of the characteristics of the instruments used in this study.

Instrument	Sample Handling	Speed	Mineralogy	Mag. Susc. or Mgt%	Density SG	Chemistry	Texture	Measurement Type
Minalyzer CS5 (Cr Tube)	On tray	50–60 m/day	-	-	Y	Y	-	Continuous/Surface
HyLogger™ 3 VNIR SWIR-TIR	On tray	130 m/day	Y	Y	-	-	-	Continuous/Surface
XCT	Transfer	12 m/day	-	-	Y	-	Y	Continuous/Volume
Geotek	Transfer	4 m/day	-	Y	Y	Y		Point/Volume/Surface
XRD	Powder	37 min	Y	Y	-	-	-	Point/Volume
XRF	Powder	N/A	-	-	-	Y	-	Point/Volume
TIMA	Polished sections	4 hrs	Y	Y		Y	Y	Surface

CONCLUSIONS

For large volumes of core, limited handling, speed, quantitative mineralogy and chemistry, hardness and density, the combination of HyLogger™ 3 (VNIR-SWIR and TIR) and Minalyzer at a scale of 1 m is optimum. The XCT is the only technique that provides full 3D, textural information and density variation at a 100 µm resolution. XCT scanning speed is fast and the subsequent data processing requires extensive skills. However, the processing can be automated once a workflow has been developed.

The Geotek provides:

- density and two types of magnetic susceptibility measurements (point and loop) ideal for detecting magnetite location and weight per cent
- resistivity, which positively correlated with stilpnomelane (strong) and orthoclase (weaker)
- P-wave velocity, which in combination with density provides domaining
- P-wave amplitude and density are negatively correlated
- Total natural γ (in particular K per cent and Th ppm) provides localisation of orthoclase (not so much stilpnomelane).

However, Geotek measurements can be time consuming and should be recommended only in specific domains already selected by the HyLogger and Minalyze analyses.

The mineralogical quantitative maps from the TIMA have provided a detailed mineralogy of the Iron Bridge deposit. The mineralogy includes 'classical' BIF mineralogy, such as quartz, magnetite, chlorite, stilpnomelane and various carbonates. However, K- feldspars (microcline/orthoclase) were more ubiquitous than initially thought. K-felspars are quite hard (Mohs scale of 6) and this should be considered during the processing of the material.

ACKNOWLEDGEMENTS

The authors thank Fortescue Metals Group Ltd for their permission to publish the results of this study.

REFERENECES

Van Kranendonk, M J, 2000. Geology of the North Shaw 1:100 000 sheet: *Western Australia Geological Survey*, 1:100 000 Geological Series Explanatory Notes.

Kisoro-Kabale iron ore project in south-western Uganda – unlocking a new high-grade iron ore region for exploration and development

L White[1], J Natukunda[2], E Tata[3], D C Ilouga[4], B Vietnieks[5], N Widmer[6] and J P Tyler[7]

1. Principal Mining/Mechanical Engineer, Kalem Group Pty Ltd, Subiaco WA 6008. Email: lee@kalemgroup.com
2. Ag. Assistant Commissioner, Directorate of Geological Survey and Mines, Entebbe, Republic of Uganda. Email: jfnatukunda@minerals.go.ug
3. Senior Exploration/Project Geologist, Kalem Group Pty Ltd, Subiaco WA 6008. Email: enerst.tata@kalemgroup.com
4. Senior Exploration/Project Geologist, Kalem Group Pty Ltd, Subiaco WA 6008. Email: charles.ilouga@kalemgroup.com
5. Managing Director, ADT Africa Limited, Kampala, Republic of Uganda. Email: ben@adtafrica.com
6. Director, ADT Africa Limited, Kampala, Republic of Uganda. Email: nathan@adtafrica.com
7. Principal Environmental Specialist, Environment Plus Pty Ltd, Subiaco WA 6008. Email: jim@environment-plus.com

ABSTRACT

The Republic of Uganda aims to exploit its' significant, un-developed iron ore resources to supply the required raw material for developing its' iron and steel industry across all the stages of the supply chain. Large prospects of iron ore exists in Uganda in the eastern (Tororo) regions and south-western (Kisoro-Kabale) regions of the country. The Kisoro-Kabale iron ore projects covers mineralisation in hematitic banded iron formation (itabirite) in the north-eastern Congo-Tanzania Craton; covering ten Iron ore prospects. The prospects include Kamena, Karukara, Nyamiringa, Kyanyamuzinda, Rugando, Katagata, Rwengongo, Kijuguta, Kihumuro and Katuna.

High-grade hematite ore prospects in south-western Uganda have known grades ranging from 55 per cent to 68 per cent iron ore content with an exploration target estimated at over 500 million tonne (Mt). Small scale production of iron ore from the south-western regions provide raw material to existing steelmaking and direct reduction iron ore plants located in the Jinja and Iganga districts of eastern Uganda. Large quantities of iron ore and other steelmaking raw material, however, are required to be imported into Uganda due to the limited exploration and development of their own iron ore mining operations.

A joint collaboration between the European Union (EU), the Department of the Geological Survey and Mines (DGSM), Ugandan Ministry of Mines in partnership with Australian drilling and mining consultancies, ADT Africa and Kalem Group as a EU funded project was undertaken during November 2022 and September 2023. Exploration and evaluation of the ten iron ore prospects in south-western Uganda was undertaken through geological mapping, drilling, assaying, and technical studies with the view to attract investment for the development of Uganda iron ore mining and steelmaking industry.

The collaboration developed mineral exploration policy, guidelines and standards, an iron ore exploration manual, a geological data management policy, and to support the sustainable development of iron ore mining in Uganda through building the capacity of local Ugandan geology and mining personnel through the transfer of knowledge and expertise from experienced iron ore experts.

The quantification of these resources is an important first step in assessing their economic potential and attracting mining investors.

This paper presents an overview of the investigation program and the preliminary results as an important first step in attracting mining investors to this exciting new iron ore province.

INTRODUCTION

Uganda's National Planning Authority (NPA) considers the iron and steel industry as one of the country's economic lifeline industries that will enable Uganda to achieve its vision of becoming an upper middle income country by 2040 (https://www.ugandainvest.go.ug/uganda-has-huge-investment-opportunities-in-iron-and-steel-sector/). According to the NPA the 2018 installed capacity of Uganda's Iron and Steel Plants was 1 000 000 tonnes per annum (tpa) with only 50.17 per cent (ie 501 700 t) utilised. Ugandan scrap iron in combination with internally produced sponge-iron (direct reduced iron) produced from Ugandan iron ore accounted for 32.89 per cent of the feed material with 67.11 per cent imported. Ugandan iron ore accounted for 10 per cent of feed materials (ie 16 500 t) despite an estimated 500 Mt of potentially available iron ore deposits. Uganda's iron ore deposits are considered as having the potential to meet the demands of an integrated iron and steel value chain ecosystem in Uganda. They in addition conclude that a developed iron and steel industry will greatly increase the current estimated 10 000 Ugandans directly employed in the sector. A similar conclusion was reached by the authors of 'Options for Improvement of the Ugandan Iron and Steel Industry' (Senfuka et al, 2011). This review concluded that the iron and steel industry in Uganda was constrained by poor capacity and a shortage of scrap iron. It also concluded that harnessing existing raw materials and increasing government significance in the exploitation of mineral resources deposits in Uganda were aspects that the iron and steel industry needed to address urgently.

This is being enabled by the European Union's decision to fund an assessment of iron ore resources outcropping in the Kisoro-Kabale region of south-western Uganda and they selected the Ugandan Department of the Geological Survey and Mines (DGSM), ADT Africa and the Kalem Group.

The scope of work includes:

- Quantification of iron ore resources in the outcropping Kamena, Karukara, Nyamiringa, Kyanyamuzinda, Rugando, Katagata, Rwengongo, Kijuguta, Kihumuro and Katuna deposits on the Kisoro-Kabale region of south-western Uganda (refer Figures 1 and 2).

- The development of an iron ore exploration manual for DGSM including mineral exploration policy, technical guidelines and geological data management.

- The transfer of knowledge and expertise to DGSM from experienced iron ore experts.

FIG 1 – Uganda iron ore prospects and steel refinery location map.

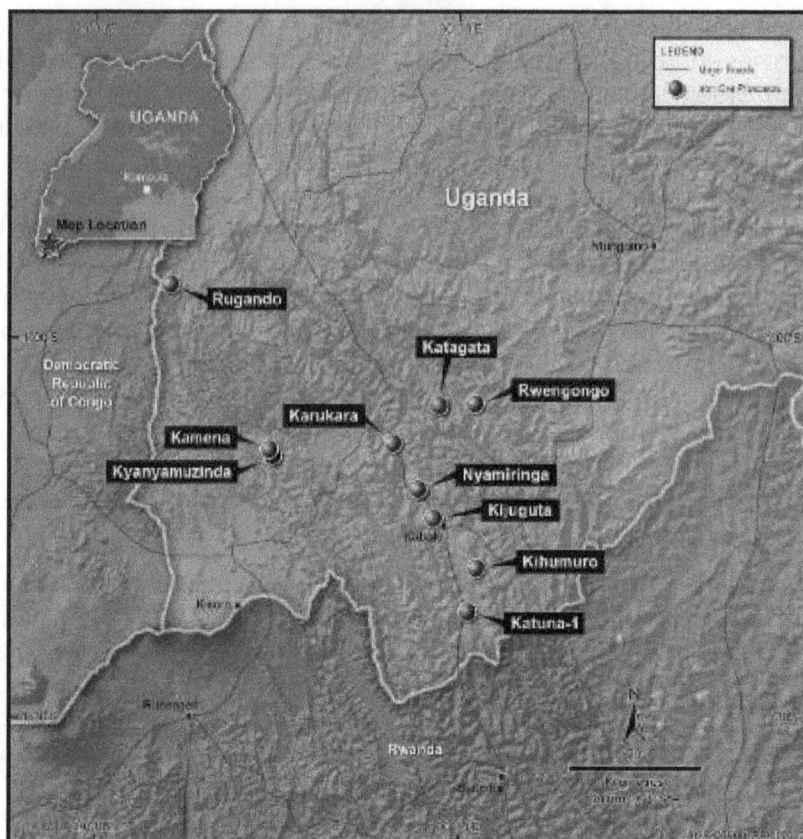

FIG 2 – Iron ore prospects south-western Uganda.

Existing knowledge

Uganda's Investment Authority states that Uganda has confirmed iron ore deposits in commercial quantities, estimated at over 500 Mt, based on the limited surveys so far undertaken. In 2014/2015, the Government of Uganda supported aerial and geological surveys which confirmed over 200 Mt (metric) of iron ore deposits in the Kigezi (ie Kabale) region of south-western Uganda alone, with huge prospects for more discoveries (https://www.ugandainvest.go.ug/uganda-has-huge-investment-opportunities-in-iron-and-steel-sector/). They conclude, however, that the surveys to date have not been extensive and further exploration is needed to determine the real extent of the iron ore deposits. Identified iron ore deposits (refer grey circles in Figure 1) largely centre on the Kabaran Rocks (ie metasediments and plutonic rock) in the south-western corner of Uganda (refer blue zone in Figure 2). The prospects for iron ore in Uganda were reviewed by Abraham *et al* (2020). This paper states that Uganda imports USD 369M worth of iron and steel products annually of which 60 per cent are raw materials for the steel processing plants. They completed a reconnaissance field survey which confirmed the existence of high-quality hematite iron ore (55 per cent to 68 per cent Fe) deposits that appear of suitable quality for iron production and in volumes sufficient to support a steel industry but they require detailed quantification and characterisation and this needs to be prioritised by the government.

REGIONAL SETTING

The geology of Uganda spans more than three billion years. It comprises Archaean lithospheric fragments, welded together, intersected or surrounded by Proterozoic fold belts. These fold belts can be related to the Eburnian (2.20–1.85 Ga), Grenvillean (1.10–0.95 Ga) and Pan-African (0.75–0.50 Ga) Orogenic Cycles (Westerhof *et al,* 2014). The south-western Uganda region shows evidence of such folding and refolding of its various rock formations. The prospects are seen to follow a regional tectonic framework characterised by a particular geodynamic episode in the evolution of this part of Africa which belongs to the northern western part of the Tanzanian Craton. Generally, the area is underlain by a weakly metamorphosed sequence of argillic and arenaceous rocks of the Kibaran System (Karagwe-Ankole rocks; Clark, 1963).

Previous mapping of the iron ore deposits in south-western Uganda indicates that there is an occurrence of high-grade (55–68 per cent Fe) ore that can easily be processed (Katto, 1997). In Uganda, iron ore occurs abundantly mainly in two areas: hematite iron ore found in Muko in Kabale and Kisoro districts of south-western Uganda and magnetite iron ore in Sukulu and Bukusu in Tororo District in eastern Uganda. Hematite iron ore is also known to occur at Mugabuzi, in Sembabule District in central Uganda (Rupiny, 2021).

The iron ore deposits in south-western Uganda are comprised of a series of hematite beds interbedded conformably with low-grade metamorphosed sedimentary horizons of pink/grey phyllites that are believed to be of mudstone, shale and sandy origin, which points towards a sedimentary source of the deposits. Banding and vuggy quartz have been observed on the Kihimuro outcrops (Figure 3) suggesting a banded iron formation (BIF) type deposit. The deposit is comprised of a series of thin beds ranging from 5 m to about 100 m wide, with a general north-west to south-east (NW-SE) strike direction and dipping at 30°SW to almost vertical.

Geologically, the mapped areas are underlain by a low-grade metamorphosed sequence of argillic and arenaceous rocks of the Kibaran System (Karagwe-Ankole rocks). The argillic rocks consist of mudstones, shales and phyllites while siltstones, sandstones, quartzites and arkoses make up the arenaceous group. The phyllites show a foliation and the commonest type of phyllites is the grey, sandy variety, which weathers to a very fine-grained silty clay and the second variety weathers to plastic finely banded red, yellow and grey clay. Mineralisation is mainly hematite which is usually metallic grey in colour and highly foliated attaining the specular texture. Like the enclosing country rocks, the hematite horizons are also marked by steep angles of dip. Occasionally, fine to medium grain octahedral magnetic crystals are found mixed with hematite.

According to data from the DGSM, the country's iron ore resources are estimated to be about 580 Mt, (Table 1) and they majorly exist in the south-western and eastern part of the country, (Figure 3).

TABLE 1

Uganda iron ore occurrences.

Type of iron ore	Location	Region in Uganda	Quantity (Mt)	Fe content
Hematite (Fe_2O_3)	Buhara		205	68
	Rubuguri		55	69
	Rugando	south-western	65	67
	Muko		155	68
	Mugabuzi Hill		12	60
	Nyaituma		5	60
Magnetite (Fe_3O_4)	Bakusu, Nakhupa, Nangawale, Surumbusa	Eastern Uganda	41	50
	Tororo		45	60
TOTAL			583	65.87

Source: Directorate of Geological Survey and Mines, 2018 as cited in Abraham *et al* (2020).

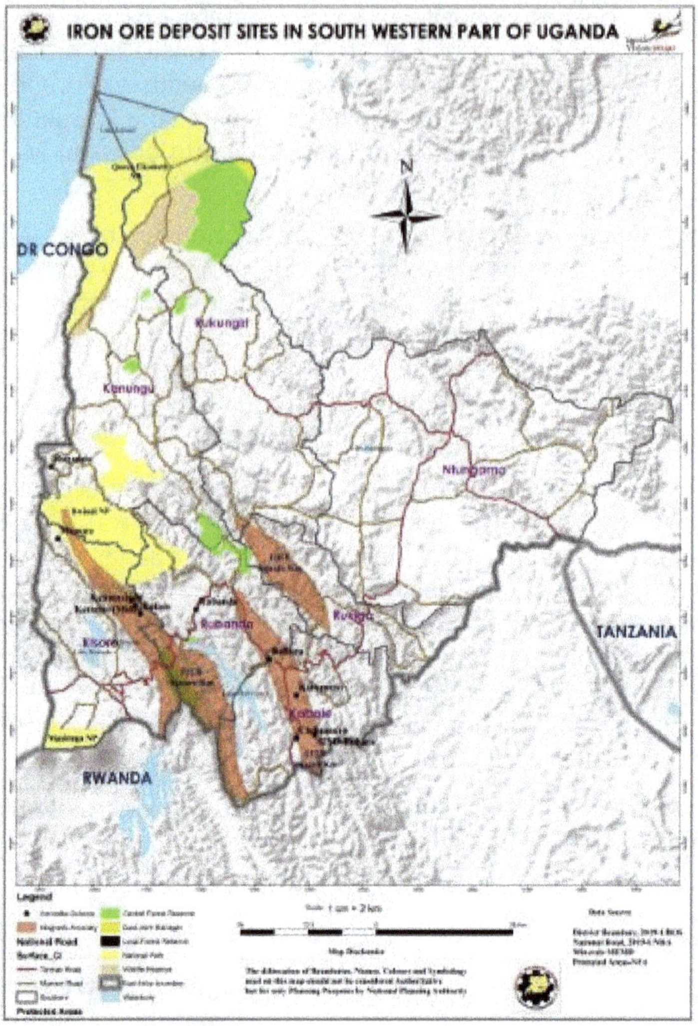

FIG 3 – DEM overlain by NW-SE trending iron ore prospects (Source: National Planning Authority).

DEPOSIT GEOLOGY

Geomorphology

Geomorphologically, south-western Uganda is characterised by numerous rolling hills and incised valleys. The topography of Kigezi highlands is similar throughout the landscape comprising mainly extensive flat-topped ridges and hills, broken by short numerous steep-sided deep subsidiary strike valleys separated by fluted spurs, usually 3–6 km. The topography is extremely rugged, consisting of narrow steep convex (20°–45°) and gentle (10°–15°) slopes. The landscape has steep slopes and deep narrow valleys with an amplitude of 600–700 m or greater with gentle pediment slopes.

Geology

The geology is composed of sedimentary rock system of Precambrian age. They are collectively grouped into the Karagwe-Ankolean system, underlain by metamorphic gneiss and granite intrusions, as well as shale and phyllite, which have given rise to clay deposits. The Kigezi highlands, which are largely non-volcanic in nature, are covered by non-volcanic soil properties.

Climate and vegetation

The climate of Kigezi highlands is warm to cool humid characterised by a bimodal rainfall pattern with annual rainfall of 1092 mm, which can be classified as moderate. Rainfall, however, increases to 1250–1540 mm or more in high-altitude areas of greater than 2000 m above sea level. The main rainfall seasons are from mid-February to May with a peak in March–April, and September to December with a peak in October/November. The vegetation cover of these highlands was until about a century ago characterised by montane forests. Human interference has led to serious degradation and in some cases depletion of vegetation cover. Presently, the common vegetation cover in the highlands is characterised by eucalyptus trees, pines, shrubs and thickets.

HOST AND ORE ROCKS CHARACTERISTICS

The various rocks identified in the prospects are described systematically in the sections below. These rocks are however, ubiquitous and found in almost all prospects within the area. These rocks range from isoclinal folds to open folds and in some cases show signs of multiple tectonic episodes with principal stress fields-oriented east–west. Common rocks encountered in close proximity to and in contact with the iron ore include: sandstone, siltstone, shale and phyllite.

Kihumuro prospect

This prospect lies between UTM 167600 to 169200 m and 9855100 to 9858000 m in Zone 36S. It is oriented in the NNW–SSE direction (Figure 4). It is elongated and bulges at the central part characterised by BIF-like deposit. The iron ore occurs as large boulders in this central part of the deposit and is also exposed in small cliffs about 1.5 m high at the eastern margin. These large boulders of low-grade iron ore are *in situ* and exposed by weathering and artisanal mining activities (Figure 5).

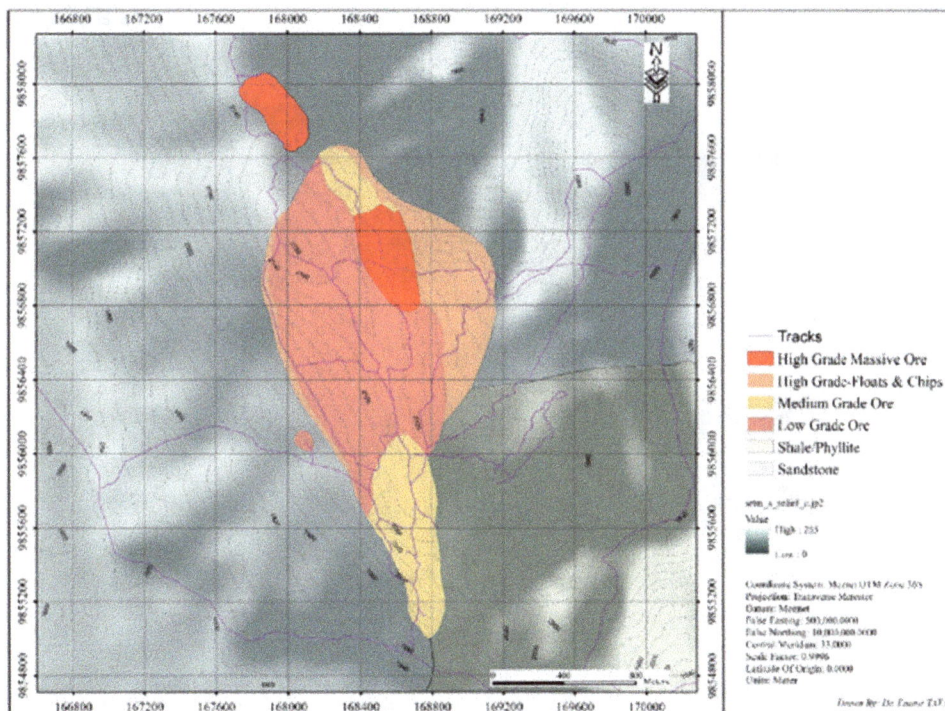

FIG 4 – Location of Kihumuro prospect.

FIG 5 – (a) Large boulders of low-grade ore exposed by artisanal mining activities in the background; (b) 1.5 m cliff of low-grade ore exposed in the eastern part of the orebody.

The surrounding rocks and orebody tend to dip steeply (> 80°) to the west in the southernmost part. It is sandwiched between red sandstones to the west, north and north-east while in the east and south-east are grey shales. The prospect is made up of three main iron ore groups characterised by field observation and grab sample description. These are:

1. *Low-grade ore*: the low-grade ore is characterised by the presence of discontinuous bands and stringers of quartz. They appear to be partially enriched BIF which occupy the central portion of the ridge. Some samples present pockets of quartz which may have formed from recrystallisation during periods of low-grade metamorphism. Some of the samples look whitish as a result of the abundance of quartz (Figure 6).

2. *Medium-grade ore:* The medium grade or is found at the south end of the prospect and towards the northern end. This ore is dark to bluish in colour and in some cases, it is whitish as a result of the presence of quartz. It is mostly shiny in lustre as a metallic appearance is imparted on the samples interpreted to have resulted from pressure during low-grade metamorphic conditions. Most often foliated structures and textures are visible in the grab samples from this domain (Figure 7).

3. *High-grade ore:* The high-grade ore occurs towards the northern end of the prospect. It occupies a smaller surface area north of the main body. A smaller concentration of high-grade ore is seen within the vicinity of the army barracks. High-grade ore occurs as floats of enriched hematite, bluish in colour and foliated. It is dominated by specularite (Figure 8).

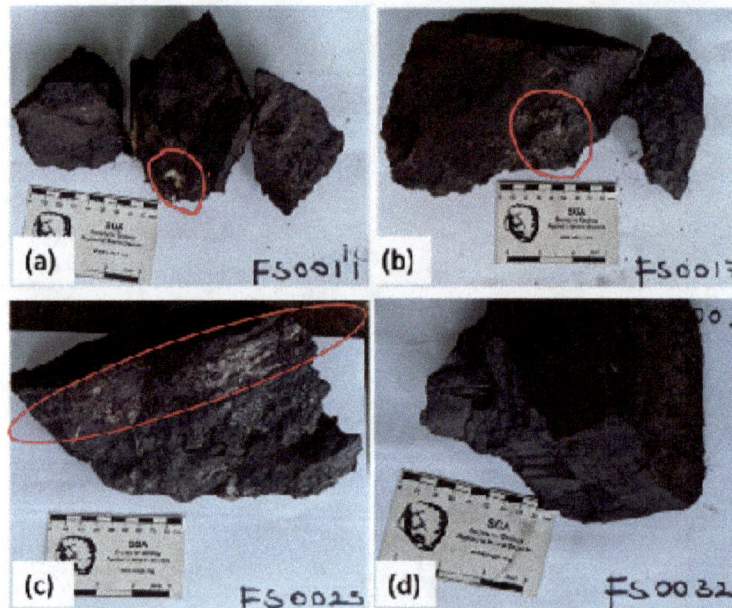

FIG 6 – Samples of 'low-grade' iron ore from Kihumuro prospect: (a) pockets of quartz in sample – red circle; (b) friable quartz – red circle; (c) discontinuous bands of quartz; (d) whitish grey ore sample.

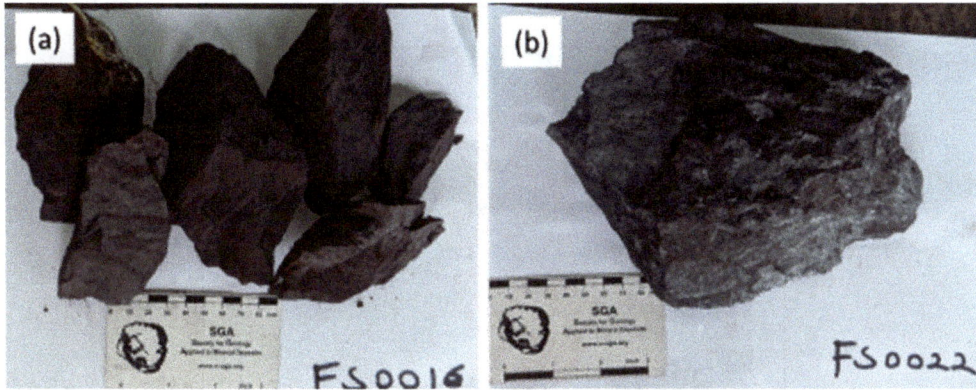

FIG 7 – (a) Dark brown partially oxidised medium grade ore; and (b) foliated bluish grey medium grade ore.

FIG 8 – (a) High-grade specularite; (b) High-grade massive ore and specularite from the same outcrop.

In the eastern margins of the orebody, there is an abundance of high-grade float and chips mixed with weathered materials or ferricrete. In some instances, it appears to form canga.

Katuna 1 prospect

The Katuna 1 prospect lies within Zone 36S and is south of the Kihumuro prospect. It is basically a north–south elongated body which has a width of about 200 m at its northern end and gradually tapers to less than 100 m at the southern end. The Katuna 1 deposit, which extends between 167300–167650 m east and 9848640–9849500 m north, has a strike length of about 800 m and dips very steeply to the west (Figure 9).

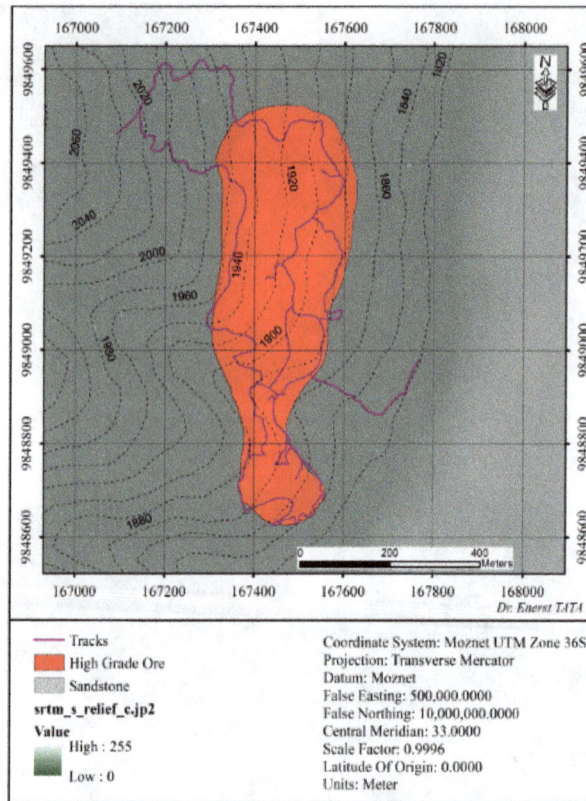

FIG 9 – Location and geology of Katuna 1 prospect.

The Katuna 1 prospect is characterised by high-grade hematite ore. The ore is massive and bluish in colour with some samples showing microplaty textures. Other massive ore types show vuggy textures probably as a result of the leaching of interstitial quartz. In some outcrops specularite textures are well developed and the ore becomes foliated (Figure 10). Within this deposit is oxidised hematite ore that is associated with duricrust and goethite. The orebody is sandwiched between sandstone layers and dips at greater than 80° to the west similar to the Kihumuro deposit to the north.

FIG 10 – (a) Massive blue hematite ore; (b) microplaty hematite sample; (c) specularite sample; and (d) massive vuggy bluish hematite.

Kamena prospect

The Kamena deposit is found at UTM Zone 35 south between 804600–805200 m east and 98772460–9873280 m north, forming a fold-shaped structure. This deposit is characterised by blue high-grade massive and specularite hematite. The ore shows evidence of tectonic folding which has caused visible structural inclination in the outcrops. The Kamena prospect is quite close to the Kyanyamuzinda prospect and would have formed a single deposit though separated by a sandstone band (Figure 11).

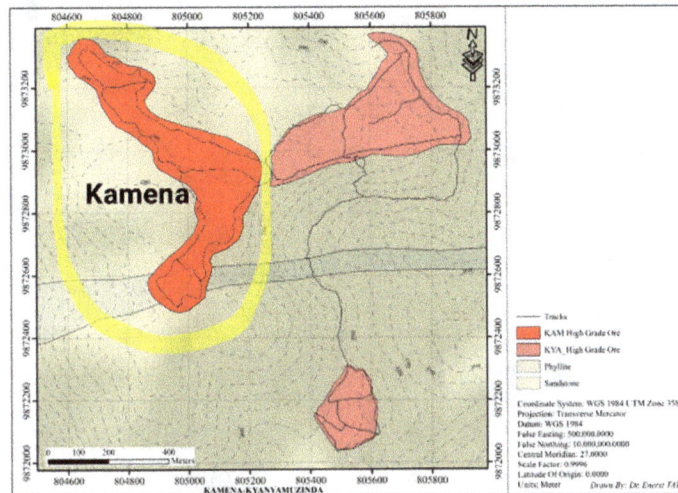

FIG 11 – Location of the Kamena iron ore prospect.

The Kamena ore samples display a variation of textures and colours resulting from oxidation. Most samples from the northern part of this deposit are massive, specularite type and occur as large outcrops. They display typical metallic lustre with minor pockets of friable quartz. Some of the outcrops have goethite stains resulting from oxidation of hematite. Massive specularite boulders occur as outcrops as well and show minor pockets of sugary quartz which are remnants of leaching. They are characteristically blue in colour typical of high-grade hematite (Figure 12).

FIG 12 – (a) Massive blue hematite ore with minor traces of leached quartz remnants; (b) specularite quartz with pockets of quartz; (c) typical Kamena high-grade specularite; (d) high-grade ore with goethite imprint from oxidation.

Massive outcrops of high-grade hematite are exposed at the central part of the prospect area. The southern portion of this deposit is densely habited with houses constructed on the orebody with visible high-grade outcrops (Figure 13).

FIG 13 – (a) massive specularite outcrop; (b) Outcrop of massive hematite ore with a NNW strike and an average dip of 65° towards the west; (c) High-grade outcrop within the habited area in Kamena; (d) Cliff of high-grade hematite ore at the centre of the deposit.

Kyanyamuzinda prospect

The Kyanyamuzinda deposit is in two parts separated by sandstones and phyllite. The two areas where high-grade hematite crops out are separated by 600 m of sandstones and phyllite. There is high-grade ore at zone (A) which is within 805400–805600 m east and 9872080–9872300 m north as well as high-grade massive and specularite ore in zone (B) which is within 805250–805900 m east and 9872900–9873380 m north in UTM 35S (Figure 14).

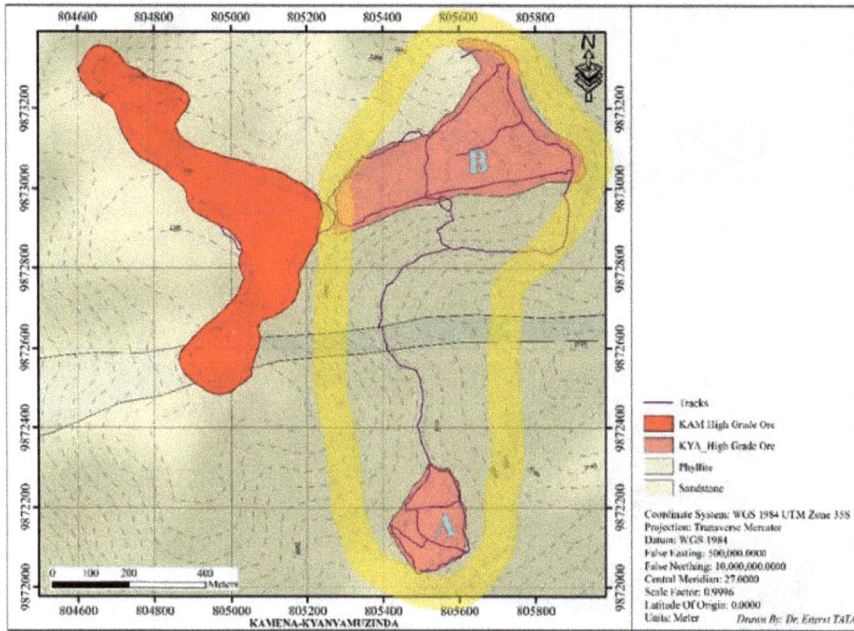

FIG 14 – Location of Kyanyamuzinda prospect zones A and B.

In zone A are massive outcrops of hematite with goethite-coated vuggy textures, specularite hematite textures and minor quartz as pockets or ribbons. Artisanal mining in the area has however destroyed some of the outcrops. Grab samples collected show variations in texture from blue to black in colour and specular to massive (Figure 15).

In zone B the ore occurs within thick layers of screes lying on top of the main body as well as exposed large outcrops of massive, specular hematite bodies. This massive ore goes up to 2500 m altitude from 2300 m. The grab ore samples are blue in colour with very little goethite imprints (Figure 16). There is a soil cover which may go up to 1 m thick in some portions of this zone.

Goethite is seen coating most of the outcrop ores while some appear as canga (fragments of high-grade hematite cemented by iron oxide; Figure 17).

FIG 15 – Zone A: (a) Massive hematite ore with pockets of remnant quartz; (b) High-grade hematite with stringers of quartz; (c) Massive specularite type hematite; (d) massive hematite with goethite coating.

FIG 16 – Zone B: (a) Massive blue hematite; (b) Specularite hematite with strong metallic lustre.

FIG 17 – (a) Scree layer capping massive hematite in north of zone B; (b) goethite coated massive boulders in zone A; (c) massive specularite boulders in zone B; (d) Massive blue Hematite boulder outcrop in zone A; and (e) Fragments of high-grade ore cemented by iron oxides in zone A (Brecciated ore).

Karukara prospect

The Karukara deposit is a narrow north–south deposit with a maximum outcrop with averaging 100 m width, and a length of 600 m separated by a sandstone band of about 20 m in the southern part of the outcrop. It is located on the slope of the hill and dips to the west at an angle greater than 80°. This deposit is located at UTM Zone 35S, 823800–823900 m east and 9873480–9873950 m north. It tends to narrow out in the V-shaped valley at 9873800 m north (Figure 18). It is completely sandwiched between sandstones.

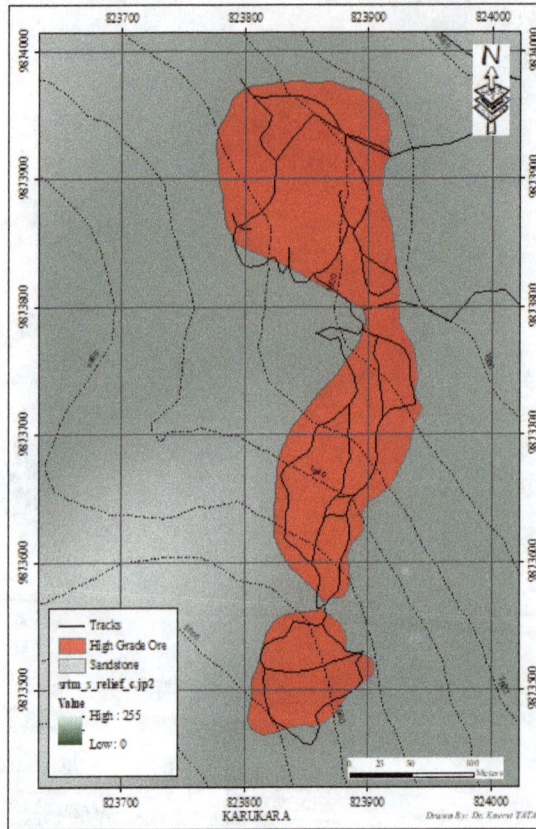

FIG 18 – Location of Karukara prospect.

Grab samples show extremely high-grade ore with minor quartz pockets or stringers as seen in the other deposits (Figure 19). Specularite hematite is also common with a strong metallic lustre.

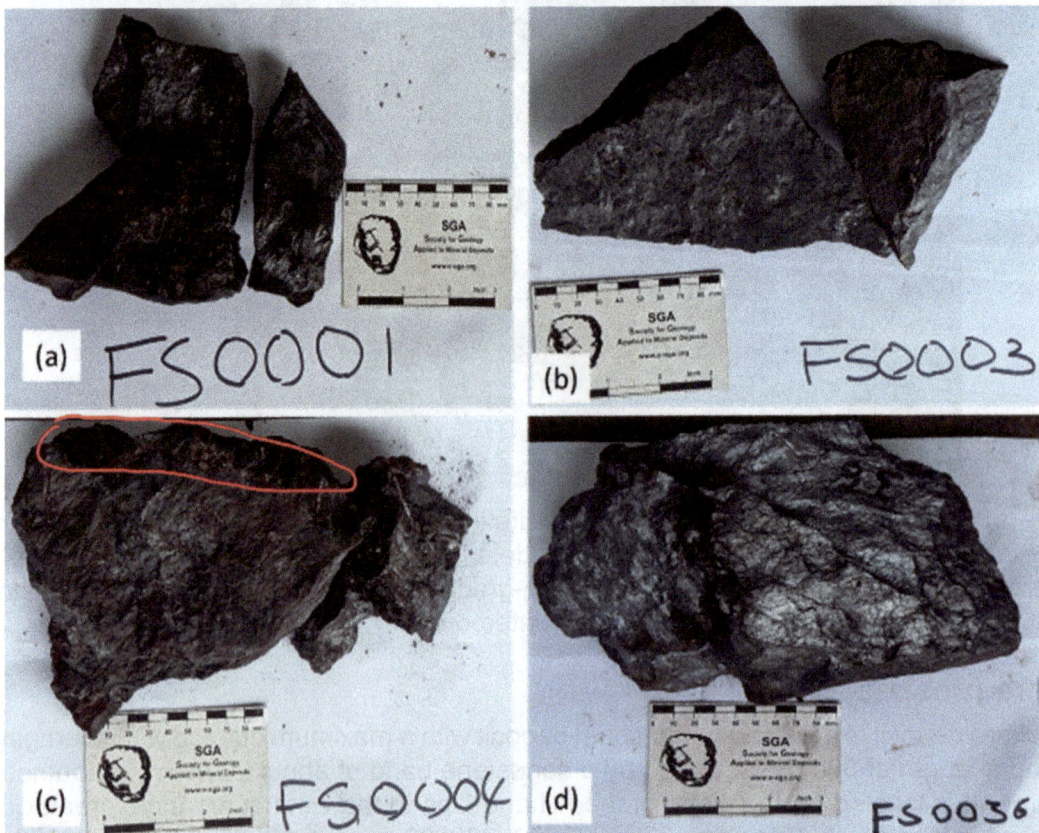

FIG 19 – (a) Massive hematite ore with blackish blue colour; (b) massive blue hematite ore; (c) black hematite ore with remnant quartz grains; and (d) blue specularite ore.

This deposit is characterised by high-grade hematite ore which is blue and massive. Large outcrops of boulders are seen cropping out both in the northern section and southern sections of the deposit. These outcrops are dispersed within the entire deposit area (Figure 20).

FIG 20 – (a) Massive outcrops of high-grade hematite ore; and (b) large boulder/outcrops of high-grade ore on the slopes of the Karukara prospect.

Kijuguta prospect

This deposit is located in UTM Zone 35 south and runs NNW–SSE and is a about 100–200 m thick and along strike for about 1 km (Figure 21). It is narrow at the northern end and slightly broader in the southern part. This deposit is in contact with sandstone in the west, shale in the north-east and quartz-vein-rich sandstone in the east. These rocks tend to dip steeply in the south-western direction.

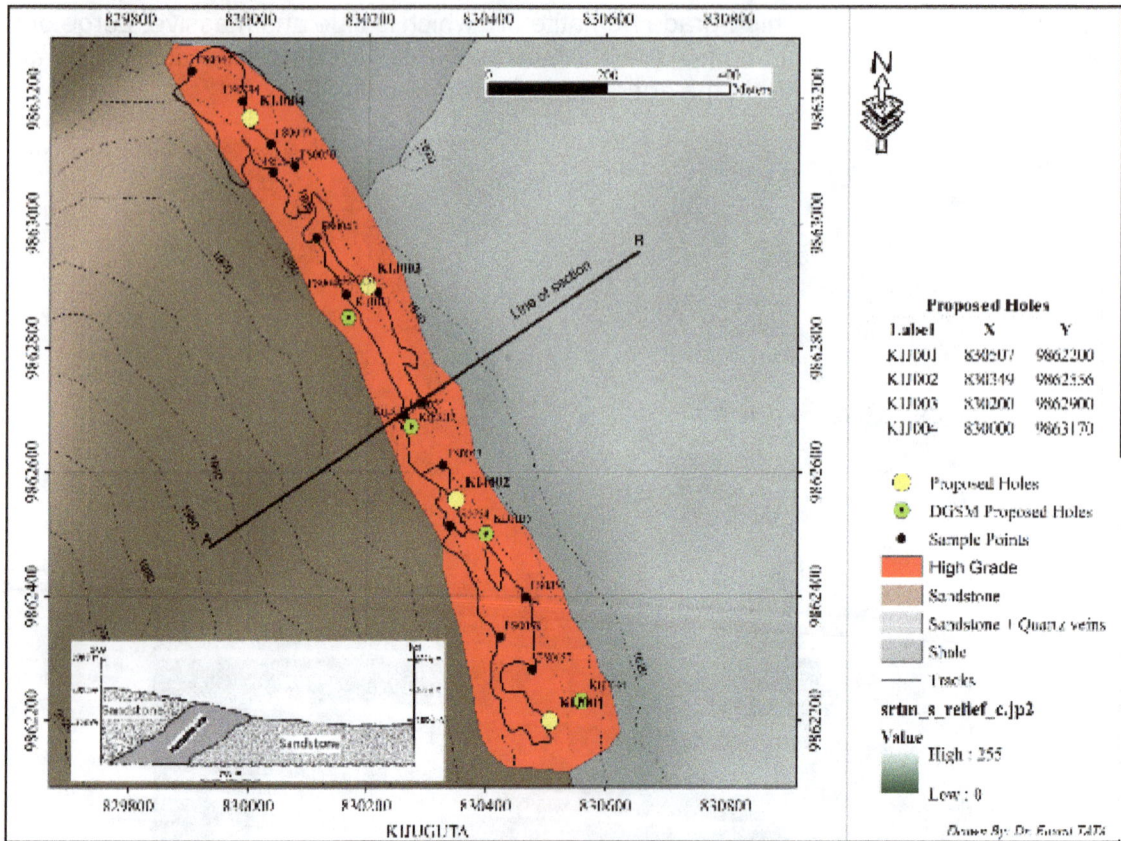

FIG 21 – Kijuguta geology and tracks.

It is composed of high-grade ore with quartz stringers and pockets of remnant quartz from leaching process. The high-grade ore in this deposit is also coated by goethite (Figure 22).

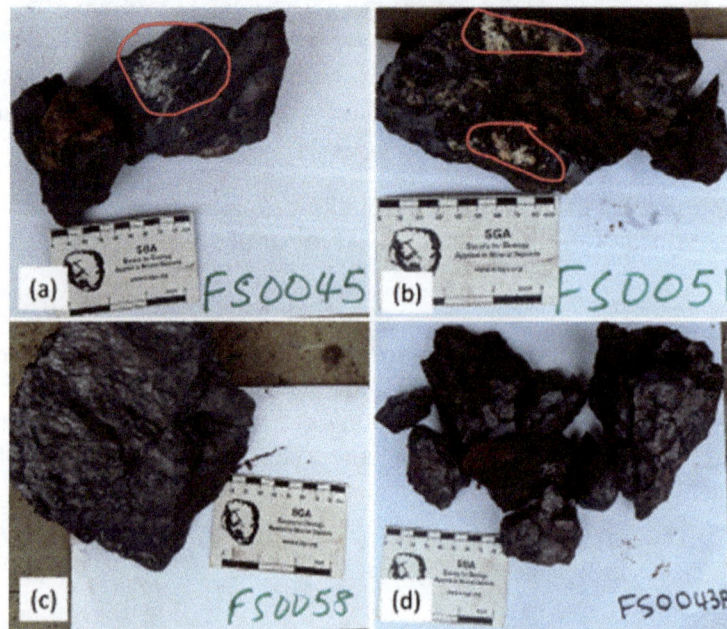

FIG 22 – (a) high-grade hematite sample with quartz; (b) massive ore with pockets of quartz; (c) massive blue hematite; and (d) botryoidal hematite indication iron remobilisation.

High-grade outcrops at Kijuguta are represented by large boulders 1.5 m below scree and soil. Artisanal mining has exposed large areas of scree indicating the extent of mineralisation. Sandstones at the base of the ore is being mined exposing structures that give an indication of the possible dip of the orebody (Figure 23).

FIG 23 – (a) Cliff of scree materials on outcrop of boulders; (b) the base of scree is represented by high-grade ore; (c) scree materials exposed by artisanal mining; and (d) mining of sandstone exposing the dip direction of the rocks in the area.

Nyamiringa prospect

The Nyamiringa prospect is an elongated body cropping-out in a NW–SE orientation. It is about 150 m in width and 700 m in length. It crops out at an average altitude of 2080 m. This deposit is sandwiched between sandstones to the east and phyllite to the west (Figure 24).

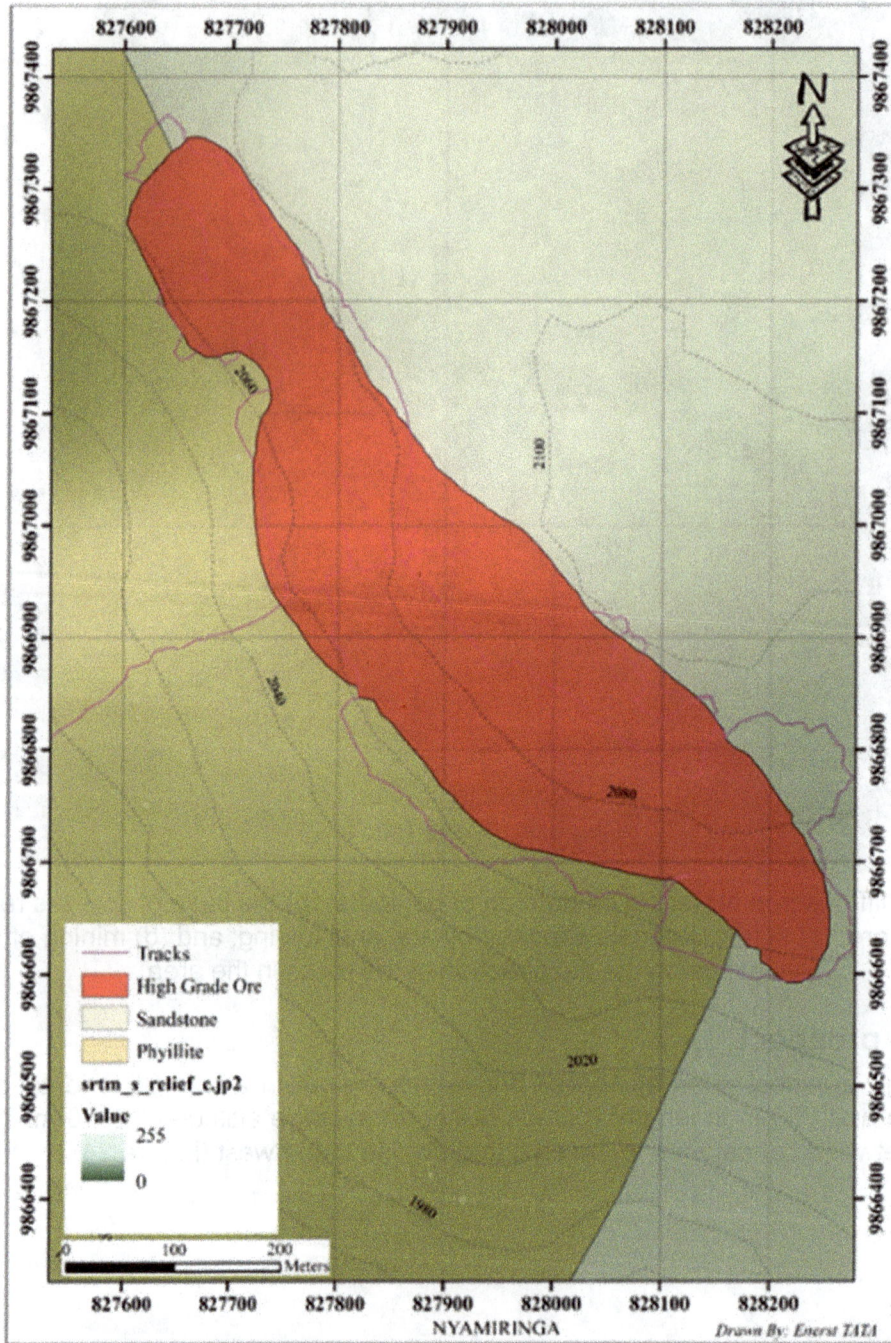

FIG 24 – Geology and orebody outcrop of the Nyamiringa prospect.

The grab samples are massive blue in colour. High-grade hematite has stringers and pockets of friable quartz which represent remnants of the leaching process. Some samples have been weathered and show coatings of goethite (Figure 25). Occasionally coarse to medium-grained magnetite is seen along with the hematite.

FIG 25 – (a) Goethite-stained high-grade samples; (b) minor quartz stringers in high-grade hematite ore; (c) high-grade specular hematite; and (d) friable quartz remnants in high-grade ore.

Rwengongo prospect

The Rwengongo prospect lies within Zone 36S. This deposit is characterised by goethite rich hematite ore. It is a low-grade deposit which is 150 m in width and 400 m long. It has a strike along the NW–SE direction (Figure 26). No outcrops are suitable for structural data collection. This deposit is at an average altitude of 1700 m. The samples recovered are dark brown to brown in colour with black spots. They show vuggy characteristics probably as a result of quartz leaching (Figure 27).

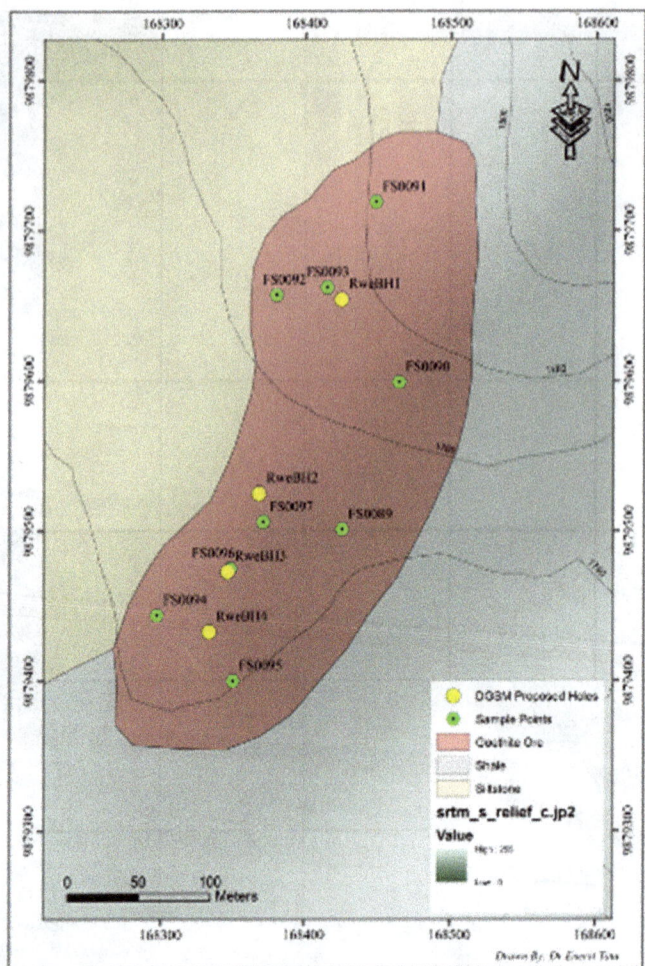

FIG 26 – Geology and location of Rwengongo prospect.

FIG 27 – Goethite samples from Rwengongo with: (a) yellowish brown stained by goethite ore; and (b) vuggy nature as a result of quartz leaching.

The area has sandstones and shales. Boulders of hematite-stained pegmatites are also encountered. The goethite crops out within siltstone and shale.

Katagata prospect

The Katagata prospect lies within UTM Zone 36S close to Rwengongo. It is characterised by goethite material and therefore of low-grade. It is sandwiched in phyllite characterised by goethitic veins (Figure 28).

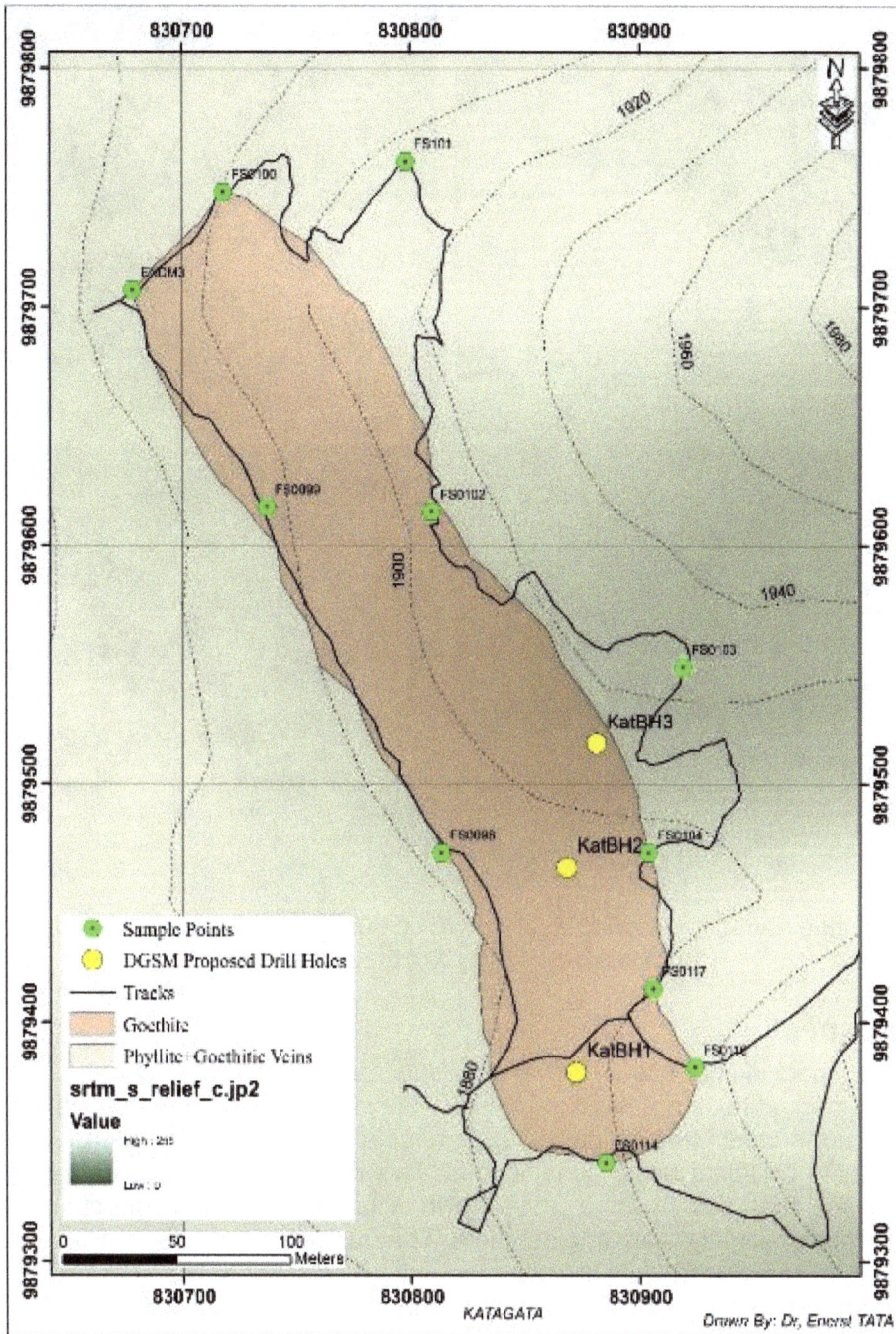

FIG 28 – Location and geology of the Katagata prospect.

The rocks are characterised by brownish colours stained by black hematitic material. Some of the grab samples are earthy with vuggy textures (Figure 29).

FIG 29 – Goethite samples showing dark brown colours with dark spots (a and c) and earthy colours with vuggy textures (b and d).

Rugando prospect

The Rugando Prospect is characterised by two hills located within UTM Zone 35S. This prospect is characterised by magnetite as opposed to the hematite in the other prospects within south-western Uganda. It has a north–south trend (Figure 30). The mineralisation extends to the summit of the hills with large magnetite boulders being conspicuous. This deposit appears to be veining which runs within the sandstone host rock. In some cases, north of the deposit, the magnetite is associated with bluish non-magnetic or slightly magnetic hematite. The magnetite samples are black in colour, blue and granular and, in some cases, specularite magnetite is encountered.

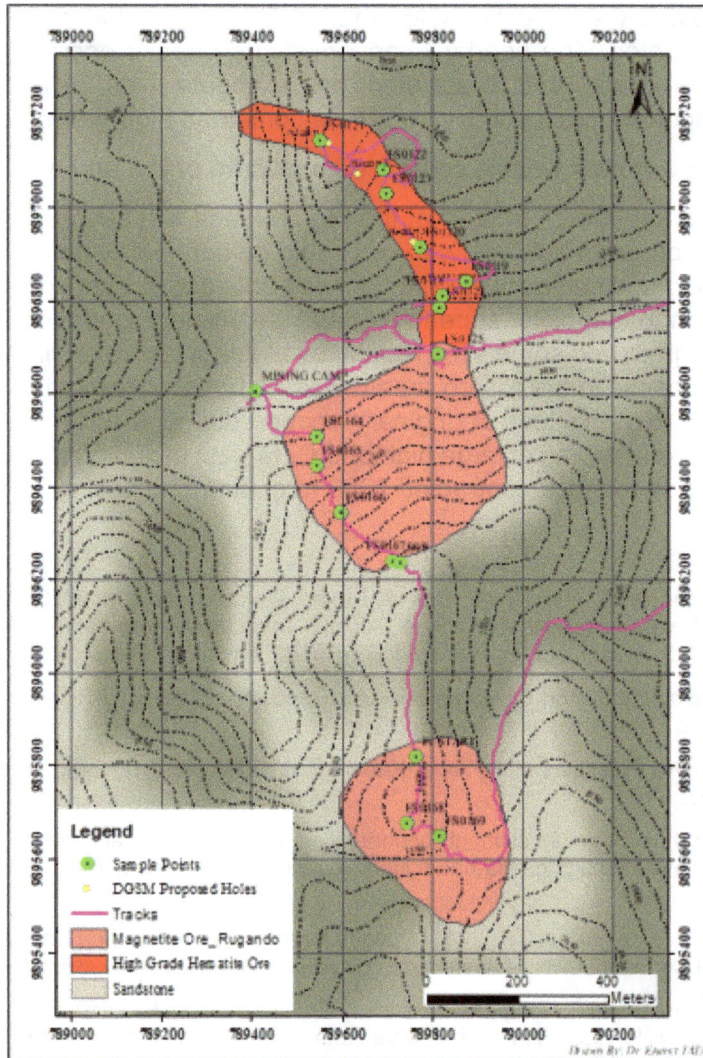

FIG 30 – Location and geology of the Rugando deposit.

CONCLUSION

The work program commenced with mapping the extent of surface mineralisation at each of the ten target sites. This work was undertaken by Dr Charles Ilouga and Dr Ernest Tata engaged by Kalem group. They were assisted by DGSM geologists capitalising on the opportunity to learn from highly experienced iron ore geologists. Dr Ilouga and Dr Tata initially worked together prior to assuming rotating rosters that ensured the presence of at least one expert iron ore geologist for the duration of the work program.

The geological survey team accompanied by representatives of DGSM and local police met with local landowners in each study area prior to commencing the non-intrusive mapping program. It was made clear to the landowners that the surveys would be non-intrusive and would enable optimum locations of drill holes to be determined. Once this was done, DGSM would return to finalise the drill hole locations, formally negotiate compensation and pay compensation to the landowners prior to the commencement of intrusive drilling.

The mapping of the orebodies was undertaken through a combination of GPS points, expert evaluation of surface rock and, the collection of representative rock samples with local land-owners engaged to help with the work program.

Mapping commenced on the Kataguta, Karukara and Kamena deposit ridgelines in the mountainous Kisoro area near the border of Rwanda and the Democratic Republic of Congo. These deposits largely comprised high-grade outcropping boulders of hematite but access to water for diamond drilling and particularly at Kamena was seen as a challenge. For these areas reverse circulation (RC). Drilling was considered more appropriate than diamond drilling for an initial assessment and in the case of Katugata it was questionable whether drilling the low-grade iron grade goethite mineralisation was warranted. The nearby Kyanyamuzinda deposit encroached on habited areas and included both high-grade and brecciated ore. Kijuguta was extensive, high-grade and considered to be an important prospect. Nyarmaringa contained high-grade scree and boulders.

The large Kihumuro deposit near Kabale comprised a mixture of low and high-grade hematite.

Mapping commenced in late March at start of the March–June wet season and heavy rain created challenges for the work program but most surveys could be completed on most of the deposits with a three to four day period and sufficient information was obtained to scope the drilling program. The Katagata and Rwengongo deposits appear to have insufficient good quality ore to justify an investigative drilling program and some sites lacking available water for a diamond drilling campaign including the Kamena deposit in the Kisoro area would be better investigated using a RC drill rig. It was proposed, however, that this loss of originally scoped diamond drill holes be compensated for with more intensive diamond drilling of large high priority targets including the Kijuguta and Kihumoro deposits with no change to the total contracted 2000 m of diamond drilling.

REFERENCES

Abraham, M J B, Gift, R, Hennery, S, Asuman, G, Rita, A, Dexter, D and Joseph, M, 2020. The Prospects of Uganda's Iron Ore Deposits in Developing the Iron and Steel Industry, *Journal of Minerals and Materials Characterization and Engineering*, 8:316–329. Available from: <https://doi.org/10.4236/jmmce.2020.84019>

Clark, D A C, 1963. A Preliminary Report on Some Iron Ore Deposits at Kashenyi Ridge, near Muko, Kigezi District, Geological Survey and Mines Department, Entebbe.

Katto, E, 1997. Butare Iron Ore Prospects Muko Area, Kabale District Southern Uganda. Geological Survey and Mines Department, Entebbe.

Rupiny, D, 2021. Uganda has huge investment opportunities in iron and steel sector [online]. Available from: <https://www.ugandainvest.go.ug/uganda-has-huge-investment-opportunities-in-iron-and-steel-sector/> [Accessed: 25 April 2023].

Senfuka, C, Kirabira, J B and Byaruhanga, J K, 2011. Options for Improvement of the Ugandan Iron and Steel Industry, Second International Conference on Advances in Engineering and Technology, Kampala.

Westerhof, A B, Härmä, P, Isabirye, E, Katto, E, Koistinen, T, Kuosmanen, E, Lehto, T, Lehtonen, M I, Mäkitie, H, Manninen, T, Mänttäri, I, Pekkala, Y, Pokki, J, Saalmann, K and Virransalo, P, 2014. Geology and Geodynamic Development of Uganda with Explanation of the 1:1,000,000 Scale Geological Map, Geological Survey of Finland, Special Paper 55, 387 p.

Greener downstream processing
– decarbonisation

How are Indian mills changing their feedstock in order to address decarbonisation goals?

D Goel[1]

1. CEO, BigMint Technologies Pvt Ltd, Raipur 492007 Chhattisgarh, India.
   Email: dhruv@steelmint.com

INDIAN STEELMAKING ROUTE-WISE

India, the second-largest steel manufacturer globally, achieved crude steel production of around 126 million tonnes (Mt) in the last financial year of April 2022 to March 2023 (FY23), against an installed capacity of 162 Mt. In its National Steel Policy, 2017, India has undertaken targets to take the crude steel production capacity to 300 Mt by 2030–2031 and production and demand to around 255 Mtpa, also by this deadline. The blast furnace-basic oxygen furnace (BF-BoF):electric arc furnace-induction furnace (EAF-IF) ratio in 2022 was at 44:56.

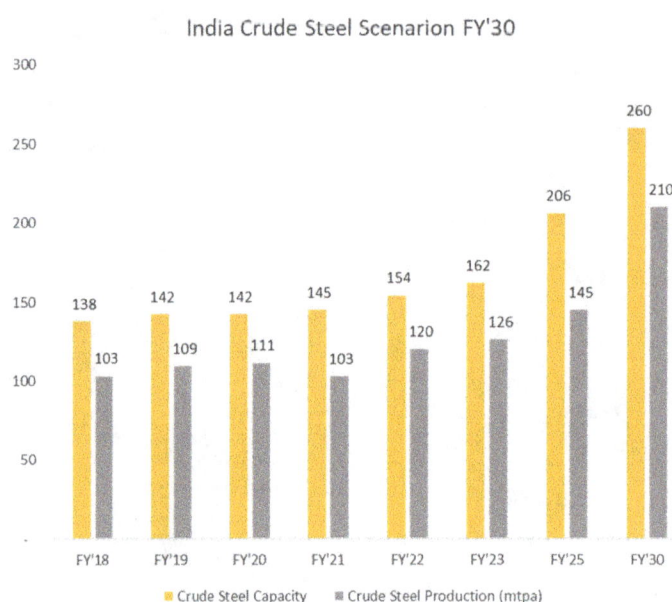

FIG 1 – India's crude steel capacity scenario by FY30.

UPCOMING STEEL CAPACITIES IN THE FORM OF BF-BOF

However, an overwhelming portion of the upcoming capacities is veered towards the blast furnace (BF) route despite the fact that mills globally are making serious efforts to transition to greener methods of steelmaking. Data maintained with SteelMint reveals that India's BF route, after dropping to its lowest point of less than 43 per cent in financial year 2011–2012, has, however, increased and has been steadily moved up post 2019–2020. There are, of course, a couple of reasons supporting BF's northward march. First, the capacities which are becoming operational since 2020 had been conceived much earlier, when decarbonisation had not gained such acceptance in India as a goal to be pursued actively. Secondly, India's steel demand by 2030 is expected to touch a substantial over-250 Mt. Mills need to achieve scalability in order to deliver on the national targets, which cannot be achieved through EAFs.

India has emerged as the fastest growing economy in the first quarter (January-March) of 2023. Its GDP grew 6.1 per cent against China's 4.5 per cent, the United States' 1.1 per cent, Japan's 1.6 per cent and Germany's -0.3 per cent. This growth, twinned with the strong infrastructure push of the government, is a pointer that the country's steel demand is likely to remain buoyant till 2030 and will warrant large volumes of supply.

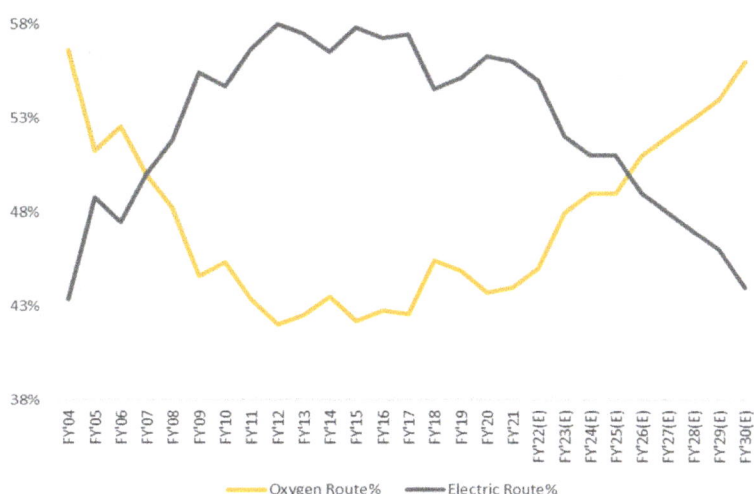

FIG 2 – India's crude steel production route-wise up to FY30.

INDIAN METALLIC MIX – CURRENT AND PROJECTED

The current metallic mix in India's crude steel production shows 78 Mt of hot metal was consumed to produce 120 Mt of crude steel in calendar year (CY) 2022, whereas DRI used was only 38 Mt and scrap, an even lower 26 Mt. But, going forward, SteelMint's data reveals that while the percentage of each will remain more or less the same in the total production, the maximum growth will happen in hot metal usage. For instance, the share of scrap in the 120 Mt of crude steel production in CY22 was 22 per cent. It is to remain stable at 22 per cent at 31 Mt in CY25, when crude steel production is expected to touch 145 Mt and climb nominally to 25 per cent in CY30, the year production will possibly hit 212 Mt. Similarly, the share of DRI will remain static at around 31 per cent over CY22–25 but dip to 25 per cent in CY30. Hot metal's share is indicated at 65–67 per cent till CY30. However, the author notices a quantum leap in the absolute share of hot metal from a moderate 78 Mt in CY22 to 143 Mt in CY30. Scrap usage will no doubt double in this period, but on a low base while DRI will show a rather toned-down 16 Mt increase in the period under review. This indicates, the share of hot metal or pig iron will increase significantly in the charge mix, going forward. DRI will not be able to make much headway because more than 70 per cent of the existing capacity is coal-based, which can add to the carbon footprint and thus must be discouraged. The Indian government is in the process of drafting policies to curb steel production through coal-based DRI. At present, the norms are still nebulous but state governments have already adopted a more proactive approach and are limiting future permits for such plants. Thus, prospects of expansion in coal-based DRI are limited till a breakthrough technology using hydrogen or any other green method is made available.

Where scrap is concerned, the government is making concerted efforts to increase scrap generation to lessen the dependency on imports. India's Vehicle Scrappage Policy, 2022 is an instance, where 15-year-old passenger and commercial vehicles are to be phased out to reduce pollution and also contribute to scrap generation. However, setting up scrappage centres along with other attendant infrastructure will take time and no immediate panacea is on the radar to take up generation to even 45 Mt by 2025. Thus, imports will have to fill in the gap.

India Crude Steel Metallic Mix in FY'30

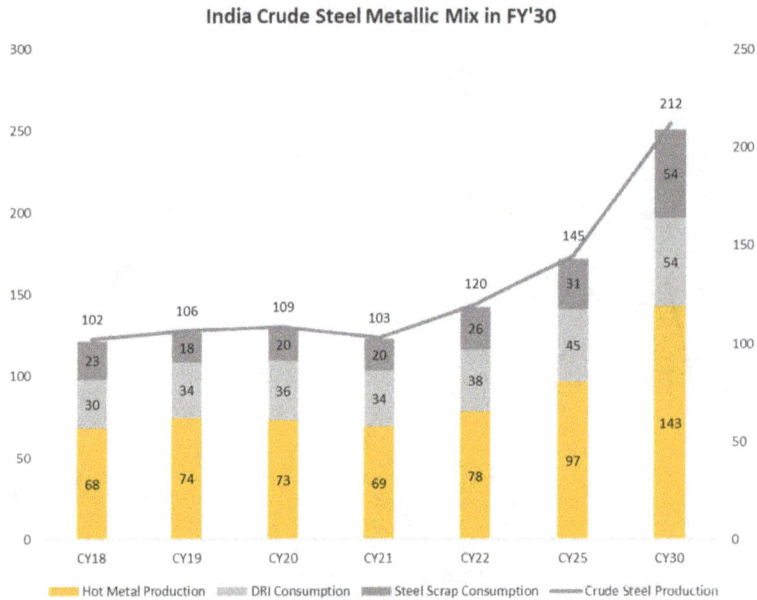

FIG 3 – Metallic share in India's crude steel production in FY30.

HOW IS THE PROPORTION OF IRON ORE/PELLET/SINTER IN IRONMAKING CHANGING?

Hot metal is perhaps the most important charge mix in the Indian context. The typical raw material feed of a blast furnace is a combination of lumps, pellets and sinter. However, the author feels that the share of pellets will increase while that of sinter, because of its polluting properties, will lessen. Pellet's share is seen expanding from a mere 17 per cent in 2018 to 30 per cent by 2030. Currently, it is at 25 per cent. One key reason for pellet's expanded role lies in the fact that the newer furnaces in India are built to accommodate more pellets, with a nod to environmental clean-up, unlike old-fangled furnaces in other steel-manufacturing countries, where pellets enjoy a mere 10–15 per cent of the charge mix.

It has been observed that higher pellet feed results in better production and helps lower the carbon emission quotient.

India Charge Mix In Hot Metal

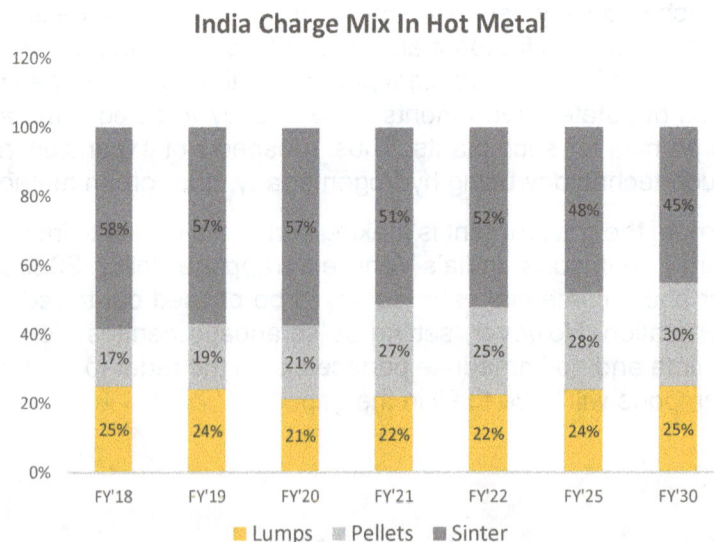

FIG 4 – Charge mix in India's hot metal production by FY30.

CHARGE MIX IN DRI

India is globally the largest producer of DRI, with an installed capacity of around 62 Mt where more than 70 per cent is derived from coal-based units and the balance from players dependent on gas as a fuel. But, with the government discouraging additional coal-based capacities, the future emphasis will be on hot metal and scrap.

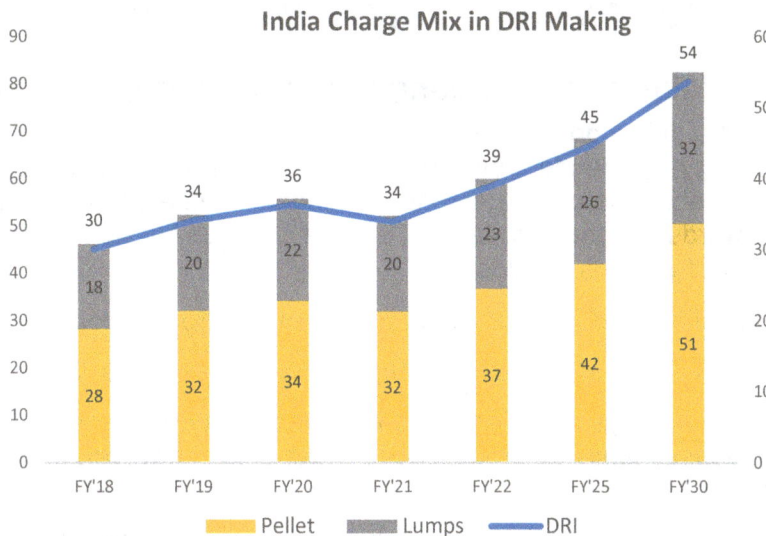

India Charge Mix in DRI Making

	FY'18	FY'19	FY'20	FY'21	FY'22	FY'25	FY'30
DRI	30	34	36	34	39	45	54
Lumps	18	20	22	20	23	26	32
Pellet	28	32	34	32	37	42	51

FIG 5 – Charge mix in India's steel production via DRI.

SCRAP IS GAINING IMPORTANCE, BUT HAS LIMITATION

India, as with all other steel-producing countries, is stressing on scrap-based production. Larger mills are increasing the proportion of scrap in their feed and some are planning to set-up EAFs. But shifting to scrap is easier said than done under current circumstances. The ecosystem for scrap generation and collection requires infrastructure which is still under installation. Till then, imports dependency will persist. But it is imperative for India to become self-sufficient in this material since other countries are becoming protectionist. As per Ukraine-based GMK Center, 43 countries have restricted exports of ferrous scrap while every third of them has banned exports of the same.

India's scrap usage is seen doubling by 2030 from the present 26 Mt but is still far lesser at 20 per cent compared to the 40–60 per cent of other developed economies.

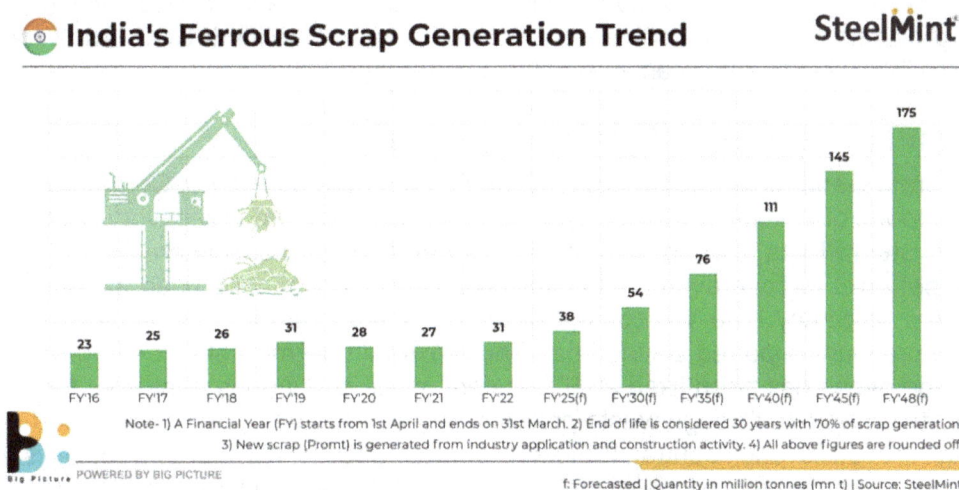

India's Ferrous Scrap Generation Trend — SteelMint

FY'16	FY'17	FY'18	FY'19	FY'20	FY'21	FY'22	FY'25(f)	FY'30(f)	FY'35(f)	FY'40(f)	FY'45(f)	FY'48(f)
23	25	26	31	28	27	31	38	54	76	111	145	175

Note- 1) A Financial Year (FY) starts from 1st April and ends on 31st March. 2) End of life is considered 30 years with 70% of scrap generation. 3) New scrap (Promt) is generated from industry application and construction activity. 4) All above figures are rounded off.

POWERED BY BIG PICTURE

f: Forecasted | Quantity in million tonnes (mn t) | Source: SteelMint

FIG 6 – India's ferrous scrap generation trend forecast by FY48.

COAL-BASED DRI WILL GRADUALLY REDUCE ON ENVIRONMENTAL CONCERNS

Sponge iron or direct reduced iron (DRI) is one of the key metallics in the Indian context. However, coal-based DRI's future looks uncertain at present. Especially since the Indian government is not too keen on its expansion, considering coal's notoriety as a carbon-spewing agent. Unless a breakthrough technology – riding on hydrogen or any other clean gas – arrives on the scene, this segment may start getting phased out beyond 2030.

GOVT INITIATIVES TO CONTROL CARBON EMISSIONS

Hydrogen mission (aimed at promoting usage of H2 in DRI making)

As per the website of the Indian government's National Green Hydrogen Mission, 'India has set its sight on becoming energy-independent by 2047 and achieving net zero by 2070. To achieve this target, increasing renewable energy use across all economic spheres is central to India's energy transition.' Working towards this end, the National Green Hydrogen Mission was approved in January 2022. Its objectives, among others, include:

- making India a leading producer and supplier of green hydrogen globally
- creating export opportunities for green hydrogen and its derivatives, amongst other.

Steel production through hydrogen involves using the gas to reduce iron pellets into sponge iron, or metallic iron that can subsequently be processed to make steel. However, in India, this technology is still at a nascent stage.

Recently, the Ministry of Steel, Government of India invited submissions of R&D proposals to promote the use of hydrogen in iron and steelmaking under the National Green Hydrogen Mission.

Coal gasification

> Gasification of coal is a process in which coal is partially oxidated by air, oxygen, steam or carbon dioxide under controlled conditions to produce a fuel gas. The hot fuel gas is cooled in heat exchangers, with the production of steam, and cleaned before combustion in a gas turbine. The offgases from the turbine are used in a boiler to produce additional steam for a steam turbine. The electrical efficiency can be around 45% with minimal impact on the environment.

as per information available in van den Berg and Boorsma (1997).

The Ministry of Coal, under the Government of India, is putting a special emphasis on surface coal gasification (SCG) projects. Key stakeholders can be classified as raw material suppliers/potential project owners; end-users/downstream players; consultants and project implementors; and technology providers. National coalminer Coal India has identified SCG as a diversification opportunity for 2030 due to significant competitive advantage.

Government allows usage of pet coke in steelmaking

India's Directorate General of Foreign Trade (DGFT), under the Ministry of Commerce and Industry, has allowed imports of low-sulfur (below 3 per cent) pet coke for use in recovery-type coke ovens of the Integrated Steel Plants (ISP) in the country, as per a recent notification. Blending of low-sulfur and low-ash pet coke in coke ovens helps improve the quality of coke used in blast furnaces and results in higher efficiency. The ash content of pet coke is only about 0.3 per cent compared to 9–10 per cent in premium imported coking coal. The maximum permissible blending ratio of pet coke in recovery-type ovens has been fixed at 10 per cent, while continuous analysers for measurement of sulfur dioxide emissions through waste/process gases will have to be installed and which will be monitored by the State and Central Pollution Control Boards (SPCB/CPCB). This measure is intended to reduce carbon emissions and also the import dependency on coking coal.

However, some amount of pet coke imports cannot be avoided if India must meet its decarbonisation goals, since the country does not produce the low-sulfur variant.

REFERENCES

van den Berg, J W and Boorsma, A, 1997. Construction Raw Materials from Coal Fired Powerstations By-products management and quality control, *Studies in Environmental Science*, 71(1997):259–268. https://doi.org/10.1016/S0166-1116(97)80209-9

Evaluation of the environmental benefits of using a conveyor belt mining system in an open pit mine

D E G Oliveira[1], T F Bravin[2], G P S Resende[3] and E A S Baeta[4]

1. Sr Mine Planning Engineer, Samarco, Belo Horizonte, Minas Gerais, 30130–141, Brazil.
   Email: douglas.goncalves@samarco.com
2. Environmental Analyst, Samarco, Belo Horizonte, Minas Gerais, 30130–141, Brazil.
   Email: thais.bravin@samarco.com
3. Junior Mining Engineer, Samarco, Belo Horizonte, Minas Gerais, 30130–141, Brazil.
   Email: gabriel.souza@samarco.com
4. Specialist Mine Planning Engineer, Samarco, Belo Horizonte, Minas Gerais, 30130–141, Brazil.
   Email: eduardo.baeta@samarco.com

ABSTRACT

There is currently in the mining industry, as well as in other industries, a movement seeking to increase investments in decarbonisation processes to help combat global warming. Among some initiatives taken is replacing diesel-powered mobile equipment in the mines with electricity-powered equipment, such as electric off-highway trucks or even belt conveyors in the mining operation. Samarco is a Brazilian mining company that since its conception has used a conveyor belt mining system in its mining operation, using only a small number of trucks to complement the mining process. This mining process that combines conveyor belts and trucks has brought environmental gains for the company since the environmental impact of conventional truck-only mining is greater than that of the combined system. To measure these gains, a medium-term mining plan was developed in which two fleet scenarios were detailed. One scenario contemplated a high level of detail of belt mining seeking its maximisation, while the other considered the operation performed 100 per cent by trucks, eliminating the use of conveyor belts. Through the comparison of these two scenarios, it was possible to make an inventory of atmospheric emissions and greenhouse gases for each one of them considering data such as types of equipment, diesel consumption, total distance travelled and energy consumption. The results showed that the fleet scenario with conveyor belts results in periods with up to 15 per cent less greenhouse gas emissions and 37 per cent less total particulates compared with the truck-only scenario. Therefore, the results confirm the environmental benefits of the method and the need to look for ways to even expand it in future operations through a study to identify the potential areas in the long-term mining plan.

INTRODUCTION

Samarco operates two open pit mines in the Alegria Complex in Minas Gerais, Brazil. These mines have been operated through the open pit mining method since their start and they'll continue to be operated in this way until the end of the project's life. The use of conveyor belts alongside trucks in the mine operation is mainly due to the characteristics of the deposit formation, which is large, thick, relatively flat, the rock is friable and because it outcrops on the surface.

The mine is operated using 197t CAT 789D trucks and wheel loaders compatible with the size of the trucks (CAT 993K and 994H), and a conveyor belt system installed directly in the mining areas, exclusively for ore haulage. The transport operation done by trucks occurs both for ore, waste and all other auxiliary movements. The conveyor belt mining operation consists of a fully truck-less operation that allows the ore to be transported from the mine to the plants.

It consists of a conveyor belt structure that is mounted inside the mine and connected to the long-distance conveyor belt system that transports the ore to the processing plants. These conveyor belts get as close as 15 m from the mining fronts and a feeder is positioned at the end of the conveyor. The ore rock is then moved by dozers close to the wheel loader, which transports the ore to the hopper, where the ore is dumped and subsequently falls into the overland conveyor belt system (Figure 1). This operation takes place at a maximum distance of approximately 100 m between the disassembled rock and the feeder, and the productivity decreases as the distance between the material and the feeder increases.

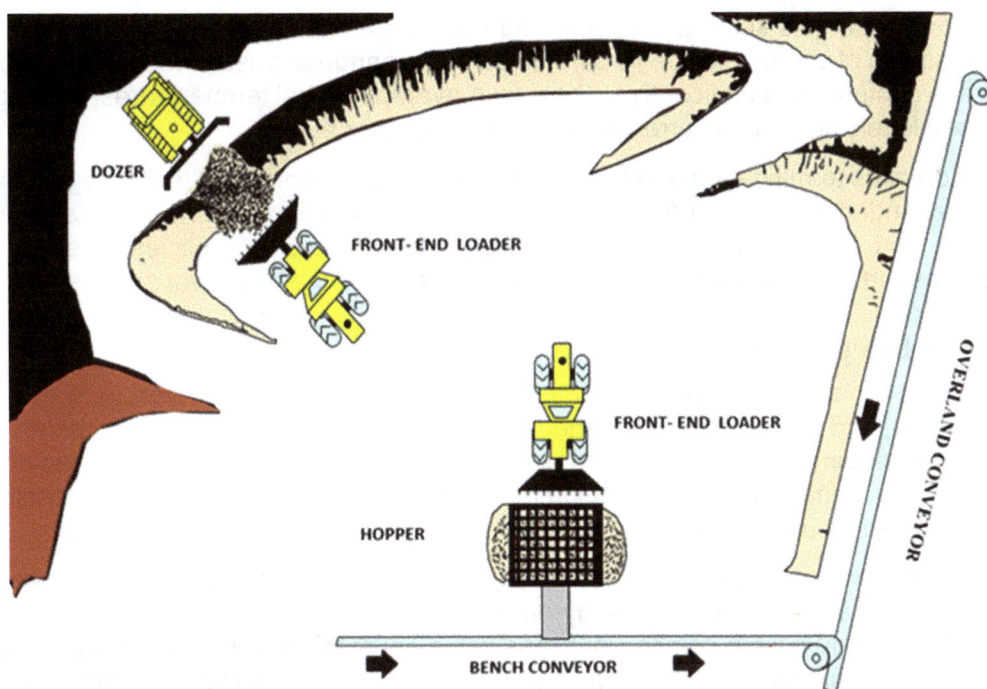

FIG 1 – Representation of the belt mining operation at Samarco.

The determination of the mine areas that can be mined with conveyor belts depends on several factors, with the main ones being:

- topographic level of the benches

- predominance of ore

- possibility of interconnecting the bench belts to the long-distance belt system

- hardness of the rock.

Conveyor belts versus off-highway trucks

Conveyor belt systems have higher CAPEX compared with off-highway trucks, but they last longer and only require replacement of wear materials. As for OPEX, it is common sense for authors like Burt and Caccetta (2014), Mohutsiwa and Musingwini (2015) and Nehring *et al* (2018) that the operating costs of conveyor belt systems are lower than off-highway trucks. According to Nehring *et al* (2018) the operating costs of off-highway trucks tend to increase over time as mines get deeper and haulage distances increase and it's a problem that occurs in most open pit mines. The main advantages of using conveyor belts in mine operations are the low operating costs and the fact that they are environmentally friendly.

On the other hand, the use of large trucks is undoubtedly the most widespread transportation method in mining worldwide, and its use provides gains of scale obtained by the large load capacity, operational flexibility, and lower acquisition cost when compared with conveyor belts in mining.

At Samarco, both modalities are used for the transport phase, combining the two operational methods according to the characteristics of the mining fronts in operation. Operating with a mixed fleet requires mine planning that considers the characteristics of both methods, with attention to the sequencing of the conveyor belt beds. Without this approach, the use of conveyor belts in mining may even become unfeasible at Samarco.

Decarbonisation

In recent years, growing concern about climate change and the need to address the environmental impacts of global warming have driven global efforts to decarbonise the global economy. Decarbonisation is a process that seeks to reduce and eliminate greenhouse gas emissions from human activities, and it emerges as an imperative to ensure a sustainable future and preserve the planet for future generations.

Dependence on fossil fuels such as coal, oil, and natural gas has been one of the main drivers of increased emissions of carbon dioxide (CO_2) and other greenhouse gases in the atmosphere. These emissions have contributed significantly to the rise in average global temperatures, melting polar ice caps, rising sea levels and the occurrence of extreme weather events.

Decarbonisation aims to mitigate these impacts by adopting solutions and technologies that reduce or eliminate dependence on fossil fuels. This requires transitioning to renewable energy sources such as solar and wind, the implementation of energy efficiency policies, the promotion of sustainable transport and the transformation of the industrial and agricultural sectors.

Statement of the problem

The mining industry should prepare to implement decarbonisation efforts to meet the Paris Agreement targets, and one solution that emerges to comply with these policies is the electrification of mine operation equipment. In this regard, electrification can be understood as replacing diesel equipment with electric batteries and replacing trucks with belt conveyors wherever possible, as this would replace fossil fuel use with electric power.

The use of diesel-powered mining equipment is a low energy-efficient operation. On the other hand, the use of electric equipment is more energy efficient and environmentally friendly in countries like Brazil, where the energy supply comes from renewable sources. Several studies carried out in the past at Samarco sought to understand the costs and operational aspects of conveyor belt mining, but the environmental benefits provided by this method were neglected for many years. It was only after the changes in the environmental policies of global industry, which occurred strongly in the last decade, that the benefits arising from the electrification of operations came into focus.

Therefore, the quantification of the environmental gains of the conveyor belt mining process at Samarco was carried out, to help better understand the impacts on particulate and greenhouse gas emissions.

METHODOLOGY

To quantify the environmental aspects of conveyor belt mining, a medium-term mine plan that simulated the period between the years 2023 and 2027 was carried out. The main purpose of this plan was to optimise the conveyor belt mining and allow a great level of detail of the conveyor belt beds and their evolution over the years, seeking to maximise the mass mined by this modal.

Two fleet sizing scenarios were carried out for this mine plan, as exemplified in Figure 2. The first one called in this work as 'mixed fleet scenario', which considered the use of conveyor belts and trucks (Samarco's current fleet) and the other considered a hypothetical scenario with 100 per cent of the material movement in the mine carried out by trucks, excluding the conveyor belt mining. This second scenario was called the 'truck-only scenario'.

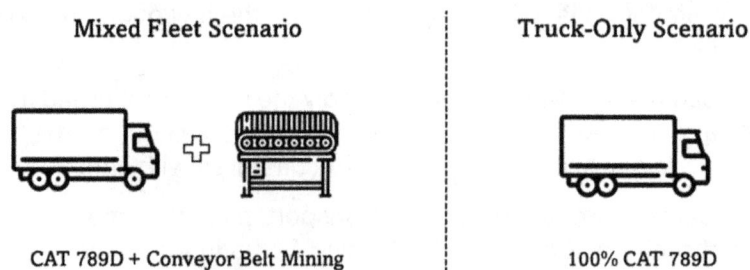

Mixed Fleet Scenario | **Truck-Only Scenario**

CAT 789D + Conveyor Belt Mining | 100% CAT 789D

FIG 2 – Fleet sizing scenarios.

Subsequently, data related to the two scenarios were collected and served as input for the Atmospheric Emissions Inventory and the Greenhouse Gas Inventory. The data collected from the two scenarios were: the total mass transported by trucks and conveyor belts; the number and the specifications of the operating equipment; the total amount of diesel consumed; the total distance travelled; the demand for electricity from mining belts; the routes to be taken by trucks and the quantity of transfers of the bench conveyor belts.

Mine movement

The mass movement is the same in both scenarios as the mine plan is the same. To quantify the tonnage mined by the conveyor belts over the years, the conveyor belt beds were considered in the mine design process, so the quantification of the tonnage mined by the conveyor belts was done in a very detailed way, which increases the level of confidence.

The total tonnage mined by conveyor belts calculated for the mixed fleet scenario was 46 Mt, while the mass mined by wheel loaders and trucks calculated for the same scenario was 187 Mt, out of a total of 233 Mt. As there is no conveyor belt mining in the truck-only scenario, the total tonnage (233 Mt) must be mined by trucks in this scenario. Figure 3 shows the mass moved by modal in both scenarios:

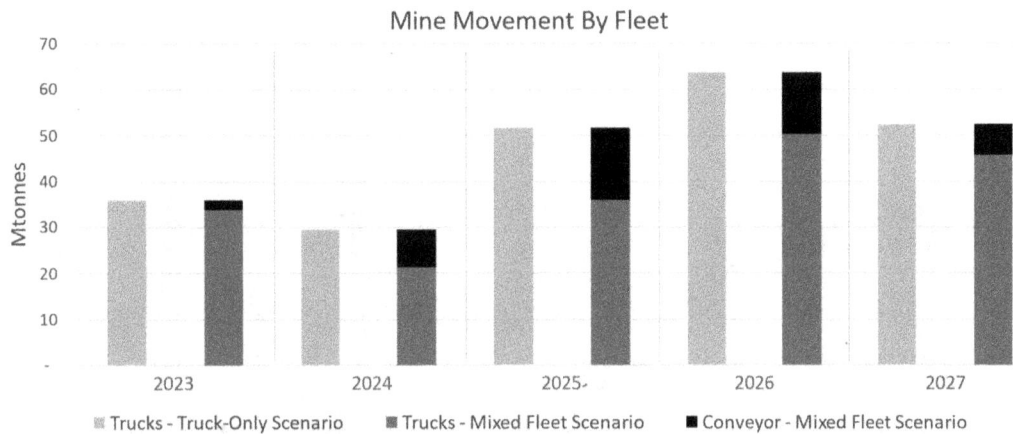

FIG 3 – Mine movement by fleet in both scenarios.

Fleet sizing

The two fleet scenarios considered all the equipment operating at Samarco's mines, differing only in the number of loaders and trucks in each of them. Table 1 shows all the equipment considered in the fleet sizing.

TABLE 1

Mobile fleet equipment that uses diesel as fuel.

Mobile mine fleet				
Equipment	Fabric	Model	Power (W)	Power (HP)
Wheel loader	Caterpillar	CAT 992K	676	907
Wheel loader	Caterpillar	CAT 993K	775	1039
Wheel loader	Caterpillar	CAT 994H	1176	1577
Off-highway truck	Caterpillar	CAT 789D	1566	2100
Dozer	Caterpillar	CAT D11T	634	850
Dozer	Caterpillar	CAT D10T	450	603
Drill	Epiroc	FlexiROC D65	403	540
Mid-size wheel loader	Caterpillar	CAT 950H	162	217
Wheel dozer	Caterpillar	CAT 834K	419	562
Motor grader	Caterpillar	CAT 16M	217	291
Tow truck	Caterpillar	CAT 777G	765	1025

Differences in the number of wheel loaders were observed between the two scenarios, which are related to the fact that conveyor belt mining requires the use of the 993K loaders for ore feeding, while in the truck-only scenario, the demand for the CAT 994H loader increases because it matches better with the loading of CAT 789D trucks. In addition, the productivity of the 993K loader is lower when feeding belts than when loading off-highway trucks, which requires more working hours.

The biggest difference between the two scenarios lies in the transportation stage, as several trucks are eliminated in the mixed fleet scenario, especially in the years when the conveyor belt mining rate is higher (Figure 4).

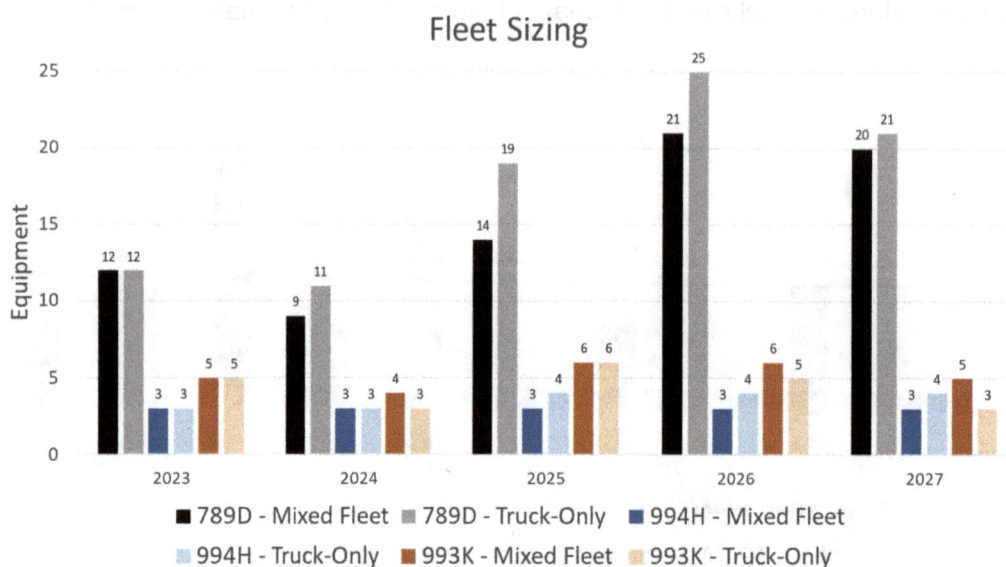

FIG 4 – Fleet sizing with focus on the equipment that differs in the fleet scenarios.

Working hours

The diesel-powered equipment working hours survey was conducted to assist in the calculation of data required for the inventories, such as annual diesel consumption and annual distance travelled in both scenarios. The mixed fleet scenario presents a total reduction of more than 76 000 worked hours of diesel-powered equipment compared with the truck-only scenario, as can be seen in Figure 5.

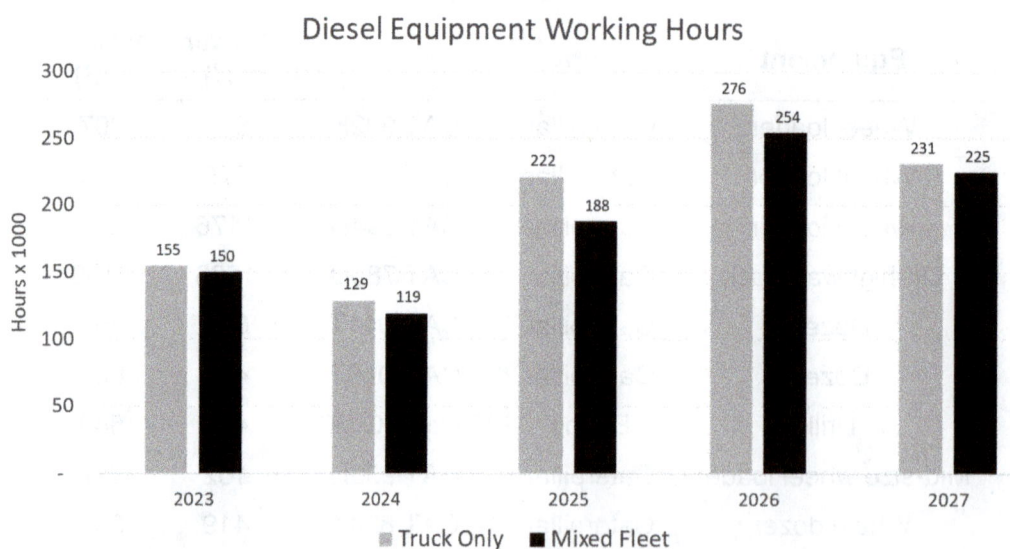

FIG 5 – Diesel equipment working hours in both scenarios.

Total distance travelled

Information such as the number of cycles per hour, the number of CAT 789D trucks and the average transport distance of each equipment for each period analysed, were obtained from the fleet sizing of the five-year plan. Based on this information and knowing the total number of hours worked by each truck, the total distance travelled per annum in each of the scenarios was calculated. The formula shown below was used to calculate the total distance travelled by the trucks:

$$Total\ Travelled\ Distance = Equipment\ Quantity \times Nr.\ of \frac{cycles}{hour} \times Working\ Hours \times (Avg.\ Haul\ Dist. \times 2)$$

The formula considers that each transportation cycle is composed of a round trip, so the average transportation distance is multiplied by two. Figure 6 shows the total distance travelled year by year in the two stipulated scenarios.

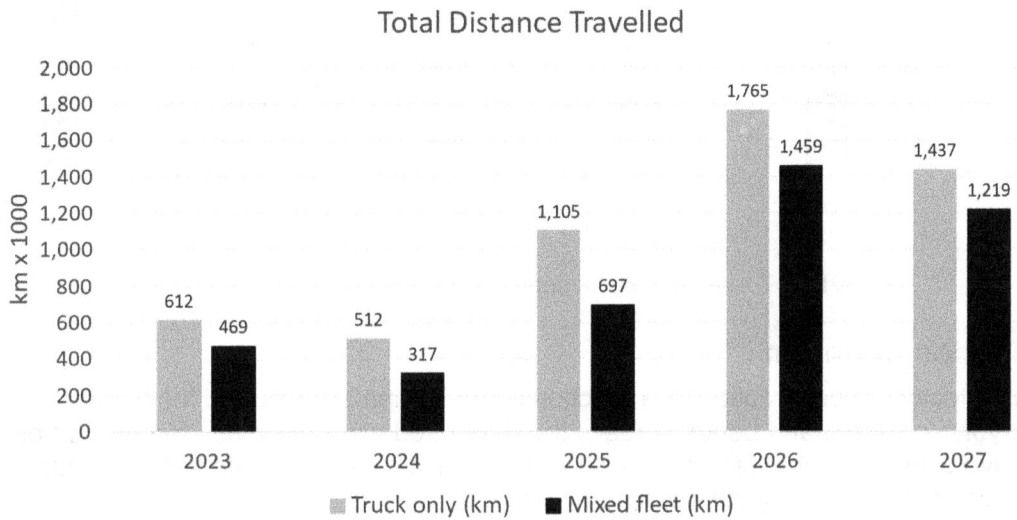

FIG 6 – Total distance travelled by year.

Diesel consumption

The specific diesel consumption of each of the large equipment was collected with the support of the operational control team. The data are presented in Table 2.

TABLE 2

Specific diesel consumption by equipment.

Equipment	Specific consumption (L/WH)
CAT 993K Operation	119.73
CAT 994H Operation	167.31
CAT 789D Operation	94.69
CAT D11T Operation	105.00
CAT D10T Operation	70.62
ROC D65 Operation	63.91
CAT 992K Mine Infrastructure	82.95
CAT D11T Mine Infrastructure	107.56
CAT 950 Mine Infrastructure	11.19
CAT D10T Mine Infrastructure	70.62
CAT 834K Mine Infrastructure	31.98
CAT 16M Mine Infrastructure	22.55
CAT 777G Mine Infrastructure	31.98

Based on the specific consumption data of each equipment and knowing the working hours per equipment in both scenarios, the diesel consumption of the equipment was calculated. By adding the individual consumption, the annual diesel consumption for each scenario was obtained, shown in Figure 7.

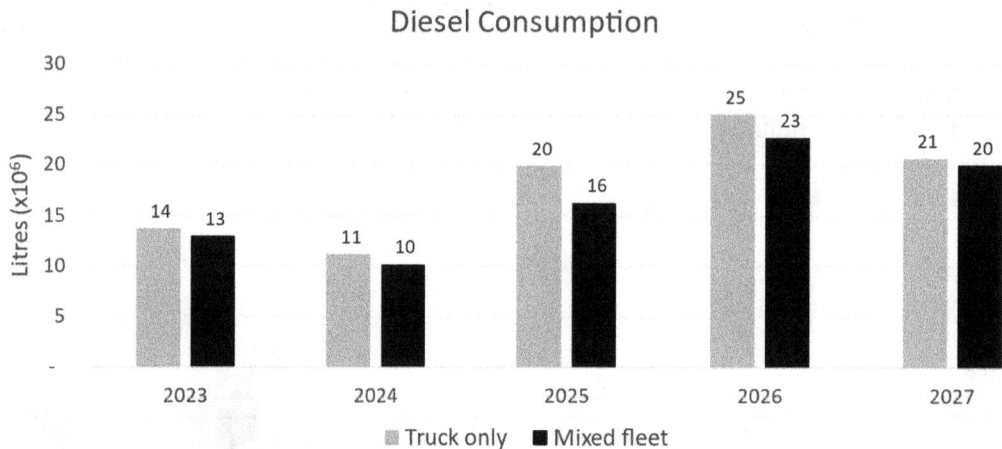

FIG 7 – Annual diesel consumption in both scenarios.

Electricity consumption

To calculate electricity consumption as a function of the equipment required for conveyor belt mining (belt conveyor, Lokotrack and bench loader), the data used were the mass fed to the belts in each year, the layout of the belt systems foreseen in the Five-Year Plan and the production capacity (nominal capacity multiplied by the operational yield).

Considering the nominal capacity of 2200 t/h (production capacity of a bench conveyor) and the operational efficiency of 70 per cent (expected efficiency for the CAT 993K wheel loader) and that when the conveyor belt system is fed via Lokotrack and hopper, 70 per cent of the mass is directed to the Lokotrack and 30 per cent to the hopper, it was possible to calculate the operating hours of each equipment and the energy demand to operate the system. Based on the average electricity consumption per hour of each equipment, the energy demand per annum was calculated, as shown in Figure 8.

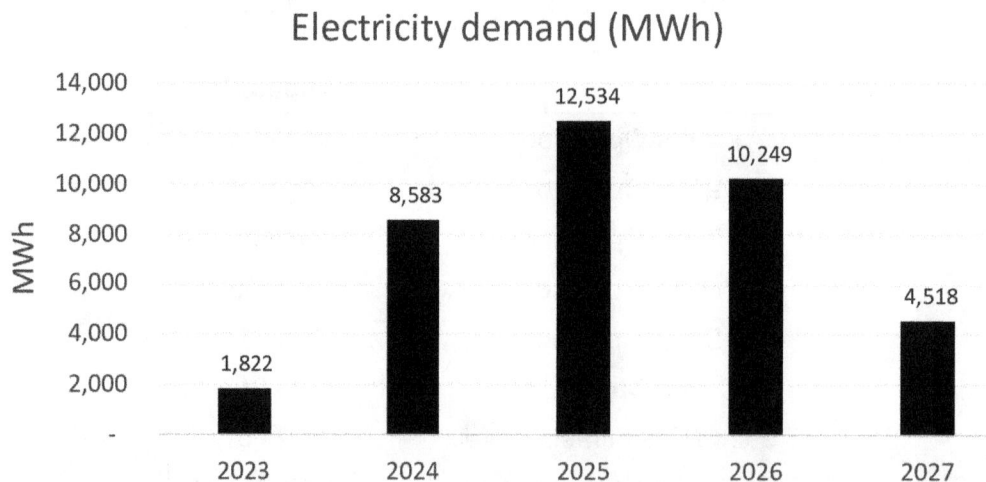

FIG 8 – Annual electricity consumption of the conveyor belts in the mixed fleet scenario.

INVENTORY OF ATMOSPHERIC EMISSIONS AND GREENHOUSE GAS

The preparation of an air emissions and greenhouse gas (GHG) inventory is done according to a methodology to quantify and evaluate the sources of air pollutant emissions at an industrial site or facility. These inventories are an essential tool for companies to understand their emissions, identify the sources and sectors with the greatest impact and set reduction targets.

Atmospheric emissions

The Atmospheric Emissions Inventory consisted of quantifying the emission rates of the main air pollutants for the study in question:

- Total Particulate Matter (TPM).

- Inhalable Particulate Matter (PM_{10})

- Fine Particulate Matter ($PM_{2.5}$)

- Sulfur Oxides, expressed as Sulfur Dioxide (SO_2).

- Nitrogen Dioxide (NO_2), expressed as NO_X.

- Carbon Monoxide (CO).

Therefore, to estimate the emissions, the main methodologies recommended by the United States Environmental Protection Agency (USEPA) and Emission Estimation Technique Manual for Combustion Engines (NPI) were used, which are guidelines for the preparation of atmospheric emission inventories worldwide. The Inventory of Atmospheric Emissions was organised and distributed in emission typologies related to the observed activities based on the following characteristics:

- Traffic lanes: the emissions of TPM, PM_{10}, $PM_{2.5}$, NO_X, SO_2 and CO coming from the activities of vehicles, trucks and trailers that circulate daily at the Alegria Complex are included.

Emissions of air pollutants from roads are generated from the burning of fossil fuels, which are emitted through vehicle exhaust. In addition to vehicle exhaust emissions, traffic roads also emit particles generated by the friction of tires on the inventoried road surface (resuspension). The description of the methodology used to estimate the emissions is presented in Table 3. The average travelled distance was calculated according to the movements made by the vehicles according to the identification of transfer routes guided by the United States Environmental Protection Agency (USEPA, 2006b) and Emission Estimation Technique Manual for Combustion Engines (NPI, 2008).

TABLE 3

Methodology used to estimate emissions from traffic lanes.

Group	Process	Formula	Parameters
Traffic lanes	Roads – exhaust (TPM, PM_{10}, $PM_{2,5}$, NO_x e CO)	$TE_i = EF_i \times \dfrac{C}{8760}$	TE_i = average emission rate of the pollutant i [kg/h] EF_i = emission factor of pollutant i [kg/L] C = Fuel consumption per annum [L/annum]
	Roads – exhaust (SO_2)	$TE_i = \dfrac{A \times \rho \times \left(\dfrac{t}{10^6}\right) \times \left(\dfrac{MM_{SO_2}}{MM_S}\right)}{8760}$	TE = average emission rate of SO_2 [kg/h] A = annual fuel consumption [L/annum] ρ = diesel density [kg/L] t = sulfur content [ppm] MM_{SO_2} = Molar mass SO_2 [g/mol] MM_S = Molar mass S [g/mol]
	Unpaved roads – particle resuspension (TPM)	$TE_i = k \times \dfrac{281,9}{1000} \times \left(\dfrac{sL}{12}\right)^{0,7} \times \left(\dfrac{W}{3}\right)^{0,45} \times DMT \times \left(\dfrac{100-EC}{100}\right)$	TE_i = average emission rate of the pollutant i [kg/h] k = particle size multiplier [lb/VMT] (k = 4.9 for TPM) sL = silt content on unpaved roads [%] W = average weight of vehicles circulating on the track [t] DMT = average haulage distance [km/h] EC = control efficiency [%]

- Operation: emissions of TPM, PM_{10}, $PM_{2,5}$, NO_x, SO_x, CO and SO_2 associated with the burning of fuel from heavy off-road machinery such as loaders, tractors and graders, generators and motor pumps.

To calculate the emission rates of this source typology, the emission factors suggested by the United States Environmental Protection Agency (USEPA, 2006b) and the Emission Estimation Technique Manual for Combustion Engines (NPI, 2008) were used. To estimate the emission of each machine/equipment, all fuel consumption of vehicles and equipment of the operational site was conservatively considered. Table 4 presents the methodology used to calculate the emission rate.

TABLE 4

Methodology used to estimate emissions from machinery and equipment.

Group	Process	Formula	Parameters
Machinery and equipment	Operation equipment (TPM, PM$_{10}$, PM$_{2,5}$, NOx e CO)	$$TE_i = EF_i \times \frac{C}{8760}$$	TE$_i$ = average emission rate of the pollutant i [kg/h] EF$_i$ = emission factor of pollutant I [kg/L] C = Fuel consumption per annum [L/annum]
	Operation equipment (SO$_2$)	$$TE_i = \frac{A \times \rho \times \left(\frac{t}{10^6}\right) \times \left(\frac{MM_{SO_2}}{MM_S}\right)}{8760}$$	TE = average emission rate of SO$_2$ [kg/h] A = annual fuel consumption [L/annum] ρ = diesel density [kg/L] t = sulfur content [ppm] $MMSO_2$ = Molar mass SO$_2$ [g/mol] MMS = Molar mass S [g/mol]

- Material transfers: emissions of TPM, PM$_{10}$, MP$_{2.5}$ associated with activities related to the transfer of all materials are included. During the transfer processes, it is possible to observe the suspension of particulate material.

For this typology, emission rates were calculated based on methodology provided by the United States Environmental Protection Agency (USEPA, 2006a). The emission rate calculation methodology used for transfers is presented in Table 5.

TABLE 5

Methodology used for material transfers.

Group	Process	Formula	Parameters
Material movement	Transfers (TPM, PM$_{10}$ e PM$_{2,5}$)	$$TE_i = k \times 0,0016 \times \frac{\left(\frac{U}{2,2}\right)^{1,3}}{\left(\frac{M}{2}\right)^{1,4}} \times A$$	TE$_i$ = average emission rate of the pollutant i [kg/h] k = particle size multiplier [dimensionless] (k = 0.74 to TPM, k = 0.35 to PM$_{10}$ and k = 0.053 to PM$_{2.5}$) U = wind average speed [m/s] M = material moisture [%] A = material handling rate [t/h]

Greenhouse gas emissions

Greenhouse Gases (GHG) are gases generated naturally or by anthropogenic action in the atmosphere. These gases have the ability to absorb infrared radiation and are responsible for retaining heat in the atmosphere. When present in high quantities, these gases contribute to the intensification of the global warming phenomenon.

The GHG inventory was developed based on the recommendations of the document 'Greenhouse Gas (GHG) Protocol Corporate Accounting and Reporting Standard – Revised Edition' of the WRI (World Resources Institute). According to the GHG Protocol, to help delineate the sources of direct and indirect emissions, improve transparency, and be useful for different types of organisations, different types of climate policies, and business objectives, two scopes were defined for the elaboration of the study.

- Scope 1 emissions are those from sources owned or controlled by Samarco. They were calculated based on the amount of fuel consumption by mobile sources, as presented in Table 6.

TABLE 6

Methodology used for stationary combustion.

Group	Emission factor	Formula	Parameters
Stationary Combustion	Fuel consumption by vehicle	$E_i = \dfrac{FE_i \times C_g}{1000}$	E_i = GHG Emissions i (t/annum) C_g = Fuel consumption g (L) FE_{ig} = GHG emission factor i applicable to the fuel g (kg GHG i/L g)

- For Scope 2, the metrics obtained from the consumption of electricity purchased or consumed by the company were calculated, purchased electricity being that which is purchased from the grid from outside to within the company's organisational boundaries (Table 7).

TABLE 7

Methodology of calculation used for electric energy consumption.

Group	Emission factor	Formula	Parameters
Electricity	Electricity consumption	$E_i = CE_g \times FE_g$	E_i = Average emission of the GHG i (t/annum) CE_g = electricity consumption associated with the origin of generation g (MWh /annum) FE_g = Emission factor associated to the origin of electricity generation g

The study considered the greenhouse gases identified and associated with its activities. Among them:

- CO_2 – Carbon Dioxide: generated from the burning of fossil fuels (diesel oil, gasoline and/or natural gas) by mobile and stationary sources.

- CH_4 – Methane: gas that is formed as a by-product of burning fuels.

- N_2O – Nitrous Oxide: generated when fuels are burned.

The Global Warming Potential (GWP) is a measure that determines how much each greenhouse gas contributes to global warming. It is a relative value that compares the warming potential of a given amount of gas with the same amount of CO_2, which by standardisation has a GWP value of 1. The GWP is always expressed in terms of CO_2 equivalence – CO_2e.

The global warming potential is calculated over a specific time interval (20, 100, or 500 years), with GHG inventories typically taking 100-year values. Its calculation is based on several factors, including the radiative efficiency and degradation rate of each gas relative to CO_2. Table 8 presents the GWP values used in the study.

TABLE 8

Global warming potentials (GWP) of some greenhouse gases (IPCC, 2014).

GHG	Chemical formulae	Global warming potential (100 years)
Carbon Dioxide	CO_2	1
Methane	CH_4	28
Nitrous Oxide	N_2O	265

RESULTS

Atmospheric emissions

The Inventory of Atmospheric Emissions was carried out for the particulate materials presented in 'Atmospheric emissions' in the previous section, from the emission sources on traffic lanes, burning of diesel and from conveyor belt transfers. The emission rates inventoried are presented below in Figure 9, considering the same graphic scale to better represent the materials with higher emission rates.

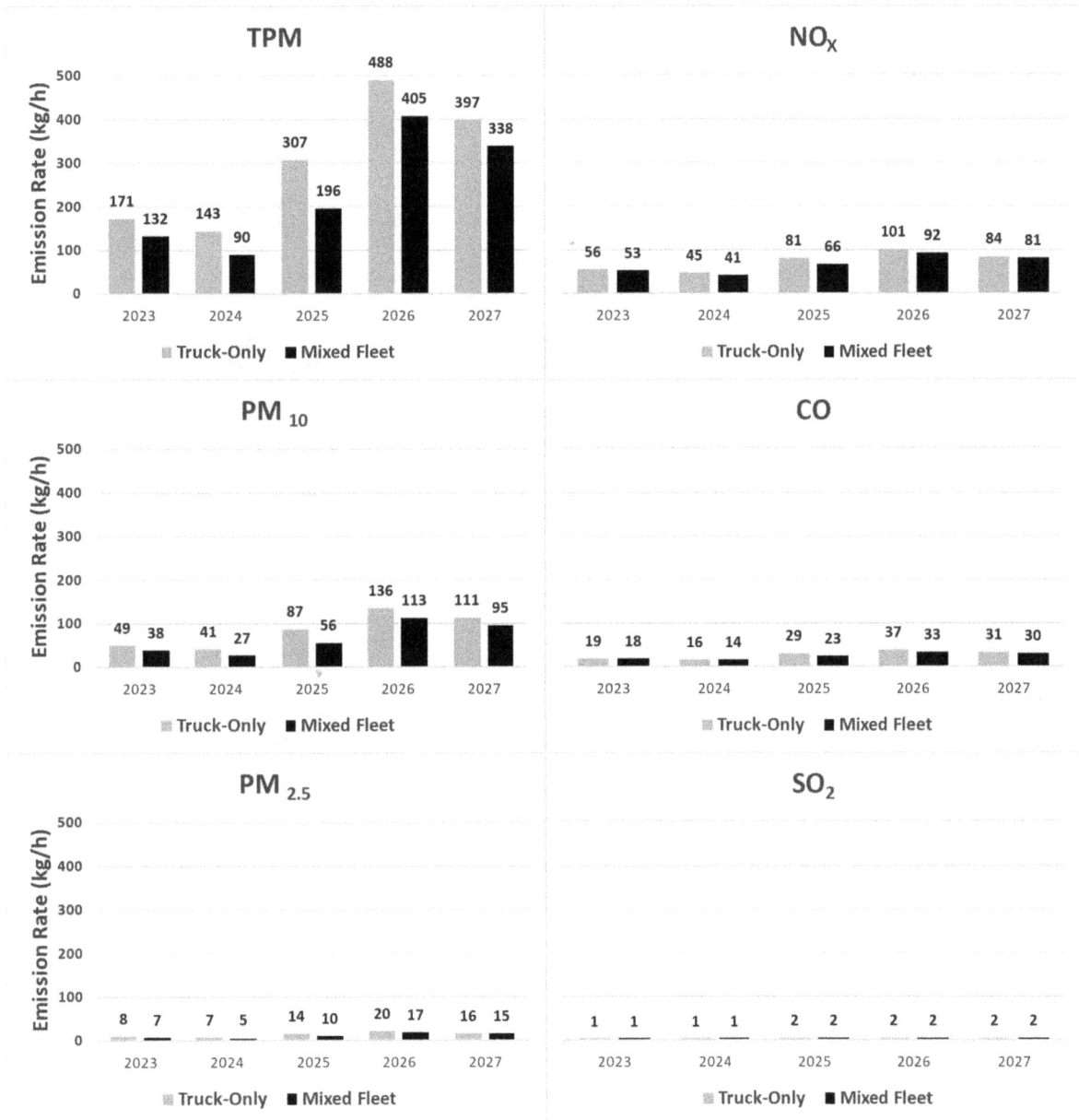

FIG 9 – Comparison of atmospheric emissions of various pollutants in both fleet scenarios.

As can be observed from the graphs, the biggest differences in calculated air pollutant emissions between the scenarios occur in the years 2024 to 2026. In these years, belt mining has been intensified in the plan due to operational factors and the sequencing of extraction.

The tonnage mined by belts is relatively low in the year 2023 because in that year, the mining development is planned to prepare for the future areas that are set to be mined by belts in the following years. In the year 2027, the mining plan shows a deepening of the pit and the need to start the development of pushbacks for the future continuity of mining that causes a reduction in belt mining. Figure 3 shows the conveyor belt mining tonnage calculated in the mine plan.

Figure 10 shows the graphs of emissions reduction (in percentage terms) for the mixed fleet scenario compared with the truck-only scenario:

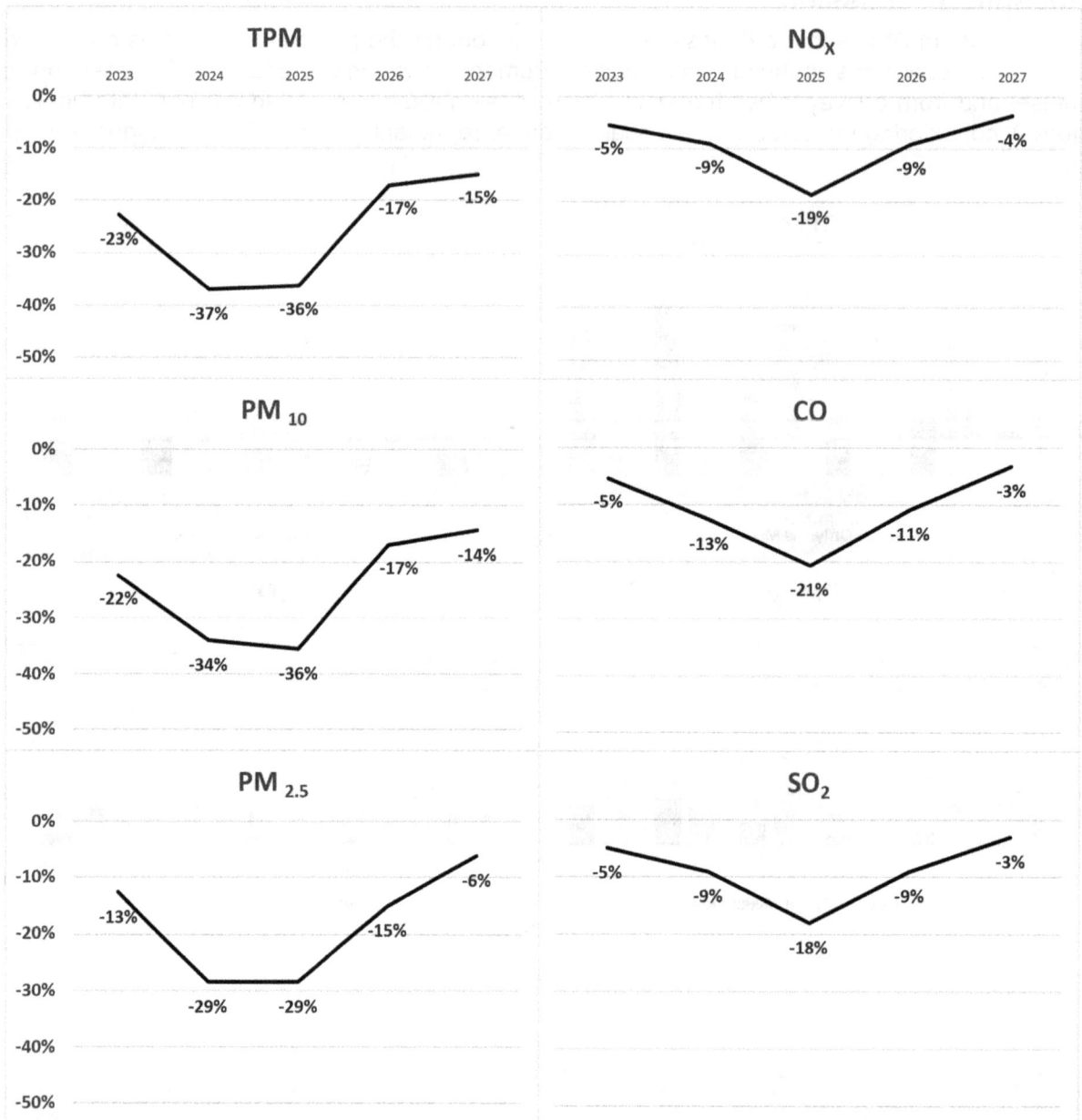

FIG 10 – Percentage reductions of pollutants from the mixed fleet scenario compared with truck-only.

The mixed fleet scenario presented a lower particulate emission rate compared with the truck-only scenario, as expected, reinforced by the fact that 98 per cent of the total particulate emissions (TPM) observed came from traffic lanes and the mixed fleet scenario presents a reduction of 1.3 million km travelled. The total distance travelled by trucks in both scenarios can also be seen in Figure 6.

It can also be seen from Figure 10 that the percentage reduction in particulate matter emitted (TPM, PM_{10} and $PM_{2.5}$) is greater than the percentage reduction in SO_2, nitrogen oxides (NO_x) and CO. As stated in 'Fleet sizing' section, although the 993K wheel loader is used more in the mixed fleet scenario than in the truck-only scenario, the final calculation shows a large reduction in truck use and hence in air emissions in this scenario because of less traffic.

Greenhouse gases

The calculations of the greenhouse gas emissions (including carbon dioxide, methane and nitrous oxide) that result from the mine operation activities that occur in both scenarios are identified as tCO_2e or tonnes of CO_2 equivalent. They are shown in Figure 11.

Greenhouse Gas Emissions

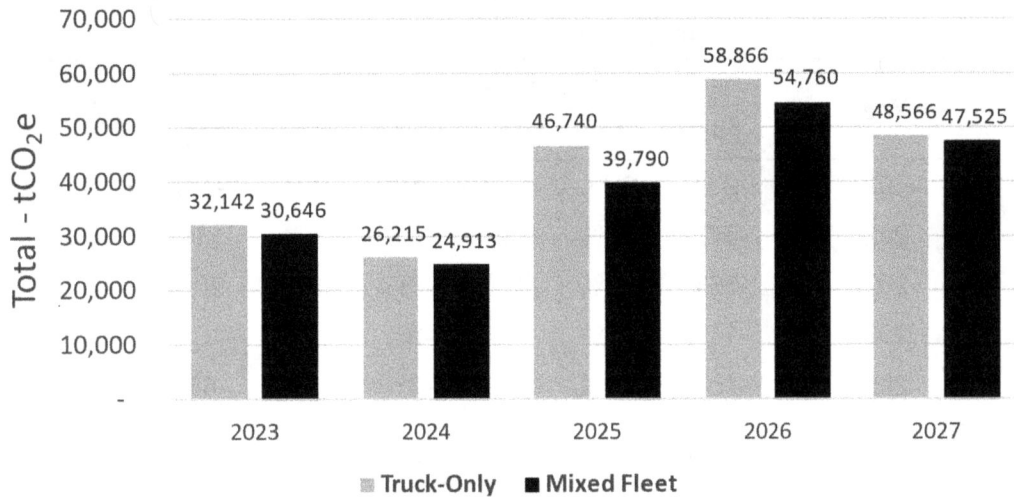

FIG 11 – GHG emissions in the scenarios.

As expected, the mixed fleet scenario results in less greenhouse gas emissions compared with the truck-only scenario. This outcome is strongly influenced by the annual diesel consumption as a direct source of emissions, which can be found in Figure 7. It also considers the electric power consumption of the belts as an indirect source of emissions at Samarco (Figure 8). The reduction of greenhouse gas emissions can be seen in percentage terms, as shown in Figure 12.

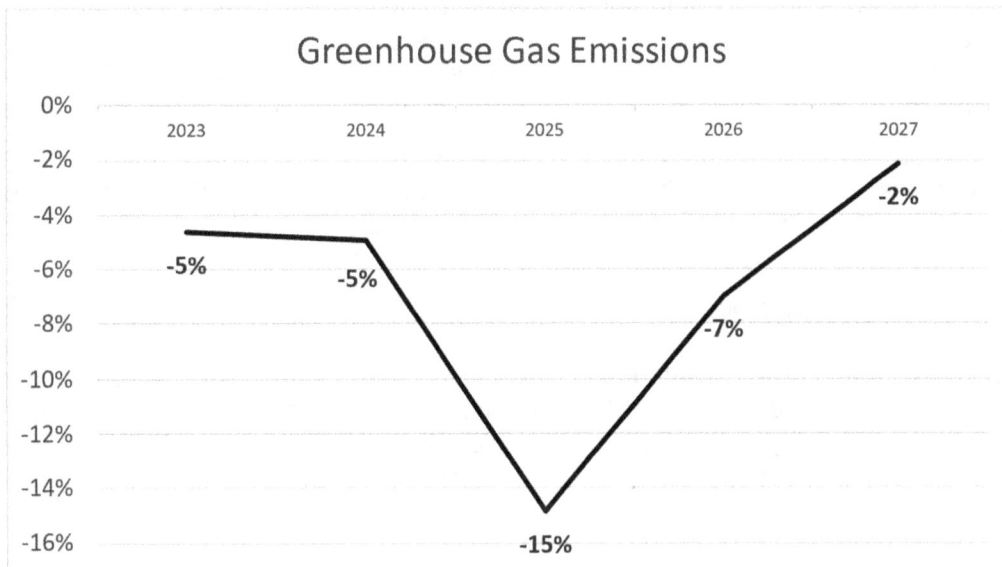

FIG 12 – The difference in greenhouse gas emissions between the mixed fleet scenario and the truck-only scenario.

The reduction in greenhouse gas emissions followed the trend of conveyor belt mining already showcased under the 'Mine movement' section, and in the 'Atmospheric emissions' section of these 'Results'. The years 2023 and 2027 have the lowest percentage of belt mining in the plan and the smallest differences between diesel consumption of the mixed fleet and truck-only scenarios.

DISCUSSION

The Atmospheric Emissions Inventory and Greenhouse Gas Inventory evaluate the environmental benefits of the conveyor belt mining combined with an off-highway truck fleet in Samarco's mine operation. Table 9 summarises all the reductions in pollutants and greenhouse gases observed in the inventory.

TABLE 9

Summary of per cent reduction in air and greenhouse gas emissions in the mixed fleet scenario compared with the truck-only fleet.

Year	Emission rate (kg/h)						tCO_2e
	TPM	PM_{10}	$PM_{2.5}$	SO_2	NO_X	CO	
2023	-23%	-22%	-14%	-5%	-5%	-2%	-5%
2024	-37%	-35%	-23%	-9%	-9%	-11%	-5%
2025	-36%	-35%	-28%	-18%	-18%	-20%	-15%
2026	-17%	-17%	-14%	-9%	-9%	-10%	-7%
2027	-15%	-14%	-11%	-3%	-3%	-4%	-2%
5 year accumulated	**-23%**	**-22%**	**-17%**	**-9%**	**-9%**	**-10%**	**-7%**

It can be observed that, in percentage terms, the greatest reductions in total particulate emissions are associated with the reduction in the travelled distance provided by the use of belts in the mine. The year 2025 is the year with the greatest rate of conveyor belt mining in the mine plan. Accordingly, 2025 is the year when the largest reduction in atmospheric and greenhouse gas emissions can be noticed.

The mixed fleet scenario resulted in a 7 per cent reduction in GHG emissions in the period between 2023 and 2027, which is equivalent to approximately 15 000 tCO_2e. This output can be considered a substantial amount, especially since it is the savings generated by a single source of emissions at Samarco, in this case the installation of conveyor belts for the mining of 45.9 Mt of ore over a period of 5 years. The company can also work on other fronts to help reduce its carbon footprint even more.

In addition, it is possible to evaluate changes that could be made in mine sequencing in the future, in order to increase the share of belts in mining, especially in the year 2027 and onwards, where less noticeable results were observed. In this year, the deepening of the pit and the need to open new pushbacks reduces the operational possibility of using belts in mining.

While significant results have been achieved in the years 2024 and 2026, belt-mined tonnage will drop considerably in 2027, indicating the need to review the sequencing of extraction from that year onwards to ensure that conveyor belt mining remains operational and can even be expanded in the future. The mine will expand to areas of less favourable topography for belt mining after 2027, but well-executed mine planning could make it possible. Given the proven environmental benefit and lower operating costs made possible by the method, this consideration should certainly be addressed.

CONCLUSION

The study presented in this paper showed with figures one of the best-known advantages of using conveyor belts in open pit mines, which is the environmental benefit of the method. However, this advantage had never been measured in Samarco's operation before and, given the current context of decarbonisation, the evaluation of this benefit is more than necessary. The results presented in this paper go beyond the analysis of operating costs between belts and trucks, already widespread in the mining industry and in the literature.

The case study demonstrated that the use of a mixed fleet of conveyor belts and off-highway trucks in mining reduces emission rates of various particulate gases and pollutants by up to 23 per cent and reduces greenhouse gas emissions by 7 per cent over the 5-year period of operation. Reductions in particulate emissions are extremely relevant considering that particulates can be harmful to workers, the community and the environment.

The reduction in greenhouse gas emissions observed over the 5-year period was 7 per cent in the mixed fleet scenario. Although this value may seem small, when it is put into perspective in relation to the total tonnage of greenhouse gases emitted over the entire period, one realises a great relevance in the amount reduced by the mixed fleet. Furthermore, when this reduction is put into perspective of the Life-of-Mine of a mineral project (Samarco has been operating with this modal since the late 1970s and will still use the belts for at least the next 15 years), a large reduction in emissions of these gases is evident. When the analysis is extrapolated to other mineral projects around the world that could use belts in the mine operation and do not, the result becomes relevant on a global scale.

The results presented in this paper help to raise the discussion for mine planning about the continued search for transportation alternatives that use electricity, mainly from renewable sources, instead of fossil fuels. It is possible to note that even small optimisations in mining sequencing that allow the reduction of the need for trucks, either by increasing the use of belts or even by reducing the average transport distance, when possible, can generate gains from an environmental point of view and, in the battle against global warming, any help has to be celebrated.

REFERENCES

Burt, C N and Caccetta, L, 2014. Equipment Selection for Surface Mining: A Review, *INFORMS Journal on Applied Analytics*, 44(2; March-April):143–162.

IPCC, 2014. Climate Change 2014: Synthesis Report, Contribution of Working Groups, I, II and III to the Fifth Assessment Report of the Intergovernmental Panel on Climate Change (eds: R K Pachauri and L A Meyer), IPCC, Geneva, Switzerland, 151 p.

Mohutsiwa, M and Musingwini, C, 2015. Parametric estimation of capital costs for establishing a coal mine: South Africa case study, *Journal of The Southern African Institute of Mining and Metallurgy*, 115(8):789–797.

National Pollutant Invention (NPI), 2008. Emission Estimation Technique Manual for Combustion Engines, ver 3.0, Commonwealth of Australia. Available from: <http://www.npi.gov.au>

Nehring, M, Knights, P F, Kizil, M S and Hay, E, 2018. A comparison of strategic mine planning approaches for in-pit crushing and conveying and truck/shovel systems, *International Journal of Mining, Science and Technology*, 28(2):205–214.

US Environmental Protection Agency (USEPA), 2006a. Revision of emission factors for AP-42, Chapter 13: miscellaneous source, Section 13.2.2: Unpaved Roads (Fugitive Dust Sources). Available from: <http://www.epa.gov/ttn/chief/ap42/index.html> [Accessed: 17 Mar 2023].

US Environmental Protection Agency (USEPA), 2006b. Revision of emission factors for AP-42, Chapter 13: Miscellaneous source, Section 13.2.4: Aggregate Handling and Storage Piles (Fugitive Dust Sources). Available from: <http://www.epa.gov/ttn/chief/ap42/index.html> [Accessed: 17 Mar 2023].

A cost-effective tool to assess iron ore products and optimise their operations in blast furnaces towards decarbonisation – state-of-the-art CFD models

X Yu[1] and Y Shen[2]

1. Postdoctoral Fellow, Shen Lab of Process Modelling and Optimisation, School of Chemical Engineering, University of New South Wales, Sydney NSW 2052.
2. Professor, Shen Lab of Process Modelling and Optimisation, School of Chemical Engineering, University of New South Wales, Sydney NSW 2052. Email: ys.shen@unsw.edu.au

ABSTRACT

The iron ore products should be assessed in terms of in-furnace behaviour and overall furnace performance, their operation should also be optimised under industry-scale blast furnace (BF) conditions, in order to be sustainable and competitive, especially in the context of decarbonisation efforts in ironmaking. While lab and pilot-scale experiments can provide useful kinetics, they are expensive and risky and cannot replicate industry-scale blast furnace conditions for proper assessment and optimisation. In this paper, we review our recent progress at Shen Lab of UNSW in using state-of-the-art Computational Fluid Dynamics (CFD) models to assess iron ore products and examine the effects of their operations on BF behaviour in basic terms, including in-furnace phenomena (flow field, temperature field, species distribution fields etc) and overall BF performance (coke rate, reduction efficiency and top gas temperature etc). We have developed state-of-the-edge CFD models for different purposes, ranging from the static-state BF model featuring unique layered burden structures for offline assessment and optimisation knowledge building-up for iron ore products, to the transient-state BF model for online assessment and optimisation of iron ore productions by describing the transient behaviours of iron ore productions under industry-scale conditions. First, we have studied several iron ore products including the iron ore products used in the present BFs and emerging iron ore products like carbon composite briquette and iron-coke. Second, several iron ore operations are studied including optimisation of conventional burden operations like oxygen enrichment, batch weight tuning and new operations like coke central charging and hydrogen injection in BFs. Third, some time-sensitive operations are also investigated including the use of wet burden materials and the drop in blast temperature on BF performance. These CFD models offer cost-effective research tools for understanding the effect of iron ore products on BF performance and serve as a reliable marketing tool for existing and emerging iron ore products under a wide range of BF conditions.

INTRODUCTION

The blast furnace serves as a dominant chemical reactor for ironmaking where iron ore products like sinter and pellets, along with metallurgical coke, are introduced alternately at the furnace top to produce liquid iron with high efficiency. To obtain 1000 kg of hot metal, more than 1500 kg of iron ore and 500 kg of carbon-containing fuel are required. Reducing the carbon footprint in this smelting practice is of paramount importance and it can be achieved by adopting new operations and new iron ore products.

The assessment of these iron ore products and the optimisation of these iron ore products can be primarily conducted through two approaches: experimental studies and modelling work. Compared to experimental studies, BF modelling is considered as a cost-effective, environmentally friendly and risk-free tool. As such, it is widely acknowledged that reliable BF modelling provides an alternative way to explore both current and emerging iron ore products, accompanied by novel operations. Multi-fluid BF modelling can be broadly divided into two branches: one utilises computational fluid dynamics (CFD) exclusively, while the other incorporates the discrete element method theory. Generally, for industrial-scale BF simulations, the former is preferred due to its high calculation efficiency. The CFD-based BF models can be categorised into three generations: first generation (1st gen): steady-state BF models; second generation (2nd gen): transient-state BF models and third generation (3rd gen): data-driven transient-state BF models. The first two gens have been achieved and the third gen is under development at Shen Lab, UNSW.

In this paper, we review the recent multi-fluid BF model developed at Shen Lab, UNSW. Our aim is to demonstrate the model's capabilities in studying different iron ore products and novel operations adopted under industrial-scale BF conditions. The paper is structured as follows: First, we briefly introduce the model framework, followed by its applications. Specifically, we report on the examination of iron ore products first, and then examine some novel operations, including burden control, oxygen level tuning in blast air and hydrogen injections. Next, we present the evolution of in-furnace phenomena over a certain time after introducing novel operations, such as hot burden charging, wet burden charging and blast condition tuning. Finally, we draw our conclusions.

MODEL DESCRIPTION

The multi-fluid BF model considers the existence of burden layer structure (Yu and Shen, 2018; Dong, 2004) and cohesive zone (Yu and Shen, 2019d; Kanbara *et al*, 1976; Sasaki *et al*, 1976; Togino *et al*, 1979) inside a BF. Besides, the interaction between two phases (gas-solid, gas-liquid and solid-liquid) and key chemical reactions (reduction of iron ore by carbon, carbon monoxide and hydrogen; water gas reaction, water gas shift reaction and carbon gasification) have been considered. The BF models are developed from a steady-state version to a transient-state version. The conservation equations for continuity, momentum, temperature and species are summarised in Table 1; and more details of the model can be found in our recent publications (Yu and Shen, 2019b, 2022; Yu, 2020).

TABLE 1
Governing equations of the CFD-based BF model.

Governing equation	Description
Continuity	$\dfrac{\partial}{\partial t}\int_{\Omega}\varepsilon\rho\,d\Omega+\int_{S}\varepsilon\rho U\bullet n\,dS=\int_{\Omega}\varepsilon S_{mass}\,d\Omega$
Momentum	$\dfrac{\partial}{\partial t}\int_{\Omega}\varepsilon\rho U\,d\Omega+\int_{S}\varepsilon\rho UU\bullet n\,dS=\int_{\Omega}\varepsilon\tau\bullet n\,dS-\int_{\Omega}\varepsilon p I\bullet n\,dS+\int_{\Omega}\varepsilon\rho g\,d\Omega+\int_{\Omega}\varepsilon S\,d\Omega$
Temperature	$\dfrac{\partial}{\partial t}\int_{\Omega}\varepsilon\rho T\,d\Omega+\int_{S}\varepsilon\rho TU\bullet n\,dS=\int_{S}\dfrac{\varepsilon\lambda}{c_p}\nabla T\bullet n\,dS+\int_{\Omega}\varepsilon\dfrac{S_h}{c_p}\,d\Omega$
Species	$\dfrac{\partial}{\partial t}\int_{\Omega}\varepsilon\rho\phi\,d\Omega+\int_{S}\varepsilon\rho\phi U\bullet n\,dS=\int_{S}\varepsilon\rho D\nabla\phi\bullet n\,dS+\int_{\Omega}\varepsilon S_\phi\,d\Omega$

MODEL APPLICATIONS

In this section, we will briefly demonstrate some typical applications of the BF model of the first two generations for the assessment of iron ore products and the optimisation of the BF operations. Firstly, we study several iron ore products, including those currently used in BFs and emerging ones like carbon composite briquettes towards carbon-neutral ironmaking. Secondly, we review our recent works of various iron ore operations, including optimising conventional burden operations such as batch weight tuning and oxygen enrichment, as well as introducing new operations like central coke charging and hydrogen injection in BFs. Lastly, we review the 2nd gen BF models – time-sensitive operations, particularly those related to the change of burden materials' properties.

Examination of iron ore products

In BF practice, iron ore and coke are alternately charged at the furnace top using a rotating chute, resulting in the existence of particle layers inside an ironmaking blast furnace. The in-furnace behaviour of iron ore and coke can be studied using the BF model, considering the burden layer structure inside a BF. It has been observed that the distributions of gas components, gas temperature and solid temperature exhibit fluctuating trends (Yu and Shen, 2018). These trends are mainly attributed to the distinct thermochemical properties between iron ore products and metallurgical coke. Moreover, these trends represent important features of an ironmaking BF, as studied by the model, which were seldom reported before this work (Yu and Shen, 2018). This way, the ordinary iron ore products can be assessed under BF conditions using this model in terms of in-furnace details and overall BF performance.

Carbon composite briquette (CCB) is an iron ore product that combines low-rank coal and iron oxides through agglomeration technology. It shows promise as a raw material for improving energy efficiency and reducing the use of high-grade coke in BF ironmaking. To study the in-furnace behaviours with CCB used as a raw material for ironmaking, the BF model is extended accordingly. (Yu and Shen, 2019a). This model allows for the detailed study of reduction behaviours for both iron ore and CCB, as well as the carbon gasification behaviours of both coke and CCB.

Figure 1 shows the reduction degree distributions of pellet and CCB, which is defined in Equation 1.

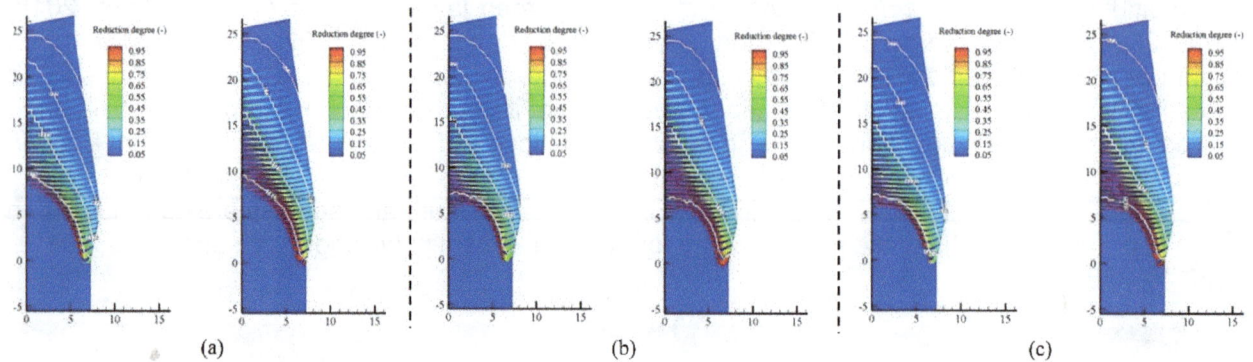

FIG 1 – Reduction degree distribution of iron oxides (for each subfigure, the left stands for iron ore while the right is for CCB) inside the BF under different CCB rates: (a) 20 kg·tHM^{-1}; (b) 40 kg·tHM^{-1}; (c) 60 kg·tHM^{-1} (Yu and Shen, 2019a).

$$R = 1 - \frac{2}{3}\frac{N_o}{N_{Fe}} \tag{1}$$

where N_o is the residual mole of oxygen element in ore, N_{Fe} is the total mole of iron element. Compared to iron ore pellet, CCB can be reduced faster under the same thermal conditions; This can be well quantified. In particular, the reduction degree of pellet is nearly 0.5 at the solid temperature of 1173 K while that of CCB is around 0.65. Additionally, it has been calculated that coke carbon can be protected through CCB charging, leading to coke saving with an increase in the amount of CCB charged.

Examination of BF operations including burden conditions, oxygen enrichment and hydrogen injection

The burden profiles and particle size control of iron ore and coke can significantly impact a BF's performance in terms of fuel rate and reducing gas utilisation efficiency. Several operations are reviewed below to demonstrate the effectiveness of the BF model.

Burden batch weight is a critical operating parameter in the blast furnace (BF) system. It directly affects solid distribution, such as coke-layer and ore-layer thickness and consequently influences gas permeability and thermal-chemical conditions inside the BF. The effects of burden batch weight on BF performance are comprehensively studied using the BF model. Through this study, it was observed that increasing the burden batch weight from 112 t to 140 t led to a nearly 1 m/s decrease in the average gas velocity in BF central regions. Additionally, it was found the height of the cohesive zone peak shortened by about 7 m, while BF reducing gas utilisation efficiency improved, lowering down the fuel rate (Yu and Shen, 2020b).

Central Coke Charging (CCC) operation, which involves admitting more coke near the furnace centre than in other regions, is a promising technology for stabilising BF operations when using low-grade materials, including low-grade iron ore, low-grade coal and nut coke (a low-grade coke). The BF model is extended to systematically study the effects of CCC patterns on the performance of an industrial-scale BF (Yu and Shen, 2019c). It is found that furnace behaviours can be quite different when adopting CCC compared to non-CCC conditions. These differences are summarised as follows. Compared to non-CCC operation:

- For CCC operation, gas permeability is improved and the cohesive zone position can also be increased.

- For CCC operation, the temperature curve of BF top gas shows a clear bell-shaped trend along the furnace radial direction, with a narrow region of high temperature, close to 1000 K, found at the furnace centre.

- For CCC operation, the reducing gas utilisation efficiency shows a slightly decreasing trend and the distribution of iron oxides inside the BF changes noticeably.

- For CCC operation, the carbon solution loss reaction of metallurgical coke becomes insignificant near the furnace central regions.

In addition, different coke charging amounts at the furnace centre can affect in-furnace phenomena and the BF fuel rate. Through a BF modelling study for an industrial-scale BF, it was observed that increasing the CCC radius from 1/64 to 1/4 of the furnace throat radius resulted in a decrease in the top gas utilisation efficiency from 47.5 per cent to 41.5 per cent. Additionally, the top gas temperature increased from approximately 500 K to around 700 K and the metallurgical coke rate to produce 1000 kg of hot metal increased by 43 kg (Yu and Shen, 2019d). However, the carbon solution loss reaction of metallurgical coke is further decreased.

Nut coke is typically considered a 'by-product' obtained from coke-making batteries and coke transportation, with a size generally less than 40 mm. Due to its fine size and the resulting decrease in furnace permeability, nut coke cannot be mixed with normal-sized coke and charged directly into BFs. However, a viable alternative is to blend nut coke with iron ore pellets, which can significantly improve furnace performance. To study the in-furnace behaviours of a BF when using a mixture of nut coke and iron ore pellets, the BF model is further extended. The relationship between the nut coke rate and BF performance is numerically examined in terms of top gas temperature, reducing gas utilisation efficiency and fuel rate. Through the study, it is calculated that utilising 20 kg of nut coke to produce 1000 kg of hot metal results in the best performance for the studied BF (Yu and Shen, 2019d).

Oxygen enrichment operation involves adding extra oxygen to the hot blast, aiming to increase the admitting rate of iron ore products at the furnace top and improve hot metal production. It is a promising technology to reduce the carbon footprint in the ironmaking process, leading to lower fuel rates. The BF model is extended to study BF performance with oxygen enrichment operation (Yu and Shen, 2020a). The simulation results align well with the measurements in terms of BF top gas temperature and productivity. Particularly, when oxygen enrichment is increased by 1 per cent, the top gas temperature decreases by 17 K. A nonlinear relationship between fuel rate and oxygen enrichment rate is also observed. By adopting the optimal oxygen enrichment rate of 7.5 per cent to produce 1000 kg of hot metal, the carbon footprint emission can be theoretically reduced by 26.1 kg.

Recently, **hydrogen** has garnered significant attention from steelmaking industries worldwide due to its large potential for substantially reducing the carbon footprint in the steelmaking process and achieving green steel production. Hydrogen can be injected into an ironmaking BF either via blast tuyeres, shaft tuyeres, or both. However, as an emerging technology, the effects of such injections on in-furnace phenomena are not yet fully understood. Therefore, the BF model is extended to study the effects of hydrogen injection on BF performance, considering different injection schemes (Yu, Hu and Shen, 2021; Zhao *et al*, 2020, 2023). For example, in the study of hydrogen shaft injection (Yu, Hu and Shen, 2021), the penetration depth of shaft-injected hydrogen is limited to 0.5 m when the injection rate is less than 5 m³/s and the thermal conditions near shaft inlets change insignificantly when the injection rate is less than 10 m³/s. Moreover, with increased hydrogen injection rates, the height of the CZ peak increases and its profile becomes more concave-like in shape. Additionally, it is calculated that a fuel rate saving of 28 kg can be achieved with a hydrogen injection rate of 30 m³/s.

Examination of iron ore productions in time-sensitive BF operations using 2nd gen BF model

It is essential to understand the evolution of in-furnace phenomena after certain operational conditions are changed in a very short window, sometimes they are unplanned eg the use of web burden after expected weather change. In these circumstances, the BF model considers the time effects on the change of burden layer structure and flow-thermal-chemical phenomena inside an ironmaking BF, ie the evolution of internal state over time. By adopting this transient BF (tBF) model, real-time *in situ* predictions, including the evolution of the cohesive zone, the formation of the deadman and the development of chemical fields, can be studied. Several time-sensitive operations tBF cases are reviewed below to demonstrate the effectiveness of the tBF model.

Hot burden refers to iron oxide sintering products with high temperatures that are not cooled to room temperature before charging into the BF. The high temperature of solid particles can provide a significant amount of thermal energy to the BF, potentially leading to fuel rate reduction. However, hot burden can also increase the flow velocity of BF reducing gas, decreasing gas residence time. The evolution of in-furnace phenomena has been investigated using the tBF model, comparing scenarios without hot burden use to those using such burden (Yu and Shen, 2022).

Figure 2 shows the distribution of Fe_3O_4 inside the BF with increasing time after hot burden deployment. It is observed that the mass fraction of Fe_3O_4 gradually increases from the furnace top to the middle shaft. Subsequently, the maximum Fe_3O_4 content in iron ore is achieved at certain heights before being further reduced to FeO. Adopting hot burden charging introduces new features in the distribution of Fe_3O_4. Particularly, the region with a high mass fraction of Fe_3O_4 gradually shifts toward the furnace top, especially in the first few hours. After nearly 10 hours, the distribution of Fe_3O_4 stabilises, indicating a new steady-state in-furnace phenomena might have been achieved. The response of thermal conditions near the furnace top to hot burden charging is more significant than the regions far from the furnace top and the model accurately captures these thermal features. Additionally, the change in furnace behaviours primarily occurs within the first 8–10 hours after deploying hot burden charging.

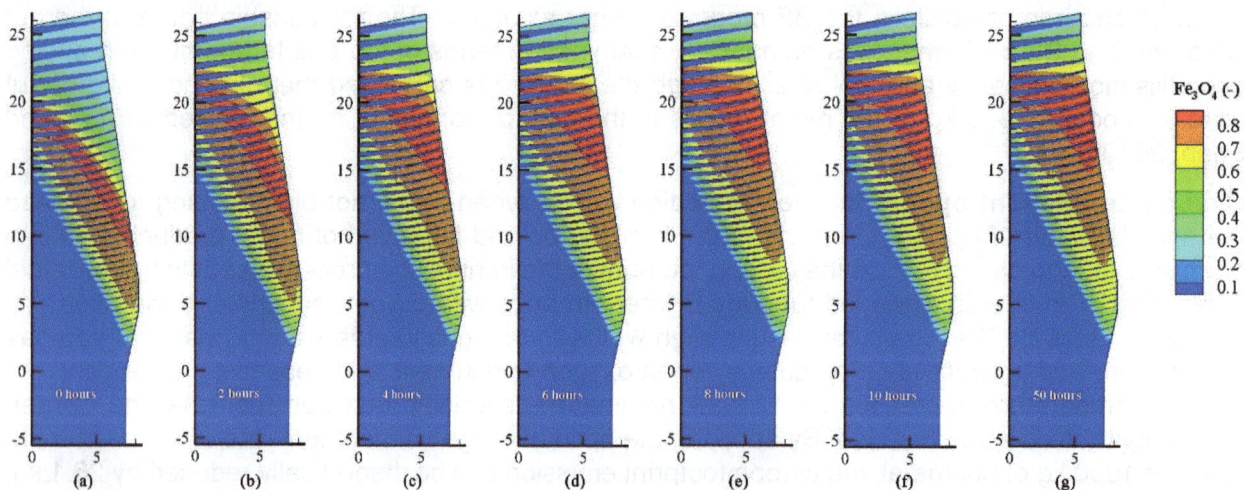

FIG 2 – Fe_3O_4 distribution snapshots at different simulation times after adopting hot burden charging: (a) 0 h; (b) 2 h; (c) 4 h; (d) 6 h; (e) 8 h; (f) 10 h; and (g) 50 h (Yu and Shen, 2022).

On the other hand, the temperature of blast air can also significantly affect the in-furnace behaviours of an ironmaking BF. The use of undesired blast temperature can result in different BF performance, with and without remedial actions taken. Therefore, the tBF model has been employed to study the thermochemical behaviours when the blast temperature fluctuates (Yu and Shen, 2023). Under the investigated conditions, it was observed that the thermal states in different regions of the BF respond to the change in quite distinct ways. Specifically, the regions near the BF gas inlet and cohesive zone showed significant responses to the blast temperature change, while the thermal conditions near the furnace top changed relatively insignificantly. By implementing certain countermeasures, such as fuel rate tuning, the thermal states can be effectively maintained even under varying blast temperatures, ensuring minimal impact on the reduction process of iron oxides. Additionally, the BF model can suggest a time window for implementing these countermeasures. The model results deepen our understanding of BF dynamic behaviours when new raw materials or operations are deployed.

Another example is the study of **charging wet burden** into a BF using the tBF model (Zhao, Yu and Shen, 2022). Since the burden moisture content varies with whether conditions, the BF in-furnace phenomena and fuel rate can also be affected when wet burden is used. Through this study, it was found that the burden can quickly dry near the BF upper shaft. However, the shaft temperature can be lowered by the evaporation process of the burden moisture. The BF top gas temperature can decrease to below 373 K within one hour after charging the wet burden under heavy rain conditions. When adopting certain countermeasures, such as changing the oxygen enrichment ratio or using dry burden, the deterioration of thermal states inside the BF can be partially or largely offset. The tBF model study provides time-related information when wet burden is used unavoidably.

CONCLUSIONS

The assessment of iron ore products in terms of in-furnace behaviours and overall furnace performance holds paramount importance in the ironmaking industry. In this paper, we review our recent progress at Shen Lab, UNSW, in using CFD-based BF models to assess iron ore products and industrial-scale ironmaking BF operations.

Our BF models feature unique layered burden structures for iron ore products, enabling systematic studies of their in-furnace behaviours. Moreover, these models allow for investigating changes in operating conditions, such as alterations in blast conditions and raw iron oxide materials. This capability facilitates the capturing of evolutions inflow field, temperature field and species distribution field. Consequently, a more detailed understanding of iron ore behaviours in an industrial-scale BF can be achieved at lower cost and risk, compared to plant tests.

Furthermore, these BF models can empower both existing and emerging iron ore products to establish their competitive edge in diverse BF conditions. By showcasing their performance and efficiency metrics, iron ore manufacturers can effectively communicate the benefits of their products to stakeholders. These CFD models pave the way for sustainable and competitive ironmaking practices, providing the iron ore industry with advanced insights and innovative marketing strategies.

REFERENCES

Dong, X, 2004. Modelling of Gas-Powder-Liquid-Solid Multiphase flow in a blast furnace, University of New South Wales.

Kanbara, K, Hagiwara, T, Shigemi, A, Kondo, S, Kanayama, Y, Wakabayashi, K and Hiramoto, N, 1976. Dissection of Blast Furnaces and Their Inside State (in Japanese), *Tetsu-to-Hagané*, 62:535–546.

Sasaki, K, Hatano, M, Watanabe, M, Shimoda, T, Yokotani, K, Ito, T and Yakoi, T, 1976. Investigation of Quenched No. 2 Blast Furnace at Kokura Works (in Japanese), *Tetsu-to-Hagané*, 62:580–591.

Togino, Y, Sugata, M, Abe, I and Nakamura, M, 1979. Investigation on the Profile of Softening-Melting Zone in the Dissected Blast Furnace, *Tetsu-to-Hagané*, 65:1526–1535.

Yu, X and Shen, Y, 2018. Modelling of Blast Furnace with Respective Chemical Reactions in Coke and Ore Burden Layers, *Metallurgical and Materials Transactions B*, 49:2370–2388.

Yu, X and Shen, Y, 2019a. Computational Fluid Dynamics Study of the Flow and Thermochemical Behaviors of a Carbon Composite Briquette in an Ironmaking Blast Furnace, *Energy and Fuels*, 33:11603–11616.

Yu, X and Shen, Y, 2019b. Model analysis of gas residence time in an ironmaking blast furnace, *Chemical Engineering Science*, 199:50–63.

Yu, X and Shen, Y, 2019c. Model Study of Blast Furnace Operation with Central Coke Charging, *Metallurgical and Materials Transactions B*, 50:2238–2250.

Yu, X and Shen, Y, 2019d. Model Study of Central Coke Charging on Ironmaking Blast Furnace Performance: Effects of Charging Pattern and Nut Coke, *Powder Technology*, 361:124–135.

Yu, X and Shen, Y, 2020a. Computational Fluid Dynamics Study of the Thermochemical Behaviors in an Ironmaking Blast Furnace with Oxygen Enrichment Operation, *Metallurgical and Materials Transactions B*, 51:1760–1772.

Yu, X and Shen, Y, 2020b. Numerical Study of the Influence of Burden Batch Weight on Blast Furnace Performance, *Metallurgical and Materials Transactions B*, 51:2079–2094.

Yu, X and Shen, Y, 2022. Transient State modeling of Industry-scale ironmaking blast furnaces, *Chemical Engineering Science*, 248(A):117185.

Yu, X and Shen, Y, 2023. Numerical study of transient thermochemical states inside an ironmaking blast furnace: Impacts of blast temperature drop and recovery, *Fuel*, 348:128471.

Yu, X, 2020. Numerical Study of Novel Operations of an Ironmaking Blast Furnace, PhD thesis, the University of NSW.

Yu, X, Hu, Z and Shen, Y, 2021. Modeling of hydrogen shaft injection in ironmaking blast furnaces, *Fuel*, 302:121092.

Zhao, Z, Yu, X and Shen, Y, 2022. Transient CFD study of wet burden charging on dynamic in-furnace phenomena in an ironmaking blast furnace: Impacts and remedies, *Powder Technology*, 408:117708.

Zhao, Z, Yu, X, Li, Y, Zhu, J and Shen, Y, 2023. CFD study of hydrogen co-injection through tuyere and shaft of an ironmaking blast furnace, *Fuel*, 348:128641.

Zhao, Z, Yu, X, Shen, Y, Li, Y, Xu, H and Hu, Z, 2020. Model Study of Shaft Injection of Reformed Coke Oven Gas in a Blast Furnace, *Energy and Fuels*, 34:15048–15060.

Greener downstream processing – DRI, HBI and pig iron production

Phosphorus distribution behaviour in hematite ore reduced in CO-H$_2$ gaseous atmosphere

E K Chiwandika[1,2], S M Masuka[3] and S-M Jung[4]

1. Lecturer, Harare Institute of Technology, 263 Harare, Zimbabwe. Email: chiwandikae@gmail.com
2. Lecturer, University of Zimbabwe, 263 Harare, Zimbabwe. Email: shebarmasuka@gmail.com
3. Lecturer, University of Zimbabwe, 263 Harare, Zimbabwe. Email: shebarmasuka@gmail.com
4. Professor, Graduate Institute of Ferrous Technology, 37673 Pohang, North Gyeongsang, South Korea. Email: smjung@postech.ac.kr

ABSTRACT

The world reserves of high-grade iron ore have depleted necessitating the need to utilise any existing natural iron ore deposits to meet the demands of the world's expanding economy using hydrogen in the reduction process to curb CO$_2$ emission. These low-grade ores include iron ores with high phosphorus content. The thermodynamic conditions in the blast furnace allow for the reduction of phosphorus that readily dissolves into the molten iron. Phosphorus in the gaseous state readily dissolves into the molten iron and therefore must be removed due to its negative influence on the mechanical properties of steel. The removal of phosphorus from molten iron is difficult causing the phosphorus to concentrate in metallic iron. The aim of this study is to provide basic knowledge on the phosphorus distribution after reduction using the CO-H$_2$ and possibly provide the optimum conditions for phosphorus partitioning into slag, increasing the probability of its removal by magnetic separation. Total porosity, pore size distribution and surface area of the pellets used were measured using a mercury pressure porosimeter. The reduction was carried out using thermogravimetric analyses (TGA). The reduction rate increased with increasing H$_2$ and was highest when 100 per cent H$_2$ was used. Metal elements distribution in the reduced pellet was clarified using electron probe microanalysis (EPMA) and showed that phosphorus was distributed in the same areas as Ca. More phosphorus was concentrated in slag when 50 per cent CO–50 per cent H$_2$ was used, increasing the probability of dephosphorization using magnetic separation. The concentration of phosphorus into slag was related to the rate of reduction. Further investigations into the reduction kinetics of this ore by CO-H$_2$ is suggested to validate these findings. Nevertheless, a comprehensive thermodynamic analysis of the reduction of phosphorus containing phases found in iron ore with carbon, CO and H$_2$ was provided to assist in selecting the optimal reduction experimental conditions.

INTRODUCTION

The global annual steel production is increasing (Pati and Vinacy, 2008) while the high-grade iron ore reserves has depleted coupled with the need to decarbonise the steel industry. The world has no choice but to utilise low-grade ore using hydrogen in the reduction process. These low-grade ore include iron ores that have high amounts of phosphorus (Delvasto *et al*, 2007). When ores that contain high levels of phosphorus are charged directly into the Blast furnace, which is still the dominant route of producing molten iron, the thermodynamic condition allows for the reduction of the phosphorus containing phases converting some P into a gaseous state which readily dissolves into molten iron. The removal of P from the molten iron is very difficult hence P is concentrated in the iron affecting the mechanical properties of the final product. It is very important to develop methods of removing the P in the ore before charging the ore into the Blast furnace.

Research has been done around world to develop processes such as bioleaching, chemical leaching, roasting and magnetic separation, selective agglomeration and reverse flotation to upgrade the quality of phosphorus-bearing iron ores (Tang, Guo and Zhao, 2010). The technologies that have been developed are still to be applied at large scale operations because they are time consuming and not cost-effective. The reduction of phosphorus containing phases in gaseous atmosphere coupled by melt separation showed that the use of high temperatures that favours high metallisation degree were not favourable for phosphorus partitioning into slag. The presence of CaO increased the partitioning ratio of the phosphorus between metal and slag (Tang, Guo and Zhao, 2010). Thermodynamic predictions showed that the carbothermic reduction of phosphorus in the apatite phase is difficult. The evaporation of phosphorus in this experiment was not significant enough to consider as dephosphorization even at temperatures of around 1200°C (Matinde and Hino, 2011).

Slag basicity was reported to influence the reduction of apatite with some portion of phosphorus being converted into a gaseous state, escaping together with CO_2, while the other part remained unreduced and distributed either in the iron rich phases or in the gangue phase (Han and Duan, 2012). The addition of about 10 wt per cent of Na_2CO_3 in the carbothermic reduction of phosphorus containing ore showed increased iron recovery and a decrease in phosphorus content in the iron (Bai *et al*, 2012). The reduction of tramp elements such as Si, P and Mn were reported to be thermodynamically slow in a hydrogen rich environment resulting in a metal of better purity as compared to carbothermal reduction but the phosphorus remained almost similar to that in carbothermic reduction.

Enhanced gaseous diffusion was reported due to the formation of some fissures in the oolitic units of the ore after the microwave pre-treatment of a phosphorus containing ore. The kinetic analysis of the ore showed that at a reduction fraction lower than 0.7, the apparent activation energy was between 20 and 40 kJ/mol which lies in mixed control region of interfacial reaction and gaseous diffusion (Tang *et al*, 2013). The application of the Hancock and Sharp method in combination with the model fitting method during isothermal reduction of an iron ore fine under hydrogen gas of between 0.25–1 atmosphere showed that the reaction was controlled by a single-phase boundary reaction in between the temperature range of 400–500°C. Some multistep mechanisms were rate controlling at temperature in the range of 400–900°C (Hessels *et al*, 2022). Enhanced iron metallisation of between 84.38–96.49 wt per cent was observed with increasing reduction time from 10 to 60 minutes in the time dependent phosphorus migration behaviour during the carbothermic reduction of a medium-phosphorus magnetite ore. Dephosphorisation was increased from 21.03 wt per cent to 33.07 wt per cent and then decreased to 31.61 wt per cent because the P in gaseous state was absorbed into metallic iron (Zhang *et al*, 2020).

A new method of upgrading a high phosphorus oolitic ore using sodium magnetisation roasting-magnetic separation-leaching was recently developed and showed an iron recovery rate of 88.7 per cent and a dephosphorisation rate of 82.7 per cent under optimal conditions (Pan *et al*, 2022). A thermodynamic prediction of phosphorus volatilisation during carbothermic reduction showed accelerated rate by the presence of SiO_2, Al_2O_3 and Fe_2O_3 (Zhang *et al*, 2018). Thus, the reduction temperature of the apatite was decreased by the presence of SiO_2, Al_2O_3 and Fe_2O_3 increasing the probability of the escape of gaseous P with the off-gases.

Research mainly focused on the phosphorus partitioning or removal methods using carbothermic reduction for the pre-treatment stage and only few research was done using gaseous pre-treatment methods especially with the use of H_2. The removal of phosphorus from the ores is further complicated by the nature of existence of phosphorus in the ore which is not clear at the moment. The global target of decarbonising the steel industry has also increased the need to investigate the fundamentals of hydrogen reduction. The aim of this study is to provide basic knowledge on the phosphorus distribution after reduction using the CO-H_2 gas combination and to provide the optimum conditions for phosphorus partitioning into slag. This might increase the probability of P removal from the pre-treated ore by magnetic separation. But, a comprehensive thermodynamic analysis of the reduction of phosphorus-containing phases with carbon, CO and H_2 will be provided to assist in the selection of the optimal experimental conditions.

A thermodynamic consideration

The state of phosphorus existence in natural iron ores is not clear at the moment (Pan *et al*, 2022; Roy, Drafall and Roy, 1978). This makes the process of developing an effective P removal process very difficult. Basic thermodynamic data was used in this experiment to provide a basic understanding of the process and to assist in selecting the experimental conditions even though there are several chemical natural interactions in the ore. Experimental evidence is of paramount important to verify the thermodynamic prediction. The following thermodynamic data was used to assist in the selection of the experimental conditions in this investigation:

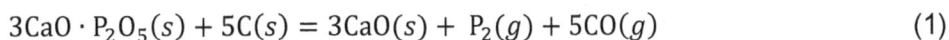

$$3CaO \cdot P_2O_5(s) + 5C(s) = 3CaO(s) + P_2(g) + 5CO(g) \qquad (1)$$

$$\Delta G° = 3,615,500 - 1,311.80T \text{ (J/mol)} \qquad (2)$$

$$Fe_2O_3(s) + 3C(s) = 2Fe(s) + 3CO(g) \tag{3}$$

$$\Delta G° = 586,390 - 508.5T \ (J/mol) \tag{4}$$

$$3CaO \cdot P_2O_5(s) + 5CO(g) = 3CaO(s) + P_2(g) + 5CO_2(g) \tag{5}$$

$$\Delta G° = 904,500 - 127.70T \ (J/mol) \tag{6}$$

$$C(s) = \underline{C} \tag{7}$$

$$\Delta G° = 24,740 - 43.51T \ (J/mol) \tag{8}$$

$$3CaO \cdot P_2O_5(s) + 5\underline{C} = 3CaO(s) + P_2(g) + 5CO(g) \tag{9}$$

$$\Delta G° = 1,625,800 - 770.95T \ (J/mol) \tag{10}$$

$$P_2O_5(s) + 5\underline{C} = 2\underline{P} + 5CO(g) \tag{11}$$

$$\Delta G° = 267,750 - 427.55T \ (J/mol) \tag{12}$$

$$P_2O_5(s) + 5C(s) = P_2(g) + 5CO(g) \tag{13}$$

$$\Delta G° = 1,044,980 - 974.9T \ (J/mol) \tag{14}$$

$$P_2(g) = 2\underline{P} \tag{15}$$

$$\Delta G° = -315,700 + 10.79T \ (J/mol) \tag{16}$$

$$3CaO \cdot P_2O_5(s) + 5H_2(g) = 3CaO(s) + P_2(g) + 5H_2O(g) \tag{17}$$

$$\Delta G° = 1,077,000 - 277T \ (J/mol) \tag{18}$$

$$\tfrac{1}{2}H_2(g) = \underline{H} \tag{19}$$

$$\Delta G° = 36,459.4 - 30.5T \ (J/mol) \tag{20}$$

$$3CaO \cdot P_2O_5(s) + 10\underline{H} = 3CaO(s) + P_2(g) + 5H_2O(g) \tag{21}$$

$$\Delta G° = 1,825,706 - 251T \ (J/mol) \tag{22}$$

$$P_2O_5(s) + 5H_2(g) = P_2(g) + 5H_2O(g) \tag{23}$$

$$\Delta G° = 371,400 - 6T \ (J/mol) \tag{24}$$

$$\tfrac{1}{5}P_2O_5(s) + 2\underline{H} = \tfrac{2}{5}P_2(g) + H_2O(g) \tag{25}$$

$$\Delta G° = -61,778.4 - 111.36T \ (J/mol) \tag{26}$$

$$\tfrac{1}{2}P_2(g) + \tfrac{3}{2}H_2(g) = PH_3(g) \tag{27}$$

$$\Delta G° = 71,500 - 108.2T (J/mol) \tag{28}$$

$$2FeO(s) + SiO_2(s) = 2FeO \cdot SiO_2(s) \tag{29}$$

$$\Delta G° = -36,200 + 21.09T \ (J/mol) \tag{30}$$

$$2FeO \cdot SiO_2(s) + 2C(s) = 2Fe(s) + SiO_2(s) + 2CO(g) \tag{31}$$

$$\Delta G° = 354,140 - 314.59T \ (J/mol) \tag{32}$$

$$FeO(s) + C(s) = Fe(s) + CO(g) \tag{33}$$

$$\Delta G° = 317,940 - 320.5T \ (J/mol) \tag{34}$$

$$3CaO \cdot P_2O_5(s) + 3SiO_2(s) + 3C(s) = 3CaO \cdot 3SiO_2(s) + P_2(g) + 5CO(g) \tag{35}$$

$$\Delta G° = 2,946,902 - 2055.14T \ (J/mol) \tag{36}$$

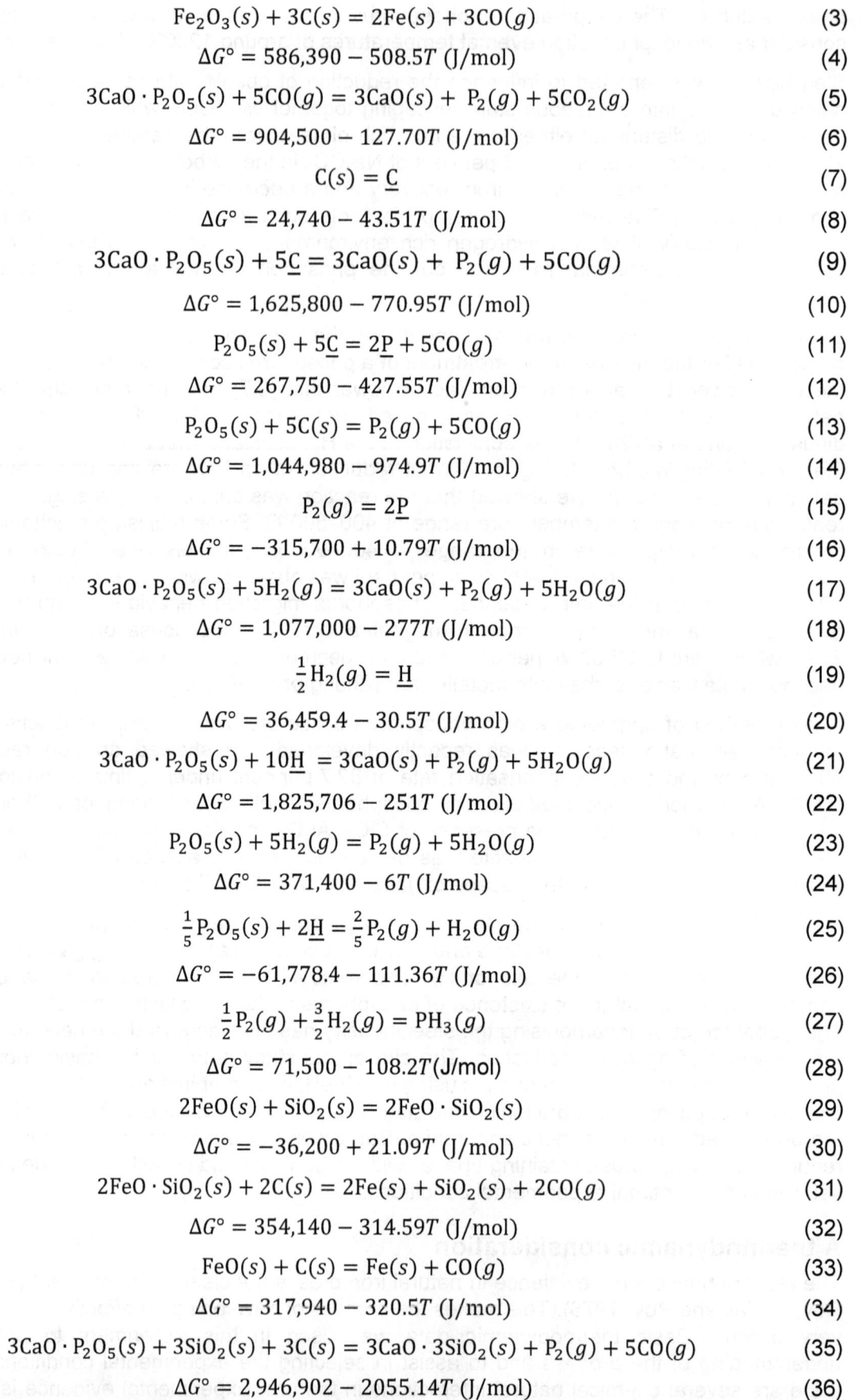

The thermodynamic data was from: (Tang, Gau and Zhao, 2010; Matinde and Hino, 2011; Kashiwaya and Hasegawa, 2012; Lee, 1999; Hino and Ito, 2010; Chen *et al*, 2022) The thermodynamic data above was used in the plot of $\Delta G°$ versus Temperature shown in Figure 1.

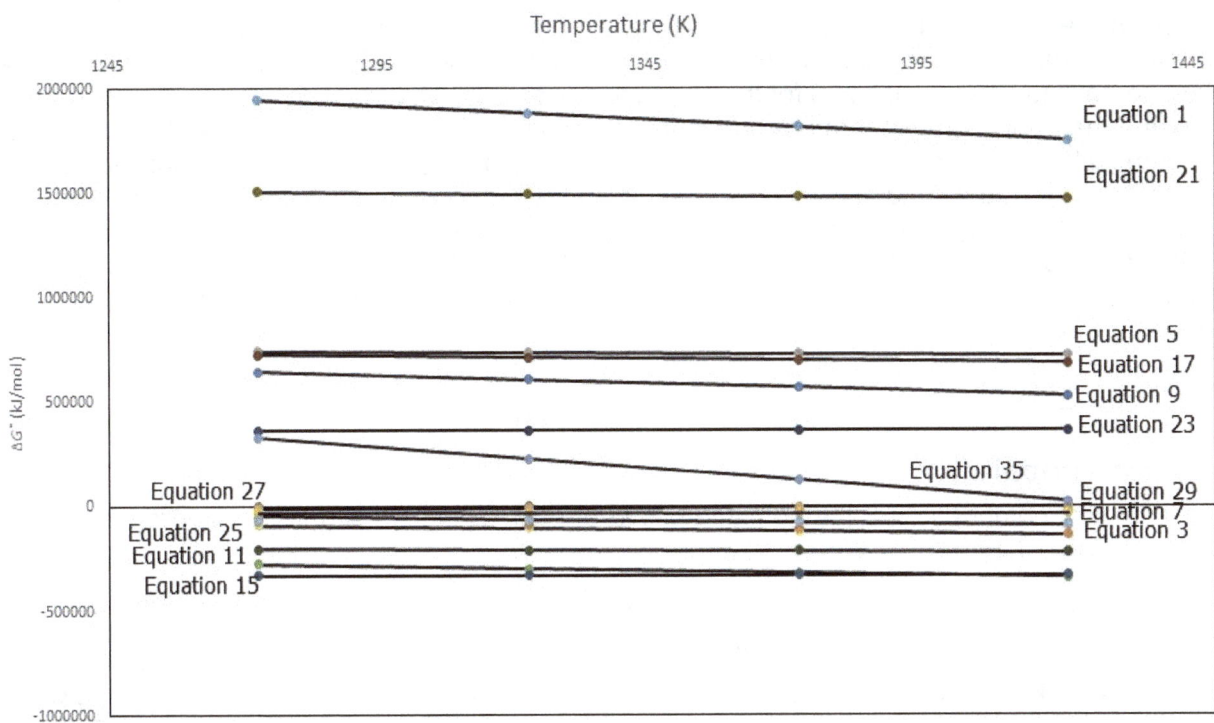

FIG 1 – Plot of $\Delta G°$ versus Temperature in the temperature range of 1000–1150°C.

Thermodynamic predictions from previous research (Matinde and Hino, 2011) reported that the carbothermic reduction of phosphorus that exists in the ore as apatite is difficult which is in agreement with the current prediction that shows a positive $\Delta G°$ value in this current research shown by Equation 1 in Figure 1. The CO which is a reaction product of Equation 1 can also act as a reducing agent for apatite according to Equation 5 but thermodynamic predictions also shows that the possibility of that reaction occurring is low since $\Delta G°$ is also positive. Some of the carbon in the system may dissolve in the metal according to Equation 7, which is highly likely since $\Delta G°$ is negative at this temperature range. The dissolution of carbon into the metal is predicted thermodynamically using Equation 7 to occur above 296°C. Reduction of the apatite with the dissolved C is also predicted by thermodynamics to be difficult as can be shown by the thermodynamic data of Equation 9. Comparison of the thermodynamic data of the direct reduction of hematite, Equation 3, with Equations 1, 5 and 9, it is clear that the reduction of hematite is highly expected as compared to that of the apatite phase at this temperature range. Research targets the removal of P through partial reduction of the iron ore containing P in the form of apatite where the P phase remain unreduced, concentrated into the slag rich phase and subject the pre-reduced ore to magnetic separation after crushing. The effectiveness of these methods highly depends on the ability to concentrate the unreduced P in the slag phase.

Most iron ores contain phosphorus, with some in acceptable levels of around 0.08 wt per cent (Delvasto *et al*, 2007; Oyama *et al*, 2009). In general, what is of major interest is that we have to find ways of using the ores with high levels of P so that we can minimise the negative impact caused by the presence of P in steel, by utilising the unused resource. Currently, most high P ores remain unused. Since the existence of P in iron ore is not yet clear, P can exist in another form like the P_2O_5. The reduction of P_2O_5 by carbon or dissolved carbon has been predicted thermodynamically to easily occur. The reduced P can easily dissolve into the iron as can be seen from the thermodynamic data of Equations 11 and 15. This will complicate P removal from the ore hence the investigation into methods where the P can be stabilised in a non-reducible phase and concentrated into slag must be done. This might provide a better chance of decreasing the amount of P that remains trapped into the pre-reduced ore by applying technologies such as magnetic separation before the application of hydrometallurgical methods.

If hydrogen is used as the main reducing gas, it can be seen that the reduction of apatite by either dissolved hydrogen or hydrogen gas is not expected to be thermodynamically easy according to Equations 17 and 21. Thermodynamics also shows that P_2O_5 reduction by H_2 has a positive $\Delta G°$ in this temperature range according to Equation 23, however the reduction of P_2O_5 by dissolved hydrogen is expected to be thermodynamically easy, Equation 25. The dissolution of H_2 into iron can be thermodynamically predicted according to Equation 20 to occur above 922°C that is well above 296°C predicted using Equation 8 for the dissolution of carbon. The presence of SiO_2 decreased the carbothermal reduction temperature of apatite Equation 35. This might complicate the chances of the removal of P because if more P is converted into a gaseous state, the probability of the dissolution of P into the metal is increased and also the formation of phases like fayalite which are difficult to reduce might be highly favoured resulting in an increase in coke rate. The careful control of the reduction experiment using H_2 or a combination of $CO-H_2$ gas might provide a solution for removing P by trapping P in an unreduced phase, concentrating P in slag followed by magnetic separation.

EXPERIMENTAL

Material preparation

The chemical composition of the hematite-based ore containing P used in this investigation is as shown in Table 1.

TABLE 1

Chemical composition of the ore.

Total Fe	FeO	Fe_2O_3	SiO_2	Al_2O_3	CaO	K_2O	MgO	P	TiO_2	LOI
55.37	1.29	77.73	10.96	4.87	0.316	0.302	0.429	0.20	0.153	3.75

The phosphorus content in the ore was analysed using inductively coupled plasma atomic emission spectroscopy (ICP–AES) and was found to be 0.20 wt per cent that is higher than 0.08 wt per cent considered to be the low phosphorus content in iron ore, and higher than 0.103 that was considered to be a high phosphorus hematite-goethite ore (Delvasto et al, 2007; Oyama et al, 2009). X-ray diffraction (XRD; Bruker AXS) was used for the identification of the major phases in this ore. The Cu tube at a scan angle of 10° to 95°, scanning rate of 2°/min, voltage of 40 kV and current of 40 mA were used during the analyses and the results are shown in Figure 2.

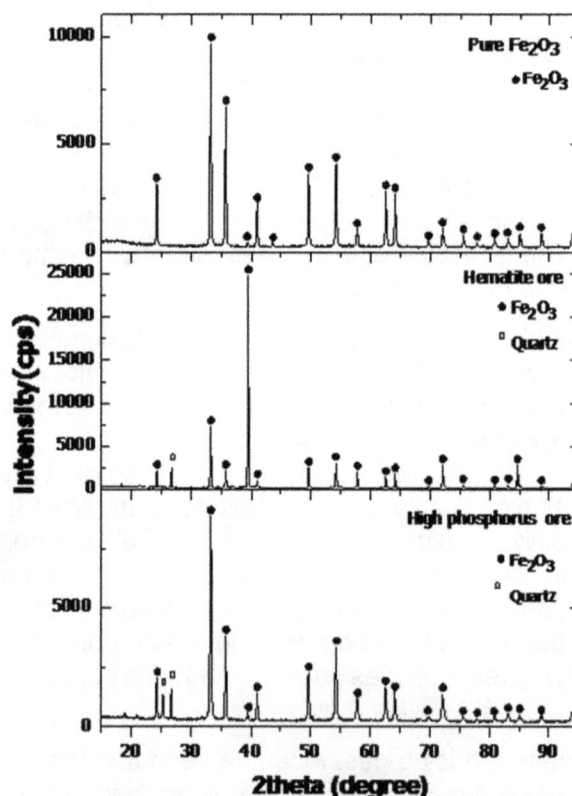

FIG 2 – Phase identification of the ore by XRD.

There were no major differences in phases identified in the high P ore and in the normal ore. The major phases identified in the both ores were Fe_2O_3 and SiO_2. The P containing phase could not be identified using the XRD because of its low amount in the ore. Distribution of elements in the ore was therefore investigated by mapping using electron probe microanalysis (EPMA; Joel (Japan)/JXA-8530F). The mapping was done on a mounted iron ore sample to specifically have an understanding of the distribution behaviour of phosphorus in the ore. This mapping was performed using a voltage of 20 keV and a beam current of 50 nA at a magnification of 400X, and the results are shown in Figure 3.

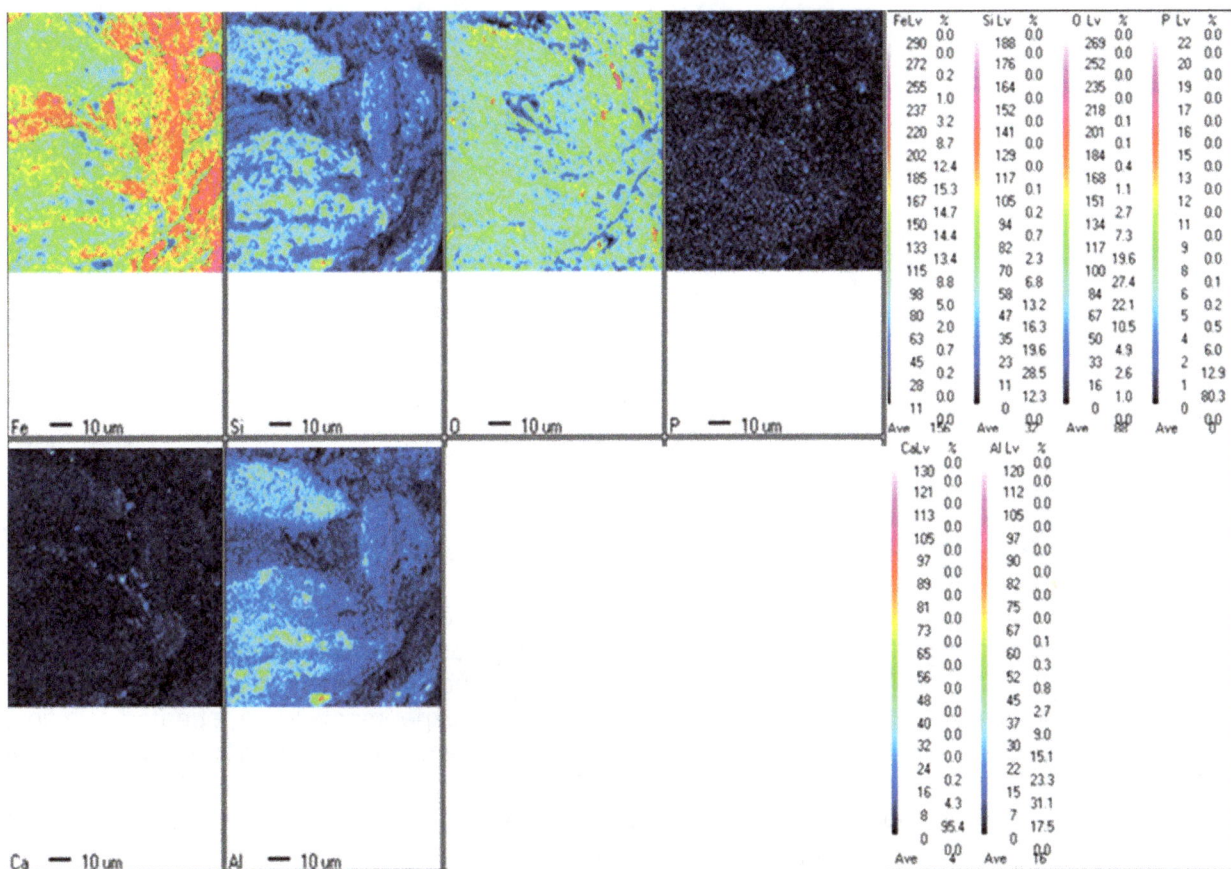

FIG 3 – Distribution of elements in the phosphorus-containing iron ore before reduction.

Figure 3 clearly shows that phosphorus was highly concentrated in the same areas as Al, Si and low amounts of Ca. This phosphorus-containing phase in the ore could not be clearly defined based on this finding.

Experimental procedure

The ore with the average particle size of 0.15 µm was formed into spherical pellets of radius 5.4 mm weighing approximately 2 g which were then cold pressed isostatically at 50 MPa for 30 minutes and sintered at 1200°C for 30 minutes in a 250 mL/min Ar atmosphere. The physical properties of the ore before and after sintering are shown in Table 2.

TABLE 2

Physical properties of the ore.

Sintering	Porosity (%)	Surface area (m^2/g)	Density (g/cm^3)
Before	36.83	0.97	2.85
After	33.06	0.92	2.60

The percentage porosity, surface area and density decreased after sintering. The CO was used as the main reducing gas hence the optimal gas flow rate was determined by placing the 2 g sintered pellet into a platinum basket that was then suspended on the balance arm of high temperature TGA, Rubotherm, and reduction experiments were performed at 1000°C using different CO flow rates at 75 minutes holding time. The results are shown in Figure 4 where the fractional reduction was evaluated as:

$$\text{Reduction degree (per cent)} = \frac{\Delta W}{W_0} \times 100 \tag{37}$$

where ΔW is the weight loss recorded by TGA and W_0 is the initial oxygen content in the pellet.

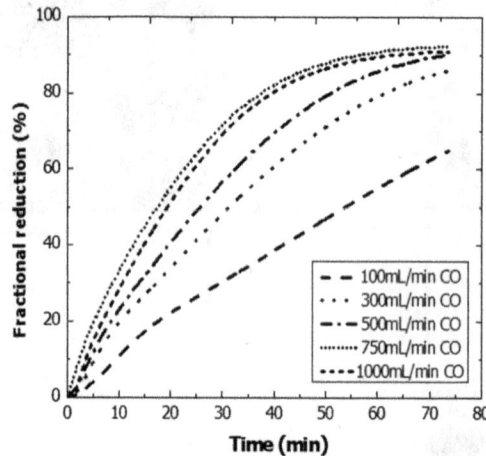

FIG 4 – Fractional reduction with change in the gas flow rate at 1000°C.

The reduction rate increased with increasing CO gas flow rate and it was believed to be caused by the increase in driving force. The optimal CO gas flow rate was therefore determined to be above 750 mL/min. The CO was then gradually replaced by increasing H_2 amounts in order to investigate the effect of H_2 on the P distribution behaviour. The reduction experiments were repeated at a flow rate of 1000 mL/min at 70 minutes holding time. The reduced samples were then taken for chemical analysis to determine the amount of metallic Fe. EPMA mappings were also performed on the reduced and mounted samples to investigate the distribution of elements after reduction to identify the positions where phosphorus was highly distributed.

RESULTS AND DISCUSSION

Effect of H_2 on the reduction behaviour of the hematite ore

The hematite with P iron ore pellets were reduced with various CO-H_2 gas compositions at 1000°C at 25 minutes holding time, and the results are shown in Figure 5.

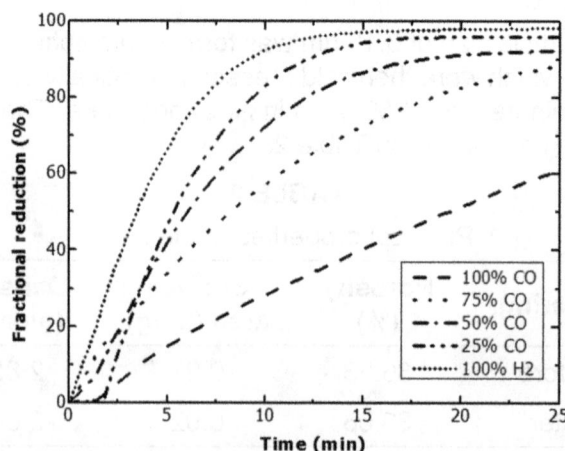

FIG 5 – Fractional reduction with various gas compositions at 1000°C.

The reduction rates were initially high up to a certain extent followed by a decreased rate to the end. The initial rate of reductions was lagging behind with 50 per cent CO–50 per cent H_2 in the early stages of reduction. This behaviour was pronounced with 25 per cent CO–75 per cent H_2 and rate of reduction increased just after the first few minutes. It has been reported in the previous research (Wagner et al, 2006) that the reduction by H_2 is endothermic, while that by CO is exothermic. As CO was replaced by H_2, it is believed that a decrease in available energy for the reaction to proceed might have occurred and might explain the lagging behind of the rate of reaction in the initial stage with 50 per cent CO and 25 per cent CO but with the progress of the reduction, the bursting of pores by CO might cause H_2 to easily diffuse into the reacting surface speeding up the rate of reduction in the process (Chiwandika and Jung, 2023). More research is required to clarify on this reduction behaviour.

The reduction rate was highest with H_2 and decreased with increasing CO. The slowing down in the rate of reduction in the last stages of reduction is believed to be caused by difficulty faced by the reducing gas to reach the reaction layer since diffusion plays a major role in the last stages of reduction since research (El-Geassy, 1986; Moon, Rhee and Min, 1998; Mousa, Babish and Senk, 2013) has shown that the reduction of iron ore starts from the outside following the topochemical receding interface model. However, more results are needed to investigate in detail the reduction kinetics of this ore. The main aim of this current research was to clarify the element distribution after reduction and to identify where phosphorus was likely to be highly located. This might provide the basic knowledge on the prospects of P removal by accumulating phosphorus in slag followed by magnetic separation.

The experiments were repeated with 70 minutes holding time to increase the amount of metallic iron in the reduced pellets. These pellets were analysed using chemical methods and ICP–AES to determine the amount of metallic Fe, Total Fe and Fe^{2+} and the results are shown in Table 3 where the Fe^{3+} was evaluated as:

$$Fe^{3+} = T.Fe - M.Fe - Fe^{2+} \qquad (38)$$

where T.Fe is the total Fe and M.Fe is the metallic Fe in the sample. The Fe^{3+} was converted into Fe_2O_3.

TABLE 3

Chemical analysis results of the pellets reduced with CO-H_2 gas at 1000°C after 70 minutes holding time.

mL/min		wt%					Fractional reduction	
CO	H_2	Total Fe	FeO	Fe_2O_3	Metallic Fe	O Remained	Calculated	TGA
1000	0	74.00	5.31	4.37	66.82	2.49	0.92	0.92
750	250	74.78	2.04	4.55	70.02	1.82	0.94	0.94
500	500	74.34	1.68	2.89	71.02	1.24	0.97	0.97
250	750	73.98	1.59	2.25	71.18	1.02	0.97	0.97
0	1000	73.55	4.14	2.86	68.34	1.78	0.94	0.99

The percentage reduction degree evaluated by TGA was based on the oxygen attached to Fe, the calculated reduction degree used for comparison purposes was evaluated as:

$$\text{Calculated Reduction degree (per cent)} = \frac{(^O/_{Fe})_i - (^O/_{Fe})_f}{(^O/_{Fe})_i} \times 100 \qquad (39)$$

where $(O/Fe)_i$ and $(O/Fe)_f$ are the initial and final ratios of oxygen in the sample respectively.

The chemical analyses results showed that metallic Fe increased with increasing H_2 up to 750 mL/min and decreased when using 1000 mL/min H_2. Previous research (Biswas, 1981) has shown that the combinations of CO and H_2 showed better reducing properties than the use of the respective gases. This was in agreement with the increase in metallic Fe with up to 750 mL/min H_2 found in the current research. Addition of H_2 to CO in iron ore reduction increases the reaction rate due to the ease by which H_2 can diffuse and react with oxygen in iron oxide more than CO. The bursting of pores by CO together with the exothermic nature of the reaction of iron oxide helps to accelerate the reaction when using 25 per cent CO–75 per cent H_2 gas. The iron product produced by H_2 reduction is dense (Biswas, 1981). This makes the diffusion of the reducing gas more difficult and may result in the low amount of metallic Fe observed when 1000 mL/min H_2 was used. The fractional reduction evaluated from the chemical analysis and that evaluated from the TGA have close similarities except for 1000 mL/min H_2. Thus, the oxygen attached to the Fe was the major contributor to the loss of weight from the sample. The weight loss contribution from other volatiles and reducible components in the system was negligible except when using 1000 mL/min H_2.

Distribution of elements in the pellets reduced by CO-H_2

The location of P after reduction was investigated using EPMA mapping on the reduced pellets which had been mounted and polished. The iron ore pellets were reduced in the CO-H_2 gaseous atmosphere at 1000°C at 70 minutes holding time. The EPMA mapping results are as shown in Figure 6.

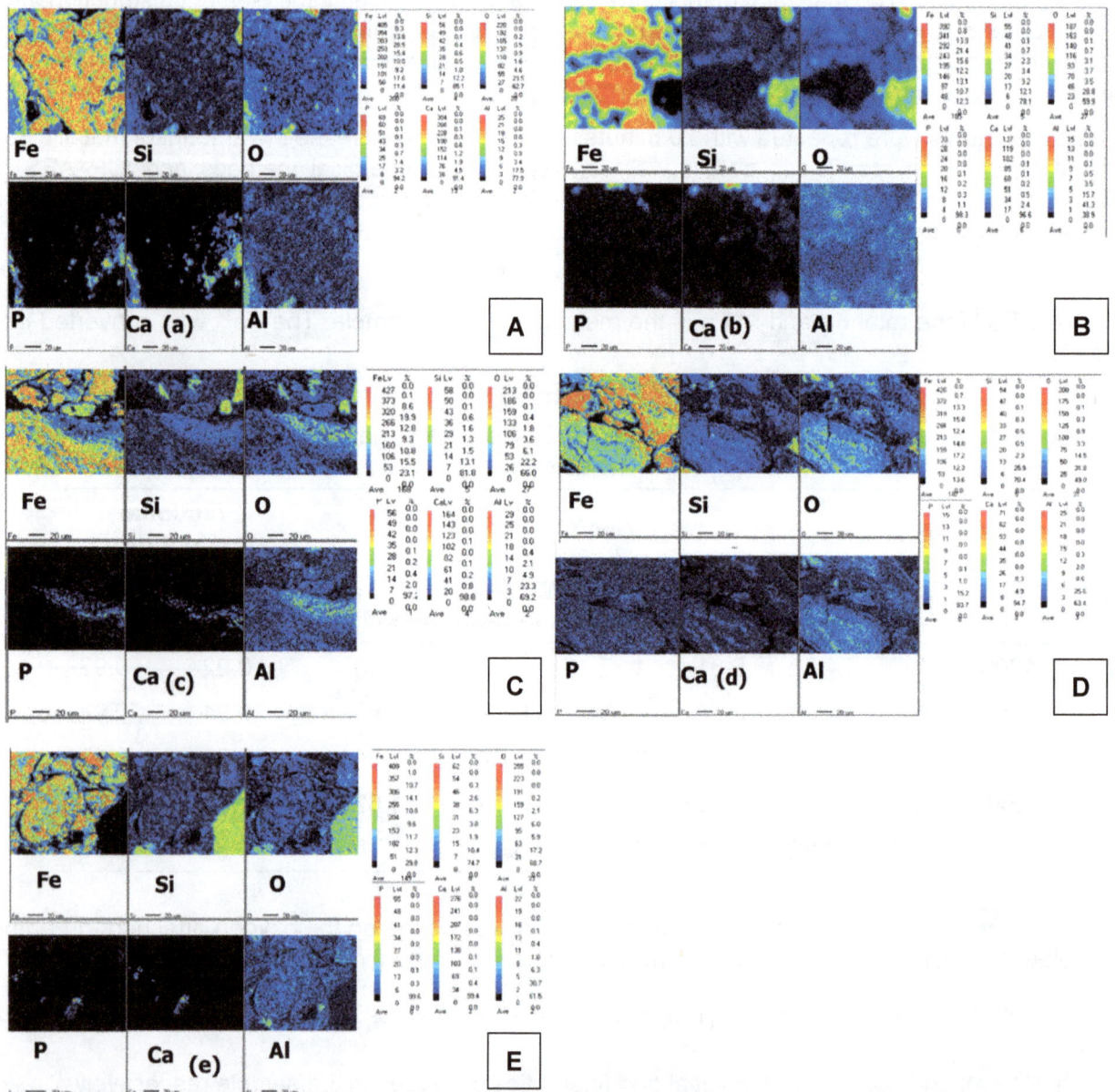

FIG 6 – Distribution of elements in the pellet reduced at 1000°C in: (a) 100 per cent CO, (b) 75 per cent CO–25 per cent H_2, (c) 50 per cent CO–50 per cent H_2, (d) 25 per cent CO–75 per cent H_2 and (e) 100 per cent H_2.

Figure 6 clearly shows the location of P that was highly distributed in the same areas as Ca and Al. The successful removal of P, especially by magnetic separation, requires its concentration in the slag phase, as shown in Figure 6c. The reduction of P containing phase may result in the formation of some gaseous P compounds which will easily dissolve into the metal as shown by the large negative thermodynamic data of Equation 15. Table 3 shows that metallic iron is available in abundance in the reduced pellet. Thus, the effective removal of P requires that P be preserved in a non-reducible phase like the apatite and concentrating this phase into the slag followed by magnetic separation before application of other methods like hydrometallurgy. In this current experiment samples that were reduced using 50 per cent CO–50 per cent H_2 gaseous atmosphere seems to have high amounts of P distributed in what appeared to be a slag phase with little amounts of Fe distributed in that same area. The concentration of the P into slag might depend on the reduction kinetics; however, more research is required for validation.

Kinetic analyses

Isothermal experiments were done in the temperature range of 1000–1150°C to investigate the kinetic effect on phosphorus distribution. The rate of conversion in a solid gas reaction is generally given by (Hessels *et al*, 2022):

$$\frac{dX}{dt} = k_{app}(T)f(X) \tag{40}$$

where $k_{app}(T)$ is the temperature dependent conversion rate constant in an isothermal experiment, $f(X)$ is the function describing the influence of the conversion extent on the conversion rate.

The rate constant k_{app} is given by the Arrhenius Equation as follows:

$$k_{app} = k_{app,0}\exp\left(\frac{Ea}{RT}\right) \tag{41}$$

where R is the universal gas constant, $k_{app,0}$ is the pre-exponent factor and E_a is the apparent activation energy.

Different mathematical expressions for $f(X)$ exist that are based on kinetic models. Integrating Equation 40 will yield:

$$g(X) = \int_0^x \frac{1}{f(X)}dx = k_{app}(T).t \tag{42}$$

When the model fitting method is applied, fitting the correct g(X) function as a function of *t* a straight line can be obtained assuming the correct kinetic model has been used. These models include but are not limited to those shown in Table 4.

TABLE 4

Different mathematical model that describe the rate limiting step.

Kinetic model	Mathematical function (g(X))
Chemical reaction (Phase boundary) control	$f_1=1-(1-X)^{1/3}$
Internal diffusion control	$f_2=1-3(1-X)^{2/3}+2(1-X)$
Nucleation and growth	$f_3=-\ln(1-X)^{1/n}$

The chemical control and the internal diffusion control models were fitted into the data obtained in the reduction of the P containing pellets obtained in the temperature range of 1000–1150°C. Figure 7 shows the plot when chemical reaction was the rate control step.

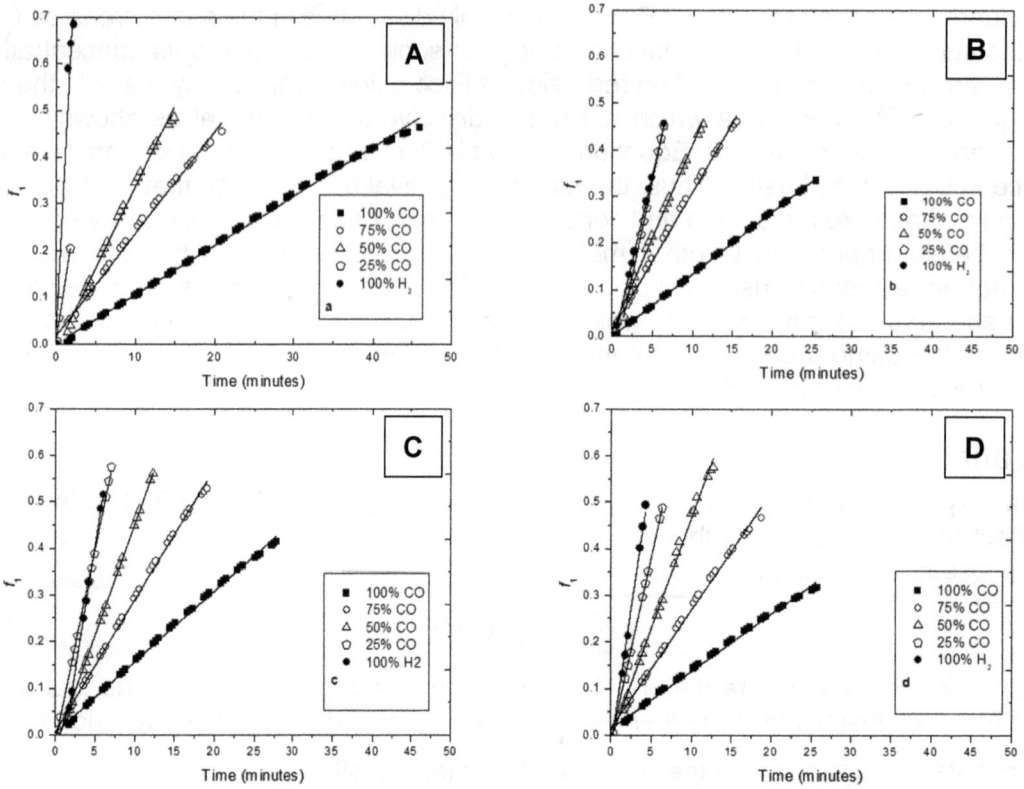

FIG 7 – Chemical reaction control plot with various CO-H_2 gas composition at: (a) 1000°C, (b) 1050°C, (c) 1100°C and (d) 1150°C.

The results from Figure 7 show that the reaction rate was controlled by the chemical reaction in the first minutes of reduction. The plot when intra-particle diffusion was the rate controlling step is shown in Figure 8.

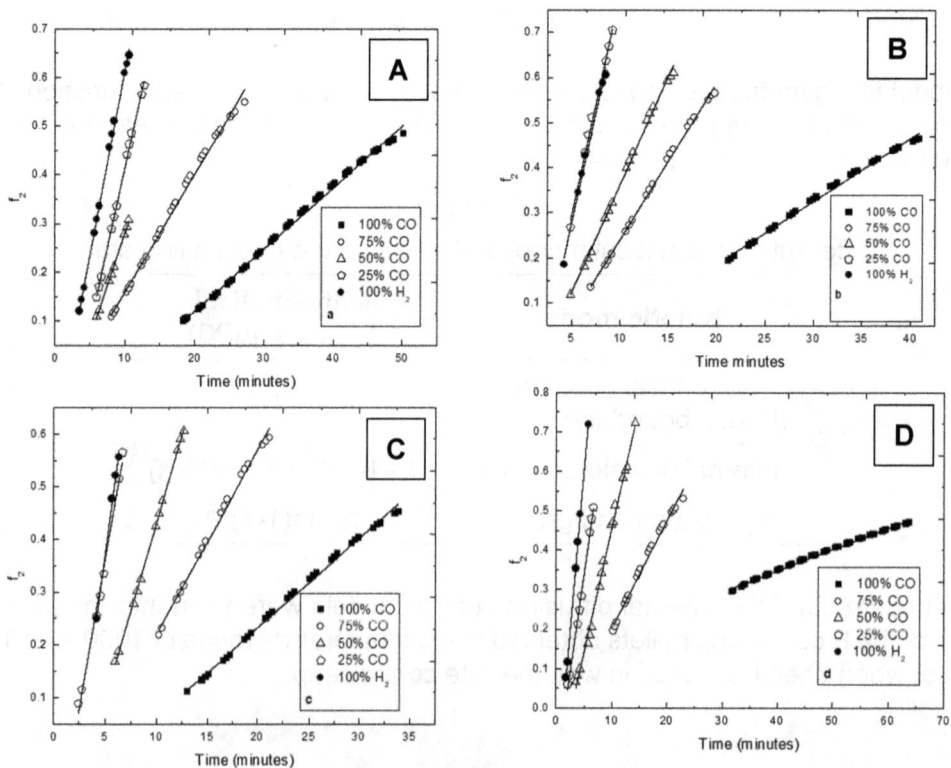

FIG 8 – Internal diffusion control plot with various CO-H_2 gas composition at: (a) 1000°C, (b) 1050°C, (c) 1100°C and (d) 1150°C.

These results indicate that the reaction could not be described by one mechanism: the reaction was initially controlled by chemical reaction and then by diffusion through the product layer in the later stage of the reduction. This can be clearly seen by checking the time axis of both the chemical and diffusion rate controlling plots. The rate parameters were calculated using the following equations (Moon, Rhee and Min, 1998):

$$k_1 = k_c(C_{A0} - C_{C0}/K_E)$$ (43)

where

k_1 is the chemical rate constant in (mol/cm^2·s)

k_c is the chemical reaction rate constant in cm/s

$C_A O$ is the bulk concentration of gas reactant in mol/cm^3

$C_C O$ is the bulk concentration of gas product

K_E is the equilibrium constant of the reaction

$$k_2 = 6D_e \left(\frac{K_E}{1+K_E}\right)(C_{A0} - C_{C0}/K_E)$$ (44)

where

k_2 is the diffusion rate constant in (mol/cm^2·s)

D_e is the effective diffusivity in cm^2/s.

The results for the evaluated kinetic parameters are shown in Table 5:

TABLE 5

Comparison of the rate parameters between CO and H$_2$.

Temp (°C)	k_c(cm/s)		D_e(cm^2/s)	
	CO	H$_2$	CO	H$_2$
1000	5.0	8.0	0.10	0.85
1050	5.5	9.7	0.97	1.15
1100	6.4	11.0	1.07	1.48
1150	7.3	13.3	0.68	1.67

Table 5 shows the calculated chemical rate constants as well as the effective diffusivity values that provides enough evidence that hydrogen has better reducing properties than CO as can be seen by the large values of k_c for H$_2$ than those for CO. H$_2$ can also diffuse much easier than CO as evidenced by the larger value of D_e. The fast H$_2$ diffusion can be as a result of its smaller atomic size. The decrease in the value of D_e at 1150 °C when CO was used shows that the diffusivity of CO is affected by partial melting that was evidenced by the sticking of the pellet on the crucible at this temperature.

Evaluation of activation energy

The activation energy values were estimated using the Arrhenius Equation.

$$k_c = k_{C0}\exp(-E_A/RT)$$ (45)

The natural logarithm of k_c against the reciprocal of absolute temperature was plotted as shown in Figure 9.

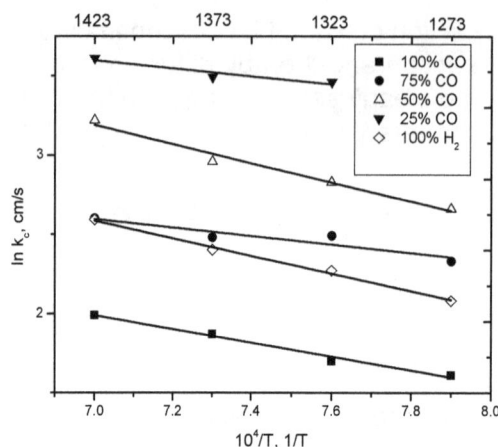

FIG 9 – Arrhenius plot of chemical rate with various CO-H$_2$ gas composition.

The slopes of the straight regression lines were used to estimate the activation energies which were found to be between 20.8–50.2 kJ/mol depending on the gas composition as shown in Table 6.

TABLE 6

Comparison of evaluated activation energies with literature.

Gas mixtures	E_A (kJ/mol) (Experiment)	E_A (kJ/mol) (Literature)	Authors
100% CO	36.3	56.5	El-Geassy, 1986
75% CO–25% H$_2$	22.2	27.4	Moon, Rhee and Min, 1998
50% CO–50% H$_2$	50.2	40.1	El-Geassy, 1986
25% CO–75% H$_2$	20.8	29.2	Moon, Rhee and Min, 1998
100% H$_2$	46.0	42.1	Moon, Rhee and Min, 1998

The evaluated activation energies showed the lowest activation energy when 25 per cent CO–75 per cent H$_2$ gaseous atmosphere was used. This was different from that by Moon, Rhee and Min (1998), findings that showed that 100 per cent CO has the lowest activation energy and that the activation energies increased with increasing the proportion of H$_2$. This difference is attributed to the different experimental conditions that were used by Moon, Rhee and Min (1998) where experiments were done at temperatures lower than 1000°C and pure hematite was used while the current experiment used high phosphorus ore at the temperatures higher than 1000°C.

The mixtures of CO and H$_2$ gases have better reducing properties (Mousa, Babich and Senk, 2013) than the use of the respective gas. This was in agreement with the current findings. The reduction rate increases with the addition of H$_2$ in the presence of CO in iron ore reduction due to the easy by which H$_2$ can be diffuse and react with oxygen in iron oxide than CO. This is mainly because of the smaller atomic size of H$_2$. The heat produced from the exothermic reaction of the reaction CO and the iron oxide together with the bursting of pores by CO helps in speeding up the reaction when using the combination of CO-H$_2$ gaseous mixtures. The lowered activation energy when using 75 per cent CO–25 per cent H$_2$ might be as a result of the catalytic effect of H$_2$ in the presence of CO (El-Geassy, 1986). Iron ore reduction by H$_2$ is thermodynamically favoured than reduction by CO above 821°C (Biswas, 1981). But the reduction by H$_2$ is endothermic while that by CO is exothermic (Wagner *et al*, 2006). This might be the reason the activation energy using 100 per cent H$_2$ is higher than that using 100 per cent CO. The high activation energy using 50 per cent CO–50 per cent H$_2$ can be as a result of attachment of CO on some sites of the oxide surface. This may lead to some areas being covered by CO molecules hindering the reduction by H$_2$ (El-Geassy, 1986). However, more research is suggested to validate this finding. The calculated activation energy values were approximately similar to those found in literature as shown in Table 6.

The general Arrhenius Equation at each CO-H_2 gaseous atmosphere were then represented by the following equations:

$$25\% \text{ CO} - k_c = 209.56 \exp(-\frac{2500}{T}) \tag{46}$$

$$75\% \text{ CO} - k_c = 86.63 \exp(-\frac{2666.7}{T}) \tag{47}$$

$$100\% \text{ CO} - k_c = 155.35 \exp(-\frac{4366.7}{T}) \tag{48}$$

$$100\% \text{ H}_2 - k_c = 637.36 \exp(-\frac{5533.3}{T}) \tag{49}$$

$$50\% \text{ CO} - k_c = 1656.28 \exp(-\frac{6033.3}{T}) \tag{50}$$

Comparison of the kinetic data with the EPMA mapping results shows that the P was concentrated into a phase that appears to be the slag phase when the general rate constant equation was given by Equation 50. If the rate of reaction is too high there will be not enough time for the unreduced P phase to accumulate into slag hence P will be trapped in the iron rich phase. A favourable rate of reduction favours the accumulation of P in the slag phase.

CONCLUSIONS

The reduction and the P distribution behaviours of pellets made using high phosphorus iron ore in CO-H_2 gaseous atmosphere were investigated at 1000°C. The following conclusions were drawn from the findings:

1. The reduction of the phosphorus-containing phase especially apatite at 1000°C with H_2 was thermodynamically predicted to be difficult compared to the reduction by carbon or CO.

2. The rate of reaction increased with increasing H_2 concentration due to the fast diffusivity of H_2. The smaller molecule size of H_2 compared to CO causes H_2 to be a better reducing agent.

3. The elemental distribution results indicated that the phosphorus was concentrated in the gangue-rich phase at 1000°C depending on the gaseous composition, which might be related to the reduction kinetics of the ore.

4. The careful control of the CO-H_2 gas compositions might assist in the production of clean steel and the kinetics of this research showed that phosphorus was concentrated in the slag when the general rate equation was given by Equation 50.

5. Approximately 50 per cent CO–50 per cent H_2 was the optimal gas composition for the concentration of phosphorus into the slag phase at 1000°C.

REFERENCES

Bai, S, Wen, S, Liu, D, Zhang, W and Cao, Q, 2012. Beneficiation of high phosphorus limonite ore by sodium carbonate-added carbothermic reduction, *ISIJ Int*, 52(10):1757–1763.

Biswas, A K, 1981. *Principles of blast furnace ironmarking*, 82 p (Cootha Publishing House).

Chen, Y, Liu, W, Chen, J and Zuo, H, 2022. Gasification behavior of phosphorus during biomass sintering of high-phosphorus iron ore, *ISIJ Int*, 62(3):474–483.

Chiwandika, E K and Jung, S-M, 2023. Effect of H_2 on the distribution of phosphorus in the gaseous reduction of hematite ore, *The Journal of The Minerals, Metals & Materials Society (JOM)*. https://doi.org/10.1007/s11837-023-05990-5

Delvasto, P, Ballester, A, Munoz, J A, Blazque, M L, 2007. Dephosphorization of an iron ore by filamentous fungus. Paper presented at the XII ENT MME/VII MSHMT-Ouro, Petro.

El-Geassy, A A, 1986. A Gaseous reduction of, Fe_2O_3 compacts at 600 to 1050°C, *J Mater Sci*, 21:3889–3900. https://doi.org/10.1007/BF02431626

Han, H and Duan, D, 2012. Dephosphorization technology of high phosphorus oolitic hematite by rotary hearth furnace iron nugget process, presented at the *6th Int Congress on the Science Technology of Ironmarking (ICSTI), 42nd Ironmaking and Raw Materials Seminar, 13th Brazilian Symposium on Iron Ore*.

Hessels, C J M, Homan, T A M, Deen, N G and Tang, Y, 2022. Reduction kinetics of combusted iron powder using hydrogen, *Powder Technology*, 407:117540.

Hino, M and Ito, K, 2010. *Thermodynamics data for steelmaking*, 264 p (Tohoku University Press Sendar).

Kashiwaya, Y and Hasegawa, M, 2012. Thermodynamics of impurities in pure iron obtained by hydrogen reduction, *ISIJ Int*, 52(8):1513–1522.

Lee, H, 1999. *Chemical thermodynamics for metals and materials*, 284 p (Imperial College Press).

Matinde, E and Hino, M, 2011. Dephosphorization treatment of high phosphorus iron ore by pre-reduction, mechanical crushing and screening methods, *ISIJ Int*, 51(2):220–227.

Moon, I J, Rhee, C H and Min, D J, 1998. Reduction of hematite compacts by H_2-CO gas mixtures, *Steel Research*, 69(8):302–306. https://doi.org/10.1002/srin.199805555

Mousa, A, Babich, A and Senk, D, 2013. Reduction behavior of iron ore pellets with simulated coke oven gas and natural gas, *Steel Research Int*, 84:1085. https://doi.org/10.1002/srin.201200333

Oyama, N, Higuchi, T, Machida, S, Sato, H and Takeda, K, 2009. Effect of high-phosphorous iron ore distribution in quasi particle on melt fluidity and sinter bed permeability during sintering, *ISIJ Inter*, 49(5):650–658.

Pan, J, Lu, S, Li, S, Zhu, D, Guo, Z, Shi, Y and Dong, T, 2022. A new route to upgrading the high-phosphorus oolitic hematite ore by sodium magnetization roasting-magnetic separation-acid and alkaline leaching process, *Minerals*, 12:568. https://doi.org/10.3390/min12050568

Pati, R and Vinacy, M, 2008. Reduction behavior of iron ore pellets, Department of Metallurgical and Materials Engineering, National Institute of Technology, Rourkela, pp 1–30.

Roy, D M, Drafall, L E and Roy, R, 1978. Crystal chemistry, crystal growth, and phase equilibria of apatite, Material Research Laboratory and Department of Material Science and Engineering, Pennsylvania State University, 285 p.

Tang, H O, Wang, J W, Guo, Z C and Ou, T, 2013. Intensifying gaseous reduction of high phosphorus iron ore fines by microwave pretreatment, *J Iron Steel Res Int*, 20:17–23.

Tang, H, Guo, Z and Zhao, Z, 2010. Phosphorus removal of high phosphorus iron ore by gas-based reduction and melt separation, *J Iron Steel Res Int*, 17:1–6. https://doi.org/10.1016/S1006-706X(10)60133-1

Wagner, D, Devisme, O, Patisson, F and Ablitzer, D, 2006. A laboratory study of the reduction of iron oxides by hydrogen, presented at the Sohn International Symposium.

Zhang, J, Luo, G, Chen, Y, Xin, W and Zhu, J, 2020. Phosphorus migration behavior of medium-phosphorus magnetite ore during carbothermic reduction, *ISIJ Int*, 60(3):442–450.

Zhang, Y, Xue, Q, Wang, G and Wang, J, 2018. Gasification and migration of phosphorus from high-phosphorus iron ore during carbothermal reduction, *ISIJ Int*, 58(12):2219–2227.

Smelting of nickel and vanadium wastes to produce green steel

N Goodman[1] and D Lalor[2]

1. Managing Director, Smelt Tech Consulting, Perth WA 6150.
   Email: neil.goodman@smelttech.com
2. Process Director, Smelt Tech Consulting, Perth WA 6150. Email: damian.lalor@smelttech.com

ABSTRACT

The increasing demand for batteries is resulting in the development of large-scale projects for the production of key battery metals such as nickel and vanadium. Several of these projects plan to extract the battery metals from the ore via chemical roasting, leaving large volumes of waste iron oxides after fuming and/or leaching of the battery metals.

These iron oxide wastes can be smelted in a HIsmelt furnace using biochar as the reductant, to produce clean, green pig iron with net zero carbon emissions. This green pig iron can then be added to basic oxygen or electric arc steelmaking furnaces to produce green, net zero carbon, steel.

This paper describes the flow sheets and the economics for potential projects producing green pig iron via HIsmelt using waste iron oxides from nickel and vanadium production plants.

INTRODUCTION

The demand for batteries is increasing as the decarbonisation of the world economy progresses. As well as lithium, a key metal in batteries is nickel, the demand for which has increased drastically, and demand is expected to accelerate further.

Future requirements for electrical power storage are expected to increase the demand for 'Redox' flow batteries that use the reduction and re-oxidation of metal oxides such as vanadium to provide and store power.

To produce nickel and vanadium from ores, large volumes of waste materials are generated after beneficiation and in many cases, the waste materials contain high levels of iron oxide.

This paper describes the technical and economic aspects of projects that will use the waste iron oxide from battery metal production processes to produce pig iron.

NICKEL PRODUCTION WASTES

Figure 1 shows a typical hydro-metallurgical flow sheet for the production of nickel from laterite via high pressure acid leaching (HPAL).

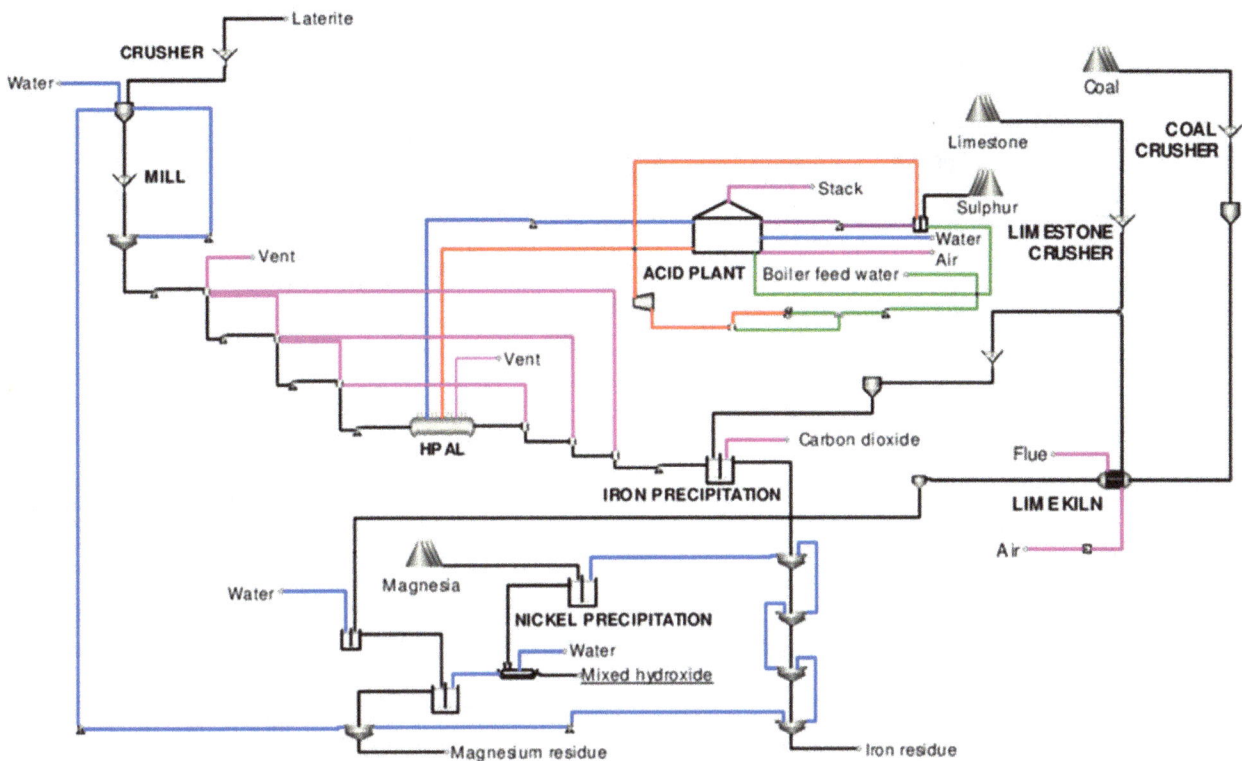

FIG 1 – Typical flow sheet for nickel extraction from laterite via HPAL.

After crushing in a gyratory crusher and grinding in a semi-autogenous grinding mill the ground laterite is preheated and fed into an autoclave (HPAL on the flow sheet) where the laterite is mixed with hot sulfuric acid which dissolves the nickel and iron. After discharge from the autoclave the iron is precipitated out via the addition of carbon dioxide and is discharged as a waste stream. The nickel rich solution is then fed into a nickel precipitator to extract the nickel. The iron residue is neutralised via the addition of lime and after thickening and dewatering, the iron residue is dumped as dry tailings.

The iron residue tailings have the typical range of analyses shown in Table 1.

TABLE 1

Typical HPAL iron residue tailings analyses.

%Fe	%CaO	%SiO$_2$	%Al$_2$O$_3$	%MgO	%S
46–52	6–10	7–14	5–9	1–2	0.2–0.6

The high level of gangue materials (lime, silica, alumina and magnesia) precludes the use of this material in a blast furnace because the gangue will form a large volume of slag blast furnace. This slag will build-up and restrict the flows of gas and liquids in the furnace, and drastically reduce production while increasing fuel consumption.

However, the HIsmelt smelting furnace can handle a large volume of slag as there are no solids in the furnace, and therefore these tailings can be smelted to produce pig iron. Compared to the benchmark 62 per cent Fe hematite, theses HPAL tailings will have a lower productivity and higher carbon consumption, resulting in a value-in-use (VIU) loss of 10–20 per cent versus the 62 per cent Fe benchmark. However, the cost of these tailings is essentially nil and as iron ore typically costs 40–60 per cent of the production cost of pig iron, the net result is that pig iron produced from these tailings will cost approximately 50 per cent less than the pig iron produced from blast furnaces.

VANADIUM PRODUCTION WASTES

Figure 2 shows a typical pyro-metallurgical flow sheet for the production of vanadium pentoxide from vanadium-titanomagnetite (VTM).

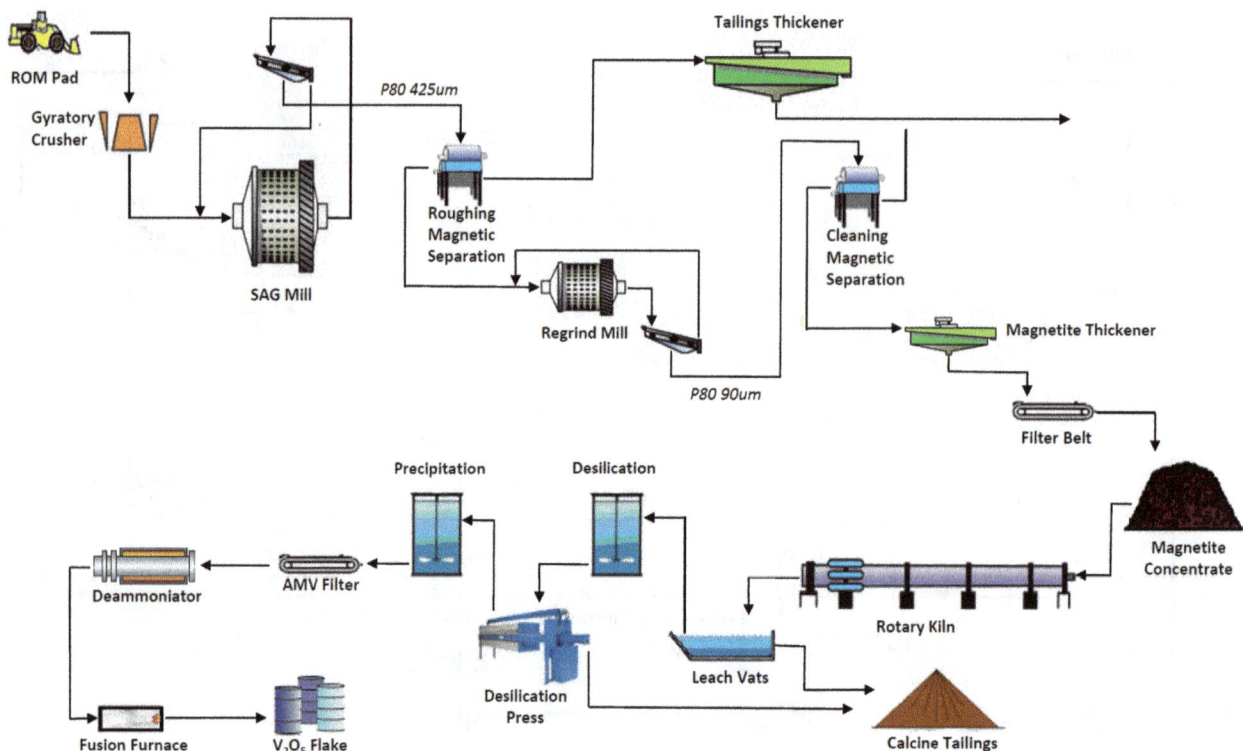

FIG 2 – Typical flow sheet for vanadium extraction from vanadium-titanomagnetite.

After crushing in a gyratory crusher and grinding in a semi-autogenous grinding mill, the VTM ore is magnetically separated. The magnetic stream is then re-ground in a ball mill and further magnetically separated. After thickening and dewatering, the magnetite concentrate is roasted with sodium bicarbonate to 800°C in a rotary kiln. The vanadium forms vanadium bicarbonate during the roasting and the magnetite is oxidised to form a calcined hematite.

The hot solids discharged from the rotary kiln are cooled and washed in water, where the vanadium bicarbonate is leached out of the hematite. The vanadium is extracted from the leachate as vanadium oxide flake via desilication, precipitation, deammoniation and fusion.

The calcined hematite tailings are a waste stream that has the typical analysis shown in Table 2.

TABLE 2

Typical calcined hematite tailings analyses.

%Fe	%TiO$_2$	%SiO$_2$	%Al$_2$O$_3$	%P	%S
52	16.5	3	3	0.001	0.01

The high level of titanium dioxide precludes the use of this material in a blast furnace because the titanium will form refractory materials in the shaft and hearth of the blast furnace. These refractory materials will build-up and restrict the flows of gas and liquids in the furnace and drastically reduce production.

However, the titanium dioxide in a HIsmelt smelting vessel will remain as a liquid in the slag and these tailings can be smelted to produce pig iron. Similar to the HPAL tailings, these calcined hematite tailings will have a lower productivity and higher carbon consumption, resulting in a value-in-use (VIU) loss of 10–20 per cent versus the 62 per cent Fe benchmark. However, the cost of these tailings is essentially nil, and as iron ore typically costs 40–60 per cent of the production cost of pig iron, the net result is that pig iron produced from these tailings will also cost approximately 50 per cent less than the pig iron produced from blast furnaces.

PIG IRON

Pig iron (or cast iron) is the highest value raw material in the steelmaking value chain and is used in increasing quantities by steelmakers who wish to produce high strength steels by melting scrap steel with electricity.

For lower strength steels such as reinforcing bar, a steelmaker can melt 100 per cent scrap steel with electricity to produce a quality product. However, for higher strength steels such as the thin flat-rolled steel used on car bodies, the levels of impurities in scrap steel are too high for the use of 100 per cent scrap. These impurities such as copper enter the scrap steel during reclamation (eg copper wiring from scrapped cars) and cannot be removed metallurgically. As more scrap steel is recycled in the steel industry, the levels of contaminants increase even further.

To reduce the level of copper in an electric furnace, 'virgin' or clean iron must be added to the furnace to dilute the copper down to an acceptable level. Steelmakers presently use direct reduced iron (DRI) containing 92 per cent Fe, 3 per cent C, 5 per cent gangue (SiO_2, Al_2O_3) and zero copper, blast furnace pig iron containing 95.5 per cent Fe, 4 per cent C, 0.5 per cent Si and zero copper, or high purity pig iron (HPPI) containing 96 per cent Fe, 4 per cent C and zero copper. For high strength steels, up to 30 per cent of the scrap will be replaced by DRI or pig iron.

This method of steel production is prevalent in developed countries such as the USA and the EU where there is large volume of steel scrap generated and recycled. In the US, approximately 70 per cent of the steel is produced via electric furnaces, and this is expected to increase further die to the lower operating cost, lower carbon dioxide emissions and lower capital cost compared to the traditional route of converting blast furnace liquid hot metal into steel by blowing with oxygen.

In developing countries, electric furnace capacity is increasing rapidly, in China more than 100 million tonnes of electric furnace capacity has been added in recent years.

Therefore, the use of pig iron in electric furnaces is expected to increase dramatically over the next decades as the developed countries decarbonise their steel production, and developing countries consume their increasing volumes of recycled scrap.

HISMELT TECHNOLOGY

Description

The HIsmelt technology smelts iron oxides into high purity pig iron and the heart of the Hismelt technology is the Smelt Reduction Vessel (SRV), see Figure 3.

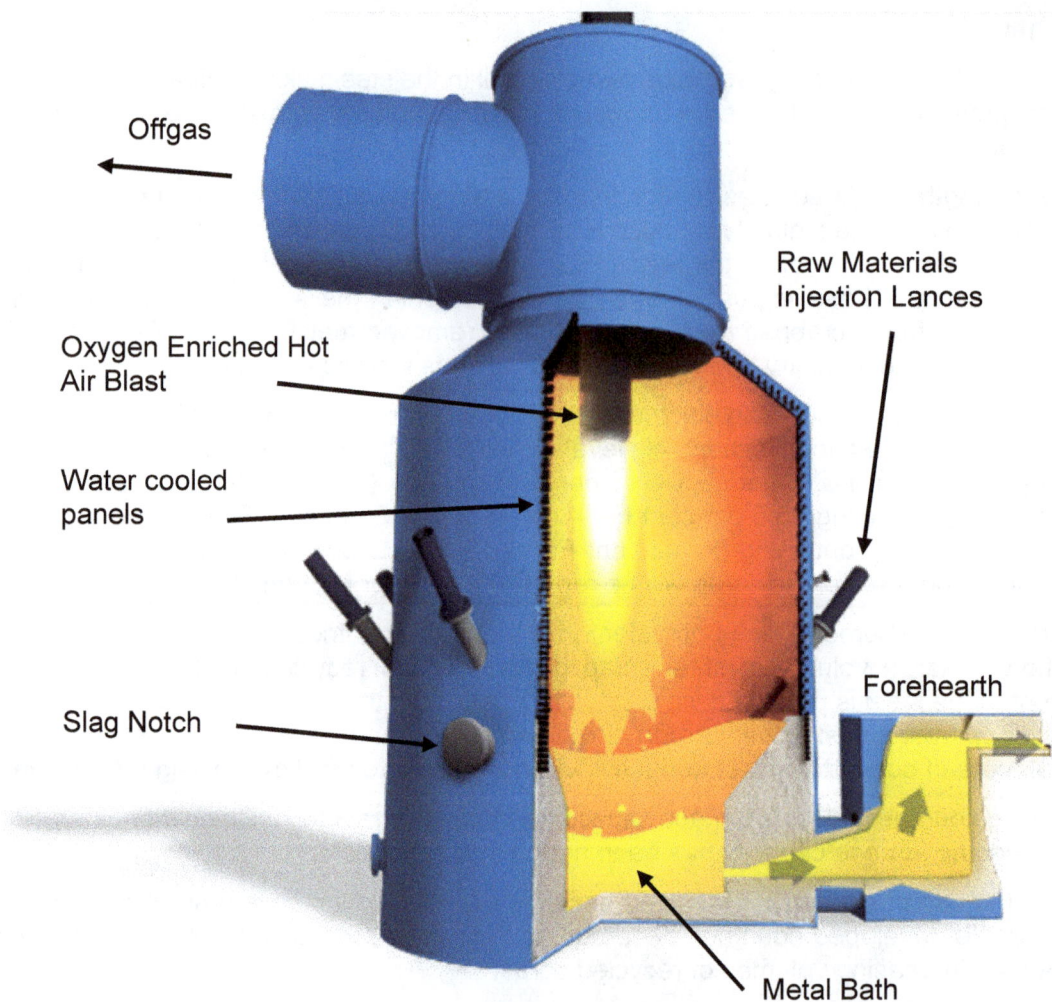

FIG 3 – Smelt Reduction Vessel (SRV).

The SRV comprises a steel pressure vessel with a refractory lined hearth, water-cooled panels around the inside shell of the barrel and roof. During operation, the refractory hearth contains a bath of liquid iron at approximately 1400°C, with a layer of molten slag on top of the hot metal. Iron ore fines, fluxes and coal are injected pneumatically downward through the slag into the hot metal bath via water cooled lances. Hot air blast (HAB) at 1200°C, enriched by oxygen to a total oxygen content of 40 per cent, is supplied by hot blast stoves and injected into the SRV via a water cooled and refractory lined lance to post combust the offgas inside the SRV.

The smelt reduction reaction between the iron oxides and carbon generates a large volume of combustible offgas that erupts from the hot metal bath and throws a large volume of slag droplets into the top space of the SRV. In this top space, the post-combustion of the offgas by the HAB generates a high temperature flame above the slag surface. The energy from the flame is transferred to the slag droplets, which in turn, transfer their heat to the endothermic smelting zone in the hot metal.

The slag layer also insulates the hot metal from the oxygen in the HAB and therefore reduces the back-oxidation of the iron to FeO.

The liquid iron (hot metal) flows continuously from the SRV via a refractory lined 'forehearth' that keeps the slag separate from the hot metal. The waste gangue from the ore and waste ash from the coal form a liquid slag that is periodically 'tapped' from the SRV via a 'slag notch'.

PATHWAYS TO SUSTAINABLE STEELMAKING USING HISMELT

Replacement of coal with Biochar

The HIsmelt technology was developed to use non-coking coal to reduce the iron oxides to iron, to provide the fuel for combustion in the top space of the SRV. The coal is injected into the SRV as a powder, and does not have any size and strength requirements, unlike the blast furnace.

Therefore, carbon-containing powders can be used in the HIsmelt technology, several different carbon-containing powders have been trialled successfully on HIsmelt pilot plants including coke breeze, blast furnace dust and charcoal derived from biomass (Biochar).

The use of biomass products such as Biochar will produce similar carbon dioxide emissions to those from coal. However, the plants that are the source of the biomass consume almost the same amount of CO_2 through photosynthesis while growing, this is released when the biomass is combusted in the SRV. Biochar can be produced from a variety of plant matter, including waste forestry waste, or energy crops, that are considered low carbon biomass sources and are deemed to emit a neutral amount of carbon in the atmosphere.

The use of biochar will comply with the current Australian regulations under the NGER Framework Requirements and the NGA accounting rules to achieve certification of carbon neutrality by Climate Active.

Carbon capture and sequestration

In 2009 the HIsmelt technology was selected by a consortium of European steelmakers to smelt liquid FeO produced by the Cyclone Converter Furnace (CCF) developed by Hoogovens (now Tata Steel). The combination of the HIsmelt SRV and the CCF is known as the 'HIsarna' process, a HIsarna pilot plant was commissioned and operated over several campaigns at Tata Steel Ijmuiden in Holland from 2010 to date.

HIsarna was developed as part of the EU Government's ULCOS (Ultra Low CO_2 Steelmaking) project, with the intention of capturing the CO_2 from the smelter offgas and pumping the CO_2 underground (sequestration). To reduce the capital and operating costs of capturing the CO_2, the nitrogen in the offgas should be reduced as much as possible, preferably to zero. To achieve this, the HIsarna SRV uses 100 per cent oxygen to post-combust the top space gas instead of hot blast.

The successful trials at Ijmuiden showed that a HIsmelt SRV can successfully use 100 per cent oxygen, therefore offers the opportunity for economically capturing and sequestrating the CO_2 in the offgas. However, the operating and capital costs of capturing and sequestrating the CO_2 are high and require favourable local geological formations such as depleted oil and gas fields to store the CO_2.

Consumption of DRI produced via hydrogen and/or biomass reduction

The flexibility of the HIsmelt process allows the SRV to smelt a wide range of ferrous feeds and the following materials have been successfully smelted in the HIsmelt plants operating in China or Australia; hematite with Fe contents of 55 to 62 per cent, magnetite concentrates with Fe contents of 57 to 68 per cent, and partially and fully pre-reduced directly reduced iron (DRI) with Fe contents of 68 to 90 per cent.

As the Fe content of the ferrous feed increases, the energy required for the reduction reactions is reduced, and the injection rate of the ferrous feed can be increased for a constant energy input. This results in an increased production of liquid iron from the SRV and a reduction in specific fuel consumption. The approximate production rates and fuel (non-coking coal or Biochar) consumptions of a 6 m SRV for various ferrous feeds are shown in Table 3.

TABLE 3

Approximate production rates and fuel (non-coking coal or Biochar) consumptions of a 6 m SRV.

Ferrous feed	Pig iron production tonnes per annum	Fuel consumption kg/t of pig iron
60% Fe Hematite	600 000	800
65% Fe Magnetite	1 000 000	700
80% Fe DRI	1 500 000	300

Directly reduced iron (DRI) can be produced in many ways with various reductants. Table 4 shows a range of possible technology routes for producing DRI.

TABLE 4

Possible technology routes for producing DRI.

Furnace type	Ferrous feed	Reductant	Technology status
Shaft	Lump ore or pellets	Natural gas	Proven, more than 100 Midrex or HYL units in operation worldwide
Shaft	Lump ore or pellets	Hydrogen	Pilot plant scale 'HYBRIT' technology
Rotary kiln	Lump ore or pellets	Coal or natural gas	Proven but uneconomic compared to shaft furnaces
Rotary hearth	Lump ore or pellets	Coal or natural gas	Proven but uneconomic compared to shaft furnaces
Fluid bed	Ore fines	Hydrogen	Proven at commercial scale at a single site – 'CircoRed' technology Pilot plant 'HYFOR' technology
Grate kiln	Ore fines	Biomass	R&D laboratory testing

The 'CircoRed' process reduces ore fines with hydrogen firstly in a circulating fluid bed (CFB) that produces DRI fines that are pre-reduced to approximately 80 per cent Fe. These pre-reduced fines are then further reduced to >90 per cent Fe in a bubbling fluid bed and then compressed into briquettes and sold as hot-briquetted iron (HBI). The 'CircoRed' process was proven at the LTV/Cliffs/Outotec 'CircoRed' plant in Trinidad from 1998 to 2002 using hydrogen generated via the steam reformation of natural gas

The combination of a HIsmelt SRV with the first stage CFB of a 'CircoRed' plant offers significant advantages for zero-carbon production of pig iron including lower cost ferrous feed (ore fines versus high-grade pellets), lower capital (no second-stage bubbling bed or briquetting required).

CONCLUSIONS

The increasing demand for batteries is resulting in the development of large-scale projects for the production of key battery metals such as nickel and vanadium. Several of these projects plan to extract the battery metals from the ore via chemical roasting, leaving large volumes of waste iron oxides after fuming and/or leaching of the battery metals.

These iron oxide wastes can be smelted in a HIsmelt furnace using biochar as the reductant, to produce clean, green pig iron with net zero carbon emissions. This green pig iron can then be added to basic oxygen or electric arc steelmaking furnaces to produce green, net zero carbon, steel.

Alternatively, the waste iron oxides could be converted to 'green' zero-carbon pig iron via pre-reduction using hydrogen followed by smelting with biochar, or the carbon dioxide can be economically captured and sequestered.

The use of waste products will reduce the working capital required to dispose of these tailings and will produce high quality pig iron for approximately half the cost of a blast furnace.

Direct reduced iron (DRI) – value of process monitoring using X-ray diffraction

U König[1] and M Pernechele[2]

1. Business Development Mining and Metals, Malvern Panalytical, Almelo, Netherlands.
   Email: uwe.koenig@malvernpanalytical.com
2. Application Specialist XRD, Malvern Panalytical, Almelo, Netherlands.
   Email: matteo.pernechele@malvernpanalytical.com

ABSTRACT

Most of the operating steelmaking capacity currently relies on conventional, coal-based steelmaking processes. Currently, direct reduced iron (DRI) is only produced in regions where energy is available at low cost. To align with net-zero emissions goals, steelmaking capacity must transition to lower-emissions steelmaking technology. Direct reduction of iron ore will get one of the preferred methods in the future due to its high technology readiness. New reduction technologies include hydrogen-based direct reduction processes. To ensure optimal efficiency during DRI production and to cope with newly developed technologies, processes need to be monitored more frequently and accurate. Traditionally quality control of raw materials and DRI has relied on time consuming physical testing, wet chemistry or the analysis of the elemental composition. The mineralogy of DRI that defines its properties is not frequently monitored. Process monitoring and quality control of DRI is possible by using modern X-ray diffraction (XRD) methods. In addition to the metallic iron content (Fe_{met}), metallisation (Met^n) and the total carbon content (C_{tot}), XRD analysis also identifies the overall mineralogical phase composition. This allows additional information about the reduction process and enables fast counteractions on changing raw material mixtures. The benefits of the quantitative determination of the phase composition and the calculation of process critical parameters to produce DRI will be demonstrated with a case study.

INTRODUCTION

Steel is the most important engineering metallic material, used across countless applications in transport, infrastructure, and energy conversion. Its production of >1.8 billion tons per annum, with the largest fraction reduced from iron oxides using carbon, leads to huge anthropogenic CO_2 emissions, accounting for ~30 per cent of all industrial CO_2 emissions (~7 per cent of the total CO_2 emissions) (Raabe, Tasan and Olivetti, 2019; Spreitzer and Schenk, 2019; Ma *et al*, 2022). Direct reduction of iron ore is a growing technology since it has a better carbonisation potential to move towards net-zero. DRI, also called sponge iron, is produced by the reduction of iron ore in the form of lumps, pellets or fines with a reducing gas.

Hydrogen-based direct reduction (HyDR) is among the most attractive solutions for green ironmaking, with high technology readiness. The underlying mechanisms governing this process are characterised by a complex interaction of several chemical (phase transformations), physical (transport) and mechanical (stresses) phenomena. In general, DRI using hydrogen proceeds two to three times faster than CO-based direct reduction. This is attributed to the physical properties of hydrogen, ie its small molecule size, low viscosity and high mobility, when diffusing as molecule through pores or as dissociated atom through solids Spreitzer and Schenk (2019).

To achieve maximum efficiency of the production process, several parameters need to be monitored, such as metallic iron content (Fe_{met}), metallisation (Met^n), total carbon content (C_{tot}) and the mineralogical phase content. Traditionally a combination of time intense wet chemical analysis and elemental analysis are the preferred methods.

X-ray diffraction (XRD) is a new, innovative, fast and precise analytical tool for rapid process control of DRI that can easily applied for process control in iron and steel production environments. Besides the common process parameter, it allows the quantitative monitoring of the mineralogical phase composition.

METHODS

A batch of 20 direct reduced iron pellets with variable chemical and mineralogical composition were used in this study to assess the precision and accuracy of the XRD method in comparison to classical wet chemistry techniques. Moreover, the NISR 691 standard for DRI was also characterised with X-ray diffraction.

The particle sizes of the pellets were first reduced with a jaw crasher and then pulverised with a Herzog HSM-100. The final particle size was below 60 microns, measured using laser diffraction with a Mastersizer 3000 analyser. The samples were pressed for 30 second at 10 kN with pressing machine; a 10 wt per cent of cellulose binder was used to assure the mechanical stability of the press pellets.

The XRD pattern were collected using the Metals Edition of Aeris compact X-ray diffractometer with a goniometer radius of 145 mm, cobalt-anode. The Bragg-Brentano measurement covered a range of 15–105° 2θ. The use of a linear PIXcel1D Medipix3 detector allows a scan acquisition time of seven minutes.

The phase identification was done with the HighScore Plus software package version 5.1 (Degen *et al*, 2014). The Rietveld analyses were also performed with HighScore Plus using the advance smart batch functionalities (Rietveld, 1969).

Chemometric models such as Partial Least Square Regression (PLSR) were applied to predict chemical composition directly from the raw diffraction data without performing phase ID, Rietveld fitting and stochiometric calculations. These models were created by the same software suite HighScore Plus.

RESULTS

Phase identification

The following 12 phases were identified in the set of 21 samples: metal iron Fe, cohenite Fe_3C, wuestite FeO, magnetite Fe_3O_4, hematite Fe_2O_3, bixbyte $(Fe_{1-x},Mn_x)_2O_3$, quartz SiO_2, pyroxene $CaMg_{0.5}Fe_{0.5}(Si_2O_6)$, forsterite Mg_2SiO_4, fayalite Fe_2SiO_4, olivine Ca_2SiO_4 and lime CaO. The peak positions of the major phases are shown in Figure 1.

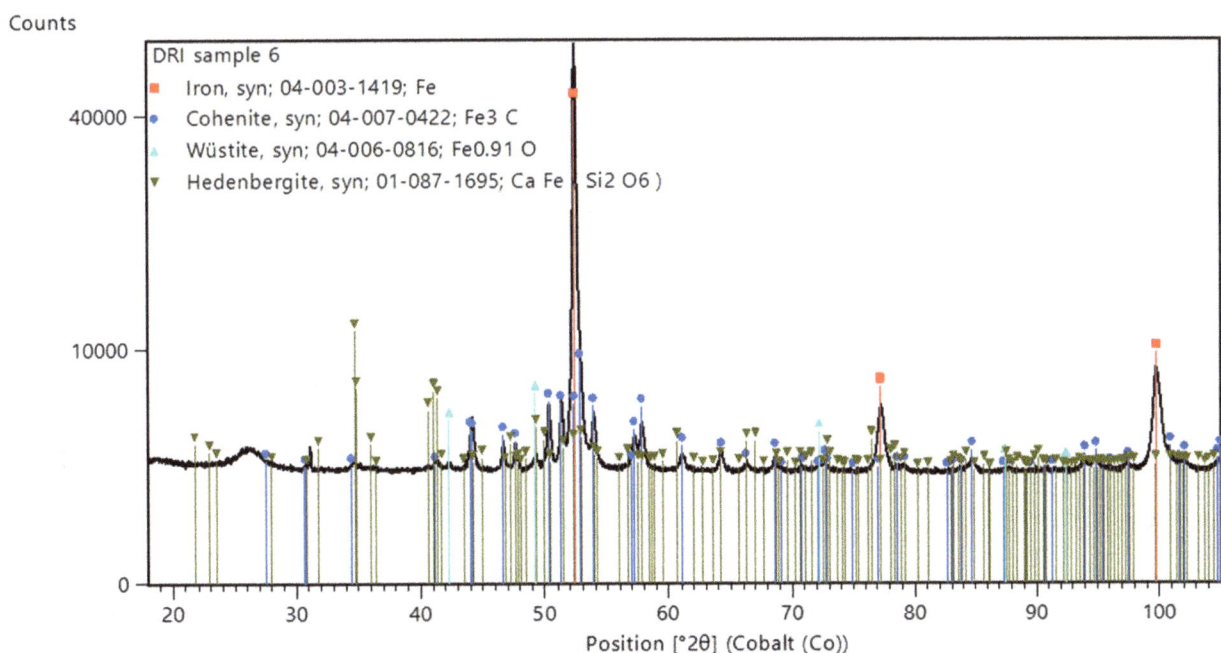

FIG 1 – Phase identification of direct reduced iron (DRI), marked are only the peak positions of the four major phases.

Phase quantification and recalculation

Figure 2 shows a typical example of a DRI analysis using XRD in combination with an automatic Rietveld refinement. The full mineralogical content is determined by providing quantitative information about the metallic iron-contributing phases (α-iron Fe, cohenite Fe_3C) as well as iron oxide phases (magnetite Fe_3O_4, hematite Fe_2O_3, wuestite FeO) and various silicatic phases (pyroxenes, olivines etc). From these phase contents the process- critical parameters Fe_{met}, Fe_{tot}, Met^n, and the C_{tot} are directly calculated as part of the automated analysis. When compared to independent reference data from wet-chemical analysis and carbon analysers the data obtained from XRD show a very good agreement over a wide range of values, see Figure 3, Tables 1 and 2.

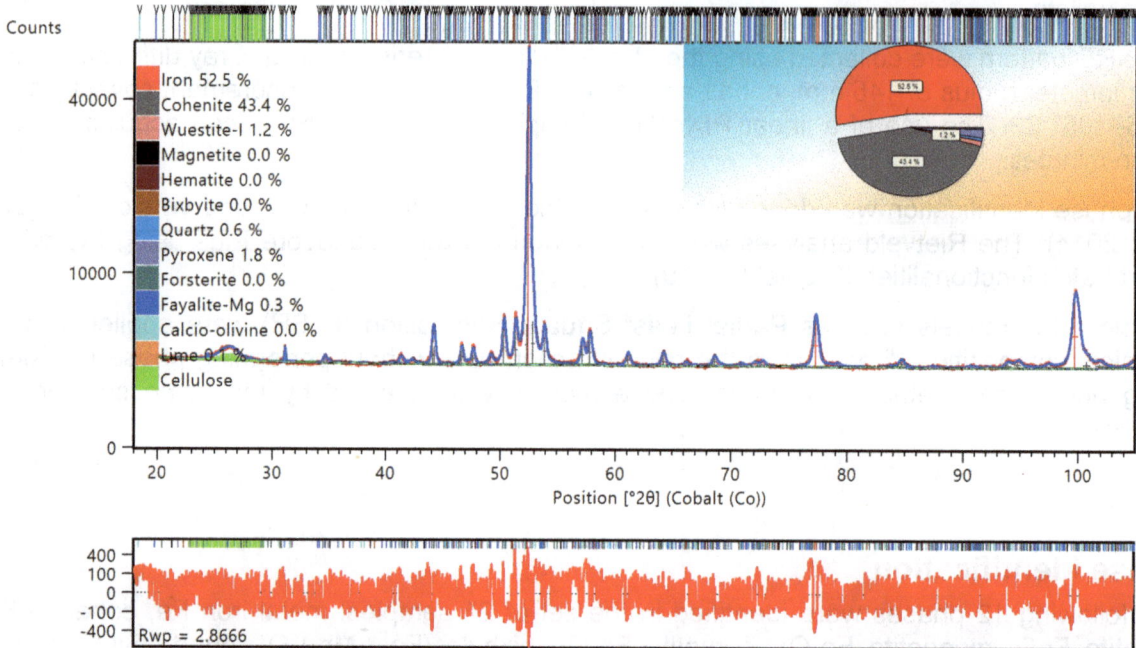

FIG 2 – Example of a quantification of 12 phases of direct reduced iron (DRI), cellulose was used as binder for sample preparation.

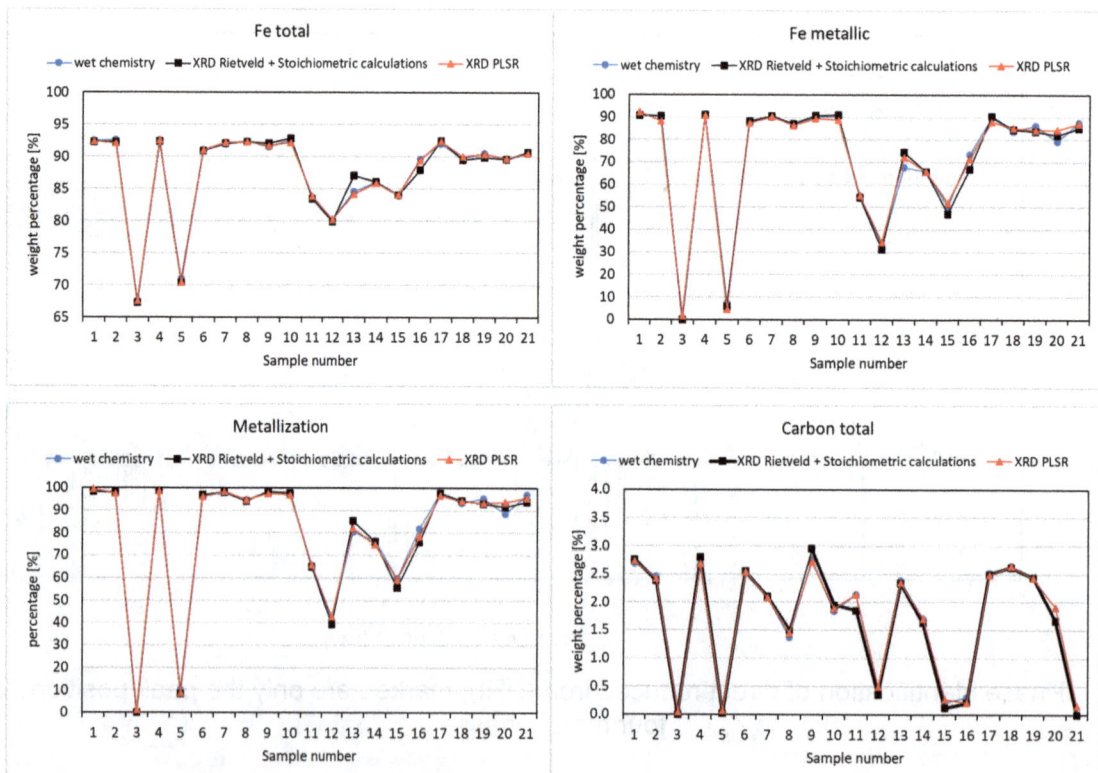

FIG 3 – Comparison of XRD based results with reference values of 21 DRI samples.

TABLE 1
Results phase composition of 21 DRI samples (Rietveld quantification).

Sample ID	Iron	Cohenite	Wuestite	Magnetite	Hematite	Bixbyite	Quartz	Pyroxene	Forsterite	Fayalite	Ca-Olivine	Lime	Rwp
1	52.4	41.2	0.0	0.0	0.3	0.2	0.7	3.9	0.5	0.7	0.0	0.2	2.9
2	57.3	35.6	0.0	0.7	0.2	0.2	1.0	3.3	0.7	0.9	0.0	0.3	2.8
3	0.0	0.1	0.1	4.1	91.6	2.6	0.0	0.0	0.6	0.3	0.0	0.7	3.0
4	52.0	41.9	0.0	0.0	0.2	0.1	0.3	4.6	0.5	0.3	0.0	0.2	3.0
5	5.7	0.2	7.4	41.6	40.8	1.2	0.5	0.1	0.9	0.4	0.0	1.2	2.5
6	52.5	38.2	2.2	0.1	0.0	0.0	2.6	3.3	0.2	0.6	0.0	0.4	2.9
7	61.0	31.4	0.6	0.0	0.0	0.0	1.1	4.2	0.6	0.6	0.0	0.4	2.9
8	66.1	22.4	5.3	0.6	0.1	0.7	0.7	1.7	1.3	0.5	0.0	0.7	3.0
9	49.5	44.1	0.2	0.5	0.0	0.3	0.8	2.9	0.9	0.6	0.0	0.2	5.4
10	63.7	29.2	1.2	0.2	0.0	0.4	1.1	2.5	0.8	0.6	0.0	0.3	3.2
11	28.4	27.7	24.8	12.5	0.0	0.6	1.5	1.8	0.0	1.0	0.0	1.8	2.8
12	26.4	5.4	57.6	3.9	0.0	0.0	2.3	1.9	0.6	0.9	0.0	0.9	2.5
13	41.8	35.0	2.8	11.7	1.5	0.2	1.2	2.4	1.9	0.8	0.0	0.7	2.7
14	42.9	24.4	7.3	13.8	5.8	0.7	0.7	1.3	1.3	0.8	0.0	1.1	2.8
15	45.3	1.7	44.2	2.7	0.0	0.2	1.3	1.0	1.7	1.0	0.0	0.7	2.7
16	64.0	3.1	17.1	9.6	0.1	0.5	0.8	0.9	1.0	1.0	0.1	1.8	3.0
17	55.6	37.4	0.8	0.3	0.6	1.0	0.0	2.3	1.0	0.5	0.0	0.7	2.9
18	47.9	39.3	0.6	4.6	0.1	0.0	2.6	2.8	0.8	1.1	0.0	0.4	2.5
19	49.7	36.4	3.0	4.1	0.1	0.0	2.1	2.0	1.1	0.9	0.0	0.5	2.5
20	58.8	24.9	6.6	1.6	0.0	0.0	1.7	3.2	1.4	1.0	0.0	0.7	2.6
21	85.0	0.0	1.3	0.0	0.0	0.0	0.0	1.5	2.5	7.9	0.3	1.4	3.5

TABLE 2
Comparison of process parameter obtained by wet chemistry, XRD Rietveld and XRD PLSR.

Sample ID	Wet chemistry				XRD Rietveld + Stoichiometric calculations							XRD PLSR				Status
	Fe total	Fe metallic	Metallization	Carbon total	Fe total	Fe metallic	Metallization	Carbon total	FeO total	Fe2+	Fe3+	Fe total	Fe metallic	Metallization	Carbon total	
1	92.5	90.9	98.4	2.7	92.3	90.9	98.4	2.8	1.9	1.2	0.2	92.4	92.0	99.6	2.7	Included
2	92.6	90.5	97.8	2.5	92.3	90.5	98.1	2.4	2.3	1.4	0.4	92.1	88.2	97.6	2.4	Included
3	67.6	0.2	0.3	0.1	67.3	0.1	0.1	0.0	86.6	1.3	66.0	67.6	1.6	1.2	0.0	Included
4	92.5	91.1	98.5	2.7	92.4	91.0	98.6	2.8	1.7	1.2	0.1	92.6	90.8	98.5	2.7	Included
5	71.0	7.2	10.1	0.1	70.5	5.9	8.4	0.0	83.1	16.0	48.5	70.4	4.3	8.8	0.1	Included
6	91.0	87.5	96.1	2.6	91.0	88.2	96.9	2.6	3.6	2.8	0.0	91.0	87.5	96.0	2.5	Validation
7	91.9	89.7	97.6	2.1	92.1	90.3	98.1	2.1	2.3	1.8	0.0	92.2	90.3	98.2	2.1	Included
8	92.4	87.2	94.3	1.4	92.3	87.0	94.3	1.5	6.8	5.0	0.4	92.3	86.3	94.8	1.4	Included
9	91.5	89.9	98.2	2.8	92.1	90.6	98.4	3.0	1.9	1.2	0.3	91.7	89.4	97.6	2.7	Included
10	92.6	90.4	97.7	1.8	92.9	90.9	97.9	2.0	2.5	1.9	0.1	92.2	88.9	97.0	1.9	Validation
11	83.6	54.9	65.6	2.1	83.5	54.3	65.0	1.9	37.6	23.2	6.0	83.8	55.0	66.0	2.1	Included
12	80.1	32.7	40.8	0.5	80.0	31.5	39.3	0.4	62.4	46.6	1.9	80.2	34.4	42.6	0.5	Included
13	84.5	67.8	80.3	2.4	87.1	74.5	85.5	2.3	16.2	6.0	6.7	84.2	72.1	82.1	2.4	Included
14	86.0	65.7	76.4	1.7	86.1	65.7	76.3	1.6	26.3	9.7	10.7	85.9	65.7	74.9	1.7	Included
15	84.0	50.1	59.6	0.2	84.1	46.9	55.8	0.1	47.8	35.8	1.3	84.1	51.9	59.4	0.3	Validation
16	89.7	73.5	81.9	0.3	88.0	66.9	76.1	0.2	27.1	16.4	4.7	89.5	71.3	78.4	0.2	Included
17	92.0	90.0	97.8	2.5	92.5	90.5	97.9	2.5	2.6	1.4	0.6	92.3	87.9	96.6	2.5	Included
18	89.7	83.6	93.1	2.6	89.5	84.5	94.4	2.6	6.5	2.8	2.2	90.0	85.2	94.2	2.6	Included
19	90.5	86.1	95.1	2.4	90.0	83.7	93.0	2.4	8.1	4.3	2.1	90.4	84.2	93.4	2.4	Included
20	89.5	79.3	88.6	1.7	89.6	82.1	91.6	1.7	9.7	6.8	0.8	89.7	84.4	93.6	1.9	Validation
21	90.6	87.7	96.9	0.0	90.7	85.1	93.8	0.0	7.3	5.7	0.0	90.5	87.2	95.6	0.2	Included

Three samples of NIST 691 were prepared as press pellet and measured with Aeris to test the accuracy of the mineral quantification (Figure 4). The NIST 691 has a certified weight percentage of metallic iron of 84.7 ± 0.5 and a certified weight percentage of total iron of 90.5 ± 0.6. The individual Aeris XRD results as well as the Aeris XRD averages are within the error bar of the certificate, proving the accuracy of the analysis.

Sample ID	Iron	Cohenite	Wuestite	Magnetite	Hematite	Bixbyite	Quartz	Pyroxene	Forsterite	Fayalite	CalcioOlivine	Lime	Fe metallic	Fe total	Metallization	FeO total	Fe2+	Fe3+	Carbon total	Rwp
Nist 1	85.0	0.0	1.3	0.0	0.0	0.0	0.0	1.5	2.5	7.9	0.3	1.4	85.0	90.7	93.8	7.3	5.7	0.0	0	3.5
Nist 2	84.5	0.0	1.6	0.0	0.0	0.0	0.0	1.8	2.8	8.0	0.0	1.4	84.5	90.5	93.4	7.7	6.0	0.0	0	3.4
Nist 3	84.9	0.0	1.1	0.1	0.0	0.0	0.0	1.5	3.1	8.2	0.0	1.2	84.9	90.6	93.7	7.4	5.7	0.1	0	3.4

	Average	84.8	90.6
	Std.dev.	0.31	0.14

FIG 4 – Results of the repeatability test for three DRI NIST standard samples (metallic iron Fe_{met} 84.8 per cent ± 0.5 per cent, total iron Fe_{tot} 90.6 per cent ±0.6 per cent).

Chemometric models

Chemometric modelling using PLSR is an innovative method to predict parameter directly from raw measurements. It is mainly used in industrial environments to correlate process parameter quantitatively. With this method no pure phases, crystal structures or modelling of peak shapes are necessary. Setting up requires a set of validation samples for creating a prediction model.

The method was tested on the same set of samples used for the Rietveld quantification. Seventeen samples were used to create the model (reference samples). Four validation samples were used to estimate the error of the method. Reference values were obtained by wet chemistry.

Figure 5 shows two examples of PLSR prediction models for the metal iron and metallisation. A clear correlation between reference values and predicted results can be seen for reference and validation samples. Increased accuracy and predictability require a larger set of samples and will be focus of future tests. However, the small sample set already indicated that PLSR on XRD raw data can be used as addition or alternative for the analysis of DRI samples.

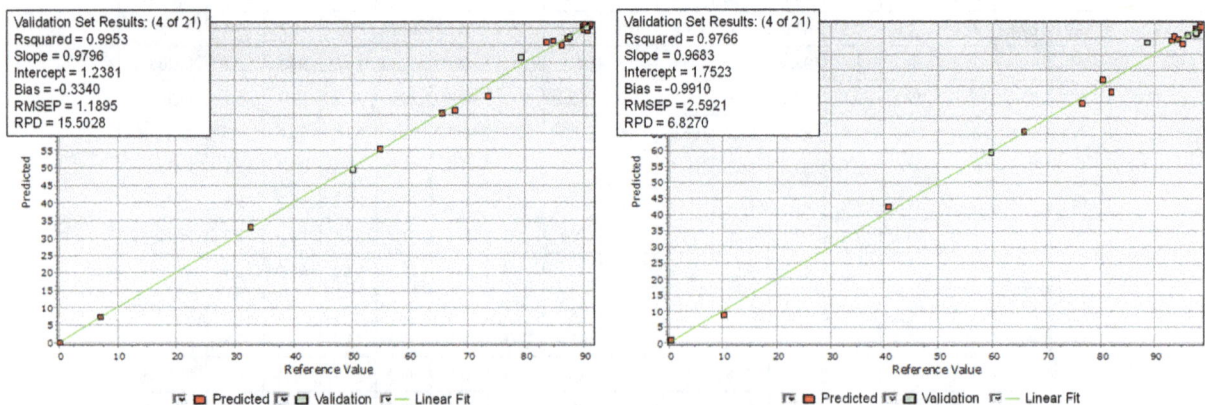

FIG 5 – Partial Least Square Regression (PLSR) analysis of the metallic iron content and metallisation.

Figure 3 and Table 2 summarise the results of the different methods to monitor process parameter for the manufacturing of DRI, wet chemistry (reference), XRD (Rietveld quantification plus re-calculation) and the use of chemometric methods (PLSR).

CONCLUSIONS

The study shows that XRD is a powerful and fast method to monitor process critical parameter and control quality of DRI within less than 10 minutes. In addition, XRD can also characterise the overall mineralogical contents in a DRI sample, which provides further important information about the reduction process. The full analysis can be automated and integrated in fully automated environments including automated sample preparation, measurements, analysis and reporting of results. The method can be applied to monitor traditional DRI manufacturing as well as direct reduction using hydrogen in the future.

REFERENCES

Degen, T, Sadki, M, Bron, E, König, U and Nénert, G, 2014. The HighScore suite, *Powder Diffraction*, 29(S2):S13–S18. doi:10.1017/S0885715614000840.

Ma, Y, Sourza Filho, I R, Bai, Y, Schenk, J, Patisson, F, Beck, A, Bokhoven, J A, Willinger, M G, Li, K, Xie, D, Ponge, D, Zaefferer, S, Gault, B, Mianroodi, J R and Raabe, D, 2022. Hierarchical nature of hydrogen-based direct reduction of iron oxides, *Scripta Materialia*, 213:114571.

Raabe, D, Tasan, C C and Olivetti, E A, 2019. Strategies for improving the sustainability of structural metals, *Nature*, 575:64–74.

Rietveld, H M, 1969. A profile refinement method for nuclear and magnetic structures, *J Appl Cryst*, 2:65–71.

Spreitzer, D and Schenk, J, 2019. Reduction of iron oxides with hydrogen – A review, *Steel Research International*, 90:1900108.

Mechanisms on the impacts of basicity on hydrogen-rich gas-based direct reduction of pellets

J Pan[1], C M Tang[2], D Q Zhu[3], Z Q Guo[4], C C Yang[5] and S W Li[6]

1. Professor, School of Minerals Processing and Bioengineering, Central South University, Changsha 410083, China. Email: pjcsu@csu.edu.cn
2. PhD candidate, School of Minerals Processing and Bioengineering, Central South University, Changsha 410083, China. Email: tcm986@csu.edu.cn
3. Professor, School of Minerals Processing and Bioengineering, Central South University, Changsha 410083, China. Email: dqzhu@csu.edu.cn
4. Associate professor, School of Minerals Processing and Bioengineering, Central South University, Changsha 410083, China. Email: guozqcsu@csu.edu.cn
5. Associate Professor, School of Minerals Processing and Bioengineering, Central South University, Changsha 410083, China. Email: smartyoung@csu.edu.cn
6. PhD, School of Minerals Processing and Bioengineering, Central South University, Changsha 410083, China. Email: swli@csu.edu.cn

ABSTRACT

The short process of electric furnace steelmaking with direct reduction iron (DRI) as raw material is an important process for smelting high-quality and special steels. In order to prepare high-quality DRI, the pellets with basicity of natural basicity and 0.4 were prepared from magnetite concentrate ($M_{R=0.09}$, $M_{R=0.4}$) and hematite concentrate ($H_{R=0.02}$, $H_{R=0.4}$). Based on the HYL method, the gas-based direct reduction experiments were conducted under the condition of rich hydrogen ($H_2/CO = 2.6$). The reduction index (RI), reduction swelling (RSI), cold compressive strength (CCS), and phase changes of the pellets were systematically studied. The results show that the RI of four kinds of pellets is greater than four in the temperature range of 800~950°C. For the magnetite pellets, at a fixed reduction temperature, the RSI increases and then gradually decreases, reaching a peak between 50 per cent and 80 per cent. Moreover, the RSI enhances with the increase of the reduction temperature. In general, the RI, RSI, and reduced CCS of the $M_{R=0.09}$ pellets are better than those of the $M_{R=0.4}$ pellets. For the hematite pellets, the maximum RSI decreases from 23 per cent to 17 per cent during hydrogen-rich reduction, while the compressive strength increases from 500 to 1070 N·pellet^{-1}. The main reason is that the glass phase generated by the increase of basicity adheres to the surface of hematite grains, hindering the contact between reducing gas and particles. As a result, the apparent activation energy is reduced, which is not conducive to the reduction process. In addition, the glassy phase attached to the hematite reduces the migration rate of Fe^{2+}/Fe^{3+}, resulting in a difference in the reduction expansion rate of hematite in the inner and outer layers of the pellet. Thus, concentric cracks are formed inside the pellets, the RSI increases and reduced CCS decreases.

INTRODUCTION

In recent years, China's ISI has been growing rapidly, and iron and steel production has continued to increase (Li, Lei and Pan, 2016; Boyle *et al*, 2021). Crude steel production in China reached up to 1033 million tons in 2021, representing over 53 per cent of global crude steel production. However, fossil energy resources, including coal, coke, and natural gas, as the main energy sources, have been used for ISI, resulting in the largest carbon emission behind the power system (Na *et al*, 2019). Global CO_2 emissions hit a record 36.3 billion tons, and China's CO_2 emissions contributed nearly one-third in 2021 (Xu and Cang, 2010; Wang, Wei and Shao, 2020). In traditional carbon reduction metallurgy, the BF ironmaking process is the main source of CO_2 and gas pollutants in the ISI (Hasanbeigi, Arens and Price, 2014). Before ironmaking, CO_2 emissions account for 88.75 per cent of the total. It is crucial to actively promote low-carbon reduction measures prior to ironmaking. Electric furnace steelmaking, in comparison to the traditional long process, offers advantages such as shorter processing time, relatively lower investment, faster construction, and energy efficiency, leading to environmental sustainability (Zhu *et al*, 2022). Therefore, achieving ultra-low carbon and zero-carbon green production in the iron and steel industry requires the substitution of fossil energy with green hydrogen and renewable electricity, alongside the development of gas-based direct reduction-electric arc furnace (DR-EAF) short processes (Sane *et al*, 2020; Voraberger *et al*, 2022).

The direct reduction process is a method that reduces iron oxides to obtain metallic iron at temperatures below the melting point. The product obtained from this process is known as direct reduced iron (DRI), which offers advantages such as high purity, environmental friendliness, low cost, and a compact process design. Moreover, DRI has a well-defined chemical composition compared to steel scrap and exhibits efficient melting properties in electric arc furnaces. As a result, there is a growing demand for direct reduced iron in electric arc steelmaking. Currently, the gas-based vertical shaft direct reduction process primarily includes the Midrex process and the HYL-III process. In comparison to the Midrex method, the HYL-III method utilises steam as a cracking agent, leading to a higher proportion of H_2 in the reducing gas and lower CO_2 emissions. This makes the HYL-III method more energy-efficient and environmentally friendly (Sui *et al*, 2017; Jain, 2009; Zhang *et al*, 2016; Oh and Noh, 2017). The primary feedstock for gas-based direct reduction is composed of pelletised ores and lumpy ores. However, with the depletion of high-grade natural lumpy ores, there has been a growing utilisation of low-grade fine-grained disseminated ores as raw materials for pellet production. To produce high-quality pellets, appropriate additives are often employed to enhance the pelletisation process. Limestone ($CaCO_3$), dolomite (Ca, Mg (CO_3)$_2$), and olivine (Mg_2SiO_4) are among the most used flux materials in the production of iron ore pellets (Dwarapudi *et al*, 2014; Kemppainen *et al*, 2015; Guo *et al*, 2019; Prakash *et al*, 2000). Extensive research has demonstrated that increasing the basicity of the sintered pellets can improve their strength by facilitating the formation of an adequate amount of silicate melt within the fluxed pellets (Wynnyckyj and Fahidy, 1974). At present, the gas-based direct reduction process is not yet implemented in China, and research on its mechanism remains limited. Specifically, there is a lack of comprehensive studies examining the impact of different raw materials used to produce gas-based reduction oxidising pellets on their reduction performance. Moreover, there is a need to further develop the technology for preparing oxidising pellets specifically for ultra-pure iron ore within the country.

In this article, a domestic super high-grade magnetite with an iron grade of 71.68 per cent and an imported hematite concentrate was selected as the iron-bearing material, with limestone used as an additive. Acidic and alkaline pellets were prepared with different ratios and basicity levels. Subsequently, under conditions simulating the HYL process, direct reduction tests were conducted on the iron ore pellets using a hydrogen-rich gas, and the performance variations and mechanisms during the reduction process of acidic and alkaline pellets were analysed. Through this study, we hope to provide reference systems and a theoretical foundation for the development of hydrogen-rich gas-based direct reduction technology for oxidised pellets.

MATERIALS AND METHODS

Raw materials

The following materials were used for preparing green pellets: an imported hematite concentrate H, super high-grade magnetite M, limestone, and bentonite. The chemical compositions of the raw materials are shown in Table 1. Their size distributions and specific surface areas are shown in Table 2. As can be seen from the table, the hematite concentrate contains 65.46 per cent Fetotal, 3.22 per cent SiO_2, and low content of other elements, which can be a good charge for shaft furnace ironmaking. The magnetite bears a very high iron grade as high as 71.68 per cent and very low gangue and detrimental impurities. In addition, iron ore concentrate has a very fine particle size with 95.5 per cent passing 0.025 mm and a high specific surface area (SSA) of 2782 cm^2/g. The size distributions of hematite concentrate are coarse with just 72.47 per cent passing -0.074 mm, and the specific surface areas are only 660 cm^2/g, which is much lower than that for conventional pelletising. Bentonite is used as the binder for pellets, and fine ground limestone is used as the flux to adjust the basicity of pellets (R=CaO/SiO_2).

TABLE 1

Chemical composition of raw materials (wt%).

Materials	TFe	FeO	SiO$_2$	Al$_2$O$_3$	CaO	MgO	Na$_2$O	K$_2$O	P	S	LOI
Hematite	65.46	1.55	3.22	1.16	0.07	0.04	0.01	0.06	0.07	0.01	1.46
Magnetite	71.68	29.49	0.16	0.15	0.035	0.038	-	-	0.002	0.005	-3.12
Limestone	0.42	-	2.32	0.38	52.69	0.78	0.014	0.15	-	0.10	42.25
Bentonite	11.81	-	49.94	25.34	2.03	3.03	2.27	0.14	0.067	0.04	13.22

TABLE 2

Size distributions and specific surface area of raw materials (wt%).

Size/mm	+0.180	0.074–0.180	0.043–0.074	0.025–0.043	-0.025	SSA (cm^2/g)
Hematite	2.83	24.70	30.88	21.45	20.14	670
Magnetite	0.00	0.14	0.62	3.74	95.50	2782
Limestone	6.54	12.94	10.80	9.46	60.26	3524
Bentonite	0.06	6.82	28.86	24.38	39.88	-

Sample preparation

In this study, the basicity of the pellets was adjusted to natural basicity and 0.40 by adding finely-ground limestone (CaCO$_3$). The iron concentrates were thoroughly mixed with 0.8 wt per cent bentonites, 9.5 wt per cent water, and the appropriate ratio of limestone. Subsequently, the mixture was pelletised in a disk pelletiser to form green balls with a diameter of 10–16 mm. The green pellets were then dried for five hours at 105°C. The resulting green pellets were placed in a corundum crucible and heated to 950°C in a muffle furnace, holding this temperature for 10 mins, followed by firing at 1250°C for 15 mins. Throughout the preheating and roasting processes, the air was continuously pumped into the reaction system at a flow rate of 5 L/min. After the roasting process, the pellets are taken out and allowed to cool naturally to room temperature in ambient air. Table 3 displays the chemical compositions of the fired pellets with different basicity. The fired pellets exhibit remarkably high iron content, along with low levels of gangue and harmful impurities. This makes them highly suitable for producing high-quality direct reduced iron, positioning them as an exceptional burden material.

TABLE 3

Chemical compositions of fired pellets with varying basicities (wt.%).

Pellets	Basicity	TFe	SiO$_2$	Al$_2$O$_3$	CaO	MgO	Na$_2$O	K$_2$O	P	S
Magnetite	0.09	69.14	0.54	0.34	0.05	0.06	0.032	0.009	0.003	0.005
	0.40	69.01	0.55	0.34	0.22	0.06	0.032	0.010	0.003	0.005
Hematite	0.02	66.15	3.55	1.32	0.09	0.05	0.025	0.057	0.072	0.010
	0.40	65.18	3.57	1.32	1.43	0.07	0.025	0.060	0.071	0.013

Experimental methods

Gas-based direct reduction experiments were conducted based on the metallurgical performance testing standards of the HYL-III method (Technologies, 2015a, 2015b, 2015c). Figure 1 shows a schematic diagram of the electric resistance furnace used to simulate a gas-based direct reduction process. The reduction experiments were carried out in the furnace with reducing gas to replicate the HYL-III shaft furnace process, as outlined in Table 4.

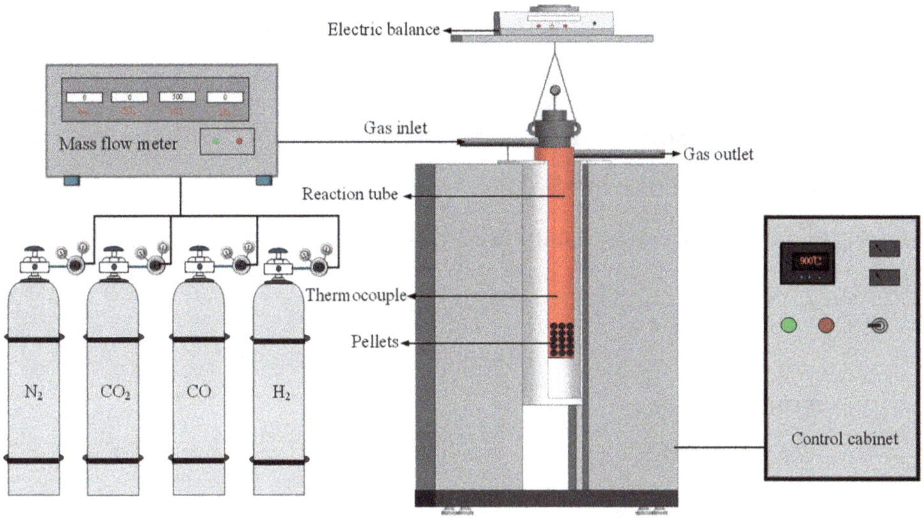

FIG 1 – Schematic diagram of the shaft furnace for gas-based direct reduction.

TABLE 4

Compositions of reducing gases.

Gas content/vol%				H_2/CO
H_2	CO	CO_2	N_2	
55	21	14	10	2.6

A 500 g sample should be in the size range of 16 mm to 10 mm, with 50 per cent of particles falling within the range of 16–12.5 mm and the other 50 per cent within the range of 12.7–9.5 mm. Place the test portion evenly in the reduction tube and close the top. Suspend the reduction tube centrally from the weighing device, ensuring that it does not touch the furnace or heating elements. During the heating process, inject N_2 as a protective gas. Continue heating and maintaining the flow of inert gas until the mass of the test portion becomes constant (mass m1) and the temperature is stable at 950 ±5°C. Then, inject the reducing gas with a flow rate of 55 ± 2 normal L/min. According to the standard of ISO 11258: 2015, the reduction degree (η) is calculated through the following formula.

$$R = \frac{O_{2rem}}{O_{2red}} \times 100 \tag{1}$$

Where:

R	=	Percentage of reduction
O_{2rem}	=	Removed oxygen from sample
O_{2red}	=	Reducible oxygen in sample

Reducible oxygen is calculated as follows:

$$O_{2\mathrm{red}} = (\frac{WO}{W_{Fe}}) \times (1.5Fe\ t - 0.5Fe_{\mathrm{II}}) \tag{2}$$

Where:

WO	=	atomic weight of oxygen
W_{Fe}	=	atomic weight of iron
Fe t	=	per cent of total iron in sample
Fe_{II}	=	per cent of iron (II) in sample

Percentage of reduction as a function of time is calculated using the recorded weight of sample and the reducible oxygen (O_{2red}) in the following formula:

$$R = \left(1 - \frac{W}{W1}\right) \times \left(\frac{10^4}{O_{2red}}\right) \qquad (3)$$

Where:

W $\qquad$ = $\qquad$ weight of sample (function of time)

W1 $\qquad$ = $\qquad$ initial weight of sample

Usually, experimental data can be fitted to a first order kinetic model. The reduction rate equation is:

$$\frac{\delta R}{\delta t} = k \times (1 - R) \qquad (4)$$

Where:

($\delta R/\delta t$) $\qquad$ = $\qquad$ reduction rate

k $\qquad$ = $\qquad$ rate constant

R $\qquad$ = $\qquad$ degree of reduction

By integration of the reduction rate equation:

$$Ln\left(\frac{1}{1-R}\right) = k \times t \qquad (5)$$

Plotting in [1/(1-R)] against time yields a straight line. The slope of this line is the value of k. A higher value of k implies a better reducibility. The Reducibility Index (RI) is defined as the value of k multiplied by 100.

$$RSI = \left(\frac{V1-V0}{V0}\right) \times 100\% \qquad (6)$$

where V1 and V0 is the volume of the reduced and fired pellets, respectively, and both of them were determined by the sand discharge method.

The schematic diagram of HYL-III low-temperature disintegration (LTD) test equipment is illustrated in Figure 2. The test sample in the size range of 9.5 mm to 15.9 mm (50 per cent + 10 mm–12.7 mm; 50 per cent + 12.7 mm–15.9 mm). Place the test portion in the reduction tube so that the surface is even. Then insert the reduction tube into the furnace. Start the heating and while heating, pass a flow of inert gas through the test portion at a flow rate of approximately 2.0 L/min N_2. When the temperature reaches 500°C, for a good homogenisation of the sample temperature, continue passing inert gas for another 20–30 mins. Then, introduce the reducing gas at a flow rate of 20 L/min to replace the inert gas and begin rotating the reduction tube. Keep reducing the sample using the reducing gas for 2 hours. After 2 hours of reduction time, stop the flow of the reducing gas and the rotation of the tube. Cool the sample to ambient temperature within the tube bypassing a flow of 5 normal L/min of inert gas. Once cooled, screen the reduced sample to determine its particle size distribution and count the number of unbroken pellets. The low-temperature disintegration (LTD) as a percentage by mass, is calculated from the following formula:

$$LTD + 6.3mm = \left(\frac{m1}{m0}\right) \times 100 \qquad (7)$$

$$LTD - 3.2mm = \left(\frac{m0-m1-m2}{m0}\right) \times 100 \qquad (8)$$

$$LTDup = \left(\frac{n2}{n1}\right) \times 100 \qquad (9)$$

Where:

m0 represents the theoretical mass in grams if the reduction is carried out to magnetite

m1 is the mass in grams of the oversize fraction retained on the 6.3 mm sieve

m2 is the mass in grams retained on the 3.2 mm sieve and passed through the 6.3 mm sieve.

Additionally, the formula requires two counts: n1, the number of initial pellets, and n2, the number of final unbroken pellets.

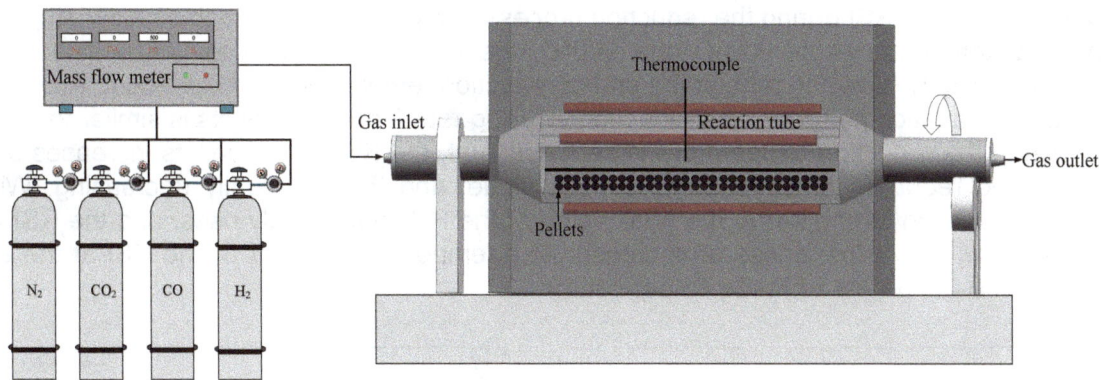

FIG 2 – Schematic diagram of the low temperature disintegration test equipment for HYL.

RESULTS AND DISCUSSION

Reduction behaviour of fired hematite pellets in hydrogen-rich gases

Reduction behaviour of fired pellets with natural basicity

The variations of the RI for $M_{R=0.09}$ pellets and $H_{R=0.02}$ pellets at different reduction temperatures are depicted in Figure 3. Figure 3a reveals that at a reduction temperature of 800°C, the RI for $M_{R=0.09}$ pellets is 7.54, which is notably higher than 4. With an increase in temperature to 950°C, the RI for $M_{R=0.09}$ pellets reaches 15.01. As the temperature rises, there is a substantial growth in RI. The fitted slope of the RI-temperature relationship for $M_{R=0.09}$ pellets is 0.04891, signifying an approximate 26 per cent increase in the reduction degree coefficient for every 50°C temperature increment. This demonstrates that a higher temperature significantly enhances the reaction rate of the reduction process for $M_{R=0.09}$ pellets. From Figure 3b, it can be observed that as the reduction temperature increases from 800°C to 950°C, the RI increases from 7.28 to 18.34. The fitted slope of the RI-temperature relationship is 0.07238, indicating an approximately 36 per cent increase in the RI for every 50°C increment in temperature. The impact of temperature on the RI of $H_{R=0.02}$ pellets is more pronounced compared to the M pellets.

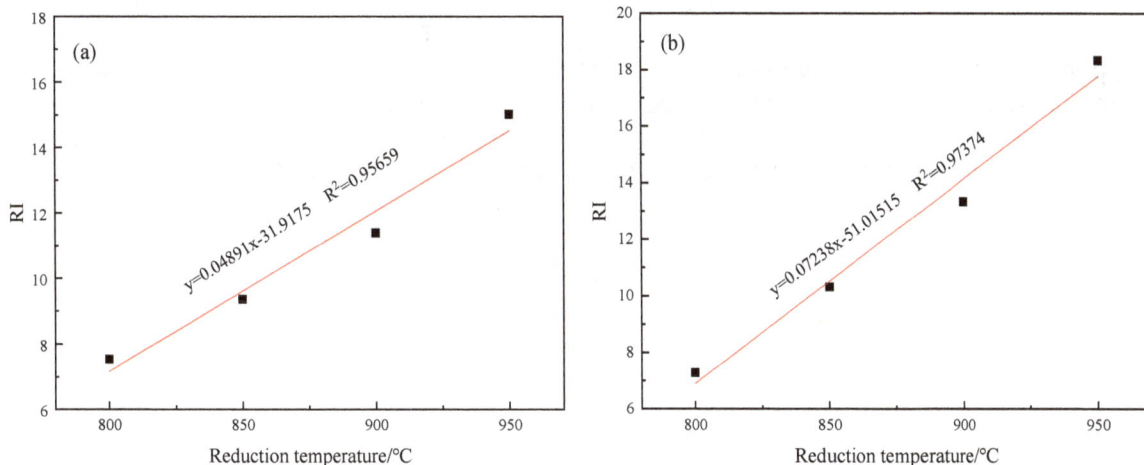

FIG 3 – Effect of reduction temperature on reduction index (RI) of (a) $M_{R=0.09}$ and (b) $H_{R=0.02}$ pellets.

The RSI of $M_{R=0.09}$ pellets and $H_{R=0.02}$ pellets at different reduction temperatures are depicted in Figure 4. From Figure 4a, it is evident that as the reduction time progresses, the RSI of the pellets undergoes an initial sharp increase, followed by a significant decrease. Towards the later stages of the reduction process, the RSI gradually decreases, reaching its peak within 10 mins. The reduction temperature has a substantial impact on the RSI of $M_{R=0.09}$ pellets. As the reduction temperature rises, the maximum swelling ratio during reduction increases significantly, and the RSI at the end of reduction exhibits a notable rise as well. This indicates that the reduction temperature significantly influences the phase transition and crystal structure of the pellets. When the temperature is below

900°C, the maximum RSI during the reduction process of $M_{R=0.09}$ pellets remains below 15 per cent. However, at temperatures of 900°C and 950°C, the maximum RSI of $M_{R=0.09}$ pellets exceeds 15 per cent, while the swelling ratio at the end of reduction remains below 15 per cent. According to Figure 4b, the variation pattern of the reduction swelling rate for $H_{R=0.02}$ pellets is similar to that of M pellets. When the reduction temperature is kept constant, the RSI of the pellets increases sharply with prolonged reduction time, reaching a peak value, and then gradually decreasing. With an increase in reduction temperature, the peak value of the RSI significantly rises, and the RSI at the end of reduction also experiences an increase. At a temperature of 950°C, the RSI at the end of reduction for $H_{R=0.02}$ pellets exceeds 15 per cent.

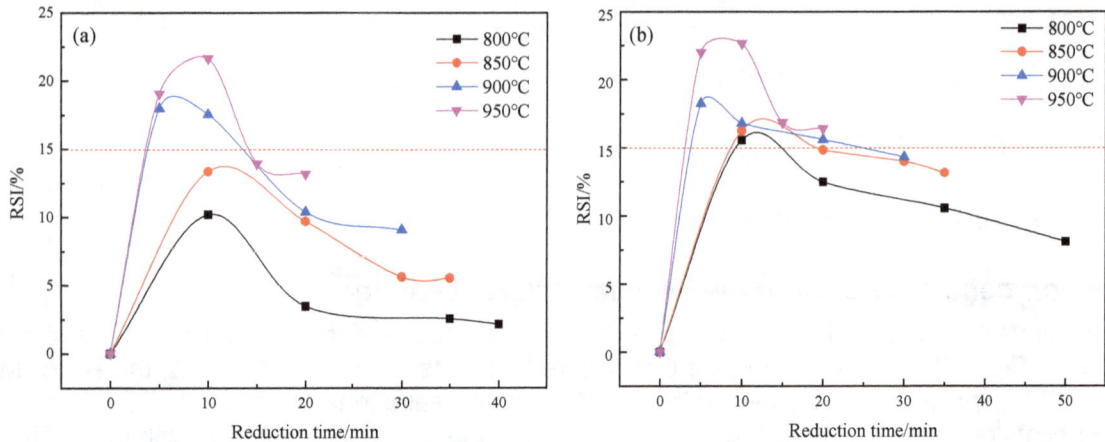

FIG 4 – Effect of reduction duration on RSI of (a) $M_{R=0.09}$ and (b) $H_{R=0.02}$ pellets under 800°C ~ 950°C.

Figure 5 illustrates the variation of cold compressive strength (CCS) of $M_{R=0.09}$ and $H_{R=0.02}$ pellets with reduction time within the temperature range of 800°C to 950°C. According to Figure 5a, at the same temperature, the CCS of the pellets initially decreases sharply with increasing reduction time, followed by a slight increase. Under identical conditions, as the reduction temperature increases, the minimum CCS of the pellets significantly decreases, and the CCS of the pellets at the reduction end point also decreases notably. At 800°C, the CCS of the pellets at the reduction end point is 1298 N·pellet[-1], whereas, at 950°C, it is only 484 N·pellet[-1]. Figure 5b shows that the CCS of the $H_{R=0.02}$ pellets decreases rapidly during the reduction process, followed by minimal changes. Under the same conditions, as the reduction temperature increases, the CCS at the reduction end point of the pellets decreases to some extent. Compared to the M pellets, the temperature has a smaller impact on the post-reduction compressive strength of the $H_{R=0.02}$ pellets.

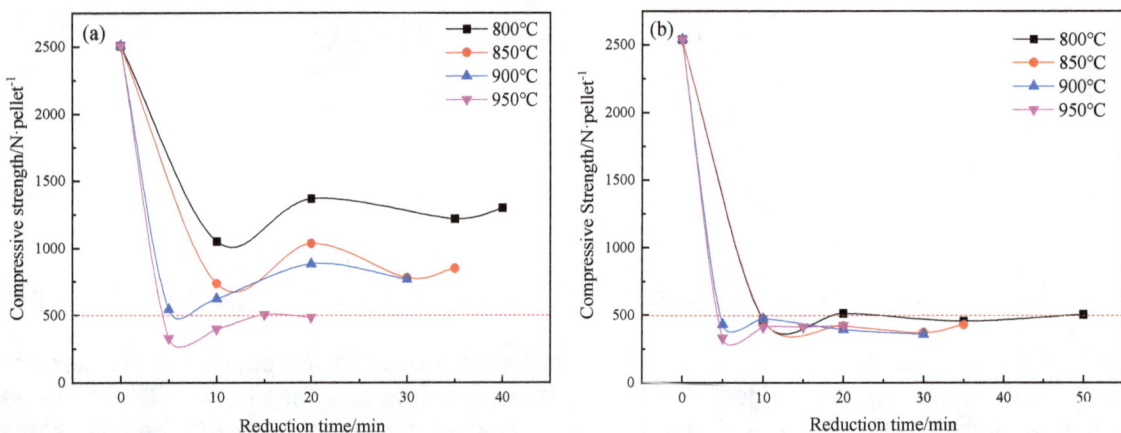

FIG 5 – Effect of reduction duration on CCS of (a) $M_{R=0.09}$ and (b) $H_{R=0.02}$ pellets under 800°C ~ 950°C.

Reduction behaviour of fired pellets with 0.4 basicity

The influence of reduction temperature on the RI of $M_{R=0.4}$ and $H_{R=0.4}$ pellets is shown in Figure 6. As depicted in Figure 6a, at a reduction temperature of 800°C, the RI for $M_{R=0.4}$ pellets is 6.58, which increases to 12.49 when the temperature is raised to 950°C. The RI of the pellets exhibits a significant increase with the rise in temperature. The fitted slope of the RI-temperature relationship is 0.03932, indicating an approximate 24 per cent increase in the RI for every 50°C increment in temperature. Figure 6b demonstrates that at a reduction temperature of 800°C, the RI for $H_{R=0.4}$ pellets is only 3.63, failing to meet the standard requirement of an RI not lower than 4. When the temperature increases to 850°C, the RI increases to 5.22. Within the temperature range of 800°C to 950°C, the fitted slope of the RI-temperature relationship is 0.05445, suggesting an approximately 48 per cent increase in the RI for every 50°C increment in temperature. Compared to $M_{R=0.4}$ pellets, the influence of temperature on the reduction degree coefficient of $H_{R=0.4}$ pellets is the most significant.

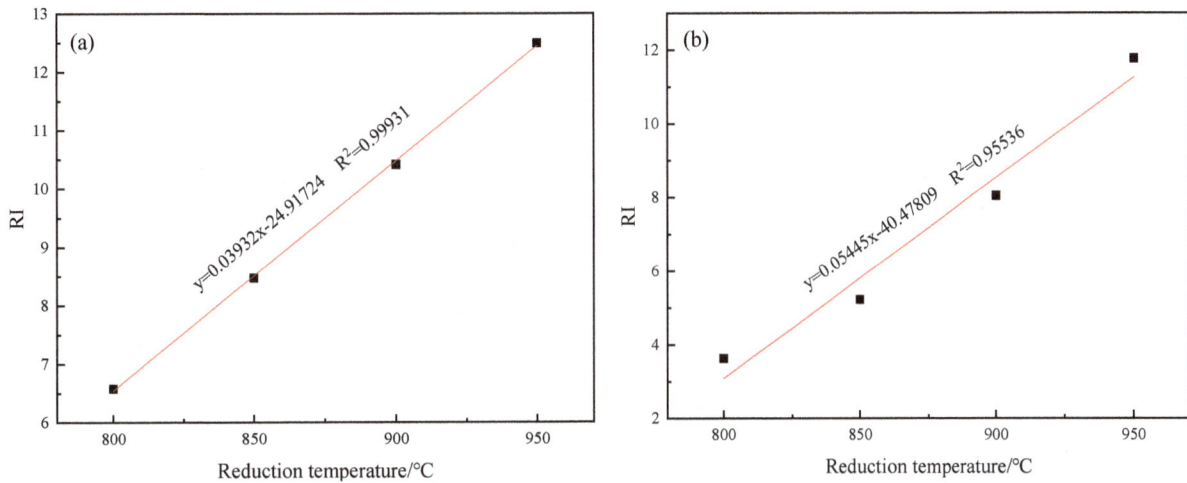

FIG 6 – Effect of reduction temperature on reduction index RI of (a) $M_{R=0.4}$ and (b) $H_{R=0.4}$ pellets.

The effect of different temperatures on the reduction swelling rate of $M_{R=0.4}$ and $H_{R=0.4}$ pellets is illustrated in Figure 7. As shown in Figure 7a, when the reduction temperature is constant, the RSI of the pellets initially increases and then decreases with increasing reduction time, reaching its peak within 10 mins. The reduction temperature has a significant impact on the RSI of the pellets, with higher temperatures leading to a notable increase in the maximum RSI and the RSI at the reduction end point. This pattern is similar to the behaviour observed in $M_{R=0.09}$ pellets. At 800°C, the maximum RSI of $M_{R=0.4}$ pellets is below 15 per cent. At 850°C and 900°C, the maximum RSI exceeds 15 per cent, while it remains below 15 per cent at the reduction end point. Figure 7b gives the transformation of RSI of $H_{R=0.4}$ pellets during the reduction process. As observed, the higher the reduction temperature, the more rapidly and dramatically increases of RSI of the pellets in the initial stage, as well as the larger peak value of RSI of pellets and the bigger RSI of the reduced products. Compared with the $H_{R=0.02}$ pellets, the $H_{R=0.4}$ pellets stronger resistance against reduction swelling, with the RSI being slightly above 15 per cent at 950°C and always far below 15 per cent between 800°C and 900°C.

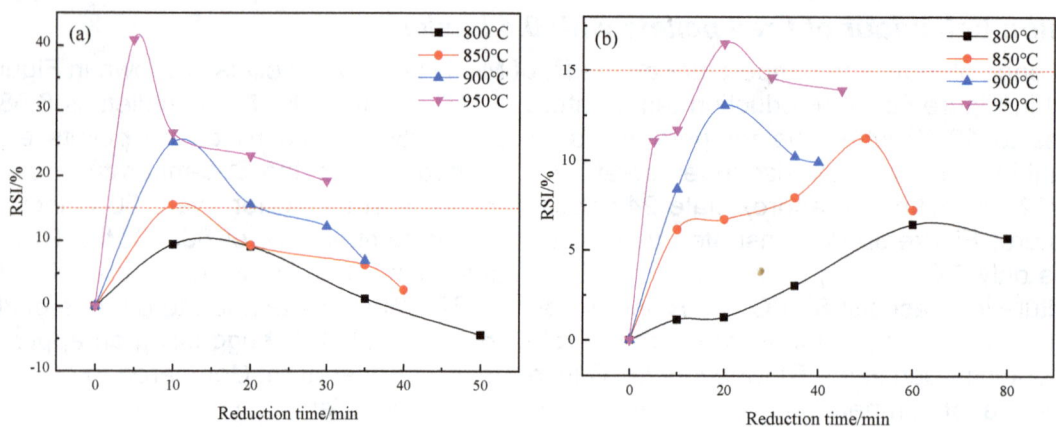

FIG 7 – Effect of reduction duration on RSI of (a) $M_{R=0.4}$ and (b) $H_{R=0.4}$ pellets under 800°C ~ 950°C.

Figure 8 illustrates the variations in compressive strength of $M_{R=0.4}$ and $H_{R=0.4}$ pellets with respect to reduction time within the temperature range of 800°C to 950°C. As shown in Figure 8a, at the same temperature, the CCS of the pellets undergoes a sharp decline during the reduction process, followed by a slight increase. Under the same conditions, with an increase in reduction temperature, the minimum CCS of the pellets significantly decreases, and the CCS of the pellets at the reduction end point also decreases noticeably as the reduction temperature rises. The CCS of $M_{R=0.4}$ pellets at the reduction end point is 1444 N·pellet⁻¹ when reduced at 800°C, but it decreases to 307 N·pellet⁻¹ at 950°C. The variation of CCS of $H_{R=0.4}$ pellets shown in Figure 8b is similar to that with natural basicity (R=0.02), and lower reduction temperature leads to a major increase in the CCS of reduced pellets. For example, the CCS of reduced products is even as high as 1500 N/pellet at a reduction temperature of 800°C. It is valuable that the CCS of $H_{R=0.4}$ pellets is always higher than the requirement for the production of gas-based direct reduction (500 N·pellet⁻¹) during the reduction process in the range of experimental temperature.

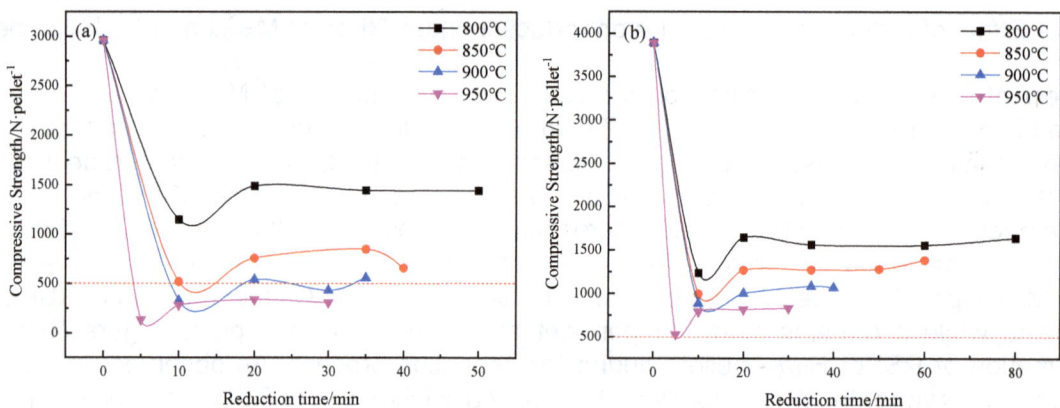

FIG 8 – Effect of reduction duration on CCS of (a) $M_{R=0.40}$ and (b) $H_{R=0.40}$ pellets under 800°C ~ 950°C.

Mechanism analysis

To further investigate the influence of basicity on the gas-based reduction mechanism of magnetite and hematite pellets, an analysis of their phase composition and microstructure after reduction was conducted. Figure 9 shows the phase changes during the reduction process of $M_{R=0.09}$ pellets and $M_{R=0.4}$ pellets, Figure 10 depicts the macroscopic morphological changes during reduction, and Figure 11 presents the SEM-EDS analysis of pellets at a specific moment during reduction.

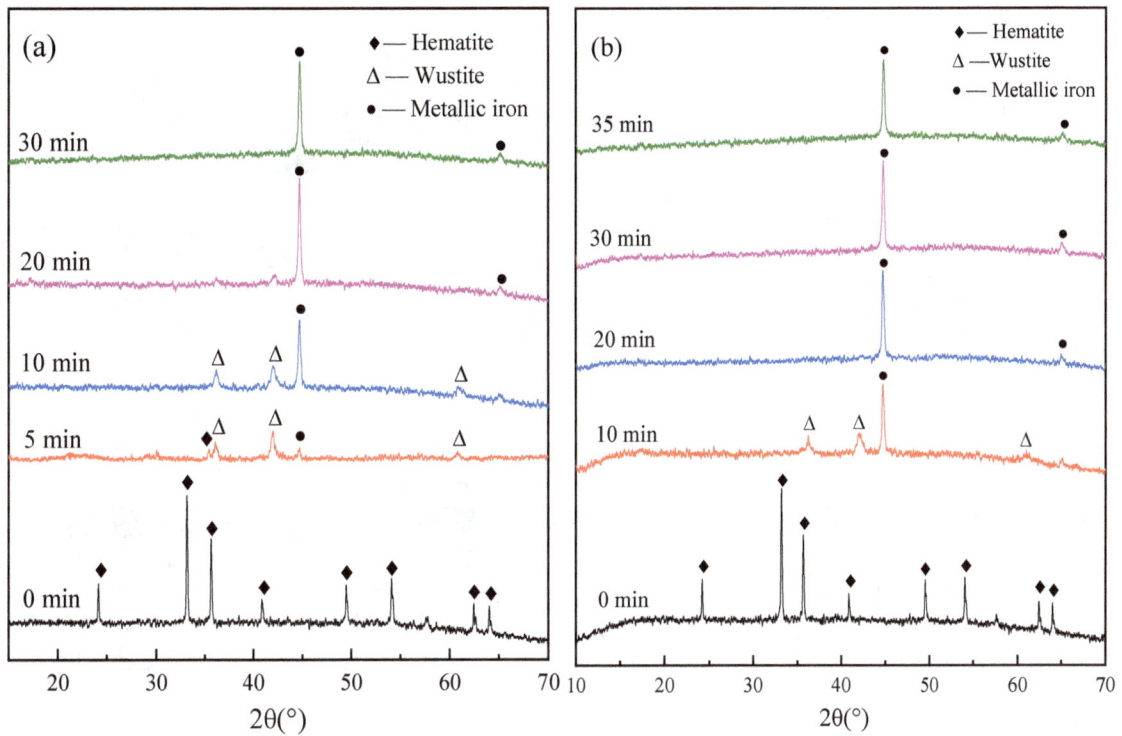

FIG 9 – XRD patterns of reduced (a) $M_{R=0.09}$ pellets (900°C) and (b) $M_{R=0.4}$ pellets (900°C).

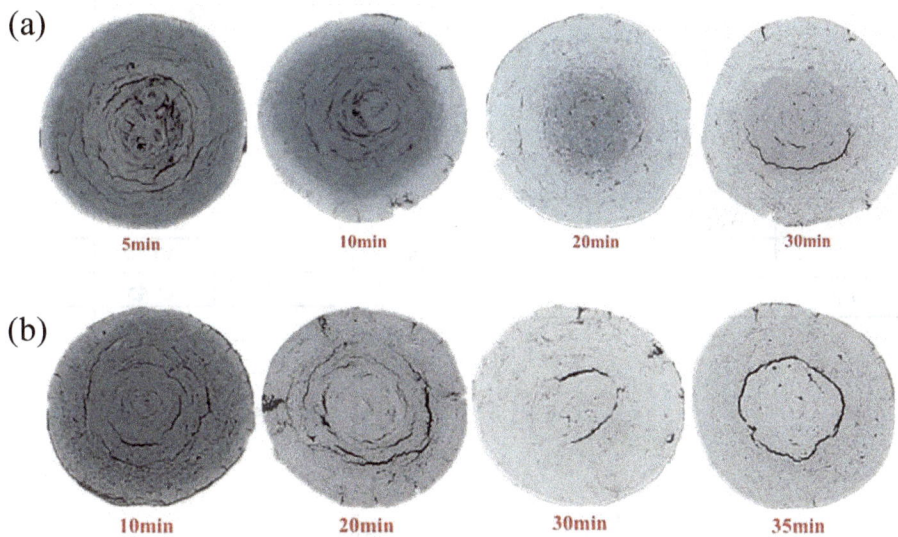

FIG 10 – Macroscopic morphology of (a) $M_{R=0.09}$ pellets and (b) $M_{R=0.4}$ pellets during reduction process (900°C).

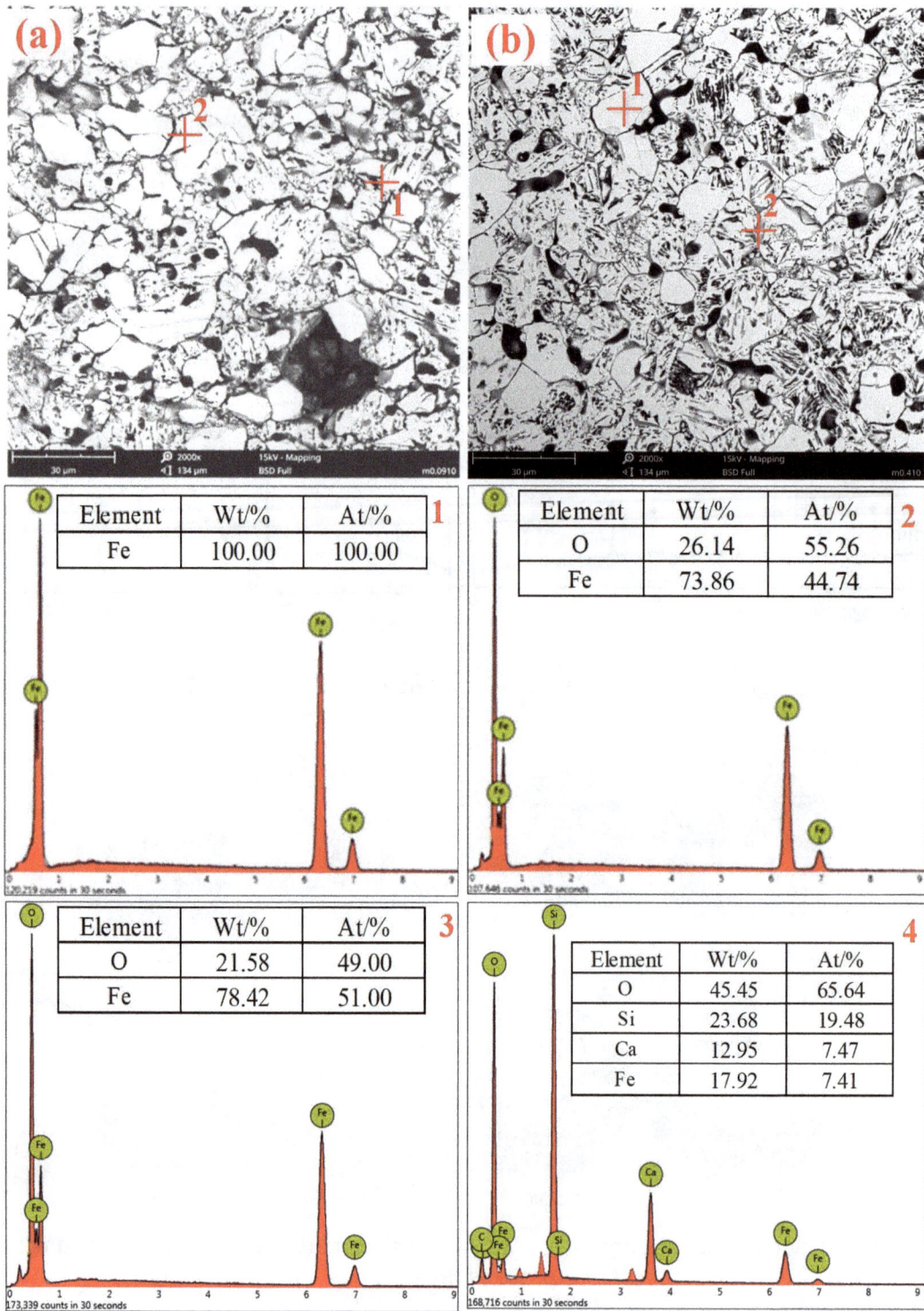

FIG 11 – SEM and EDS of (a) $M_{R=0.09}$ pellets and (b) $M_{R=0.4}$ pellets reduced at 900°C for 10 min (2000X).

From Figure 9a, it can be observed that the initial phase composition of the $M_{R=0.09}$ pellets is solely Fe_2O_3 before reduction. After 5 mins of reduction, the $M_{R=0.09}$ pellets predominantly consist of wüstite (Fe_xO) with a small amount of unreduced Fe_2O_3, along with the formation of some metallic iron. As shown in Figure 10a, after 5 mins of reduction, the $M_{R=0.09}$ pellets exhibit a thin white outer shell of metallic iron in its macroscopic state, while the internal structure appears greyish-black macroscopically, mainly consisting of ferrous oxide with a small amount of Fe_2O_3. After 10 mins of reduction, the $M_{R=0.09}$ pellets consist only of wüstite (Fe_xO) and metallic iron, with the phase peak of metallic iron significantly stronger than that observed after 5 mins of reduction. Figure 10a reveals that after 10 mins of reduction, the outer layer of the pellet exhibits a higher degree of reduction, with only the core region of large grains remaining unreduced, containing some ferrous oxide. The internal region of the pellets is predominantly composed of wüstite, with a small amount of dispersed metallic iron particles. The outer layer exhibits a significantly higher porosity, while the inner layer has a very low porosity, with grains tightly packed together. After 20 mins of reduction, it can be observed from Figure 9a that the $M_{R=0.09}$ pellets contain only metallic iron, and the peaks corresponding to a metallic iron phase are significantly enhanced, while the ferrite phase content is too low to be detected. At this stage, macroscopically, the $M_{R=0.09}$ pellets have only a small portion of grey-black areas, while the rest appears bright white. After 30 mins of reduction, XRD analysis at this point shows no significant changes compared to the 20 min reduction, indicating a single metallic iron phase.

For the $M_{R=0.4}$ pellet, after 10 mins of reduction, no Fe_2O_3 phase is detected in its interior. The phase transitions from a single Fe_2O_3 to wüstite and metallic iron. As shown in Figure 10b, after 10 mins of reduction, the cross-section of the $M_{R=0.4}$ pellet appears predominantly grey-black, with a thin white shell at the edges. Referring to the microstructure in Figure 11b of the $M_{R=0.4}$ pellet after 10 mins of reduction, the white region observed in the macroscopic cross-section corresponds mainly to metallic iron, while the grey region is primarily composed of wüstite. The outer layer of the pellets is undergoing hematite reduction to metallic iron. The middle layer of the pellets mainly consists of hematite, with dispersed metallic iron particles present. The inner layer of the pellets is primarily composed of hematite. After reduction for 20 mins and 30 mins, the XRD spectra exhibit metallic iron phases. The structural differences between the inner and outer layers of the pellets are insignificant. The dominant phase is metallic iron, with a small amount of hematite present, all located in the middle of larger grains. This is because the reducing gas diffuses from the outside to the inside during the reduction process. Due to encapsulation, gas diffusion is limited, resulting in an incomplete reduction of the core of larger grains. Figure 11b also indicates that the outer layer of the pellets is mainly composed of metallic iron, with a small amount of unreduced hematite in the core of the metallic iron grains. The porosity is relatively high, forming a honeycomb-like structure. The inner layer is predominantly hematite, with traces of calcium ferrite olivine visible at 2000 times magnification.

Figure 12 illustrates the phase transformation during the reduction process of $H_{R=0.02}$ pellets and $H_{R=0.4}$ pellets. Figure 13 depicts the macroscopic morphological changes during the reduction process, while Figure 14 demonstrates the SEM-EDS analysis of the pellets at a specific moment during reduction. From Figure 12a and Figure 13a, it can be observed that under the condition of 800°C, after 10 mins of reduction, ferrite and metallic iron are formed within the $H_{R=0.02}$ pellets, and a small amount of Fe_2O_3 remains unreduced. The SEM-EDS analysis of the pellets in Figure 14a also indicates that during a 10 mins reduction, the outermost layer of the pellets is primarily reduced to metallic iron with a small amount of impurities. The inner layer consists of ferrite and bonded impurities, with ferrite being the main component. At this stage, the outer layer of the pellets already exhibits a honeycomb-like structure with numerous pores, while the inner layer shows significantly fewer pores, and the grain structure remains intact. After a 20 mins reduction, the $H_{R=0.02}$ pellets exhibit a significant enhancement in the metallic iron phase peak, with only a weak ferrite phase peak remaining. The cross-sectional view of the pellets shows a visibly thicker white shell, and a smaller and lighter grey-black central area, indicating further progression of the reduction reaction. After a 35 mins reduction, only the metal iron phase is present, and its phase peak is significantly enhanced. At this point, the $H_{R=0.02}$ pellets appear uniformly coloured at a macroscopic level. After a 50 min reduction, the pellets reach a high reduction degree. The XRD analysis and macroscopic morphology show no significant changes compared to the 35 mins reduction stage.

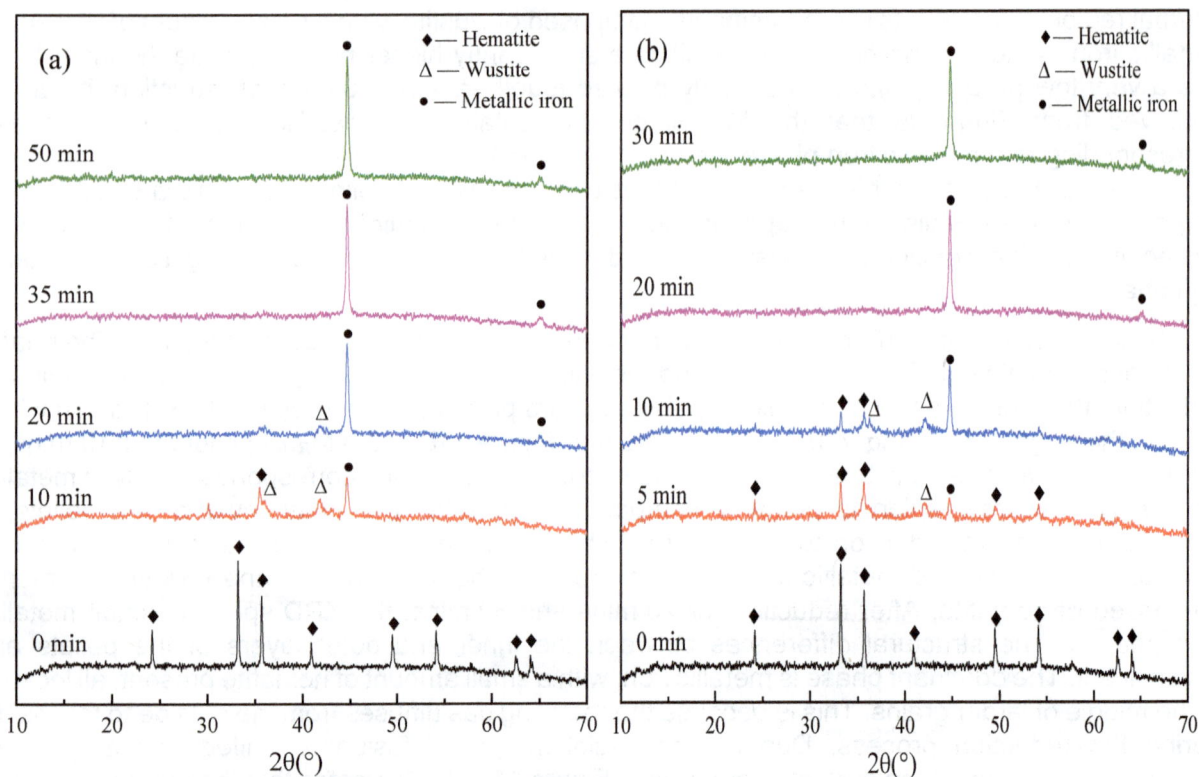

FIG 12 – XRD patterns of reduced (a) $H_{R=0.02}$ pellets (800°C) and (b) $H_{R=0.4}$ pellets.

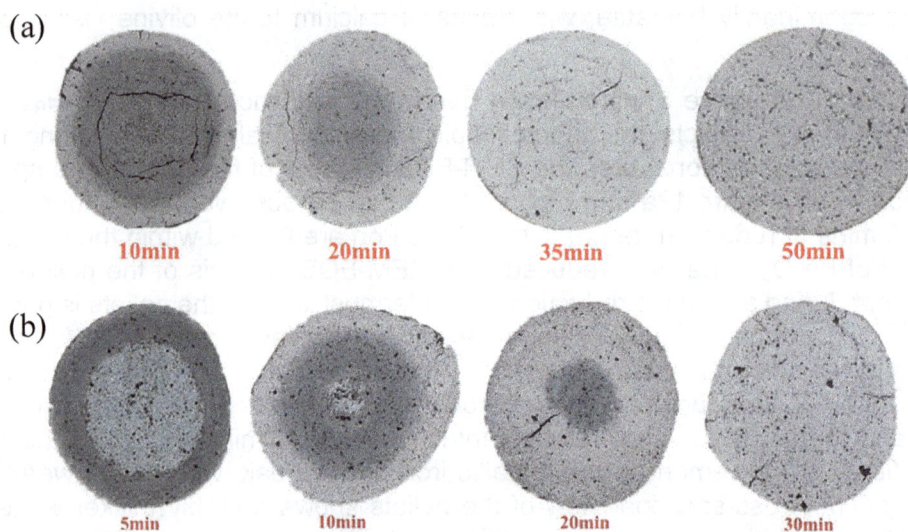

FIG 13 – Macroscopic morphology of (a) $H_{R=0.02}$ pellets during reduction process (800°C) and (b) $H_{R=0.4}$ pellets during reduction process (950°C).

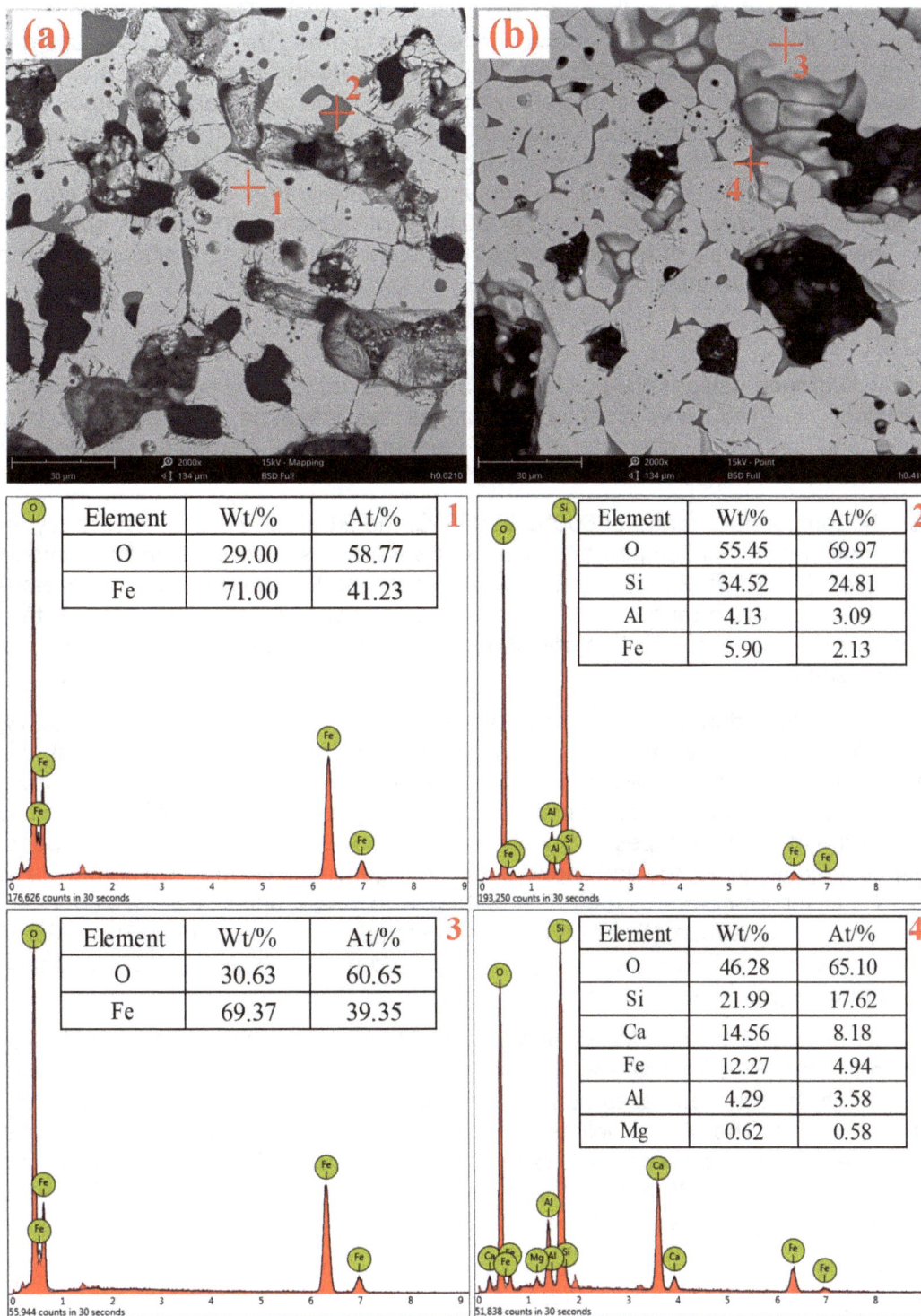

FIG 14 – SEM and EDS of (a) $H_{R=0.02}$ pellets reduced at 800°C for 10 min (2000X) and (b) $H_{R=0.4}$ pellets reduced at 950°C for 10 min.

From Figures 12b and 13b, it can be observed that the predominant phases in $H_{R=0.4}$ pellets are Fe_2O_3, wüstite, and metallic iron after 5 mins of reduction. The brown region in the inner layer of the pellets corresponds to Fe_2O_3, while the grey-black region represents wüstite. There is a thin layer of metallic iron on the outermost surface of the pellets, although its content is low and not clearly visible in the macroscopic cross-sectional image. After 10 mins of reduction, the macroscopic profile of $H_{R=0.4}$ pellets shows three distinct layers: outer white-grey, middle grey-black, and inner brown. Compared to the 5 mins reduction, the core region of Fe_2O_3 undergoes significant shrinkage, the grey-black wüstite region progresses towards the inner layer, and the outer layer exhibits the presence of metallic iron. Figure 12b indicates a noticeable increase in the peak intensity of the metallic iron phase and a significant decrease in the Fe_2O_3 phase peak. Figure 14b shows that there is a small amount of liquid phase both inside and outside the pellets. The inner layer of the pellets is mainly composed of Fe_2O_3, with intact grains and few micropores. The middle layer of the pellets is dominated by wüstite, with the appearance of micropores and the beginning of grain destruction. The outer layer of the pellets is mainly metallic iron, with no large grains and numerous micropores, forming a honeycomb-like structure. After 20 mins of reduction, the $H_{R=0.4}$ pellets appear predominantly grey-white, with a small amount of grey-black regions in the middle. At this point, the XRD spectrum of the pellets only shows the presence of metallic iron, and the peak corresponding to metallic iron is significantly enhanced. After 30 mins of reduction, the pellets appear uniformly grey-white. The XRD analysis of the pellets does not show any significant difference compared to the results obtained after 20 mins of reduction.

CONCLUSIONS

To improve the hydrogen-rich gas-based shaft furnace direct reduction of hematite pellets to promote green ironmaking, the reduction process as well as the influence mechanism of basicity on consolidation and reduction characteristics of magnetite and hematite concentrates were investigated. The main conclusions are as follows:

- Within the temperature range of 800°C to 950°C, the RI of the pellets significantly increases with the rise in reduction temperature. The temperature has the most significant impact on the $H_{R=0.4}$ pellet. The RI of $M_{R=0.09}$ pellets, $M_{R=0.4}$ pellets, and $H_{R=0.02}$ pellets is all greater than 4 within the temperature range of 800°C to 950°C. For the pellets with an $H_{R=0.4}$, the RI at a reduction temperature of 800°C is only 3.63. However, when the temperature rises to 850°C, the RI increases to 5.22.

- The RSI variations during the reduction process are relatively similar for the $M_{R=0.09}$ pellets, $M_{R=0.4}$ pellets, and $H_{R=0.02}$ pellets. They initially increase sharply, reaching a peak value, and then gradually decrease. Moreover, RSI increases with higher reduction temperatures. For the $H_{R=0.4}$ pellets, at higher reduction temperatures (900°C to 950°C), the RSI variation pattern is similar to the other three types of pellets. It reaches a peak value between 70 per cent and 90 per cent reduction degree and shows a relatively gentle increase in RSI during the reduction process at lower temperatures (800°C to 850°C). The $M_{R=0.09}$ pellets and $H_{R=0.4}$ pellets have an RSI end point of less than 15 per cent within the temperature range of 800°C to 950°C, while the $M_{R=0.4}$ pellets and $H_{R=0.02}$ pellets have an RSI end point of less than 15 per cent within the temperature range of 800°C to 900°C.

- The CCS of $M_{R=0.09}$ pellets, $M_{R=0.4}$ pellets, and $H_{R=0.4}$ pellets decreases significantly at the beginning of the reduction process and then slightly increases. On the other hand, the CCS of $H_{R=0.02}$ pellets sharply decreases during the initial stages of reduction and then stabilises with minimal changes. The $M_{R=0.09}$ pellets and $M_{R=0.4}$ pellets have a CCS end point greater than 500 N·pellet^{-1} within the temperature range of 800°C to 900°C. The $H_{R=0.02}$ pellets have a CCS end point greater than 500 N·pellet^{-1} at a reduction temperature of 800°C. The $H_{R=0.4}$ pellets maintain a CCS higher than 500 N·pellet^{-1} throughout the experimental temperature range.

- The gas-based direct reduction reaction of pellets follows the unreduced core model, proceeding from the outer to inner layers. During the reduction process, the large grains within the pellets are gradually disrupted, resulting in a higher porosity in the reduced pellets. Compared to the magnetite pellets, the hematite pellets grains are smaller. The purity of the pellets of $M_{R=0.09}$ and $M_{R=0.4}$ is high, as even under 2000 times magnification, it is difficult to observe many impurities. However, after reduction, $H_{R=0.02}$ pellets and $H_{R=0.4}$ pellets show significantly more visible impurities.

ACKNOWLEDGEMENTS

The authors want to express their gratitude for the financial support from the National Natural Science Foundation of China (No. 52174329), the Youth Natural Science Foundation of China (No. 51904347), Natural Science Foundation China (No. 52274343), Youth Natural Science Foundation China (NO.51904347) and China Baowu Low Carbon Metallurgy Innovation Foundation (BWLCF202102).

REFERENCES

Boyle, A D, Leggat, G, Morikawa, L, Pappas, Y and Stephens, J C, 2021. Green new deal proposals: comparing emerging transformational climate policies at multiple scales, *Enegy Res Soc Sci*, 81:102259.

Dwarapudi, S, Banerjee, P K, Chaudhary, P, Sinha, S, Chakraborty, U, Sekhar, C, Venugopalan, T and Venugopal, R, 2014. Effect of fluxing agents on the swelling behaviour of hematite pellets, *Int J Miner Process*, 126:76–89.

Guo, H, Jiang, X, Shen, F M, Zheng, H Y, Gao, Q J and Zhang, X, 2019. Influence of SiO_2 on the compressive strength and reduction-melting of pellets, *Metals*, 9(8):852–870.

Hasanbeigi, A, Arens, M and Price, L, 2014. Alternative emerging ironmaking technologies for energy-efficiency and carbon dioxide emissions reduction: a technical review, *Renewable Sustainable Energy Rev*, 33:645–658.

Jain, I P, 2009. Hydrogen: the fuel for 21st century, *Int J Hydrog Energy*, 34(17):7368–7378.

Kemppainen, A, Ohno, K, Iljana, M, Mattila, O, Paananen, T, Heikkinen, E, Maeda, T, Kunitomo, K and Fabritius, T, 2015. Softening behaviours of acid and olivine fluxed iron ore pellets in the cohesive zone of a blast furnace, *ISIJ Int*, 55(10):2039–2046.

Li, L, Lei, Y L and Pan, D Y, 2016. Study of CO_2 emissions in China's iron and steel industry based on economic input–output life cycle assessment, *Natural Hazards*, 81(2):957–970.

Na, H M, Du, T, Sun, W Q, He, J F, Sun, J C, Yuan, Y X and Qiu, Z Y, 2019. Review of evaluation methodologies and influencing factors for energy efficiency of the iron and steel industry, *Int J Energy Res*, 43(11):5659–5677.

Oh, J and Noh, D, 2017. The reduction kinetics of hematite particles in H_2 and CO atmospheres, *Fuel*, 196:144–153.

Prakash, S, Goswami, M C, Mahapatra, A K S, Ghosh, K C, Das, S K, Sinha, A N and Mishra, K K, 2000. Morphology and reduction kinetics of fluxed iron ore pellets, *Ironmak Steelmak*, 27(3):194–201.

Sane, A, Buragino, G, Makwana, A and He, X Y, 2020. *Enhancing Direct Reduced Iron (DRI) for Use in Electric Steelmaking* (Air Products and Chemicals, Inc.: Allentown).

Sui, Y L, Guo, Y F, Jiang, T and Qiu, G Z, 2017. Reduction kinetics of oxidized vanadium titano-magnetite pellets using carbon monoxide and hydrogen, *J Alloys Comp*, 706:546–553.

Technologies, HYL, 2015a. Iron ores for direct reduction, Reducibility Test 3–9.

Technologies, HYL, 2015b. Iron ores for direct reduction, Swell Test 10–12.

Technologies, HYL, 2015c. Iron ores for direct reduction, Low Temperature Disintegration Test.

Voraberger, B, Wimmer, G, Dieguez, S U, Wimmer, E, Pastucha, K and Fleischanderl, A, 2022. Green LD (BOF) Steelmaking—Reduced CO_2 Emissions via Increased Scrap Rate, *Metals*, 12(3):466.

Wang, X L, Wei, Y W and Shao, Q L, 2020. Decomposing the decoupling of CO_2 emissions and economic growth in China's iron and steel industry, *Resour Conserv Recycl*, 152:104509.

Wynnyckyj, J R and Fahidy, T Z, 1974. Solid state sintering in the of iron ore pellets induration, *Metall Trans A*, 5:991–1000.

Xu, C B and Cang, D Q, 2010. A brief overview of low CO_2 emission technologies for iron and steel making, *J Iron Steel Res Int*, 17(3):1–7.

Zhang, F, Zhao, P C, Niu, M and Maddy, J, 2016. The survey of key technologies in hydrogen energy storage, *Int J Hydrog Energy*, 41(33):14535–14552.

Zhu, D Q, Xue, Y X, Pan, J, Tang, Z C, Yang, C C and Jiang, R C, 2022. Research progress and development thinking of gas-based direct reduction process (in Chinese), *Sinter and Pellet*, 47(01):1–9+86.

Influence of iron ore chemistry and mineralogy on sticking and breakage of pellets inside shaft furnace

S Purohit[1,4], M I Pownceby[2,4] and A Guiraud[3,4]

1. Research Scientist, CSIRO Mineral Resources, Clayton South Vic 3169.
   Email: suneeti.purohit@csiro.au
2. Principal Research Scientist, CSIRO Mineral Resources, Clayton South Vic 3169.
   Email: mark.pownceby@csiro.au
3. Principal Research Consultant, CSIRO Mineral Resources, Clayton South Vic 3169.
   Email: adrien.guiraud@csiro.au
4. Heavy Industry Low-carbon Transition Cooperative Research Centre, Adelaide SA 5000.

ABSTRACT

Sticking and breakage of iron ore pellets during high temperature reduction can result in several operational challenges during shaft furnace ironmaking. With the shift to H_2-based DRI processes and decreasing iron ore quality, this issue becomes more prominent. It is therefore important to understand how the iron ore chemistry and mineralogy affects the sticking and breakage of pellets inside the shaft furnace. This study provides a review of the mechanisms that lead to sticking and breakage of pellets inside the shaft furnace. The sticking index of pellets is related to the rate of metallic iron production as well as the precipitation morphology of product iron. Sticking of pellets is also enhanced by the formation of low-melting slag phases. The breakage of pellets inside the shaft furnace is often associated with the mechanical load of burden and pellet swelling, which is caused due to iron whisker or protruded iron formation. This paper investigates how the composition of iron ore, particularly the presence of different impurity oxides such as SiO_2, Al_2O_3, MgO, CaO and alkali oxides in the pellets, influences the sticking and swelling indices. The effect of ore mineralogy on these adverse behaviours is also discussed.

INTRODUCTION

Sticking and swelling of pellets inside shaft furnace ironmaking processes can result in severe operational challenges along with reduced productivity and increased energy consumption. As hydrogen DRI processes become more prevalent, the significance of these issues is expected to grow. Pellet sticking is usually observed in the lower part of the reduction shaft and as the sticking becomes more pronounced, it results in the formation of larger clusters. Clustering, through sticking, can have several critical detrimental effects in the shaft furnace, especially in the Direct Reduced Iron (DRI) shaft. These include:

- Disrupting the upward gas movement, thereby forming gas channels and decreasing gas utilisation efficiency.

- Lowering the percentage reduction and consequently, the metallisation of the burden.

- Disrupting the smooth downward movement of burden, thereby causing sudden slips.

- Damaging the nozzles and other furnace parts.

- In severe cases, completely preventing the discharge of direct reduced iron (DRI), which can result in a plant shutdown.

Excessive swelling of pellets on the other hand results in pellet disintegration and thereby lowers burden permeability and increases burden hanging and flue dust formation.

Researchers worldwide have extensively investigated these issues, leading to numerous theories on the sticking and swelling mechanisms and the development of preventive measures. This paper undertakes a comprehensive review of the previous research works on pellet sticking and breakage due to swelling to gain a deeper understanding of the underlying causes and the influence of iron ore chemistry and mineralogy on these phenomena. The authors also aim to provide a detailed understanding of how the diminishing quality and grade of iron ores will impact the sticking and swelling characteristics of pellets in future hydrogen-based shaft furnace ironmaking processes.

PRIMARY CAUSES OF PELLET STICKING AND BREAKAGE

Sticking of pellets inside the reduction shaft furnace is reported to be caused by: (a) sintering of the iron whiskers; (b) freshly formed dense metallic iron; and (c) low-melting slag phases. During the reduction of wüstite to metallic iron, based on the reduction conditions, the precipitation morphology of iron can be porous, dense, or fibrous (Gudenau *et al*, 2005). The latter two textures appear to cause higher sticking in pellets. For example, several researchers have reported the formation of iron whiskers as the primary cause of pellet sticking. The iron whiskers developed in neighbouring pellets hook together and form mechanical bonds between the adjacent pellets (Wang *et al*, 2019). As the reduction proceeds the iron whiskers grow and thicken, thereby enhancing the sticking strength. Freshly precipitated dense metallic iron is also attributed as another cause of pellet sticking. The sticking strength is enhanced with a rise in reduction temperature as the diffusion, crystallisation and growth of newly formed iron proceeds. In addition, some researchers have reported the formation of low-melting eutectic phases as the cause of sticking (Mandal and Sinha, 2017; Yi, Huang and Jiang, 2013). Yi, Huang and Jiang observed bonding of wüstite phase at the pellet interface at a lower reduction degree however with the progress of reduction, low-melting eutectic phase formation within the $CaO-SiO_2-FeO$ system became the dominant factor of pellet.

Pellet breakage inside the shaft furnace can happen due to excessive or uneven burden distribution as well as excessive swelling. Volumetric swelling of the pellets is a common phenomenon accompanying the reduction of iron oxide. During the initial hematite to magnetite reduction, pellets undergo a volume change of about 20–24 per cent and this is considered as normal swelling (Wright and Morrison, 1980). Theoretically hematite to magnetite reduction should result in 4.9 per cent volume rise, however the crystallographic transformation of hexagonal to cubic structure disrupts the original hematite structure and results in anisotropic growth of magnetite. Lu (1974) suggested that the directional growth of magnetite results in microcracks and causes this normal swelling. The latter stages of reduction involving the conversion of wüstite to metallic iron however can result in a catastrophic volume increase of a few hundred percent. This abnormal swelling can lead to disintegration and breakage of pellets, thereby lowering gas permeability and hence, productivity. The abnormal swelling has been widely studied by researchers and their findings strongly suggest iron whisker formation as the primary cause. The growth of iron fibres or whiskers pushes the formerly adjacent particles in the pellet apart, resulting in a significant increase in the external volume of the pellets (Man and Feng, 2016).

Since iron whisker formation has been linked to both pellet sticking and swelling it is therefore important to understand the mechanism of the iron whisker formation.

MECHANISMS OF IRON WHISKER FORMATION

The formation of iron whiskers from the wüstite was extensively investigated in the late 20th century and several notable theories were proposed to elucidate the mechanism underlying their formation. Some of the most prominent theories are mentioned further.

Nicolle and Rist's theory

Nicolle and Rist (1979) proposed a model to explain the mechanism of iron whisker formation in the early stage of wüstite reduction. The theory was based on Wagner's mechanism of reduction of non-stoichiometric wüstite (Wagner, 1952). According to Wagner's mechanism, the reduction of wüstite results in the transfer of oxygen ions to the gas phase and conversion of Fe(III) ions to Fe(II) ions. This leads to a higher Fe/O ratio on the wüstite surface, creating a concentration gradient that results in the inward diffusion of iron ions and outward diffusion of a vacancy. The diffused Fe ions supersaturate the wüstite layer below the surface. As the reduction proceeds the activity of iron at the surface progressively increases and eventually exceed that of metallic iron (unity). Upon reaching a critical value, high enough for nucleation, the first iron nucleus forms at the surface. The newly formed nucleus then starts to grow by assimilating iron ions diffused from the supersaturated wüstite layer. This process persists until the completion of reduction and the depleted supersaturated wüstite and virgin wüstite both continue to receive iron.

Based on this mechanism, Nicolle and Rist (1979) suggested that at any stage of the wüstite reduction, the reaction is controlled by the production of Fe ions and their subsequent diffusion. When the rate of diffusion of Fe ions is negligible compared to their production (diffusion-controlled reaction), iron accumulation is restricted to the wüstite surface (as shown in Figure 1a). The slow diffusion conditions in all directions ensures the formation of multiple nuclei across the surface. As there is no iron build-up in the bulk, the growth of nuclei can only take place via oxygen removal around it, hence the nuclei can grow only radially. Under this diffusion-controlled reaction, the nuclei grow radially at their periphery in the shape of plates and finally merge to form a layer of sponge iron.

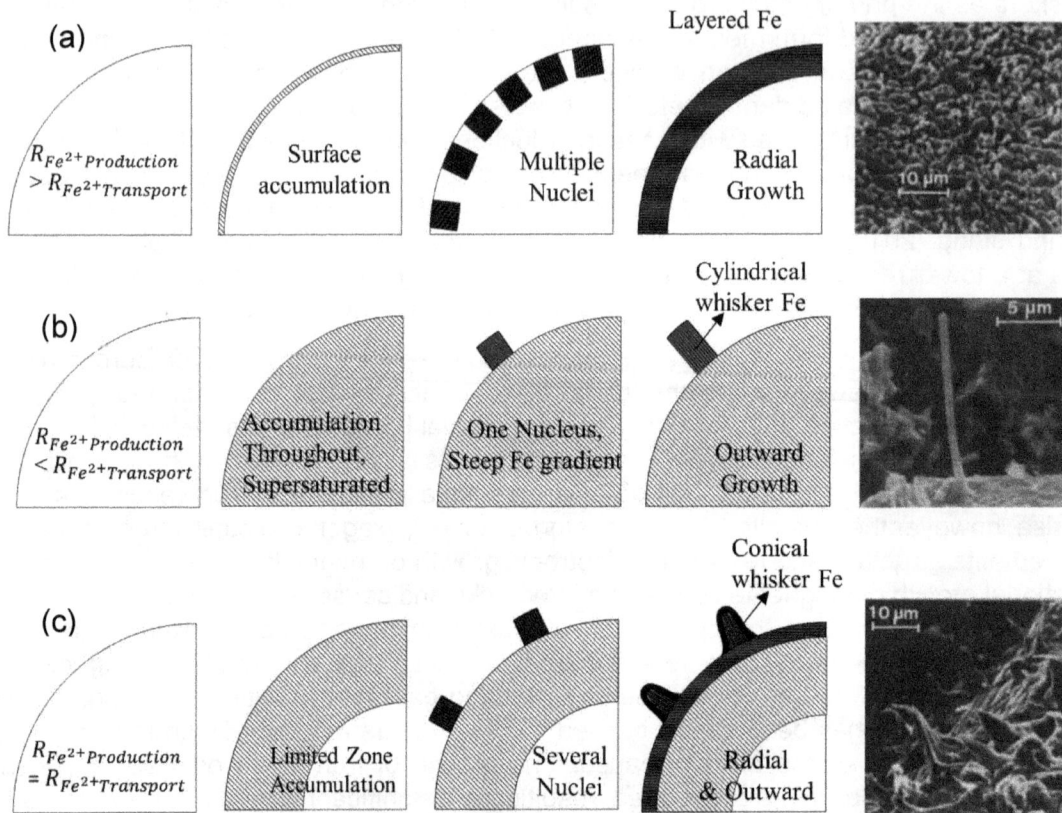

FIG 1 – Illustration of the precipitation morphology of metallic iron under: (a) diffusion control; (b) chemical reaction control; and (c) mixed control (recreated from Nicolle and Rist, 1979).

When the rate of production of Fe ions is negligible compared to their diffusion (ie in chemical reaction control), iron starts to build-up uniformly throughout the volume of wüstite, as shown in Figure 1b. This supersaturated wüstite reservoir feeds iron to the newly formed iron nucleus down a steep concentration gradient. This condition facilitates the outward growth of the first iron nucleus without any significant oxygen removal. The bulk wüstite then gets depleted of its excess iron and the first nucleus remains the only one on the particle. Thus, a cylindrical whisker is formed. Wüstite reduction is seldom so close to chemical control to form only one nucleus (Nicolle and Rist, 1979).

When the production rate of Fe ions at the surface is of same order as the inward diffusion rate (ie under mixed control), iron build-up takes place in a limited zone. This condition results in the formation of several nuclei that grow both radially by the continuous oxygen removal and outwardly by the Fe ion diffusion from the supersaturated zone (shown in Figure 1c). The nuclei take the shape of conical whiskers.

According to this theory of Nicolle and Rist (1979), replacing CO gas by H_2 gas lowers the chance of iron whisker formation as the chemical rate constant is increased by a factor of 40, thereby, displacing the control towards diffusion and thereby forming layered iron (Figure 1a).

Bleifuss' theory

Bleifuss' theory (Bleifuss, 1971) of iron whisker formation is based on the restricted growth of iron nuclei. According to this theory, the oxygen removal process in calciferous wüstite reduction leads to an elevated concentration of CaO near the surface. As the reduction progresses to a depth of 300–400 Angstroms, it is anticipated that the surface layer will be enveloped by a monolayer of lime-saturated wüstite. With the continuation of chemical reaction, the iron-rich lime layer becomes more prominent and continuous, thereby limiting the potential sites for iron nucleation. In addition, the lateral growth of newly formed iron nuclei is prevented and the restricted growth results in iron whisker formation.

vom Ende et al's theory

The theory of iron whisker formation suggested by vom Ende, Grebe and Thomalla (1971) is based on iron nucleation. The theory suggests that the uneven distribution of lime in wüstite promotes the nucleation of iron whiskers. The larger calcium ions cause local distortion in the wüstite lattice and create high-energy spots due to lattice strain. This allows preferential nucleation of metallic iron on the wüstite surface. This theory does not explain why the calcium-promoted nuclei results in whisker formation after nucleation.

Lu's theory

Lu (1974) critically analysed the iron whisker formation theory by Bleifuss (1971) and vom Ende, Grebe and Thomalla (1971). In this analysis, Lu identified several limitations and shortcomings in these theories and put forth his own alternative explanation, which was supported with experimental data. A schematic representation of Lu's theory showing the steps of iron whisker formation is shown in Figure 2. According to Lu's theory, when the pellet containing both hematite and magnetite (Figure 2a) is reduced, magnetite absorbs more calcium oxide compared to hematite and forms lime containing dirty wüstite (shown as W_{Ca} in Figure 2). Lime lowers the activity of dirty wüstite and limits the chemical reaction. However, the oxygen removal reaction continues from the pure wüstite (from hematite) and increases the Fe/O ratio, thereby favouring ion nucleation (Figure 2c). This results in the formation of a sponge iron layer. With time, the reduction of pure wüstite finishes or slows down and reduction of the dirty wüstite begins (Figure 2d). During the reduction process, there is a generation of surplus Fe(II) ions in the dirty wüstite. However, the accumulation of these ions to a level sufficient for the iron nucleation may not be possible as the presence of CaO accelerates the diffusion of ferrous ions to the nearby metallic iron sink (Figures 2e and 2f). The diffusion of Fe(II) ions from the sides and the relatively stationary oxygen beneath the metal phase ensures the outward growth of metallic iron as whiskers.

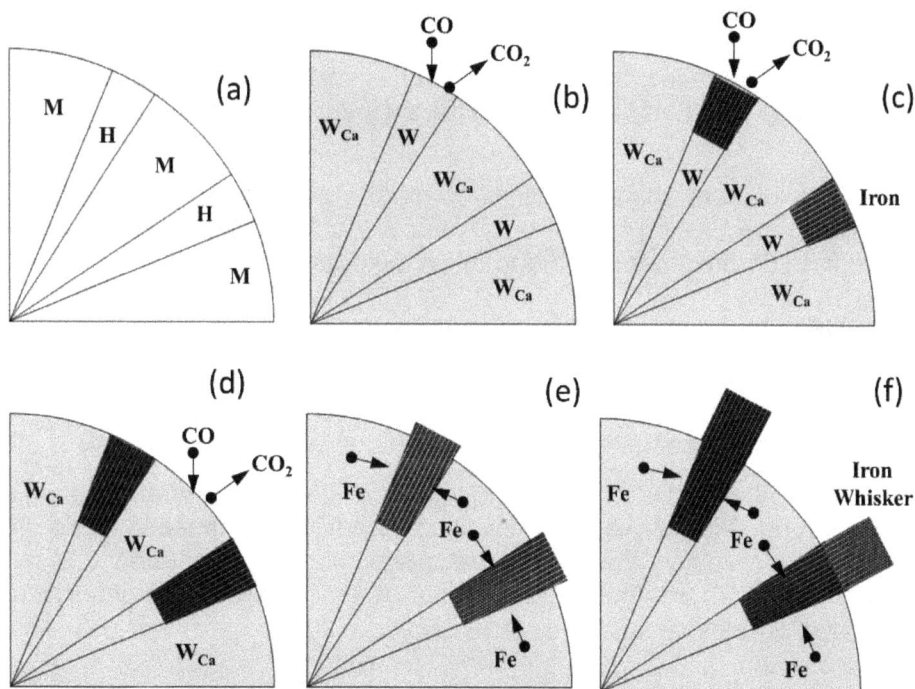

FIG 2 – Schematic representation of Lu's theory showing the steps involved in iron whisker formation (Recreated from Lu, 1974).

According to Lu's theory two important conditions that lead to iron whisker formation are high pellet porosity and uneven distribution of impurities that are soluble in wüstite.

FACTORS AFFECTING PELLET STICKING AND SWELLING

The whisker formation mechanisms described above are influenced by several physio-chemical factors. According to Nicolle and Rist's theory, the precipitation of iron as whiskers occurs when there is a significant accumulation of Fe ions and a limited number of favourable nucleation sites on the surface. In contrast, Lu's theory emphasizes the uneven distribution of impurities in wüstite as the primary cause of whisker formation. These conditions for whisker formation are governed by the chemical rate constant and the diffusion coefficient, which are affected by several factors such as pellet characteristics, reduction temperature, reducing gas and presence of foreign cations in wüstite. This section reviews the effects of iron ore chemistry and mineralogy on pellet sticking and breakage. The effects of different impurity oxides, such as SiO_2, Al_2O_3, MgO, CaO, Na_2O and K_2O, in the iron ore pellets are thoroughly investigated to understand their impact on the sticking and swelling behaviour of pellets in the shaft furnace ironmaking. It is important to differentiate the influence of these impurity oxides as intrinsic components of iron ore pellets rather than as external coating materials. The effect of coating materials in preventing sticking will be discussed in the forthcoming publication.

Iron

The existing literature provides evidence that high-grade iron ore pellets containing lower gangue content exhibit elevated sticking index (SI) and increased reduction swelling index (RSI). In their study, Sharma, Gupta and Prakash (1990) investigated the swelling behaviour of pellets made from two different grade Indian iron ores and pure hematite. They observed a highest swelling of 119 per cent for pure iron oxide, 103 per cent for 68 wt. per cent Fe iron ore and 42 per cent for 64 wt. per cent Fe iron ore. They expanded their study to include a range of gangue contents (2 to 10 wt. per cent) in the iron ore and pure hematite pellets by adding pure reagent grade SiO_2 and Al_2O_3. The swelling tests for all the pellets suggested a significant decrease in the RSI from about 80 per cent at 2 wt. per cent gangue to less than 20 wt. per cent at 8 wt. per cent gangue (Sharma, Gupta and Prakash 1990).

Sharma, Gupta and Prakash (1990) observed long whiskers in the pellets made from pure hematite and 68 wt. per cent Fe iron ore, while the pellets from 64 wt. per cent Fe iron ore comprised few whiskers of shorter size. They suggested that the increased presence of gangue materials in the iron ore pellets led to the formation of low melting slag phases, which in turn enhanced the bonding strength of the pellets. This phenomenon prevented the iron whiskers from exerting mechanical pressure on the adjacent surfaces, thereby reducing the swelling effect. Granse (1971) conducted swelling tests on ten kinds of pellets, made from different individual concentrates and corelated larger swelling with higher iron content.

The use of high-grade iron ore has a similar effect on the pellet sticking. Basdağ and Arol (2002) reported an increase in the sticking index from 5 per cent to over 30 per cent when the iron content of pellet increased from 65.5 wt. per cent Fe to 67 wt. per cent Fe.

Silica and alumina

As mentioned, a higher gangue content is reported to be advantageous for the sticking and swelling behaviour of iron ore pellets during the shaft furnace reduction. The authors suggest that SiO_2 forms fused layer of glassy silicates that inhibits the growth of iron whiskers (Lu, 1974).

Granse (1971) conducted a study to assess the impact of gangue minerals containing silica, such as quartz, quartzite and feldspar, on the swelling behaviour of pellets made from magnetite concentrate. The study revealed that when the quartz content exceeded 2.2 wt. per cent, the reduction swelling index (RSI) was below 20 per cent. A similar effect was observed with quartzite. However, in the case of feldspar addition, an RSI below 20 per cent was only achieved with an 8 per cent feldspar addition.

According to Lu (1974), North American magnetite concentrates with high silica contents do not exhibit abnormal swelling, unlike Marcona (Peru) ores and many other Swedish (Kiruna) ores. Lu proposed that the presence of silica and alumina in the concentrates enhances slag bonding and improves resistance against the force exerted by growing whiskers. Sharma, Gupta and Prakash (1993) observed a lower RSI of 28 per cent for pellets with 2 wt. per cent SiO_2 and 3.6 wt. per cent

Al_2O_3 compared to the 100 per cent RSI for pure hematite pellets. They also reported the pellets made from iron ore with 64 wt. per cent Fe had the RSI of 34 per cent at 2 wt. per cent Al_2O_3 and 12 per cent at 8 wt. per cent Al_2O_3. However, a pellet from pure hematite and 68 wt. per cent Fe iron ore showed an RSI of over 60 per cent even at 8 wt. per cent Al_2O_3.

Qing *et al* (2018) observed an RSI of 16.5 per cent with a 4.8 wt. per cent SiO_2 containing pellet. By lowering the SiO_2 to 2.8 wt. per cent and 1.8 wt. per cent, the RSI increased to 34.5 per cent and 55.8 per cent, respectively. Wang *et al* (2021) reported a decrease in the RSI of fired briquettes from 18 per cent to 5.7 per cent with an increase in the SiO_2 content from o to 8 wt. per cent. Wang *et al*, observed that as the Al_2O_3 content increased, the RSI initially decreased, followed by an increase and eventually reached a plateau without further changes.

The lowered SI of pellets at higher gangue contents has also been reported by several other researchers. Basdağ and Arol (2002) reported a decrease in the SI from 30 per cent to 40 per cent at 1.25 wt. per cent SiO_2 and from 20 per cent to 25 per cent at 1.55 wt. per cent SiO_2 in commercial HYL pellets. Bahgat, Hanafy and Lakdawala (2016) reported about a 20 per cent reduction in SI by increasing the SiO_2 content of iron ore from 1.2 to 1.5 wt. per cent. They also reported a similar ~20 per cent reduction in SI by increasing the Al_2O_3 content of iron ore from 0.55 to 0.85 wt. per cent.

While higher gangue contents may reduce sticking and swelling, it has negative implications for ironmaking operations, such as decreased productivity and increased fuel and flux costs. Moreover, elevated gangue levels can adversely affect the strength of fired pellets. Guo *et al* (2019) proposed that pellets with high SiO_2 contents led to a higher proportion of liquid phase (mainly $2FeO.SiO_2$) formation during the induration process. The liquid shrinks during the cooling stage, resulting in the formation of numerous small cracks thereby reducing the compressive strength of the pellets. With an increase in SiO_2 content from 2.19 to 8.13 wt. per cent, the compressive strength decreased from 3.39 kN to 2.20 kN. In addition, there are also studies that report the formation of iron whiskers in silica containing pellets. Abdel-Halim *et al* (2009) observed both whisker and platelet structures developed in precipitated iron from 2 wt. per cent SiO_2 doped wüstite.

Basic oxides – lime

The use of CaO-based coating materials is a common practise in industrial operations to prevent sticking. However, when CaO is present as an impurity within the iron ore pellets, the sticking and swelling of the pellets are enhanced. The theories of iron whisker formation mentioned earlier suggest the presence of lime to be a catalyst to iron whisker formation. The impact of lime on the sticking and swelling of iron ore pellets has been extensively studied and most research findings support these theories. Wright (1978) suggests that the presence of CaO in contaminated wüstite reduces the thermodynamic activity of iron, causing iron to nucleate at uncontaminated surfaces and subsequently grow outward. Elkasabgy and Lu (1980) suggested that the nucleation and growth of iron whiskers occur at sites of calcio-wüstite surface that is higher in CaO.

Additionally, lime is reported to enhance the sticking by reacting with other impurities such as Al_2O_3 and SiO_2 present in the iron ore, forming low melting calcium-alumino-silicate slag phases. Dwarapudi *et al* (2014) suggested that high CaO generates more calcium silicate melt phase, resulting in high strength, less porosity and enhanced sticking. Wang *et al* (2021) found the RSI of 40 per cent in fired briquettes containing 0.25 wt. per cent CaO and the samples also were characterised by large number of slender iron whiskers. With a rise in CaO content to 2 wt. per cent the RSI reached 62.2 per cent. The porosity of these CaO-containing briquettes was very high, typically between 20 per cent to 40 per cent. Iljana *et al* (2015) observed an increasing trend of the RSI for Russian Karelian pellets when the limestone content was increased from 0.5 wt. per cent to 3.2 wt. per cent.

There are also some contradictory theories that suggest a better performance of pellets with the presence of CaO. Basic pellets are known to have higher reducibility and upon reduction they form a highly porous Fe structure with many macro and micro pores (Biswas, 1981). In comparison, acidic pellets form dense Fe with only large macropores. Odo and Nwoke (2019) suggested that the layers of Fe metal hinder the gas flow to the core of pellets. In the unreduced core, a low melting slag phase of FeO forms and blocks the pores, thereby enhancing the sticking and clustering. Sharma, Gupta and Prakash (1993) suggested that addition of CaO forms low melting slag phases during firing. The slag phase envelops the iron oxide surface and restricts the nucleation and growth of iron whiskers. Bai *et al* (2021) suggested that the CaO forms calcio-ferrite phases that promotes liquid slag formation and eventually fills the porosity and lowers reduction expansion. The calcio-ferrite phases suppress the migration of iron atoms, thereby delaying the whisker formation and optimising the whisker morphology.

Granse (1971) reported complete elimination of abnormal swelling of magnetite concentrate with just 2 wt. per cent limestone addition. Further addition only had a minor effect. Addition of dolomite and slaked lime has a similar positive effect in lowering any abnormal swelling. Mandal and Sinha (2015) observed no sticking of pellets with 4 wt. per cent CaO at 1300°C. Meyer (1980) reported that the compressive strength of pellets rises with 0.5 per cent $Ca(OH)_2$ addition, initially increasing with up to 5 per cent addition and with the strength diminishing afterwards due to the formation of glassy structures. With $CaCO_3$ addition, the compressive strength initially increased, reaching a maximum at 6 per cent at 1150°C and 8 per cent at 1200°C firing temperature and thereafter the strength decreased.

Basic oxides – MgO

MgO is commonly considered as a desirable material for iron ore pellets. Elkasabgy and Lu (1980) suggested that the size of the Mg^{2+} ions (0.6 Å) is smaller than Fe^{2+} ions (0.75 Å) and Ca^{2+} ions (0.99 Å). This enables easier and more uniform distribution of MgO in the wüstite lattice compared to CaO distribution. Hence, the chance of iron whisker formation is lowered when MgO is present as impurity.

According to Basdağ and Arol (2002), the addition of MgO resulted in a decrease in the SI from approximately 40 per cent at 0.55 wt. per cent MgO to around 20 per cent at 0.85 wt. per cent MgO. This decrease in SI was attributed to the reaction between MgO and Fe_2O_3 during the induration process, which leads to the formation of a high melting point $MgFe_2O_4$ spinel phase. In contrast, in the absence of MgO, calcium diferrite ($CaO.2FeO$) and other silicate compounds form an intergranular slag phase, contributing to an increased sticking tendency.

Lu (1974) suggest that the higher mobility of Mg^{2+} ions and their higher solubilities in wüstite fails to initiate abnormal swelling. Sugiyama *et al* (1983) observed lower swelling of pellets with increasing MgO/SiO_2 ratio. Sharma, Gupta and Prakash (1993) observed a decrease in pellet swelling with increased MgO addition and suggested that the high melting compounds of MgO ($MgO.Al_2O_3$, $MgO.Fe_2O_3$) restrict the growth of iron whiskers. Wang *et al* (2021) reported that the relative RSI of fired briquettes containing MgO exhibited an initial increase, reaching a maximum value of 17 per cent when the MgO content was 0.5 wt. per cent and subsequently decreased with increasing MgO. The surface of the briquette containing MgO was mostly granular metallic iron and exhibited low porosity of between 13 per cent to 16 per cent. Gao *et al* (2013) found that the addition of MgO resulted in reduced swelling and reduction degradation of the pellets. This can be attributed to the stabilising effect of MgO on magnetite, which helps to decrease volume changes during the process. However, these researchers also noted a decrease in compressive strength, from 3100 N at 0 wt. per cent MgO to 2700 N at 2 wt. per cent MgO. They proposed that the introduction of MgO caused a transformation of uniformly distributed compact hematite crystallites to fragile magnetite-intermingled hematite crystallites, which may have contributed to the observed decrease in compressive strength.

Iljana *et al* (2013) reported lower swelling of acid pellets containing 5.34 wt. per cent SiO_2 compared to olivine pellets, under simulated blast furnace conditions. They proposed that this could be attributed to the formation of cohesive slag bonds resulting from the higher gangue content present in the acid pellets. According to Shen *et al* (2014), pellets containing MgO exhibited higher reduction compressive strength, which is considered an indicator of the reduction swelling index.

Basicity (CaO/SiO₂)

As mentioned in the previous sections, the influence of gangue and basic oxides on pellet sticking and swelling is a complex subject with contradictory findings in the literature. Due to this complexity, it becomes crucial to examine the impact of basicity, specifically the ratio of CaO to SiO_2, on optimising the pellet sticking and swelling. The interaction between CaO and SiO_2 can result in the formation of various calcium silicate compounds. At higher CaO/SiO_2, more stable and low-melting slag phases form and these can enhance the pellet strength and lower swelling tendency. Conversely, lower basicities may result in the formation of more reactive and glassy slag phases and increase pellet swelling (Friel and Erickson, 1980).

Wright and Morrison (1980) conducted isothermal reduction experiments using single high-grade pellets of different CaO/SiO_2 ratios to understand the effect of basicity on pellet swelling. Their findings are shown in Figure 3a, which suggests maxium swelling ocurs at a pellet basicity of 0.49 (corresponding to 0.67 wt. per cent CaO in the feed). Later, Morrison and Wright (1980) conducted pot-grate tests on high-grade hematite pellets from the Pilbara and found similar trends of pellet swelling with increasing basicity.

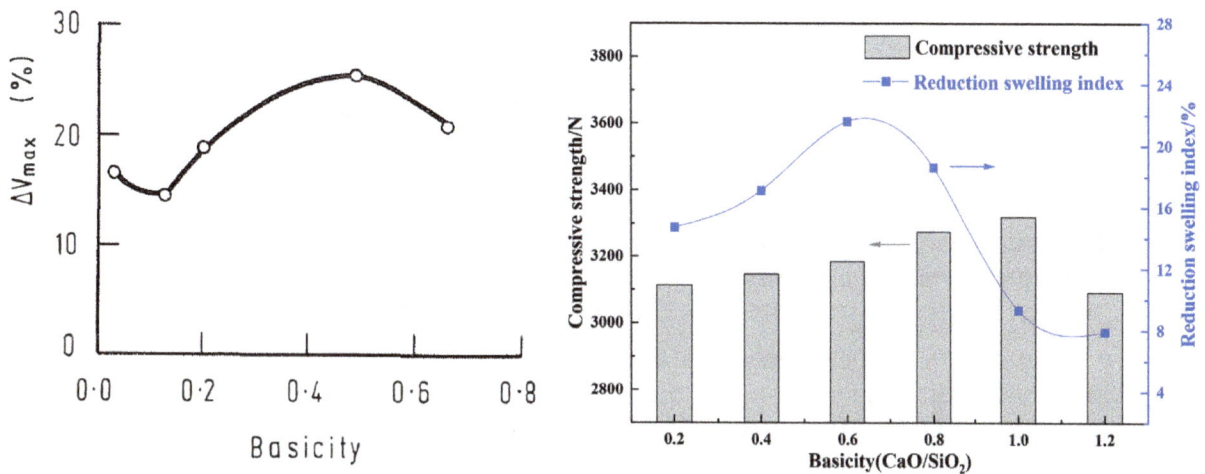

FIG 3 – (a) Effect of basicity (CaO/SiO_2) on volume change of single pellets (Wright and Morrison 1980); (b) Effect of basicity on RSI and compressive strength (Guo *et al*, 2021).

In a recent study conducted by Guo *et al* (2021), similar trends of pellet swelling were observed in relation to pellet basicity, as depicted in Figure 3b. The RSI increased rapidly with increasing basicity from 0.2 to 0.6, reaching a maximum value of 21.6 per cent at a basicity of 0.6, followed by a subsequent decrease. The lowest RSI value of 7.9 per cent was observed for pellets with a basicity of 1.2.

According to Guo *et al* (2021), for pellets with lower basicity, the main bonding phase was calcium iron silicate, which enveloped the hematite particles. The presence of this silicate phase led to slower reduction rates compared to hematite, resulting in an uneven expansion and internal stresses within the pellets. On the other hand, for pellets with a higher basicity, the main bonding phase identified was SFCA (silico-ferrite of calcium and aluminium). SFCA exhibited excellent reducibility similar to hematite, which effectively mitigated internal stresses and reduced swelling tendencies.

Shi *et al* (2022) investigated the swelling behaviour of fired hematite pellets with 0.02 to 1 CaO/SiO_2 ratios under H_2-rich reduction. They observed an initial upward trend in the RSI, reaching its peak at a basicity range of 0.6 to 0.8, followed by a subsequent decrease. They suggested that higher basicity promotes slag formation in the fired hematite pellets. The slag closes the pores within the pellets and absorbs the volume expansion from the wüstite to iron phase transition and lower reduction swelling. Lu (1973) mentions some other studies where maximum swelling had occurred between 0.5 to 0.7 basicity range. Granse (1971) reported that the basicity range over which the swelling was maximised was from 0.5 to 1.0. They suggested that at high basicities, the metallisation of wüstite grains that are contaminated with an abundance of basic oxides is delayed, thereby preventing whisker growth.

Abdel-Halim, Nasr and El-Geassy (2011) found lower sticking after reduction of basic pellets of 1.62 basicity $\left(\frac{CaO+MgO}{SiO2+Al2O3}\right)$ compared to acidic pellets of 0.37 basicity. The basic pellets after reduction resulted in a highly porous structure with large number of macropores between iron grains and micropores inside grains whereas the acidic pellets showed denser and relatively large, sintered iron grains with only large macropores in between Fe grains. Fayalite phase was also seen in the reduced acidic pellets and was considered to enhance clustering.

Alkali oxides

Alkalis such as sodium and potassium oxides (Na_2O and K_2O) are introduced to the iron ores during their formation through various geological processes. Alkalis are detrimental for the ironmaking processes, especially the blast furnace. Alkalis have been linked to several detrimental effects, including causing swelling in iron ore feedstocks, forming scaffolds, corroding the refractory lining, increasing coke consumption and negatively impacting the physical and chemical properties of the slag (Maharshi, Bitan and Mitra, 2018). Numerous studies have been conducted to investigate the impact of alkalis during the reduction process of iron ore pellets.

Elkasabgy (1984) found that when alkalis react with other compounds present in the iron ore, they form low-melting point slag phases and enhance sticking and swelling of pellets. Elkasabgy observed normal swelling in high-purity pellets (made from reagent-grade hematite) with no whisker formation. However, when the pellets were doped with 0.58 wt. per cent Na_2O, they showed abnormal swelling of 57 per cent and 138 per cent at induration temperatures of 1200°C and 1300°C, respectively. Similarly, pellets doped with 0.68 wt. per cent K_2O showed swelling of 39 per cent and 172 per cent, respectively. In both cases, iron whiskers were detected under scanning electron microscopy (SEM). Elkasabgy suggested that the alkali ferrites, formed during induration, are incorporated into wüstite as a solid solution component in an irregular pattern which leads to whisker formation.

Elkasabgy (1984) also conducted swelling tests with alkali-doped commercial acid pellets and observed liquid alkali iron silicates. They suggested the swelling and cracking of the alkali-containing acid pellets was because of the uneven distribution of the liquid slag and not because of the formation of iron whiskers. In normal acid pellets, silicate gangue bridges between the iron oxide grains and these bridges bear the stress from subsequent phase transformations of the Fe oxide. Alkalis flux the silicate binding bridges and then react with ferrous iron oxide forming low melting point slag, which weakens the pellets and causes them to degrade during reduction (Elkasabgy, 1984).

Nicolle and Rist (1979) suggested that the addition of alkali oxides to the wüstite solid solution causes a lattice expansion as the diameter of Na^+ (0.97 Å) and K^+ (1.33 Å) ions are larger than the Fe^{2+} (0.75 Å) ions. The expanded lattice promotes the diffusion of vacancies to the surface and helps the accumulation of large amounts of iron. The surface defects also provide nucleation sites which promotes whisker formation.

There are several other experimental studies that detail the adverse effect of alkalis on pellet sticking and disintegration. For example, Pichler et al (2014) observed the formation of a dense iron layer on alkali-soaked iron ore lumps that resulted in higher sticking and degradation. vom Ende, Grebe and Thomalla (1971) and Lu (1973) also reported an adverse effect of sodium content and its uneven distribution on reduction swelling.

Ore mineralogy

The effect of ore mineralogy on the pellet sticking and swelling is critically important, especially in the current and future scenarios of degrading global iron ore quality. Traditional pellets are made using magnetite concentrates or high-grade hematite but currently there is very limited research on the effect of ore mineralogy on the sticking and swelling behaviour of pellets.

Meyer (1980) suggested that pellets from magnetite concentrates have a higher compressive strength (>1000 N/pellet) compared to hematite (200 N/pellet) because of the higher reactivity of freshly formed hematite. As a result, pellets from magnetite can bear a higher burden load and have a correspondingly lower degradation. Morrison and Wright (1986) found that Swedish pellets with 2.81 wt. per cent gangue (1.02 basicity) and Brazilian pellets with 3.22 wt. per cent gangue (0.56 basicity) both showed a decrease in compressive strength after 650°C reduction. However, Australian acid pellets with 3.4 wt. per cent gangue (0.07 basicity) showed a decrease in compressive strength after 800°C, suggesting that the pellet strength after reduction is sensitive to ore type. Furthermore, de Moraes and Ribeiro (2019) suggested that the presence of porous goethite and martite constituents in the iron ore negatively impact the mechanical strength of the pellets due to the higher amount of retained magnetite in the core of the indurated pellets.

Pichler *et al* (2014) treated limonitic (63 per cent Fe) and hematitic ores (67 per cent Fe) with alkali-bearing aqueous solutions and observed higher sticking with the limonitic ore. de Alencar *et al* (2021) suggested that a microcrystalline porous hematite ore is more likely to stick compared to a compact granular and lamellar hematite ore. They suggested the higher moisture content in the microporous iron ore results in it exhibiting higher sticking.

SUMMARY

A detailed literature review was carried out to gain a comprehensive understanding of the sticking and swelling behaviour of pellets inside the shaft furnace. This review aimed to identify the underlying causes and assess the impact of iron ore chemistry and mineralogy on pellet swelling and sticking. While iron whisker formation was identified as the primary factor, it is crucial to note that it is not the sole cause of pellet sticking and abnormal swelling. Several mechanisms of iron whisker formation were presented and these highlighted how the operating conditions and the presence of impurities can affect iron whisker formation. The study emphasizes how the impurity oxides of iron ores can affect the swelling and sticking behaviour of pellets as it has commonly been observed that the presence of Al_2O_3, SiO_2 and MgO all have a beneficial effect in reducing these adverse effects. However, the role of CaO content of pellets in this regard remains controversial. Nonetheless, a consensus was reached regarding the maximum swelling of pellets occurring within an intermediate basicity (CaO/SiO_2) range of 0.4–0.8. The presence of alkali oxides was found to have a highly detrimental impact on the sticking and swelling of pellets during the shaft furnace. Research investigating the influence of iron ore mineralogy on these adverse effects remains limited, with only a few studies suggesting that certain mineralogical compositions such as moisture-rich iron ores, can lead to increased sticking and reduced strength.

ACKNOWLEDGEMENTS

This work has been supported by the Heavy Industry Low-carbon Transition Cooperative Research Centre (HILT CRC) whose activities are funded by its industry, research and government Partners along with the Australian Government's Cooperative Research Centre Program. This is HILT CRC Document 23–012. The authors would like to express gratitude to Associate Professor Tom Honeyands and Dr. Rou Wang from the University of Newcastle for their invaluable contributions and insightful discussions, which greatly enhanced the quality of this paper.

REFERENCES

Abdel-Halim, K S, Bahgat, M, El-Kelesh, H A and Nasr, M I, 2009. Metallic iron whisker formation and growth during iron oxide reduction: Basicity effect, *Ironmaking and Steelmaking*, 36(8):631–640. https://doi.org/10.1179/174328109 X463020

Abdel-Halim, K S, Nasr, M I and El-Geassy, A A, 2011. Developed model for reduction mechanism of iron ore pellets under load, *Ironmaking and Steelmaking*, 38(3):189–196. https://doi.org/10.1179/030192310X12816231892305

Bahgat, M, Hanafy, A and Lakdawala, S, 2016. Influence of Iron Ore Mineralogy on Cluster Formation inside the Shaft Furnace, *International Journal of Materials and Metallurgical Engineering*, 10(5):614–619.

Bai, K, Liu, L, Pan, Y, Zuo, H, Wang, J and Xue, Q, 2021. A review: research progress of flux pellets and their application in China, *Ironmaking and Steelmaking*, 48(9):1048–1063. https://doi.org/10.1080/03019233.2021.1911770

Basdağ, A and Arol, A I, 2002. Coating of iron oxide pellets for direct reduction, *Scandinavian Journal of Metallurgy*, 31(3):229–233. https://doi.org/10.1034/j.1600-0692.2002.310310.x

Biswas, A K, 1981. *Principles of Blast Furnace Ironmaking : Theory and Practice*, 528 p.

Bleifuss, R L, 1971. Volumetric Changes in the Hematite-Magnetite Transition Related to the Swelling of Iron Ores and Pellets during Reduction, *ICSTIS*, pp 52–56.

de Alencar, J P S G, do Carmo, G S, Silva, N L A, Bastos, N D and Figueiredo, P S, 2021. Avaliação da influência da tipologia do minério em testes de sticking, *Tecnologia Em Metalurgia, Materiais e Mineração*, 18(January):e2373. https://doi.org/10.4322/2176-1523.20212373

de Moraes, S L and Ribeiro, T R, 2019. Brazilian Iron Ore and Production of Pellets, *Mineral Processing and Extractive Metallurgy Review*, 40(1):16–23. https://doi.org/10.1080/08827508.2018.1481056

Dwarapudi, S, Banerjee, P K, Chaudhary, P, Sinha, S, Chakraborty, U, Sekhar, C, Venugopalan, T and Venugopal, R, 2014. Effect of fluxing agents on the swelling behavior of hematite pellets, *International Journal of Mineral Processing*, 126:76–89. https://doi.org/10.1016/j.minpro.2013.11.012

Elkasabgy, T El and Lu, W K, 1980. The Influence of calcia and magnesia in wustite on the kinetics of metallization and iron whisker formation, *Metallurgical Transactions B*, 11(3):409–414. https://doi.org/10.1007/BF02676884

Elkasabgy, T, 1984. Effect of alkalis on reduction behavior of acid iron ore pellets, *Transactions of the Iron and Steel Institute of Japan*, 24(8):612–621. https://doi.org/10.2355/isijinternational1966.24.612

Friel, J J and Erickson, E S, 1980. Chemistry, microstructure and reduction characteristics of dolomite-fluxed magnetite pellets, *Metallurgical Transactions B*, 11(2):233–243. https://doi.org/10.1007/BF02668407

Gao, Q-J, Shen, F-M, Wei, G, Jiang, X and Zheng, H-Y, 2013. Effects of MgO containing additive on low-temperature metallurgical properties of oxidized pellet, *Journal of Iron and Steel Research International*, 20(7):25–28. https://doi.org/10.1016/S1006-706X(13)60121-1

Granse, L, 1971. The influence of Slag forming Additions on the Swelling of Pellets from Very Rich Magnetite Concentrates, *ISIJ*, 11:45–51.

Gudenau, H W, Senk, D, Wang, S, De Melo Martins, K and Stephany, C, 2005. Research in the reduction of iron ore agglomerates including coal and C-containing dust, *ISIJ International*, 45(4):603–608. https://doi.org/10.2355/isijinternational.45.603

Guo, H, Jiang, X, Shen, F, Zheng, H, Gao, Q and Zhang, X, 2019. Influence of SiO$_2$ on the Compressive Strength and Reduction-Melting of Pellets, *Metals*, 9(852):1–18.

Guo, Y, Liu, K, Chen, F, Wang, S, Zheng, F, Yang, L and Liu, Y, 2021. Effect of basicity on the reduction swelling behavior and mechanism of limestone fluxed iron ore pellets, *Powder Technology*, 393:291–300. https://doi.org/10.1016/j.powtec.2021.07.057

Iljana, M, Kemppainen, A, Paananen, T, Mattila, O, Pisilä, E, Kondrakov, M and Fabritius, T, 2015. Effect of adding limestone on the metallurgical properties of iron ore pellets, *International Journal of Mineral Processing*, 141:34–43. https://doi.org/10.1016/j.minpro.2015.06.004

Iljana, M, Mattila, O, Alatarvas, T, Kurikkala, J, Paananen, T and Fabritius, T, 2013. Effect of circulating elements on the dynamic reduction swelling behaviour of olivine and acid iron ore pellets under simulated blast furnace shaft conditions, *ISIJ International*, 53(3):419–426. https://doi.org/10.2355/isijinternational.53.419

Lu, W K, 1973. The Effect of Sodium on the Swelling of Hematite during Reduction, *Scandinavian Journal of Metallurgy*, 2:65–67.

Lu, W K, 1973. The Suppression and Revival of Swelling of Iron Ore Pellets during Reduction, *Scandinavian Journal of Metallurgy*, 2:169–172.

Lu, W K, 1974. The Cause and Control of Abnormal Swelling of Iron Ore Pellets During Reduction, *Ironmaking Conference*, pp 61–72.

Maharshi, G D, Bitan, K S and Mitra, M K, 2018. Effect of alkali on different iron making processes Optimisation of Sinter Microstructure View project PhD Work View project, *Material Science and Engineering International Journal*, 2(6):304–313.

Man, Y and Feng, J-X, 2016. Effect of iron ore-coal pellets during reduction with hydrogen and carbon monoxide, *Powder Technology*, 301:1213–1217. https://doi.org/10.1016/j.powtec.2016.07.057

Mandal, A K and Sinha, O P, 2015. Characterization of Fluxed Iron Ore Pellets as Compared to Feed Material for Blast Furnace, *Journal of Progressive Research in Chemistry*, 2(1). https://www.researchgate.net/publication/279997875_Characterization_of_Fluxed_Iron_Ore_Pellets_as_Compared_to_Feed_Material_for_Blast_Furnace

Mandal, A K and Sinha, O P, 2017. Effective utilisation of waste fines in preparation of high-basicity double-layer DRI pellets for minimisation of sticking, *Transactions of the Institutions of Mining and Metallurgy, Section C: Mineral Processing and Extractive Metallurgy*, 126(3):182–190. https://doi.org/10.1080/03719553.2016.1210936

Meyer, K, 1980. *Pelletizing of Iron Ores* (Springer-Verlag).

Morrison, A L and Wright, J K, 1980. The making and evaluation of directly reduced iron pellets, *The AusIMM Conference*, pp 225–236.

Morrison, A L and Wright, J K, 1986. Evaluation of Raw Materials By Simulation of Direct Reduction in the Shaft Furnace, *Transactions of the Iron and Steel Institute of Japan*, 26(10):858–864. https://doi.org/10.2355/isijinternational1966.26.858

Nicolle, R and Rist, A, 1979. The mechanism of whisker growth in the reduction of wüstite, *Metallurgical Transactions B*, 10(3):429–438. https://doi.org/10.1007/BF02652516

Odo, J and Nwoke, V, 2019. Effect of Core Diameter on the Compressive Strength and Porosity of Itakpe Iron Ore Pellets, in *10th International Symposium on High-Temperature Metallurgical Processing* (eds: T Jiang, J-Y Hwang, D Gregurek, Z Peng, Z Baojun, J P Downey, O Yücel, E Keskinkilic and R Padilla) (Springer International Publishing).

Pichler, A, Schenk, J L, Hanel, M B, Mali, H, Hauzenberger, F, Thaler, C and Stocker, H, 2014. Influence of alkalis on mechanical properties of lumpy iron carriers during reduction, in *METAL 2014, 23rd International Conference on Metallurgy and Materials Conference Proceedings*, pp 87–92.

Qing, G, Wu, K, Tian, Y, An, G, Yuan, X, Xu, D and Huang, W, 2018. Effect of the firing temperature and the added MgO on the reduction swelling index of the pellet with low SiO_2 content, *Ironmaking and Steelmaking*, 45(1):83–89. https://doi.org/10.1080/03019233.2016.1242248

Sharma, C, Gupta, T R and Prakash, B, 1993. Effect of firing condition and ingredients on the swelling behaviour of iron ore pellets, *ISIJ International*, 33(1):446–453.

Sharma, T, Gupta, R C and Prakash, B, 1990. Effect of gangue content on the swelling behaviour of iron ore pellets, *Minerals Engineering*, 3(5):509–516. https://doi.org/10.1016/0892-6875(90)90043-B

Shen, F M, Gao, Q J, Jiang, X, Wei, G and Zheng, H Y, 2014. Effect of magnesia on the compressive strength of pellets, *International Journal of Minerals, Metallurgy and Materials*, 21(5):431–437. https://doi.org/10.1007/s12613-014-0926-5

Shi, Y, Zhu, D, Pan, J, Guo, Z, Lu, S and Xu, M, 2022. Improving hydrogen-rich gas-based shaft furnace direct reduction of fired hematite pellets by modifying basicity, *Powder Technology*, 408(August):117782. https://doi.org/10.1016/j.powtec.2022.117782

Sugiyama, T, Shirouchi, S, Tsuchiya, O, Onoda, M and Fujita, I, 1983. Effect of magnesite on the properties of pellets at room and low (900 degree C) temperatures, *Transactions of the Iron and Steel Institute of Japan*, 23(2):146–152. https://doi.org/10.2355/isijinternational1966.23.146

Vom Ende, H, Grebe, K and Thomalla, S, 1971. Anomalous Swelling During the Reduction of Iron Ore Pellets with Lime Additions, *Stahl Und Eisen*, 91(14):815–824.

Wagner, C, 1952. Mechanism of the Reduction of Oxides and Sulphides to Metals, *JOM*, 4:214–216.

Wang, G, Zhang, J, Li, Y, Xu, C and Liu, Z, 2019. Study on sticking mechanism of pellets under low H_2/CO conditions, *Powder Technology*, 352:25–31. https://doi.org/10.1016/j.powtec.2019.04.026

Wang, P, Dai, M, Chun, T, Long, H and Wei, J, 2021. Influence of the Gangue Compositions on the Reduction Swelling Index of Hematite Briquettes, *Metallurgical and Materials Transactions B: Process Metallurgy and Materials Processing Science*, 52(4):2139–2150. https://doi.org/10.1007/s11663-021-02177-8

Wright, J K and Morrison, A L, 1980. Volume and Structural Changes Occurring During the Isothermal and Non-isothermal Reduction of High-Grade Hematite Pellets, *Australia Japan Extractive Metallurgy Symposium*, pp 167–178.

Wright, J K, 1978. Swelling of High-Grade Iron Ore Pellets Reduced by Hydrogen in a Fixed Bed, *Australasian Institute of Mining and Metallurgy*, pp 1–7.

Yi, L, Huang, Z and Jiang, T, 2013. Sticking of iron ore pellets during reduction with hydrogen and carbon monoxide mixtures: Behavior and mechanism, *Powder Technology*, 235:1001–1007. https://doi.org/10.1016/j.powtec.2012.11.043

The development of DRI pellets for green steelmaking

C R Ure[1]

1. FAusIMM, Principal Process Engineer, Worley, Perth WA 6000. Email: chris.ure@worley.com

ABSTRACT

To substantially reduce, if not eliminate, carbon dioxide (CO_2) emissions from ironmaking it is necessary to convert from blast furnace (BF) ironmaking using coking coal as the reductant to direct reduction (DR) ironmaking using hydrogen (H_2) and green electricity.

Currently, almost all natural gas (NG) based direct reduced iron (DRI) production utilises iron oxide pellets in shaft furnaces, as opposed to fines in fluidised beds. Also, these pellets are ordinarily produced from high-grade magnetite concentrate, rather than hematite and goethite with typically higher gangue contents.

The use of high-grade magnetite concentrate is driven by the preference to melt DRI in conventional electric arc furnaces (EAFs), thereby avoiding the basic oxygen furnace (BOF) for making crude steel (CS). However, in replacing BF with DR ironmaking it is not economically feasible to source enough high-grade magnetite to replace the lower grade hematite and goethite currently consumed.

That is not to say that hematite and goethite, or even low-grade magnetite, cannot be used in DR ironmaking. However, their higher gangue content, resulting in higher slag production, requires that the EAF producing CS be replaced with a submerged arc furnace (SAF) that will produce pig iron (PI) for the BOF as does the BF.

Here, critical issues to be addressed are, what are the technicalities in replacing high-grade magnetite concentrate with hematite and goethite and NG with H_2 in DRI pellet production and melting?

This paper summarises certain relevant published information on DRI pellet research, test work and production together with relevant information on Pilbara ores. It attempts to illustrate the inter-relationship of composition and processing conditions on DRI pellet properties. It is intended to serve as a rudimentary guide to understanding the complexities of formulation, testing, evaluation, production and utilisation of DRI pellets using hematite and goethite and offer insight into this challenge.

INTRODUCTION

Decarbonising the steel industry will require several aspects:

- Replace BF ironmaking with DR and electric furnace melting ironmaking.
- Transition from using DRI pellets predominantly produced from high-grade magnetite (Fe_3O_4) concentrate to lower grade hematite (Fe_2O_3) and goethite (FeOOH) iron ores.
- Transition DR from using reformed NG to using 100 per cent H_2 reductant.
- Transition from DRI melting in the conventional EAF to the SAF operated in open-bath mode.
- Carburise the melting DRI to PI in the SAF, preferably with a sustainable carbon source.
- Retain the BOF converter but capture and preferably reuse its carbon emissions.

PILBARA ORES

To understand the potential to use low-grade Pilbara ore fines for DR ironmaking it is useful to compare Pilbara ore fines compositions to those of DR and BF grade pellets.

Comparison of pellet and Pilbara ore fines compositions

A range of historical (Gerstenberg *et al*, 1979; Poveromo and Swanson, 1999) and recent (LKAB, 2019) DR and BF indurated pellet compositions have been compiled and compared with Pilbara ore fines compositions (Park *et al*, 2022).

To achieve an unbiased comparison the following adjustments were applied.

- As the indurated pellet compositions are moisture free then the Pilbara ore fines compositions (Park *et al*, 2022) have been adjusted to reflect that of an indurated pellet produced therefrom.

- As many of the pellet compositions were reported with per cent Al_2O_3 and per cent TiO_2 combined then the same here.

The relevant compositions values are given in Table 1 and displayed in Figure 1.

TABLE 1

Compositions of various DR and BF grade pellets and Pilbara ore fines.

Pellets			Pellets (cont.)		
%SiO$_2$	%Al$_2$O$_3$ + %TiO$_2$	%Σ	%SiO$_2$	%Al$_2$O$_3$ + %TiO$_2$	%Σ
1.70	0.35	2.05	2.11	0.38	2.49
1.63	0.45	2.08	1.45	0.72	2.17
1.15	0.50	1.65	2.12	1.42	3.54
1.40	0.30	1.70	0.80	0.34	1.14
1.32	0.24	1.56	1.85	0.67	2.52
1.40	0.23	1.63	2.20	0.45	2.65
2.75	0.33	3.08	2.60	0.48	3.08
1.31	0.77	2.08	**Pilbara low Al$_2$O$_3$ fines**		
1.23	0.76	1.99	%SiO$_2$	%Al$_2$O$_3$ + %TiO$_2$	%Σ
1.28	0.62	1.90	5.59	1.44	7.03
1.23	0.62	1.85	6.77	1.82	8.59
2.64	0.51	3.15	**Pilbara high Al$_2$O$_3$ fines**		
3.85	0.76	4.61	%SiO$_2$	%Al$_2$O$_3$ + %TiO$_2$	%Σ
1.60	0.71	2.31	4.38	2.81	7.19
2.61	1.46	4.07	4.36	2.83	7.19
4.60	1.13	5.73	5.77	2.79	8.56
1.09	0.35	1.44	3.54	2.01	5.55
3.19	0.40	3.59	4.76	2.62	7.38
1.22	1.04	2.26	6.22	3.30	9.52

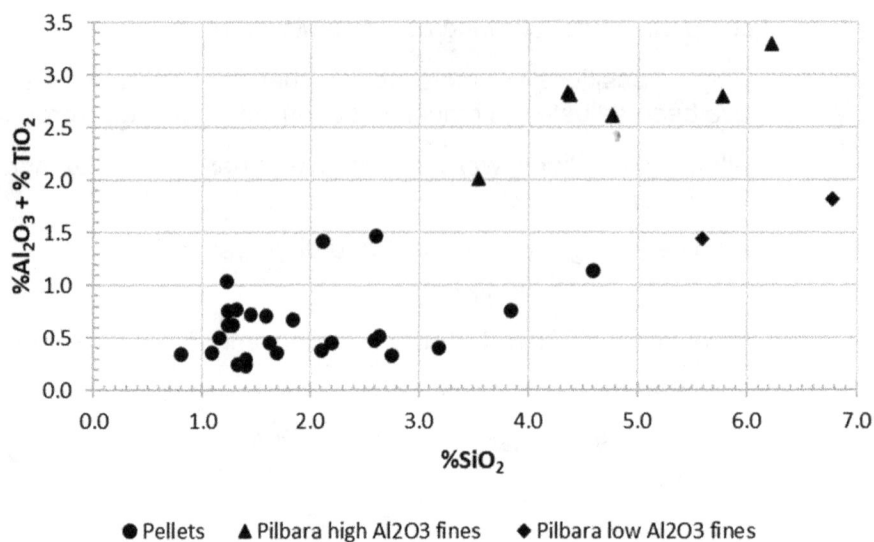

FIG 1 – Compositions of various DR and BF grade pellets and Pilbara ore fines.

It can be seen in the graph of Figure 1 that the compositions can be broadly grouped into three categories.

1. The cluster of pellets at lower left that would comprise classical high-grade DR pellets ideally suited to being processed via the DR-EAF route.

2. A lower branch of several pellets extending to the two Pilbara low per cent Al_2O_3 ore fines.

3. Three pellets extending to the upper branch of several Pilbara high per cent Al_2O_3 ore fines, which comprise the bulk of ore produced in the Pilbara.

ORE PREPARATION

Grinding

For pellet making the iron ore fines need to be further ground to a sufficient fineness in order that both the granular particles making up a pellet have a high total surface area and there is sufficient intergranular porosity to both facilitate exposure to and allow the reducing and product gases to permeate the pellet respectively.

Blaine value

The measure for this purpose is the Blaine value, having dimensions surface area per mass, cm^2/g. Unlike the dimensions suggest, it is not possible to equate it to a specific particle size, since grinding results in a particle size distribution range. Instead, it is measured by a standardised test (ASTM, 1946).

There is no ideal Blaine value for DRI pellets and an optimum value needs to be determined for each specific ore, or ore blend, typically ±2000 g/cm^2 up to 3000 g/cm^2.

Note that certain ores, such as goethite, even when finely ground may retain a degree of intragranular porosity that is further beneficial to gas diffusion.

Beneficiation

Beneficiation of Pilbara ores is challenging and the results are dependent on ore type.

Hematite

The microplaty hematite ores, such as Newman high-grade (HG) fines, can get to above the conventional DR grade of 67 per cent Fe with coarse grinding and beneficiation, but with low mass yield (Warnock and Bensley, 1996; Bensley, Fenton and Turner, 1999). Newman HG fines were used for BHP's Boodarie Iron hot briquetted iron (HBI) operation.

Hematite-Goethite

A few hematite-rich high-P Brockman ores ground to an 80 per cent passing size (P_{80}) of 45 µm can get to about 65 per cent Fe but with less than 80 per cent mass yield. Many hematite-goethite ores tested struggle to reach 65 per cent Fe, even at a P_{80} of 20 µm to 45 µm. Generally, the hematite-goethite ores must be ground to less than 10 µm to achieve liberation and then have poor mass yield (Nunna *et al*, 2021).

Goethite-Hematite

Goethite dominant iron ores generally won't get to greater than 60 per cent Fe since pure goethite is only 63.8 per cent Fe. Also, typically their gangue SiO_2 and Al_2O_3 are tied up in the crystal lattice (Pownceby *et al*, 2019). Thus, the main known issue is that with solid-state direct reduction the SiO_2 and Al_2O_3 report as sub-micron inclusions in the DRI. Moreover, goethite ground to below 20 µm is notoriously difficult to recover.

PELLET PREPARATION

Green pellets

Green pellets are made from various materials in addition to the finely ground iron ore and water:

- Binder agent (eg bentonite and kaolinite)
- Fluxes (eg CaO, MgO and SiO_2).

Pellet producers may add carbon to hematite pellets to replicate the shorter magnetite pellet induration cycle, increasing production rate (Cribbes, July 2023, personal communication). Another advantage is that it significantly increases the indurated pellet porosity (Firth, Douglas and Roy, 2011), aiding reduction. However, carbon cannot be added to goethite containing pellets as it disrupts the removal of the combined moisture by too rapid heating (Cribbes, July 2023, personal communication).

Pellets may be formed using a drum or disc balling unit. DR pellet size ranges are (Astier, 1979):

- preferred – 7 mm to 15 mm
- tolerable – 5 mm to 18 mm.

However, within these limits there may be reasons to want to utilise pellets that are at the smaller or larger size end of the ranges, due to their reduction or/and melting behaviour.

Induration

Once formed green pellets are dried and hardened by induration (firing), which is conducted in one of four types of furnaces:

1. Grate-kiln (eg Whyalla, South Australia).
2. Straight-grate or travelling grate (eg see below).
3. Circular-grate (from Primetals Technologies).
4. Shaft furnace (eg Port Latta, Tasmania).

Two induration plants of the straight-grate type have previously operated at locations in the Pilbara, utilising hematite-goethite ores and fired with fuel-oil, producing BF grade pellets for export to Japan:

1. Robe River, owned and operated by Cliffs Iron Ore Associates, a JV between Cleveland Cliffs and Mitsui, between 1972 and the mid-1980s. The plant was shutdown in the mid-1980s when OPEC formed and drove up the oil price rendering the pellet production uneconomic. It was eventually scrapped in the late 1990s (Cribbes, July 2023, personal communication).

2. Dampier, owned and operated by Hamersley Iron Pty. Ltd., started in 1968 (Baker *et al*, 1973). It was also shutdown around the same time as Robe River for the same reason.

Note that these events occurred prior to the availability of NG from the offshore gas resources of NW Western Australia.

Since pellets may be made from any type of iron ore, a few primary reactions take place during firing:

- Free moisture is evaporated.

- In the case of magnetite, it is oxidised to hematite.

- In the case of goethite, the crystal bound moisture is removed converting it to hematite.

- In the case where carbon has been added to hematite pellets the carbon reduces the hematite to magnetite, which is subsequently re-oxidised to hematite (Firth, Douglas and Roy, 2011).

The main issue with Pilbara goethite-containing ores is their high free and combined moisture contents. This restricts the heating rate for the initial drying phase, requiring that the induration cycle be extended to drive off the free and bound moisture in a timely manner so as not to result in the disintegration of the pellets because of rapid steam formation (Baker *et al*, 1973).

Advantageously, Pilbara hematite-goethite ores can contain gangue material that acts as a natural binder. For instance, the Hamersley ore that was used at the Dampier pellet plant contained 6 per cent kaolinite and the only further addition was 1.5 per cent of limesand (Baker *et al*, 1973). Similarly in the case of the Robe River pellet plant (Cribbes, July 2023, personal communication).

There is a critical fusion temperature, being approximately 1340°C (Cribbes, July 2023, personal communication), with respect to forming the pellets primary interparticle bond. Overheating results in softening and deformation of the pellets.

In all cases, the binder agent and any fluxing agents are melted and react with a proportion of the ore's gangue oxides, forming an inter-particle slag that further bonds the ore particles together when it solidifies upon cooling.

Cold compressive strength

The pellet cold compressive strength (CCS) after induration is also important after reduction.

After induration

In Figure 2a CCS increases with induration (firing) temperature and can be positively impacted by a lower Blaine value (Mathisson, 1979). Note that the graph is only plotted up to 1360°C. This is consistent with the previous mentioned critical fusion temperature of approximately 1340°C and that overheating causes the pellet to soften and deform.

In Figure 2b CCS further increases as the slag amount increases (Mathisson, 1979). As well, the clustering and scattering of the data points indicate that for a given SiO_2 the CCS decreases with increasing CaO/SiO_2, suggesting that the benefit of slag to increasing CCS may be maximised when CaO/SiO_2 is less than 1.

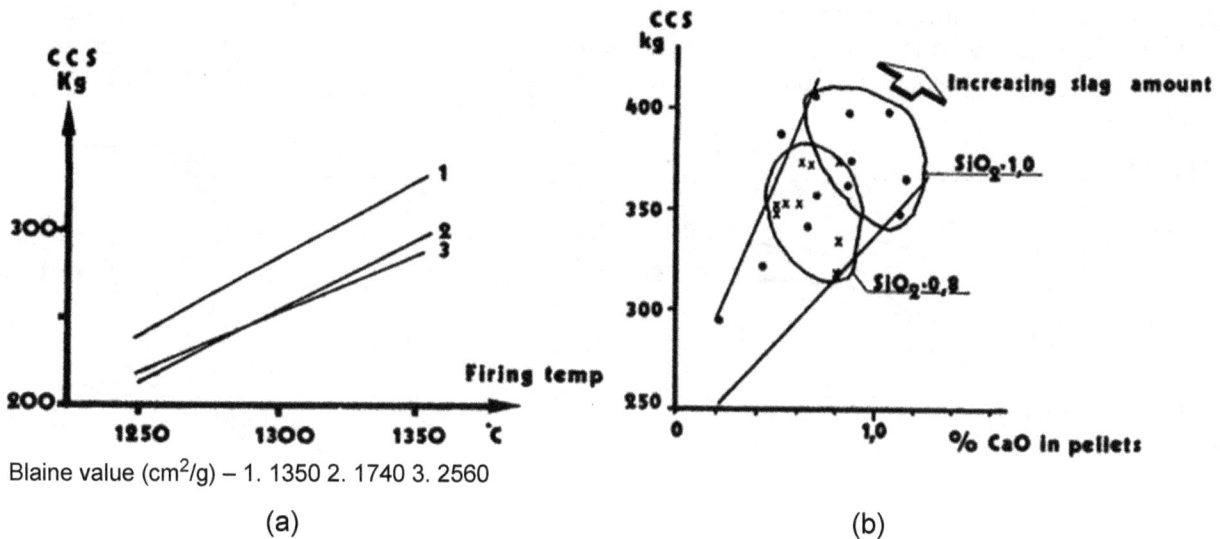

Blaine value (cm²/g) – 1. 1350 2. 1740 3. 2560

(a)   (b)

FIG 2 – (a) CCS of acid pellets as a function of firing temperature for various Blaine values; (b) increasing CCS of pellets fluxed with CaO as a function of slag content (Mathisson, 1979).

After reduction

The CCS after reduction is linearly proportional to and, to a first approximation, one-third of that after induration (Mathisson, 1979).

Hot briquetted iron

If the DRI pellets are subsequently to be compressed into HBI or like, then the CCS after reduction is of interest since the force and specific energy of compression would be proportional to it.

PELLET REDUCTION

Pellet reduction is affected by several factors.

- reducibility
- rate of reduction
- maximum reduction temperature
- sticking, clustering and cluster strength
- swelling and disintegration.

Reducibility

Reducibility is the maximum degree of reduction with respect to removal of O and is not the same as the degree of metallisation. For instance, if Fe_2O_3 is reduced to wüstite (FeO), as 2FeO, then 33 per cent reduction has occurred but 0 per cent metallisation, since no metallic iron (Fe) has been produced. If half of 2FeO is then reduced to Fe then 66 per cent reduction and 50 per cent metallisation has occurred. Of course, 100 per cent reduction equates to 100 per cent metallisation.

Maximising reducibility is important since SAF ironmaking slag, like the BF, has a naturally low FeO content of 1 per cent to 3 per cent (Ure, 2000). Therefore, any direct reduction below the optimally high value means that the balance of reduction must be completed in the SAF (Ure, 2000), increasing the amount of carbon (C) otherwise added to the SAF to carburise the PI, as well as increasing the furnaces specific electric power consumption.

Reducibility is influenced by a few factors, including ore and pellet composition. In the graph of Figure 3a, a pellet CaO/SiO_2 of less than approximately 1.0 is required to achieve nearly 100 per cent reduction. In the case of the MPR pellet the CaO/SiO_2 is 0.17, although the $(CaO+MgO)/(SiO_2+Al_2O_3)$ is 0.31 (Mathisson, 1979). Note also that while the MPR pellet achieves nearly 100 per cent reduction it does so at a slower rate, which is elaborated upon next. The graph of Figure 3b shows numerous examples of reducibility of commercial pellets and lump ores by pot-grate test (Burghardt and Kortmann, 1979). Notably, the Kakogawa pellet (■) only achieves a maximum degree of reduction of just over 80 per cent.

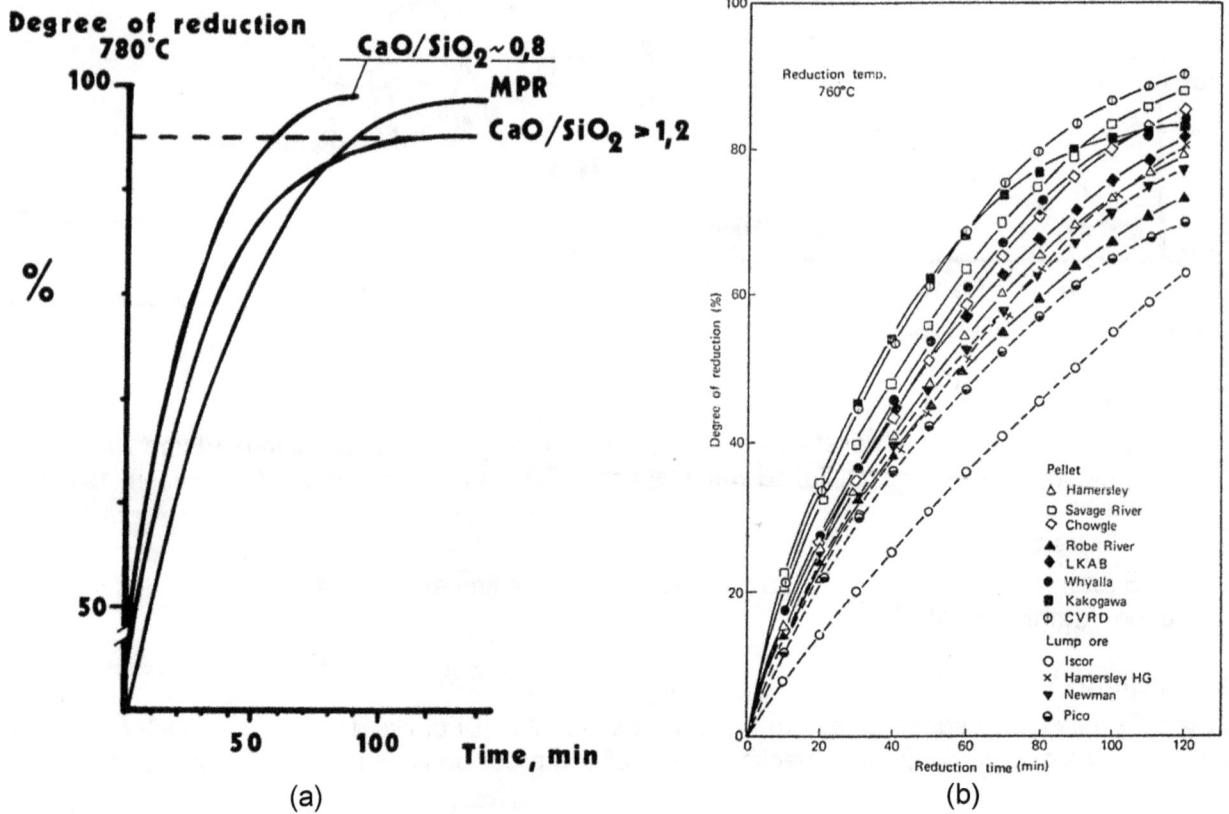

FIG 3 – (a) Examples of degree of pellet reducibility as a function of CaO/SiO_2 (Mathisson, 1979); (b) Examples of commercial pellets and lump ore in static bed reduction testing (Burghardt and Kortmann, 1979).

Rate of reduction

The achievable rate of pellet reduction is important in that it in part determines the DRI production rate.

The rate of reduction can vary significantly, as can be seen in the graph of Figure 3b. Here the rate of reduction, as indicated by the varying slope of the curves, for the Savage River pellets (□) is substantially higher than for the Robe River pellets (▲), while the Hamersley pellets (△) fall between. It can also be seen in these examples that pellets typically outperform lump ores.

Several factors influence the rate of reduction:

- porosity
- composition
- reduction temperature
- reducing gas composition and supply.

Porosity

Grain

Pilbara goethite-hematite ores are inherently porous and so even when finely ground any remnant micropores in the grains will benefit internal gaseous diffusion and reduction.

Pellet

Pellets have a more uniform intra-granular porosity, but total porosity will vary with ground grain size distribution and can be further reduced by flux additions.

Cracking

During reduction it is normal for pellets to swell and this often results in cracking and macropores.

Composition and temperature

In lieu of the variable slopes of the degree of reduction curves of Figure 3, a more practical measure and indicator of the rate of reduction is the time to achieve 95 per cent reduction (t_{95}).

For example, in the graph of Figure 4 it can be seen that t_{95} is substantially reduced with increasing reduction temperature, while also being significantly influenced by the CaO/SiO_2 ratio, here exhibiting minima at ~0.8 over the temperature range 780°C to 1000°C. However, it can also be observed that the CaO/SiO_2 is less significant a factor the higher the reducing gas temperature.

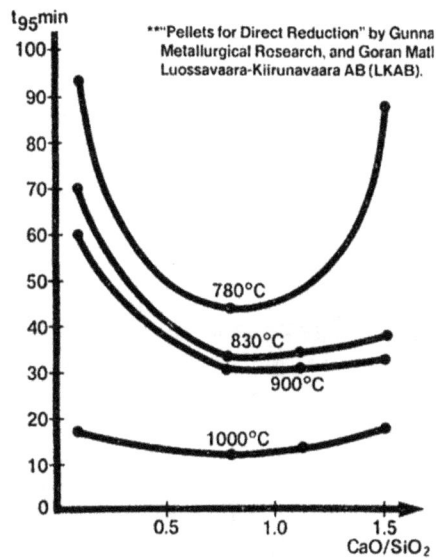

FIG 4 – t_{95} versus CaO/SiO_2 at various reduction temperatures (Burghardt and Kortmann, 1979).

Reducing gas composition and temperature

Reduction rate is also influenced by the reducing gas composition. For example, in the graph in Figure 5 the reduction rate not only increases with increasing temperature but also with increasing hydrogen (H_2) content of the reducing gas, while it is negatively impacted by the addition of water vapour (H_2O) or methane (CH_4).

The influence of H_2O is of concern in so far as that it is the product of the H_2 reacting with the iron oxide and its proportion increases as the reducing gas mixture travels up through the descending pellet bed in the reactor shaft.

In the case of CH_4, it is of concern where it is introduced as a means of DRI carburising via cracking.

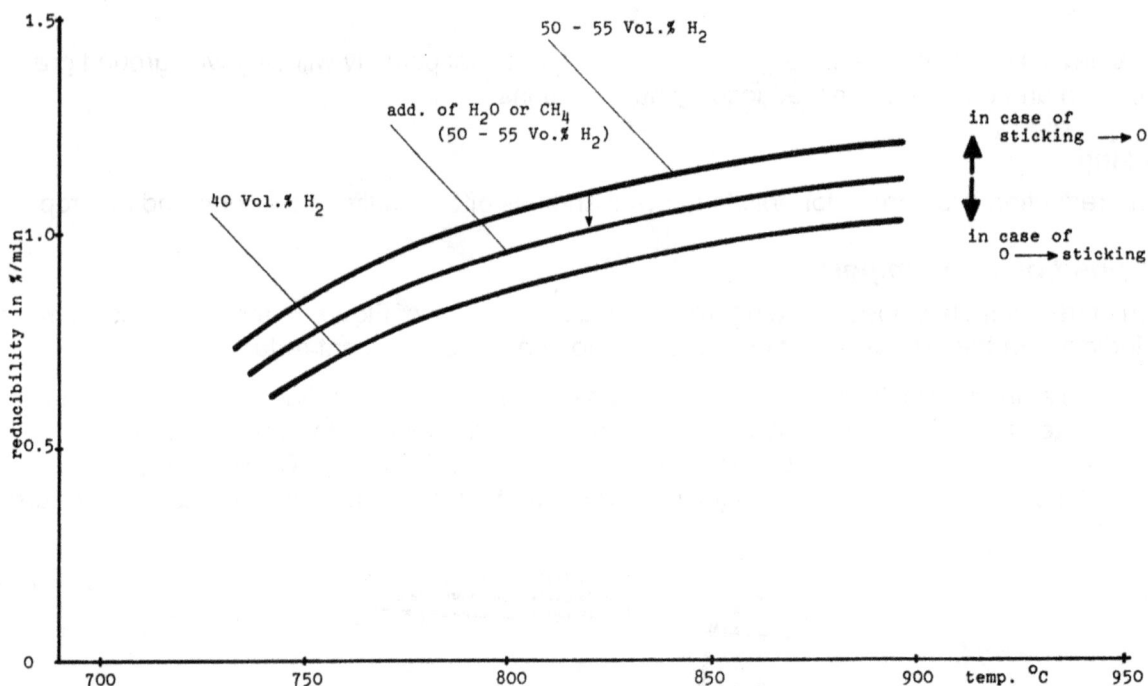

FIG 5 – Reduction rate versus temperature for various reducing gas H_2 contents (Burghardt and Kortmann, 1979).

Reducing gas supply

Irrespective of the previous, the overall rate of reduction may still be limited by the DR reducing gas supply rate. For instance, a DR vessel operating at 1 atm gauge pressure will require a gas volume flow rate five times that of a vessel operating at 5 atm.

Maximum reduction temperature

Given that the rate of reduction is substantially influenced by the reduction temperature, then what factors govern or limit the reduction temperature?

The graph in Figure 6 indicates that the maximum processing temperature is dependent upon the total gangue oxide content of the pellets, as well as the density and smoothness of the pellet surface. The primary reason for there being a maximum reduction temperature is associated with the tendency for pellets to stick together, forming clusters.

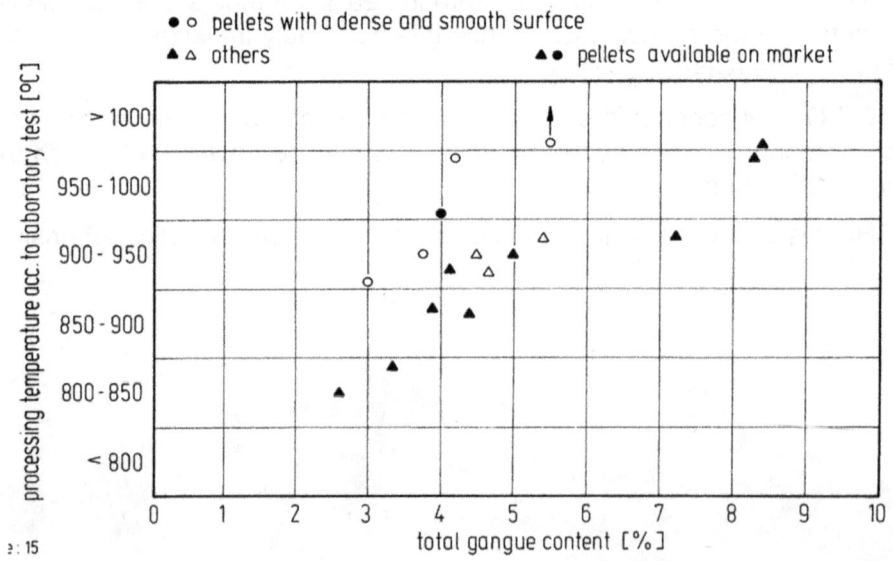

FIG 6 – Maximum reduction temperature versus total gangue content and pellet surface smoothness (Gerstenberg *et al*, 1979).

Sticking and clustering

Pellet sticking, forming clusters, occurs in the lower part of the DR module. The main threat of cluster formation is the potential for bridging at the bottom of the DR module, preventing the DRI from discharging. To ensure that the DRI pellets discharge readily then it is necessary to avoid or minimise sticking and clustering.

The degree of sticking is influenced by a few factors.

- coating
- reduction temperature
- reducing gas composition.

Coating

A common method of preventing sticking is to coat pellets with an anti-sticking agent, such as lime or cement.

Reduction temperature and gas composition

It can be seen from the graph in Figure 7 that as well as the reduction temperature the reducing gas composition significantly influences sticking tendency.

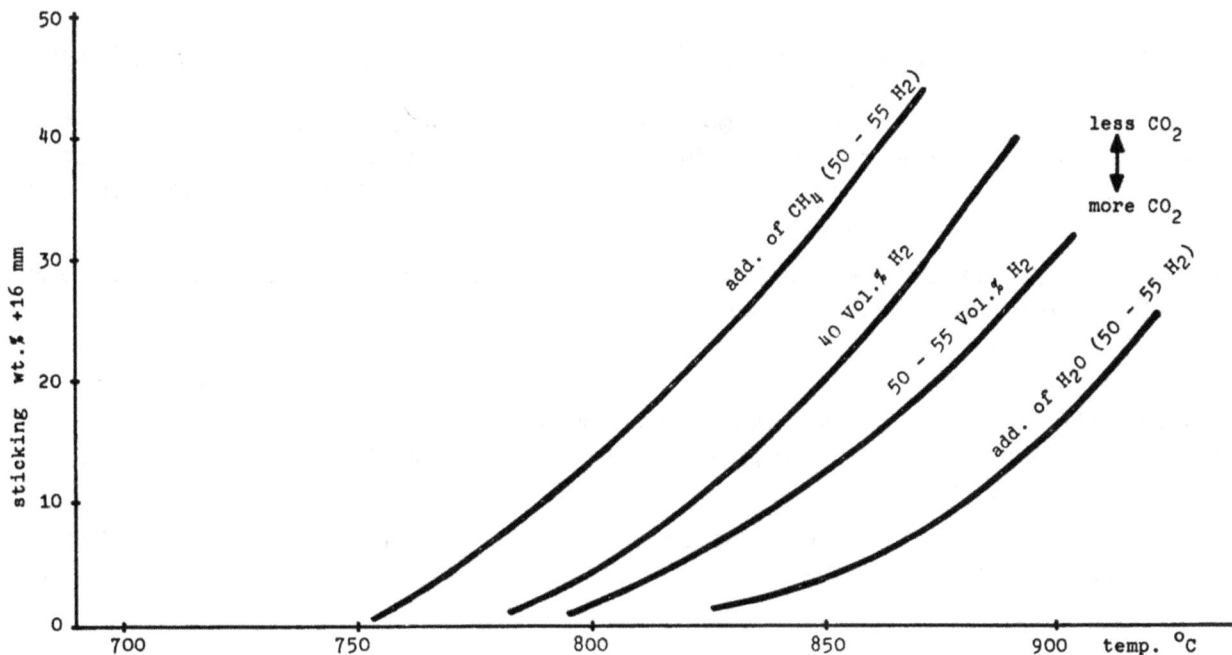

FIG 7 – Sticking versus reduction temperature for various reducing gas compositions (Burghardt and Kortmann, 1979).

In particular, the tendency to stick reduces with increasing H_2 content (H_2/CO), which clearly favours a transition to 100 per cent H_2 rather than reformed NG as the reductant gas. Furthermore, the addition of CH_4 promotes sticking whilst addition of H_2O decreases sticking.

Cluster strength

Once a cluster has formed then there is the matter of cluster strength and robustness. For instance, in a static scenario the cluster can readily remain intact, but in a dynamic scenario, such as during discharge from the DR module and transfer to the melting furnace, the cluster could be broken up.

Accordingly, cluster strength is assessed by means of a tumble test and determined to be the percentage of clustering remaining after tumbling for 5 minutes. It can be seen from the graph of Figure 8 that test work has determined that the cluster strength reduces to zero when the Fe content of the indurated pellets is sufficiently low, here <65 per cent Fe (Maarouf *et al*, 1979).

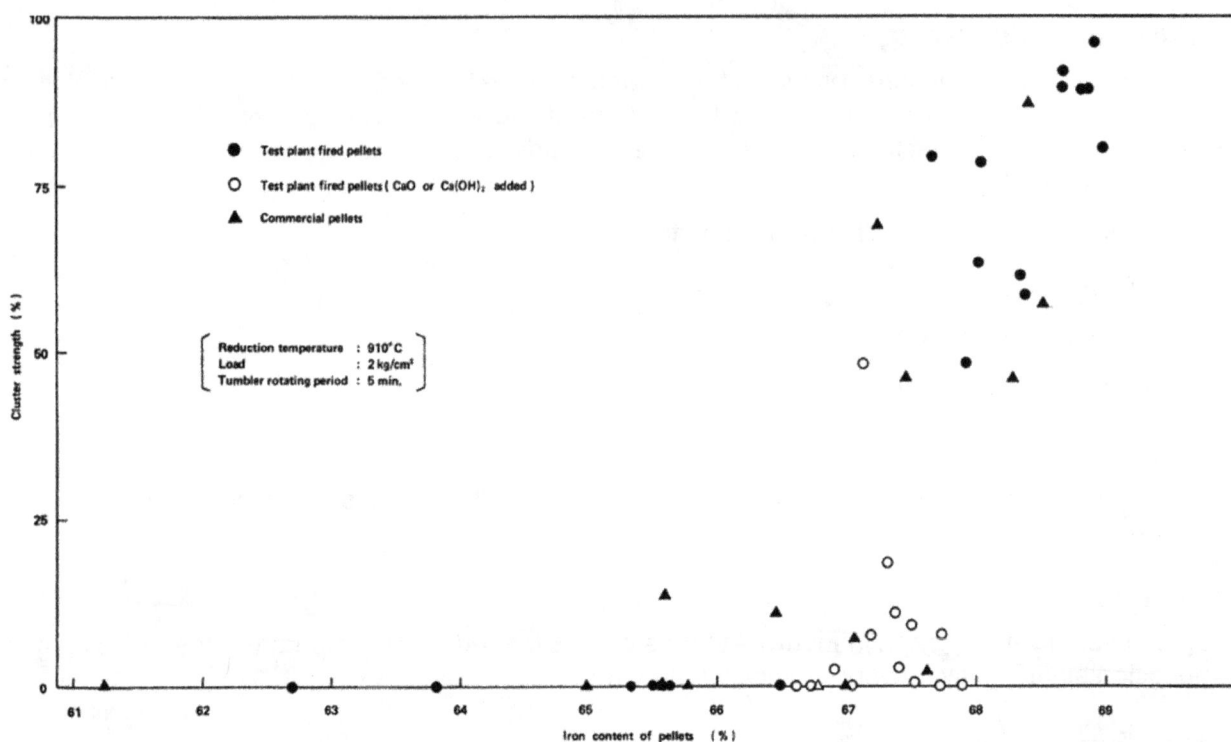

FIG 8 – DRI pellet cluster strength versus indurated pellet Fe content (Maarouf *et al*, 1979).

Assuming that the Fe content of the indurated pellets is present as Fe_2O_3, then this would correspond to gangue contents of >7 per cent. Even without fluxing, this would be the case for pellets produced from all the Pilbara fines ores shown in Figure 1, except for the one case of high $(Al_2O_3+TiO_2)/SiO_2$ fines having the lowest total gangue content of ~5.5 per cent $Al_2O_3+TiO_2+SiO_2$.

Thus, even for those fines, adequate fluxing of the pellets with CaO could be expected to result in DRI pellets with a zero, or near zero, cluster strength.

Swelling and disintegration

Swelling and disintegration of pellets during reduction is a major problem. It leads to a restriction of gas flow due to an increase in the pressure differential from inlet to outlet and consequential reduction in production rate.

In one study (Kortmann *et al*, 1973), the effects of fluxing of pellets with lime on swelling and disintegration was analysed in detail.

Swelling

Swelling is a consequence of multiple factors, but two are of primary importance:

1. Oxide changes.
2. Flux additions.

Oxide changes

When rhombohedral hematite is reduced to cubic magnetite, swelling occurs due to the increase in the crystal volume.

Flux addition

It is common practice to flux pellets with lime (CaO), usually by addition of limestone ($CaCO_3$). In some cases, dolomite ($CaMg(CO_3)_2$) is added where MgO is desired to improve certain pellet properties. The effects of fluxing pellets with lime on swelling can be summarised as follows:

- Swelling is observed in the temperature range 900°C to 1100°C.

- When CaO content is absent, the lower the gangue silica content the greater the swelling (10 per cent to 30 per cent).

- The addition of ±0.3 per cent CaO massively increases swelling (40 per cent to 80 per cent) when the gangue silica content is less than approximately 3 per cent.

- The addition of 1 per cent to 2 per cent CaO increases swelling moderately when the gangue silica content is greater than approximately 3 per cent (20 per cent to 35 per cent). Beneficially, it raises the CCS and abrasive strength of indurated pellets and reduces disintegration in the low-temperature reduction range.

- The optimal range for minimising swelling is CaO/SiO$_2$ less than 0.7 for gangue content up to approximately 10 per cent.

- Swelling behaviour is adversely affected in the higher basicity range.

Disintegration

In the same study (Kortmann *et al*, 1973) it was determined that a complex set of reactions occur in the low-temperature range 400°C to 600°C under weakly reducing conditions as a function of the amount and composition of the gangue, causing pellet disintegration when the CaO/SiO$_2$ is outside the range 0.1 to 0.8.

DRI MELTING IN AN ELECTRIC FURNACE

DRI in an EAF

Melting of DRI pellets, or HBI, is typically performed in an EAF, sometimes in combination with steel scrap (SS), to produce crude steel (CS) directly, circumventing the basic oxygen furnace (BOF).

A criterion for melting DRI or HBI and any requisite fluxes in an EAF is that the amount of slag generated relative to CS produced is preferably not more than about 5 per cent.

Where the use of DRI or HBI alone does not meet this or another preferred limit, the addition of SS is beneficial in reducing the slag percentage. Of course, in a fully integrated steel mill there is a natural uprising of SS, from the hot and cold rolling mills. This SS can be beneficially recycled through the EAF to dilute slag production up to a point.

In the case of melting Pilbara low-grade goethite-hematite derived DRI plus fluxes, the amount of slag generated would likely be more than can be readily handled in an EAF, at least not without slagging-off part way through producing a CS heat. Alternatively, a requisitely large amount of SS might be employed, but this would exceed the typical uprising in an integrated steel mill.

Note that slagging-off part way through producing a CS heat was practiced at New Zealand Steel when its vanadiferous titanomagnetite (VTM) DRI was originally melted in an EAF (Evans, 1993, personal communication).

DRI in a SAF

Alternatively, if high gangue content DRI is melted in a SAF then the molten iron and slag accumulate in the bath and can be intermittently tapped out separately as required, while the SAF melts on a continuous basis.

Unfortunately, the SAF is not equipped or constructed to handle the high temperatures of CS and attendant slag so the molten iron must be carburised to PI.

In the case where the DRI is produced with 100 per cent H$_2$ as reductant, carburising to PI can only be achieved by addition of carbon and any fluxes, to the SAF.

Mode of operation

When DRI is melted in a SAF, it cannot be operated in submerged mode. This is because submerging the electrodes in electrically conductive DRI results in short-circuiting, circumventing electrical heating of the slag. Instead, the SAF is operated in 'open-bath' mode so that the DRI enters the slag bath without contacting the electrodes, as practiced at New Zealand Steel (Ure, 2000).

Pellet melting

When fed to a SAF, upon entering the slag bath [S] as shown in Figure 9a, a DRI pellet [A] melts into its component metallic iron and gangue oxides, including residual FeO. The gangue oxides are incorporated into the molten slag, while the metallic iron particles melt and coalesce into raindrop-like prills and descend into the PI bath [Fe]. Any residual FeO may also be subsequently reduced by the PI carburising agent added.

Melting time of the pellet is governed by its slag composition as shown in Figure 9b and pellet size.

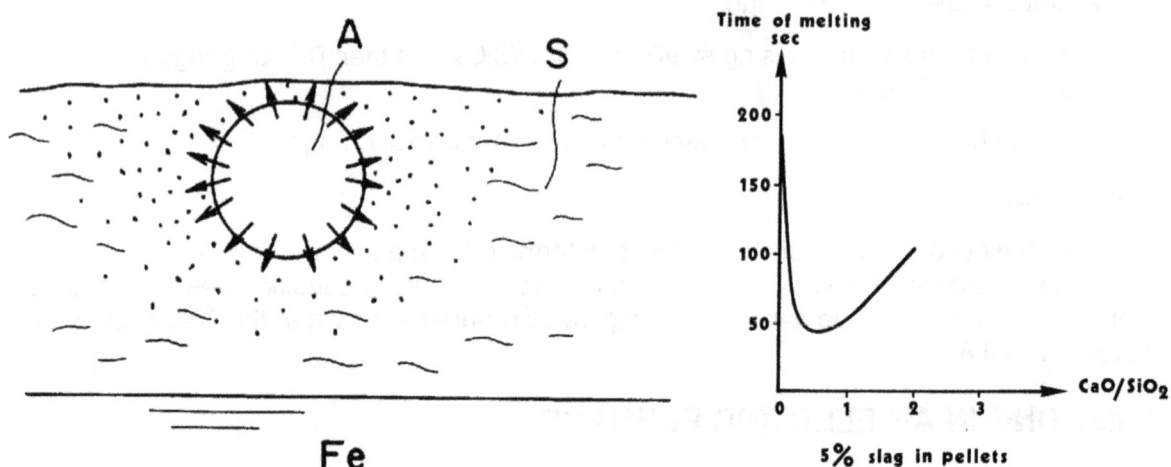

FIG 9 – (a) A DRI pellet melting in the slag bath (Kobe Steel, Ltd, 2000); (b) Example of melting time versus CaO/SiO_2 for DRI pellets containing 5 per cent slag (Mathisson, 1979).

Unlike the CS produced in an EAF, the PI produced in a SAF subsequently needs to be converted to CS in a BOF. The drawback to this is that the BOF produces carbon emissions because of having to carburise to PI in the SAF, although the carburising agent may be renewably sourced.

SAF slag

As with the BF, there may be potential for SAF slag to be formulated to be suitable for subsequent use, such as for slag cement production.

CONCLUSIONS

Pilbara hematite-goethite iron ore fines fall into two distinct composition groups, one having low $(Al_2O_3+TiO_2)/SiO_2$ ratio and one having high $(Al_2O_3+TiO_2)/SiO_2$ ratio.

They have both previously been exploited to produce BF pellets and should be suitable to produce DRI pellets.

Their variable hematite-goethite constitution and porous nature require that the pellet induration cycle be adapted to drive off free and combined crystalline moisture in a timely manner so as not result in the disintegration of the pellets because of rapid steam formation.

Carbon additions must not be made to the hematite-goethite green pellets as this disrupts crystalline moisture removal from the goethite during induration.

Their high gangue oxide contents impart certain benefits to the pellets in the DR process, such as minimising or avoiding sticking and clustering and enabling higher reduction temperatures.

Higher reduction temperatures enable higher reduction rates, subject to reducing gas supply rate.

Direct reduction using 100 per cent H_2, instead of reformed NG, will increase reduction rate and further reduce the propensity for pellets to stick.

The higher gangue content of Pilbara DRI pellets makes them less suitable for melting in an EAF and in such case would likely require that they are blended with a requisite proportion of scrap steel. They would be better suited for melting in a SAF operating in open-bath mode.

Pilbara DRI pellets would likely benefit from blending of the high and low ratio Al_2O_3 fines, to optimise the SAF ironmaking slag composition and minimise additional slag fluxing requirements.

Pilbara pellets will likely require fluxing with CaO in the range 0.5 to 0.7 CaO/SiO_2, to optimise both reduction and melting performance, subject to pot grate testing in the laboratory.

Carburising agent will need to be added to the SAF to produce pig iron from 100 per cent H_2 reduced DRI pellets. If the FeO content in the DRI pellets fed to the SAF is more than a couple of percent then additional carburising agent will likely be needed to achieve the equilibrium slag FeO content.

The pig iron produced in the SAF requires conversion to crude steel in a BOF.

ACKNOWLEDGEMENTS

The traditional custodians of the land we are meeting on, the Whadjuk Nyoongar people.

REFERENCES

Astier, J, 1979. Iron ore feed required by direct reduction processes, in DR Pellet Symposium, (LKAB: Kiruna-Malmberget).

ASTM, 1946. Standard Test Methods for Fineness of Hydraulic Cement by Air-Permeability Apparatus, American Society for Testing of Materials publication ASTM C204.

Baker, L A, Thomas, C G, Cornelius, R J, Lynch, K S and Armstrong, G J, 1973. Effect of goethite on production rate in a travelling grate pellet plant, *Transactions of the Society of Mining Engineers,* 254:270–278.

Bensley, C N, Fenton, K L and Turner, J, 1999. Iron ore beneficiation by BHP Iron Ore in Western Australia, (eds: J T Woodcock, and J K Hamilton), Monograph Series 19 (The Australasian Institute of Mining and Metallurgy: Melbourne).

Burghardt, O and Kortmann, H, 1979. Test methods for feed materials used in direct reduction, in DR Pellet Symposium, (LKAB: Kiruna-Malmberget).

Firth, A R, Douglas, J D and Roy, D, 2011. Understanding reactions in iron ore pellets, in *Proceedings Iron Ore Conference 2011*, pp 413–423 (The Australasian Institute of Mining and Metallurgy: Melbourne).

Gerstenberg, B, Knop, K, Kropla, H-W and Huschens, B, 1979. Requirements of ore properties on the basis of results of PUROFER shaft furnaces and methods for testing these properties, in DR Pellet Symposium, (LKAB: Kiruna-Malmberget).

Kabushiki Kaisha Kobe Seiko Sho Inventions (Kobe Steel, Ltd), 2000. Method of making iron and steel, United States Patent Number 6,149,709. https://patents.justia.com/patent/6149709

Kortmann, H A, Burghardt, O P, Grover, B M and Koch, K, 1973. Effect of lime addition upon the behaviour during the reduction of iron ore pellets, in *Transactions of the Society of Mining Engineers,* 254:184–192.

LKAB, 2019. Pellets, in LKAB products 2019 (LKAB: Kiruna-Malmberget).

Maarouf, E, Miki, O, Katsumata, Y, Narita, K and Kaneko, D, 1979. Starting up of direct reduction plant in Qatar Steel Co., in *DR Pellet Symposium*, p 15 (LKAB: Kiruna-Malmberget).

Mathisson, G, 1979. LKAB DR pellets – research and development, in DR Pellet Symposium, (LKAB: Kiruna-Malmberget).

Nunna, V, Suthers, S P, Pownceby, M I and Sparrow, G J, 2021. Processing options for removal of silica and alumina from low grade hematite-goethite iron ores, in *Proceedings Iron Ore Conference 2021*, pp 490–506 (The Australasian Institute of Mining and Metallurgy: Melbourne).

Park, J, Kim, E, Suh, I-k and Lee, J, 2022. A short review of the effect of iron ore selection on mineral phases of iron ore sinter, in *Minerals*, 12:35.

Poveromo, J J and Swanson, A W, 1999. Iron-bearing raw materials for direct reduction, Chapter 5 in *Direct Reduced Iron – Technology and Economics of Production and Use*, pp 59–79 (The Iron & Steel Society: Warrendale).

Pownceby, M I, Hapugoda, S, Manuel, J, Webster, N A S and MacRae, C M, 2019. Characterisation of phosphorus and other impurities in goethite-rich iron ores – possible P incorporation mechanisms, in *Minerals Engineering*, 143:106022.

Ure, C R, 2000. Alternative ironmaking at BHP New Zealand Steel, in *58th Electric Furnace Conference and 17th Process Technology Conference Proceedings 2000,* pp 535–546 (The Iron & Steel Society: Warrendale).

Warnock, W W and Bensley, C, 1996. The Port Hedland hot briquetted iron project, Perth, in *Proceedings of the Australasian Institute of Mining and Metallurgy Annual Conference 1996,* pp 197–202 (The Australasian Institute of Mining and Metallurgy: Melbourne).

Greener downstream processing – green iron and steel production

The challenges of removing sulfur from magnetite ores to meet direct reduction specifications

D Connelly[1]

1. Principal Consulting Engineer, METS Engineering Group, Perth WA 6004.
   Email: damian.connelly@metsengineering.com

ABSTRACT

Magnetite ores can be used to produce very high-grade concentrates that are suitable feed for the direct reduction process route but the high sulfur content of some concentrates can cause substantial environmental issues. Processing of high sulfur magnetite ore is therefore increasingly problematic for the industry. Generally, the industry demands a concentrate with <0.05 per cent sulfur and <2 per cent silica. This paper reports the outcomes of a study to produce a concentrate with the requisite low sulfur content from a sulfur-rich magnetite ore.

The main findings may be summarised as follows: if the sulfur is present in magnetite as nonmagnetic pyrrhotite or pyrite, it can be beneficiated using magnetic separation and hence, monoclinic pyrrhotite is magnetic and is magnetically recovered with magnetite leading to a concentrate that is too high in sulfur. X-ray diffraction and QEMSCAN analyses revealed that the magnetite concentrate was liberated at P_{80} 30 microns and was associated with maghemite, pyrrhotite (magnetic), carbonates and hematite with very low amounts of chalcopyrite, sphalerite and arsenopyrite. Sulfur removal using flotation was the first stage of investigation and this required flash flotation in the grinding stage followed by column flotation of the initial concentrate to generate the final concentrate. At the same time, it was observed that oxidation of the pyrrhotite had a negative effect on its flotation requiring copper sulfate to activate it. Bioleaching and concentrate leaching were also considered and achieved over 90 per cent sulfur removal but were deemed unsuitable owing to the production of highly reactive wastes and high CAPEX. Reference is made to a number of operating magnetite producers and how they manage the sulfur in feed.

INTRODUCTION

Approximately, 7–9 per cent of global greenhouse gas (GHG) emissions are attributable to steel production. This is largely owing to the widely used practice of using a blast furnace (BF) to reduce the iron ore chemically by feeding mixed burdens of lump ore, sinter and pellets with metallurgical coke acting as a fuel, reductant and metallurgical feed. Similarly, sinter plants and pellet plants contribute to the carbon footprint. This is the current nature of steelmaking, which has evolved and grown over the last 200 years.

With growing concern about global warming, which is largely attributed to carbon dioxide emissions, the steel industry is now faced with a huge challenge of how to produce green steel in a relatively short time frame. Efforts have been made to address these emissions, emphasising the significance of alternative technologies. Direct reduction (DR) processes, which use high-grade magnetite as the feed include the MIDREX (Midland-Ross Direct Iron Reduction) process which is one of the most promising and developed methods. Ultimately, the long-term goal is a switch from integrated blast furnace and basic oxygen furnace (BF/BOF) steelmaking to DR ironmaking followed by electric arc furnace (EAF) steelmaking.

Decarbonisation presents a huge opportunity for the direct reduction industry throughout its entire value chain, whether it is used as EAF feedstock to enable the circular economy for steel by reducing harmful metallic impurities in steel scrap, or as BF burden feedstock to enable lower CO_2 emissions. As an interim step, the above processes will use natural gas (methane, CH_4) and gas reformers to produce hydrogen but, in the longer term, the industry needs to wean off natural gas and convert to hydrogen reduction to produce hot briquetted iron (HBI), which can be fed into EAF and produce green steel assuming the electricity has been sourced from solar, wind or nuclear sources.

Magnetite ores can be used to produce very high-grade concentrates that are suitable for DR processes, but it is not uncommon for fresh magnetite orebodies to contain pyrite (FeS_2) and pyrrhotite (Fe_{1-x}, where x = 0 to 0.2). Whilst the pyrite is not recovered during magnetic separation, the pyrrhotite is and thus, the concentrate will not meet the specifications imposed by the direct reduction (DR) plants with pyrrhotite still present. The target specification for a green magnetite concentrate is 68 per cent Fe and <0.05 per cent sulfur. This paper discusses work conducted to produce high-grade magnetite concentrates with the requisite low sulfur content.

Magnetite is suitable for DR production and can use either FINMET, Circored, Midrex, HYL or Pered technologies. The units will need to be purged with nitrogen on start-up and shutdown to eliminate the risk of explosions when using hydrogen or reformed methane.

ORE FEED COMPOSITION AND MINERALOGY

The compositions of the ore feeds used in this study is summarised in Table 1. In addition to the bulk sample, two variability samples were collated from existing drill cores: one to represent high sulfur mineral content and one to represent high manganese content respectively. These were included to simulate ores that may be mined in the future. In addition, 10 per cent mining dilution was included in the test samples to reflect mining practice.

TABLE 1

Summary of ore feed compositions.

Element %	Composite
SiO_2	18.0
Al_2O_3	2.68
Fe2O3	46.4
MgO	6.2
CaO	13.1
Na_2O	0.15
K_2O	0.39
TiO_2	0.08
P_2O_5	0.02
MnO	2.54
Cr_2O_3	<0.01
V_2O_5	<0.01
LOI	9.58
Sum	99.1
S	0.14
Fe[1]	32.5
Sat[2]	39.4
Fe (mag)	28.5

Fe[1] grade calculted from the Fe_2O_3 result. [2]Satmagan grade expressed as Fe_3O_4.

The results of QEMSCAN analysis of the composite sample used in this work is summarised in Figure 1. The magnetite orebodies are strata bound in dolomitic units. The magnetite has fine inclusions of pyrrhotite and the manganese association is with the siliceous gangue (calcium aluminosilicates). Optical microscopy revealed no major iron ore minerals other than magnetite. There was a large amount of other iron-bearing minerals but they were rejected in the magnetic separation process for magnetite beneficiation. The exceptions were monoclinic pyrrhotite and maghemite. The QEMSCAN analysis indicated that it would be a greater challenge to reach targeted concentrate Fe and S levels because of the presence of pyrrhotite.

The evaluation was primarily based on the pyrrhotite content and its ratio to pyrite.

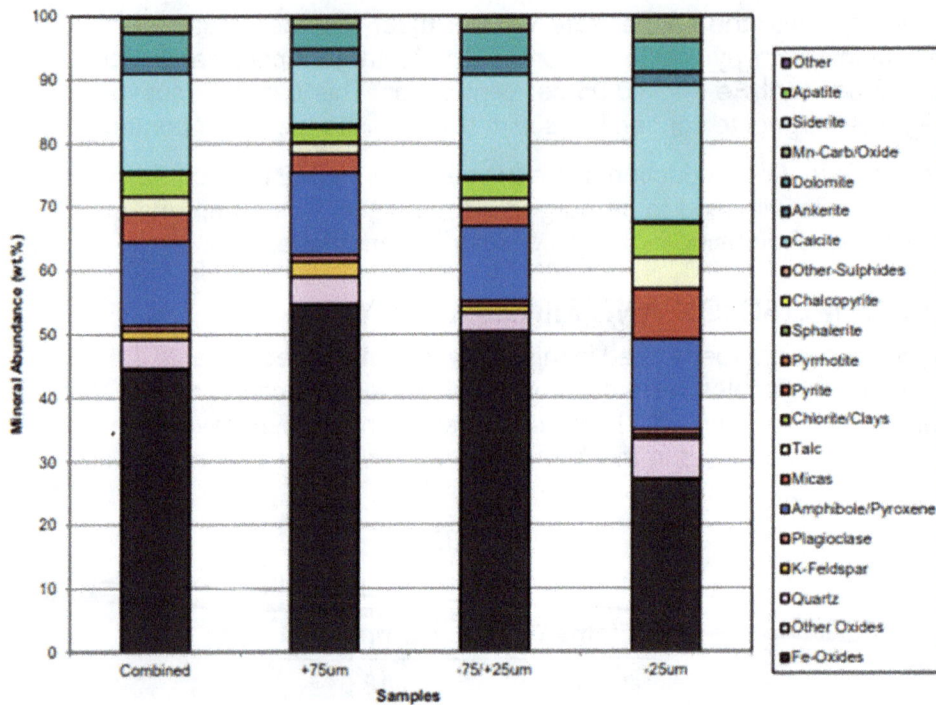

FIG 1 – QEMSCAN analysis of the nature and occurrence of minerals in the (Arvidson *et al*, 2013).

SULFIDE REMOVAL OPTIONS

A literature survey of processes for removing sulfides from magnetite concentrates revealed very few technical papers on the subject, which is surprising given that the presence of sulfides in fresh magnetite ores is not uncommon. Thus, the following options to reduce the sulfur level in the final magnetite concentrate were considered:

- Flotation to discard the sulfide concentrate.

- Magnetic separation for pyrite removal but not pyrrhotite.

- Bacterial leaching of the concentrate.

- Chemical leaching.

- Electrochemical leaching.

It was decided, based on a literature survey of industry practice, that flotation should be evaluated as the first step as the ore was ground and flotation is a widely practised technique. Helpful references were found relating to Swedish magnetite ores, where it is not an uncommon occurrence to find sulfur and apatite (Arvidson *et al*, 2013).

Bacterial oxidation in tanks was also considered but was deemed inferior owing to acid waste issues and much larger plant CAPEX due to long leaching times in excess of 96 hours. The same applies to other leaching methods.

FLOTATION TEST WORK

It is likely that excessive oxidation of the pyrrhotite happened in the time lag between obtaining the final magnetic concentrate and start of the flotation tests. It was confirmed several weeks after the testing that indeed a very long time elapsed from the time the magnetite concentrate was available and the start of flotation. Such oxidation is known to deactivate the sulfide surface, especially for pyrrhotite, and, consequently, interfere with the flotation performance (Hodgson and Agar, 1989; Buswell and Nicol, 2002).

The sulfide is liberated at relatively coarse sizes and hence, the use of two stages of flotation, flash flotation and column flotation. The magnetite however required fine grinding to achieve the high-grade required for DR (Figure 2).

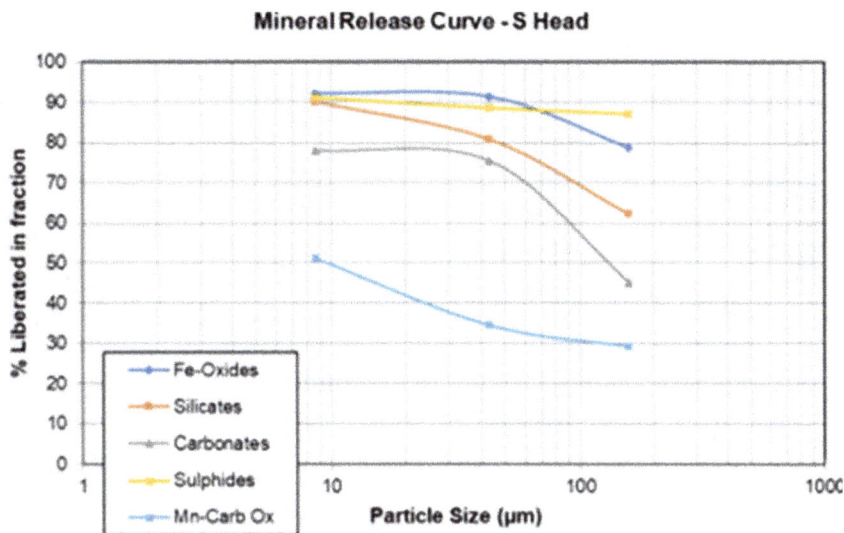

Mineral Release Curve - S Head

FIG 2 – Release curves for sulfur, based on composite sample QEMSCAN at three test sizes.

FLASH FLOTATION

Even though many ores require fine grinding to achieve the maximum recovery, most ores liberate a large percentage of the mineral values as they pass through the grinding unit. By placing a flash flotation cell between the grinding unit and classifier, it was found to be possible to quickly save the free mineral values from the coarse gangue at a low cost.

Given that the QEMSCAN on the composite sample studies established that over 80 per cent of the sulfides are liberated at a coarse size, (>150 microns), flash flotation provided a good option for processing the ore. The flash flotation technology has a low initial cost, low reagent consumption, prevents over grinding and increases the overall recovery.

Metso SkimAir is a flotation cell designed to be installed in a flash flotation role in the grinding circuit (Figure 3). The cell is typically installed on the hydrocyclone underflow stream to capture any sufficiently liberated, fast floating particles that may have reported here.

FIG 3 – Metso SkimAir flash flotation cell (Metso.com).

Due to the higher SG of the sulfide and magnetite there is a strong likelihood that they will report to the hydrocyclone underflow stream despite being in a suitable size range for flotation. By recovering these particles from the circulating load, it prevents them from reporting back to the mill and being overground. Excess grinding of these particles can result in slimes, which are difficult to recover in downstream flotation circuits. Recovering these particles from the circulating load not only boosts overall plant recovery but also has an impact on mill capacity. It can also lead to improved concentrate dewatering characteristics.

COLUMN FLOTATION

The first flotation machine design to use the column concept was developed by Town and Flynn in 1919 (Cheng and Liu, 2015, p 540), a column was used with an injection of air at the bottom through a porous plate. In subsequent decades, this type of device was substituted by the impeller-type flotation machines in mineral processing plants due to the absence of effective and reliable air spargers for fine bubble generation and the lack of automatic control systems on the early columns (Rubenstein, 2018). Column flotation devices were re-introduced for mineral processing by Boutin and Wheeler (1967) in Canada, when wash water addition to the froth was used to eliminate entrainment of hydrophilic materials to the float product. By the late-1980s column flotation had become a proven industrial technology in the mineral processing industry. The column cells were pioneered and developed in Canada.

The column is fed with ore pulp in its upper third portion and, at the lower section, air is injected at high velocities. This causes the pulp to flow down against a swarm of rising air bubbles. This counter-current flow promotes the suspension of particles in the pulp. In addition, the energetic injection of air provides for the generation of small bubbles and promotes the contact of these bubbles with the ore, leading to the collection of hydrophobic particles.

The loaded bubbles ascend and form a thick froth layer at the column top, which is favoured by the shape of the device, which has a smaller diameter than its height. Just over the column top, a system gently distributes water over the froth, which washes most of the entrained hydrophilic material back to the pulp. The froth thickness and this washing process enable a higher enrichment of ore in the froth, improving the quality of concentration and the recovery. The froth that is rich in hydrophobic material is discharged in the launders. In direct flotation, this froth corresponds to the concentrate. Hydrophilic particles flow down and leave the column through a barometric leg, which includes a level control system. This corresponds to the underflow and, in the case of direct flotation, to the tailings. Column cells (Figures 4) are considered to have two distinct zones; the collection zone and the froth zone. The deep froth zone coupled with mist wash water was found to have the ability to wash the froth, which is considered advantageous in removing only sulfides.

Column flotation was first used in Australia at Harbour Lights (Subramanian, Connelly and Wong, 1988). Flotation columns and mechanical flotation cells respond in a similar way to the chemical variables listed here, and many of the physical ones:

- air rate
- froth depth
- wash water and positive bias
- design of column cells
- air spargers
- level control system.

Feed characteristics will have a similar effect on both machines. They also respond to reagents in much the same way.

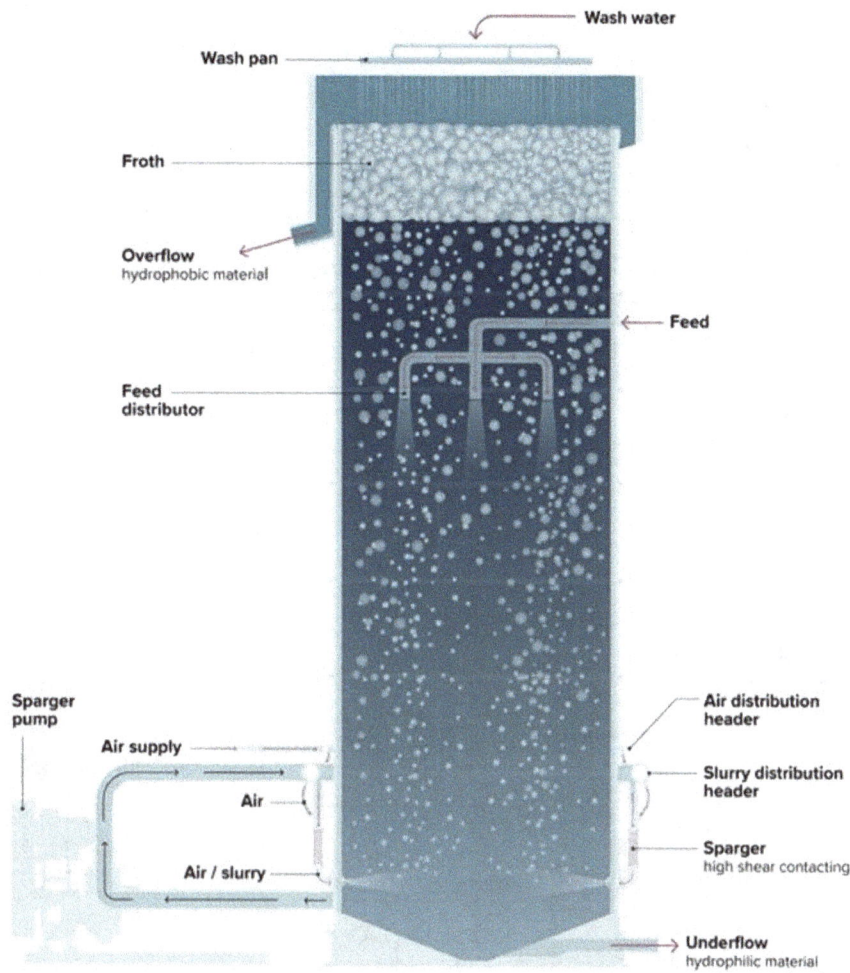

FIG 4 – Column flotation cell (source: Metso.com).

PROPOSED PROCESS FLOW SHEET

The proposed process flow sheet, designed as a result of this study, is shown in Figure 5. Some 10 Mtpa of ore is to be processed annually to produce 3.3 Mtpa of 68 per cent Fe concentrate with <0.05 per cent sulfur to qualify as a green magnetite feed. The process consists of a primary gyratory crusher with the crushed ore stockpiled. Thereafter, a two-stage cone crusher is used, close circuited by a screen and a final stage of high-pressure grinding rolls (HPGR). Thereafter, coarse dry cobbing is used to reject 20 to 30 per cent of the ore as a waste.

The ore is then wet ground in a conventional ball mill close circuited by cyclones and wet magnetic separation (LIMS) of the cyclone overflow to remove waste. The magnetic concentrate is then subjected to flash flotation to remove the sulfides, followed by secondary magnetic separation and regrind to P_{80} 30 microns in a tower mill using ceramic media. Thereafter, primary and secondary magnetic separation is used to recover clean magnetite followed by a further stage of flotation to further remove some sulfides. The column flotation cell is particularly suited to removing a small weight fraction from the concentrate and the ability to wash the froth. The tailings are pumped to a tailings dam. The concentrate is pumped via a slurry pipeline and filtered at the port before being dried and placed in the hold of a ship for transport overseas.

METALLURGICAL BALANCE

The overall metal balance is summarised in Table 2. The level of silica is just above the commonly accepted 2 per cent and this could be reduced by including reverse flotation to remove silica. It is possible that, in the laboratory, excessive oxidation of the pyrrhotite happened in the time lag between obtaining the final magnetic concentrate and start of the flotation tests. Piloting would be required to probe this flow sheet and optimise as necessary to produce a concentrate that meets the DR specifications with minimal GHG emissions.

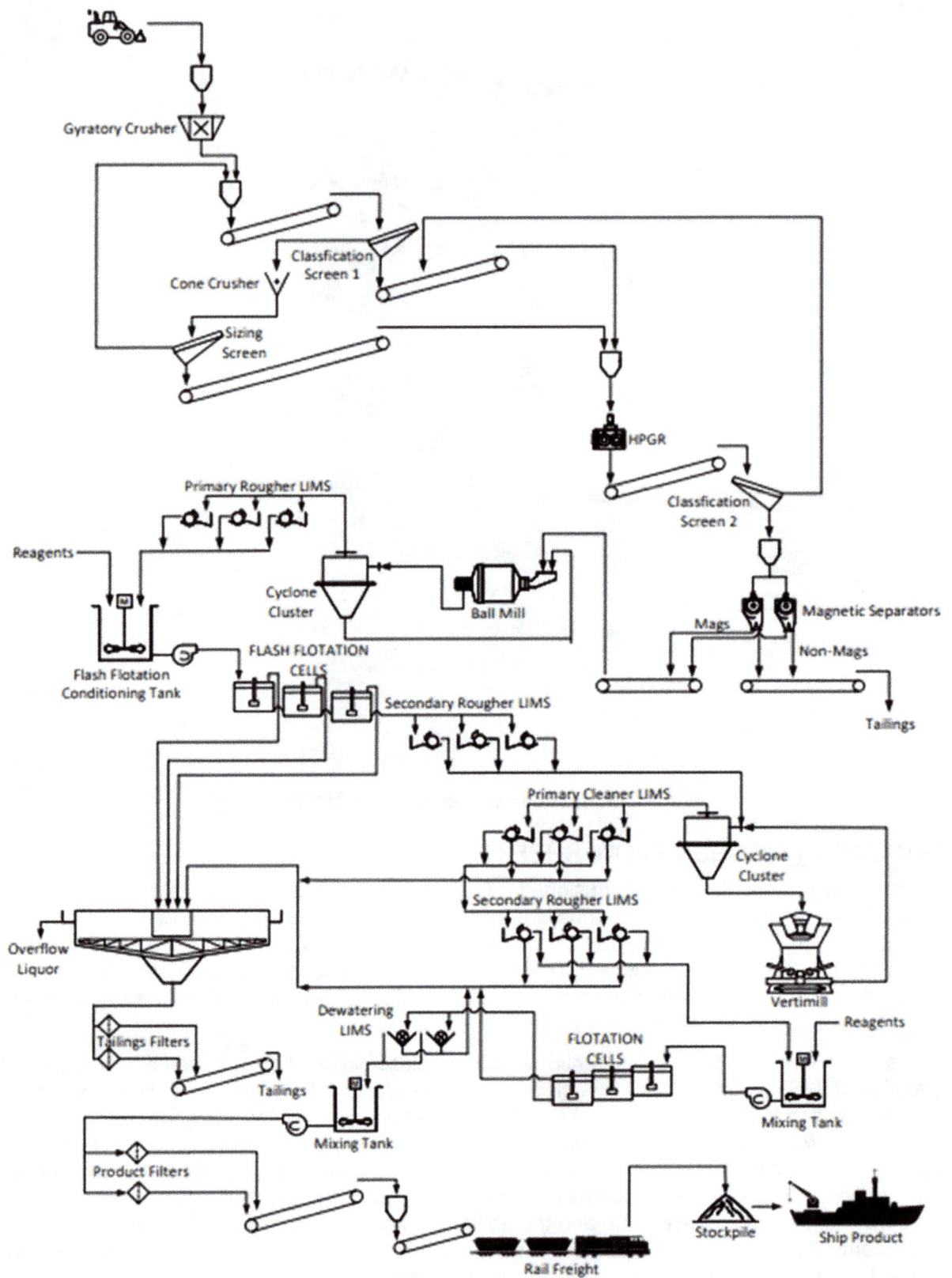

FIG 5 – Proposed process flow sheet.

TABLE 2

Overall metal balance.

Stream	Weight	Assays,%												
	%	SiO₂	Al₂O₃	Fe₂O₃	MgO	CaO	Na₂O	K₂O	TiO₂	P₂O₅	MnO	S	Fe¹	Fe(Mag)
Final Concentrate	45.18	2.15	0.28	96.70	1.08	0.69	< 0.01	0.01	0.02	0.02	1.56	0.08	67.63	n/a
Dewatering N-Mags	0.15	32.40	2.63	19.70	13.30	12.20	0.16	0.39	0.08	0.07	7.67	0.26	13.78	n/a
S Concentrate	2.43	6.80	0.44	83.40	3.45	2.54	< 0.01	0.02	0.02	0.02	2.13	1.25	58.33	57.26
Finisher N-Mags	9.40	31.90	2.11	11.60	15.70	16.00	0.10	0.18	0.05	0.02	6.03	0.32	8.11	n/a
Rougher N-Mags	24.34	30.10	4.23	10.70	11.10	20.40	0.36	0.46	0.22	0.05	3.60	0.51	7.48	1.33
Cobber N-Mags	18.50	42.40	9.27	13.70	6.32	14.00	0.94	1.15	0.39	0.07	2.32	0.57	9.58	2.60
Calc. Head	100.00	19.36	3.08	51.97	5.94	9.45	-	0.35	0.14	0.04	2.64	0.33	36.35	2.20
Direct Head	-	20.70	3.35	51.30	6.24	9.92	0.30	0.41	0.12	0.03	2.71	0.35	35.88	30.84
1 Fe grade calculated from the Fe₂O₃ WRA result														

The final concentrate did not achieve the specification of <0.05 per cent sulfur. This was believed to be due to oxidation in the laboratory due to the time taken to complete the test work. Future work will freeze the samples in between flotation stages.

MAGNETITE PROJECTS IN AUSTRALIA

The presence of sulfides in magnetite is an added complication for what are usually difficult projects. The significant infrastructure costs for magnetite projects such as power, water, accommodation and ports for greenfield projects can be a significant impediment for the development of magnetite projects. Existing projects like Savage River had existing hydro power, water and was also depreciated many years ago while Project Magnet in South Australia is an integrated project built off other infrastructure and has been successful, although margins are tight. Karara has struggled and survived because of Ansteel solid support but the project went over budget and has been problematic. Citic Pacific in the Pilbara has struggled and has been a very difficult project from a delivery and operation perspective (Dowson, Connelly and Yan, 2009).

FMG's Iron Bridge $3.5B magnetite project has gone over budget and was late, highlighting the difficulties faced by these projects. There are a number of other projects such as Ridley, Razorback, Iron Road and Southdown that are currently on hold.

Some projects in Australia contain sulfur levels that are problematic.

OTHER PROJECTS

The occurrence of pyrite and pyrrhotite in Swedish magnetite ores is common. The new Kaunis mine is undergoing an expansion partly to solve the problem with off spec magnetite concentrate due to high sulphur levels. Flotation to remove sulfides is being included in the upgraded circuit.

Northland Resources is developing several magnetite mineral resources in northern Europe. The Tapuli, Sahavaara and Pellivuoma mineral resources are in Sweden and the Hannukainen resource in Finland. Three of these resources (Sahavaara, Pellivuoma and Hannukainen) require flotation to remove more than 98 per cent by mass of the sulfur in the feedstock to produce a saleable magnetite concentrate with a sulfur level below 0.05 per cent (Australian Financial Review, 9th Dec 2022).

CONCLUSIONS

Magnetite projects in Australia have experienced a troubled history and could not compete with hematite projects because of the high capital cost and higher operating cost (Connelly, 2015). However, going forward, magnetite is projected to sell at a premium and traditional hematite could face diminishing markets. This will have a significant impact on the miners of the Pilbara ores as these ores are mainly lower-grade hematite, which cannot be fed directly to the technologies that are currently being developed to address greenhouse gas emissions.

The use of flash flotation and conventional flotation cells will allow the DR specification to be met for sulfur. The oxidation of pyrrhotite during test work is a trap to be mindful of. Piloting would be recommended to process fresh sample not subject to oxidation.

The challenges of meeting the DR specification with respect to silica is one issue but reducing the sulfur level to <0.05 per cent adds another level of complexity.

According to Wood Mackenzie, as much as US$1.4 trillion needs to be spent on new mines, renewable energy, new steel mills, new hydrogen generators, new EAFs to achieve 90 per cent of the worlds steel net zero carbon emissions by 2050. Green premiums for iron ores (magnetite) are already emerging in the market place. For this to happen green hydrogen needs to be produced for sub US$2/kg. Australia's FMG is leading the charge to hydrogen production in Australia.

ACKNOWLEDGEMENTS

The author would like to thank the AusIMM and METS Engineering Group for permission to publish this paper and all colleagues and engineers at various sites, METS staff and other consultants for their contribution and the management of METS for their permission and constructive criticism of various drafts of this presentation. Thanks also to various vendors for providing technical input.

REFERENCES

Arvidson, B, Klemetti, M, Knuutinen, T, Kuusisto, M, Man, Y and Hughes-Narborough, C, 2013. Flotation of pyrrhotite to produce magnetite concentrates with a sulfur level below 0.05% w/w, *Miner Eng*, 50:4–12.

Boutin, P and Wheeler, D A, 1967. Column Flotation Development using an 18 in Pilot Unit, *Canadian Mining Journal*, 88(3):94–10.

Buswell, A M and Nicol, M J, 2002. Some aspects of the electrochemistry of the flotation of pyrrhotite, *Journal of Applied Electrochemistry*, 32(12):1321–1329.

Cheng, G P and Liuc, J T, 2015. Development of Column Flotation Technology, *Journal of Chemical and Pharmaceutical Research*, vol 7. https://api.semanticscholar.org/CorpusID:137947633

Connelly, D, 2015. What is the future for magnetite projects in Australia and why have some of the new projects been so problematic?, in *Proceedings Iron Ore 2015*, pp 107–114 (The Australasian Institute of Mining and Metallurgy: Melbourne).

Dowson, N, Connelly, D and Yan, D, 2009. Trends in magnetite ore processing and testwork, in *Proceedings Iron Ore 2009*, pp 231–241 (The Australasian Institute of Mining and Metallurgy: Melbourne).

Hodgson, M and Agar, G E, 1989. Electrochemical investigations into the flotation chemistry of pentlandite and pyrrhotite: process water and xanthate interactions, *Canadian Metallurgical Quarterly: The Canadian Journal of Metallurgy and Materials Science*, 28(3):189–198. https://doi.org/10.1179/cmq.1989.28.3.189

Rubenstein, Y, 2018. *Column Flotation: Theory and Practice*, Instituto De Ingenieros De Minas Del Peru.

Subramanian, K, Connelly, D and Wong, K, 1988. Commercialisation of A Column Flotation Circuit for Gold Sulfide Ore, *Column Flotation 88*, SME Annual Meeting.

The role of steelmaking desulfurisation and EAF-smelter in hydrogen-based steelmaking route of low-grade iron ores

Q Fan[1]

1. QRF Consultant, Melbourne Vic 3000. Email: quanrongfan@gmail.com

ABSTRACT

The steelmaking desulfurisation plays an important role in coal-based steelmaking route of blast furnace and BOF converter within the flow sheet of BF-deS-BOF. With the advent of H_2-based ironmaking and steelmaking process, the transition flow sheet of DRI-Smelter-BOF has been assessed and developed by the steelmaking industry. What is the role of deS in H_2-based route of DRI-smelter-BOF? This paper will present the influence of sulfur-containing iron ore on steelmaking desulfurisation and the leading role of EAF-smelter played in H_2-based steelmaking route. The sulfur contents of iron ore vary considerably around the world and lower-grade iron ores with high sulfur are likely to be used for H_2-based route, a considerable amount of sulfur that comes from DRI will report to the melt of EAF-smelter and the operating conditions of EAF-smelter also impact the sulfur content of the hot melt for steelmaking deS. The EAF-smelter of H_2-based route is the focus of the steelmaking industry at moment, unfortunately, an industrial-scale EAF-smelter has been the missing link of the flow sheet of DRI-Smelter-deS-BOF, despite similar EAF-smelters being used for various applications of Fe-Ti and Fe-Ni processes. The latest progress is that several pilot-scale and industrial-scale EAF-smelters with reducing capability are planned for construction within next several years and low-grade Australia iron ores are certainly to be employed as basic feedstocks for H_2-based route of DRI-Smelter-deS-BOF.

INTRODUCTION

The current attentions and activities of steelmaking industry have been placed on the carbon-neutral transition and H_2-based steelmaking route, the technology of steelmaking process will be altered profoundly with carbon being replaced by hydrogen and renewable energy. However, the raw materials of iron ores cannot be substituted and remain the same feedstocks used for the existing BF-BOF route. The H_2-based routes face the ongoing issues of lower grade iron ores with higher impurities of phosphorus and sulfur. With limited amount of coal consumed in H_2-based flow sheet, the sulfur content of hot metal of EAF-smelter will be lower than that of coal-based blast furnace. Nevertheless, a certain amount of sulfur can be captured by the hot melt of H_2-based EAF-smelter when distribution ratio of sulfur decreases due to the higher oxygen potential of EAF-smelter and deS stations currently used for carbon-based route can be integrated into H_2-based flow sheet in the order of DRI-Smelter-deS-BOF.

With respect to H_2-based DRI-Smelter-deS-BOF, the industrial-scale EAF-smelter between DRI and deS is the missing link of the flow sheet despite comparable industrial-scale EAF- smelters being used for various applications of Fe-Ti and Fe-Ni smelting processes. There is no doubt that industrial-scale EAF-smelter, around one million tonnes per annum (1 Mtpa), will play the decisive role in H_2-based transition and the steelmaking industry in cooperation with iron ore major players is accelerating the implementation of several pilot and industrial-scale EAF-smelters for low-grade iron ores. Fundamental research and investigations are also required to understand and clarify the performance of EAF-smelters such as the reduction mechanism of FeO. With pilot and industrial-scale EAF-smelters having advanced into the engineering design stage and intensified activities in research and development, the positive outcome is anticipated for the industrial-scale applications of EAF-smelters in H_2-based steelmaking transition.

The role of deS in H_2-based steelmaking route

The H_2-based route of DRI-Smelter-deS-BOF wishes for the least amount of coal to be used and total sulfur entering the H_2-based route will be reduced significantly. The sulfur captured by the hot melt of H_2-based EAF-smelter mainly depends on the sulfur content of DRI plus some sulfur from the required amount of coal and lime fluxes used in the EAF-smelter. Globally, the sulfur contents of iron ores diverge considerably from 0.001 wt per cent to 0.1 wt per cent and sulfur content of

Australia iron ores is in the range of 0.01–0.04 wt per cent with an average content of 0.025 per cent (Park *et al*, 2022). With an assumption of DRI with sulfur content of 0.025 per cent being charged into EAF-smelter and 10 per cent of sulfur is reported to the hot melt and 90 per cent to the slag phase. The sulfur content of the molten iron is estimated to be around 50 ppm. With this range of sulfur in the hot melt, the hot melt is not necessary for deS refining and can be delivered directly to BOF station for oxygen blowing.

The scenario would be changed when 20 per cent of DRI sulfur remains in the hot melt of EAF-smelter in addition to the sulfur coming from the coal-char and lime fluxes used in the EAF-smelter. The sulfur content of hot metal may approach 100 ppm or higher, therefore, deS refining is necessary for H_2-based flow sheet to be extended into DRI-Smelter-deS-BOF and existing deS station of coal-based route can be assimilated into H_2-based steelmaking route.

It is well known that around 10 per cent of sulfur in coal-based blast furnace reports to hot melt due to lower oxygen potential of 1–3 per cent of FeO in slag phase. The oxygen potential of H_2-based EAF-smelter is expected to be higher with 3–10 per cent of FeO in slag phase. More than 10 per cent of sulfur of DRI charged into EAF-smelter could be arrested by the hot melt, deS refining of molten iron from EAF-smelter may be required with changed characteristics as compared to that of coal-based route as displayed in Table 1.

TABLE 1

DeS of hot melt for H_2-based and coal-based route.

	Coal-based	H_2-based
Melt temperature	lower	high – prefer CaO reagent
Melt silicon, carbon	high	low – higher oxygen potential
Melt sulfur	high	low – less reagents used
Blowing time	long	short – less temperature drop of the melt
Slag amount	more	less – less slag to hold melt droplets
Slag viscosity	high	low – less retaining of melt droplets
Mg reagent efficiency	high	low – escaping of Mg from the bath

The temperature of hot metal from H_2-based EAF-smelter could be around 1450–1550°C, higher than the temperature of hot melt from the coal-based blast furnace. The higher temperatures of the melt prefer CaO reagent and co-injected magnesium could be compromised in efficiency due to the quick vapourisation and escapement from the molten bath. The blowing time and reagents will be adjusted for the changed characteristic of deS of H_2-based route. Continuous improvement of deS refining has been the theme of steelmaking industry for decades and that endeavour will continue to move into the H_2-based steelmaking route.

The latest development of steelmaking deS refining has been the invention of Dynamic Free Lance, abbreviated as DFL. The DFL injection lance was originally proposed as a top-inserted injection lance for direct iron smelting reduction (Fan, 2008) and not anticipated for steelmaking deS refining. Nevertheless, the developments either for iron smelting reduction or deS refining follow the same procedure of a small-scale modelling, fundamental research and industrial trial tests. DFL injection lance inherits the advantage of easy handling and operation of a top-inserted lance and possesses the intrinsic capability of mixing the entire bath efficiently. The journey started with theoretical analysis and physical modelling, throughout rigorous assessments and safety checks by the steelmaking industry, DFL injection lance has been successfully conducted for deS blowing of co-injection of CaO-Mg in a 70 t ladle of molten iron.

The breakthrough technology of DFL was originated with a physical phenomenon observed during experimentations in 2005. It was surprising to find that the lance could move naturally in connection with a flexible joint without a mechanical driving force as shown in Figure 1. The lance was named 'Dynamic Free Lance' with two implications: (1) liberate the stationary lance and change the stationary lance into a free-movement lance; (2) random movement of DFL under the opposite reaction force from the blowing injection jet. The first impression of DFL has been the beautiful movement of the injection lance with improved blowing dynamics as the high frequency vibration of the stationary lance transformed into lower frequency movement of DFL by energy conservation.

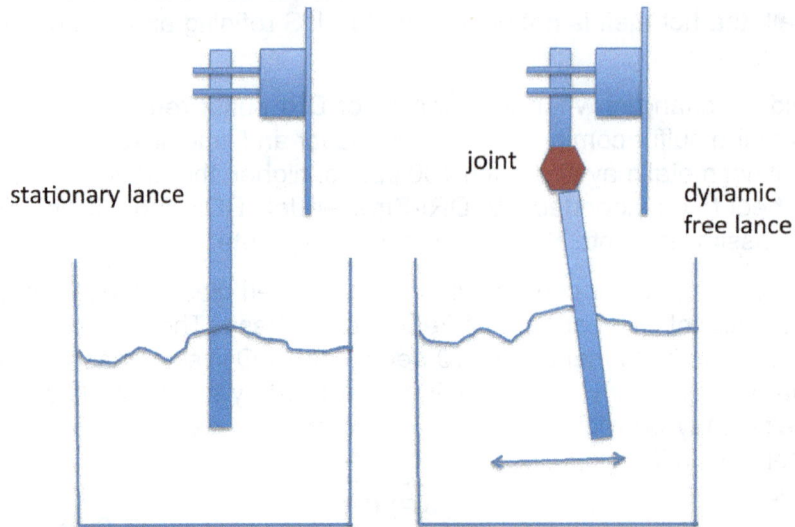

FIG 1 – Comparison of dynamic free lance with stationary lance injection.

Figure 2 displays the water modelling of submerged injection of DFL, the lance moved towards the right side at that moment and the injection jet under the lance was pushed towards the left side of the lance with bubbles floating into the wake of the lance. Figure 3 shows the lance moving towards the right side nearly approaching the maximum angled position. If the lance had been placed stationary in that angled position, the jet trajectory should be located on the right side of the lance, however the jet plume shifts to the left side of DFL under the drag force from the liquid bath. The drag force overpowers the buoyancy force of the bubbles and causes forceful mixing of the injection gas with the liquid phase and the mixing intensity of the injection jet with the liquid phase may be expressed as:

$$R = M_L U_L / (M_g U_g + M_a U_a) \tag{1}$$

Where M_L is the mass of the liquid phase impacted on the blowing jet with impact velocity of U_L, which is equal to the lance movement speed. M_g and U_g is the mass and velocity of the injection jet and M_a and U_a the mass and velocity of the reagent carried by the injection gas.

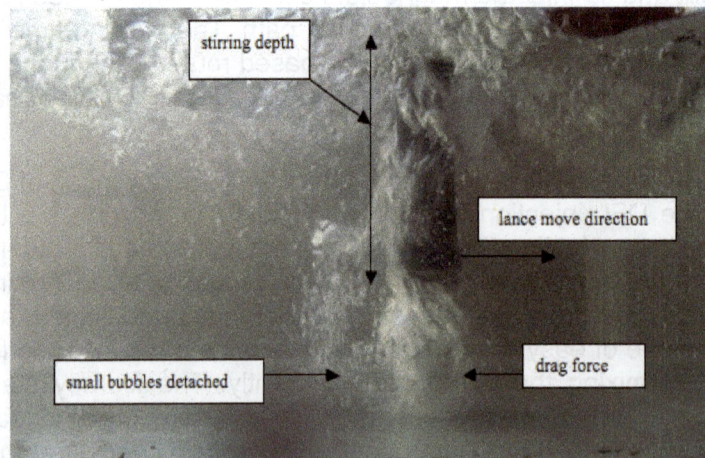

FIG 2 – Bubble trajectory of DFL injection.

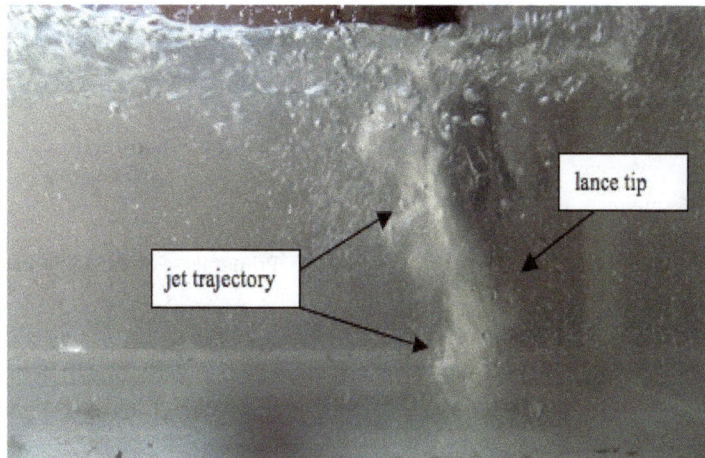

FIG 3 – Bubble trajectory of angled DFL.

The DFL injection lance of 700 kg in weight has been employed for steelmaking desulfurisation of 70 t molten iron for co-injection of Mg-CaO reagents. It was tested to pass the feasibility study and rigorous safety checks without drawbacks identified. The DFL was submerged about 2.0 m into the molten bath for deS refining of 5–9 minutes. Figure 4 displays DFL injection lance in different positions in molten bath during the injection blowing. The dynamic injection of DFL improves the entire bath mixing with smooth blowing and reduced splashing accordingly. It was estimated that the splash intensity of DFL was reduced by 30–40 kg/min in comparison to the stationary lance injection under the same injection conditions.

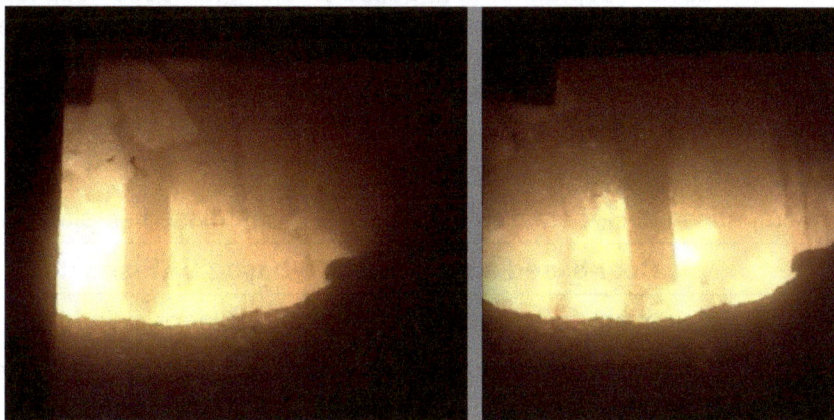

FIG 4 – Images of DFL for steelmaking desulfurisation injection.

The role of EAF-smelter in H₂-based route

The transition into H_2-based route leads to the extensive investigations and development to respond to the missing link of industrial-scale EAF-smelter of DRI-Smelter-deS-BOF. Several pilot and industrial-scale EAF-smelters have been queued for engineering design and constructions, the latest statement of Rio Tinto EAF-smelter with Baosteel, the EAF-smelter of Hy4Smelt by Primetals Technologies in cooperation with Fortescue and demonstration of EAF-smelter by Hatch-BHP in Australia. Three major players of Australia iron ores are actively involved in the development of three different EAF-smelters respectively. It is expected that EAF-smelters with different engineering design for lower grade iron ores will be developed and installed globally within next several years and these industrial activities and plant commissions of H_2-based EAF-smelters repeat the scenario of coal-based direct smelting furnaces pursued 40 years ago as displayed in Figure 5.

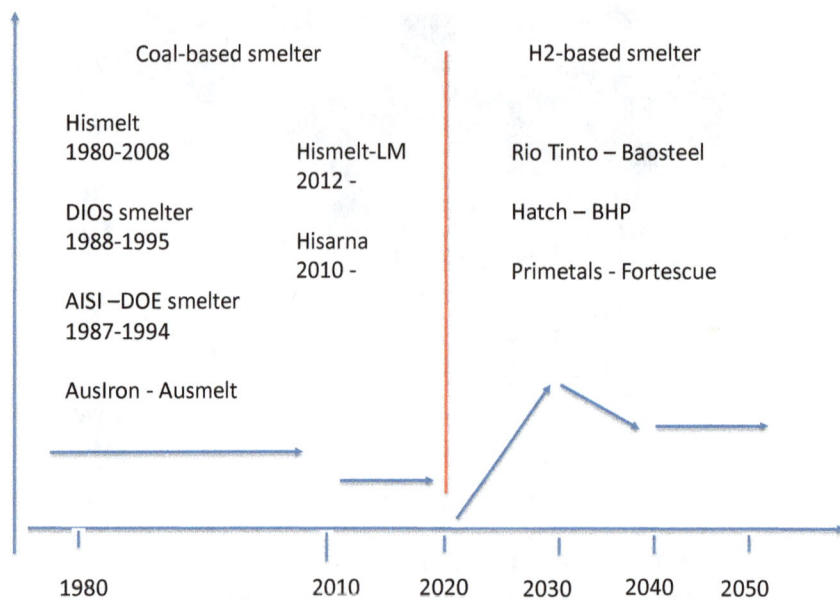

FIG 5 – Development of coal-based smelter and H_2-based EAF-smelter.

The coal-based smelters of both pilot and industrial-scale have been broadly investigated and developed for 30 years with intention to replace the blast furnace and up-to-date industrial activities are declined towards Hismelt-LM and Hisarna smelter. The coal-based smelters were designed with a post combustion of CO to supply the heat energy for FeO reduction and the heat energy supplied from the electrical power was not adopted for coal-based smelting furnaces despite well-known EAF furnaces to be used by the industry. With the renewable energy transition, the electric electrodes can be used to supply the heat energy in place of $CO-O_2$ post combustion in EAF-smelter and the energy shift consequently changes the features and design of EAF-smelters compared to the coal-based smelter as presented in Table 2.

TABLE 2

Comparison of H_2-based EAF-smelter and coal-based smelter.

	Coal-based smelter	**H_2-based EAF-smelter**
Shape	mainly cylindrical	rectangular or cylindrical
Heat energy	post-combustion of CO	electric arc from green energy
Bath size	small	large
Hot spot	flame above the bath	electrical arc in the slag phase
Off-gas volume	high	less
Furnace height	high	short

The H_2-based EAF-smelter may need a larger smelting bath than the coal-based one for equivalent smelting capacity. The electrodes of about 1.0 m in diameter occupy a certain bath area in a six-in-line rectangular EAF-smelter, also considering the distance between the electrodes and electrode to the bath wall. H_2-based EAF-smelter might enlarge the smelting bath above 50 m^2 to approach the productivity range of 500 000 t/a that was accomplished by the coal-based Hismelt smelting furnace with a bath diameter of 30 m^2.

The heat energy transportations in the coal-based smelter and H_2-based EAF-smelter are also carried out in different modes. The post combustion flame of the coal-based smelter is located above the molten bath and heat transfer from the combustion flame to the molten bath is conducted by radiation and convection with the splashed droplets also carrying the heat energy back to the bath. The heat transfer efficiency is constrained due to the gap between the combustion flame and the bulk bath. With a H_2-based EAF-smelter, the electrical electrodes and arcs are submerged into the slag layer or in contact with the slag foam. The slag phase near the electrodes is in direct contact with the hot arcs and the slag streams can carry the heat energy throughout the entire bath. The heat energy generated by the electric arcs and energy consumption by FeO reduction take place in the same slag phase for efficient energy utilisation.

It is well-known that the heat absorbing reduction of FeO in slag phase can be expressed by the following two equations:

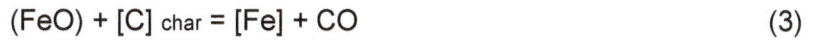

$$(FeO) + [C]_{\text{melt droplet}} = [Fe] + CO \tag{2}$$

$$(FeO) + [C]_{\text{char}} = [Fe] + CO \tag{3}$$

Where the carbon consumptions from carbon of the melt droplets and coal char are indispensable for the final reduction of FeO inside H_2-based EAF-smelter, also considering the following FeO reduction with the carbon of melt bath via the slag-melt interfacial area calculated by the bath diameter:

$$(FeO) + [C]_{\text{melt bath}} = [Fe] + CO \tag{4}$$

It has been acclaimed from the investigations of coal-based smelters that FeO reduction in the smelting furnace mainly occurs in the slag phase via Equations 2 and 3; and reduction (Equation 4) contributes marginally for the overall reduction of FeO. The latest investigations suggest that the reduction mechanism of FeO could be more complicated than the suggested slag phase as the dominant reduction site. The slag-melt interface across the entire bath could be highly unstable due to FeO reduction and escaping of CO bubbles as shown in Figure 6. The special measurements conducted by Ruuska et al (2006) suggested that the slag droplets may be emulsified considerably into the melt bath in the steelmaking BOF converter, based on a radio-wave interferometry and slight change in the receiving signals when waves passing through the interface of two layers of different characteristics.

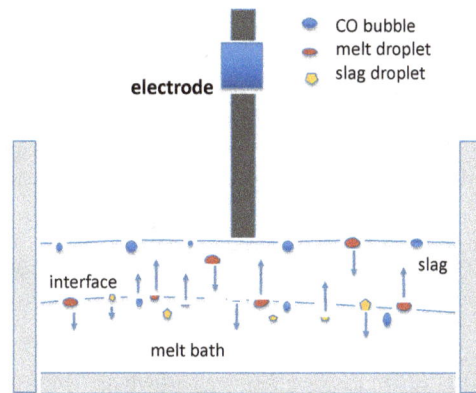

FIG 6 – Two-site dynamic model of FeO reduction in H_2-based EAF-smelter with slag droplets emulsified into melt emulsion.

The radio-wave measurements implied that the melt emulsion with slag droplet could be formed in BOF converter and that phenomena could happen again near the melt-slag interface in H_2-based EAF-smelter, where two reduction sites may be dynamically formed with the melt emulsion of slag droplets located under the slag emulsion as shown in Figure 6. Across the melt-slag interface of the two reduction sites, CO bubbles carry the melt droplets from the melt emulsion into the slag phase. At the same time, the melt droplets descend from the slag phase to the melt emulsion and entrain some slag droplets into the melt emulsion. The rise of CO bubbles and up-down transportations of melt and slag droplets cause strong perturbation near the slag-melt interface and the slag droplets could be emulsified considerably into the melt bath, that entail the following reduction of FeO carried by the emulsified slag droplets with carbon of the melt bath.

$$(FeO)_{\text{slag droplet}} + [C]_{\text{melt bath}} = [Fe] + CO \tag{5}$$

The emulsified slag droplets are supposed to be small in micro or millimetre diameter and generate large specific surface area for FeO reduction with carbon of the melt bath. It is not a surprise if the interfacial area generated by the emulsified slag droplets is larger in one or two magnitudes than slag-melt interface calculated by bath diameter. The FeO reduction of Equation 4 might be incorporated into the FeO reduction of Equation 5 with an acceptable deviation and total reduction rate of FeO for ironmaking inside EAF-smelter may be expressed as:

$$R_{total} = R_{melt\ droplet} + R_{char} + R_{slag\ droplet} \qquad (6)$$

Where the overall FeO reduction is composed of Equations 2 and 3 in the slag emulsion in addition to Equation 5 of slag droplets emulsified in the melt emulsion. In general, the reduction mechanism of FeO will never change from a coal-based smelter to H_2-based EAF-smelter. FeO reduction for ironmaking is accomplished on the interfacial area either of the slag droplets emulsified in the melt bath or the melt droplets and char particles emulsified inside the slag phase. Considerable research and investigations have been conducted to understand FeO reduction via the melt droplets and char particles in the slag phase, whereas, understanding of the emulsified slag droplets in the melt bath is limited and literature is scarce or nothing there on FeO reduction of emulsified slag droplets in melt emulsion. Preliminary analysis and investigations suggest that the slag droplets emulsified in the melt bath should not be overlooked for FeO reduction and the emulsified slag droplets could contribute to more than 50 per cent of total reduction of FeO in EAF-smelter:

$$R_{slag\ droplet} > R_{melt\ droplet} + R_{char} \qquad (7)$$

The melt emulsion with the slag droplets could be the dominant reduction site, overtaking the site of slag emulsion for FeO reduction in two-site reduction model of H_2-based EAF-smelter. More research and explorations are required in the future when H_2-based pilot and industrial EAF-smelters are ready to provide more information to clarify the reduction mechanism and the leading reduction site of H_2-based EAF-smelter for low-grade iron ores.

CONCLUSION

- The lower-grade iron ores facing H_2-based steelmaking route may lead to the flow sheet of DRI-Smelter-deS-BOF when sulfur content of the melt from EAF-smelter approaches 100 ppm or higher.

- The improvement of deS refining has been the long-term theme of the industry and will continue to improve with new technology such as DFL injection lance for steelmaking deS with reduced splashing of 30–40 kg/min as compared to the stationary lance.

- H_2-based EAF-smelter for low-grade iron ores will be developed globally for the coming decades, the repeating scenarios of coal-based smelters chased 40 years ago to replace the ironmaking blast furnace.

- Without the post combustion of CO, the characteristics of EAF-smelters is changed with heat energy generated inside the slag phase by the electric arcs and consumed locally by FeO reduction.

- The mechanism of FeO reduction never changes from coal-based smelter to H_2-based EAF-smelter, via the interface of slag droplets emulsified in the melt bath and melt droplets and char particles distributed in the slag phase.

- Two-site Dynamic Model of FeO reduction is proposed for H_2-based EAF-smelter with the site of melt emulsion under the site of slag emulsion. The site of melt emulsion could contribute more than 50 per cent of total reduction of FeO.

- Several pilot and industrial-scale EAF-smelters for low-grade iron ores have advanced into engineering design stage and demonstration of full-scale EAF-smelter is anticipated to play the decisive role in H_2-based steelmaking transition.

REFERENCE

Fan, Q, 2008. Dynamic lance furnace for direct iron smelting reduction. in 2008 AISTech Conference Proceedings.

Park, J, Kim, E, Suh, I-K and Lee, J, 2022. A short review of the effect of iron ore selection on mineral phases of iron ore sinter, *Minerals*, 12:35.

Ruuska, J, Ollila, S, Bååth, L and Leiviskä, K, 2006. Possibilities to use new measurements to control LD-KG-converter, in *Proceedings of the 5th European Oxygen Steelmaking Conference*, pp 210–217.

Phosphorus association with goethite – effects of Fe(II)-catalysed recrystallisation

N Karimian[1,2], M I Pownceby[3] and A Frierdich[4]

1. CSIRO, Mineral Resources, Clayton South Vic 3169. Email: niloofar.karimian@csiro.au
2. School of Earth, Atmosphere & Environment, Monash University, Clayton Vic 3800.
3. CSIRO, Mineral Resources, Clayton South Vic 3169. Email: mark.pownceby@csiro.au
4. School of Earth, Atmosphere & Environment, Monash University, Clayton Vic 3800. Email: andrew.frierdich@monash.edu

ABSTRACT

Naturally occurring goethite is typically associated with impurities such as SiO_2, Al_2O_3 and P_2O_5. These may be found with goethite through mechanisms including direct substitution or coprecipitation of nanocrystalline phases. Understanding P incorporation in goethite has significant implications for Australia's iron ore industry as it lowers their commodity value. The current study tests a novel low-temperature technique to extract P impurities from synthetic goethite phases under O_2-free ambient conditions by utilising Fe(II)-catalysed recrystallisation. The objective of this study was to examine the effect of the Fe(II)-catalysed recrystallisation of goethite on the geochemical behaviour (mobility and speciation) of co-associated P. To achieve this, we tracked changes in labile P (aqueous and 1M NaOH extractable) concentrations over time in goethite suspensions which were reacted with $Fe(II)_{aq}$ for 14 days under circumneutral pH conditions and room temperature conditions. Temporal changes in Fe mineralogy were investigated via synchrotron-based X-ray absorption spectroscopy. Iron K-edge extended X-ray absorption fine structure (EXAFS) spectroscopy confirmed goethite as the only mineral phase in P-free treatments while goethite (~86 per cent) and feroxyhyte (~14 per cent) were the main phases in 1.0 per cent P-containing treatments after a 14-day reaction with $Fe(II)_{aq}$. According to our findings, the treatment of P-goethite with Fe(II) for a duration of 7–14 days, followed by a 24-hour extraction of the reacted solids using 1M NaOH, results in the removal of approximately 79.5 per cent of the initial P content. This removal encompasses both the aqueous fraction (~4.5 per cent) and the surface adsorbed fraction (~75 per cent) from the 1 per cent P-bearing mineral phase. Notably, this approach demonstrates an improvement in P extractability, with approximately a 10 per cent increase compared to the 1 per cent P-goethite sample that was not reacted with Fe(II).

INTRODUCTION

Goethite is one of the most prevalent iron(III) oxides in soils and sediments and is generally found in association with several types of mineral deposits including iron ore, nickel laterites, manganese nodules, and bauxite ores. Naturally occurring goethite is typically associated with impurities, such as Al_2O_3, P_2O_5, and SiO_2. These may be found with goethite through mechanisms including direct substitution (Cornell and Schwertmann, 2003), adsorption (Hsu et al, 2020), or coprecipitation of nanocrystalline phases (eg apatite, $Ca_5[PO_4]_3(OH)$; Rasmussen et al, 2021). Understanding the nature of P incorporation in goethite has significant implications for Australia's iron ore industry. Iron ore producers are increasingly reliant on goethite-rich materials that often contain variable quantities of critical impurities P, Al, and Si which can reach a collective level of 5 per cent (Pownceby et al, 2019). These lower-grade ores command a lower market price as the contaminants significantly increase the energy intensity of the ironmaking process through the increased volumes of slag generated in ironmaking and additional production costs associated with dephosphorization at the hot metal stage (American Public Health Association (APHA), 1998).

The redox-induced phase transformation and recrystallisation of Fe(III) oxide/hydroxides represents a fundamental geochemical process of great importance. This process has been proven to expedite the conversion of thermodynamically less stable Fe(III) oxides, such as ferrihydrite, into more stable phases like goethite (Burton, Hockmann and Karimian, 2020; Frierdich and Catalano, 2012; Frierdich et al, 2019a). It operates through electron transfer and atom exchange mechanisms between dissolved Fe(II) and structural Fe(III). The catalysed mineralogical transformation of Fe(III) oxides can have profound implications, leading to significant alterations in the number of vacant sites available for surface complexation and the reactive surface area of the resulting mineral phase (Burton, Hockmann and Karimian, 2020).

Contrary to metastable Fe(III) oxides like ferrihydrite, the reaction of Fe(II) with goethite does not typically result in significant mineralogical transformations. However, it is now well-established that aqueous Fe(II) can act as a catalyst for the recrystallisation of goethite (Burton, Hockmann and Karimian, 2020; Frierdich and Catalano, 2012; Frierdich et al, 2019a, 2019b). The exchange between structural Fe atoms in goethite and the aqueous Fe(II) phase facilitates the uptake or release of co-associated species. This phenomenon shows promise for the leaching of nickel from goethite-bearing lateritic ores (Frierdich et al, 2019a) but has not been explored in previous studies concerning phosphorus. Therefore, utilising this technique may hold potential for the removal of phosphorus from goethite-rich sources, offering a novel approach that has not been previously investigated.

The objective of this study was to examine the effect of the Fe(II)-catalysed recrystallisation of goethite on the geochemical behaviour (mobility and partitioning) of co-associated P and the mineralogy of goethite. To achieve this objective, we tracked changes in Fe mineralogy and P aqueous phase and surface adsorbed concentration over time in goethite suspensions which were reacted with $Fe(II)_{aq}$ for 14 days under circumneutral pH conditions and ambient temperature. The extent of goethite recrystallisation was monitored using isotopically labelled aqueous Fe(II) to quantify the Fe atom exchange into and out of the goethite structure. Dynamics in P aqueous concentrations during Fe(II)-catalysed goethite recrystallisation were examined by monitoring aqueous and 1M NaOH extractable P concentrations, while the corresponding changes in Fe mineralogy were investigated via synchrotron-based X-ray absorption spectroscopy (XAS). This research on mechanistic studies will provide valuable insights into impurities substitution in oxyhydroxide phases, enhancing our understanding of the recrystallisation process. These findings will also contribute to the development of strategies for producing high-quality, environmentally friendly iron ores.

METHODS

Two series of goethites were synthesized from a green rust precursor using a modification of the procedure described by Cornell and Schwertmann (2003) and Fey and Dixon (1981) which involves oxidative hydrolysis of mixed $P/FeCl_2$ -bicarbonate solutions buffered at pH 6.5–8.2: one pure goethite (used as a control); and one P-goethite (1.0 mol per cent P). The initial synthetic phases were analysed for composition and mineralogy using bulk X-ray fluorescence spectroscopy (XRF), Inductively Coupled Plasma Mass spectroscopy (ICP-MS) and X-ray powder diffraction (XRD). For preliminary qualitative phase identification, random powder diffraction patterns were collected using a Philips X-Pert diffractometer using: $CoK\alpha$ radiation.

X-ray absorption spectroscopy (XAS) was also performed on these samples to confirm the mineralogy and speciation of Fe, at the XAS beamline, Australian Synchrotron. X-ray absorption spectra at the Fe K-edge were collected in transmission mode at room-temperature to K = 12 Å^{-1}. For standard background subtraction and edge-height normalisation, we utilised the ATHENA program (Ravel and Newville, 2005) and linear combination analysis (LCF). We used the acquired Fe K-edge EXAFS spectra to quantify the abundance of distinct Fe phases in the synthetic, Fe(II)-reacted and non-reacted goethite samples.

The determination of the aqueous-phase Fe isotope composition was carried out using ICP-MS with a Thermo iCAP-Q ICPMS instrument. For isotope analysis, samples were diluted to approximately 100 µg L^{-1} Fe and measured in the reaction cell mode with the use of He gas to eliminate polyatomic interferences. To calculate the iron isotope mole fractions (f), the counts per second (cps) of the Fe isotope (n) were divided by the sum of the total Fe isotope cps.

$$f^n\text{Fe} = \frac{ncps}{54cps + 56\,cps + 57cps + 58cps} \tag{1}$$

The f^{57}Fe values of the aqueous Fe(II) were utilised to estimate the proportion of Fe atoms in goethite that undergo exchange with the aqueous Fe(II). This estimation was performed based on the following equation where N(solid)$_{total,}$ and N(solid)$_{exchangeable}$ are the total moles of Fe in the goethite, and the moles of Fe in the solid phase that have exchanged with the aqueous phase, respectively (Handler *et al*, 2014):

$$Fe\ exchange\ (\%) = \frac{N\ solid\ ^{exchange}}{N\ solid\ ^{total}} \times 100$$

$$= \frac{NFe(II)}{N\ solid\ ^{total}} \times \frac{f^0 Fe(II) - f^t Fe(II)}{f^t Fe(II) - f^0 solid} \times 100 \qquad (2)$$

Recrystallisation experiments

Experiments were based on reacting P-free and P-bearing goethite samples with Fe(II)$_{aq}$. Reactions were conducted inside an anoxic chamber (~2–3 per cent H$_2$, balance N$_2$) with an O$_2$ content maintained at <1 ppm. All labware, reagents, and mineral powders were equilibrated with the chamber atmosphere for >48 h before use. MQ water and all the solutions will be sparged with N$_2$ prior to entry into the chamber and were sparged within the chamber atmosphere for an additional 30 min once inside.

Reaction batches were prepared by adding 20 mg of goethite mineral powder to 15 mL polypropylene tubes. After transferring into the anoxic chamber, the mineral was suspended into 8.9 mL of water, 1 mL of 100 mM MOPS-Na salt (3-(N-Morpholino propane sulfonic acid sodium salt) (adjusted to ~ pH 7.60 with HCl or NaOH), and 0.048 mL of 100 mM NaOH. The reaction was then initiated by adding 0.05 mL of 100 mM ^{57}Fe-enriched Fe(II)$_{aq}$ (prepared by dissolution of ^{57}Fe0, Isoflex, in 1 M HCl) and was continued for 14 days. Samples were collected following 1h, 1 day, 2 days, 7 days, and 14 days reaction by removing the entire suspension with a syringe and immediately filtering (0.2 µm Polyethersulfone (PES)) the aliquot to remove the iron oxide particles and stopping the reaction. The filtrate was then acidified (HCl, trace metal grade) inside the anaerobic chamber. The surface adsorbed fraction of P was measured following a 24 h extraction of the Fe(II)-reacted and non-reacted 1 per cent P-goethite phases with 1 M NaOH. The Fe and P concentrations in the filtrate were measured by ICP-MS (Agilent 8900 ICP-MS).

RESULTS AND DISCUSSIONS

Changes in mineralogy

X-ray diffraction analysis (Figure 1) confirmed that the initial P-free mineral phase was pure goethite. The XRD data also indicated that the synthetic mineral phase with 1 per cent P was indeed goethite but with the presence of one minor peak attributable to lepidocrocite phase (Figure 1). The composition analysis of the samples revealed that the P-free goethite contained ~56 per cent Fe and the 1 per cent P-goethite phase consisted of ~55 per cent Fe and ~0.32 per cent P.

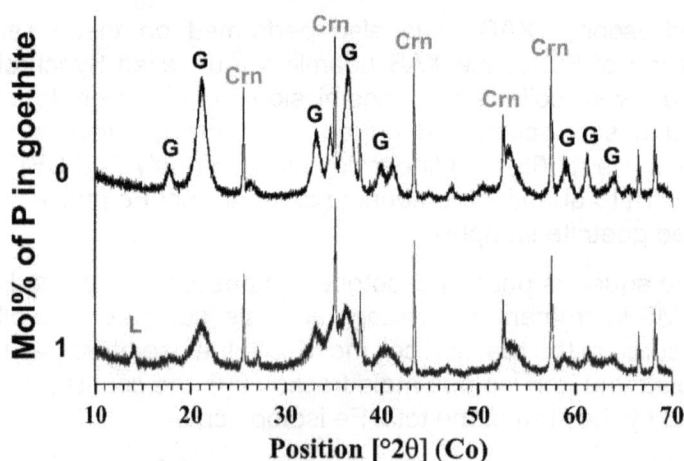

FIG 1 – X-ray diffraction pattern for initial synthetic P-free – and 1 per cent P-goethite. Crn refers to corundum used as an internal standard.

The quantitative results from Fe K-edge EXAFS spectroscopy (Figure 2 and Table 1) showed no significant changes in mineralogy over 14 days in the P-free/Fe(II)-reacted treatment. However, according to the LCF analysis of the Fe K-edge for the 1 per cent P-goethite sample collected after

1 hr following the reaction with $Fe(II)_{aq}$ the sample contained ~65.2 per cent goethite, ~28.5 per cent ferrihydrite, and ~6.3 per cent feroxyhyte. This material then completely transformed to goethite (86 per cent) and feroxyhyte (14 per cent) by the end of the 14 days reaction period.

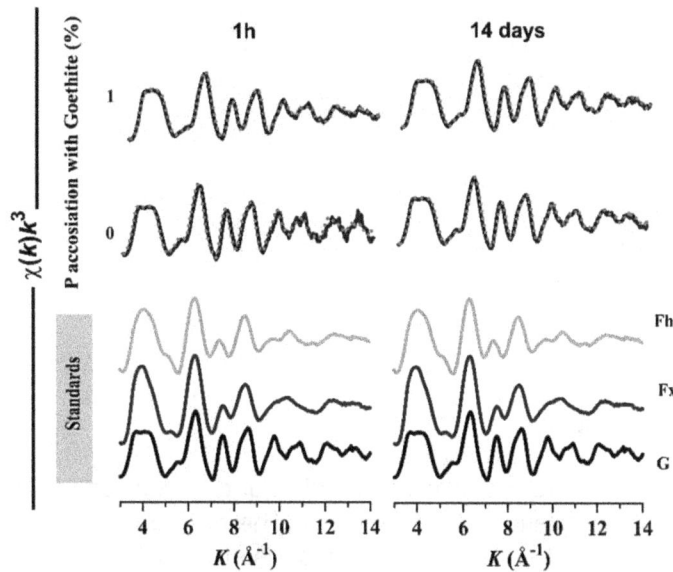

FIG 2 – k^3-weighted Fe K-edge EXAFS spectra of reference compounds and reacted samples after 1 hr and after 14 days total reaction time.

TABLE 1

Fe Speciation data expressed as a proportion of total Fe and based on LCF fits of Fe K-edge EXAFS spectra.

Sample	Goethite%	Ferrihydrite%	Feroxyhyte%	R^2
G + Fe(II) 1h	100	0	0	0.006
1% P-G + Fe(II) 1h	65.2	28.5	6.3	0.006
G + Fe(II) 14d	100	0	0	0.001
1% P-G + Fe(II) 14d	86	0	14	0.001

The presence of poorly crystalline phases such as ferrihydrite and feroxyhyte in the initial 1 per cent P-goethite mineral is challenging to confirm based solely on the XRD data. However, it is evident from the Fe XAS LCF results that the incorporation of P impacts the Fe mineralogy of the solid phase, potentially influencing the rates and extents of goethite formation from a green rust precursor, as outlined in our mineral synthesis procedure. The observed transformation of less thermodynamically stable ferrihydrite to goethite induced by Fe(II) in this study aligns with the findings of previous experiments (Boland *et al*, 2014; Bolanz *et al*, 2013; Hansel, Benner and Fendorf, 2005).

Phosphorous dynamics during the Fe(II)-catalysed recrystallisation of goethite

Figure 3 depicts the changes in P (both aqueous and 1M NaOH-extractable) and $Fe(II)_{aq}$ concentrations over a period of 14 days. Figure 3b shows that the concentration of aqueous Fe(II) decreases rapidly within the initial 1–24 hours and remains relatively constant throughout the rest of the experiment duration. The aqueous-phase Fe isotope analysis showed that the $f^{57}Fe$ values of aqueous Fe(II) decreased from an initial value of ~0.954 to final values of ~0.0474, and 0.0606 for 1 per cent P-goethite and P-free goethite phase by day 14 in the treatments, respectively. These trends reflect the exchange of Fe atoms between aqueous Fe(II) enriched in ^{57}Fe and Fe(III) within goethite. Plugging the fraction of ^{57}Fe values of $Fe(II)_{aq}$ into Equation 2 demonstrates that

~86 per cent in the 1 per cent P-goethite and ~60 per cent of the Fe atoms in the P-free goethite sample have exchanged with $Fe(II)_{aq}$ over the 14 day reaction period.

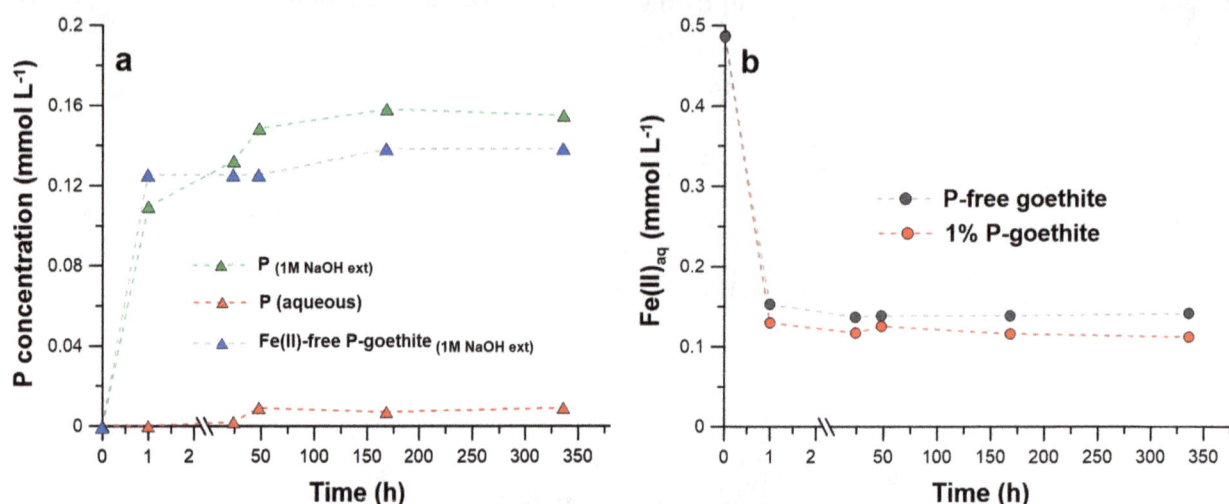

FIG 3 – (a) Changes in 1M NaOH extractable and aqueous concentrations of P over 14 days; (b) changes in $Fe(II)_{aq}$ concentrations over 14 days (The data points on the graph represent the average of two replicates, and the standard deviation is smaller than the size of the symbol used for each data point).

The stability of goethite in the P-free treatment suggests that the observed exchange of Fe atoms can be attributed to recrystallisation, wherein there is an exchange of Fe atoms between goethite and the aqueous phase without any noticeable alteration in mineralogy which was observed and reported in previous studies (Burton, Hockmann and Karimian, 2020).

On the other hand, in the presence of Fe(II), the reaction with 1 per cent P-goethite leads to some mineralogical changes, as evidenced by the LCF results. The reaction induces the transformation of ferrihydrite and feroxyhyte into goethite and, after 14 days, only goethite and feroxyhyte were detected in this treatment.

Figure 3a illustrates the variations observed in the labile P fractions over time. Upon the addition of Fe(II) and following the 14 day reaction period, a slight increase of approximately 0.009 mmol L^{-1} was observed in the concentrations of aqueous P (Figure 3a). The aqueous P concentration in the control treatment (ie Fe(II)-free 1 per cent P-goethite) was below the detection limit (<10 µgL^{-1}) at all sampling time points.

Additionally, the concentrations of 1M NaOH-extractable P showed an increase following the introduction of aqueous Fe(II) to the 1 per cent P-goethite system with the 1M NaOH-extractable P concentration reaching approximately 0.15 mmol L^{-1} over the 14-day duration of the experiment. This increase in NaOH-extractable P resulted in a nearly 5 per cent rise in the NaOH extractability of P, reaching approximately 75 per cent compared to the control non-Fe(II) reacted 1 per cent P-goethite sample, which exhibited a NaOH extractability of approximately ~70 per cent.

CONCLUSIONS

Our results indicate that treating 1 per cent P-goethite with Fe(II) for a period of 7–14 days at room temperature, followed by a 24-hour extraction of the reacted solids with 1M NaOH, leads to the removal of approximately 79.5 per cent of the initial P content (including both the aqueous (~4.5 per cent) and surface adsorbed fractions (~75 per cent). This shows about a 10 per cent increase in the extractability of P compared to a non-Fe(II) reacted 1 per cent P-goethite sample. The reaction with an enriched aqueous $^{57}Fe(II)$ tracer provided direct evidence of Fe(II)-catalysed goethite recrystallisation. Iron K-edge EXAFS spectroscopy and XRD results confirmed goethite as the dominant mineral phase after the 14-day reaction with $Fe(II)_{aq}$. Goethite recrystallisation caused a slight increase in the total amount of aqueous P. Our results also confirmed that surface-bound P (based on 1M NaOH extraction on the Fe(II)-reacted mineral phases) was the dominant fraction of available P with (~75 per cent of the total) removed by the end of the experiment.

ACKNOWLEDGEMENTS

The authors gratefully acknowledge the financial support provided by CSIRO through the CERC postdoctoral fellowship awarded to Dr. Niloofar Karimian. The assistance rendered by the staff of ANSTO at the Australian Synchrotron is also acknowledged, with special recognition given to Dr. Jessica Hamilton, the XAS beamline scientist who contributed significantly to the data collection process. Furthermore, the authors express their appreciation to Dr. Nick Owen and Dr. Nathan Webster, the XRD staff at CSIRO, for their valuable contributions to this research endeavour.

REFERENCES

American Public Health Association (APHA), 1998. *Standard Methods for the Examination of Water and Wastewater*, 20th edition, APHA, American Water Works Association and Water Environmental Federation, Washington DC.

Boland, D D, Collins, R N, Miller, C J, Glover, C J and Waite, T D, 2014. Effect of Solution and Solid-Phase Conditions on the Fe(II)-Accelerated Transformation of Ferrihydrite to Lepidocrocite and Goethite, *Environmental Science and Technology*, 48(10):5477–5485.

Bolanz, R M, Bläss, U, Ackermann, S, Ciobotă, V, Rösch, P, Tarcea, N, Popp, J and Majzlan, J, 2013. The Effect of Antimonate, Arsenate, and Phosphate on the Transformation of Ferrihydrite to Goethite, Hematite, Feroxyhyte, and Tripuhyite, *Clays and Clay Minerals*, 61(1):11–25.

Burton, E D, Hockmann, K and Karimian, N, 2020. Antimony Sorption to Goethite: Effects of Fe(II)-Catalyzed Recrystallization, *ACS Earth and Space Chemistry*, 4:476–487.

Cornell, R M and Schwertmann, U, 2003. *The Iron Oxides: Structure, Properties, Reactions, Occurrences and Uses* (Wiley VHC: Weinheim, Germany).

Fey, M V and Dixon, J B, 1981. Synthesis and Properties of Poorly Crystalline Hydrated Aluminous Goethites, *Clays and Clay Minerals*, 29:91–100.

Frierdich, A J and Catalano, J G, 2012. Controls on Fe(II)-Activated Trace Element Release from Goethite and Hematite, *Environmental Science and Technology*, 46:1519–1526.

Frierdich, A J, McBride, A, Tomkinson, S and Southall, S C, 2019a. Nickel Cycling and Negative Feedback on Fe(II)-Catalyzed Recrystallization of Goethite, *ACS Earth and Space Chemistry*, 3:1932–1941.

Frierdich, A J, Saxey, D W, Adineh, V R, Fougerouse, D, Reddy, S M, Rickard, W D A, Sadek, A Z and Southall, S C, 2019b. Direct Observation of Nanoparticulate Goethite Recrystallization by Atom Probe Analysis of Isotopic Tracers, *Environmental Science and Technology*, 53:13126–13135.

Handler, R M, Frierdich, A J, Johnson, C M, Rosso, K M, Beard, B L, Wang, C, Latta, D E, Neumann, A, Pasakarnis, T, Jeewantha Premaratne, W A P and Scherer, M M, 2014. Fe(II)-catalyzed recrystallization of goethite revisited. *Environmental Science & Technology,* 48(19):11302–11311. https://doi.org/10.1021/es503084u

Hansel, C M, Benner, S G and Fendorf, S, 2005. Competing Fe(II)-induced mineralization pathways of ferrihydrite, *Environmental Science and Technology*, 39:7147–7153.

Hsu, L-C, Tzou, Y-M, Ho, M-S, Sivakumar, C, Cho, Y-L, Li, W-H, Chiang, P-N, Teah, H Y and Liu, Y-T, 2020. Preferential phosphate sorption and Al substitution on goethite, *Environmental Science: Nano*, 7:3497–3508.

Pownceby, M I, Hapugoda, S, Manuel, J, Webster, N A S and Macrae, C M, 2019. Characterisation of phosphorus and other impurities in goethite-rich iron ores – Possible P incorporation mechanisms, *Minerals Engineering*, 143:106022.

Rasmussen, B, Muhling, J R, Suvorova, A and Fischer, W W, 2021. Apatite nanoparticles in 3.46–2.46 Ga iron formations: Evidence for phosphorus-rich hydrothermal plumes on early Earth, *Geology*, 49:647–651.

Ravel, B and Newville, M, 2005. ATHENA, ARTEMIS, HEPHAESTUS: data analysis for X-ray absorption spectroscopy using IFEFFIT, *Journal of Synchrotron Radiation*, 12:537–541.

Hydrogen reduction of Australia iron ores

L Lu[1], Y Tang[2] and S Hapugoda[3]

1. Senior principal scientist, CSIRO Mineral Resources, Pullenvale Qld 4069.
   Email: liming.lu@csiro.au
2. Experimental scientist, CSIRO Mineral Resources, Pullenvale Qld 4069.
   Email: yajun.tang@csiro.au
3. Experimental scientist, CSIRO Mineral Resources, Pullenvale Qld 4069.
   Email: sarath.hapugoda@csiro.au

ABSTRACT

The steel industry is one of the global leading CO_2 emitters, accounting for approximately 7 per cent of global anthropogenic CO_2 emissions. In 2022, the world produced 1878.5 Mt of crude steel. About 70.8 per cent of the world's crude steel was produced by integrated steel mills through the blast furnace and basic oxygen furnace (BF-BOF) process, while the remainder was produced largely by mini mills using electrical arc furnaces (EAF). Compared with gas-based DRI and scrap-based EAF processes, the direct CO_2 emissions from the BF-BOF process and the coal based DRI and EAF process are significantly more, as both the processes rely heavily on fossil fuels, such as coke and pulverised coal, to provide energy and reduce iron oxides. Hydrogen is very reactive and reduces iron oxides to metallic iron without generating CO_2. Therefore, substitution of hydrogen for fossil fuels in these processes presents a good opportunity to significantly decarbonise the steel industry. This paper will first discuss the thermodynamic and kinetic characteristics of hydrogen and carbothermic reductions of iron oxides and then examine the behaviours of different ore types present in Australia iron ores during hydrogen reduction.

INTRODUCTION

The Paris agreement sets out a global framework to avoid dangerous climate change by limiting global warming to well below 2°C, preferably 1.5°C, compared to pre-industrial levels. To achieve this long-term temperature goal, countries need to reach a global peak of greenhouse gas emissions as soon as possible and achieve a climate neutral world by mid-century. However global CO_2 emissions from energy combustion and industry reached their highest ever annual level to 36.3 gigatonnes (Gt) in 2021, which was 6 per cent higher than 2020. Clearly the international community is falling far short of the Paris goals, with no credible pathway to 1.5°C in place. Therefore, an urgent system-wide transformation is urgently needed to avoid climate disaster.

Manufacturing of iron and steel is one of the most energy intensive industries and contributes about 7.2 per cent greenhouse gas (GHG) emissions from the energy related sector. The blast furnace and basic oxygen furnace (BF/BOF) route is currently the dominant ironmaking process and produced about 70 per cent world crude steel. The process relies on coke and pulverised coal as fuel and reducing agents to produce hot metal with consistent quality for the BOF process, which inevitably results in a large amount of GHG emissions for every tonne of crude steel produced (about 1.9 t-CO_2/t crude steel). The global crude steel production was about 1878.5 Mt in 2022, which corresponds to a large quantity of GHG emissions. Therefore, fundamental technology and process changes combined with a reduction of material demand and increased recycling are need for the steel industry to meet the Paris target. One opportunity to avoid or reduce GHG emissions is substituting hydrogen for carbon as an energy source and reducing agent. This paper first discusses the thermodynamic and kinetic characteristics of hydrogen and carbothermic reductions of iron oxides. The behaviour of different ore types present in Australian iron ores during hydrogen reduction is further examined.

CHARACTERISTICS OF HYDROGEN AND CARBOTHERMIC REDUCTIONS

Thermodynamic characteristics

Figure 1 shows the stability areas of different iron oxides in the presence of CO/CO_2 and H_2/H_2O gas mixtures of varying gas oxidation degrees (COD). COD is defined as the ratio of oxidised gas components over the sum of oxidised and oxidisable gas components, it is a good indicator of the reducing potential of the gas mixture. Clearly the stability of iron oxides depends heavily on the temperature and GOD of the gas atmosphere. When temperature is over 570°C, reduction of iron oxides is expected to take place in three consecutive steps, $Fe_2O_3->Fe_3O_4->FeO->Fe$, through the following reactions (1–3 for reduction by CO and 4–6 for reduction by H_2):

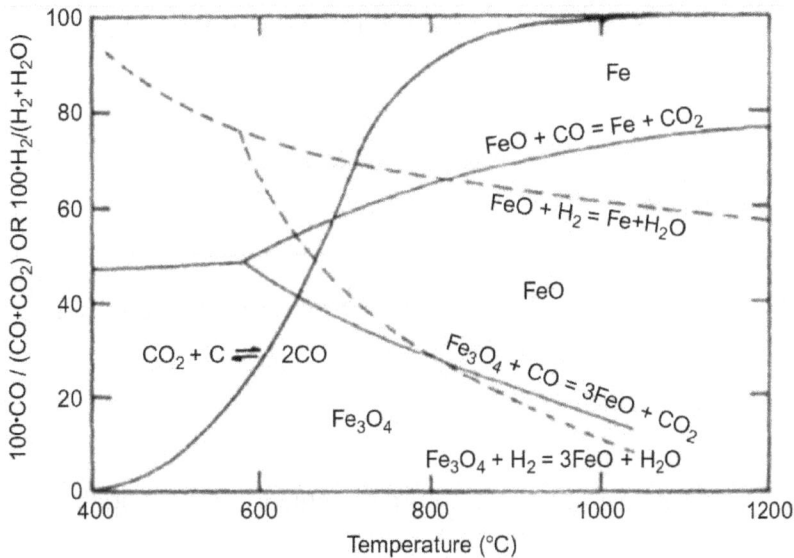

$$3Fe_2O_3+CO = 2Fe_3O_4+CO_2 \tag{1}$$

$$Fe_3O_4+CO = 3FeO+CO_2 \tag{2}$$

$$FeO+CO = Fe+CO_2 \tag{3}$$

$$3Fe_2O_3+H_2 = 2Fe_3O_4+H_2O \tag{4}$$

$$Fe_3O_4+H_2 = 3FeO+H_2O \tag{5}$$

$$FeO+H_2 = Fe+H_2O \tag{6}$$

FIG 1 – Stability diagram of iron oxides in the presence of CO/CO_2 and H_2/H_2O gas mixtures of varying reducing potentials (Yang, Raipala and Holappa, 2014).

However, at temperatures below 570°C, as wüstite is meta-stable as evidenced in the Fe-O phase diagram, magnetite is directly reduced to metallic iron via the following reactions without first being converted to wüstite:

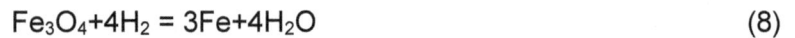

$$Fe_3O_4+4CO = 3Fe+4CO_2 \tag{7}$$

$$Fe_3O_4+4H_2 = 3Fe+4H_2O \tag{8}$$

Therefore, reduction of iron oxides takes place in two steps, $Fe_2O_3->Fe_3O_4->Fe$, at temperatures below 570°C.

As shown in Figure 1, at about 1073 K (800°C), the equilibrium ratios of $CO/(CO+CO_2)$ and $H_2/(H_2+H_2O)$ are very close to each other. In fact, the hydrogen and CO reduction lines cross at 821°C; below this temperature, CO is a stronger reducing agent and above this temperature, H_2 is a stronger reducing agent (Lu and Lu, 2022). This is true for both Fe_3O_4 and FeO reductions. In addition, unlike carbothermic reduction which is exothermic, the overall reduction with hydrogen is endothermic, which means that for reduction with hydrogen to proceed, additional energy must be added to the system to ensure the reduction temperature maintained.

Finally, Figure 1 also shows the equilibrium line of the Boudouard reaction (Equation 9) at a gas pressure of 1 bar and a carbon activity of 1. If a CO/CO_2-containing gas mixture has a temperature and composition below the equilibrium line, carbon deposition will take place. On the other hand, there is no such an issue for the hydrogen reduction.

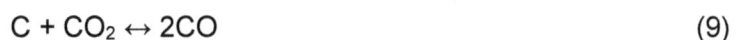

$$C + CO_2 \leftrightarrow 2CO \tag{9}$$

Kinetic characteristics

The kinetics of reactions between gases and solids are typically modelled using shrinking-core theoretical frameworks, in which the formation of a uniform solid product layer covering the entire solid surface with a sharp interface between the solid reactant and the product is assumed. Based on the shrinking-core model in Figure 2, reduction of non-porous iron ore particles is believed to consist of the following process steps (Spreitzer and Schenk, 2019):

- mass transfer of the reducing gases, such as H_2 and CO, through a laminar gas film from the bulk gas stream to the solid/gas interface

- diffusion of reducing gases through the solid layer of reduction product to the reactant

- reduction reaction at the interface between the solid reactant and the product

- diffusion of gaseous reduction products through the solid layer of reduction product back to the laminar gas film

- mass transfer of gaseous reduction products through the laminar gas film back to the bulk gas stream.

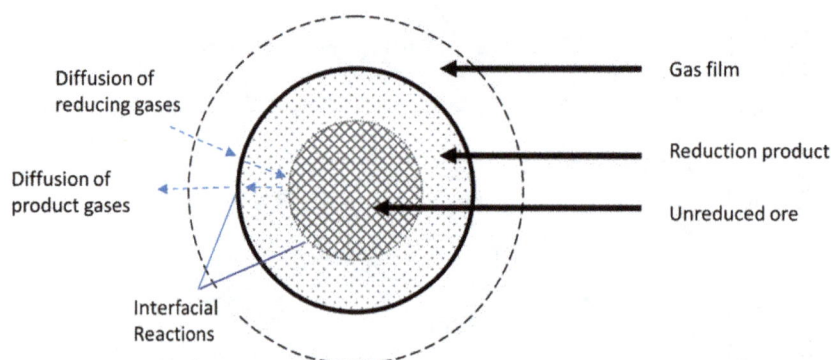

FIG 2 – Schematic diagram of shrinking core model.

As the iron ore particles in the shrinking-core model are nonporous, the reducing gases are not able to penetrate the particles and after adsorbing on the surface of particles, react with the solid reactant, leading to the formation of a layer of solid reduction product on the solid reactant with a clear boundary or interface between them. As the reduction reaction proceeds, the product/reactant interface moves inwards and the layer thickness of the reduction product increases. Due to the formation of the product layer, both the gaseous reduction products and reactants need to diffuse through the solid product layer away from and towards the interface. In the case of porous iron ore particles, the reducing gas can penetrate into the particles partially and reduction reaction hence takes place across the interface into the solid reactant, leading to formation of a reaction zone between the solid product and reactant, through which the concentrations of reactants and products varies.

The overall reduction rate of iron ore particles is determined by the slowest step of the above process steps, or the rate limiting step. In general, at low temperatures, the interfacial gas/solid reactions are considerably slower than the transport of gaseous reactants and products and become the rate limiting step. The interfacial gas/solid reaction rate increases exponentially with temperature according to the Arrhenius equation. At high temperatures, the interfacial gas/solid reactions occur faster than the transport of gaseous reactants and products, hence mass transport is rate limiting. At intermediate temperatures such as the conditions occurring in blast furnaces, shaft furnaces and fluidised bed reactors, the mass transport and interfacial gas/solid reactions take place at a similar speed, a mixed interfacial reaction and diffusion mechanism is often in operation.

Compared with CO, the hydrogen molecule is much smaller and shows better diffusion behaviour. Therefore hydrogen molecules can diffuse further into smaller pores of porous iron ore particles, which is expected to increase the reaction area and hence enhance reaction kinetics. However, water vapour is formed during reduction with hydrogen and shows a slow diffusion and desorption which may deter the reduction progress. In case of reduction with CO, deposition of carbon under the Boudouard reaction equilibrium line can also impede the reduction reaction.

EXPERIMENTAL

Raw materials

An Australian ore consisting of mixed hematite/martite and goethite ore types was sourced for the present study. The as received sample was first dried at 105°C and screened at 1 and 0.5 mm. The -1+0.5 mm size fraction obtained was then used in reduction tests. The sample was analysed by XRF and thermogravimetric analyses and had an Fe grade of 58.83 per cent and total LOI of 6.7 per cent. Based on qualitative observation using an optical microscope, the sample contained predominantly earthy, vitreous goethite, martite-goethite and hydrated hematite particles.

Experimental set-up

Reduction tests were carried out in a laboratory scale fixed bed reactor (Figure 3) at a temperature of 750°C in a flow of 55 per cent H_2 + 45 per cent N_2 gas mixture at a total flow rate of 1 L/min. For each reduction test, a subsample of about 9 g was weighed out and loaded onto a porous supporting disc of the inner reaction tube which was then sealed with an external quartz tube. After being purged with a N_2 flow of 1 L/min for 15 min, the reactor was inserted into the hot zone of an electrically heated furnace preheated at a temperature to target a sample temperature of 750°C. The sample was purged continuously with N_2 until the target temperature was achieved. After being held at the target temperature for a further 30 min, the reaction gas mixture was introduced to replace the purging gas. The reduction reaction was interrupted at varying reaction times of 5, 15 and 40 min by switching the reaction gas back to purging gas and withdrawing the reactor from the furnace hot zone.

FIG 3 – Experimental set-up used in reduction tests.

Sample characterisation

The partially reduced samples were weighed after being fully cooled and subsamples were prepared for characterisation by quantitative XRD and optical microscopical analyses.

RESULTS

Change in sample mass during reduction

Figure 4 shows the total mass loss and the mass loss due to reduction as the reduction was progressed up to 40 min. Clearly as the reduction proceeded, the reduction became slower, especially, in the later stage of reduction. Compared with the mass losses in the first 5 min and from the fifth to 15th min, the mass loss in the last 25 min from the 15th to 40th min was considerably lower at only 4.4 per cent. This is due to the formation of dense layers of reduced products, such as reduced iron, around the reactants, which deterred the diffusion of H_2 into and H_2O out of the interfaces where the reduction reaction was taking place.

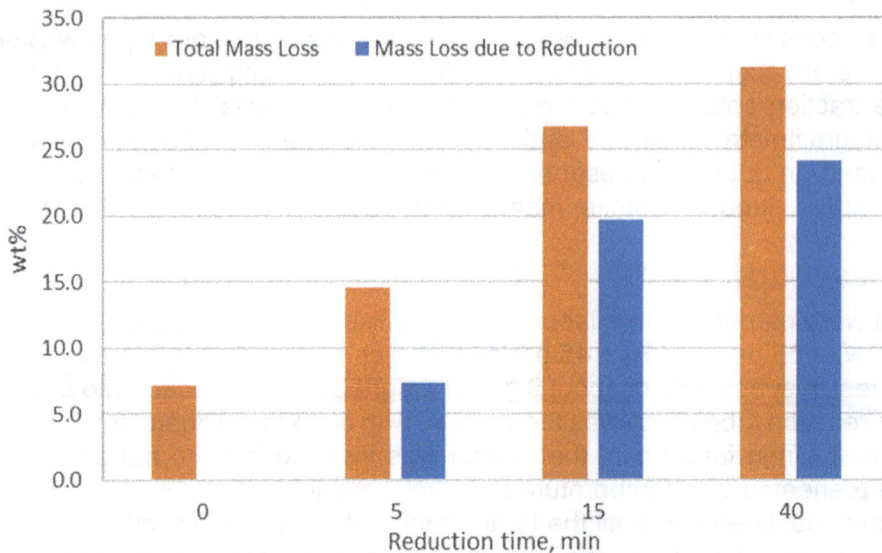

FIG 4 – Experimental set-up used in reduction tests.

Evolution of iron bearing mineral phases during reduction

Figure 5 presents the XRD quantitative results of Fe bearing mineral phases present in the partially reduced samples. As expected, the raw sample consisted predominantly of hematite/martite and goethite minerals, which agreed well with the total LOI measured for the sample by thermogravimetric analysis. As the sample was heated up to the target temperature, dehydration of goethite took place to form proto-hematite. The goethite to proto-hematite conversion occurred very fast and was completed well before introduction of the reaction gas. Therefore, the calcined sample was dominant with hematite and proto-hematite.

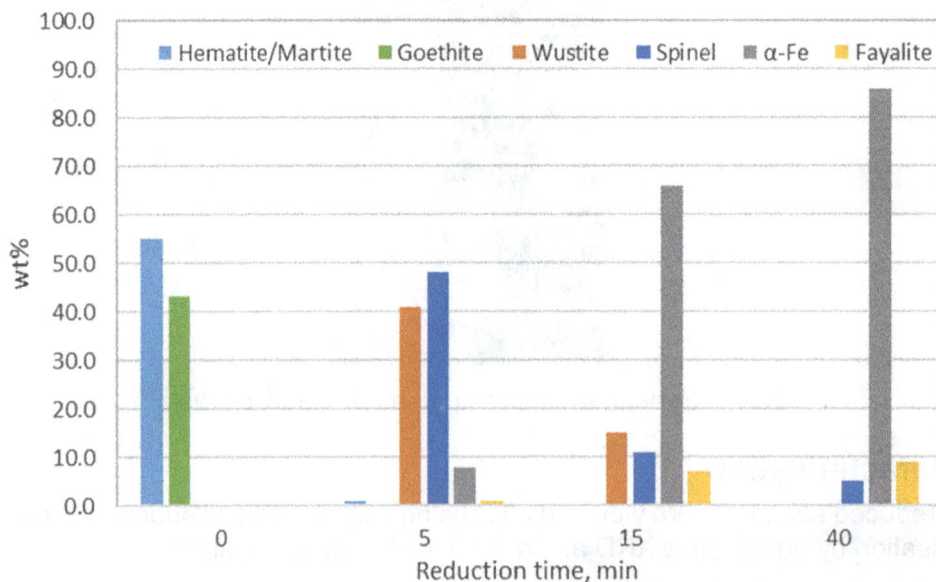

FIG 5 – Iron-bearing mineral phases present in partially reduced samples.

Initial reduction was very fast as evidenced by the mineral composition of the 5 min reduction sample. The hematite and geothite phases in the sample almost disappeared within 5 min of reduction and were replaced by magnetite (Fe_3O_4) and wustite (FeO) phases. There was also a small amount of reduced Fe in the sample. The dominance of magnetite (Fe_3O_4) and wustite (FeO) phases in the sample was consistent with the mass loss measured for the sample due to reduction, 7.4 per cent. A mass loss of 7.4 per cent corresponded to a reduction degree of 28.9 per cent, less than a reduction degree of 33.3 per cent, estimated for a complete reduction of Fe_3O_4 to FeO. Therefore, Fe_3O_4 was not expected to be fully reduced to FeO, which explained the presence of spinel in the sample. Furthermore, it was possible that some H_2 molecules diffused to the FeO phase area. Therefore, part of FeO in the sample was reduced to Fe simultaneously, leading to a lower-than-expected FeO content in the 5 min sample.

As reduction time proceeded from 5 to 15 min, Fe_3O_4 and FeO were continuously reduced. Therefore, the contents of Fe_3O_4 and FeO in the 15 min reduction sample decreased while the proportion of reduced Fe increased considerably. Again, this sample achieved a mass loss due to reduction of 19.7 per cent, corresponding to a reduction degree of 77.7 per cent, suggesting most of FeO in the sample was reduced to Fe. Consequently, the 15 min reduction sample was dominant with reduced iron with a small proportion of Fe_3O_4 and FeO phases.

Further reduction from 15 to 40 min led to a small decrease of Fe_3O_4 phase and disappearance of FeO phase from the sample. The reduction degree was increased only from 77.7 to 95.2 per cent, confirming the deterrence of reduction at the late stage of reduction. A small amount of unreduced magnetite was likely encapsuled by the reduced iron and kept unreduced. In the presence of SiO_2 in the sample, it can readily react with FeO and Fe_3O_4 at the temperature of interest, leading to formation of fayalite.

Therefore, the reduction steps from hematite to Fe were partially overlapping instead of a step wise process. The iron bearing phases coexisted in the partially reduced samples. The reduction took place quickly in the beginning and slowed down gradually due to the longer diffusion distance and slower reduction kinetics of FeO to Fe.

Behaviour of iron bearing minerals during H2 reduction

Figure 6 shows the cross-sectional optical micrographs of the martite-goethite and vitreous goethite particles after being partially reduced for 15 min. The fine bright phase in the micrographs was believed to be reduced iron particles, which were surrounded by a light grey wustite phase. As expected, reduction of these particles occurred in consecutive steps from Fe_2O_3 to Fe.

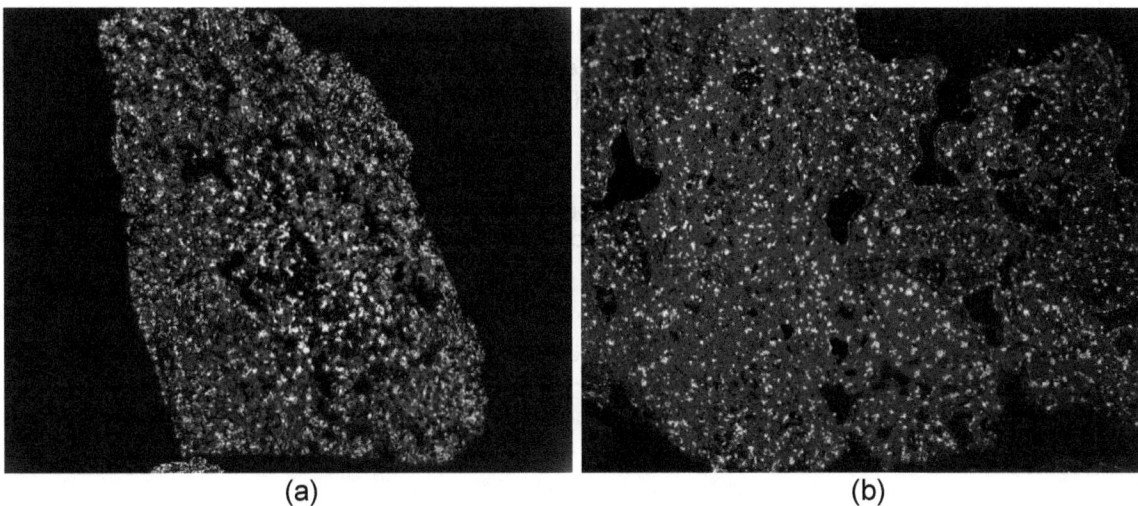

(a)   (b)

FIG 6 – Cross-sectional optical micrographs of partially reduced particles observed in the 15 min reduction sample (a) Partially reduced martite-goethite particle; (b) Partially reduced vitreous goethite particle.

The bright phase was found to be distributed evenly across the particles, suggesting reduction occurred evenly across the particles. Due to the porous nature of these particles, H_2 molecules were readily available for both the external and internal pore surfaces of the particles and reacted with the Fe bearing mineral phases across the particles, leading to a rapid and uniform reduction. Therefore, the primary porous nature of dominant ore types in Australia ores and the secondary porosity resulting from dehydration of goethite will no doubt promote the reduction kinetics of iron ore with H_2.

As evidenced in Figure 6a, the bright reduced iron phase is apparently coincident with the martite mineral phase. This may suggest the hematitic minerals in the sample, such as martite and microplaty hematite, is more reducible than the goethitic minerals. Also comparing the two ore types present in Figure 6, the martite-goethite particle appeared more open in pore structure and hence reduced more than the goethite particle.

CONCLUSIONS

H_2 is a stronger reducing agent only at temperatures above 821°C. Unlike carbothermic reduction which is exothermic, the overall hydrogen reduction is endothermic. Therefore, additional energy must be added to the system to proceed the reduction of iron ore with hydrogen.

At intermediate temperatures such as the conditions occurring in blast furnaces, shaft furnaces and fluidised bed reactors, the overall reaction is controlled by both the mass transport and interfacial gas/solid reactions.

As reduction with H_2 proceeded, it became slower, in particular, in the later stage of reduction, due to the formation of dense layers of reduced products, such as reduced iron, around the reactants, which deterred the diffusion of H_2 into and H_2O out of the interfaces where reduction reaction was taking place.

Within 5 min reduction, the hematite and protohematite phases in the sample almost disappeared completely and turned to first magnetite and then wustite. As reduction proceeded, the proportions of magnetite and wustite phases started to decrease after reaching their maximum, while the proportion of reduced iron increased. The reduction steps from Fe_2O_3 to Fe was partially overlapped. The iron bearing phases often coexisted in the partially reduced samples.

The reduction products were found to be distributed evenly across the particles, suggesting reduction occurred evenly across the particles. The porous nature of the dominant ore types in Australia ores will no doubt promote the reduction kinetics of iron ore with H_2. Comparing the ore types present in the sample, the hematitic particle, such as martite and microplaty hematite, appeared more open in pore structure and hence reduced more than the goethite particle.

ACKNOWLEDGEMENTS

The authors would like to thank CSIRO Mineral Resources Characterisation Team Nathan Webster and Rong Fan for carrying out he quantitative XRD analysis.

REFERENCES

Lu, L and Lu, Y, 2022. Blast furnace ironmaking and its ferrous burden requirements, in *Iron Ore: Mineralogy, Processing and Environmental Sustainability* (ed: L Lu), second edn, ch 17, pp 579–604 (Woodhead Publishing). https://doi.org/10.1016/B978-0-12-820226-5.00019-7

Spreitzer, D and Schenk, J, 2019. Reduction of Iron Oxides with Hydrogen — A Review, *Steel Research Int*, 90:1900108.

Yang, Y, Raipala, K and Holappa, L, 2014. Ironmaking, in *Treatise on Process Metallurgy* (eds: S Seetharaman, A McLean, R Guthrie and S Sridhar), pp 2–88. doi:10.1016/b978-0-08-096988-6.00017-1

Hydrogen reduction of as-received and pre-oxidised NZ titanomagnetite ironsands in a small-scale high-temperature fluidised bed

B Maisuria[1], S Prabowo[2], D Del Puerto[3], R J Longbottom[4], B J Monaghan[5] and C W Bumby[6]

1. PhD Candidate, Robinson Research Institute, Lower Hutt, Wellington 5010, New Zealand. Email: bavinesh.maisuria@vuw.ac.nz
2. Researcher, Robinson Research Institute, Lower Hutt, Wellington 5010, New Zealand. Email: sigit.prabowo@vuw.ac.nz
3. Research Scientist, Callaghan Innovation, Lower Hutt, Wellington 5010, New Zealand. Email: diegodelpuerto@callaghaninnovation.govt.nz
4. Research Fellow, Pyrometallurgy Group, School of Mechanical, Materials, Mechatronic and Biomedical Engineering, University of Wollongong, Wollongong NSW 2522. Email: raymond_longbottom@uow.edu.au
5. Professor, Pyrometallurgy Group, School of Mechanical, Materials, Mechatronic and Biomedical Engineering, University of Wollongong, Wollongong NSW 2522. Email: brian_monaghan@uow.edu.au
6. Professor, Robinson Research Institute, Lower Hutt, Wellington 5010, New Zealand. Email: chris.bumby@vuw.ac.nz

ABSTRACT

The use of hydrogen as a reducing agent can substantially reduce carbon dioxide (CO_2) emissions from the ironmaking process. Iron ore fines can be reduced in a fluidised bed (FB) using hydrogen (H_2) at high temperatures (>800°C), although several studies have reported the sticking of particles which causes the bed to defluidise, effectively shutting down the process. Here, we report results from the reduction of New Zealand (NZ) titanomagnetite (TTM) ironsands in a small-scale laboratory FB (100 g) at temperatures between 800–1000°C using H_2 flow rates up to 5 standard L/min. No sticking phenomena are observed under any of these conditions, which is attributed to the formation of a stable titanium-bearing oxide layer on each particle's exterior, preventing iron-iron contact at the particle surfaces. Pre-oxidised NZ TTM ironsands have also been studied and shown not to stick. Interestingly, at lower reaction temperatures the reduction kinetics for pre-oxidised ironsands is faster than for as-received ironsands. This increased kinetic rate can be attributed to the pre-oxidation stage inducing micro-fractures that create a void for the hydrogen to diffuse into the inner regions of the particle. In addition, oxidation of TTM appears to segregate the particle into two distinct phases, a high Aluminium-Magnesium (Al-Mg) phase and a high Titanium (Ti) phase. The findings are important as increasing the operating temperature of the FB reactor results in a faster reaction rate and higher gas utilisation, making the process more economically attractive. Furthermore, oxidation of NZ TTM ironsands results in a faster reduction rate at lower temperatures (800°C– 900°C) opening the opportunity to reduce the reaction temperature which might be beneficial in a final commercial-scale process.

INTRODUCTION

In recent times, the global focus on reducing CO_2 emissions in the steelmaking process has led to increased interest in hydrogen direct reduction of iron (Hessling, Tottie and Sichen, 2021; Shahabuddin, Brooks and Rhamdhani, 2023; Spreitzer and Schenk, 2019b; Vogl, Åhman and Nilsson, 2018). This method, among various gas-based processes, specifically utilises FB processing, which offers advantages such as the use of pure hydrogen gas and iron ore fines with minimal additional steps (Schenk, 2011; Spreitzer and Schenk, 2019a; Wolfinger, Spreitzer and Schenk, 2022; Zhang, Lei and Zhu, 2014). However, when it comes to conventional hematite ores, a significant challenge arises due to the loss of fluidisation caused by ore particles sticking at temperatures around 800°C (Guo *et al*, 2020; Hayashi and Iguchl, 1992; Lu *et al*, 2023). This issue has been observed during the commissioning of several pilot commercial-scale processes, resulting in their discontinuation (Goodman, 2019; Hillisch and Zirngast, 2001; Plaul, Böhm and Schenk, 2008; Schenk, 2011). One notable exception is the FINEX (POSCO) process, which employs a syn-gas FB process at temperatures below 800°C for a pre-reduction stage, achieving only 60–70 per cent metallisation (Yi, Cho and Yi, 2018). To complete the metallisation process, the partially reduced powder is then introduced into a smelting-reduction furnace (melter gasifier) for further carbothermic reduction. However, recent research indicates that NZ ironsand, unlike other iron ores, does not experience sticking during hydrogen reduction, even at temperatures up to 1000°C (Prabowo et al, 2019).

NZ ironsand serves as the primary source of iron for domestic steel production in NZ. A notable distinction between NZ ironsand and conventional magnetite or hematite ores is its higher titanium content, which stands at approximately 8 per cent by weight TiO_2 (Brathwaite, Gazley and Christie, 2017; Longbottom et al, 2019; Zhang et al, 2020). The composition of NZ ironsand mainly consists of single-phase grains of TTM, which is a spinel solid solution formed by the combination of magnetite (Fe_3O_4) and ulvö¨spinel (Fe_2TiO_4). The presence of titanium in the ore poses challenges in the conventional Blast Furnace (BF) ironmaking process due to the formation of high melting point phases such as titanium carbides or carbo-nitrides and perovskite. These phases contribute to increased slag viscosity and accumulate in the hearth, resulting in reduced productivity (Park et al, 2004; Templeton, 2006). In contrast, NZ ironsand undergoes processing in a coal-fired rotary kiln instead of the traditional blast furnace route (Evans, 1986).

Extensive research has focused on TTM ores as a promising and cost-effective alternative as-received material for steel production (Longbottom, Monaghan and Mathieson, 2013; She et al, 2013; Sun et al, 2013; Wang et al, 2016; Yu et al, 2021). Previous studies have indicated that TTM ore exhibits a slower reduction rate compared to magnetite ore when exposed to either H_2 or carbon monoxide (CO) gas. This phenomenon is generally attributed to the stabilising effect of Ti^{4+} occupying the octahedral sites within the TTM spinel crystal structure (Longbottom et al, 2019; Park and Ostrovski, 2003; Sun et al, 2013).

Prabowo (Prabowo, 2020; Prabowo et al, 2019, 2022) presented findings indicating that the hydrogen-FB reduction of NZ ironsand avoids particles sticking through the formation and the continuous presence of a Ti-rich oxide shell around each particle. This protective shell is established early in the reaction and remains intact throughout, completely enclosing the reduced iron metal and preventing particle adhesion. Other research studies have also observed the non-sticking behaviour of TTM ores during FB reduction using H_2 gas at temperatures up to 950°C (Shannon, Kitt and Marshall, 1960; Yu et al, 2021, 2022). However, the specific mechanism responsible for preventing sticking was not identified in these studies. Collectively, these investigations highlight the significant advantage of utilising TTM ore as a feed material for a FB system, offering superior performance compared to conventional hematite ores.

However, it has been demonstrated that pre-oxidising TTM to titanohematite (TTH) can enhance the rate of gaseous reduction (Longbottom, Ostrovski and Park, 2006; Park et al, 2004; Sui et al, 2017; Sun et al, 2017; Wang et al, 2017; Wolfinger et al, 2022). The accelerated reduction rate is commonly attributed to the formation of microcracks resulting from internal stresses caused by the volume expansion during the structural transformation from rhombohedral TTH to cubic TTM (Longbottom, Ostrovski and Park, 2006; Park et al, 2004; She et al, 2013; Yu et al, 2021; Zheng et al, 2023). These microcracks facilitate improved gas penetration to and from the reaction sites within the solid material.

The predominant forms of TTM ores are typically pellets or packed beds (Longbottom, Ostrovski and Park, 2006; Park et al, 2004; Sui et al, 2017; Wang et al, 2017; Zhang et al, 2020). Previous studies have investigated the reduction of pre-oxidised South African TTM ores using CO gas in a FB system. These studies observed that TTM particles maintained fluidisation without any adhesion issues during both oxidation and reduction processes, even at temperatures reaching 950°C. In another study, Shannon, Kitt and Marshall (1960) examined the reduction of pre-oxidised NZ ironsand using hydrogen gas within a spouted bed reactor. They also observed a more rapid reduction reaction of pre-oxidised ironsand without any particle sticking problems at temperatures up to 900°C. However, this research did not provide any data regarding the phase transformation behaviour and microstructural analysis of the ironsand during the reduction process. There have been some studies looking into the phase evolution during the oxidation of magnetite using hot-stage X-ray diffraction (XRD), (Zheng et al, 2023) however there is little work on TTM.

This research paper presents a comparison between the reduction process of as-received and pre-oxidised NZ ironsand utilising H_2 gas within an experimental FB system. The primary objective of this study was to analyse the differences in the phase development and microstructural changes occurring in the ironsand during hydrogen reduction within the FB system at different temperatures.

EXPERIMENTAL

Materials

Ironsand supplied from the Waikato North Head mine was used in the oxidation and reduction experiments and washed and sieved to a particle size fraction of 106–125 microns. The sieved material was prepared following the standard 'fused bead' method of oxidising in air at 1000°C for 1 hr. X-ray fluorescence (XRF) was performed on the oxidised material and the chemical composition is given in Table 1. The mass change from oxidative roasting is expressed as the Loss of Ignition (LOI) (given as negative in Table 1 due to an increase in mass from oxidation). The ironsand was then mixed with lithium tetraborate melt as a standard calibrated matrix and calibrated XRF analysis was performed using an XRF Bruker S8-Tiger.

TABLE 1

Chemical composition of the ironsand sample after full oxidation as measured by XRF.

Composition (equivalent mass%)									
Fe	TiO_2	Al_2O_3	MgO	MnO	P_2O_5	V_2O_5	CaO	SiO_2	LOI
60.7	8.1	3.4	2.6	~0.7	~0.6	~0.6	~0.2	~0.8	-3.1

XRD spectra of the as-received ironsand sample is presented in Figure 1 and will be discussed in detail in the results and discussion section.

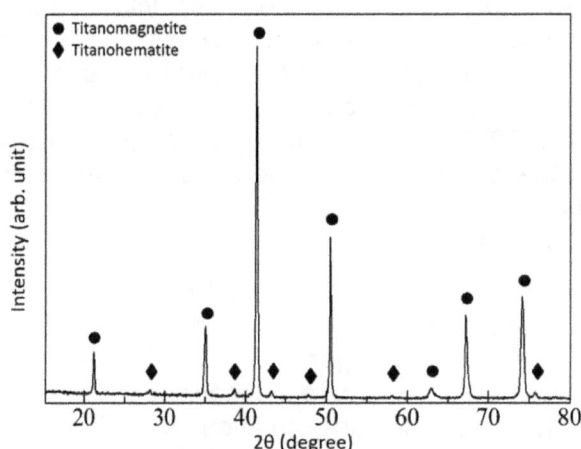

FIG 1 – XRD patterns of the as-received ironsand concentrate, 106 to 125 µm size fraction.

The as-received ironsand concentrate used in this study mainly comprises TTM with a small proportion of TTH. Note that the atomic ratio of titanium (Ti) calculated from the XRF data is 0.26 at. per cent, which is consistent with a stoichiometric Ti content of x ~ 0.27 ± 0.02 for TTM ($Fe_{3-x}Ti_xO_4$), as previously reported for NZ ironsand (Park and Ostrovski, 2003, 2004).

The morphology of naturally occurring particles in NZ ironsand concentrate can be categorised into two main classes. TTM comprises the majority of particles in the form of uniform grains (Figure 2a). Less than 10 per cent of the total particles consist of non-uniform particles that contain exsolved lamellae structures within a TTM matrix (Figure 2b). The previously identified lamellae structures within these particles have been classified as TTH (Park and Ostrovski, 2003). The findings presented in Table 2 support this observation, as the EDS point analysis reveals that the O/(Fe + Ti) ratio in the lamellae (points 2 and 3 in Figure 2b) is higher compared to the uniform TTM grain in Figure 2a (point 1) and the TTM matrix in Figure 2b (point 4). Furthermore, the Ti:Fe ratio in the lamellae is also higher than that of the TTM matrix.

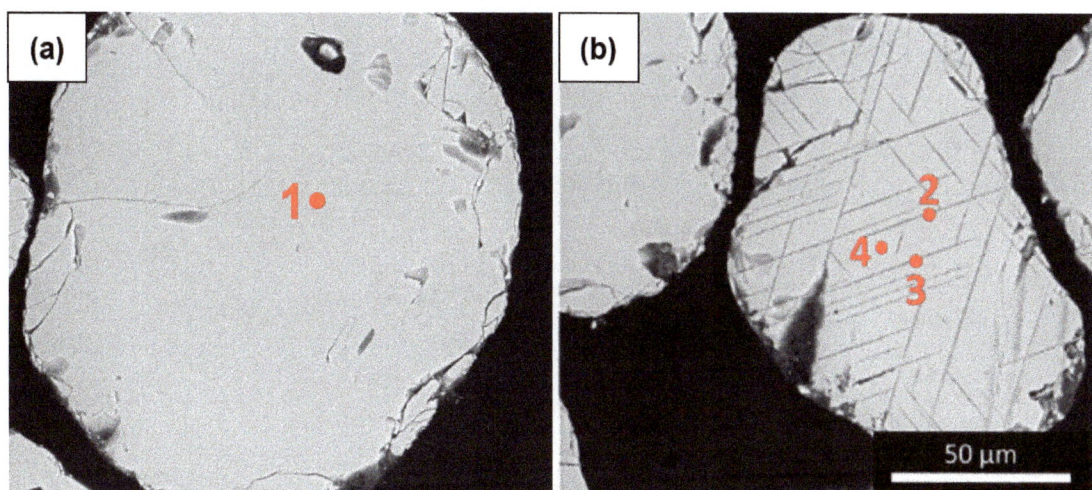

FIG 2 – Back scatter SEM image of two types of particles in the as-received NZ ironsand; (a) is the uniform TTM grain; (b) is the non-uniform particle containing TTH lamellae.

TABLE 2

Elemental spot EDS analysis of the points shown in Figure 2.

	Point (at.%)			
Element	**1**	**2**	**3**	**4**
O	53.8	61.6	61.9	58.7
Mg	3.2	-	-	1.0
Al	2.2	0.6	0.5	1.3
Ti	3.9	6.2	7.4	3.2
Mn	0.5	-	1.6	0.4
Fe	36.3	31.2	31.3	37.7
O/(Fe+Ti)	1.3	1.6	1.6	1.4

Experimental reactor

Figure 3 depicts the schematic diagram of the laboratory-scale FB reactor utilised in this study, with comprehensive details available elsewhere (Prabowo *et al*, 2019). In summary, the system consists of a vertical quartz tube housed within a radiant furnace. A gas distributor in the form of a fused quartz frit (pore size: 40–90 μm) is embedded within the tube. The temperature of the bed was monitored using a thermocouple inserted within the bed (referred to as the 'bed thermocouple' in Figure 3). This thermocouple served as a feedback input for the furnace temperature controller, regulating the heating power output. The maximum fluctuation in bed temperature observed during any experimental reduction was ±4°C, primarily occurring right after introducing H_2 gas into the reactor. To prevent water condensation at the reactor outlet, the top section of the quartz reactor area is insulated with alumina fibre insulation and the outlet flange temperature is controlled using a band heater. Notably, the experimental reactor features a custom-made venturi-sipper sampling system, facilitating repeated bed material sampling throughout a single experimental run without interrupting the FB operation. Real-time measurement of the water vapour partial pressure (PH_2O) in the outlet gas stream was achieved using a Vaisala HMT337 humidity sensor. A small cyclone was installed between the reactor and the sensor to collect any fine ironsand particles (<106 μm) that could potentially harm the sensor. The outlet gas line (indicated by the red line in Figure 2), the cyclone and the sensor were heated using heating tape to maintain a temperature of approximately 120°C, preventing water condensation within the pipeline connecting the reactor and the sensor. The sensor underwent *in situ* calibration at the operating temperature of 120°C by introducing controlled H_2-H_2O gas mixtures with varying H_2O mol per cent (ranging from 5 per cent to 50 per cent) into the reactor.

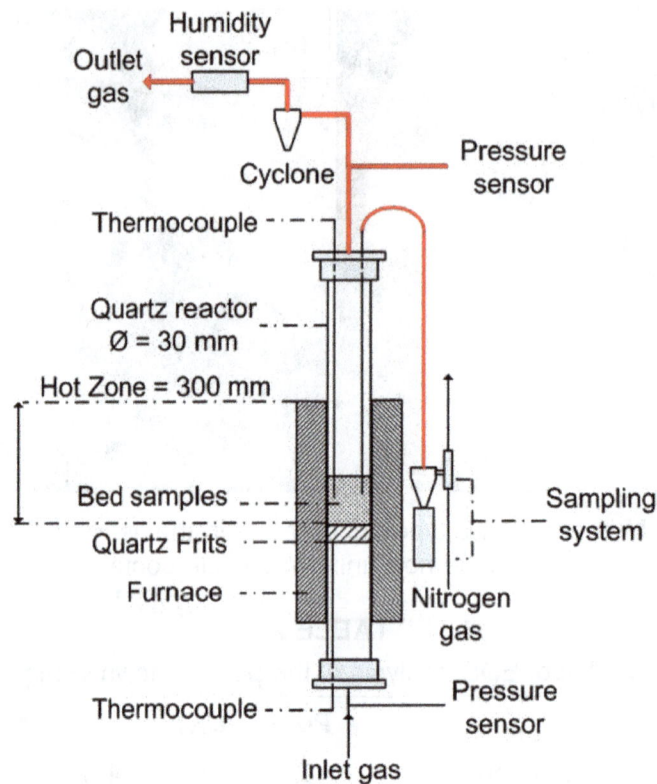

FIG 3 – Schematic diagram of the experimental FB reactor used in this work (taken from Prabowo, 2020).

Experimental procedure

Oxidation

To begin each oxidation run, an initial charge of 100 g of as-received ironsand was introduced into the reactor at room temperature. The bed was subsequently purged using argon gas (99.99 per cent purity) flowing at a rate of 3 L/min, while gradually heating the bed at a rate of 10°C/min until reaching the desired oxidation temperature. The oxidation temperatures investigated ranged from 800°C to 1000°C, with increments of 50°C. Temperatures exceeding 1000°C were not explored further due to observed issues of particle sticking and loss of fluidisation. Once the bed reached the target temperature, the fluidising gas was switched to clean, dry air sourced from a pressurised cylinder at a flow rate of 3 L/min. At predetermined sampling intervals, samples weighing approximately 2 g were extracted into a small sample container and rapidly cooled by water quenching. Upon completion of the oxidation experiment, the entire bed sample was extracted while still hot into the sample container and similarly cooled by quenching.

Reduction

To conduct the reduction experiments, we utilised 100 g of the pre-oxidised ironsand sample for each trial. Initially, the sample was placed into the reactor at room temperature. Pure hydrogen (99.98 per cent purity) was employed as the reducing gas, with a flow rate of 5 L/min. The reduction temperatures explored ranged from 800°C to 1000°C, with increments of 50°C. The procedures for charging the sample, heating, switching gas and collecting samples were akin to those employed in the oxidation experiments.

Humidity sensor

Real-time measurements of the PH_2O were conducted during the reduction reaction using a Vaisala HMT337 humidity sensor, manufactured by Vaisala in Finland. The sensor is designed for high-temperature applications up to 180°C and was installed downstream from the reactor in the gas outlet line, as depicted in Figure 3. To safeguard the sensor and prevent gas line blockage, a small cyclone separator was positioned between the reactor and the sensor to collect fine ironsand particles smaller than 106 μm. Heating tape was employed to raise the temperature of the outlet gas line (highlighted in red in Figure 3), the cyclone and the sensor to a minimum of 120°C, ensuring that water condensation would not occur anywhere along the pipeline. *In situ* calibration of the sensor was performed at an operating temperature of approximately 120°C. Controlled gas mixtures of H_2-H_2O with varying H_2O mol. per cent ratios (ranging from 5 per cent to 50 per cent) were fed into the reactor to establish a continuous linear calibration across the entire range of gas compositions. The sensor exhibited a response delay of less than 10 sec to changes in moisture levels. The calibrated Td values were then utilised to calculate PH_2O using Equation 1.

Sample characterisation

The crystal phases present in each partially reduced ironsand sample were examined using XRD analysis conducted on a Bruker D8 Advance instrument. Co Kα radiation source was utilised, with a step size of 0.05° and a collection time of 2 sec per step, covering an angular range of $15° \leq 2\theta \leq 80$. Comprehensive XRD pattern fittings were performed using TOPAS 4.2 software from Bruker. This enabled quantitative XRD (q-XRD) analysis of both the weight percentage (wt per cent) and lattice parameter of each crystalline phase within every sample. The degree of metallisation (per cent met) was determined from the q-XRD data using Equation 1.

$$\%met = \frac{Fe^o}{Fe_{Total}} \times 100 \tag{1}$$

Fe_{Total} represents the total iron content across all iron-containing phases, while $Fe°$ specifically refers to the metallic iron content. To validate the accuracy of the q-XRD analysis, a standard chemical titration method outlined in ISO 16878:2016 (ISO, 2016) was employed independently and the results from both methods were compared. The comparison revealed a remarkable agreement within 3 per cent met for the entire range of achievable metallisation degrees, reinforcing the reliability of the q-XRD analysis (Prabowo *et al*, 2019).

Microstructural characterisation of the samples was performed using a field-emission scanning electron microscope (FEG-SEM, FEI, Nova 450) operating at 20 kV. An integrated energy dispersive spectroscopy (EDS) detector at 15 kV facilitated elemental spot analysis and mapping. For SEM analysis, the powder samples were initially embedded in an epoxy resin mold. To prepare the resin-mounted samples, a grinding process was employed using SiC paper ranging from #220 to #4000 grade. Subsequently, the samples underwent polishing with 3 μm and 1 μm diamond paste. To enhance conductivity and minimise charging effects, the polished samples were carbon-coated prior to being placed in the SEM for imaging and analysis.

RESULTS AND DISCUSSION

Oxidation of TTM

Figure 4 displays the oxidation degree (per cent) of TTM plotted against time. The data reveals that the oxidation rate exhibits a substantial increase as the temperature rises, with the most rapid oxidation occurring at 1000°C. After 1 hr of exposure at 1000°C, complete oxidation is achieved. Notably, there is an absence of any sticking during the oxidation process. However, at temperatures below 950°C, incomplete oxidation persists even after a duration of 2 hr.

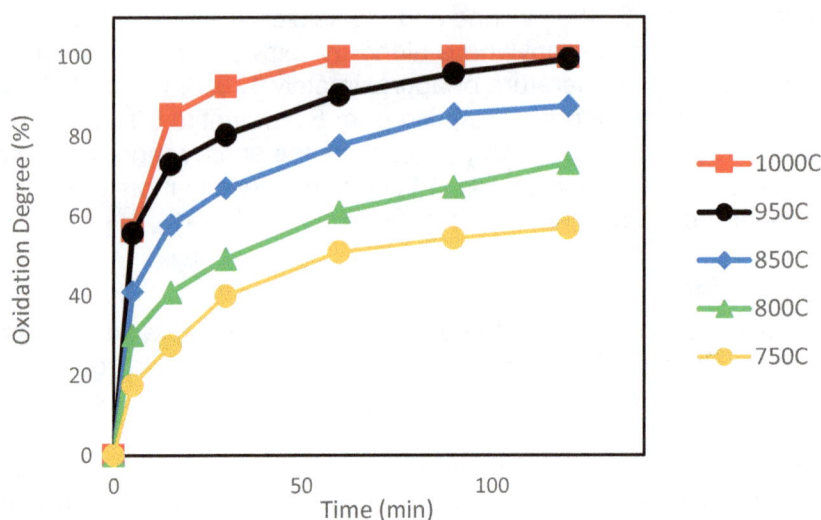

FIG 4 – Plot of per cent oxidation degree during oxidation in flowing air (3 L/min) at various bed temperatures.

Figure 5 provides the changes in phases during the oxidation process. The concentration of magnetite falls sharply during the first 15 min as the oxidation to hematite occurs after which it slows down. The formation of pseudobrookite starts to then increase indicating that Ti is diffusing and segregating away from the TTH phase.

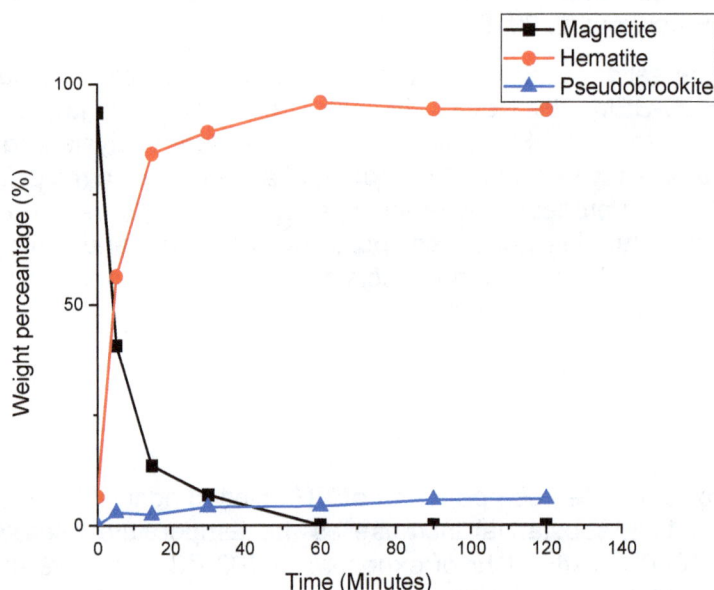

FIG 5 – Plot of weight percentage of magnetite, hematite and pseudobrookite during the oxidation at 1000°C in a FB using 3 L/min of air.

Morphological and microstructural changes during the oxidation of TTM

Figure 6 shows backscatter electron (BSE) images of the cross-section of particles and provides a comparison of the morphology and microstructure between an as-received ironsand particle (Figure 6a and 6c) and a fully oxidised ironsand particle (Figure 6b and 6d). In Figure 6a, the as-received ironsand particle exhibits a uniform structure throughout. EDS analysis conducted on point 7 confirms the presence of iron (Fe), Ti, magnesium (Mg) and aluminium (Al) within the particle.

Figure 6b depicts a fully oxidised ironsand particle after being subjected to 60 min of exposure at 1000°C with a flow rate of 3 L/min of air in the FB system. Notably, Figure 6b reveals a non-uniform particle with two distinct phases: a light region and a dark region. EDS analysis (Figure 7) conducted on the dark regions (points 1, 2 and 3) indicates a high concentration of Mg and Al and a low concentration of Ti. Conversely, the lighter regions (points 4, 5 and 6) display a low concentration of

Mg and Al but a high concentration of Ti. Furthermore, the concentration of Fe in the dark regions appears to be lower. Additionally, the oxidised particle exhibits voids, which can be attributed to the crystal expansion from magnetite to hematite.

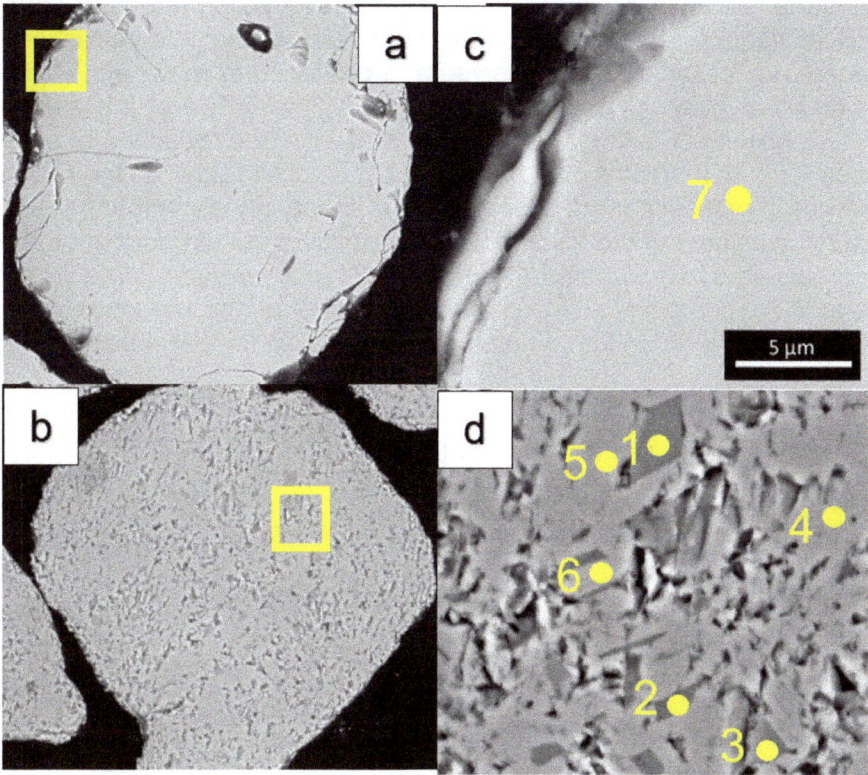

FIG 6 – BSE-SEM image of as-received ironsand particle (a) and a fully oxidised ironsand particle (b). Images (c) and (d) are high magnification of the as-received and oxidised particles of the yellow outlined area.

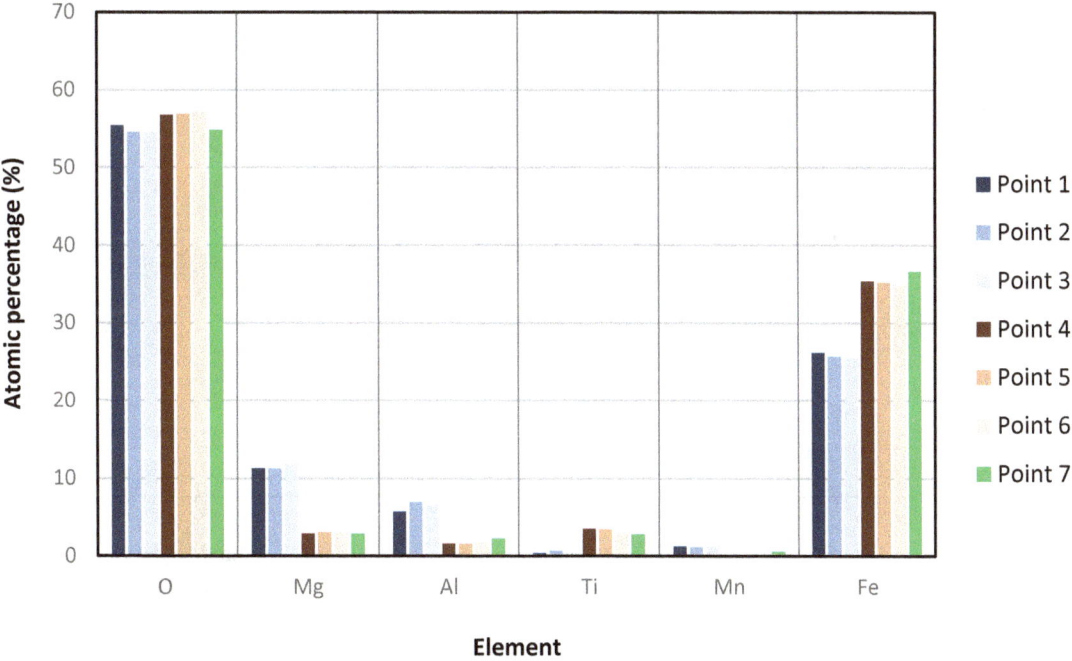

FIG 7 – Elemental spot EDS analysis of points shown in Figure 5.

These observations provide valuable insights into the structural and elemental changes that occur during the oxidation process of ironsand particles, highlighting the formation of distinct phases and the presence of voids resulting from crystal expansion.

Reduction of TTM and TTH

Effect of temperature on the oxidation degree of TTM and TTH

Figure 8 illustrates the relationship between the degree of reduction and time during the reduction process in a FB at various temperatures. The data reveals that the reduction rate rises with increasing temperatures for both the as-received and pre-oxidised materials. Particularly at 800°C (Figure 8a), a significant disparity in the total reduction time between the as-received and pre-oxidised materials is observed. Initially, during the first 10 min, the reduction rate shows a relatively close resemblance for both materials, but subsequently, it accelerates. As the reaction temperature climbs, the difference in reduction rates diminishes and eventually converges. At 950°C (Figure 8d), the time required to complete the reduction process is relatively similar for both the as-received and pre-oxidised materials. It is noteworthy that despite the higher oxygen content in the pre-oxidised material, known as TTH, compared to the as-received material, the reduction process is faster for TTH, necessitating more hydrogen for complete reduction at all temperatures.

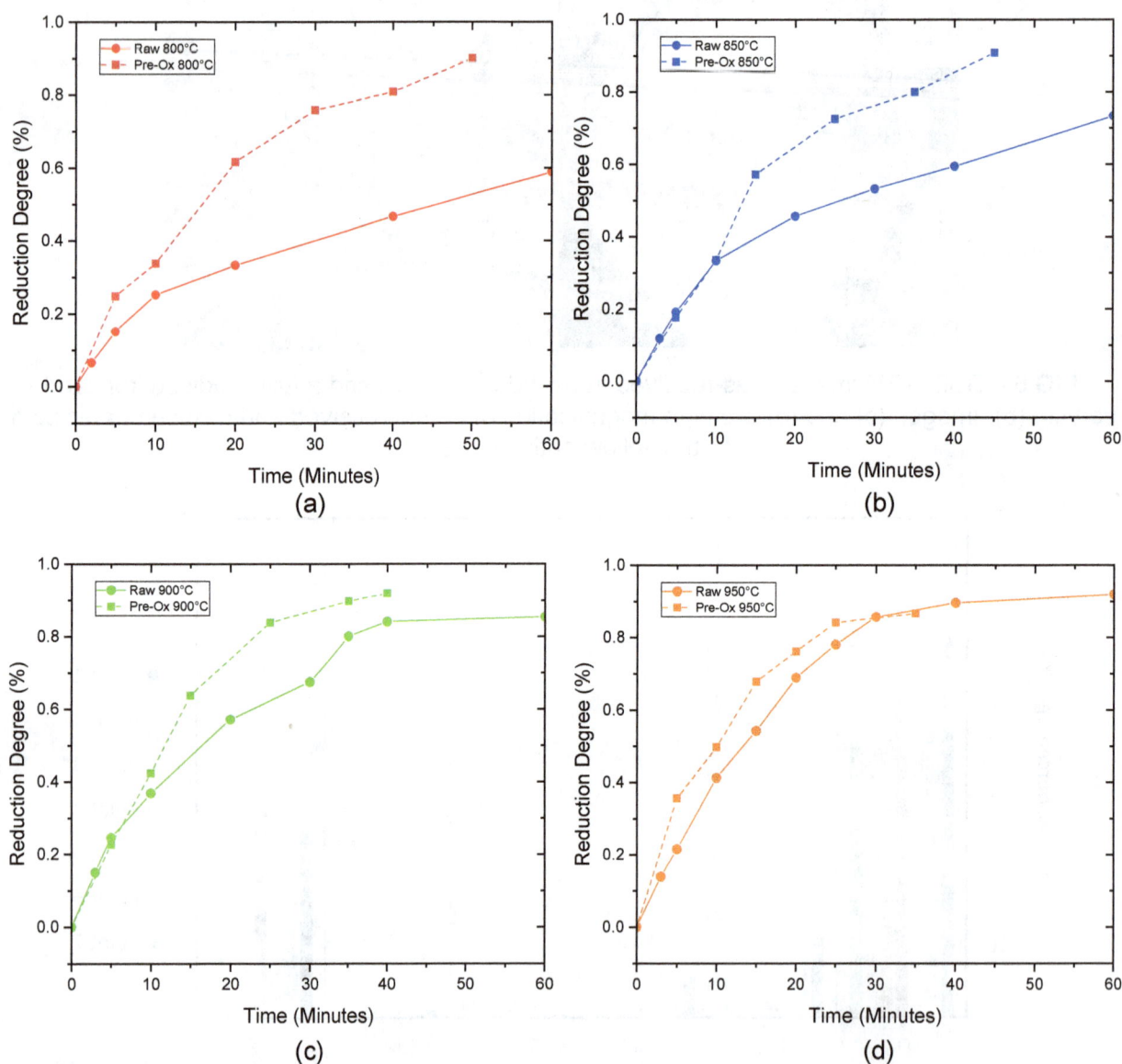

FIG 8 – Plot of reduction degree (per cent) as a function of time during the reduction of pre-oxidised ironsand in 5 L/min H_2 at different temperatures: 800°C (a), 850°C (b), 900°C (c) and 950°C (d).

Comparison of the phase evolution during the reduction of TTM and TTH

Figure 9 presents the concentration profile of TTM and wustite during the reduction process in a FB flowing 5 L/min of hydrogen. It is important to note that the complete reduction of all hematite has been excluded from this report as it occurs within 5 min in each experiment.

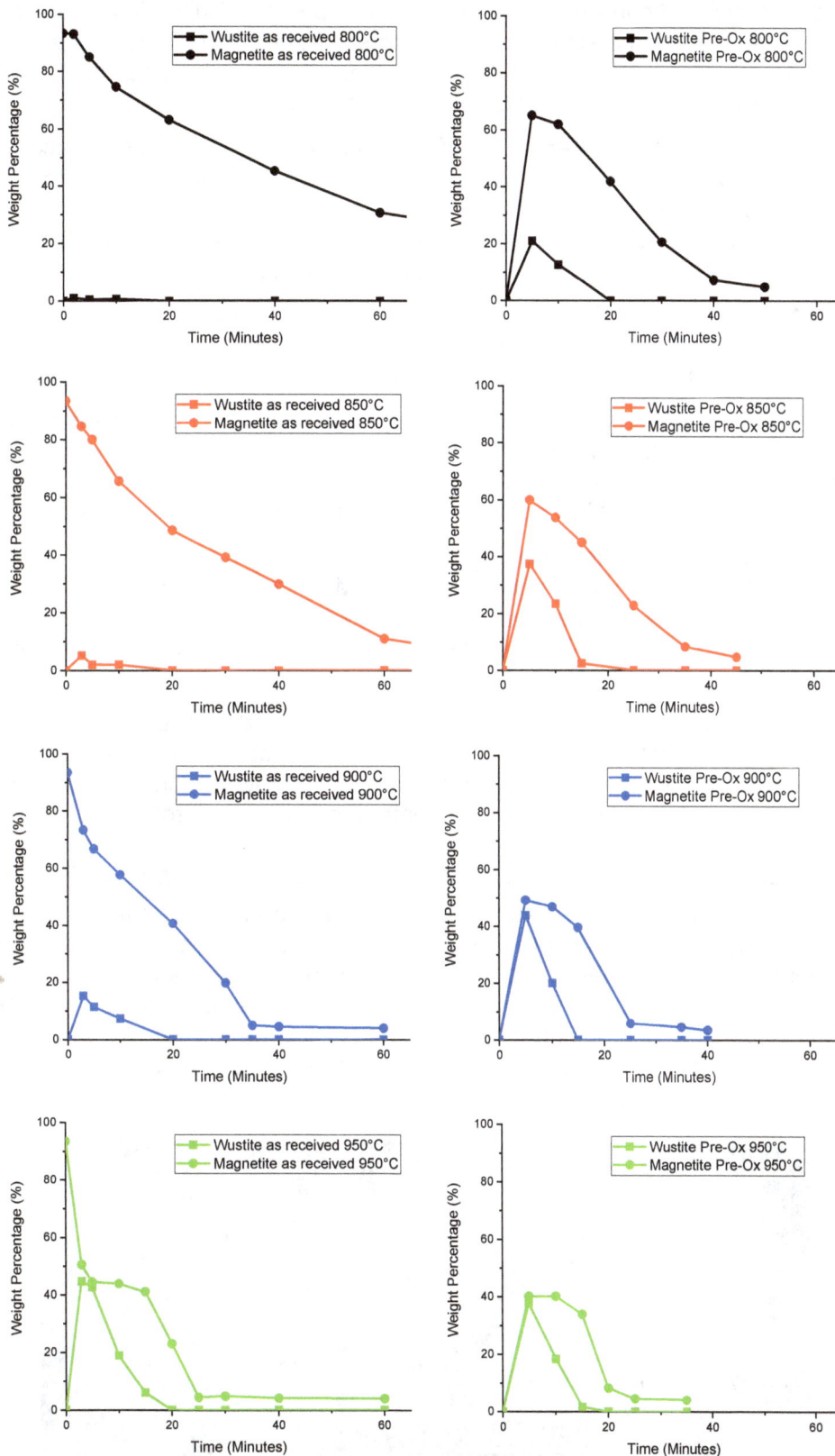

FIG 9 – Plots to show the phase evolution of magnetite and wustite as a function of time during the reduction process using 5 L/min of hydrogen at various temperatures; 800°C (black), 850°C (red), 900°C (blue) and 950°C (green).

Across all temperatures, the reduction of magnetite is observed to be faster for the oxidised material compared to the as-received material. This suggests that the magnetite produced is more readily reducible when applying a pre-oxidation step. One possible explanation for this phenomenon is the separation of aluminium (Al) from the TTM (as depicted in Figures 6 and 7), which has been found to enhance the reduction rate of TTM. Previous studies (Kapelyushin *et al*, 2015, 2017) have shown that high concentrations of Al can slow down the reduction process. This is in addition to an increase in surface area due to the volume expansion of the crystal from TTM to TTH creating voids allowing the diffusion of gas to move active sites. The formation of pseudobrookite during oxidation may remove Ti from the TTH phase which could enhance the reduction rate.

At a temperature of 950°C, the concentration of TTM using as-received and pre-oxidised ironsand reaches a plateau between approximately 5 to 10 min. This plateau is attributed to the enrichment of titanium (Ti) in the TTM due to the formation of wustite. Ti is known to be insoluble in wustite, leading to an increased concentration of Ti in the TTM. This higher concentration makes the reduction of high Ti-containing TTM challenging, resulting in a shift towards reducing wustite instead.

At temperatures of 800°C and 850°C when utilising as-received ironsand, minimal wustite production is observed. This can be attributed to the relatively slower reduction rate of TTM compared to the reduction rate of wustite at those temperatures, resulting in no significant accumulation of wustite.

However, when oxidised ironsand is employed, a distinct peak in wustite concentration is observed, providing further evidence that the oxidation step significantly increases the reduction rate of TTM to a level comparable to or higher than the rate of wustite reduction. This suggests that the oxidation process enhances the reactivity of TTM, allowing it to be reduced at a similar rate as wustite.

Furthermore, after 15 min of reduction at all temperatures, the concentration of wustite gradually decreases implying the reduction rate is quicker than TTM. This can be explained by the enrichment of Ti in the TTM, as discussed earlier, which slows down the reduction process. As a result, the reduction of TTM becomes the rate-limiting reaction, leading to a reduction in the concentration of wustite over time.

These findings highlight the influence of oxidation and temperature on the production and concentration of wustite during the reduction process. The oxidation step enhances the reduction rate of TTM, while the presence of Ti in the TTM can affect the overall kinetics and dynamics of the reduction reactions, ultimately impacting the concentration of wustite throughout the process.

Morphology and microstructural changes during the reduction of TTM and TTH

Figure 10 presents BSE images of an as-received ironsand particle after full reduction at 950°C, compared to a pre-oxidised ironsand particle subjected to the same conditions. The most significant difference between the two particles is the formation of a shell (yellow arrow) on the exterior of the as-received particle. This shell primarily consists of a Ti-rich oxide layer, along with other impurities such as Mg and Al oxides. This shell formation is believed to be the reason why the as-received particle does not experience sticking during the reduction process, as previously reported by Prabowo (2020) and Prabowo *et al* (2019).

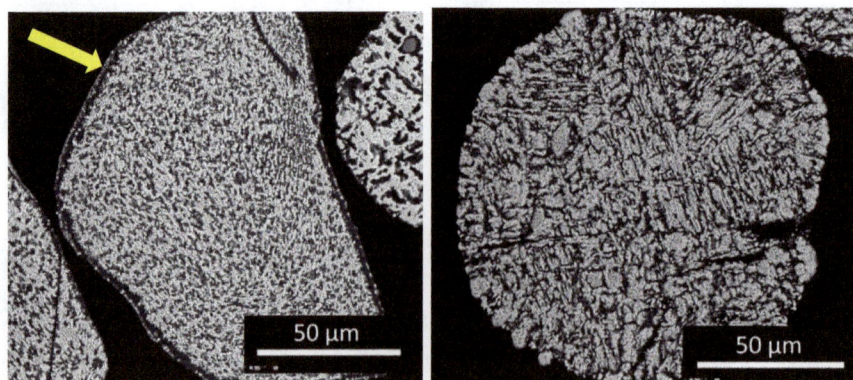

FIG 10 – BSE images of fully reduced as-received (left) and pre-oxidised (right) NZ ironsand at 950°C using 5 L/min of hydrogen.

In contrast, the BSE image on the right displays the pre-oxidised particle, which does not exhibit any shell formation. Nevertheless, it is noteworthy that this pre-oxidised particle also does not experience sticking at 950°C. The exact cause for the formation of the shell in the as-received ironsand particle has not yet been definitively determined and requires further investigation.

The observation of the shell formation in the as-received particle provides valuable insights into the mechanisms underlying sticking prevention during the reduction process. The presence of a Ti-rich oxide layer and other impurities within the shell likely plays a crucial role in inhibiting sticking behaviour. Further research is needed to fully understand the factors contributing to the formation of the shell and its impact on the reduction dynamics.

CONCLUSIONS

The FB reduction of NZ TTM and TTH ironsand using H_2 gas has been successfully conducted without encountering sticking issues within the temperature range of 800°C to 950°C. When TTM is utilised, a Ti-rich oxide shell forms, effectively preventing direct Fe-Fe contact and eliminating the sticking mechanism. In contrast, no such shell forms during the reduction of TTH, resulting in the presence of an evenly distributed matrix of Fe and Ti-rich oxides containing Mg and Al oxides.

Notably, the reduction of TTH exhibits different kinetic behaviour compared to TTM, particularly at lower temperatures (800°C, 850°C and 900°C). This behaviour can be attributed to the higher reduction rate of magnetite, leading to larger concentrations of wustite in the TTH particles. These differences arise from changes in the morphology and microstructure following oxidation at 1000°C. Two significant changes are observed in the particles. First, the oxidation of TTM to TTH causes particle expansion, resulting in microcracks throughout the structure. These microcracks facilitate deeper penetration of hydrogen into the particle and increase the number of active sites for the reduction reaction, without the need for solid-state diffusion. Second, a two-phase segregation occurs within the originally homogeneous TTM matrix, where a phase with high Mg and Al content coexists with a phase containing low Ti, while a high Ti phase is dispersed among regions with low Mg and Al content. It is hypothesised that the high Mg and Al phases hinder the reduction process, while the segregation may aid in the diffusion of Ti through the particle.

This study highlights the potential advantages of implementing a pre-oxidation process on NZ ironsands to enhance the reduction rate at temperatures below 950°C. This, in turn, can lower the required process temperature, offering potential energy savings and process efficiency improvements.

ACKNOWLEDGEMENTS

This research was supported by funding received from the Endeavour Fund of the New Zealand Ministry of Business Innovation and Employment (Grant No. RTVU1907).

REFERENCES

Brathwaite, R L, Gazley, M F and Christie, A B, 2017. Provenance of titanomagnetite in ironsands on the west coast of the North Island, New Zealand, *Journal of Geochemical Exploration*, 178:23–34. https://doi.org/10.1016/j.gexplo.2017.03.013

Evans, N, 1986. Direct reduction at New Zealand's Glenbrook Works, *Steel Times International*, 38.

Goodman, N J, 2019. The HIsmelt technology: from Australia to China... and back again?, in *Proceedings Iron Ore 2019*, pp 313 (The Australasian Institute of Mining and Metallurgy: Melbourne).

Guo, L, Bao, Q, Gao, J, Zhu, Q and Guo, Z, 2020. A review on prevention of sticking during fluidized bed reduction of fine iron ore, *ISIJ International*, 60(1):1–17. https://doi.org/10.2355/isijinternational.ISIJINT-2019-392

Hayashi, S and Iguchi, Y, 1992. Factors Affecting Bed Reduction the Sticking of Fine Iron Ores during Fluidized, *ISIJ International*, 32(1).

Hessling, O, Tottie, M and Sichen, D, 2021. Experimental study on hydrogen reduction of industrial fines in fluidized bed, *Ironmaking and Steelmaking*, 48(8):936–943. https://doi.org/10.1080/03019233.2020.1848232

Hillisch, W and Zirngast, J, 2001. Status of Finmet plant operation at BHP DRI, Australia, *Steel Times International*, 25:20.

International Organization for Standardization (ISO), 2016. ISO 16878:2016 – Iron ores – Determination of metallic iron content – Iron(III) chloride titrimetric method.

Kapelyushin, Y, Sasaki, Y, Zhang, J, Jeong, S and Ostrovski, O, 2017. Formation of a Network Structure in the Gaseous Reduction of Magnetite Doped with Alumina, *Metallurgical and Materials Transactions B: Process Metallurgy and Materials Processing Science*, 48(2):889–899. https://doi.org/10.1007/s11663-016-0897-1

Kapelyushin, Y, Xing, X, Zhang, J, Jeong, S, Sasaki, Y and Ostrovski, O, 2015. Effect of Alumina on the Gaseous Reduction of Magnetite in CO/CO_2 Gas Mixtures, *Metallurgical and Materials Transactions B: Process Metallurgy and Materials Processing Science*, 46(3):1175–1185. https://doi.org/10.1007/s11663-015-0316-z

Longbottom, R J, Ingham, B, Reid, M H, Studer, A J, Bumby, C W and Monaghan, B J, 2019. In situ neutron diffraction study of the reduction of New Zealand ironsands in dilute hydrogen mixtures, *Mineral Processing and Extractive Metallurgy: Transactions of the Institute of Mining and Metallurgy*, 128(3):183–192. https://doi.org/10.1080/03719553.2017.1412877

Longbottom, R J, Monaghan, B J and Mathieson, J G, 2013. Development of a bonding phase within titanomagnetite-coal compacts, *ISIJ International*, 53(7):1152–1160. https://doi.org/10.2355/isijinternational.53.1152

Longbottom, R J, Ostrovski, O and Park, E, 2006. Formation of Cementite from Titanomagnetite Ore, *ISIJ International*, 46(5):641–646. https://doi.org/10.2355/isijinternational.46.641

Lu, F, Zhong, H, Liu, B, Xu, J, Zhang, S, and Wen, L, 2023. Particle agglomeration behavior in fluidized bed during direct reduction of iron oxide by CO/H2 mixtures, *Journal of Iron and Steel Research International*. https://doi.org/10.1007/s42243-022-00882-5

Park, E and Ostrovski, O, 2003. Reduction of Titania-Ferrous Ore by Carbon Monoxide, *ISIJ International*, 43(9):1316–1325. https://doi.org/10.2355/isijinternational.43.1316

Park, E and Ostrovski, O, 2004. Reduction of Titania-Ferrous Ore by Hydrogen, *ISIJ International*, 44(6):999–1005. https://doi.org/10.2355/isijinternational.44.999

Park, E, Lee, S-B, Ostrovski, O, Dong-Jun, M and Rhee, C-H, 2004. Reduction of the Mixture of Titanomagnetite Ironsand and Hematite Iron Ore Fines by Carbon Monoxide, *ISIJ International*.

Plaul, F J, Böhm, C and Schenk, J, 2009. Fluidized-bed technology for the production of iron products for steelmaking, *Journal of The South African Institute of Mining and Metallurgy*, 109(2009):121–128.

Prabowo, S, 2020. Reduction of New Zealand Titanomagnetite Ironsand by Hydrogen Gas in a Fluidised Bed System, PhD thesis, Te Herenga Waka-Victoria University of Wellington. https://doi.org/10.26686/wgtn.17151755

Prabowo, S W, Longbottom, R J, Monaghan, B J, del Puerto, D, Ryan, M J and Bumby, C W, 2019. Sticking-Free Reduction of Titanomagnetite Ironsand in a Fluidized Bed Reactor, *Metallurgical and Materials Transactions B: Process Metallurgy and Materials Processing Science*, 50(4). https://doi.org/10.1007/s11663-019-01625-w

Prabowo, S W, Longbottom, R J, Monaghan, B J, del Puerto, D, Ryan, M J and Bumby, C W, 2022. Phase transformations during fluidized bed reduction of New Zealand titanomagnetite ironsand in hydrogen gas, *Powder Technology*, 398. https://doi.org/10.1016/j.powtec.2021.117032

Schenk, J L, 2011. Recent status of fluidized bed technologies for producing iron input materials for steelmaking, *Particuology*, 9(1):14–23. https://doi.org/10.1016/j.partic.2010.08.011

Shahabuddin, M, Brooks, G and Rhamdhani, M A, 2023. Decarbonisation and hydrogen integration of steel industries: Recent development, challenges and technoeconomic analysis, *Journal of Cleaner Production*, vol 395. https://doi.org/10.1016/j.jclepro.2023.136391

Shannon, W, Kitt, W and Marshall, T, 1960. Experimental fluidized-bed reduction of New Zealand ironsands, *New Zealand Journal of Science*, 3:74–90.

She, X F, Sun, H Y, Dong, X J, Xue, Q G and Wang, J S, 2013. Reduction mechanism of titanomagnetite concentrate by carbon monoxide, *Journal of Mining and Metallurgy, Section B: Metallurgy*, 49(3):263–270. https://doi.org/10.2298/JMMB121001020S

Spreitzer, D and Schenk, J, 2019a. Iron Ore Reduction by Hydrogen Using a Laboratory Scale Fluidized Bed Reactor: Kinetic Investigation—Experimental Setup and Method for Determination, *Metallurgical and Materials Transactions B: Process Metallurgy and Materials Processing Science*, 50(5):2471–2484. https://doi.org/10.1007/s11663-019-01650-9

Spreitzer, D and Schenk, J, 2019b. Reduction of Iron Oxides with Hydrogen—A Review, *Steel Research International*, 90(10). https://doi.org/10.1002/srin.201900108

Sui, Y-L, Guo, Y-F, Jiang, T, Xie, X-L, Wang, S and Zheng, F-Q, 2017. Gas-based reduction of vanadium titano-magnetite concentrate: behavior and mechanisms, *International Journal of Minerals, Metallurgy and Materials*, 24(1):10–17. https://doi.org/10.1007/s12613-017-1373-x

Sun, H, Adetoro, A A, Pan, F, Wang, Z and Zhu, Q, 2017. Effects of High-Temperature Preoxidation on the Titanomagnetite Ore Structure and Reduction Behaviors in Fluidized Bed, *Metallurgical and Materials Transactions B: Process Metallurgy and Materials Processing Science*, 48(3):1898–1907. https://doi.org/10.1007/s11663-017-0925-9

Sun, H, Wang, J, Han, Y, She, X and Xue, Q, 2013. Reduction mechanism of titanomagnetite concentrate by hydrogen, *International Journal of Mineral Processing*, 125:122–128. https://doi.org/10.1016/j.minpro.2013.08.006

Templeton, F, 2006. Iron and steel – Attempts to extract iron, *Te Ara – the Encyclopedia of New Zealand*. https://teara.govt.nz/mi/iron-and-steel/page-2

Vogl, V, Åhman, M and Nilsson, L J, 2018. Assessment of hydrogen direct reduction for fossil-free steelmaking, *Journal of Cleaner Production*, 203:736–745. https://doi.org/10.1016/j.jclepro.2018.08.279

Wang, Z, Pinson, D, Chew, S, Rogers, H, Monaghan, B J, Pownceby, M I, Webster, N A S and Zhang, G, 2016. Behavior of New Zealand Ironsand During Iron Ore Sintering, *Metallurgical and Materials Transactions B: Process Metallurgy and Materials Processing Science*, 47(1):330–343. https://doi.org/10.1007/s11663-015-0519-3

Wang, Z, Zhang, J, Jiao, K, Liu, Z and Barati, M, 2017. Effect of pre-oxidation on the kinetics of reduction of ironsand, *Journal of Alloys and Compounds*, 729:874–883. https://doi.org/10.1016/j.jallcom.2017.08.293

Wolfinger, T, Spreitzer, D and Schenk, J, 2022. Analysis of the Usability of Iron Ore Ultra-Fines for Hydrogen-Based Fluidized Bed Direct Reduction — A Review, *Materials,* 15(7). https://doi.org/10.3390/ma15072687

Wolfinger, T, Spreitzer, D, Zheng, H and Schenk, J, 2022. Influence of a Prior Oxidation on the Reduction Behavior of Magnetite Iron Ore Ultra-Fines Using Hydrogen, *Metallurgical and Materials Transactions B: Process Metallurgy and Materials Processing Science*, 53(1):14–28. https://doi.org/10.1007/s11663-021-02378-1

Yi, S H, Cho, M Y and Yi, J G, 2018. Footprints for the future of commercial FINEX plants, in *Proceedings of the 8th International Congress of Science and Technology of Ironmaking*, pp 25–28.

Yu, J, Hu, N, Xiao, H, Gao, P and Sun, Y, 2021. Reduction behaviors of vanadium-titanium magnetite with H2 via a fluidized bed, *Powder Technology*, 385:83–91. https://doi.org/10.1016/j.powtec.2021.02.038

Yu, J, Ou, Y, Sun, Y, Li, Y and Han, Y, 2022. Hydrogen reduction behaviors and mechanisms of vanadium titanomagnetite ore under fluidized bed conditions, *Powder Technology*, 402. https://doi.org/10.1016/j.powtec.2022.117340

Zhang, A, Monaghan, B J, Longbottom, R J, Nusheh, M and Bumby, C W, 2020. Reduction Kinetics of Oxidized New Zealand Ironsand Pellets in H2 at Temperatures up to 1443 K, *Metallurgical and Materials Transactions B: Process Metallurgy and Materials Processing Science*, 51(2):492–504. https://doi.org/10.1007/s11663-020-01790-3

Zhang, T, Lei, C and Zhu, Q, 2014. Reduction of fine iron ore via a two-step fluidized bed direct reduction process, *Powder Technology*, 254:1–11. https://doi.org/10.1016/j.powtec.2014.01.004

Zheng, H, Daghagheleh, O, Ma, Y, Taferner, B, Schenk, J and Kapelyushin, Y, 2023. Phase transition of magnetite ore fines during oxidation probed by in situ high-temperature X-ray diffraction, *Metallurgical and Materials Transactions B: Process Metallurgy and Materials Processing Science*. https://doi.org/10.1007/s11663-023-02754-z

Driving the transformation to hydrogen ironmaking – an experimental vertical shaft H$_2$-DRI reactor facility in NZ

B H Yin[1], S Mendoza[2], B Rumsey[3], Y Iwasaki[4], M Lynch[5], M McCurdy[6] and C W Bumby[7]

1. Senior Scientist, Robinson Research Institute, Faculty of Engineering, Victoria University of Wellington, Wellington 5010, New Zealand. Email: ben.yin@vuw.ac.nz
2. PhD candidate, Robinson Research Institute, Faculty of Engineering, Victoria University of Wellington, Wellington 5010, New Zealand. Email: shaira.mendoza@vuw.ac.nz
3. Senior Technician, Robinson Research Institute, Faculty of Engineering, Victoria University of Wellington, Wellington 5010, New Zealand. Email: ben.rumsey@vuw.ac.nz
4. Research Assistant, Robinson Research Institute, Faculty of Engineering, Victoria University of Wellington, Wellington 5010, New Zealand. Email: yukinori.iwasaki@vuw.ac.nz
5. Junior Engineer, Robinson Research Institute, Faculty of Engineering, Victoria University of Wellington, Wellington 5010, New Zealand. Email: matt.lynch@vuw.ac.nz
6. Energy Materials Scientist, GNS Science Wellington 5010, New Zealand. Email: m.mccurdy@gns.cri.nz
7. Chief Scientist, Robinson Research Institute, Faculty of Engineering, Victoria University of Wellington, Wellington 5010, New Zealand. Email: chris.bumby@vuw.ac.nz

ABSTRACT

Iron and steel production is responsible for approximately 6.3 per cent of global anthropogenic CO$_2$ emissions. Direct reduction of iron with hydrogen is a potential route for decarbonising the iron and steel industry. Extensive small-scale laboratory experiments have been performed to test the feasibility of hydrogen-direct reduction of iron ores sourced from diverse geographical locations. However, test results from the batch reduction of small-sample do not necessarily translate to continuous commercial-scale processes. Establishing a small-scale continuous test reactor for H$_2$-DRI process is challenging, requiring substantial resources and investment. The limited availability of small-scale reactors that enable testing under relevant conditions has become a major obstacle for H$_2$-DRI development, leaving a big gap between laboratory success and proof of commercial feasibility. Here, we report on the recent commissioning of a continuous counter-flow vertical shaft reactor, which has a processing capacity of 6 kg/hr iron ore pellets at up to 100 NL/min H$_2$ flow. We believe this to currently be the largest H$_2$-DRI facility in the Southern Hemisphere. In this talk, we will discuss the design and construction journey for this reactor, starting from the initial system design to the initial demonstration of H$_2$ reduction of titanomagnetite ironsand pellets. This reactor now enables a range of research capacities, including the validation of laboratory results on a larger scale with practical operational variables, feasibility assessment of hydrogen direct reduction of different iron ores on a commercial scale; identification of scalability, sticking, and other potential issues before building a full-scale plant; and a viable solution for producing small-quantity DRIs for downstream melting investigations. Overall, this new research facility offers significant benefits for assisting the transformation of iron and steel production to a more sustainable and green industry.

INTRODUCTION

The steelmaking industry is one of the largest contributors to global anthropogenic CO$_2$ emissions, accounting for approximately 6.3 per cent of the total emissions. This has led to increased pressure on the industry to transition to more sustainable and environmentally friendly practices (Vogl, Åhman and Nilson, 2018).

Hydrogen direct reduction of iron (H$_2$-DRI) has emerged as a potential solution to decarbonise the steelmaking industry. H$_2$-DRI is a process that involves reducing iron ore using hydrogen gas instead of carbon monoxide, which is the primary reducing agent in traditional blast furnaces. Rather than generating CO$_2$, the only by-product of an H$_2$-DRI process is water vapour. The carbon footprint is substantially reduced if the hydrogen used in an H$_2$-DRI process is generated from renewable energy sources such as wind and solar power (Vogl, Olsson, Nykvist, 2021; Midrex, 2013; van Vuuren et al, 2022). Although there have been extensive laboratory experiments demonstrating the feasibility of the H$_2$-DRI processes at the gram scale, these small batch samples are not necessarily representative of a large-scale continuous process (Midrex, 2013). It is therefore desirable to establish a small-scale test reactor that can provide a test platform to study the behaviour of different iron ores under practical continuous-reduction conditions. However, such research facilities are scarce due to their high cost and complexity. This leaves significant uncertainties regarding the scalability of the process, which is a major barrier to promoting green hydrogen steelmaking.

To address these challenges, we have designed and constructed a laboratory-scale continuous vertical shaft H_2-DRI reactor capable of processing up to ~20 kg of pelletised iron ore fines in a single run. This paper documents the journey of designing, assembling, and commissioning this reactor. The reactor now enables the validation of laboratory results on a larger scale, replicating the operating conditions of commercial-scale processes. It will provide insights into the behaviour of different iron ores during H_2-DRI processing and also allows the identification of potential challenges that may arise during future scale-up of the process. In addition, it enables kg-quantities of reduced H_2-DRI pellets to be produced for use in subsequent investigations of behaviour during melting and slag separation. This new facility aims to offer practical and reliable information to both iron ore suppliers and the steelmaking industry, in order to facilitate the successful commercial development of green hydrogen steelmaking.

DESIGN AND CONSTRUCTION OF THE REACTOR

Hydrogen is a highly flammable and potentially dangerous gas, and the operation of an H_2-DRI reactor involves a sequence of gas-solid reactions that must be performed at high temperatures. As a result, utmost attention to health and safety (H&S) considerations are paramount in the design and construction of an experimental test reactor. Reactor construction requires detailed knowledge and expertise in process and control system design, reaction engineering, handling hazardous materials, and managing complex engineering projects. This project was meticulously planned and executed, adhering to a systematic step-wise approach, as shown in Figure 1.

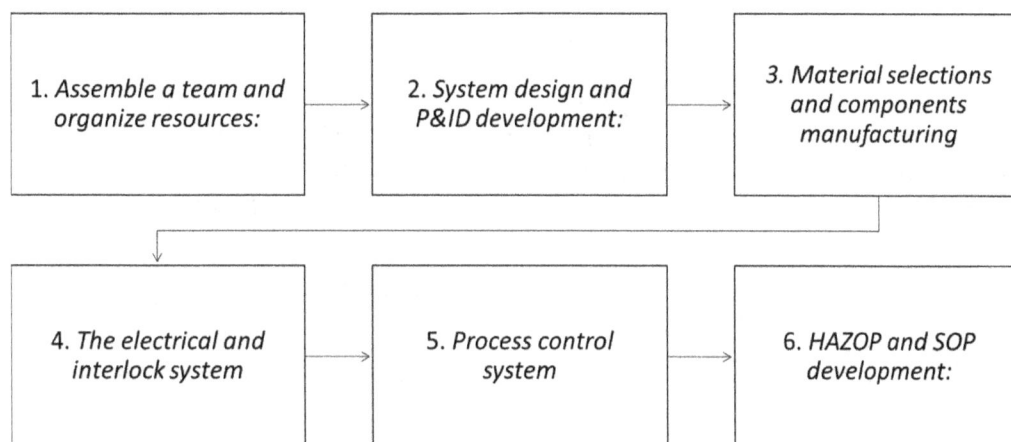

FIG 1 – A flow of critical steps.

Assemble a team and organise resources

This reactor forms part of a Research Program funded by New Zealand (NZ) government the MBIE Endeavour scheme. A project team with complementary expertise was assembled, consisting of professionals with diverse expertise in chemical and process engineering, metallurgical engineering, electrical engineering, applied physics, and materials science. The selection of an appropriate site for the reactor system is a crucial consideration. The site must meet specific criteria, including ample space to accommodate the reactor and associated infrastructure, as well as providing easy access for iron ore handling, milling, and palletisation activities. Moreover, it is essential to comply with local regulations and safety standards in order to ensure the protection of both personnel and the surrounding environment. The former site of the NZ Coal Research Laboratory in Lower Hutt, Wellington fulfils all these requirements. This site also benefits from its strategic location within a hydrogen research cluster in NZ, encompassing the Robinson Research Institute, Callaghan Innovation, and GNS Science. These organisations house specialist expertise and advanced characterisation instruments.

A test bunker measuring 4.8 m (width) × 3.8 m (length) × 5.4 m (height), was designated for the installation of the H_2-DRI reactor. The generous height of the bunker allows for the installation of the reactor without any space constraints. This bunker offers the advantage of good air circulation, which is crucial for maintaining a safe and well-ventilated environment within the reactor area. To further enhance air exchange, a high-capacity extraction system and additional louver ventilation were implemented, to eliminate the possibility of a gas leak leading to hydrogen accumulation within the bunker during reactor operation.

System design and P&ID development

The design parameters for the H_2-DRI system (Table 1) were established from the outset, prioritising essential test requirements.

TABLE 1

The design parameter of the H_2-DRI reactor.

Design parameters	
Temperature	Room – 1200°C
Pressure	Ambient – 2 Bar (relief pressure)
Iron ore capacity	Up to 20 kg per batch
Production rate	Up to 12 kg/hr
Heating zone	400 mm
Total hydrogen volume	25 m^3
Hydrogen flow rate	0–250 SLPM
Nitrogen flow rate	0–100 SLPM

A Process and Instrumentation Diagram (P&ID) was created to depict the system design. The P&ID provides a visual representation of the various process components, equipment, and interconnections of the vertical shaft H_2-DRI reactor. As illustrated in Figure 2, the system can be broadly divided into three main functional components, namely the gas distribution unit, the solid distribution and reaction unit, and the support unit. A summary breakdown of each unit is as follows:

1. The gas distribution unit is responsible for distributing and controlling the gas flow in the reactor system: It comprises three upstream gas lines:

   o *Reactant H_2* is sourced from external gas cylinders, and the flow rate is regulated by a mass flow controller (MFC).

   o *Process N_2* is used for system conditioning and as balance N_2 for the reactor. It is supplied from gas cylinders and the flow rate was controlled by an MFC.

   o *Emergency purge N_2* is a high-flow emergency line that can be rapidly introduced to purge the reactor and furnace hot zone in the event of the emergency interlock being activated (see next section).

   o *A downstream exhaust line* is responsible for discharging the end gases of the reaction outdoors.

2. The solid distribution and reaction unit plays a crucial role in distributing the iron ore feedstock and facilitating the chemical reactions within the reactor. It comprises several parts arranged from bottom to top:

   o *Pellet collector* collects the ironsand pellets after the reduction process and holds these under an inert atmosphere. It is equipped with a load cell that measures the weight of discharged samples.

   o *Auger* controls the movement of the solid materials through the reactor and regulates the residence time of ironsand pellets within the reaction zone.

- Reactor is a silicon carbide (SiC) tube within which the reduction process occurs.
- Electrical Furnace heats the SiC tube in order to maintain the desired reaction temperature.
- Cooling jacket surrounds specific regions of the reactor and helps lower the temperature of overheated areas, preventing any potential damage to the reactor.
- Hopper stores the ironsand pellets (at an elevated temperature) before they are transported downwards through the reactor. This enables a continuous supply of feedstock to the system.
- Cyclone is employed to remove entrained fine particles from the exhausted gas. This prevents damage to the humidity sensor or blockage of the exhaust line.
- Condenser collects and removes water vapour produced by the reaction.

3. The support unit consists of auxiliary systems and devices that provide the necessary support for the overall operation of the H_2-DRI system, which comprises:

- Cooling water system is designed to provide continuous cooling water to maintain the required operating temperatures within the desired range throughout the system. It can also help to fine-tune the temperature of the system.
- Temperature (×10), humidity (×1), and pressure (×3) probes are strategically placed throughout the system to monitor and measure the temperature, humidity, and pressure levels. This data is crucial for maintaining system control and ensuring safe operating conditions.
- Heating jacket is installed to maintain the temperature of the downstream components and prevents any moisture from condensing before reaching the humidity sensor.
- Laboratory ventilation system ensures proper ventilation in the laboratory area at all times that the reactor is operating.
- CCTV and atmosphere sensors add another layer of detection on any abnormal changes in the bunker, enabling prompt notification and response to any potential issues.

The system operates with a gas-solid counter-current flow configuration, where the reduction takes place within the SiC tube, which is heated by the electrical furnace to facilitate the reaction. During system operation, reactant gas H_2 is introduced into the reactor above the auger and flows upwards through the SiC tube. Simultaneously, pre-pelletised iron ore feed is stored in the hopper and move downwards at a rate controlled by the auger. The resulting reduced pellets are discharged into a pellet collector for further handling and processing. Unreacted H_2 and water vapour produced from the reaction are discharged and exhausted externally. It should be noted that in a large commercial-scale plant, the hydrogen in the exhaust gas would be collected and recycled using a condensing heat exchanger to maximise gas utilisation. However, this level of complexity is not necessary for a small-scale experimental test reactor.

The P&ID allows for a comprehensive understanding of the system's functionality and precise placement of components, piping, and instrumentation. Importantly, it enables efficient communication and collaboration among team members and with external stakeholders, focusing on effective discussions and decision-making during the construction project.

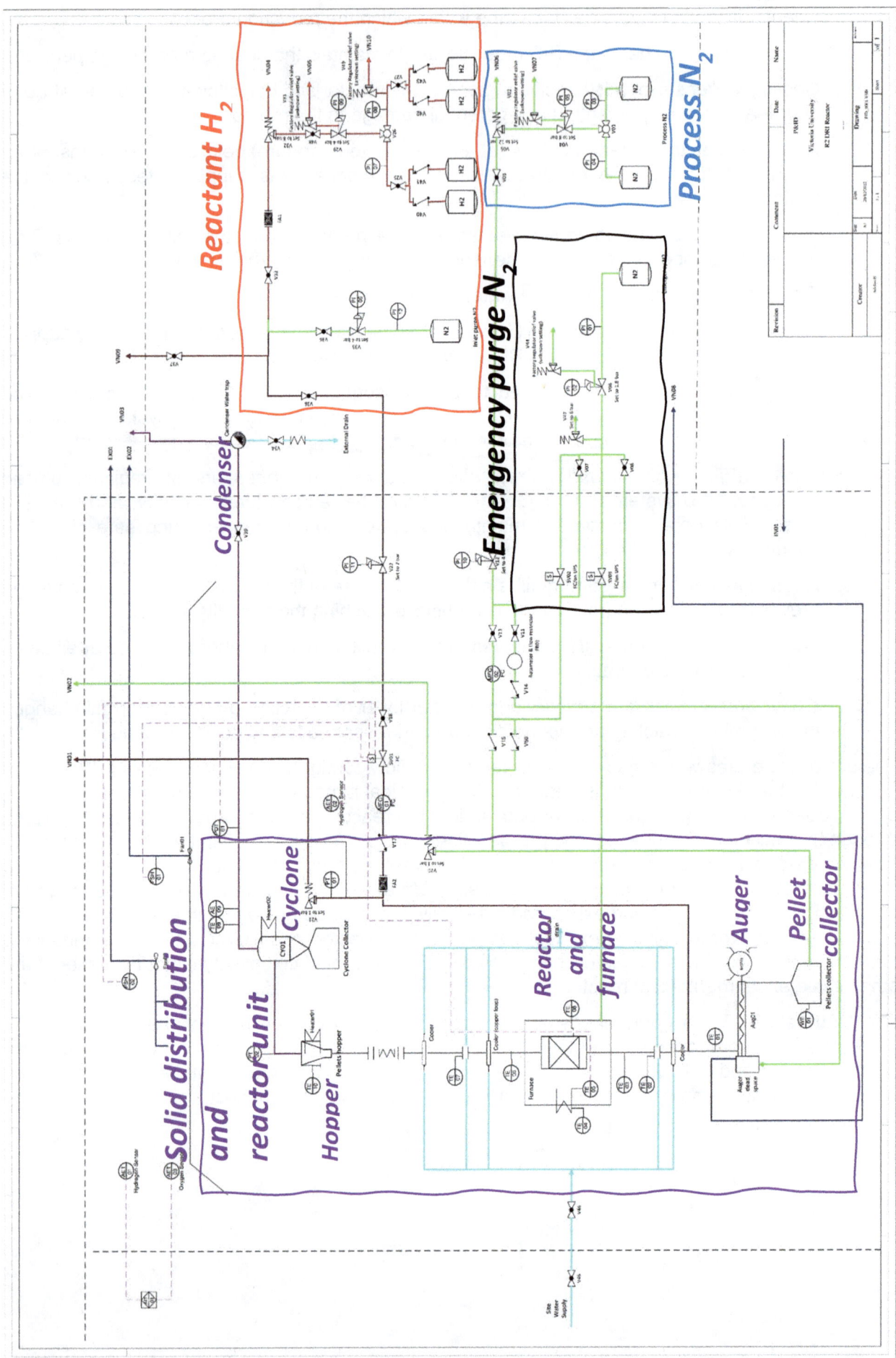

FIG 2 – P&ID diagram of the vertical shaft H₂-DRI reactor system.

Material selections and components manufacturing

The selection of appropriate materials and the manufacturing of components play a crucial role in ensuring the safe and efficient operation of the H_2-DRI reactor system.

An extensive evaluation of potential reactor construction materials was conducted, specifically focusing on their resistance to high temperatures, hydrogen embrittlement, mechanical strength, and toughness. From the available material options, silicon carbide (SiC) was selected for the reactor tube within which the high-temperature reaction takes place. SiC offers exceptional thermal stability, hydrogen tightness, and mechanical properties. The feasibility of utilising SiC tubes was verified through a series of pre-construction tests, which confirmed negligible hydrogen leakage and pressure tolerance at various evaluated temperatures. Stainless steel 316 is the primary material used for other reactor components operating at lower temperatures, as well as the connecting gas and exhaust lines. Tri-clover clamps with Viton seals, which can withstand temperatures up to 250°C, were utilised to connect the reactor components. A cooling water circulation system was installed at those areas at risk of exceeding the polymer seal temperature limitations.

Materials choices, manufacturing processes, and quality assurance measures have been fully documented to ensure traceability and compliance with relevant standards and regulations. This documentation will be essential for future reference, maintenance, and potential modifications/upgrades to the reactor system.

The electrical and interlock system

Hydrogen is known for its highly reactive and hazardous nature, and this demands heightened attention to precautionary safety measures. Regardless of the variations in H&S regulations across different international jurisdictions, it is widely recognised that hydrogen-related processes must meet stringent H&S standards.

A key component of the safety system is an electrical interlock that will immediately shutdown hydrogen in the bunker and purge the reactor if it detects an unsafe operating condition. This interlock is designed to be fail-safe and operates as a watchdog that is electrically isolated and independent from the remote-operated automated process control system.

The design of the electrical control system and wiring are required to comply with stringent hazardous area electrical standards, specified within AS/NZS 60079.14 and 60079.13 (which largely follow the IECEx standards). All the electrical components installed in the hazardous zone were selected as 'explosion-proof rated'. External engineering consultants were engaged to facilitate compliance with these requirements, including hazardous area assessment and identification, electrical and interlock system design, and the selection of key electrical components. These external consultants also provided an external review of the over-arching H&S design strategy for the project.

A key lesson from the design and building of this hydrogen reactor within a research-organisation setting is the importance of engaging professional consultancy services at an early stage. This proactive approach allowed the project team a comprehensive understanding of the regulatory requirements and H&S standards, enabling potential challenges and compliance issues to be identified ahead of time, mitigating time and technical risks, and facilitating smooth project execution. However, it is imperative to recognise that adherence to a construction standard does not inherently ensure the process is safe.

Process control system

Personnel cannot be within the reactor bunker when H_2 is present in the reactor, meaning that all process control occurs through a remote-operated electronic control system, that allows for all parameters to be monitored and adjusted if necessary. The control system is designed in such a way that all operating parameters must be designated before the process begins, in order to reduce the chance of human error during the process. However, it also allows for continuous adjustment to these parameters if necessary to maintain the desired conditions during a run. In addition, it also provides an additional layer of safety. Key parameter values are compared against pre-defined safe operating ranges, and the operator is alerted to these risks before these reach the threshold for interlock activation. This allows the operators to adjust parameters or initiate a soft shutdown of the system before the interlock halts the reactor. As the control system is entirely separate from the interlock there is no interference with the correct and safe function of the interlock.

HAZOP and SOP development

The development of Hazard and Operability (HAZOP) analysis and Standard Operating Procedures (SOP) are established practices for identifying and assessing potential hazards and ensuring consistent and safe operation of a hazardous gas process. Significant effort has been invested to create comprehensive HAZOP and SOP documentation for this H_2-DRI reactor, taking into account the complex nature of the reactor system and the potential hazards associated with hydrogen and high-temperature processes.

Three primary sub-systems were identified: gas distribution unit, solid distribution and reaction unit, and support unit. The sub-systems were further broken down into 11 study nodes, which were thoroughly analysed through the HAZOP process. Every component within each study node was examined under various scenarios, including abnormal process conditions, equipment failure, and human error, to ensure a comprehensive understanding of the potential hazards and their consequences. Mitigation measures were developed to address each identified hazard, aiming to minimise the associated risks and enhance the safety of the reactor system. The SOP was developed in conjunction with the control system, providing a detailed set of instructions and precautions for equipment set-up, start-up and shutdown procedures, emergency response protocols, and maintenance activities.

The HAZOP and SOP documents were reviewed by a senior external auditor team. To maintain the relevance and accuracy of these documents, regular reviews and updates of the HAZOP and SOP documentation are conducted. Any modifications or improvements in the reactor system are incorporated to ensure that the analysis remains up-to-date throughout the project's life cycle, promoting ongoing safety and operational excellence.

COMPLETION AND COMMISSIONING

Reactor completion

Following the design and build phase, the H_2-DRI reactor was successfully completed and commissioned in early 2023, as depicted in Figure 3. This significant milestone marked the start of a new phase of operation, where its functionality and performance can be rigorously evaluated and fine-tuned during commissioning.

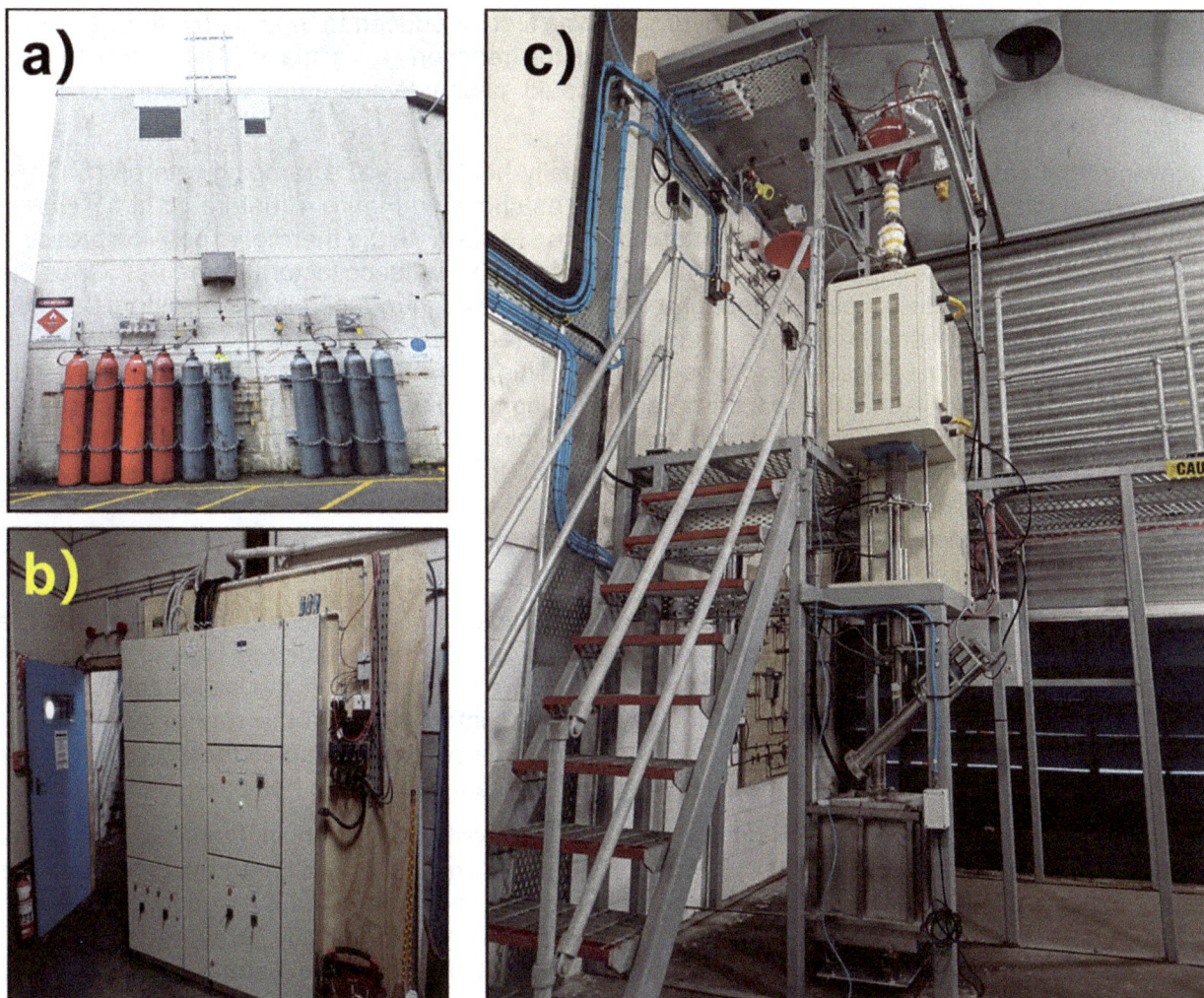

FIG 3 – (a) The gas distribution unit, the condenser, and the exhaust lines located outdoors; (b) The entrance of the bunker where the electrical and interlock system is installed; and (c) The solid distribution and reactor unit located in the bunker.

Commissioning with H_2 reduction of NZ TTM ore

Initial commissioning of the reactor was conducted using NZ titanomagnetite (TTM) ore as the feedstock. NZ TTM ironsand is abundantly available along the west coast of the North Island of New Zealand and is a significant source of iron for domestic steel production. Current commercial processing of NZ TTM ore relies on a coal-fired rotary kiln process that is carbon-intensive (Longbottom *et al*, 2017). The behaviour of NZ TTM ore during H_2 reduction has been studied in gram-scale laboratory experiments in our lab, and we have investigated its reaction kinetics and sticking characteristics (Zhang *et al*, 2020; Longbottom *et al*, 2017). In small-scale fluidised bed experiments, it has been observed that a protective titanium-rich oxide shell forms around each particle during the initial reduction stage, preventing the ironsand particles from sticking together at temperatures up to 1000°C (Prabowo *et al*, 2022a, 2022b, 2022c). The vertical shaft H_2-DRI reactor offers an excellent opportunity to assess the performance of NZ TTM ore during hydrogen reduction on a larger scale, validating the previous lab results and identifying any potential issues that may arise during large-scale production.

In preparation for feeding into the reactor, the raw NZ TTM ores were milled and pelletised to form spherical pellets with a diameter of approximately 5 mm. These pellets are required to meet specific compressive strength criteria, 500 N in our case, to ensure their integrity during transportation and handling. Other work within our group has investigated the effect of pelletisation process parameters on the strength of the resulting pellets (Mendoza *et al*, 2022). Prior to loading the TTM pellets into the reactor hopper, alumina balls were charged throughout the reactor from the auger to the hopper. This served two purposes: first, during the heating and stabilisation process, the discharge of some alumina balls facilitated sample flow and accommodated thermal expansion; second, it ensured that at the beginning of the experiment when hydrogen was first introduced to the reactor, all of the TTM pellets were positioned above the hot zone where the reaction takes place. This precautionary measure prevented rapid initial reaction, which could cause temperature excursions or other transient 'shocks' during start-up.

Following final safety checks, the reactor was gradually heated up at a ramping rate of 15°C per minute while nitrogen was used to flush the reactor, as shown in Figure 4 (before point a). During the heat-up process, the auger was intermittently rotated to alleviate the thermal expansion pressure in the hot zone, while the alumina balls were discharged to the pellet collector (Figure 4a). Once the target reaction temperature was reached, process nitrogen was introduced into the system at a flow rate equivalent to the hydrogen flow rate required for the reaction. This was set to 100 SLPM for our initial commissioning runs (Figure 4, from point a to b). This step was taken to achieve a steady state temperature distribution within the reactor before initiating the reaction. The hot zone was observed to move upwards as the purging gas distributed heat within the reactor, resulting in the temperature increase in the hopper and the upper flange (Figure 4c). Simultaneously, there was a slight increase in pressure due to the high flow rate of the purging gas (Figure 4a). The pressures throughout the reactor are carefully monitored during each run to detect any potential blockages in the reactor during the reaction.

Once the system had reached a steady state, hydrogen gas was introduced into the reactor, and the auger was rotated at a predetermined speed to ensure the desired residence time of the pellets within the consistent hot zone (Figure 4, point b). The weight change in the pellet collector was measured over time, for two purposes: Firstly, it provides information on the rate of discharge of pellets, and hence their residence time within the reactor. Secondly, it enables the end point of the experiment to be determined, once the desired quantity of reduced pellets has been discharged.

Key reactor parameters during an initial commissioning test are shown in Figure 4. The initial reaction temperature was set at 900°C (from point b to c). The act of switching hydrogen into the reactor led to a slight adjustment in temperature and pressure distributions throughout the reactor, due to differences in gas properties such as thermal conductivity and viscosity compared to nitrogen. Importantly, a sharp increase in water vapour concentration within the exhaust line was also observed, indicating the onset of the water-producing reaction (Figure 4b). As additional TTM pellets were transferred into the hot zone, the water vapour levels gradually rose until a new steady state was achieved, determined by the thermodynamics and kinetics of the reaction. Following stabilisation at each prior temperature, the reaction temperature was then subsequently increased in stepped increments to 950°C (from point c to d), 1000°C (from point d to e), and finally 1050°C (from point e to f). The experiment was terminated upon observing the occurrence of sticking behaviour after the reactor temperature was raised to 1050°C. The identification of sticking within the reactor tube was based on several parameters, which included a decrease in humidity and pressure due to reduced water production, as well as temperature ramping in the upper section of the reactor (hopper and upper flange) caused by the absorption of additional heat by the jammed pellets, leading to their elevated temperature.

Another event occurred after the onset of sticking at 1050°C, which can be observed (between points e and f) in Figure 4b. A distinct and sharp increase in humidity can be observed in this plot, which is believed to arise from previously jammed pellets suddenly becoming dislodged and falling into a cavity created by the earlier jamming. This rapid rise in water vapour level was attributed to the swift reaction of these dropped pellets within the hot zone. However, the reaction then quickly stopped again indicating further sticking in the reactor tube and necessitating that the experiment be halted. Subsequently, the reactor was gradually cooled down to room temperature to conclude the experiment. The plots shown in Figure 4 demonstrate the value of real-time monitoring of the measured system parameters, which are highly sensitive to changes in conditions within the reactor.

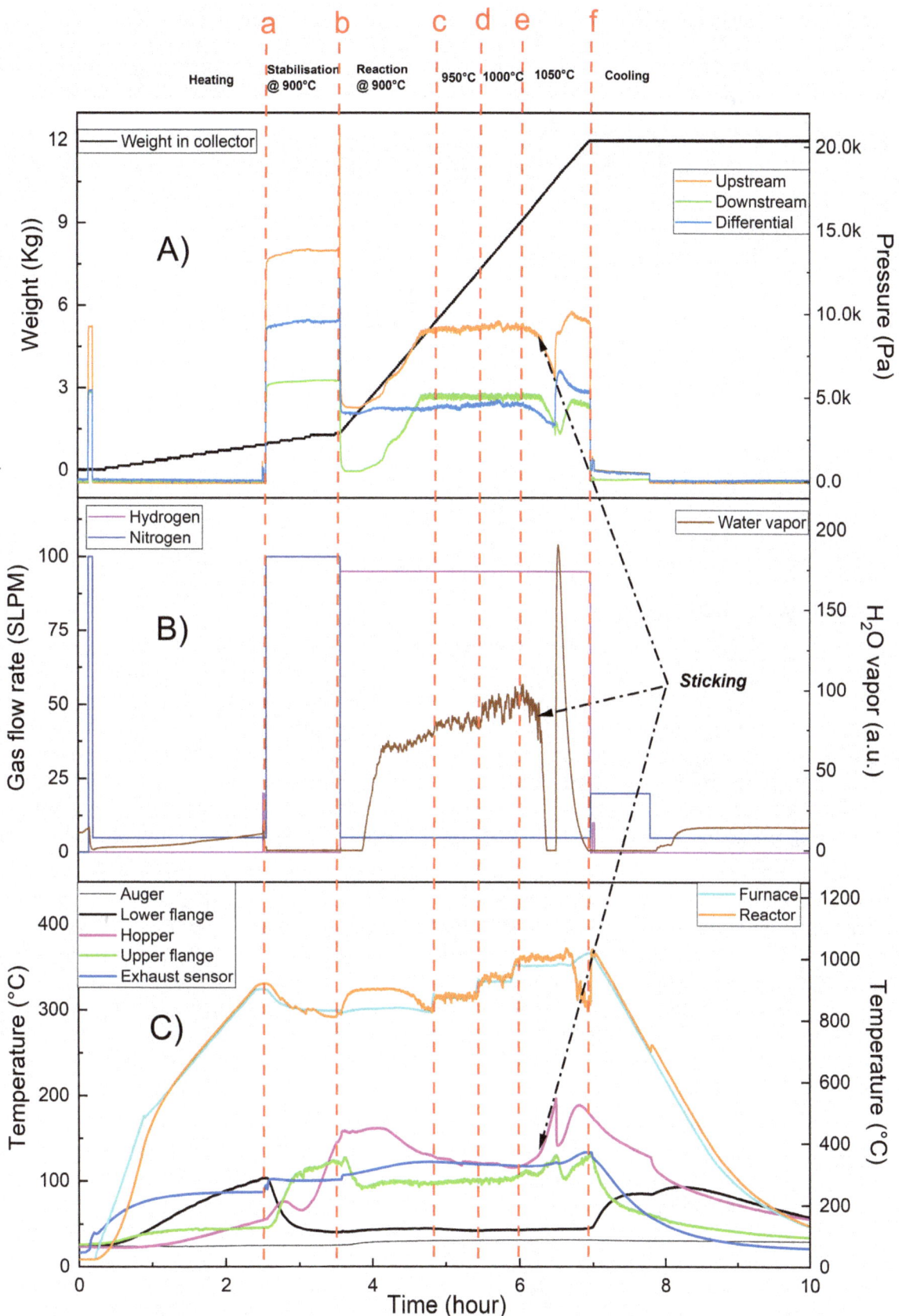

FIG 4 – Data collected from the commissioning test with NZ TTM pellets. (a) Pressure and weight change, (b) Gas flow rates and humidity reading, and (c) the temperatures of the system.

The reduced DRI pellets produced at various temperatures were collected and subjected to a re-oxidation process at 1000°C in air. The mass gain on ignition (GOI%) could be obtained, which provides an indication of the reduction degree of DRI samples. Figure 5a shows that the GOIs of the DRIs produced at 900°C, 950°C and 1000°C, are all above 32 per cent, indicating a reduction degree close to 100 per cent. The pellets which became 'stuck' at 1050°C, were less reduced with a GOI of approximately 30.89, whereas the pellets collected above the sticking point were not reduced at all because they never entered the hot zone. Powder XRD results also confirm (Figure 5b) the absence of iron oxide in the reduced DRI pellets. Characteristic peaks of metallic Fe are dominant in these pellets and $FeTiO_3$ was also observed at 950°C.

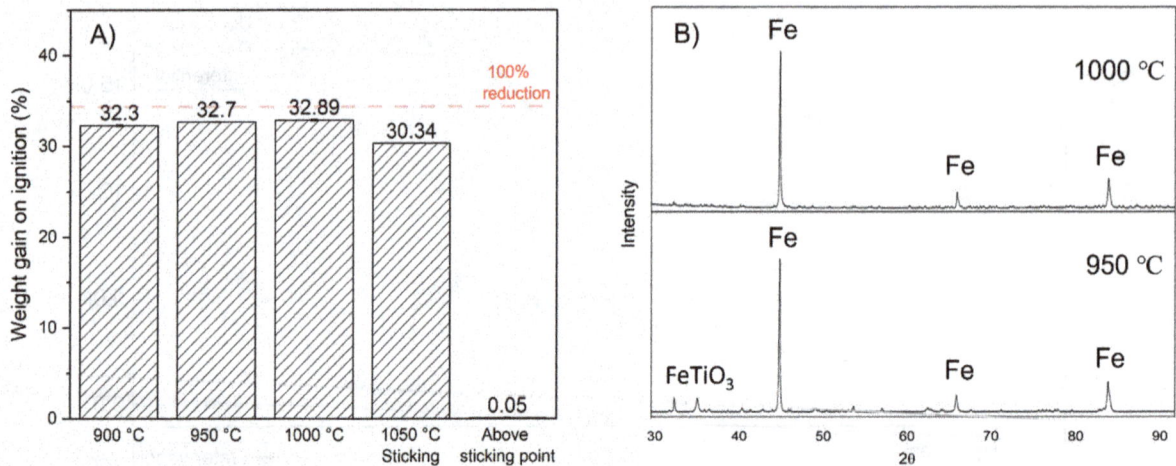

FIG 5 – (a) Mass gain of ignition, and (b) PXRD patterns of DRI samples.

SUMMARY AND PROSPECTS

A 20 kg-scale continuous vertical shaft H_2-DRI reactor has been constructed in this work, marking a major milestone in capability for conducting feasibility studies on H_2 reduction of various iron ores at the Robinson Research Institute in Lower Hutt, New Zealand. With the ability to manipulate temperature, residence time, and gas flow rate and composition, the H_2-DRI reactor now offers a flexible test bed for in-depth studies on thermodynamics, kinetics, and sticking within the counter-flow pellet DRI process. Furthermore, the reactor can produce sufficient quantities of H_2-DRI to enable subsequent studies of melting and slag separation at the bulk multi-kg scale. This will enable improved insights into the impact of introducing H_2-DRI pellets into the overall steel production chain.

To enhance the reactor's capabilities and provide more comprehensive information, several further improvements are planned. Firstly, a multi-point thermal probe will be installed within the SiC reactor to provide accurate data on the spatial temperature profile throughout the hot zone during the DRI reaction. In addition, the water vapour sensor will be calibrated under a hydrogen atmosphere to provide accurate quantification of the absolute moisture content within the exhaust gas. This will facilitate the determination of gas utilisation as a function of reactor operating parameters. Furthermore, materials characterisation techniques such as XRD and SEM-EDX-EBSD will be utilised to gain insights into the morphology, elemental composition, and distribution of the reduced DRI pellets produced by this reactor.

Building upon the experience and confidence gained from the successful commissioning of this reactor, our team is now pursuing a program of research targeting the use of NZ TTM ironsand as a feedstock. In addition, this reactor is now available to undertake commercial testing for industry partners from Australia (or elsewhere) who wish to test specific iron ore materials under continuous H_2-DRI conditions. It is envisaged that results obtained from this reactor will provide improved understanding of the process and ore requirements that must be met in order to economically produce H_2-DRI from the many lower-grade iron ores that are found within New Zealand and Australia. This will facilitate the future development and commercial implementation of H_2-DRI processes utilising these ores.

REFERENCE

Longbottom, R J, Ingham, B, Reid, M H, Studer, A J, Bumby, C W and Monaghan, B J, 2017. In situ neutron diffraction study of the reduction of New Zealand ironsands in dilute hydrogen mixtures, *Miner Process Extr Metall*, 128:183–192.

Mendoza, S, Yin, B H, Zhang, A and Bumby, C W, 2022. Pelletization and sintering of New Zealand titanomagnetite ironsand, *Adv Powder Technol*, 33(12):103837.

Midrex, 2013. The MIDREX ® Process, Midrex Technol. Inc.

Prabowo, S W, Longbottom, R J, Monaghan, B J, del Puerto, D, Ryan, M J and Bumby, C W, 2022a. Sticking-Free Reduction of Titanomagnetite Ironsand in a Fluidized Bed Reactor, *Metall Mater Trans B*, 50:1729–1744.

Prabowo, S W, Longbottom, R J, Monaghan, B J, del Puerto, D, Ryan, M J and Bumby, C W, 2022b. Phase transformations during fluidized bed reduction of New Zealand titanomagnetite ironsand in hydrogen gas, *Powder Technol* 398:117032.

Prabowo, S W, Longbottom, R J, Monaghan, B J, del Puerto, D, Ryan, M J and Bumby, C W, 2022c. Hydrogen Reduction of Pre-oxidized New Zealand Titanomagnetite Ironsand in a Fluidized Bed Reactor, *JOM*, 74:885–898.

van Vuuren, C, Zhang, A, Hinkley, J T, Bumby, C W and Watson, M J, 2022. The potential for hydrogen ironmaking in New Zealand, *Cleaner Chemical Engineering*, 4:100075.

Vogl, V, Åhman, M and Nilsson, L J, 2018. Assessment of hydrogen direct reduction for fossil-free steelmaking, *J Clean Prod*, 203:736–745.

Vogl, V, Olsson, O and Nykvist, B, 2021, Phasing out the blast furnace to meet global climate targets, *Joule*, 5:2646–2662.

Zhang, A, Monaghan, B J, Longbottom, R J, Nusheh, M and Bumby, C W, 2020. Reduction Kinetics of Oxidized New Zealand Ironsand Pellets in H_2 at Temperatures up to 1443 K, *Metall Mater Trans B*, 51:492–504.

Greener downstream processing – sintering and pelletising

Sintering characteristics of iron ore blends containing iron ore concentrates

H Han[1], L Lu[2] and S Hapugoda[3]

1. Research Scientist, CSIRO Mineral Resources, Technology Court, Pullenvale Qld 4069. Email: hongliang.han@csiro.au
2. Senior Principal Research Scientist, CSIRO Mineral Resources, Technology Court, Pullenvale Qld 4069. Email: liming.lu@csiro.au
3. Experiment Scientist, CSIRO Mineral Resources, Technology Court, Pullenvale Qld 4069. Email: sarath.hapugoda@csiro.au

ABSTRACT

With the recent rapid development of Chinese steel industry, the world iron ore resources of good quality are depleting rapidly. As a result, iron ore deposits of lower quality have been or are being developed to meet the market demand and are becoming increasingly important to iron and steel companies. As concentrates have advantages of high Fe grade and less harmful impurities, they are ideal sweeteners to maintain sinter grade and quality when high proportions of low-grade ores are used in the sinter blends. However, due to their fine nature, addition of concentrates will inevitably increase the proportion of adhering fines and reduced the proportion of nucleus particles of the sinter blends, which can adversely affect the green bed permeability and sintering productivity. In the present study, lab scale sintering tests were carried out under the conditions similar to the sintering process to examine the effect of concentrates on the high temperature sintering characteristics of sinter blends and the structure and strength of the resultant sinter analogues.

INTRODUCTION

Sintering is the most widely used agglomeration process for iron ore fines for blast furnace use. In modern blast furnaces, particularly those operating in East Asia, iron ore sinter constitutes more than 60 per cent of blast furnace burdens. Good sinter quality and high sintering productivity are thus required to sustain the extreme operating conditions and productivity of modern blast furnaces. While sinter quality is influenced by many factors, the mineralogy and physical structure of sinter undoubtedly play an important role. As the sinter mineralogy and physical structure are formed upon solidification of the initial melt generated in the sintering process, the formation and fluidity of the initial sinter melt of adhering fines, penetration and assimilation of nucleus particles by the initial melt, which at fixed sintering conditions depend mainly on the characteristics of iron ore fines in the blend, are particularly important in determining sinter quality (Lu and Manuel, 2021).

With the rapid development of the steel industry, the iron ore resources of good quality have been depleted all over the world in recent years. As a result, iron ore deposits of lower quality have been or are being developed to meet the market demand and these types of iron ores have become increasingly important to most iron and steel companies. As concentrates have advantages of high Fe grade and less harmful impurities, they are ideal sweeteners to maintain sinter grade and quality when high proportions of low-grade ores are used in the sinter blends (Higuchi and Lu, 2017). However, due to their fine nature, addition of concentrates will inevitably increase the proportion of adhering fines and reduced the proportion of nucleus particles of the sinter blends, the quantity of adhering fines increases and the basicity of adhering fines decreases. which can adversely affect the green bed permeability and sintering productivity (Han, Lu and Hapugoda, 2023).

Many researchers widely investigated the melt formation and fluidity, assimilation and penetration behaviours of adhering fines with different composition (Kasai, Wu and Omori, 1991; Debrincat, Loo and Hutchens, 2004; Scarlett et al, 2004; Hida and Nosaka, 2013; Jeon et al, 2014; Liu et al, 2016; Andrews, 2018). However, these studies have not comprehensively investigated the sintering behaviour of sintering blends containing concentrate ores, little information has been obtained concerning the sintering characteristics of initial melt formation and fluidity as well as initial melts penetration and assimilation of sintering blends containing different concentrates. Hence understanding of these high temperature characteristics of ore blends containing different concentrates is urgently needed.

In the present study, lab scale sintering tests were carried out under the conditions similar to the sintering process to examine the effect of concentrates on the high temperature sintering characteristics of sinter blends and the structure and strength of the resultant sinter analogues.

EXPERIMENTAL

Raw materials

In the present study, three ore blends were constituted from the selected ores. Blend 1 was the base blend, which consisted of nine selected ores but did not contain concentrate ores. Blend 2 consisted of 91.4 per cent base blend and 8.6 per cent hematite concentrate. While Blend 3 consisted of 91.4 per cent base blend and 8.6 per cent magnetite concentrate. In addition, all the ore blends mixed with fluxes to target the sinter basicity and chemical composition.

Ore material of -0.15 mm particle size was selected for the adhering fines. Hematite concentrate and magnetite concentrate used in the present study were very fine, with 99.22 per cent and 99.59 per cent below 0.15 mm respectively, therefore the CaO content decreased in the adhering fines when concentrates were added into the sintering blends.

Experimental methods

Formation and fluidity of initial melt

Firing experiments of adhering fines tablets were used to evaluate the initial melt formation and fluidity ability of different sintering blends, as initial melts were formed in the adhering fines layer of pseudo-particles, mainly by the reaction of iron ore with CaO.

A schematic diagram of the experiment for initial melt formation and fluidity ability is shown in Figure 1. The adhering fines from different sintering blends compacted into a cylindrical tablet sample. The tablet sample was produced using an iron mould under a pressure of 800 N. The sample was placed on Ni foil and positioned in a heating furnace to simulate the sintering thermal profile under an air atmosphere. The sample was heated from room temperature to 400°C within 1 minute, from 400°C to 850°C within 1 minute and from 850°C to the set temperature (1240°C, 1280°C and 1320°C) within 1 minute and then held for 4 minutes. Subsequently, the sample was removed from the furnace and cooled down to room temperature in air. After each experiment, the initial melt formation and fluidity ability of adhering fines was quantitatively evaluated based on the change in tablet shape and the vertical projected area of the sample was measured to calculate the fluidity index. High fluidity index means high initial melt formation and fluidity ability.

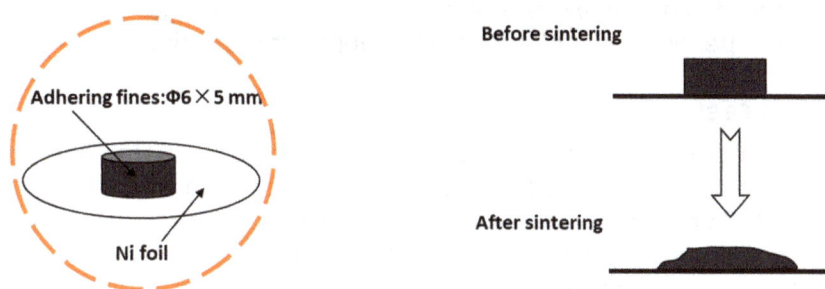

FIG 1 – Schematic diagram of initial melt formation and fluidity ability experiment.

The fluidity index = (the projected area of the sample after sintering – the projected area of the sample before sintering)/the projected area of the sample before sintering.

Penetration and assimilation of nucleus particles with initial melt

To simulate the penetration and assimilation behaviour of the melt from fines within the adhering layer in pseudo-particles, sintering experiments on two-layer tablets composed of iron ore substrates and adhering fines were carried out. Figure 2 shows the schematic diagram of the experiment for penetration and assimilation ability. The iron ore substrate was made from mixtures of two classes of iron ore sizes, ie -0.15 mm and 0.15–0.25 mm, in equivalent mass ratios to simulate coarse particles. The mixed iron ore was pressed into a cylindrical tablet whose diameter and height were 25 mm and 5 mm under a pressure of 5000 N. As an initial melt source, a cylindrical tablet of adhering fines was used whose diameter and height were 6 mm and 5 mm respectively under a pressure of 800 N.

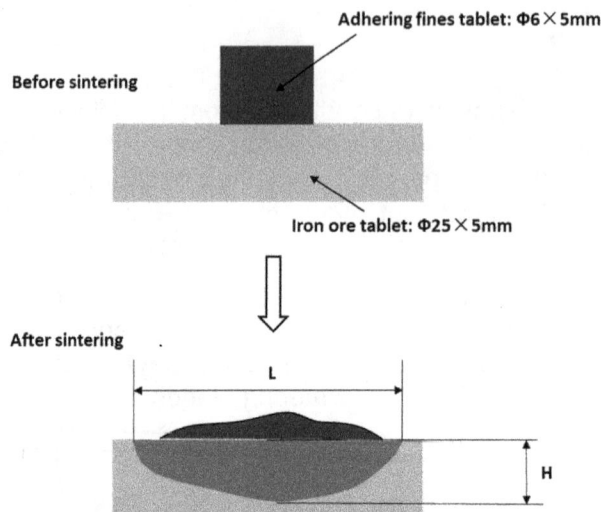

FIG 2 – Schematic diagram of penetration and assimilation ability experiment.

The sample was placed on Ni foil and heated in a furnace with a horizontal alumina tube to simulate the sintering thermal profile under an air atmosphere (the same conditions described in previous section). After the experiment, the sample was dissected perpendicular to the interface, mounted in epoxy resin and polished. Then representative cross-section images of the sample were obtained using an optical microscope.

During the sintering process, the initial melt first forms within the adhering layer of pseudo-particles and then spreads out through the adhering layer to react with surrounding adhering fines and nucleus particles to generate the primary sinter melt. Figure 2 also shows a schematic diagram of the interaction zone present in the substrate after the penetration and assimilation test. To quantify the penetration and assimilation behaviour of the initial melt, the volume of interaction zone (VOR) was proposed.

$$\text{VOR} = \pi \times \left(\frac{L}{2}\right)^2 \times \frac{H}{2}$$

Where H and L are the depth and width of the melt and substrate interaction zone. A large VOR value indicates a higher penetration and assimilation ability of the initial sinter melt.

Lab-scale sintering test

To simulate sintering behaviour of adhering fines and nucleus particles, lab-scale sintering test was conducted. A schematic diagram of this test is shown in Figure 3. In this test, the selected iron ore was screened to -1.0+0.85 mm and used as nucleus particles. Adhering fines were coated onto the nucleus particles by rolling particles with a spatula after the nucleus particles were wetted with water. The granulated particles were then pressed into a tablet whose diameter was 12 mm under a pressure of 800 N. The total weight of nucleus particles and adhering fines was set at 2.5 g for the purpose of controlling the tablet height and the weight ratio of nucleus particles to adhering fines was set according to the actual ratio of adhering fines and nucleus particles in different blends.

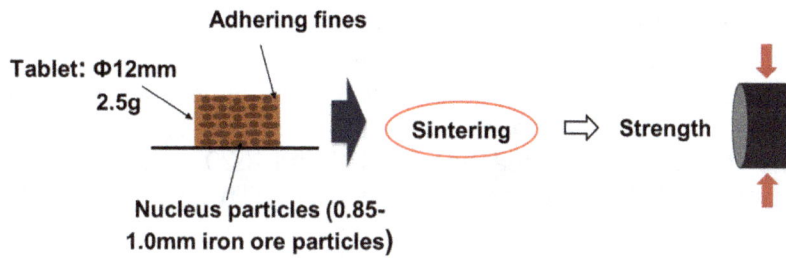

FIG 3 – Schematic diagram of lab-scale sintering test.

The sample was placed on a Ni foil and heated in a horizontal heating furnace to simulate the sintering thermal profile under an air atmosphere (the same conditions described in previous section). Five samples were tested under the same conditions and then were placed vertically on a compressive strength measuring machine to test their strength.

RESULTS AND DISCUSSION

Formation and fluidity ability of initial melt

The melt formation and fluidity of different blends adhering fines were evaluated at 1240°C, 1280°C and 1320°C. Figure 4 shows the top view of the solidified tablets after the formation and fluidity experiment while Table 1 displays their corresponding fluidity indices. The adhering fines followed a similar fluidity–temperature profile. As the temperature increased, both the viscosity and surface tension of initial melt decreased, the projection area and fluidity index of the initial melt increased.

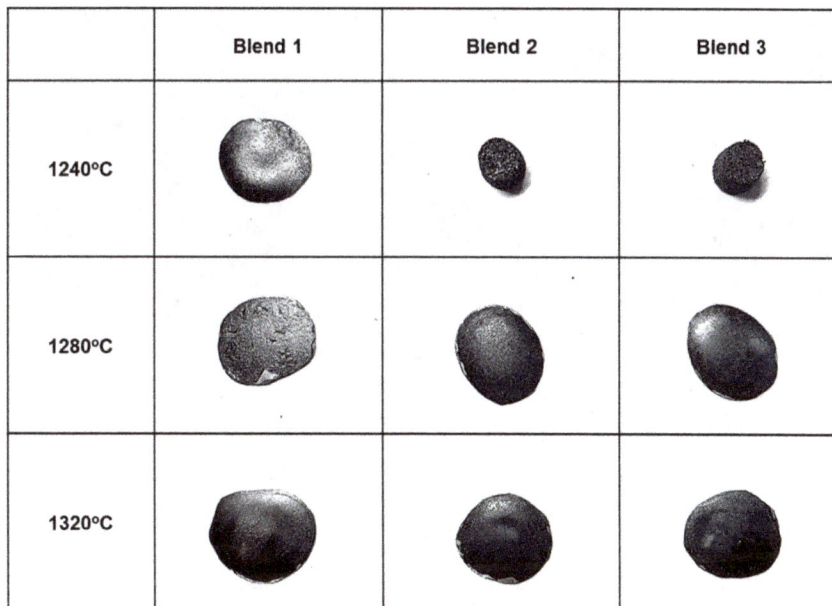

FIG 4 – Change in the projection area with different temperature of tablets prepared using different sintering blends adhering fines.

TABLE 1

Fluidity index of different blends adhering fines with various temperatures.

	Blend 1	Blend 2	Blend 3
1240°C	3.75	0	0
1280°C	4.48	3.95	3.97
1320°C	5.24	4.09	4.26

Different adhering fines had different initial melt formation and fluidity ability and different adhering fines required distinct temperature to form the initial melt and spread out. When the temperature was

1240°C, the initial melt from Blend 1 adhering fines was formed and started to spread out, initial melt formation and fluidity ability was high and the fluidity index reached 3.75. However, addition the hematite and magnetite concentrates decreased the basicity of adhering fines which increased the liquidus temperature of the melts formed, so no melt was formed for Blend 2 and Blend 3 adhering fines at 1240°C. When the temperature reached 1280°C and 1320°C, all these three blends adhering fines showed very similar initial melt formation and fluidity ability, the fluidity index of the initial melt was all more than 3.95.

Overall, addition of the hematite concentrate and magnetite concentrate in sintering blends obtained very similar initial melt formation and fluidity ability as the sintering blend without concentrate when the temperature was more than 1280°C. But their initial melt formation and fluidity ability was much lower than the sintering blend without concentrate when the temperature was 1240°C.

Penetration and assimilation of initial sinter melt

The cross-sectional images of the sample assemblies with initial sinter melts of different blend adhering fines on the same iron ore substrate after the penetration and assimilation experiments at 1240°C, 1280°C and 1320°C are shown in Figure 5. Table 2 shows the measured results of the spread width and penetration depth.

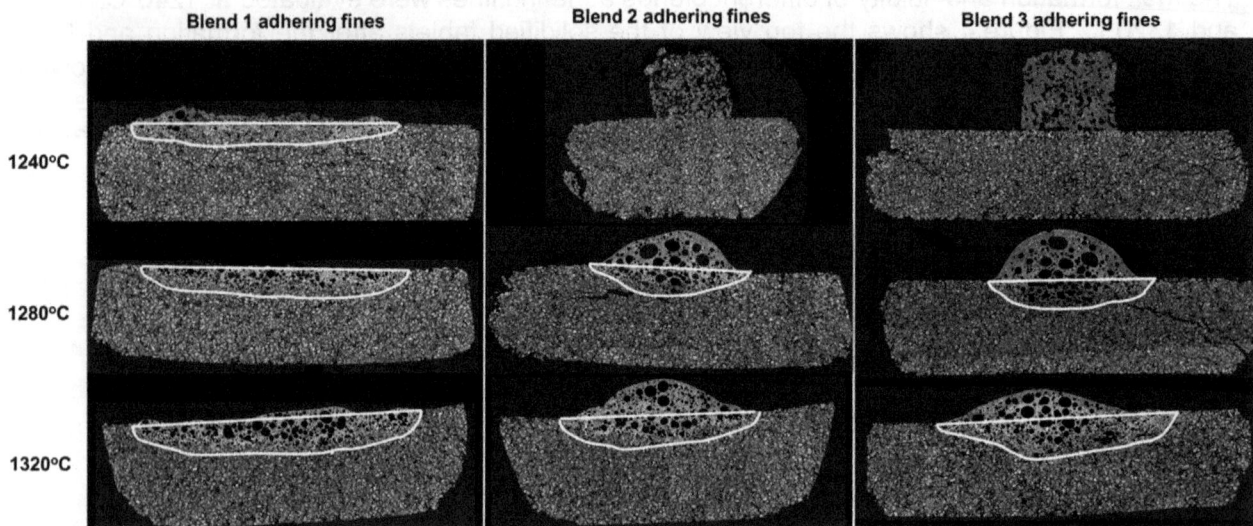

FIG 5 – Cross-sectional images of the sample assemblies with initial sinter melt of different blend adhering fines on iron ore substrate after the penetration and assimilation experiments.

TABLE 2

Calculated volume of interaction zone between initial melts from different adhering fines and iron ore substrate/mm^3.

	Blend 1	Blend 2	Blend 3
1240°C	98.83	0.00	0.00
1280°C	138.17	62.43	70.33
1320°C	220.98	89.27	147.29

For all the adhering fines tested, the penetration width and depth of melt both increased with temperature consistently. As the temperature increased, both the viscosity and surface tension of initial melt decreased, promoting the penetration of the melt into the substrate.

Different adhering fines had different penetration and assimilation ability. At the same temperature, Blend 1 adhering fines generally showed larger interaction volumes than Blend 2 and 3, indicating higher penetration and assimilation ability. Addition of the hematite concentrate and magnetite concentrate in sintering blends also showed different penetration and assimilation ability, compare the blend containing hematite concentrate (Blend 2), the blend containing magnetite concentrate

(Blend 3) possessed higher penetration and assimilation ability. Complete melting was observed at 1240°C for Blend 1 adhering fines, 1280°C for Blend 2 and Blend 3 adhering fines, which again suggests that Blend 1 adhering fines required low temperature to form a melt required for penetration and assimilation, compared with Blend 2 and Blend 3 adhering fines.

In addition, Blend 1 adhering fines initial melt appeared to spread more and Blend 1 and Blend 3 adhering fines initial melt appeared to spread less, but very similar penetration depth was obtained by all blends adhering fines initial melt. Addition of concentrates in the sintering blends, the quantity of adhering fines increases and the basicity of adhering fines decreases, large penetration depth of initial melt can improve the sinter quality.

Overall, the sintering blend without concentrate had higher penetration and assimilation ability than the sintering blends containing concentrates. But relatively large penetration depth can be obtained by the sintering blends containing concentrates, which can positively affect the quality of sinter. In addition, the blend containing magnetite concentrate possessed higher penetration and assimilation ability than the blend containing hematite concentrate.

Compressive strength of the sintered samples

Table 3 shows the results of the compressive strength of sintered samples. For all the samples tested, the compressive strength increased with increasing temperature. As the temperature increased, formation and fluidity ability as well as penetration and assimilation ability of initial melt increased, promoting the strength of sintered sample increased.

TABLE 3

Compressive strength of sintered samples/N.

	Blend 1	Blend 2	Blend 3
1240°C	86.49	34.67	37.78
1280°C	116.17	122.64	215.08
1320°C	395.49	394.47	498.49

Different adhering fines had different compressive strength of sintered samples. When the temperature was 1240°C, the strength of Blend 1 adhering fines samples was higher than the strength of Blend 2 and Blend 3 adhering fines samples. When the temperature reached 1280°C and 1320°C, all these three blends adhering fines samples showed very similar strength, even the strength of samples containing concentrates higher than the strength of samples without concentrates. At lower temperature, no melt was formed and no penetration and assimilation behaviour was occurred for the adhering fines containing concentrates, so the strength of Blend 2 and Blend 3 adhering fines samples was lower. While when the temperature was more than 1280°C, very similar initial melt formation and fluidity ability, similar penetration depth obtained for the adhering fines with and without concentrates, all of these made a similar or higher strength of samples containing concentrates.

CONCLUSIONS

Sintering blends containing concentrates obtained very similar initial melt formation and fluidity ability as the sintering blend without concentrate at higher temperature. But their initial melt formation and fluidity ability was much lower than the sintering blend without concentrate at lower temperature.

Addition the concentrates in sintering blends decreased penetration and assimilation ability of initial melt, but relatively large penetration depth can be obtained by the sintering blends containing concentrates, which can positively affect the quality of sinter.

Addition the concentrates in sintering blends could not affect the strength of sinter at higher temperature. But the strength of sinter was much lower than the sintering blend without concentrate at lower temperature.

Addition the concentrates in sintering blends, the quantity of adhering fines and nucleus particles, the basicity of adhering fines, the high temperature characteristics of blends all changed, some measures should be adopted for ensuring productivity and sinter quality.

ACKNOWLEDGEMENTS

The authors would like to thank CSIRO Mineral Resources for permission to publish the paper and for the financial support of this work.

REFERENCES

Andrews, L, 2018. Experimental model systems to investigate factors driving iron ore sintering coalescence, The University of Newcastle, Australia.

Debrincat, D, Loo, C E and Hutchens, M F, 2004. Effect of iron ore particle assimilation on Sinter structure, *ISIJ International,* 44:1308–1317.

Han, H, Lu, L and Hapugoda, S, 2023. Effect of ore types on high temperature sintering characteristics of iron ore fines and concentrate, *Minerals Engineering*, vol 107.

Hida, Y and Nosaka, N, 2013. Evaluation of iron ore fines from the viewpoint of their metallurgical properties in the sintering process, *Mineral Processing and Extractive Metallurgy,* 116:101–107.

Higuchi, T and Lu, L, 2017. Evaluation of sintering behaviour based on ore characteristics, in *Proceedings Iron Ore 2017*, pp 303–308 (The Australasian Institute of Mining and Metallurgy: Melbourne).

Jeon, J-W, Kim, S-W, Suh, I-K and Jung, S-M, 2014. Assimilation Behavior of Quasi-particle Comprising High Alumina Pisolitic Ore, *ISIJ International,* 54:2713–2720.

Kasai, E, Wu, S and Omori, Y, 1991. Factors Governing the Strength of Agglomerated Granules after Sintering, *ISIJ International,* 31:17–23.

Liu, D H, Zhang, J L, Xue, X, Wang, G W, Li, K J and Liu, Z J, 2016. Infiltration behavior of sintering liquid on nuclei ores during low-titanium ore sintering process, *International Journal of Minerals, Metallurgy, and Materials,* 23:618–626.

Lu, L and Manuel, J, 2021. Sintering Characteristics of Iron Ore Blends Containing High Proportions of Goethitic Ores, *JOM,* 73:306–315.

Scarlett, N V Y, Pownceby, M I, Madsen, I C and Christensen, A N, 2004. Reaction sequences in the formation of silico-ferrites of calcium and aluminum in iron ore sinter, *Metall Mater Trans B,* 35:929–936.

Development of an iron ore analysis method applying assimilation reaction to coarse ore

Y Hong[1], Y M Lee[2] and B C Kim[3]

1. Research Engineer, Hyundai Steel R&D Center, Chungnam 31719, Republic of Korea. Email: yulhong@hyundai-steel.com
2. Senior Researcher, Hyundai Steel R&D Center, Chungnam 31719, Republic of Korea. Email: thirdhand81@hyundai-steel.com
3. Team Leader, Hyundai Steel R&D Center, Chungnam 31719, Republic of Korea. Email: bckim@hyundai-steel.com

ABSTRACT

Since sintered ore is the primary source for producing molten iron in the blast furnace process, maintaining stable quality despite the utilisation of low-grade iron ore is crucial. However, conventional analysis methods, such as chemical composition, ore size distribution and types of iron ore, have limitations in developing the utilisation of new low-grade iron ore and predicting sintered ore properties. Sintered ores are formed through a complex melting process of granule particles consisting of iron ore, flux and fuel. Consequently, it has been investigated that analysis methods which apply the reactive properties of iron ore to predict sintered ore characteristics. In this study, we estimated the properties of iron ore by simulating the assimilation process of coarse ore, which influences the bonding strength of sintered ores and quantifying the degree of assimilation reaction. Pseudo particles were formed by granulating coarse ore (1–3 mm) with adhering fines, which act as primary melts and sintered at 1623 K. The bonding strength of individual sintered ores, formed from 11 different types of iron ore, was measured using a simplified Tumbler Index (TI) test and each value was indexed as the Iron ore Reactivity Index (IRI). By investigating the relationship between the degree of assimilation depending on iron ore properties and the characteristics of sintered ore, we demonstrated that sintered ores formed from more reactive iron ore exhibited stronger bonding strength. Moreover, when blending conditions resulted in a higher Iron Blending ore Reactivity Index (IBRI), the physical properties of sintered ore improved. The development of this analysis method is expected to not only enable careful consideration of iron ore usability but also allow for the advance prediction of sintered ore qualities by deriving the IBRI.

INTRODUCTION

Sintered ore is produced through a complex melting process involving a mixture of iron ores, fuel and flux at high temperatures. It serves as the primary source of molten iron in the blast furnace process. Since sintered ore consists mainly of iron ore, the properties of the iron ores significantly impact the sinter quality. Therefore, it is crucial to understand the characteristics of the ores through accurate analysis. The physico-chemical properties of sintered ore that influence the blast furnace include reactivity, mechanical strength and reduction disintegration. These properties are influenced by the chemical composition and the formation of bonding phases during the sintering process.

However, sinter quality is significantly difficult to predict due to its production process involving the melting reaction of a granule mixture. As the characteristics of primarily used iron ore deteriorate, it becomes necessary to investigate strategies for maintaining the physico-chemical properties of sintered ore, even when utilising low-grade iron ore. Conventional analysis methods such as composition analysis, size distribution and microstructure, which have been used to assess the suitability of iron ores for sintering, do not provide an intuitive reflection of the melting behaviour during the bonding phase formation in sintered ores. Thus, these analysis methods have limitations when it comes to diversifying the range of usable iron ore types.

The assimilation of adhering fines and nucleus particles during sintering is attributed to the formation of the bonding phase in sintered ore, leading to changes in its chemical and physical properties (Kenichi, Jun and Seiji, 2020). The behaviour of melt formation varies depending on the characteristics of the iron ore, such as gangue contents, combined water and ore types (Jun *et al*, 2003; Guo-liang *et al*, 2014, 2015; Sheng-li *et al*, 2014). Consequently, researchers have investigated new analysis methods to simulate the assimilation properties of iron ore, aiming to overcome the limitations of conventional analysis methods. The assimilation mechanism involves the formation of primary melt through the reaction between fine ores and fluxes, as well as the penetration of primary melt into coarse ores. Notably, the assimilation of coarse ores influences the surface properties and bonding strength of the sintered ore (Kenichi, Jun and Seiji, 2020).

Therefore, the objective of this study was to assess the utilisation of iron ores by developing and implementing a new analysis method that simulates the melting process of coarse ores. The aim was to establish a correlation between the bonding strength of sintered ore and the assimilation properties of iron ore. The degree of assimilation of individual iron ores was examined using the Tumbler Index (TI) test, which utilised small sintered ore samples formed through simulated melting reactions. It was observed that sintered ore produced from more reactive iron ores exhibited stronger bonding strength. This simple reactivity test was easily applicable to various types of iron ore source used as a sinter feed. Furthermore, analysis of the pot test results demonstrated that the physical characteristics of sintered ores formed from blended ores with superior reactivity were improved. Factors influencing the reactivity of iron ore were investigated and the applicability of new low-grade ores with high SiO_2 content was assessed through the assimilation test. By incorporating this new analysis method into the iron ore evaluation process, it is expected that the usability of low-grade iron ore can be reevaluated and the quality of sintered ores can be predicted in advance by deriving IBRI.

EXPERIMENTAL METHODS

In this study, the assimilation of a total of 11 different types of iron ore was measured. Pseudo particles were utiflised to simulate the sintering process. The coarse ore from each iron ore type had a size range of 1–3 mm, serving as nuclei particles for the pseudo particles. These coarse ore particles were mixed with 26 per cent CaO and 74 per cent Fe_2O_3 fines to form a primary melt at the sintering temperature. The weights of each component were 307.7 g, 50 g and 142.3 g, respectively. After the addition of water, the materials were thoroughly blended and the mixture was then compressed into tablets with a diameter of 10 mm and a height of 20 mm under a pressure of 7.5 MPa. The pseudo particles were then sintered at a temperature of 1623 K using a vertical tube furnace. The resulting small sintered ores were then evaluated for their strength using a simplified tumbler test. The degree of assimilation was quantified by calculating the IRI using this equation:

$$IRI = (\frac{\text{Over 2.8 mm sintered ore weight after the TI test}}{\text{Total sintered ore weight}} \times 10) - 5$$

The chemical composition of the iron ores was analysed using X-Ray Fluorescence (XRF) and the volatile content in iron ores was determined through the Loss On Ignition (LOI) test using Thermogravimetric Analyzers (TGA). Sinter pot tests were conducted to investigate the relationship between the qualities of sintered ores formed from blended iron ores and the IBRI. An 80 kg mixture of iron ores, fluxes and fuels was granulated and sintered in a pot. The properties of the resulting sintered ores, such as Shatter Index (SI), Reduction Disintegration Index (RDI) and Mean Size (MS), were measured following the respective standard test methods. IBRI for each blending condition was calculated using a weighted mean formula that considered the IRI values and the usage of each iron ore. The relationship between IBRI and the qualities of the sintered ores was then investigated. Additionally, an optical microscope was used to examine the microstructure of the sections of the sintered ores.

RESULTS

Reactivity characteristics of each iron ore and additional iron ore source

As indicated in Table 1, each individual iron ore source exhibits distinct characteristics. Upon evaluating the degree of assimilation for each iron ore source, varying values emerge based on their respective properties. When sufficient assimilation occurs between the initial melt and iron ore, strong bonding phases are formed, which improve mechanical strength of sintered ore, resulting in an enhanced TI for the sintered ore. In other words, the IRI values corresponds to 5 when the surface of coarse ore fully react with initial melt and 0 when the bonding which formed by the melting reaction between coarse ore and initial melt is weakened. Since the chemical compositions and physical properties of iron ores vary, they have a diverse impact on the formation of sintered ore. Comparing these properties using various analysis methods made it complex to make quick decisions regarding usability. However, the reactivity of iron ores was easily able to be quantified and compared by the assimilation test. It has been discovered that several iron ores display superior reactivity despite being classified as low-grade ore types. This study demonstrates that the analysis method, which incorporates assimilation reactions, not only overcomes the limitations of simple composition analysis but also enables the re-evaluation of iron ore.

TABLE 1

Chemical compositions of iron ore source and IRI.

Iron ore source		T.Fe (%)	Al_2O_3 (%)	SiO_2 (%)	LOI (%)	IRI
	A	66.51	0.19	3.81	0	5.00
	B	65.72	1.32	1.43	2.96	5.00
	C	59.9	1.51	4.68	10.26	4.54
	D	63.66	1.31	4.74	3.08	4.50
	E	59.68	1.98	4.99	11.02	4.38
Iron ores	F	61.54	2.85	4.45	5.91	4.27
	G	62.21	0.61	8.54	3.63	4.06
	H	61.62	2.45	4.34	6.64	4.01
	I	59.55	1.82	5.29	10.67	3.78
	J	59.09	2.74	5.41	7.86	2.50
	K	58.18	3.31	5.02	11.20	1.84
	L	72.57	0.23	0.80	0	5.00
Additional iron ore source	M	64.43	2.00	3.94	7.80	5.00
	N	64.73	1.65	5.15	0.60	5.00
	O	65.77	1.70	4.02	0.70	4.02

Relationship between IBRI and qualities of sintered ore

To investigate the relationship between the reactivity of blending ores and the qualities of sintered ore, pot tests were conducted. The sinter feed was granulated from a mixture of iron ore, flux and fuel. Four different blending conditions were established with specific IBRI value ranges, 3.68, 3.88, 4.08 and 4.28 using eight kinds of iron ores. The values for basicity, fuel ratio, slag volume and burnt lime ratio were fixed to minimise the influence of other factors. The obtain values were normalised for sinter quality. The impact of IBRI on sintered ore qualities, including SI, RDI and MS, is illustrated in Figure 1. The results indicate there are trends that the properties of sintered ores improved as IBRI increased. When the sinter feed is composed of iron ores with strong assimilation properties, the melt that connects the iron ores and forms the bonding phase changes the iron ore surface, improving the sintering quality. The availability of IBRI was able to be verified by applying the simply evaluated IRI result of a single iron ore to analyse the tendency of qualities of sintered ore blended with various iron ores. This result demonstrates the possibility of predicting the quality of sintered ore by deriving and comparing the IBRI for each blending condition.

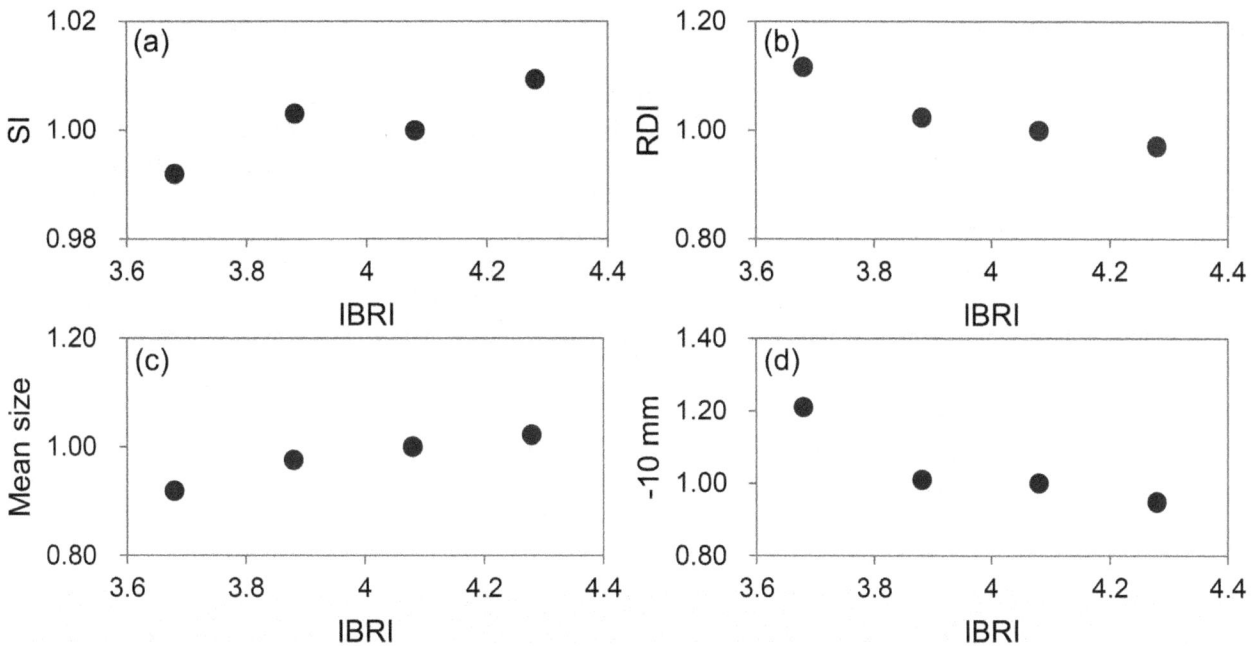

FIG 1 – Change in (a) SI, (b) RDI, (c) under 10 mm ratio of sintered ores and (d) MS with IBRI.

Influence of characteristics of iron ore on assimilation reaction

The factors that influence the reactivity of iron ores are representatively known as the mineral components of iron ores, ore types and contents of combined water (Jun *et al*, 2003; Guo-liang *et al*, 2014, 2015; Sheng-li *et al*, 2014). Among them, three factors on the reactivity of iron ore were analysed. The most significant factor is the influence of Al_2O_3 on the melting properties (Figure 2a). Al_2O_3 raises the melting point of the initial melt and increases its viscosity. This results in a reduced amount of melt being produced and makes it difficult to form the bonding phase (Xiaobo *et al*, 2017). The bonding phase containing a high amount of Al_2O_3 generates a relatively weak phase, known as columnar calcium ferrite, which deteriorates the bonding strength of the sintered ore (Rui-Feng, Yu and Xing-Min, 2022). The second factor is the impact of volatiles in ores (Figure 2b). Ores with high volatile content tend to create pores as the volatile matter evaporates during the sintering process, which increases the surface area with which the initial melt needs to react. The gangue in the surface area of the ore decreases the permeability of the melts, making it challenging to form the bonding phase (Kenichi, Jun and Seiji, 2020). However, in our research, the correlation between the LOI content of the ore and the IRI was not significantly high. Increasing the SiO_2 content in iron ore is known to lead to increased viscosity of the liquid phase, which decreases the melting region and the bonding area. However, through the analysis of the influence factors on assimilation, it was proven that the influence of SiO_2 was relatively low (Figure 2c). This implies that the usability of high SiO_2 iron ore, which exhibits superior assimilation properties, could be re-evaluated. Moreover, the SiO_2 present in the iron ore could potentially replace the SiO_2 added as flux.

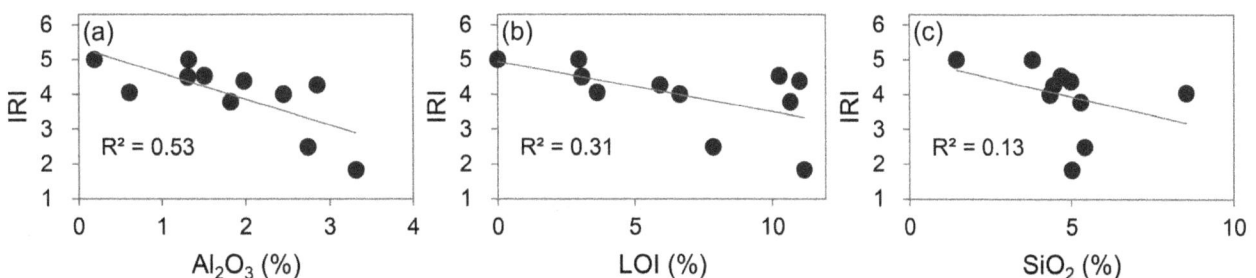

FIG 2 – Relationship between the IRI and (a) LOI, (b) Al_2O_3 and (c) SiO_2 content in the iron ore.

Application of assimilation analysis method

The usability of a new iron ore with a high SiO_2 content, referred to as iron ore X, was investigated through assimilation analysis. The chemical compositions and the IRI of iron ore X and middle-grade ore D are presented in Table 2. It can be observed that iron ore X exhibits superior reactivity despite having a significant SiO_2 content. Therefore, the potential replacement of iron ore D with iron ore X was evaluated through a pot test. Since the IRI of ore X is higher than that of ore D, the IBRI slightly increases with the increase in the replacement ratio, suggesting an improvement in the properties of sintered ores. The pot tests confirmed that as the replacement rate increased, the qualities of sintered ores, such as SI, RDI and MS, were either similar or slightly improved, as predicted. Therefore, this application test demonstrates that the assimilation evaluation analysis of iron ores can accurately estimate the usability of new iron ores in the sintering process, which cannot be determined solely through composition analysis. Furthermore, it was proven that the change in sintered ore qualities can be predicted by deriving the IBRI for each blending condition in advance.

TABLE 2

The usability evaluation results of a new iron ore X through assimilation test and pot tests.

Comparison of iron ore characteristics and IRI					
Iron ores	T.Fe (%)	Al_2O_3 (%)	SiO_2 (%)	LOI (%)	IRI
X	60.40	0.81	9.81	2.91	4.57
D	63.66	1.31	4.74	3.08	4.50

	The replacement ratio of ore D (%)				
	$D \rightarrow X$	0	25	75	100
	IBRI	4.504	4.506	4.520	4.520
	SI	1.00	0.99	1.00	1.00
Normalised sinter quality	RDI	1.00	0.94	0.90	0.90
	-10 mm	1.00	0.99	0.95	0.92
	MS	1.00	1.02	1.02	1.05

CONCLUSIONS

This study aimed to investigate a new analysis method that utilises the assimilation reaction of coarse ores to reassess the potential of low-grade iron ore and predict the qualities of sintered ores. Several conclusions were drawn based on various studies conducted:

1. Through laboratory-scale experiments that simulated the assimilation process of coarse ore and evaluated different types of iron ore, it was possible to rank various iron ore source. This iron ore analysis method allows for the re-evaluation of the value of iron ores with overcoming the limitation of chemical composition analysis.

2. The influence of SiO_2 on assimilation, which can lead to the formation of high-viscosity melt, was found to be relatively low. However, iron ores with high Al_2O_3 content exhibited an increase in the temperature required for melt formation and the formation of weak bonding phases during the sintering process. The effect of Al_2O_3 content on assimilation was found to be significant.

3. Increasing the IBRI resulted in improved sinter qualities. Through simple pot tests, we verified the possibility of predicting the tendency of sinter qualities by deriving the IBRI for each blending condition. However, it is necessary to further demonstrate the correlation between the IBRI and sinter qualities by applying the various data obtained from pot tests and sintering plants to improve the reliability of the IBRI evaluation method.

Overall, this study highlights the potential of the assimilation evaluation analysis method for reassessing low-grade iron ore and predicting the qualities of sintered ores.

ACKNOWLEDGEMENTS

This paper was prepared with the approval of Hyundai Steel.

REFERENCES

Guo-liang, Z, Sheng-li, W, Bo, S, Zhi-gang, Q, Chao-gang, H and Yao, J, 2015. Influencing factor of sinter body strength and its effects on iron ore sintering indexes, *International Journal of Minerals, Metallurgy and Materials*, 22(6):553–560 (University of Science and Technology: Beijing and Springer-Verlag: Berlin Heidelberg).

Guo-liang, Z, Sheng-li, W, Shao-guo, C, Bo, S, Zhi-gang, Q and Chao-gang, H, 2014. Influence of gangue existing states in iron ores on the formation and flow of liquid phase during sintering, *International Journal of Minerals, Metallurgy and Materials*, 21(10):962–968 (University of Science and Technology: Beijing and Springer-Verlag: Berlin Heidelberg).

Jun, O, Kenichi, H, Yohzoh, H and Kazuyuki, S, 2003. Influence of iron ore characteristics on penetrating behavior of melt into ore layer, *ISIJ International*, 43(9):1384–1392.

Kenichi, H, Jun, O and Seiji, N, 2020. Influence of melting characteristics of iron ores on strength of sintered ores, *ISIJ International*, 60(4):674–681.

Rui-Feng, X, Yu, D and Xing-Min, G, 2022. Effect of alumina on crystallization behavior of calcium ferrite in Fe_2O_3-CaO-SiO_2-Al_2O_3 system, *Materials*, 15(15):5257–5275.

Sheng-li, W, Guo-liang, Z, Bo, S and Chao-gang, H, 2014. Influencing factors and effects of assimilation characteristic of iron ores in sintering process, *ISIJ International*, 54(3):582–588.

Xiaobo, Z, Shengli, W, Heng, Z, Lixin, S and Xvdong, M, 2017. Flow and penetration behaviours of liquid phase on iron ore substrate and their effects on bonding strength of sinter, *Ironmaking & Steelmaking*, 47(7):405–416.

Mineralogy and reducibility of sinter analogues in the Fe_3O_4-CaO-SiO_2 (FCS) ternary system under hydrogen atmosphere

I R Ignacio[1,2], G Brooks[3], M I Pownceby[4], M A Rhamdhani[5] and W J Rankin[6]

1. (Former) PhD Candidate, Swinburne University of Technology, Hawthorn Vic 3122.
   Email: irosaignacio@swin.edu.au
2. (Current) Postdoctoral Researcher, CSIRO Mineral Resources, Pullenvale Qld 4069.
   Email: isis.rosaignacio@csiro.au
3. Professor, Swinburne University of Technology, Hawthorn Vic 3122.
   Email: gbrooks@swin.edu.au
4. Senior Principal Research Scientist, CSIRO Mineral Resources, Clayton South Vic 3169.
   Email: mark.pownceby@csiro.au
5. Professor, Swinburne University of Technology, Hawthorn Vic 3122.
   Email: arhamdhani@swin.edu.au
6. Adjunct Professor, Swinburne University of Technology, Hawthorn Vic 3122.
   Email: wjrankin@outlook.com

ABSTRACT

Steelmaking has a significant contribution in producing greenhouse gases and there is a worldwide push towards decarbonising the process. One of the approaches to reduce CO_2 emissions is to use hydrogen in the blast furnace (BF) as the reductant gas. However, hydrogen reduction is an endothermic reaction, bringing changes to the temperature distribution in the BF. Understanding iron ore sinter mineralogy and reducibility under a hydrogen atmosphere is essential to optimise the process. In this study, sinter analogues in the Fe_3O_4-CaO-SiO_2 (FCS) ternary system together with industrial sinters were reduced in H_2-rich and CO atmospheres and the resulting reducibility and associated mineralogical changes were observed. Results show that magnetite-lime-silica (MLS) sinter analogues had similar reduction rates under H_2 and CO, and also higher reducibility than most of the industrial sinters, suggesting that magnetite sinters could be used efficiently in a conventional or hydrogen blast furnace.

INTRODUCTION

Iron and steelmakers currently face two major challenges. The first is dealing with the changes in the iron ore grade. High-grade ores are being consumed rapidly to meet the high demand from steelmakers and low-grade ores with high levels of gangue are the new reality. In the last decades, magnetite ores have been considered as a viable source of iron. Although magnetite ores require intense processing, contributing to higher productions costs, after beneficiation they typically reach 65–70 per cent iron content with low levels of impurities (eg phosphorous and sulfur) and steelmakers report lower environmental impacts when using higher grade magnetite ores (Davies and Twining, 2018). It is anticipated that either stricter sinter grade standards or the availability of raw materials will lead to an increase in the use of magnetite concentrates in sinter blends (Han and Lu, 2018). However, in contrast to hematite ores, the fundamental behaviour of magnetite concentrates in sinter blends is poorly known (Clout and Manuel, 2003). Therefore, understanding of how magnetite ores behave in the overall ironmaking process and, especially in the sintering process, is fundamental.

The second challenge is reducing the environment impacts coming from the conventional high-intensity carbon-based ironmaking process. However, any change to non-carbon-based processing routes could have unknown implications for complex feed materials such as iron ore sinter. As well, the quality of iron produced through non-carbon-based methods may not be the same as that produced through the existing blast furnace process.

Significant effort is being made by researchers and practitioners globally to decarbonise the ironmaking process through development of new technologies or optimising old ones, so as to align the industry with the new sustainable global emission guidelines. These range from optimising the blast furnace feed materials, including iron ore sinter, by using cleaner reduction agents, ideally produced from renewable energy sources. The use of hydrogen in blast furnace ironmaking is a promising strategy to replace coke and therefore reduce CO_2 emissions since hydrogen reduction produces water instead of carbon dioxide. However, hydrogen reduction of iron oxide is an endothermic reaction and the temperature distribution in the blast furnace is significantly impacted with the introduction of hydrogen. The temperature in the shaft region decreases as the hydrogen concentration rises and therefore, at higher hydrogen conditions, better reducibility of iron ore burden at lower temperatures is needed (Murakami *et al*, 2020). However, at lower temperatures, the reduction disintegration behaviour of sinters is expected to change, altering the permeability of the blast furnace (Murakami, Kodaira and Kasai, 2015). It is necessary in this situation to optimise the quality of iron ore sinter and investigate the replacement of hydrogen from thermodynamic and mineralogical points of view (Angalakuditi *et al,* 2022). Although the reduction of iron oxides under hydrogen is well understood (Monazam, Breault and Siriwardane, 2014; Spreitzer and Schenk, 2019; Abolpour, Afsahi and Azizkarimi, 2021), the mineralogical changes that take place in the iron ore sinter during reduction under hydrogen have not yet been thoroughly investigated (Angalakuditi *et al,* 2022).

Considering the above, the aims of this research were to:

- Create synthetic sinter analogues in the Fe_3O_4-CaO-SiO_2 (FCS) ternary system under different sintering conditions of temperature and holding time.

- Reduce the synthetic sinter analogues and industrial sinters under H_2-rich and CO atmosphere and compared their behaviours.

- Analyse the effect of sintering conditions on the reducibility of the various sinter types.

- Characterise the microstructural changes occurring during reduction of the sinter analogues and industrial sinters under H_2-rich and CO atmospheres.

EXPERIMENTAL

Materials

Sinter analogues were created from synthetic powders of magnetite (Fe_3O_4), lime (CaO), and silica (SiO_2) with the purity and particle size of each component shown in Table 1. Compact pellets were prepared by combining the powdered oxides in the compositional amounts listed in Table 2. The basicity (the ratio between CaO and SiO_2) was kept constant at 2.0, which is a typical basicity used in industrial iron ore sintering. The use of synthetic materials is justified in this study since their composition is closely controlled and easily reproduced. The powders were weighed using an electronic precision scale and then mixed manually in a mortar and pestle under acetone for homogenisation. The amount of acetone used was enough to form a slurry and to ensure non-segregation of material. This slurry was then dried in an oven for 1 hour at 110°C. The dry powder blends were then compacted into pellets of 2.5 g (±0.05 g), with dimensions of 13 mm diameter by approximately 6.5 mm height.

TABLE 1

Purity and particle size of the synthetic powder oxides used in this study.

Chemical	Purity	Particle size
Fe_3O_4	95%	<5 μm
CaO	98%	<0.16 μm
SiO_2	99.8%	<0.014 μm

TABLE 2

Bulk composition of the sinter pellets before sintering (wt%).

Tablet	Fe_3O_4	CaO	SiO_2	Basicity*
MLS	85.0	10	5	2.0

MLS = magnetite + lime + silica (Fe_3O_4 + CaO + SiO_2); *Basicity = CaO/SiO_2

Five industrial sinters were used for comparison. The samples were identified with the prefix IS (Industrial Sinter) and their chemical composition is given in Table 3.

TABLE 3

Bulk chemical compositions of the industrial sinter samples (wt%) determined via X-ray fluorescence (XRF) spectroscopy.

Sample	Total Fe	FeO	CaO	SiO_2	Al_2O_3	MgO	Mn	TiO_2	CaO/SiO_2	Porosity
IS1	57.2	7.7	10.3	5.1	1.8	0.9	0.5	0.1	2.0	9.5
IS2	56.4	7.2	11.0	5.5	1.9	1.0	0.3	0.1	2.0	6.1
IS3	56.7	6.6	9.2	6.1	1.9	1.3	0.4	0.3	1.5	8.2
IS6	56.5	7.2	10.0	6.1	1.8	1.3	0.4	0.1	1.6	6.6
IS7	56.7	7.7	9.8	6.2	1.7	1.3	0.3	0.1	1.6	4.3

Sintering test

The compacted tablets were sintered in duplicate in a horizontal tube resistance furnace (Nabertherm RHTH 120–300/18) under a controlled atmosphere and temperature conditions. An alumina rod with an alumina boat attached to the tip was used to insert and remove the two samples simultaneously into the hot zone. The samples were held for a specific time (t_{hold}) and then rapidly cooled, simulating the conditions of the flame front in actual sintering. Table 4 shows a summary of the experimental conditions. For cooling, the sample holder was removed from the high-temperature zone to the cool end of the furnace in open air, where the flange is constantly cooled by circulating water.

TABLE 4

Samples identification and experimental conditions used in the sintering experiments.

Sample	T_{max} (°C)	t_{hold} (min)
MLS001	1350	4
MLS003	1250	4
MLS004	1300	2
MLS005	1300	6

To control the furnace atmosphere, a pre-mixed bottled gas containing 99.5 per cent N_2 and 0.5 per cent O_2 was used, injecting into the furnace a low oxygen partial pressure of $pO_2 = 5 \times 10^{-3}$ atm, typical for laboratory-based iron ore sintering experiments (Clout, 1994; Pownceby and Clout, 2003; Harvey, 2020) at a flow rate of 100 L/h, controlled by a gas flowmeter. This value for oxygen partial pressure is generally chosen for laboratory studies because it generates mineral assemblages and microstructures close to those observed in industrial iron ore sinters (Hsieh and Whiteman, 1989).

For this work, four separate samples were analysed (see Table 4); however, a total of 30 experiments were performed with broader range of compositions and sintering conditions. The additional samples had their mineralogy, strength, porosity and reducibility analysed in previous publications (Ignacio *et al*, 2022a, 2022b, 2023) and a complete study will soon be published as a PhD thesis.

Mineralogical characterisation

Samples were prepared for image analysis and phase characterisation by optical microscopy. The sintered pellets were roughly crushed, mounting a chip from each pellet in epoxy resin, before being plane ground and polished using an automated polisher. Optical microscopy was carried out by placing the polished samples in a moving stepping stage fixed onto an optical microscope (Olympus

BX61). Photomicrographs were obtained using a high-resolution camera, allowing clear images of the cross-section surface of the sample at 200× magnification.

X-ray diffraction analysis (XRD) was used to identify and quantify the mineral phases in the sintered samples formed before and after reduction. XRD patterns were collected with a PANalytical X'Pert Pro Multi-purpose Diffractometer using Fe filtered Co Ka radiation, automatic divergence slit, 2° anti-scatter slit and fast X'Celerator Si strip detector. Patterns were collected from 4 to 80° in steps of 0.017° 2θ with a counting time of 0.5 sec per step, for an overall counting time of approximately 35 min. Phase identification was performed using PANalytical Highscore Plus© software (V4.8) which interfaces with the International Centre for Diffraction Data (ICDD) PDF 4+ 2021 database. Quantitative phase analysis (QPA) was carried out *via* the Rietveld method using TOPAS V6. The internal standard method was used for calculation of amorphous/unidentified phase content, using highly crystalline corundum as the standard.

Reducibility test

A thermogravimetric analysis (TGA) set-up was used in this study to measure the reducibility of the samples. The set-up consisted of a vertical tube furnace (Eurotherm) with a control system, and a precision balance (FX-300i with 1 mg accuracy) placed at the top of the furnace measuring the weight loss of the sample during the reaction. This set-up was used in previous work by Purohit (2019) and Ignacio *et al* (2022b) and simulates a scaled-down version of the ISO 7215:2015 standard technique.

The sintered sample was placed inside an alumina crucible held by a platinum wire basket and then placed in the hot zone of the furnace. The furnace was sealed, and argon gas was purged to maintain a neutral atmosphere before and after each run for 5 min. The reducing atmosphere was created by purging a mixture of H_2/N_2 (15 per cent and 85 per cent, respectively) at 0.4 L/min. This flow rate was found to be the optimum flow rate in previous studies (Kuila, Chatterjee and Ghosh, 2016). The same tests were repeated under a CO reducing atmosphere for comparison. For the CO experiments a mixture of CO/N_2 (40 per cent and 60 per cent, respectively) was used. In all tests, the temperature was kept constant at 950°C, following the ISO 7215:2015 standard. The weight loss was recorded continuously every 0.125 sec using WinCT software for 150 min.

RESULTS AND DISCUSSION

After isothermal reduction in the H_2-rich atmosphere, the surface of some samples appeared marginally modified. Figure 1 shows typical industrial sinter and sinter analogues before and after reduction. For most of the sintered magnetite-containing the samples, no significant difference in surface appearance was noted after reduction. The samples had slightly shrunk and the hardness was maintained (not easily breakable). Cracks were noted in the surface of a few samples as shown in Figure 1b (right). In comparison, the industrial sinters initially presented as slightly reddish in colour and with bright patches randomly distributed across the surface of the unreduced samples. After reduction, the samples had a greyish and dull surface as shown in Figure 1a. A slight shrinkage in the overall dimensions was also noted for the industrial samples.

FIG 1 – (a) Industrial sinter before reduction (left) and after reduction (right); (b) Sinter analogues before reduction (left) and after reduction (right).

Figure 2 shows the weight loss versus time curves for the reduction of the sinter analogues containing magnetite-lime-silica (MLS001, MLS003, MLS004 and MLS005) under H_2-rich and CO atmospheres reduced at 950°C for 150 min. The reaction between Fe_3O_4 and the reductant gases (H_2 and CO) generates FeO and α-Fe and releases other gases (H_2O, in the case of H_2, and CO_2, in the case of using CO as the reductant gas), leading to shrinkage of the sample. Therefore, the weight loss of the sample after reduction is an indication of the reducibility. Angalakuditi *et al* (2022) confirmed the correlation of weight loss (%) with degree of reduction in their work. The W_{loss} (%) for the samples analysed reached 15 to 18 per cent, for H_2 reduction and 13 to 21 per cent, for CO reduction. From the graphs it can be noted that the rate of reduction was very similar for the first hour for all samples but slightly changed after the initial period, except for the MLS005 sample, which remained very similar throughout the entire reduction period. The reduction of MLS001 was improved when H_2 was used, while the opposite behaviour was observed for the MLS003 sample. Sample MLS004, fired to 1300°C for 2 min presented a higher rate of reduction compared to the others in both reducing atmospheres. Note that the curve for MLS004 – H_2 was interrupted after one and a half hours due to a technical issue in the recording software, but it is still, possible to observe the trend.

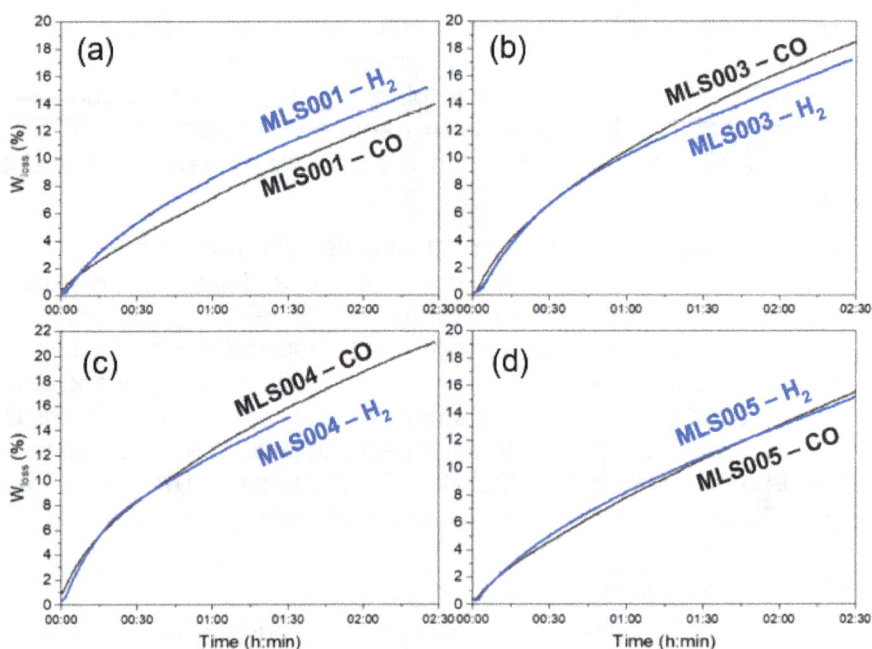

FIG 2 – Weight loss W_{loss} (%) versus time (h:min) for sinter analogues under both H_2 and CO reducing conditions.

Figure 3 shows the weight loss versus time curves for the reduction of the industrial sinters IS2, IS3, IS6 and IS7 under H_2-rich and CO atmospheres at 950°C for 150 min. The W_{loss} (%) for the samples analysed reached 13 to 19 per cent, for H_2 reduction and 10 to 16 per cent, for CO reduction. It can be noted that, for industrial samples, the rate of reduction for H_2 reduction was much faster than for CO reduction from the beginning of reaction. The reduction rate was improved by using H_2 as reductant gas for all the industrial samples, indicating a higher kinetic efficiency when compared to CO, confirming previous investigations by Kuila, Chatterjee and Gosh (2016). The results on the industrial sinters were different to the sinter analogue experiments and we attribute this to be due to the chemistry of the natural versus synthetic samples eg industrial sinters use natural ores which contain additional impurities such as alumina. Alumina is important in forming the sinter matrix bonding phase(s) SFCA which may contribute to the differences in reducibility between the industrial sinters (containing SFCA) and the synthetic sinter analogues (SFCA-free).

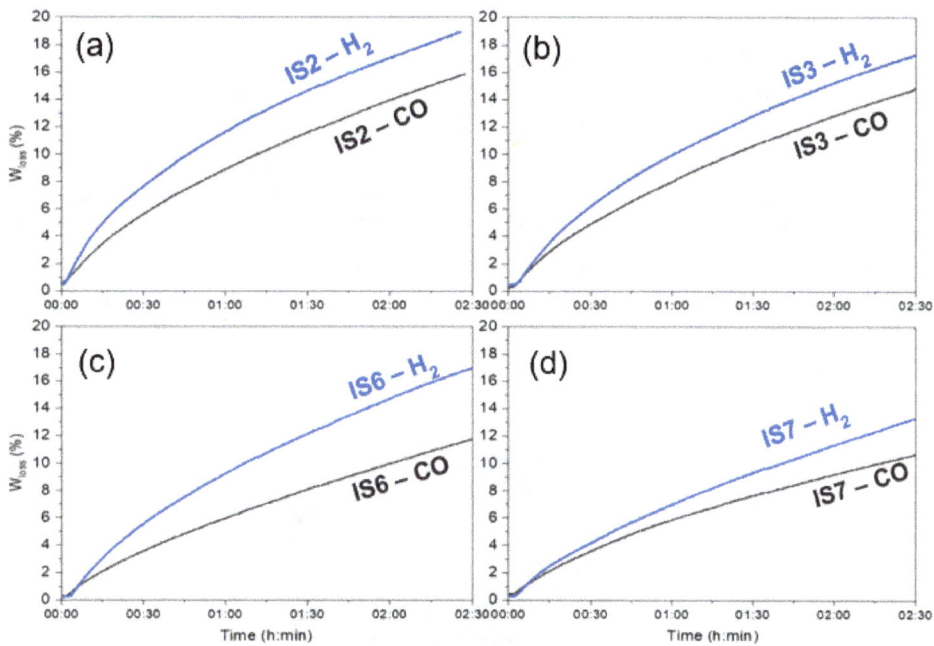

FIG 3 – Weight loss W_{loss} (%) versus time (h:min) for industrial sinters under both H_2 and CO reducing conditions.

Figure 4 compares the most and the least reducible sinter analogues (MLS004 and MLS004, respectively) with the most and least reducible industrial sinters (IS2 and IS7, respectively) examined in this work. Interestingly, the magnetite-lime-silica (MLS) samples performed significantly better than the industrial sinters in both the CO and H_2-rich reducing atmospheres. Similar observations were made in previous publications by Ignacio *et al* (2022b) where the reducibility of hematite-lime-silica samples, fired under similar conditions, were investigated. The similarity in reducibility with hematite-based sinters suggests that magnetite sinters could potentially be used in the blast furnace when sintered under appropriate conditions.

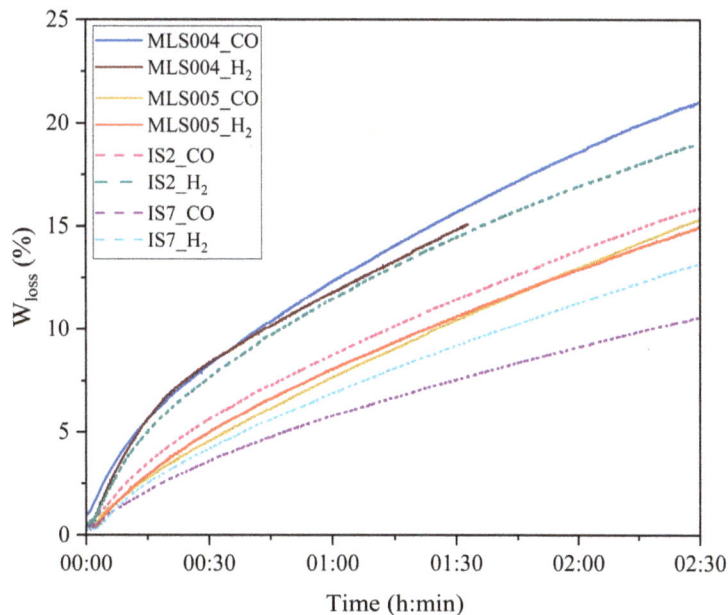

FIG 4 – Weight loss W_{loss} (%) versus time (h:min) comparing the most and the least reducible samples among the sinter analogues and industrial sinters.

Figure 5 shows the mineralogical changes for the MLS004 and MLS005 sinter analogue samples, which presented the highest and lowest weight losses on reduction respectively. Figures 5a and 5d show the samples before reduction, Figures 5b and 5e, show the samples after reduction under CO atmosphere and Figures 5c and 5f, show the samples after H_2 reduction. The bright phase in the reduced runs indicates the presence of metallic iron. The samples showed variations in the total porosity after being sintered under different conditions of temperature and holding times. MLS004 was sintered at 1300°C for 2 min, which created smaller pores but a large number (23.0 per cent). With the increase in holding time to 6 min for the MLS005 samples, the coalescence of the pores was favoured, leading to pore enlargement but reduction of the total porosity (8.50 per cent). Porosity was measured using image analysis techniques described in previous investigation (Ignacio *et al*, 2023). Increasing the temperature (from 1250°C to 1350°C) and holding time (from 2 to 6 min) caused a reduction in total porosity (Ignacio *et al*, 2022b). The predominant phases found in the MLS samples in this study were magnetite (61 to 77 wt per cent), hematite (10 to 21 wt per cent), larnite (8 to 10 wt per cent), γ-CFF (3 to 6 wt per cent), β-CFF (<1 wt per cent), with the remaining material being amorphous/unidentified phases.

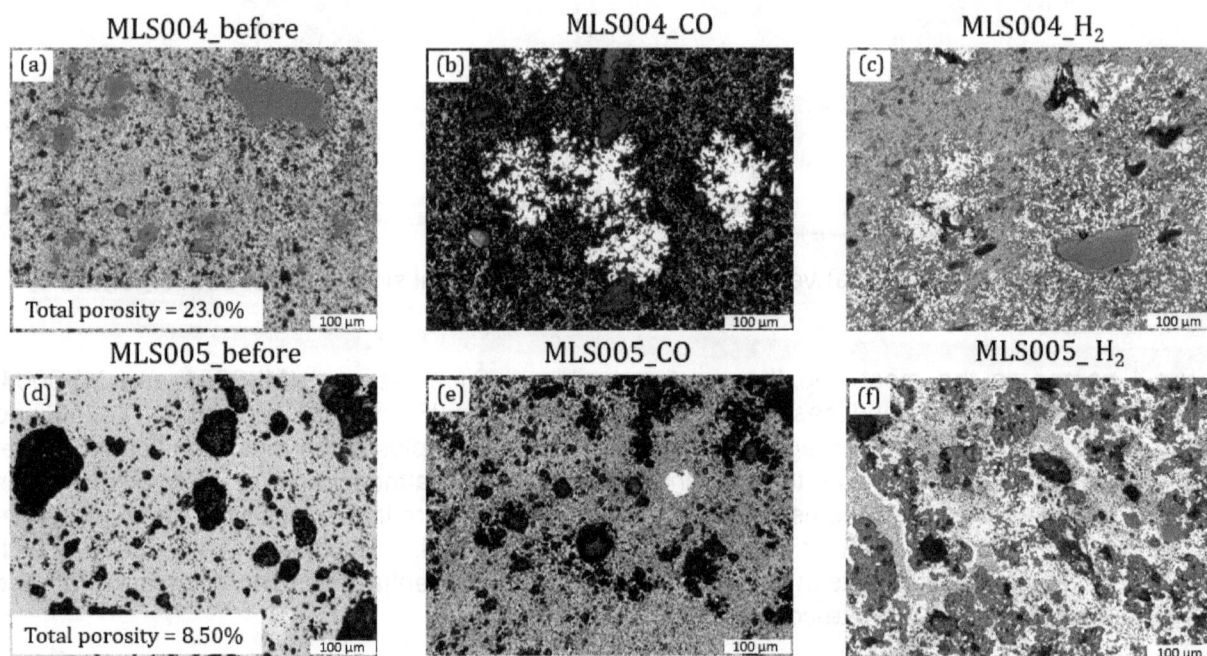

FIG 5 – Reflected light photomicrographs showing the mineralogical changes within samples MLS004 (highest W_{loss}%) and MLS005 (lowest W_{loss}%) under CO and H_2 reduction conditions.

After reduction under H_2, it was observed the mineralogy comprised the coexistence of metallic iron, wüstite, magnetite and hematite (in a much smaller fraction). The reduction reactions occurring in the MLS samples in this work followed the two-step sequence $Fe_3O_4 \rightarrow FeO \rightarrow Fe$, described by Kuila, Chatterjee and Gosh (2016) in their study on the kinetics of hydrogen reduction of magnetite ore fines. During the reduction reaction, magnetite is converted initially to wüstite (Fe_3O_4 to FeO), at an intermediate rate and finally the conversion of wüstite to metallic iron (FeO to Fe), at a slower reduction rate (He *et al*, 2023).

Figure 6 shows the mineralogical changes for the IS2 and IS7 samples, which presented the highest and lowest weight losses respectively, among all the industrial sinter samples in this study. Figures 6a and 6d show the samples before reduction, Figures 6b and 6e, show the samples after reduction under CO atmosphere and Figures 6c and 6f, show the samples after H_2 reduction. The slight difference in porosity between the two samples (6.1 per cent for IS2 and 4.3 per cent for IS7) is presumed have acted towards improving reducibility. The main phases observed in the industrial sinter samples were skeletal hematite, calcium ferrite and silico-ferrite of calcium and aluminium (SCFA), as indicated in Figure 6a.

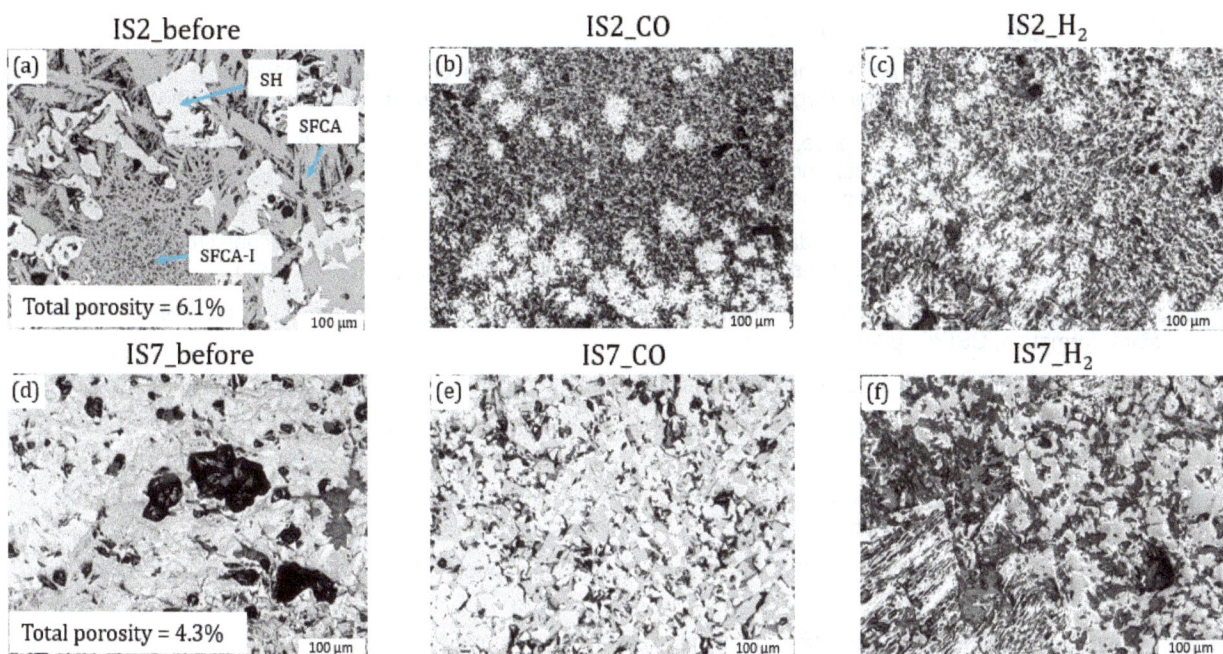

FIG 6 – Reflected light photomicrographs showing the mineralogical changes within samples IS2 (highest W_{loss}%) and IS7 (lowest W_{loss}%) under CO and H_2 reduction conditions. SH = skeletal hematite, CF = calcium ferrite, SCFA = silico-ferrite of calcium and aluminium.

CONCLUSIONS

The mineralogy and reduction behaviour of magnetite-based sinter analogues and industrial sinters under CO and H_2 were investigated. Synthetic iron ore sinter analogues from the Fe_3O_4-CaO-SiO_2 (FCS) ternary system were created under controlled laboratory tests where the maximum temperatures and holding times were varied. The sinter analogues as well as a range of industrial sinters were reduced under isothermal and controlled conditions in a TGA furnace under H_2-rich and CO-rich atmospheres. From this work it can be concluded that:

- Magnetite-lime-silica (MLS) sinter analogues had similar reduction rates under H_2 and CO; while the reduction rates of industrial sinters was improved under H_2, showing a higher reduction efficiency when compared to CO.

- The synthetic magnetite-lime-silica (MLS) sinter analogues had higher reducibilities than most of the industrial sinters examined in this study, in both of the reductant atmospheres, suggesting that magnetite sinters could be used in a conventional blast furnace or in a hydrogen blast furnace.

- Reducibility was strongly connected to porosity for both the industrial sinters and MLS sinter analogues.

Although the results found are promising when considering the hydrogen reduction of sinter analogues using magnetite concentrates, it is still too early to assume that this method could replace the conventional blast furnace in the near future. A valuable continuation of this work would come from further developing synthetic sinter analogues with expanded compositional ranges, including the effect of alumina for example, as well as the use of natural hematite and magnetite ores, which would be more in line with industrial sinters. Also, extending the sintering conditions examined to include a range of temperatures, holding times, and reducing atmosphere conditions will better inform future developments.

ACKNOWLEDGEMENTS

The authors acknowledge the Swinburne University of Technology and The Commonwealth Scientific and Industrial Research Organisation (CSIRO) for providing financial support for this research through an Australian Government Research Training Program (RTP) Scholarship and a CSIRO PhD Top-up Scholarship.

REFERENCES

Abolpour, B, Afsahi, M M and Azizkarimi, M, 2021. Hydrogen reduction of magnetite concentrate particles, *Mineral Processing and Extractive Metallurgy: Transactions of the Institute of Mining and Metallurgy*, 130(1):59–72.

Angalakuditi, V B, Bhadravathi, P, Gujare, R, Ayyappan, G, Singh, L R and Baral, S S, 2022. Mineralogical aspects of reducing lump iron ore, pellets and sinter with hydrogen, *Metallurgical and Materials Transactions B: Process Metallurgy and Materials Processing Science*, 53B:1036–1065.

Clout, J M F and Manuel, J R, 2003. Fundamental investigations of differences in bonding mechanisms in iron ore sinter formed from magnetite concentrates and hematite ores, *Powder Technology*, 130:393–399.

Clout, J M F, 1994. Formation of key iron ore sinter phases from Hamersley fines: implications for ultra fines chemistry and sinter formation, CSIRO, p 107.

Davies, M and Twining, M, 2018. Magnetite: South Australia's resource potencial, *MESA Journal*, 86(1):30–44.

Han, H and Lu, L, 2018. Recent advances in sintering with high proportions of magnetite concentrates. *Mineral Porcessing and Extractive Metallurgy Review*, 39(4):217-230.

Harvey, T, 2020. Influence of mineralogy and pore structure on the reducibility and strength of iron ore sinter, PhD Thesis, University of Newcastle, Newcastle.

He, J, Li, K, Zhang, J and Conejo, A N, 2023. Reduction kinetics of compact hematite with hydrogen from 600 to 1050°C, *Metals*, 13(464).

Hsieh, L H and Whiteman, J A, 1989. Effect of oxygen potential on mineral formation in lime-fluxed iron ore sinter, *ISIJ International*, 29(8):625–634.

Ignacio, I R, Brooks, G, Pownceby, M I, Rhamdhani, M A and Rankin, W J, 2022a. Effects of sintering conditions on porosity, strength and reducibility of magnetite tablets, in *Proceedings International Mineral Processing Conference Asia and Pacific, IMPC 2022*, pp 1580–1587.

Ignacio, I R, Brooks, G, Pownceby, M I, Rhamdhani, M A and Rankin, W J, 2023. Porosity in iron ore sinter analogues, *Iron & Steel Technology*, pp 70–77.

Ignacio, I R, Brooks, G, Pownceby, M I, Rhamdhani, M A, Rankin, W J and Webster, N A S, 2022b. Porosity, mineralogy, strength and reducibility of sinter analogues from the $Fe_2O_3(Fe_3O_4)$-CaO-SiO_2 (FCS) ternary system, *Minerals*, 12(1253):1–14.

International Organisation for Standardisation, 2015. ISO 7215:2015 Iron ores for blast furnace feedstocks – Determination of the reducibility by the final degree of reduction index, August 2015.

Kuila, S, Chatterjee, R and Ghosh, D, 2016. Kinetics of hydrogen reduction of magnetite ore fines, *International Journal of Hydrogen Energy*, 41(22):9256–9266.

Monazam, E R, Breault, R W and Siriwardane, R, 2014. Kinetics of hematite to wüstite by hydrogen for chemical looping combustion, *Energy and Fuels*, 28(8):5406–5414.

Murakami, T, Kodaira, T and Kasai, E, 2015. Reduction and disintegration behavior of sinter under N_2-CO-CO_2-H_2-H_2O gas at 773 K, *ISIJ International*, 55(6):1181–1187.

Murakami, T, Wakabayashi, H, Maruoka, D and Kasai, E, 2020. Effect of hydrogen concentration in reducing gas on the changes in mineral phases during reduction of iron ore sinter, *ISIJ International*, 60(12):2678–2685.

Pownceby, M I and Clout, J M F, 2003. Importance of fine ore chemical composition and high temperature phase relations: applications to iron ore sintering and pelletising, *Mineral Processing and Extractive Metallurgy*, 112(1):44–51.

Purohit, S, 2019. Alternative processing route for magnetite ores, PhD thesis, Swinburne University of Technology, Melbourne.

Spreitzer, D and Schenk, J, 2019. Reduction of iron oxides with hydrogen – A review, *Steel Research International*, 90(10).

Pelletisation of New Zealand titanomagnetite ironsand for hydrogen direct reduction

S Mendoza[1], B H Yin[1,2], A Zhang[1], M Nusheh[3] and C W Bumby[1,2]

1. Robinson Research Institute, Faculty of Engineering, Victoria University of Wellington, Wellington, New Zealand. Email: shaira.mendoza@vuw.ac.nz
2. The MacDiarmid Institute for Advanced Materials and Nanotechnology, Victoria University of Wellington, Wellington, New Zealand.
3. Hot Lime Labs, Lower Hutt, New Zealand.

ABSTRACT

Hydrogen-direct-reduction of iron ore in a vertical shaft furnace is a potential process that could significantly reduce CO_2 emissions from the steel industry. This approach requires the pelletisation of iron ore fines, which then undergo high temperature induration to strengthen the pellets before they can be fed into the shaft furnace. Here we report on investigations into the disc-pelletisation behaviour of New Zealand titanomagnetite ironsand, and its subsequent induration and hydrogen-reduction behaviour. Initial studies focused on the effect on pellet strength of varying ironsand particle size and induration conditions using bentonite and a commercially sourced carboxymethyl-cellulose binder. Green pellets formed with ironsand of average particle size of 65 μm exhibited optimal strength. The compressive strength of these pellets after induration at 1200°C for 2 hrs in air was measured to be 976 N. Interparticle bonding within the indurated pellets occurs due to a combination of titanohematite recrystallisation (from oxidation of titanomagnetite), and the formation of a liquid bonding phase due to interdiffusion with the bentonite and subsequent melting. Further investigation with various binders showed that a combination of both organic and inorganic binders was essential to achieve optimal pellet strength. Carboxymethyl-cellulose binders provides green strength, whilst inorganic binders such as bentonite promoted high indurated pellet strength. Hydrogen reduction tests in a thermogravimetric furnace showed that all the pellets could achieve a maximum reduction degree of 97 per cent at 1100°C, indicating that the different pellet preparation procedures had no significant effect on reducibility. Pellet recipes developed in this work are now being used to investigate vertical shaft hydrogen-DRI processing of NZ ironsand in a kg-scale laboratory reactor.

INTRODUCTION

Steel is one of the most essential materials for modern construction and manufacturing, and its production continues to see an increasing trend. However, the steel industry emits significant quantities of carbon dioxide (CO_2), contributing ~7 per cent of all anthropogenic CO_2 emissions worldwide (Kueppers *et al*, 2022). The major reason for this is that the initial ironmaking process traditionally uses coal as both a chemical reactant, and a source of heat. In New Zealand (NZ), a carbothermic approach is used for the reduction of the local iron ore source – titanomagnetite (TTM) ironsand. NZ ironsand contains high levels of titanium (equivalent to ~8 wt. per cent TiO_2) and as a result, it cannot be processed in a conventional blast furnace process (Brathwaite, Gazley and Christie, 2017). Instead, direct reduction (DR) of NZ ironsand occurs at ~1000°C in rotary kilns, followed by smelting in an arc furnace to produce pig iron (Templeton, 2006). This two-step process is extremely CO_2 and energy intensive, meaning that the iron and steel sector in NZ accounts for ~5 per cent of gross national CO_2 emissions (Ministry for the Environment, 2021).

Direct reduction (DR) of iron ore with hydrogen (H_2) is a potential alternative approach to 'near-zero CO_2' ironmaking (van Vuuren *et al*, 2022; Vogl, Åhman and Nilsson, 2018). The only by-product of H_2-DR is water vapour. In addition, if green H_2 is sourced and the entire process utilises a renewable form of energy, then the CO_2 emissions could drop to almost zero. There have been promising advancements worldwide in demonstrating H_2-DR in a vertical shaft furnace process, a technology that is already well-established for gas-based DR using fossil-derived syngas (Pei *et al*, 2020; Ripke and Kopfle, 2017). In this process, iron ore pellets or lumps are charged from the top of the shaft furnace and reduced by a counter-flowing gas. Free-flowing mass transfer of the solids stream through the reactor is important, thus raw iron ore fines are pelletised before they can be fed into a shaft furnace. Previous studies have already shown that NZ ironsand can be reduced with H_2, which opens a potential for NZ to adapt H_2 ironmaking to its domestic steel industry (McAdam, Dall and Marshall, 1969; Prabowo et al, 2022; Zhang et al, 2020). However, the optimisation of a pelletisation process for NZ ironsand is a key prerequisite for the development of a vertical shaft H_2-DR process using this material.

In a typical pelletisation process, a mixture of iron ore fines, binder and moisture are agglomerated into 'green' pellets in a disc or drum pelletiser. The green pellets are subsequently strengthened via induration at high temperatures of up to 1300°C (De Moraes, De Lima and Ribeiro, 2018; Zhu et al, 2015). Throughout the pelletisation process, certain strength criteria must be met to ensure pellet quality and minimise loss of material. In general, green pellets must pass a minimum of four 'drops' and exhibit a compressive strength of at least 22 N after drying (Lu, Pan and Zhu, 2015). This would ensure that the pellets remain intact when taken through the induration furnace. The indurated pellets have a higher strength criterion as they would experience higher crushing forces in a shaft furnace. For example, the MIDREX process requires a minimum compressive strength of 2450 N for pellets with diameters between 9 mm to 16 mm (Lu, Pan and Zhu, 2015). To achieve these strength criteria, different pelletising parameters are required for different types of iron ores. This is because various types of iron ore fines have distinct pelletisation behaviour, due to their unique properties such as particle morphology, surface characteristics, and gangue compositions (Eisele and Kawatra, 2003; Kawatra and Claremboux, 2022a).

The pelletisation behaviour of conventional hematite and magnetite ores has been extensively documented, and the major parameters affecting pellet quality are well understood (Eisele and Kawatra, 2003; Kawatra and Claremboux, 2022a). For example, the particle size distribution of iron ore fines is important because green pellets adhere together through capillary and viscous forces from the liquid trapped between solid particles (Qiu et al, 2004). Patra, Kumar and Rayasam (2017) reported that green pellets with coarse sized hematite particles were unable to meet the strength criteria due to high interstitial spaces. Iron ore particles are typically milled to a finer size distribution (<75 μm) to facilitate sufficient bonding in green pellets (Umadevi et al, 2008). In addition, the use of binders is necessary to enhance the strength of both green and indurated pellets. Bentonite clay is the most widely used binder for iron ore pelletising as it can achieve both functions (Kawatra and Claremboux, 2022b). However, it is an inorganic compound that primarily contains silica and alumina, which contribute to the gangue constituents that must be removed as slag in downstream processes (Sivrikaya and Arol, 2014). Therefore, non-contaminating organic binders such as starch and cellulose derivatives have also previously been investigated (Halt and Kawatra, 2014; Srivastava, Kawatra and Eisele, 2013). Commercially manufactured carboxymethyl cellulose (CMC) binders such as Peridur® by Nouryon are among the most useful organic binders, as these can deliver a comparable green strength to bentonite at significantly lower dosage levels. It should be noted however, that organic binders fully decompose into volatile gases upon induration, which can result in the loss of strength in the indurated pellets (Claremboux and Kawatra, 2022).

The pelletising behaviour of NZ TTM ironsand has not been comprehensively investigated in the literature, as the current industrial process does not require a pelletised feedstock. NZ TTM ironsand particles have a natural particle size distribution in the range of 100 to 300 μm (Prabowo et al, 2019). These occur as smooth spheroidal particles, a shape which can present challenges for pellet agglomeration (Xing et al, 2021). Prior work has investigated the H_2 reduction behaviour of NZ TTM ironsand pellets, but these studies were not focused on optimising and understanding the pelletising process (McAdam, Dall and Marshall, 1969; Zhang, 2020). This study, therefore, aims to fill this gap of knowledge. Firstly, the effects on the pellet strength of different induration temperatures (800–1200°C), induration times (0.5–3 hrs), and average particle size of ironsand (17–81 μm) were investigated with pellets formed with 1 wt. per cent bentonite and 0.1 wt. per cent organic binder (Peridur®). Thereafter, the performance of alternative organic and inorganic binders was studied to identify a binder combination that aids in the strength of green pellets as well as the final indurated pellet. Minimum target criteria for the strength of 5 mm pellets in this work were calculated based on the strength requirement for 9–16 mm industrial-sized pellets. Lastly, all pellet samples were reduced with H_2 gas in a thermogravimetric analysis (TGA) furnace to study the effect of various pelletising conditions on the reduction behaviour of NZ ironsand pellets.

MATERIALS AND METHODS

Materials

NZ TTM ironsand concentrate was obtained from Waikato North Head mine in New Zealand, with a median average particle diameter (D50) of 130 μm (Figure 1a). SEM imaging showed that most of the as-received NZ TTM ironsand particles are spheroidal in shape with smooth surfaces (Figure 1b). XRF analysis of the chemical composition of the as-received NZ TTM ironsand (Figure 1c) revealed a total Fe content of ~60 wt. per cent.

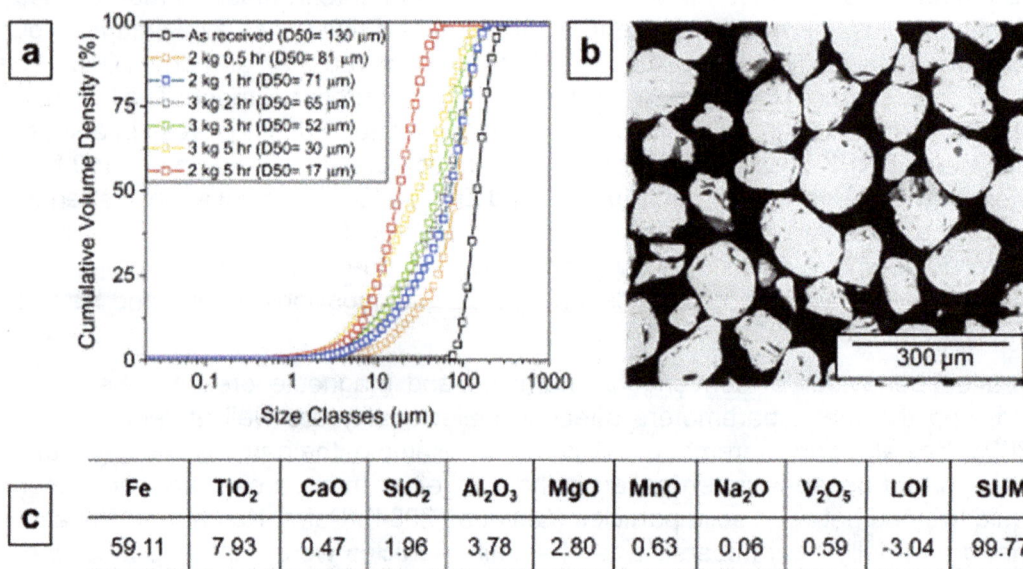

	Fe	TiO$_2$	CaO	SiO$_2$	Al$_2$O$_3$	MgO	MnO	Na$_2$O	V$_2$O$_5$	LOI	SUM
c	59.11	7.93	0.47	1.96	3.78	2.80	0.63	0.06	0.59	-3.04	99.77

FIG 1 – (a) Particle size distributions measured by laser scattering for as received ironsand and after various milling conditions; (b) Back-scatter electron SEM image of as-received NZ TTM ironsand; (c) XRF oxide analysis of ironsand after heating in air for 1 hr at 1000°C. Note that LOI = mass 'loss on ignition'. A negative value indicates mass gain due to oxidation (Mendoza *et al*, 2022).

All the binders used in this study were sourced as solid powders. The organic binders consisted of reagent grade carboxymethylcellulose (BDH CMC) from BDH Chemicals Ltd., and commercial CMC binders from Nouryon such as Peridur®, Finnfix10000, and Finnxfix30000. The inorganic binder used in this study was sodium bentonite from Federal.

Ironsand pelletisation

As received NZ TTM ironsand particles were firstly milled at different mass loadings and milling times to obtain a range of different particle sizes (Figure 1a). Wet ball milling was conducted using a 5 L stainless steel jar on a roller mill equipped with a Teco E2 T-Verter speed drive set at 90 rev/min. The wet milled ironsand was completely dried (80°C, >8 hrs) then thoroughly mixed with the binders and 1 wt. per cent water prior to pelletising. A bench-top Lurgi disc pelletiser from Mars Mineral US was used to make the green pellets at a disc-inclination angle of 45° and rotation speed of 35 rev/min. To facilitate the agglomeration of particles into spherical pellets, the mixed powders were added alternatingly with finely sprayed water. A moisture content of 7–8 wt. per cent was consistently used throughout this study. This was determined in a preliminary work to be the optimum moisture content for forming green pellets of acceptable strength. Once the green pellets had grown to the target size of ~5 mm, they were collected and dried at 80°C for >8 hrs, and then indurated in a muffle furnace which was heated at a rate of 10°C/min in air.

Table 1 shows the pellet composition and induration conditions for each set of samples. The naming convention for each individual sample is '*sample set–varied parameter*' (ie *P-65 μm* for a sample from set *P* produced using 65 μm ironsand particles.

TABLE 1

List of samples and pelletisation parameters.

Sample set	Variable	Inorganic binder (1 wt.%)	Organic binder (0.1 wt.%)	Particle size (μm)	Induration temperature (°C)	Induration time (hr)
	Base Case	Bentonite	Peridur	-	-	-
S	Sintering Condition	Bentonite	Peridur	65	800–1200	0.5–3
P	Particle Size	Bentonite	Peridur	17–81	1200	2
B	Organic Binder	Bentonite	Peridur, Finnfix10000, Finnfix30000, BDH CMC	65	1200	2

H₂ reduction of ironsand pellets

H_2 reduction of indurated pellets was performed using a TGA furnace (Mettler USA TA1 Thermobalance). Figure 2 shows a diagram depicting the experimental set-up. A sheathed B-type thermocouple is propped on top of a weighing balance located below the furnace hot zone. This thermocouple holds an alumina crucible on top where the pellet sample is placed. Next to the crucible is a capillary tube where the process gases are flowed through. For each test, an individual pellet of ~5.2 mm diameter was reduced at 1100°C with 100 vol. per cent H_2 gas at a flow rate of 500 mL min⁻¹. This flow rate was previously determined to be above the critical flow rate of 380 mL min⁻¹ in which the reaction rate is not affected by the gas transport rate (Zhang *et al*, 2020).

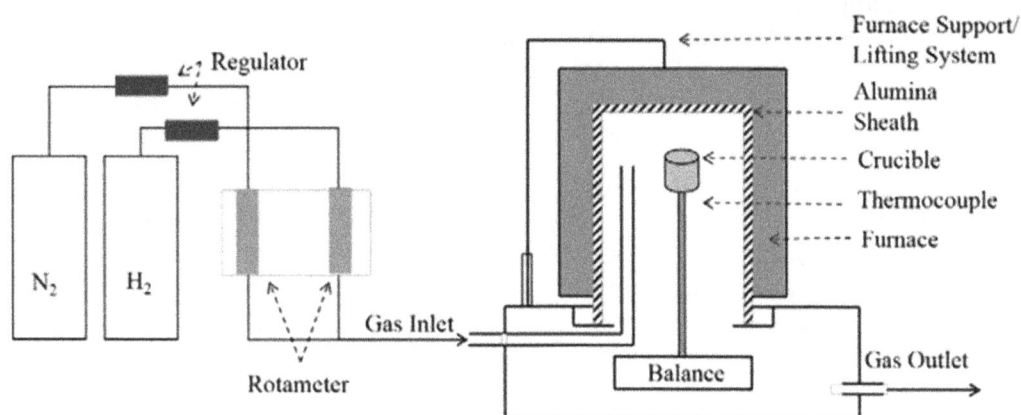

FIG 2 – A schematic of the experimental set-up for H_2 reduction in a TGA furnace (Zhang *et al*, 2020).

The reduction degree (R) was calculated based on the weight loss resulting from the removal of oxygen in the sample using the equation below.

$$R = (w_0 - w_t)/(X * w_0) \times 100\%$$

In the equation, w_0 is the original weight of the pellet and w_t is the weight of pellet after time t of reduction. The factor X stands for the weight fraction of the removable oxygen from the pellets. For fully oxidised pellets X was calculated to be 0.25. For partially oxidised pellets, X was calculated by first determining the Fe(II)/Fe(III) ratio through a titration method (ISO 9035: 1989E).

Characterisation methods

Materials characterisation

The particle size distribution of raw powders was determined by laser diffraction (Mastersizer 3000), and their chemical composition was measured by X-ray Fluorescence (XRF) spectrometry after pre-oxidation at 1000°C. Backscattered scanning electron microscopy (SEM) images were taken with an FEI Nova NanoSEM. Elemental spot analysis and elemental mapping were conducted using an integrated energy dispersive spectrometer (EDS) detector. X-ray diffraction (XRD) was conducted with a Bruker D8 Advance instrument fitted with a Co Kα radiation source.

Evaluation of pellet strength

The strength of green pellets was characterised by performing a 'drop test'. Here, multiple green pellets were dropped repeatedly from a height of 45 cm onto a steel plate, and the average number of drops to fracture was recorded as the 'drop number'. Meanwhile, the compressive strength of dry and indurated pellets was tested using a Universal Testing Machine (Tinius Olsen, UK H10KT). Parallel steel plates were set at a crosshead speed of 1 mm/min and the maximum load to fracture was recorded.

Minimum target criteria for pellet strength

The minimum target criteria (MTC) for the strength of 5 mm pellets are: a drop number of four for green pellets (Lu, Pan and Zhu, 2015), a compressive strength of 5 N for dry pellets, and 550 N for indurated pellets. The next section explains how the compressive strength criteria were set.

RESULTS AND DISCUSSION

Determining the minimum target criteria for pellet strength

In this study, ~5 mm diameter pellets were produced for H_2 reduction in a laboratory reactor. These pellets are smaller compared to the 9–16 mm diameter pellets typically used in large-scale industrial processes like MIDREX. Compressive strength criteria for 9–16 mm pellets are 22 N for dry pellets and 2450 N for indurated pellets (Lu, Pan and Zhu, 2015). This means that at a similar force criterion, a 5 mm pellet would experience a significantly higher stress than a 9–16 mm pellet (as stress(σ) = force (F)/area (A)). Thus, an appropriate MTC for 5 mm pellets was identified.

Figure 3 shows the compressive strength of dried green pellets with different sizes formed with the base case binders and 65 µm ironsand. There is a linear fit with an R^2 value of 0.97 between the mid-plane cross-sectional area of the pellet and its compressive strength. This confirms that there is a linearly proportional correlation between these two parameters. As a result, the MTC for the compressive strength of 5 mm pellets was calculated by scaling from the 16 mm case, to be: a compressive strength of 5 N for dry pellets; and a compressive strength of 550 N for indurated pellets.

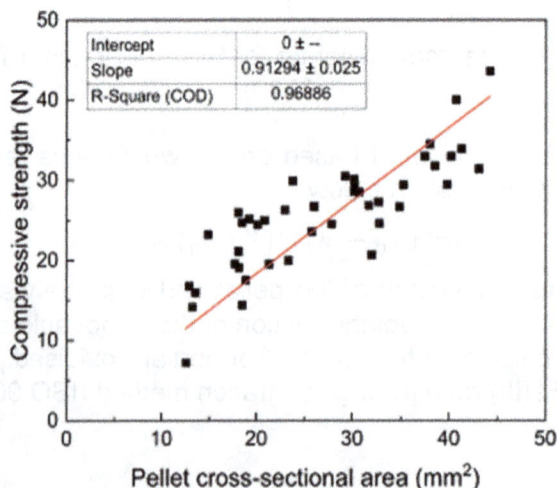

Intercept	0 ± --
Slope	0.91294 ± 0.025
R-Square (COD)	0.96886

FIG 3 – Relationship between green pellet size and the compressive strength.

Effect of induration condition on the strength of pellets

The optimum induration condition for NZ TTM ironsand pellets were first determined by indurating pellets formed with the base case binder recipe and 65 µm ironsand (see Table 1) at different temperatures and durations. Figure 4a shows that the compressive strength of indurated pellets was enhanced by increasing the temperature, more than doubling from 380 N to 976 N when the temperature was increased from 1100°C to 1200°C. Similarly, Figure 4b shows that the compressive strength of indurated pellets was improved by prolonging the induration time at 1200°C, with an approximately linear increase of compressive strength as the time was extended from 0.5 to 3 hrs.

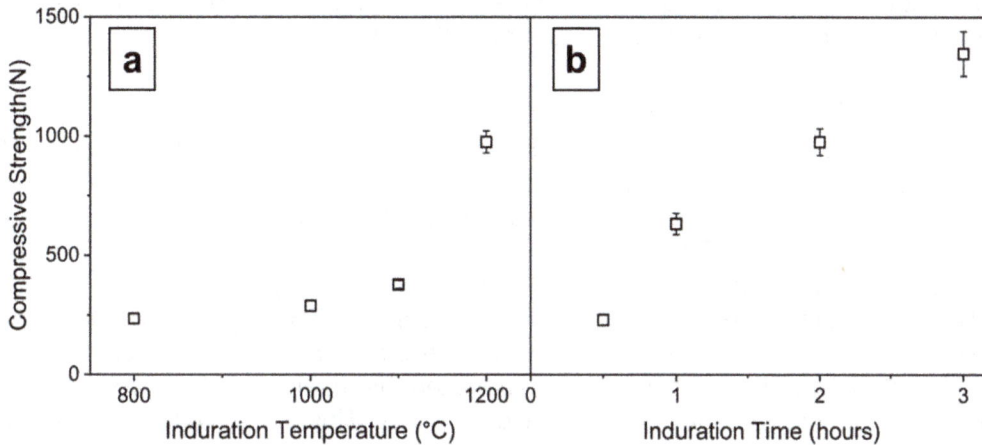

FIG 4 – Compressive strength of indurated ironsand pellets at different (a) induration temperatures (all for 2 hrs), and (b) induration times (all at 1200°C) (from Mendoza *et al*, 2022). Error bars represent the standard error for a sample size of 10–15 pellets.

Induration of NZ ironsand at a higher temperature and a longer time can promote increased oxidation of TTM. The subsequent recrystallisation of TTH phase can improve the interparticle bonding, leading to a higher compressive strength of the indurated pellets (Forsmo *et al*, 2008; Prusti, Nayak and Biswal, 2017). XRD patterns in Figure 4 show the crystalline phase changes that occur within the pellet at different induration conditions. At a relatively low temperature of 800°C for 2 hrs (Figure 5a) and a shorter induration time of 0.5 hr at 1200°C (Figure 5b), unoxidised peaks of TTM and pseudobrookite (PSB) were detected. The intensity of these TTM and PSB peaks decreased as the induration temperature and time increased. After induration at 1200°C for at least 2 hrs, the pellets were fully oxidised with only the TTH phase present.

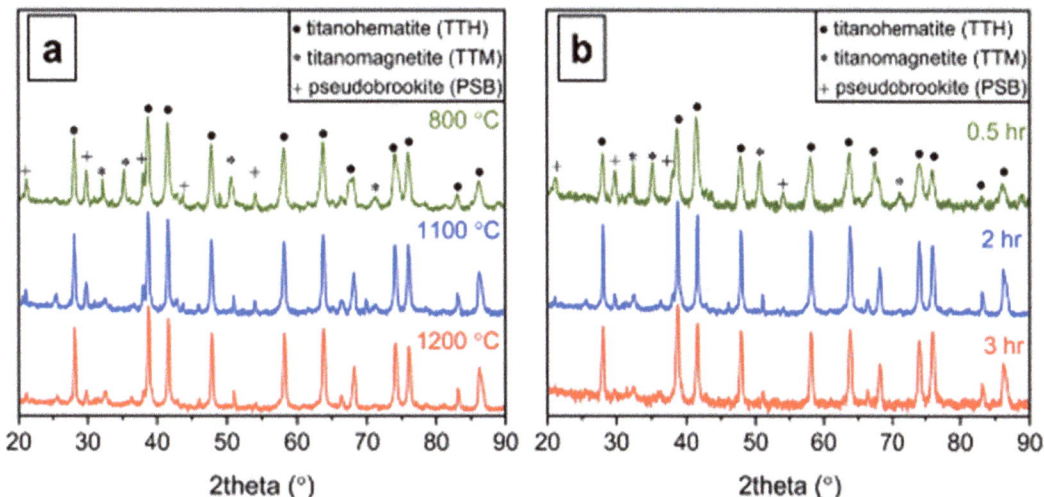

FIG 5 – XRD patterns of indurated pellets at different (a) induration temperature (all for 2 hrs), and (b) induration times (all at 1200°C) (from Mendoza *et al*, 2022).

In addition, the binder also affects the compressive strength of indurated pellets. Previous work has presented SEM/EDS element distribution maps of typical interparticle bonding regions in indurated ironsand pellets (Mendoza *et al*, 2022). The maps showed that after induration at 800°C for 2 hrs, the interparticle bonding region was identified to mainly comprise the calcined bentonite binder. In contrast, samples indurated at a higher temperature of 1200°C for 2 hrs showed interparticle bonding regions that were denser and consolidated by a liquid bonding phase. Whilst the melting temperature of bentonite is above 1200°C (Sivrikaya and Arol, 2018), it is likely that interdiffusion of gangue component within ironsand occurs within the bonding region during induration at 1200°C. This would lead to a slight decrease in the local melting point. The onset of liquid bonding formation may explain the significant strength difference observed between the induration temperatures of 1100°C and 1200°C (see Figure 4). Overall, these observations suggest that formation of a liquid bonding phase is the key mechanism for achieving high compressive strength in indurated pellets. As such, all further samples within this study were indurated at 1200°C for 2 hrs.

Effect of ironsand particle size on the strength of pellets

Initial pelletisation trials on as received ironsand particles was unsuccessful because of the large particle size distribution of the unmilled ironsand (D50 = 130 µm) which led to insufficient agglomeration during disc pelletisation. The ironsand was then milled to a D50 below 85 µm to allow the formation of green pellets. Figure 6 shows the strength of pellets containing different milled ironsand particle sizes prepared using the base case binder recipe and the induration condition outlined in Table 1. Coarse milled ironsand particles in sample *P-81 µm* formed green pellets with a drop number of 8 and dry pellets with a compressive strength of 18 N. In comparison, sample *P-65 µm* which contains relatively finer particles exhibited the maximum drop number of 18 for green pellets and a compressive strength of 30 N for dry pellets. This is because finer particles have a higher surface area which can enhance the interparticle adhesion forces from the binders (Abazarpoor *et al*, 2020). A similar behaviour was reported for hematite ore, wherein green pellets bonded with bentonite showed higher drop numbers when the particles were milled from 106 µm to 44 µm (Patra, Kumar and Rayasam, 2017). However, in the current study, further decreasing the particle size of ironsand below 65 µm (see samples *P-48 µm* to *P-17 µm*) resulted in a decrease in strength for both green and dry pellets. The additional surface area from excessively fine particles have possibly surpassed the available coverage from the binders, leading to weaker interparticle bonding.

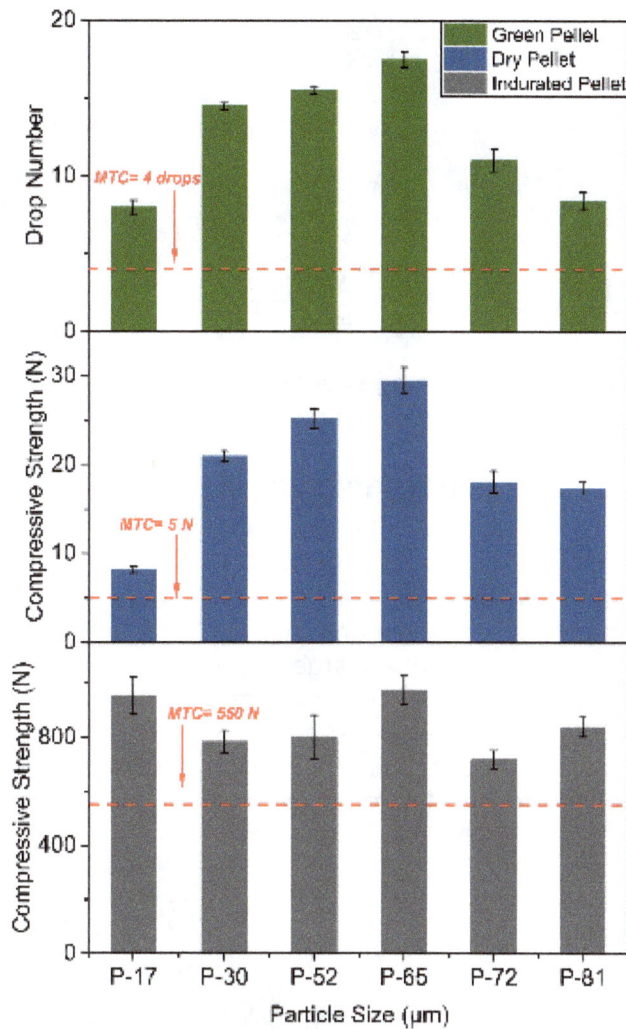

FIG 6 – Effect of ironsand particle size on the strength of green, dry, and indurated pellets (Mendoza *et al*, 2022).

The particle size did not have a significant effect on the strength of indurated pellets. Figure 6 shows that the compressive strength of all indurated pellets exceeded the minimum target criteria of 550 N, and the variation of strength between samples can be attributed to experimental variability. Figure 7 shows the backscattered electron SEM images of cross-sectioned indurated pellets under high magnification. Small particles in sample *P-17 μm* (Figure 7a) exhibit significantly more interparticle 'necking' compared to the coarse particles in sample *P-81 μm* (Figure 7b). This is because smaller particles have a higher local curvature and hence a higher surface energy, which promotes interparticle diffusion and 'necking'. However, both samples also exhibit a liquid bonding phase (dark grey region) between the ironsand particles. This liquid bonding phase was previously discussed in section 3.1 as a primary contributor to the strength of indurated pellets, and hence explains why there is an insignificant difference in the compressive strength of samples *P-17 μm* and *P-81 μm*. Based on these results, it was decided that an intermediate ironsand particle size of D50 = 65 μm represented an optimum size for pelletising and this was used for all further studies in this report.

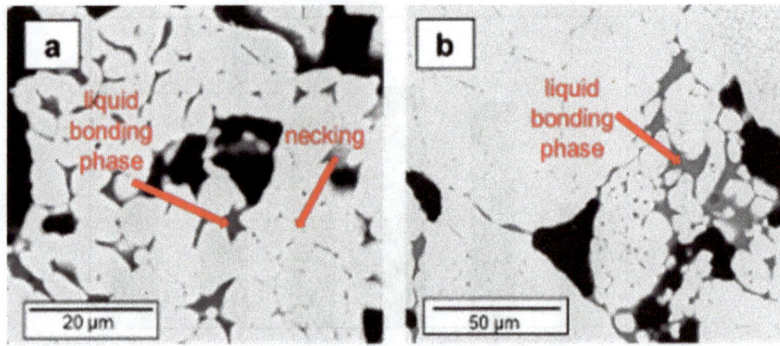

FIG 7 – Back-scatter electron SEM images of polished cross-sections of indurated pellets: (a) sample *P-17 µm*; and (b) sample *P-81 µm*.

Effect of organic binders and inorganic additives

A binder dosage of 1 wt. per cent bentonite has been reported to be effective for pelletising conventional ores like hematite (Eisele and Kawatra, 2003; Sivrikaya and Arol, 2014; Srivastava, Kawatra and Eisele, 2013). However, pelletisation of NZ TTM ironsand with only 1 wt. per cent bentonite resulted in green pellets with a low drop number of approximately two drops (Figure 8a). This drop number does not meet the minimum target criteria and would be problematic for materials handling in a large-scale pelletising process. The low strength of green pellets is likely due to the smooth surface and spheroidal shape of ironsand particles (Figure 1b). These particles require stronger adhesive and cohesive forces for successful agglomeration (Abazarpoor *et al*, 2020). Thus, different binder combinations were investigated to overcome these issues.

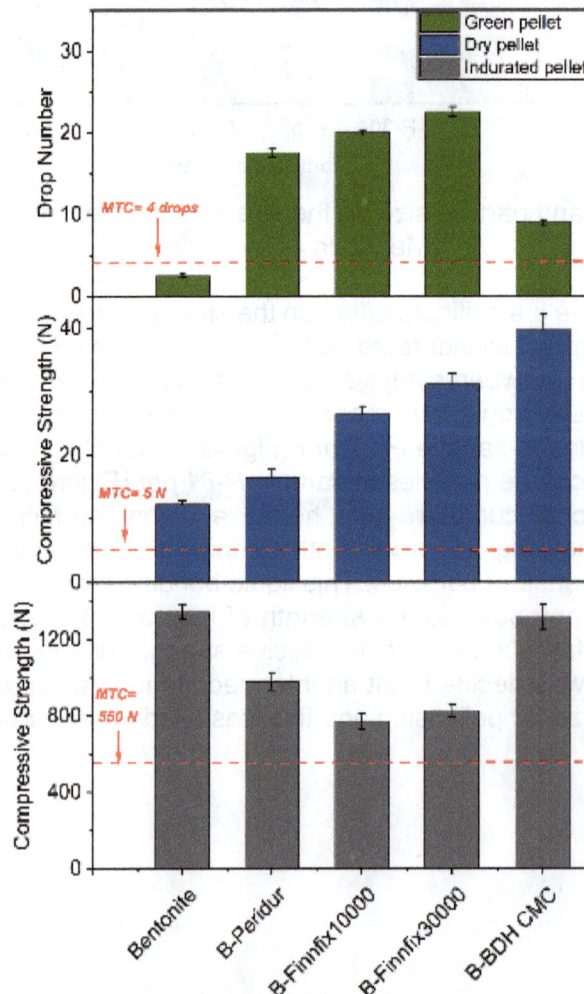

FIG 8 – Effect of organic CMC binder (0.1 wt. per cent) addition with bentonite (1 wt. per cent) on the strength of ironsand pellets (sample set B) (Mendoza *et al*, 2022).

Figure 8 shows the drop number of green pellets that were formed with 1 wt. per cent bentonite and an additional 0.1 wt. per cent of different CMC binders. Green pellets containing commercial CMC binders such as B-Peridur®, B-Finnfix10000, and B-Finnfix3000 were found to have the highest drop numbers of 18, 20, and 23, respectively. This is because organic binders increase the viscosity of the liquid trapped between particles during pelletisation, which reinforces the strength in green pellets (Qiu *et al*, 2004). Furthermore, Figure 8 shows that the organic binder contributes to the strength of dry pellets as all the samples doped with an organic binder (regardless of type) had significantly higher strengths compared to the dry pellets with only bentonite. In contrast, the addition of an organic binder appeared to have little effect on the compressive strength of indurated pellets, with only small variations between the different types of CMC.

Interestingly, indurated pellets formed with solely bentonite showed a significantly higher compressive strength of over 1200 N compared to the pellets formed with organic binders. It is possible that during pelletising, the organic binder competes with bentonite in moisture absorption. This might hinder the hydration of bentonite and its dispersion between the ironsand particles (Eisele and Kawatra, 2003), affecting the liquid phase formation in the subsequent induration process. It should also be noted that organic binders fully decompose during the induration process, meaning that it cannot be used as a sole binder as it would compromise the strength of the indurated pellet (Devasahayam, 2018; Srivastava, Kawatra and Eisele, 2013). Overall, these results suggest that a combination of organic and inorganic binder can result in the optimised strength of green, dry, and indurated pellets, but that dosage ratios may need to be optimised to minimise competing reactions between the different additives.

Effect of pelletising conditions on the reduction behaviour of pellets

Figure 9 shows the reduction degree of various pellet samples against the reduction time. Regardless of the pelletising parameter, each sample exhibited a reduction degree of ~97 per cent. Sample *P-81 µm* achieved 90 per cent reduction within 6 min which was slightly slower compared to the 4.5 min required for 90 per cent reduction of samples *P-65 µm* and *P-17 µm*. This suggests that a particle-scale mechanism may be the rate-limiting step in the reduction of sample *P-81 µm*. However, for samples with an ironsand particle size of ≤65 µm, particle-scale effects no longer have an influence on the reduction rate. This is consistent with previous observations of a pellet-scale shrinking core reaction in these type of pellets (Zhang *et al*, 2020). In comparison, Figure 9b shows that the induration condition and varying inorganic additives had little effect on the reduction rate of pellets. It should be noted that samples with different types of organic binders were excluded from the reduction test, because the binders were expected to be fully decomposed after induration.

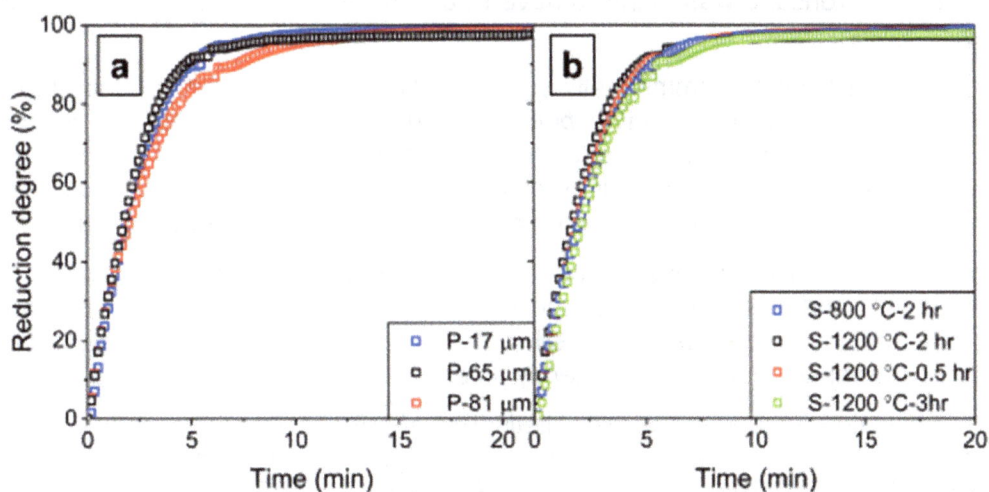

FIG 9 – Reduction degree of pellets in 500 mL min⁻¹ of H₂ gas at 1100°C: (a) shows the effect of ironsand particle size (sample set P), and (b) shows the effect of induration temperature and time (sample set S).

Backscattered electron SEM images in Figure 10 show a typical morphology of a fully reduced ironsand pellet in 500 mL min⁻¹ H₂ at 1100°C. This representative sample is *S-1200°C-2 hrs* which contains the base case binder recipe and 65 μm ironsand, indurated at 1200°C for 2 hrs in air. The metallic Fe can be observed as the bright region in bulbous clusters, whereas the impure oxides are seen as the dark grey region which coexist as fine islands within the metallic Fe structure. This morphology was homogeneous throughout the whole pellet and did not vary for any samples despite the different pelletising procedures.

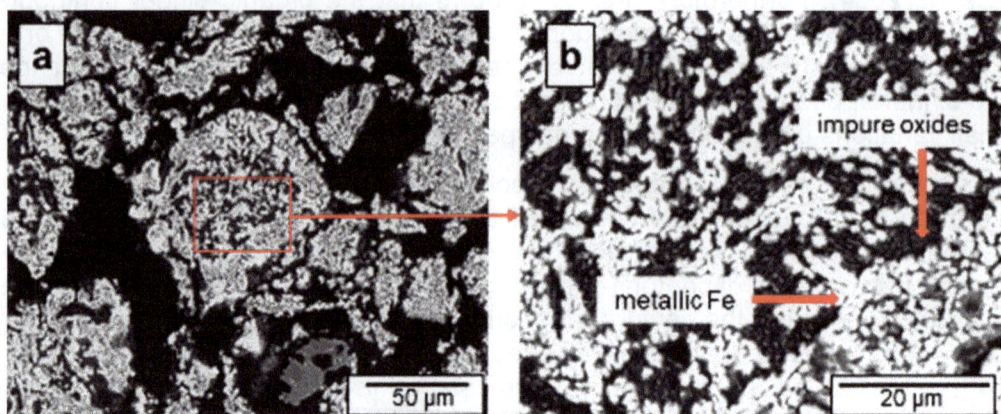

FIG 10 – Backscattered electron SEM images of reduced pellet sample formed from the base case binders with 65 μm ironsand, indurated at 1200°C for 2 hrs in air (Sample S-1200°C-2 hrs).

CONCLUSIONS

In this work, NZ TTM ironsand pellets were successfully produced by disc pelletisation. The effects of different induration conditions, ironsand particle size, and type of binders on the strength and reducibility of pellets in H_2 were analysed. The following conclusions can be drawn from this study:

- Induration of ironsand pellets at higher temperatures and longer durations increased the compressive strength of indurated pellets. For temperatures from 800°C to 1100°C, interparticle 'necking' occurred through oxidation of TTM and the subsequent recrystallisation of TTH grains. However, significant additional strength was imparted by indurating at a temperature of 1200°C, due to the formation of a liquid bonding phase.

- Milling of ironsand from 130 μm to <85 μm enabled the formation of green pellets that meet the target strength criteria. A maximum green drop number of 18, and a dry compressive strength of 30 N were obtained using milled ironsand with a D50 of 65 μm. However, the particle size of ironsand was found to have relatively little effect on the strength of indurated pellets.

- Organic binders such as commercially sourced CMC can effectively enhance the strength of green and dry pellets compared to pellets produced with only bentonite. However, these organic binders deliver no additional strength to indurated pellets due to its decomposition at high temperatures. In comparison, inorganic binders such as bentonite promote high strength in indurated pellets through the formation of a liquid bonding phase between particles.

- All pellet samples with an ironsand particle size of ≤65 μm exhibited a maximum reduction degree of ~97, and closely similar reduction rates in flowing H_2. This indicates that different pellet preparation procedures described in this study have little effect on the H_2 reduction behaviour of ironsand pellets, if sufficiently small ironsand particles are used.

These results show that future large-scale pelletisation of NZ TTM ironsand for a H_2-DR process is feasible, and that pelletisation must include pre-milling of ironsand, addition of both an organic binder and inorganic additive, and induration at ~1200°C. Future work will focus on characterising other key parameters for the H_2-DR process of these pellets, such as the disintegration and potential sticking behaviour in a continuous laboratory-scale vertical shaft reactor that has recently been commissioned within our laboratory.

ACKNOWLEDGEMENTS

This work was funded through New Zealand MBIE Endeavour Grant No. RTVU1907.

REFERENCES

Abazarpoor, A, Halali, M, Hejazi, R, Saghaeian, M and Zadeh, V S, 2020. Investigation of iron ore particle size and shape on green pellet quality, *Canadian Metallurgical Quarterly*, 59:242–250. https://doi.org/10.1080/00084433.2020.1730116

Brathwaite, R, Gazley, M and Christie, A, 2017. Provenance of titanomagnetite in ironsands on the west coast of the North Island, New Zealand, *J Geochem Explor*, 178:23–34. https://doi.org/10.1016/j.gexplo.2017.03.013

Claremboux, V and Kawatra, S K, 2022. Iron Ore Pelletization: Part III. Organic Binders, *Mineral Processing and Extractive Metallurgy Review*, 00:1–17. https://doi.org/10.1080/08827508.2022.2029431

De Moraes, S L, De Lima, J R and Ribeiro, T R, 2018. Iron Ore Pelletizing Process: An Overview, *Iron Ores and Iron Oxide Materials*, pp 41–59 (IntechOpen). https://doi.org/10.5772/intechopen.73164

Devasahayam, S, 2018. A novel iron ore pelletization for increased strength under ambient conditions, *Sustainable Materials and Technologies*, 17:e00069. https://doi.org/10.1016/j.susmat.2018.e00069

Eisele, T C and Kawatra, S K, 2003. A review of binders in iron ore pelletisation, *Mineral Processing and Extractive Metallurgy Review*, 24:1–90. https://doi.org/10.1080/08827500306896

Forsmo, S P E, Forsmo, S E, Samskog, P O and Björkman, B M T, 2008. Mechanisms in oxidation and sintering of magnetite iron ore green pellets, *Powder Technol*, 183:247–259. https://doi.org/10.1016/j.powtec.2007.07.032

Halt, J A and Kawatra, S K, 2014. Review of Organic Binders for Iron Ore Concentrate Agglomeration, *Minerals and Metallurgical Processing*, 31:73–94. https://doi.org/10.1007/bf03402417

Kawatra, S K and Claremboux, V, 2022a. Iron Ore Pelletization: Part I, Fundamentals, *Mineral Processing and Extractive Metallurgy Review*, 43:529–544. https://doi.org/10.1080/08827508.2021.1897586

Kawatra, S K and Claremboux, V, 2022b. Iron Ore Pelletization: Part II, Inorganic Binders, *Mineral Processing and Extractive Metallurgy Review*, 43:813–832. https://doi.org/10.1080/08827508.2021.1947269

Kueppers, M, Hall, W, Levi, P, Simon, R and Vass, T, 2023. Iron and Steel, International Energy Agency (IEA), Paris. Available from: <https://www.iea.org/reports/iron-and-steel>

Lu, L, Pan, J and Zhu, D, 2015. Quality requirements of iron ore for iron production, *Iron Ore: Mineralogy, Processing and Environmental Sustainability*, pp 475–504 (Woodhead Publishing). https://doi.org/10.1016/B978-1-78242-156-6.00016-2

McAdam, G D, Dall, R E A and Marshall, T, 1969. Direct gas reduction of NZ Ironsands, *New Zealand Journal of Science*.

Mendoza, S, Yin, B H, Zhang, A and Bumby, C W, 2022. Pelletization and sintering of New Zealand titanomagnetite ironsand, *Advanced Powder Technology*, 33. https://doi.org/10.1016/j.apt.2022.103837

Ministry for the Environment, 2021. New Zealand's Greenhouse Gas Inventory 1990–2019. Available from: <https://environment.govt.nz/assets/Publications/Greenhouse-Gas-Inventory-1990-2019/New-Zealands-Greenhouse-Gas-Inventory-1990-2019-Volume-1-Chapters-1-15.pdf>

Patra, S, Kumar, A and Rayasam, V, 2017. The effect of particle size on green pellet properties of iron ore fines, *Journal of Mining and Metallurgy A: Mining*, 53A:31–41. https://doi.org/10.5937/jmma1701031s

Pei, M, Petäjäniemi, M, Regnell, A and Wijk, O, 2020. Toward a fossil free future with hybrit: Development of iron and steelmaking technology in Sweden and Finland, *Metals (Basel)*, 10:1–11. https://doi.org/10.3390/met10070972

Prabowo, S W, Longbottom, R J, Monaghan, B J, del Puerto, D, Ryan, M J and Bumby, C W, 2022. Hydrogen Reduction of Pre-oxidized New Zealand Titanomagnetite Ironsand in a Fluidized Bed Reactor, *Journal of The Minerals, Metals & Materials Society (JOM)*, 74:885–898. https://doi.org/10.1007/s11837-021-05095-x

Prabowo, S W, Longbottom, R J, Monaghan, B J, del Puerto, D, Ryan, M J and Bumby, C W, 2019. Sticking-Free Reduction of Titanomagnetite Ironsand in a Fluidized Bed Reactor, *Metallurgical and Materials Transactions B: Process Metallurgy and Materials Processing Science*, 50:1729–1744. https://doi.org/10.1007/s11663-019-01625-w

Prusti, P, Nayak, B K and Biswal, S K, 2017. Study of Temperature Profile in the Induration of Magnetite Iron Ore Pellets, *Transactions of the Indian Institute of Metals*, 70:453–462. https://doi.org/10.1007/s12666-016-1011-8

Qiu, G, Jiang, T, Fan, X, Zhu, D and Huang, Z, 2004. Effects of binders on balling behaviors of iron ore concentrates, *Scandinavian Journal of Metallurgy*, 33:39–46. https://doi.org/10.1111/j.1600-0692.2004.00668.x

Ripke, J and Kopfle, J, 2017. MIDREX H_2: Ultimate Low CO_2 Ironmaking and its place in the new Hydrogen Economy, Direct from MIDREX.

Sivrikaya, O and Arol, A I, 2014. Alternative Binders to Bentonite for Iron Ore Pelletizing – Part I: Effects on Physical and Mechanical Properties, *Holos*, 3:94–103. https://doi.org/10.15628/holos.2014.1758

Sivrikaya, O and Arol, A I, 2018. Thermal Investigation of Some Potential Binders for Iron Ore Pelletizing, in *Proceedings of the 16th International Mineral Processing Symposium (IMPS 2018)*, pp 550–557.

Srivastava, U, Kawatra, S K and Eisele, T C, 2013. Study of Organic and Inorganic Binders on Strength of Iron Oxide Pellets, *Metallurgical and Materials Transactions B: Process Metallurgy and Materials Processing Science,* 44:1000–1009. https://doi.org/10.1007/s11663-013-9838-4

Templeton, F, 2006. Iron and steel – Attempts to extract iron, Te Ara – the Encyclopedia of New Zealand.

Umadevi, T, Kumar, M G S, Kumar, S, Prasad, C S G and Ranjan, M, 2008. Influence of raw material particle size on quality of pellets, *Ironmaking and Steelmaking*, 35:327–337. https://doi.org/10.1179/174328108X287928

van Vuuren, C, Zhang, A, Hinkley, J T, Bumby, C W and Watson, M J, 2022. The potential for hydrogen ironmaking in New Zealand, *Cleaner Chemical Engineering,* 4:100075. https://doi.org/10.1016/j.clce.2022.100075

Vogl, V, Åhman, M and Nilsson, L J, 2018. Assessment of hydrogen direct reduction for fossil-free steelmaking, *J Clean Prod*, 203:736–745. https://doi.org/10.1016/j.jclepro.2018.08.279

Xing, Z, Cheng, G, Gao, Z, Yang, H, Xue, X, 2021. Effect of Incremental Utilization of Unground Sea Sand Ore on the Consolidation and Reduction Behavior of Vanadia–Titania Magnetite Pellets, *Metals (Basel)*, 11:269. https://doi.org/10.3390/met11020269

Zhang, A, 2020. Reduction of New Zealand Titanomagnetite Ironsand Pellets in H_2 Gas at High Temperatures, Phd thesis, University of Wollongong. Available from: <https://ro.uow.edu.au/theses1/1036>

Zhang, A, Monaghan, B J, Longbottom, R J, Nusheh, M and Bumby, C W, 2020. Reduction Kinetics of Oxidized New Zealand Ironsand Pellets in H_2 at Temperatures up to 1443 K, *Metallurgical and Materials Transactions B: Process Metallurgy and Materials Processing Science*, 51:492–504. https://doi.org/10.1007/s11663-020-01790-3

Zhu, D, Pan, J, Lu, L and Holmes, R J, 2015. Iron ore pelletization, in *Iron Ore: Mineralogy, Processing and Environmental Sustainability*, pp 435–473 (Woodhead Publishing). https://doi.org/10.1016/B978-1-78242-156-6.00015-0

QPM hematite by-product as an ironmaking feed material

G Reynolds[1] and L Lu[2]

1. Operations Manager, Queensland Pacific Metals, Brisbane Qld 4000.
   Email: greynolds@qpmetals.com.au
2. Senior Principal Scientist, CSIRO Mineral Resources, Pullenvale Qld 4069.
   Email: liming.lu@csiro.au

ABSTRACT

Queensland Pacific Metals (QPM) is developing its flagship Townsville Energy Chemicals Hub (TECH) facility to produce nickel and cobalt sulfates from imported high-grade nickel laterite ore for nickel-cobalt-aluminium and nickel-cobalt-manganese Li-ion batteries. The TECH facility is expected to use a patented leaching, recovery and recycling process called the DNi Process™. It uses the nitric acid leaching process for extracting nickel, cobalt and other valuable metals from laterite ore. Apart from the nickel and cobalt sulfates, the TECH Project will also produce high purity alumina, hematite (~640 ktpa) and other by-products. The hematite by-product precipitated from the DNi Process™ (hematite precipitate hereafter), is high in Fe and particularly low in impurities. QPM is working to identify potential applications for these high-purity by-products to generate additional revenue and to achieve its zero-waste ambition. As such, QPM has engaged CSIRO to evaluate various agglomeration options to produce low-cost, hematitic products suitable as ironmaking feed materials. This paper will first reveal the unique characteristics of the hematite precipitate arising from the TECH Project and then examine the sintering and pelletising processes for preparing the hematite precipitate as an ironmaking feed.

INTRODUCTION

Queensland Pacific Metals (QPM) plans to import high-grade nickel laterite ore from New Caledonia and process the ore through its proposed Townsville Energy Chemicals Hub (TECH) facility to produce nickel and cobalt sulfates for nickel-cobalt-aluminium and nickel-cobalt-manganese Li-ion batteries. Given the significant benefits entailed, the TECH project was recently declared by the Qld government as 'a prescribed project'.

The TECH facility is expected to use a patented recovery and recycling process called the DNi Process™. It uses the nitric acid leaching process for extracting nickel, cobalt and other valuable metals from laterite ore. As more than 98 per cent of the nitric acid in the DNi Process is recycled, the process is environmentally friendly with no requirement for tailing dams and generates minimal waste products. Apart from the nickel and cobalt sulfates, the process also generates high purity alumina, hematite and other by-products. QPM is working actively to identify potential applications of these high purity by-products to achieve its zero-waste ambition. QPM estimates to produce about 640 kt of hematite by-product a year.

The hematite by-product precipitated from the DNi Process™ is high in Fe and particularly low in impurities. However, it is highly crystallised, very fine with at an estimated top size of 20 microns and contains a high moisture content of about 15 per cent. As the material is considerably different from natural hematite ore, preparation methods are required to allow its utilisation in sintering and pelletising while maintaining production productivity and product quality. This paper will first reveal the unique characteristics of the hematite by-product arising from the TECH Project and then examine and optimise the sintering and pelletising processes for preparing this material as an ironmaking feed.

EXPERIMENTAL

Materials

Table 1 shows the chemical composition of the initial hematite by-product sample received. The hematite by-product contained a high LOI and consequently a low Fe grade. The dehydration process occurring during sintering and pelletising is expected to raise the sample Fe grade to 64.55 per cent. The sample was particularly low in Si and P and can be blended and utilised to counteract the rising trend in the SiO_2 and P contents of ores commercially available. The high LOI content of the sample is due to the presence of hydrated hematite obtained through atmospheric hydrolysis. The future sample will be prepared through a pressurised hydrolysis process and will be less hydrated.

TABLE 1

Chemical composition of QPM hematite by-product (wt per cent).

	Fe	SiO_2	Al_2O_3	P	S	CaO	Mn	MgO	Cr	Ni	LOI425	LOI650	LOI1000
Units	%	%	%	%	%	%	%	%	%	%	%	%	%
PV1269	58.45	0.22	1.94	0.005	0.21	0.01	1.28	0.41	1.23	0.169	8.25	8.83	9.45

Figure 1 shows the particle size distribution and SEM image of the hematite by-product sample received. The as received QPM hematite by-product is characterised by many micron size spheres of a cauliflower surface structure. It has a narrow size distribution with a P_{80} of 16.4 μm, which is considerably finer than typical pellet feeds available. The sample showed a very high Blaine Index of 2345 cm^2/g and a high capacity of absorbing and holding water.

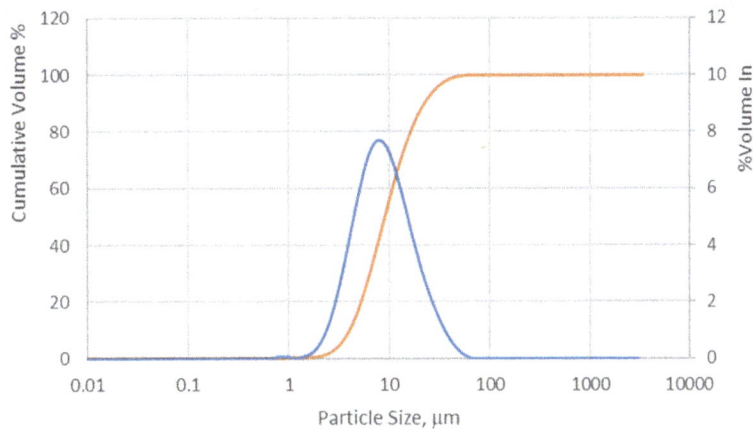

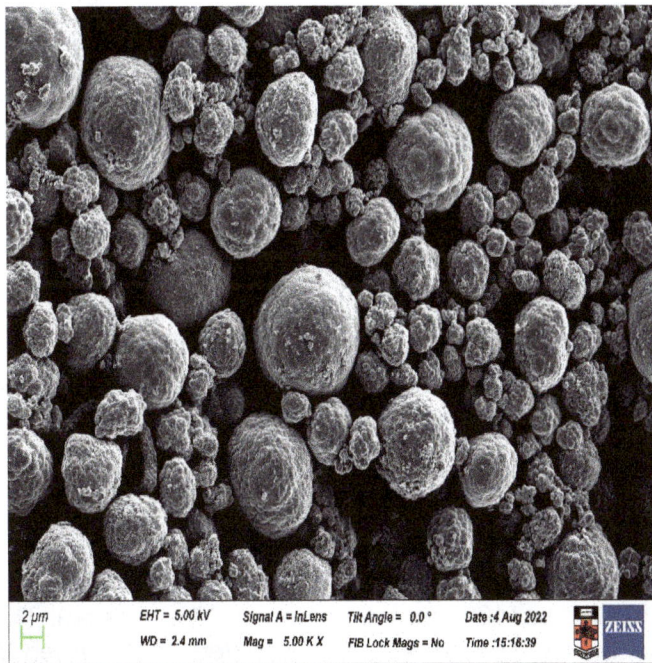

FIG 1 – Particle size distribution and SEM image of QPM hematite by-product.

Sintering test method

The granulating and sintering performance of the hematite by-product was evaluated at 15 wt per cent to substitute 10 per cent of magnetite concentrates and 5 per cent of Australian ores in a reference industrial sinter blend. The reference blend comprised of typical global iron ores from Australian (65 per cent) and Brazilian (25 per cent) iron ore fines as well as concentrates from China (10 per cent). Table 2 summarises the sintering test conditions applied and sinter chemistry targets, which were achieved by addition of limestone, dolomite and silica sand. Granulation was carried out by making up a number of sinter mixes with different water additions from the component raw materials and measuring the permeability of a packed green granule bed. These granulated sinter mixes of varying moisture contents are expected to have different granule size distributions, strength and morphologies and therefore different resistance to air flow when packed in the sinter pot. The effect of mix moisture on the granulating characteristics was then investigated. During sintering, the granulated sinter mix was charged onto a layer of approximately 30 mm thick hearth material, consisting of sinter particles of -16+10 mm supported on a thick stainless steel grate bar. The coke breeze at the top surface of the sinter mixture was ignited for 90 sec at a constant suction of 8 kPa by multiple propane burners. At the end of the 90 sec ignition cycle, the propane burners were switched off and the suction across the sinter bed was ramped up and maintained at 16 kPa to move the flame front (the layer of burning coke breeze) downwards. The sintering process finished when the flame front reached the layer of hearth material. After being cooled, the fired sinter cake was processed using the CSIRO combined drop tower/jaw crusher procedure and the size distribution of the sinter particles was determined. Based on the data recorded, the sintering productivity and fuel rate, sinter yield and return sinter fines balance for each firing test were then calculated. The sinter particles were further sampled for determination of sinter chemistry and quality indices, such as TI, RI and RDI. The sinter quality was measured in accordance with the relevant ISO standards.

TABLE 2

Sintering test conditions and sinter chemistry targets.

Parameters	Units	Sintering conditions
Bed height (total)	mm	600
Return fines (dry mix basis)	%	20
Return fines sizing	mm	-5
Hearth layer depth	mm	30
Hearth layer size	mm	-16 + 10
Ignition flame temperature	°C	1100
Ignition time	s	90
Ignition suction	kPa	8
Sintering/cooling suction	kPa	16
Sinter basicity (CaO/SiO_2)	kg/kg	1.90
Sinter SiO_2 level	%	5.20
Sinter MgO level	%	1.20

Pelletising test method

The pelletising performance of the hematite by-product was evaluated at 100 per cent with or without pre-treatment. The mixture containing the hematite by-product, binders and water was calculated to target the required mix moisture and binder contents and then mixed manually. Green balls were then produced using a balling disc (1000 mm diameter × 180 mm deep). The compression strength (average kg/pellet) and drop strength (average drops/pellet) of the green balls were measured to gauge the ability of the green balls to withstand handling in an operating pellet plant. The green balls produced were then preheated and indurated in a laboratory-scale pellet induration simulator in an air atmosphere to establish the preheating and firing profile necessary to achieve the required strength of the fired pellets. The preheated and fired pellets were characterised in terms of cold compressive strength. Selected preheated and fired pellets were mounted in epoxy resin and polished to a surface finish of 1 μm. The polished blocks then underwent qualitative microscopic examination to determine the extent of consolidation and bridging/crystal growth.

UTILISATION OF QPM HEMATITE BY-PRODUCT IN SINTERING

Figure 2 compares the average sinter quality and sintering performance of the blends without (Reference blend) and with 15 per cent QPM hematite by-product (Blend 4), both using 1.5 per cent hydrated lime (HLM) as binder. Compared with the reference blend at the same HLM addition, Blend 4 containing 15 per cent QPM hematite by-product:

- Required a higher fuel rate to achieve a balanced sintering outcome.

- Showed a lower flame front speed resulting in a lower sintering productivity.

- Achieved a better sinter quality as reflected by a higher sinter TI, mean size and reducibility but a slightly inferior RDI. This was unexpected as strong sinter often has a better (lower) RDI, but in this case, the higher reducibility of Blend 4 sinter is probably responsible for its higher than expected RDI.

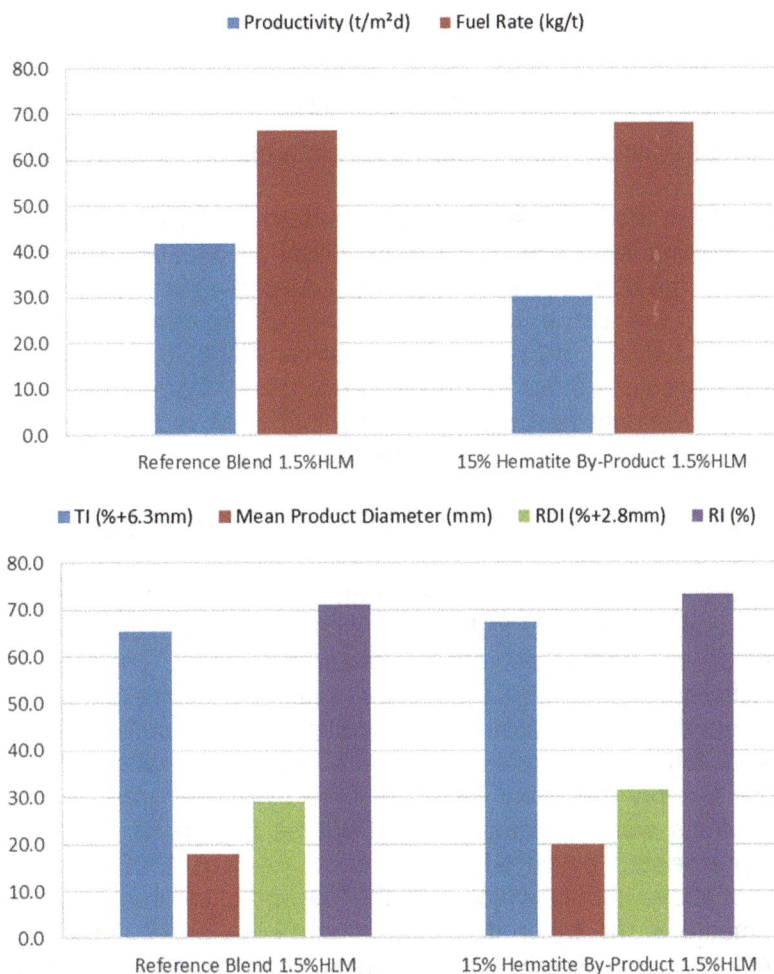

FIG 2 – Average sinter quality (top) and sintering performance (bottom).

Clearly, the most significant impact of introducing the QPM hematite by-product into the reference blend is on the sintering productivity. Strategies are hence required to recover the productivity loss due to the substitution of QPM hematite by-product for magnetite concentrates and Australian iron ore fines.

The effect of hydrated lime (HLM) on the sintering performance and sinter quality, in particular, sintering productivity, of the blend containing 15 per cent QPM hematite by-product (Blend 4), was investigated. As shown in Figure 3, the addition of hydrated lime was found to be very effective in improving the sintering productivity for the blend containing 15 per cent QPM hematite by-product. This is largely due to increased flame front speed. At 4 per cent HLM addition, the blend containing

15 per cent QPM hematite by-produced achieved an excellent productivity of 40 t/m²d, close to that of the reference blend. However, the sinter TI strength deteriorated slightly with the increased addition of hydrated lime for the blend due to reduced bulk density and consolidation time as a result of increased flame front speed.

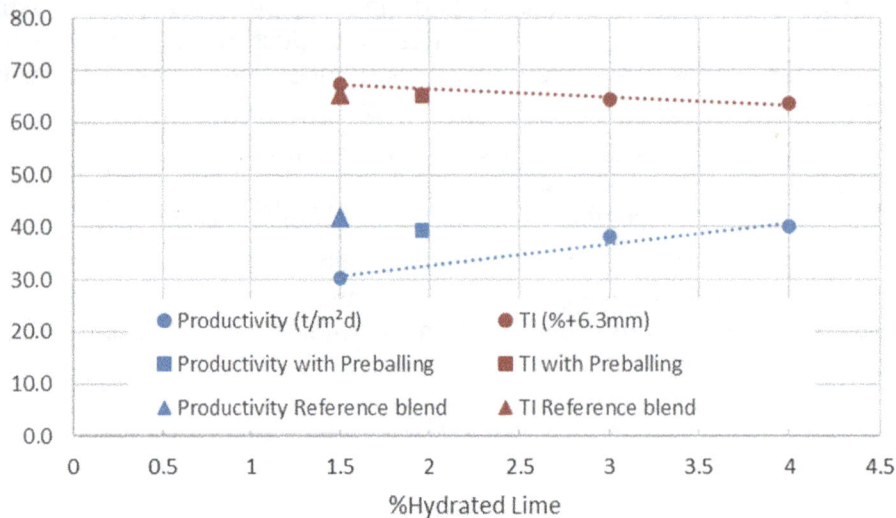

FIG 3 – Effect of HLM addition and pre-balling on sintering productivity and sinter strength.

Other material pre-treatment technologies, such as pre-balling, have also been tested. Green balls of -4 mm in diameter were made from the QPM hematite by-product using an Eirich intensive mixer, then dried and mixed with the other blend components. As shown in Figure 3, using the pre-balling pre-treatment technology, the same blend containing 15 per cent QPM hematite by-product achieved an excellent productivity of 39.2 t/m²d and sinter TI of 65.9 approximately at 2 per cent HLM, which is quite comparable to those for the reference blend. Comparing the case with only HLM addition, the pre-balling treatment reduced the amount of HLM required to achieve the sintering productivity and slightly improved the sinter strength.

UTILISATION OF QPM HEMATITE BY-PRODUCT IN PELLETISING

Balling tests were first trialled with the as received material. As it was extremely fine, acidic and had a high capacity of absorbing water, green ball formation was very difficult. The green balls made were not in a good shape and hard to achieve the required strength. Then the as received material was pre-treated by different single step processes, including neutralisation with HLM, roasting, HPGR (high pressure grinding roll) treatment and addition of other pellet feed. Some of these pre-treatment methods were found to improve the balling ability of the material, however most of these methods except for HPGR pre-treatment were not able to lift the green ball quality. Preliminary laboratory scale firing tests were therefore carried out on the green balls made from the QPM hematite by-product pre-treated by HPGR. The pellets fired at 1300°C for 20 min achieved a compression strength of above 2000N, which is on the low end of the acceptable range. It is believed that the hydrated hematite in the by-product has affected the pellet consolidation and therefore strength of fired pellets adversely. To confirm this, lab scale firing tests were also carried out on the green balls made from the QPM hematite by-product roasted at 650°C. Under the same firing conditions of 1300°C for 20 min, the pellets achieved an average compression strength of 3975N. However as mentioned earlier, the green balls from the roast only material was not up to the acceptable strength. Therefore, a combined roasting and HPGR treatment was proposed.

Figure 4 shows the compression strength of pellets preheated and fired from green balls made from the combined roasting and HPGR pre-treatment. Clearly the pellets preheated at 1000°C for 10 min achieved an excellent compression strength of above 500N at both addition levels of limestone slurry. Figure 5 shows an optical micrograph of the pellets fired at 1200°C for 10 min. The hematite grains are well connected and fused together to form a strong matrix which explains the good pellet strength as shown in Figure 4. All the pellets required a lower firing temperature to achieve the required strength. The fired pellets contained 62.69 per cent Fe, 1.8 per cent SiO_2 and 2.39 per cent Al_2O_3 and 0.87 per cent CaO at 1.5 per cent limestone slurry, which is suitable for the blast furnace process. Reducing the addition of limestone slurry from 1.5 to 1.0 per cent is expected to improve the Fe grade and reduce the CaO content of the fired pellets slightly.

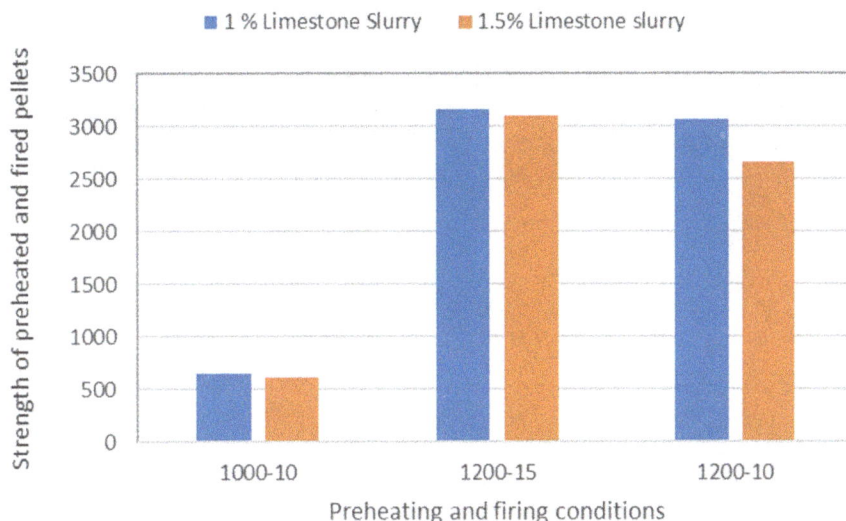

FIG 4 – Strength of pellets preheated and fired from green balls made from roasted and HPGR treated QPM hematite by-product.

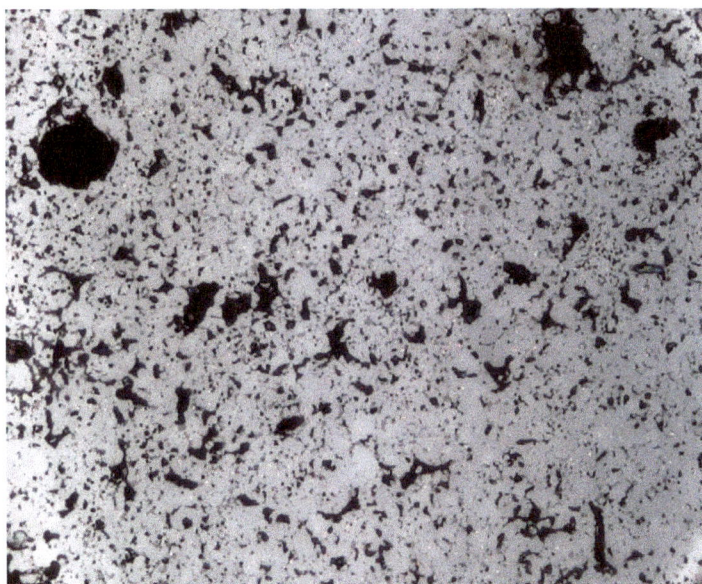

FIG 5 – Optical micrograph of the cross-sectional area of the pellets fried at 1200°C for 10 min (1.5 per cent limestone slurry).

CONCLUSIONS

The as received QPM hematite by-product is unique and very different from conventional ironmaking raw materials, such as sinter fines and pellet feeds. It is characterised by many micron size spheres of a cauliflower surface structure with a narrow size distribution, high specific surface area and strong capability of adsorbing and holding water.

Introducing QPM hematite by-product into the reference blend has an adverse impact on the sintering productivity. However, the addition of HLM and pre-balling treatment were found to be effective in recovering the loss in productivity resulting from addition of QPM hematite by-product. Comparing the case with only HLM addition, the pre-balling treatment reduced the amount of HLM lime required to achieve the sintering productivity and slightly improved the sinter strength.

Due to its unique characteristics, the as received QPM hematite by-product is hard to form green balls and achieve the preheated and fired pellet strengths. A novel multiple step pre-treatment process including roasting and HPGR was proposed and demonstrated to be effective in achieving the quality of green and fired pellets.

ACKNOWLEDGEMENTS

The authors would like to thank Queensland Pacific Metals for their financial support to conduct the test work and permission to publish this manuscript. The authors also wish to thank CSIRO Mineral Resources Sintering Team for the technical support and resources for completing the test work.

Effect of -1 mm particle size fraction on JPU permeability and granulation characterisation

T Singh[1], S Mitra[2], O A Aladejebi[3], D O'Dea[4] and T Honeyands[5]

1. AAusIMM, Research Associate, Centre for Ironmaking Materials Research, The University of Newcastle, Callaghan NSW 2308. Email: tejbir.singh@newcastle.edu.au
2. MAusIMM, Research Academic, Centre for Ironmaking Materials Research, The University of Newcastle, Callaghan NSW 2308. Email: subhasish.mitra@newcastle.edu.au
3. Research Associate, Centre for Ironmaking Materials Research, The University of Newcastle, Callaghan NSW 2308. Email: tosin.aladejebi@newcastle.edu.au
4. Principal Technical Marketing, BHP Marketing Iron Ore, Brisbane Qld 4000. Email: damien.p.odea@bhp.com
5. FAusIMM, Director, Centre for Ironmaking Materials Research, The University of Newcastle, Callaghan NSW 2308. Email: tom.a.honeyands@newcastle.edu.au

ABSTRACT

Granulation of the iron ore blend and the additives is the foremost step in the agglomeration of iron ore fines to achieve a permeable bed during sintering, which is crucial for sinter productivity. This study investigated the granulation of four types of blends to determine the impact of granulation on bed permeability. The blends were granulated in a rotating drum at various moisture contents and their JPU permeability was measured at each moisture content in a permeability pot (diameter 95 mm and height 500 mm) at various flow rates. Several independent experiments were carried out to investigate the changes in permeability when mixtures of different particle size range (-1 mm and +1 mm) were introduced into a packed bed with a larger particle size range (-6.3 mm to +4.0 mm). The results revealed that the addition of particles in the -1 mm size range had a more pronounced negative impact on permeability compared to particles in the +1 mm size range. This can be attributed to the relatively higher energy losses (mainly inertial) associated with the presence of -1 mm particles in the mixture. The findings were used to quantify the granulation process based on the -1 mm particle size fraction in the granulated mix and their utilisation (denoted by $UI_{-1\,mm}$). For achieving optimal permeability in a specific blend, it was found that a $UI_{-1\,mm}$ (utility index of -1 mm particle size) ranging from 0.82 to 0.92 was required. However, exceeding a $UI_{-1\,mm}$ value of 0.95 resulted in weaker granules prone to deformation and led to decreased bed permeability. The study also showed that the $UI_{-1\,mm}$ was highly dependent on the water addition per kg of -1 mm in dry feed and suggested adding ~80 to 110 mL of water per kg of -1 mm in dry feed to attain the best possible permeability. Additionally, the research proposed a mechanism involving the melting of granules at high temperatures and highlighted the influence of -1 mm particles on bed permeability during the sintering process.

INTRODUCTION

Granulation is the first step in sintering and plays a crucial role in the sintering of iron ore where the granules are formed from raw materials such as iron ore blend, coke, fluxes and return fines. The principal mechanism in raw feed granulation involves the layering of fine particles onto larger nuclei particles, resulting in a coarser particle size distribution (Litster, Waters and Nicol, 1986; Litster and Waters, 1988; Litster, 1990). The granulation output has a direct impact on the permeability of the packed bed during sintering, influencing the sintering reaction and bed permeability (Singh *et al*, 2019, 2022; Ellis, Loo and Witchard, 2007).

Many researchers have extensively studied iron ore granulation (Khosa and Manuel, 2007; Litster, 2003; Iveson *et al*, 2001a, 2001b; Niu *et al*, 2023; Yang *et al*, 2018; Takehara *et al*, 2022; Formoso *et al*, 2003; Li *et al*, 2019; Litster, Waters and Nicol, 1986; Litster and Waters, 1988; Litster, 1990). Litster, Waters and Nicol (1986) investigated the kinetics and mechanisms of particle growth during granulation and proposed a model to predict particle size distribution. Additionally, Khosa and Manual (2007) examined the effect of various ore types on granulation to predict the optimum moisture levels and JPU permeability, though the effect of various size fractions of the particles on the permeability was not studied. In a series of past studies (Standish and Collins, 1983; Yu and

Standish, 1987), the effect of different particle size groups in mixtures on bed permeability and packed bed voidage using spherical glass beads have been shown.

Maintaining an appropriate particle size distribution is crucial for preserving the permeability of the packed bed before and during sintering. The presence of excessive proportions of small-sized particles in the mixture can disrupt the bed permeability (Ellis, Loo and Witchard, 2007; Niu *et al*, 2023; Kasai, Rankin and Gannon, 1989; Zhou *et al*, 2020). From a quantitative perspective, these studies have indicated that the moisture available for granulation (total moisture minus the moisture absorbed by feed materials) and the size distribution of raw materials are the primary factors influencing the process. On the other hand, properties like particle shape and surface characteristics of the ore are considered secondary factors, as their influence is primarily mediated through their impact on moisture absorption capability (Higuchi, Lu and Kasai, 2017; Litster and Waters, 1988).

It is noted that existing studies in the literature largely report particle size distribution measurement as a primary measure to quantify the granulation process. However, they often fail to establish a direct link with the permeability of the granules (Linhares *et al*, 2020; Que *et al*, 2019; Maldonado *et al*, 2011). The present study aims to address this knowledge gap by experimentally investigating the influence of granulation on the permeability of the green bed prior to sintering.

METHOD AND MATERIALS

Raw material

In this study, three types of Australian iron ores were used and blended with a generic Asia Pacific base blend and magnetite concentrate to obtain final ore blends for sintering experiments. The dry sizing of each ore is given in Table 1. Ore 1 and ore 2 were similar except for their water-holding capacity (WHC) with ore 2 having a higher WHC than ore 1.

TABLE 1

Dry size distribution and water holding capacity of components of the raw mixture.

Ore	Type	%+6. mm	%-6.3+ mm	%- mm	WHC
Base blend	Mixed	15.3	39.9	44.9	6.07
Magnetite concentrates	Magnetite	0.0	0.0	100.0	5.45
Aus ore 1	Brockman Marra-mamba	8.6	41.1	50.3	5.37
Aus ore 2		8.8	43.0	48.2	9.15
Aus ore 3	Pisolitic-Goethite	38.4	42.5	19.1	6.81

Three blends were formulated by combining three distinct types of ores. The blending process involved maintaining a consistent proportion of 70 per cent base blend and 10 per cent magnetite while allocating 20 per cent of the blend to each of the test ores (Aus ore 1, 2 and 3).

Each iron ore blend was mixed with additives to achieve a 7 per cent coke rate, 1.9 basicity and 20 per cent return fines and named as blend 1, blend 2 and blend 3 representing the sinter raw mix. The particle size distribution of each blend used is shown in Figure 1. The P-G type ore (ore 3) was coarser with around 20 per cent of -1 mm particles compared to BK-MM (ore 1 and 2) which has around 50 per cent of -1 mm particles and hence the PSD for blend 1 and blend 2 was similar compared to blend 3. Blend 3 was relatively coarser with ~38 per cent of -1 mm particle fraction compared to ~43 per cent in blends 1 and 2.

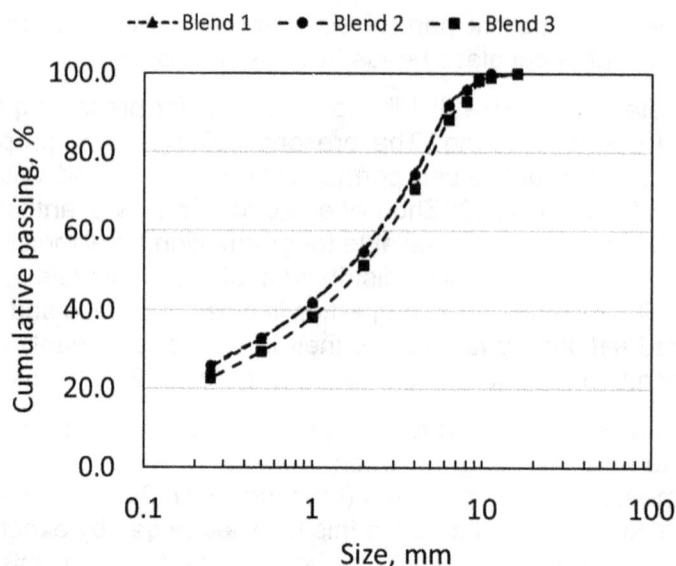

FIG 1 – Frozen particle size distribution of blends 1, 2 and 3 (representing sinter raw mix) before granulation (blend 1 and 2 overlapping in the figure).

To examine the impact of different particle sizes in a packed bed, Aus ore 2 was subjected to screening, resulting in the segregation of particles into four distinct size groups: +6.3 -4.0 mm (PS1), -2.8 +2.0 mm (PS2), -2.0 +1.0 mm (PS3) and -1.0 +0.5 mm (PS4). The packed bed of screened particles was formed by taking a proportion of smaller particle size (PS2, PS3 and PS4) in larger particle size (PS1) at three different proportions ie 10 per cent, 20 per cent and 30 per cent.

Apparatus and methods

Granulation and JPU permeability

The granulation process was performed in a rotating drum with a depth of 310 mm and a diameter of 300 mm. The process included a dry mixing stage for 2 minutes followed by a wet mixing stage for 3 minutes, with the drum rotating at a speed of 20 rev/min to achieve a target Froude number of 0.005. A calculated amount of water was added by sprinkler to the raw mix during granulation to obtain the target moisture content. After granulation, a sub-sample of the raw mix was taken to measure the experimental moisture content, apparent density and particle size distribution (Li *et al*, 2019; Ellis, Loo and Witchard, 2007).

To measure the bed permeability, airflow rates through a packed bed (diameter: 95 mm, height: 500 mm) of granulated raw mix and a mixture of screened particles were measured at different suction pressures. The JPU permeability of the granulated raw mix was calculated using Equation 1 as follows:

$$JPU = u * \left(\frac{L}{\Delta P}\right)^{0.6} \tag{1}$$

where u is the superficial velocity in m/min, L is the length of the packed bed in mm and ΔP is the pressure drop across the bed in mm H_2O.

RESULTS AND DISCUSSION

JPU permeability

Figure 2 presents the JPU permeability of the three blends with moisture content ranging from 6 per cent to 11 per cent. The moisture content at which the maximum permeability occurred was considered the optimum moisture content (shown with dotted lines). The permeability of the blends follows a parabolic pattern exhibiting an increase in the moisture content below the optimum moisture level, followed by a decrease thereafter. Although the granule size increases with increasing moisture content (discussed in the next section), the granules become weaker after optimum moisture, resulting in their deformation in the packed bed and reduced permeability (Ellis, Loo and

Witchard, 2007). Blend 2 exhibits the highest optimum moisture, followed by blend 3 and blend 1, with values of 8.9, 8.7 and 8.4, respectively due to variations in their WHC.

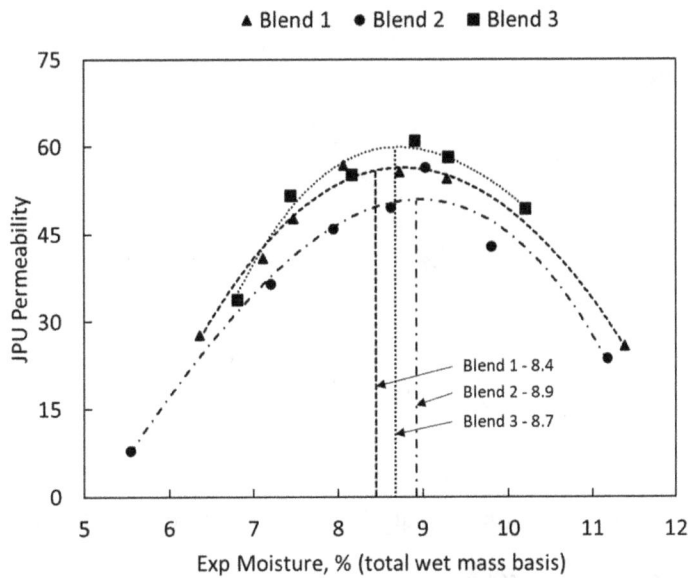

FIG 2 – JPU permeability of each blend at various moisture levels.

To investigate the effect of smaller particle size fraction on the bed permeability, PS1 was mixed with PS2, PS3 and PS4 at various proportions and the corresponding JPU permeability trends are shown in Figure 3. Figure 4 shows the measured pressure drop per unit length (DP/L) versus velocity curve for each mixture. The graph clearly illustrates that the packed bed containing an increasing fraction of -1 mm particles exhibits a higher DP/L compared to the packed bed with +1 mm particles.

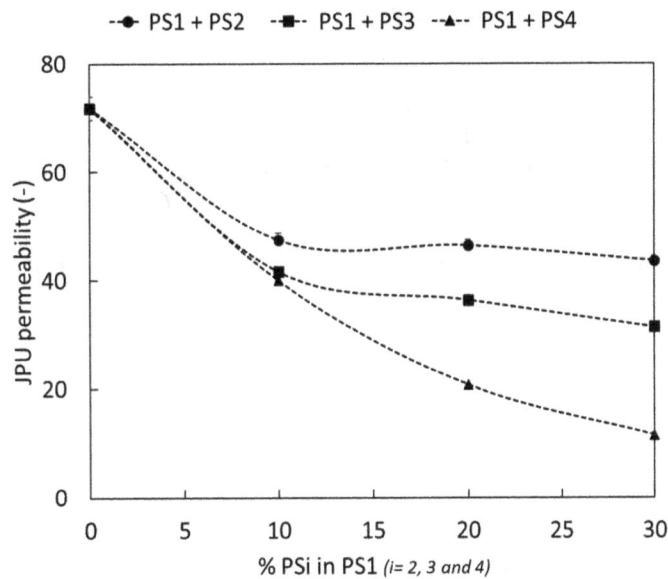

FIG 3 – Variation in JPU permeability in a packed bed of -6. mm+4. mm particles size with smaller particle size at 10 per cent, 20 per cent and 30 per cent proportions.

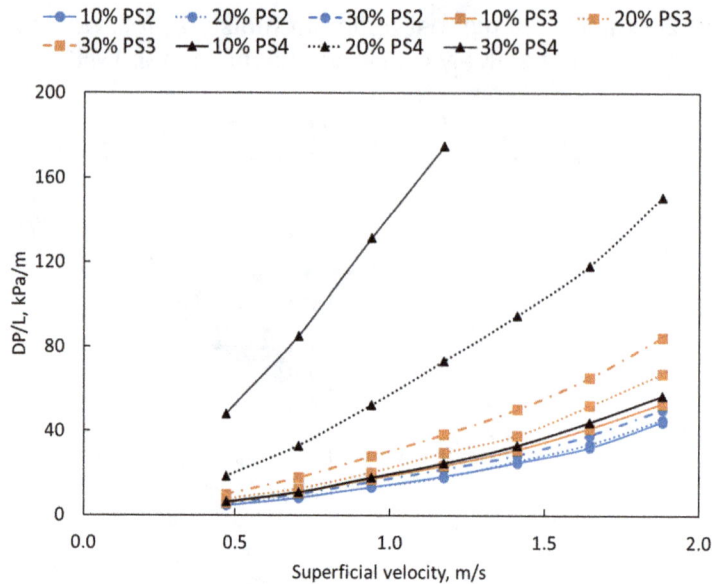

FIG 4 – DP/L versus velocity for various packed bed mixtures.

It is evident from Figure 3 that the effect of -1 mm size fraction particles in the packed bed is more detrimental than the +1 mm particle fraction in decreasing the bed permeability. The measured decrease in the JPU permeability can be attributed to the higher energy losses (mainly inertial) upon adding -1 mm particle in the bed compared to +1 mm particles which can be concluded from the pressure drop gradient versus superficial velocity plot shown in Figure 4.

Granulation

Fundamentally, the granulation of raw mix is a process where the particle size distribution becomes coarser after the layering of adhering fine particles (usually -0.25 mm particles) onto the larger nuclei particles (usually +1 mm particles) (Litster, Waters and Nicol, 1986).

In Figure 5, with an increase in moisture content, an increasing trend is observed in the Sauter mean diameter (SMD) of each blend at different moisture contents (Figure 5a). The differences in the level of SMD for various blends are normalised by plotting SMD against water addition for granulation per kg of dry feed (Figure 5b).

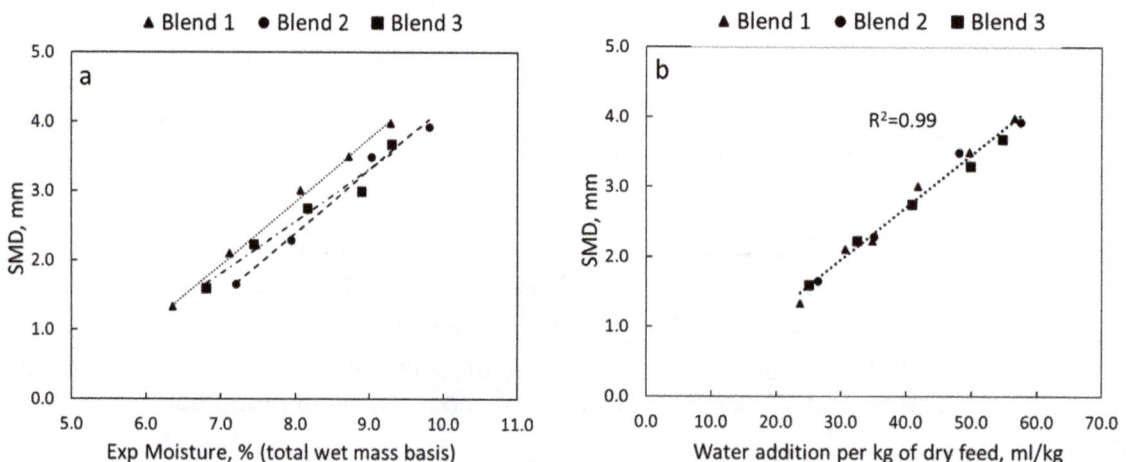

FIG 5 – SMD for each blend at various moisture levels after granulation.

It is understood from previous studies that during granulation, -0.25 mm particles (or a fraction) adhere to +1 mm particles to form granules (Khosa and Manuel, 2007; Litster, Waters and Nicol, 1986). The intermediate size range particles (-1 mm and +0.25 mm) take part in granulation as nuclei or as adhering fines though some fraction of these particles remains free. It was shown in the previous section that the -1 mm particles have a notable detrimental effect on the permeability of the bed. Hence, the granulation performance of various blends was evaluated by determining the utility of -1 mm particles during granulation, defined as the utility index of -1 mm ($UI_{-1\,mm}$). The $UI_{-1\,mm}$ was calculated by taking the ratio of the difference between -1 mm particles in the dry feed and granulated feed to the -1 mm particle fraction in the dry feed as follows:

$$UI_{-1mm} = \frac{(-1\,mm\,fraction\,in\,dry\,feed) - (-1\,mm\,fraction\,in\,granulated\,feed)}{(-1\,mm\,fraction\,in\,dry\,feed)} \qquad (2)$$

It is noted that the $UI_{-1\,mm}$ increases with increasing water addition per kg of -1 mm in dry feed as shown in Figure 6. The $UI_{-1\,mm}$ value increases rapidly to 0.9 with water addition and then stabilises. A similar level of $UI_{-1\,mm}$ was achieved at a certain amount of water addition per kg of -1 mm in dry feed for different blends.

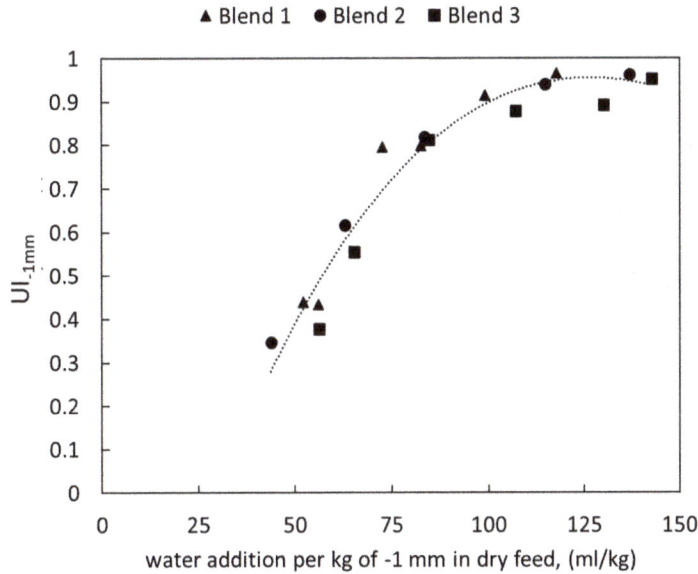

FIG 6 – Utility index of -1 mm particles ($UI_{-1\,mm}$) of each blend at various water addition per kg of -1 mm in dry feed (mL/kg).

The effects of $UI_{-1\,mm}$ were investigated on the permeability of the bed. Figure 7 shows the change in the ratio of experimental moisture to optimum moisture which represents the variation in JPU permeability (ratio = 1, max JPU) with $UI_{-1\,mm}$. A $UI_{-1\,mm}$ value ranging from 0.82 to 0.92 was identified as necessary to achieve optimal permeability. This finding is useful for determining the appropriate amount of water added during the granulation process, which is recommended to be within the range of 80 to 110 mL/kg of -1 mm in dry feed.

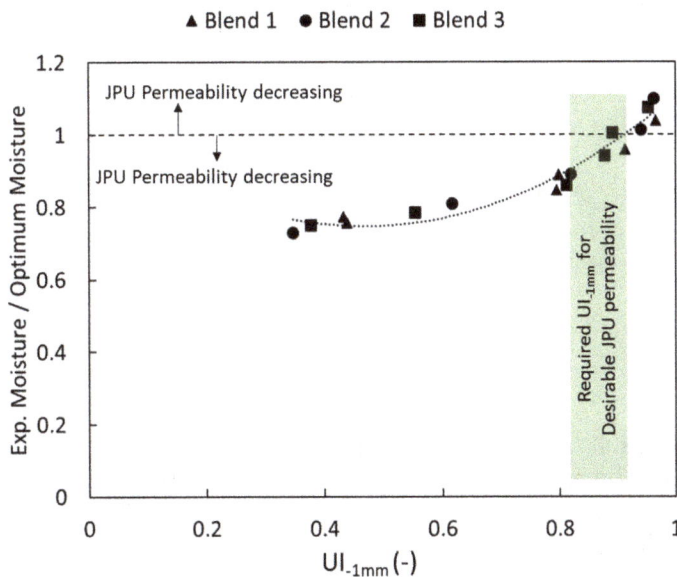

FIG 7 – Variation in ratio of experimental moisture to optimum moisture with the utility index of -1 mm particles.

The granulation performance of blends containing different types of ores can be improved simply by increasing the water addition per kg of -1 mm in dry feed, though theoretically an extra amount of fuel (coke) would be required to compensate for the large latent heat (2250 kJ/kg) required to evaporate the added water.

Pathway of melting and -1 mm particles

The pathway of the melting of granules during sintering is proposed based on the literature findings (Loo and Leaney, 2002). Figure 8 shows the mechanism of melting of a group of particles for two cases: low $UI_{-1\,mm}$ and high $UI_{-1\,mm}$ with similar adhering fines to nuclei ratio. For low $UI_{-1\,mm}$, the amount of free -1 mm particles that are not part of the adhering fines (shown in red) is higher than in a high $UI_{-1\,mm}$ scenario. A green bed with a high $UI_{-1\,mm}$ is expected to have a large channel flow diameter and will be more permeable as shown in Figure 7. In the beginning, the layer of adhering fines (AF) and free -1 mm particles begin to react and form a melt phase. This aids in bond formation with other granules, followed by the melting of nuclei, either partially or fully. A high amount of AF and free -1 mm particles will produce comparatively more melt which will hinder airflow in the bed. During heating, the fuel (coke) present in the bed combusts and the flux starts to react which results in formation of the vacant sites (Yang and Standish, 1991). The pore generation process by nuclei during melting depends on the intra-porosity of the particle. Melting of porous nuclei results in comparatively high porosity generation and lower resistance to airflow. This melt formation phenomenon during sintering explains the variation in airflow resistance for a varying $UI_{-1\,mm}$ and AF/nuclei ratio. A higher $UI_{-1\,mm}$ is desirable to achieve a relatively higher bed permeability before and during sintering. It is expected that more work will be conducted to quantify the effect of $UI_{-1\,mm}$ on the sintering permeability in future.

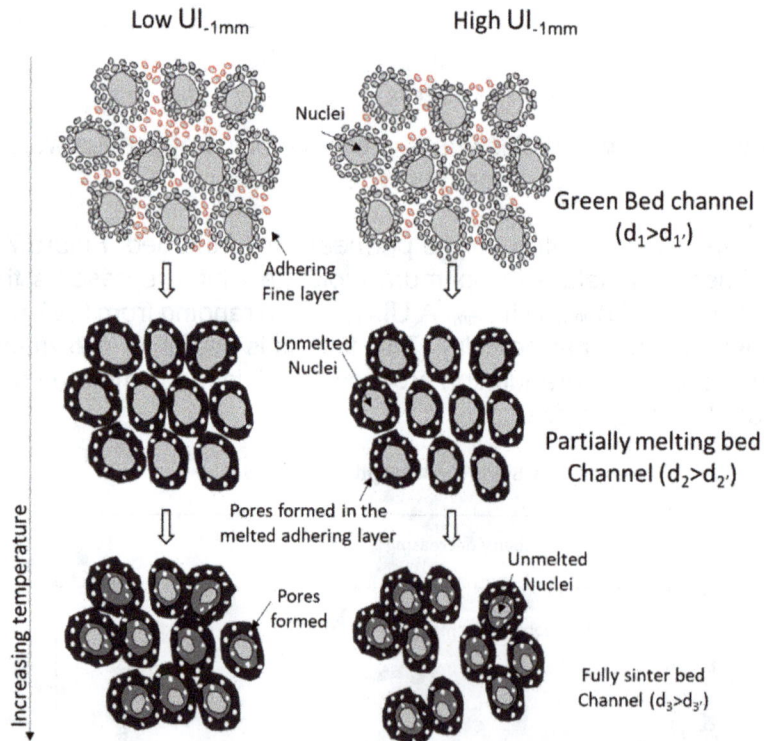

FIG 8 – Pathways of bonding of two granules with increasing temperature during sintering.

CONCLUSIONS

This study emphasizes the significance of particle size and $UI_{-1\,mm}$ in achieving a permeable bed during sintering, thereby enhancing productivity. The granulation and permeability for three types of blends containing three different ores, two BK-MM ore and one P-G ore, were investigated and the following conclusions were drawn:

1. The presence of free -1 mm particle size fraction in a packed bed has a more pronounced detrimental effect on bed permeability compared to +1 mm particles. This can be attributed to

the higher energy losses experienced in beds with a relatively higher proportion of -1 mm particles.

2. A similar level of utility index of -1 mm particles can be obtained for the same water addition per kg of -1 mm size fraction in dry feed. A $UI_{-1\ mm}$ from 0.82 to 0.92 resulted in the generation of granules which contributed to optimal packed bed permeability through the addition of 80 to 110 mL water per kg of -1 mm size fraction in dry feed.

3. A granule melting mechanism was proposed at high temperatures for low and high $UI_{-1\ mm}$. It is concluded that for a similar AF/nuclei ratio, a high $UI_{-1\ mm}$ results in a lower airflow resistance during sintering. It is possible to achieve a higher $UI_{-1\ mm}$ by increasing the aforesaid water addition requirement.

ACKNOWLEDGEMENTS

The authors gratefully acknowledge BHP for their financial support for the Centre for Ironmaking Materials Research and permission to publish this work.

REFERENCES

Ellis, B G, Loo, C E and Witchard, D, 2007. Effect of ore properties on sinter bed permeability and strength, *Ironmaking and Steelmaking*, pp 99–108.

Formoso, A, Moro, A, Fernández Pello, G, Menéndez, J L, Muñiz, M and Cores, A, 2003. Influence of nature and particle size distribution on granulation of iron ore mixtures used in a sinter strand, *Ironmaking and Steelmaking*, pp 447–460.

Higuchi, T, Lu, L and Kasai, E, 2017. Intra–particle water migration dynamics during iron ore granulation process, *ISIJ International*, pp 1384–1393.

Iveson, S M, Litster, J D, Hapgood, K and Ennis, B J, 2001a. Nucleation, growth and breakage phenomena in agitated wet granulation processes: a review, *Powder Technology*, pp 3–39.

Iveson, S M, Wauters, P A, Forrest, S, Litster, J D, Meesters, G M and Scarlett, B, 2001b. Growth regime map for liquid-bound granules: further development and experimental validation, *Powder Technology*, pp 83–97.

Kasai, E, Rankin, W J and Gannon, J F, 1989. The effect of raw mixture properties on bed permeability during sintering, *ISIJ International*, pp 33–42.

Khosa, J and Manuel, J, 2007. Predicting granulating behaviour of iron ores based on size distribution and composition, *ISIJ International*, pp 965–972.

Li, C, Moreno-Atanasio, R, O'Dea, D and Honeyands, T, 2019. Experimental Study on the Physical Properties of Iron Ore Granules Made from Australian Iron Ores, *ISIJ International*, pp 253–262.

Linhares, F M, Victor, C C F, Lemos, L R and Bagatini, M C, 2020. Effect of three different binders and pellet feed on granulation behaviour of sintering mixtures, *Ironmaking and Steelmaking*, pp 991–997.

Litster, J D and Waters, A G, 1988. Influence of the material properties of iron ore sinter feed on granulation effectiveness, *Powder Technology*, pp 141–151.

Litster, J D, 1990. Kinetics of iron ore sinter feed granulation, *Powder Technol*, pp 125–134.

Litster, J D, Waters, A G and Nicol, S K, 1986. A model for predicting the size distribution of product from a granulation drum, *Trans ISIJ*, pp 1036–1044.

Litster, J, 2003. Scaleup of wet granulation processes: science not art, *Powder Technology*, pp 35–40.

Loo, C E and Leaney, J C M, 2002. Characterizing the contribution of the high – temperature zone to iron ore sinter bed permeability, *Mineral Processing and Extractive Metallurgy*, pp 11–17.

Maldonado, R D, Haehnel, S, Drain, P and Heslin, J, 2011. Investigation of Mixing and Rolling Drum Performance at Port Kembla's Sinter Machine Through Full-Scale Sampling and Laboratory Scale Experiments, in *Proceedings of the Iron Ore Conference 2011*, pp 475–483 (The Australasian Institute of Mining and Metallurgy: Melbourne).

Niu, L, Zhang, J, Wang, Y, Kang, J, Li, S, Shan, C, Li, Z and Liu, Z, 2023. Iron ore granulation for sinter production: Developments, progress and challenges, *ISIJ International*, pp 601–612.

Que, Z, Wu, S, Zhai, X and Li, K, 2019. Effect of characteristics of coarse iron ores on the granulation behaviour of concentrates in the sintering process, *Ironmaking and Steelmaking*, pp 246–252.

Singh, T, Honeyands, T, Mitra, S, Evans, G and O'Dea, D, 2019. Measured and Modelled Air Flow Rates during the Iron Ore Sintering Process: Green and Sintered Beds, in *Proceedings of the Iron Ore Conference 2019*, pp 156–165 (The Australasian Institute of Mining and Metallurgy: Melbourne).

Singh, T, Mitra, S, O'Dea, D, Knuefing, L and Honeyands, T, 2022. Quantification of Resistance and Pressure Drop at High Temperature for Various Suction Pressures During Iron Ore Sintering, *ISIJ International*, pp 1768–1776.

Standish, N and Collins, D, 1983. The permeability of ternary particulate mixtures for laminar flow, *Powder Technology*, pp 55–60.

Takehara, K, Higuchi, T, Hirosawa, T and Yamamoto, T, 2022. Effect of Granule's Moisture on Granulation Rate of Iron Ore, *ISIJ International*, pp 1363–1370.

Yang, C, Zhu, D, Pan, J and Lu, L, 2018. Granulation effectiveness of iron ore sinter feeds: Effect of ore properties, *ISIJ International*, pp 1427–1436.

Yang, Y and Standish, N, 1991. Fundamental mechanisms of pore formation in iron ore sinter and pellets, *ISIJ International*, pp 468–477.

Yu, A and Standish, N, 1987. Porosity calculations of multi-component mixtures of spherical particles, *Powder Technology*, pp 233–241.

Zhou, H, Lai, Z, Lv, L, Fang, H, Meng, H, Zhou, M and Cen, K, 2020. Improvement in the permeability of sintering beds by drying treatment after granulating sinter raw materials containing concentrates, *Advanced Powder Technology*, pp 3297–3306.

Effect of high oxygen enrichment on sintered ore quality and sintering phenomena

K Takehara[1], T Higuchi[2] and T Yamamoto[3]

1. Senior Researcher, JFE Steel Corp., Fukuyama Hiroshima 721–8510, Japan.
   Email: ke-takehara@jfesteel.co.jp
2. Senior Researcher, JFE Steel Corp., Fukuyama Hiroshima 721–8510, Japan.
   Email: ta-higuchi@jfesteel.co.jp
3. General manager, JFE Steel Corp., Fukuyama Hiroshima 721–8510, Japan.
   Email: tets-yamamoto@jfesteel.co.jp

ABSTRACT

Iron ore fines are used as raw materials for sintered ore for blast furnaces in the world. Since the production and quality of sintered ore have a great influence on the stable operation of a blast furnace, many technologies were developed for controlling the heat profile in the sintering machine, which largely affect the sintered ore properties. In particular, the influence of oxygen in the suction gas, which plays an important role in combustion, was investigated over the years. It is known that increasing the oxygen ratio by about 30 per cent not only significantly improves the combustion rate, but it also increases the strength and yield of sintered ore. However, the details of the mechanism for improving these qualities are unknown. In this study, we investigated the influence of oxygen enrichment at higher ratios (O_2: 21–40 per cent) and tried to understand the sintering behaviour enhanced by oxygen. The effect of oxygen enrichment on sintered ore quality and the heat profile were evaluated by sintering tests using a quartz glass pot. From the results, sintered ore yield had a maximum value for O_2 = 30 per cent, but strength was increased for O_2 = 40 per cent. From heat pattern analysis, it is clear that the high temperature holding time also reached a maximum at O_2 = 30 per cent. Furthermore, we found that with increased oxygen ratio, the combustion speed (flame front speed) was increased, but the area of combustion was enlarged. It is considered that there could be a maximum oxygen ratio value for the yield from the balance of these positive and negative effects on high temperature holding time. To clarify the reason for strength improvement, thermodynamic calculations were performed and it was found that melting behaviour could be improved by changing the oxygen ratio.

INTRODUCTION

Sintered ore, which is the main raw material for blast furnaces, is produced mainly using a Dwight Lloyd (DL) type sintering machine that sucks air and gas from top to bottom. The DL type sintering machine uses a process that melts and agglomerates iron ore, limestone and auxiliary materials using bonding agents such as coke breeze. For sintering operations, it is important to produce high-strength sintered ore and achieve a high production rate in order to stabilise the blast furnace (Sasaki and Hida, 1982). Therefore, various studies have been conducted on the effect of oxygen ratio, which is the main factor for combustion, on the sintering heat pattern and the properties of sintered ore.

The effect of oxygen ratio has been studied mainly by sintering pot tests and simulations and it has been reported that the sintering speed increases when enriched with oxygen. Voice and Wild (1957) conducted a pot test with varying oxygen ratios in the suction gas from 21 vol per cent to 100 vol per cent and found that the flame front speed (FFS) increased under high oxygen enrichment conditions. Also, they pointed out that with higher oxygen ratio, the difference of the movement speed of the heat pattern and the flame front increased and the maximum temperature in the layer decreased. On the other hand, regarding quality, in the oxygen ratio range of 21 to 35 vol per cent, a case (Ishimitsu *et al*, 1963) where the strength and yield did not change significantly, a case (Sugawara *et al*, 1973) where the yield improved and a case (Rajak *et al*, 2021) in which both strength and yield increased regardless of the shortening of the sintering time have been reported. These qualities are said to be inversely related to sintering time, but tend to improve or change less with oxygen enrichment. The reason for the improvement of strength and yield is discussed from two viewpoints. The first point is that coke combustion is promoted by oxygen enrichment, the temperature in the combustion zone increases and the high temperature holding time is extended. It is believed that this effect increases the amount of melt and contributes to quality improvement. The second point is that calcium ferrite (CF), which is considered to have high strength, increases due to oxygen enrichment (Rajak *et al*, 2021), which is said to be the cause of high strength. On the other hand, under the oxygen-enriched conditions referred to by Voice and Wild (1957) where the heat pattern movement speed deviates from the combustion speed, the maximum temperature begins to decrease and the quality is expected to deteriorate. Thus, there seemed to be an optimum oxygen ratio for improving the sinter quality, but the amount and mechanism for the optimisation are not clear. Also, the effect of oxygen on liquid phase formation in iron ore sintering has not been discussed.

Therefore, in this study, we conducted sintering pot tests where the oxygen ratio of the suction gas was increased to a maximum of 40 vol per cent and we analysed the heat pattern in the combustion zone to clarify the effect of oxygen enrichment on the yield. In addition, the effect of oxygen partial pressure on the liquid fraction was analysed thermodynamically and the effect of high oxygen enrichment on sinter strength is discussed.

EXPERIMENTAL PROCEDURE

Method of sintering pot test

Table 1 shows the mixing ratio of the sintering raw materials used in the experiment and Table 2 shows the chemical composition of the ore. Ore A is a pisolite ore from Australia and ores B and C are hematite ores from South America. The mixing ratio of the raw materials was 1.0 mass per cent of quicklime and 20 mass per cent of return fines compounded (Takehara, Higuchi and Yamamoto, 2023).

TABLE 1

Blending composition of sinter mixture (mass per cent).

Ore A	34.4
Ore B	18.9
Ore C	11.5
Return fine	19.1
Silica sand	0.1
Limestone	10.7
Burnt lime	1.0
Coke Breeze	4.3

TABLE 2

Chemical composition of iron ores (mass per cent).

	T. Fe	SiO_2	Al_2O_3	MgO	LOI*
Ore A	57.6	3.4	1.9	0.1	10.6
Ore B	65.5	1.3	1.2	0.0	2.4
Ore C	62.6	6.9	1.2	0.1	0.6

* LOI: Loss on Ignition

Figure 1 shows a schematic image of the test equipment. We adopted a quartz glass pot that allows observation of the combustion state during sintering. Firstly, all raw materials were charged into a concrete mixer, mixed for 180 s at a rotation speed of 20 rev/min and then granulated for 300 s at 12 rev/min. The raw material after granulation was then charged into a quartz glass pot of 300 mmφ × 400 mmH and sintering tests were performed under a constant suction pressure of 5.0 kPa. In order to adjust the oxygen content in the suction gas blown into the raw material layer to

21, 30 and 40 vol per cent, pure oxygen was added. At the same time, the oxygen content of the air was adjusted to 21 vol per cent. Oxygen was introduced 30 s after ignition and was stopped when the exhaust gas temperature reached the maximum temperature and began to fall.

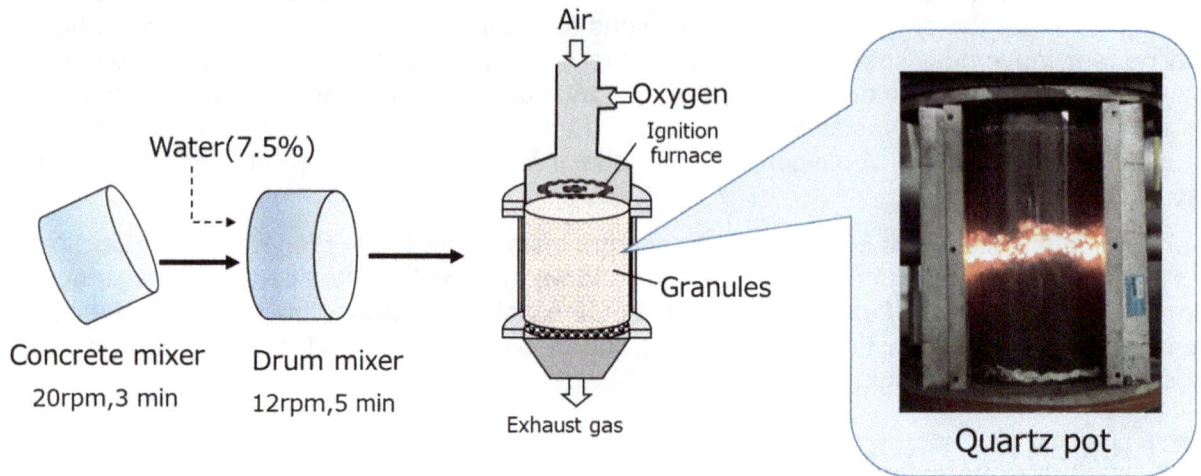

FIG 1 – Schematic image of sinter pot test equipment.

In the sintering tests, evaluation of productivity and quality were conducted under the same sintering conditions as well as heat pattern measurement to eliminate the influence of the thermocouple on quality. For temperature measurement, two R-type thermocouples were fixed at positions 130 mm (lower layer) and 260 mm (upper layer) from the bottom and the raw material after granulation was charged to form a packed bed.

Thermodynamic equilibrium calculation

Equilibrium calculations of the Fe_2O_3-FeO-CaO-SiO_2-Al_2O_3-MgO system were performed using FactSage 7.1. FToxide and FACTPS were used for the solid-phase and liquid-phase databases of oxides. Regarding the oxygen partial pressure, calculations for approximately 10^{-1} kPa simulating the reducing atmosphere of the combustion zone and for approximately 10 kPa which is close to atmospheric pressure are reported. Considering the non-uniformity of the oxygen partial pressure in the combustion zone, this study also set the oxygen partial pressure to 2.0×10^{-1} to 4.1×10 kPa (2.0×10^{-3} to 4.0×10^{-1} atm) so as to include a reducing atmosphere condition and the suction gas atmosphere and the liquid fraction and composition were investigated. The initial composition was set to be the same as the sintered ore described in above section, ie 83 mass per cent-Fe_2O_3, 5.0 mass per cent-CaO, 10.0 mass per cent-SiO_2, 1.6 mass per cent-Al_2O_3 and 0.4 mass per cent-MgO. In this calculation, the composition after calculation was different from the initial composition, because oxygen exchange is performed freely so that the conversion of FeO and Fe_2O_3 proceeds when the oxygen partial pressure is changed.

EXPERIMENTAL RESULTS

Sintering pot test results

Figure 2 shows the pot test results (Takehara, Higuchi and Yamamoto, 2023). The sintering time decreased with increasing oxygen ratio and the yield reached its maximum value at 30 vol per cent oxygen ratio. The suction air volume did not change significantly with respect to the oxygen ratio. The sintering productivity increased with shortening of the sintering time and TI (Tumbler index: Strength) also increased with increasing oxygen ratio.

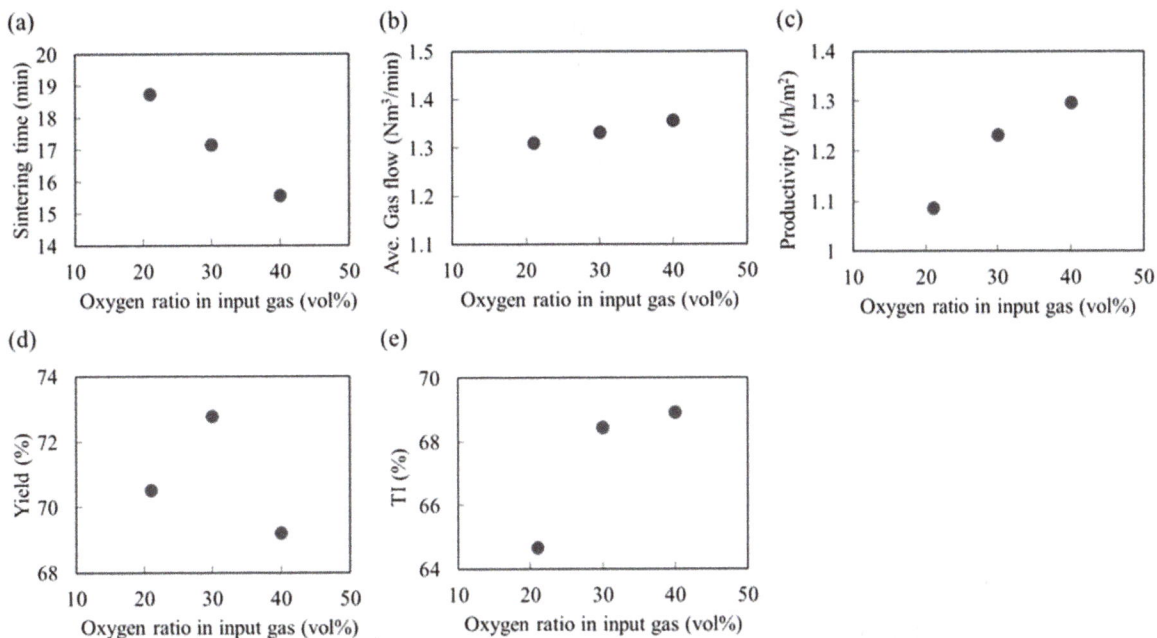

FIG 2 – Results of sinter pot tests with oxygen ratio in the input gas of 21~40 vol per cent.

Figure 3 shows the observed results (Takehara, Higuchi and Yamamoto, 2023) for the sintering area when the combustion zone reached the middle layer (height of about 200 mm from the bottom). The high temperature area observed through the quartz glass pot showed that the range increased with oxygen enrichment. In the white combustion zone where coke combustion is active and the red heat zone where the temperature is relatively low, the width of both the combustion zone and the red heat zone increased as the ratio of oxygen in the suction gas increased and the red heat zone in particular widened more than the combustion zone.

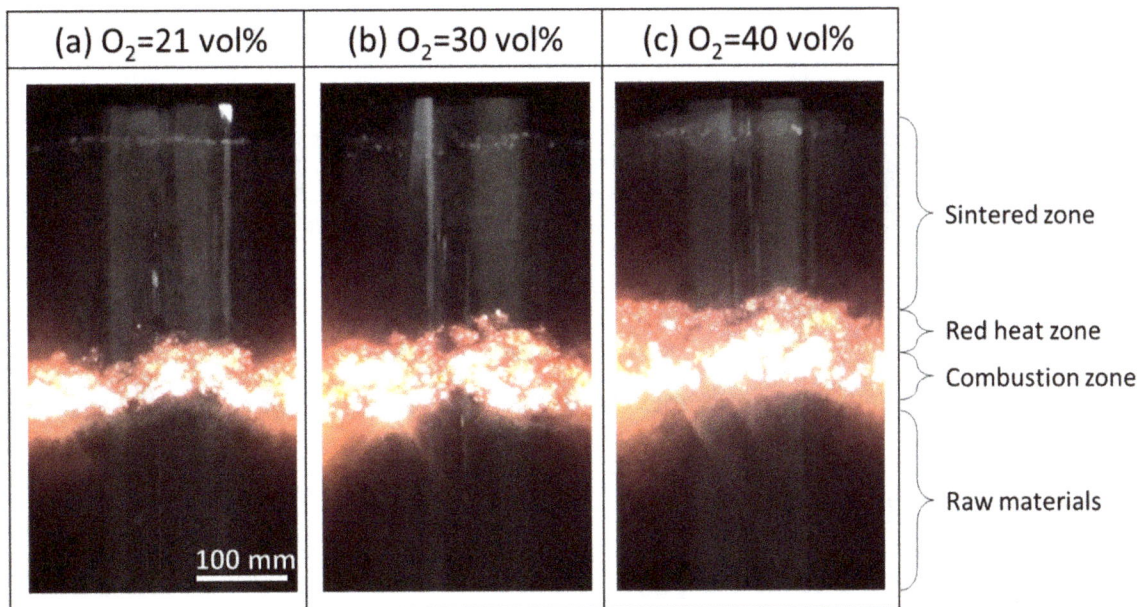

FIG 3 – Appearance of sintering process during sinter pot tests with oxygen ratio in the input gas of 21 to 40 vol per cent.

Figure 4 shows the high temperature holding time above 1200°C measured by the thermocouples for the upper and lower layers. In both cases, the high temperature holding time was extended when the oxygen ratio was 30 vol per cent. In the upper layer, the maximum temperature at 40 vol per cent oxygen ratio was lower than that at 21 vol per cent oxygen ratio and the high temperature holding time was shorter. In the lower layer, the maximum temperature at 40 vol per cent oxygen ratio was slightly lower than that at 21 vol per cent and the high temperature holding time was significantly shorter than that at 30 vol per cent. These trends corresponded to the sinter yield in Figure 2.

(a) Upper layer

(b) Lower layer

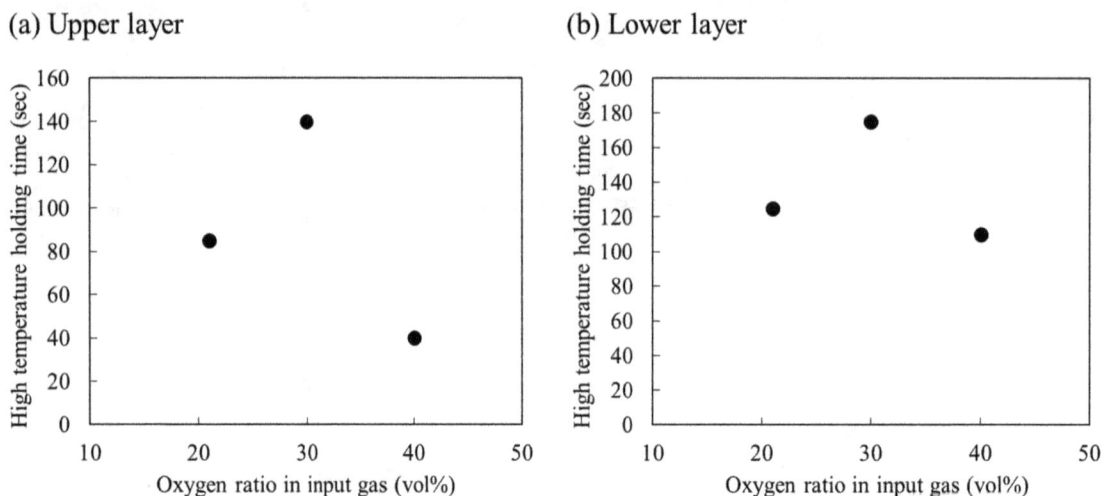

FIG 4 – High temperature holding time above 1200°C in the (a) upper layer and (b) lower layer for each oxygen ratio in the input gas.

Previous reports have indicated that the high-temperature holding time is extended with oxygen enrichment in the range up to 35 vol per cent, but at 40 vol per cent both the maximum temperature and the high-temperature holding time tend to decrease.

Figure 5 shows the cumulative pore volume (10–1000 μm) of sintered ore obtained by the mercury intrusion method. The total pore volume decreased with increasing oxygen ratio. This result confirms that the strength increases as the oxygen ratio increases.

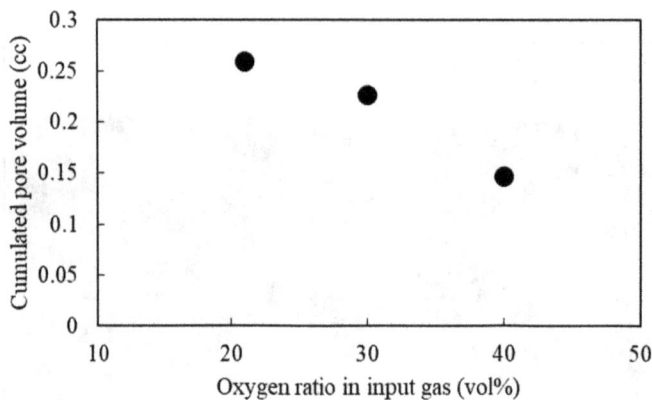

FIG 5 – High temperature holding time above 1200°C in the (a) upper layer and (b) lower layer for each oxygen ratio in the input gas.

Analysis of quartz glass pot tests

Image analysis was performed to examine the heat pattern from the results observed in the quartz glass pot tests. Figure 6 shows the analysis flow. First, the obtained image was converted to 8-bit grey scale and the luminance distribution in the height direction was obtained. After that, the luminance at each height was divided by the 8-bit maximum value (= 255) to convert to the Relative Intensity (-) (Takehara, Higuchi and Yamamoto, 2023).

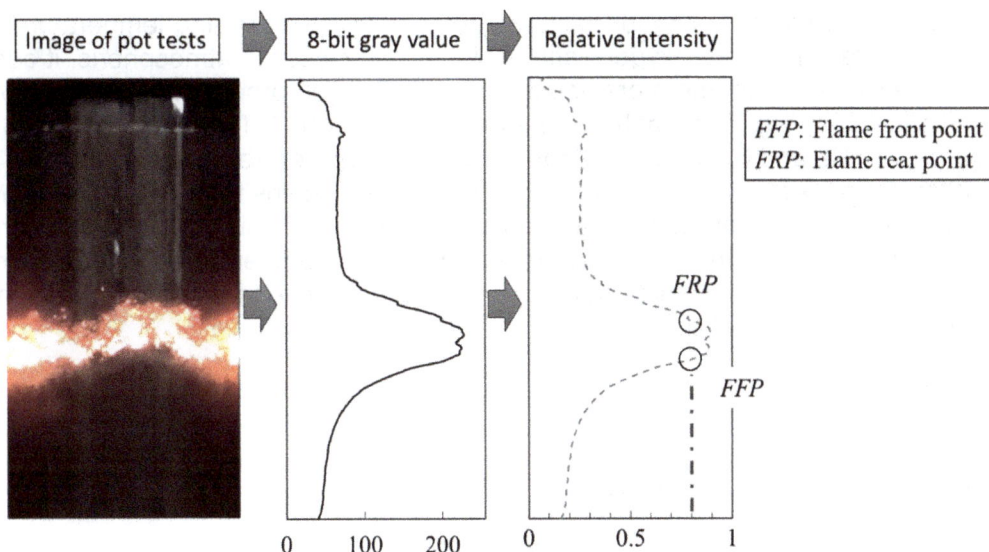

FIG 6 – Flow analysis of flame rear and front points acquired from pot test images.

Figure 7 shows the results of analysing the heat pattern at each point in time in the pot test using relative intensity. The observed results were image-processed every minute and the remarkable data extracted to evaluate sintering phenomena. Focusing on the shape of the heat pattern, it can be seen that the narrow peak shape in the upper layer after ignition changes to a wider width over time. At an oxygen ratio of 21 vol per cent, the peak width expanded after around 13 minutes. At an oxygen ratio of 30 vol per cent, the peak width tended to expand clearly after 9 minutes. At 40 vol per cent oxygen, expansion occurred even earlier. This indicates that the positions of the combustion zone and the heat pattern are separated.

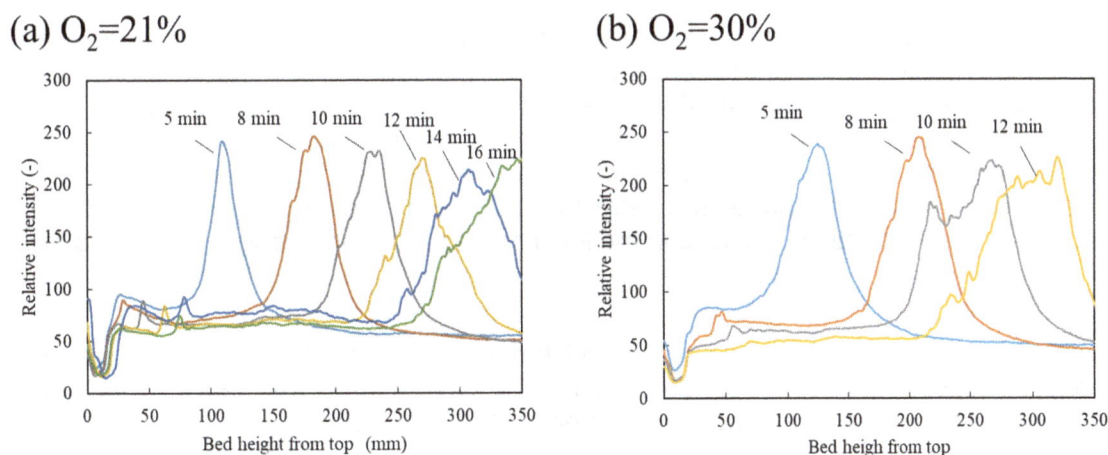

FIG 7 – Flow analysis of flame rear and front point acquired from pot test images.

With oxygen enrichment, the high-temperature holding time increases due to the expansion of the high-temperature zone and at 21 vol per cent O_2 and above, it is possible to obtain improved yield and strength effects. On the other hand, in the combustion zone, when the oxygen ratio exceeds an appropriate value, the combustion rate exceeds the expansion effect and it is assumed that the expansion of the heat pattern results in a decrease in the maximum temperature and a decrease in the high temperature holding time. This mechanism seems to be the reason that there is optimum oxygen ratio at 30 vol per cent for sinter yield.

THERMODYNAMIC EQUILIBRIUM CALCULATION RESULTS

The sintering process is a heterogeneous reaction and it is assumed that the temperatures in the vicinity of the powdered coke are wide ranging. In the present study, therefore, the liquid phase fraction at 1200–1400°C was determined at an oxygen partial pressure of 2.0×10^{-1} ~40 kPa.

Figure 8 shows the calculated results for each oxygen partial pressure and temperature. Focusing on 20 kPa and 40 kPa, where the oxygen ratio is higher and closer to atmospheric, it can be seen that the higher the oxygen ratio, the more liquid phase likely to be produced. The effect of a higher oxygen ratio is particularly significant at high temperatures of 1400°C. The same applies to the lower oxygen ratios of 2.0 kPa and 0.2 kPa, where the higher the oxygen ratio, the more likely the liquid phase is to be produced. At low oxygen ratios, the liquid phase begins to appear at temperatures as low as 1250°C, but the effect of oxygen ratio on the increase in liquid phase in that temperature range is small and it is thought that the effect appears at higher temperatures. It also appears to be possible to achieve oxygen enrichment effects over a relatively wide temperature range in low ratio conditions than in high ratio conditions.

(a) O_2: 20 kPa and 40 kPa case

(b) O_2: 0.2 kPa and 2 kPa case

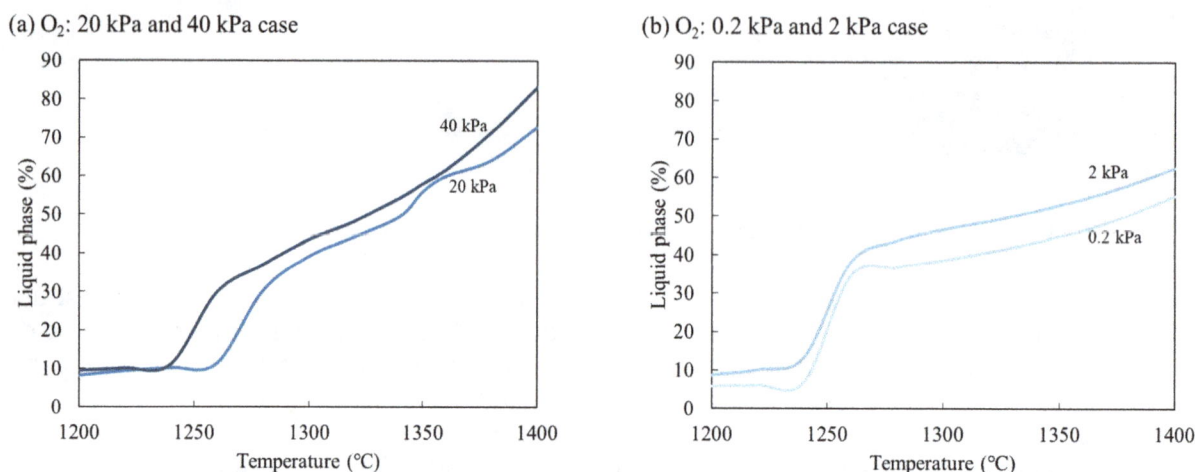

FIG 8 – Calculation results for liquid phase ratio at O_2 partial pressures of 2.0×10^{-1} to 4.1×10 kPa.

From the above, the effect of oxygen enrichment increases the liquid phase fraction as expected in the conditions of high temperatures around 1400°C and low oxygen partial pressures. These conditions correspond to high-temperature combustion zones in the sintering layer. The effect of increased high temperature holding time due to oxygen enrichment is described above, while the effect of the increase in the metallurgical liquid phase is described in this section. Thus, oxygen enrichment is expected to have two effects: improving the heat pattern and promoting thermodynamic melt formation, resulting in the low porosity and high strength of sintered ore.

CONCLUSIONS

In this study, the effect on quality of oxygen enrichment in the sintering process was evaluated at higher oxygen ratios of 21~40 vol per cent than in the past and the mechanism was investigated using heat pattern analysis and thermodynamic equilibrium calculations.

1. Sintering time was reduced and sintering productivity improved with increasing oxygen enrichment. The yield showed a maximum value at an oxygen ratio of 30 vol per cent and the strength increased with oxygen enrichment.

2. At an oxygen ratio of 30 vol per cent, the high temperature holding time increased due to the expansion of the combustion zone. On the other hand, as the oxygen ratio increases, the combustion zone and heat pattern become separated, suggesting that the effect of extending the high temperature holding time decreases or disappears.

3. Thermodynamic calculations showed that the liquid phase content increased at high temperatures or under low oxygen partial pressures. It is considered that under oxygen-enriched conditions, the two effects of the extending high-temperature holding time in the combustion zone and promoting the formation of the liquid phase resulted in the formation of high-strength sintered ore with low porosity.

ACKNOWLEDGEMENTS

The authors wish to express their sincere gratitude to Dr Hidetoshi Matsuno of JFE Techno-Research Corporation for thoughtful advice on this research.

REFERENCES

Ishimitsu, A, Wakayama, S, Tokuma, S and Sato, K, 1963. Use of Preheated and Oxygen-Enriched Air for Sintering Iron Ore, *Tetsu-to-Hagané*, 49:342–350.

Rajak, D K, Ballal, N B, Viswanathan, N N and Singhai, M, 2021. Effect of Oxygen Enrichment on Top Layer Sinter Properties, *ISIJ Int*, 61:79–85.

Sasaki, M and Hida, Y, 1982. Consideration on the Properties of Sinter from the Point of Sintering Reaction, *Tetsu-to-Hagané*, 68:563–571.

Sugawara, M, Koitabashi, T, Hamada, T, Okabe, K, Saino, M and Sato, Y, 1973. Effect of Heat Supply on the Iron Orre Sintering Operation, *Kawasaki Steel Giho*, 5:30–41.

Takehara, K, Higuchi, T and Yamamoto, T, 2023. Effect of Oxygen Enrichment on Melting Behavior, *Sintering Process*, 109:235–244.

Voice, E W and Wild, R, 1957. Importance of Heat Transfer and Combustion in Sintering, *Iron Coal Trad Rev*, 175(10):841–850.

Health, safety, environment and community – dust mitigation and reduction of water consumption

Supply chain approach to iron ore dust monitoring and remediation – Pilbara, WA

P Hapugoda[1], S Hapugoda[2], G Zhao[3], E R Ramanaidou[4], G O'Brien[5] and Y Jin[6]

1. Senior Experimental Scientist, CSIRO Mineral Resources, Pullenvale Qld 4069.
   Email: priyanthi.hapugoda@csiro.au
2. Senior Experimental Scientist, CSIRO Mineral Resources, Pullenvale Qld 4069.
   Email: sarath.hapugoda@csiro.au
3. Research Scientist, CSIRO Mineral Resources, Pullenvale Qld 4069.
   Email: guangyu.zhao@csiro.au
4. Team Leader, CSIRO Mineral Resources, Pullenvale Qld 4069.
   Email: ramanaidou.eric@csiro.au
5. Principal research Consultant, CSIRO Mineral Resources, Pullenvale Qld 4069.
   Email: graham.o'brien@csiro.au
6. Team Leader, CSIRO Mineral Resources, Pullenvale Qld 4069. Email: yonggang.jin@csiro.au

ABSTRACT

In 2021 the ports of Port Hedland and Dampier in the Pilbara Region of Western Australia handled 900 Mt of iron ore. This ore was produced in the Hamersley's mines and transported by trains to the ports for subsequent export. Dust, which is generally considered to be particles less than 100 μm in diameter, is generated at all steps in the supply chain. Whilst all dust may have potential impacts on human health and mangroves and buildings a proportion of the dust is classed as respirable (PM_{10}) dust, which consists of particles <10 μm in diameter.

In 2021, in response to the community concern with iron ore dust, CSIRO commissioned an internal study to characterise and benchmark the dustiness of the different ores and to develop targeted strategies to minimise dust impacts, without affecting other iron properties, such as stickiness during transportation. A specific concern is that that the dust in the PM_{10} fraction may contain significant amounts of particulates, such as free silicates which may have health implications for the workers and residents.

This study provided detailed mineral and chemical characterisation of the dust generated from the railed fine ore products for five generic Australian ore types: Australian Channel iron Deposit (ACID), Australian Brockman (AB), Australian Marra Mamba (AMM), Australian Pilbara (AP) and Australian magnetite (AM). The dust from these fine ore products were compared to that generated from a sample of Brazilian Hematite (BH) type ore fines.

The ores were evaluated for their physical, chemical and mineralogy properties, dust generating potential and understandings of the relationship between the dustiness of iron ores and their physiochemical properties. The size distributions of the railed fine ore products were determined and a suite of analyses were conducted on subsamples of total (-106 μm) and fine (-38 μm) dust. For all the ores, the total dust fraction contains less Fe and more contaminants (SiO_2 and Al_2O_3) than did the bulk ore. The fine dust fraction contained even less Fe and more contaminants (SiO_2 and Al_2O_3) than the total dust fraction. This suggests that the iron ore dust contains more clay and potentially free silica than does the bulk ore. These findings were supported by optical and Scanning electron microscopy analyses of the samples.

Dustiness measurements were made for each of the different ores (-106 μm) to investigate the relationship of dustiness of various iron ores with their physiochemical properties and assess their responses to dust suppression by the application of water sprays.

INTRODUCTION

In 2021 the ports of Port Hedland and Dampier in the Pilbara Region of Western Australia handled 900 Mt of iron ore (Pilbara Ports Authority, 2021). This ore was produced in the Hamersley's mines and transported by trains to the ports for subsequent export (Darby, 2005; Ashby, 2012; Australasian Railway Association, 2013). Dust, which is generally considered to be particles less than 100 μm in diameter, is generated at all steps in the supply chain. Whilst all dust may have potential impacts on human health and mangroves and buildings a proportion of the dust is classed as respirable (PM_{10}) dust, which consists of particles <10 µm in diameter.

In 2021, in response to the community concern with iron ore dust, CSIRO commissioned an internal study to characterise and benchmark the dustiness of the different ores and to develop targeted strategies to minimise dust impacts, without affecting other iron properties, such as stickiness during transportation. A specific concern is that that the dust in the PM_{10} fraction may contain significant amounts of particulates, such as free silicates which may have health implications for the workers and residents.

Current approach to managing and regulating dust in Port Hedland.

- All port operator's licence conditions include extensive controls about managing dust emissions from the iron ore companies' operations and DWER has recently released its Port Hedland Regulatory Strategy.

- As outlined in the Strategy, industry must demonstrate that expansion in operations does not result in an increase of dust emissions.

- This is conditioned through the operating licence, with modelling and validation monitoring required to demonstrate that dust does not increase.

- Additionally, collection of dust samples and dust characterisation is a condition of the companies operating licence.

There are two main initiatives as the holistic approach for iron ore dust management:

1. Long-term initiative: Majority of Australian iron ore fines are martite-goethite types and upgrading these types of ores to high quality ore (high Fe and low Al, Si and P) as a feed for green steelmaking production. This could be achieved by upgrading M-G ore, or only the low quality/finer goethitic type at the mine by dehydration and or agglomeration processes. This will result in a reduction of iron ore dust including the reduction in PM_{10} and $PM_{2.5}$ particles emitted during transport and at the ports.

2. Short/Medium term initiative: Supply chain approach to iron ore dust monitoring and remediation.

This paper shares the preliminary findings from the characterisation of iron ore dust generated by screening the -106 µm size fractions from selected ores.

Source of iron ore dust

All the iron ore mines in Australia are open cut and the ore is usually extracted by conventional drilling and blasting methods (Lu, 2015; Hagemann *et al*, 2008; Ramanaidou and Morris, 2010; Morris, 1985). These methods generate a lot of fugitive dust a portion of which is respirable (PM_{10}) metalliferous dust. These ores are transported to the port by rail and dust is produced at all stages of the supply chain. For example, mine site processes such as haul road transportation, crushing and screening produce lump and sinter fines products, stockpiling, conveyor belt transfers veneering, and. their transport from pit to port generates a significant amount of dust.

In addition, the new more complex ore types, which have entered the market often require additional processing to achieve the required grade which requires various nonmagnetic physical separation methods such as scrubbers, jigs, spirals, hydrocyclones, cyclones and heavy medium drums (Adams, 2021). All these processes add to the complexity of benchmarking the dust generating potential of the different ores at all steps in the supply chain.

Dust research and development

CSIRO Environment and Sustainability Research Team has developed functional nanomaterials and novel processes (Figure 1).

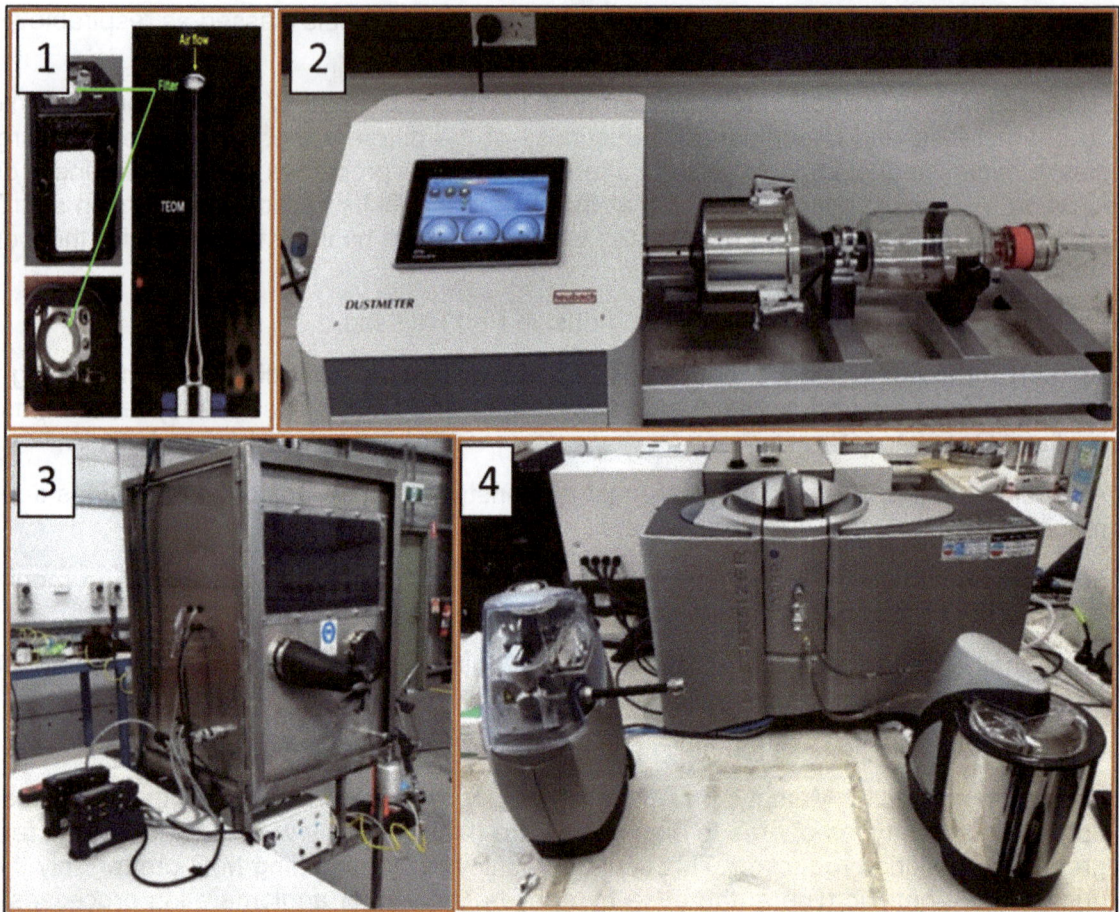

FIG 1 – (1) Monitoring of personal dust exposure (respirable dust and crystalline silica), (2) Dustiness measurement of iron ores, (3) Laboratory dust testing chambers, (4) Facility for dust characterisation.

Quantifying dust generated from the different iron ore types

CSIRO Mineral Resources have previously done extensive work around coal shipping ports and open cut coalmines using combined optical and SEM characterisation methods (Warren *et al*, 2015; Warren, Krahenbuhl and O'Brien, 2013). We believed that the same approach could be applied to iron ore ports and mines, to benchmark mine generated dust from different ore types, produce different amounts of dust and are there differences in the corresponding chemical properties.

For example, Figure 2 shows some images of different types of iron ore samples which have different amounts of hematite-goethite contents and potentially different dust generating potential. This project undertook a comprehensive assessment of the mineralogical and chemical characterisation of the entire range of particulates that could be produced from the different ore types as the first step to obtaining an understanding of their potential impacts on the local environment.

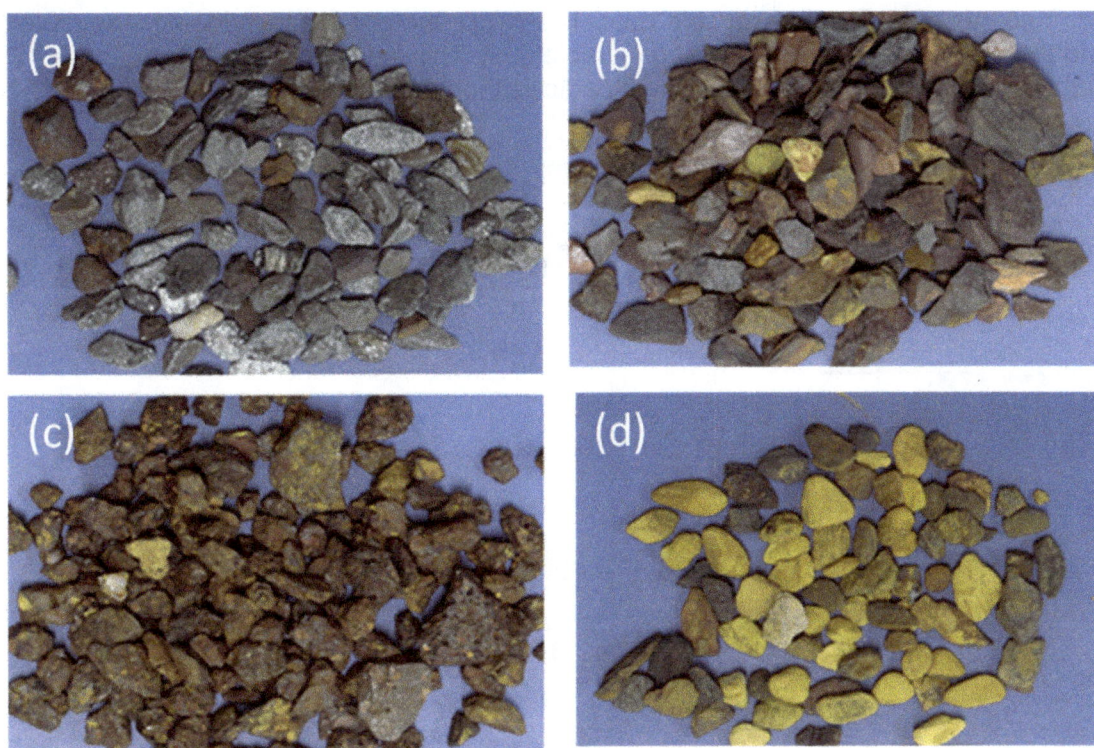

FIG 2 – Different types of iron ores with different amounts of hematite-goethite contents. These iron ore types display different dust generating potential: (a) dense-medium porous hematite ore, (b) dense-medium porous martite (hematite)-goethite ore, (c) medium to high porous channel iron-goethite ore, (d) porous-high porous ochreous goethite ore.

METHODOLOGY

For this study five generic Australian ore types: Australian Channel iron Deposit (ACID), Australian Brockman (AB), Australian Marra Mamba (AMM), Australian Pilbara (AP) and Australian magnetite concentrates (AM) were selected to investigate the dust generating potential of the different ores and determine the mineral and chemical characterisation of the dust generated from their potential railed fine ore products. The dust from these fine ore products were compared to that generated from a sample of Brazilian Hematite (BH) type ore fines.

The ores were evaluated for their physical, chemical and mineralogy properties, dust generating potential and understandings of the relationship between the dustiness of iron ores and their physiochemical properties. The following microscopic and chemical methods were used:

- The dust characterisation was carried out to identify the dust from different iron ore fines, using optical microscopy and SEM imaging methods to obtain size and composition detail for the individual particles.

- The size distributions of the railed fine ore products were determined and a suite of analyses were conducted on subsamples of total (-106 µm) and fine (-38 µm) dust to investigate the amounts of total and respirable dust from different ore types.

- Chemical properties of the dust samples were determined for the total ore and for all dust (-106 µm) and fine dust (-38 µm) fractions for each ore type.

- The dustiness of iron ores with and without water spray application was evaluated by rotating drum method using a Heubach Dustmeter.

RESULTS

Size distributions

The size distributions of the railed fine ore products were first determined by combined physical sizing and laser sizing results (Figure 3). These results established that the e different ores produced different amounts of total and respirable dust. The ACID fine ore produced least respirable (PM$_{10}$) dust and the BH fine ore produced most respirable dust. This indicate that, although the ACID ore is dominantly goethite but contained less finest particles than the hematite dominant Brazilian hematite ore.

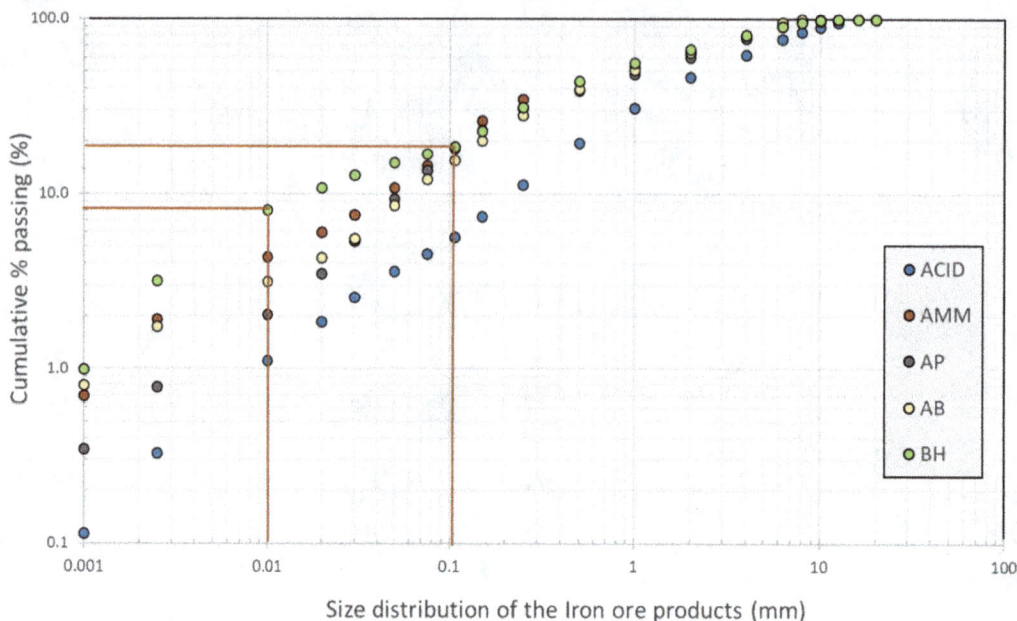

FIG 3 – Size distribution of the iron ore railed products.

Chemical properties

XRD and XRF tests were then conducted on samples of the fine ore product, the total dust fraction (-106 µm) and the fine dust fraction (-38 µm) for each of the different ores. These results presented in Table 1 showed that the total dust for all the ores, the total dust fraction contains less Fe and more contaminants (SiO$_2$ and Al$_2$O$_3$) than did the bulk ore. The fine dust fraction contained even less Fe and more contaminants (SiO$_2$ and Al$_2$O$_3$) than the total dust fraction.

TABLE 1

Chemical properties of the dust samples was determined for the total ore and for all dust (-106 µm) and fine dust (-38 µm) fractions for each ore type.

	ACID			AB			BH			AP			AMM		
	Fe	SiO$_2$	Al$_2$O$_3$	Fe	SiO$_2$	Al$_2$O$_3$	Fe	SiO$_2$	Al$_2$O$_3$	Fe	SiO$_2$	Al$_2$O$_3$	Fe	SiO$_2$	Al$_2$O$_3$
Total ore	57.9	5.4	1.4	62.9	4.2	2.2	65.1	1.7	1.4	63.8	2.6	1.3	61.8	3.2	2.0
-106 µm	55.0	7.1	3.1	63.0	4.3	2.4	64.2	1.8	2.2	60.8	4.1	2.9	60.4	4.2	2.9
-38 µm	54.3	7.4	3.7	61.9	5.0	2.6	64.0	1.7	2.3	59.8	4.4	3.2	57.1	5.8	3.9

This suggests that the iron ore dust particularly Australian ores, contains more clay and potentially free silica than does the bulk ore. These findings were supported by optical and Scanning electron microscopy analyses of the samples.

Dust characterisation using optical and SEM microscopy

Optical reflected light microscopy and Scanning Electron Microscopy (SEM) methods were used to obtain quantitative information of the different constituents for the different iron and particulates in each of the dust samples and to provide indicative size distribution for the different constituents.

Coal Grain Analysis (CGA) optical imaging system

CGA is a reflected light microscopy system which has been used successfully to obtain quantitative information on the different particulates in dust samples collected around Australian coal ports (Koval *et al*, 2018). The raw optical and the characterised image for a subset of the analyses conducted for one of the ores (ACID) are shown in Figure 4. Whilst this method could identify the different iron ore particles of hematite and goethite, it could not distinguish between the clays and quartz particles which were grouped into the overarching class of dark minerals. For each of the four different ore types studied the proportion of goethite, hematite and dark minerals (quartz plus clays) present in the total dust sample and in the PM_{10} fraction of the sample are shown in Table 2.

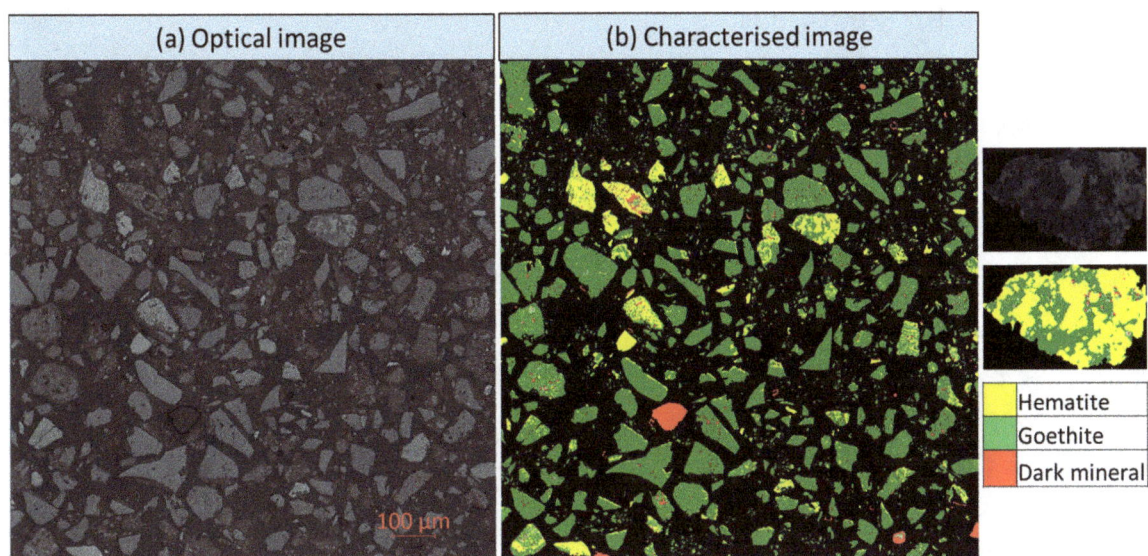

| (a) Optical image | (b) Characterised image |

Hematite
Goethite
Dark mineral

FIG 4 – CGA images from optical microscope for the ACID total dust (-106 µm) sample, (a) Original image, (b) Characterised image.

TABLE 2

Summary results from CGA analysis.

Ore type	Dust size (microns)	Goethite	Hematite	Dark minerals (clays and quartz)
ACID	all dust (-106 µm)	66.6	16.4	17.1
	PM_{10} (-10 µm)	61.8	2.6	35.6
AB	all dust (-106 µm)	27.2	60.7	12.0
	PM_{10} (-10 µm)	39.4	20.9	39.8
AP	all dust (-106 µm)	35.9	46.4	17.8
	PM_{10} (-10 µm)	25.7	3.6	70.7
BH	all dust (-106 µm)	66.6	16.4	17.1
	PM_{10} (-10 µm)	61.8	2.6	35.6

Summary observations from CGA analysis data as showed in Table 2 are:

- CGA cannot currently provide separate information on quartz and clays.

- Majority of quartz and clays in the overall dust sample appear to preferentially report to the finer (PM_{10}) fraction.

- Hematite does not appear to preferentially report to the PM_{10} fraction.

- Goethite shows approximately the same amount in both the overall dust and PM_{10} fractions.

SEM analysis

SEM was used to investigate whether additional information could be obtained for the quartz particulates in the dust samples. The raw and characterised image for one of these samples (AB) are shown in Figure 5. The summary results obtained for this analysis were:

- Iron ore dust (-106 micron) were observed under SEM.

- Samples, resin mounted and sprinkled on filter paper and on conductive tape were prepared.

- Secondary electron (SE) and Backscattered (BSE) images were taken.

- Under SEM backscattered conditions hematite display bright white, goethite is light-mid grey and quartz is dark grey colours.

- ACC, Voltage (Kv), aperture, contrast brightness need optimisation to get better images.

- EDS mapping will be an advantage for the identification of magnetite, hematite, goethite and gangue minerals.

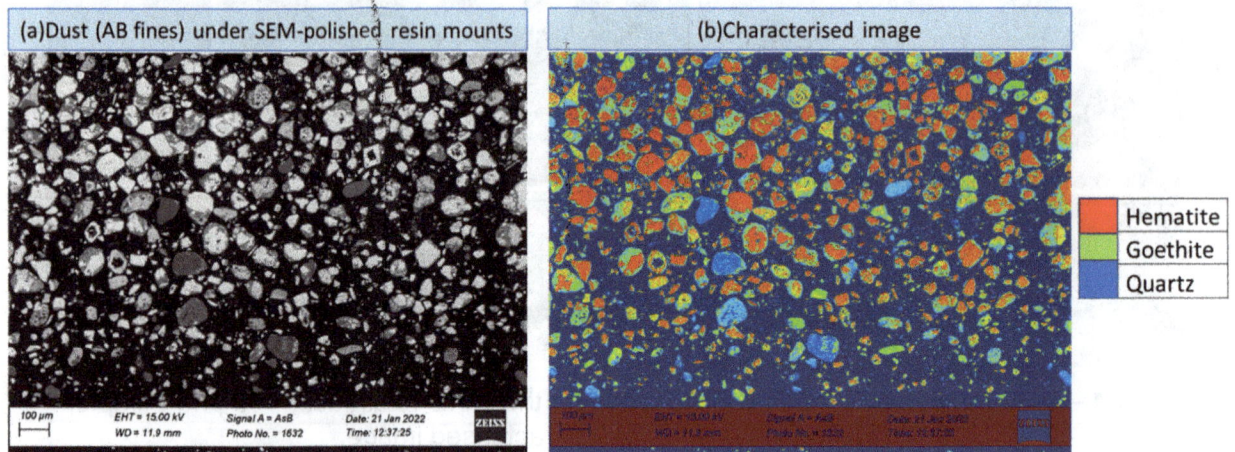

FIG 5 – SEM analysis of Dust (AB fines): (a) BSE image, (b) Characterised image.

Dustiness evaluation

The dustiness of the iron ores (-106 μm) was evaluated using a Heubach Dustmeter and the relationship of dustiness of various iron ores with their physiochemical properties was studied. As shown in Figure 6, while there is no obvious correlation between dustiness levels and particle sizes, a strong correlation between the dustiness and goethite content was observed.

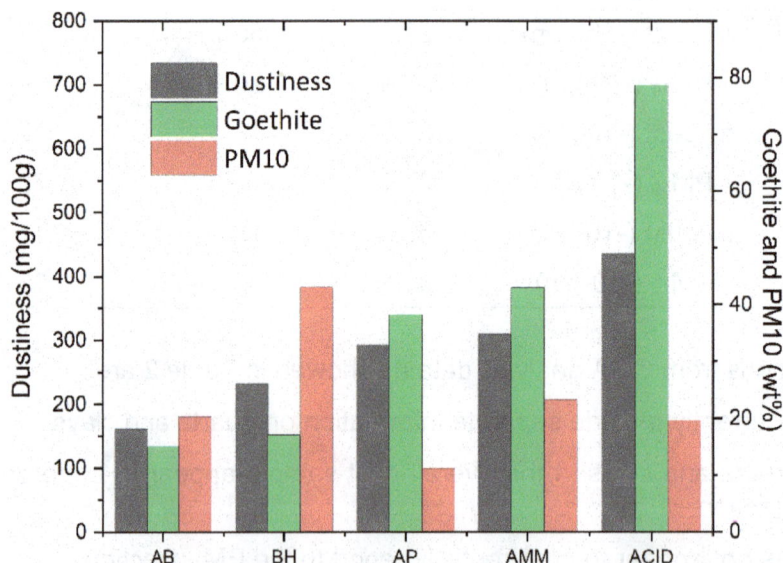

FIG 6 – Correlation of dustiness with PM_{10} and goethite contents (based on XRF analysis results) of the iron ores (-106 μm).

Effect of water spray on dust suppression was investigated by measuring the dustiness of ACID iron ore with pre-treatment of water spray (5 wt per cent). As shown in Table 3, water spray (5 wt per cent) can reduce dustiness value of ACID iron ore sample by 65 per cent with 5 min cure time. The dustiness value increases as increasing cure time likely due to water evaporation or variation of water interaction with the ore during the cure period.

TABLE 3

Dust suppression for ACID iron ore sample (-106 µm).

Suppressant applied	Cure time	Sample weight (g)	Dust value (mg/100 g)	Dust value reduction (%)
No treatment	N/A	30	429.8	N/A
water	5 min	30	148.5	65.4
water	60 min	30	175.3	59.2
water	180 min	30	278.3	35.2

CONCLUSIONS

For the different fine ore samples studied it was found that they produced different amounts of total dust and PM_{10} dust. These results suggest that between 1 per cent (ACID) and 8 per cent (BH) of the railed fine ore products were respirable (PM_{10}) dust.

Chemical (XRF and XRD) tests and optical imaging (CGA) indicate that the fine dust contained more clays and quartz than did the overall dust. CGA cannot currently discriminate between clay and quartz particles.

CGA provides size detail (particle, length, width, area) on dust particles >1 micron. Information is generally obtained on between 10 000 and 100 000 particles, which enabled particles to be partitioned into nuisance >10 µm and respirable <10 µm (PM_{10}) size fractions.

SEM does provide discrimination between quartz and clays and different ore types in dust particles.

Further SEM processing development is required to also obtain particle size and abundance information on a large number (10 000 to 100 000) of dust particles.

Preliminary testing suggests that characterised SEM images can be imported into the CGA software for processing to obtain this information.

This suggests that the iron ore dust contains more clay and potentially free silica than does the bulk ore. These findings were supported by optical and Scanning electron microscopy analyses of the samples.

Dustiness measurements were made for each of the different ores (-106 µm) and a strong correlation between the dustiness and goethite content of the iron ores was observed. Water spray is proved to be an effective way for iron ore dust suppression.

ACKNOWLEDGEMENTS

The authors sincerely acknowledge the support provided by CSIRO Mineral Resources for supporting this research. The external reviewers are thanked for their valuable comments and improvements to the paper.

REFERENCES

Adams, T, 2021, 3 Feb. Iron ore mining in Australia-history, top locations and companies, Global Road Technology International Holdings (HK) Limited, Industry Articles. Available from: <https://globalroadtechnology.com/iron-ore-mining-in-australia/>

Ashby, I, 2012. BHP Billiton Iron Ore – Growth and Outlook, Presentation to Australian Journal of Mining (AJM) Conference.

Australasian Railway Association, 2013. Australian Rail Industry Report 2011–12, Report prepared for the ARA by the Apelbaum Consulting Group, Melbourne. Available from: <www.ara.net.au/publications>

Darby, M, 2005. BHP Billiton Iron Ore Railroad, Presentation to Analysts June 2005, BHP Billiton. Available from: <www.bhpbilliton.com/home/investors/reports/Documents/2005/MikeDarbyRailOperations.pdf> [Accessed: November 2013].

Hagemann, S, Rosière, C A, Gutzmer, J and Beukes, N J, 2008. *Banded Iron Formation-Related High-Grade Iron Ore*, Reviews in Economic Geology series, vol 15. https://doi.org/10.5382/Rev.15

Koval, S, Krahenbuhl, G, Warren, K and O'Brien, G, 2018. Optical microscopy as a new approach for characterising dust particulates in urban environment, *Journal of Environmental Management*, 223:196–202.

Lu, L, 2015. *Iron Ore: Mineralogy, Processing and Environmental Sustainability*, 666 p (Elsevier).

Morris, R C, 1985. Genesis of iron ore in banded iron-formation by supergene and supergene-metamorphic processes — a conceptual model, in *Handbook of Strata-bound and Stratiform Ore Deposits* (ed: K H Wolf), 13:73–235.

Pilbara Ports Authority, 2021. Port Statistics and Reports, 2020-2021, Port of Dampier and Port of Headland, Available from: <https://www.pilbaraports.com.au/about-ppa/news,-media-and-statistics/port-statistics>

Ports Australia, 2012. Trade Statistics for 2011–12, Ports Australia. Available from: <www.portsaustralia.com.au/tradestats [Accessed: March 2013].

Ramanaidou, E R and Morris, R C, 2010. Comparison of supergene mimetic and supergene lateritic iron ore deposits, *Applied Earth Science*, 119:35–39.

Warren, K, Krahenbuhl, G and O'Brien, G, 2013. Quantitative Analysis of Coal in Dust Samples Using CSIRO's Reflected Light Optical Imaging System, 10th Australian Coal Science Conference 2013.

Warren, K, Krahenbuhl, G, Mahoney, M, O'Brien, G and Hapugoda, P, 2015. Estimating the fusible content of individual coal grains and its application in coke making, *Int J Coal Geol*, 152(2015):3–9.

Development and implementation of low-cost sensors and LiDAR for managing dust emissions due to mining activity

D Ilic[1], A Lavrinec[2], L Sutton[3] and J Holdsworth[4]

1. Senior Research Associate, University of Newcastle, Callaghan NSW 2308.
   Email: dusan.ilic@newcastle.edu.au
2. Research Associate, University of Newcastle, Callaghan NSW 2308.
   Email: aleksej.lavrinec@newcastle.edu.au
3. PhD Candidate, University of Newcastle, Callaghan NSW 2308. Email: liam.sutton@uon.edu.au
4. Associate Professor, University of Newcastle, Callaghan NSW 2308.
   Email: john.holdsworth@newcastle.edu.au

ABSTRACT

Air quality is a big topic of discussion in regional mining communities as well as urban environments situated close to port terminals handling bulk solids. Community perception of poor air quality can significantly affect the reputation of the mining operation and handling facilities.

Dust monitoring for regulatory compliance relies on using a handful of accurate but very expensive sensors, usually situated at the boundary, relatively far from the activities that generate the airborne dust. These reference monitors measure dust leaving the site but dust mitigation and management on-site largely relies on operator judgement.

At a fraction of the cost, small wireless solar powered units can be used to supplement reference monitors in real-time. Due to their compact nature, portability and low price, these units can be implemented on a mass scale to increase spatial resolution of dust measurements and, more importantly, in much closer proximity to both the dust generating activity and the sources of dust. This can provide valuable insights into the origins and timing of dust generation thus producing data-driven decision-making while minimising subjectivity. Similarly, light detection and ranging (LiDAR) is another supplemental technology that can be used to increase the spatial resolution of dust measurements in near real-time, however, the cost of commercial systems even exceeds that of regulatory monitors. LiDAR also presents difficulties in quantifying dust plume concentrations.

This study presents learnings of implementing both technologies for dust emissions management. Several low-cost sensor units and a fast-scanning custom-built 3D LiDAR system operating at a range of 1.5 km with a resolution of 1.5 m were deployed on a surface mine to monitor dust emissions from different activities. The project explores the use of real-time sensor data to quantify LiDAR measurements in parallel with integration into mine planning. The approach can assist in prioritising dust mitigation activities thereby reducing instances of emissions exceedance and minimising water use through databased and targeted application. Demonstrated is an opportunity for a step-change in industry best practice as well as commercialisation and operationalisation of the method across different industries, including future mineral extraction.

INTRODUCTION

The Upper Hunter in NSW is situated along a north-west/south-east direction through the Great Dividing Range and contains townships of Muswellbrook and Singleton. The region has many coalmines and powerplants covering 19 000 km² that is populated by over 50 000, of whom 30 000 are employed (Hunter Joint Organisation of Councils, 2018). Coal mining is of significant economic benefit to the region, state and nationally, providing employment with annual exports of over $20 billion, accompanied by over 7000 mining related businesses (Coal Services, 2018; NSW Minerals Council, 2023a). This revenue supports government services and infrastructure delivered by regional councils (Mills, 2019) and in turn the industry and local communities are supported by the government (NSW Government, 2020) through a range of initiatives including sponsorships and funds. High economic benefits of mining positively impact regional communities and can stimulate additional growth that can contribute to greater sustainability. On the other hand, negative impacts can include, rising living costs, reduced housing affordability, high unemployment, poverty in disadvantaged local communities and pollution (Cottle and Keys, 2014).

The topography, meteorology, mining, and power generation all influence air quality in the region (Climate and Health Alliance, 2015; Hibberd *et al*, 2013) with main sources of emissions being smoke, soil, petrol and diesel vehicles, sea spray and secondary sulfate (Hibberd *et al*, 2013). A map of the area is provided in Figure 1, showing regions of resource extraction. Air quality in the Upper Hunter is impacted by different sources however coal mining contributes almost 90 per cent and 70 per cent of particulate matter (PM) below 10 and 2.5 μm, PM_{10} and $PM_{2.5}$ respectively, while 90–100 per cent of sulfur and nitrogen dioxide emissions are primarily due to electricity generation (NSW Government, 2017; Campbell, 2014). The meteorology and landscape also result in high and prolonged cumulative emissions. In 2019, this was exacerbated by low rainfall, dry weather patterns and drought with poorest air quality in the previous decade, exceeding state, and national standards (NSW Government, 2019).

FIG 1 – Scatter of resource extraction along the valley in the Upper Hunter (left) while the Upper Hunter Air Quality Monitoring Network (UHAQMN) is limited to a handful of locations (right) (Google Maps; NSW DPIE, 2023).

Significant progress has been made, such as implementation of pollution reduction programs and development of best practice guidelines for controlling dust emissions, as well as, planning for investment renewables, gradual decarbonisation, and closure of powerplants (NSW Government, 2011, 2014; NSW Minerals Council, 2023a; AGL, 2015). However, there is ongoing concern in the region (eg McCarthy, 2018; Goetze, 2019) as well as increased pressures on both industry and government by various organisations (eg Climate and Health Alliance, Doctors for the Environment Australia, Lock the Gate Alliance). This has contributed to the administration of the Upper Hunter Mining Dialogue (UHMD) by the NSW Minerals Council and the Upper Hunter Air Quality Monitoring Network (UHAQMN) by the NSW state government. The location of the UHAQMN monitors is also shown in Figure 1. Comparable trends have promulgated at other locations around Australia leading to similar initiatives (eg Port Hedland Industries Council monitoring network and WA guidelines on dust emissions) where resolution has involved state regulator investigation and interventions that include industry led buybacks of private land (Wilkie, 2019; Connick, 2023; Government of Western Australia, 2021a).

Some of the operational initiatives to minimise the impact of mining on air pollution include postponing blasting in adverse weather, limiting haul road traffic, wetting dry/dusty areas, reducing vehicle speeds, and halting operations (NSW Minerals Council, 2023b). During mining activity, haul trucks, drills and mobile machinery emit dust during their operation, and predominantly, transport from their tyres, bodies, and trays. Consequently, major dust generating activities include tyre generated dust and wind erosion, drilling/blasting and loading/unloading.

The approach to dust mitigation during mining is controlled and implemented via what is known on-site as a Trigger Action Response Plan (TARP). In most cases, the methods for controlling dust employed are predominantly reactive and it is the operators of the heavy vehicles on-site themselves that trigger an action control response, such as application of water carts based on their visual observation during loading and transport activities. Typically, water application can involve haul roads and stockpiles, although it is also used in the preparation plant to reduce dust, with some remaining in the processed product and during vehicle wash-down. Water is often used as a dust suppressant on an as-needed basis around the facility, either caused by visual observation or an exceedance of the any of the regulatory boundary monitors. This method is representative of typical mine site dust mitigation methods, which effectively rely on methods based on operator perceptions of observable nuisance dust. The fine particles of concern on the other hand are almost invisible to the human eye.

Due to the significant dust suspension time and travel distance, the surrounding regional communities are vulnerable, with detrimental health impacts directly correlating to proximity of mining operations (Aneja, Isherwood, and Morgan, 2012). Increased particulate retention and travel time exceeding 15 mins has been reported (Gautam, Prusty and Patra, 2015), indicating high particulate concentrations remain following cessation of mining activity. As these fine particles are difficult to see, this shows that the reactive nature of the operators to trigger an event to mitigate the dust only helps when the dust is visible and does not trigger an event for PM of the order that is damaging to the health of workers and communities. Localised elevated emissions at the mine site and in proximity have also been observed to exceed background average concentrations by up to 15 times during peak production periods (Sahu, Patra and Kolluru, 2018). To comply with operating conditions and regulations, some operations must cease all dust generating activities during adverse weather and/or adverse PM_{10} concentrations measured at the regulatory or UHAQMN monitors, with the definition of 'adverse' determined during the granting of the Environmental Protection License (EPL).

BACKGROUND

The general approach to managing dust due to mining activity and resource handling operations includes measuring air quality dust concentration, deposition, as well as use of factors and meteorological data to predict emissions. Standard, reference grade or equivalent methods such as high-volume air sampling (Hi-Vol), Tapered Element Oscillating Microbalance (TEOM) and Beta Attenuation Monitor (BAM) are expensive to install and operate. These devices involve drawing a sample of air through an inlet in a specific location and measuring PM concentration. However, typically, dust monitoring is limited to only a handful of locations, limiting spatial resolution and these methods are expensive both with regards to cost and labour.

Australian air quality regulations generally align with US and European environmental agencies, with the basis generally referencing US EPA classification. A three-tiered air quality monitoring approach involves reference methods (regulatory monitoring, 10 per cent precision and bias error) as the regulation standard, followed by near-reference (supplemental/personal exposure, 20–30 per cent precision and bias error) and indicative (education, hot spot identification, 30–50 per cent precision and bias error) (Williams *et al*, 2014a). Recent advancement of technology, coupled with affordability, has seen increasing use of low-cost sensors (LCS) to monitor local air quality in urban settings and by individual businesses. However, unlike expensive reference methods, an application, standard use, and certification framework for the use of LCS in air quality monitoring is not yet in place. This has exacerbated uncertainty regarding LCS accuracy, purpose, operation, and calibration (Williams *et al*, 2018).

LCS may be viewed as affordable for personal use (A$150 to A$2000) sensors that use light scattering technology which results in lower accuracy compared to more established versions employing the same or similar technology such as the DustTrak for example (Williams *et al*, 2014b). However, despite lower reported accuracy, both the US EPA and the European Committee for Standardization (CEN) have studied and promoted LCS in an effort towards developing a framework for their application as supplemental, for research and education, while accepting that their intended application is unlikely to rival or replace standard or equivalent reference methods (Williams *et al*, 2014a, 2018). Neither organisation provides a specific definition of what a LCS is defined as, however, both infer these are emerging, low prices sensors (Williams *et al*, 2014b; Karagulian *et al*, 2019).

In contrast to reference methods that involve drawing air in and measuring the concentration of material that adheres to the measuring filter/surface, similarly to light scattering measurement of the LCS, Light Detection and Ranging (LiDAR) systems detect an object by sending out a pulse of electro-magnetic radiation that subsequently scatters off, or reflects from, the object and back to a detection device. The period between transmitting the signal and receiving the backscattered light gives a time-of-flight duration. This duration and the speed of light determine the range of the LiDAR. When light scatters by particles that are approximately of identical size to the wavelength of light (Mie Scattering), it can be calculated according to principles of planar light scatter by assuming a homogenous sphere (Musa and Paul, 2017; Sutton *et al*, 2022).

Unlike standard reference methods, and similarly to LCS, a standardised framework does not exist for LiDAR with respect to application to air quality monitoring and visualisation. Despite this, several commercial LiDAR systems exist which can be used for aerosol detection (eg MiniMPL and WindCube) and enable much greater spatial range (ie few kilometres) compared to single point aerosol detection devices, with resolutions of 20 to 40 m and some systems having capability of measuring wind speed and direction. However, both acquisition rate and resolution can be slow compared to the dynamics of the atmosphere and dust plumes and upfront costs range approximately A\$400 000 to A\$800 000 (Sutton *et al*, 2022; Hill, 2018). These commercial systems have been used in several LiDAR studies, generally producing 2D visualisation/planar plots (CRC Care, 2014; Hashmonay *et al*, 2020) and high-profile instances (Government of Western Australia, 2018a, 2018b) however, these reports show evidence of a considerable expertise gap in configuring LiDAR reconnaissance and data interpretation necessary to enable science-based conclusions from the studies. This is a barrier to progress in standardisation and establishing novel practical measurement outcomes for both industry and regulation. The secondary effect of this lack of expertise is on societal trust and expectations, with a consequence that opportunities to use this equipment to address current air quality concerns are missed.

Meanwhile, the environmental and social consequences of ineffective dust management in the mining and resource industry include reduced air quality, physical impairment, loss of psychological value and function of amenities, chronic illness, and reduced life expectancy. Typically, emissions are controlled by adding water, which is a scarce resource and ensuring its use is effectively integrated into dust management is crucial in minimising environmental impact. Understanding the impact of water addition is necessary to maintain a balance between dust emissions (insufficient water) and operational interruptions due to handling problems (excess water).

Excessive water is also economically costly and can result in significant revenue loss to businesses in the supply chain due to operational bottlenecks. The influence of water addition and the content required to minimise dust emissions can be classified through the Dust Extinction Moisture (DEM). DEM is more common in the iron ore industry and presently a condition of maintaining handled product in accordance with (or above) the DEM has been integrated into the EPL for some operations (Government of Western Australia, 2021b). While the standard method for DEM (Standards Australia, 2013) is specific to coal, its use in the coal industry is not widespread, nor has it been used as a regulatory tool. Additionally, wind tunnel testing can also be used to assess the amount of moisture that is needed to prevent dust lifting off from the surface of materials such as those on mining haul roads. However, again, a standard method or framework for water application based on activities that produce specific emissions does not exist. These methods of assessing moisture required to suppress dust and integrating together with means of identifying dust generating activities during mining present an opportunity for a step-change in practice.

With funding from both government and industry partners, we have developed both LCS units and a LiDAR system for supplementing standard air quality monitoring procedure. Over the past three years, we have developed a system for air quality monitoring that combines optical laser scanning using LiDAR and stand alone, solar and/or battery powered and remotely operated LCS units. Testing and development in different workplaces, field trials and mine site campaigns, we have demonstrated successful use of a number of LCS units in parallel with a fast-scanning 3D LiDAR system that visualises dust contours over a distance of 1.5 km and spatial resolution of 1.5 m. Following is an overview of our advancements, with a focus on application for managing dust emissions due to mining activity and an example that can act to promote improvement in dust emission management in the mining and resources industry.

DEVELOPMENT OF LOW-COST SENSOR UNITS

While LCS units are not accurate enough to replace standard reference methods at present, their low cost in providing real-time air quality data enhances their appeal in increasing spatial resolution on a mine site, even if qualitatively. This can broaden understanding of activities generating dust and provide real-time alerts that can be used to control dust.

Unit components and workflow

While there are many commercially available LCS, some like the Plantower PMS5003 stand out as common, widely studied, and utilised, with high correlation to reference monitors and readily available (US EPA, 2019). The units were initially powered off the mains supply but were replaced with a 15 W solar panel and a 17 Wh LiPo battery for the field and mine site trials. The workplace trials consisted of first building LCS units that consisted of the following:

- Plantower PMS5003 sensor

- ESP32 board

- Charger with a charging cable

- 3D printed enclosure.

The total of these components is in the order of A$50. Using learnings from workplace trials, new units were assembled for field trials, with the additional components for outdoor use requiring:

- weatherproof case

- solar panel

- battery

- solar battery charger

- long range antenna

- battery voltage monitoring system

- mounting system

- 4G modem for an autonomous gateway to facilitate connection to the local internet source.

The total cost of unit components for a field ready model is in the order of A$300, while the gateway cost is in the order of A$500. The sensors work by taking a reading every 2 seconds (customisable down to 1 second). Taking readings less frequently would result in some power savings, but they would be significant only if the readings are so infrequent that shutting the fan off is appropriate. After accumulating 60 seconds of readings, sensor data are averaged and then sent using a low power, 2.4 GHz, long range, star topology ESP-NOW protocol to the gateway. The star topology means all units send data to a centralised gateway without communicating with each other like in a mesh network. The system was tested for range of up to 2 km line of sight.

After receiving the readings from the sensor units, the gateway stores them locally in a csv file and sends the data to a cloud-based database. It is only the gateway that is connected to the internet. Data is visualised as line plots, maps and diagrams using Grafana, which is a web-based browser running on a cloud server. While it is possible to set-up alerts (eg email or text message notification) this has not been explored yet. A line style visualisation is illustrated in Figure 2.

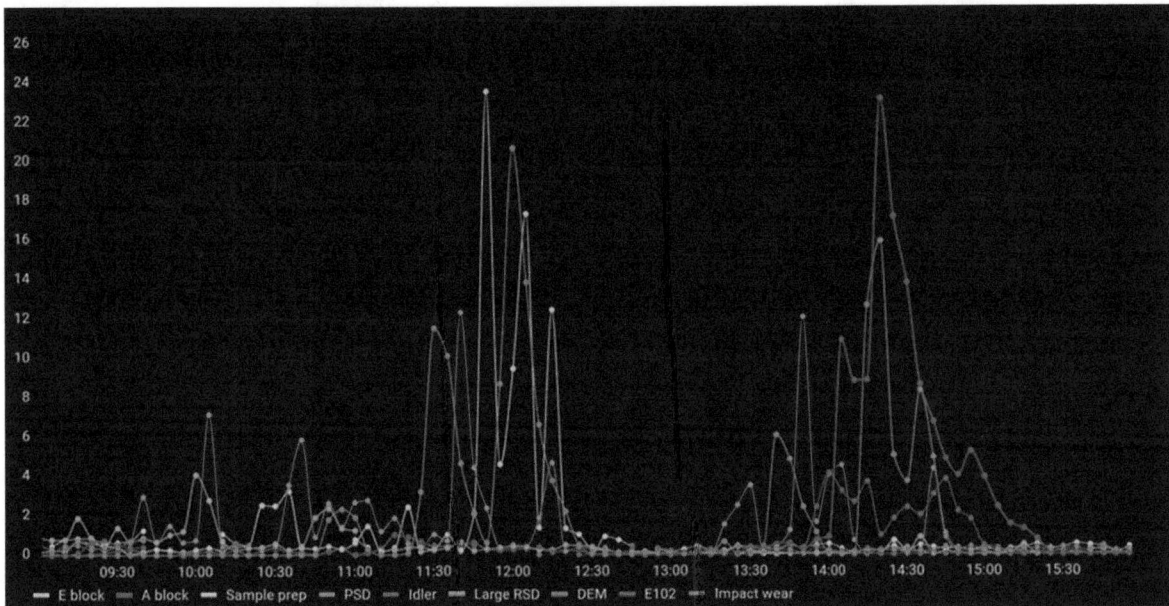

FIG 2 – Illustration of real-time dust measurement visualisation, with y-axis showing PM concentration in (µg/m³) and x-axis, the time of the day.

In addition to the LCS, it is possible to query, download, store, and visualise the state government monitoring data. This can be useful in assessing data relative to the local airshed.

Workplace trials

The LCS units were first implemented in a local industrial workplace, where activities involve sample preparation and testing of bulk solid and granular materials, with some activities generating dust. Dust in the workplace is managed using dust extraction and air scrubbers and implementing a network of these sensors next to what are perceived as dust generating activities allowed assessment of the effectiveness of these dust management practices. The image in Figure 3 shows a plan view of the industrial workplace, located at the Newcastle Institute for Energy and Resources (NIER), with the monitors located in different locations. During workplace trials, all sensors were powered directly from the mains and data was monitored remotely for a period of 4–5 months during which no real issues were experienced. Using a similar methodology and parts, the workplace trials resulted in the implementation of several in-house assembled and installed bespoke LCS units for real-time monitoring, although slightly different in design (eg most had LCD warning lights above a programmed dust concentration threshold).

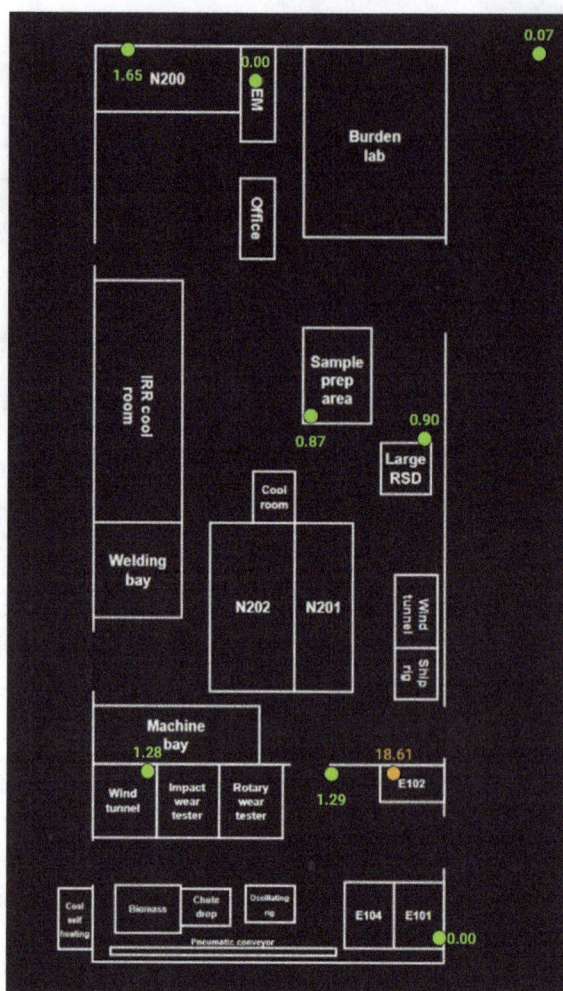

FIG 3 – Floor map of industrial workplace used in the trials and real-time LCS readings.

Field trials

Following successful workplace trials, the sensor units were weatherproofed and deployed in the field. Initial development involved deployment at a local sand quarry, however, this only involved collocation with Hi-Vol gravimetric samplers and still required power from the mains supply from the same location. Subsequently, a single LCS unit was trialled, located approximately 200 m away from the mains powered units and Hi-Vol, mounted on a tripod, and powered by a battery and solar panel (Ilic and Lavrinec, 2022). During these field trials, despite prolonged extremely wet weather, some packet loss was observed, however, the units remained on-site for months during which data was monitored remotely. These trials were used to fine-tune the size of the battery and solar panel needed for the portable units.

Mine site campaigns

Several portable LCS monitors were deployed on an open cut mine site in the Upper Hunter in different stages. The first stage involved collocating three units used in the workshop and three portable units adjacent to a calibrated regulatory monitor. The sensors were deployed in this location to enable comparison of the data between the LCS and a regulatory monitor as well as a calibration procedure that is under development.

The sensors co-located with the regulatory monitor were calibrated using a simple slope and offset. This is a simple way to calibrate the sensors and is a pathway suggested by the Australian Standard to prove sensor equivalency (AS 3580.9.17). While there is no intention to prove that these sensors are indeed equivalent to reference method monitors, it is useful to assess how they compare. Several such studies have already been undertaken in field environments (eg Karagulian *et al,* 2019; Kosmopoulos *et al,* 2020; Sayahi, Butterfield and Kelly, 2019), but it will be useful to re-assess them in the context of dust emissions at an Australian surface mine with a high baseline dust concentration adjacent to surrounding community, and in close proximity to government owned and controlled monitoring network that is used to inform the community.

There is a more complex way of calibrating the sensors involving the correction factor to account for humidity effects (in the absence of a heated inlet, the particles grow slightly larger when saturated by water). However, for simplicity and brevity, it has not been applied in this case, but it does provide a slightly better correlation with the regulatory monitor. This kind of correction is currently being examined and will be applied in future once more data is collected. The correction applied using humidity data is recommended by US EPA, however, the correction factors that the US EPA suggests are taken from studies that are based on smoke and not coal/overburden dust. A comparison between LCS and a regulatory reference monitor is shown in Figure 4.

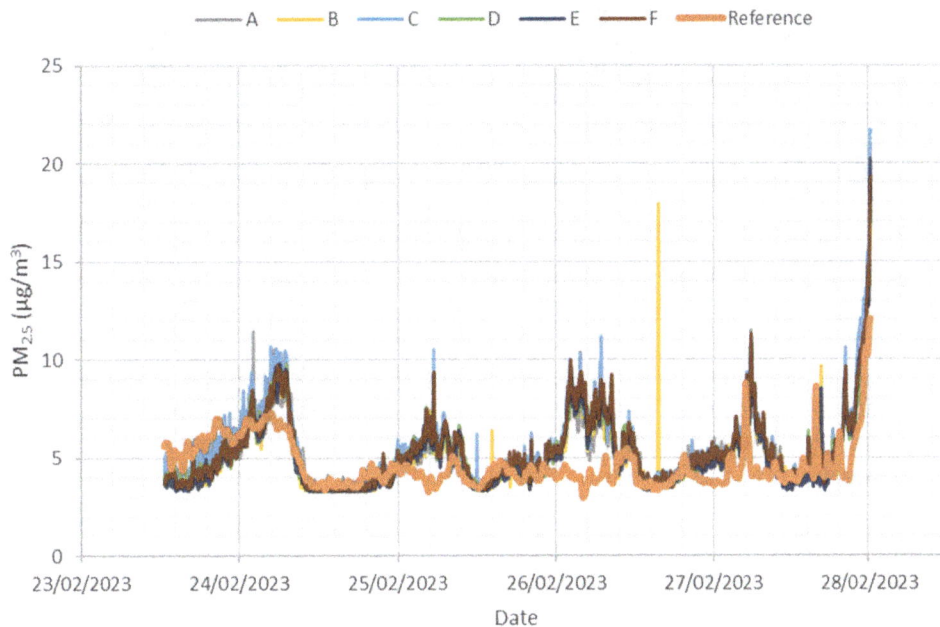

FIG 4 – Illustration of comparison between LCS units A to F and the regulatory reference monitor.

The figure above demonstrates that the LCS units show very low intra-sensor variability and compare well with the regulatory sensor. More results will need to be gathered to make definitive conclusions, but in principle good agreement is observed.

The second stage of deployment occurred approximately one month later, where additionally an autonomous solar battery-powered gateway and total of 16 sensors were deployed in and around the active areas of the mine site. While clear communication, planning and coordination with site stakeholders (eg technical services manager, environmental superintendent, mine services supervisor) are required leading into deployment, once a location was selected, each solar-powered unit, mounted on a star-picket, took approximately 10–15 minutes to install in any active area of the mine. This includes the time needed to travel to a specific location, park the vehicle, safely walk to, and drive a star-picket into the soil/overburden. The gateway is installed first, nominally in an area that is easiest to access, and in line-of-sight with mining activity being monitored. Communication over the radio is required as active mine sites include significant hazards such as moving heavy machinery.

Each LCS unit, housed in a 3D printed enclosure with solar panel attached is then slipped over the top of the star picket. Air quality in the area is measured, and data is then sent to the cloud facilitating access via a browser dashboard in real-time. The real advantage of this set-up is that each unit is already measuring data during the actual deployment and as such, data transmission to the gateway and the dashboard can be tested immediately during the installation. As any active mine site is extremely dynamic in its topography, to locate the exact location of each sensor, either a GPS can be used (trade-off being more power required) or location identified using a map application on the mobile phone (eg Google Maps). The sensor and gateway set-up are shown in Figure 5.

FIG 5 – Gateway and LCS units installed in the active area of a surface mine.

Over two months of field use, a few issues occurred. Namely, two sensors had to be relocated due to mining activity. More importantly, and impacting outcomes, the gateway itself had to be moved twice due to mining activity. The second time, the gateway was moved without clear communication and was placed by mine personnel outside of the line-of-sight, leading to a loss of data. This, as well as spiders crawling inside a few LCS units resulted in data loss. However, a few weeks of data was obtained. The sensors and gateway were removed from site and redeployed again one week later. They are currently on-site and in future it is planned that they will be employed in special tasks such as assessing air quality over the duration of a single shift for specific workers that are outside of cabins on-site (eg shotfirers).

DEVELOPMENT OF LIDAR SYSTEM

Over several years, a LiDAR system has been developed, with the aim of increasing acquisition time and spatial resolution of measurements. The coaxial system utilises 3D printed mountings for both optical and mechanical components. The transmission path consists of 355 nm wavelength laser, with a 25 µJ pulse energy and 1000 Hz a pulse rate (Sutton *et al*, 2022). Energy output and pulse time are used to measure the laser output energy and trigger the time-of-flight which in turn corresponds to a range measurement. The laser beam is expanded 40 times to increase the spot size necessary for eye safety (Holdsworth, 2020). A 355 mm, F4.6 parabolic Newtonian telescope, that sits on a Dobsonian mount is used to receive the signal. Several lenses then reduce the received light into an image size sufficiently smaller than the size of the photomultiplier aperture. The visible light that is captured by the telescope is separated from the signal by using a 355 nm dichroic reflector and 10 nm bandwidth filter (Sutton *et al*, 2022). The backscattered dust plume light is converted into a photocurrent via transmission through a photocathode photomultiplier, where the voltage is measured using a 100 MHZ analogue to digital converter prior to producing a range resolution of 1.5 m. Photomultiplier voltage, laser energy, trigger, and telescope azimuth and elevation movement is controlled using a National Instruments PXI crate, using fast analogue-digital conversion, and store data at ~35 Mb/sec. This system compares well to the commercial units and was assembled using a make-shift budget of under A\$75 000 (Sutton *et al*, 2022).

Field trials

Field trials were initially undertaken in a public area located at Kooragang Island, NSW, in clear view of water vapour plumes of local industry (Sutton *et al*, 2022). The first field trial involved transporting the system out of a laboratory setting, powering by battery and operating in the field for the first time. These trials involved fine-tuning signal acquisition and general performance of the unit, as well as fully developing its portability and eliminating reliance on mains power. Further development included investigating normalisation of the laser pulse energy and algorithms to subtract background noise, as well as range correction to account for scatter. A methodology was developed to graph colour contour plots that can be produced from averaging scans. The system demonstrated 2D planar plots as observed by commercial units, however, with greater resolution and faster acquisition (Sutton *et al*, 2022). These trials showed dust plumes can be measured with an averaged single line scan of less than 1 sec (~700 pulses) and a range of 2.2 km with a resolution of 1.5 m.

Mine site campaigns

The LiDAR has been deployed in several campaigns over a three-year period at a surface mining operation located in the Upper Hunter. Visualisation has included assessing different mining activities such as heavy machinery digging, excavation and loading haul trucks, truck movements, drilling and blasting operations. During deployment a set-up/pack-up and store procedure and a calibration method to align the laser and optics were developed. Data is processed using laser normalisation and correction for background and range.

The field ready LiDAR system requires adequate space for transport to site (eg large, enclosed trailer or hi-top roof van) and is generally located on the periphery overlooking the active mining areas. The system is battery operated, completely autonomous and is typically positioned on a platform, thus enabling manoeuvrability, although the telescope base is fixed during data acquisition. The system positioned on a platform and ready to acquire data on-site is shown in Figure 6.

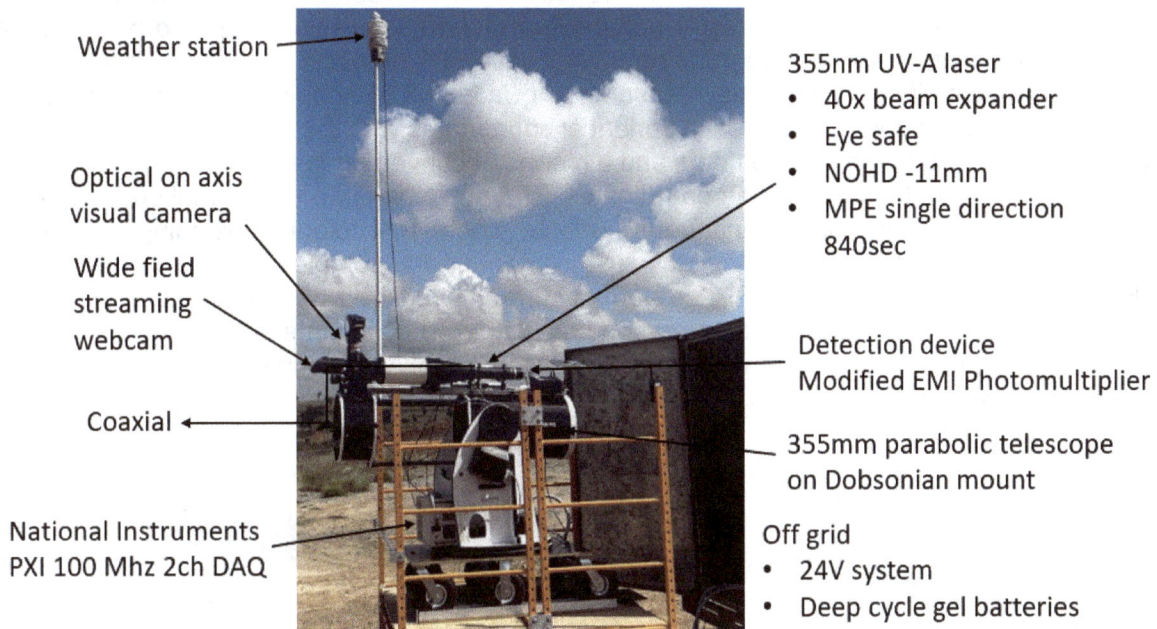

FIG 6 – LiDAR system ready on-site.

Due to telescope grade optics, transporting the LiDAR system from Newcastle and setting-up on mine site located 130 km away in the Upper Hunter can occur by lunchtime on the same day. This does not include carefully dis-assembling individual fragile components and packing up the day before. The system is typically located over-looking the active mining areas, on the bunds, and generally in the zones that only require a visitor induction and driving licence. Due to fast scanning speed, plume visualisation occurs in near real-time, although data processing is necessary and takes some time. Figure 7 illustrates LiDAR data for a volume of the atmosphere at a range of 0.25 km, located at a specific area of a haul road that has just had a cart apply water. The increased concentration shown in yellow in the lower plot is an indication of an increase in dust emissions, demonstrating our ability to assess the duration of water application effectiveness.

immediately after water cart

5-6 minutes after water cart

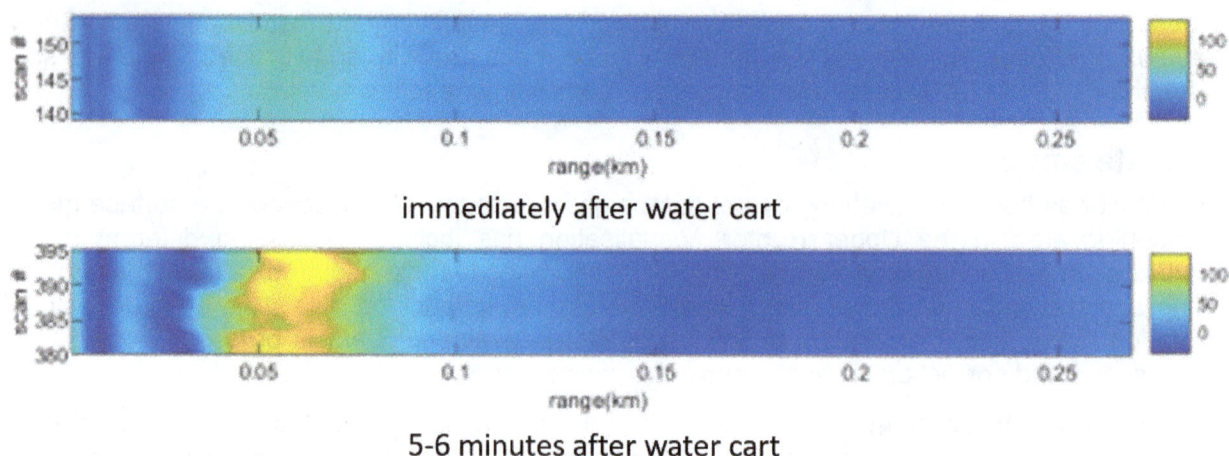

FIG 7 – Illustration of near-real time LiDAR sensing (close range, <0.25 km), yellow indicates increased signal intensity (ie increased concentration of dust particles).

At present, a method is being developed to reconstruct in near real-time a dust plume in 3D. The process involves a pe-programmed raster pattern during data acquisition, (eg a horizontal scan of 10 degrees and vertical for 0.5 degree, repeating several times and then returning to starting position). Combining LiDAR with Inertial Measurement Units (IMUs) enables the collection of precise 3D spatial data. LiDAR data is acquired in spherical coordinates with the additional information from IMUs, providing orientation measurements and a means of converting the LiDAR raster pattern from IMU-spherical coordinates to cartesian coordinates. The outcome is an iso-normal surface plot that represents the LiDAR data as a continuous surface, where the colour or shading indicates the intensity of the backscattered light. By plotting in this manner, the complex 3D structure captured by LiDAR is effectively visualised. The 3D iso-normal surface overlayed on top of a 2D mine planning map is illustrated in Figure 8, with the corresponding 2D plot showing intensity of signal in Figure 9. The visualised plume corresponds well to the location of heavy machinery working in the areas indicated by the red circles in Figure 9.

FIG 8 – Illustration of near-real time LiDAR sensing 3D construction (yellow, red, and green mine planning markings correspond to those shown in the image of Figure 9).

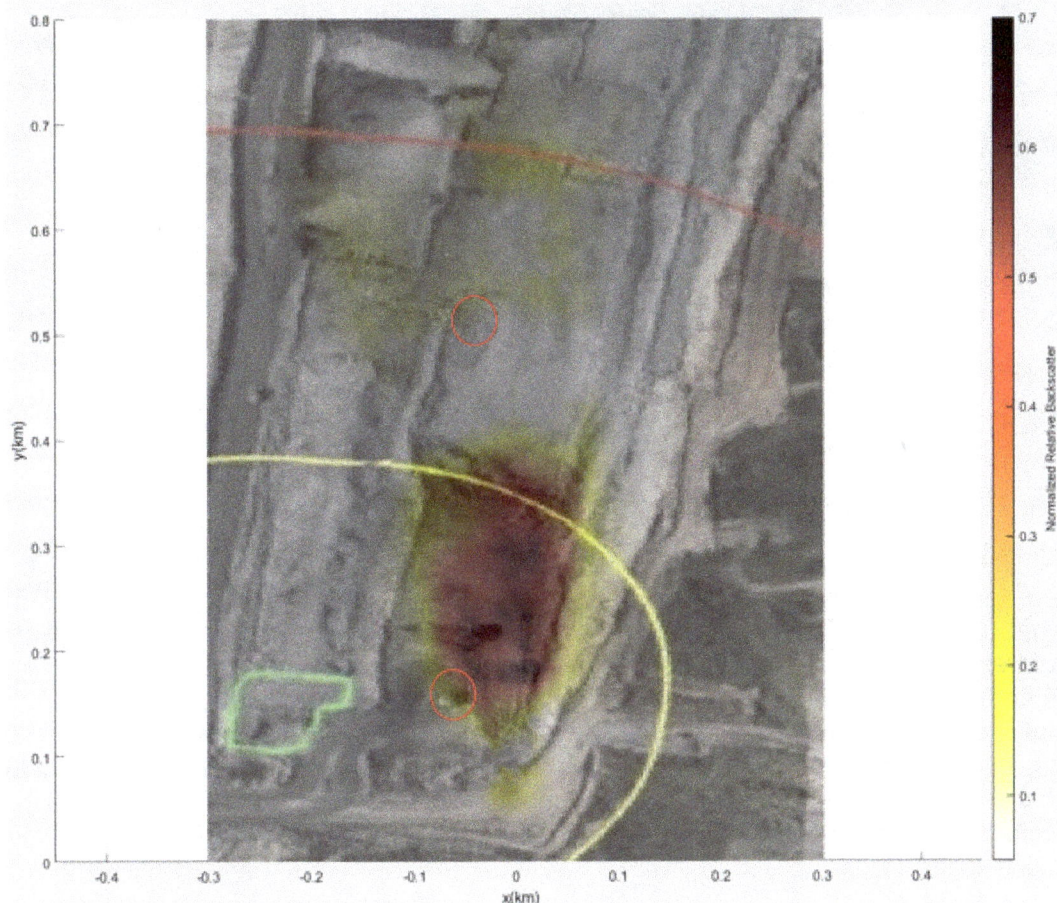

FIG 9 – Illustration of near-real time LiDAR sensing (increased range, plot size 0.6 km × 0.8 km).

CONCLUSIONS

For mining communities, the resources and minerals industry and government, air quality is a contentious topic. While the previous few years have seen a reduction in prominence due to the global pandemic and wet weather, air quality is anticipated to decrease with a return to the dry weather patterns and bushfire season. However, air quality monitoring is also limited to only a handful of locations, typically used for regulatory compliance. This is usually close to mine boundary, measures dust moving across the site and provides only limited understanding of what activities generated that dust, let alone how to best prevent, or mitigate the emissions.

We have demonstrated how expensive regulatory-grade sensors can be supplemented with small, autonomous solar-powered units at a fraction of the cost. LCS portability and easy implementation enables rapid deployment at mass scale, much closer to sources of dust thus enabling a broader capture of the mining activities that contribute to site emissions and affect local air quality. Similarly, LiDAR systems are emerging in addressing air quality in terms of increased spatial resolution; however, commercial units require significant upfront costs and involve qualified data inference. It is possible to increase the spatial resolution of PM measurements across entire sites, therefore covering a much greater field of view, different locations, and activities (ie drill pad area, different pits, haul roads and mobile machinery). Of particular interest, from a worker safety perspective is the exposure of shotfirers to hazardous dust levels due to adjoining mining activity. These workers are perhaps the only staff on-site not protected in cabins. Further interest also includes spatial distribution of dust plumes associated with blasting and the effectiveness of water application.

Figure 10 shows examples of LCS units installed on-site (left), real-time visualisation map (centre) and 3D dust plumes generated using LiDAR (right). In the latter, we have integrated a 2D map used in weekly mine planning. As part of future work, we intend to combine LiDAR dust plume data into the 3D weekly mine planning open cut terrain.

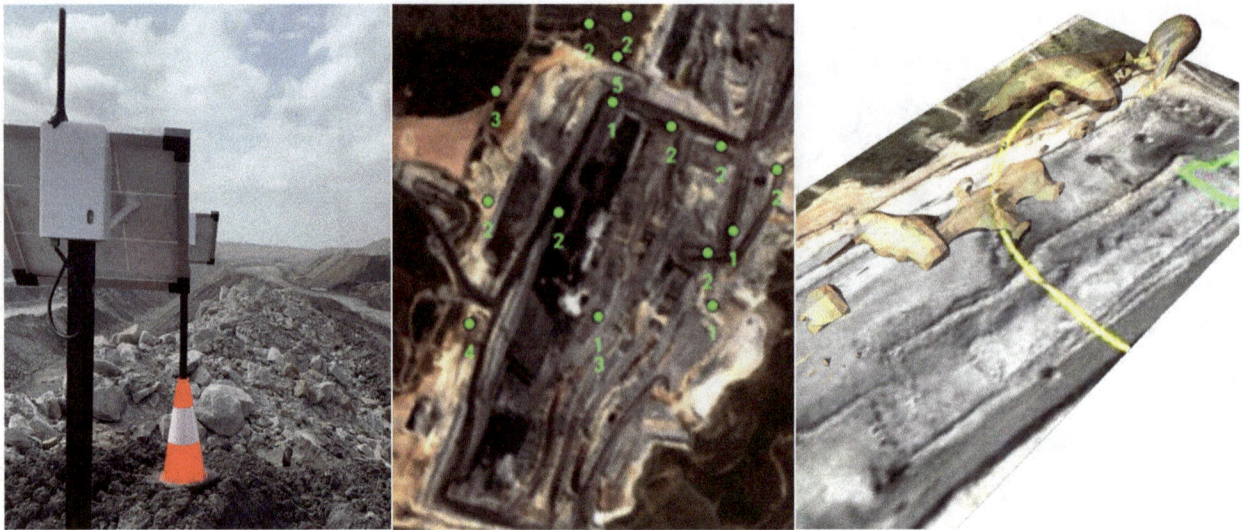

FIG 10 – Overview/illustration of system capability.

The use of LCS and LiDAR in the demonstrated method of managing dust emissions due to mining activity addresses the present lack of spatial resolution of dust measurement. Despite our work thus far concentrating on the coal industry, the unit components, general workflow, and mode of operation are unlikely to change for both the LCS units and the LiDAR when applied to iron ore or other resources. The approach can be further enhanced by implementing an even larger array of LCS that is complemented by a high-range and resolution 3D LiDAR. The sensors can be calibrated to regulatory boundary monitors, and then used to calibrate the LiDAR. Our innovative system and method can:

- improve understanding of dust sources, necessary controls to mitigate emissions and environmental impacts

- result in accurate identification of hazardous dust levels and increased coverage of mining activities at the mine site

- improve databased and targeted water application.

Use of water on-site can further be optimised by understanding the DEM and moisture required to suppress dust for the range of different materials handled. Automated, data-based decision-making has the capability of enhancing the effectiveness of the TARP, thus improving operational decisions by enabling accurate prioritisation of site activities, selection of risk-reduction and mitigating measures. The approach can also assist in eliminating disruption to equipment and operations by minimising regulatory exceedance. The outcome is an improved dust minimisation approach that can replace a regulatory shutdown requirement and lead to operational best practice in dust management for the mining and resources industry.

ACKNOWLEDGEMENTS

The authors acknowledge VGT, the Australian Government (AusIndustry Innovation Connections), Coal Services, MACH Energy and the University of Newcastle for financial assistance and resources in developing this work.

REFERENCES

AGL, 2015. AGL Greenhouse Gas Policy, Available from: <https://www.agl.com.au/>, April 2015. (Accessed: 25/07/2023).

Aneja, V P, Isherwood, A and Morgan, P, 2012. Characterisation of Particulate Matter (PM_{10}) related to surface coal mining operations in Appalachia, *Atmospheric Environment*, 54:496–501.

Campbell, R, 2014. Seeing through the dust: Coal in the Hunter Valley economy, *The Australia Institute*, Policy Brief No. 62.

Climate and Health Alliance, 2015. Coal and health in the Hunter: Lessons from one valley for the world, Summary for Policymakers, *Environmental Justice Australia*. Available from: <https://envirojustice.org.au/blog/publications/coal-and-health-in-the-hunter-lessons-from-one-valley-for-the-world/> (Accessed: 25/07/2023).

Coal Services, 2018. May 2018 – NSW coal industry statistics [online], *Industry newsletters*. Available from: <https://www.coalservices.com.au/mining/news-and-events/industry-newsletters/may-2018-nsw-coal-industry-statistics/> (Accessed: 25/05/2023).

Connick, F, 2023. Anxiety stirs in NSW community as Cadia mine investigated over heavy metals in blood tests, *The Guardian, Rural Network*. Available from: <https://www.theguardian.com/australia-news/2023/may/25/anxiety-stirs-in-nsw-community-as-cadia-mine-investigated-over-heavy-metals-in-blood-tests> (Accessed: 25/05/2023).

Cottle, D and Keys, A, 2014. Open-cut coal mining in Australia's Hunter valley: Sustainability and the Industry's economic, ecological and social implications, *International Journal of Rural Law and Policy*, University of Technology Sydney ePress, 1:1–7.

CRC Care, 2014. Advanced Lidar Port Hedland dust study: Broadscale, real-time dust tracking and measurement, *CRC CARE Technical Report* #33, CRC for Contamination Assessment and Remediation of the Environment, Newcastle, Australia.

Gautam, S, Prusty, B K and Patra, A K, 2015. Dispersion of Respirable Particles from the Workplace in Opencast Iron Ore Mines, *Environment Technology and Innovation*, 4:137–149.

Goetze, E, 2019, 25 Oct. 'Our pool is black': Residents in NSW's Upper Hunter vent air-pollution fears, ABC News, Upper Hunter Region. Available from: <https://www.abc.net.au/news/2019–10–25/air-quality-stokes-community-pollution-fears-in-nsw-upper-hunter/11638418> (Accessed: 25/05/2023).

Government of Western Australia, 2018a. Mapping Dust Plumes at Point Samson – Cape Lambert using a LiDAR, Department of Water and Environmental Regulation, January 2018.

Government of Western Australia, 2018b. Mapping Dust Plumes at Port Hedland using a LiDAR, Department of Water and Environmental Regulation, February 2018.

Government of Western Australia, 2021a. Guideline: Dust emissions, Department of Water and Environmental Regulation, July 2021.

Government of Western Australia, 2021b. Application for Licence Amendment, L4513/1969/18, Department of Water and Environmental Regulation, Amendment Report. September 2021.

Hashmonay, R, Roddis, D, Lewis, M and Eastwood, J, 2020. Open Path Boundary Monitoring for Operational Dust Control, Australian Coal Research Programme (ACARP), Report C27061.

Hibberd, M F, Selleck, P W, Keywood, M D, Cohen, D D, Stelcer, E and Atanacio, A J, 2013. Upper Hunter Particle Characterisation Study, CSIRO, Australia.

Hill, C, 2018. Coherent Focused Lidars for Doppler Sensing of Aerosols and Wind, *Remote Sensing MDPI,* pp 1–24.

Holdsworth, J, 2020. LiDAR Safety calculations, Safety review form for UON activities, University of Newcastle, Australia, 59.2019.

Hunter Joint Organisation of Councils, 2018. Upper Hunter Diversification Action Plan: Implementation Priorities, July 2018.

Ilic, D and Lavrinec, A, 2022. Evaluation and development of low-cost sensor units for continuous, real-time PM_{10} and $PM_{2.5}$ monitoring, in CASANZ22, 26th International Clean Air and Environment Conference, Adelaide, South Australia.

Karagulian, F, Barbiere, M, Kotsev, A, Spinelle, L, Gerboles, M, Lagler, F, Redon, N, Crunaire, S and Borowiak, A, 2019. Review of the performance of low-cost sensors for air quality monitoring, *Atmosphere*, 10:506.

Kosmopoulos, G, Salamalikis, V, Pandis, S N, Yannopoulos, P, Bloutsos, A A and Kazantzidis, A, 2020. Low-cost sensors for measuring airborne particulate matter: Field evaluation and calibration at a South-Eastern European site, *Science of the Total Environment*, 748:141396.

McCarthy, J, 2018, 4 July. The Independent Planning Commission is in Muswellbrook today considering a Mount Pleasant coal mine application, *Newcastle Herald*. Available from: <https://www.newcastleherald.com.au/story/5505849/muswellbrook-coal-mine-hearing-considers-serious-air-quality-concerns/> (Accessed: 25/05/2023).

Mills, C, 2019. Strategic Review of the Resources for Regions Program – Inputs Summary Report, University of Technology Sydney, October 2019.

Musa, A and Paul, B S, 2017. Prediction of electromagnetic wave attenuation in dust storms using Mie scattering, 2017 IEEE AFRICON, South Africa, pp 603–608.

NSW Department of Planning and Environment (NSW DPIE), 2023. Upper Hunter map. Available from: <https://www.dpie.nsw.gov.au/air-quality/air-quality-maps/live-air-quality-data-upper-hunter> (Accessed: 25/07/2023).

NSW Government, 2011. Coal Mining Benchmarking Study: International Best Practice Measurers to Prevent and/or Minimise Emissions of Particulate Matter from Coal Mining, Office of the Environment and Heritage, Katestone Environmental Pty Ltd.

NSW Government, 2014. Pollution Reduction Programs, Operating Procedure, *NSW Environmental Protection Authority*.

NSW Government, 2017. Better evidence, stronger networks, healthy communities, Five year review of the Upper Hunter Air quality Monitoring Network, *Office of Environment and Heritage*.

NSW Government, 2019. Air quality monitoring network, Upper Hunter, Spring 2019.

NSW Government, 2020. Regional Growth Fund, Resources for Regions. Available from: <https://www.nsw.gov.au/improving-nsw/regional-nsw/regional-growth-fund/resources-for-regions/> (Accessed: 25/05/2023).

NSW Minerals Council, 2023a. Air Quality in the Upper Hunter, Actions to improve air quality and manage impacts, Upper Hunter Mining Dialogue. Available from: <www.nswmining.com.au/mining-factsheets> (Accessed: 25/05/2023).

NSW Minerals Council, 2023b. Environmental Management. Available from: <https://www.nswmining.com.au/environmental-management> (Accessed: 25/05/2023).

Sahu, S P, Patra, A K and Kolluru, S S R, 2018. Spatial and temporal variation of respirable particles around a surface coal mine in India, *Atmospheric Pollution Research*, 9:662–679.

Sayahi, T, Butterfield, A and Kelly, K E, 2019. Long-term field evaluation of the Plantower PMS low-cost particulate matter sensors, *Environmental Pollution*, 245:932–940.

Standards Australia, 2013. AS4156.6 – Coal Preparation – Part 6: Determination of dust/moisture relationship for coal.

Sutton, L, Ilic, D, Holdsworth, J, Williams, K and Wheeler, C, 2022. Development of a 3D, fast-scanning LiDAR system for particulate matter visualisation, in CASANZ22, 26th International Clean Air and Environment Conference, Adelaide, South Australia.

US EPA, 2019. List of designated reference and equivalent methods, United States Environmental Protection Agency, December 15, 2019, pp 1–73.

Wilkie, D, 2019, 9 August. Industry to wear costs of Port Hedland buy backs, *Business News*. Available from: <https://www.businessnews.com.au/article/Industry-to-wear-costs-of-Port-Hedland-buy-backs> (Accessed: 25/05/2023).

Williams, R, Kaufman, A, Hanley, T, Rice, J and Garvey, S, 2014b. Evaluation of field-deployed low-cost PM sensors, *US Environmental Protection Agency*, EPA/600/R-14/464, December 2014, pp 1–76.

Williams, R, Kilaru, V, Snyder, E, Kaufman, A, Dye, T, Rutter, A, Russell, A and Hafner, H, 2014a. Air sensor guidebook, *US Environmental Protection Agency*, EPA 600/R-14/159, June 2014, pp 1–73.

Williams, R, Nash, D, Hagler, G, Benedict, K, MacGregor, I C, Seay, B A, Battelle, M L and Dye, T, 2018. Peer review and supporting literature review of air sensor technology performance targets, US Environmental Protection Agency, EPA/600/R-18/324, September 2018, pp 1–83.

BHP Port Hedland wind fence – innovations in dust control

L van Wyk[1] and R Peiffer[2]

1. Principal Studies, BHP, Perth WA 6000. Email: lizelle.vanwyk@bhp.com
2. Senior Air Quality Scientist, BHP, Perth WA 6000. Email: ruth.peiffer1@bhp.com

ABSTRACT

The Port of Port Hedland is the largest bulk export terminal in the world and is one of Australia's most important pieces of economic infrastructure. BHP is the largest Port user to support the export of iron ore products from mines located within the Pilbara region of Western Australia (WA). BHP's priority is the safety and well-being of its workforce and the communities where it operates. In a recent regulatory application to increase export capacity, BHP committed to the deployment of wind fences to reduce wind erosion from stockpiled ore.

Wind erosion from stockpiles was identified through inverse modelling as the most significant dust source compared to other sources such as stackers, reclaimers, shiploaders and transfer stations. The primary cause of dust emissions from stockpiles is dust lift-off during high wind conditions. BHP proposed the installation of wind fences across its Port Hedland operations, targeted at reducing emissions from wind dependent sources, primarily stockpiles, by reducing the wind speed downwind of the wind fence. Modelling predicts that installation of the wind fence will significantly reduce emissions from stockpiles, particularly at BHP's western operations (Finucane Island).

To be effective in shielding stockyards and reducing wind speeds, the wind fences are located up-wind and as close as practical to the open stockyards. BHP's wind fence design incorporates learnings from other existing installations and has been tailored for Port Hedland's cyclonic conditions. The wind fence will be higher than the stockpile height, with a mesh porosity of approximately 50 per cent. The mesh porosity allows some of the wind to pass through the mesh. This reduces the wind turbulence downstream of the wind fence, as opposed to a solid wind fence, which increases downwind turbulence, resulting in increased dust.

INTRODUCTION

The Port of Port Hedland is one of the largest bulk export terminals in the world, making this one of Australia's most important economic infrastructure assets. BHP is the largest Port user to support the export of iron ore products from mines located within the Pilbara region of WA. BHP operates two iron ore export facilities at Port Hedland: Nelson Point (NP), south of Port Hedland and Finucane Island (FI), west of town (Figure 1).

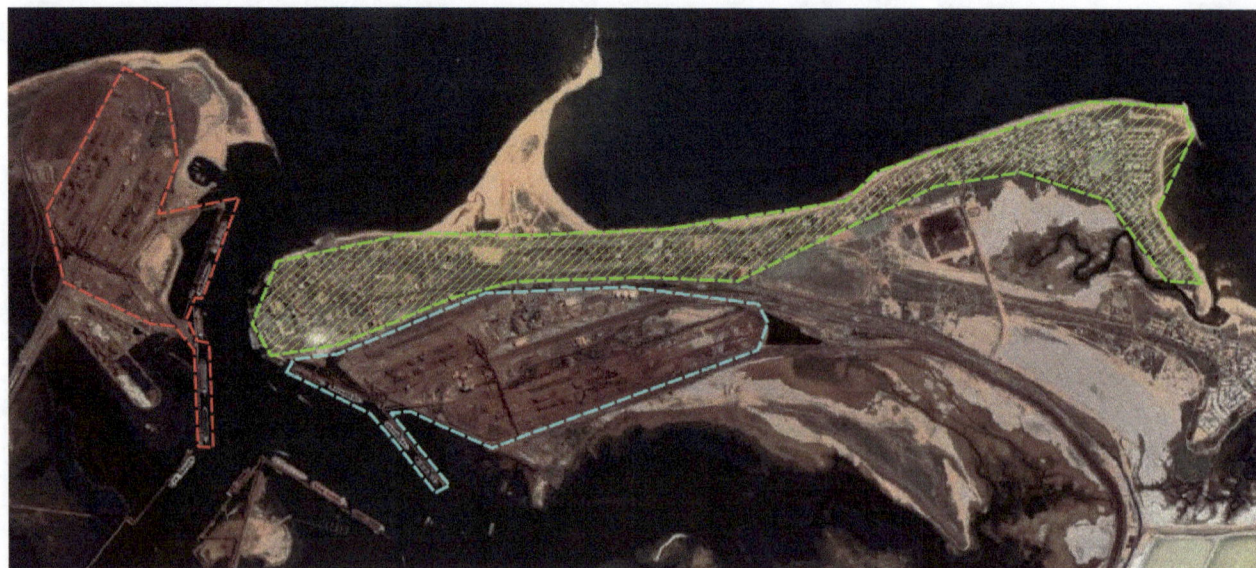

FIG 1 – Residential Port Hedland area (green outline), BHP FI site (red outline) and BHP NP site (blue outline).

The iron ore is unloaded, stockpiled and subsequently loaded onto ships via a network of rail car dumpers, conveyor belts, stackers, reclaimers and shiploaders. These operations produce fugitive dust emissions, which BHP seeks to minimise through emission control strategies and monitor with a large network of particulate (PM10) sensors across the two sites.

To better understand the contribution of specific groups of equipment or processes to the PM10 concentrations measured across the BHP monitoring network and to quantify the emissions from different equipment at the operations, BHP adopted a novel 'inverse modelling' approach in combination with the use of probabilistic programming. Inverse modelling is the process of using air quality measurements to derive the emissions that caused them. The modelling is called 'inverse' as it reverses the usual order of air quality modelling where prior emission estimates are input to an air quality model to estimate pollutant concentrations. The results of this study prompted BHP's commitment to install wind fences at the Port Hedland operations to reduce windblown dust emissions.

METHODOLOGY

Inverse modelling

The Source Attribution and Magnitude Model (SAMM), a statistical model developed by Ramboll (2020), was utilised for this study. Detailed operational information about the different emission sources at BHP's Port Hedland operations (material movement data) was collected, combined and analysed, along with 10-minute average ambient PM10 data recorded over a 5-year period across BHP's monitoring network. The measured background concentration was subtracted from the monitoring data. Particulate transport was then modelled with two common dispersion models AERMOD and CALPUFF, using unit emission rates to generate dispersion coefficients for use in the inverse modelling.

The inverse model was formulated to solve for the set of emission rates that best satisfied the relationship between the matrices of measured PM10 concentrations, forward modelling dispersion coefficients incorporating the material movement data, as well as the vector of emission rates by source type. Hourly emission rates produced from the inverse model were input into AERMOD and CALPUFF to confirm the source attribution modelling results.

RESULTS

Inverse model performance

Evaluation of model performance involved a two-step process; the inverse model was run using a sub-set of all the hours in the 5-year data set (the 'fitting' data set) and secondly the resulting emission factors are applied to all hours of the 5-year data set to estimate concentrations at each measurement location. The measured concentrations were then filtered for wind direction and for a valid background concentration and compared to the modelled concentrations to assess overall model performance.

Both the AERMOD and CALPUFF SAMM configurations predict long-term average PM10 concentrations most accurately at receptors located on Nelson Point and in Port Hedland (right of yellow dotted line, as illustrated in Figure 2). These receptors likely see contributions from all included source types, indicating both the AERMOD-SAMM and CALPUFF-SAMM configurations predict source contributions within Port Hedland relatively well.

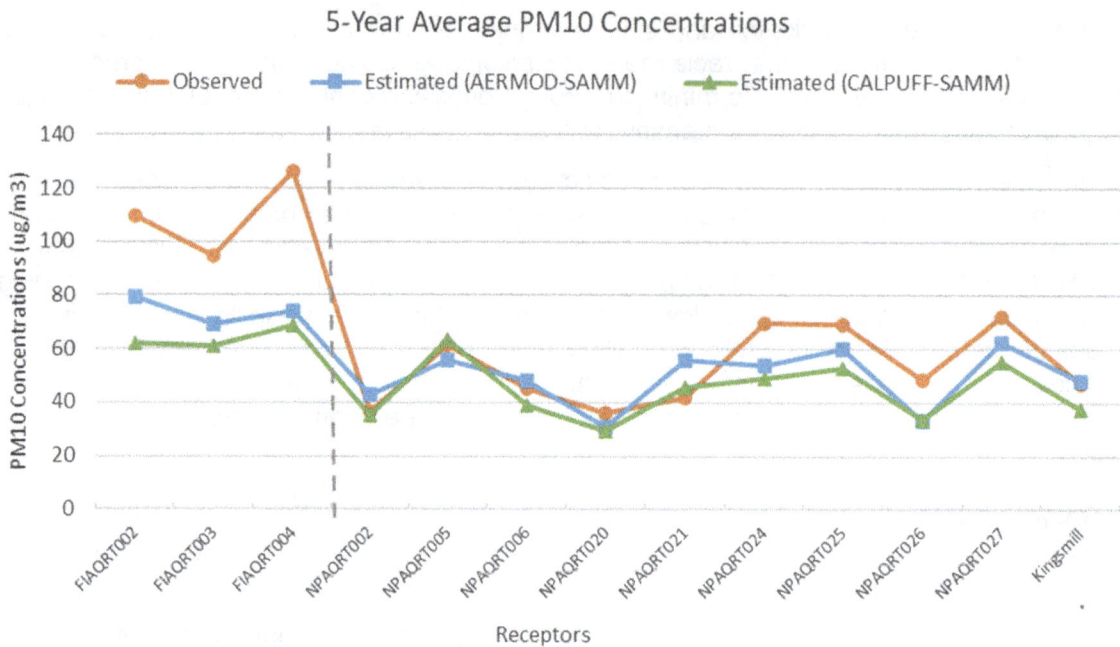

5-Year Average PM10 Concentrations

FIG 2 – Comparison of observed (orange), AERMOD-SAMM predicted (blue) and CALPUFF-SAMM predicted (green) 5-year average PM10 concentrations.

Both approaches underpredict concentrations at receptors on Finucane Island (left of grey dotted line, as illustrated in Figure 2), suggesting an emission-source near these receptors is not included in the model. Comparison of the estimated and observed concentrations on an annual and hourly basis indicates similar patterns across all receptors.

Source identification and dust control

The emission factors derived by the inverse model configurations were compared against conventional emission factor-based estimates for BHP's Port Hedland operations. The results indicated wind erosion from stockpiles at BHP's operations was the most significant potential dust source compared to other sources such as stackers, reclaimers, shiploaders and transfer stations. The primary cause of dust emissions from stockpiles is dust lift off during high wind conditions. BHP proposed the installation of wind fences across its Port Hedland operations, targeted at reducing emissions from wind dependent sources, primarily stockpiles, by reducing the wind speed downwind of the wind fence.

Wind fences are installed to control stockpile dust emissions in other parts of the world, including iron ore ports in Brazil and importing ports in Asia. BHP is developing the wind fences to withstand extreme cyclonic (Category D) weather events experienced in Port Hedland. To be effective in shielding stockyards and reducing wind speeds, the wind fences are located up-wind and as close as practical to the open stockyards. BHP's wind fence design incorporates learnings from existing installations and has been tailored for Port Hedland's cyclonic conditions. The wind fence will be higher than the stockpile height, with a mesh porosity of approximately 50 per cent. The mesh porosity allows some of the wind to pass through the mesh. This reduces the wind turbulence downstream of the wind fence, as opposed to a solid wind fence, which increases downwind turbulence, resulting in increased dust.

Estimated emission reductions

Modelling predicts that installation of the wind fence will significantly reduce emissions from stockpiles, particularly at BHP's western operations (Finucane Island). Installation of the wind fences was proposed as part of the BHP Port Licence Amendment (BHP, 2020) to increase throughput capacity from 290 million tonnes per annum (Mtpa) to 330 Mtpa. Two additional stockyard rows are included in the 330 Mtpa Licence Amendment.

The wind fences are designed to target the reduction in PM10 emissions that could potentially be directed towards the West End of Port Hedland. Figure 3 shows the estimated average PM10 emission rates for stockyards for the 290 Mtpa and 330 Mtpa scenarios, showing percentage change in average emission rate for all wind directions (ie not only emissions that could potentially be directed towards the West End of Port Hedland). Significant reduction in average estimated emissions is demonstrated for stockpiles, especially considering two additional stockpiles rows will be added at Nelson Point (NP).

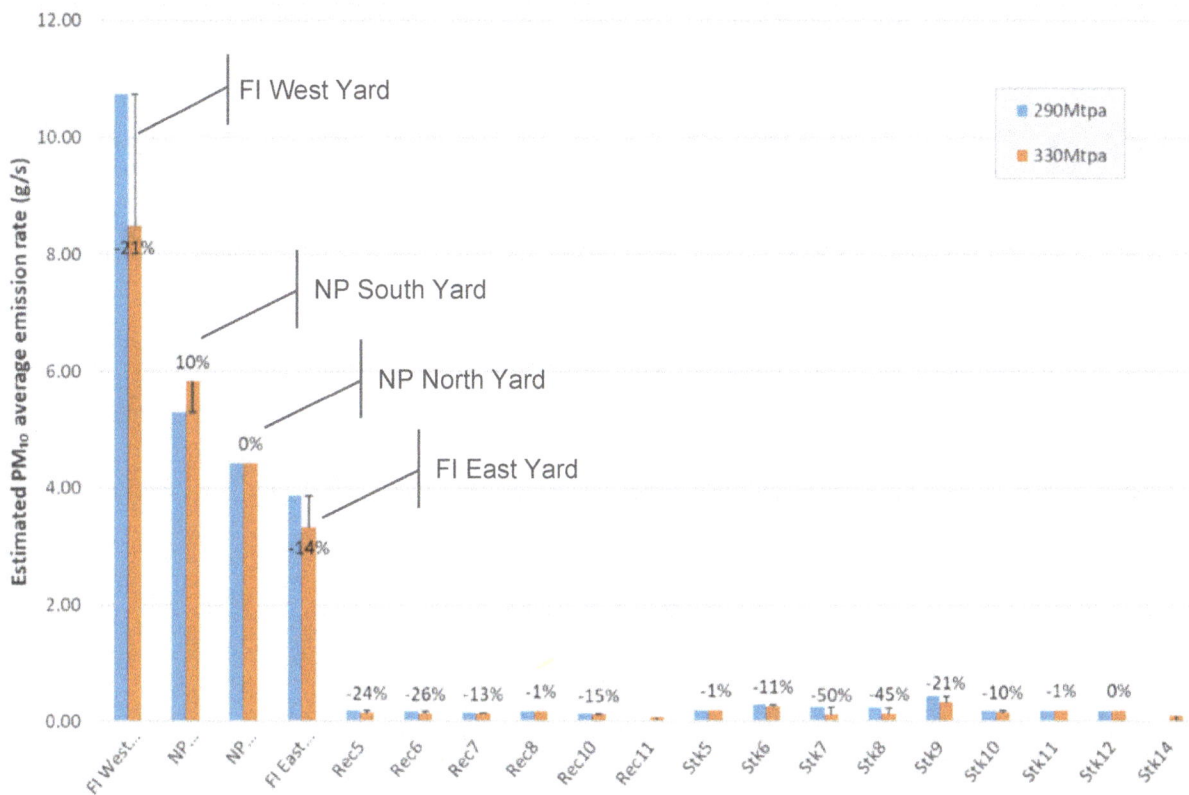

FIG 3 – Estimated average PM10 emission rates with dust abatement for stockyards, stackers and reclaimers for 290 Mtpa and 330 Mtpa, showing percentage change in average emission rate.

CONCLUSIONS

Inverse modelling was used to develop a fugitive dust emission inventory for BHP's Port Hedland operations. The strength of this approach is that it utilises ambient PM10 monitoring data recorded at the BHP site, with the magnitude of uncertainty within the emissions inventory being constrained by the site-specific ambient concentration data. The inverse modelling study identified wind erosion from stockpiles as the most significant dust source compared to other sources such as stackers, reclaimers, shiploaders and transfer stations. The primary cause of dust emissions from stockpiles is dust lift-off during high wind conditions. BHP proposed the installation of wind fences across its Port Hedland operations, targeted at reducing emissions from wind dependent sources, primarily stockpiles, by reducing the wind speed downwind of the wind fence. Modelling predicts that installation of the wind fence will significantly reduce emissions from stockpiles, particularly at BHP's western operations (Finucane Island) (GHD, 2020).

ACKNOWLEDGEMENTS

BHP would like to acknowledge Ramboll US Consulting, Ramboll Australia and GHD Pty Ltd for their input to the supporting analyses for the Port Hedland wind fence project.

REFERENCES

BHP, 2020. Environmental Protection Act 1986 Licence (L4513/1969/18) Amendment – BHP Port Operations 330 Mtpa Application.

GHD, 2020. Port Hedland Port Operations – 330 Mtpa Licence Application Air Quality Assessment.

Ramboll US Consulting and Ramboll Australia (Ramboll), 2020. BHP Port Hedland Windblown Dust Calculations. Report prepared for BHP, February 2020.

Health, safety, environment and community – health and safety practices

Advancements, innovations and initiatives for optimising the treatment of haul roads

H Haghighi[1]

1. Quality Team Leader/R&D Sector Lead, Downer EDI Works, Somerton Vic 3062. Email: hamed.haghighi@downergroup.com

ABSTRACT

Haul roads are often permanent unsealed access roads used by heavy haul trucks to transfer minerals. These roads can deteriorate rapidly which results in potholes, deformations and dust. Downer EDI Works has successfully provided efficient solutions to the mining industry to maintain their haul roads. HaulPac as a solution package is a process in which a bituminous emulsion is used to create a stable and durable pavement. The emulsion penetrates the road and binds the particles together, creating a stable base. It also seals the surface and prevents water from penetrating the underneath layers. Bituminous emulsion can increase the strength of the pavement material and decrease the road surface degradation which results in dust suppression. However, the ever-increasing price of commodities, the necessity of reducing carbon emissions and the increasing use of autonomous haul trucks have made it vital to invest in novel maintenance solutions that can meet the needs of the mining industry in a cost-effective manner. This paper outlines the benefits of bituminous emulsion in terms of productivity, safety and environment followed by Downer's approach to evaluate the suitability and to optimise the amount of product. The potential for changing the colour of the treated surface is explored and an initiative that incorporates end-of-life tyres of haul trucks into the pavement is introduced. And finally, an innovative method using artificial intelligence to estimate dust emission is presented.

INTRODUCTION

Haul roads, generally unsealed, play a crucial role in the transportation of goods, materials and equipment within mining operations. These specially designed roads are built to withstand heavy loads and provide efficient access to various areas within a mine. Haul roads serve as vital arteries, enabling the smooth flow of resources and contributing to the overall productivity and profitability of the mining operations.

Dust emission from haul roads in mines is a significant environmental concern and can have various impacts on the surrounding ecosystem and human health. The movement of vehicles, particularly heavy trucks, on unsealed or poorly maintained haul roads can generate large amounts of dust particles. These particles, known as fugitive dust, can become airborne and spread over considerable distances, affecting both the immediate vicinity of the mine and nearby communities.

The primary sources of dust on haul roads are the fine particles present in the road surface materials, such as gravel or crushed rock, and the materials being transported, such as ores or minerals. Factors such as road conditions, vehicle speed, traffic volume, and weather conditions can influence the amount of dust generated.

Excessive dust emission can have several adverse effects. Firstly, it can impair air quality, leading to respiratory problems and other health issues for workers. Inhalation of fine dust particles can cause respiratory irritation, coughing and even more severe conditions such as chronic bronchitis or silicosis. Dust can also settle on vegetation, impacting plant health and productivity.

To mitigate dust emissions from haul roads, several measures can be implemented. These include watering, surface stabilisation, speed control, road maintenance and dust control technologies. Sealing or bound paving is proven to have the maximum efficiency. The products available to control dust by bonding and sealing the pavement include surfactants (wetting agents), salts (water attracting chemicals), adhesives (such as lignin sulfonates), petroleum-based binders (waste oil, bituminous emulsions and tars), Electro-chemicals (enzymes), Synthetic polymers (PVS, PVA PAM) and microbiological binders (ARRB, 2020).

HaulPac is a dust control technology which uses an anionic bituminous emulsion at three stages of mixing with the road materials, capping the surface of the treated road followed by regular surface maintenance spraying. Each stage requires certain dosage of HaulPac to deliver the target residual binder which is determined by assessing the road material for performance and compatibility in laboratory.

BENEFITS

The benefits of bituminous treatment of haul roads are outlined for productivity, safety, environment and performance.

Productivity

Keeping haul roads operational particularly after rain events is crucial. Wet running can be greatly improved with using bituminous emulsion. It is proven that the addition of bitumen reduces moisture susceptibility which can result in less permanent deformation under traffic and better integrity of the road with less defects.

Rolling resistance which is defined as the force required to maintain a vehicle at a steady speed (Thompson and Visser, 1999) can be improved because the smooth-running surface reduces the tyre friction. It can also contribute to increased tyre life and improved fuel efficiency.

The less frequent need for regular maintenance results in reduction of watercarts and graders which helps with equipment and workforce savings.

Safety

As of the main benefit of HaulPac treatment, it greatly reduces the surface dust which results in better visibility and less respiratory problems.

The smoother ride conditions of the treated haul road improve driver comfort and reduce fatigue. As a result of capped surface, the braking distance in both dry and wet conditions can be reduced which lowers the likelihood of collision incidents.

Environment

The treated haul road reduces the need to water and can benefit water saving significantly. Generally, 2 L of water per sqm per hour is used to supress the dust on haul roads (Greenbase Pty Ltd, 2021). HaulPac as a water-based bituminous emulsion can reduce the addition of water up to 100 per cent depending on the requirements.

Less watering results in less equipment which lowers CO_2 emissions. A generic diesel engine can consume up to 263 L of fuel per hour (Caterpillar, 2017). According to National Greenhouse and Energy Reporting website (https://www.cleanenergyregulator.gov.au/NGER/Forms-and-resources/Calculators), 1000 L of diesel oil can produce 3 t of CO_2 equivalence (CO_2-e). Therefore, every one hour of less fuel consumption per equipment can save up to 0.79 t of CO_2-e.

Less dust results in better growth of native flora allowing an improved and more stable habitat for local fauna. The use of HaulPac is considered to have a low risk profile and sensitive receptors can be protected with the implementation of controls in line with published guidance commonly applied in the industry (WSP Australia, 2020).

Performance

A recent study at Monash University has investigated the permanent deformation (rutting potential) of HaulPac-treated crushed rock by using the extra-large wheel tracker at Australian Road Research Board (ARRB). It was found that the increase of bituminous emulsion results in reduction of deformation and increase of rut resistance. The outcomes of this research will be published in early 2024.

SUITABILITY EVALUATION

It is recommended that suitability of the treatment is assessed in the laboratory by looking into the performance and compatibility of the pavement materials. The process should include determination of the main engineering properties of the road materials such as particle size distribution, dry density/moisture relationship, Atterberg limits and pH. A range of HaulPac percentages are added to the road material to achieve a residual binder up to 3.0 per cent (typically). The treated material is compacted by Marshall hammer and cured in a drying oven at 40±5°C for 72±2 hours to replicate the medium-term condition of pavement after construction (Leek and Jameson, 2011). At completion of curing, the samples are tested for strength in terms of resilient modulus. The samples are also soaked in water bath under vacuum until fully saturated. Upon completion of saturation, the samples are examined for material loss (fine fraction) which is considered as an indication of efficacy of the treatment: the less the material loss the more efficient is the treatment. Figure 1 shows a typical example of relationship between material loss and residual binder content. It also includes photos of an untreated sample, an untreated soaked sample and a soaked sample when treated with 3 per cent residual binder HaulPac.

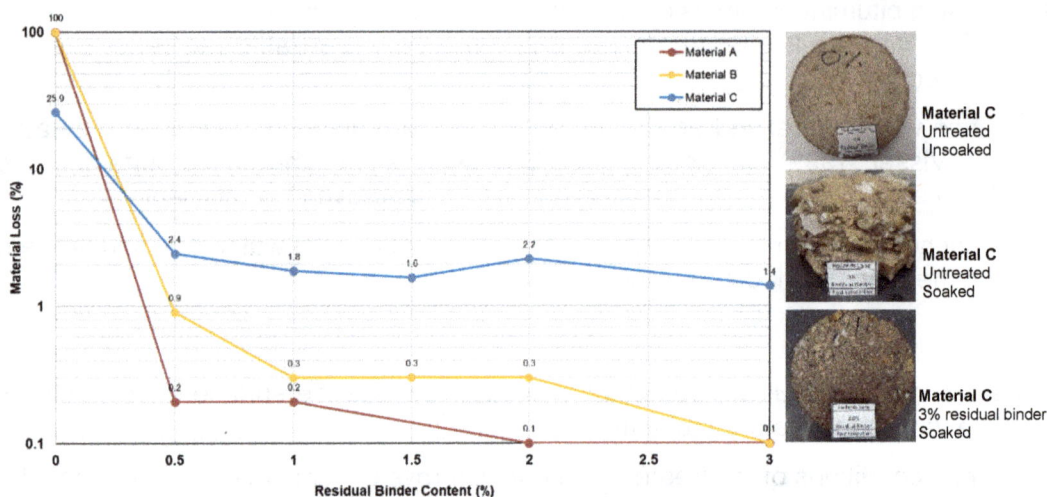

FIG 1 – Typical relationship between material loss and residual binder content.

The efficiency of capping layer can be assessed by applying HaulPac at various dilution ratios to the surface of the compacted samples at a typical rate of 0.3 L/m². The capped surface is examined for consistency, set time, run-off, appearance and tackiness. Figure 2 shows examples of different pavement materials as capped with various diluted HaulPac (not capped, 25 per cent H: 25 per cent HaulPac + 75 per cent water; 50 per cent H: 50 per cent HaulPac + 50 per cent water; and 100 per cent H: 100 per cent HaulPac). The importance of the suitability evaluation can be seen on the two sets of the samples as different dilution ratios can react differently on various types and conditions of the haul road materials. For example, in the second set, not diluting HaulPac has resulted in partial coverage of the surface due to the dry condition of the surface and consequently premature breaking of the product. Figure 2 also includes an example of mine haul road as adequately capped with HaulPac.

FIG 2 – Examples of capping layer assessment (left), and HaulPac-capped mine haul road.

AUTONOMOUS HAUL TRUCKS AND SURFACE REFLECTION

Autonomous mine haul trucks are equipped with advanced sensor and perception systems that allow them to navigate and operate safely in various conditions, including different road surface reflections. These systems enable the trucks to detect and react to changes in the road environment, including reflections, to ensure optimal performance and safety.

They use a combination of sensors and cameras to perceive their surroundings. These sensors can detect changes in the road surface, including reflective surfaces, by measuring the properties of the reflected light or radio waves.

Depending on the characteristics of the road surface reflection and the truck's programming, the autonomous truck may adjust its speed, trajectory, or behaviour to maintain safe and efficient operations. For example, if a reflection is detected that could potentially interfere with the truck's sensors or cause confusion, the truck may slow down, change lanes, or take alternative routes.

Different autonomous truck systems may have varying approaches to handling road surface reflections, as they can be influenced by factors such as the specific sensors used, the algorithms employed, and the level of programming and customisation implemented by the mining company or manufacturer.

As a result of collaboration with mine operations, the effect of various matting agents and colour pigments on HaulPac as sprayed on the surface has been investigated. The laboratory assessment could confirm a glossiness reduction of Glossiness Units (GU) up to 80.0 per cent. However, field trials are recommended to measure the efficiency.

HAUL ROADS AND RECYCLED RUBBER

The management of end-of-life tyres at mines can vary depending on the type of operation, location and applicable regulations. One of the efficient practices is tyre recycling and repurposing. By establishing partnerships with specialised recycling companies, the separated rubber component can be repurposed in construction of haul roads.

A feasibility study was conducted at Downer's National Research and Development Laboratory which indicated that the performance of HaulPac-treated road material is not compromised when crumb rubber is introduced. The positive outcome of initial assessment can make it possible to introduce crumb rubber as a component of HaulPac formulation or as a separate component as mixed directly with HaulPac-treated material.

Diverting the end-of-life tyres from landfill and repurposing them onto the roads, has a huge environmental benefit which cannot be ignored if the engineering properties are assessed in laboratory and tested under real conditions in the field.

DUST QUANTIFICATION USING ARTIFICIAL INTELLIGENCE

A recent study at Monash University (De Silva *et al*, 2022, 2023) has used machine learning models to quantify dust emissions from the images taken from vehicle-induced dust clouds. This new approach was developed as part of an advanced assessment of HaulPac and its impact on unsealed roads which can result in reduction of deformation and increase of rut resistance.

A new vision data set with the goal of advancing semantic segmentation was generated to identify and quantify vehicle-induced dust clouds from images. Field experiments were conducted on ten unsealed road segments with different types of road surface materials in varying climatic conditions to capture vehicle-induced road dust. A digital single-lens reflex (DSLR) camera was used to capture the dust clouds generated due to a utility vehicle travelling at different speeds. A research-grade dust monitor was used to measure the dust emissions due to traffic. A total of ~210 000 images were photographed and refined to obtain ~7000 images. These images were manually annotated to generate masks for dust segmentation. The baseline performance of a truncated sample of ~900 images from the data set is evaluated for U-Net architecture which is used to validate the quality and trainability of the images.

Based on static image performance, the selected models have been used to identify and segment dust clouds in video recordings of road dust emissions. These are used to visually analyse the segmentation quality of the best-performing machine learning models and determine whether the development of novel machine learning models for segmentation of dust is necessary.

ACKNOWLEDGEMENTS

The author would like to thank the dedicated team at Downer's National Research and Development Laboratory and Blended Products business, without whom the continuous technical support, problem solving and the delivery of the products could not be achieved. The author would also like to thank the research team at Monash University's SPARC HUB for the partnership and collaboration. And special thanks to Fortescue Metals Group for providing me with the access to the relevant parts of their Pilbara's operations.

REFERENCES

Australian Road Research Board (ARRB), 2020. *Road Materials, Best Practice Guide*, vol 1, ARRB, May 2020. Available at: https://www.arrb.com.au/bestpracticeguides

Caterpillar, 2017. Cat® C32 Diesel Generator Sets. Available at: https://s7d2.scene7.com/is/content/Caterpillar/CM20170920–31126–63881

De Silva, A, Ranasinghe, R, Sounthararajah, A, Haghighi, H and Kodikara, J, 2022. Semantic Segmentation Model Performance on Vehicle-induced Dust Cloud Identification on Unsealed Roads. 10.21203/rs.3.rs-2239765/v1.

De Silva, A, Ranasinghe, R, Sounthararajah, A, Haghighi, H and Kodikara, J, 2023. A benchmark dataset for binary segmentation and quantification of dust emissions from unsealed roads, *Scientific Data,* vol 10. Available at: https://www.nature.com/articles/s41597–022–01918-x

Greenbase Pty Ltd, 2021. Emissions Study – Havieron Project Stage 2, Greenhouse Gas Assessment, Technical Report, ver 2.0. Available at: https://www.epa.wa.gov.au/sites/default/files/Referral_Documentation/Appendix%20L%20Greenhouse%20Gas%20Emissions%20Report.pdf

Leek, C and Jameson, G, 2011. Review of Foamed Bitumen Stabilisation Mix Design Methods, Austroads Publication No. AP–T178/11.

Thompson, R J and Visser, A T, 1999. Designing and Managing Unpaved Opencast Mine Haul Roads for Optimum Performance, SME Annual Meeting, March, Colorado.

WSP Australia, 2020. HaulPac – Environmental Consideration (Desktop Risk Review) – Mining Haul Road Application, Available upon request.

Health, safety, environment and community – licence to operate

A guide to ground disturbance permitting and managing ESG risk

B Spence[1], H Arvidson[2] and S Helm[3]

1. Industry Solutions – NRG Manager, K2fly, Subiaco WA 6008. Email: brian.s@k2fly.com
2. Chief Geoscientist, K2fly, Subiaco WA 6008. Email: heath.a@K2fly.com
3. General Manager Industry Solutions, K2fly, Subiaco WA 6008. Email: sean.h@k2fly.com

ABSTRACT

Mining is the art of disturbing ground. All phases of a mine's life cycle from exploration to building and operating a mine, through closure, has a physical impact on the earth. There is currently increasing public and stakeholder awareness owing to a wide range of issues including heritage protection, environmental stewardship, and social impacts therefore ground disturbance must be conducted in a way that honours and respects all agreements, regulations, legislation, and social expectations.

A comprehensive ground disturbance permitting process has been developed through consultation with major mining companies and Traditional Owner groups. It operates from strategic business planning through to closure and rehabilitation. Adherence to the principles and process of permitting helps to minimise risk by ensuring that mining companies conduct their business in an inclusive, professional, reliable, and consistent manner that stakeholders reasonably expect, and that mining companies can be proud of. Built on a geographical information system, supplemented with a specialist database, this process allows request, review, and approval by subject matter experts, to operate in a common environment that is transparent, facilitates collaboration, improves efficiencies, and reduces the risk of error or omission. The application of new technology can effectively highlight changes that naturally occur in today's dynamic ESG domain and provides the perfect platform to record actual disturbance against authorised and planned allowances.

The social license to operate demands that strong governance of the ground disturbance process, described herein, can be demonstrated to any interested stakeholder, and mining companies recognise that this principle is central to the betterment of the mining industry.

INTRODUCTION

This paper outlines years of learning that has shaped K2fly's understanding of what is required to systematically approach the complex problem of managing Environmental, Social and Governance (ESG) risk in a dynamically evolving field of standards, legislation, and community expectations. Unlike disciplines that are more transactionally oriented or have been practiced and standardised for decades, ESG is relatively new, and requirements continue to evolve. Complexity is compounded by the fact that, much of what needs to be known is not yet known! By that we mean that discovery via formal survey or inspection is required to gather data, information and knowledge that precedes risk analysis and decision-making. This extends timelines for development, increases project risk, operational pressure, and total cost.

The sheer number of stakeholders, internal and external, who are and need to be involved is also growing. Broad areas are being sub-divided into many individual specialised parts. What once required one or two signoffs now requires potentially a dozen different perspectives to be reviewed and approved. Requirement for multiple signoffs adds to issues with tracking activity, visibility of overall process, cross functional communication, and the need to collaborate across such a broad range of stakeholders and subject matter experts.

We see every mining organisation grappling with these issues. In many cases well intentioned business areas, are each trying to solve the problem independently, ensuring their part of governance is in order and their area is not holding up approvals. Internal systems extend into areas they were never intended or designed to functionally cover. Few are truly integrated, Geographical Information Systems (GIS) sit outside application boundaries.

This situation is similar to where many mining organisations were in the late 1990s in Australia, with the introduction of integrated Enterprise Resource Planning systems (eg SAP) requiring independent functional areas (Finance, HR, Supply, Maintenance) to interoperate in a common system. The biggest challenges were not the system itself but internally addressing the need for process ownership, common practice, standardisation, and change management. ESG is facing similar challenges today.

Working with proponents and Traditional Owner groups, over nearly ten years, has identified a systematic, comprehensive approach to these challenges.

DEFINITION OF RESOURCE GOVERNANCE

Our definition of Resource Governance is based on a view that the 'Resource' is anything existing on a parcel of land that influences the value of a mineral project. Value is defined as revenue minus cost, therefore anything that is sold to generate revenue (eg ore, by-products) and anything that adds a cost (eg protecting cultural heritage, carbon mitigation, waste management, protecting plants and animals etc) influences value.

In terms of governance, we adopt the following definition from Standards Australia (2003):

> Governance refers to the framework of rules, relationships, systems, and processes by which an enterprise is directed, controlled and held to account and whereby authority within an organisation is exercised and maintained. It encompasses authority, accountability, stewardship and leadership, and direction and control exercised in any organisation.

THE FOUNDATION

Companies such as K2fly spent years defining and refining the underlying data model for Resource Governance. More recently newer graph database technology has been added to traditional SQL relational databases. Graph technology provides a greater ability to represent complex relationships in data, allowing users to perform 'traversal queries' based on connections and apply graph algorithms to find patterns, paths, communities, influencers, single points of failure, and other relationships. More efficient analysis is thus enabled at scale against large amounts of data. The power of graphs is in analytics, the insights they provide, and their ability to link disparate data sources. Users don't need to execute countless joins and the data can more easily be used for analysis, data science, and machine learning (Oracle Australia, 2023).

Data consolidation is fundamental to efficiently and effectively supporting the ground disturbance permitting process. The creation of an operational data store with integration to the sources of truth, including spatial data, is at the foundation of providing a comprehensive environment for governing the ground disturbance process and managing ESG risk. This enterprise operational repository allows for all stakeholders to collaborate, interact, and make informed decisions in a structured, systematic, transparent, and auditable manner. The use of Graph technology allows for extended relationships amongst this complex data set to be created either automatically or manually.

Figure 1 represents the data that is required to support the process of managing ground disturbance. It shows the breadth, domains, and the interrelated nature of the data.

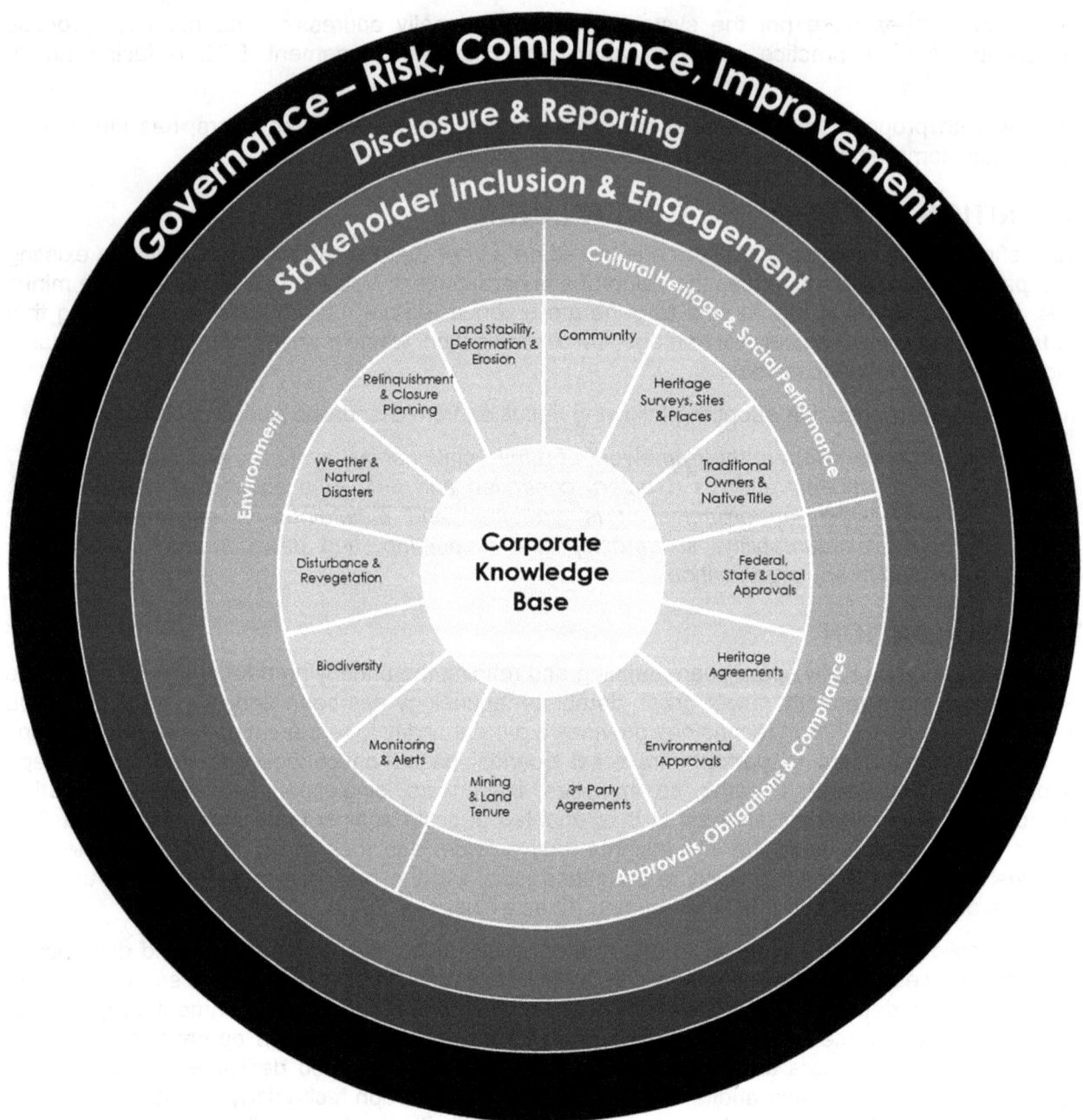

FIG 1 – Ground disturbance knowledge base.

Data is sourced from both internal and external sources. It is consolidated in one integrated operational data store at an enterprise level. Consolidating the data and building the complex interrelationships into a shared model, provides an extremely powerful foundation for work management, analytics, and improvement opportunities.

A COMPREHENSIVE APPROVAL PROCESS

One of the key requirements for comprehensive ground disturbance governance is to embed the process as part of operational business. It is essential that the process involves operational and specialist teams, working together, in a common environment. Siloed, independent approvals are less efficient.

Figure 2 provides a high-level representation of a comprehensive ground disturbance approval process.

It operates from strategic business planning through to closure and disturbance reconciliation.

The process can be supported by information technology to allow for systematic control, data quality assurance, transparency, and analysis.

Using information technology, enables the many internal stakeholders involved in the process to collaborate, communicate, and operate in a common environment. This improves work management – visibility, control, prioritisation, efficiency, and effectiveness.

External Approval Instruments

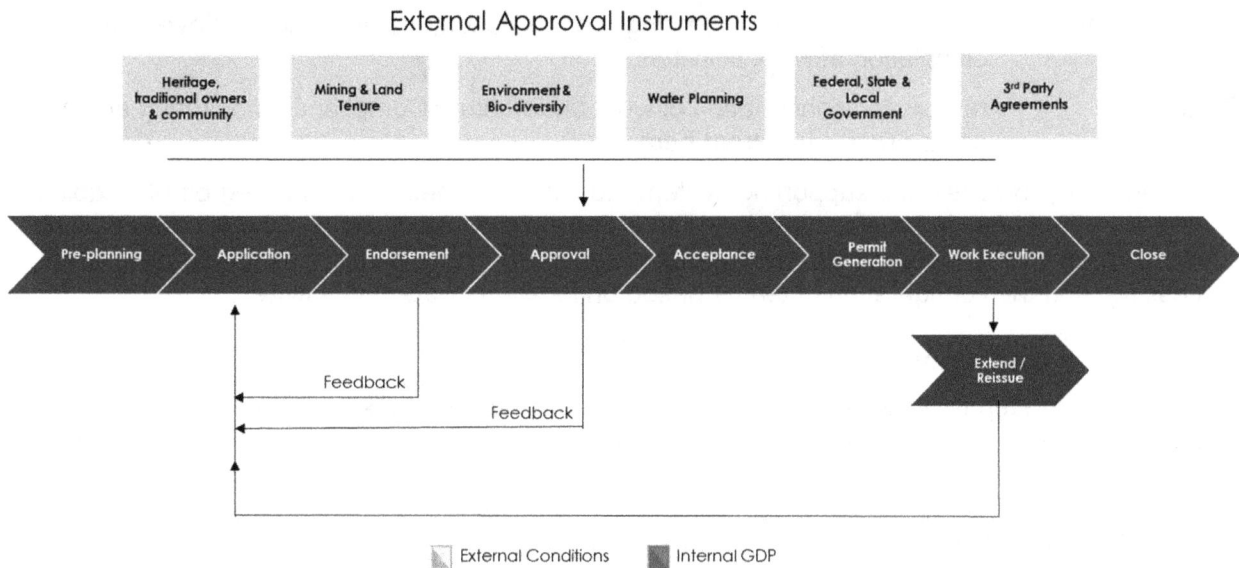

FIG 2 – Ground disturbance approval process.

Process initiation

The process begins with the outputs of the strategic business planning process, whereby planned activity can undergo initial assessment against known public and internal data. This allows for analysis and recognition of the potential long lead times required to gain relevant approvals. The system can automatically assess planned activity against existing tenure, cultural heritage, social, environmental, and stakeholder information.

The external approval instruments are representative of the data that is required to be integrated into the common data model for governance of the ground disturbance process. These approvals describe the rights granted, and the obligations imposed in order to systematically assess and approve all submitted work programs. The process allows for the easy selection and application of relevant obligations to each work activity being proposed.

The process depicted in Figure 2 follows a common workflow:

Pre-planning   the collection of relevant data to prepare an application, pre-analysis of location and area boundaries against existing registered public and internal activity in that area, relevant stakeholders, cultural heritage sites and places, tenure, environmentally sensitive areas etc.

Application   front end loaded to assist applicants to do as much as possible prior to submission, including spatial sensitivity analysis that automates the process of identifying and highlighting potential issues. The spatial analysis is driven by configurable business rules established by the specialist approval teams. The applicant can then determine how to address these issues.

Endorsement   internal coordination and assessment prior to specialists needing to be involved. This is essentially a quality assurance step.

Approval   specialist review, consultation, and application of conditions relevant to the activities proposed to be undertaken. (The final approved area or shape may differ to that which is initially lodged due to many factors such as outcomes of heritage survey, land assessment etc. also, the system will maintain a history of areas applied for, area approved for disturbance and actual disturbance undertaken, which is very useful for forensic analysis if required.)

Acceptance	applicant understands and agrees to the conditions applied to the area approved for disturbance.
Permit generation	flexibility in format, order of presentation and generation of the final permit.
Work execution	recording of actual clearing, rehabilitation, compliance achievement during completion of work activities.
Close	review and confirmation of work undertaken. Compliance monitoring and audits can be performed from here.

The governance process and supporting system must be designed to allow a degree of flexibility to handle the differences in organisational structure of proponent companies. The process described above, allows any number of approval groups, configuration of sequenced or parallel approval processing, and the configuration of teams based on location or area of activity.

Preparation and application

The preparation step is designed to provide the operational group making the application with the guidelines they need to collect and prepare the necessary information to streamline their application through approval.

Automated spatial analysis

An important early step is to spatially analyse the location of the work. Today's technology allows the applicant to access the spatial area they propose to work in against existing known information in that location. This is a powerful feature to facilitate an efficient and effective ground disturbance process. As an example, this analysis could include intersecting:

- company tenements and tenure
- federal, state and local government agreements
- commercial arrangements
- other land holdings, ownership, and tenure
- native title and protected areas
- aboriginal cultural heritage stakeholders and organisations
- cultural heritage sites and places
- environmental approvals, landforms, and environmentally sensitive features
- infrastructure.

Mapping functionality allows the applicant to clip, cut, or adjust the area they propose to work in based on the spatial analysis of the area in question. This saves a significant amount of interaction with GIS specialists to re-draft polygon boundaries. The technology also allows for shape files to be directly uploaded if available. All these early adjustments can be made by the applicant without needing to involve specialist approval or GIS teams.

Endorsement and approval

Endorsement and approval steps both follow similar workflow. As mentioned, endorsement is a pre-approval quality assurance step(s) to ensure the applicant has correctly completed the application and considers all the issues before specialist approvers need to get involved.

Approval is the more formal process of assessing the application, providing feedback and suggestions on adjustment if required, assigning the relevant obligations and conditions to the scope of work being proposed, and approving the application if appropriate.

Smaller organisations tend to require fewer approval groups whereas larger companies can have ten or more involved in the process.

Managing the approval instruments and the obligations they impose is another essential element of the process. To understand an organisation's ability to access/disturb ground, the following list captures aspects that must be known.

The organisation's obligations, with respect to:

- federal, state and local government regulations
- access agreements
- easements
- landholders
- Traditional Owners
- company obligations (social compliance).

Complexity arises in the process, with consideration of:

- Where are the obligations stored?
- Who manages them?
- Do obligations always apply?
- Did we comply?
- How can compliance be demonstrated, and with what evidence?

The ground disturbance process must have ready access to the relevant obligations for the proposed activity and its spatial area. This improves the efficiency and effectiveness of the process. A supporting information system must provide the ability to manage all forms of external approvals and related obligations. Flexibility is required for the approval specialists to manage these obligations for easy application and understanding when assigned to a permit.

The approval specialists determine if the existing instruments/approvals provide the necessary authorisation to perform the activities that are proposed.

This can be quite a complex area as the rights are often cascaded from related approvals. That is, to clear ground (for example) you would need mining tenure (or another approval instrument) initially before obtaining the secondary approvals required for the actual clearing. The flexibility to associate the permit with the appropriate approval instrument is important to accurately reflect the remaining clearing allowance.

This is where the graph (data model) delivers its value, facilitating easy and transparent analysis of all the relationships associated with the correlated instruments and the potential hierarchies that exist. This allows users to clearly identify which instruments should apply and the nature of the relevant obligations.

Applicant acceptance and permit generation

Once all approvals have been completed the applicant is notified and must review and accept the conditions as stated before the permit is considered finally approved and active. Negotiation can occur between the applicant and the relevant approval teams as required, with a prescribed feedback loop to capture any resultant amendments.

The permit is then generated based on a flexible template that is pre-configured and can be printed and saved as a PDF document.

Work execution and closure

As work progresses under the active permit the applicant can update progress against compliance requirements defined in the approval process. This can include progressive clearance and revegetation tracking, or uploading required documentation as work is completed.

Once all work is completed and all necessary information provided, the permit can be requested to be closed. At this point a further compliance check can be completed and history is maintained.

Reissue and extension

The ground disturbance process also allows for the reissue of a permit, when new information has become available, and deemed significant enough to warrant review and subsequent re-approval by the various approval teams.

In the case of a reissue, the original permit may also be placed on hold meaning no further work can occur until the permit can be reassessed.

Extension can occur when the permit expiry date is the only element that needs to be changed and relevant approvals are not exceeded.

Changed conditions

Technology allows for the monitoring of active permits to continually check that there are no new spatial area features and/or changes to the prevailing obligations. If changes are identified, then system notifications raise these to the attention of relevant internal stakeholders for reassessment and subsequent action.

The application of technology in this area is very important for larger organisations that often manage many active permits.

Compliance and auditability

Having a common, enterprise approach to ground disturbance permitting, that is supported by information technology, demonstrates a proactive approach to managing ESG risk and ensuring compliance to relevant social expectations and legal obligations.

Every step of the ground disturbance process is transparent, monitored, repeatable, and auditable. The process ensures a level of confidence that agreed process steps are being carried out every time. This reduces the risk of missing information or individuals not executing their work to an agreed standard.

This level of transparency also means reports showing the performance of both the quality of the outcomes and the process itself can be reviewed and improved where necessary over time.

The ground disturbance process implements separation of duties, where teams or individuals have defined roles and are only allowed to perform those duties.

Governance ensures work on the ground should not commence until the permit is approved, accepted, generated and is active.

More formal compliance checks or audit steps can be added as required. This is an area many proponents grapple with. The close out and compliance steps are difficult to internally resource when specialist teams are normally fully engaged dealing with the volume of new approvals being sought.

Stakeholder interaction and inclusion

Throughout the ground disturbance process stakeholders and the interactions that occur can be recorded and importantly related to the relevant applications, permits, tenements, external approvals, or activity areas as they occur.

Internally stakeholders can see, track, and collaborate to improve process efficiency and effectiveness.

CONCLUSIONS

ESG and ground disturbance permitting are complex domains, bringing together many stakeholders, from different specialist areas, all with vested interests in Country and the resources within it.

Ground disturbance must be conducted in a way that honours and respects all agreements, regulations, legislation, and social expectations. All parties' interests must be respectfully understood, considered, and recorded.

This requires a comprehensive permitting process embedded within an organisation's business model: A process that spans the extent of operations from strategic business planning through to closure and rehabilitation, which is described herein. With so many stakeholders and such a broad spectrum of non-transactional information to manage, it becomes almost essential that a supporting information system is used. Information technology helps to minimise risk by ensuring that mining companies conduct their ground disturbing activities in an inclusive, professional, reliable, and consistent manner that stakeholders reasonably expect.

Built on a geographical information system, supplemented with a specialist database, this system captures the process to allow request, review, and approval by subject matter experts, to operate in a common environment that is transparent, facilitates collaboration, improves efficiencies, and reduces the risk of error or omission.

ACKNOWLEDGEMENTS

We would like to acknowledge and extend our gratitude to all individuals and organisations that have contributed to the completion of this technical paper. Their support and assistance have been invaluable. Notable amongst those have been Bradley Brown, Chief Executive Officer of The Keeping Place, and the team at K2fly whose guidance, feedback, and encouragement throughout has been invaluable.

We apologize for any unintentional omissions in this acknowledgement, and we sincerely appreciate the contributions of all involved.

REFERENCES

Oracle Australia, 2023. Graph Databases Defined. Available from: <https://www.oracle.com/au/autonomous-database/what-is-graph-database/> [Accessed: 23 May 2023].

Standards Australia, 2003. AS 8000–2003, Good Governance Principles, Australian Standard.

Mining and processing – new equipment

Latest fifth generation (Mk V) gyratory crushers with focus on safety and maintenance features

J Garrett[1] and W Malone[2]

1. Global Product Line Manager – Crushing and Screening, FLSmidth, Midvale Utah 84047, USA. Email: josh.garrett@flsmidth.com
2. Global Product Group Manager (VP) – Crushing and Screening, FLSmidth, Midvale Utah 84047, USA. Email: bill.malone@flsmidth.com

ABSTRACT

FLSmidth (FLS) has spent the last 15 years, working with customers and engineering houses to deliver innovative improvements in crusher safety, coupled with the latest, maintenance friendly, crusher design. The key focus being to prevent service personnel from working underneath the crusher during maintenance and obviate the need for personnel to be exposed to suspended loads when performing tasks such as, main shaft installation, for example.

Improvements include: full top-service capability to ensure the safest and fastest way to disassemble the crusher for maintenance purposes, which also removes the requirement for a service cart; a self-aligning main shaft which ensures remote assembly is now possible, keeping service personnel well out of harm's way; And rotable shells, or the ability to perform relines off-station, allows for quick shell 'swap-over' without the downtime associated with removing, refitting and curing time associated with epoxy backing.

The high flow spider improves large material entry into the crusher, reducing blockages and downtime and therefore maintains crusher capacity. Each size of crusher in the range comes with an option to increase the standard feed opening to a larger one if desired to compensate for 'blocky' or 'elongated' feed material, a common problem for Pilbara hematite iron ores.

Power capability has also been increased with a maximum of 1500 kW, to match the higher throughput capacities being achieved. This due to uprated eccentric speeds and larger discharge annulus. Increased eccentric throws and mantle diameters are also now a reality, when compared to more traditional designed gyratory crushers.

To keep up with the improved design features of the Mk V, digitalisation has also been upgraded to maximise performance capabilities to ensure that the most efficient reduction and throughput capacities are achieved, whilst utilising the higher powers now available as standard.

Condition monitoring ensures that the crusher is at its optimum operating capability by monitoring and adjusting the crusher settings via feedback from additional sensors, ie main shaft spin, temperature and vibration. Camera monitoring capabilities are available as is an advanced oil contamination monitoring system.

As a result of the above design improvements, it will be shown that the Mk V range offers benefits such as: enhanced safety features, increased availability, faster maintenance turnarounds, higher power utilisation, equivalent product size at higher throughput capacities and with Remote operation capabilities. All of which result in a gyratory crusher with higher levels of safety and a lower per tonne operating cost.

LATEST ADVANCEMENTS IN THE CRUSHING INDUSTRY

Gyratory crushers have come a long way from the original machines that were designed around the turn of the last century. Low capacity, low power, high maintenance, low reduction and no regard for safety, but still considered marvels in their day.

Nowadays there are more powerful machines and equally much more stringent regulations to provide not only crushers that can deliver the sort of throughput capacities our original machine designers would deem unbelievable, but also do it in such a way as to comply with current requirements regarding safety standards for operations and maintenance personnel.

The latest primary gyratory crushers therefore incorporate many new features that are designed to ensure operational and maintenance practices are currently the safest available. Couple these with the ever-increasing demands from today's operators for higher throughput capacity and power handling capabilities, this new generation of crushers can now satisfy all these needs.

Opening lower grade iron ore deposits that until recently were not economical due to the high capacities required are now viable propositions. Swapping out old crushers with new generation crushers can improve throughputs anywhere between 50–100 per cent. Reduced dynamic loads ensure that the old structures can still be used and careful attention to design allows for the same, or near same, footprint to be utilised.

Indeed, such are the advancements in today's technology, that a traditional-type primary gyratory crusher can now literally be swapped-out with one of the new generation crushers, which can potentially double the power rating capability and the throughput capacity of the old crusher. All this with the same or near same footprint, same or less installation height and with the benefit of reduced dynamic loads, despite the higher motor powers and increased throughputs. In other words, retrofitting older machines into existing structures to avoid costly civil works and increase capacity at the same time, is now a reality.

Figure 1 perfectly illustrates the significant advancements that have been made in gyratory crushers over the last 118 years (since 1905), from a single company point of view, and shows the progression in improvements over time such as: capacity, power handling capability, reduction ratio (increased strength) and maintenance strip down times.

Crusher Model	First Built	Designation	Feed Opening (in)	Capacity (mtph)	Power (kW)	Power to Weight Ratio	Reduction Ratio	Strip Down Time (hrs)	Notes
"Bulldog" (Mk I)	1905	Mk I	18	200	150	1 : 1	2.5 : 1	48	Side Discharge - Belt Driven (Flat, later V-type)
	1910		48	1,200	225				Maintenance Intense (Multi-Daily)
	1919		60	2,000	330				Grew over 50 years. Became basis for all other designs
	1950	Superceeded............							Eventually, over 2,400 Sold
TC (Mk II)	1950	Mk II	60	3,000	375	1 : 1.5	3.5 : 1	30	Circular Discharge - Belt Driven (Flat, later V-type)
	1969		72	3,400	525				Hydraulic Adjustment Introduced in Early 60's
	1990	Superceeded............							
NT/UD (Mk III)	1990	MK III	60 x 113	4,500	750	1 : 2.5	4.0 : 1	17	Circular Discharge - Shaft Driven
	2005		60 x 113	6,000	1,000				Incorporation of FEA and Modern Controls
	2013	Still Current.............							Modulraized and Simplified Components
TSU (Mk IV)	2006	Mk IV	63 x 114	8,000	750	1 : 2.8	4.5 : 1	5	Circular Discharge - Shaft Driven
	2009		63 x 118	10,000	1,200				Incorporation of FEA and Modern Controls
	2013	Still Current.............							Eccentric Serviced from Top
TSUV (Mk V)	2019	Mk V	72 x 130	15,000	1,500	1 : 2.7	4.5 : 1	4	As TSU but more enhancements...!
									Rotable Shells, Self-Aligning Mainshaft
		Latest Offering - And coming to an area near you soon...!							1,500 kW and capacities up to 15,000 mtph

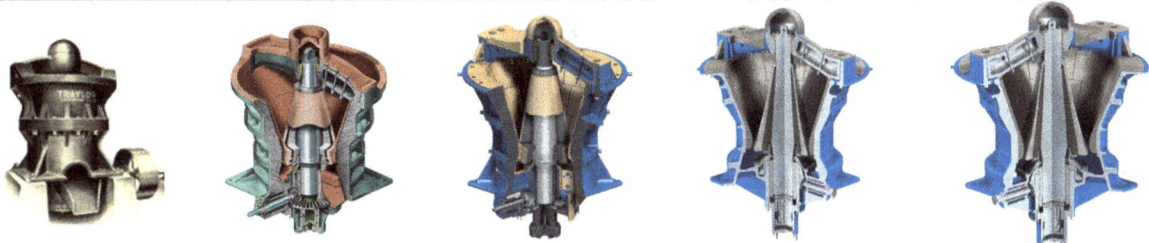

FIG 1 – Evolution of today's modern primary gyratory crusher.

Top service crushers – what's the difference? The key difference is that when disassembling traditional designed crushers, the eccentric assembly and hydraulic piston assembly had to be taken out from the underside of the crusher. Figure 2 illustrates this concept of disassembly from the top, rather than traditionally from the bottom.

This required not only a trolley-type maintenance cart that could run underneath the crusher to accept the reasonably large assemblies, but also designing enough height in the discharge chamber underneath the crusher to allow this to be performed. The extra height required added significantly to the capital cost of the excavation or the run-of-mine (ROM) pad height as well as additional concrete and steel during construction.

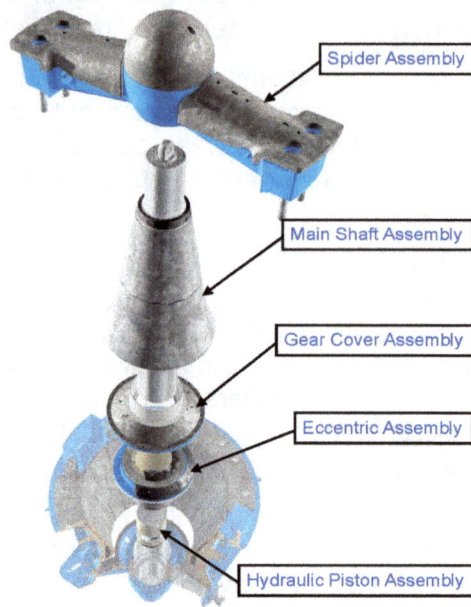

FIG 2 – Exploded view of top-service major components showing accessibility from the top.

The need for service personnel to be under the crusher whilst this was taking place was also a requirement and exposed them to all manner of hazards, ie falling rocks from above, working under heavy suspended loads, working in a confined space potentially harbouring noxious gases/chemicals/dust etc.

The top-service design renders all the above hazards obsolete, as it allows the crusher to be fully dismantled and re-assembled from the top.

Table 1 highlights the service time savings that can be expected from the top service range of crushers.

TABLE 1

Service and maintenance time comparisons.

	FLS Top Service Crusher			FLS Bottom Service Crusher		
Task Crusher Shutdown – Top Service						
Remove Shaft – Inspect Inner Ecc Bush…A1						
Remove Shaft – Inspect Outer Ecc Bush…A2						
Complete overhaul…A3	A1	A2	A3	B1	B2	B3
Task Crusher Shutdown – Bottom Service						
Remove Shaft – Inspect Inner Ecc Bush…B1						
Remove Shaft – Inspect Outer Ecc Bush…B2						
Complete overhaul…B3						
Primary Crusher Shutdown – Total Duration (hrs)	15	22	67	15	64	112
Primary Crusher Shutdown – Total Duration (days)	0.6	0.9	2.8	0.6	2.7	4.7
Clean Material from Dump Pocket	3.5	3.5	3.5	3.5	3.5	3.5
Removing Main Shaft	7	7	7	7	7	7
Removing Eccentric	4.5	11.5	16.5	4.5	53.5	53.5
Remove countershaft Assembly			40			48
Strip and Refit Concaves	78.5	78.5	78.5	78.5	78.5	78.5
Swap over of Rotable Shells	16	16	16	▪	▪	▪

DESIGN CAPACITY REQUIREMENTS FOR NEW GENERATION CRUSHERS

One of the key design features embraced was increased throughput capacity capabilities. Today's modern iron ore mines are demanding higher and higher throughputs, this in part to having to deal with deposits that demand such higher throughputs to compensate for the lower grades being processed or the improved economics of larger tonnage hematite iron ore mines. Traditionally this would have to be done with several primary gyratory crushers, and the cost of doing multiple crushers, multiple stations and enlarged truck fleets mostly rendered this option as unviable.

Today two machines can meet the throughput capacity requirements, where once four would have been required. This reduces capital expenditure in both equipment supply and structural requirements. Fewer and larger trucks can be utilised making mine planning and truck routing simpler and maintenance costs and service times can be greatly reduced also.

Table 2 illustrates the significant increased throughput capacity handling capabilities of the new generation machines.

TABLE 2

Capacity ranges for Mk V gyratory crushers.

					TSUV Mk V – Capacity (mtph)			
OSS (in)		5 in	6 in	7 in	8 in	9 in	10 in	11 in
OSS (mm)		127 mm	152 mm	178 mm	203 mm	229 mm	254 mm	279 mm
Feed opening (mm)	**Mantle diameter (mm)**							
1100 1300	1900 TSUV	1474–2272	2254–3383	2800–4058	3396–4766	-	-	-
1400 1600	2200 TSUV	-	2351–3791	2886–4519	3472–5300	-	-	-
1600 1800	2600 TSUV	-	2926–4842	4142–6733	4913–7821	5436–8494	-	-
1600 1800	3000 TSUV	-	-	4800–6050	5550–7450	6700–8750	7500–9650	8250–10 750
1600 1800	3300 TSUV	-	-	5000–6100	7250–8750	8600–10 100	10 500–12 550	12 400–15 000

Similarly, Table 3 illustrates the significant power increases made to the range.

TABLE 3

Power ranges for Mk V gyratory crushers.

TSUV Mk V – power increases (kW)				
Bottom service model	**(kW)**		**Top service model**	**(kW)**
42 × 65 NT	375	→	1900 TSUV	600
54 × 75 NT	450	→	2200 TSUV	750
60 × 89 NT	600	→	2600 TSUV	1000
60 × 113 NT	750	→	3000 TSUV	1200
60 × 113 UD	1000	→	3300 TSUV	1500

As can be seen from the increase in power, when comparing traditional machines with next generation machines, the possibilities exist to suddenly consider, not being forced to fully rebuild a plant, but simply change out the existing crusher with a newer and more efficient model.

This obviously cuts down on civil and installation costs, which as we all know, are a drain on capital investment and provide no returns or even re-sale values down the line. This leaves perhaps the only other points that should be considered, ie are my current discharge feeder and conveyor capable of handling the extra capacity or do they need to be upgraded and is my electrical distribution system also capable of handling the extra power?

DESIGN CONSIDERATIONS OF MAIN BEARINGS

Special attention was paid during the design phase of the crushers with respect to the increase in power each model would see and the impact this would have on the main bearing components.

Consequently, all bearings were redesigned to reflect this and indeed key design limits such as: design factor of safety, increased from ~2.0 to ~2.4 and bearing surface areas increased by ~33 per cent, were also upgraded.

Table 4 illustrates key differences between the bearings in traditional type gyratory crushers and the latest generation gyratory crushers (4).

TABLE 4

Bearing comparison chart – traditional versus new generation (4) gyratory crushers.

	Eccentric bushing comparison			
	(1) Other 60–89	(2) Other 63–89	(3) FLS 60–89 NT	(4) FLS 16 × 24 TSU
Installed power (kW)	600	1000	600	750
Actual power (kW)	600	750	600	750
Bushing ID (mm)	864	890	918	954
Bushing height (mm)	1250	1200	1155	1675
Design factor of safety	2	2	2	2.4
Oil film thickness (0.001")	3.1	2.5	3	3.7
Bearing surface total (in^2)	5258	5200	5162	7780
Bearing surface effective (in^2)	876	867	860	1297
Load transmitted to bearing (psi)	273	345	278	230
Conclusion	Generous	Marginal	Generous	Very generous

The above is an example of various design parameters that have been measured/calculated for a typical 60' × 89' (1600 mm × 2400 mm) gyratory crusher, which is still one of the most popular size machines in operation today. It is worth noting however, that the differences in the above results generally hold true when making the same comparisons on larger or smaller crushers in the gyratory crusher range.

Key points on the top service gyratory crushers (#4 in Table 4) are:

- largest main shaft diameter in its class
- tallest eccentric bush in its class
- highest factor of safety in its class
- thickest oil film in its class
- largest effective bearing surface in its class.
- lowest load transmitted to bearing in its class.

MECHANICAL ADVANCEMENTS – GYRATORY CRUSHERS

The overall operating principles of the gyratory crusher have remained the same as they were 100 years ago. ROM is fed into the top of the gyratory crusher, via trucks (direct tip feed) or apron feeder (indirect feed). The ROM, assisted by gravity, falls through the crushing chamber which contains a main shaft that has an eccentric motion between the mantle liner on the shaft and the concave liners on a stationary circular inner wall of the crusher shells. This 'eccentric motion' is enough to reduce the feed material in manageable stages as it passes through the crusher to the final desired product size that exits.

It seems a simple enough procedure, but to keep this process running smoothly, there are several challenges that must be overcome:

- Oversize ROM – larger than recommended feed being presented to the crusher and potentially blocking or 'bridging' the crusher feed opening, either fully or partially. The results are unplanned downtime, loss of throughput and lost revenue.

- Liner replacement – Traditional methods for replacing wear parts on a gyratory crusher could shut the crusher down for days and if a problem is encountered during this time, then even more downtime is required.

- Unplanned maintenance – top service capability ensures that, for example, should an eccentric bush require inspection, this can be actioned within 4–6 hours compared to 2–3 days with a traditional bottom service crusher. With no need to clean out a discharge hopper underneath the crusher, this again not only contributes greatly to the reduction in downtime but ensures no potential operator or maintenance team exposure to safety hazards is required.

- Increased personnel safety – additional hazards that are introduced when working *in situ* eg main shaft removal and replacement with the new generation machines can now safely be carried out without the need for personnel being under suspended heavy loads or exposed to any falling rock hazards. By utilising rotable shells, maintenance can now be carried out in a more structured, non-time critical way that a workshop environment offers.

Oversize Ore – Potential of blockages and bridging at the crusher inlet

Constant material flow-through the crusher is paramount to the process in terms of overall throughput capacity handled and indeed expected by the plant. If this process is interrupted once per shift for any significant length of time or there are regular shorter stoppages every shift, these can have a very detrimental effect on the overall plant throughput and operational viability.

Traditionally, the relationship between the feed opening gape and the maximum allowable feed particle has been limited to ~80 per cent, ie for a 1600 × 3300 size gyratory, the recommended maximum feed size would be, 1600 mm × 80 per cent = 1280 mm.

We all know that this figure is not realistic and, oversize material is regularly presented to the gyratory crushers. This could be up to/over 2 m size in the longest dimension. Rock breakers have been the answer to a certain extent in reducing these blockages when they occur, but it does require operator time and although they reduce lost time, they do not eliminate it entirely. The availability of a rock breaker in many gyratory station set-ups does also tend to make the truck loaders at the face less diligent in assessing oversize material.

The Mk V was designed with the above in mind and additionally is offered with a choice of two sizes of top shell. The standard shell with a feed opening that corresponds to traditional feed openings and the alternative top shell which has an enlarged feed opening to cope with elongated, slabby and in general oversize material that may be characteristic to a particular site. Table 5 details the standard and alternative feed openings available for each crusher in the range.

TABLE 5

Gyratory crusher feed openings.

Traditional gyratory crusher feed opening	TSUV gyratory options
1100 mm – (43 inches)	1100 or 1300 mm – (43 or 51 inches)
1400 mm – (55 inches)	1400 or 1600 mm – (55 or 63 inches)
1600 mm – (63 inches)	1600 or 1800 mm – (63 or 71 inches)

High-flow spider concept

Additional design considerations on the MK V crushers included the 'high-flow spider' concept. This design improvement allows more free space under the spider which in turn aids entry of larger size material and so helps to reduce blockages even more.

As can be seen from the comparison in Figure 3, the high-flow spider (bottom picture) on the left, offers a much better entry point in terms of height as measured from the underneath of the spider arm to the entry point of the crusher. This helps prevent larger, blockier material wedging as it occasionally gets presented at this point or can be pushed towards this point by additional incoming material. The use of a rock breaker is also enhanced by the high-flow design of the spider as material nudged over by the rock breaker will have less chance of becoming wedged as the sloping feature of the more traditional design now plays a less significant role.

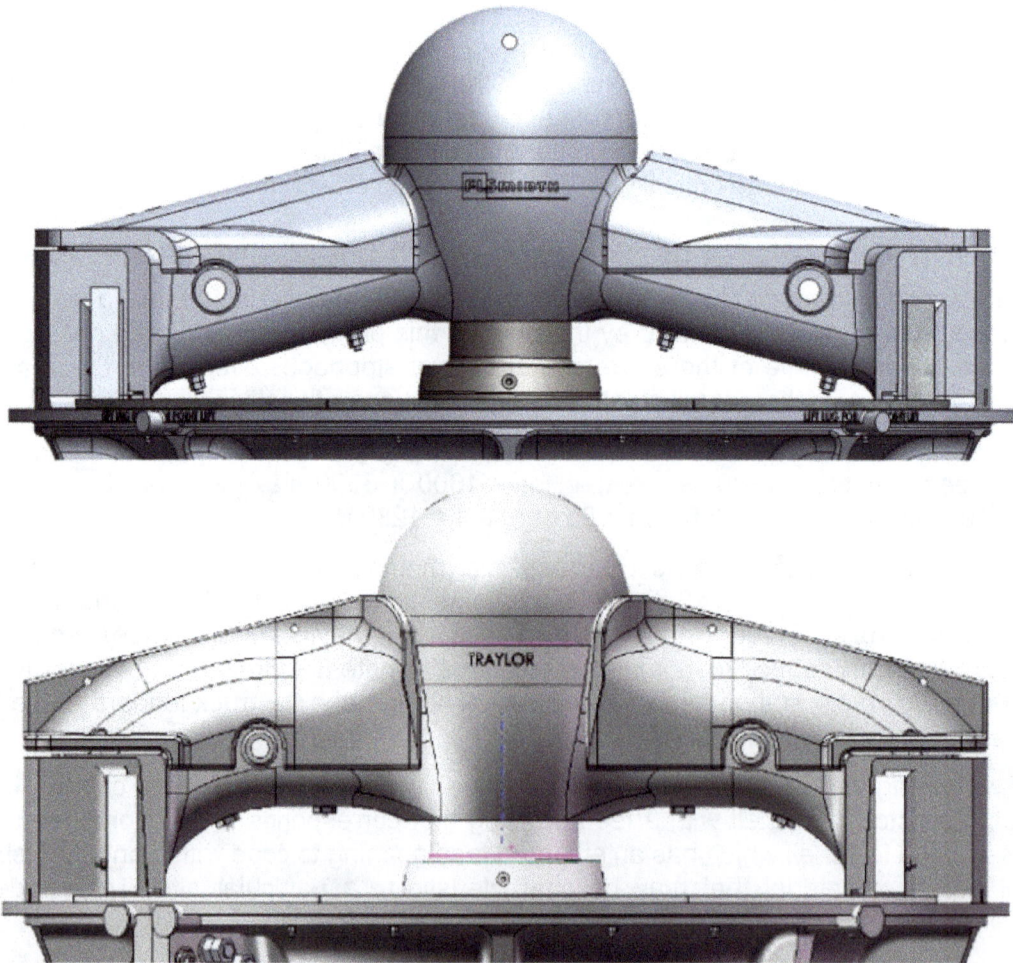

FIG 3 – Comparison between normal (upper) and high-flow spiders (lower) showing feed opening benefit under the arms.

Additionally, the COG (centre of gravity) of the high-flow spider is higher and more in-line with the lifting points than that of the standard spider and lifting in and out for maintenance purposes is much more stable as a result.

Self-aligning main shaft – service and safety benefits

The concept of the self-aligning main shaft was a major challenge in that with an ever-increasing focus on safety, operators were becoming more and more vocal with their concerns that service personnel were being exposed to unacceptable hazards each time they were obliged to position themselves under the suspended main shaft to facilitate the entry of the main shaft into the eccentric assembly when rebuilding the crusher.

Normally when dismantling the crusher, for any reason, the spider is removed first and thereafter the main shaft is removed, so allowing access to the concaves, if a liner change is required, or access to the dust collar and then the eccentric assembly if this is the goal. Similarly, even if the mantle liner is to be renewed, then the main shaft must first be removed. Upon performing the desired work, the crusher is then reassembled and put back into production, and this is where service personnel are exposed to the major hazard of having to go underneath the suspended main shaft and guide it into the eccentric assembly.

This, up until now, is a procedure that for the reasons stated above, generally is carried out anywhere from three to over five times a year for nearly every gyratory crusher in operation in the world today. Obviously, exposing service personnel this number of times a year under a suspended load that normally weighs anywhere between 40 t and 110 t, is to be avoided, if possible.

The self-aligning main shaft was specifically developed to stay the need for any personnel to be under the suspended main shaft during the rebuilding process. The solution is simple, allow the main shaft to be lowered into the crusher and by ensuring there is a generous amount of 'lead-in' allowed, then there is no need for service personnel to assist with the entry of the main shaft into the eccentric assembly directly, they can stand back, guide the crane operator from a safe distance and perform the task completely out of harm's way.

As can be seen in Figure 4, the main shaft is lowered into a specially designed dust-collar that has a generous lead-in allowance that allows the bottom of the main shaft to then make contact with tapered guides that begin to 'self-centre' the main shaft as it moves further down.

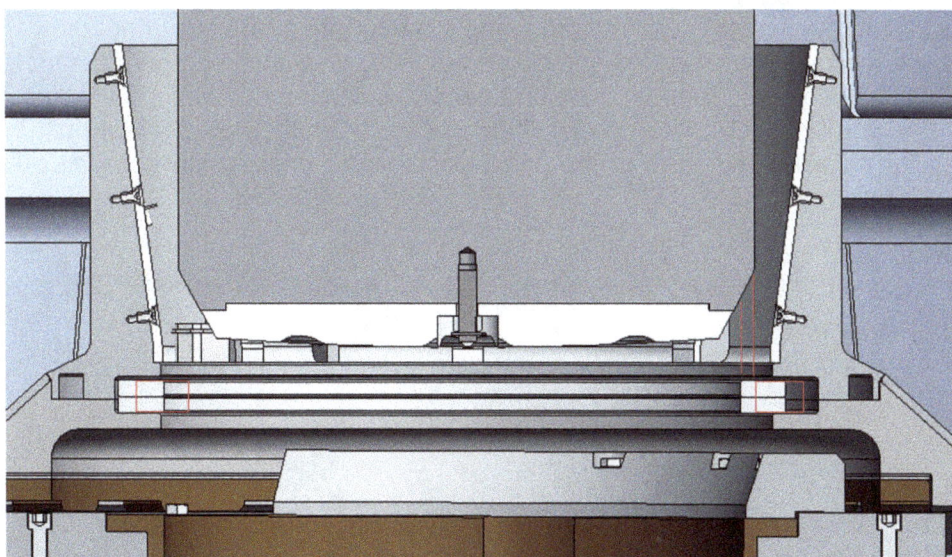

FIG 4 – Self-aligning main shaft concept.

The chamfered bottom of the main shaft automatically centres the dust seal as the shaft locates more and removes the need to centre this part as accurately as possible before installation to avoid damage and subsequent oil leakage.

Further, cameras can now be installed in the dust bonnet and utilised by service personnel and the crane operators to provide a visual of where the bottom of the main shaft is in relation to the eccentric bushing. These cameras are connected to a USB port with spring loaded dust and watertight caps. Once the main shaft is lowered into position, a quick tug on the USB cable and the crusher is ready to run and more the system will be ready to be used again during the next shutdown.

Rotable shell philosophy

The rotable shells concept is particularly useful in applications where production is required constantly and with minimum stoppages related to normal maintenance work, primarily the replacement of worn liners in the gyratory crusher.

By additional investment, normally a spare top shell and a spare middle shell, re-lining the crusher with new liners can be achieved much more efficiently and allow an increase in uptime and therefore production.

Weather is no longer a deciding factor in how well or fast a liner change-out will go or how long a period must be allowed for epoxy to cure due to performing this task in the wintertime etc.

Shells can now be re-lined in the workshop, in a more stable environment and in a proper manner as having this capability allows the work to be carried out in a normal controlled manner, free from the constraints of machine downtime and lost production, which are large driving forces when performing this task *in situ*. This leads to a better job with less chance of problems occurring and even if they do, there is time to re-work in a methodical manner and ensure there is no compromise in workmanship.

The rotable shell concept generally can take standard concave replacement time from 3 to 5 days down to 1½ to 2 days. This allows for more uptime or production to ensure monthly or yearly production quotas are easier to achieve, frees up maintenance personnel for other critical maintenance tasks within the plant and in a new 'greenfield' application, could point to a smaller intermediate stockpile being required along with a smaller stockpile conveyor being required. Smaller plant layout and cheaper stockpile conveyor could be considered 'knock-on' benefits.

The safety aspects of performing the relining tasks in a remote, controlled environment are very much enhanced and ensures a first-class job is carried out every time, without compromise.

Figure 5, in cutaway format for better clarity, illustrates the rotable shell concept. Each shell, normally the top shell and the middle shell are the concave bearing shells, is configured to be individually pre-lined with concaves without having to rely on being assembled as a complete crushing unit. The shells are supplied with their own maintenance/transportation frames that allow them to be assembled in the workshop and poured and cured with epoxy backing compound. Once ready, they are simply transported on their frames to the crusher and swapped out with the shells containing the worn liners. The crusher shells also can be fitted with a hydraulic jack system built into the shell flanges to facilitate quick separation of the shells, ensuring the swap-out goes as quickly and as smoothly as possible.

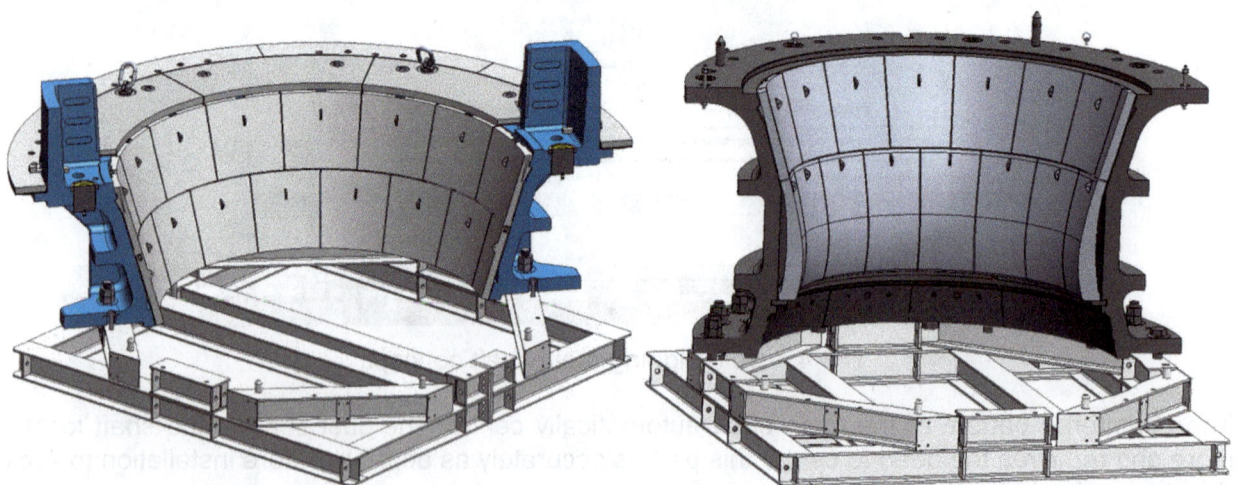

FIG 5 – Cutaway showing the rotable shell concept.

DIGITALISATION ADVANCEMENTS – GYRATORY CRUSHERS

Fully modernised control system

Along with the mechanical advancements made on the crushers in bringing the maintenance and safety practices in line with current expectations and demands. Operations and service staff need to communicate and understand the crushers current operating conditions. The ability to also capture and interpolate historical data assists in future operations by allowing more accurate planned maintenance schedules to be realised and ensuring sufficient replacement parts are identified and pre-ordered in a timely manner.

The latest control system that has been developed meets the above requirements and is also designed to extract the maximum performance from the new generation crushers.

Over the past few decades, a minimum requirement has been expected from the industry on what a control system should include. This has become an unofficial 'standard' and indeed most gyratory crusher manufacturers offer all or most of these features, eg crusher main lube oil temperature, crusher auxiliary lube oil temperature, oil reservoir levels, hydraulic pressure under the main shaft, differential pressure over filters, motor power draw and differential speed between motor and crusher.

The features are shared with the operator in a simple format on the touch screen display of the newly designed control panel. Keeping all information relative to the operating conditions of the crusher visible as a default and then being able to quickly jump to sub-levels, should the requirement for more detailed information or trending data be required for example, as can be seen in Figure 6.

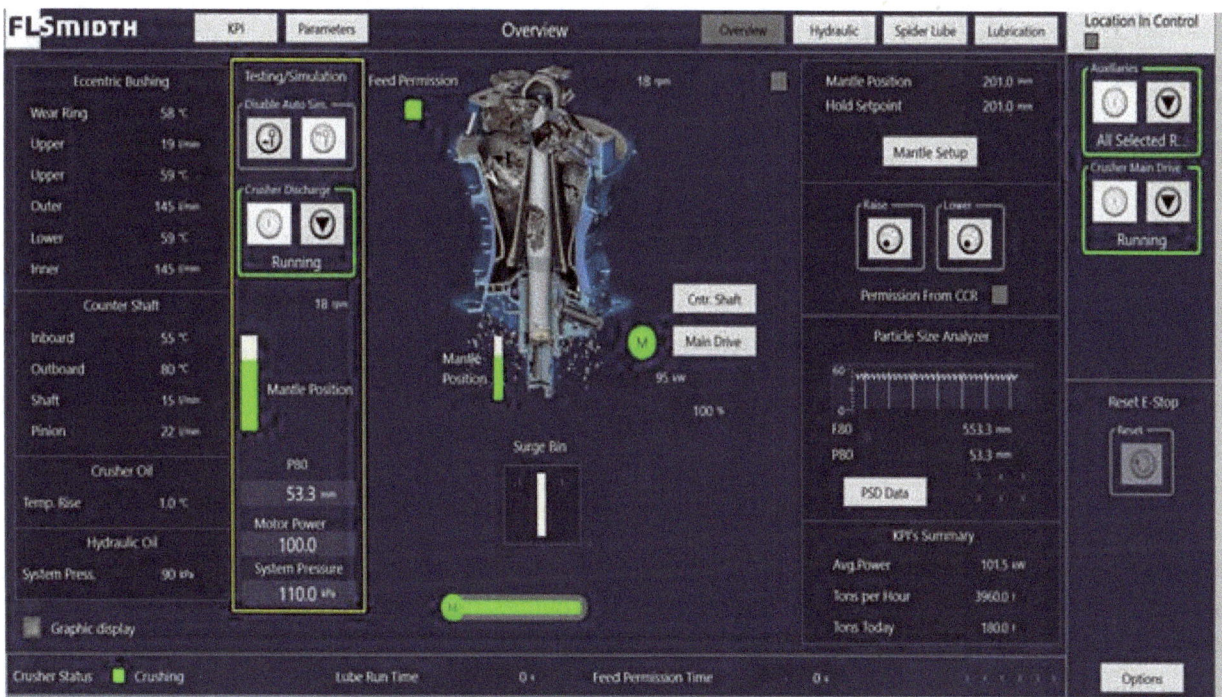

FIG 6 – Image of the gyratory Control Systems homepage, showing key data.

All data recorded is automatically stored in the control system or in the Cloud. This can then be accessed for reference and allow generation of data that might point to an unplanned incident and help explain what occurred. It can also be used for generating trends, leading to a better understanding of the whole process and allowing more accurate and confident fine-tuning of the operation of the crusher to operate, in a safe manner, at its optimum performance.

Additionally, added features now allow for the lifetime prediction of major wear components, allowing more accurate maintenance schedules to be implemented. Again, this has many benefits in terms of forward planning of production rates and forward planning of maintenance and service downtimes, to name but a few.

The latest crushing control systems are designed to have a 'plug-n-play' format and developed with future upgrading in mind being, consequently, a simple task.

All sensors are wired to a local junction box on the lubrication/hydraulic unit and in the field. These boxes are designed to plug directly into the PLC box with HMI touch screen via a single connector. The control system then interacts with the plant DCS system via the preferred communication protocol either EtherNet/IP, Profinet or Modbus TCP/IP, allowing the crusher control system to act as the primary control of the crusher and auxiliary equipment, or be used as a monitoring system with indirect control of the motor control centre. Either way, the control system will provide advance control options allowing the operator to maximise crusher performance be it through controlling the power with the auto power option, which maintains a constant power being delivered to the ROM in the crusher or maintaining a constant gap setting (OSS normally).

One newer development requested by service and operations personnel, is the main shaft spin rate sensor. It has long been known that a change in the spin rate of a main shaft, which (can only be observed when the crusher is running empty, reflects a change in the internal parts of the crusher. Generally, this information was never accurately recorded, being merely passed on via word of mouth or written in the daily log and frequently this information was lost or ignored and seldom reported to the correct team. A sensor has now been placed internally on the spider assembly that can monitor the main shaft rotation and provide data, that over time, can show a gradual increase in the rev/min spin speed of the main shaft (Figure 7a). This can point to a potential problem with not only the spider bushings but also act as an indicator that the main bushings may be experiencing issues that could lead to a more severe situation, sometime down the line (Figure 7b). Maintenance team can now forward plan this remedial action well in advance.

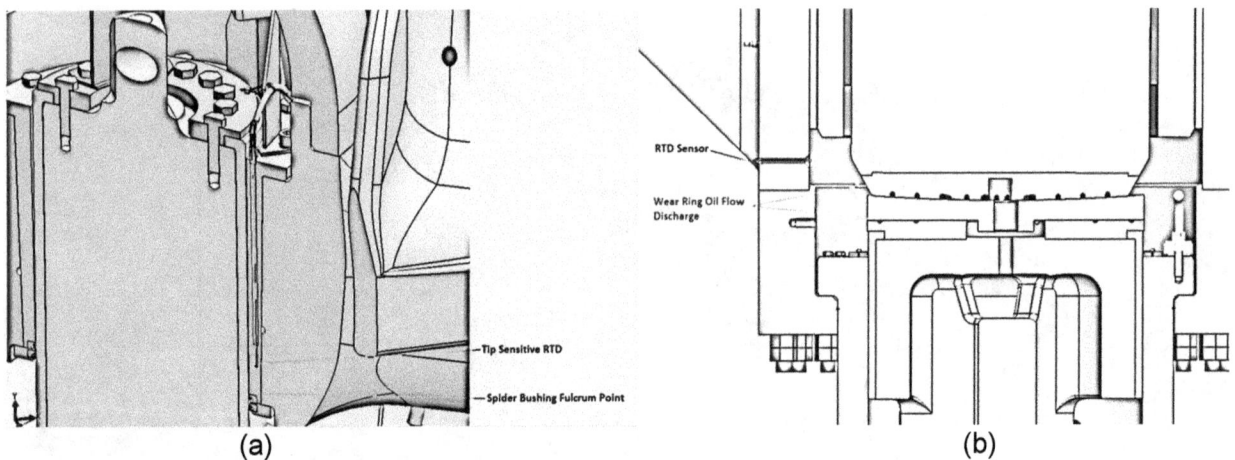

FIG 7 – (a) Spider spin sensor; (b) step bearing sensor.

Another system that is gaining traction is a torque sensor built into the torque limiting coupler that is placed between the drive motor and the crusher. It provides real-time data on how hard the crusher is working and how frequent and severe overloads are.

Integrating camera systems into the control process is now becoming more common with crushing applications. Placed at the crusher inlet, they monitor the haul trucks for oversize feed material and give the crusher operator the option to have the truck dump the oversize material away from the crusher, so avoiding a potential blockage situation.

Cameras are also being used on the crusher discharge belts to provide product size distribution (PSD) analysis. Not only can the operator see the PSD exiting the crusher, but the control system now has the capability to act on this information and adjust the OSS setting of the crusher to maintain a more constant product being sent to the next stage of the comminution process.

Condition monitoring system (CMS)

As the control system is now capable of providing the operator with more valuable data, it is important to organise this information in a way that can be understood and subsequently actioned upon.

Condition monitoring provides frequent reports that are generated via data sets collected by the system. New learning algorithms can detect and report potential issues in advance, allowing a plan to be developed, parts purchased and labour scheduled in a timely manner. This ensures maximum operational availability of the equipment and contributes towards the yearly crushing metrics being achieved.

As mentioned above the sensor package for crushers has been defined and utilised to great success, but to further the capabilities of a condition monitoring system additional sensors have been added. Localised heat generation inside the crusher has predicted bearing and gearing issues. In order to measure this, additional temperature sensors have been added to the crushing system. Oil temperature sensor in the step-bushing area and temperature sensor located at the spider bushing, give early indication and warning of any sudden temperature increases.

Three vibration sensors have been added to the crusher countershaft assembly. One vibration sensor per countershaft bearing in the radial direction with one additional vibration sensor monitoring the axial direction. These provide information on both the roller bearing health and the pre-loaded wave spring.

There are several third party companies that can also provide add-on systems to help better understand the crusher auxiliaries. One such system is an oil contamination system that can be added to the outside of the lubrication/hydraulic reservoir and on a defined schedule run an analysis on the oil. It provides three cleanliness levels and a moisture sensor to detect water in the system. The cleanliness rating can provide an indication when a sample should be to be sent to a lab for more detailed verification of what contaminates are polluting the oil, ie bronze, iron, or good old fashioned dirt.

CONCLUSION

Finally, safety in the industry is taking a 'front seat' in the mechanical engineering design of process equipment and more and more attention is being sought, nay demanded, in this long overdue theatre. With today's emphasis on safe working practices and, to summarise, 'sending people home safely in the 'same condition they reported for work', albeit a little more tired. That is our ultimate goal, not only as equipment suppliers, but by working in partnership with our responsible end-users to achieve this goal.

When we couple the above with the demand for increased throughput capacities and power handling capabilities, this gives us an ever-evolving task that has to be constantly re-looked at and balanced on a day-to-day basis. The potential is there to design machines that can comply with today's demands, ie increased throughput, power, reliability and efficiency and satisfy the requirements for safety and minimal downtime etc. Only time will tell if we head in this direction, but early indications would suggest that industry is seriously considering these latest, more environmentally friendly, safety conscious and more efficient and powerful machines and that they are taking them seriously as the next step towards long-term sustainability.

Analysing impact forces and overcoming speed, heat and pressure issues in high capacity belt support applications

C T Portelli[1]

1. Senior Mechanical Engineer, Kinder Australia, Braeside Vic 3195.
   Email: cameronp@kinder.com.au

ABSTRACT

Belt conveyor transfers are the most likely location for high wear rates and failures. Belt transfers are necessary to change the direction of conveyed material and will remain a part of belt conveyor systems into the future.

Burden being accelerated due to fall and changes in direction from one system to the next, prevents steady state flow which introduces component fatigue. The conveyor belt is considered the greatest cost item over the life of a belt conveyor system and consumes considerable downtime to replace, therefore a financial incentive exists to preserve this high-cost item. Other issues created at the transfer include health risks of uncontained dust and product losses due to spillage.

Additional consideration to support the belt is one way to improve the life of components and contain dust within the transfer chute. The humble impact cradle/bed has changed little over the years whilst conveyor systems have achieved ever greater flow rates. Further development of the impact cradle is an opportunity to reduce maintenance costs and increase uptime.

The peak technology on the market for conventional impact belt support is the dynamic impact cradle, which ensures the belt support area under the chute allows for some dynamic travel, whilst also maintaining a consistent skirt board area. The impact energy at a transfer when installed with Kinder Australia's dynamic impact belt support system was measured to further understand the forces involved and how they compare with static belt support systems. Adding further dynamic capacity to the load zone has been shown to increase belt life by at least 30 per cent. Other components also benefit from reducing impact energy in the transfer and the incorporation of polyurethane bushes at roller supports has been employed further increase roller and frame life.

Kinder Australia has developed a unique and innovative range of belt support technologies that further promote dynamic travel whilst maintaining a consistent skirt board area without a significant increase in belt-friction tension. This technology combines slider rails with rollers that can absorb impact independently within the support system and/or utilising exotic slider materials to overcome high belt speeds and/or system capacities. Kinder Australia has a vast library of documented case studies and application data to ensure the future systems being offered will survive, provide better chute sealing and protect the conveyor belt.

INTRODUCTION

Given the high failure rate to components in the transfer above all other system components along with significant wear to the conveyor belt caused at the transfer, why have transfers?

- Avoiding environmental features or roadways. Finding the cheapest path.

- Avoiding vertical/horizontal curves which allows for a simpler conveyor design and better conveyor belt tracking.

- Reducing risk in the event of a belt puncture that may propagate through the full tape length of the conveyor belt. A smaller system will damage a lesser quantity of conveyor belt in this event.

- Running bulk material through other processes such as crushing, screening, washing and stockpiling requires that a transfer be used to collect the product for transport to the next stage.

For these reasons, the conveyor belt transfer will remain a part of conveyor belt systems into the future and engineering solutions have been developed and will continue to be further developed to solve issues around the transfer. Such issues include:

- Equipment and maintenance costs:
  - belt cover and carcass impact damage (large lump)
  - belt top cover groove lines from material entrapment beneath the skirts
  - idler and structure impact damage
  - belt top cover wear induced by accelerating the bulk material from one system to the next
  - carry back induced mis-tracking causing damage to the belt
  - maintaining the chute itself (liners/steelwork).
- Health effects:
  - dust creation
  - noise pollution.
- Lost productivity:
  - chute blockages and hang-up
  - carry back induced mis-tracking causing spillage.

To prevent belt sag induced gaps at the transfer and to offer a more reliable belt support solution, the impact cradle was developed. Typically, a composite bar made up of a low friction UHMWPE (Ultra High Molecular Weight Polyethylene) top, an aluminium rail for rigidity and fastening plus, a low durometer rubber or polyurethane base to offer some dynamic travel to lessen the impact loading on the belt. The relative rigidity along the length of the bar offers a constant flat surface to prevent gaps opening between the belt and skirting system.

The impact cradle is a simple, effective and reliable solution for many belt conveyor systems, particularly at the primary end of small to medium size ore operations. They are also generally easy to maintain given drop-down wing set-ups or slide out retractability (Figure 1).

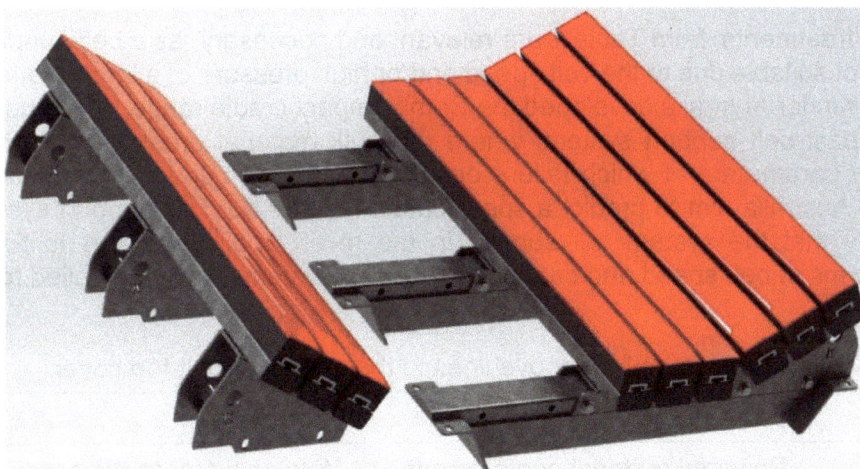

FIG 1 – K-Shield Impact Belt Support System retractability demonstration.

For lighter duty applications where the lump size, system capacity or drop height is significantly lower, a belt support system can be retrofitted to the existing frames and rollers. The K-Sure® Support (Figure 2), whilst strictly not an impact cradle, does perform the role of a constant flat surface upon which the belt to skirt gap is kept to a minimum.

FIG 2 – K-Sure® support retrofit belt support system.

Keeping the product within the transfer has obvious advantages, such as spillage and dust reduction, but there is a more sinister issue caused in the form of groove lines in the top cover of the conveyor belt on systems that do not have adequate transfer support. Entrapped material can make its way between the belt and skirt, usually between roller sets as a gap is formed by the belt sag. As the material makes its way to the next idler set it gets compressed between the belt and skirt. This forces top cover wear that can be seen as two groove lines right where the skirting exists on the belt. These groove lines cannot be easily cleaned and lead to an early replacement of the belt, as further wear at this location starts to penetrate the belt carcass.

Solutions to combat this issue are the use of spray bars within a contact skirt to ensure no product is allowed to get caught in this area, the bonding of polyurea in the groove line to extend the belt life or the tapering out of the chute to spread the concentration of wear over a greater area. Given these options have downsides and are treatments that do not address the root cause, a belt support option is always a more preferred method.

In some cases, 'treatments' from Table 1 are relevant and necessary, as a belt support system may be considered not suitable due to the belt speed and/or high pressure of a large material throughput. To counter this, Kinder Australia developed the Combi Impact cradle range, which uses rollers in the centre of an impact belt support system, where the bulk material pressure is highest (Figure 3). Limitation to the parameters at which failure occurs have been found experimentally in the past, however Kinder Australia aim to predict a specification limit for our belt support systems to ensure most importantly that the system will survive in the given application and to further push the boundaries in terms of belt speed and capacity that these systems will be installed to.

TABLE 1

Methods to combat groove lines in the conveyor belt top cover.

Solution	Pro	Con
K-Hydra Belt® Shield (Figure 3)	Powerful material containment without wear to the top cover.	Adding water to the product stream is undesirable for some product types.
Polyurea groove line fill (Figure 4)	Extends the already installed belt life and can offer a greater wear resistant area at the skirt/belt interface.	A patented, cost-effective process that is solving the symptom, not the root cause.
Tapered chute (Figure 5)	Chute design allows the potential for top cover wear to occur over a greater area, reducing groove concentration.	Slightly more complex chute and skirting design (when belt is troughed). Difficult or impossible to implement on longer transfer points or where multiple transfers exist along a system.

FIG 3 – K-Shield Combi Impact Cradle.

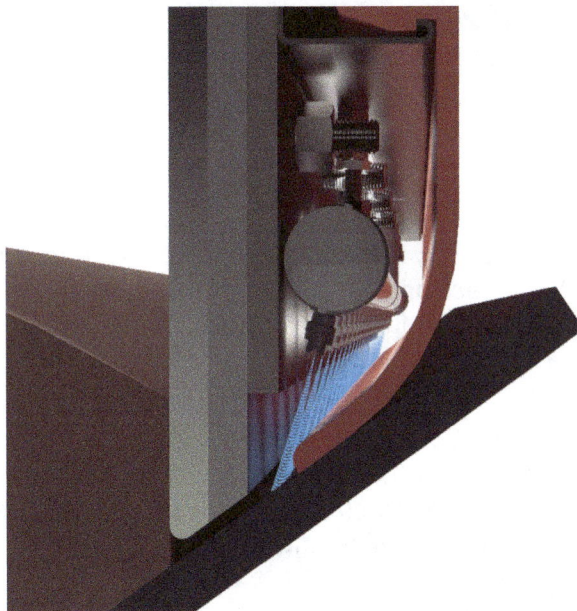

FIG 4 – K-Hydra Belt® Shield, high pressure water spray behind a lay in skirt.

FIG 5 – Belt cover repair system shown here to fill in grooves caused at skirt zones (source: C Uchtman, Wear Systems Solutions).

FIG 6 – Tapered chute on a flat belt conveyor system.

TEMPERATURE ANALYSIS

A failure regarding insufficient support is unheard of within Kinder Australia, as all our 'impact' rated systems have fully supported impact bars or fully supported rails on springs. Only the K-Sure® Support uses a slider surface that is not fully supported. However this system is not considered for severe impact loading, only as a slider surface for relatively free flowing bulk material, usually no more than 1000 TPH and lumps size no larger than 50 mm.

Ensuring the slider material will not exceed the allowable temperature requires a look at the actual system being used. We will first consider the dynamic impact cradle, as we have managed to collect some data on a system of this type in service (Figure 7). This installation is considered a success as the wear rate on the rails is to date negligible and the client is pleased with the improvements to spillage and reduced maintenance required in the area.

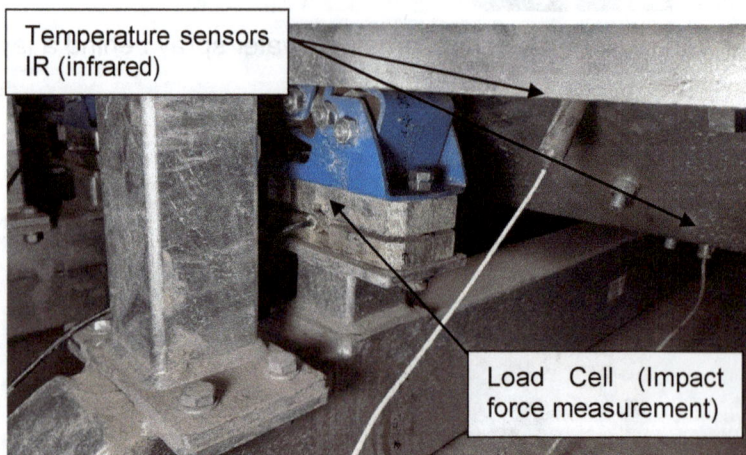

FIG 7 – K-Shield Dynamax® Impact Belt Support System with temperature and impact force data measurement devices.

This type of belt support system consists of UHMWPE slider rails, fully supported by steel plate (Figure 8). The centre section is dynamic via six torsion springs to absorb impact and the wings are static to ensure a consistent belt to skirt gap. As each slider rail is a rather large surface being acted upon by friction induced heating, we will assume the problem to be of one dimensional heat conduction transfer.

FIG 8 – K-Shield Dynamax® Impact Belt Support System.

The rate of heat energy generated at the friction boundary can be quantified by the following, with data from Table 2:

$$P = \text{Force}_{\text{Friction}} \times v \tag{1}$$

$$\text{Force}_{\text{Friction}} = \mu_{\text{UHMWPE}} \times (N_{\text{Burden}} + N_{\text{Belt}})\, L \tag{2}$$

$$N_{\text{Burden}} = \frac{TPH \times 9.81}{3.6 \times v} = 1090 \text{ N/m}$$

$$N_{\text{Belt}} = 274 \text{ N/m}$$

$$\text{Force}_{\text{Friction}} = 0.22 \times (1090 + 274) \times 3 = 900 \text{N}$$

$$P = 900 \times 1.5 = 1350 \text{ Watts}$$

Current values were measured at the drive motor before and after impact cradle installation. This allows us to confirm the accuracy of this value:

$$P = V \times \Delta I = 415 \times 3 = 1245 \text{ Watts} \tag{3}$$

TABLE 2

Data for full steel supported slider system.

Case Study 1 – K-Shield Dynamax® quarry install (hornfels 90%, granite 10%)				
Data	**Symbol**	**Qty**	**Unit**	**Notes**
System capacity	TPH	600	$Tonnes/Hour$	Nominal
Belt width	BW	1.4	Metres	Equal three roll trough
Belt speed	v	1.5	$Metres/Second$	Measured
Slider length	L	3	Metres	2 × 1.5 metre systems
Lump size		250	mm	Note lump size greater than allowable for K-Sure® Belt Support System
Friction coefficient	μ_{UHMWPE}	0.22	----	Dynamic against rubber, proprietary formula test
Thermal conductivity	K_{Rubber}	0.13	$W/m.K$	Soft Vulcanised (Çengel, 2006)
	K_{UHMWPE}	0.41		OK 1000 (Okulen, 2013)
	K_{Steel}	~60		AISI 1010 (Çengel, 2006)
Temperature ambient	T_{Ambient}	20	°C	Measured
Max allowable slider temperature	$T_{\text{Allowable}}$	80	°C	(Okulen, 2013)

The current difference was an average when fully loaded and shows an absorbed power at the drive due to an increase in friction within an acceptable error can be quantified using Equation 1. At Kinder Australia, when quoting the extra drive power required, we use a friction coefficient of 0.3 as a conservative figure and for static cases (start-up).

If we assume all the friction power generated goes into heating the slider rail, we need to ensure the allowable temperature for the material is not exceeded:

$$Q_{In} - QO_{ut} = \frac{dE_{Wall}}{dt} \tag{4}$$

$$\text{Conduction Heat Transfer Rate } Q = k. A \frac{\Delta T}{L} \tag{5}$$

Equation 4 is a one dimensional conduction in plane walls (Çengel, 2006) and is illustrated in Figure 9. Equation 5 is from Çengel (2006).

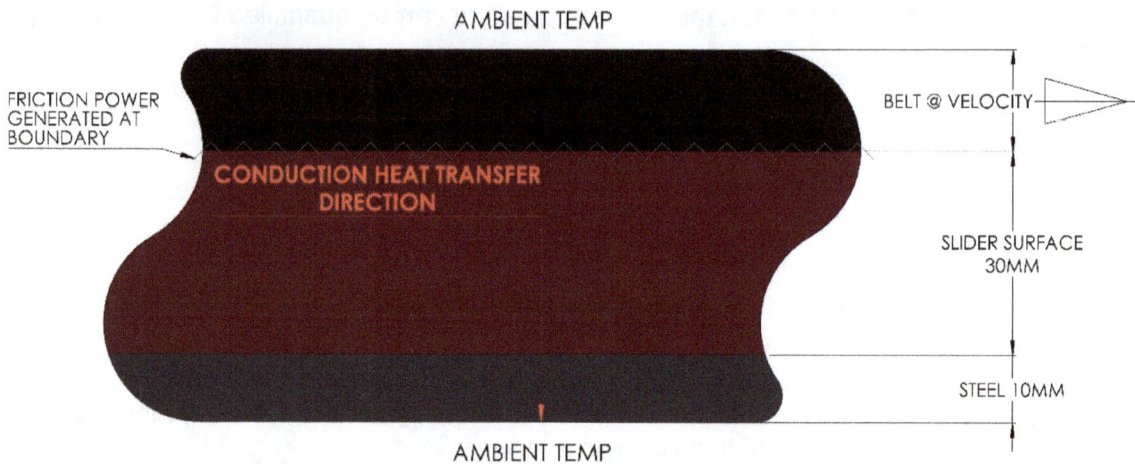

AMBIENT TEMP

FRICTION POWER GENERATED AT BOUNDARY

BELT @ VELOCITY

CONDUCTION HEAT TRANSFER DIRECTION

SLIDER SURFACE 30MM

STEEL 10MM

AMBIENT TEMP

FIG 9 – Heat transfer schematic for full steel supported slider system.

Given the belt is predominantly made of rubber which has a very low thermal conductivity value (poor conductor or good insulator) and that the steel support is a very good conductor, we can assume no heat will transfer to the conveyor belt and any temperature variation across the slider surface will be quickly conducted out by the steel support, keeping the underside of the slider at ambient or close to ambient. The steel components are large and have a large surface area for convection to then take place. We are considering this as a steady state conduction so, $\frac{dE_{Wall}}{dt} = 0$.

We want to consider the slider area that is under the greatest pressure. This area, given the trough layout is the centre of the belt where most of the product is piled onto. Typically, the mass distributed to the centre roller of an equal 3 roll idler is 66 per cent. Given this is a transfer point, we will assume 70 per cent is applied to the centre as a chute with vertical skirts typically narrows and steepens the burden profile than it would naturally sit in a regular carry trough along the system.

$$P_{Centre} = P \times 0.7 = 945 \text{ Watts}$$

Using Equation 5 and a centre slider rail effective width of 450 mm, we obtain a maximum heat transfer rate possible to avoid exceeding the maximum allowable temperature for the slider material.

$$Q_{Centre} = \frac{0.41 \times 0.45 \times 3 \times (80-20)}{0.03} = 1107 \text{ Watts}$$

The above is an allowable maximum to not exceed the temperature of 80°C when in a 20°C ambient temperature environment. Given we are only producing 945 watts of heat generation at the centre of the slider surface, theoretically 80°C will never be reached.

It is however understandable why some designs have emerged whereby the centre section of the impact cradle uses rollers instead of a slider bed surface (Figure 9). This negates an estimated 70 per cent of friction force generation and associated potential for slider material failure and risk of insufficient drive power. It does however require a very heavy roller be installed in this location so that it can resist the impact loading. Kinder Australia have also found a roller to be less reliable when considered as yet another moving part in the system and less accessible for maintenance than a slider rail system.

The effect of a raised ambient temperature is clear if we consider a limit being where P is equal to Q for the centre of the impact cradle. This puts a limit on ambient temperature at 32.1°, entirely plausible given install occurred in winter, Melbourne experienced a relatively mild summer (no days over 40°C, (Australian Government Bureau of Meteorology, 2021) and that the system is installed surrounded by a concrete structure under a jaw crusher that is shielding the system from direct sunlight and associated temperature rises.

The sensor data showed a 7°C difference measured over 20 mm (Figure 10, sensor 3), which equates to 194 watts of conduction for the centre section. This is much less than the expected 945 watts. We know our friction factor is accurate, therefore there are other rates of energy lost from the slider material. Such losses may be attributed to convection from the sides of the slider rails, quite feasible given the 450 mm effective width in the centre is made up of three 150 mm wide rails, allowing for much surface area that can interact with some potentially turbulent air due to the moving conveyor belt (Figure 11). Additionally, conduction heat transfer to the conveyor belt may be playing a part, though given its poor conduction value, this is likely playing a lesser role. Further improvements will also be made to temperature measurement device placement on future testing systems, to increase accuracy and further quantify this phenomenon.

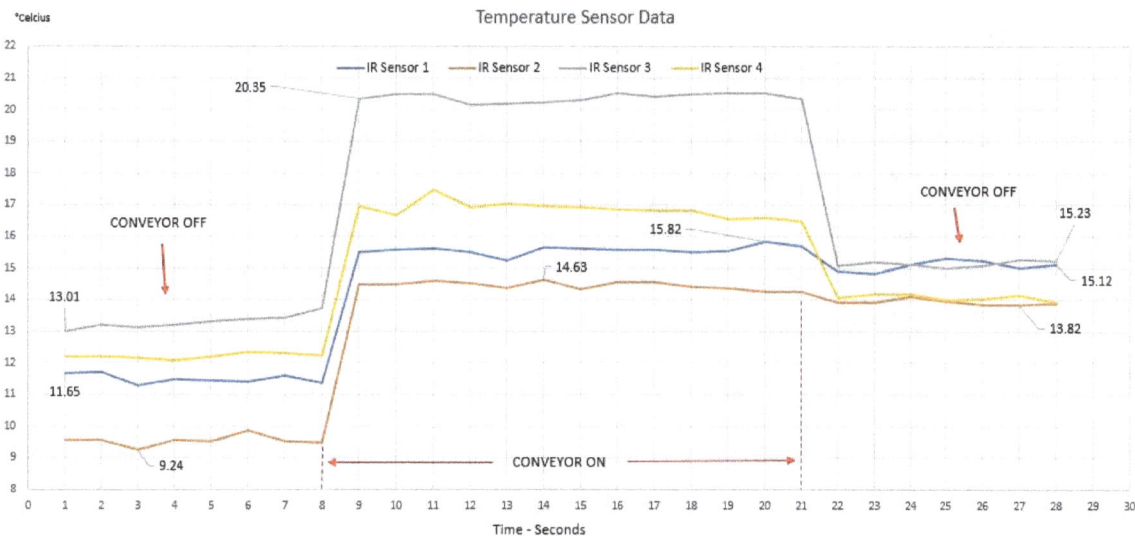

FIG 10 – Typical temperature sensor data log.

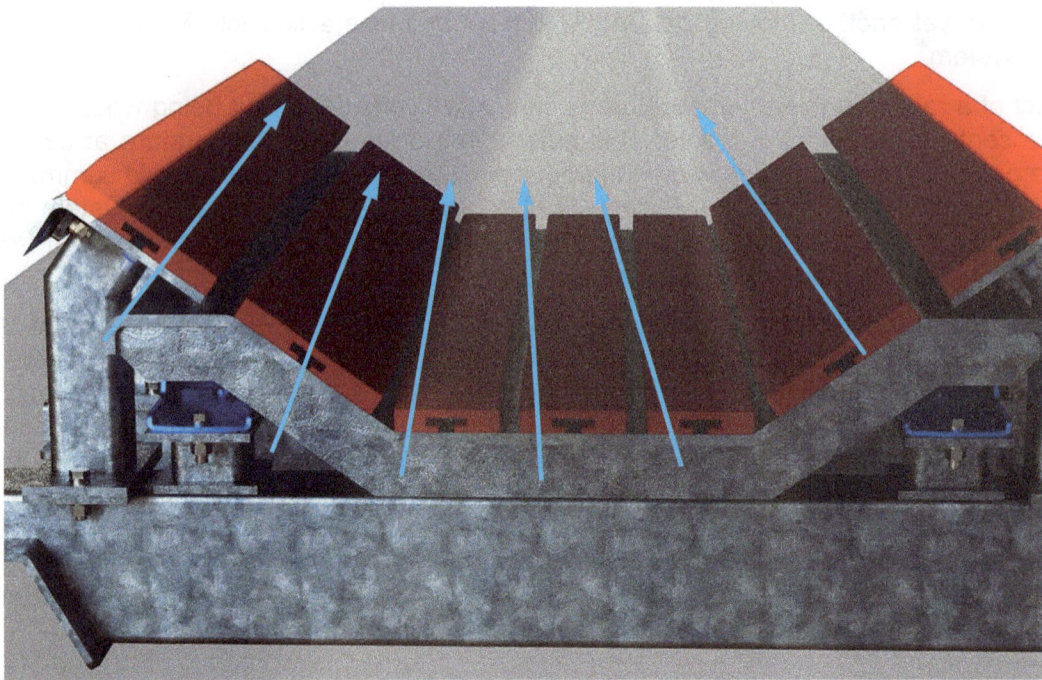

FIG 11 – Convection heat transfer illustration, arrows indicate air being drawn into slider rail gaps by the conveyor belt.

Further doubt to conduction only heat transfer analysis and assumptions is introduced when considering an impact bar made up of a low durometer rubber base with aluminium track and the same UHMWPE material slider surface, albeit a lesser 10 mm of thickness. If the same assumptions are used here, the heat energy has nowhere to go, with rubber acting as 'insulation', both top and bottom of the rail (Figure 12). For completeness, we can work out the maximum permissible friction power for this system (Çengel, 2006):

$$R_{Total} = \frac{L_1}{K_1 A} + \frac{L_2}{K_2 A} \tag{6}$$

$$Q = \frac{\Delta T}{R_{Total}} \tag{7}$$

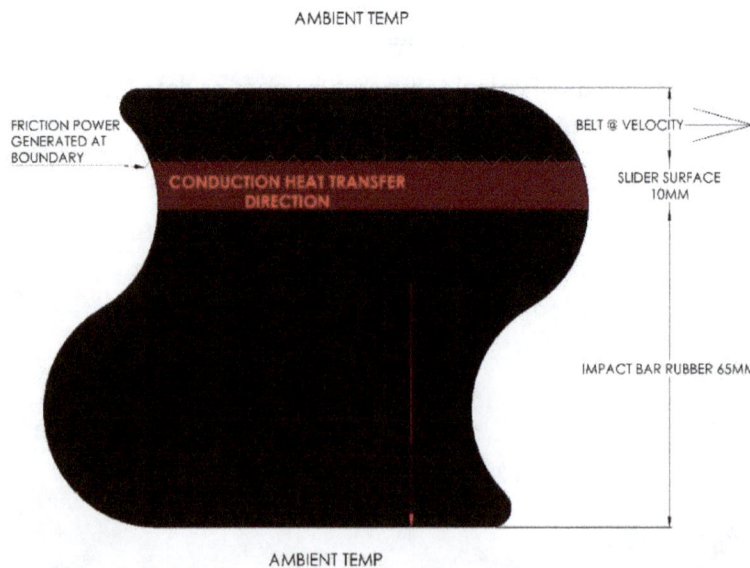

FIG 12 – Heat transfer schematic for a typical rubber based impact bar.

Using the previous equations and ignoring the quantity of heat transferred to the belt (if any at all), shows an allowable friction power of 171 watts for the centre of an impact cradle in the same application as discussed previously using data where necessary from Table 2. This is further proof that more research is required to properly ascertain for the materials and support system type, the friction power limit and deriving a maximum heat flux with unit 'watts per square metre'. This method will be a practical shortcut to avoid numerical solutions for unknown properties such as convection heat transfer, which would be heavily dependent on the amount of turbulent air in the immediate area. We can also account for 'reasonable' levels of allowable wear in the slider rail material using this method. For example, a slider material may not be reaching its temperature limit, though still wearing out in an unacceptable time frame. Other variables include thermal contact conductance (conduction value for two disjointed surfaces due to microscopic air particles) and the fact that temperature and contact pressure significantly affect the kinetic friction coefficient (Ptak, Taciak and Wieleba, 2021).

$$P = \text{Force}_{\text{Friction}} \times v \tag{1}$$

$$P_{\text{Max}} = \dot{q} \times A, \left(\frac{W}{m^2} \times m^2 \right) \tag{8}$$

$$\frac{\text{Force}_{\text{Friction}} \times v}{A} < \dot{q} \tag{9}$$

$$\frac{\text{Pressure} \times v}{A} < \dot{p} \tag{10}$$

Equation 9 (the Material based limitation factor) and Equation 10 (the Application based limitation factor) do not consider ambient temperature. We can either take a conservative approach and limit all relevant systems to the worst-case scenario, or perhaps temperature dependant levels of $\dot{q}$ and p could be derived. Additionally, $\dot{q}$ will be material dependant as the friction coefficients vary. Equation 10 was developed so all material types could be compared against each other, removing the variable material friction factors, to realise the output as an application specific factor. Kinder Australia has a vast library of documented success and failure case study specifications which can be analysed. Preliminary data shows the K-Sure Support has a much lower allowable heat flux, which may point to a sagging of the rails after some heating, thus placing greater pressure and overheating the remaining supported sections of the slider (Figures 13, 14 and 15).

FIG 13 – Possible failure mode under high ambient temperature or high belt speed and capacity conditions, placing higher loads on smaller areas due to sagged slider rails once the softening temperature is reached.

FIG 14 – Variable wear thickness along slider rail.

FIG 15 – Signs of friction generated burning on the slider rail.

SLIDER MATERIAL OPTIONS

A superior slider material has been observed in a Kinder installed application since 2011 at a lignite fired power station (Case Study 2). An increase in belt speed or capacity insists that we either look at a Combi design, use a slider material with better properties or a combination of both methods to reduce friction in the transfer. This can also be necessary in slow moving feeder belt applications, where the pressure across the transfer is very high. For Case Study 2, the belt speed is high at 5.1 m/s and moderately high capacity of 2500 TPH. To ensure there were no drive power issues or overheating of slider materials, Kinder Australia opted to use both a centre roller and proprietary K-Glideshield® product at the wings as slider surfaces.

K-Glideshield® is a unique composite engineered plastic that has far superior mechanical and thermal properties when compared to all grades of UHMWPE which lends itself to be the preferred and to date only suitable material for higher capacity and or speed belt support applications, such as Figure 16.

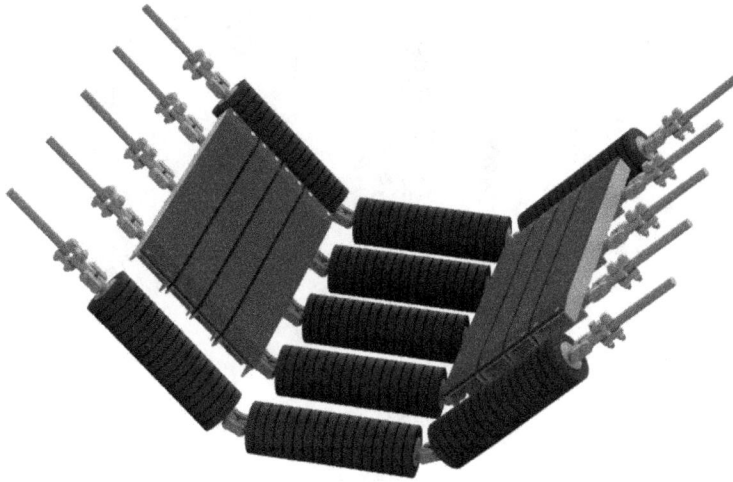

FIG 16 – Lignite fired power station high belt speed support system designed for Case Study 2.

From Table 3, comparisons between K-Glideshield® and UHMWPE materials can easily be made:

- Over 50 per cent less friction induced heat will be produced due to the lower friction coefficient.

- Heat that is generated will be able to move through the entire section of K-Glideshield® rail over 50 per cent faster.

- Any heat that must be stored by the K-Glideshield® rail will be permitted due to the much greater service temperature (assuming a similar specific heat).

TABLE 3

Slider material comparison data.

Material	Dynamic friction coefficient	Thermal conductivity $W/m.K$	Service temperature °C	Notes
K-Glideshield®	0.09	0.64	250	Internal Test
UHMWPE	0.22	0.41	80	As per Table 1
Nylacast Nylube	0.075	0.25	110	Nylacast (2019)

Therefore, both belt speed and capacity of conveyor systems can be pushed much further if a system warrants a slider bed solution. Both much lower friction heat will be generated and the slider materials capacity for temperature is much higher.

Nylacast Nylube is another material that Kinder Australia has used on occasion without success. Whilst having the smallest friction coefficient of all three, it is likely that heat generated at the friction boundary is unable to propagate through the material relying on convection over a much smaller area (sides of rails, typically 10–30 mm). The relatively minor increase in service temperature is another limit for use of this material in slider bed applications.

Whilst the Case Study 2 slider material sees much higher capacity and belt speed, the client has been impressed by the seemingly negligible wear over the past 10 years. A check performed in November 2021 showed low-generated temperatures (Figure 17) and wear of no more than 4 mm on the 10 mm thick wear surface. This system analysed as per Figure 1, shows a capacity of the K-Glideshield® material well above the actual heat it is likely to be experiencing. If UHMWPE were installed, it would likely be borderline as to whether the customer would consider the installation a success, even given there are further heat losses in these slider systems yet to be understood.

FIG 17 – Temperature data showed it was unlikely to exceed the allowable for the K-Glideshield® material.

Kinder Australia have collaborated with clients that wish to push boundaries installing belt support systems in ever greater belt speed and capacity conveyor systems. Case studies that exceed the allowable temperature of the slider material has given us invaluable data which we can apply limits for systems, perhaps using Equation 9 or Equation 10. The above calculations show the heat transfer properties of a system depends on the system being analysed, whether a conventional impact bar system, dynamic slider system, K-Sure® Belt Support System or a Combi system.

Using Equation 10, which omits the friction factor to keep the limitation factor (p value) comparable across all materials, will allow us to impose limits for the different materials and system type combination. This is assuming a steady lead in and lead out idler set run the belt onto and off the belt support system. Without this, additional pressure will be applied to the slider material.

Some interesting K-Sure® Belt Support case studies arose when comparing this subset. Figure 18 shows some successful case studies in green and failures that occurred in red.

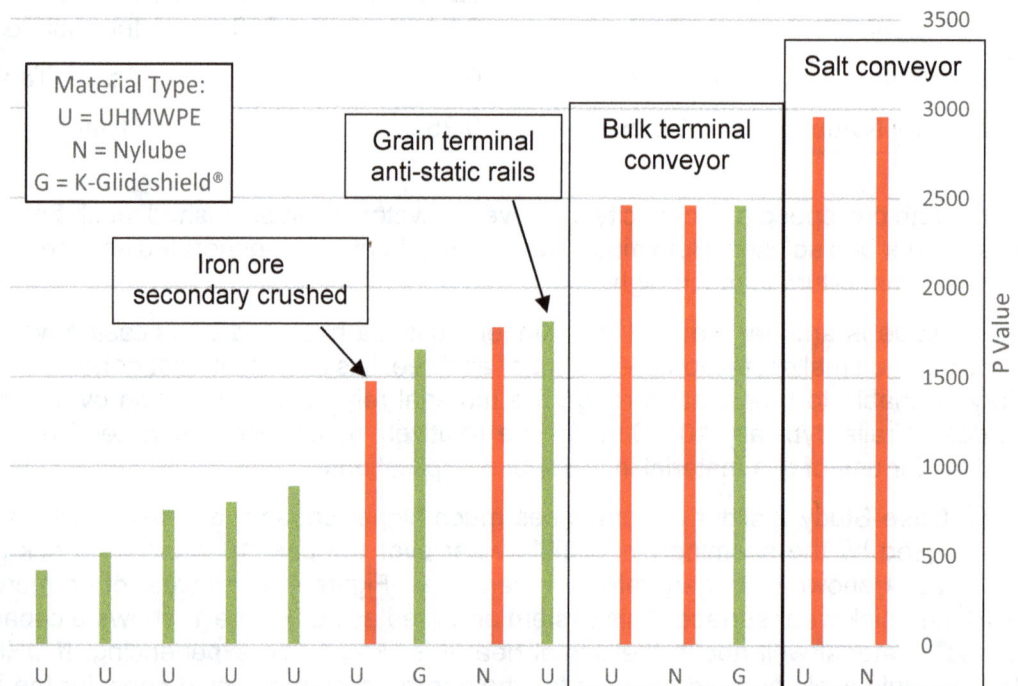

FIG 18 – Case study application factor comparisons (K-Sure® support only).

Case Study 3 – Bulk terminal K-Sure® Support (mineral sand)

- Belt Width: 850 mm

- Belt Speed: 4.2 m/s

- System Capacity: 1150 TPH

- Drop Height: 2 metres

This bulk terminal provided us with invaluable data, given all three material types were tried. It is known for sure that to use anything less than the K-Glideshield® in this application results in a failure.

Case Study 4 – Iron Ore Secondary Crushed Conveyor K-Sure® Support

- Belt Width: 900 mm

- Belt Speed: 3.4 m/s

- System Capacity: 700 TPH

- Drop Height: 2 metres

Our lowest application factor UHMWPE failure occurred in the Pilbara (North-West Australia) where temperatures regularly exceed 45°C (Department of Primary Industries and Regional Development, 2021). This shows a need to consider further limiting the application factor when ambient temperatures are high (Figures 14 and 15). This was also a consideration for the salt conveyor, though with such a high application factor, it was no wonder these material types failed, however it's very likely both these applications would have succeeded using K-Glideshield®.

Case Study 5 – Grain terminal, K-Sure® Support with anti-static UHMWPE rails

- Belt Width: 1050 mm

- Belt Speed: 4.3 m/s

- System Capacity: 1000 TPH

- Drop Height: 2 metres

Our highest successful application factor using UHMWPE stands out, having outlasted lesser duty applications. Further investigation is required as the use of anti-static material may have played a part. This conveyor is located in a milder climate than case study 4.

Kinder Australia plan to apply this method to our other product types and rolling this out to the wider sales and engineering teams to ensure clients will only be supplied a successful system type and material combination. Further data is required to see the effect of the drop height. Initial theories are that a flow rate force is small enough to be excluded in applications where the K-Sure® Support is considered, however with a greater lump size and capacity warranting a more robust system, this may become a factor. For chutes where flow is interrupted with the use of ledges and rock boxes and other soft flow chute applications may allow for a complete exclusion of this force in playing a part.

Case Study 6 – Primary Crushed Copper Ore K-Shield Dynamax® Impact Belt Support

- Belt Width: 2600 mm

- Belt Speed: 4.5 m/s

- System Capacity: 11 000 TPH

- Drop Height/Lump Size: 7.3 metres/600 mm

- p value: 22 600

Another notable system installation for its ability to resist friction induced heat (Figure 19). This heavy-duty dynamic impact cradle with rubber based impact bars topped with K-Glideshield® was custom built to replace the standard UHMWPE rubber based impact bars supplied by Kinder's competitor, which started to smoke upon the early commissioning of the plant. This client has since asked for further cradle installations as the plant ramps up further.

FIG 19 – Case study 6 – belt support system.

A p value of around 12 000 is where Kinder Australia has started to see UHMWPE topped rubber based impact bar systems begin to fail. At 22 600, it is no wonder the copper ore client was witnessing an overheating of the originally installed slider surface, with which they attempted to keep cool by constantly hosing with water until the K-Glideshield® solution arrived.

IMPACT FORCE ANALYSIS

Whilst Kinder Australia have not experienced a structure failure, there are opportunities to decrease the impact forces to protect the belt. One such solution developed by Kinder Australia was the dynamic impact idler, the first of which was commissioned originally for belt protection then went on to increase the conveyor belt system uptime, requiring fewer unscheduled shutdowns to replace failed rollers and idler frames. The case study was covered in more detail in an earlier submitted white paper (Portelli, 2019).

Case Study 7 – Primary Crushed Iron Ore K-Shield Dynamax® Impact Idler

- Belt Width: 1800 mm
- Belt Speed: 4.0 m/s
- System Capacity: 5500 TPH
- Drop Height/Lump Size: 6.0 metres/500+mm

The client would not entertain the use of slider beds in any form as a focus on maintainability of smaller lightweight components was within their scope. Thus, the client was willing to put up with the potential for further spillage. The heavy conveyor belt used on this system was not likely to sag enough between roller sets, keeping these gaps small enough to prevent spillage in the original layout, however the idler added dynamic travel and hence would further open skirting gaps, with the potential for further spillage (Figure 20). With a large lay in skirt panel effectively able to soak up some of the belt sag upon impact travel, the client was willing to bear any potential disadvantages for to increase conveyor belt life and to promote system uptime. Regarding the issue of potential skirting gaps opening upon impact, Kinder Australia have a polyurethane lay in skirt designed to follow the belt due to its pre-tensioned installation location (Figure 21).

FIG 20 – K-Shield Dynamax® Idlers installed beneath a primary sizer.

FIG 21 – K-Snap Loc® lay in belt skirting system can self-adjust to belt sag and skirting wear.

Since a site trial and further implementation across both the primary sizer collection belts, this belt support system has been commissioned to another of this client's sites. Kinder Australia have also made improvements to these systems based on client feedback:

- Risk of torsion springs to fill with bulk material dust. This build-up has the potential to cause the torsion spring to lock up and eventually crack its mount plates as has been the case on other types of equipment where these are used (Figures 22 and 24).

- Roller shaft compressing the support plates due to severe impact (Figures 23 and 25).

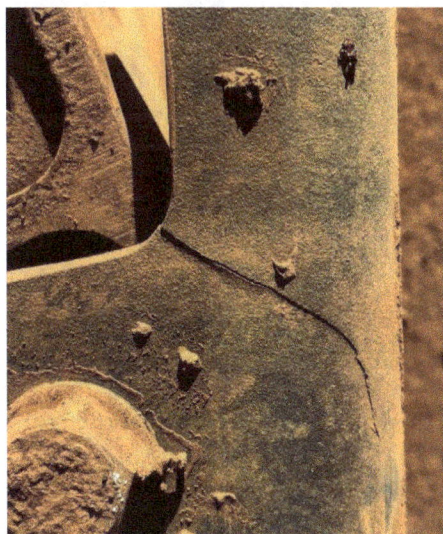

FIG 22 – Crack detail.

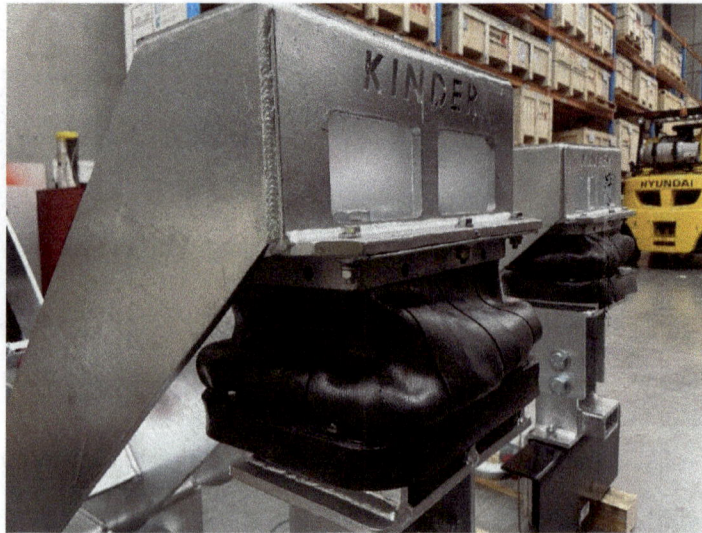

FIG 23 – Torsion spring boot added to prevent ingress.

FIG 24 – Roller shaft support plate steel compression.

FIG 25 – Roller bush to minimise steel on steel contact (roller not shown as not supplied).

This site has seen shutdown cycles for belt replacement move from 6–9 months out to 12 months. This minimum 30 per cent belt life improvement makes savings on a belt change that is reportedly a $250 k exercise (as of 2019, likely much more now), not to mention the opportunity to allocate the labour elsewhere on-site for shutdown works. With roller and idler frame sets lasting 4–6 weeks prior to the Dynamax® Idler install, requiring unscheduled shutdowns of the conveyor for replacement, they are now having no such issues and able to complete multiple shut cycles before offering any preventable maintenance on these items.

Kinder Australia have seen an improvement to belt life when incorporating dynamic travel in the belt support system as well as an overall more reliable transfer. Dynamic travel may be in the form of torsion springs like those used on our K-Dynamax® Idler and dynamic impact cradle, or it may be simply in the form of rubber based impact bars as this low durometer base provides for some dynamic travel. Where dynamic travel is great enough, it should be limited to the load area of the support system where practical and hence why the dynamic impact cradle was developed with only the centre portion able to travel.

Impact force measurements were taken from our equipment installed to Case Study 1. An attempt was made to take baseline data from a basic impact idler system (Figure 26) prior to installation of the dynamic cradle solution. Use of the same load cells and housings to be used on the dynamic solution proved to be a wrong decision. To cover a comparable length of the transfer as would be covered by the dynamic idler, the load cells were placed under and bolted to a rail made from steel angle section. Once all bolted together, the load cells were unable to act independently and deliver much change in force data, even when the system was clearly delivering heavy lump material flow. Future development work is required to ensure the next project can obtain quality baseline data, to see what improvement (decrease in force) if any a dynamic impact cradle makes compared with an impact idler system.

FIG 26 – Baseline impact data recording system.

We can observe the recorded data from when installed to the dynamic cradle, as one load cell was used per spring (independent). It was also possible to observe the impact loading live on-site via the control box (Figure 27).

FIG 27 – Control box for live data view and logging.

With so much data taken over many weeks, it was necessary to seek data of interest, for us this was the peak readings. On many occasions, we noted that higher force readings were recorded on the lead in and lead out torsion springs. Due to the six spring layout (see Figure 28), this leads us to believe that under lump impact the trough panel is rotating about the centre springs (springs three and four, see Figure 29).

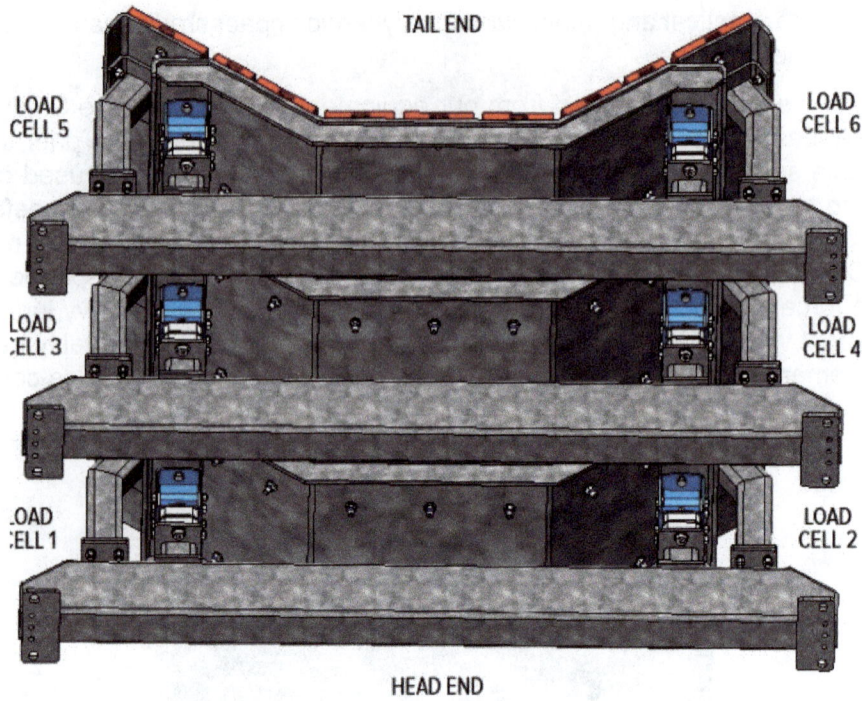

FIG 28 – Cell numbering layout.

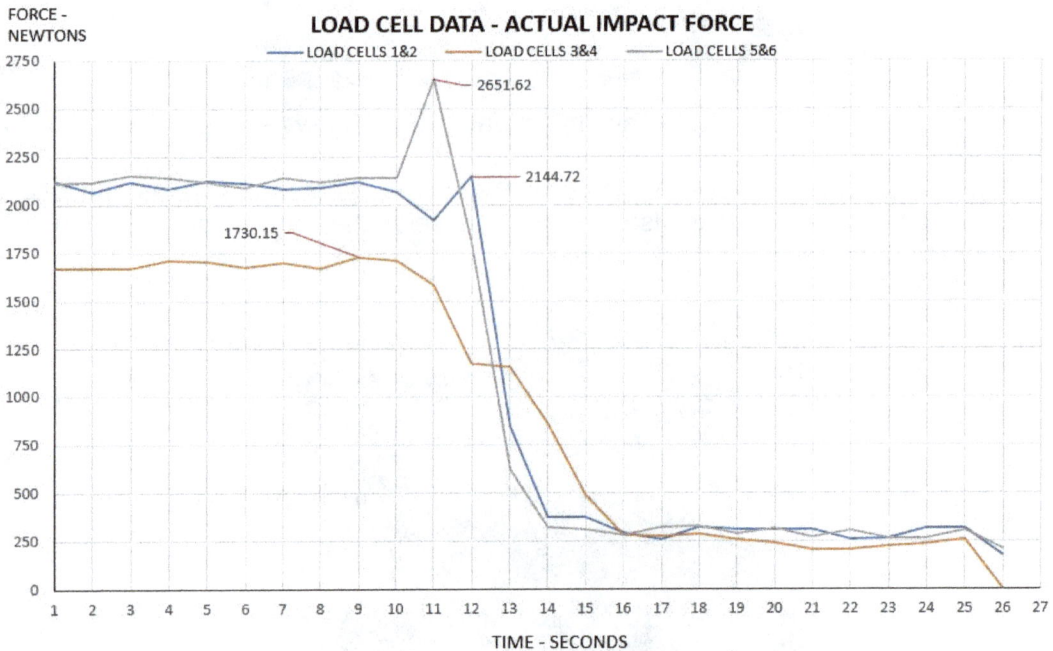

FIG 29 – Typical peak load cell force data.

OPPORTUNITIES TO IMPLEMENT FINDINGS (CONCLUSION)

Other spring layout configurations are an opportunity for further analysis for our dynamic impact cradles. For example, a stiffer four spring configuration may work better than the current six spring layout, potentially allowing a more dynamic capacity at the location of impact. As shown in, as a lump impacts the tail end of the load zone, the cradle is responding by taking up all the force on springs located at five and six, whilst load is coming off the springs elsewhere. Whilst this shows that at least the cradle is semi-independent at the location of impact, there is further scope to allow the dynamic section to respond even better to impact by reducing the quantity of springs as there will be fewer non-impact loaded springs working to unnecessarily stiffen the system.

As this was our first attempt at data capture in a load zone application, further improvement will be made to the data collection system for another attempt at baseline data and subsequent belt support system installation. A summary of these improvements are:

- Ensure baseline data can be obtained accurately, perhaps using two cells per idler frame rather than trying to spread multiple frames over fewer load cells.

- Consider other load cell types and data capture that can produce higher frequency data collection to ensure peaks are accurately captured.

- Improve temperature capture to achieve more reliable data closer to the heat source.

- Allow load cell readings to be independent, rather than averaged across a set of two.

- Installation to a harder/denser rock or ore carrying system, given the hornfels rock from the initial data collection system is quite brittle, lessening the impact intensity.

Our final case study is one that pushes the innovation boundaries to further soften, control and allow adjustment of the spring configuration, as well as incorporating some other improvements from other belt support systems. All credit to our client for opting to incorporate a tapered chute on this troughed system (Figures 30 and 31). They realised the benefits for spreading skirting related wear and opted to spread this over a greater area. This further pushed Kinder Australia to come up with a completely bespoke system to ensure the dynamic section too followed the tapered shape of the chute.

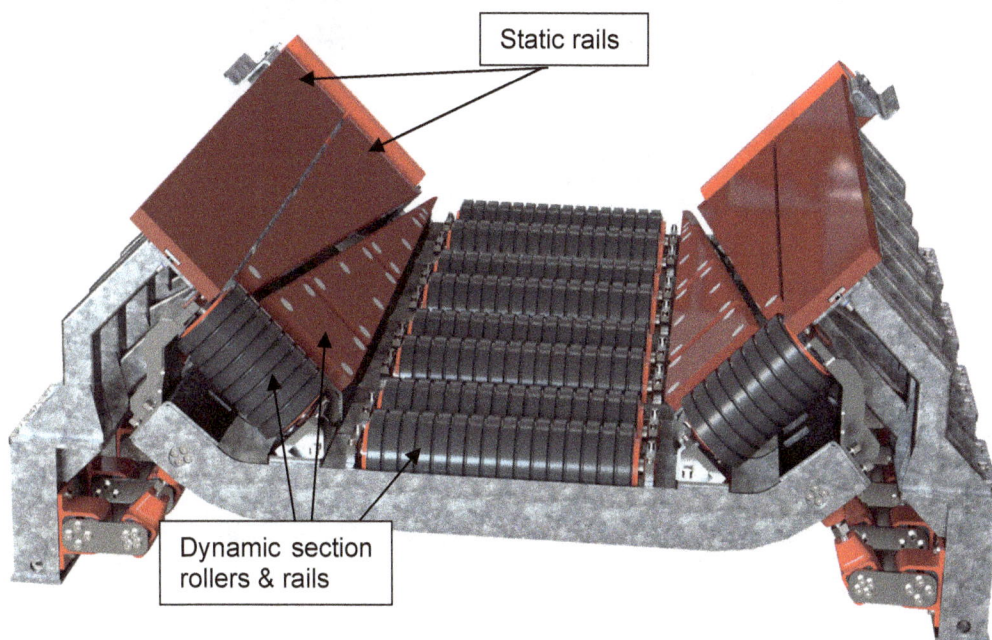

FIG 30 – K-Shield Dynamax® Combi Impact Bed.

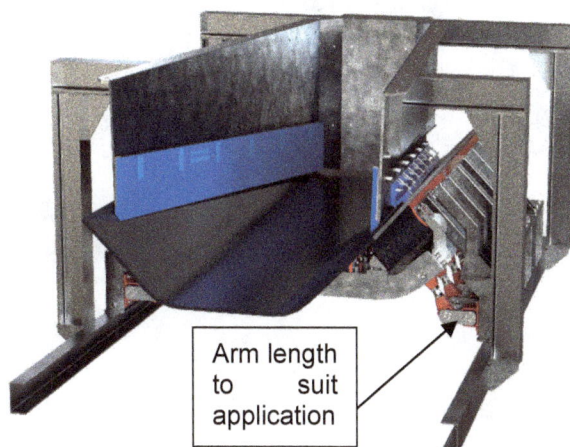

FIG 31 – Tapered chute assembly.

Moving away from 'off the shelf' torsion spring assemblies allowed us to control the distance from pivot to pivot points, thus optimising dynamic travel for a given impact force. This will allow for any future application to be fine-tuned, within the limits that the packaging envelope allows (Figure 32). It also allowed us to incorporate an adjustment rod, that will allow the ride height of the dynamic section to be maintained over the life of the torsion springs, an issue encountered with the K-Shield Dynamax® Impact Idler (Portelli, 2019). Design incorporations from lessons learnt in other transfer applications include:

- Combi design to take the largest friction forces out of the equation.
- Polyurethane isolation bushes incorporated to all impact loaded rollers.
- Pre-tension of one torsion spring to further dial in the dynamic system behaviour.
- Completely independent dynamic sections to ensure springs are less likely to work together producing a higher (harder) spring rate.
- Spring assembly is more open, making build-up issues less likely.
- Lead in rollers ensures excessive pressure on slider rails will not occur.

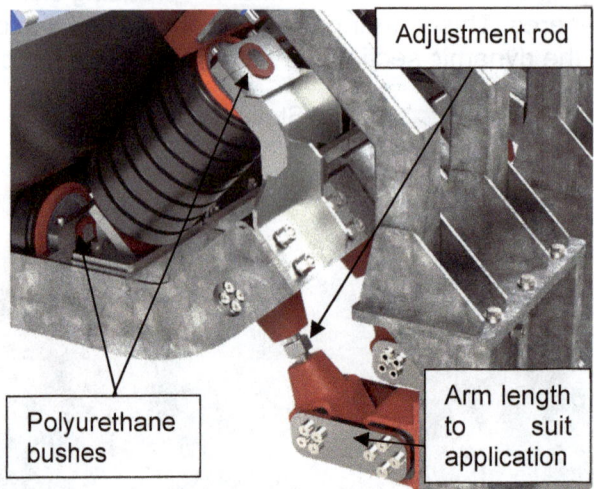

FIG 32 – New torsion spring configuration.

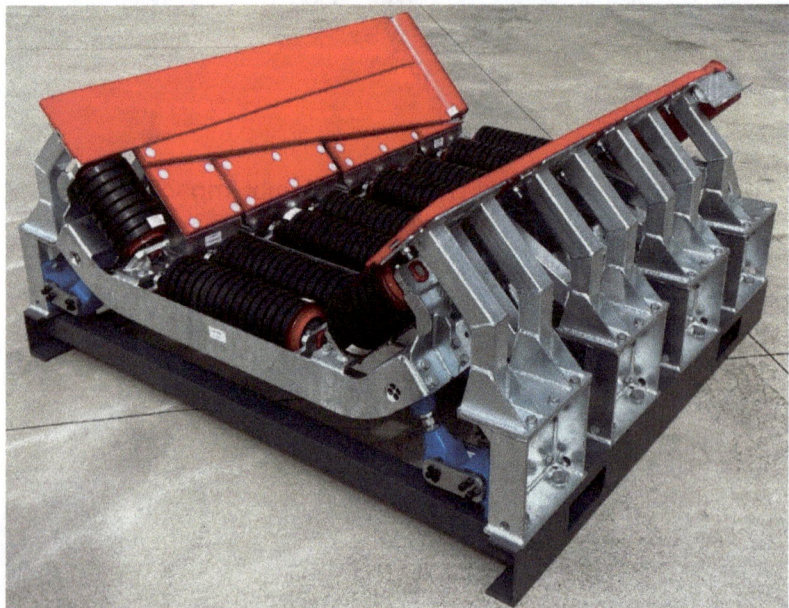

FIG 33 – K-Dynamax® Combi Impact Cradle ready for shipping.

REFERENCES

Australian Government Bureau of Meteorology, 2021, December. Scoresby, Victoria, Daily Weather Observations, Available from: <http://www.bom.gov.au/climate/dwo/202112/html/IDCJDW3072.202112.shtml>

Çengel, Y A, 2006. *Heat and Mass Transfer* (McGraw-Hill: New York).

Department of Primary Industries and Regional Development, 2021, May 3. Climate in the Pilbara region of Western Australia, Agriculture and Food. Available from: <https://www.agric.wa.gov.au/climate-change/climate-pilbara-region-western-australia#:~:text=Temperature%20in%20the%20Pilbara&text=In%20northern%20inland%20areas%2C%20such,%C2%B0C%20across%20the%20region.

Nylacast, 2019, February 01. Nylacast Nylube. Available from: <https://cdn-nylacast.s3.eu-west-2.amazonaws.com/nylacast/uploads/2015/08/NylacastNylubeTechnicalData19.pdf

Okulen, 2013. OK 1000 Technical Data, Ahaus-Ottenstein: Ottensteiner Kunststoff GmBH & Co.

Portelli, C T, 2019. Reducing Belt Conveyor Transfer Impact Energy Using a Dynamic Idlei, in *Proceedings of the 13th International Conference on Bulk Materials Storage, Handling and Transportation (ICBMH 2019)*, pp 272–290.

Ptak, A, Taciak, P and Wieleba, W, 2021. Effect of Temperature on the Tribological Properties of Selected Thermoplastic Materials Cooperating with Aluminium Alloy. *Materials*, 14(2021).

Economic recovery of low-grade fine iron ore using the Optima Classifier®

F F van de Venter, T Kale and J Monama

1. Process Manager, Gravitas Minerals, Benoni, Gauteng 1501, South Africa.
   Email: franco.vandeventer@gravitasminerals.com
2. Chief Executive, Gravitas Minerals, Benoni, Gauteng 1501, South Africa.
   Email: tebogo.kale@gravitasminerals.com
3. Associate – Process Technologies, Gravitas Minerals, Benoni, Gauteng 1501, South Africa.
   Email: jones.monama@gravitasminerals.com

ABSTRACT

For a multitude of years fine iron ore, -1 mm, across the world has been deposited as discard material in tailings facilities. This contributes to environmental problems, decreases plant yield and increases the cost of processing. Fine iron ore recovery processes are often deemed unfeasible or uneconomical, due to the complexity of the reclamation thereof in comparison with high-grade ores. However, recent advances in the recovery of fine iron ore using the Optima Classifier® has shown great promise. The Optima Classifier® is a new hindered settling separator that makes use of an upward current of water for density separation. It's unique design combines three different separating zones for increased efficiency and recovery. In this study the potential of the Optima Classifier® to produce a +63 per cent Fe concentrate from -1 mm feed material was evaluated from a standalone and process perspective. It was found that a +63 per cent Fe product can be produced from a single stage, using the Optima Classifier®. Fine iron ore feed with an average head grade of 50 per cent Fe was upgraded to average concentrate grades of 63.2 per cent Fe. Displaying upgrade ratios of up to 1.26 in a single stage. A three-stage, rougher, cleaner and scavenger system did however show the best performance with recoveries of up to 70 per cent, yielding 56 per cent of a 63 per cent Fe product. Furthermore, the dewatering of the product and tailings was evaluated using the UltraG ultra-fine dewatering screen ensuring dry, stackable material. The tailings could be used in co-disposal, while loads on thickeners and filter presses are significantly reduced. The high throughput of the Optima Classifier® ensures a small plant footprint, minimal civil and structural requirements and low operating utility consumption such as electricity, minimising the carbon footprint of a fines beneficiation plant. The Optima Classifier® in combination with the UltraG dewatering technology has provided a platform for economical recovery of fine iron ore, both current arising and historical. Providing opportunity to lower plant operating costs, reduce environmental liability of tailings facilities and increasing immediate process water recovery.

INTRODUCTION

With the increased steel production in recent years, the production of iron ore rose to above 2 million metric tonnes. In 2021 western Australia was the biggest supplier of iron ore in the world, with 918 Mt, accounting for 37 per cent of the global supply (Government of Western Australia, 2022). With the gradual depletion of high-grade lumpy ore, emphasis is put on the utilisation of low-grade and fine iron ore recovery. In South Africa a substantial future resource for iron ore production is the Banded Iron Formation (BIF), which is considered a low-grade resource, with head grades at 44 per cent Fe. An amenability study on the gravity processing of South African BIF fines was done utilising a combination of milling and shaking table tests, where a series of three different processing routes was evaluated (Da Corte, Singh and Letsoalo, 2023). These processing routes are displayed in Figure 1.

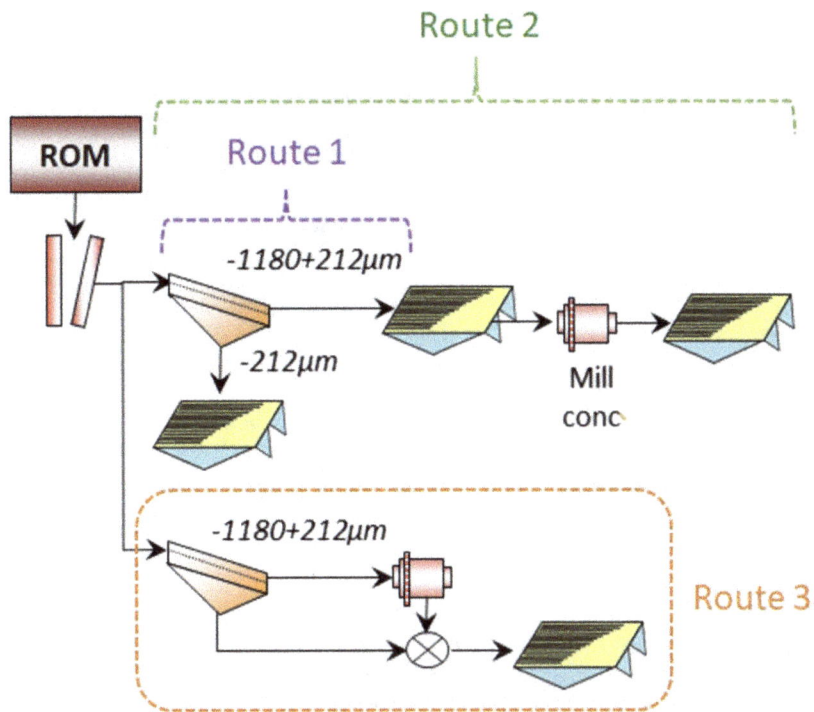

FIG 1 – Gravity processing routes considered in South African BIF processing (Da Corte, Singh and Letsoalo, 2023).

Route 2 posed the economical advantage of reduced slime production and decreased milling costs. It also displayed the best recovery of a 63 per cent Fe product at 45.1 per cent. The study has successfully demonstrated that there is beneficiation potential of low-grade/fine iron ore, with yield and recoveries similar to that of Indian BIF ore (Da Corte, Singh and Letsoalo, 2023). With many available processing technologies, the future of low-grade/fine iron ore beneficiation is evaluated in this paper, with the integration thereof in the environmental, social and governance (ESG) framework.

BACKGROUND

In the advances of low-grade/fine iron ore beneficiation emphasis is put on the cost-effectiveness of the process. This grounds substantial research going into gravity concentration, seeing that the operating costs is generally less compared to processes that needs additives such as flotation or heavy media separation. The most applied technologies in gravity concentration of fine iron ore are jigging and spirals (Sahin, 2020). Although jigging is regarded as a cost-effective fine beneficiation solution, it makes use of moving parts for pulsation, with a decrease in throughput at finer particle sizes in production units (Das *et al,* 2007). Spirals on the other hand, does not make use of any moving parts, but is limited to low upgrade ratios per beneficiation stage, misplacement of hematite and addition of wash water, resulting in the need of various stages in the recovery of fine iron ore (Sadeghi and Bazin, 2020). This is generally CAPEX and OPEX intensive seeing that with increased processing stages come increased use of structures, civils, pumps, piping etc. A new technology, the Optima Classifier® (OPC), addresses these technical and operational restrictions in the processing of fine iron ore.

Optima Classifier®

Design and operating principle

The OPC is a gravity concentration technology that exploits hindered settling and the properties of minerals to separate particles based on density and size. The OPC is comprised of three main sections namely the fluidisation chamber, fluidised bed chamber and the recovery chamber, as seen in Figure 2.

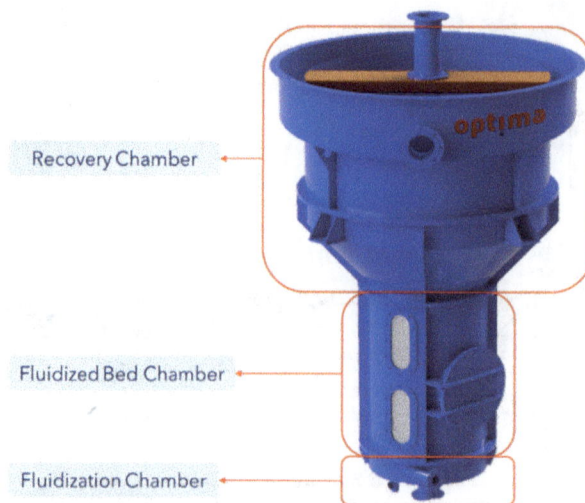

FIG 2 – Optima Classifier® design.

Slurry feed is introduced to the top of the OPC, through a feed pipe that extends to the middle of the tapered extension in the recovery chamber. Fluidisation water is introduced through the fluidisation chamber, where a uniform column of water is created using equally spaced nozzles. With an accumulation of material in the fluidised bed chamber, hindered settling is created where dense material sink and less dense material is carried upward in the unit with the fluidisation water. The fluidised bed chamber therefore account for the primary separation section of the OPC. In the tapered extension of the recovery zone an accumulation of fine dense particles and less dense coarse particles is observed, resulting in the creation of an autogenous dens media. This aids in separation from the moment feed is introduced. The recovery chamber has a bigger diameter than the fluidised bed chamber, which reduces the upward velocity of particles. This ensures the minimisation of misplacing dense particles to the overflow. In the recovery chamber a second hindered settling zone is created, leading to the use of a third separating mechanisms in the OPC. The OPC is controlled using a density set point that is calculated using a differential pressure in the fluidised bed chamber. The dense material, in this case, hematite is discharged through the underflow of the unit, using a control valve. Less dense material, typically silica, is discharged through the overflow of the OPC. The OPC is capable of processing material with a top size of up to 6 mm, with effective separation in a top to bottom particle size ratio of 1:4 to 1:6.

Optima Classifier® applications

The OPC has been tested in various commodities including but not limited to coal, chrome, iron ore, mineral sands, copper/cobalt ore, manganese and diamonds. In fine coal applications the OPC has been implemented in an export duff process, where -6 mm coal is beneficiated from a 20 MJ/kg; 35 per cent ash feed to a +27 MJ/kg; -15 per cent ash product. In the -6 + 1 mm fraction, capacities of more than 35 t/h/m^2 has been observed, where the -1 mm fraction was processed up to 25 t/h/m^2. These capacities allowed for the construction of a 60 t/h fine coal beneficiation plant built on a 12 m × 12 m area, including a feed preparation and dewatering section. In another application the OPC was used to produce a chemical grade chrome product with -0.8 per cent SiO_2 and + 46 per cent Cr_2O_3 content. In one month the OPC produced 4000 t of product with an average chrome and silica grades of 46.7 per cent Cr_2O_3 and 0.6 per cent SiO_2. Due to the high capacity per installed square metre the OPC replaced two processing technologies in the plant that had the same function. With so much promise the OPC was tested in a fine, low-grade iron ore application where a case study was done on the technical and economic feasibility of this application.

CASE STUDY

Feed material

A -1 mm South African BIF sample with a head grade of approximately 50 per cent Fe was used in this study with the aim of producing a 63 per cent Fe product, with maximum recovery of iron ore. A 30 per cent Fe tailings grade was thus also targeted.

Experimental set-up

A continuous lab scale OPC150 system was used to accumulate material for down steam processing. The experimental set-up of the OPC system is shown in Figure 3.

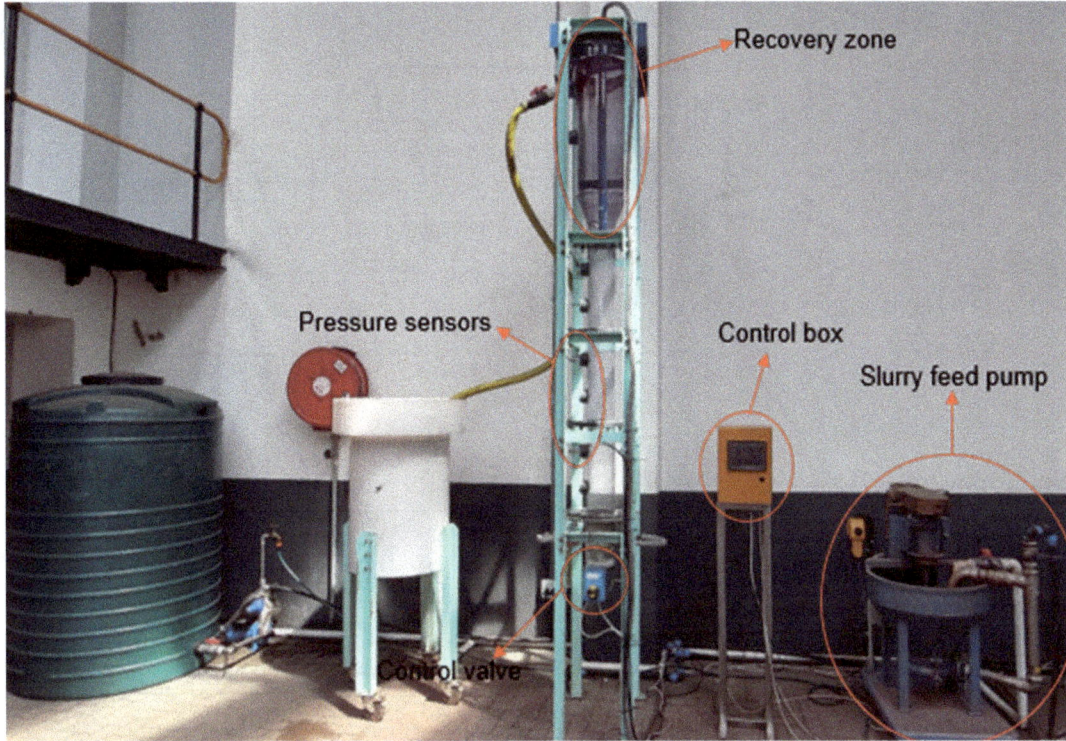

FIG 3 – OPC system lab scale experimental set-up.

The slurry feed pump was used to introduce feed to the unit at approximately 20 per cent solids. In iron ore, the lab scale OPC has a throughput of up to 300 kg/h depending on the beneficiation stage. Once a steady slurry feed rate was obtained, fluidisation water was introduced at the bottom of the OPC150. The overflow of the OPC150 flowed into a conical tank, where the solids settled and the process water overflowed back to the slurry feed tank. The OPC150 was filled with sample and a relative density set point was obtained and used as the cut point. A fully automated experiment was conducted where the control box would open the underflow valve when the set point was reached. The overflow and underflow were continuously sampled while feed was introduced to the unit. Various test scenarios were investigated using different PSD feeds, process configurations and processing parameters. A summary of this is seen in Table 1.

TABLE 1

Different scenarios investigated.

Scenario 1	Scenario 2	Scenario 3
-1 mm pre-concetrator	-1 + 0.3 mm rougher	-1 + 0.1 mm rougher
-1 mm cleaner	-1 + 0.3 mm cleaner	-1 + 0.1 mm cleaner
-1 mm scavenger	-0.3 + 0.075 mm rougher	-1 + 0.1 mm scavenger
-1 mm scavenger cleaner	-0.3 + 0.075 mm cleaner	

Across the three scenarios a total of 11 different test runs were conducted. In all three scenarios the aim of the tests was achieved where a 63 per cent Fe product was achieved. In the pursuit of balancing recovery and economical implementation to a production scale application, scenario 3 was found to be the best. A simplified process flow of scenario 3 can be seen in Figure 4.

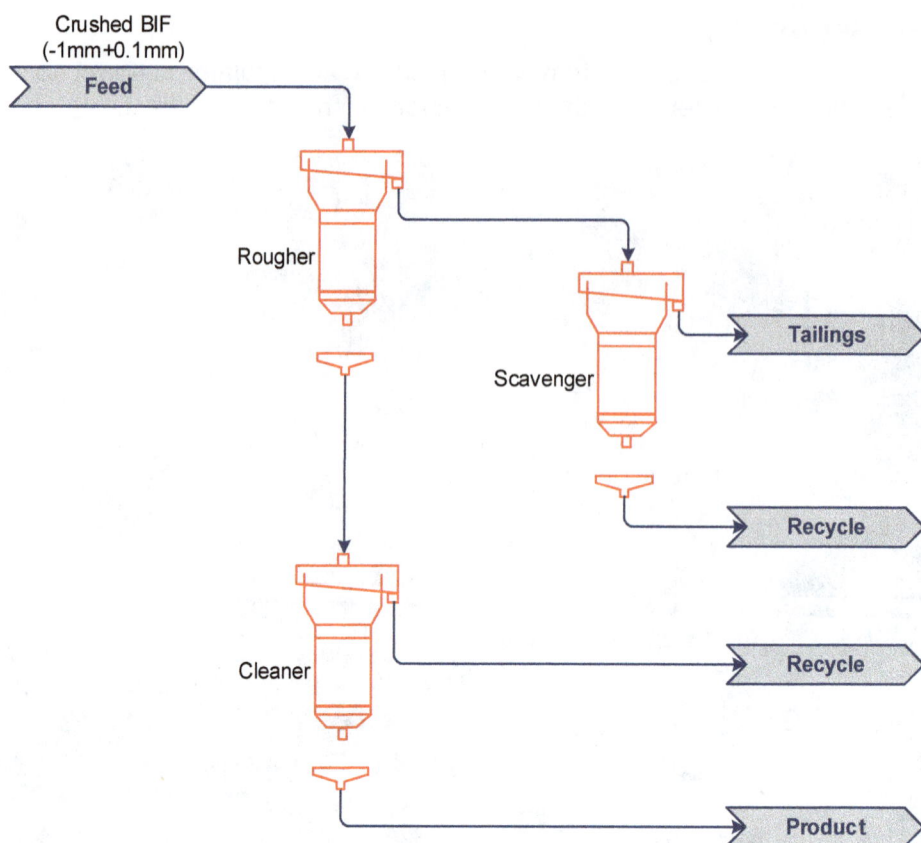

FIG 4 – Test flow sheet.

In the flow sheet shown in Figure 4 a rougher stage was used to pre concentrate the feed. The overflow reported to a scavenger unit where the aim was to produce a 30 per cent Fe tailings. The underflow of the scavenger unit was considered a recycle stream. The underflow of the rougher reported to a cleaner unit where the aim was to produce a 63 per cent Fe product. The overflow of the cleaner was considered a recycle stream.

Results

The relative density cut points for the rougher, cleaner and scavenger was 1750 kg/m^3, 1750 kg/m^3 and 1500 kg/m^3 respectively. The relative density inside the unit is based on several factors such as bed height and fluidisation velocity and is not the true density of the material. The fluidisation was set to 600 L/h, 800 L/h and 500 L/h for the rougher cleaner and scavenger stages. The performance of each stage is summarised in Table 2.

TABLE 2

Three-stage test results.

	Stage		
Parameter	**Rougher**	**Cleaner**	**Scavenger**
Feed grade (%Fe)	51.95	62.30	36.96
Overflow grade (%Fe)	36.96	57.30	33.50
Underflow grade (%Fe)	62.30	63.60	45.60
Underflow mass yield (%$_{wt}$)	54.10	75.50	30.00
Recovery (%)	64.87	77.02	37.08
Upgrade ratio	1.20	1.03	1.23

By considering the cleaner overflow and the scavenger underflow as a recycle streams, the final yield and recovery was simulated using the behaviour of the OPC's with regards to yield and upgrade ratios. The process yield and recovery were thus calculated to be 56.0 per cent and 70.7 per cent respectively. To confirm the accuracy of the test work, a metal balance was done over each stage as well as over the three-stage process. The overall experimental error was calculated to be approximately 5 per cent and is contributed to mass loss materials handling such as sub-sampling. A visual representation of the feed, tailings and product is shown in Figure 5.

FIG 5 – Feed, tailings and product.

Although the focus of the study was based on the recovery of iron ore, a secondary study was done on the immediate recovery of process water. For this study an ultra-fine dewatering screen was used to dewater at a 50 micron aperture. Desliming upfront at 0.1 mm minimises the amount of slimes in both the product and tailings dewatering effluent, when dewatering at a 50 micron aperture. A mass balance over the three-stage system indicated that the tailings had 20 per cent solids, where the product at 60 per cent solids. The tailings were dewatered to a 22 per cent moisture content and the concentrate to 15 per cent moisture. Both these oversizes were conveyable and stackable, with the effluent recyclable as process water.

IMPLEMENTATION

A high-level feasibility study was conducted on a 30 t/h pilot plant that includes a feed preparation, beneficiation and dewatering section. A simplified process flow diagram can be seen in Figure 6.

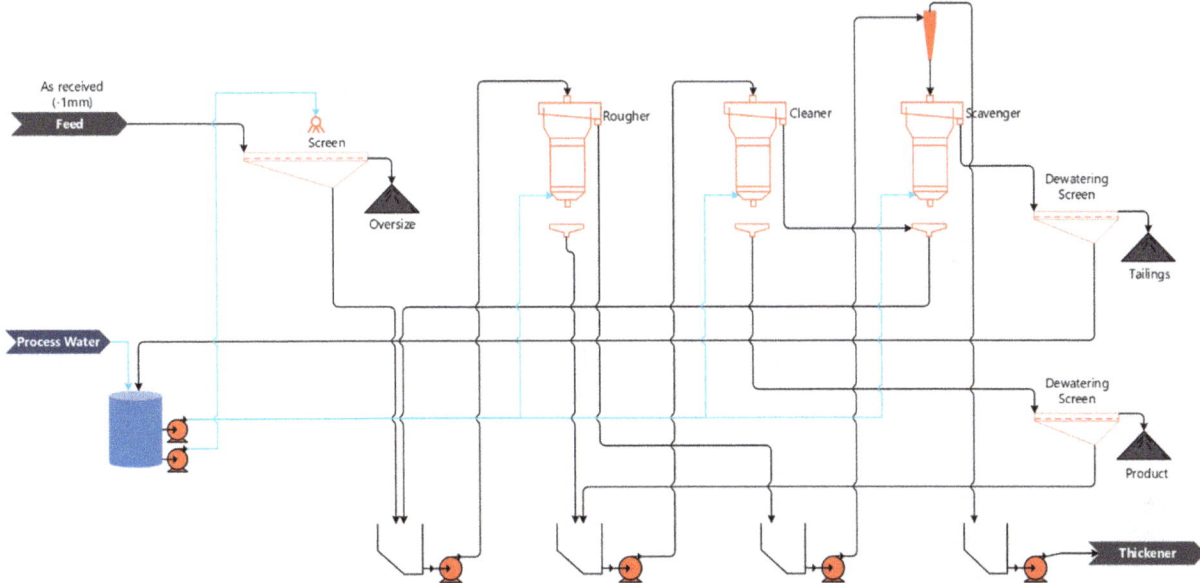

FIG 6 – The 30 t/h pilot implementation.

From Figure 6 the only water that exists the process is that of the desliming cyclone overflow. Under conventional circumstances the total water load to the thickener was calculated to be 224.8 m³/h.

However, with the effluent of the screens being recyclable the water load to the thickener was reduced to 133 m³/h as seen in Table 3.

TABLE 3

Recycled water comparison.

	Water in product deposition	Water in tailings deposition	Water recycled	Water to thickener
Conventional	2.8	2.7	84.2	224.8
With dewatering screens	2.8	2.7	176.0	133.0

Aside from the water saving of 40 per cent, the solids load to the thickener is also reduced from 13.2 t/h to 5.2 t/h. This will result in a smaller thickener needed which reduces the CAPX and OPEX of the plant. In the economic evaluation of the pilot plant a theoretical selling price of iron was assumed at USD50/t (half of the current selling price), as shown in Table 4.

TABLE 4

Revenue data for the 30 t/h pilot plant.

Process data	
Feed (tons per month)	16 500
Product yield (%)	56
Product (tons per month)	9 240
Cost Data	
CAPEX (USD)	1 850 000
Revenue	
Sales price per ton (USD)	50
Monthly revenue (USD)	462 000
12 Month revenue (USD)	5 544 000

On an assumed plant availability of approximately 75 per cent the pilot plant is forecasted to produce 9240 tons of sellable product per month. This leads to a monthly revenue of USD462 000 and a CAPEX payback period of four months. This shows that the plant is not only economically feasible, but with immediate recycled water and co-disposable tailings, the environmental liability associated with fine iron ore beneficiation is also minimised. This process is not only integrate-able with current arising fines but can also be used in the reprocessing of historical fines captured in tailing facilities.

CONCLUSION

The Optima Classifier® can be used in a multitude of commodities. In iron ore it addresses conventional technology's limitations such as particle size, low upgrade ratios, low throughputs and misplacement of hematite. The OPC can produce a 63 per cent Fe product, with upgrade ratios of 1.03 to 1.23. A three-stage system resulted in a 56 per cent product yield, with a 70 per cent overall recovery. The use of ultra-fine dewatering screens can result in a 40 per cent process water saving, with reduced loads to a plant thickener. The economic evaluation showed that a 30 t/h pilot plant is economically feasible, with current arising and historical reclamation processing capabilities. This process does not generate revenue but also reduces environmental liability and materials handling cost.

ACKNOWLEDGEMENTS

A sincere thanks to the authors of the paper as well as Gravitas Minerals® for endorsing this study and paper.

REFERENCES

Da Corte, C, Singh, A and Letsoalo, K, 2023. Amenability of South African Banded Iron Formation (BIF) to Fines Gravity Processing, *Mining, Metallurgy and Exploration*, 40:885–891. https://doi.org/10.1007/s42461–023–00758–6

Das, B, Prakash, S, Das, S K and Reddy, P S R, 2007. Effective Beneficiation of Low Grade Iron Ore Through Jigging Operation, *Journal of Minerals and Materials Characterization and Engineering*, 7(1):27–37.

Government of Western Australia, 2022. Western Australia Iron ore profile, Department of Jobs, Tourism, Science, and Innovation.

Sadeghi, M and Bazin, C, 2020. The Use of Process Analysis and Simulation to Identify Paths to Improve the Operation of an Iron Ore Gravity Concentration Circuit, *Advances in Chemical Engineering and Science*, 10:149–170. https://doi.org/10.4236/aces.2020.103011

Sahin, R, 2020. Beneficiation of low/off grade iron ore: a review, *International Journal of Research – Granthaalayah*, 8(8):328–335. https://doi.org/10.29121/granthaalayah.v8.i8.2020.934

Mining and processing – ore processing and beneficiation

Removal of phosphorus and gangue minerals from iron ore by a combined approach of alkaline drying and reverse flotation

A Batnasan[1], H Takeuchi[2], K Haga[3], K Higuchi[4] and A Shibayama[5]

1. Project lecturer, Akita University, Akita, 010–8502, Japan. Email: altansukh@gipc.akita-u.ac.jp
2. Graduate student, Akita University, Akita, 010–8502, Japan. Email: m6021211@s.akita-u.ac.jp
3. Associate professor, Akita University, Akita, 010–8502, Japan. Email: khaga@gipc.akita-u.ac.jp
4. Process Research Laboratories, Research and Development, Nippon Steel Corporation, Japan. Email: higuchi.t9g.kenichi@jp.nipponsteel.com
5. Professor, Akita University, Akita, 010–8502, Japan. Email: sibayama@gipc.akita-u.ac.jp

ABSTRACT

High silica (SiO_2), alumina (Al_2O_3), and phosphorus (P) contents in iron ore adversely affect the quality of iron products. The removal of those impurities from iron ore is an essential point in further iron ore processing. In this paper, the removal of P, SiO_2 and Al_2O_3 impurities from iron ore by alkaline drying followed by reverse cationic flotation is discussed.

The reverse cationic flotation experiments are performed using the iron ore samples without and with alkaline drying pretreatment. All other flotation conditions except pulp pH (4–12) and particle size distributions (-45 to -250 µm) are held constant. A typical collector and frother, such as dodecylamine acetate (DAA) and methylisobutylcarbinol (MIBC), are used in the flotation experiments. Alkaline treatment is conducted before flotation using sodium hydroxide at fixed conditions. The impact of alkaline drying on the performance of iron ore flotation is evaluated based on the removal rates of P and $SiO_2 + Al_2O_3$, iron grade, and iron recovery compared with the iron ore flotation without alkaline drying (direct flotation). X-ray diffraction (XRD) analysis indicates hematite and goethite are the main minerals in the iron ore, which assays 62.8 mass per cent Fe, 0.232 mass per cent P, 7.1 mass per cent $SiO_2 + Al_2O_3$ (total) determined by X-ray fluorescence analysis (XRF).

Results indicate that a combined approach is more effective for removing P and $SiO_2 + Al_2O_3$ impurities from iron ore than the direct processing technique because the maximum removal rates of P and $SiO_2 + Al_2O_3$ are 48.2 per cent and 66.5 per cent from the alkaline dried iron ore when 8.4 per cent and 13.1 per cent from the non-alkaline dried iron ore.

INTRODUCTION

Iron is widely distributed in the earth's crust, occurring in the form of various minerals. However, the primary sources of iron ore minerals are magnetite (Fe_3O_4), hematite (Fe_2O_3), goethite (FeO(OH)), ochreous goethite (FeO(OH)·n(H_2O)), siderite ($FeCO_3$) and pyrite (FeS_2). Among them, the important commercial minerals are magnetite and hematite because of their high iron content, fewer impurity and lower processing cost for beneficiation before being used for steel production by the integrated blast furnace (BF) and basic oxygen furnace (BOF) process (Clout and Manuel, 2015; Kildahl *et al*, 2023; Nikolaeva *et al*, 2021).

Due to long-time mining activity and the enormous increase in steel production, iron content in iron ores is reducing with the simultaneous increase in the level of impurity elements, mainly silica, alumina and phosphorus. On the other hand, marketable iron ores containing over 60 per cent Fe possess a high amount of impurity elements (Pownceby *et al*, 2019; Rocha, Cancado and Peres, 2010; Roy, Nayak and Rath, 2020).

Silica and alumina in iron ore are mostly associated with gangue minerals such as quartz, alumina (corundum), clay minerals, gibbsite, illite, diaspore and lateritic minerals, while some other remained portion is combined with the iron minerals. Phosphorus occurs as apatite and is also included in iron hydroxide minerals such as goethite, ochreous goethite and ferrihydrite (Clout and Manuel, 2015; Han, Lu and Hapugoda, 2023; Park *et al*, 2022; Krumina *et al*, 2016).

The high amounts of iron ore impurities adversely affect the quality and productivity of steel. Thus, removing impurities from high-impurity iron ore is more critical to obtaining a marketable product. The beneficiation techniques vary from screening and size classification to complex processing (gravity separation, magnetic separation and froth flotation), depending on the nature and mineralogical characteristics of the iron ore. Many studies have focused on removing silica, alumina, and phosphorus from iron ores by flotation based on the differences in the surface hydrophobicity of dispersed particles (Das and Rath, 2020; Haga *et al*, 2020; Nakhaei and Irannajad, 2017; Tohry *et al*, 2021; Zhang *et al*, 2019a). However, the flotation route is more effective in removing quartz and alumina minerals from iron ore but much less effective for phosphorus removal because of phosphorus occurrence in iron ore minerals. The combination of thermal treatment and chemical leaching is favourable to removing phosphorus from high-phosphorus iron ore (Fisher-White, Lovel and Sparrow, 2012; Mochizuki and Tsubouchi, 2019; Zhang *et al*, 2019b).

In this study, the removal of silicon, alumina and phosphorus from high-impurity iron ore is investigated using a combined method consisting of alkaline drying and reverse cationic flotation to produce a qualified iron product. Alkaline drying, slurry pH and particle size distribution are examined to remove the impurities from the iron ore.

EXPERIMENTAL

Materials and equipment

An iron ore received from a mine in Australia is used in this study. Various size fractions ranging from -45 μm and -500 μm are prepared from the received sample by crushing, milling and sieving (JIS Z 8801 sieves, IIDA Manufacturing Co, Ltd).

Reagent-grade chemicals used in this study are sodium hydroxide (NaOH) as an alkaline additive in the roasting process and as a slurry pH regulator during the reverse flotation experiments, hydrochloric acid (HCl) as an adjustment of slurry pH, dodecylamine acetate (DAA, $C_{12}H_{27}N \cdot C_2H_4O_2$) as a cationic collector and methyl isobutyl carbinol (MIBC, $C_6H_{14}O$) as a frother. All aqueous solutions used in this study are prepared using distilled water. The pH of the slurry of flotation experiments is determined by a pH/ORP metre (Toko, TPX-999 Si).

An electric muffle furnace (Advantec, KM-160) and laboratory flotation test machine with a 1000 mL cell volume are used in roasting and flotation experiments.

Methods

The research process consists of two steps, alkaline drying and reverse froth flotation, as shown in Figure 1. The iron ore sample with an appropriate amount of NaOH solution is heated at 60°C for 24 hrsrs (Batnasan *et al*, 2022). The dried sample is cooled in a desiccator to room temperature and ground with a pestle and mortar. Then the ground sample is used in flotation experiments. The reverse flotation experiments are carried out using non-alkaline roasted and alkaline roasted iron ore samples under the same conditions. The effects of alkaline roasting, slurry pH (4–12) and particle size distributions (-45 to -500 μm) on the impurities removal are investigated. Pulp density (5 per cent), impeller speed (1200 rev/min), collector dosage (1000 g/t DAA) and frother dosage (200 g/t) are held constant throughout the flotation experiments. After each flotation experiment, froth and tailing are filtered and washed with distilled water. In a drying oven, froths and tailings from reverse flotation experiments are dried at 60°C for 1 day for further analysis.

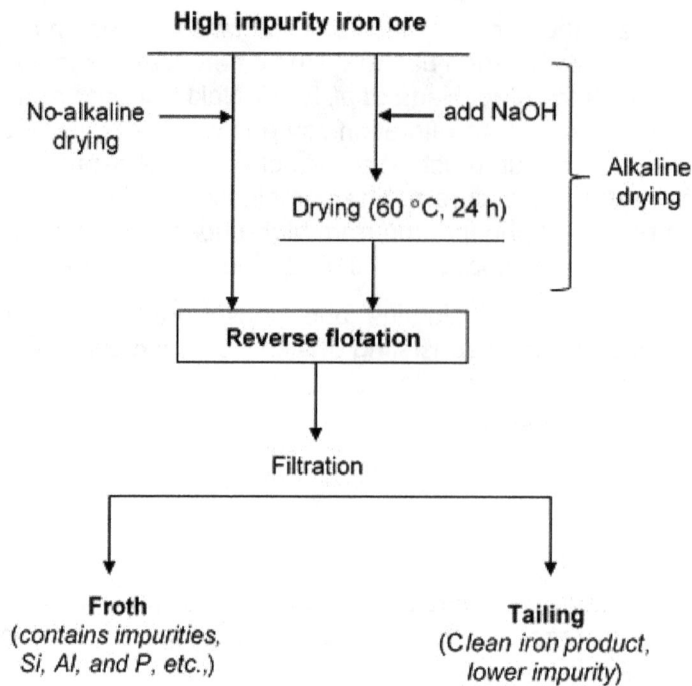

FIG 1 – A process flow sheet for the removal of impurities from high-impurity iron ore.

Analysis and evaluation

XRF (Primus-II, Rigaku), XRD (RINT–2200/PC, Rigaku) and scanning electron microscopy with energy dispersion spectroscopy, SEM-EDS (SU-70, Hitachi Hightech) are occupied with characterising iron ore samples and solid samples from roasting and flotation experiments.

The removal rate ($R_{removal}$, per cent) of quartz (SiO_2), alumina (Al_2O_3) and phosphorus (P) as impurities, iron loss ($R_{Fe, loss}$, per cent) and iron recovery ($R_{Fe, Rec}$, per cent) through the flotation are estimated using the chemical content of each element determined using XRF, as follows.

The removal rate of impurities and iron loss, respectively, per cent:

$$R_{Removal/Fe\ loss}(\%) = 1 - \frac{C_T * W_T}{C_C * W_C} * 100 \tag{1}$$

Iron recovery, per cent:

$$R_{Fe,Recovery}(\%) = \frac{C_T * W_T}{C_C * W_C} * 100 \tag{2}$$

Where:

C_C is for the SiO_2, Al_2O_3, P and Fe content in the original iron ore sample used in flotation

C_T is for the SiO_2, Al_2O_3, P and Fe content in the products from each flotation

W_C stands for the mass of the iron ore used in each flotation experiment

W_T stands for the mass of the product from each flotation experiment

RESULTS AND DISCUSSIONS

High-impurity iron ore characterisation

The chemical compositions of the prepared iron ore samples with various size distributions are summarised in Table 1, which shows that the iron ore contains an acceptable level of Fe. However, the contents of P and SiO_2 + Al_2O_3 are several times greater than their required levels because the typical specifications for iron ores in Japanese steelmaking are over 60 per cent Fe, lower than 0.1 per cent P, below 5 per cent SiO_2 + Al_2O_3 and less than 0.08 per cent for S (Clout and Manuel, 2015). It indicates that iron ore is a high-impurity iron ore due to its higher P and high SiO_2 + Al_2O_3 contents.

TABLE 1

Chemical contents of the iron ore samples with various size distributions, mass per cent.

Particle size, (µm)	T.Fe	P	SiO$_2$	Al$_2$O$_3$	CaO	MgO	Na	S
-250 (feed)	62.6	0.250	5.06	2.86	0.05	0.09	0.008	0.013
-500+250	60.4	0.250	4.72	2.80	0.03	0.07	0.004	0.011
-250+125	61.2	0.250	4.89	2.88	0.02	0.03	0.004	0.008
-125+90	62.6	0.224	4.63	2.73	0.03	0.02	0.004	0.008
-90+75	63.5	0.232	4.52	2.52	0.02	0.10	0.004	0.008
-75+45	62.7	0.242	5.00	2.74	0.03	0.04	0.004	0.008
-45	63.3	0.253	4.38	2.34	0.03	0.10	0.003	0.008

Figure 2 shows the mineral composition of the iron ore as a received sample in which hematite (Fe$_2$O$_3$) and goethite (FeO(OH)) are the main constituents. However, the iron ore is referred to as high-impurity iron ore, discrete minerals containing Si, Al and P are not identified by XRD analysis. It could be related to the amorphous form of the individual impurity minerals or the inclusions of the impurity elements on the iron oxide minerals.

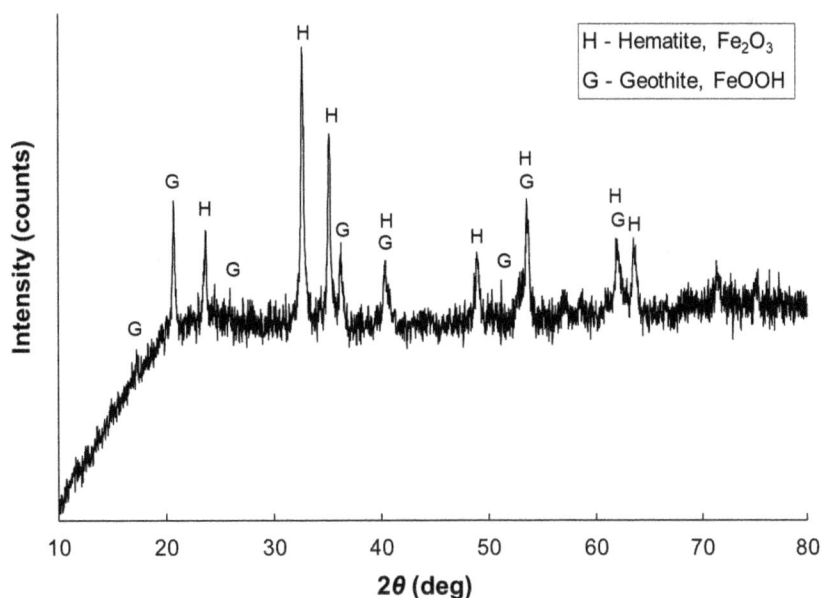

FIG 2 – XRD pattern of Australian iron ore sample.

The SEM image of a polished sample of the iron ore revealed many micropores distributions on the surface of the iron ore particles (Figure 3). SEM-EDS colour mapping analysis for all elements showed that Si (yellow colour), Al (apple green colour) and P (dark red colour) maps are overlapped with Fe distribution (blue colour) due to its high-grade in the ore. It implies that the impurity elements such as Si, Al and P are distributed at the interface of iron oxide/hydroxide or gangue minerals region into the iron minerals. On the other hand, Si and Al occur as discrete grains in the iron ore.

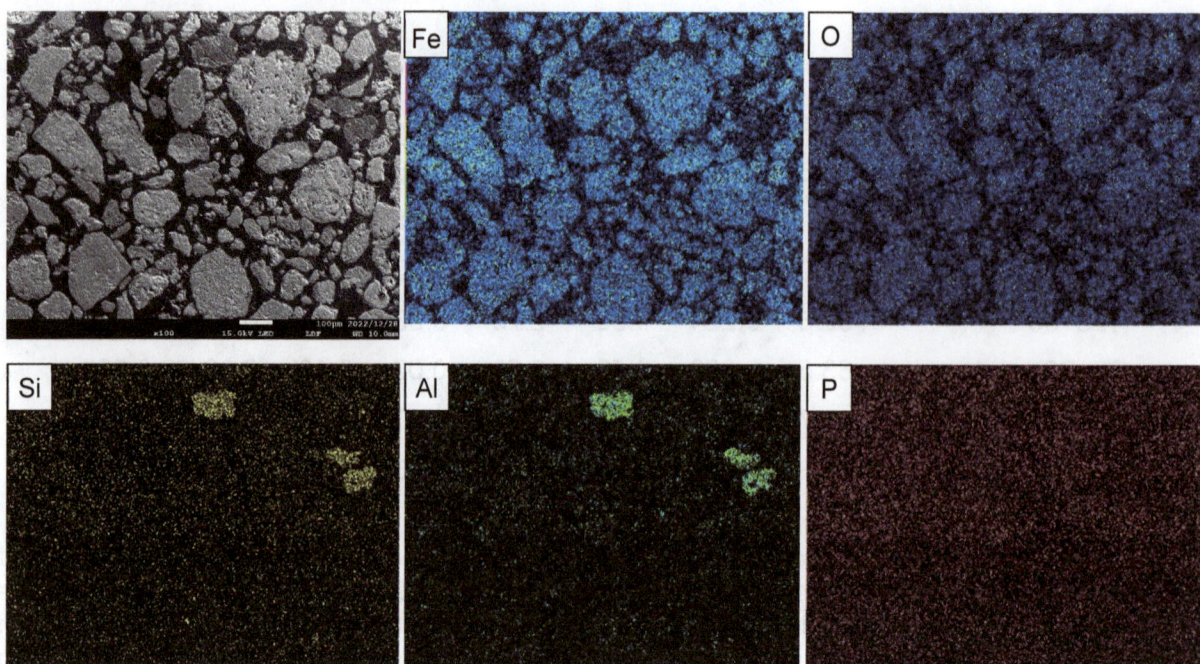

FIG 3 – SEM-EDS images of the iron ore sample.

Reverse cationic flotation

The impurity elements such as Si, Al and P in iron ore are removed using reverse cationic flotation at various conditions. The results obtained are shown below.

Effect of alkaline roasting and slurry pH

Figure 4 compares the removal rates of $SiO_2 + Al_2O_3$ and P from the initial ore as a feed and alkaline-dried iron ore samples by reverse cationic flotations conducted at the same conditions, which are a particle size of <250 µm, pulp density of 5 per cent, DAA dosage of 1000 g/t, MIBC dosage of 200 g/t and flotation time of 10 minutes. The initial pH of the slurries prepared using feed and alkaline-dried iron ore samples is 6 and 12, respectively.

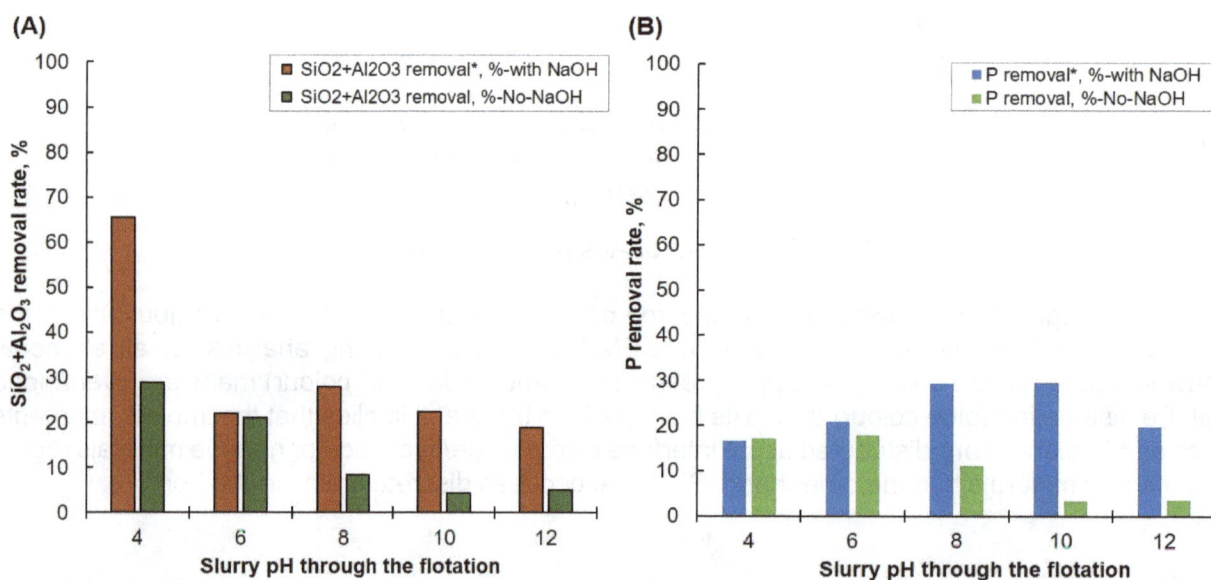

FIG 4 – Comparison of $SiO_2 + Al_2O_3$ (A) and P (B) removal from the iron ore feed (non-alkaline dried) and alkaline-dried iron ore (50 g/kg-ore NaOH, 60°C, 24 hrs); Flotation: 5 per cent pulp density, 1000 g/t DAA, 200 g/t MIBC, 10 min flotation).

During the flotation experiments, the slurry pH is adjusted between 4 and 12 using 1 M HCl and 5 M NaOH solutions.

The removal rates of impurities such as $SiO_2 + Al_2O_3$ and P, iron recovery and iron loss from the high-impurity iron ore are estimated by Equations 1 and 2 using their contents in initial iron ore and tailings from the reverse cationic flotation of non-alkaline and alkaline dried iron ore sample, respectively. It can be seen that the removal rate of $SiO_2 + Al_2O_3$ from both feed and alkaline dried samples is decreased with the slurry pH from 4 to 12. It indicates that besides the removal rates, the $SiO_2 + Al_2O_3$ removal from both samples has the same tendency between the broad pH ranges. Moreover, the phosphorus removal tendency from both samples differs between the pH ranges (pH 4–12). The P removal from the iron ore feed decreases from 17.1 per cent to 3.5 per cent with the increasing slurry pH from 4 to 12.

In comparison, the P removal from alkaline-dried iron ore is first raised from 16.6 per cent to 29.4 per cent in the pH range between 4 and 10 and then decreases to 19.8 per cent at pH 12. Results indicate that the removal rates of $SiO_2 + Al_2O_3$ and P from alkaline dried iron ore at the alkaline medium (pH 8–12) are several times higher than the rates from the feed sample. The finding suggests that alkaline drying pre-treatment effectively enhances the removal of impurities from high-impurity iron ore. However, there are drawbacks, such as a lower iron recovery and higher iron loss from the alkaline dried iron ore in the pH range between 6 and 10 compared to the direct flotation with feed sample due to the float of a considerable amount of iron oxides with impurity minerals as shown in Figure 5. It might be related to the hetero-coagulation of P, Si and Al-containing minerals as clay minerals with iron oxide particles in the alkaline dried iron ore. The result suggests using a depressant such as starch in the iron ore flotation at alkaline media (pH 8–10) to reduce iron loss. On the other hand, de-sliming and wet sieving of the iron ore sample before flotation is necessary to minimise iron loss.

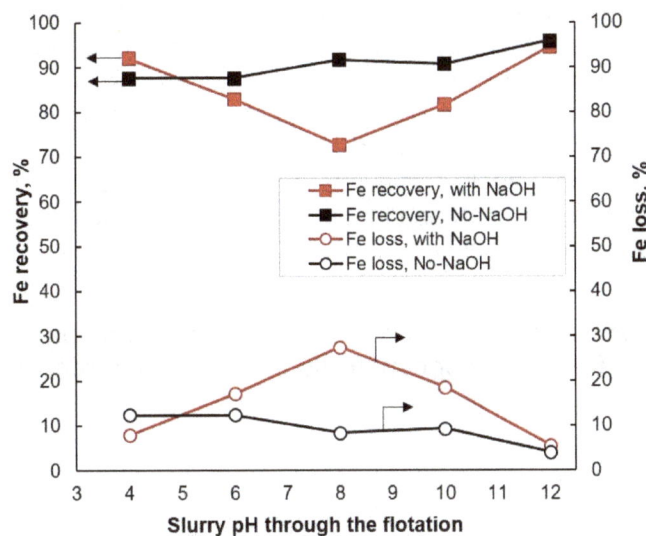

FIG 5 – Effect of slurry pH on the Fe recovery and Fe loss from the iron ore feed (non-alkaline dried) and alkaline dried iron ore (50 g/kg-ore NaOH, 60°C, 24 hrsrs); Flotation: 5 per cent pulp density, 1000 g/t DAA, 200 g/t MIBC, 10 min flotation).

At pH 12, 94.5 per cent of the highest Fe recovery and 5.5 per cent of the lowest iron loss are obtained, but impurities removal rates are lower than that from pH 8 and 10 (Figure 5). Due to the relatively lower impurities removal rates and operational problems caused by the filtration of a slurry from the flotation at pH 12, further experiments are conducted at pH 10, where the iron recovery and iron loss are 82 per cent and 18.4 per cent, respectively.

Effect of particle size distribution

Figure 6 depicts the removal of $SiO_2 + Al_2O_3$ and P from the alkaline dried iron ore with the different particle size distributions ranging from -500 + 250 µm to -45 µm. All flotation experiments are performed at the same conditions as described above. Decreasing the particle size distribution from -500 + 250 µm to -90 + 75 µm, the removal rates of $SiO_2 + Al_2O_3$ and P from the alkaline dried iron ore increase from 27 per cent to 66.5 per cent and 37.5 per cent to 48.2 per cent, respectively.

Nevertheless, their removal rates decrease to 42.8 per cent for $SiO_2 + Al_2O_3$ and 35.6 per cent for P when the particle size becomes finer (-45 µm).

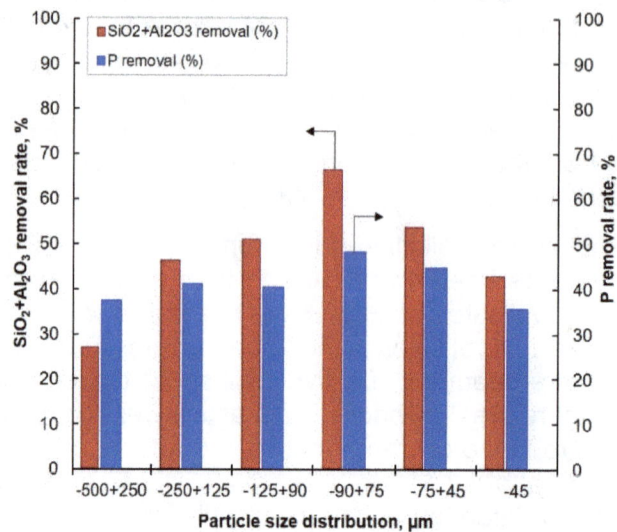

FIG 6 – Effect of particle size distribution on the removal of impurities from alkaline dried iron ore (Drying: 50 g/kg-ore NaOH, 60°C, 24 hrs; Flotation: 5 per cent pulp density, pH 10, 1000 g/t DAA, 200 g/t MIBC, 10 min flotation).

The results indicate that the removal efficiencies of impurities from medium particles fraction (-90 + 75 µm) are the highest because of the maximum removal rates of 66.5 per cent for $SiO_2 + Al_2O_3$ and 48.2 per cent for P. A higher impurity removal from the medium particles fraction may be due to the sufficient liberation of the minerals containing the impurity elements. In contrast, a lower removal of impurities from coarse and fine particles could be related to insufficient liberation of the gangue minerals from the iron oxide minerals. In particular, the main reasons for the lower removal efficiency through flotation are low bubble collision and insufficient kinetics to colloid particles and bubbles and initiate the particle-bubble attachment in the slurry (Ran *et al*, 2019; Farrokhpay, Filippov and Fornasiero, 2020).

Figure 7 shows that by reducing the particle size from -500 + 250 µm to -75 + 45 µm, the iron recovery decreases slowly from 89.1 per cent to 85.0 per cent and reaches 75.0 per cent with the finer size fraction (-45 µm).

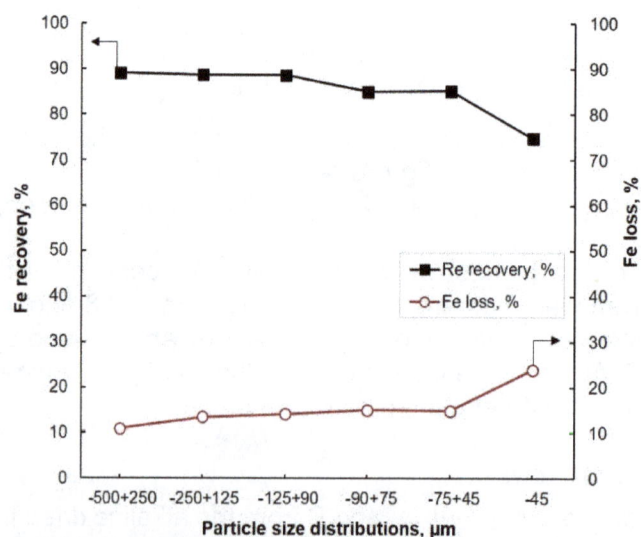

FIG 7 – Effect of particle size distribution on the Fe recovery and Fe loss from alkaline dried iron ore (Drying: 50 g/kg-ore NaOH, 60°C, 24 hrs; Flotation: 5 per cent pulp density, pH 10, 1000 g/t DAA, 200 g/t MIBC, 10 min).

It is observed that the lower iron recovery (74.5 per cent) and higher iron loss (24 per cent) from the fine particles of iron ore might be related to the floatability of finer iron ore particles caused by the adsorption of the amine collector (DAA) on their surfaces and the increase in froth stability through the flotation. This finding is consistent with other authors who concluded that the iron loss of ultrafine particles in reverse cationic flotation is higher than that in reverse anionic flotation because of the entrainment of ultrafine hematite particles (Rocha, Cancado and Peres, 2010; Ma, Marques and Gontijo, 2011; Safari *et al*, 2020). Due to high impurities removal rates, the medium particles of -90 + 75 μm are selected as optimum particle size fraction through the flotation.

Comparison of XRD patterns of iron ore feed, heated sample and products obtained by reverse cationic flotation at pH 10 are shown in Figure 8. It can be seen that the new diffraction peaks corresponding to clay minerals such as albite and kaolinite appeared in the froth from the flotation. At the same time, the mineral composition of tailing is similar to the feed and alkaline dried samples, indicating it consists of hematite and goethite.

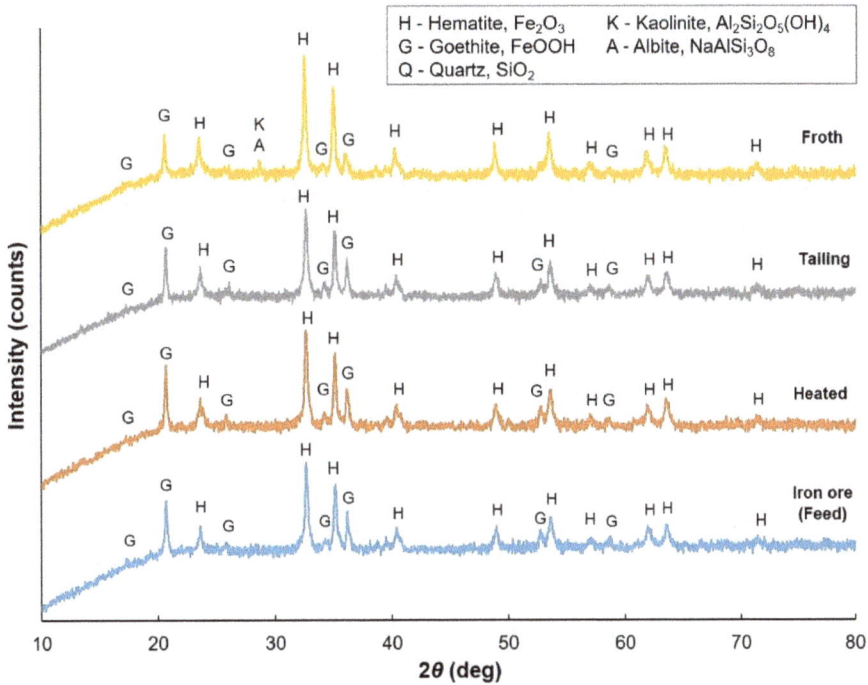

FIG 8 – XRD patterns of iron ore samples (feed and heated) and products from reverse cationic flotation (Particle size: -90 + 75 μm, Drying: 50 g/kg-ore NaOH, 60°C, 24 hrs; Flotation: 5 per cent pulp density, slurry pH 10, 1000 g/t DAA, 200 g/t MIBC, 10 min flotation)`.

The result shows that the kaolinite/albite clay minerals containing amorphous silica float with the DAA collector through reverse cationic flotation.

The mass balance of the processing of the high-impurity iron ore by alkaline drying and reverse cationic flotation is shown in Figure 9. It is worth noting that the SiO_2 + Al_2O_3 content achieved in tailing as a clean product from flotation at pH 10 is 2.8 mass per cent, while the P content reached 0.144 per cent. It indicates that the SiO_2 + Al_2O_3 and P content in the high-impurity iron ore decrease from 7.04 mass per cent to 2.8 mass per cent and from 0.232 mass per cent to 0.144 mass per cent, accounting for 66.5 per cent and 48.2 per cent removal rates, respectively, estimated by Equation 1. The iron content in the tailing (clean iron product) increases slightly compared to the iron ore feed. In comparison, Fe, SiO_2 + Al_2O_3 and P contents in the froth containing gangue minerals are 60.4 mass per cent, 7.84 mass per cent and 0.184 mass per cent, respectively. Through the flotation at pH 10, 42.5 g of the charged iron ore feed (50 g) into the flotation cell was settled down at the bottom, while 3.2 g minerals in the feed were attached to the froth. In addition, the loss in weight of charged feed was 4.3 g, accounting for 8.5 per cent distribution due to iron ore minerals agglomeration in the slurry. Notably, about 15 per cent iron loss through the reverse flotation at the condition is attributed to the entrainment of finer iron ore particles and iron minerals agglomeration in slurry.

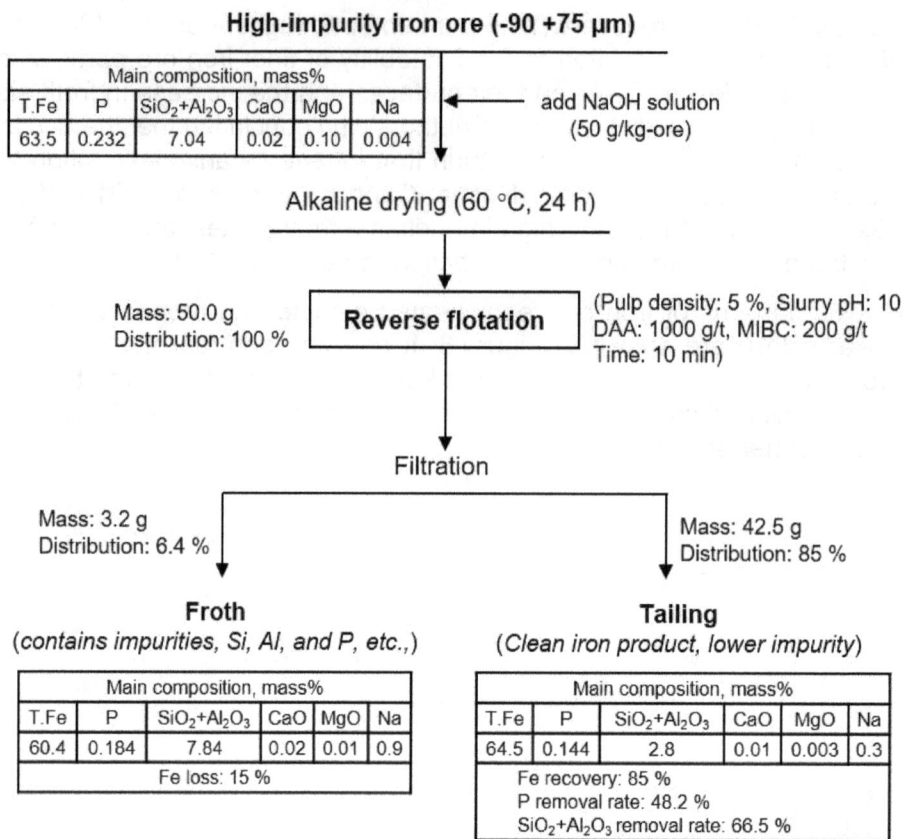

FIG 9 – Mass balance of the beneficiation of high-phosphorus iron ore by a combined method of alkaline drying and reverse cationic flotation.

CONCLUSIONS

The removal of impurities such as phosphorus, silica and alumina in high-impurity iron ore during the reverse cationic flotation in response to alkaline drying, different slurry pH and various particle sizes is investigated in this study. The results have shown that alkaline drying can promote the removal of SiO_2, Al_2O_3 and P as impurities from the iron ore because their removal rates from alkaline dried iron ore were several times higher than that from the initial iron ore sample at the same alkaline media (pH 8–12). However, a lower iron recovery and higher iron loss occur from the alkaline dried iron ore flotation than the direct iron ore flotation. For instance, at pH 10, the removal rates of P and $SiO_2 + Al_2O_3$ from alkaline dried iron ore are eight and five times higher than that from the initial iron ore sample, while the iron recovery is about 9 per cent lower due to the hetero-coagulation of P, Si and Al-containing minerals with iron oxide particles.

Besides the alkaline drying, the particle size is critical in the removal of impurities and flotation performance of iron ore because of low removal rates of P and $SiO_2 + Al_2O_3$ from the course (-500 + 125 µm) and fine particle (-45 µm) size fractions in comparison to the medium size fractions (-90 + 75 µm). The iron recovery decreases to 75 per cent by reducing the iron ore particle size from -500 µm to -45 µm because of the floatability or froth stability of finer iron ore particles.

The maximum removal rate of 48.2 per cent for P and 66.5 per cent for $SiO_2 + Al_2O_3$ are achieved from the alkaline dried iron ore with medium particle fraction (-90 + 75 µm), resulting in the decrease of P and $SiO_2 + Al_2O_3$ contents in the high-impurity iron ore from 0.232 mass per cent to 0.144 mass per cent and 7.04 mass per cent to 2.8 mass per cent, respectively. At this condition, the iron content in the final product is 64.5 mass per cent, with a recovery of 85 per cent.

The findings suggest that the alkaline-drying pre-treatment could improve the removal of P, SiO_2 and Al_2O_3 from high-impurity iron ore by reverse cationic flotation.

ACKNOWLEDGEMENTS

This study is funded by the New Energy and Industrial Technology Development Organization (NEDO) project (№19101845-0). The authors acknowledge the Nippon Steel Corporation in Japan for providing iron ore samples and technical assistance in sample analysis and collaborative research.

REFERENCES

Batnasan, A, Takeuchi, H, Haga, K, Shibayama, A, Mizutani, M and Higuchi, K, 2022. Effect of alkaline roasting conditions on the dephosphorization of high-phosphorus iron ore, in *Proceedings of the IMPC Asia Pacific 2022 Conference*, pp 2–12 (The Australasian Institute of Mining and Metallurgy: Melbourne).

Clout, J M F and Manuel, J R, 2015. Mineralogical, chemical, and physical characteristics of iron ore, in *Iron ore mineralogy, Processing and Environmental Sustainability* (ed: Liming Lu), pp 45–84 (Woodhead Publishing).

Das, B and Rath, S S, 2020. Existing and new processes for beneficiation of Indian iron ores, *Trans Indian Inst Met*, 73(3):505–514. doi:10.1007/s12666-020-01878-z

Farrokhpay, S, Filippov, L and Fornasiero, D, 2020. Flotation of fine particles: A review, *Miner Process Extr Metall Rev*, pp 1–11. doi:10.1080/08827508.2020.1793140

Fisher-White, M J, Lovel, R R and Sparrow, G J, 2012. Phosphorus removal from goethitic iron ore with a low temperature heat treatment and a caustic leach, *ISIJ International*, 52(5):797–803.

Haga, K, Batnasan, A, Mizutani, M, Higuchi, K and Shibayama, A, 2020. Upgrading the iron from low-grade and high-silicate iron ores by magnetic separation, *World of Mining, Surface & Underground*, 72(3):157–161.

Han, H, Lu, L and Hapugoda S, 2023. Effect of ore types on high temperature sintering characteristics of iron ore fines and concentrate, *Miner Eng*, 197:108062. doi: 10.1016/j.mineng.2023.108062

Kildahl, H, Wang, L, Tong, L and Yulong Ding, Y, 2023. Cost effective decarbonisation of blast furnace – basic oxygen furnace steel production through thermochemical sector coupling, *J Clean Prod*, 389:135963. doi: 10.1016/j.jclepro.2023.135963

Krumina, L, Kenney, J P L, Loring, J S and Persson, P, 2016. Desorption mechanisms of phosphate from ferrihydrite and goethite surfaces, *Chem Geol*, 427:54–64. doi: 10.1016/j.chemgeo.2016.02.016

Ma, M, Marques, M and Gontijo, C, 2011. Comparative studies of reverse cationic/anionic flotation of Vale iron ore, *Inter J Miner Process*, 100:179–183. doi: 10.1016/j.minpro.2011.07.001

Mochizuki, Y and Tsubouchi, N, 2019. Upgrading low-grade iron ore through gangue removal by a combined alkali roasting and hydrothermal treatment, *ACS Omega*, 4:19723–19734. doi: 10.1021/acsomega.9b02480

Nakhaei, F and Irannajad, M, 2017. Reagents types in flotation of iron oxide minerals: A review, *Miner Process Extr Metall Rev*, pp 1–36, doi: 10.1080/08827508.2017.1391245

Nikolaeva, N V, Aleksandrova, T N, Chanturiya, E L and Afanasova, A, 2021. Mineral and technological features of magnetite–hematite ores and their influence on the choice of processing technology, *ACS Omega*, 6(13):9077–9085. doi: 10.1021/acsomega.1c00129

Park, J, Kim, E, Suh, I-K and Lee, J, 2022. A short review of the effect of iron ore selection on mineral phases of iron ore sinter, *Minerals*, 12:35. doi: 10.3390/min12010035

Pownceby, M I, Hapugoda, S, Manuel, J, Webster, N A S and MacRae, C M, 2019. Characterisation of phosphorus and other impurities in goethite-rich iron ores – Possible P incorporation mechanisms, *Miner Eng*, 143:106022. doi: 10.1016/j.mineng.2019.106022

Ran, J-C, Qiu, X-Y, Hu, Z, Liu, Q-J, Song, B-X and Yao, Y-Q, 2019. Effects of particle size on flotation performance in the separation of copper, gold and lead, *Powder Technology*, 344:654–664. doi: 10.1016/j.powtec.2018.12.045

Rocha, L, Cancado, R Z L and Peres, A E C, 2010. Iron ore slimes flotation, *Miner Eng*, 23:842–845. doi: 10.1016/j.mineng.2010.03.009

Roy, S K, Nayak, D and Rath, S S, 2020. A review on the enrichment of iron values of low-grade iron ore resources using reduction roasting-magnetic separation, *Powder Technol*, 367:796–808. doi: 10.1016/j.powtec.2020.04.047

Safari, M, Hoseinian, F S, Deglon, D, Leal Filho, L S and Souza Pinto, T C, 2020. Investigation of the reverse flotation of iron ore in three different flotation cells: Mechanical, oscillating grid and pneumatic, *Miner Eng*, 150:106283. doi: 10.1016/j.mineng.2020.106283

Tohry, A, Dehghan, R, Zarei, M and Chehreh Chelgani, S, 2021. Mechanism of humic acid adsorption as a flotation separation depressant on the complex silicates and hematite, *Miner Eng*, 162:106736. doi: 10.1016/j.mineng.2020.106736

Zhang, L, Machiela, R, Das, P, Zhang, M and Eisele, T, 2019b. Dephosphorization of unroasted oolitic ores through alkaline leaching at low temperature, *Hydrometallurgy*, 184:95–102. doi: 10.1016/j.hydromet.2018.12.023

Zhang, X, Gu, X, Han, Y, Parra-Álvarez, N, Claremboux, V and Kawatra, S K, 2019a. Flotation of iron ores: A review, *Miner Process Extr Metall Rev*, pp 1–29. doi: 10.1080/08827508.2019.1689494

Design and operation of large drum scrubbers

G S Beros[1] and K M Edwards[2]

1. Technical Marketing Manager, Queensland Pacific Metals, Brisbane Qld 4000.
   Email: gberos@qpmetals.com.au
2. Principal Process Engineer, Calibre, Perth WA 6000.
   Email: kristine.edwards@calibregroup.com

ABSTRACT

Over the past 20 years drum scrubbers have become a widely adopted technology within iron ore beneficiation plants. Generally, the drum scrubbers in use are considered high-capacity machines capable of accepting larger-sized feed material. Together with a water addition to a desired pulp density and a mild energy input, these large rotating machines effect an attritioning action that removes water soluble clays and other deleterious materials. Within the Western Australian iron ore industry, drum scrubbers have been successfully applied to Channel Iron, Marra Mamba, Bedded and Detrital orebodies. The introduction of drum scrubbers has enabled Western Australian producers to satisfy the substantial increase in demand for iron ore from Chinese steelmakers by making the lower-grade sections of the above orebodies economic. However, despite being a relatively simple unit process, the requirements for successful design and operation of these machines are not fully understood. This paper draws on the collective experience of both authors to provide some points to consider for incorporating drum scrubbers into iron ore beneficiation plant flow sheets.

INTRODUCTION

A large drum scrubber is a specialised piece of equipment used in many iron ore processing operations for the purpose of separating and cleaning materials such as rocks, ores and minerals. It is designed to remove impurities and contaminants from the surface of the material by subjecting it to mechanical agitation and a scrubbing action in the presence of water. Drum scrubbers are especially effective for processing materials with high clay content or sticky characteristics where dry processing methods may not be suitable.

The scrubber consists of a rotating drum, typically made of durable materials such as steel and fitted with shell liners and material lifters which are generally manufactured from rubber. The drum is mounted at a slight angle, allowing the rock slurry to flow into one end and progress through the scrubbing process. As the drum rotates, the material is lifted and tumbled by the movement of the drum and the flow of water.

The wet drum scrubber utilises the combined effects of abrasion, attrition and impact to dislodge and break down particles adhering to rocks. The scrubbing action helps to remove clay, mud and other fine materials that may be present, improving the quality and purity of the final product. Additionally, the water serves as a medium for transporting the dislodged impurities away from the rocks.

The lifters on the interior of the drum enhance the scrubbing action. Together with the speed of rotation, water flow rate, and residence time of the material inside the scrubber these can be adjusted to optimise the scrubbing efficiency and achieve the desired cleaning results.

For this paper, a large drum scrubber is defined as having the ability to process rocks of up to 300 mm in size at a solids feed rate of greater than 1500 t/h and be rotated utilising a pinion gear from a motor with a variable speed drive. Figure 1 shows a series of drum scrubbers in operation with the photograph being taken from the feed end looking down to the discharge end of the drums. These drums have a diameter of 5.0 m with an effective grinding length (EGL) of 11.75 m. The EGL is the length of the mill cylinder where grinding, or scrubbing in this case, can occur. EGL is generally measured as the length from the inside of the drum scrubber feed end liner to the scrubber discharge weir.

FIG 1 – Drum scrubbers in operation at a Pilbara mine site.

PROCESS TESTING AND DESIGN

Scrubbers are used for two different reasons in the treatment of iron ore. The first is to break down clay particles found within a deposit. Clay contains up to 40 per cent alumina (Al_2O_3) and 45 per cent silica (SiO_2) and when subjected to enough time, water and energy, will break down into its natural particle size of less than 20 micrometres (μm). Removal of these clays is integral to improving the grade of the ore and meeting the desired alumina and silica product quality targets. The residence time to process these types of materials tends to lie between 60 sec and 2 min.

The second reason is to provide a robust method to mix sticky, ultrafine material with water and remove them downstream to aid with materials handling issues. The residence time of these scrubbers is often 30 sec or less.

The key to success with any drum scrubbing process is to apply enough energy, time and water to the material while being cognisant of the same functions to minimise attrition losses. This is best addressed in the testing and more specifically, the design phase. Note that for all tests hereafter 'residence time', unless specified, refers to the *average* residence time. Fine material, that which flows with the water, is assumed to be less than 0.5 mm in size.

Testing

Testing for determination of the important parameters of a drum scrubber can be highly questionable. The issue being that the overall requirement for determination of residence time and energy input is likely to be undertaken on a sample which is not easy to obtain and is not completely representative from an ore characterisation perspective. For large drum scrubbers, the sample should also contain large particles of up to 300 mm and be statistically representative of the particle size distribution. However, this is problematic as generally the samples are obtained from PQ diamond drill core which has a diameter of 85 mm. This means that the process engineer will have to make assumptions for material that is 200 mm to 300 mm in size.

Commonly in industry, two methods of testing have been observed. The first is a batch drum test and was utilised by Beros (2000) at Mesa J. This is where the sample is added to a drum with water provided up to the desired pulp density. The drum is then closed and rotated for a set time to mimic the residence time. After the set time period, the drum hatch is then opened and the drum rotated to allow the sample to be emptied either into a 200 L drum or dropped onto a screen and finally, hosed out to remove any remaining fines. The question here is, how much time and energy does all of that add to the desired residence time and energy input? If the ore requires a very short residence time this will not yield a satisfying result. However for many ore types, running this test over increasing lengths of time usually shows an increase in alumina rejection from the coarse material. This often will plateau indicating that a possible optimum residence time has been achieved. Figure 2 shows a typical batch drum test arrangement together with the resultant residence time and screening results.

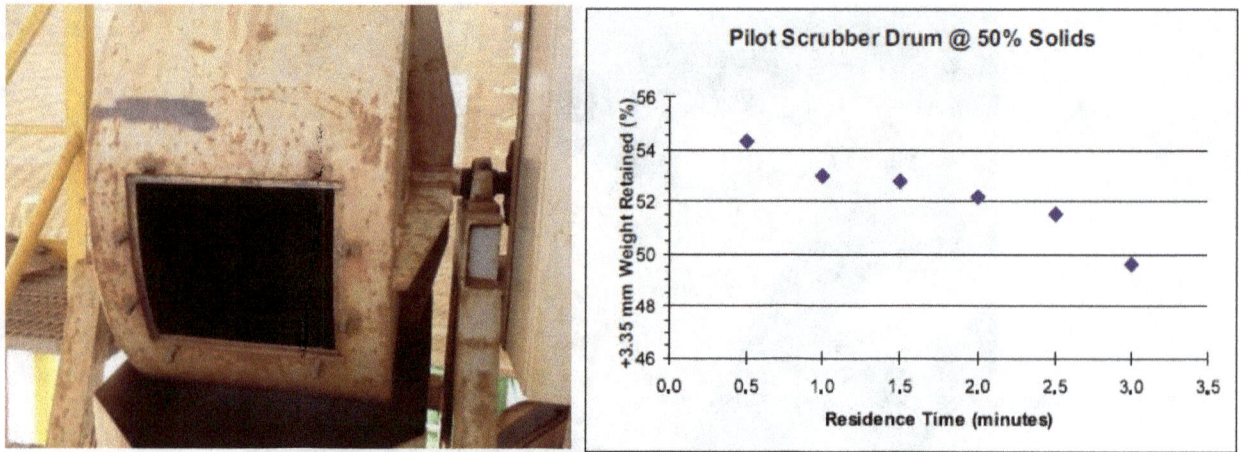

FIG 2 – Batch drum testing arrangement and results.

The second type of testing is a pilot scrubber. Pilot scrubbers can potentially be several metres in length, capable of handling a significant sample tonnage and generally accommodating a larger particle size. The scrubber is fed with solids at a fixed rate with a water flow set to achieve the required pulp density. The pilot scrubber has a discharge lip that can be adjusted to vary the residence time within the unit. After reaching steady state, the scrubber is sealed and the charge measured to ensure the time was set. This will also provide a graph of residence time versus silica and alumina rejection. Figure 3 shows a typical pilot scrubber testing arrangement together with results for alumina rejection from a particular series of tests. In this test work there was no availability to alter the motor speed of either the batch or pilot units so various energy consumptions could not be tested.

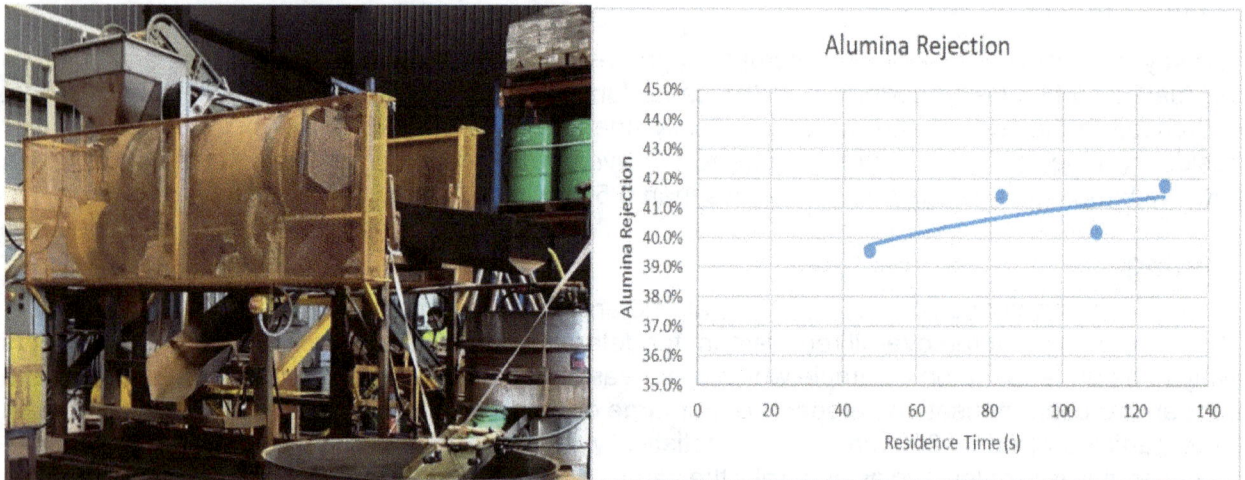

FIG 3 – Pilot drum scrubbing testing arrangement and alumina rejection results.

The most readily available calculation for determining a scrubber size is presented by Miller (2004). The paper proposed a calculation for scrubber sizing (see the paper for more details) and any unknown values were assumed to be as indicated in the 'Miller' model. As the pilot scrubbing can provide the actual residence time based on the load, the data was used to determine the variables associated with the Miller calculation. Retention time is a function of feed rates together with the length and the fall (gradient) of the scrubber. The feed inlet is higher than the discharge lip and the resultant gradient will contain a particular volume of material. The average retention time is determined using the calculated volume divided by the volumetric flow rate of solids and water. It is accepted that the water will carry the fine material out faster than the coarse material which is moved down the scrubber by rotation and the lifters. The rule of thumb says the coarse solids retention time is 2.5 times longer than the average, whereas water and fine particles travel at 0.6 times the average. As the pilot scrubbing can provide the actual residence time based on the load, Miller demonstrated how the data can be used to determine the variables associated with the calculation proposed in Equation 1 and depicted in Figure 4.

$$Q = C_d a D_0^b H_{tot}^c \qquad (1)$$

where:

Q   = Slurry flow

C_d = Weir Coefficient, assumed to be 0.63

Coefficients a, b and c

H_{tot} = The distance between the weir and the operating level, calculated from Equation 1

The H_{tot} is then used to give the total scrubber charge.

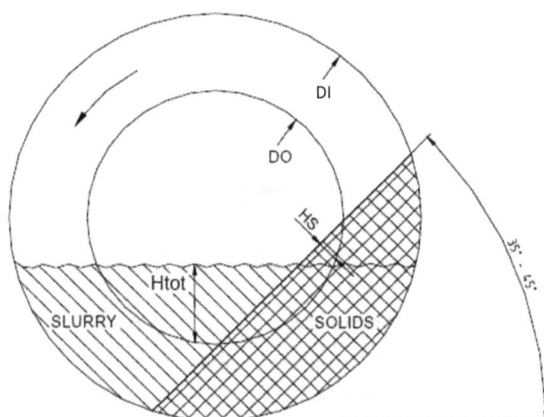

FIG 4 – Miller scrubber residence time geometry model.

Given that a = 0.0002, b = 0.5 and c = 2, the pilot scrubber data for residence time based on the measured solids was compared to the value determined using the Miller calculation (Equation 1). This data, presented in Table 1, gives a good correlation and is used for the initial drum scrubber sizing and power requirements for the design.

TABLE 1

Comparison of residence time, actual versus calculated from the Miller model.

Residence time based on volume (sec)	Miller model residence time (sec)
47.8	52.3
82.8	81.6
108.2	105.8
127.3	131.2

Process design

Scrubber sizing

After testing a range of residence times, it was established that there exists satisfactory contaminant rejection from as little as 45 sec and substantially greater rejection at 85 sec. From a rejection perspective it seems prudent to design a scrubber for a range of residence times. This is achieved by fixing the nominated residence time at 85 sec and then letting the physical dimensions of any given scrubber dictate the actual range possible. In this case, the Original Equipment Manufacturers (OEM's) were asked to provide machines capable of providing a range of residence times from 60 sec to 120 sec if possible. Generally, this is as much information a process engineer will receive to design and purchase equipment.

Given the desired feed rate, density, water content, bulk density and estimated residence time, the size of the scrubber can be determined. The inlet diameter (Df), as shown in Figure 5, is often a function of the maximum feed particle size. Nominally Df is set at 2.5 times the largest particle size to prevent blockages at the chute angle. The EGL is the inside length which will provide the residence time along with the discharge opening diameter (Do).

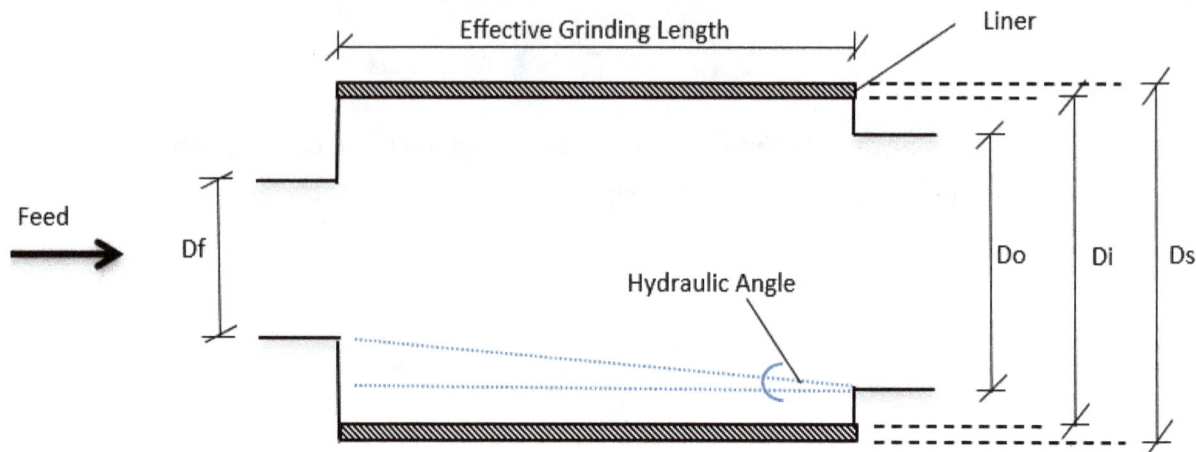

FIG 5 – Drum scrubber dimensions.

During the design phase, the OEM's are consulted and given the same data to determine the ultimate physical design of their machines. Four different discharge weir dimensions were provided and as calculations from OEM's are considered intellectual property, only the general arrangement and total calculated residence time and energy were provided which were then compared to the calculations determined from the Miller calculation using the estimated variables from the pilot data. This data is provided in Table 2.

TABLE 2

Comparison of OEM residence time to the Miller model.

	OEM residence time (s)	Miller model residence time (s)
Weir 1	65	69.1
Weir 2	80	83.5
Weir 3	95	99.0
Weir 4	111	115.0

As it is very difficult to determine loads in a full-scale drum scrubber, it is positive that the dimensions and values generated from the model aligns well to that provided by the OEM's.

System design

Historically, it has always been difficult to size scrubbers perfectly due to the variable nature of the orebodies and the difficulty in obtaining representative samples at large sizes. As such it is prudent to allow as much flexibility in the scrubber and the system supporting the scrubber. The key elements to a design are the:

- scrubber outlet diameter
- maximum feed rate
- power consumption
- water addition.

With careful planning, the scrubber outlet diameter can be designed with some flexibility in actual operation. A weir (fixed ring) is located at the end of the scrubber and different sized weirs can be inserted during maintenance shutdowns to either restrict the outlet for a longer residence time or open the outlet for a shorter residence time. In a recent project, the original design included four different options to outlet sizes and were available at site during commissioning and operations. The weirs themselves offered a range in residence times of between 65 sec and 111 sec.

The scrubber feed rate is generally set as a design value by the project. Because this can highly influence the residence time it is preferable to size the scrubber at the desired rate because adding a design figure to this can potentially lead to longer residence times. However, should the residence time be shorter in the plant operation, an increase in solids flow rate can be applied. As this is a possibility, it is prudent in the design of the feed conveyor and chutes to provide a significant design factor. Of course, this is also dependent on the downstream process capability. For best control practices, the feed to the scrubber should have a weightometer to provide controlled flow into the scrubber and to be used to calculate the water addition in tandem with a water flowmeter.

Scrubbers are usually fitted with a variable speed drive (VSD) motor control. This allows for a variation in critical speed that directly affects the energy input. As with all mechanical equipment there will always be a limit to the range that can be effectively applied. Typical scrubber design would allow for a range between 30 per cent and 70 per cent of critical speed. In general, this will relate to a VSD speed of between 50 per cent and 100 per cent, which is more often the reported value in the control system. This set point would commonly be an engineer-controlled setting and its optimum would be developed by plant testing and production data. Definite care should be taken when assessing the starting torque on the VSD. It is inevitable that at some point the scrubber will crash fully loaded and need to be started again.

As the feed density is of importance in the operation of the scrubber, the design of the water system for dilution is critical. For example, the project mentioned above initially attempted to design between 50 per cent solids and 70 per cent solids with a nominal set point of 60 per cent solids. Generally, this leads to a very large valve being required and since the control of the water flow is important, a globe valve is recommended. This requires proper planning as these valves are large and heavy and require space for maintenance access. In this case, availability of these types of valves ended up limiting the design envelope for the water addition, thus ending in the lowest density at 53 per cent, while maintaining solids rate.

Other key considerations in the design of a scrubber, is the outlet chute. Typically, the scrubber will feed a screen directly and the chute above the screen needs to allow for the biased scrubber outflow. A good design allows for a chute to redirect and spread the material across the entire screen width. To allow for flexibility due to possible changes in scrubber speed, dilution etc, the chute can be made up of three sections with the central redirection chute able to be adjust laterally to ensure the even feed to the screen.

GUIDELINES FOR OPERATING

Operating large wet drum scrubbers to process up to 300 mm rocks above 1500 t/h at the appropriate critical speed requires adherence to certain guidelines. A brief discussion of some salient general operating guidelines will cover the following points:

- feed rate control
- water flow and spray system
- retention time
- scrubbing media
- drum speed
- discharge system
- monitoring and maintenance.

Feed rate control

It is important to maintain a consistent feed rate of rocks into the drum scrubber to ensure optimal processing. Many operations utilise run-of-mine (ROM) blending stockpiles so that the correct ratio of harder, more competent rock is fed to achieve the required scrubbing action. This will promote rock on rock cascading which is desirable to remove and breakdown the softer, clayey and friable material. As the nominated feed rate of solids in this paper is above 1500 t/h, it is important that all feeders, conveyors and chutes have been appropriately designed to effectively handle rock sizes of 250 mm to 300 mm. The main feed belt conveyor, or apron feeder, which delivers ore into the scrubbers must be fitted with an appropriate and accurate tonnage weightometer so that measurements can be used to achieve the required water addition and therefore the correct operating pulp density.

Water flow and spray system

Water addition plays a crucial role in a drum scrubber and assists to soften the material being processed. The addition of water initially creates a slurry or suspension that facilitates the mechanical action of the scrubber in breaking down the softened material. As the drum rotates, the water flow helps dislodge and remove impurities ensuring a cleaner final product. Water addition also allows operators to control the drum scrubber operating pulp density. Correct water flow settings and adjustments maintains the separation efficiency of the equipment. As mentioned above, the water addition is controlled in tandem with the feed weightometer and highlights the importance to ensure that equipment used in control loops is regularly maintained and calibrated.

Retention time

Retention time refers to the duration that the scrubbing charge spends inside the drum scrubber. The appropriate retention time to ensure thorough cleaning of the rocks is determined in the test work and engineering design phase. Adequate retention time allows for sufficient agitation and contact between the rocks and water to ensure thorough removal of clayey and friable material.

Generally during the design phase an outlet weir, which is essentially an internal ring of a nominated height, is incorporated into the machine. The main function of the outlet is to ensure the appropriate total filling of the drum which is then factored into the scrubber mass. For large drum scrubbers the total mass can be in the order of 150 t together with a charge of a similar weight too. The possibility of installing additional weirs does exist which has the effect of increasing the residence time.

Should the characteristics of the feed material change from that used in the overall design then installing additional weirs or altering the weir height will be necessary. These are all potential options but are still constrained by the total mass of the drum and the charge as it is not possible to change the physical dimensions of the drum itself.

Scrubbing media

The selection of suitable scrubbing media to facilitate the drum scrubbing process is important. Common options include rubber or polyurethane liners and lifters, which help prevent wear and tear of the drum and improve scrubbing efficiency. Regular inspection is required and replacement of worn liners and lifters to maintain optimal performance is recommended. This will also identify 'hotspots', generally towards the feed end of the drum, where wear rates are higher and may have a more frequent replacement cycle than media further down toward the discharge end of the drum.

Many manufacturers of liners and lifters have the capability to undertake discrete element modelling (DEM) and can be used to gain a better understanding of the particle flow within a drum scrubber. Figure 6 depicts a cross-section profile of a drum scrubber charge that achieves the desired metallurgical performance. The diameter of the scrubber drums depicted in all figures below is 5.0 m.

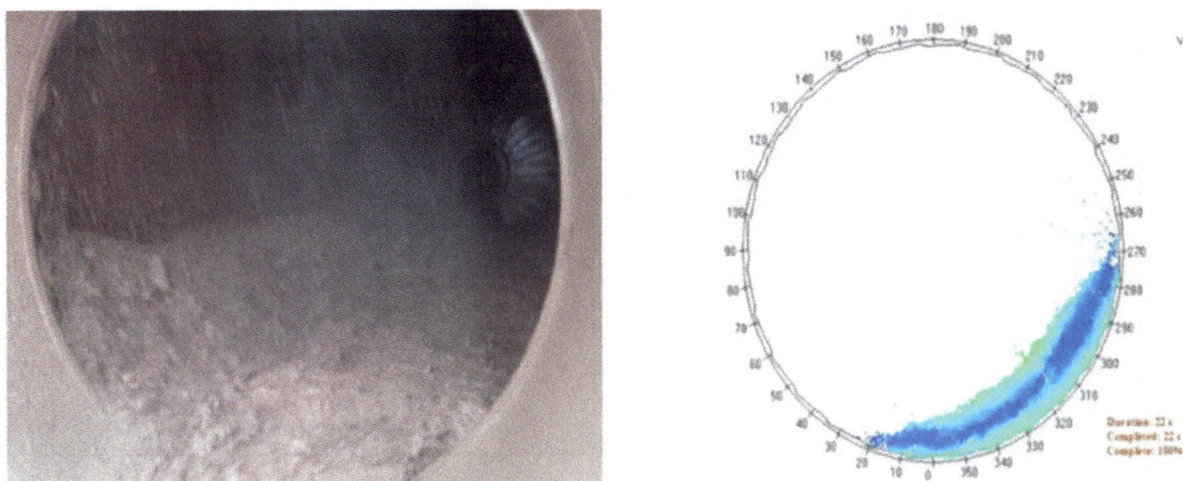

FIG 6 – Optimum drum scrubber charge profile.

There is good agreement with the modelled profile to the actual profile of the drum scrubber, pictured operating with a feed rate of 2100 t/h at a density of 55 per cent w/w solids. However, the lifters were modelled at the 50 per cent worn condition. Generally new lifters require some time to be 'worn' into operation. Following installation, as can be seen in Figure 7, new lifters generate some excessive rise of the bed which lowers scrubbing efficiency. At this site, the lifter manufacturer was engaged and a new profile was designed which provided the desired metallurgical condition immediately following installation. This design still maintained the same number of lifter rows inside the drum but had a longer width on the base to accommodate a lower lifter face angle which decreased from 70° to 55° from the horizontal. The longer base width is likely to increase the number of operating hours achieved between installations. The DEM profiles are presented in Figure 7.

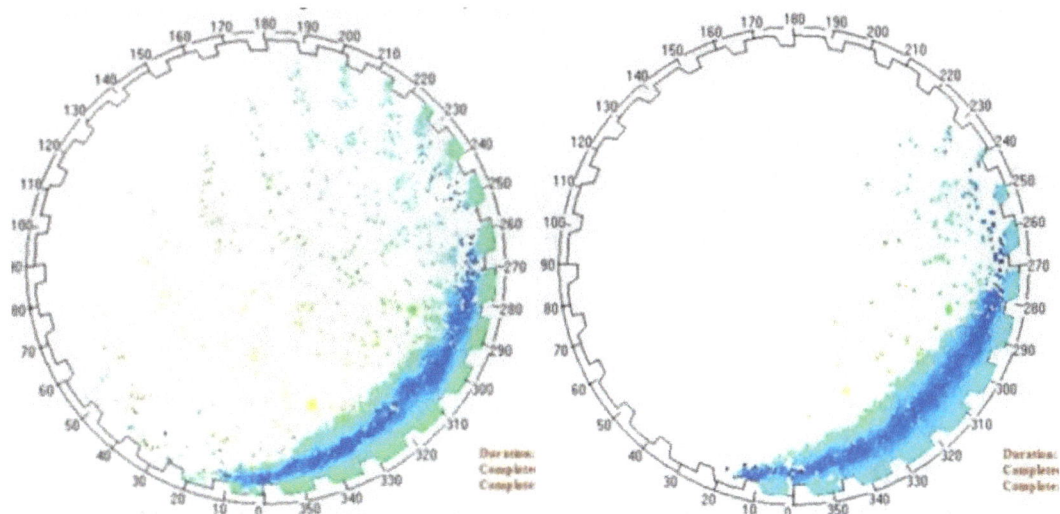

FIG 7 – Profiles of existing (left) and revised (right) lifter designs after installation (Source: Metso Minerals Wear Protection).

Drum speed

The appropriate scrubber drum speed is determined from test work in the engineering design phase. Generally, most scrubbers operate within a range of 40 per cent to 60 per cent of critical speed with the design criteria typically having a wider critical speed range of say 30 per cent to 70 per cent. A large drum scrubber is likely to be installed with a girth gear drive which requires extensive oil cooling and filtering systems. The drive motors are fitted with a variable speed drive (VSD) and it is recommended that drum speed, in revolutions per minute (rev/min), is determined at various VSD set points to build-up operating charts similar to the following graphs highlighted in Figure 8.

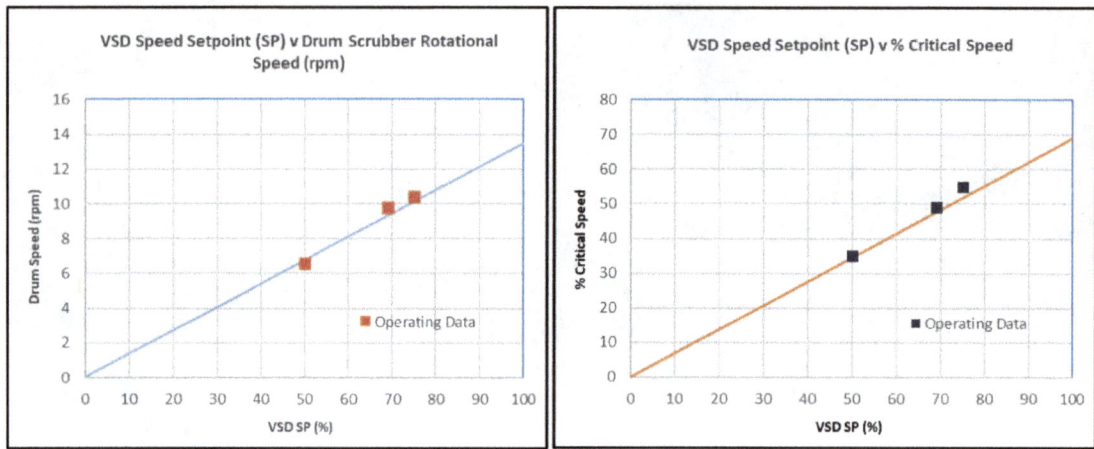

FIG 8 – Drum speed and critical speed as VSD set points.

This correct drum speed setting maximises the scrubbing action and enhances the separation of impurities from the rocks. Drum speed should be adjusted as required to ensure the desired product quality is achieved. In Figure 9, the wet screen pictured makes a separation at 1.0 mm. The screen oversize is conveyed directly to final product and the undersize is pumped to a concentration (or desand) plant. However, the photograph on the left shows that there is excessive free water still contained in the product. To overcome this, it was advised to reduce the drum speed from 9.7 rev/min to 8.3 rev/min with no change to any other operating variable. The image on the right of Figure 9 shows the improvement achieved and resulted in overall product moisture being reduced by 2 per cent; this equates to a 4 per cent reduction in screen oversize moisture.

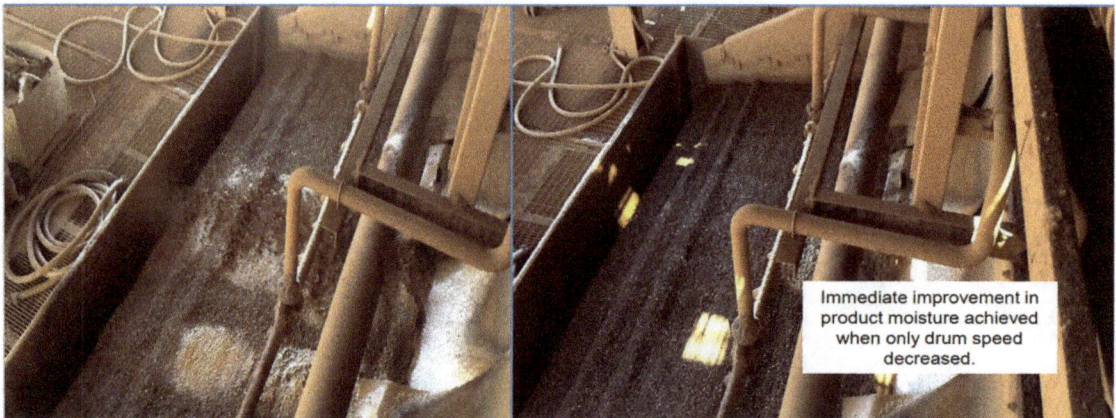

FIG 9 – Improvement in wet screen operation following a reduction in drum speed.

Discharge system

Product from large drum scrubbers is typically presented to a series of wet vibrating screens. It is important that the design of chutes presenting material promotes full width flow along the screen. Further to this, sprays must be installed which delivers water at a suitable volume and pressure to enable maximum removal of all the clay and friable material broken down in the scrubbing process from the rock. It is desirable to for the rock to be as clean as possible after wet screening.

Monitoring and maintenance

A drum scrubber's performance must be regularly monitored. Reference should be made to the equipment's operating and maintenance manuals for any specific requirements. It is recommended that as a minimum monitoring includes the feed rate, water flow, scrubber speed and overall efficiency, including downstream equipment which process the scrubbed product. A comprehensive maintenance program to inspect and service the scrubber components, replace worn parts and address any issues promptly must be implemented.

BENEFITS OF DRUM SCRUBBING TO SINTERING

The average quality of iron ore has been in decline for years as higher grades have been mined out and supply significantly expanded to meet rapidly growing Chinese demand. The deterioration in product quality of the average grade of sinter fines, especially for gangue content between 2010 and 2019, can be seen in Table 3. The trend is continuing.

TABLE 3

Average global sinter fines grades (1998–2019).

Year	Fe (%)	SiO_2 (%)	Al_2O_3 (%)	P (%)
1998	63.9	4.11	1.70	0.048
2010	62.9	4.10	1.73	0.056
2019	61.9	5.16	1.87	0.067

In the Pilbara context this means that resource and reserve tonnages of lower grade goethite, martite-goethite, bedded, detrital and channel iron deposits is increasing. It can be demonstrated that wet drum scrubbers assist to not only improve the grade of lower grade ore but also enhance the textural characteristics of the sinter fines product. Using the goethite classification proposed by Manuel and Clout (2017), it is the softer and friable ochreous goethite that breaks down in the scrubbing process.

A mineralogical survey of a set of primary hydrocyclones presented by Beros (2019), confirms the effect created by the operation of drum scrubbers on overall mineralogy. These hydrocyclones are the first downstream step for the treatment of wet screen underflow from the scrubbing circuit. Using a coloured textural coding, as highlighted in Figure 10, the browns and dark blue colours represent the harder hematite/goethite material. The lighter colours from the fawn to the greyish colours on the chart represent friable ochreous goethite plus alumino-silicates and other gangue minerals. Depending on the ore type presented to the primary cyclones, generally 35 per cent to 45 per cent of mass is rejected to the overflow. The survey clearly indicates that there is a concentration of harder hematite/goethite material to the underflow and rejection of the alumino-silicate minerals and ochreous/friable goethite to the overflow. The hydrocyclones also have a cut-point of 50 μm which marginally coarsens the sinter fines product and provide additional benefits for sintering.

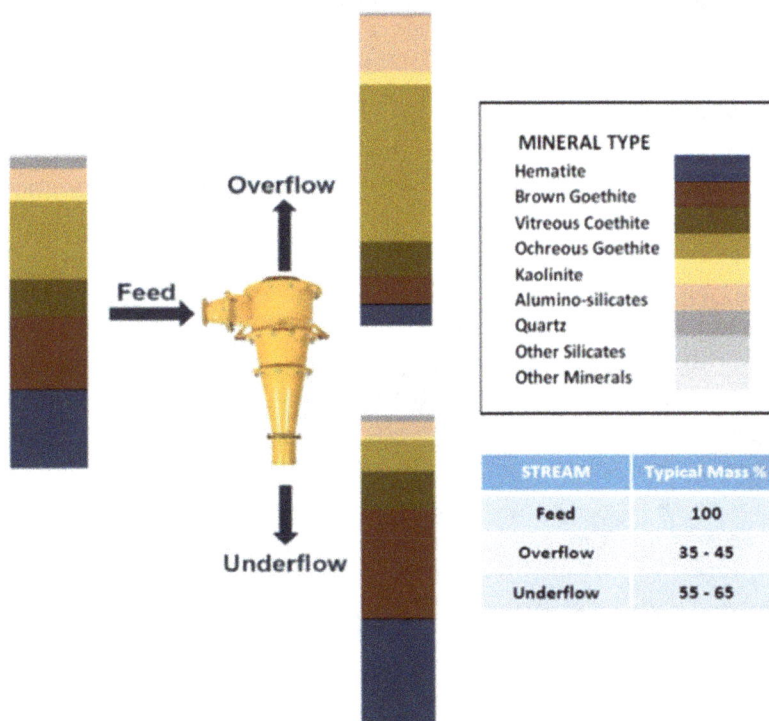

FIG 10 – Hydrocyclone mineralogical survey.

From a sintering and ironmaking perspective, Clout (2002) highlights that when ochreous goethite is present in sinter fines, it can affect these processes in several ways:

- Reducibility: Ochreous goethite has a relatively low reducibility, which means it is less easily reduced to metallic iron during the blast furnace process. However, this depends on the interaction of the other sintering factors listed below.

- Melting behaviour: Ochreous goethite has a higher melting point compared to other iron oxide minerals. This can affect the ironmaking process by increasing the energy required for achieving the desired softening and melting temperatures. It can also lead to higher fuel consumption and lower sintering productivity.

- Porosity and bonding: Ochreous goethite has a higher porosity compared to other iron ore minerals. This can result in an uneven flame front propagation creating more open pores in the sinter bed, affecting the sinter strength and reducing the overall mechanical strength of the sintered product.

- Permeability: The presence of ochreous goethite can reduce permeability and potentially hinder the proper flow of gases, affecting the efficiency of the ironmaking process.

The above clearly demonstrates that the inclusion of a drum scrubber circuit into a flow sheet can mitigate the negative effects of ochreous goethite on downstream ironmaking processes. It can also prolong resource and reserve volumes and assists in extending the operating mine life. Companies also benefit from reducing Scope 3 emissions with their customers.

CONCLUSION

The drum scrubber is a common piece of equipment for the beneficiation and upgrading of iron ore. The requirement to improve grade and reduce contaminants will continue to increase in the coming years to support the future of 'green' steel and reductions in emissions. As such, the design and operation of these drum scrubbers will become of increasing importance in the iron ore industry.

It is important that the knowledge of how to operate drum scrubbers needs to be applied to the design of the scrubber and the systems around the scrubber to allow for the flexibility required to address the unknowns that may arise between the testing phase and the operation of the full-scale plants. Flexibility is often equated to expense and over-design in the engineering field. However, the modest expense in the beginning will lead to a more robust and long-term process that will allow for optimisation, increased plant productivity and the ability to process ore-types that were not foreseen in the testing and design phase.

ACKNOWLEDGEMENTS

The authors would like to acknowledge Calibre Group for permission to publish this paper and the numerous people who have provided support over many years in several projects.

REFERENCES

Beros, G S, 2000. Processing of pisolitic iron ore at Mesa J, in *Proceedings of the Seventh Mill Operators Conference*, pp 85–92 (The Australasian Institute of Mining and Metallurgy: Melbourne).

Beros G S, 2019. Not all sub-60 per cent Fe ores are the same: the link between processing and sintering of Fortescue ores, in *Proceedings Iron Ore 2019*, pp 853–861 (The Australasian Institute of Mining and Metallurgy: Melbourne).

Clout, J M F, 2002. Upgrading processes in BIF-derived iron ore deposits, in *Proceedings Iron Ore 2002*, pp 237-242 (The Australasian Institute of Mining and Metallurgy: Melbourne).

Manuel, J R and Clout J M F, 2017. Goethite classification, distribution and properties with reference to Australian iron deposits, in *Proceedings Iron Ore 2017*, pp 567–574 (The Australasian Institute of Mining and Metallurgy: Melbourne).

Miller, G, 2004. Drum scrubber design and selection, in *Proceedings Metallurgical Plant Design and Operating Strategies, MetPlant 2004*, pp 529–540 (The Australasian Institute of Mining and Metallurgy: Melbourne).

The use of Zimbabwean limonite-coal composite pellet as a sustainable feed for pig iron production

T Chisahwira[1], S M Masuka[2], S Maritsa[3] and E K Chiwandika[4]

1. Metallurgical Engineering Student, University of Zimbabwe, Harare 263, Zimbabwe. Email: chisahwirathemba@gmail.com
2. Lecturer, University of Zimbabwe, Harare 263, Zimbabwe. Email: shebarmasuka@gmail.com
3. Metallurgical Engineering Student, University of Zimbabwe, Harare 263, Zimbabwe; Lecturer, Harare Institute of Technology, Harare 263, Zimbabwe.
4. Lecturer, Harare Institute of Technology, Harare 263, Zimbabwe. Email: chiwandikae@gmail.com

ABSTRACT

The challenges facing the iron and steel industry on a global scale include depletion of resources, excessive energy consumption and CO_2 emissions. The demand for iron products in Zimbabwe is increasing due to massive infrastructural development like the expansion and improvement of mines, transportation systems and water reticulation systems. In Zimbabwe, cast iron is mainly produced from recycling scrap metals. However, there is currently a shortage of scrap, which has resulted in the vandalism of state-owned properties such as bus terminals and railway wagons and lines. Nevertheless, Zimbabwe is rich in limonite ores, which are idle, and it is the aim of this research that this ore be effectively used to produce feed for the blast furnace. The Blast furnace is still the dominant route for producing cast iron where sintering and coking processes are the main routes for preparing the Blast furnace feed. Sintering limonite ore fines produce a product of poor grade because of the high value of loss on ignition that has a negative impact on Blast furnace efficiency. Furthermore, the sinter plant and the coke making plant are a source of huge environmental pollutants. Since low-grade Zimbabwean coal fines are also available in abundance, they were used in the preparation of the composite materials ensuring maximum utilisation of this resource. Results showed that the limonite ore-carbon composite pellets sample with the binary basicity ratio (CaO/SiO_2) of about 0.27 and coal addition of 2.5 wt per cent achieved above the minimum physical quality requirements for pellet of drop number of at least 4, a dry compression strength greater than 2.2 kg/pellet and an indurated compression strength greater than 250 kg/pellet. Limestone was used as the binder avoiding the addition of more impurities from bentonite since the limonite ore has a high SiO_2 content of about 8.78 wt per cent.

INTRODUCTION

Global steel production has been increasing gradually in the past decade due to an increase in demand for steel which is predicted to continue. The iron and steel industry for many countries, including Zimbabwe, is considered the pillar for infrastructural development and economic advancement, employing more than 6.0 million people globally, with approximately 900 billion US dollars turnover per annum; contributing immensely to global economic development (World Steel Association, 2018). The continued expansion of the iron and steel industry presents a challenge for producers in terms of the cost of raw materials, carbon dioxide emissions and energy usage. The iron and steel industry is highly energy intensive, reportedly the second largest global energy consumer after the chemical and petrochemical industry and with most of the energy used in the blast furnace in ore based steel production (Energy Technology Perspectives, 2010). Apart from being energy intensive, the iron and steel industry is responsible for about 5 per cent of global carbon dioxide emissions according to the International Energy Agency (IEA, 2023). This puts a considerable amount of pressure on producers to develop more sustainable, energy-efficient and environmentally sound methods for iron ore processing and steel production (Holmes and Li, 2015).

Even though many alternative ironmaking technologies have been developed globally, the blast furnace remains the most cost-effective and highly productive process for hot metals, responsible for approximately 70 per cent of all global steel production, and is expected to remain so in the future (Mousa, 2019).

As alluded to earlier, the high cost of raw materials as well as the depletion of high-grade ores is a challenge facing the iron and steel industry, necessitating the need for innovative and alternative ways for processing poor-grade ores to supply the ever-growing industry (Geerdes, 2009). Zimbabwe has not been spared in this global crisis and has in the past relied on scrap metal for cast iron production. However, the price of scrap metals has been increasing because of an acute shortage of scrap, which has resulted in the closure of many foundries in the country. Another problem associated with the dependency of Zimbabwe on scrap metal is the vandalism of state-owned property such as rail wagons (Phiri, 2022), as suppliers seek to provide the much-needed commodity to cast iron producers. Meanwhile, the country has vast amounts of limonite ores that are not being utilised. Ripple Creek Deposits, located in the Midlands Province of central Zimbabwe, is estimated to have 111 million t of iron ore, 59 per cent of which meet blast furnace requirements and the remainder being soft limonite ores that generate fines during mining and processing.

Sintering is an easy, energy-efficient way of processing iron ores. However, sintering limonite ores results in more pores in the sinter due to the high water of crystallisation content and the high loss on ignition of the same affects the sinter strength, making sinter made from these ores unsuitable for blast furnace feed. In terms of carbon dioxide emissions, the production of high-quality sinter using the limonite ore would require high coke rates and subsequently high carbon dioxide emissions (Clout and Manuel, 2015). Limonite ores can therefore be effectively utilised through the preparation of limonite–carbon composite pellets, which will not only utilise this idle resource but also improve the properties of the iron-bearing burden, increasing reduction efficiency by improving reactivity and lower carbon dioxide emissions thus contributing towards Zimbabwe's aim to reduce emissions by 40 per cent per capita by 2030. The source of the carbon is coal fines, which have not found use in the country lying idle in dumps in coal mining companies and as such, this will provide a way of waste management for the coal fines (IEA, 2022).

The use of iron–carbon composite pellets in iron and steelmaking has been studied in recent years as researchers seek environmentally sound ways of ironmaking and a way of utilising fines generated in coal and iron ore mining processes. The depletion of high-grade iron ores globally necessitates the use of lower grade ores, including limonite ores such as those used in this study. According to Mishra (2020), using composite pellets offers a way of fines utilisation without high-temperature burden preparation, increases productivity due to an increased rate of reduction and allows for the use of non-coking coal as a reductant. The increased contact between the carbon and the iron oxide increases carbon monoxide utilisation efficiency and lowers the reducing agent rate. Factors that influence the reduction kinetics of the composite pellets include pellet composition, pellet geometry and particle size according to Huang (2018).

Somerville (2016) investigated the properties of composite briquettes made with Australian hematite ore and charcoal in order to determine the effect of heating time and carbon-to-iron ratio (C/Fe) on the strength, density and iron metallisation of the fired briquettes. It is reported that briquette strength increased as the C/Fe was decreased and at longer heating time, strength increased. Density increased slightly as heating time was increased but further increase did not have an effect on the briquette density while the degree of iron metallisation was reported to be 45 per cent after 3 minutes of heating and increased to 80 per cent after 5 minutes. Further heating did not result in an increase in metallisation. Briquette density after heating decreases with increasing C/Fe ratio in general while increasing C/Fe ratio increases the degree of metallisation in the heated composite briquettes.

Pal (2018) also reports on the replacement of bentonite as a binder in pellet formation because of the high levels of alumina and silica, with CaO. Apart from the impurities that are in bentonite that cause a reduction in the pellet quality, the increase in demand for bentonite clay in crude oil exploration has resulted in the prices of bentonite increasing (Zhu, 2022). When CaO is used, it promoted primary bonding between particles and imparted strength to the green pellets as a result of the formation of $Ca(OH)_2$. Upon treatment with carbon dioxide, calcium hydroxide is converted to calcium carbonate which is responsible for cold crushing strength and this transformation is also associated with a 1.52 per cent volume expansion. This transformation improves porosity but in excess results in a breakdown of the pellets (Pal, 2018).

Park et al (2018) reported on the reduction behaviour of magnetite – coke composite pellets flued with dolomite where the C/O ratio and dolomite content influenced the rate of reaction. Higher dolomite content and high C/O resulted in a high reduction degree. However, dolomite levels higher than 5 wt per cent compromised the physical strength of the pellets. The physical strength of pellets, particularly those used in the blast furnace, is important as they should be able to withstand the compressive pressure on the pellet bed from the furnace load. Increased reducibility of iron – carbon composite is significant in terms of increased efficiency of the blast furnace which subsequently leads to lower emissions as the iron and steel industry is aiming to be greener. The aim of this study is to investigate the possibility of using Zimbabwean limonite ore–carbon composite pellets as an alternative feed for the blast furnace. This is expected to provide the basic knowledge of this ore and to provide a way of effectively using the limonite ore and waste coal fines that are lying idle. Apart from contributing towards the lowering of carbon dioxide emissions as aforementioned, this study also aims to increase the resource base for the iron and steel industry since high-grade ores are depleting.

EXPERIMENTAL

Materials

The iron oxide used in this study is a limonite ore obtained from Kwekwe, in the Midlands province of central Zimbabwe. The coal was obtained from waste dumps at Hwange Colliery Company (HCC) in the Matabeleland North province of Zimbabwe. Chemical analysis was done for the limonite and limestone as shown in Table 1 while proximate analysis for the coal fines was done according to the Indian Standard IS:1350 (Part-I)-1984 (Table 2).

TABLE 1

Analysis of Zimbabwean Limonite ores and limestone

Analyte	T-Fe	SiO_2	CaO	Al_2O_3	MgO	Mn	S	P	LOI
Zimbabwean Limonite Ore	52.51	8.78	1.32	1.30	0.59	2.1	0.005	0.04	11.87
Limestone	1.34	7.10	45.70	1.19	4.22	0.31	0.17	0.01	39.96

TABLE 2

Coal proximate analysis.

Fixed carbon	Volatile matter	Ash content	Moisture content
64.28	24.12	10.40	1.20

The determination of the phases in the iron ore was done using X-ray diffraction at a scanning speed of 0.5 with increments of 0.02 at a 2θ range of 10–80 degrees and was determined to be hematite, goethite and silica as shown in Figure 1.

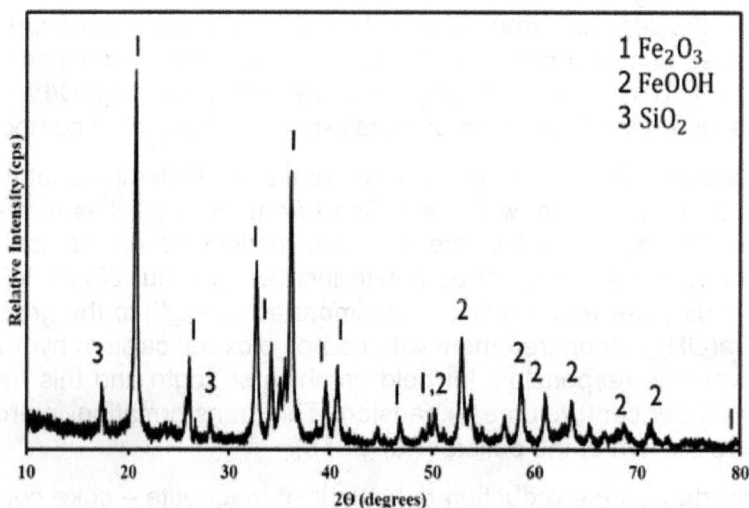

FIG 1 – Phase identification of limonite ore.

Preparation of raw materials

Moisture was removed from limestone and limonite by oven drying at 383 K for 24 hours. Size reduction for the ore and the limestone was done using a laboratory jaw crusher to an average of 10 mm and a riffle splitter was used to ensure the homogeneity of the samples. Coal fines, which had an average size of 2 mm, limestone and limonite ore, were further reduced in size to 80 per cent passing 75 μm using a pulveriser. The samples were sieved on a 75-micron sieve and the oversize material was re-pulverised until the target size of less than 75 μm was achieved to ensure uniform mixing of the raw materials for efficient pelletising.

Optimisation of limestone and coal

Five blends were prepared each with varying amounts of limestone, 0, 2.5, 5, 7.5 and 10 wt per cent to attain the calculated chemical composition and a basicity ratio (CaO/SiO_2) illustrated in Table 3.

TABLE 3

Evaluated chemical composition of pellets.

Analyte	T-Fe	SiO_2	CaO	Al_2O_3	MgO	CaO/SiO_2
Blend 1	52.51	8.78	1.32	1.30	0.59	0.15
Blend 2	51.26	8.73	2.40	1.30	0.67	0.27
Blend 3	50.07	8.70	3.43	1.29	0.76	0.39
Blend 4	48.94	8.66	4.42	1.29	0.84	0.51
Blend 5	47.86	8.62	5.35	1.29	0.92	0.62

After homogenisation, a pelletising disc of 610 mm diameter, rotating at 28 rev/min and inclined at 40° was then used to make green pellets. The necessary amount of water was added to the rolling pellet feed for green pellets formation which were removed once the target size (12 mm) was attained. The green pellets were air-dried for 48 hours. Drop tests and moisture content tests determined the optimum level of limestone to be 2.5 wt per cent. Drop tests involved the random selection of ten pellets from each batch which were dropped on a steel plate at a height of 45 cm. The number of drops that were required to break an individual pellet was noted as the drop number. The average of the ten pellets was then considered as the drop number of that batch.

$$Drop\ number_{batch} = \frac{\sum drop\ number\ of\ each\ pellet}{10}$$

Optimisation of coal then followed, where varying amounts of coal (0, 2.5, 5, 7.5 and 10 wt per cent) were added to pellets with 2.5 wt per cent limestone. The pellets were taken for drop tests and compression tests and moisture content tests after air drying for 48 hours. Composite pellets between 12–16 mm diameter were placed in a ceramic crucible of 80 mm internal diameter and 110 mm height and indurated at 1373 K in a muffle furnace for 25 minutes. Cooling was done naturally and further tests, morphological analyses using the optical microscope and compression tests were done on the indurated composite pellets.

Dry compressive strength was done using the Electronic Universal Testing Machine (ADW-50S) on 20 pellets. The average maximum load a pellet was able to crack was recorded as compressive strength in units of kg/pellet (kg/p). Induration compression tests were also done using the Electronic Universal Testing Machine (ADW-50S) where 20 indurated pellets from each batch were compressed and the average was recorded as the compression strength. A schematic representation of the composite pellet preparation is shown in Figure 2.

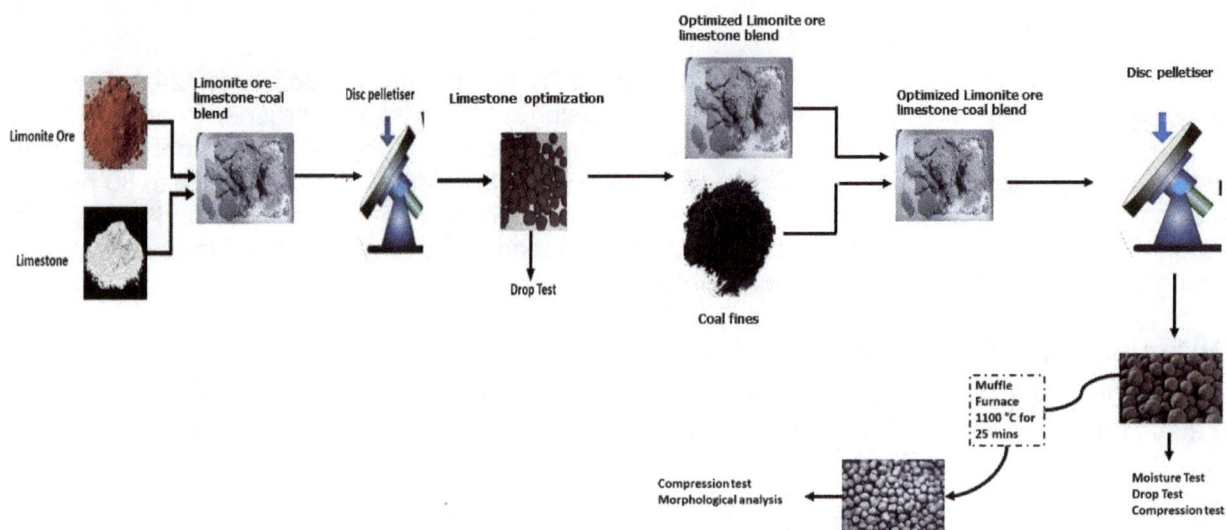

FIG 2 – Schematic presentation of pellet preparation.

RESULTS AND DISCUSSION

Optimisation of limestone

Bentonite is traditionally used as a binder in pelletising iron ores, but it has the disadvantage of adding impurities such as alumina and silica. Results of the chemical analysis of limonite ores showed a high amount of silica (8.78 wt per cent) and so limestone was used and the amount added had to be optimised. Apart from acting as a binder, limestone is used as an additive to produce pellets with high reducibility and good melting and softening properties (Iljana *et al*, 2022). The optimum amount of limestone was determined through drop tests of the green pellets. Drop tests are done to ensure that pellets do not deteriorate and generate fines during handling and transportation prior to induration, therefore the drop test is a measure of the ability of the composite to maintain its integrity (Geerdes, 2009). The addition of limestone resulted in an increase in the drop number from 2 to 6 at 0 wt per cent and 2.5 wt per cent limestone, respectively. However, further addition resulted in a decrease in the drop number as shown in Figure 3.

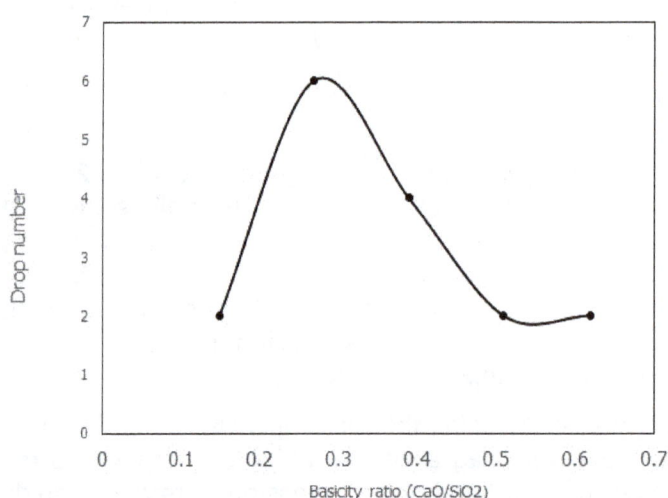

FIG 3 – Drop test results of the blends.

The drop test determined the optimal binary basicity ratio (CaO/SiO_2) to be 0.27 as the drop number was 6, which is above the minimum required drop number of 4 (Sivrikaya and Ali, 2012). The trend for the drop number against basicity shows a peak at 2.5 wt per cent limestone, after which a decrease in the drop number is observed. The initial increase in the drop number may be because of the binding properties of limestone (Iljana *et al*, 2022). Blend 5 which had the largest amount of limestone with a binary basicity ratio (CaO/SiO_2) of 0.62 had a drop number of 2. Increase in the amount of limestone to above 2.5 wt per cent lowers the drop number probably due to the increase in the distance between the components responsible for forming the bonding phase in the blend. However, further investigations are required for clarification.

Optimisation of coal

The optimum binary basicity ratio of 0.27 was maintained for the optimisation of the coal. Moisture content in the composite pellets was analysed for uniformity purposes as a quality control measure. Results are shown in Figure 4.

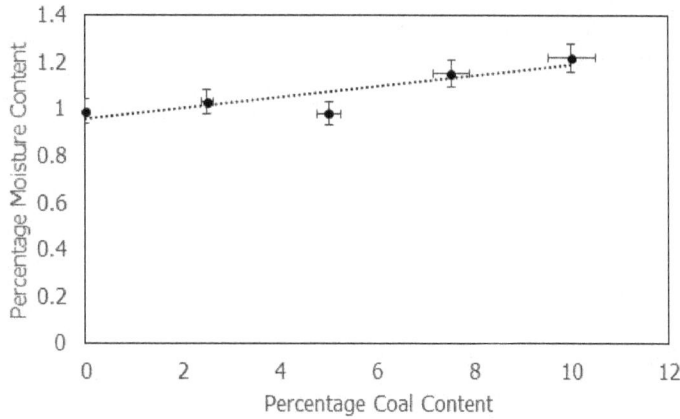

FIG 4 – Moisture content against percentage coal.

Moisture tests on the composite pellets indicated an increase in the moisture content of the pellets as the amount of coal was increased. The difference is about 0.2 per cent and is negligible, confirming uniformity in the composite blends. The small increase in the moisture content in the air-dried pellets with increasing coal content may be from the inherent coal moisture.

Drop tests on the composite pellets indicated a decrease in the drop number with an increase in the amount of coal added. There was an anomaly at 10 wt per cent coal addition, 12 wt per cent coal was added to get a general trend with increasing coal. The decreasing trend with increasing coal addition was clarified. The first blend with no coal added had a drop index of 6, while the fifth blend with the highest amount of coal at 10 wt per cent had a drop index of 3 as shown in Figure 5.

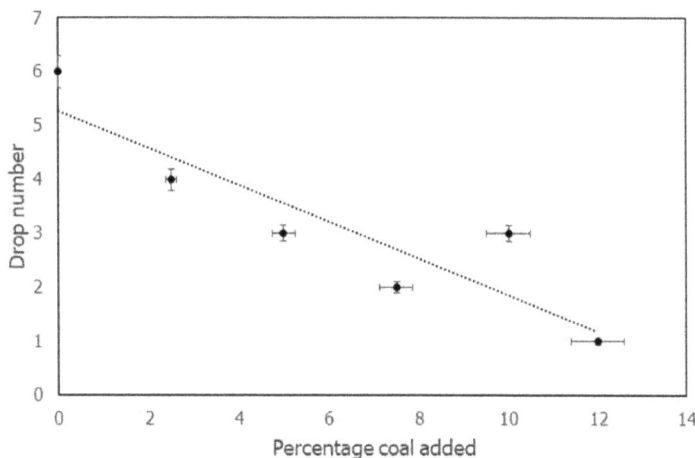

FIG 5 – Drop number of green pellets versus coal percentage.

The composite blend with 2.5 wt per cent coal gave a drop number of 4 and therefore less than 2.5 wt percent coal addition was determined as the optimum amount of coal. The decrease in the drop number may be a result of the hydrophobic nature of the coal which causes a decrease in the amount of water absorbed during palletisation resulting in weaker bonds (Deepak *et al*, 2021). Furthermore, an increase in the percentage of coal increases the iron-to-iron distance such that there are weaker bonding bridges during pellet formation.

Dry compression strength

The dry compressive strength index shows the ability of green pellets to resist disintegration when loaded by other green pellets on the conveyer belts and grates (Shaik *et al*, 2020). Pellets containing no carbon had a dry compressive strength value below the standard industrial requirement of 2.2 kg/pellet (Sivrikaya and Ali, 2012). The dry compressive strength increased from 1.4 kg/pellet at 0 per cent coal content to 2.6 kg/pellet at 2.5 per cent coal. However, further increase resulted in much lower compressive strength as indicated by the third blend which had a coal content of 5 wt per cent and a compressive strength of 1.3 kg/pellet. The blends with 7.5 wt per cent and 10 wt per cent coal added had a compressive strength of 2 kg/pellet and 1.9 kg/pellet respectively. Results are illustrated in Figure 6.

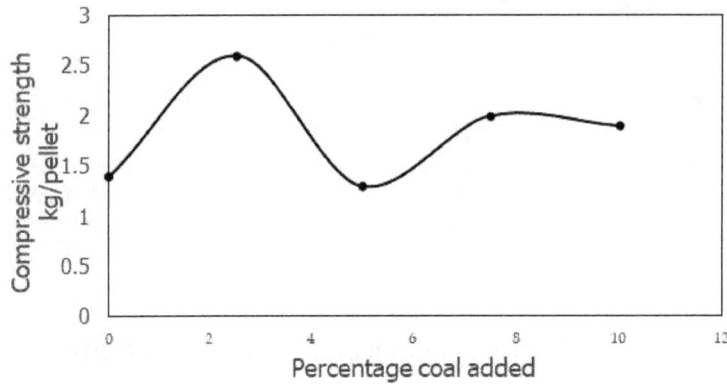

FIG 6 – Dry compressive strength of composite pellets.

Indurated compressive strength

Indurated compressive strength of pellets is one of the key quality indicators for blast furnace ironmaking as the process has high strength demand to minimise generation of fines during processing. The minimum requirement for industrial pellets is 250 kg/pellet (Sivrikaya and Arol, 2014) and the compression strength of carbon containing pellets had a value above the minimum as shown in Figure 7.

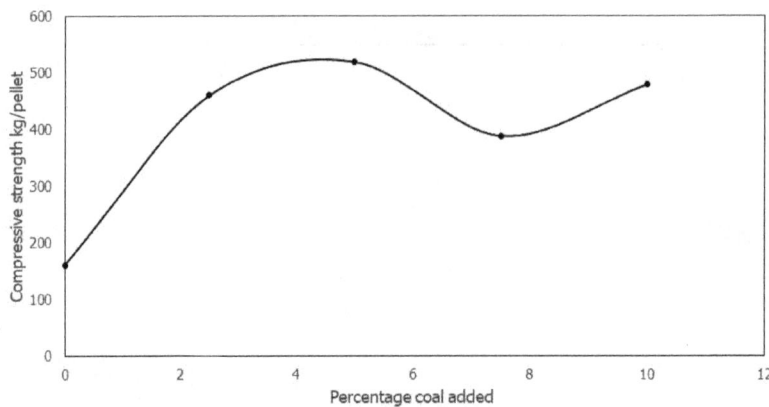

FIG 7 – Compression strength of indurated pellets.

The first blend with no coal had an indurated strength of 160 kg/pellet and upon the addition of 2.5 wt per cent carbon, the compression strength increased to 460 kg/pellet. Figure 7 illustrates that there was a general increase in the compression strength of the pellets with an increase in the amount of carbon added. There was however a drop in compressive strength at 7.5 wt per cent carbon which might be a result of increased porosity, however, further investigations are required to ascertain this reasoning. Decreasing the distance between iron oxide particles by increasing the iron-to-carbon ratio in the compact mix or by decreasing the particle size was also found to increase the strength of the pellets after induration (Somerville, 2016). In this study, the size was maintained while carbon–iron ratio was varied where the trend indicated an overall increase in strength. This might be related to the type and amount of bonding phases developed during the induration process.

A thermodynamic analysis was done as well as morphological analysis in order to provide more evidence.

Thermodynamic analysis

Prediction of the probable phases that might develop during the induration of the pellet was done using basic thermodynamics relying on the chemical composition of the blends. It should be noted that the thermodynamic analysis was done to assist with the theoretical understating of the phases even though natural iron ore with diverse chemical interactions was used. The limonite phase is converted into hematite in the dehydration temperature range around 453 to 673 K according to Equation 1.

$$2FeOOH(s) = Fe_2O_3(s) + H_2O(g) \tag{1}$$

Research has shown that the lowered water content in the goethite enhances the specific area of the iron ore, according to Zulkania $et\ al$ (2022). Research by Zhao $et\ al$ (2019), however, indicated that goethite could not be fully decomposed to hematite at temperatures between 458 and 673 K. Since limestone ($CaCO_3$) was used to control the basicity ratio and as a binder the decomposition temperature of $CaCO_3$ was estimated using the Equation 2.

$$CaCO_3(s) = CaO(s) + CO_2(g) \tag{2}$$

$$\Delta G° = 168\ 400 - 143.94\ T \text{ J/mol (298–1112 K)} \tag{3}$$

The decomposition temperature was estimated using Equation 3 to occur above 1170 K. With no coal addition, the bonding phases that are likely to be formed based on the chemical composition of the blend can be predicted thermodynamically as follows:

$$CaO(s) + Al_2O_3(s) + SiO_2(s) = CaO \cdot Al_2O_3 \cdot SiO_2(s) \tag{4}$$

$$\Delta G° = -105\ 900 + 14.2\ T \text{ J/mol (298–1673K)} \tag{5}$$

$$CaO(s) + Al_2O_3(s) = CaO \cdot Al_2O_3(s) \tag{6}$$

$$\Delta G° = -18\ 000 - 19\ T \text{ J/mol (773–1878K)} \tag{7}$$

$$2CaO(s) + SiO_2(s) = 2CaO \cdot SiO_2(s) \tag{8}$$

$$\Delta G° = -118\ 800 - 11.3\ T \text{ J/mol (298–2403K)} \tag{9}$$

$$CaO(s) + Fe_2O_3(s) = CaO \cdot Fe_2O_3(s) \tag{10}$$

$$\Delta G° = -30\ 000 - 4.8\ T \text{ J/mol} \tag{11}$$

$$2CaO(s) + Fe_2O_3(s) = 2CaO \cdot Fe_2O_3(s) \tag{12}$$

$$\Delta G° = -53\ 100 - 2.5T \text{ J/mol} \tag{13}$$

When coal is added, the direct reduction of the hematite to magnetite is expected as well as the indirect reduction where the Boudouard equation is of great importance, Equation 14.

$$CO_2(g) + C(s) = 2CO(g) \tag{14}$$

$$\Delta G° = 169\ 000 - 172.16\ T \text{ J/mol (298–2500 K)} \tag{15}$$

The Boudouard equation is thermodynamically estimated using Equation 15 to occur at temperatures above 981 K. The presence of CO will result in the formation of magnetite according to Equation 16:

$$3Fe_2O_3(s) + CO(g) = 2Fe_3O_4(s) + CO_2(g) \tag{16}$$

$$\Delta G° = -43\ 100 - 52.84\ T \text{ J/mol (298–2500 K)} \tag{17}$$

The presence of magnetite in the sample stabilises the fayalite phase (Chiwandika and Jung, 2020) according to Equation 18.

$$2FeO \cdot Fe_2O_3(s) + SiO_2(s) = 2FeO \cdot SiO_2(s) + Fe_2O_3(s) \tag{18}$$

The fayalite phase can also be formed according to Equation 19.

$$2FeO(s) + SiO_2(s) = 2FeO \cdot SiO_2(s) \tag{19}$$

$$\Delta G° = -36,200 + 21.09T \text{ (J/mol)} \tag{20}$$

The thermodynamic data of Equations 2 to 20 was obtained from Lee (1999). Using Equations 2 to 20, a $\Delta G°$ versus temperature is given in Figure 8.

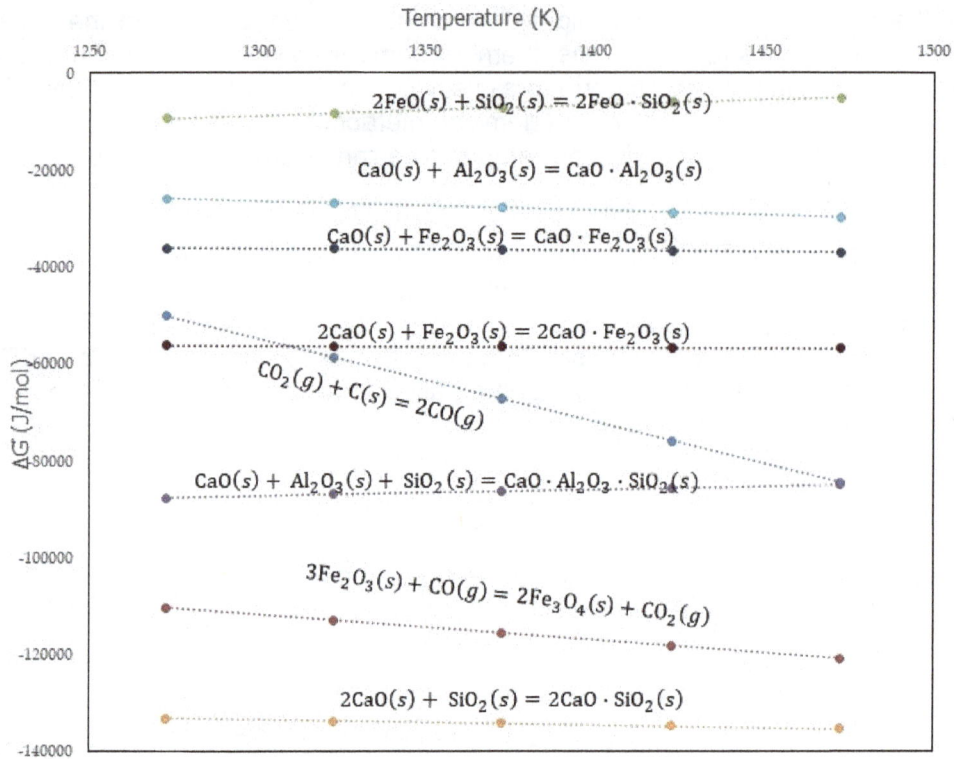

FIG 8 – Plot of $\Delta G°$ versus Temperature in the temperature range of 1250–1500 K.

The induration of the pellets was performed at 1373 K well above the estimated decomposition temperature of $CaCO_3$. The CaO from the decomposition of $CaCO_3$ was therefore used in the formation of bonding phases like the thermodynamically stable $2CaO \cdot SiO_2$, $CaO \cdot Al_2O_3 \cdot SiO_2$, $2CaO \cdot Fe_2O_3$, $CaO \cdot Fe_2O_3$ and the $CaO \cdot Al_2O_3$. However, based on the low amount of CaO available in the blend, low amounts of bonding phases are predicted to be found. The addition of coal allowed for the reduction of hematite to magnetite facilitating the stabilisation of the low melting temperature fayalite phase. Figure 8 shows that the fayalite phase is less likely to be the preferred phase thermodynamically. However, silica is available in abundance and will react with magnetite forming the fayalite phase (Equation 18), increasing the number of bonding phases available. This might be the main reason for the increased compressive strength in the indurated limonite–coal composite pellets. However, more results are required from XRD to validate this hypothesis. Pellet reducibility studies also need to be investigated.

Research has also shown that in the absence of CaO, bonding phases with high melting temperatures like the mullite may be formed. The presence of CaO results in the formation of feldspar ($CaAlSi_3O_8$) which has a lower melting point of about 1866 K than mullite ($3Al_2O_3 \cdot 2SiO_2(s)$) which has melting points of 2183 K (Prusti *et al*, 2020). Prakash (1996), reports that the formation of calcium diferrite bonding phase ($CaO \cdot 2Fe_2O_3$), improves the reduction properties of the composites. However, the amount of CaO in our blends was limited resulting in the formation of low amounts of such a phase. Fayalite has also a lower melting point of about 1473 K allowing more bonding phases to be formed in the limonite–coal composite pellets. Studies by Somerville (2016) indicated that samples that were close to the fayalite phase had a melting temperature of 1473 K and when cooled, the slag structures held the composite briquettes together and imparted some strength (Somerville, 2016) which was in agreement with the current findings basing on the thermodynamic predictions.

Morphological analysis

Results of the morphological analysis is shown in Figure 9.

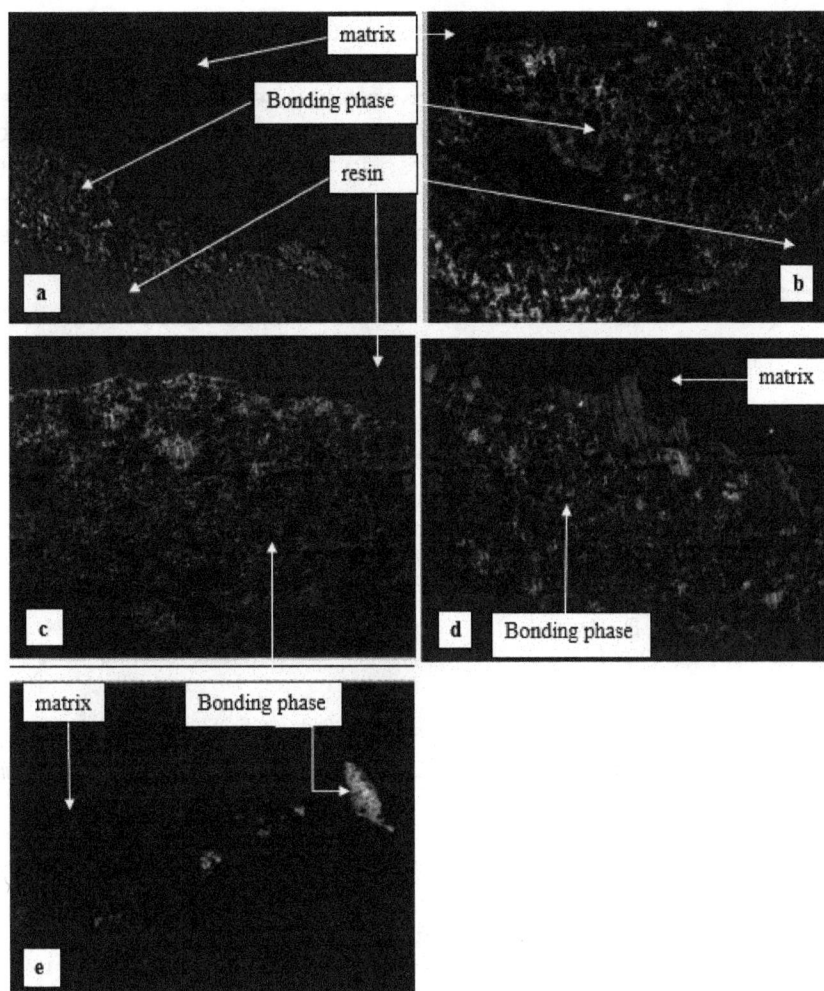

FIG 9 – Morphological changes with increasing coal.

Morphology of the indurated composite pellets was investigated at coal percentages of 0 per cent, 2.5 per cent, 5 per cent, 7.5 per cent and 10 per cent through the use of a metallurgical optical microscope as shown in Figure 9a–9e. The formation of a slag bonding phase was observed in all the varieties of the indurated composite pellets. The nucleation and following growth of the bonding phase formed at the surface of the pellets as from Figure 9a–9e, the bonding phase is located in contact with the resin that encapsulates the pellet. The intensity of the bonding phase formed increased from the pellet containing coal at 0 wt per cent, 2.5 wt per cent up to 5 wt per cent as shown in Figure 9a–9c. This is the reason why the indurated crushing strength increased from 160 kg/pellet at coal addition of 0 wt per cent to 460 kg/pellet at coal addition of 2.5 wt per cent and lastly to 520 kg/pellet at 5 wt per cent coal addition. The intensity of the bonding phase increased from 0 wt per cent coal pellet to 5 wt per cent coal pellet due to the subsequent increase in the quantity of coal therefore enhanced reduction of the iron oxides and possibly the formation of fayalite as one of the bonding phase. The drop in the intensity of the bonding phase observed in Figure 9d and 9e may be due to the increased distances between the iron oxides to be reduced and silicon dioxide. High concentrations of carbon, that is 7.5 wt per cent and 10 wt per cent, may be responsible for increasing the distance between the active species responsible for forming the bonding phase.

CONCLUSIONS

It has been shown that the use of Zimbabwe limonite–carbon composite pellets could potentially be an alternative raw material for the cast iron production as the optimisation of limestone and carbon achieved physical properties that are above the minimum requirements for blast furnace operations. Analysis of the ore indicated a significantly high amount of silica 8.78 wt per cent, which then made the use of bentonite as a binder unfavourable and resulted in the use of limestone as an alternative. The addition of 2.5 wt per cent limestone that resulted in a basicity ratio (CaO/SiO_2) of 0.27 was selected as the optimum since the green pellets achieved a drop number greater than 4. Composite pellets with 2.5 wt per cent coal were selected as the optimum since a drop number and a compressive strength of 4 and 2.5 kg/pellet was obtained respectively. Indurated pellets had a compressive strength above 250 kg/pellet, which is above the minimum allowable compressive strength. Morphology analysis showed an increase in the bond phase with an increase in coal content. Overall, it is recommended that more research be done to optimise and test other parameters such as swelling, softening and melting properties and reduction disintegration.

ACKNOWLEDGEMENTS

This research was funded by the University of Zimbabwe Research and Innovation Directorate.

REFERENCES

Chiwandika, E K and Jung, S-M, 2020. Effects of ilmenite ore on phase development of hematite ore sinter, *Metallurgical and Materials Transaction B*, 51, pp 1469–1484.

Clout, J M F and Manuel, J R, 2015. Mineralogical, Chemical and physical characteristics of iron ore, in *Iron Ore: Mineralogy, Processing and Environmental Sustainability* (ed: L Lu), pp 45–84 (Woodhead Publishing). <https://doi.org/10.1016/B978-1-78242-156-6.00002-2>

Deepak, N, Nigamananda, R, Nilima, D, Swagat, S R, Soobhankar, P and Partha, S D, 2021. An Optimal Route for Preparation of Metallised Composite Pellets from Ilmeite Concentrate, *Journal of Sustainable Metallurgy*, 7(1):1102–1115.

Energy Technology Perspectives, 2010. *Scenarios and Strategies to 2050* (OECD/IEA).

Geerdes, M, 2009. The Ore Burden: Sinter, Pellets, Lump Ore, in *Modern Blast Furnace Iron Making: An Introduction*, pp 31–35 (IOS Press: Amsterdam).

Holmes, R J and Li, L, 2015. Introduction: Overview of the global iron ore industry, in *Iron Ore* (ed: L Lu), *Mineralogy, Processing and Environmental Sustainability*, pp 1–42 (Woodhead Publishing) <https://doi.org/10.1016/B978-1-78242-156-6.00001-0>

Huang, Z, 2018. Reduction Enhancement Mechanisms of a Low-Grade Iron Ore–Coal Composite by NaCl, *Metallurgical and Materials Transactions B: Process Metallurgy and Materials Processing Science*, 49(1)411–422.

Iljana, M, Paananen, T, Mattila, O, Kondrakov, M and Fabritius, O, 2022. Effect of Iron Ore Pellet Size on Metallurgical Properties, *Metals*, 12(2):302.

Indian Standard, 1984. IS:1350-1 (1984) Methods of test for coal and coke, Part 1: Proximate analysis, second revision [online]. Available at: <https://law.resource.org/pub/in/bis/S11/is.1350.1.1984.pdf>

International Energy Agency (IEA), 2022. Coal 2022 [online]. Available at: <https://www.iea.org/reports/coal-2022> [Accessed: 06/02/2023].

International Energy Agency (IEA), 2023. Iron and Steel [online]. Available at: <https://www.iea.org/energy-system/industry/steel> [Accessed: 06/02/2023].

Lee, H, 1999. Chemical thermodynamics for metals and materials (Imperial College Press).

Mishra, S, 2020. Review on reduction kinetics of iron ore – coal composite pellet in alternative and sustainable iron making, *Journal of Sustainable Metallurgy*, 6:541–556.

Mousa, E A, 2019. Morden Blast Furnace Ironmaking Technology: Potential to Meet the Demand of High Hot Metal Production and Lower Energy Consumption, *Metallurgical and Materials Engineering*, 25(2):69–104.

Pal, J, 2018. Innovative development on agglomeration of iron ore fines and Iron oxides wastes, *Mineral Processing and Extractive Metallurgy Review*, 40(4):248–264.

Park, H, Sohn, I, Tsalapatis, J and Sahajwalia, V, 2018. Reduction behaviour of dolomite-fluxed magnetite: coke composite pellets at 1573K, *Metallurgical and Materials Transactions*, 49(B):1109–1118.

Phiri, M, 2022. Zimbabwe's scrap metal rush creates a circular economy, and headaches for authorities, China Dialogue [online]. Available at: <https://chinadialogue.net/en/business/zimbabwes-scrap-metal-rush-creates-a-circular-economy-and-headaches-for-authorities/> [Accessed: 10 February 2023].

Prakash, S, 1996. Reduction and sintering of fluxed iron ore pellets – a comprehensive review, *The Journal of the Southern African Institute of Mining and Metallurgy*, pp 3–15.

Prusti, P, Barik, K, Dash, N, Biswal, S K and Meikap, B C, 2020. Effect of limestone and dolomite flux on the quality of pellets using high LOI iron ore, *Powder Technology*. doi:10.1016/j.powtec.2020.10.063

Shaik, M B, Sekhar, C, Dwarapudi, S, Gupta, N, Paul, I, Patel, A K, Tudu, S and Kumar, A, 2020. Characterization of Colemanite and Its Effect on Cold Compressive Strength and Swelling Index of Iron Ore Pellets, *Mining, Metallurgy and Exploration*, 38(1)217–231. doi:10.1007/s42461-020-00331-5

Sivrikaya, O and Ali, I, 2012. The bonding / strengthening mechanism of colemanite added organic binders in iron ore pelletisation, *International Journal of Mineral Processing*, pp 90–100.

Sivrikaya, O and Arol, A, 2014. Alternative binders to bentonite for iron ore pelletizing – part i: Effects on physical and mechanical properties, *Holos*, 3:94–103.

Somerville, M, 2016. The Strength and Density of Green and Reduced Briquettes Made with Iron Ore and Charcoal, *Journal of Sustainable Metallurgy*, 2(3):228–238.

World Steel Association, 2018. World steel in figures [online]. Available at: <https://worldsteel.org/wp-content/uploads/2018-World-Steel-in-Figures.pdf> [Accessed: 11 May 2022].

Zhao, H, Li, Y, Song, Q, Liu, S, Ma, Q, Ma, L and Shu, X, 2019. Catalytic reforming of volatiles from co-pyrolysis of lignite blended with corn straw over three different structures if iron ores, *Journal of Analytical and Applied Pyrolysis*, vol 144.

Zhu, D, 2022. Utilization of Hydrated Lime as Binder and Fluxing Agent for the Production of High Basicity Magnesium Fluxed Pellets, *ISIJ International*, 62(4):632–641. doi:10.2355/isijinternational.ISIJINT-2021-157

Zulkania, A, Rochmadi, R, Hidayat, M and Cahyono, R B, 2022. Reduction reactivity of low grade iron ore – biomass pellets for a sustainable iron making process, *Energies*, 15(1):137.

Iron ore transfer chutes – directions after 30 years of scale modelling

P Donecker[1]

1. Director, Bulk Solids Modelling, Capel WA 6271. Email: peter@bulksolidsmodelling.com.au

ABSTRACT

Like all chains, iron ore values chains are only as strong as their weakest link, which in ore processing is often the humble chute. While mechanically the simplest element in the iron ore flow sheet, chutes are often the most problematic and regularly the major cause of unscheduled downtime.

Chute designs have remained static for decades, despite the extra demands being placed on them by grade reduction, which impacts chutes in three ways *viz*:

1. Higher volumetric rates as infeed increases to meet production targets.
2. Higher moisture content as mining goes below water table.
3. Stickier properties as the dilution phase is often clay based.

While there is no shortage of ideas and the geometry options are infinite, the fact is that chute design has not kept pace with grade decline and has remained essentially unchanged.

One reason cited for this lack of innovation is the difficulty evaluating and de-risking designs. While the power of computers running Discrete Element Method (DEM) models has advanced, the models themselves still need to be 'calibrated' with the dynamic properties of the ore. Flow properties commonly measured for bin design are useful for static situations only and were never intended to predict dynamic behaviour as in chutes, a critical qualification lost in the mists of time. So while they always produce a colourful and reassuring animation, DEM models informed by static ore properties often result in wildly inaccurate real world predictions.

Responding to this low success rate, many operators are now utilising physical modelling to help reduce risk and avoid the consequences of having an underperforming chute in their value chain. Physical testing, in a dimensionless scale-down of the proposed system (bulk material and chute geometry), addresses the shortcoming of mathematical models designs and when used in combination with DEM is the best option.

In the ideal world, the focus of chute modelling should be innovative design concepts aimed at addressing future needs, however in reality the focus is nearly always on modifications to installed chutes that are underperforming even today.

This situation can be traced back to design procedures which if informed at all, are based static ore properties.

As an industry, we need to break this cycle and become more proactive, which means addressing flaws in the current design process. In 30 years of providing modelling support to the industry, I have looking at hundreds of underperforming chutes in the iron ore industry. This practical experience has led to a unique and unbiased perspective on chute design and their performance in the real world. The knowledge presented in this paper can not only be used to help avoid baked-in problems now, but also help guide and evaluate designs so we can deal with the challenges of our low-grade future.

INTRODUCTION

Physical scale modelling of transfer chutes was first conceived and developed in 1993 in the laboratories of CRA Advanced Technical Development in Bentley in Western Australia, in the Materials Science Section. It evolved from a study of material performance across the group companies which included Hamersley Iron, Argyle Diamonds and Comalco Mineral Products. Former Rio Tinto Chief Group Scientist Robin Batterham followed the development and mentions it in a paper on corporate innovation (Batterham, 1995). Since that time the technology has been developed and improved at Bulk Solids Modelling (BSM). That development has allowed a unique insight into chute design in the iron ore industry and the physics of granular flow.

Part of the materials performance study involved the systematic monitoring of several transfer chutes in Hamersley Iron with the aim of correlating materials performance with laboratory wear tests. This proved to be more difficult than anticipated due partly to the misalignment of the wear plates that had happened over time. A view of the chute being monitored at this stage is shown in Figure 1.

FIG 1 – Hamersley Iron chute from 1993 study.

To resolve this problem a request was made for the lower section of one of the chutes to be re-built. This was done, but although the wear plates were now well aligned, they were 100 mm further apart that in the original. On the next monthly inspection, the wear liners had developed holes and the chute was surrounded by spillage.

Clearly, there had been a change in the flow pattern in the chute and the influence of this was far greater than the influence of the wear materials. Searching for a way to understand this change, the idea of constructing a scale model was conceived. Such scale modelling is well known in Chemical, Aeronautical and Marine Engineering and relies on the principle of dimensional similarity. It had come into public consciousness in Australia a decade earlier with the Americas Cup races and the work of Ben Lexcen testing scale models a test tank in the Netherlands (Lexcen, 1982).

Following the first test circuit set-up in the laboratory at CRA in 1993, the technique has evolved continuously over the last 30 years to a point where all aspects of a chute can be accurately modelled, including conveyor transitions, troughing, belt widths, head pulley diameter, belt speeds and volumetric throughput. Parameters are controlled and logged in real time using a dedicated computer system. Materials have been developed to simulate a wide variety of ore types. Cohesion is controlled by moisture level; this is monitored with a near-infrared meter. A view from the control console showing monitor with circuit mimic and a model with laser profile scanner in the background is shown in Figure 2.

FIG 2 – View of a modelling circuit from the control desk.

SCALING THEORY

Dimensional similarity can be expressed in terms of dimensionless numbers. If these are held constant when the system is scaled, then dimensional similarity ensues. Going through the standard theory (Buckingham, 1914) it was determined that the key dimensionless group in the case of non-cohesive ores was the Froude Number (Ang and Trewella, 1994). By coincidence, this a number that Lexcen relied on in his tests.

A knowledge of the Froude Number is important if the scale modelling process is to be understood. It can be expressed as:

$$V^2/2Lg$$

Where:

V	is a characteristic velocity
L	is a characteristic length
g	is the gravitational constant

Flow is Independent of density

The first thing to be noted is that there is no density term in the expression for the Froude Number. This leads to the important conclusion that the modelling process is independent of density. This independence was explored and verified in the laboratory with two undergraduate thesis projects (Ang and Trewella, 1994; Tonkin, 1995).

Armed with this knowledge, a model of the Hamersley Iron chute was constructed, the flow patterns were verified by tracing out the wear marks on the liner and matching them to the wear marks on the acrylic surfaces of the model. The effect of the change in the separation distance of the liner plates was duplicated in the model at the flow rates corresponding to the real-world case. The test material used in these experiments was cracked wheat.

The great advantage of density independence in the modelling process is that it is not necessary to use some scaled-down version of the actual ore. This has allowed modelling of chutes from countries around the world without the problems of shipping samples. An example of this is the Vale S11D project in Brazil for which the whole process was modelled in Australia using low density test materials. Other chutes for iron ore mines in Canada have also been modelled here.

The University of Para in Brazil has an experimental test loop (Mesquita and Braga, 2015), built with guidance from an earlier paper on this subject (Donecker, 2011) where they have validated density independence. In Australia however, this concept seems to be slow to take hold (Chen *et al*, 2019).

Density independence applies to discrete element model (DEM) simulations as well, Hilton and Cleary (2011) discuss Froude Number scaling in hopper discharge over several orders of magnitude.

The independence from density applies not only to models, but to the full-scale chutes. Chutes are volumetric devices; they handle volume not tonnage. If coal with an equivalent sizing was run through a chute designed for lump iron ore, at the same volumetric throughput, the flow patterns would be essentially the same.

Flow within a typical iron ore transfer chute can be divided into five main categories each of which is independent of density.

1. Discharge from the feed conveyor:

   A summary of conveyor discharge trajectory calculation methods is provided by Hastie and Wypych (2006). None of the formulae presented contain a density term.

2. Flow along the impact plate:

   The differential equations presented by Roberts (1969) for flow on curved surfaces such as impact plates do not contain a density term.

3. Free fall:

   For free-falling particles, the density independence was famously demonstrated by Galileo in the 16th century (Viviani, 1717).

4. Free surface flow:

   Free surface flow down an incline has been studied by many researchers and shown to be independent of density. The most famous of these is Bagnold (1954).

5. Flow-through horizontal and vertical orifice:

   At the bottom of a chute the ore generally moves downwards and forwards through the loading section under the influence of both gravity and the receiving belt. Beverloo, Leniger and Van de Velde (1961) developed an equation for this that is generally known as the Beverloo equation. Similar expressions have been developed for different orientations. The equation is usually expressed in terms of mass flow, but dividing both sides by density provides and expression for volume flow rate that is independent of density.

Initially, volumetric throughput in the modelling circuit was done by conducting a crash stop and a belt cut. More recently, an in-house a laser profile technique has been developed that provides real-time information about volumetric throughput. This has sped up the process considerably. The volumetric data can be converted to tonnage throughput at any bulk density of interest. There is no need to run tests with materials of different densities.

Other parameters

Angle of repose is another dimensionless parameter that is relevant to the modelling process. This number captures features related to both shape and internal friction. It has been commonly used as a means of setting DEM parameters and this creates a perception that some measurements such as coefficient of restitution, Youngs modulus, rolling resistance and so forth are necessary. The reality is that in the physical realm many quite different materials that are the result of crushing processes adopt an angle of repose like that of iron ore. The scale modelling process does not seek to make spheres behave in a way that simulates angular particles, or to increase the convergence speed of equations of motion. Neither do contact physics need to be defined in the real world. Over the years, a range of angular test materials have been used with good results. More recently, particles have been specially fabricated for the role.

FLOW REGIMES

To understand transfer chute flow, it is essential to understand granular flow regimes. These can be divided into three categories.

1. Quasi-static flow. This occurs when material that is at rest first begins to move. Velocity at this stage is negligible and particle packing is at a maximum. In this regime inertial forces are negligible in relation to gravitational forces.

2. Dense granular flow occurs when the ore has increased in velocity and the particles have moved further apart. This is the principal mode of flow in a transfer chute. In this regime the inertial forces are significant in relation to the gravitational forces and Froude Number scaling applies.

3. Dispersed flow occurs when the particle velocity has increased further and the particles have moved further apart, behaving rather like molecules in a gas. This type of flow occurs in fluid beds. Inertial forces are dominant in this regime.

Focusing on dense granular flow, there are two key things to understand.

1. The theory of quasi-static flow does not apply to dense granular flow. Jenike (1961, p 217) explains that his theory only applies where inertial forces are negligible in relation to gravitational forces.

2. There exists a range of angles through which a non-cohesive granular material such as lump iron ore will flow with fixed, not increasing velocity. This is sometimes known as the Bagnold range. Operation within this range allows the energy change associated with the change in gravitational potential to be matched by frictional forces. Scale modelling has been successfully used many times to assist in the design of chutes that operate within this range. These chutes that can operate with ore-on-ore flow, minimal belt impact and close to zero wear.

   Silbert *et al*, (2001) discuss this with the help of DEM simulations for chute flows on rough surfaces. Figure 3 illustrates the type of relationship they observed between bed depth, angle of inclination and flow type. The absolute values of the axes and positions of the boundaries will depend on the material type.

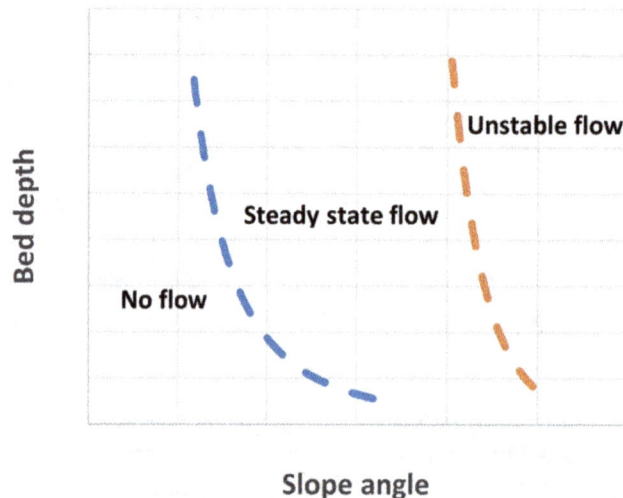

FIG 3 – Flow regimes on an inclined plane.

COHESIVE ORES

Much of the previous discussion has been based on non-cohesive flow. It can however be extended to cohesive ores such as iron ore fines. Froude Number scaling still applies, so independence from density is maintained, but in addition the ratio of the cohesive forces to the gravitational forces in the system needs to be considered. This ratio is known as the Granular Bond Number. If we are to maintain dimensional similarity in a scale model, we must maintain a constant Granular Bond Number; the cohesive forces in the test material must be scaled in the same proportion as the other forces. This means that they must typically be reduced by a factor of ten. As mentioned by Carr (2019, p 255), in practice, it is not possible to measure cohesion or adhesion with any degree of confidence even in the original ore.

Some people think that it is essential to use some form of scaled down iron ore fines to achieve dimensional similarity for cohesive materials. This is not the case. Johanson (1972) discusses this in relation to scale models of hoppers where he points out that crushed-down samples of the actual ore can cause hang-up where it does not occur in the full-scale installations. Although this is a quasi-static situation, similar concepts apply in the dense granular flow regime. This can be explained with the aid of Figure 4.

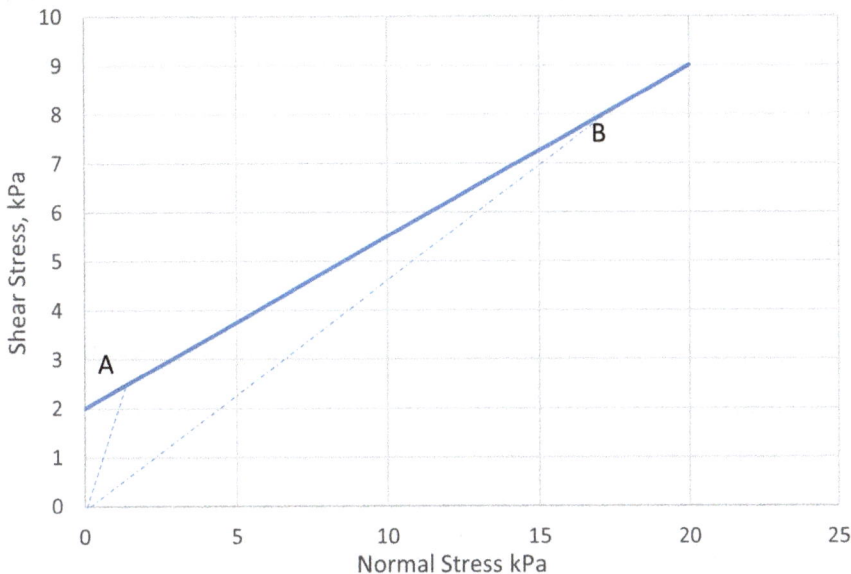

FIG 4 – Shear cell data.

Figure 4 shows the classic form of the results of a set of shear cell tests as a graph of the relationship between the shear stress at failure and the normal stress. The y-axis intercept is the cohesion. According to Jenike (1961) and Roberts (1998), this relationship is set by the shear in the fines and the large particles have but little influence on the result. That is why the tests are conducted on fines. So, almost any size distribution could be used, so long as the fines were the same. One approach might for example be to scalp the material to a top size that is 1/10th of the actual ore.

Firstly, consider a full scale quasi-static situation, where the normal stress is perhaps 17 kPa below a free surface. The effective angle of internal friction can be found by taking a line from the origin to the point on the graph corresponding to 17 on the x-axis (B).

Consider a 1/10 scale model of this situation. If the scaling is correct, the normal stress will, like the other forces in the system, be reduced by a factor of 10. So, it will be 1.7 kPa. Now construct the same line and determine the effective angle of internal friction. This gives a line from the origin to point A. It has a much higher slope. This simple exercise serves to illustrate that it is not valid to simply remove some of the coarse material and use the rest in a scale model test. The resultant effective angle of internal friction will always be exaggerated and flow will be inhibited. Neither would such a move represent a scaling of the size distribution.

Some propose not only using a scaled version of iron ore but maintaining the moisture content at the same level as the full-scale material. A scaled-down material will have a much higher specific surface area and a different water holding capacity. Moisture content is not a scalable quantity. As the material is circulated, it will dry out so water additions will need to be made. It is well recognised that iron ore suffers from wetting hysteresis and delayed response to added water, so this represents further problems.

Any laboratory procedure that involves multiple handling of cohesive iron ore samples will to lead to agglomeration. Such agglomeration affects experimental outcomes (Guo, 2014, p 68; Carr, 2019, p 127). A scale modelling circuit is just such a situation. Any attempt to use iron ore in such a system will inevitably lead to progressive, irreversible agglomeration and the degree of agglomeration will affect the experimental results. Carr (2019) has even gone so far as to suggest using agglomeration to combat sticky ore issues.

The first simulation of cohesive ores using synthetic material dates to 2001. Since that time, these materials have been the subject of ongoing development and refinement, producing test materials for simulating cohesive iron ore that do not undergo agglomeration and do not have delayed wetting response. These materials can be recirculated for hours in a test circuit. The level of cohesion can be simply varied by altering the water content. In the same way as iron ore fines, density decreases as we increase the moisture content above the dust extinction moisture level so the volume of material in the circuit must be adjusted. This is easily done with the aid of real-time volume measurement. These materials have been shown in hundreds of test runs to simulate full-scale chutes with excellent accuracy.

THE ADVANTAGES OF PHYSICAL SCALE MODELLING

Validation of trajectory

One of the first requirements of transfer chute design is the correct prediction of trajectory. In a scale model this relies on the correct head pulley diameter, belt speed and transition geometry. Over the course of many years, an excellent correspondence has been found with the upper and lower trajectory predictions using the method of Huque (2004). The prediction of lateral spread is strongly reliant on the correct transition geometry, a parameter that is often overlooked in some models and numerical predictions.

Simulation speed and duration

A scale model simulation runs faster than the full-scale case. This is because simulation time varies with the square root of the scale. One hour of real time can be covered by a 20-minute run. This kind of speed and duration cannot be approached by DEM. It allows consideration of temporal effects.

Cover of complete range of throughputs.

Scale modelling allows steady-state simulation of any throughput rate between zero and the maximum, in a matter of minutes.

Real-time variation of geometry

While a model is running, it is possible to adjust features such as impact plate position and inclination in real time. Other adjustments are often made using 3D software to generate new geometry that can be rapidly created using a laser cutter. Modifications can be made and the simulation re-run over the whole range of conditions in a time scale of an hour. This is much faster than DEM.

Variations in belt speeds

A common requirement is to investigate the effect of change in belt speeds. This applies particularly in the case of stackers. Such changes can be easily made via control software in real time.

Real time variation in cohesion

By varying the moisture content, a range of cohesion levels between the dust extinction moisture level and the worst handling condition can be covered.

Reality check

A DEM simulation is always subject to the possibility that the input parameters are not correct, or the time steps are too ambitious. The concept of garbage in, garbage out applies. An output is possible that cannot be achieved in the real world. As a cross-check, end users often request an independent validation. Using DEM for the validation does little to mitigate this risk. A physical scale model, on the other hand, cannot produce a result that is physically impossible and it uses a completely different technology. It provides a very effective way to validate DEM results.

Cost and timing

As an outsourced service particularly in the case of independent design validation, physical scale modelling costs are on a par with, or less than, DEM simulation costs. There is a time delay amounting typically to three weeks, but the output that can be generated once the model is in operation can represent many years of DEM simulation time. The use of in-house test materials eliminates costs of shipping samples, the use of which has no basis in theory in any case.

TEST METHODOLOGY

Modelling a chute that handles cohesive ores is usually conducted in the following manner:

- Using a material that is slightly above the dust extinction moisture, the circulating load is increased to the desired operating capacity, as measured by the laser profile scanner.

- If the chute can handle this volume, the circulating load is slowly increased until blockage occurs, or the conveyors cannot handle the load. This sets the maximum capacity point. It also provides information about flow behaviour at start-up. This process typically takes 20 minutes.

- Returning to the design operating point, the moisture is slowly increased to simulate an increase is cohesive properties. At the same time, material must be removed from the circuit to accommodate the decrease in bulk density that occurs. Moisture is increased until the chute either blocks or it reaches a point corresponding to the maximum cohesion expected. Beyond this point, the bulk density increases with increase in moisture content and material must be added to the circuit to maintain capacity. Such adjustments are only made possible with real time monitoring of throughput.

- The process might be repeated using test material that is at a moisture level corresponding to the design specifications and increasing the flow to the design operating throughput.

- Throughout all these tests, moisture must be adjusted continually to allow for drying. This can only be done with the help of real-time monitoring.

- Having gone through this process many hundreds of times, some common observations can be made. The first is that any gross design issues will become apparent in the first minute, sometimes even in the first seconds. Such problems include trajectory issues at low throughput such at the ore missing an impact plate, excessive dead zones and extreme side loading at low throughput and even direct impact on the belt without any chute contact.

SOME OBSERVATIONS AND REFLECTIONS ON SCALE MODELLING

Continuous innovation

Scale modelling was an innovation in 1993. In the past 30 years at BSM this has been added to. Model making has gone from hand cutting to lasers. 3D computer software is used now. The innovative development of materials to simulate cohesive ores that started in 2001 has continued. Instrumentation to measure volumetric throughput has been specifically developed for the application after working through some initial failures. Continuous monitoring of moisture levels belt speeds and other variables has been added. Belt materials have been trialled and optimised. Methods to ensure exact replication of belt transitions have been conceived trialled and introduced. Head pulleys fabrication has gone through more iterations than can be counted. Conveyors have been designed and built and scraped. Radical designs have been tested, refined and re-tested. Experiments with computer vision and image recognition software have been conducted, analysed and re-conducted.

This is the way innovative development goes. A constant hunger for new ways to do things.

Build-up – what is it?

Observation of the build-up in model chutes over many years reveals that it occurs in two stages. The first stage is the deposition of material with a composition the same as the bulk ore. This occurs very quickly, usually on ledges or similar features within the chute or in areas of constriction. It seldom occurs by direct deposition of sticky ore on open, inclined walls at the slopes usually specified.

The second stage is the slow accretion of ultra-fines on this surface. It may also occur on other surfaces. This happens over an extended period. Build-up within a chute is typically composed of a combination of these or by ultra-fines alone.

Figure 5 shows a cross-section through a piece of this build-up. The layered structure is clear. All of this material is less than 150 microns, and the size range extends down to at least 0.04 microns.

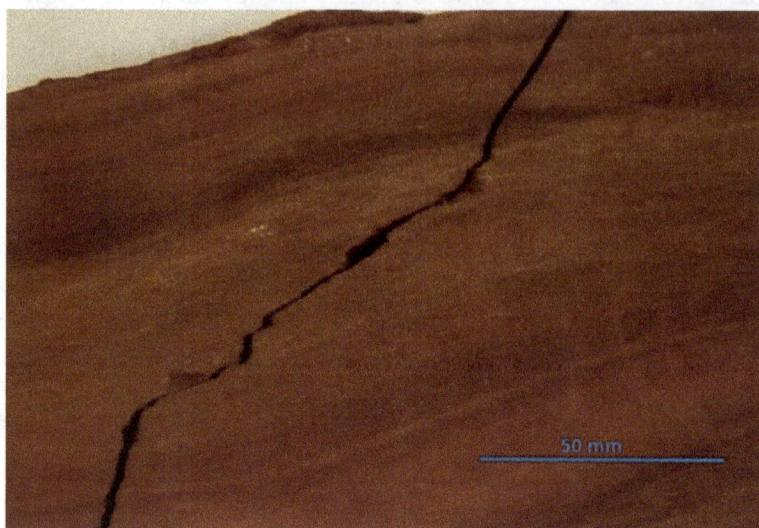

FIG 5 – Sample of build-up.

Although this phenomenon is observed in physical scale modelling, achieving accurate replication remains challenging. In a circulating system, fines accretion will occur at multiple locations, causing removal of fines from the circulating load. Even the extended simulation times available with physical scale modelling do not match the time scale of usually associated with these accretions.

Accretion in this context can be understood as a mass transfer process that relies on the mobilisation of ultra-fine particles through the presence of water. The occurrence of this build-up varies across different locations. It is crucial to compare these variations to gain insights into the underlying process and explore potential control measures.

The process of fines accretion is not related to the flow properties of the bulk ore. Current tests of ore properties will not predict it. It is more likely to be related to design issues and water control.

Wall friction – does it matter?

Coefficient of friction between chute walls and the ore is often though to play a vital role in chute capacity. Trials conducted at BSM over many years have demonstrated however that increasing the surface friction by applying sandpaper to the model chute surfaces or spraying them with adhesive so that they capture a layer of particles, has minimal effect on chute capacity. It is worth examining wall friction more closely.

Investigations of iron ore samples falling onto inclined surfaces reveal that there is a critical release angle. This angle is not influenced by the nature of the surface but is instead related to material properties (Carr, 2019, p 190). This release angle can be compared with the left-hand boundary in Figure 3.

When the inclined surface is set at less than the release angle, the ore creates its own autogenous surface at the minimum release angle. Carr (2019) points out that this angle is constant for one ore type. This is also referred to as the stall angle (Benjamin, 2015). The same result has been observed many times the course of scale modelling over the last 20 years. The flowing ore interacts with this surface in just the same was as any other material. Although it is in one sense build-up, it is in essence the same as any other surface. There is no more driving force for further build-up than there is with any other surface. The angles observed in the scale models have been found to match the full-scale chute and fall within a surprisingly small range. Both the modelling and the full-scale designs are achieved with a minimum of property measurements and no shear cell or similar test results.

The great advantage of an autogenous surface is that it is not subject to wear. Chutes designed to utilise this type of surface have been shown to operate for extended periods with no liner wear. Some component of frictional drag is also a key to providing a centring action on chute flow.

Knowledge of the angles of the autogenous surfaces formed by any given iron ore are crucial to adopting a design approach that uses these surfaces. This comes firstly from scale modelling and subsequently from field observations. Such experience is not gained without accepting risk. Scale modelling can mitigate this risk. Appropriate design can also reduce risk by reducing susceptibility to variations in the autogenous release angle.

Shear cell test data – is it necessary?

The measurement of ore properties using shear cell tests has been universally declared to be an essential first step in any chute design endeavour. This, even though, as mentioned above, they do not apply to the flow regime that dominates in iron ore transfer chutes. Such tests have been available to the industry since its inception. Yet there has been little real progress in chute design based on these tests. It could even be said that the single-minded focus on the hopper design concepts developed by Jenike (1961, 1964) have hindered progress. Perusal of the literature generated in relation to chute design reveals that it contains no mention of concepts like the Granular Bond number or of the work of Bagnold and many others on the dense granular flow regime. Froude Number appears to have been a recent discovery.

It is encouraging to some degree to see that laboratory reports now incorporate some *ad hoc* tests and that recommendations for chute design based directly on shear cell results have diminished in extent. Tacit acceptance of the limitations of this approach.

Over the last 30 years, the modelling process has been successfully applied to hundreds of chutes without any attempt to apply shear cell data to the test materials. In a similar way, many of the chutes that have been modelled have not used this kind of data in their design. Designs have been developed based on qualitative information about ore behaviour and some basic information such as size distribution and moisture. These designs depart significantly from convention, yet they have been successfully installed and operated in large numbers. They incorporate autogenous surfaces that significantly extend liner life.

If we are to make real progress, then perhaps it is time to look beyond shear cell tests and begin to take onboard some broader perspectives that recognise the effects of inertia, the real nature of build-up and wall friction and the importance of capturing field observations compared to laboratory tests.

Risky business

Over the years it is noticeable that chute specifications which have evolved from hopper design mentality have become more detailed and restrictive. As this happens it becomes more difficult to innovate. Even small, rational changes to these designs that have been verified by both DEM and physical scale modelling have been rejected due to this mindset. Yet the ongoing problems with these chutes are evident.

There is a saying about doing the same thing and expecting a different result. Yet this is exactly what is happening with chute design. There is an overwhelming impression that it is considered safer to follow the path of proven failure than to risk something new. There is some aspect of cultural attitudes involved here since Brazil and other countries have been far more open to innovations than has Australia.

Somehow this wall must be breached. This can only come from the highest levels. There will be no gain without risk. No new designs. A component of innovative risk must be written into corporate policy and separated from personal liability.

CONCLUSIONS

Physical scale modelling began as an innovation and has now developed into a mature technology. This has happened through a process of continuous innovation. Despite this, some of the basic principles are not well understood within the industry. In particular, the concept that use of some form of scaled ore ensures similarity is flawed.

The theories of hopper flow dating back to the 1960s are past their use-by date. It is time for fresh thinking. There are some small moves in the right direction, but it needs a shake-up.

Scale models provide a perspective that is independent of other laboratory tests. A process plant is a full-scale model. Every effort should be made to extract data directly from that source.

Innovation comes with risk. The risk of adopting innovative designs must be accepted, even mandated, and written into corporate policy.

REFERENCES

Ang, J and Trewella, L, 1994. Dimensional Analysis of Granular Solids Flow in a Model Conveyor Transfer Station, Honours thesis, Department of Mechanical Engineering, Curtin University of Technology.

Bagnold, R A, 1954. Proceedings of the Royal Society of London, Series A, Mathematical and Physical Sciences, 225(1160):61.

Batterham, R J, 1995. The Innovation Process – A Corporate Perspective, in *Nurturing Creativity in Research Conference*, p 6 (Australian National University: Canberra).

Benjamin, C W, 2015. Transfer chute, AU2013302325B2.

Beverloo, W A, Leniger, H A and Van de Velde, J, 1961. The flow of granular material through orifices, *Journal of Chemical Engineering Science*, 15:260–296.

Buckingham, E, 1914. On Physically Similar Systems; Illustrations of the Use of Dimensional Equations, *Physical Review*, 4(4):345–376. doi:10.1103/PhysRev.4.345.

Carr, M J, 2019. Identification, Characterisation and Modelling of Dynamic Adhesion for Optimised Transfer System Design, Doctoral dissertation, The University of Newcastle.

Chen, B, Zhao, X, Lu, M, Qiao, G, Wu, H and Roberts, A, 2019. Transfer Chute Analysis using Continuum Method, DEM and Scale Modelling, paper presented at the *13th International Conference on Bulk Materials Storage, Handling and Transportation*, Gold Coast, Australia.

Donecker, P, 2011. Dynamic Scale Modelling (DSM) of Transfer Chutes, paper presented at the *Successfully Managing Wet, Sticky and High Clay Ores Conference*, Perth, WA.

Guo, J, 2014. Investigation of arching behaviour under surcharge pressure in mass-flow bins and stress states at hopper/feeder interface, Doctoral dissertation, The University of Newcastle.

Hastie, D and Wypych, P, 2006. Conveyor Trajectory Prediction Methods – A Review, in *Bulk Materials Handling Conference, Bulkex 2006*, Melbourne, Australia.

Hilton, J E and Cleary, P W, 2011. Granular flow during hopper discharge, *Physical Review E*, 84(1):011307. doi:10.1103/PhysRevE.84.011307.

Huque, S T, 2004. Analytical and Numerical Investigation into Belt Conveyors Transfers, Doctoral dissertation, The University of Wollongong.

Jenike, A W, 1961. Gravity flow of bulk solids, *Bulletin of the University of Utah: Utah Engineering Experiment Station*, 52(29), Bulletin No. 108, October.

Jenike, A W, 1964. Storage and flow of solids. *Bulletin of the University of Utah: Utah Engineering Experiment Station*, 53(26), Bulletin No. 123, November.

Johanson, J R, 1972. Modelling flow of bulk solids, *Powder Technology*, 5(2):96.

Lexcen, B, 1982. Yacht keel with fins near tip, Patent Application No. AU8566882.

Mesquita, A L A and Braga, E M, 2015. Desenvolvimento de μm Laboratório de Transportadores de Correia em Escala (Development of a Small-Scale Conveyor Belt Laboratory), Trabalho de Conclusão de Curso – Graduação em Engenharia Mecânica (Undergraduate Thesis in Mechanical Engineering), Universidad Federal do Pará.

Roberts, A W, 1969. An Investigation of the Gravity Flow of Non-cohesive Granular Materials through Discharge Chutes, *Transactions of ASME Journal of Engineering in Industry*, 91(2:Series B):373–381.

Roberts, A W, 1998. Basic Principles of Bulk Solids, Storage, Flow and Handling, Centre for Bulk Solids and Particulate Technologies, The University of Newcastle.

Silbert, L E, Ertaş, D, Grest, G S, Halsey, T C, Levine, D and Plimpton, S J, 2001. Granular flow down an inclined plane: Bagnold scaling and rheology, *Physical Review E*, 64(5):051302.

Tonkin, M, 1995. Design of Impact Plates used in Belt Conveyor Transfer Stations, Honours Thesis, Department of Mechanical Engineering, Curtin University of Technology.

Viviani, V, 1717. Racconto istorico della vita di Galileo Galilei (Historical Account of the Life of Galileo Galilei), 606 p.

Developing dephosphorisation technique for iron ore with reduction process

O Ishiyama[1], K Higuchi[2] and K Saito[3]

1. Senior Researcher, Ironmaking Research Laboratory, Process Research Laboratories, Research and Development, Nippon Steel Corporation, Futtsu Chiba 293-8511, Japan. Email: ishiyama.7n9.osamu@jp.nipponsteel.com
2. General Manager, Ironmaking Research Laboratory, Process Research Laboratories, Research and Development, Nippon Steel Corporation, Futtsu Chiba 293-8511, Japan. Email: higuchi.t9g.kenichi@jp.nipponsteel.com
3. Senior Fellow, Nippon Steel Research Institute Corporation, Chiyoda Tokyo 100-0005, Japan. Email: saito.k.hwh@nsri.nipponsteel.com

ABSTRACT

Iron ore in Australia is predicted to increase in its phosphorus content in the near future, which is detrimental to properties of steel product. Steelmakers are concerned by the content of phosphorus in iron ore will exceed limits for steelmaking and increase the converter slag ratio to molten steel, which will finally lead to a decline in competitiveness in high-grade steel production. In this paper, fundamental development of a dephosphorisation technique by reduction gasification is considered. Basic reduction examination was carried out using Australian and a synthetic model iron ore sample with phosphorus compound reagent, varying reaction time and composition of substances to find out suitable process condition for dephosphorisation. With successful lowering of phosphorus concentrations by gasification reduction of the iron ores, dephosphorisation was assumed and further elucidation works were carried out.

More than 10 mass per cent of dephosphorisation ratio is obtained from Australian iron ore through batch-type 25 g furnace experiment using hydrogen as a part of reducible mixed gas. Lowering of phosphorus content occurred preferentially either in wustite or in another reducing zones, suggesting a relationship between reducing conditions to efficient dephosphorisation of iron ores. Another result of the fundamental experiment using a model sample with reagent is different dephosphorisation occurred with different reagents contributing to with unified reduction model for phosphorus, providing positive possibility for comparative evaluation, even if new morphologies of phosphorus in iron ore are clarified in future. According to these approaches, it is important to reduce iron ore efficiently and to avoid gasified phosphorus retaining and reattaching to adjacent iron ore particle at the same time. A kiln-type furnace is suggested to overcome previous technical issue, and additional experiments using larger rotating furnace were carried out for verification.

INTRODUCTION

Australian iron ore is one of the nearest natural resource for Japanese steel industry, but deterioration in its grade is ongoing. Total Fe content is predicted to decrease down to 59.7 per cent in the average of imports to Japan by 2030 due to increase of SiO_2, which is one main component of gangue minerals (Figure 1). Phosphorus, which degrades properties of the steel product, is also predicted to show a boost up to 0.085 per cent in average simultaneously, and is a concern of overloading the steelmaking process and higher converter slag ratio, which finally leads to a decline in competitiveness in high-grade steel production. This current state of supply is related to historical demand and depletion of relatively larger iron ore deposits in Australia, from low phosphorus Brockman ore to either Marra Mamba ore which is relatively hard to agglomerate in sinter process, or high combined water content pisolithic ore, and eventually to high phosphorus Brockman ore. With this background, fundamental development of dephosphorisation technique is considered.

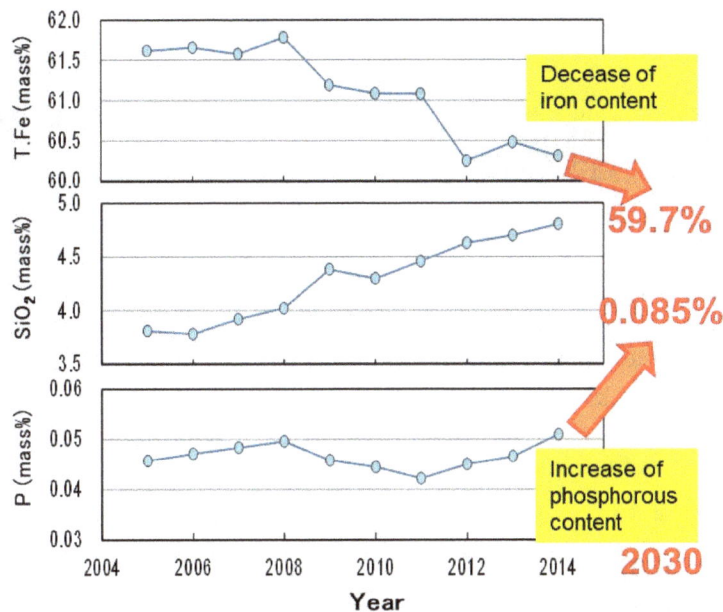

FIG 1 – Trends and future predictions of iron ore grades (average of imports in Japan).

Many technical studies on the possibility of removing phosphorus in iron ore is reported (Anyakwo and Obot, 2011; Jin *et al*, 2006; Fisher-White, Lovel and Sparrow, 2012; Lonkov *et al*, 2012; Matinde and Hino, 2011; Nunes *et al*, 2012; Pandey, Sinha and Raj, 2010; Xu *et al*, 2012; Zhu *et al*, 2013). Although some of them report significant dephosphorisation, technical issues still remain including losses of mass of iron ore because of its wet processing, or installing of another process such as crushing in advance. On another front, a study focused on reaction between gas and solid to realise efficient processing has been reported (Sasabe, Iida and Yokoo, 2014). The fundamental work suggest feasibility of dephosphorisation by gasification, following reduction of the iron ores.

EXPERIMENTAL

It is understood that separation of phosphorus is difficult from phosphates form or following selective adsorption due to its stability, so reduction reaction is necessary to decompose and to gasify phosphorus in iron ore. But on the other hand, if the reduction has progressed completely, there is a concern that phosphorus once detached from the iron ore will be taken up again because of a large affinity with hot metal formed through the reduction. So, it is significant to find out process conditions such as reduction temperature, reduction time, gas composition, gas flow rate, particle size and so on for phosphorus to gasify and discharge outside the ore and in this study high phosphorus content iron ore fine from Western Australia is reduced to verify the hypothesis.

Specifically, to develop a dephosphorisation process for 10 mass per cent dephosphorisation ratio from a variety of phosphorus morphologies and distributions, a fundamental batch-type 25 g furnace experiment was carried out using iron ore from Western Australia and a synthetic model sample with phosphorus compound reagent, under controlled reduction gasification and affinity of phosphorus to metallic iron conditions. Additional experiment using a larger rotating kiln-type furnace was carried out for efficient continuous process verification.

Dephosphorisation of high phosphorus iron ore from Western Australia

Figure 2 shows a schematic diagram and heat profile of iron ore dephosphorisation using high phosphorus Brockman iron ore and its chemical contents are also shown in Table 1. The sample containing a high (0.219 mass per cent) amount of phosphorus is rinsed and screened into 5–10 mm in size and then 25 g is charged in a stainless steel mesh basket. Reduction using an electric furnace, with its reaction chamber inner diameter of 73 mm, by reducible gas composition of 70 per centN_2–30 per centH_2 and 70 per centN_2–15 per centCO–15 per centCO_2, at the total gas flow rate of 15 NL/min. The furnace temperature is increased by 10°C/min from room temperature to 1000°C with N_2 and then the furnace is held 30 min at the temperature to be stable. N_2 gas is switched to reducible gas for a predetermined time period (3–60 min), finally after lowering the furnace temperature the samples are recovered and assayed.

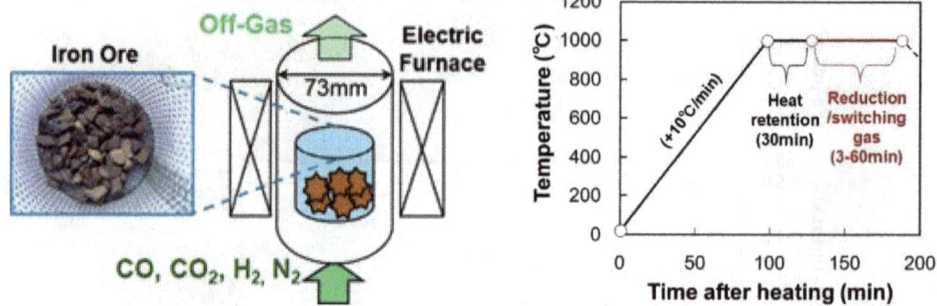

FIG 2 – Schematic diagram and heat profile of iron ore dephosphorisation.

TABLE 1

Chemical contents of the high phosphorus Brockman iron ore sample, mass per cent.

T.Fe	FeO	M.Fe	P	SiO_2	Al_2O_3	MgO	CaO	CW	LOI
60.34	0.27	0.07	0.219	2.40	3.01	0.09	0.04	6.89	7.51

Dephosphorisation retrial with condition controlling deviation of phosphorus content among iron ore particles

From large phosphorus content deviation among natural iron ore particles in previous experiment, the importance of sample preparation was determined. As shown in Figure 3, sample preparation was re-examined, specifically, 2 kg of particles are extracted from a 5 kg bulk sample and the entire amount is finely pulverised to $D_{50} = 45\ \mu m$ and then thoroughly mixed to ensure the representativeness of the sample. Then, 0.8 g of pulverised fine ore is compacted into a tablet model iron ore sample with a diameter of 8 mm and a height of 7 mm (compacting pressure 3 MPa), these tablets are 25 g (about 31 pieces), and are charged into a stainless steel mesh basket and then reduced for dephosphorisation retrial.

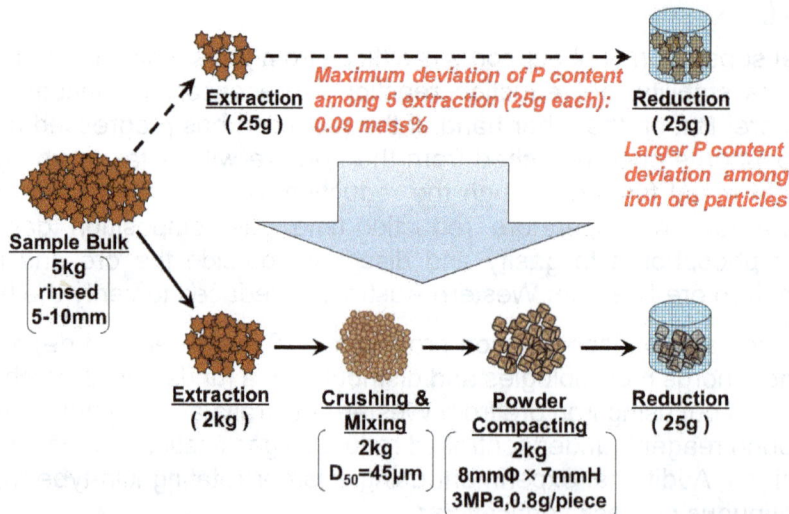

FIG 3 – Re-examining sample preparation to avoid influences of deviation of phosphorus content among iron ore particles.

Dephosphorisation of synthetic model iron ore using phosphorus compound reagent

Sasabe, Iida and Yokoo (2014) works on gasifying phosphorus from different phosphorus compounds and phosphates, suggest a different chemical reaction for each substance. It is necessary to understand the effect of various form and association of phosphorus on the dephosphorisation ratio to develop an efficient process, reduction on synthetic model iron ore using several phosphorus compound reagent was carried out. Hydroxyapatite ($Ca_5(OH)(PO_4)_3$), calcium triphosphate ($Ca_3(PO_4)_2$) and ferric phosphate ($FePO_4$) are selected individually to design composition P = 0.35 mass per cent and P/Fe = 0.005 constant respectively, by blending with hematite (Fe_2O_3) or goethite (FeO(OH)) reagents. In this experiment, reduction conditions are as follows: gas composition of N_2–30 per centH_2; gas flow rate of 15 NL/min; reduction temperature of 1000°C and reduction time of 5 min.

Continuous dephosphorisation processing of high phosphorus iron ore from Western Australia with larger rotating kiln-type furnace

It is important to reduce iron ore efficiently and to avoid gasified phosphorus retaining and reattaching to adjacent iron ore particles at the same time. To overcome the technical issues a kiln-type furnace is suggested due to rolling performance of iron ore particles in the chamber and additional experiments using a larger rotating furnace are carried out for efficient continuous process verification. Figure 4 shows a schematic diagram of iron ore dephosphorisation using another high phosphorus Brockman iron ore and its chemical contents are also shown in Table 2. The sample containing a high (0.284 mass per cent) amount of phosphorus is screened into 5–10 mm in size, it is charged into a rotating chamber with the rate of 1 kg/h by both screw feeder (SF) and vibration feeder (VF) from hopper for 1 h. The reaction chamber inner diameter is 92 mm rotating at a rate of 2 rev/min and at 1 deg tilt angle, its central 500 mm portion out of total length 2000 mm is heated at least up to 700°C internally with outer electric resistance heating element furnace. Pre-heated reducible gas composition is 50 per cent N_2–25 per cent H_2–25 per cent CO_2 and is injected at the total gas flow rate of 60 NL/min. All the processed iron ores are recovered at kiln outlet in rotating carousel boxes, on a time series basis and are assayed to evaluate the dephosphorisation ratio after lowering the furnace temperature.

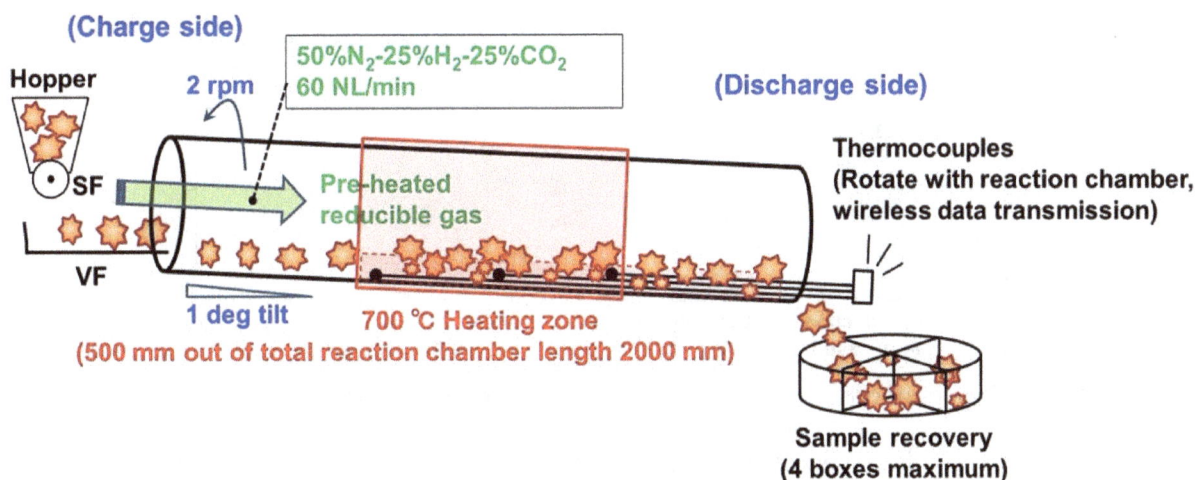

FIG 4 – Schematic diagram of iron ore continuous dephosphorisation.

TABLE 2

Chemical contents of high phosphorus Brockman iron ore used in continuous dephosphorisation verification, mass per cent.

T.Fe	FeO	M.Fe	P	SiO$_2$	Al$_2$O$_3$	MgO	CaO	CW	LOI
62.59	0.33	0.02	0.284	2.66	2.18	0.04	0.01	4.54	5.00

RESULTS AND DISCUSSIONS

Dephosphorisation of high phosphorus iron ore from Western Australia

The trend of reduction weight loss in process and appearance of the recovered samples are shown in Figure 5. Figure 6a shows the dephosphorisation ratio results from average of four results under H_2 reduction and seven results under CO–CO_2 reduction. Dephosphorisation ratio (R_{dephos}, mass per cent) through reduction is estimated with results of chemical analysis, as follows.

$$R_{dephos} (mass\%) = (1 - \frac{P_B/Fe_B}{P_A/Fe_A}) * 100$$

Where:

P_A is for the P content in the original iron ore sample used in reduction

P_B is for the P content in the dephosphorisation processed iron ore sample

Fe_A is for the Fe content in the original iron ore sample used in reduction

Fe_B is for the Fe content in the dephosphorisation processed iron ore sample

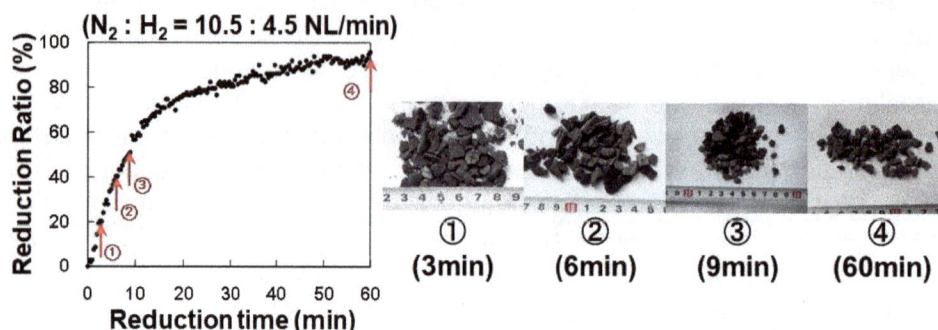

FIG 5 – Trend of reduction ratio and appearance of the recovered samples.

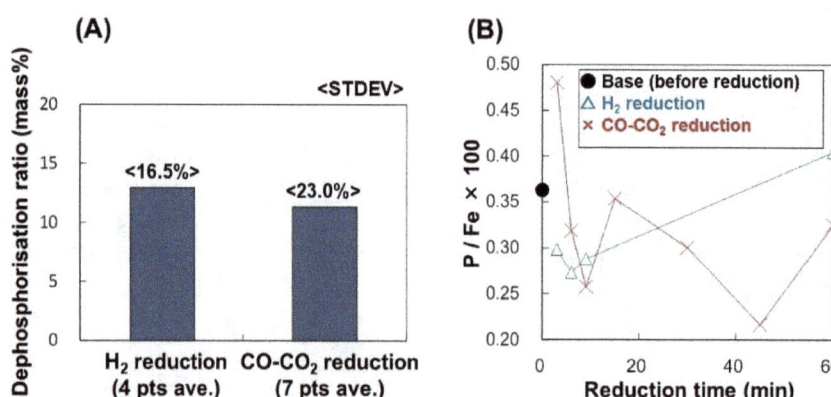

FIG 6 – Dephosphorisation results in average (a) and its breakdown by assay (b).

Dephosphorisation ratio is based on the premise that any losses of iron content in iron ore during gasification and removal reaction of phosphorus would not occur. From the results, it is found that both H_2 and CO–CO_2 reduction is effective. The obtained dephosphorisation ratio was more than 10 per cent from the average of multiple samples, but there were a large standard deviation. Figure 6b shows the breakdown of results by assay, here seen a variation in the amount of residual phosphorus (P/Fe) after reduction. During the course of the experimentation, difficulty in evaluating dephosphorisation behaviour under different reduction time was revealed. This may be due to heterogeneity of the phosphorus distribution within a particle and between particles and to experimental accuracy.

Dephosphorisation retrial with condition controlling deviation of phosphorus content among iron ore particles

In order to overcome the previous issue faced, dephosphorisation retrial was carried out with a new sample preparation method described above. Results of reduction and dephosphorisation ratio is shown in Figure 7. Good repeatability is obtained at each reduction disruption point (3, 6, 9, 60 min). The dephosphorisation ratio is over 10 mass per cent (from an average of eight samples: 14.9 mass per cent start mass). Since no significant change is confirmed in the dephosphorisation ratio after the achievement of reduction ratio 25 per cent, the relationship between reduction time and dephosphorisation ratio is unclear in this experiment. It is considered that iron oxide is reduced somewhat to wustite, based on reduction ratio increasing up to 25 per cent within 3 min of reduction time and since the dephosphorisation progression is saturated at relatively low reduction ratio. It is plausible that dephosphorisation ratio does not progress even if the reduction ratio is increased. From the observation, limiting the reduction of iron ore to wustite may be significant in developing an efficient dephosphorisation process.

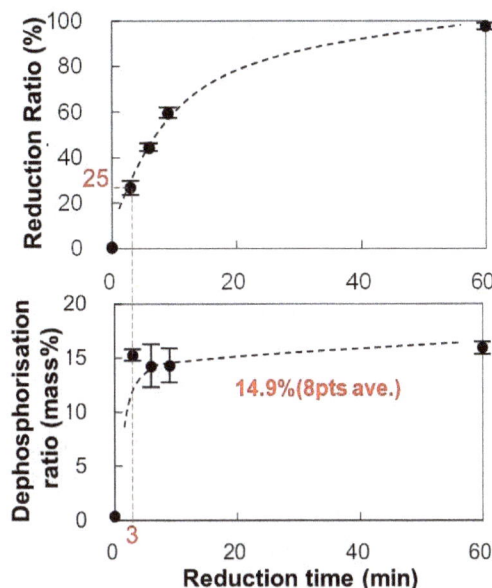

FIG 7 – Results of reduction and dephosphorisation ratio after deviation of phosphorus content controlled.

Dephosphorisation of synthetic model iron ore using phosphorus compound reagent

Figure 8 shows the effect of various forms of phosphorus on dephosphorisation using a synthetic model iron ore. From these result, different phosphorus compounds indicate different dephosphorisation ratio under similar unified reduction condition, allowing comparative evaluation of mechanistic pathway for dephosphorisation even if new forms of phosphorus in iron ore are found in future. In addition, when mixed with hematite reagent compared to goethite, even higher dephosphorisation ratio is obtained. This may be due to once gasified, the phosphorus might be trapped into minute pores generated after goethite dehydration but yet clarified.

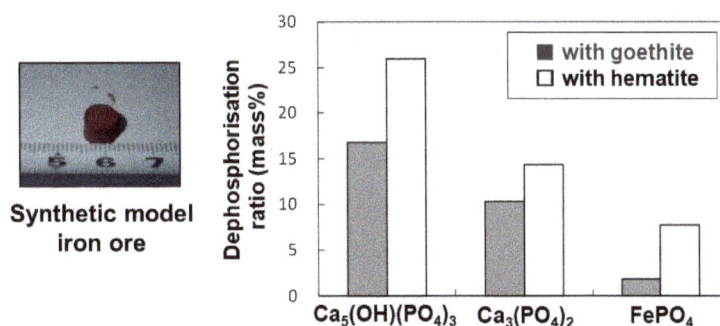

FIG 8 – Effect of various morphologies of phosphorus on dephosphorisation using synthetic model iron ore.

Continuous dephosphorisation processing of high phosphorus iron ore from Western Australia with larger rotating kiln-type furnace

Results of continuous dephosphorisation with a rotating kiln-type furnace is shown in Table 3, with 855.3 g total amount of continuously dephosphorisation processed samples recovered. The maximum dephosphorisation ratio in portion is 26.5 mass per cent, while the average for total samples obtained is 16.8 mass per cent. With this experiment, continuous dephosphorisation by rotating kiln-type furnace is verified as an efficient mass treatment process.

TABLE 3
Results of continuous dephosphorisation with rotating kiln-type furnace.

Box no.	Recovery (g)	T.Fe (mass%)	FeO (mass%)	M.Fe (mass%)	Removed P (g/t-ore)	Reduction ratio (mass%)	Dephosph-orisation ratio (mass%)
1	480.0	70.21	61.07	1.33	2271	15.7	15.9
2	321.9	70.85	65.36	0.69	2249	15.4	16.7
3	53.4	72.38	68.30	1.99	1985	16.2	26.5
Total /avg	855.3	70.58	63.13	1.13	2246	15.6	16.8

Removal behaviour of phosphorus in iron ore during reduction

Figure 9 shows schematic diagram of removal behaviour hypothesis of phosphorus in iron ore during reduction. Before reduction, phosphorus is adsorbed on the surface or inside iron ore grain, then phosphate begins to decompose with reduction gas. When reduction progresses, phosphorus separation is promoted and decomposition of phosphorus compounds to elemental phosphorus occurs. There is a concern that some of detached phosphorus may be re-adsorbed on adjacent particles depending on its arrangement, or re-bonded during the cooling process.

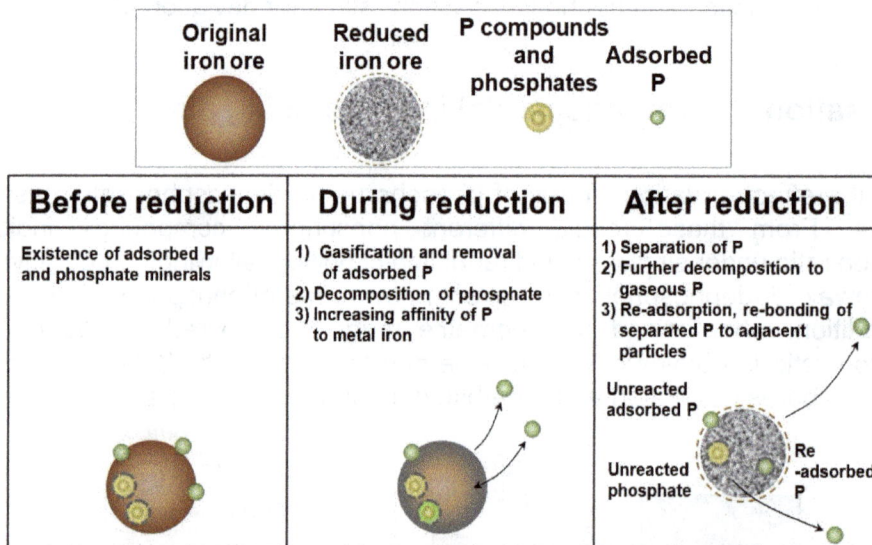

FIG 9 – Schematic diagram of removal behaviour of phosphorus in iron ore during reduction.

As previously mentioned, it is significant to reduce iron ore efficiently and to avoid gasified phosphorus being retained and reattaching to adjacent iron ore particle at the same time. In order to realise efficient reduction, it is suggested to overcome previous technical issues using a kiln-type furnace with a rolling action of iron ore particles in the chamber. In addition, this method would be able to obtain higher dephosphorisation ratio by combining with other treatment methods such as grinding, beneficiation, preheating and so on before and after the process. This study and verification needed to be continued and progressed in the future, to solve worldwide natural resource issue for sustainable steelmaking activities. A schematic diagram of dephosphorisation process conceivable for high phosphorus content iron ore in Australia is shown in Figure 10.

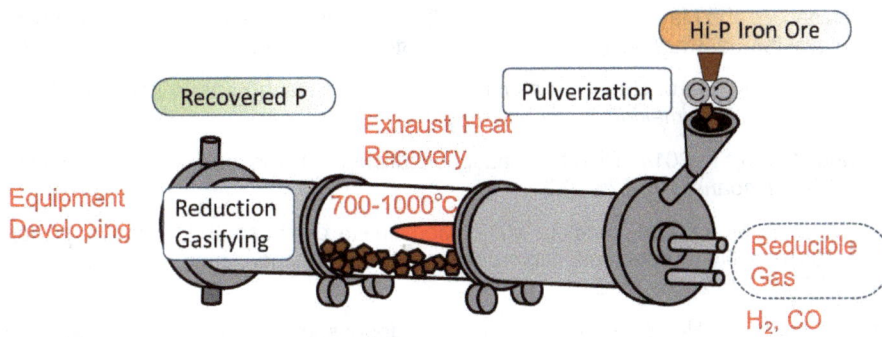

FIG 10 – Schematic diagram of dephosphorisation process conceivable for high phosphorus content iron ore in Australia.

CONCLUSIONS

Dephosphorisation ratio of 14.9 mass per cent in average is obtained through a reduction experiment using H_2 gas, the process condition to achieve over 10 mass per cent is provided. Iron oxide was partially reduced to wustite, consistent with reduction ratio increase up to 25 per cent within 3 min of reduction time. Since the dephosphorisation progression is saturated at relatively low reduction ratio, it is confirmed that the dephosphorisation ratio does not progress even if the reduction ratio is promoted higher. From the observation, limiting the reduction of iron ore to wustite may be significant in developing an efficient dephosphorisation process.

Different phosphorus compounds show a different dephosphorisation ratios under similar reduction condition, allowing comparative evaluation of mechanistic pathway for dephosphorisation, even if new forms of phosphorus in iron ore are found in future. In addition, when mixed with a hematite reagent compared to goethite, an even higher dephosphorisation ratio is obtained. This may be due to dispelling of gasified phosphorus trapped into minute pores generated after goethite dehydration.

From the results of continuous dephosphorisation with rotating kiln-type furnace, the maximum dephosphorisation ratio in portion is 26.5 mass per cent, while the average for the total samples obtained is 16.8 mass per cent. With this experiment, continuous dephosphorisation by rotating kiln-type furnace is verified as an efficient mass treatment process.

It is proposed for lowering of phosphorus in Australian iron ores, the efficient reduction of and avoiding gasified phosphorus from retaining and reattaching to adjacent iron ore particle at the same time are significant. In order to realise efficient reduction, kiln-type furnace is suggested to overcome previous technical issue due to rolling action of iron ore particles in the chamber. Finally, the work suggests conceivable reduction in phosphorus from high phosphorus content iron ores in Australia, using the dephosphorisation process.

ACKNOWLEDGEMENTS

This paper is based on results obtained from a project commissioned by the New Energy and Industrial Technology Development Organization (NEDO).

REFERENCES

Anyakwo, C and Obot, O, 2011. Laboratory studies on phosphorus removal from Nigeria's Agbaja iron ore by Bacillus Subtilis, *Journal of Minerals & Materials Characterization & Engineering*, pp 817–825.

Fisher-White, M, Lovel, R and Sparrow, G, 2012. Phosphorus removal from goethitic iron ore with a low temperature heat treatment and a caustic leach, *ISIJ Intl*, 52:797–803.

Jin, Y, Jiang, T, Yang, Y, Li, Q, Li, G and Guo, Y, 2006. Removal of phosphorus from iron ores by chemical leaching, *Journal of Central South University of Technology*, 13:673–677.

Lonkov, K, Gaydardzhiev, S, Bastin, D, Correa de Araujo, A and Lacoste, M, 2012. Removal of phosphorus through roasting of oolitic iron ore with alkaline earth additives, *IMPC XXVI Intl Mineral Processing Congress Proceedings*, pp 2194–2205.

Matinde, E and Hino, M, 2011. Dephosphorization treatment of high phosphorus iron ore by pre-reduction, mechanical crushing and screening methods, *ISIJ Intl*, 51:220–227.

Nunes, A, Pinto, C, Valadao, G and Magalhaes Viana, P, 2012. Floatability studies of wavellite and preliminary results on phosphorus removal from a Brazilian iron ore by froth flotation, *Minerals Engineering*, 39:206–212.

Pandey, J, Sinha, M and Raj, M, 2010. Reducing alumina, silica and phosphorous in iron ore by high intensity power ultrasound, *Ironmaking & Steelmaking*, 37(8):583–589.

Sasabe, M, Iida, Y and Yokoo, T, 2014. Direct dephosphorization from iron ore containing higher concentration of phosphorus, *Tetsu-to-Hagané*, 100:325–330.

Xu, C, Sun, T, Kou, J, Li, Y, Mo, X and Tang, L, 2012. Mechanism of phosphorus removal in beneficiation of high phosphorous oolitic hematite by direct reduction roasting with dephosphorization agent, *Transactions of Nonferrous Metals Society of China*, pp 2806–2812.

Zhu, D, Chun, T, Pan, J, Lu, L and He, Z, 2013. Upgrading and dephosphorization of Western Australian iron ore using reduction roasting by adding sodium carbonate, *International Journal of Minerals, Metallurgy and Materials*, 20(6):505–513.

Improved iron ore processing through digitalising conveyed ore flows using representative real time multi-elemental analysis

H Kurth[1]

1. Chief Marketing Officer, Minerals Consultant, Scantech International Pty Ltd, Camden Park SA 5038. Email: h.kurth@scantech.com.au

ABSTRACT

High specification Prompt Gamma Neutron Activation Analysis (PGNAA) has been successfully applied to conveyed flows at iron ore operations since 2002 to representatively measure ore quality for improved quality management and process control. While the PGNAA technique has changed little since that time, significant improvements have been made in measurement capability. Elements previously unable to be measured well or at all using this technique are now measurable and one of these is phosphorus. Aluminium and magnesium are also key elements in addition to calcium and silicon in determining sinter basicity in ironmaking. Other elements measured routinely in iron ores include iron, carbon, potassium, sulfur and manganese.

The benefits of representative measurement cannot be overstated as iron ore mining operations in South Africa, Europe, Australia, Asia and the Americas have discovered. The technology is applicable to any conveyed flow where elemental analysis in real time assists with quality management, from raw materials in ironmaking through to scrap steel in steelmaking and coal quality for coke-making. Major benefits are realised when raw material processing is optimised and this often includes diversion of material meeting a required quality so it can bypass unnecessary processing. This includes waste rock which can be diverted from a plant feed stream in 30 sec increments or ore that meets product quality being diverted to bypass beneficiation. Analysers are successfully utilised to identify ore types that assist with upgrade optimisation in jig plants; measuring feed, product and discard flows to also account for metal content. Product flows to stockpiles help manage production flows and stockpile qualities for different products and assist in blending ores into trains or ships to confirm specification compliance. Technologies can be used in magnetite processing but are predominantly used in hematite and goethite ore processing operations. The paper details examples of the above applications at various operational sites providing estimates of economic benefit as well as operational improvements.

INTRODUCTION

As better quality resources are exhausted, there is an increasing need to manage quality so that specifications can be met and lower quality iron ores can be sufficiently upgraded for use in ironmaking. Magnetite ores are effectively upgraded through magnetic separation and produce reasonably pure concentrates while hematite and goethite ores may require more intense processing to produce a saleable quality. The need for representative measurement to manage the process performance is therefore increasing as the quality of iron ore resources available for mining generally decline. Measurement data is used in different ways to generate benefits for iron ore mining and processing operations. This paper explains technology that has been successfully applied to many of these situations and discusses some of the outcomes achieved.

PROMPT GAMMA NEUTRON ACTIVATION ANALYSIS (PGNAA)

Prompt Gamma Neutron Activation Analysis (PGNAA) is a well-proven technique used to representatively measure conveyed flows in real time with successful implementation in coal and cement sectors since 1990s and in iron ore since the early 2000s (Matthews and du Toit, 2011). PGNAA has proven successful in sinter basicity control in ironmaking and effective for scrap steel measurement in steelmaking (Balzan, 2022; Kurth and Kalicinski, 2023).

PGNAA utilises a neutron source, typically Californium-252, housed within the analyser and underneath the conveyor that generates a zone of neutron flux that conveyed load passes through continuously. Elemental nuclei absorb neutrons (hence Neutron Activation) and instantaneously release gamma energy with signatures unique to each element (hence Prompt Gamma). An array of detectors, commonly sodium iodide, or bismuth germanium oxide in high specification systems, determine the gamma energies emitted and enable accumulation of spectra for a parcel of material. Customised signal processing algorithms resolve the spectral response into proportions of each element present.

Analysers are calibrated for a range of elements and those useful for quality monitoring and management are reported to a plant control system in real time. Calibrations in high specification systems are customised based on responsiveness of elements to the technique (as low as carbon in atomic number), expected composition ranges for each element and belt load variability inputs. A PGNAA system is shown in Figure 1. The high specification system can accommodate steel corded and chlorine containing Fire Resistant Anti-Static specification (FRAS) conveyor belts. Analysers are successfully utilised in both surface and underground operations on such belts.

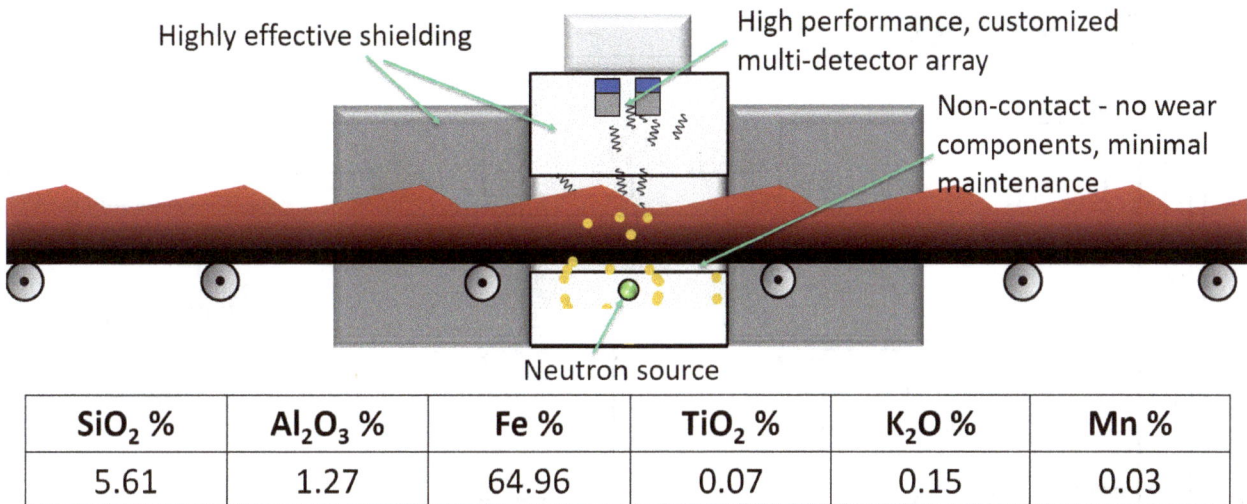

SiO$_2$ %	Al$_2$O$_3$ %	Fe %	TiO$_2$ %	K$_2$O %	Mn %
5.61	1.27	64.96	0.07	0.15	0.03

FIG 1 – Cross-section through GEOSCAN high specification PGNAA analyser showing main components and an example of elemental results for a two minutes measurement increment (Scantech).

The advantage of this technique over others is that measurement is fully penetrative of conveyed flows (through 500 mm) irrespective of particle size, belt speed, composition, moisture content, dust, segregation or layering, or surface coatings. The full conveyed width and depth passing through the analyser is sensed continuously and a tonnage weighted average measurement of multiple elements is generated for each measurement period. The measured parcel size reflects the throughput rate (input from a belt scale) and measurement accumulation time (increment). Increments of two minutes are standard for high performance PGNAA systems and increment size can be reduced to 30 sec at similar precisions using a higher specification configuration. A timeframe of 30 sec is useful in bulk diversion applications as it provides an optimal balance between high measurement precision and small increment size to optimise selectivity to provide the best parcel segregation outcomes possible. This optimises metal recovery in the 'accept' stream compared to poorer measurement systems (Kurth, 2022). A longer measurement time can be used when monitoring quality or controlling blending of materials into a process.

PGNAA is an unmatched analysis technique for representative multi-elemental analysis of conveyed bulk flows in real time and is of particular benefit in the measurement of contaminants at low concentrations. Elements commonly measured in iron ores include iron, silicon, aluminium, titanium, manganese, calcium, magnesium, sulfur, phosphorus and carbon. High specification PGNAA has proven to measure phosphorus in iron ores to below 0.01 weight percent. It should also be noted that different configurations and lower specification PGNAA systems are normally unable to provide timely or precise measurements. Lower specification systems may have smaller neutron sources, fewer or lower specification detectors and digital multi-channel analysers, longer measurement time requirements, poorer precisions, inability to measure common elements well or at all and price.

Based on feedback given to the author from mining companies over recent years some 70–80 per cent of neutron activation (including thermal activation) based elemental analysers installed in the Western Australian iron ore market from multiple vendors have failed to achieve expected measurement performances or produced any usable data since installation. Successful implementation has predominantly occurred using higher specification systems which can also measure the phosphorus and aluminium well.

APPLICATIONS

Digitalising the conveyed flow quality in real time allows operations to make real time decisions that can influence quality and operational activities. Benefits will vary depending on the location of the measurement. Analysis can occur as soon as the iron ore is on a conveyor. Figure 2 shows an example of multiple installations in a single iron ore operation. Individual applications and benefits are discussed in detail in the following sections.

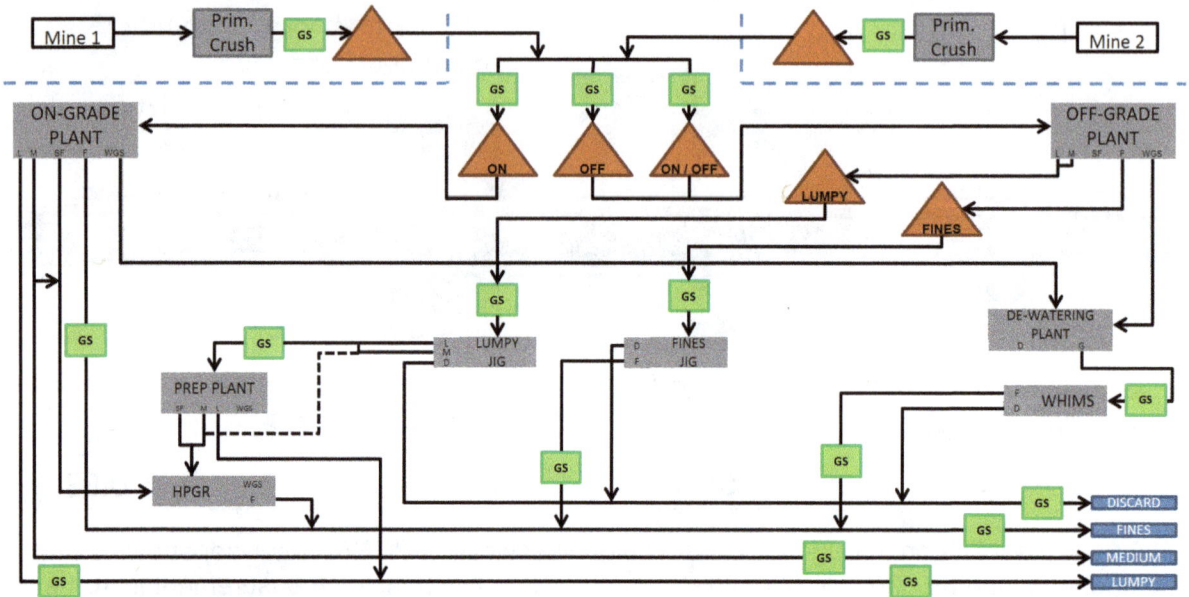

FIG 2 – Iron ore plant flow sheet at Assmang Khumani in South Africa showing GEOSCAN high specification PGNAA analysers (green boxes) used throughout the operations (after Matthews and Du Toit, 2011).

Run-of-mine ore

Measuring run-of-mine ore is challenging from a calibration perspective due to the difficulties in collecting representative samples on primary crushed conveyed material. Analyser data is compared to expected ore quality from blasthole and trucking data over longer time frames in such cases.

Figure 3 shows a comparison of analyser measurements to averaged laboratory assay data calculated for each equivalent batch of ore. The main application of this data is to ensure plant feed quality is on target and that product quality is likely to be achieved. Measurements are used to reconcile expected mined grade with actual grades and this confirms confidence in the mining schedule and geological resources and reserves modelling. One site also recovered from mining a poorly defined zone assumed to be ore that proved to be below the acceptable quality and ceased haulage from that area once the lower quality was detected to prevent a full shift of product being diluted to below acceptable specification levels. This action saved the operation many hundreds of thousands of Australian dollars.

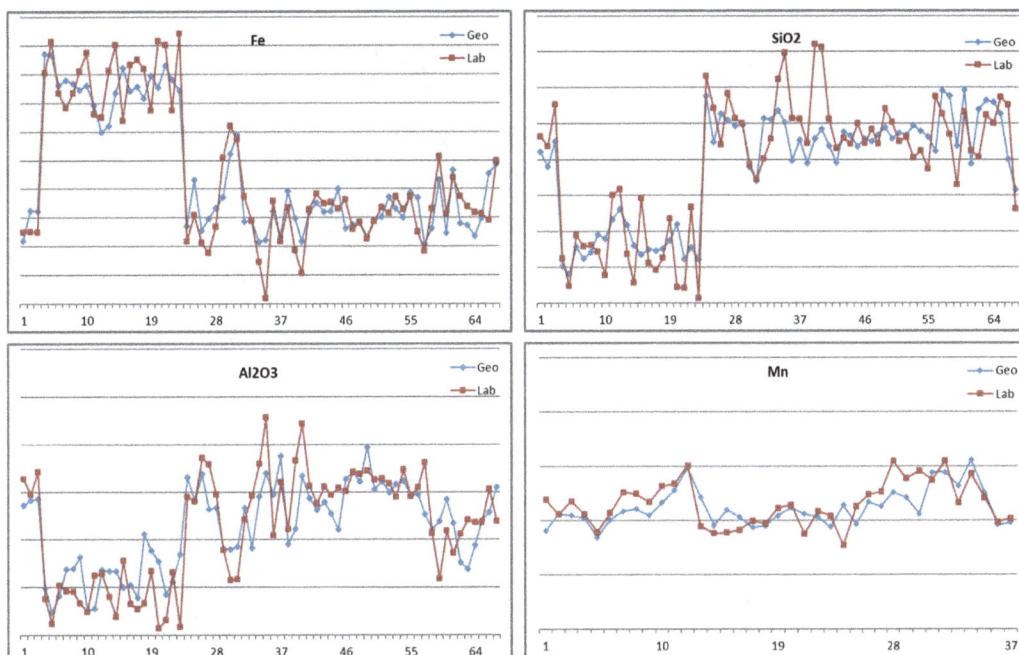

FIG 3 – Iron ore calibration data for run-of-mine ore at an iron ore mine in Western Australia. Vertical axis gridlines are at 0.5 per cent intervals (after Balzan, Beven and Harris, 2015).

Changes in ore type detected through continuous monitoring of run-of-mine ore enables beneficiation plant operators to optimise responses and ensure maximum process recovery for each ore type. This is a feed forward application that increase plant efficiencies as ore changes are detected before they reach the process. Ore quality variability affects process efficiency (Goodall, 2021) and using real time measurements to control ore blend results in significant recovery improvements. At Assmang Khumani operations in South Africa the measurement of ore quality from each mine is incorporated in a site wide elemental balance process to ensure continuous operational optimisation.

Increment diversion

Measured parcels of conveyed flow are diverted when they meet predetermined quality criteria. There is an opportunity to divert material that is uneconomic to process or does not require processing as this may incur unnecessary costs if processed for no additional value. Material below an economic cut-off can be diverted to a low-grade or waste destination and discarded or processed at a later date if economics permit. A major benefit of diverting poorer quality material is that higher quality material can replace it and increase product output relative to process plant capacity.

Product quality bypass is a common application in bulk commodities where further processing cost can be avoided for mined ore that meets product quality requirements. Table 1 shows the criteria considered in diverting flows between multiple mines and a process plant at Assmang Khumani on overland conveyors (shown centre top of Figure 2).

TABLE 1

Algorithms used for run-of-mine classification (after Matthews and du Toit, 2011).

	%Fe	%Al$_2$O$_3$	%K$_2$O + %Na$_2$O	%SiO$_2$	% Estimated yield
On-grade	>=65.3	<=1.8	<=0.25	<=2.5	>91
Just barely off spec	>=65.3	>1.8 and <=2.2	>0.25 and <=0.35	>2.5 and <=4.5	>80
Off-grade high	<65.3 and >60	>2.2 and <=3.2	>0.35 and <=0.40	>4.5 and <=6.5	>60 and <80
Off-grade low	<65.3 and >60	>3.2	>0.40 and <=0.54	>6.5 and <=12	>45 and <60
Waste1	<60 and >=39		>0.54	>12	
Waste2	<39				

Approximately one third of the mined ore meets the product quality and is diverted to bypass the jig plant. This provides an annual saving in processing costs of some A\$6–8 million. GHG emissions savings in avoiding beneficiation can also save approximately 8 kg CO_2 e/t which for 5 Mtpa of iron ore bypassing a jig plant can save 40 000 t CO_2 e/a.

Process optimisation and plant design

Differences between ore types can affect beneficiation performance and metallurgical test work on ore types can advise process operators of optimal upgrade factors expected during processing. In the case of Assmang Khumani, the jig feed and product qualities are measured in real time to assist operators to maximise iron ore recovery. Ore type is identified by the multi-elemental analysis and the upgrade factor determined to ensure it is close to that expected. Discard flows are also measured to ensure all iron ore losses are accounted for in reviewing plant performance.

Reducing feed quality variation improves predictability of plant performance and reduces process upsets. This ensure consumables including energy, water, reagents and equipment wear are minimised and this in turn reduces GHG emissions to reduce environmental impact.

Assmang had successful experiences with PGNAA prior to designing and building the Khumani operations. The detailed knowledge of the ore types and processing characteristics and confidence in the analyser's measurement capabilities enabled the plant design to incorporate the bypass application. This reduced required jig plant size and cost (CAPEX) as they knew a large proportion of the ore to be product quality and this did not need processing. Utilising the analysers provided an opportunity to save CAPEX in designing a smaller throughput jig plant in addition to ongoing operational savings (OPEX).

Kumba Iron Ore in South Africa, like Assmang, has adopted high specification PGNAA analysers for quality monitoring throughout their operations (Balzan *et al*, 2019) as part of their process optimisation strategy and achieved significant operational improvements from the mines to the port.

An iron ore operation in Western Australia used a similar philosophy to minimise the number of sample towers required for their quality monitoring. Sample towers (priced at some A\$15–20 million each at the time of construction) were installed on product flows only and analysers specified for ore and intermediate conveyed flows. At least three sample towers initially included in the flow sheet design were replaced with analysers, achieving major CAPEX and OPEX savings.

Product quality

Real time elemental measurement of product streams increases confidence of specification compliance through the ability to respond to product quality in real time. Much iron ore produced is direct shipping quality (DSO) and there may be limited opportunity to blend product stockpiles at ports or during ship loading. Improved management of product quality at the mine results in fewer quality related problems during downstream handling of the iron ore.

At Assmang, product quality of each beneficiation stage is measured so that the overall contribution of products to each stockpile and an overall quality for a stockpile can be determined. Not shown in Figure 2 is the reclaim analyser after the stockpiles that measures the quality removed from each stockpile and loaded onto each train. The deduction of tonnes and grade from a stockpile allows the stocks in each pile to be continually updated and a 'live' quantity and average quality to be determined.

Figure 4 shows comparisons of analyser data with assay data from a full stream sampling system (typically used on product streams) at an iron ore mine in Western Australia. Analyser calibrations on product streams are therefore easier to manage due to availability of data and better sampling capabilities.

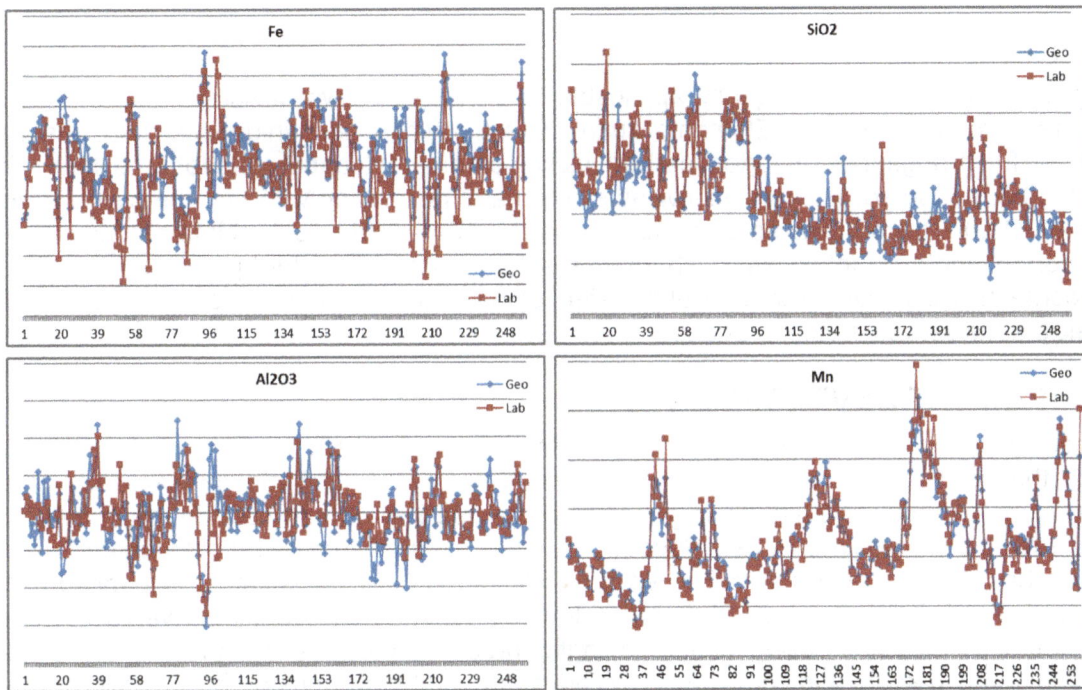

FIG 4 – Iron ore calibration data for product at an iron ore mine in Western Australia. Vertical axis gridlines are at 0.5 per cent intervals, except SiO_2 is 1 per cent (after Balzan, Beven and Harris, 2015).

Ore quality loaded onto trains at a mine is communicated to the port operations to allow each train to be unloaded and blended according to desired stockpiles and subsequent shipment quality. Advanced stockyard management systems can track the quantity and quality of iron ore being delivered, stored and reclaimed to optimise port logistics. Using elemental analysis at the port enables stockpile mapping in fine detail so that reclaim can be carefully controlled to ensure shipment qualities are optimised. Analysers are currently being implemented in these applications at a port in South Africa.

Sinter chemistry control in ironmaking

Iron ore quality remains a concern for downstream processing and the chemistry of blast furnace feed is one example. Sinter feed quality is a specific application where PGNAA has been proven effective and beneficial (Balzan, Harris and Bauk, 2016). Sinter basicity is measured and controlled, most commonly through limestone addition, to ensure the blast furnace feed chemistry minimises energy consumption and unplanned downtime. Furnace 'freezes' can shutdown the ironmaking process and result in many millions of dollars in lost iron production for a single event.

High specification PGNAA enables elements such as iron, silicon, aluminium, calcium and magnesium to be measured with high precision to determine sinter basicity to optimise limestone addition by a controller to achieve a desired set point. Such controllers are well established at numerous sites (Di Giorgio, Pinson and Noble, 2019). Most existing PGNAA installations produce only a simplified basicity calculation of CaO/SiO_2 which provides only an approximation of basicity.

A high specification PGNAA system enables more elements (eg aluminium and magnesium) to be precisely measured so that a more accurate calculation of basicity can be determined according to $(CaO + MgO) / (SiO_2 + Al_2O_3)$. Accurate measurement of these elements in turn leads to a greater level of control and better outcomes for the plant (Balzan, 2022). One example indicates a sinter basicity precision of 0.05 for a target ratio of approximately 1.25 using laboratory sample data for 310 corresponding periods of analyser measurement. During the comparison the composition of the elements measured varied significantly (7.5–11.5 per cent CaO, 2.9–4.0 per cent MgO, 7.5–10 per cent SiO_2 and 1.8–2.6 per cent Al_2O_3) but the basicity remained consistent (1.2–1.3).

Application of the PGNAA technology has also assisted in coal blending to optimise coke-making operations with the resultant coke added with sinter to the blast furnace to produce pig iron.

Scrap steel

High specification PGNAA is equally applicable to conveyed flows of scrap steel in electric arc furnace feed for steelmaking. Elements such as iron, silicon, aluminium, calcium, magnesium, titanium, sulfur, manganese, nitrogen, carbon, nickel, chromium, copper, phosphorus, lead, zinc and cobalt can be measured at levels as low as single parts per million with high precision (Kurth and Kalicinski, 2023). This enables scrap to be bulk sorted by measured parcel into batches for smelter feed to control contaminant levels which affect steel quality and therefore potential value and uses. Higher contaminant concentrations results in steel of lower quality and therefore only suitable for use as rebar, the lowest value steel product.

MEASUREMENT DEVELOPMENTS

High specification PGNAA systems are characterised by proven performances in plant applications. Systems that measure more elements to lower detection limits at better precisions over shorter measurement accumulation times will generate more useful data and confidence in these results enables effective process responses. Successful implementation of over 50 of these units in the iron ore sector has demonstrated the value available from such data. Benefits are enhanced as further measurement capabilities are added. A major development has been the improvements in spectral measurement accumulation using optimised configurations and arrays of high performance bismuth germanium oxide detectors to compliment the advantages of highly customisable calibrations.

Incorporating these developments has enabled measurement times to be reduced from the initial two minutes to 30 sec at similar measurement precisions. Additional elements have been included through spectral characterisation work to allow reporting of elements once thought impossible to quantify from the combined spectral response. Phosphorus is one such element in iron ore that was initially expected to be unmeasurable at levels as low as 0.01 per cent P. Figure 5 shows analyser phosphorus results compared with laboratory assays at multiple iron ore sites. Calibrations have been retrofitted as part of normal product support agreement calibration reviews to all existing iron ore users interested in receiving phosphorus results.

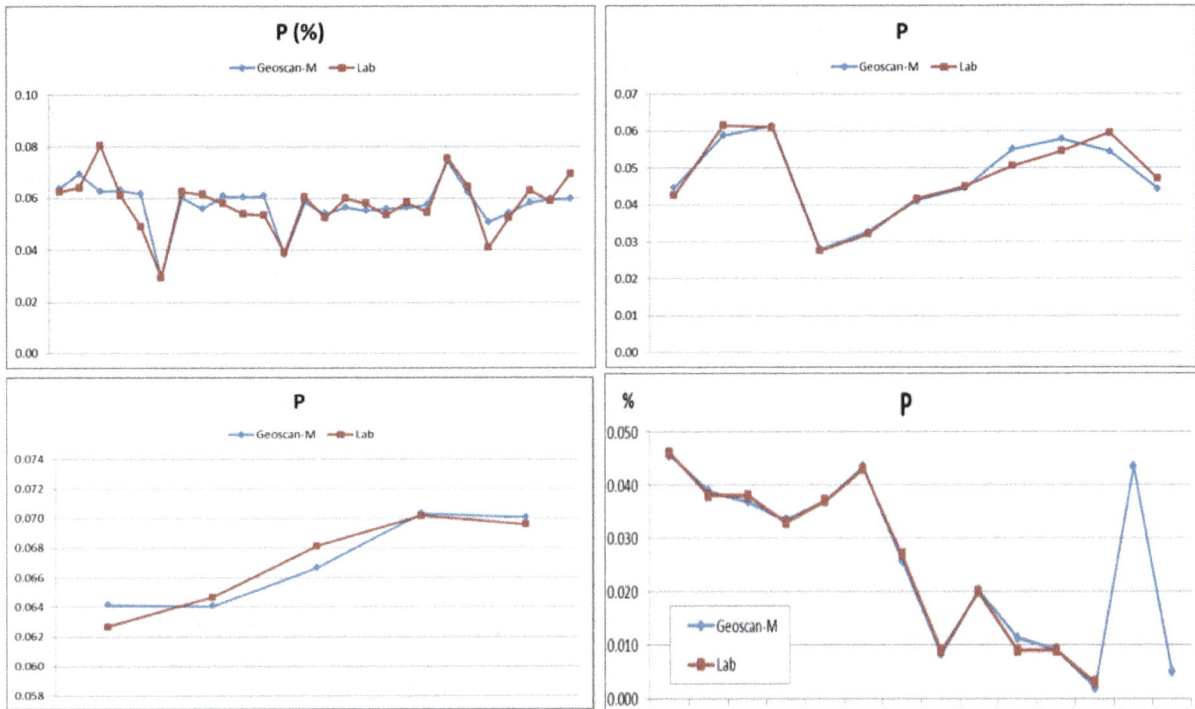

FIG 5 – Iron ore calibration data for phosphorus (after Balzan and Nieuwenhuys, 2021).

Another element which has previously been assumed to be difficult to reliably measure in PGNAA systems is carbon. Carbon spectral peaks are used for detector spectral alignment. The application of analysers on conveyed flows resulted in constant carbon response from the rubber conveyor belt and components in the analyser; including carbon steel in the frame and carbon from the high density polyethylene (HDPE) shielding material. The improvements in spectral resolution and background measurement have resulted in the ability to clearly define additional carbon from within the conveyed material. In Figure 6 the carbon measurement in iron ore sinter feed material demonstrates excellent performance which has been successfully duplicated in other analyser applications.

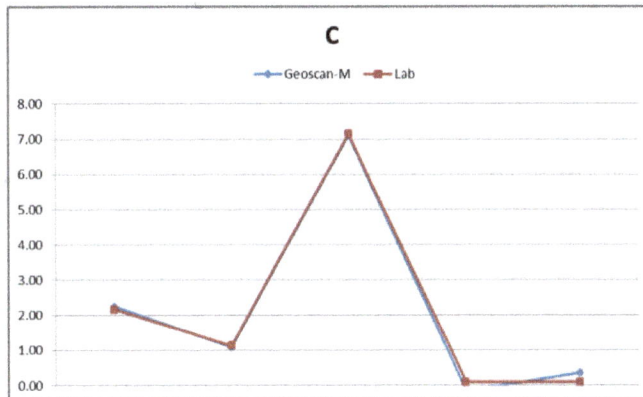

FIG 6 – Iron ore calibration data example for carbon (Scantech).

An analyser is successfully utilised in iron ore to measure mercury content over a 1–20 ppm Hg range at 0.5 ppm precision. This assists the user to monitor and control mercury levels prior to ironmaking. Further developments have included the ability to measure conveyed materials on conveyors containing chlorine, which previously reduced measurement performance capability due to chlorine's strong spectral response relative to other elements. High performance PGNAA analysers have consequently been successfully used in underground and enclosed gallery conveying systems using FRAS rated conveyor belts.

CONCLUSIONS

The paper summarises many successful applications of high specification PGNAA systems and where possible shows economic benefits where end users have communicated these. Many more benefits have not been shared due to confidentiality and competitive reasons. High specification PGNAA analysers have proven to representatively measure more elements to lower levels at better precisions over shorter time increments than any other available on-conveyor technology. Faster measurement at higher precisions has created bulk sensing data suitable for bulk sorting which significantly reduces operational costs and emissions. Capabilities in measuring additional elements such as phosphorus, carbon, magnesium, mercury and aluminium at high precisions has increased potential benefits for ore, product, scrap and downstream process control. Plant designs utilising analyser capabilities have allowed CAPEX savings of tens of millions of dollars in addition to ongoing operational cost benefits. Analyser paybacks have been as short as weeks or days depending on value of benefits derived. While few analysers in the market can be considered high specification, there is obvious value in performing effective due diligence to ensure technical and operational risk is minimised in selecting the correct analyser specification for each application.

ACKNOWLEDGEMENTS

The author acknowledges the work done by Scantech research and development, calibration and service engineers in continuing to develop analyser capabilities and their published and unpublished contributions. It is also appropriate to acknowledge collaboration of iron ore customers who have encouraged and supported further analyser developments. Scantech approval to publish is acknowledged given the competitive nature of technology suppliers, hence publishing proven capabilities and performance is preferred over aspirational and unachieved performance claims.

REFERENCES

Balzan, L, 2022. Improved Stability in Control of Sinter Feed Basicity by Using GEOSCAN Real-Time On-Belt Analysis, in *Proceedings of the Iron & Steel Technology Conference, AISTech 2022*, pp 1388–1394. doi: 10.33313/386/161

Balzan, L and Nieuwenhuys, F, 2021. Real time phosphorus analysis using GEOSCAN on belt analysers at Assmang Khumani Mine in the Northern Cape, in *Proceedings Iron Ore Conference 2021*, pp 342–350 (The Australasian Institute of Mining and Metallurgy: Melbourne).

Balzan, L, Beven, B J and Harris, A, 2015. GEOSCAN online analyser use for process control at Fortescue Metals Group sites in Western Australia, in *Proceedings Iron Ore Conference 2015*, pp 99–105 (The Australasian Institute of Mining and Metallurgy: Melbourne).

Balzan, L, Harris, A and Bauk, Z, 2016. Process improvement through real-time analysis, in *Proceedings of the 46th Ironmaking, 17th Iron Ore and 4th Agglomeration Conference 2016*, Associação Brasileira de Metalurgia, Materiais e Mineração.

Balzan, L, Swart, E, Gray, L and Kalicinski, M, 2019. Process improvement at Kumba Iron Ore Sishen and Kolomela mines through the use of GEOSCAN on belt analysis equipment, in *Proceedings of Iron Ore Conference 2019*, pp 649–659 (The Australasian Institute of Mining and Metallurgy: Melbourne).

Di Giorgio, N, Pinson, D and Noble, G, 2019. Optimising the sintering process at Port Kembla using real time feed composition, in *Proceedings Iron Ore Conference 2019*, pp 68–76 (The Australasian Institute of Mining and Metallurgy: Melbourne).

Goodall, W, 2021. Understanding what is feeding your process: how ore variability costs money, *Process Mineralogy Today*, March 10, 2021. Available from: <https://minassist.com.au/understanding-what-is-feeding-your-process-how-ore-variability-costs-money/>

Kurth, H and Kalicinski, M, 2023. Real-Time On-Line Elemental Analysis of Scrap for Steelmaking, in *Proceedings of the Iron & Steel Technology Conference (AISTech 2023)* doi: 10.33313/387/074

Kurth, H, 2022. Ore quality measurement and control using Geoscan-M PGNAA real time elemental analysis, in *Proceedings International Mining Geology Conference 2022*, pp 338–345 (The Australasian Institute of Mining and Metallurgy: Melbourne).

Matthews, D and du Toit, T, 2011. Validation of material stockpiles and roll out for overall elemental balance as observed in the Khumani iron ore mine, South Africa, in *Proceedings Iron Ore Conference 2011*, pp 297–305 (The Australasian Institute of Mining and Metallurgy: Melbourne).

Minimising cost of rheology control additives for iron ore slurries/tailings for enhancing processing performance

Y K Leong[1]

1. Professor, University of Western Australia, Crawley WA 6009.
   Email: yeekwong.leong@uwa.edu.au

ABSTRACT

Occasionally additives are used in the wet processing of iron ore tailings and slurries, for examples, in the thickening of tailings with flocculant to recover process water, and in the smooth pumping of concentrated slurries. They can mitigate slurry behavioural issues that could significantly affect the plant operation and production. The nature of iron ore business requiring the plant to have a high throughput to be economical, large amounts of slurries are therefore processed every day. The quantity of additives needed will correspondingly be large and so are the associated costs. This could have a significant impact on the plant profitability. If additives are used they must be very cost-effective. NaOH is among the lowest cost chemicals available, at ~US$400 per ton. In this paper we will look at use of composite additives involving phosphate compounds with NaOH added to reduce the cost, for the control of iron ore slurries rheology. The NaOH worked synergistically with the phosphate compounds. The effectiveness of these composite additives in viscosity reduction can be neutralised by another cheap additive such as lime or a Ca(II)-based salt turning the slurries back to being a paste. Tailings paste is an ideal material for safe storage in the dam. These composite additives can also be used to facilitate the recovery of iron ore at commercial grade from tailings (Leong, 2020a, 2021a).

INTRODUCTION

Iron ore mineral processing plants do not often use chemical additives to aid the wet beneficiation processes due to the product low price and high cost associated with the high throughput of the ore materials. The use of polymeric flocculants to hasten the thickening process of the iron ore tailings in the thickener is unavoidable. However, the dosage needed is quite small and hence it is cost-effective. At times, the thickened tailings acquired a yield stress much higher than can be handled by the pump hampering the flow of the thickened tailings to the dam. The use of water to reduce the yield stress and viscosity defeats the purposes of reducing the process water usage and disposing high density tailings. With current price of iron ore at a historically high level, the use of cheap additives to control the tailings rheology (Leong, Drewitt and Bensley, 2019; Leong *et al*, 2023a; Leong, 2018, 2020b, 2021b) and the nature of particle interactions in the iron ore slurries will greatly improve the efficiency of the plant in terms increased throughput, concentrated tailings being disposed and less process water being used (Leong, 2021b). In this paper the results of rheological effects of some cheap bulk chemical additives on iron ore tailings are presented.

For safe iron ore tailings storage, the tailings should develop a paste-like consistency in the tailing storage facilities (TSF) or dams. It will also be demonstrated in this paper that with the correct chemical additives combinations, a low viscosity concentrated tailings was first produced so it can be pumped effortlessly to the TSF and then at the dam a cheap neutralising additive can be used to convert it back to the paste-like consistency (Leong, Drewitt and Bensley, 2019; Leong *et al*, 2023a; Leong, 2021b).

The main components of iron ore tailings are ocherous goethite (OG) and clay, in particular, kaolin. Time-dependent behaviour may be another rheological issue posed by the tailings due to the presence of clay. Thixotropy has been incorporated in the rheological model used in the modelling of tailings flow (Talmon *et al*, 2023). In this study, the time-dependent behaviour of the tailings were also discussed.

MATERIALS AND METHODS

Three iron ore tailings were sourced from iron ore mines in the Pilbara, Western Australia. They are designated as Pilbara mine A, B and C. The main components of these tailings were goethite, hematite, kaolin and quartz. AR grade reagents, NaOH, pyrophosphate, Na polyphosphate or Na (polyPO₄) and others were used. The yield stress was measured with Brookfield vane viscometers. The stepdown shear rate was performed using the Anton Paar cone-and-plate rheometer. The shear stress-shear rate behaviour was characterised with a Haake VT 550 cone-and-plate viscometer. The cones used in both rheometers have an angle of 1° and so the shear rate is constant everywhere in the tailings in the gap. In the ageing test, the gel (50–100 g) was mixed or stirred thoroughly for 2 mins with a spatula or pH/conductivity probes. After that the yield stress was measured at function ageing time. The gels were left undisturbed during ageing and the vane yield stress measurements were conducted in regions not disturbed by a previous measurement. Orion 3 and 4 stars pH and conductivity metres were used to measure the tailings pH and conductivity. The pH and conductivity of the supernatant water of Pilbara mine A and B tailings were 7.25 and 0.56 mS/cm, and 7.2 and 1.39 mS/cm respectively. The Pilbara mine C tailings was used to demonstrate that safer tailings storage practice is possible. The viscosity or yield stress of this tailings was reduced to a very low level by the composite NaOH:Na (polyPO₄) additives and then shown to be converted back into a paste with a neutralising calcium solution.

RESULTS AND DISCUSSIONS

The cheap additives used to reduce the yield stress or viscosity of concentrated tailings is a UWA-patented composite additives of NaOH and sodium polyphosphate. The cost of NaOH is ~US$400 per ton and of sodium phosphate about US$1000 per ton. The effective of this composite additives is dependent upon the surface chemistry of the tailings such as pH and, salt chemicals and concentration. The relationship of yield stress with additive dosage is shown in Figure 1(a) and with pH in Figure 1(b).

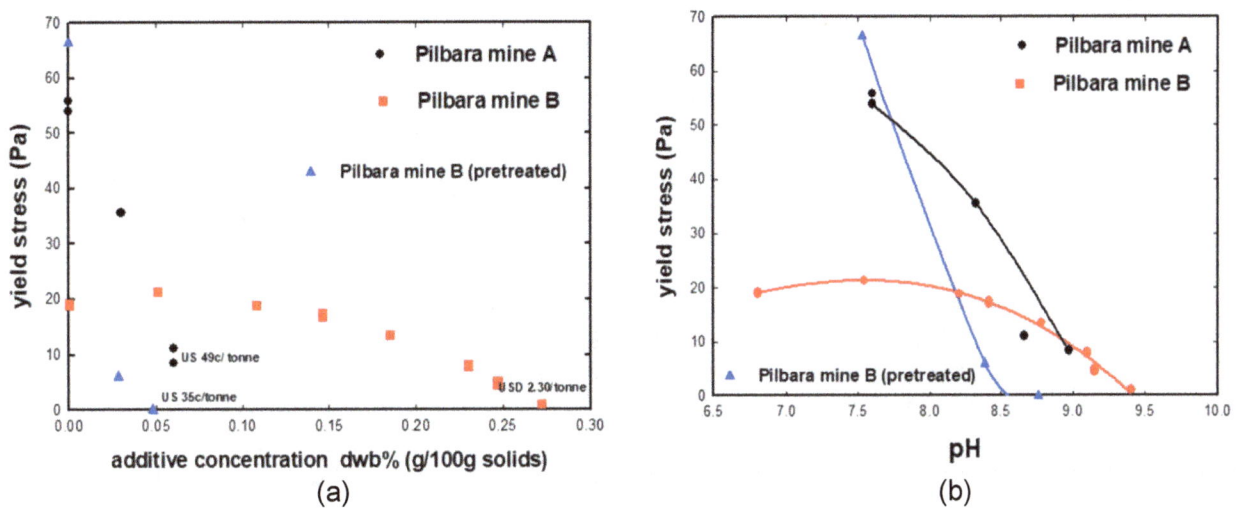

FIG 1 – The relationship between yield stress and (a) composite additive dosage and (b) the pH (changed by the additive addition) for Pilbara mine A and B tailings.

The reduction is yield stress with additive dosage is very sharp or dramatic for tailings A. The yield stress was reduced from 55 Pa to 10 Pa needing a dosage of only 0.06 g/100 g tailings solids. The cost is only US$0.49 per ton of tailings solids. For Pilbara mine B tailings the dosage required is much higher, 0.27 g/100 g tailings solids to reduce the yield stress from 20 Pa to ca. 0 Pa. The cost is prohibitive, US$2.30 per ton of tailings solids. However, an appropriate pretreatment can reduced the cost of US$0.36 per ton tailings. The reduction in yield stress from 67 Pa to 0 Pa required a dosage of 0.05 g per 100 g tailings solids. The concentration of this tailings is 72 wt. per cent solids compared to the untreated sample of 49 wt. per cent solids. The pretreatment cost is prohibitive. However, the understanding gained can be used to design changes to the management of the tailings to reduce the cost of the additives. The pH of the treated tailings were all less than 10, the upper limit of tailings disposal pH depending upon the location of the tailings dam.

Figure 2a showed the yield stress of the tailings mine C being reduced to ~20 Pa from an initial value of 1000 Pa by the composite additives so it can be handled and pumped over long distances to TSF. A similar result was obtained with the same composite additives but with a higher content of polyphosphate (see Figure 2b). The pH at which the yield stress reduction to 20 Pa occurred at a slightly lower pH. The dosage used for both composite additives was approximately the same. So the use of the lower phosphate content composite additives would therefore be more cost-effective. At the TSF it can be converted back to a paste of ~1000 Pa with a neutralising agent $CaCl_2$ or lime CaO (Leong, Drewitt and Bensley, 2019; Leong, 2021b). Paste storage is considered to be much safer in the event of a dam failure.

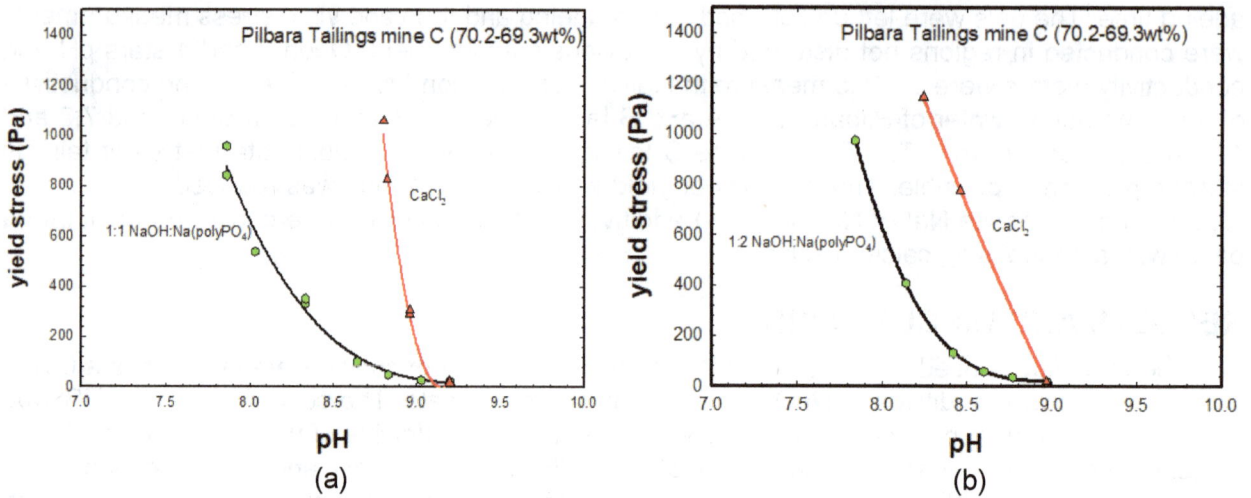

FIG 2 – Demonstration with Tailings mine C showing a dramatic reduction of yield stress by the composite additives and its conversion back to its paste state with a neutralising additives keeping pH below 10.

Tailings with a significant component of smectite clay such as bentonite (or sodium montmorillonite NaMnt) and hectorite generally posed a significant processing problem due to a drastic change in rheological behaviour and filterability (Carman, 1938). They will display significant time-dependent behaviour such as thixotropy and rheopexy (Leong *et al*, 2021a). Thixotropy is depicted by a decreasing viscosity or shear stress with time at a constant shear rate. Rheopexy represents the opposite effect. The time-dependent behaviour was studied under the stepdown shear rate and ageing modes. In the stepdown shear rate mode, ideally the gel should sheared to an equilibrium state at $1000s^{-1}$ prior to its abrupt stepdown to $10\ s^{-1}$. If the gel is time-dependent it will show an immediate increase the shear stress with at the low shear rate as shown in Figure 3 for 5.04 wt. per cent NaMnt gel. The gel was prepared 26 days earlier. It has an aged yield stress (left undisturbed for 26 days) of 28.6 Pa. After vigorous stirring with spatula, the yield stress decreased to 17.7 Pa. At this state, the gel was subjected to stepdown shear rate characterisation. As a result the decrease in shear stress with time is quite small at $1000\ s^{-1}$. This characterises the thixotropic behaviour of the gel. Upon stepdown the shear stress increases with time and this characterises the rheopectic behaviour of the gel. The aligned platelets structure at high shear rates is being disrupted by the strong electric double layer (EDL) repulsive forces acting between platelets as a result of a much weaker hydrodynamic shear force at $10\ s^{-1}$ (Leong *et al*, 2021b). We discovered that this EDL repulsive force is responsible for the time-dependent behaviour of clay gels as it governed the development of the structure.

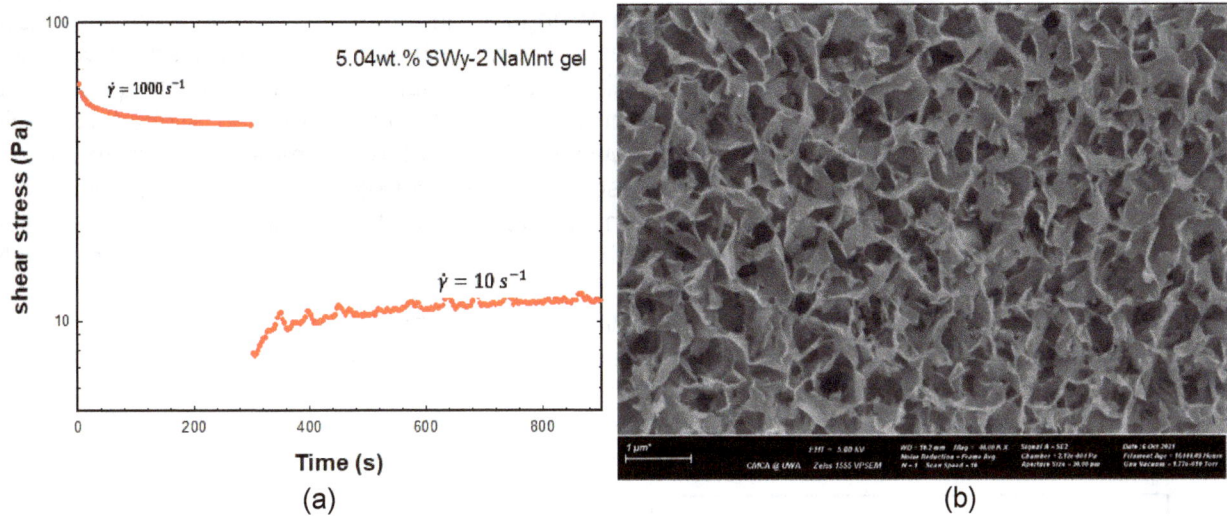

FIG 3 – (a) The stepdown shear rate result showed thixotropic behaviour (shear stress or viscosity reducing with shearing time at constant shear rate of 1000 s^{-1}) and rheopectic behaviour at low shear rate of 10 s^{-1} with shear stress or viscosity increasing with shearing time of 5 wt. per cent NaMnt gel. (b) The microstructure of 5 wt. per cent NaMnt gel.

The microstructure of a 5 wt. per cent NaMnt gel in Figure 3b showed an open structure formed by highly flexible platelets that are small and very thin. Even at such a low solids concentration, there were already enough particles to form a relatively strong 3D network structure that occupied the whole gel volume. The EDL repulsive forces between the faces of the platelets open up the structure and also is responsible for the time-dependent behaviour as it breaks bonds, move and orientate the platelets to form stronger bond (Du *et al*, 2018; Leong *et al*, 2018). The continued spontaneous increase in the ageing yield stress spanning several months is a reflection of this behaviour (Leong *et al*, 2021a). As the gel gets stronger its free energy state goes lower. The gel strengthening process is slow and a reflection of the gel driven thermodynamically to reach a minimum free energy state. The platelet-platelet bond is formed by positive edge-negative face attraction and the configuration of this interaction can range from 90 to 0 degree. At high salt concentration, the ageing yield stress did not show an immediate increase with wait time. High ionic strength greatly diminished the EDL force (Leong *et al*, 2021a) – see Figure 4. So attractive forces such as van der Waals dominated the particle-particle in the gel. Tailings with a high content of NaMnt or bentonite was anticipated to pose significant consolidation and filtration issues (Carman, 1938).

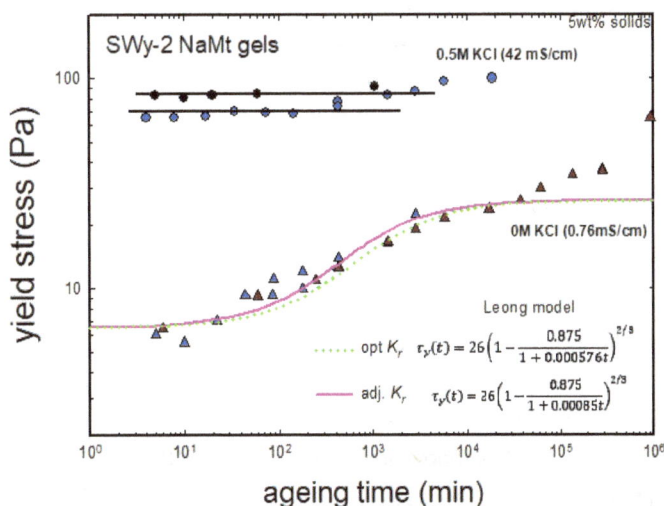

FIG 4 – The ageing behaviour of 5 wt. per cent NaMnt gels at low and high salt concentrations (Leong *et al*, 2021a).

The stepdown shear rate behaviour of Pilbara tailings C in Figure 5a showed no significant immediate increase in the stepdown shear stress. A similar behaviour was observed with the 39 wt. per cent kaolin suspension. The mine C tailings at 54 wt. per cent solids and with a zero yield stress after being treated with 0.1 dwb per cent composite additives showed slight thixotropic behaviour at 1000 s⁻¹. The shear stress at stepdown were below the rheometer measurement limit and showed no significant increase with time. These treated and untreated iron ore tailings displayed weak time-dependent behaviour despite their high solids loadings. This stepdown shear rate result also confirmed the dominant clay component in the tailings is kaolin. Note that it takes a relatively small critical amount of NaMnt to bring significant change to the suspension rheological behaviour. This critical concentration can be just a small fraction of the kaolin present. If the NaMnt concentration reached more than 3 wt. per cent based on NaMnt and water only, the rheology could alter markedly (Leong *et al*, 2023b). Note that the resistance to filtration for thixotropic or time-dependent mud is at least an order magnitude larger than colloidal clay (Carman, 1938).

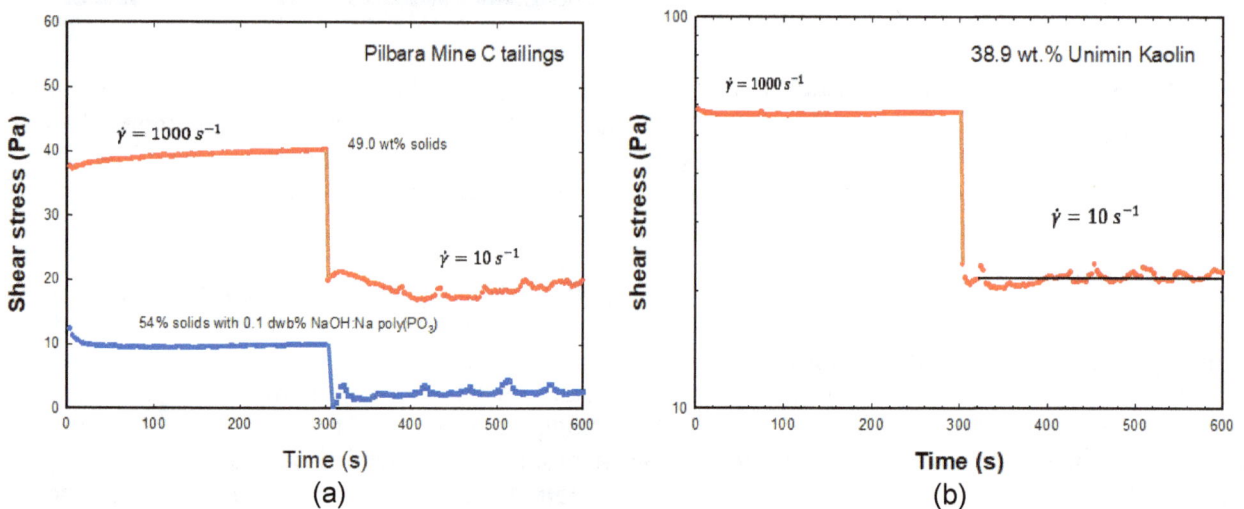

FIG 5 – The stepdown shear rate behaviour of (a) Pilbara Mine C tailings and (b) kaolin suspension.

CONCLUSIONS

Depending upon the chemistry of the tailings it is possible to reduce the yield stress to a manageable or almost free flowing consistency at a low cost with NaOH and polyphosphate composite additives, US$0.50 per ton. With some tailings, the cost of these composite additives can be reduced further by increasing the NaOH content for the same yield stress reduction and without significantly impacting on the pH of the tailings. The use of these additives provides the following benefits: (i) an increased iron ore throughput via the minimisation of downtown, (ii) a higher recovery of process water for reuse, (iii) a smaller environmental foot print and (iv) a safer storage of tailings in dams.

Time-dependent behaviour is not very significant with iron ore tailings confirming the lack of smectite clay present. The composite additives appeared to impart a small degree of thixotropy to the iron ore tailings.

REFERENCES

Carman, P C, 1938. Fundamental principles of industrial filtration, *Transaction of the Institute of Chemical Engineers*, 16.

Du, M, Liu, J, Clode, P L and Leong, Y K, 2018. Surface chemistry, rheology and microstructure of purified natural and synthetic hectorite suspensions, *Physical Chemistry Chemical Physics,* 20:19221–19233.

Leong, Y K, 2018. *Method of Rheology Control*, Innovation Patent No: 2018100304.

Leong, Y K, 2020a. Improved Mineral Separation of Tailings, Innovation Patent No: 2020900408.

Leong, Y K, 2020b. Controlling Rheology of Iron Ore Tailings Slurries, Innovation Patent No: 2020103937.

Leong, Y K, 2021a. Mining iron ore from tailings with minimal use of process water, in *Proceedings of the Iron Ore Conference 2021*, pp 654–660 (The Australasian Institute of Mining and Metallurgy: Melbourne).

Leong, Y K, 2021b. Controlling the rheology of iron ore slurries and tailings with surface chemistry for enhanced beneficiation performance and output, reduced pumping cost and safer tailings storage in dam, *Minerals Engineering*, 166:106874.

Leong, Y K, Drewitt, J and Bensley, S, 2019. Reducing the viscosity of concentrated iron ore slurries with composite additives for quality upgrade, reduced power, safer tailings storage and smaller environmental footprint, in *Proceedings of the Iron Ore Conference 2019*, pp 738–743 (The Australasian Institute of Mining and Metallurgy: Melbourne).

Leong, Y K, Liu, P, Au, P I, Clode, P and Liu, J, 2021b. Microstructure and Time-Dependent Behavior of STx-1b Calcium Montmorillonite Suspensions, *Clays and Clay Minerals*, 69:787–796.

Leong, Y K, Liu, P, Clode, P and Liu, J, 2021a. Ageing behaviour spanning months of NaMt, hectorite and Laponite gels: Surface forces and microstructure – a comprehensive analysis, *Colloids and Surfaces A: Physicochemical and Engineering Aspects,* 630:127543.

Leong, Y K, Liu, P, Liu, J, Clode, P and Huang, W, 2023b. Microstructure and time-dependent behavior of composite NaMnt-kaolin gels: Rigid-flexible platelet interactions and configurations, *Colloids and Surfaces A: Physicochemical and Engineering Aspects*, 656:130476.

Leong, Y K, Du, M, Au, P I, Clode, P and Liu J, 2018. Microstructure of sodium montmorillonite gels with long aging time scale, *Langmuir*, 34:9673–9682.

Leong, Y K, Bensley, S J, Drewett, J and Burkett, S, 2023a. A cost-effective tailings solution to rheology issues while meeting the environmental constraints using inexpensive additives, in *Proceedings of Paste 2023: 25th International Conference on Paste, Thickened and Filtered Tailings,* G W Wilson, N A Beier, D C Sego, A B Fourie and D Reid (eds), pp 580–587 (University of Alberta, Edmonton and Australian Centre for Geomechanics, Perth). https://doi.org/10.36487/ACG_repo/2355_43

Talmon, A M, Meshkati, E, Nabi, M, Simms, P and Nik, R M, 2023. Comparing various thixotropic models and their performance in predicting flow behavior of treated tailings, in *Proceedings of Paste 2023: 25th International Conference on Paste, Thickened and Filtered Tailings,* G W Wilson, N A Beier, D C Sego, A B Fourie and D Reid (eds), pp 588–600 (University of Alberta, Edmonton and Australian Centre for Geomechanics, Perth). https://doi.org/10.36487/ACG_repo/2355_44

A comparative study of two laboratory crushing techniques for a South African banded iron formation (BIF)

N Maistry[1] and A Singh[2]

1. Senior Engineer, Mintek, South Africa. Email: nicholem@mintek.co.za
2. MAusIMM, Senior Technical Specialist: Physical Separation, Mintek, South Africa.
   Email: ashmas@mintek.co.za

ABSTRACT

It is estimated that in less than a decade, high-grade iron ore reserves in South Africa will be depleted, leaving only low-grade ores with a dominant lithology of banded iron formation (BIF) hematite. BIF is a complex ore reserve as it is typically liberated below 1 mm, resulting in high energy costs. However, due to its abundance, its beneficiation potential should be evaluated to maximise recovery at an acceptable product grade. To develop a beneficiation program for BIF hematite, the first step is to determine the effect of two laboratory crushing techniques on mineralogical characteristics, specifically liberation and shape. For this study, the cone crusher and high pressure grinding rolls (HPGR) were used.

The Electron MicroProbe Analysis (EMPA) and Mineral Liberation Analyser (MLA) were used to compare the mineralogical characteristics. The liberation characteristics for hematite are similar for both crushing techniques; however, HPGR preferentially liberates quartz. HPGR achieved 8–12 per cent additional quartz liberation in the greater than 80 per cent and 100 per cent liberation classes respectively and specifically in +75 µm size range. Improved quartz liberation is associated with lower energy requirements for downstream milling and improved separation efficiencies for beneficiation units that operate in this size range. Both samples displayed similar shape characteristics; most particles were elongated smooth and the rest were elongated angular. Good hematite liberation is expected at grinds of 212 µm to 150 µm. Further consideration should be given to throughputs, energy efficiency and downstream equipment selection prior to selecting a crushing technique.

INTRODUCTION

In recent years, due to stricter environmental laws, the steel industry has been focused on reducing carbon emissions. Thus, there has been an increased demand for high-grade iron ores as they lower coke required during processing. These high-grade iron ores are usually produced by Australia, Brazil and South Africa (Creamer Media Reporter, 2017).

Australia suppliers (Rio Tinto, BHP Billiton and Fortescue Metals) are the lowest cost producers, followed by Brazil (Vale) as they supply higher grade ores of simple mineralogy that are easier and cheaper to process. South Africa (SA) is significantly higher on the cost curve; however, SA iron ore has a unique trace element mix that complements the concentrates from the major producers. Thus, it is highly sought after at a premium price due to:

- high iron content, with an average of 64.10 per cent Fe (Creamer Media Reporter, 2018)
- contains less deleterious elements
- more competent and less friable resulting in less product breakdown and fines generation.

Unfortunately, it is forecasted that South Africa's high-grade ores will be exhausted by 2033. The remaining low-grade ores have a major lithology of banded iron formation (BIF) hematite, which is typically liberated at 1 mm. Thus, BIF hematite is associated with high energy costs, which lowers the demand for this ore. However, the beneficiation potential of BIF hematite should be investigated as it is available in abundance. This is an opportunity to maximise recovery and produce an acceptable product grade as well as increase life-of-mine (LOM). This first step in development of a beneficiation program for BIF hematite is to determine the effect of two laboratory crushing techniques on mineralogical characteristics, specifically liberation and shape. The crushing techniques used for this study were the cone crusher and high pressure grinding rolls (HPGR).

LITERATURE REVIEW

Types of crushers

Crushing, which is the first stage in comminution, is achieved through compression or impact of ore against a rigid surface to ensure liberation of valuable minerals. It is typically a dry process, which consists of two to three stages in an open, or closed circuit. The requirements of the crushing circuit are determined by the ore characteristics (strength, hardness etc), handling capacity or required ratio for downstream beneficiation. The three stages of crushing are primary, secondary and tertiary. Common crushers used for primary crushing are jaw and gyratory crushers, whereas cone, roll and impact crushers are used for secondary and tertiary crushing. Although secondary and tertiary crushing use the same types of crushers, the two differences are that a closer set is used and tertiary crushers are used in closed circuits (Haldar, 2018).

Cone crusher

A cone crusher (Figure 1) is a gyratory crusher, which has a conical crushing head with the smaller diameter at the top. The crushing bowl has a similar shape to the crushing head, but it is inverted to create a crushing chamber that is wider at the top than at the bottom (Wescone Distribution, 2010).

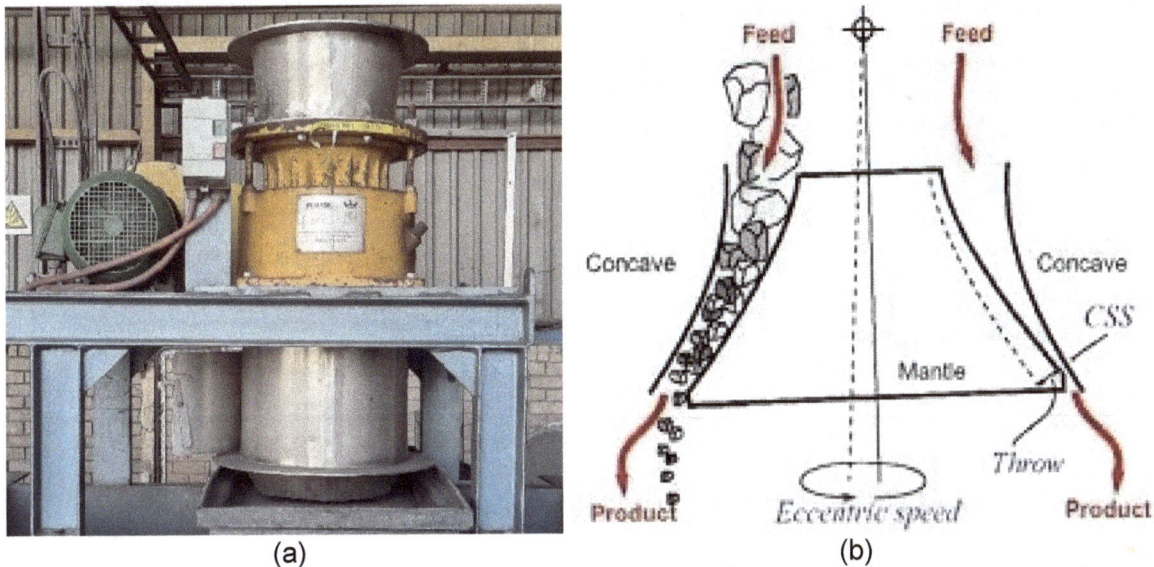

FIG 1 – Cone crusher: (a) at Mintek (Wescone Model: 300/2); (b) operating principle (Svedensten, 2016).

The oscillating motion of the head provides the crushing action. The main shaft of the crusher is rotated by the drive about the vertical axis, causing the head to gyrate in the crushing chamber (Wescone Distribution, 2010). Simultaneous crushing occurs on opposite sides – top left and bottom right. The operating principle is shown in Figure 1(b).

High pressure grinding rolls (HPGR)

The HPGR (Figure 2) consists of two counter-rotating rolls that are mounted on a frame. One roll floats on the rails perpendicular to the roll axis and it is positioned using pneumatic hydraulic springs, whereas the other roll is fixed, as shown in Figure 2(c). The non-liner stiffness response of the spring provides the force required in the compressed bed, and it is determined by various parameters such as initial gap setting, initial nitrogen pressure in the pneumatic accumulators and the initial hydraulic oil pressure.

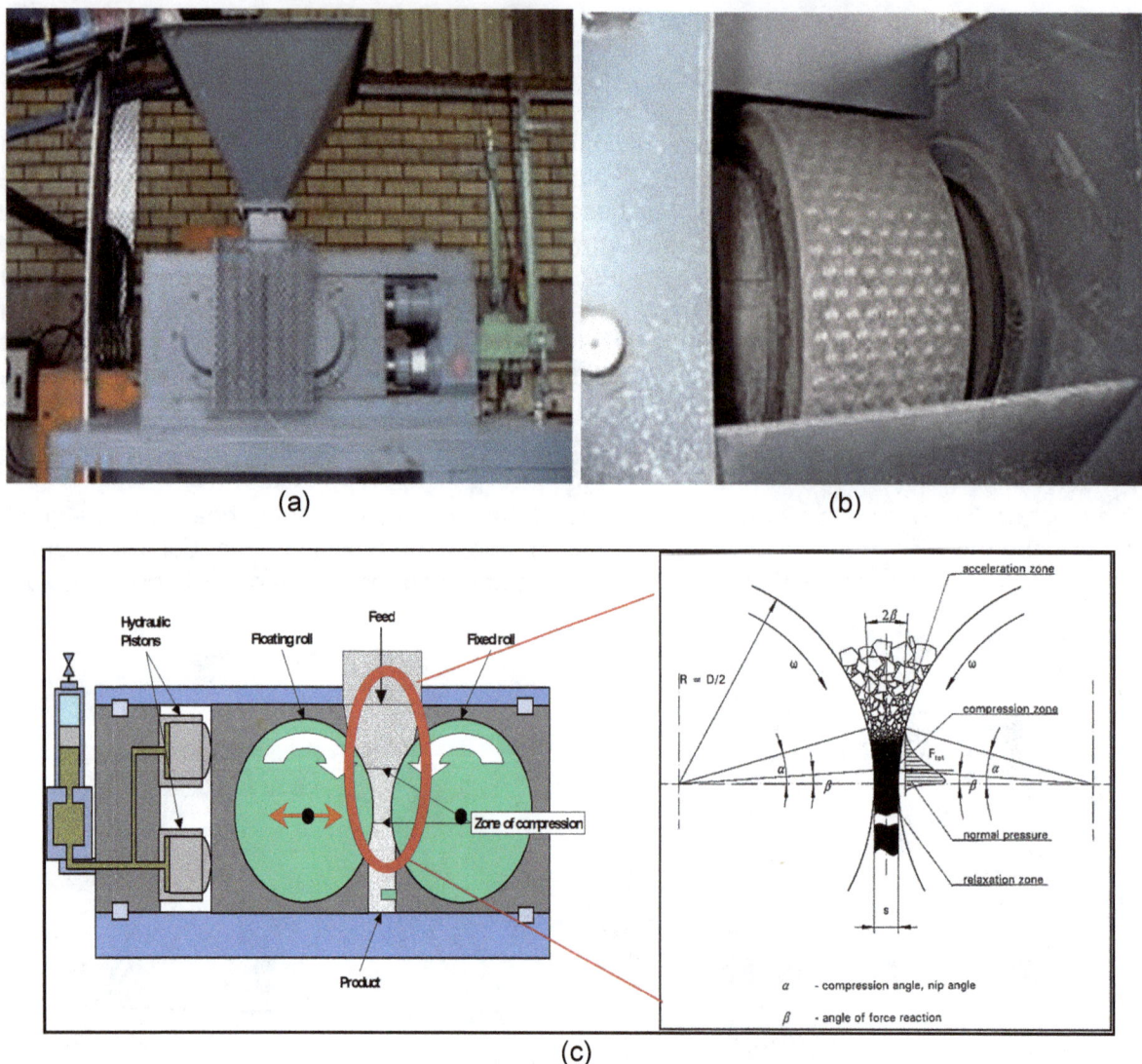

FIG 2 – Laboratory HPGR (a) and (b) at Mintek; (c) operating principle.

HPGR and preferential liberation

Daniel (2007) evaluated the energy efficiency of a HPGR/ball milling circuit compared to a conventional SAG/ball milling circuit for three different ore types, namely lead/zinc, bauxite/aluminium and platinum/chromite. Evaluation of the circuit products, via Mineral Liberation Analysis, showed no conclusive proof that preferential liberation is achieved with HPGR. Palm (2012) made a similar finding, as there were minor differences observed in the liberation characteristics of a sphalerite ore crushed by HPGR versus a cone crusher. Conversely, Chapman et al (2013) found that HPGR significantly improved percentage liberation of Platinum Group Minerals (PGM). However, there is no conclusive evidence to confirm that HPGR improve flotation kinetics and increase metal recovery for sphalerite and sulfides. Therefore, the benefits of HPGR, in terms of material fracture along grain boundaries to improve liberation, is inconclusive.

However, Daniel (2007) observed extensive particle micro-cracking in the HPGR product, which was not present in conventionally crushed products. The presence of micro-cracks may weaken particles, resulting in a reduced energy requirement for the ball mill (Daniel, 2007; Barani and Balochi, 2016; Liu *et al*, 2017; Van der Meer and Maphosa, 2012). The benefits of micro-cracks are not fully understood, but the creation of more porous material is expected to improve leaching of gold, silver and zinc ores (Drosdiak, 2011; Ghorbani et al, 2013). Figure 3 shows an example of a micro-crack generated by HPGR crushing a siliceous brecciated hematite sample. Chile's Los Colorades iron ore operation used HPGR to generate -7 mm material for coarse magnetite pre-concentration, with the pre-concentration subjected to ball milling for further magnetic separation. Test work conducted by Van der Meer and Maphosa (2012) showed an 18 per cent increase in Bond Ball mill Work Index

(BBWI) and a 30 per cent increase in ball mill capacity, as well as a higher quality magnetic pre-concentrate when compared to the cone crushed product.

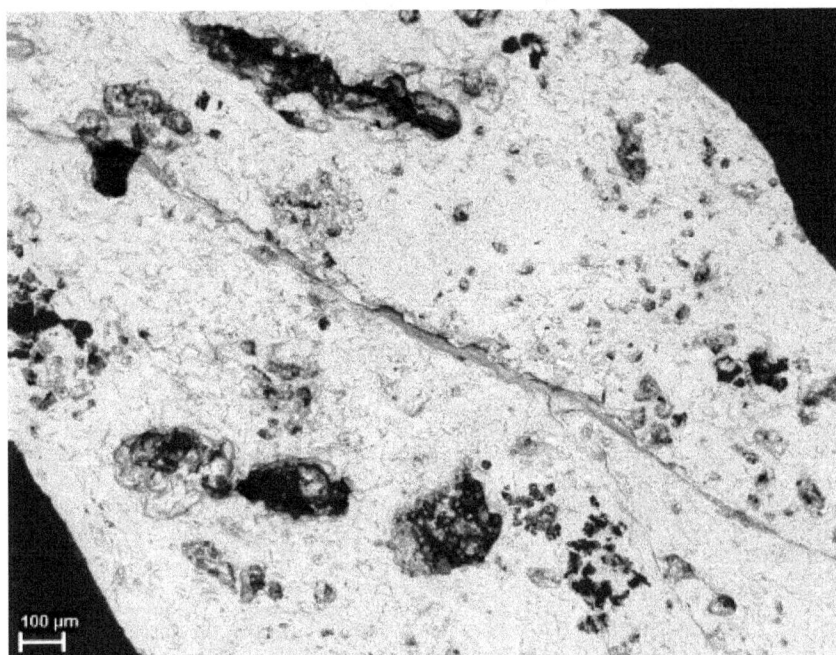

FIG 3 – Micro-crack generated by HPGR crushing a siliceous brecciated hematite sample.

Literature comparing the liberation characteristics of HPGR and cone crushing, specifically for hematite could not be found. Therefore, to develop a beneficiation plan for BIF hematite, it is important to establish if there are any mineralogical benefits of HPGR crushing. Thus, the intention of this paper is to provide a comparative assessment of two crushing methods for BIF hematite.

EXPERIMENTAL

Physical characteristics of the ore

The BIF hematite sample originated from a deposit in the Limpopo Province, South Africa. The sample was dry screened at 8 mm and 1.18 mm to determine the particle size distribution. Table 1 shows that 86.8 per cent of particles are greater than 1.18 mm and 49.2 per cent reports to the -8+1.18 mm size fraction, indicating that the sample is predominantly coarse. The sample has a top size of 40 mm, thus, coarse comminution bench scale testing (eg CWI) was not possible as -75+50 mm particles are required for testing. This would have established the energy requirement for a given crushing operation.

TABLE 1

Particle size distribution of sample.

Size fraction (mm)	Mass (%)
+8	37.5
-8+1.18	49.2
-1.18	13.2
Total (calculated)	**100.0**

Figure 4 shows the texture of the BIF hematite sample in the -8+1.18 mm size fraction. Due to the laminated and intergrown nature of the material, it is not suitable for coarse beneficiation.

FIG 4 – Image of -8+1.18 mm BIF hematite from Limpopo.

Sample preparation

Each size fraction was separated into half and both +8 mm size fractions were individually staged crushed to -12.5 mm using a jaw crusher. Each crusher product was composited with the respective halves of -8+1.18 mm and -1.18 mm size fractions, subjected to either laboratory cone or HPGR crushing. The crusher products were individually blended and subsampled for head assay, size by assay and mineralogical characterisation. A subsample of the feed was removed for mineralogical analysis prior to separating the sample.

Cone crushing

Approximately 131 kg of material with a top size of 12.5 mm was crushed to -1.18 mm using the laboratory Wescone crusher (Model: W300/2).

HPGR crushing

Approximately 131 kg of material with a top size of 12.5 mm was crushed to -1.18 mm using the Polysius LABWAL laboratory HPGR. The material was fed through a hopper to maintain a constant flow of feed material. The rollers are 250 mm in diameter and 100 mm in width. The unit is fitted with studded rolls, each driven by a separate motor that delivers a speed of 0.68 m/s.

The working gap for HPGR was set to 3 mm. The hydraulic oil pressure was set at 90 bar based on the fine grind required, with the nitrogen pressure set at two-thirds of the hydraulic oil pressure, at 60 bar.

Size by assay of the crusher products

The HPGR and cone crushing -1.18 mm products were independently subjected to a standard root 2 sizing analysis (850 µm, 600 µm, 425 µm, 300 µm, 212 µm, 150 µm, 75 µm, 53 µm and 38 µm) to determine the particle size distribution. Each size fraction produced was prepared for chemical analysis to determine the Fe and gangue deportment.

Mineralogy

Electron MicroProbe Analysis (EMPA)

A representative normal mount polished section of the feed was analysed on the Cameca SX 50 electron microprobe using wavelength dispersive spectrometry. The system was calibrated using pure oxide reference standards and was cross-referenced with the hematite standards during measurement to ensure reproducibility and quality of results. Oxygen content was calculated by stoichiometry. An accelerating voltage of 20 kV was used with a beam current of 30 nA. Two types of analysis were performed at the same coordinate as follows:

- Point analysis was conducted with a maximum 5 μm spot size in what appeared to be clean hematite. Low Fe totals in point analysis results indicates impure composition.

- Area analysis was conducted with a raster scan over an area of 100 μm × 75 μm. Area analysis with high Fe and low totals indicate the presence of porosity or fracturing between hematite and other grains (if present). Low Fe totals in area analysis results indicates mixed composition of particles (ie presence of other phases).

Particle Tracking Analysis (PTA)

Based on the size distribution, three size fractions of the HPGR and cone crushing products were selected for mineralogical characterisation via Particle Tracking Analysis (PTA). Each size fraction was characterised in terms of mineral, density and size distribution of the sample. The liberation characteristics of each mineral by size class were also reported.

PTA was conducted on the +850 μm, -850+212 μm and -212 μm size fractions. The size fractions were prepared into sections, carbon coated and analysed using autoSEM Mineral Liberation Analysis (MLA). Each size fraction had 1–2 sections for statistical confidence.

The results obtained from MLA were verified using the particle size distribution and chemical analysis of the feed.

RESULTS

Size by assay of crusher products

Figure 5 shows the discrete Fe grade across size for HPGR and cone crushing products. This graph indicates that the Fe grade remains relatively constant across all the size ranges, with the cone crushing product having a slightly higher head grade – 46.1 per cent versus 44.0 per cent – however, still within a 5 per cent error.

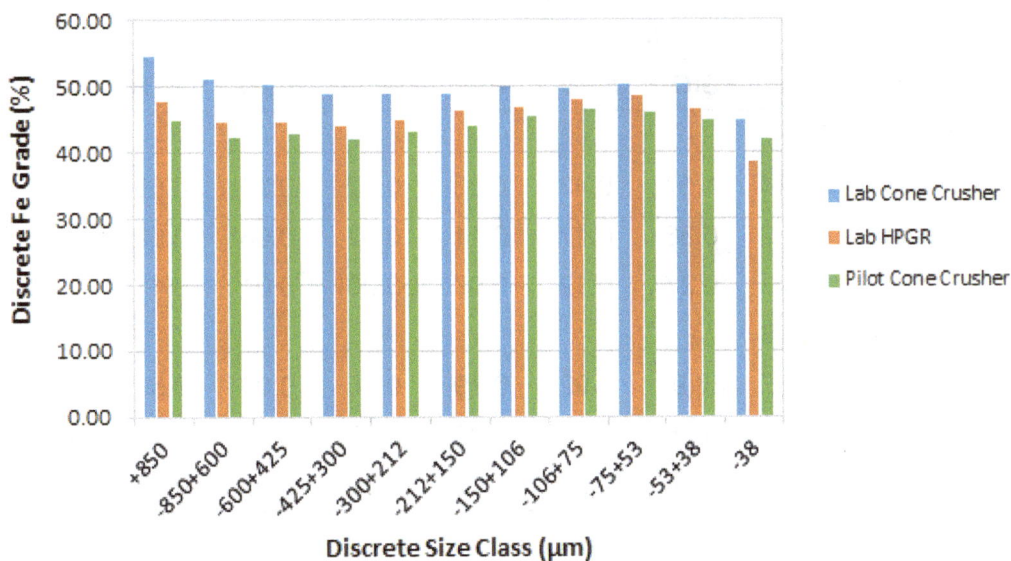

FIG 5 – Discrete Fe grade across size for laboratory crushing products.

Figure 6 shows the discrete Fe deportment closely follows mall pull due to the relatively constant Fe grade across the size ranges. It indicates a bimodal mass distribution for both crusher products, with peaks in the -850+600 μm and 38 μm size fractions. The cone crusher product has 8.3 per cent more material reporting to the -850+300 μm size range; however, this can be offset by the 11.1 per cent additional material reporting to the +850 μm size fraction for the HPGR product. This indicates that HPGR produces coarser material, especially in the +425 μm size range. Furthermore, approximately 18 per cent of Fe reports to the slimes, thus, magnetic separation is required to improve overall Fe recovery.

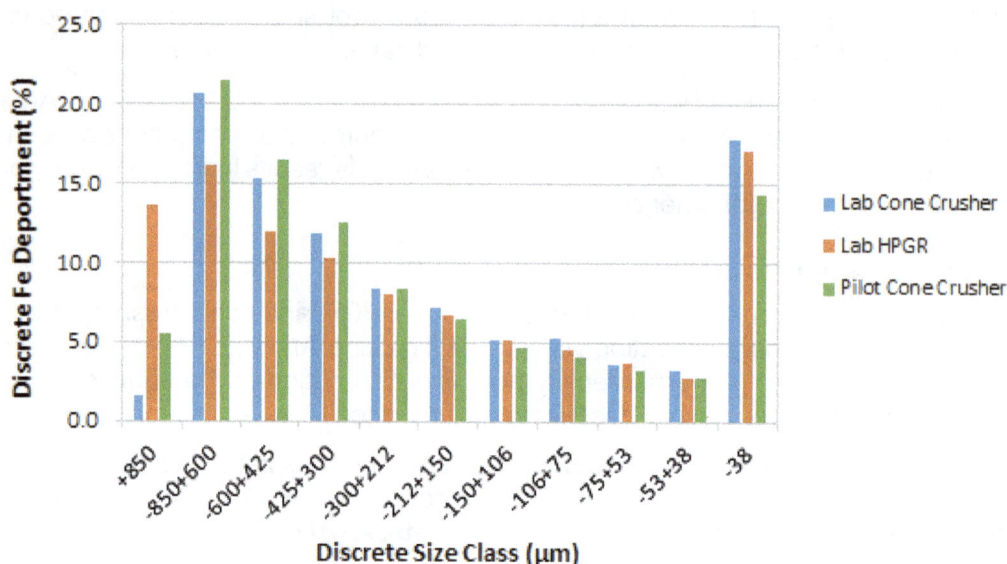

FIG 6 – Discrete mass and Fe deportment across size for laboratory crushing products.

Mineralogical characterisation

Electron MicroProbe Analysis (EMPA)

EMPA was conducted on the -1.18 mm feed in order to classify the hematite phases correctly and then reliably assign Fe-grade and density to each phase identified. The results are presented in Table 2. The point analysis indicates the average Fe content for hematite is 69.23 per cent and the area analysis shows the Fe grade is significantly lower at 62.44 per cent.

TABLE 2

EMPA determined chemistry.

	N = 92	Fe	SiO$_2$	Al$_2$O$_3$	MgO	MnO	CaO	TiO$_2$	K$_2$O	Total
Point Analysis	Minimum	67.87	-	-	-	-	-	-	-	98.61
	Maximum	70.14	1.57	0.43	0.33	0.27	0.20	0.09	0.03	100.78
	Average	69.23	0.36	0.17	0.11	0.07	0.06	0.09	0.03	88.44
	Std Dev.	0.39	0.27	0.12	0.07	0.06	0.04	0.11	0.00	0.46
Area Analysis	Minimum	13.03	-	-	-	-	-	-	-	98.61
	Maximum	70.05	68.96	0.92	0.19	0.65	0.09	0.50	0.22	100.78
	Average	62.44	8.85	0.33	0.06	0.09	0.05	0.11	0.09	88.44
	Std Dev.	12.08	17.52	0.25	0.03	0.10	0.02	0.15	0.09	0.46

This indicates that hematite is not clean and includes disseminated silicates in the -100+75 µm size range as shown in Figure 7. Low totals can also be attributed to the presence of pores, which adversely affect gravity separation processes due to reduced particle density. The area analysis indicates the achievable limit of the concentrate Fe grade in the -100+75 µm size range for this BIF hematite sample.

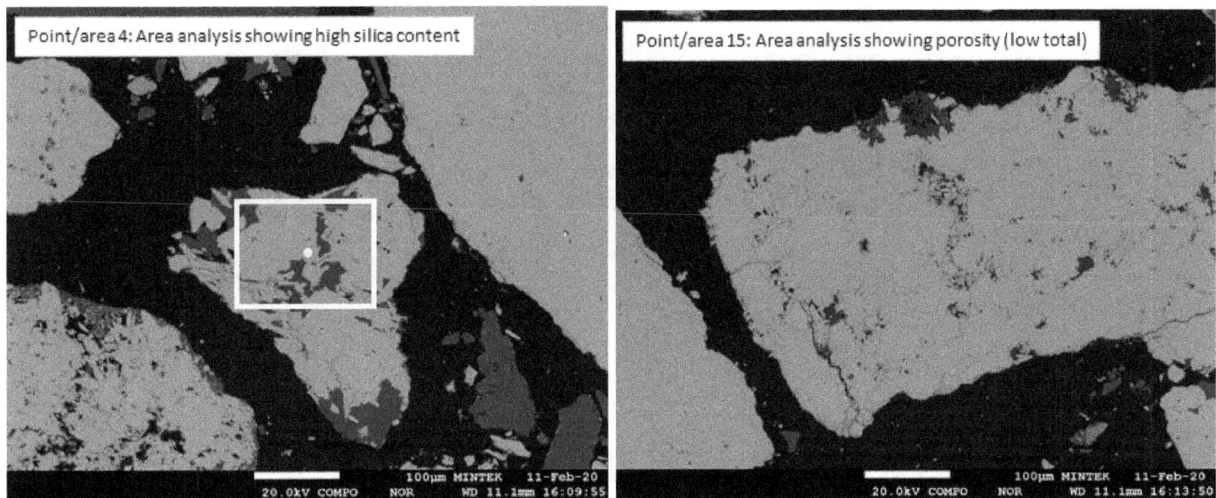

FIG 7 – EMPA point and area analysis showing porous hematite and high silica content.

The Fe grade for hematite, obtained via EMPA, varied significantly, thus, results were evaluated in terms of density and chemistry of the different hematite phases present in the sample, as shown in Table 3.

TABLE 3

Three defined hematite phases.

Mineral	Density	Fe (%)	Si (%)	Al (%)	Mg (%)	Mn (%)	Ca (%)	K (%)	O (%)	C (%)	H (%)
Hematite phase A	5.15	69.44	0.33	0.20	0.00	0.00	0.07	0.06	30.34	0.00	0.00
Hematite phase B	5.12	67.22	0.53	0.45	0.00	0.00	0.00	0.00	29.90	0.00	0.00
Hematite phase C	4.85	62.44	4.65	0.18	0.04	0.07	0.04	0.07	28.88	0.00	0.00

Particle Tracking Analysis (PTA)

Feed mineral abundances and liberation characteristics

The mineral abundance and liberation characteristics of the -1.18 mm crushing products are shown in Table 4. The results indicate that hematite phase C is the dominant hematite phase, with a mineral mass yield of 64.8–72.0 per cent, followed by hematite phase B, with a ~3.2 per cent mineral mass yield. The difference in head grade may be due to varying composition and size of the bands in the coarser size fraction of the feed. The differences are more apparent for the gangue as it is present in smaller amounts. As shown in Table 3, the hematite product grade for the sample with a top size of 1.18 mm will not exceed 62.44 per cent Fe.

The cone crusher product has a higher measured Fe head grade, thus it contains more hematite than the HPGR product. Magnetite is present in small quantities (1.5–1.8 per cent) and quartz is the dominant gangue mineral (23.2–30 per cent).

Table 4 also shows two liberation classes for each sample namely 100 per cent completely liberated (100 per cent of the particle area constitutes the mineral) and greater than or equal to 80 per cent liberated (80 per cent or more of the particle's area is consumed by the mineral). Both samples have similar liberation characteristics with 46–47 per cent completely liberated and 80–81 per cent for the ≥80 per cent liberated class. The two main gangue minerals (quartz and carbonates) follow a similar trend. However, the HPGR product shows improved quartz liberation of 8–12 per cent for both liberation classes, indicating that this method of crushing offers superior quartz liberation.

TABLE 4

Mineral abundance and liberation of the laboratory crushing products.

Mineral Name	Density (g/cm³)	Cone crusher product			HPGR crusher product		
		Mass (%)	Mass (%) in liberation class		Mass (%)	Mass (%) in liberation class	
			≥80%	100%		≥80%	100%
Hematite A	5.15	0.0			0.0		
Hematite B	5.12	3.2	80.8	46.2	3.2	79.9	47.0
Hematite C	4.85	72.1	47.9	20.1	64.8	60.6	29.0
Quartz	2.65	23.2			30.0		
Magnetite	5.15	1.4			1.8		
Others	-	0.1			0.2		
Total	4.07	100.0			100.0		

As shown in Table 5, the liberation characteristics of hematite phase C by size are similar for both crusher products. Approximately 80 per cent of hematite phase C is liberated in the 80 per cent liberated class for both samples, with the +212 µm size fraction being less liberated than the total. Furthermore, for ~80 per cent of the hematite phase C to be 80 per cent, 90 per cent and 100 per cent liberated classes, grind sizes of 212/150 µm; 106 µm and 38 µm would be required, respectively.

TABLE 5

Hematite phase C liberation by size of laboratory crushing products.

Size class (µm)	Cone crusher product Cumulative mass% in liberation class			HPGR crusher product Cumulative mass% in liberation class		
	80	90	100	80	90	100
-300+212	79.1	66.3	40.6	77.4	67.5	48.0
-212+150	85.6	74.7	48.8	77.3	70.9	45.2
-150+106	82.3	75.4	46.5	81.9	74.5	46.6
-106+75	84.5	80.0	59.7	84.0	80.1	57.7
-75+53	88.3	81.4	65.8	88.0	79.9	60.6
-53+38	91.4	86.7	77.1	88.1	82.2	72.6
-38+25	94.3	91.1	86.4	90.4	86.7	79.4

The liberation characteristics of quartz by size for both crusher products, presented in Table 6, indicates the quartz liberation is greater in the HPGR product. For the 80 per cent liberation class, HPGR crushing improves the quartz liberation in particles coarser than 75 µm. For the -75 µm size fraction, both crushing methods provide similar quartz liberation characteristics. Preferential liberation at coarser sizes using the HPGR will reduce energy requirements for downstream milling.

TABLE 6

Quartz liberation by size of the laboratory crushing products.

Size class (µm)	% Mass reporting to ≥80% liberation class	
	Cone crusher product	HPGR product
-1180+850	24.21	38.37
-850+600	31.23	48.06
-600+425	35.34	52.79
-425+300	42.16	57.41
-300+212	44.34	60.80
-212+150	53.09	47.88
-150+106	45.28	60.04
-106+75	52.36	59.41
-75+53	67.11	68.98
-53+38	71.11	73.11
-38+25	84.43	81.06
-25+18	89.71	88.88
-18+13	93.59	93.74
-13+9	98.13	97.63
-9+6	99.15	98.99
-6+1	99.71	99.61
Total	**47.92**	**60.64**

Density distribution

The density by size distribution for the laboratory cone crusher and HPGR products presented in Figure 8. These figures are a visual representation of the contribution of each size class of the sample, highlighting the mass percent and amount of near-density fractions present. Both products have a similar profile, but the cone crusher product has higher proportions of material in the intermediate SGs of 3.40–4.75 for the coarser size fractions. This further demonstrates quartz dilution of the coarser particles for cone crusher product.

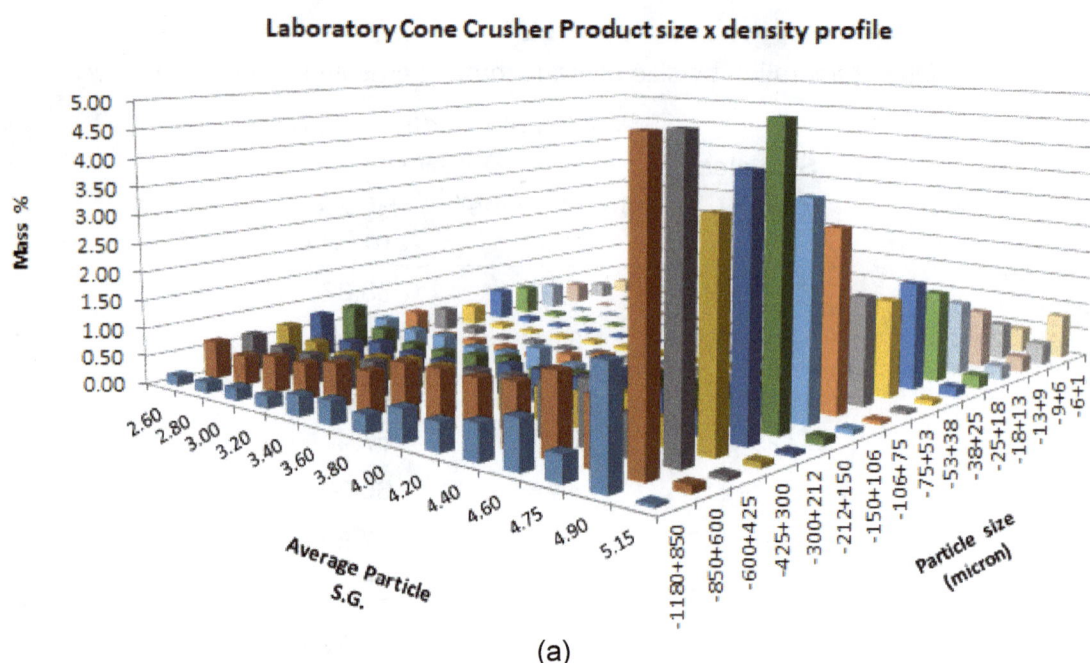

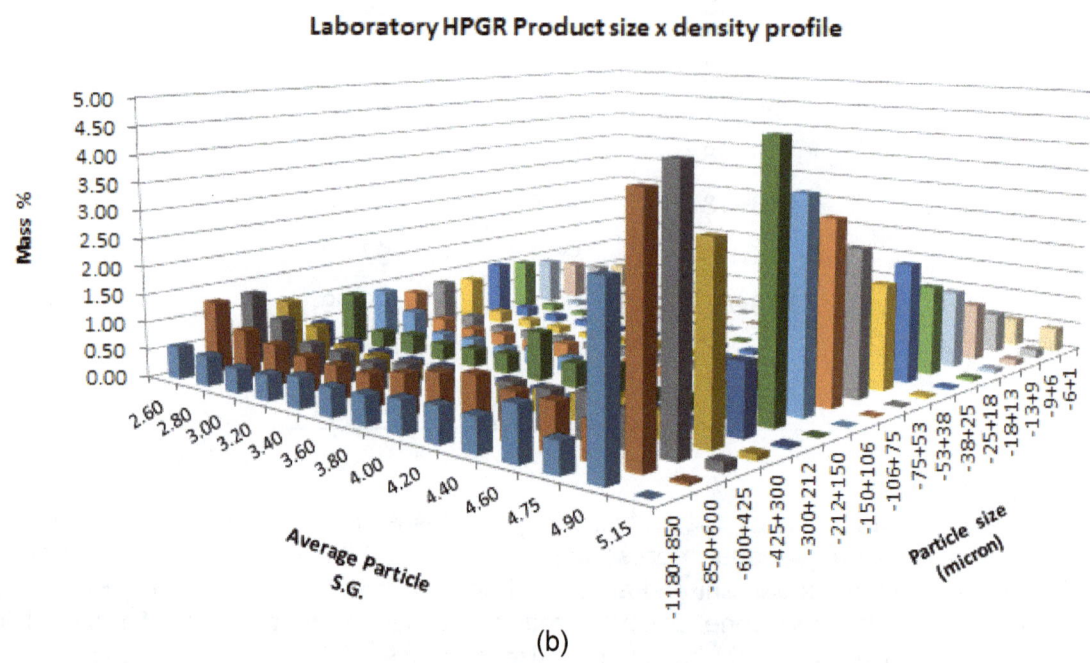

FIG 8 – Feed size by density for (a) cone crusher product (b) HPGR product.

Shape characteristics

The shape of the laboratory cone crusher and HPGR products were described based on roundness and aspect ratio.

Roundness is the ratio of the cross-sectional area of the particle to that of a circle or sphere with a diameter equivalent to the maximum Feret diameter (Femax) of the particle (Little *et al*, 2017), as shown in Equation 1.

$$Roundness = \frac{4.Area}{\pi.Fe_{max}^2} \qquad (1)$$

An aspect ratio is a proportional relationship between an image's width and height. Aspect ratios are given as a formula of width to height, eg 3:2 (0.6).

Five different shapes classes were characterised based on the roundness and aspect ratio and is presented in Table 7.

TABLE 7

Classification of sizes classes based on roundness and aspect ratio.

Shape description	Roundness	Aspect ratio
Angular	<0.4	>0.5
Elongated angular	<0.4	<0.5
Elongated smooth	0.4–0.75	>0.5
Equant	0.4–0.75	<0.5
Round	>0.75	>0.5

Figure 9, which gives the shape characteristics for cone crusher and HPGR products, indicates both products have similar shape characteristics with the dominant shape being elongated smooth (71–72 per cent) followed by elongated angular (16–19 per cent).

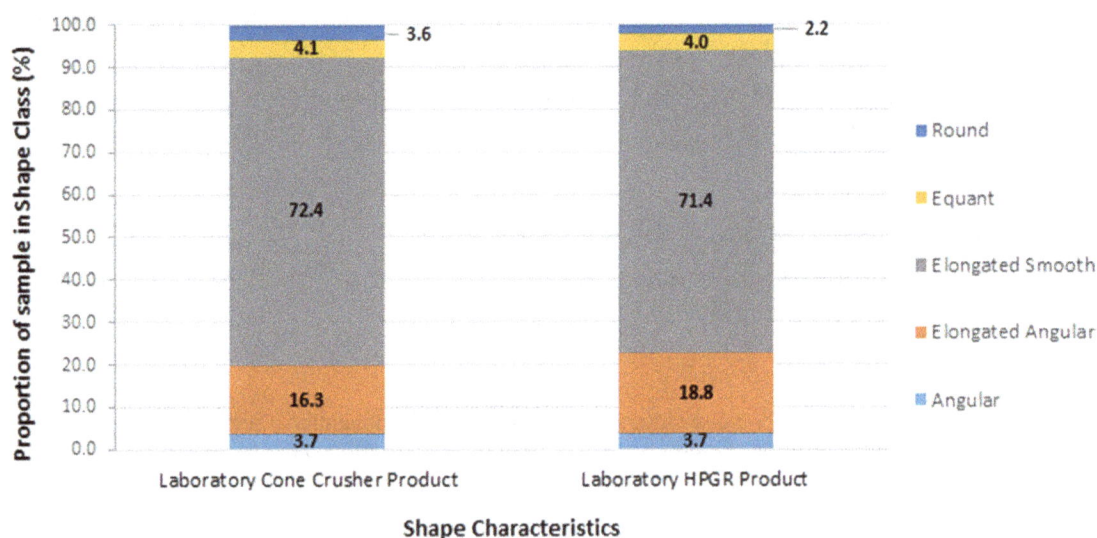

FIG 9 – Shape characteristics of the laboratory crushing products.

DISCUSSION

Preferential liberation

Improved quartz liberation is observed for the 80 per cent liberation class in particles coarser than 75 µm, indicating that HPGR preferentially liberates quartz when compared to cone crushing. 95 per cent of the sample contains hematite and quartz, thus, an improvement in quartz liberation was expected to be associated with an improvement in hematite liberation in the same liberation class. However, this trend was only observed for hematite phase C in the +212 µm fraction for the 90 per cent liberation class. This may be due to a smaller proportion of quartz than hematite, resulting in reduced effect on hematite's liberation characteristics. Improved quartz liberation is expected to improve separation efficiencies for beneficiation units that operate in this size range, eg gravity separation and magnetic separation.

Approximately 80 per cent of the hematite reports to the 212/150 µm liberation class, which corresponds to a study conducted by Vidyadhar *et al* (2019) on the beneficiation of BIF hematite from India. It revealed that a grind of -150 µm is required to liberate the gangue from the iron oxides. During the study, a sample with a grade of 38.97 per cent underwent gravity separation, magnetic separation and flotation. The resulting product grade was 63.7 per cent, with a yield and recovery of 28.0 per cent and 45.8 per cent, respectively. The low recovery indicates that there may be Fe losses to slimes. However, the results and literature indicate that there is potential for BIF hematite to be beneficiated to produce a reasonable product grade to maximise resource utilisation.

Scale-up and other considerations

Drosdiak (2011) states that pilot testing is the only method to ensure reliable scale-up, sizing and selection of a HPGR. According to Daniel (2007), for a laboratory HPGR to relate to a pilot unit, the ratio of the diameter of the coarse particle to the working gap expected for a full scale unit should be targeted in the laboratory. This prevents a weak particle bed and compression zone, which may result in different particle interactions and interparticle comminution in laboratory and pilot units. However, since this study is not focused on design and sizing of crushers, the trend of preferential silicate liberation is expected to be observed on a pilot scale.

The effects of throughputs, energy efficiency and equipment wear rates regarding HPGR crushing should be investigated.

CONCLUSIONS AND RECOMMENDATIONS

Comparative laboratory crushing revealed no preferential hematite liberation characteristics. However, HPGR shows 8–12 per cent additional quartz liberation, specifically in the +75 µm size fraction, which indicates a potential decrease in energy consumption in downstream milling. Both crushing products have similar shape characteristics, with the dominant shape being elongated smooth, followed by elongated angular.

It is recommend that the selection of crushing techniques should also consider throughputs, energy efficiency and equipment wear rates.

ACKNOWLEDGEMENTS

The authors would like to acknowledge Mintek for funding the research project and providing the facilities and resources required to complete this study. The authors would also like to thank Mintek's mineralogical division for their contribution to this research.

REFERENCES

Barani, K and Balochi, H, 2016. A comparative study of the effect of using conventional and high pressure grinding rolls crushing on ball mill grinding kinetics of an iron ore, *Physiochemical Problems of Mineral Processing*, 52(2):920–931.

Chapman, N A, Shackleton, N J, Malysiak, V and O'Connor, C T, 2013. Comparative study of the use of HPGR and conventional wet and dry grinding methods on flotation of base metal sulphides and PGMs, *The Journal of Southern African Institute of Mining and Metallurgy*, 113(5):407–413.

Creamer Media Reporter, 2017, June 23. Real Economy Insight 2017: Iron-Ore, Creamer Media's Research Channel Africa. Available from: <https://www.researchchannel.co.za/login.php?lir=1&url=/article/rei-2017-iron-ore-2017-06-23> [Accessed: July 27, 2017].

Creamer Media Reporter, 2018, May 2. Iron-Ore 2018: A review of the iron-ore sector, Creamer Media's Research Channel Africa. Available from: <http://www.researchchannel.co.za/login.php?&url=/article/iron-ore-2018-a-review-of-the-iron-ore-sector-2018-05-02/searchString:iron%20ore%20may%202018> [Accessed: May 15, 2018].

Daniel, M, 2007. Energy efficient mineral liberation using HPGR technology, PhD thesis, School of Engineering, University of Queensland.

Drosdiak, J, 2011. A pilot-scale examination of a novel high pressure grinding roll/stirred mill comminution circuit for hard-rock mining applications, MSc thesis, The University of British Columbia.

Ghorbani, A, Mainza, A, Petersen, J, Becker, M, Franzidis, J-P and Kalala, J, 2013. Investigation of particles with high crack density produced by HPGR and its effect on the redistribution of particle size fraction in heaps, *Minerals Engineering*, 43–44:44–51.

Haldar, S K, 2018. *Mineral Exploration: Principles and Applications,* second edition (Elsevier: Amsterdam).

Little, L, Mainza, A, Becker, M and Wiese, J, 2017. Fine grinding: how mill type affects particle shape characteristics and mineral liberation, *Minerals Engineering*, 111:148–157.

Liu, L, Tan, Q, Liu, L, Li, W and Lv, L, 2017. Comparison of grinding characteristics in high-pressure grinding roller (HPGR) and cone crusher (CC), *Physiochemical Problems of Minerals Processing*, 53(2):1009–1022.

Palm, N S, 2012. The effect of using different comminution procedure on the flotation of sphelarite, *Minerals Engineering*, 24(8):731–736.

Svedensten, P, 2016. *Principles of Mechanical Crushing* (Quarry Academy).

Van der Meer, F and Maphosa, W, 2012. High pressure grinding moving ahead in copper, iron and gold processing, *The Journal of the Southern African Institute of Mining and Metallurgy*, 112:637–647.

Vidyadhar, A, Singh, A K, Srivastava, A, Nayak, B, Rao, K V and Das, A, 2019. Beneficiation of banded hematite quartzite from Meghatuburu mine, eastern India, in *Proceedings of the XI International Seminar on Minerals Processing Technology (MPT-2010)* (eds: R Singh, A Das, P K Banerjee, K K Bhattacharyya and N G Goswami), pp 583–589 (NML Jamshedpur).

Wescone Distribution, 2010. Wescone W300 series 2 crushers: operating and maintenance instruction manual, Bentley, Australia: Wescone Distribution Pty Ltd.

Comparison between ball mill and vertical stirred mill for the fine grinding of a low-grade iron ore

J Mesquita[1], E Kleiderer[2], L Southavy[3], R Belissont[4], Y Basselin[5], H Turrer[6], M Badawi[7] and Y Foucaud[8]

1. Mining and Mineral Processing Senior Research Engineer, ArcelorMittal Maizières Research, Maizières Lès Metz 57208, France. Email: josue.mesquita@arcelormittal.com
2. Mineral Processing Technician, ArcelorMittal Maizières Research, Maizières Lès Metz 57208, France. Email: elena.kleiderer@arcelormittal.com
3. Mineral Processing Technician, ArcelorMittal Maizières Research, Maizières Lès Metz 57208, France. Email: leslie.southavy@arcelormittal.com
4. Senior Characterization Research Engineer, ArcelorMittal Maizières Research, Maizières Lès Metz 57208, France. Email: remi.belissont@arcelormittal.com
5. Characterization Technician, ArcelorMittal Maizières Research, Maizières Lès Metz 57208, France. Email: yves.basselin@arcelormittal.com
6. Head of Mineral Processing Service, ArcelorMittal Maizières Research, Maizières Lès Metz 57208, France. Email: henrique.turrer@arcelormittal.com
7. Associate Professor, Laboratoire de Physique et Chimie Théoriques – Université de Lorraine, 57500 Saint-Avold, France. Email: michael.badawi@univ-lorraine.fr
8. Associate Professor, Laboratoire GeoRessources – Université de Lorraine, Vandoeuvre-les Nancy 54505, France. Email: yann.foucaud@univ-lorraine.fr

ABSTRACT

The Mont Reed iron ore deposit, owned by ArcelorMittal, is situated approximately 140 km southwest of the Mont-Wright mine in the southern region of the Labrador Trough, Canada. Acquired by Québec Cartier Mining (QCM) in the 1960s, the Mont Reed mining property has remained unexploited due to the complex nature of the ore in the area. However, during the last years ArcelorMittal's R&D team has been dedicated to developing a process to beneficiate the Mont Reed ore and produce high-quality pellet feed suitable for direct reduction furnaces. Extensive research revealed that achieving a good liberation degree of particles required grinding the material below 75 µm (P_{80} of 53 µm). To find the most effective technology for this task, a detailed comparison between ball mill and vertical stirred mill grinding was conducted. Both grinding technologies underwent detailed testing in bench-scale experiments. The ground products were comprehensively characterised, including size distribution analysis and chemical analysis per size fraction. Additionally, liberation analysis was performed using an automated scanning electron microscope (SEM). To determine the grinding specific energy, a method previously developed at Samarco (Brazil) for fine grinding of iron ores was employed. Furthermore, magnetic separation and flotation tests were conducted on the ground products. After a thorough evaluation of all the results, it was observed that both ball mill and vertical stirred mill grinding produced samples with similar physical characteristics. However, the vertical mill demonstrated higher efficiency in grinding the material. To conclude the study, an economic evaluation was conducted for both options considering various factors. Overall, ArcelorMittal's R&D team's dedicated efforts have brought them closer to unlocking the potential of the Mont Reed ore deposit and paving the way for its future utilisation to the production of high-quality pellet feed.

INTRODUCTION

The Mont Reed iron deposit is situated in the southern section of the Labrador Trough (Greenville tectonic province), Québec, Canada (Figure 1). The mining property was initially acquired by Québec Cartier Mining (QCM now ArcelorMittal) during the 1960s and remains unexploited since (Québec Cartier Mining Company, 1977).

The Mont Reed iron deposit is situated in the southern region of the Labrador Trough, specifically in the Greenville tectonic province of Québec, Canada (Figure 1). Back in the 1960s, the mining property was acquired by Québec Cartier Mining (now ArcelorMittal) and since then, the deposit has remained unexploited (Québec Cartier Mining Company, 1977).

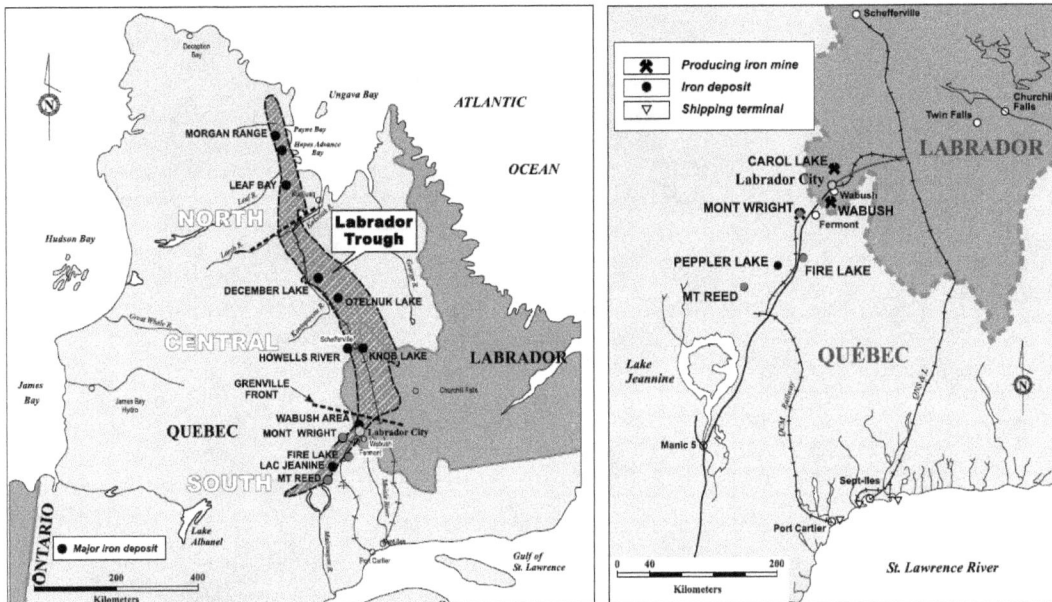

FIG 1 – Mont Reed location (ArcelorMittal assets in red).

The Labrador Trough contains world-class Palaeoproterozoic, Lake Superior-type, banded iron formations (BIFs, 1.88 Ga), that were weakly to intensely metamorphosed during the late Precambrian orogeny (ca. 1.0 Ga). In this context, the iron formation at Mont Reed is composed of fine to medium grained magnetite-hematite-quartz units, associated with quartzite, marble, gneiss and silicate-carbonate facies. The orebody extends over an area of about 2 × 3 km, with a thickness ranging from 60 to 100 m. Geological interpretations describe Mont Reed as primarily folded in an antiform shape that was later refolded orthogonally (Ibrango, 2013). In addition to iron oxides, most of mineral assemblages of the iron formation consist of quartz, Mg-Fe orthopyroxenes (enstatite–ferrosilite series), Ca-clinopyroxenes to a lesser extent (eg diopside–hedenbergite series) and carbonates (Klein, 1978).

One particularity of the Mont Reed deposit when compared to the other ArcelorMittal Mines Canada (AMMC) operations lies in the fact that the ore contains both magnetite and hematite-rich ores, with an overall magnetite to hematite ratio of about 2:1, together with the occurrence of a Mn-rich layer that contaminates the orebody (Belissont, 2021). Furthermore, the liberation degree of the particles in the Mont Reed deposit is lower compared to other deposits in the region. Additionally, the mineralogy of the gangue is much more complex, characterised by the presence of pyroxenes, carbonates, silicates and amphiboles.

Hence, in addition to geological and resource modelling, one key issue for developing the project corresponds to find an optimal process route to concentrate the ore (recovering both hematite and magnetite) and avoiding the presence of gangue minerals in the concentrate. Due to that, since 2020 ArcelorMittal Mining and Mineral Processing Team (MMP) has been studying Mont Reed ore aiming at proposing a feasible process route to concentrate the material and produce a high-quality pellet feed (SiO_2 <1.8 per cent) which can be fed into a direct reduction furnace.

Throughout the research process, a comprehensive characterisation study was conducted using representative samples from the Mont Reed deposit (Belissont, 2021). Additionally, a thorough series of bench-scale tests was carried out to evaluate various methodologies for ore concentration (Mesquita, 2021). Building upon the findings from these investigations, the initial flow sheet, as depicted in Figure 2, was developed.

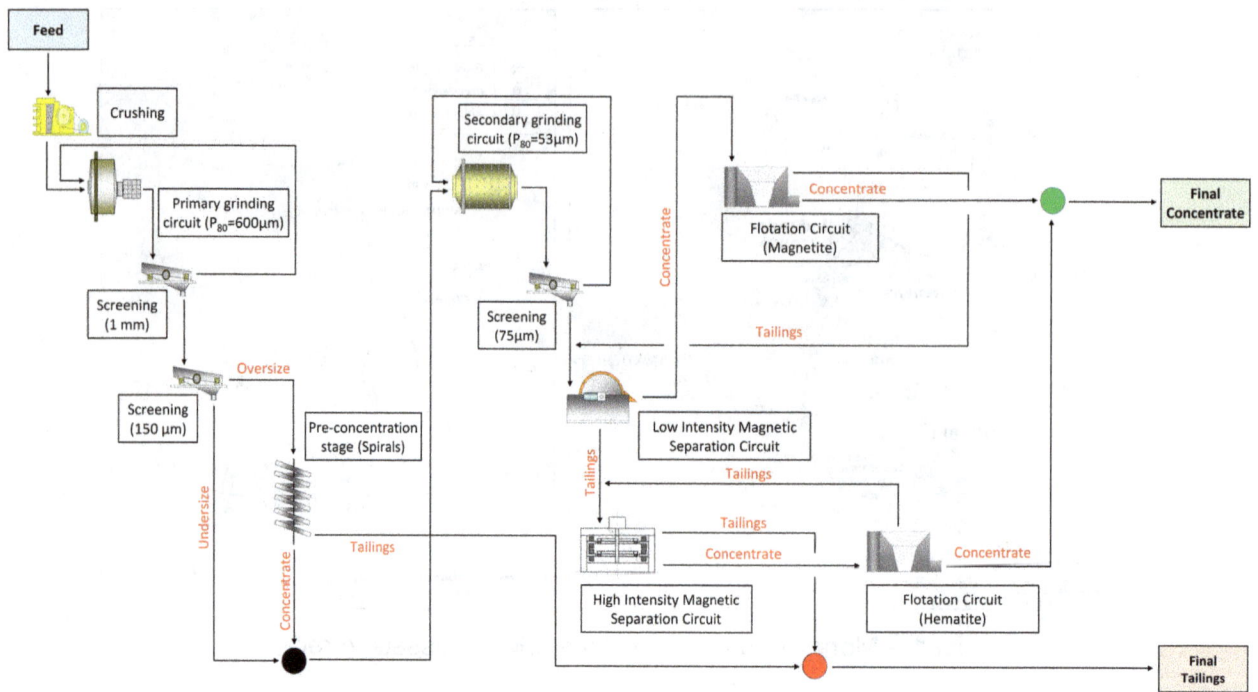

FIG 2 – Mont Reed preliminary flow sheet.

The flow sheet comprises several stages, beginning with a primary crushing step to adjust the run-of-mine (ROM) material before proceeding to the SAG mill circuit. The main objective of the SAG circuit is to liberate a portion of the gangue, producing a product with a P_{80} of 600 µm, which is then subjected to classification at 150 µm.

The oversize material from the screening operation is directed to a pre-concentration stage, utilising spirals. Importantly, the classification is performed before the pre-concentration study to prevent any loss of fine iron oxide particles in the spirals. The concentrate obtained from the spirals and the undersize material from the 150 µm classification are then fed into the secondary grinding circuit, ensuring further liberation of the particles.

The magnetite present in the ground product is recovered through a low-intensity magnetic separation stage (LIMS) and after this step, a reverse flotation operation is conducted. On the other hand, to recover the hematite from the LIMS stage tailings, a different circuit is utilised, consisting of wet high-intensity magnetic separation, followed by reverse flotation.

In this context, the characterisation study and metallurgical tests revealed a significant finding about the iron oxide particles from the Mont Reed deposit: approximately 80 per cent of them are already liberated below 150 µm. However, to attain high-grade products with low silica levels, it was concluded that additional grinding would be necessary (Belissont, 2021). As a result, it was determined that further grinding below 75 µm (P_{80} of 53 µm) would be necessary to meet the desired specifications (Mesquita, 2021).

This fact holds immense significance for the project's development, considering that grinding operations are widely recognised as energetically inefficient, accounting for approximately 34 to 44 per cent of the energy consumption in a mineral processing plant (Esteves *et al*, 2021). Given this context, the pursuit of energy-efficient technologies for comminution processes remains a constant focus in mineral processing studies. Notably, the use of stirred mills has been on the rise, particularly for ultrafine grinding.

Against this backdrop, engineers at MMP decided to conduct a comparison between the conventional tubular ball mill option and the vertical stirred mills for performing the secondary grinding in the Mont Reed flow sheet. The study's outcomes, as outlined in this paper, shed light on the results obtained from this investigation, offering valuable insights into the choice of grinding technology for this specific application.

FINE GRINDING AND STIRRED VERTICAL MILLS

The primary objective of comminution in mineral processing is to release the constituent minerals present in an ore, enabling the separation of valuable components from the gangue (Xiao *et al*, 2012). Achieving effective liberation of particles pose a significant challenge in modern ore processing because the increasing complexity of ores, coupled with finer grain sizes, necessitates finer grinding to attain the required degree of liberation (Jankovic, 2000).

As particle becomes smaller, the probability of flaws decreases and their strength increases. Hence, for very small particles, higher applied forces may be needed and, additionally, relatively high probabilities for successful application of the stress on these particles (Jankovic, 2000). Such high probabilities do not exist in the conventional ball mills as they use relatively large grinding media (Xiao *et al*, 2012). By the other side, the vertical mills are loaded with smaller balls, which have higher specific surface and, consequently, increase the grinding action, especially for smaller particles (Esteves *et al*, 2021).

In this sense, as the particle size decreases, it is necessary to reduce the size of the grinding media while increasing the media velocity to generate the necessary energy for particle breakage. Stirred mills have the capability to achieve exceptionally high media velocities and energy densities by vigourously stirring the media at a rapid rate (Jankovic, 2000). Morrison, Cleary and Sinnott (2009) showed that the basic difference between vertical and tubular ball mills arises from modifications in the frequency and energy of balls collisions. The higher efficiency of the vertical mills is due to the higher frequency of lower energy impacts and, by the same token, smaller frequency of higher energy impacts when compared to conventional ball mills.

This observation is supported by Esteves *et al* (2021), who noted that the breakage mechanisms occurring in vertical mills are like those observed in tubular ball mills, with the distinction that the rates of breakage in vertical mills are higher, given the same ball charge and energy inputs.

Due to that, the use of vertical stirred mills in the mining industry has increased remarkably over the past few decades, mainly due to the growing demand for finer ore grinding. In several applications, the vertical mill demonstrated the same ability to achieve the products specifications of the ball mills while proved to be able to reduce the average specific energy consumption in about 40 per cent (Rosa, Oliveira and Donda, 2014; Xiao *et al*, 2012).

Thus, the utilisation of vertical mills for regrinding operations has become a well-established practice, with its application being incorporated into most newly designed process flow sheets. Regarding secondary grinding circuits, the vertical mill has increasingly emerged as a viable alternative to traditional ball mills as in the case of Mont Reed deposit (Esteves *et al*, 2021).

MATERIALS AND METHODS

The methodology used in the work is presented in Figure 3.

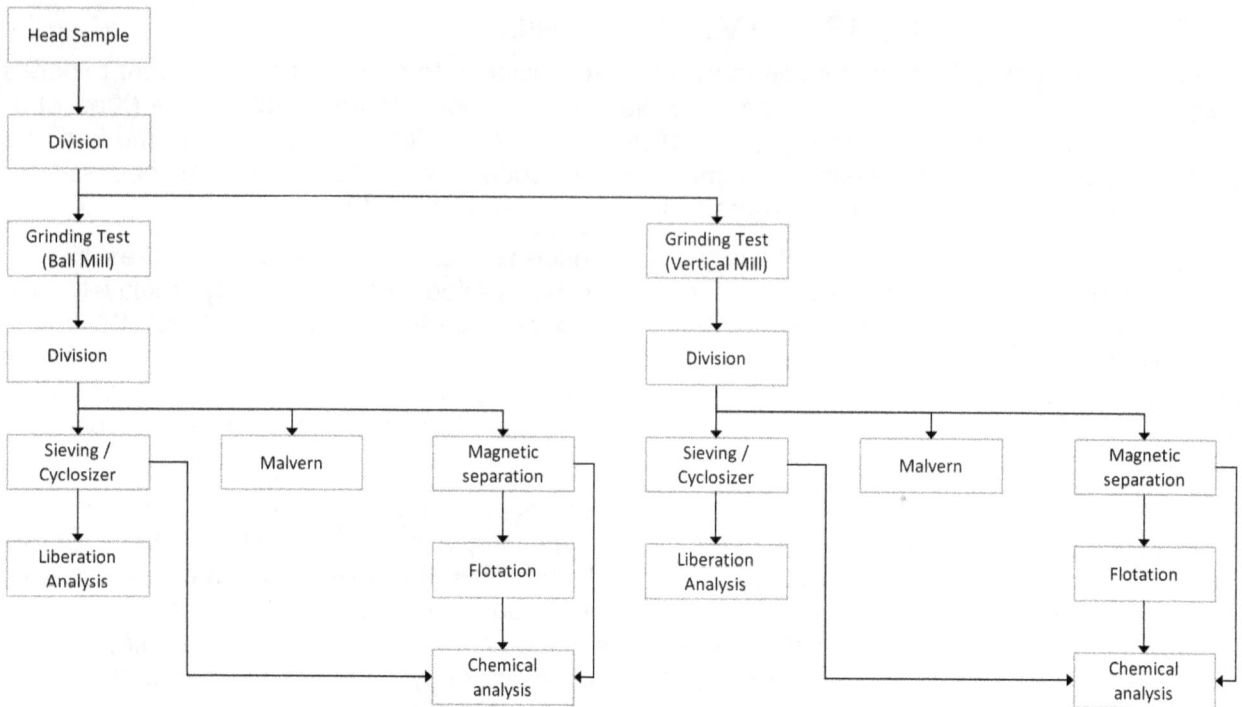

FIG 3 – Project methodology.

The same sample underwent testing in two distinct grinding processes and the resulting ground products were comprehensively characterised. Furthermore, both magnetic separation and flotation processes were employed to evaluate the behaviour of the ore within the Mont Reed flow sheet, based on the specific grinding method utilised. This approach allowed for a thorough assessment of how the grinding method influenced the performance and characteristics of the products during subsequent processing steps.

Sample

The sample selected for the study corresponded to the fresh feed of the secondary grinding stage pilot test performed with Mont Reed ore. The material is a blend of pre-concentration product generated using spirals and the undersize of 150 µm screening. The chemical analysis of the sample is presented in Table 1 and its mineralogical composition is presented in Table 2.

TABLE 1

Grinding feed sample chemical analysis.

Grade (%)												
Fe	FeO	SiO$_2$	Al$_2$O$_3$	CaO	MgO	TiO$_2$	P	Mn	S	Na$_2$O	K$_2$O	LOI
39.78	14.10	34.90	0.10	2.84	2.47	0.01	0.012	0.769	0.004	0.057	0.013	1.350

TABLE 2

Grinding feed sample mineralogical composition.

Mineral	Mineral class	Chemical formula	Proportion (%)
Magnetite	Oxides	Fe_3O_4	39.7
Quartz	Tectosilicate	SiO_2	36.7
Hematite	Oxides	Fe_2O_3	10.6
Ferrosilite	Orthopyroxenes	$Fe_2Si_2O_6$	5.5
Dolomite	Carbonates	$CaMg(CO_3)_2$	2.8
Diopside	Clinopyroxenes	$CaMgSi_2O_6$	2.8
Calcite	Carbonates	$CaCO_3$	1.2
Hornblende	Carbonates	$Ca_2Fe_5Si_7AlO_{22}(OH)_2$	0.3
Talc	Phyllosilicate	$Mg_3Si_4O_{10}(OH)_2$	0.3

Grinding tests

As mentioned before, grinding test were performed in bench scale aiming at evaluating two different comminution methodologies: tubular ball mill and vertical stirred mill. The methodology selected to evaluate both scenarios was proposed by Donda (2003).

The Donda's methodology was initially developed by engineers at Samarco (Brazil) with the primary objective of estimating the energy consumption during the grinding operations of the company (Donda, 2003). This approach stemmed from the application of a well-established practice utilised since 1995 to estimate the specific energy involved in primary grinding operations using ball mills. Engineer Joaquim Donda adapted and formalised this method, customising it to determine the energy consumption of the grinding stages in Samarco's operations. The methodology showed a strong correlation with industrial results, as reported by Rosa and Donda (2014).

According to Donda (2003), the specific energy needed to reduce the size of the ores in the pinion shaft of a ball mill can be determined by Equation 1.

$$E = \frac{1}{K} \times ln\frac{F}{P}$$

(1)

Where:

E   = Specific energy in kWh/t, in tubular mills, on the pinion shaft of the motor

K   = Characteristic parameter of the ore for a given mesh

F   = Retained mass of particles in a specific mesh in the circuit feed (usually P_{80})

P   = Retained mass of particles in a specific mesh in the circuit product (usually P_{80})

The application of Donda's methodology aims to calculate the characteristic parameter 'K' of the ore using a standard laboratory test This 'K' value is specific to the material being studied for a given mesh size and under specific grinding conditions (Bottosso et al, 2023). The parameter 'K' represents the slope of the grinding curve, which essentially indicates how easily the material can be ground and also indicates its performance in a grinding circuit. A higher 'K' value implies better grindability or improved performance in the grinding process (Rosa, Oliveira and Donda, 2014).

The significant contribution of Donda's methodology lies in two aspects. Firstly, it introduced a laboratory test to determine the 'K' parameter, providing a practical and efficient way to quantify the grindability of the material. Secondly, Donda's methodology demonstrated its industrial applicability and reliability by successfully applying it in the grinding facilities of SAMARCO (Donda, 2003). These accomplishments validated the methodology's effectiveness, solidifying its value as a useful tool in the assessment and optimisation of grinding processes.

The Donda's test procedure consists in grinding four aliquots of the same sample using different energies (grinding times) and considering standardised operational parameters. The cumulative oversize mass in the interest mesh of the different products (feed and ground products) are determined and the information is used to define the value of 'K' through an exponential regression curve, as shown in Figure 4 and Equation (2).

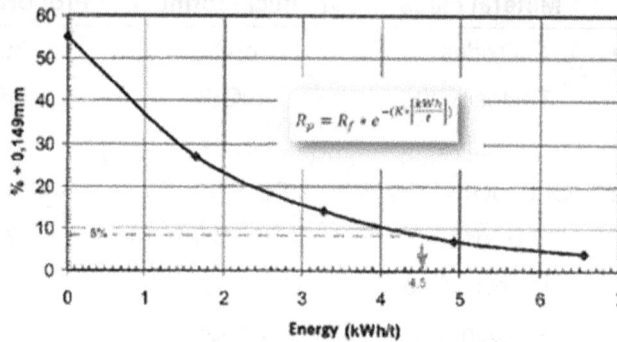

FIG 4 – K parameter determination.

$$Rp = Rf * e^{-(K*E)} \tag{2}$$

Where:

E = specific energy applied in the comminution process (kWh/t) – 'x' value

K = parameter related to the analysed screen aperture, within a specific milling condition (t/kWh)

Rf = % cumulative retained at analysed screen aperture, of feed material

Rp = % cumulative retained at analysed screen aperture, of products material – 'y' value

The grinding energies to be evaluated in the laboratory tests are suggested by Rosa and Donda (2014) according to the comminution stage of the process (primary grinding or regrinding), being converted into grinding times using *Rowlands'* Equation (3) applied to small diameter mills (Rowland, 1984).

$$Kwb = 6.3 * \sin\left(51 - 22 * \left(\frac{2.44 - D}{2.44}\right)\right) * (3.2 - 3 * V_p) * C_s * \left(1 - \frac{0,1}{2^{(9 - 10 * C_s)}}\right) \tag{3}$$

Where:

Kwb = Kilowatt hour per ton of balls, in the mill pinion axis

D = Mill diameter in metres, inner part of the coating

V_p = Fraction of the mill volume filled by balls charge

C_s = Fraction of the critical speed of the mill

The power applied on the pinion axis of a laboratory mill (in kWh) is obtained by multiplying the calculated *Kwb* by the mass of the ball charge used in the test (in tonnes). This power can be transformed into kWh/t dividing the original number by the mass of ore fed into the mill (in tonnes). Considering the energy consumption of this system running during one our (which will be have the same value of the power calculated before but in kWh/t) it is easy to determine a conversion factor between energy and grinding time dividing the energy consumption by 60 (kW/t/min). Using the conversion factor, it is possible to transform the evaluated energies in grinding times. The 'K' value couple with the formula proposed by Donda (2003) allows to estimate the specific energy spent in a comminution process.

Regarding the vertical stirred mills, in an ideal scenario, a laboratory-scale machine could be employed for scaling up purposes, like batch tubular ball mills are adopted for scaling up industrial operations. However, in contrast to ball mills, where laboratory machines allow for the use of balls with diameters like or slightly smaller than those used in industrial operations, vertical mills require a proportional scaling-down of the ball size based on the mill dimensions (this ensures that the ratio between the diameter of the largest ball, as well as the gap between the screw and the inner wall of the mill is preserved). Unfortunately, it turns out that the maximum particle size that can be tested in a batch vertical mill is very small, not reflecting the desired feed size of industrial applications (Esteves *et al*, 2021).

To solve this problem, the design of vertical stirred mills for specific applications are generally based on laboratory tubular ball mill tests. Accordingly, Metso, which is one of the main manufacturers of vertical stirred mills, adopts a standard batch grinding test with smaller ball diameters (Bergerman and Delboni, 2019).

Such a test is very similar to Donda's one and provides a relationship between specific energy consumed and product particle size (P_{80}). The only requirement is the implementation of an efficiency factor of 0.65 in the specific energy obtained in the laboratory, that accounts for the increased rate of breakage observed in vertical stirred mills compared to ball mills (Bergerman and Delboni, 2019). By incorporating this factor, the performance of vertical mills can be appropriately adjusted and optimised.

In this way, the two grinding methodologies were evaluated using Donda's methodology and followed the information presented above. The grinding tests were conducted using a Retsch-supplied ball mill model TM300 and the operational parameters are presented in Table 3.

TABLE 3
Grinding tests parameters.

Parameter	Value			
Mill internal diameter (m)	0.3048			
Mill internal length (m)	0.3048			
Mill load (%)	30			
Relation V pulp/V voids	1.00			
Percentage of solids in weight	80			
Percentage of critical speed	70			
Ore weight (kg)	5.000			
Evaluated Energies (kWh/t)	3	6	12	18
	Diameter (mm)	**No. of balls**	**Weight (kg)**	**Proportion (%)**
	63.5	9	10.70	35
Ball charge distribution (Ball Mill Tests)	50.8	27	12.56	41
	38.1	27	5.42	18
	25.4	32	1.84	6
	Total	**95**	**30.52**	**100**
Ball charge distribution (Vertical Mill Tests)	10.0	7863	32.24	100
	Total	**7863**	**32.24**	**100**

Liberation analysis

Liberation analyses were performed on sized samples using Zeiss's Mineralogic system, which is based on a Zeiss Sigma 300 scanning electron microscope with a field emission gun (FEG-SEM). This system allows to use backscattered electrons (BSE) images and energy dispersive X-ray analyses (EDS) to create digital mineral images and extract various quantitative mineralogical information like mineral liberation, association and morphological parameters of grains and particles.

The methodology used in Mineralogic analyses starts with the processing of BSE images using the grey scale values of minerals to, first, distinguish particles from the resin and then, to identify iron oxides (high brightness) on the one hand and gangue minerals (low brightness) on the other hand.

The second step consists in classifying each particles' grains into mineral classes, using both IP outputs and EDS analyses. Given the known complexity of Mont Reed gangue silicates, to expedite the analyses, a simplified mineral classification and the spot centroid analysis mode was used.

Each particle is classified into three liberation categories, depending on the degree of liberation of the interest mineral: above 90 per cent, the particle is liberated, between 30 per cent and 90 per cent, the particle is a middling and below 30 per cent, the particle is locked. The overall liberation of the sample is then computed as the weight of a specific mineral in liberated particles, over the weight of the same mineral in all particles, as Equation (4).

$$\text{Liberation} = \frac{W^{liberated}}{W^{liberated} + W^{middling} + W^{locked}} * 100 \qquad (4)$$

Where the weight (W) is calculated as the area of mineral multiplied by their specific gravity.

Laser particle size analysis

During the project the size distribution of several samples were evaluated using a laser analyser model Mastersizer 3000 distributed by Malvern Panalytical.

Powder X-ray diffraction

Powder X-ray diffraction (XRD) analyses were performed using a Bruker D2 Phaser diffractometer, which features a goniometer in Bragg-Brentano geometry, a Co Kα (1.789 Å) source and a Lynxeye 1D silicon-drift detector. Prior to the acquisition, the samples are ground to a top size of 63 µm using a ring mill. Diffractograms were acquired as coupled $2\theta/\theta$ scans on the 7–80°2θ range, with a step size of 0.02°2θ, a dwell time of 1.5 sec per step.

Phase identification was performed using Bruker's EVA software, which features the ICDD PDF-2 database. Phase quantification was performed using the Rietveld refinement method in TOPAS software. Refinement parameters included a background fitting with a four to six order Chebyshev polynomial, sample displacement correction, crystal structure refinement, ie cell parameters and crystallite size, preferential orientations correction (on major phases only) and peak fitting using Bruker-designed fundamental parameters (FP).

Sieving

Sieving's were performed in wet via using 30 cm screens mounted in a Sinex AT 450 vibrating sieve-shaker. The screens apertures were selected considering the top sizes of the samples and adopting Tyler mesh series.

Cyclosizer

Cyclosizer tests were performed in a machine M16 supplied by MARC Technologies Pty Ltd. The cutting sizes of the cyclones were adjusted using the operational parameters of the tests as suggested by MARC Technologies Pty (2014).

Magnetic separation tests

The magnetic separation tests were performed in a bench scale low intensity magnetic separator model WD(20)111–15 supplied by Outotec. The tests were performed using samples of 2 kg, 20 per cent of solids, pulp flow rate of 1.5 L/min and 600 G magnetic field. The objective of the tests was to evaluate the influence of grinding methodology into magnetite particles recovery.

Flotation tests

Flotation tests were performed using a Denver Metso Flotation Cell model D12 equipped with 1 L cell. The operation was performed in pH 10.5 with a pulp of 30 per cent of solids. Flotigan EDA (supplied by Clariant) was used as collector and corn starch as depressor. The depressor dosage was fixed in 300 g/t and three different collector dosages were tested: 100 g/t, 300 g/t and 500 g/t. The conditioning stage of the reagents was performed using an agitation of 1200 rev/min (with 50 per cent of solids in pulp) while the flotation operation was performed with a lower speed (900 rev/min). An air flow rate of 4 L/min was adopted and the froth was collected until exhaustion.

To prepare the depressor, 2 g of starch was first diluted in 20 mL of water and simultaneously, 0.5 g of NaOH was diluted in the same amount of water using a separate container. Once both reagents were fully dissolved, they were combined in a beaker and thoroughly mixed using a magnetic agitator until the starch completely gelatinised. The resulting gelatinised solution was then further diluted and utilised for conducting the flotation tests. It is crucial to highlight that demineralised water was employed in the preparation of the reagents and throughout the flotation tests to ensure accurate and consistent results.

RESULTS

The particle size distributions of Donda's tests products for each grinding methodology are presented in Figure 5.

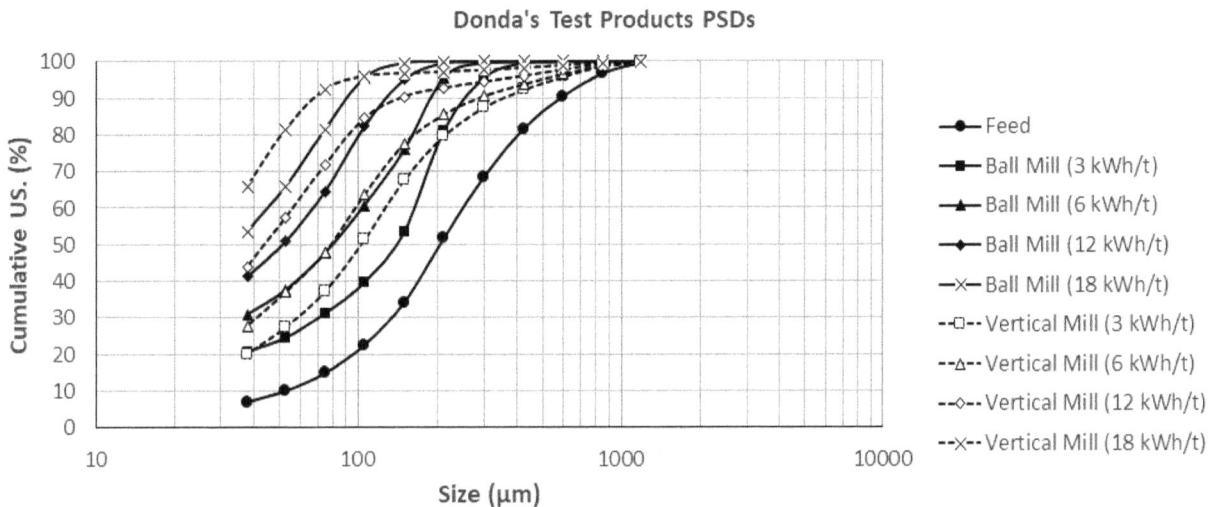

FIG 5 – Donda test products PSDs.

The graph illustrates that the vertical mill tends to generate a higher proportion of fine particles overall, particularly at higher energy levels. However, it is not as efficient as the tubular ball mill in handling coarser particles, particularly those with larger top sizes. This difference in efficiency can likely be attributed to the effect of lower energy impacts on both types of particles in each grinding method.

To determine the K value for both grinding conditions, the cumulative oversize masses at 53 µm were plotted against the grinding energy, as demonstrated in Figure 6. This approach allowed for a quantitative assessment of the grindability of the ore under the two grinding conditions, providing valuable insights into their respective performances.

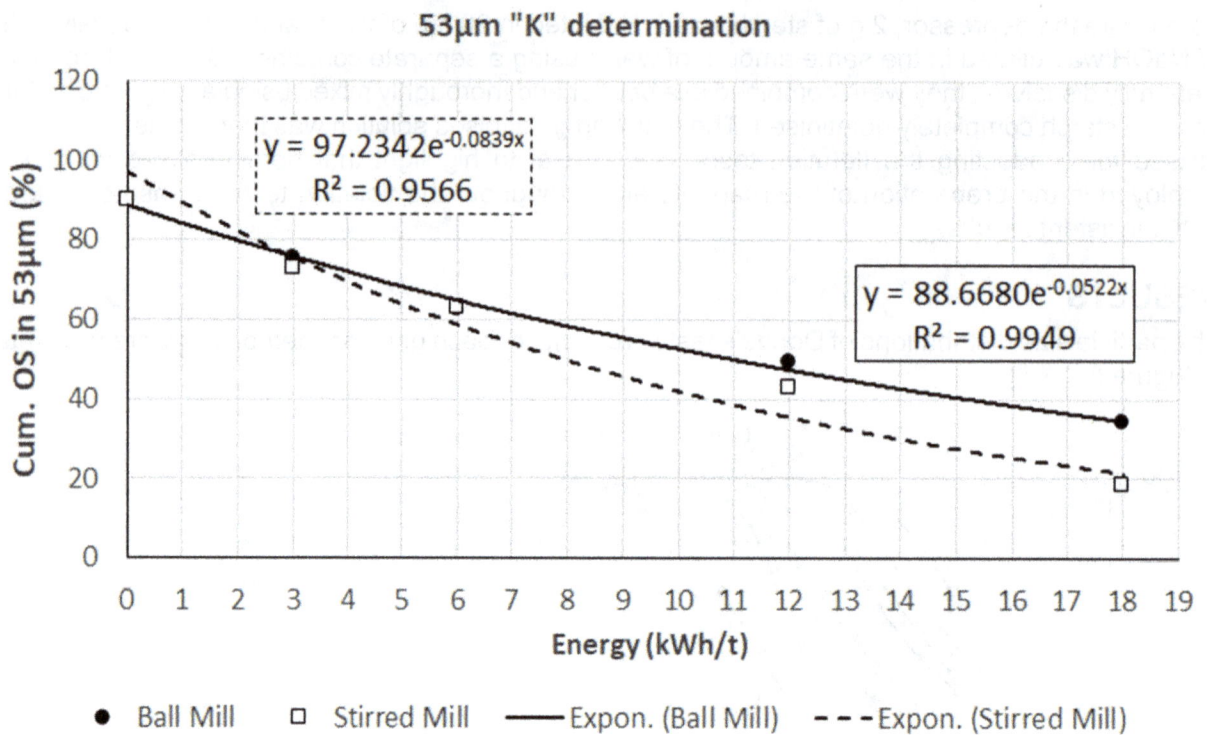

FIG 6 – 'K' parameter determination.

The K values for both cases were defined by employing the exponential regressions displayed in the graph. Subsequently, the corresponding grinding energies were calculated using the equation proposed by Donda (2003). The outcomes obtained from these calculations are summarised and presented in Table 4. This table provides a comprehensive overview of the K values and the respective grinding energies for each grinding condition, offering valuable data for further analysis and comparison.

TABLE 4

Grinding energy calculation.

Option	K (t/kWh)	53 µm oversize (%) Feed	53 µm oversize (%) Product	Energy (kWh/t)	Blaine (cm²/g)
Ball Mill	0.0522	90	20	28.8	2278
Vertical Mill*	0.0839	90	20	11.7	2914

* The 0.65 factor suggested by METSO is already considered in the calculated energy.

The K parameter obtained from the vertical mill test is significantly higher than that of the ball mill, indicating a higher efficiency in the grinding process for the former. This observation is further reinforced when analysing the calculated grinding energies. Notably, to produce a similar product, the ball mill consumed 2.46 times more energy than the vertical mill. While it is possible to optimise the ball mill energy industrially by manipulating the ball charge and operational parameters, even in such cases, the vertical mill would remain substantially more efficient. Furthermore, the Blaine value measurements for both samples demonstrated that the vertical mill could produce a much larger superficial area compared to the ball mill.

In continuation, new samples were subjected to grinding using the calculated energies and the size distribution of the resulting products is presented in Figure 7. These ground samples were later utilised in beneficiation tests.

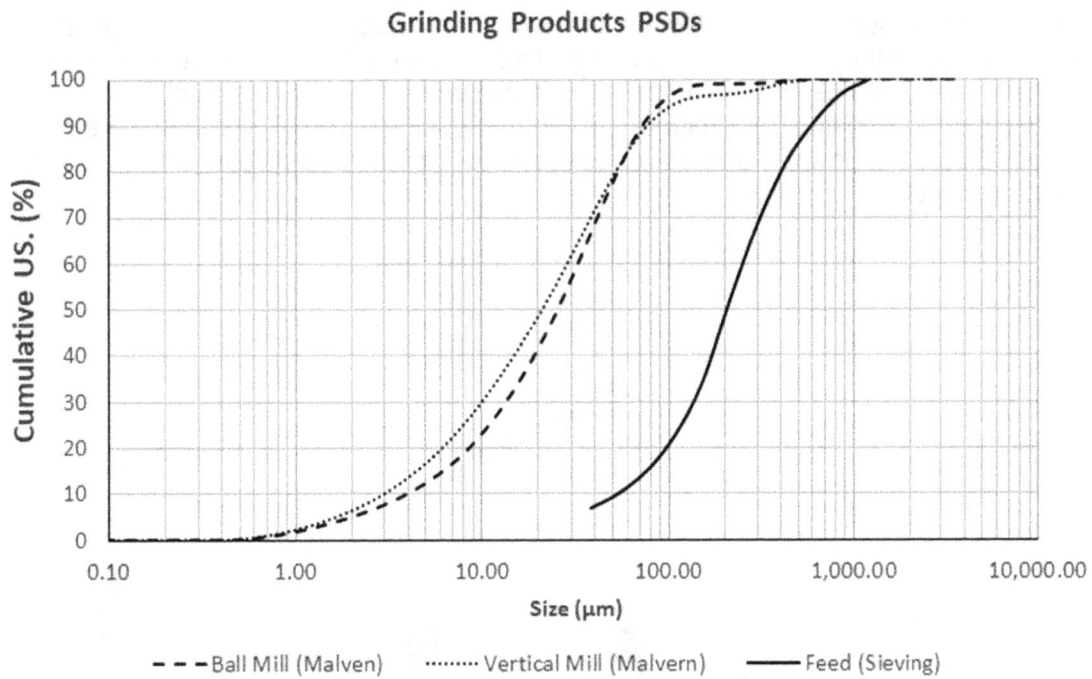

FIG 7 – Grinding products PSDs.

Both samples presented similar P_{80} values, but the vertical mill was more efficient to produce particles in the range between 40 µm and 2 µm (Table 5). Again, a bigger top size was observed in the vertical mill product.

TABLE 5

Ground products information.

Stream	P_{10} (µm)	P_{50} (µm)	P_{80} (µm)
Grinding Feed	50	200	400
Ball Mill Product	4	25	50
Vertical Mill Product	3	20	50

Regarding the distribution of elements per size fraction, the size-by-size results of both ground products are presented in Table 6.

TABLE 6

Ground products size-by-size information.

Size (µm)	Mass (%)		Fe		FeO		SiO$_2$		Al$_2$O$_3$		CaO		MgO	
	BM	VM	BM	VM	BM	VM	BM	VM	BM	VM	BM	VM	BM	VM
106	0.7	4.6	45.61	58.61	13.35	10.69	27.90	12.84	0.84	0.17	3.42	1.30	1.57	1.80
75	4.2	3.3	44.29	50.59	13.35	9.69	32.00	25.80	0.11	0.09	1.50	0.77	2.00	1.30
53	11.0	10.5	43.86	41.62	14.40	12.50	32.90	36.60	0.11	0.09	1.65	1.08	1.93	1.67
35	22.9	21.0	51.30	47.47	17.35	16.65	23.47	29.16	0.11	0.16	1.58	1.41	1.66	1.60
24	9.4	9.2	41.41	44.65	15.05	17.05	32.30	28.20	0.11	0.12	2.88	2.96	2.62	2.64
15	12.6	13.1	32.52	31.50	12.00	12.40	44.00	45.20	0.13	0.12	3.73	3.68	2.54	2.51
12	8.8	8.8	32.74	30.93	12.00	12.65	44.80	44.00	0.12	0.14	3.24	4.13	2.44	2.73
8	5.3	5.4	31.63	32.02	11.85	12.65	42.50	41.10	0.14	0.16	4.17	4.80	2.65	2.87
<8	25.0	24.1	30.27	31.50	11.11	12.65	40.60	37.98	0.20	0.23	5.21	5.92	3.84	3.99

There are minimal discrepancies between the two methodologies, with more significant grade variations observed in the coarser size fractions. This discrepancy likely results from challenges in breaking coarse particles using the vertical mill. In this context, it appears that the richer particles are more resistant to grinding, as illustrated in Figure 8. Additionally, Fe grades were found to be higher above 24 μm, while SiO_2 grades increased below this size, indicating the presence of two distinct particle populations in the ore.

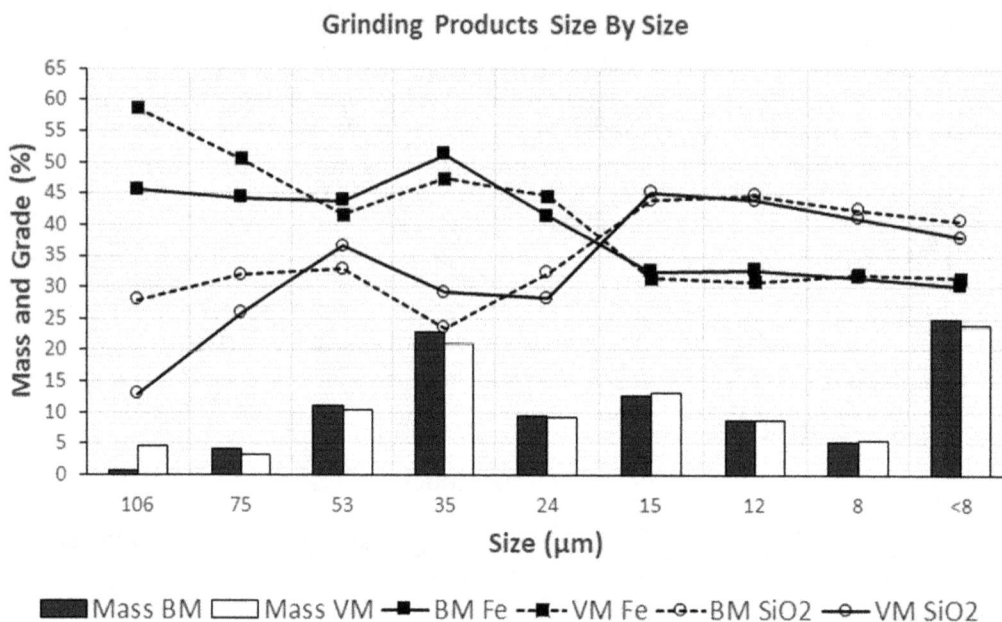

FIG 8 – Grinding products size-by-size.

The average liberation degree of particles in the>38 μm size fraction of the ground products is provided in Table 7. It is important to note that the <38 μm fraction of the samples was removed prior to liberation analysis to prevent any interference from ultrafine particles in the quality of SEM images. Nevertheless, it is assumed that the particles are fully liberated within this size range. By focusing on particles bigger than 38 μm, the liberation analysis yields valuable insights into the extent of particle liberation, which is crucial for assessing the efficiency of the grinding process and its potential impact on downstream operations.

TABLE 7

Ground products liberation study.

Particle type	Iron oxides		Quartz		Ferrosilite		Dolomite		Diopside	
	Ball mill	Vertical mill	Ball mill	Vertical mill	Ball mill	Vertical mill	Ball mill	Vertical mill	Ball mill	Vertical mill
Liberated (%)	96.2	96.9	96.3	96.4	72.8	92.2	87.6	92.7	91.3	88.3
Middlings (%)	2.8	2.6	3.2	3.3	23.5	6.9	2.1	6.1	7.5	10.0
Locked (%)	1.0	0.5	0.5	0.4	3.7	0.9	10.4	1.2	1.2	1.7

In general, the minerals are well liberated in both cases and, despite the very similar values observed for iron oxides and quartz, some interesting differences were found for Ferrosilite and Dolomite. In these cases, the vertical mill was more efficient to liberate the gangue minerals. In addition, a slightly higher quantity of locked particles was observed in ball mill product.

Overall, the minerals show good liberation in both cases and while iron oxides and quartz exhibit very similar liberation values, intriguing differences are observed for Ferrosilite and Dolomite. Notably, the vertical mill demonstrated greater efficiency in liberating the gangue minerals in these instances. Additionally, a slightly higher quantity of locked particles was observed in the ball mill product. These findings underscore the significance of the grinding method in influencing mineral liberation and underscore the potential advantages offered by the vertical mill in terms of gangue mineral liberation.

When analysing the SEM images (Figure 9), no significant visual differences were observed in terms of particle shapes. Both grinding methods appear to produce particles with relatively similar shapes. This finding suggests that the primary grinding mechanism may not have a significant impact on the overall particle morphology. However, it is essential to note that this visual assessment is preliminary and a more detailed quantitative analysis may be required to gain deeper insights into particle shape variations between the two grinding methods.

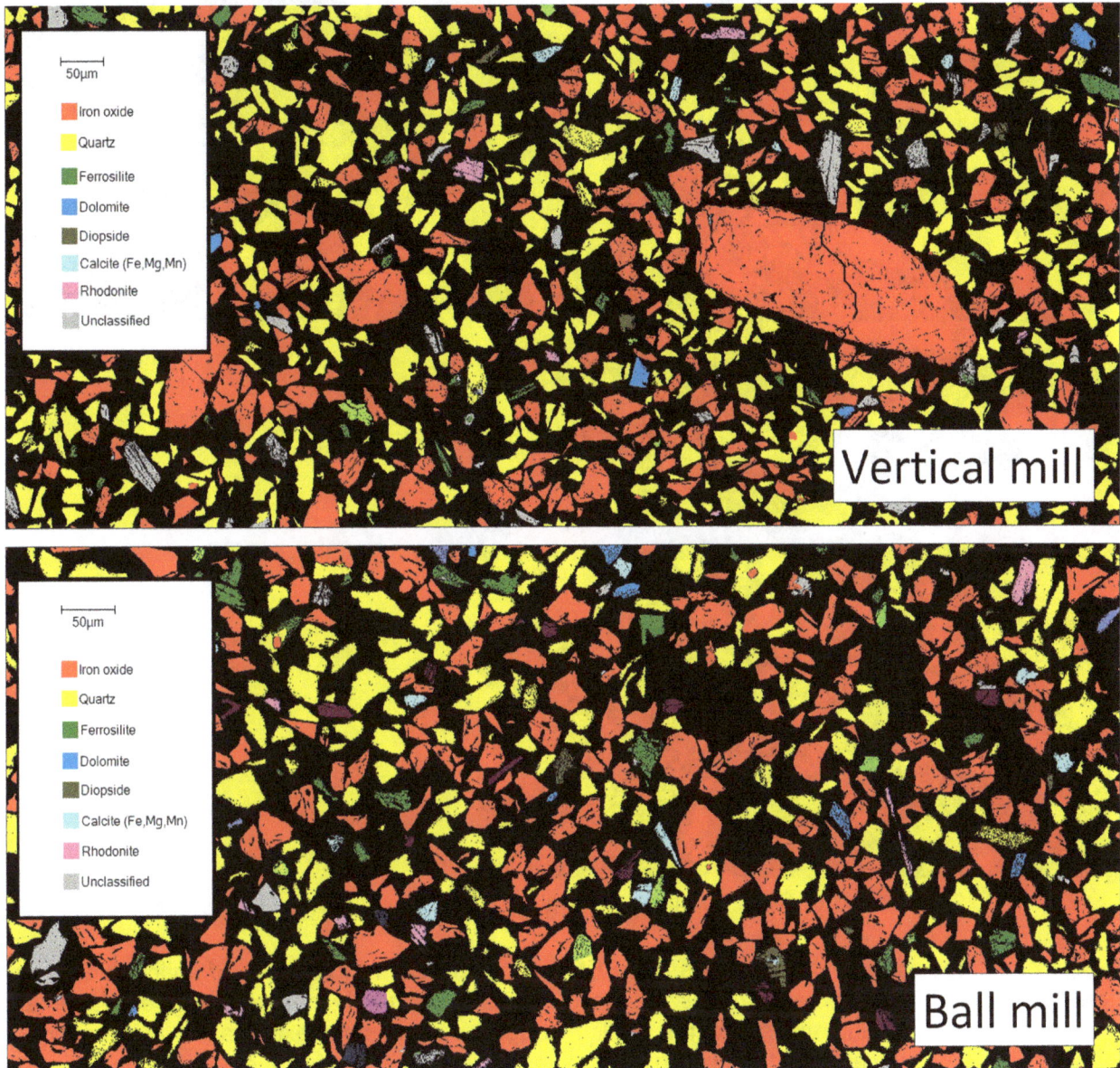

FIG 9 – SEM images (liberation analysis).

The SEM images of both grinding products reveal a noteworthy observation, concerning the presence of coarse particles in the vertical mill product (Figure 10).

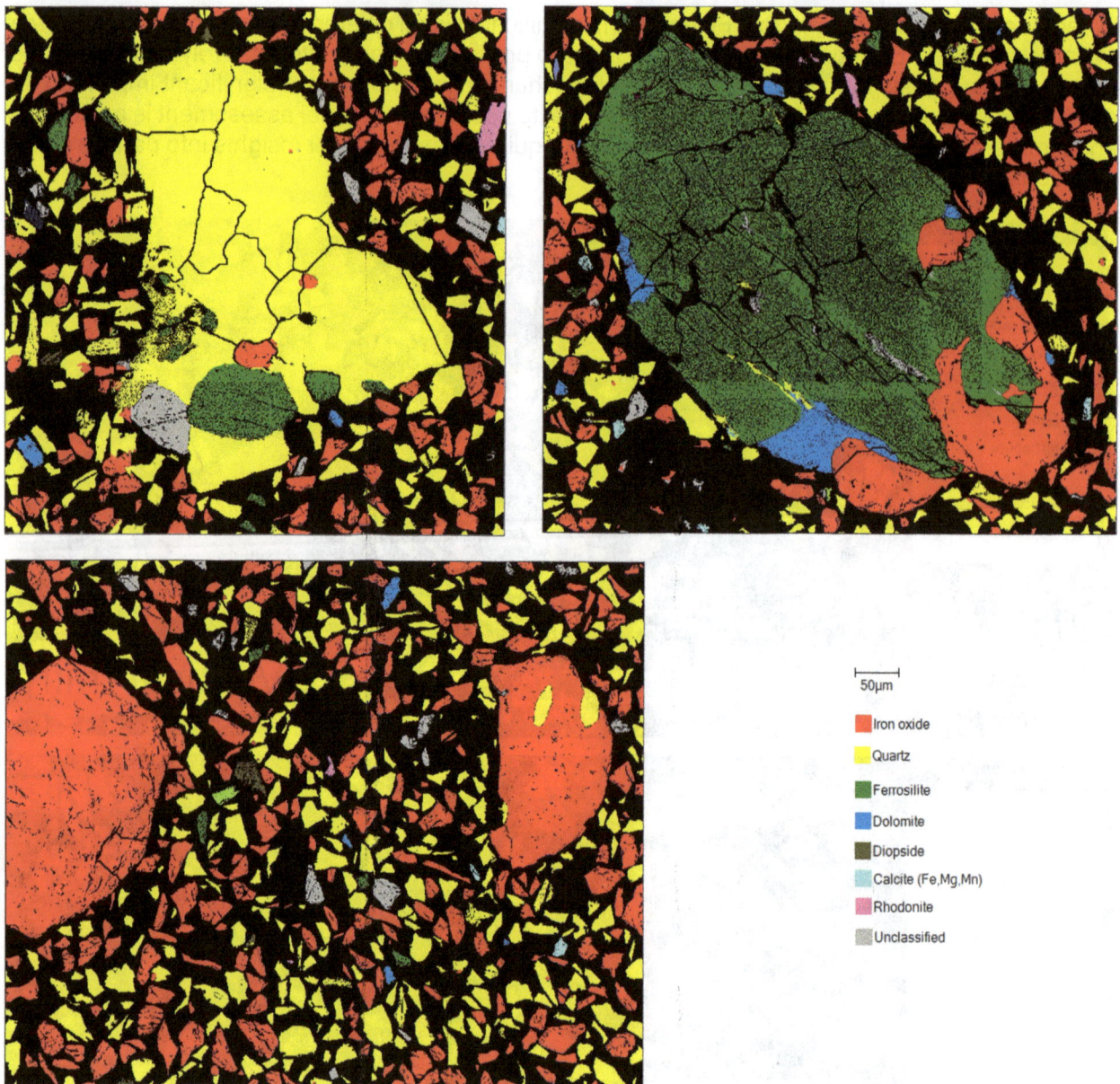

FIG 10 – Vertical mill product coarse particles.

Despite the larger sizes of the particles, the SEM images reveal the presence of noticeable cracks generated during the grinding process. These cracks are frequently observed in the attachment zones between different minerals. This finding suggests a possible explanation for the higher efficiency of the vertical mill in liberating the gangue minerals. The lower energy impacts in the vertical mill may have a greater tendency to concentrate on the weaker points of the particles, prioritising the detachment of individual grains over the complete breakage of particles belonging to the same mineral. This mechanism seems to allow the vertical mill to selectively target and release the gangue minerals more effectively, potentially leading to the observed differences in mineral liberation between the two grinding methods.

Following the grinding process, the ground samples were then introduced into a low-intensity magnetic separation drum to analyse their behaviour in this specific process. The key findings and outcomes from the magnetic separation tests are illustrated in Figure 11.

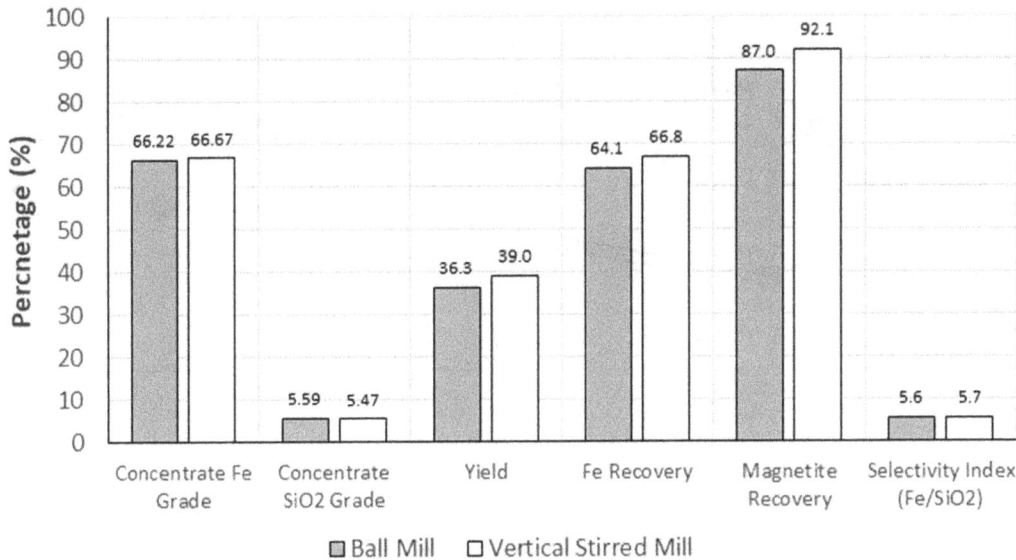

Low Intensity Magnetic Separation Tests Comparison

□ Ball Mill   □ Vertical Stirred Mill

FIG 11 – Low intensity magnetic separation results.

Despite of the similar concentrate qualities and selectivity index observed, the vertical mill product presented higher recoveries in the magnetic separation process. This is probably due to the lower quantity of Fe being lost to the tailings in the form of mixed particles (higher liberation degree of gangue minerals). In addition, the magnetite recovery indicates that the additional iron recovered in the concentrate of vertical mill product is really contained in magnetite structure. In this case, the higher presence of fine particle does not seem to be harmful to the process and the silica levels of the concentrate corroborates the necessity of using an additional flotation step to produce a pellet feed adapted to direct reduction furnaces.

Moreover, the magnetite recovery data suggests that the additional iron recovered in the concentrate of the vertical mill product is genuinely contained within the magnetite structure. This finding indicates that the higher presence of fine particles in the vertical mill product does not appear to have a detrimental effect on the magnetic separation process.

Furthermore, the silica levels of the concentrate align with the need for an additional flotation step to produce a pellet feed suitable for direct reduction furnaces.

In the next stage of the research, the tailings of magnetic separation tests (21.2 per cent Fe and 56.0 per cent SiO_2 in both cases) were fed into a flotation process to evaluate the recovery of the hematite contained in the sample. The results obtained are presented in Figure 12.

In contrast to the previous concentration stage, the presence of ultrafine particles played a significant role in the separation process during flotation. The vertical mill product required a higher dosage of collector to effectively float the gangue, likely due to its larger superficial area resulting from the abundance of ultrafine particles. However, this higher dosage of collector had an adverse effect, leading to increased drag of iron to the tailings and subsequently reducing the overall Fe recovery.

Generally, the tailings of the ball mill products exhibited poorer quality compared to the vertical mill ones, but the recoveries were higher, except for the lowest collector dosage. This observation highlights the challenges faced by the flotation process when dealing with ultrafine particles, necessitating the evaluation of a desliming operation before flotation, with a cut near 3–5 µm, to effectively handle these fine particles.

Furthermore, the results underscore the need for multiple stages of concentration to recover the hematite contained in the Mont Reed ore fully. This finding suggests that a multi-step approach to concentration may be necessary to achieve optimal recovery rates and high-quality products from the flotation process. The implications of ultrafine particles in the flotation process and the potential benefits of desliming and multi-stage concentration are essential considerations in designing an effective beneficiation strategy for the Mont Reed ore.

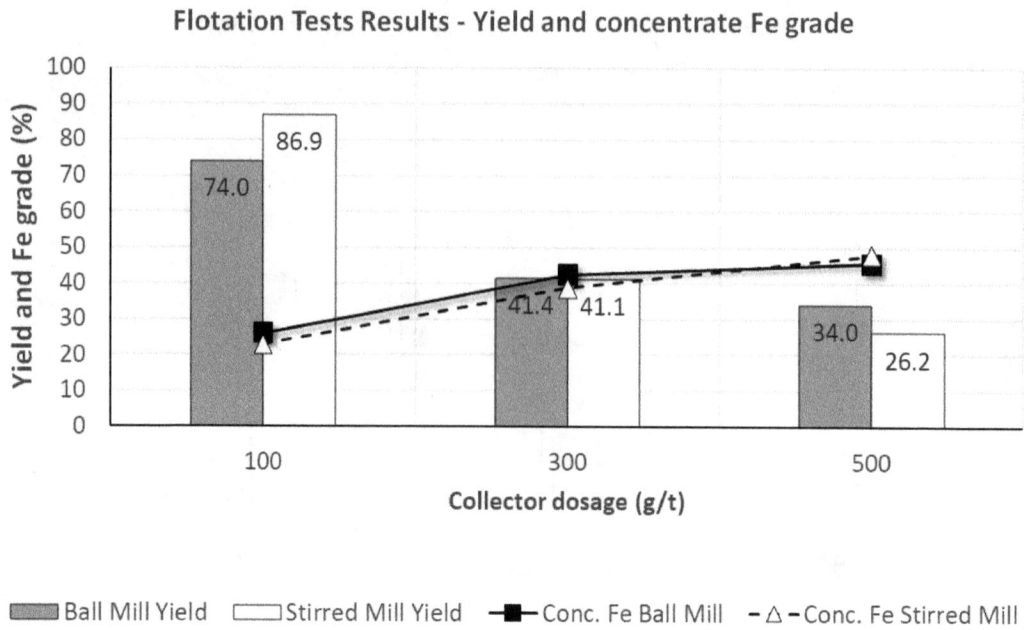

Flotation Tests Results - Yield and concentrate Fe grade

Ball Mill Yield   Stirred Mill Yield   Conc. Fe Ball Mill   Conc. Fe Stirred Mill

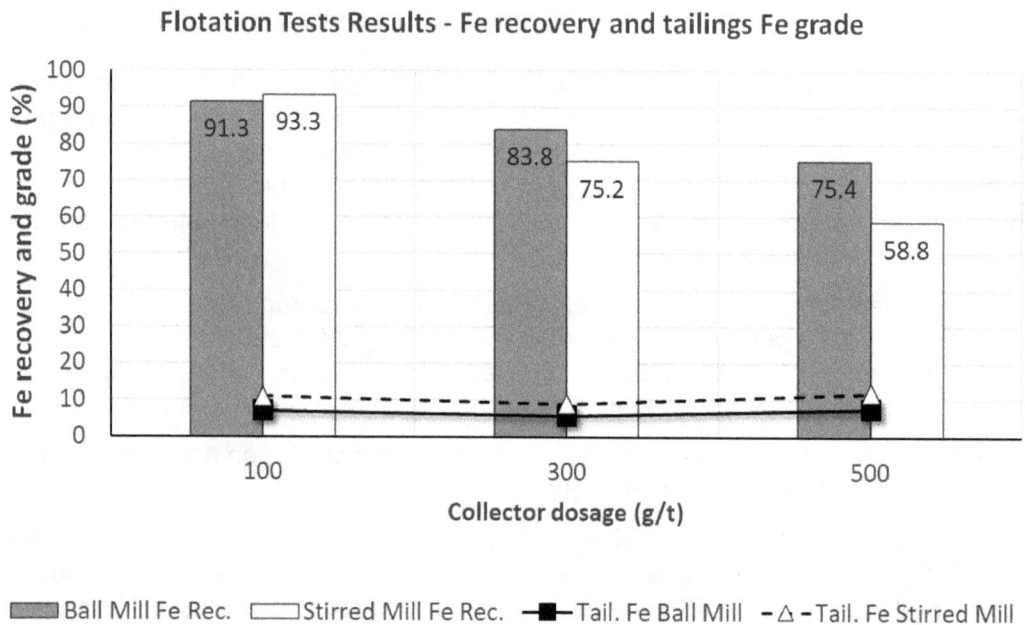

Flotation Tests Results - Fe recovery and tailings Fe grade

Ball Mill Fe Rec.   Stirred Mill Fe Rec.   Tail. Fe Ball Mill   Tail. Fe Stirred Mill

FIG 12 – Flotation tests results.

Economic analysis

In the preceding sections, a detailed technical comparison between tubular ball mills and vertical stirred mills was performed. To further augment the information, a simplified economic evaluation comparing both machines is presented. The initial aspect assessed pertains to the prices of the two types of mills. For illustration purposes, an example from Samarco is used to highlight the differences (Rosa, Oliveira and Donda, 2014). By conducting this economic evaluation, a comprehensive understanding of the cost implications associated with each grinding technology can be gained.

Samarco is a Brazilian mining company with shareholding equally divided between Vale and BHP Biliton. Samarco acts in the international iron ore pellets market producing 24.0 million tons of iron ore concentrate (Germano concentrators) per annum and 21.7 million tons of pellets (Rosa, Oliveira and Donda, 2014).

The basic process flow sheet of the second concentrator of Germano includes crushing, grinding, desliming, two stages of flotation, regrinding and thickening. This concentrator started up in July

2008 and achieved a capacity of 8.5 million tons of concentrate per annum in 2010 (Rosa, Oliveira and Donda, 2014).

In this context, a vertical mill was installed as an alternative to increase the energy available in the regrind circuit of Germano concentrator. This model of equipment was chosen due to layout requirements: the difficulty to install a new ball mill in an operational plant with little footprint available was a key decision factor (Rosa, Oliveira and Donda, 2014).

Thus, the chosen vertical mill (VTM1500, 1500cv) was set-up in parallel with the two existing regrinding ball mills (18 × 33 ft, 6400 kW, each), being the system designed to split the same percentage of the total feed rate between the three equipment (Rosa, Oliveira and Donda, 2014).

Considering this information, it is possible to state that the 18 × 33 ft, 6400 kW ball mill performs the same task than the VTM1500 mill in Samarco operation, being the flow rate and the requested reduction ratio of the machines similar in both cases. That said, a direct comparison between the prices of the equipment was done with the help of the mine and mill equipment estimator's guide published by Info Mine USA Inc (2010). The prices of similar equipment were collected in the document and are presented in Table 8 (2010 price basis).

TABLE 8
Comparison between mills prices.

Mill type	Model	Power (KW)	Price (USD)	Price (%)
Ball Mill	18 × 33 ft	6400	4 139 000	100
Vertical Mill	72" (183 cm) mill radius × 44'2" (13.5 m) mill height	1100	2 383 000	58
	Difference		1 756 000	42

From the table, it is possible to conclude that in 2010 the price of a vertical mill corresponded to 58 per cent of the price of a ball mill, considering that both machines were able to perform the same task.

Aiming at comparing part of the operational costs of both machines, the energies calculated for Mont Reed ore were transformed into operational costs considering the cost of the kWh practiced in ArcelorMittal Canadian operations (0.1 USD/kWh). The comparison between both values is presented in Table 9.

TABLE 9
Comparison between energy costs.

Mill type	Specific energy (kWh/t)	Cost (USD/t)	Cost (%)
Ball Mill	25.8	2.58	100
Vertical Mill	10.8	1.08	42
Difference		1.50	58

Once again, the vertical mill showed to be more advantageous to perform the required fine grinding. An 58 per cent lower energy cost was observed for this type of machine.

Finally, to conclude the study, a qualitative analysis was conducted, focusing on the wear of the balls charge during the laboratory tests with Mont Reed ore. The results of this analysis are presented in Table 10, shedding light on the wear patterns and behaviour observed during the grinding tests.

TABLE 10

Consumption of charge of balls.

Mill type	Ball charge mass loss (g)	Ball charge consumption (kg/t)	Ball charge consumption (%)
Ball Mill	13	2.6	100
Vertical Mill	7	1.4	54
Difference		1.2	46

In this case, the vertical mill demonstrated a 46 per cent lower energy consumption compared to the ball mill. Based on the evaluation of various aspects, it is evident that the vertical mill emerges as a superior choice for performing fine grinding operations.

Vertical stirred mills constrain

While the results presented above are promising and favour the use of vertical mills in the mineral industry, there are certain challenges that need to be addressed by the suppliers. One such challenge is to focus on making the maintenance of vertical mills more accessible and streamlined. Simplifying maintenance procedures would help in consolidating the use of this technology in various mineral processing operations (Rosa, Oliveira and Donda, 2014).

Another drawback associated with vertical mills is the limited availability of high-capacity machines in the market. This limitation poses a challenge when dealing with large-scale operations that require high flow rates. In such cases, multiple mills may be necessary to achieve the desired comminution capacity. Addressing this limitation and expanding the availability of high-capacity vertical mills could facilitate their wider adoption in industrial-scale applications.

In conclusion, while the results favour the use of vertical mills, addressing maintenance issues and expanding the availability of high-capacity machines will be crucial for maximising the potential of this technology in the mineral sector.

CONCLUSIONS

In this study, a comparison between tubular ball mills and vertical stirred mills was conducted to achieve a material with a P_{80} of 53 µm, using the feed from the secondary grinding stage of Mont Reed circuit.

The analysis revealed that the ball mill consumed 2.46 times more energy than the vertical mill while producing a material with a Blaine value 21 per cent smaller. Both ground products exhibited similar P_{80} and P_{10} values, but the vertical mill was more efficient in producing particles within the range of 40 µm to 2 µm. However, the vertical mill encountered challenges in breaking coarse particles.

In terms of liberation degree, both grinding methods resulted in high values, but the vertical mill proved more efficient in liberating the gangue present in the ore. This was likely due to the vertical mill's ability to detach mineral grains from mixed particles more effectively, primarily through lower energy impacts.

When evaluated within a concentration process, the higher liberation degree of the vertical mill product proved advantageous for the recovery of magnetite during the low intensity magnetic separation stage. However, the increased quantity of ultrafine particles in the vertical mill product posed difficulties for the flotation efficiency.

In a simplified economic analysis, the vertical mill emerged as an interesting option for fine grinding processes. Specifically, the price of the vertical mill was found to be 42 per cent lower than the ball mill and the energy costs were 58 per cent lower for the same machine.

Overall, these findings suggest that the vertical mill offers significant advantages in terms of energy efficiency, liberation degree and economic viability for fine grinding operations, making it a compelling choice for Mont Reed project.

REFERENCES

Belissont, R, 2021. Mont Reed Process Route Development Phase I – Samples Characterisation, ArcelorMittal Maizières Research Centre technical report.

Bergerman, M G and Delboni, H, 2019. Development and validation of a simplified laboratory test to design vertical stirred mills, *KONA Powder and Particle Journal*, 10.14356.

Bottosso, N, Mesquita, J, Araujo, A C and Guimarães, 2023. Effect of mineralogy on grinding energy of Mont-Wright iron ore, paper presented to the 13th International Comminution Symposium (Comminution '23).

Donda, J D, 2003. Um método para prever o consumo específico de energia na (re)moagem de concentrados de minérios de ferro em moinho de bolas, Phd thesis, Universidade Federal de Minas Gerais, Belo Horizonte.

Esteves, P M, Mazzinghy, D B, Galéry, R and Machado, L C R, 2021. Industrial vertical stirred mills screw liner wear profile compared to discrete element method simulations, *Minerals*, 11:397.

Ibrango, S, 2013. Technical note, 2010–061 Mount-Reed resource modelling 2012, Met-Chem technical report for ArcelorMittal Mines Canada.

Info Mine USA Inc, 2010. *Mine and mill equipment costs – An estimator's guide*, pp 44–48 (InfoMine USA Inc.; Aventurine Engineering, Inc.: Washington).

Jankovic, A, 2000. Fine grinding in Australian minerals industry, *Journal of Mining and Metallurgy*, 36(1–2)A:51–61.

Klein, C, 1978. Regional metamorphism of Proterozoic iron-formation, Labrador Trough, Canada, *Am. Mineral*, 63:898–912.

MARC Technologies Pty, 2014. Cyclosizer model M16 operational and maintenance manual, MARC Technologies Pty manual.

Mesquita, J, 2021. Mont Reed Process Route Development Phase II, ArcelorMittal Maizières Research Centre technical report.

Morrison, R D, Cleary, P W and Sinnott, M D, 2009. Using DEM to compare the energy efficiency of pilot scale ball and tower mills, *Minerals Engineering*, 22:665–672.

Québec Cartier Mining Company, 1977. Mt Reed Iron Ore Project, Québec Cartier Mining Company technical report.

Rosa, A C and Donda, J D, 2014. A Lei de Moagem – Comprovação para minério de ferro, 1st ed (Livraria and Editora Graphar: Ouro Preto).

Rosa, C A, Oliveira, P S and Donda, J D, 2014. Comparing ball and vertical mills performance: an industrial case study, in *Proceedings XXVII International Mineral Processing Congress 2014*, ch 8 (Gecamin: Santiago).

Rowland, C A, 1984. Testing for selection of comminution circuits to prepare concentration feed, in Proceedings of the Australasian Institute of Mining and Metallurgy 1984.

Xiao, X, Zhang, G, Feng, Q, Xiao, S, Huang, L, Zhao, X and Li, Z, 2012. The liberation effect of magnetite fine ground by vertical stirred mill and ball mill, *Minerals Engineering*, 34(2012):63–69.

Effect of hydrodynamic parameters on iron ore reverse flotation

M Safari[1], F S Hoseinian[2], D Deglon[3], L Leal Filhoc[4] and T C Souza Pinto[5]

1. Minerals Processing Division, Mintek, Randburg 2125, South Africa.
   Email: MehdiS@Mintek.co.za
2. Department of Mining and Metallurgical Engineering, Amirkabir University of Technology, Tehran, Iran.
3. Centre for Minerals Research, Department of Chemical Engineering, University of Cape Town, Cape Town 7700, South Africa.
4. Mining and Petroleum Department, Polytechnic Engineering School, University of São Paulo, São Paulo, Brazil.
5. Mineral Development Center – CDM/Vale, BR381 KM450, 33040–900 Santa Luzia, MG, Brazil.

ABSTRACT

In this study, the effect of hydrodynamic parameters on the iron ore reverse was investigated in a mechanical flotation cell. The work carries out using the case study of Iron ore from Vale in Brazil. The influence of hydrodynamic factors such as froth height, impeller speed, superficial gas velocity and solids percentage were evaluated on the quartz flotation. The results showed that the hydrodynamic conditions have a great effect on the flotation performance, especially on the fine and coarse quartz. The size-by-size analysis of the froth products showed that froth height, impeller speed, superficial gas velocity and solids percentage have a significant effect on true flotation quartz particles. The increase in the froth height decreased the recovery of quartz particles. The increase in the impeller speed increases the recovery of quartz particles to an optimum. The increase in the superficial gas velocity improved the flotation of coarse quartz particles. The increase in the solids percentage increases the recovery of quartz particles.

INTRODUCTION

Flotation is a selective separation process for separating different minerals from each other. The flotation separation efficiency directly depends on recoveries of the mineral species result in true flotation and entrainment (Hoseinian *et al*, 2021; Neethling and Cilliers, 2009; Wang *et al*, 2015). The chemical variables and hydrodynamic condition within the flotation cells effect on the recovery and selectivity of the flotation process (Hoseinian *et al*, 2018, 2019b; Nazari *et al*, 2023). Cationic reverse flotation of quartz is a common method for selective separation of quartz from iron ores to product the pellet feed (Safari *et al*, 2018). There are two major challenges in the reverse flotation of quartz: the present of coarse quartz particles in the tailing and fine iron oxide particles in the concentrate (Safari *et al*, 2021; Vieira and Peres, 2007).

Flotation is a size dependent process which each particle, ie fine, intermediate and coarse particles has different flotation behaviour during the flotation process (Farrokhpay and Fornasiero, 2017; Safari *et al*, 2022). The flotation recovery versus particle size curves are the useful technique to determine the maximum size of particle for adequate flotation in various chemical and hydrodynamic conditions. The flotation efficiency of coarse particles is low due to high collision probability and low adhesion probability during the flotation process in the high turbulent regions of cells (Farrokhpay and Fornasiero, 2017; Safari and Deglon, 2020). The adhesion probability increases with decreasing particle size. A high content of fine particle size in the reverse cationic flotation of iron ores increases the required collector dosage and reduces the selectivity (Filippov, Severov and Filippova, 2014). Vieira *et al* (2007) studied the effect of type and concentration of collector, pH and particle size on the flotation performance of quartz particles in a mechanical cell. They reported that the recovery of quartz particles with different size can be effectively floated using appropriate collectors, so that the collector of ether diamine for the flotation of medium and coarse quartz and ether monoamine for flotation of fine quartz are more effective (Vieira and Peres, 2007). Ma, Marques and Gontijo (2011) compared the reverse cationic/anionic flotation for iron ore in batch scale flotation tests. They reported that the metal loss of ultrafine particles in reverse cationic flotation is higher than that in reverse anionic flotation due to the entrainment of hematite ultrafine particles. While, in the coarse particle size range, the performance of reverse anionic flotation is slightly less than reverse cationic flotation (Ma, Marques and Gontijo, 2011).

The recovery of fine particles due to the entrainment can be related to the water recovery (Cilek, 2009). It can be explained by the mechanisms of the hydrodynamic boundary layer, the wake and swarm effect of rising bubbles (Hoseinian *et al*, 2020). The size and rising velocity of bubbles are depended to operational factors including the type and concentration of frother, the froth removal rate, froth residence time, airflow rate and impeller speed (Hoseinian, Irannajad and Safari, 2017; Hoseinian *et al*, 2019a). In addition, the flotation cell hydrodynamics determined the probability of bubble particle collision, as well as the probability of bubble particle detachment, which is strongly depend on the particle size, bubble size and the flotation cell turbulence (Cilek and Yılmazer, 2003; Safari *et al*, 2016). Therefore, it is apparent that the entrainment and true flotation strongly depends on the flotation cell hydrodynamic conditions. By far the majority of iron flotation research, particularly with respect to chemical condition such as reagent selection for investigate the flotation performance, is carried out using mechanical cell (Araujo, Viana and Peres, 2005; Mowla, Karimi and Ostadnezhad, 2008; Pavlovic and Brandão, 2003; Yuhua and Jianwei, 2005).

Previous test work done by other researchers on a number of Brazilian iron ores (Lima *et al*, 2016; Lima, Valadão and Peres, 2013; Ma, Marques and Gontijo, 2011) indicates that while the mechanical flotation cell was efficient at cleaning quartz particles between 10 and 150 μm, the collection efficiency decreased significantly for larger particles (Hassanzadeh *et al*, 2022). In this study, the effects of various hydrodynamic operating parameters including solid percentage, particle size, impeller speed, froth height and superficial gas velocity on the hematite recovery with a view to evaluate the entrainment during the process were investigated.

MATERIAL AND METHODS

The iron sample was obtained from the rougher feed from Vale, Brazil. To ensure that all samples were representative, feed sample were blended and divided into individual feed samples for the flotation tests using two different rotary sample dividers in two different steps. The elemental composition of the iron sample was determined by XRD and ICP analysis (64.4 per cent Fe_2O_3 and 34.9 per cent SiO_2). Ether dodecyl amine (EDA) was received in liquid form and used as collector. MIBC (methyl isobutyl carbinol) was used as a frother. Starch was received in powder form and used as a depressant. Reagents were prepared on a daily basis using deionised water and adjusted to the appropriate pH by NaOH and HCl, if required. All reagents were of the analytical grade quality. The conditions for the flotation experiments are presented in Table 1.

Flotation tests were carried out using a Leeds laboratory flotation cell. Several concentrates were collected at the specified times during the flotation experiments. At the end of the flotation experiment, the concentrate samples and tail were separately filtrated, dried and weighed for mineral recovery calculation, screening in seven different size fractions and sizing using a Malvern Mastersizer™ 2000. The flotation products were subjected to a chemical analysis for the determination of the content of total Fe and SiO_2.

TABLE 1

Experimental conditions for flotation experiments.

Condition/Parameter	Value
Feed mass (Hematite/Quartz Ore)	1–3 kg
Froth height (Relative to cell height)	5.60 to 33.30%
Impeller speed	600 to 1500 rev/min
pH	10.5

RESULT AND DISCUSSION

The particle size and density of hematite and quartz particles have a significant influence on their flotation. In order to evaluate the effect of particle size on the hematite flotation efficiency, flotation experiments were carried out in different conditions in which some samples of the concentration were collected during different times for chemical and size distributions analyses. The result of one flotation experiment has been shown in Figure 1. The mass recovery and water recovery increased with increasing time during flotation. The particle size has a main effect on flotation efficiency. In this regard, the effect of hydrodynamic conditions on process was evaluated with emphasize on particle size using a recovery-size curve.

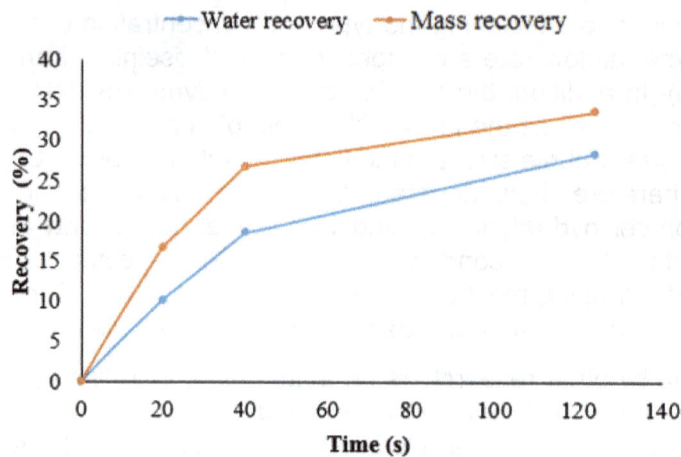

FIG 1 – Effect of particle size on the entrainment of hematite.

Forth height

To determine the effect of froth height on recovery of quartz particles, flotation experiment were performed as a function of froth height at a fixed impeller speed, flotation time, collector dosage, depressant dosage and frother dosage. The results are shown in Figure 2. The mass recovery and water recovery decreased almost linearly with increasing froth height due to increasing froth residence time which promotes the drainage of both particles and water from the froth phase back to the pulp phase (Wang *et al*, 2015). The recovery of quartz particles for different particle sizes and froth height shows that maximum peaks for the particle size of 31 µm can be attributed to the entrainment phenomena. Furthermore, maximum peaks for particle size of 126 µm can be due to both of the increasing attachment efficiency and decreasing detachment efficiency in this particle size fraction. The coarse quartz particle recovery also decreases almost linearly with increasing froth height. The increase in the particle size led to a more decrease in the coarse particle recovery rather than the decrease in the fine particle recovery. The increase of bubble drainages causes the detachment of coarse quartz particles with higher density that have less chance to be transported to the concentrate and settled more quickly and backed to the pulp phase.

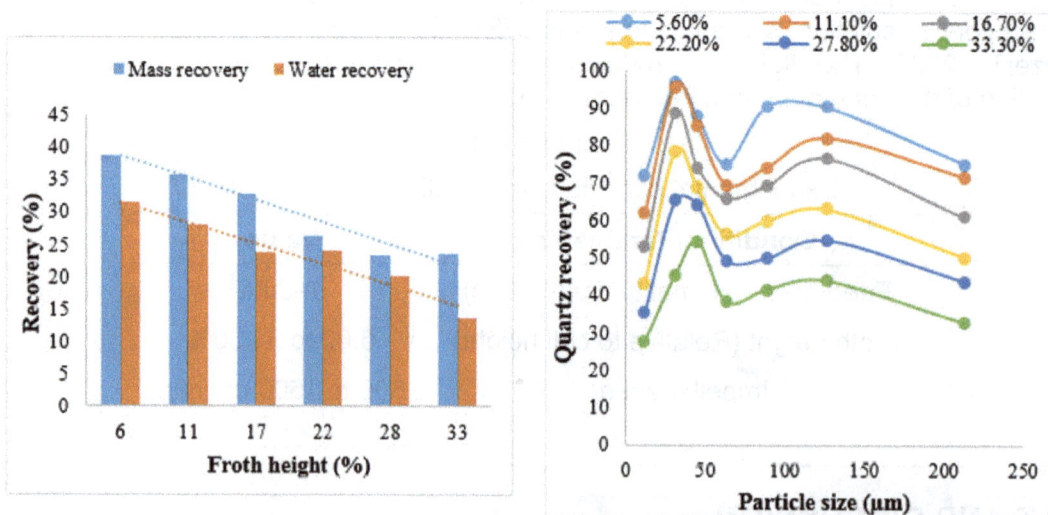

FIG 2 – Effect of froth height on the recovery of quartz particles (impeller speed = 1200 rev/min, Jg = 0.6 cm/s, solid percentage = 25 per cent, depressant dosage = 1000 g/t, collector dosage = 100 g/t).

Impeller speed

The impeller speed relate directly to sub-processes of the air dispersion into bubbles, particle suspensions in the pulp, the probabilities of collision, attachment and detachment of particle–bubble. To determine the effect of impeller speed on the process, flotation experiment were performed as a function of impeller speed at a fixed froth height, flotation time, collector dosage, depressant dosage and frother dosage. The results are shown in Figure 3. The results show that the impeller speed has a strong effect on the recovery of quartz particles. The increase of impeller speed significantly increased particle–bubble detachment of coarser particles but has a lesser effect on fine particles. The pulp phase detachment of solid particles from rising particle-bubble aggregates plays an important role in the reduction in recovery of coarse quartz particles. Mass recovery and water recovery increased with the increase in impeller speed from 600 rev/min to 1200 rev/min, while it decreased at impeller speed of 1350 rev/min. The recovery of coarse quartz particles increased with increasing impeller speed to an optimum, after which they decrease. It is clear from this figure that increasing impeller speed beyond the optimum leads to very large decreases in coarse quartz recovery (Wang et al, 2015). It is clear from this figure that there is an optimum impeller speed range for coarse quartz particle recovery, beyond which recovery is very poor (<10–20 per cent).

The higher impeller speed enhances the attachment efficiency by increasing the overcoming of the energy barrier to particle-bubble attachment that led to increase of recovery. However, the high impeller speed causes an increase in detachment and disruption of bubble-particle aggregates, especially for coarser particles and also may create more turbulence to the froth and affect the froth stability. It leads to decrease of recovery of coarse quartz particles at impeller speed higher than 1200 rev/min. The results demonstrated that increasing impeller speed generally increases the recovery of coarse quartz particle to an optimum.

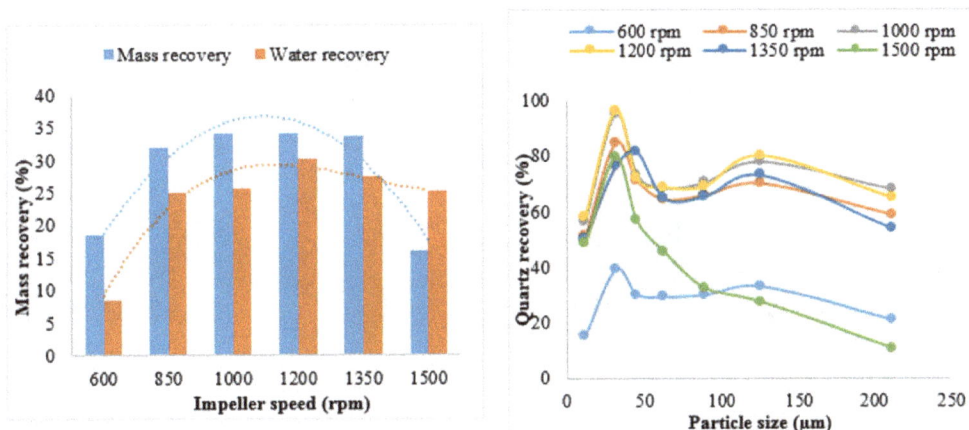

FIG 3 – Effect of impeller speed on the recovery of coarse quartz and fine ore particles (Froth height = 16.7 per cent, Jg = 0.6 cm/s, solid percentage = 25 per cent, depressant dosage = 1000 g/t, collector dosage = 100 g/t).

Superficial gas velocity

The effect of superficial gas velocity on the coarse quartz recovery shown in Figure 4 reveals that at 0.9 cm/s, the maximum coarse quartz recovery with fraction size of 126 µm is 83.76 per cent. It is seen in Figure 4 that change in the superficial gas velocity from 0.4 cm/s to 0.9 cm/s increased the expected value of the mass recovery but the recoveries of water also increased. The high superficial gas velocity increased the recovery of the coarse particles of quartz. Coarse particles are heavy, high superficial gas velocity is preferable in achieving the high coarse quartz recovery. A larger number of bubbles were produced with an increase in the superficial gas velocity that carry more water into the froth phase from the pulp phase. Furthermore, superficial gas velocity also affects the extent of water drainage during the process.

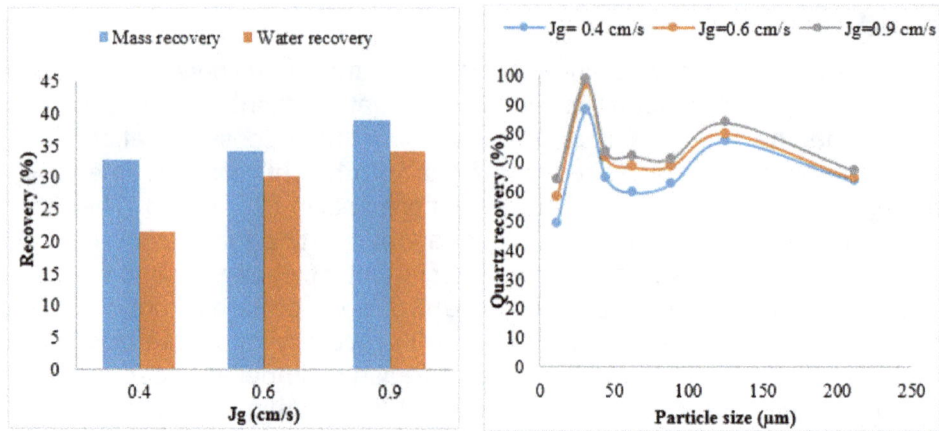

FIG 4 – Effect of superficial gas velocity on the recovery of coarse quartz and fine ore particles (impeller speed = 1200 rev/min, Froth height = 16.7 per cent, solid percentage = 25 per cent, depressant dosage = 1000 g/t, collector dosage = 100 g/t).

Solid percentage

Solids percentage in the pulp has a significant effect on the mass recovery and water recovery, as shown in Figure 5 which demonstrates the effect of solids percentage, froth height and particle size on the recovery by true flotation. The solid dispersion in the pulp below the pulp/froth interface depends on some parameters such as the particle size and impeller speed that determine the amount of solids entrained to the concentrate. The recovery of quartz particles larger than 126 decreased, which indicates that the coarse quartz particles tend to settle in flotation cell. Also, coarse hematite particles are less entrained to the concentrate.

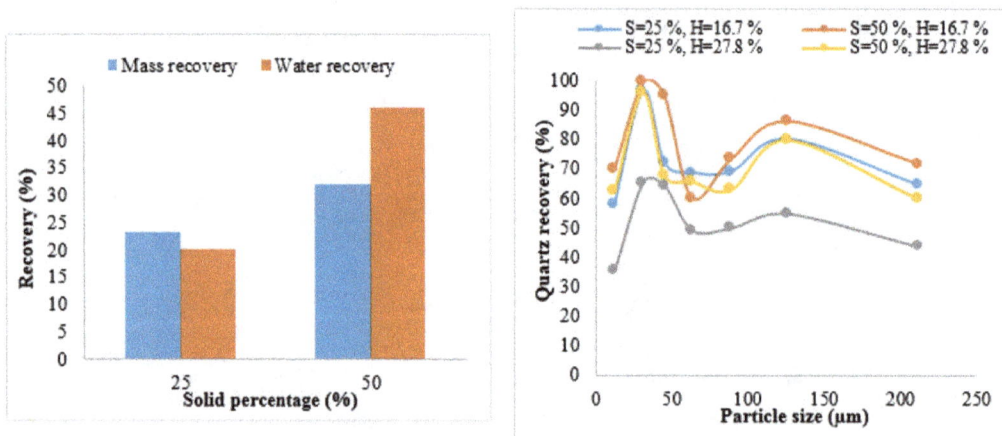

FIG 5 – Effect of solids percentage on the recovery of coarse quartz and fine ore particles (impeller speed = 1200 rev/min, Jg = 0.6 cm/s, depressant dosage = 1000 g/t, collector dosage = 100 g/t).

CONCLUSION

The effects of particle size on the true flotation during the reverse flotation of hematite were performed to assess the influence of hydrodynamic operating parameters such as froth height, impeller speed, superficial gas velocity and solids percentage. The results showed that the hydrodynamic parameters have high effect on the efficiency of reverse flotation of hematite. The increase in the froth height decreased the mass and water recovery due to the increasing froth residence time and water drainage. The recovery of coarse quartz particles decreased with increasing the froth height due to the increase of bubble drainages and detachment of coarse quartz particle. In the mechanical flotation cell, the impeller speed adjusts the dispersion and mixing of air and solids in the pulp phase and probabilities of collision, attachment and detachment of particle–bubble. The effect of impeller speed indicated that increasing impeller speed increases the recovery of coarse quartz particle to an optimum. With increasing the superficial gas velocity and solid percentage, the recovery of quartz particles increased. The results showed that the hydrodynamic conditions have a great effect on the reverse flotation of quartz, especially on the coarse quartz flotation.

REFERENCES

Araujo, A, Viana, P and Peres, A, 2005. Reagents in iron ores flotation, *Minerals Engineering*, 18(2):219–224.

Cilek, E and Yılmazer, B, 2003. Effects of hydrodynamic parameters on entrainment and flotation performance, *Minerals Engineering*, 16(8):745–756.

Cilek, E C, 2009. The effect of hydrodynamic conditions on true flotation and entrainment in flotation of a complex sulphide ore, *International Journal of Mineral Processing*, 90(1–4):35–44.

Farrokhpay, S and Fornasiero, D, 2017. Flotation of coarse composite particles: Effect of mineral liberation and phase distribution, *Advanced Powder Technology*, 28(8):1849–1854.

Filippov, L, Severov, V and Filippova, I, 2014. An overview of the beneficiation of iron ores via reverse cationic flotation, *International Journal of Mineral Processing*, 127:62–69.

Hassanzadeh, A, Safari, M, Khoshdast, H, Güner, M K, Hoang, D H, Sambrook, T and Kowalczuk, P B, 2022. Introducing key advantages of intensified flotation cells over conventionally used mechanical and column cells, *Physicochemical Problems of Mineral Processing*, 58.

Hoseinian, F S, Irannajad, M and Safari, M, 2017. Effective factors and kinetics study of zinc ion removal from synthetic wastewater by ion flotation, *Separation Science and Technology*, 52(5):892–902.

Hoseinian, F S, Rezai, B, Kowsari, E and Safari, M, 2018. Kinetic study of Ni (II) removal using ion flotation: Effect of chemical interactions, *Minerals Engineering*, 119:212–221.

Hoseinian, F S, Rezai, B, Kowsari, E and Safari, M, 2019a. Effect of impeller speed on the Ni (II) ion flotation, *Geosystem Engineering*, 22(3):161–168.

Hoseinian, F S, Rezai, B, Kowsari, E and Safari, M, 2020. The effect of water recovery on the ion flotation process efficiency, *Physicochemical Problems of Mineral Processing*, 56.

Hoseinian, F S, Rezai, B, Safari, M, Deglon, D and Kowsari, E, 2019b. Effect of hydrodynamic parameters on nickel removal rate from wastewater by ion flotation, *Journal of Environmental Management*, 244:408–414.

Hoseinian, F S, Rezai, B, Safari, M, Deglon, D and Kowsari, E, 2021. Separation of nickel and zinc from aqueous solution using triethylenetetramine, *Hydrometallurgy*, 202:105609.

Lima, N P, de Souza Pinto, T C, Tavares, A C and Sweet, J, 2016. The entrainment effect on the performance of iron ore reverse flotation, *Minerals Engineering*, 96:53–58.

Lima, N P, Valadão, G E S and Peres, A E C, 2013. Effect of particles size range on iron ore flotation, *Rem: Revista Escola de Minas*, 66(2):251–256.

Ma, X, Marques, M and Gontijo, C, 2011. Comparative studies of reverse cationic/anionic flotation of Vale iron ore, *International Journal of Mineral Processing*, 100(3–4):179–183.

Mowla, D, Karimi, G and Ostadnezhad, K, 2008. Removal of hematite from silica sand ore by reverse flotation technique, *Separation and Purification Technology*, 58(3):419–423.

Nazari, S, Hoseinian, F S, Li, J, Safari, M, Khoshdast, H, Li, J and He, Y, 2023. Synergistic effect of grinding time and submicron (nano) bubbles on the zeta potential state of spent lithium-ion batteries: A gene expression programming approach, *Journal of Energy Storage*, 70:107942.

Neethling, S and Cilliers, J, 2009. The entrainment factor in froth flotation: Model for particle size and other operating parameter effects, *International Journal of Mineral Processing*, 93(2):141–148.

Pavlovic, S and Brandão, P R G, 2003. Adsorption of starch, amylose, amylopectin and glucose monomer and their effect on the flotation of hematite and quartz, *Minerals Engineering*, 16(11):1117–1122.

Safari, M and Deglon, D, 2020. Evaluation of an attachment–detachment kinetic model for flotation, *Minerals*, 10(11):978.

Safari, M, Harris, M, Deglon, D, Leal Filho, L and Testa, F, 2016. The effect of energy input on flotation kinetics, *International Journal of Mineral Processing*, 156:108–115.

Safari, M, Hoseinian, F, Deglon, D A, Leal Filho, L d S and Pinto, T C d S, 2021. Investigating the effect of operational parameters on the flotation of itabirite iron ore in a pneumatic flotation cell, in Proceedings IMPC2020.

Safari, M, Hoseinian, F, Deglon, D A, Leal Filho, L d S and Pinto, T C d S, 2018. Investigation of the reverse flotation of hematite in three different types of laboratory flotation cells, in Proceedings IMPC 2018.

Safari, M, Hoseinian, F, Deglon, D, Leal Filho, L and Pinto, T S, 2022. Impact of flotation operational parameters on the optimization of fine and coarse Itabirite iron ore beneficiation, *Powder Technology*, 408:117772.

Vieira, A M and Peres, A E, 2007. The effect of amine type, pH and size range in the flotation of quartz, *Minerals Engineering*, 20(10):1008–1013.

Wang, L, Peng, Y, Runge, K and Bradshaw, D, 2015. A review of entrainment: Mechanisms, contributing factors and modelling in flotation, *Minerals Engineering*, 70:77–91.

Yuhua, W and Jianwei, R, 2005. The flotation of quartz from iron minerals with a combined quaternary ammonium salt, *International Journal of Mineral Processing*, 77(2):116–122.

Coarse processing of South African banded iron formation (BIF)

A Singh[1], N Maistry[2] and C Carelse[3]

1. MAusIMM, Senior Technical Specialist: Physical Separation, Mintek, South Africa. Email: ashmas@mintek.co.za
2. Senior Engineer, Mintek, South Africa. Email: nicholem@mintek.co.za
3. Senior Scientist: Mineralogy, Mintek, South Africa, Email: candicec@mintek.co.za

ABSTRACT

Banded Iron Formations or more commonly known as BIF are sedimentary rock formations with alternating silica-rich layers and iron-rich layers that are mainly composed of iron oxides (hematite and magnetite), iron-rich carbonates, and/or iron-rich silicates. Considering the depletion of high-grade iron ore reserves in South Africa, there is a need to maximise resource utilisation through the exploitation of low-grade/uneconomic resources particularly BIF material, which makes up more than 65 per cent of current reserves. Previous test work conducted by alternative BIF sources globally indicates the need for fines processing in order to produce a concentrate. South African BIF however is amenable to coarse and fines processing producing an on-grade product or blended feedstock.

C-Type material within South Africa's Northern Cape Province grading at approximately 33 per cent Fe was supplied for research purposes. Top size optimisation via gravity Heavy Liquid Separation (HLS) tests was conducted. The laboratory results being perfect separation (Ep=0) were modelled to predict pilot performance, the results of which were tested on a pilot scale for the finest size fraction (-1 mm). It was observed that a top size of -2 mm provided optimum results in terms of upgrade and recovery, with crushing finer not necessarily producing superior results. At a target grade of 60 per cent Fe, recoveries of 65.6 per cent were achieved on the laboratory scale, whilst recoveries of approximately 58 per cent were achieved on the pilot scale. Targeting higher upgrades of 63 per cent would inevitably result in lower recoveries of 57.2 per cent on a laboratory scale and approximately 54 per cent on a pilot scale. This is in line with grades and recoveries achieved globally for BIF material albeit at finer grind sizes.

INTRODUCTION

Ferrous minerals (eg iron ore, manganese, chrome etc) are one of the four most important commodity sectors for South Africa's economy with iron ore alone accounting for about 15 per cent of the total mineral sales.

Kumba Iron Ore (KIO) a subsidiary of global miner Anglo American and Assmang a JV of Assore and African Rainbow Minerals (ARM) are South Africa's primary iron ore producers accounting for greater than 80 per cent of iron ore production within the country. Their life-of-mine (LOM) forecast suggests that their high-grade lumpy material will be depleted within the next two decades (average life of approximately 12–13 years) in the absence of further exploration and discoveries (Anglo American, 2011).

There exists potential to unlock value from low-grade/uneconomic resources namely Banded Iron Formation (BIF) lithology which would further extend mining operations for another two decades. Seeing that BIF is projected to make up more than 65 per cent of current reserves, the study undertaken focused on exploiting this resource.

The steady rise in mining costs and escalating electricity prices have resulted in the higher production cost curves and necessitated South Africa to explore innovative ways of mining and downstream beneficiation to resolve the aforementioned challenges with the objective of growing and sustaining the iron ore reserves aimed at reduced operational and capital costs, improved energy efficiency and utilisation of unexploited reserves (Media, 2019).

BANDED IRON FORMATION (BIF)

BIF are distinctive units of sedimentary rock consisting of alternating layers of iron oxides and iron-poor chert/gangue, which are often red in colour of similar thickness as represented in Figure 1.

Almost all of these formations are of Precambrian age, with most deposits dating to the late Archean period (2500–2800 Ma) (Gross, 1965; Condie, 2015).

FIG 1 – Image of mixed BIF material tested at Mintek.

BIF are thought to have formed in sea water, although initially dissolved iron accumulated during earlier times of oxygen deficiency in the Earth's atmosphere, with iron oxides subsequently deposited as a result of oxygen production by photosynthetic cyanobacteria. This oxygen combined with the dissolved iron in the ocean to form insoluble iron oxides, which precipitated out, forming a thin layer on the ocean floor. BIF's tend to be extremely hard, tough and dense thus making them resistant to erosion (Trendall, 2002). BIF account for more than 65 per cent of primary iron ore reserves in South Africa, thus representing the future of iron ore mining. The most important BIF ore district in South Africa is the Northern Cape Province with C-grade (40–50 per cent Fe) material being utilised for testing.

GEOLOGY OF NORTHERN CAPE C-GRADE MATERIAL

C-grade (40–48 per cent Fe) material from Kolomela located approximately km from Postmasburg in the Northern Cape Province of South Africa was utilised for research purposes. This material is a younger source containing more contaminants than the massive Sishen deposit and hence is an ideal base case for similar ore resources within the region. The Transvaal Supergroup or Griqualand West Supergroup as it is referred to where it occurs in the Northern Cape Province of South Africa is the largest known resource of high-grade hematite (Eriksson, Hattingh and Altermann, 1995; Carney and Mienie, 2013). The Transvaal sedimentation began with predominantly fragments of sedimentary rocks followed by carbonate bearing rocks and BIF (Eriksson, Hattingh and Altermann, 1995). The BIF ore within this region is classified primarily into four ore types namely: i) laminated with or without specular hematite, ii) massive/clastic-textured hematite, iii) brecciated hematite, iv) conglomeratic hematite (Basson *et al*, 2018).

Hematite ore is currently being beneficiated at Sishen, Khumani, Kolomela and Beeshoek. The two latter mines represent the Sishen South deposits which are located approximately 65 km south of the Sishen mine. Sishen and Kolomela are owned by Kumba Resources Limited (a subsidiary of Anglo American) with Khumani and Beeshoek being owned by Assmang (a joint venture between African Rainbow Minerals and Assore). Figure 2 represents the regional geological map of the Kolomela mine reflecting the BIF zones (Basson *et al*, 2018) with Figure 3 being a simplified stratigraphic column depicting the lithology types.

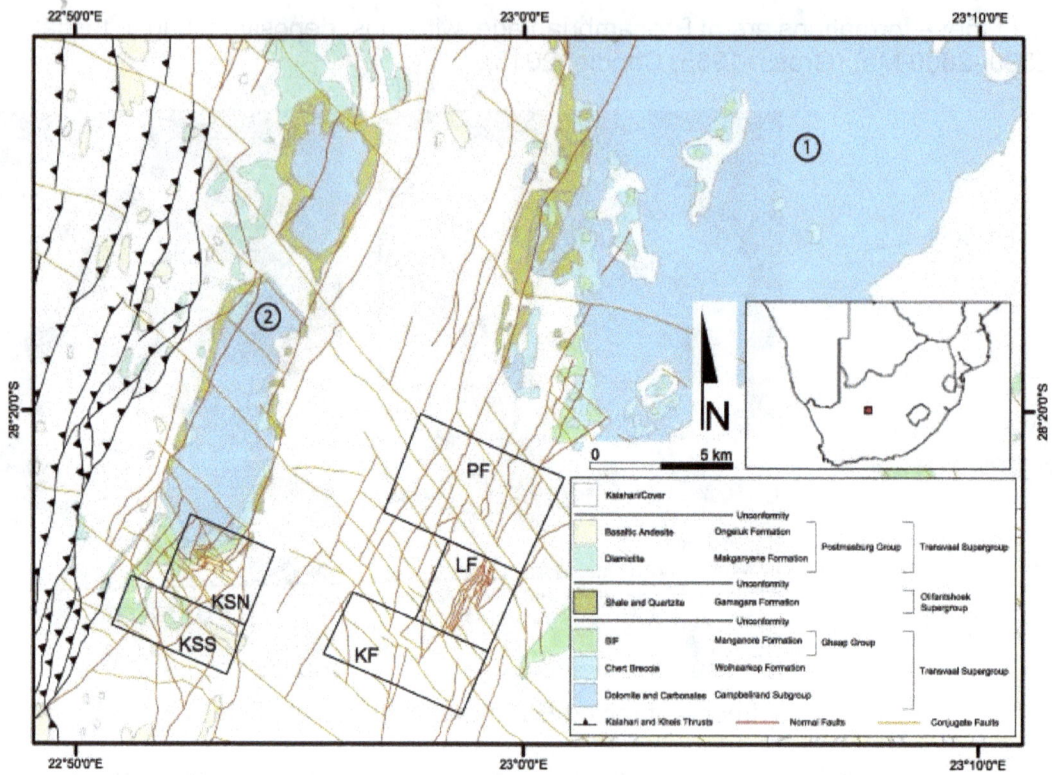

FIG 2 – The structural setting of mineralisation at Kolomela Mine showing BIF ore zones.

FIG 3 – Simplified stratigraphic column depicting the Kolomela local geology.

METHODOLOGY

Samples

Approximately 17 t of moist C-grade material from the Northern Cape Province in South Africa (Kolomela mine) was received, containing both coarse rocks exceeding 21 mm and fines as shown in Figure 4. The 'as received' material was dried, hand sorted to remove the +10 mm rocks which were manually crushed to -10 mm.

FIG 4 – 'As received' C-grade BIF.

Sample preparation

The pre-prepared Run-of-mine (ROM) to -10 mm was screened at 3 mm with the oversize crushed to -3 mm. A 500 kg representative sub-sample was removed for sizing analysis followed by test work. Two alternative representative splits of 500 kg each was removed and prepared to top sizes of -2 mm and -1 mm for comparative top-size optimisation tests to determine the top size that achieves the best Fe product grades and recoveries with the lowest contaminant levels.

The -30+1 mm, -20+1 mm, -12+1 mm size fractions generated for each sample was subjected to Heavy Liquid Separation (HLS) in discrete fractions, the combined overall performance of which is presented within the report. The natural -1.18 mm size fractions generated has been chemically assayed allowing for an overall mass balance to be generated. Gravity separation was considered for these BIF samples in order to ensure low OPEX costs are attained. HLS test work provides a benchmark of what can be achieved on dense media separators. In addition, a representative sub-sample of the feed was crushed and prepared to -1.1 mm for mineralogical evaluation.

Mineralogical evaluation of feed

Feed material was prepared to a top size of -1.1 mm for mineralogical evaluation. The aim of the study is to determine the bulk modal mineralogy, hematite chemistry, as well as hematite grain size, liberation and mineral association. Representative polished sections were prepared for electron microprobe analysis. The feed was screened into different size fractions and mounted into polished sections for analysis by automated scanning electron microscopy, QEMSCAN.

Heavy liquid separation

HLS utilises the differences in particle density to concentrate material into various density classes. The ore is introduced to the fluid medium and the density gradient results in particles that are heavier than the medium to sink, while particles lighter than the medium float. Separation is facilitated by a separation medium as follows:

- For densities less than 2.96 g/cm³ Tetrabromo-Ethane (TBE) and Acetone are used to prepare the mixture.

- For densities between 2.96 g/cm³ and 3.7 g/cm³, TBE and Atomised Ferrosilicon are used.

- For densities at and greater than 3.8 g/cm³, TBE and Tungsten Carbide are used.

For a top size of -3 mm, densimetric (HLS) test work was conducted on -30+2 mm, -20+ 1 mm, -12+ mm, -8+3.3 mm and -3.35+ mm discrete fractions the results of which were mathematically combined to represent a -30+ mm size fraction. The combined headgrade at -30+ mm reported at 32.3 per cent Fe. Similarly for a top size of -2 mm, HLS test work was conducted on -20+1 mm, -12+ mm, -8+3.3 mm and -3.35+ mm discrete fractions the results of which were mathematically combined to represent a -20+ mm size fraction which also reported at 32.3 per cent Fe. Lastly for a top size of -1 mm, HLS test work was conducted on -12+ mm, -8+ 3.3 mm and -3.35+ mm discrete fractions the results of which were mathematically combined to represent a -12+ mm size fraction. The headgrade at -12+ mm reported marginally higher at 32.7 per cent Fe. This was further compared to treating a wide -12+ mm fraction to compare the effect on overall efficiency (upgrade and recovery).

The combined effect for each top size is presented within this paper. Approximately 30–50 kg of each size fraction was separated in the HLS laboratory at ten density cut points namely 3.8, 3.9, 4.0, 4.1, 4.2, 4.3, 4.4, 4.5, 4.6 and 4.7 g/cm³ as per medium combinations discussed above. The HLS procedure employed for this study is outlined in Figure 5.

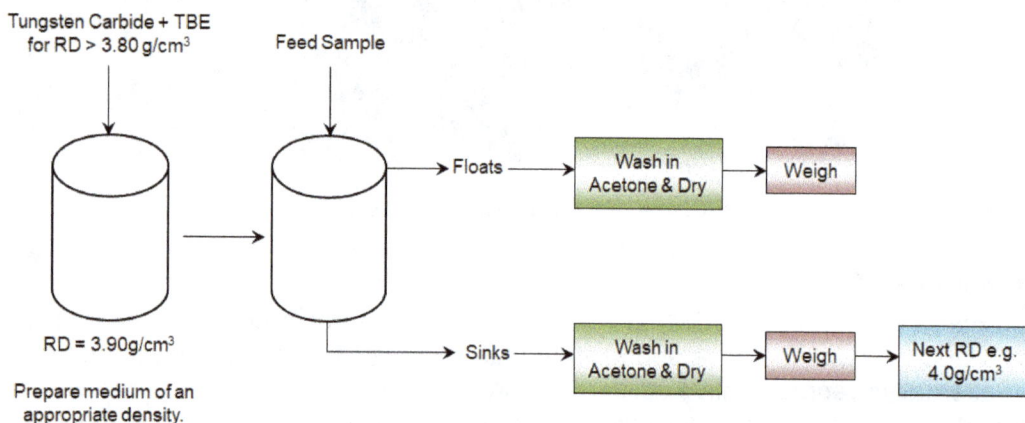

FIG 5 – Overview of the HLS procedure.

The resulting densimetric fractions (ten floats and one sink) for each size fraction was dried, weighed and a sub-sample removed for quantitative chemical analysis to generate washability data and determine the presence of near density material at the cut point that achieves the target grades of 60 per cent Fe and 63 per cent Fe.

Predicted partition curves for dense media separation

Utilising the washability characteristics obtained during HLS test work, partition curves were constructed to predict dense media separation (DMS) performance. An empirical Weibull function, derived by Rao (2004), is used to represent the partition surface of gravity concentrators in terms of size and density attributes as illustrated in Equation 1.

$$Y = 100\left(1 - exp\left(-\left(\tfrac{1}{1-Y_P}\right)\right)\left(\tfrac{\rho}{\rho_P}\right)^{pd^q}\right) \tag{1}$$

Where:

Y	is the Partition number, a function of particle size and density
Y_P	is the Pivot partition number, representing the fraction of bypass in gravity concentrators
ρ_p	is the Pivot density (in kg/m³)
ρ	is the Particle density (in kg/m³)
p	captures viscosity effects
q	represents flow conditions (turbulence) of the separator
d	represents particle size (in mm)

The Weibull model is used in the Mintek Model to predict the performance of DMS operations. The effect of particle size and difference in performance of DMS are built into the Mintek Model. The Mintek Model is therefore able to predict efficiency of separation (Ecart Probable, Ep) at various density cut points for DMS plants. The predicted product yield, iron grade (and main contaminant grades silica and aluminium), iron recovery with the predicted Ep at several density cut points were determined.

The parameters used in the Mintek Model include the geometric mean size, Y_P, ρ_p, p and q. Each of these parameters is related to the particle size, density cutpoint and Ep of the separation. Sets of values for these parameters were derived based on test work done in the past on DMS operations. These parameters are used as the basis on which the Mintek model predicts the performance in DMS plants.

RESULTS AND DISCUSSION

Mineralogical evaluation of feed

Chemical analysis revealed that the average headgrade of the C-grade material reported at 32.9 per cent Fe, 17.2 per cent Si and 4.8 per cent Al. Phosphorus and manganese was found to be negligible within the sample.

Based on mineralogical analysis it was confirmed that the feed sample consists mainly of hematite (49.6 mass per cent) associated with illite and quartz as the major gangue minerals present. Hematite has an average Fe concentration of 68.59 wt per cent. Hematite particles are coarse with 58 per cent of the particles being >106 µm in size.

Liberation classification is based on area percent of the mineral of interest (eg hematite) over the total area of a particle. Discrete liberation data shows that 67.2 mass per cent of hematite is liberated with 6.1 mass per cent being completely locked in gangue. The rest (26.7 mass per cent) falls into the low and high middlings classes (Figure 6).

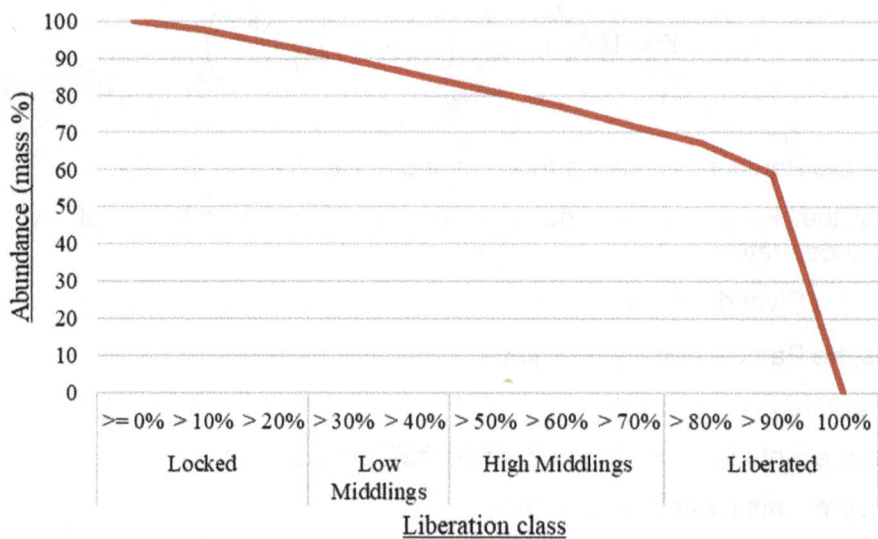

FIG 6 – Cumulative liberation of hematite.

Heavy liquid separation

The unit cumulative iron grade-recovery curve for the mathematically combined individual top size fractions for the C-grade sample is presented in Figure 7. At iron grades of 60 per cent and above all the top size fractions have a similar recovery profile with the coarser size fractions achieving higher recoveries.

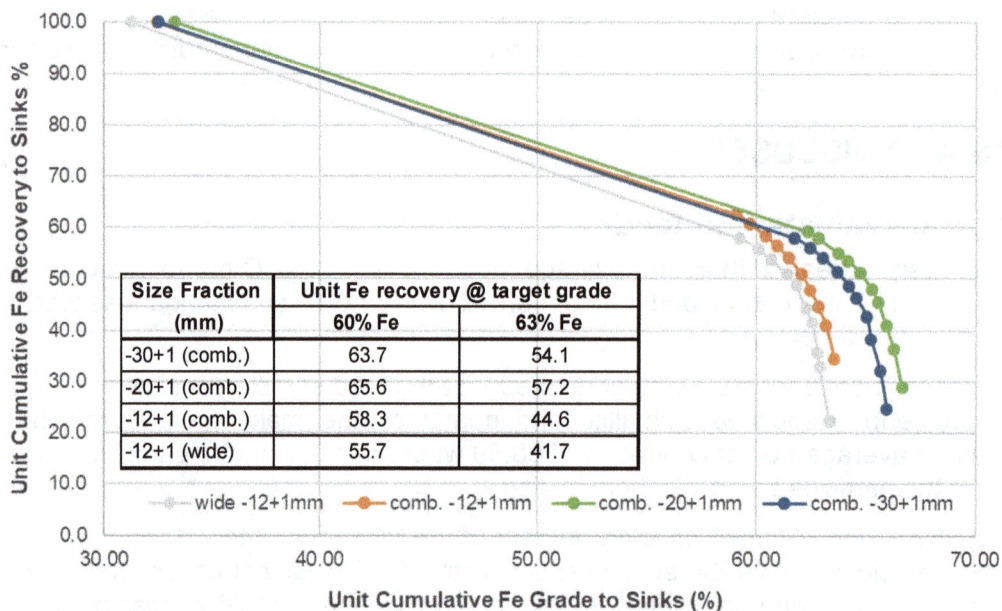

Size Fraction	Unit Fe recovery @ target grade	
(mm)	60% Fe	63% Fe
-30+1 (comb.)	63.7	54.1
-20+1 (comb.)	65.6	57.2
-12+1 (comb.)	58.3	44.6
-12+1 (wide)	55.7	41.7

FIG 7 – Iron grade-recovery curve for C-grade BIF material.

Chemical analysis was conducted to confirm if the contaminant grade specifications for silica, and alumina at the coarser size fractions were met at the target Fe grade specification(s) as represented in Table 1. It can be seen that alumina target is met for both grades however, silica specification is not achieved for the coarsest size class of -30+ mm targeting a grade of 60 per cent Fe. Interesting to note was the -20+ mm top size which produced the highest recovery at 65.6 per cent Fe and 57.2 per cent Fe for grades of 60 and 63 per cent Fe respectively, meeting contaminant target specifications. Crushing coarser seems not to liberate as much Fe from gangue and crushing finer results in Fe being displaced to gangue.

Furthermore, the indicated cut density to achieve a concentrate at 60 and 63 per cent Fe indicates the need for use of an Ultra High Density Dense Medium Separator (UHDDMS) as cut densities range from lowest of 3.74 g/cm³ to highest of 4.6 g/cm³ which is high for a conventional DMS application.

TABLE 1

Cutpoint and contaminant specification at specified target Fe grade.

Size Fraction (mm)	@60%Fe target grade				@ 63%Fe target grade			
	Grade Spec		<3.5%	<1.5%			<3.5%	<1.5%
	Fe Rec (%)	D_{50}	Si (%)	Al (%)	Fe Rec (%)	D_{50}	Si (%)	Al (%)
-30+1mm	63.7	3.75	3.91	0.97	54.10	4.00	2.24	0.72
-20+1mm	65.6	3.74	3.34	0.83	57.20	3.92	2.11	0.57
-12+1mm	58.3	3.94	2.62	0.72	44.60	4.55	1.04	0.59
-12+1mm (wide)	55.7	3.89	3.52	0.96	41.70	4.63	1.53	0.88

Predicted partition curves for dense media separation

The predicted DMS iron grade recovery curve for each top size is compared to the HLS results in Figure 8 which indicates similar responses were achieved, with the DMS results relatively superimposed on the HLS results. Comparing the laboratory HLS results, it can be seen that at a higher target grades a significant drop (10 per cent to 14 per cent) in the iron unit recovery is observed. The same is observed for the modelled DMS data.

The highest upgrade and recovery are obtained at a -20+ mm size class, increasing the top size could have a significant impact on the overall Si content within the final product as demonstrated by the HLS results.

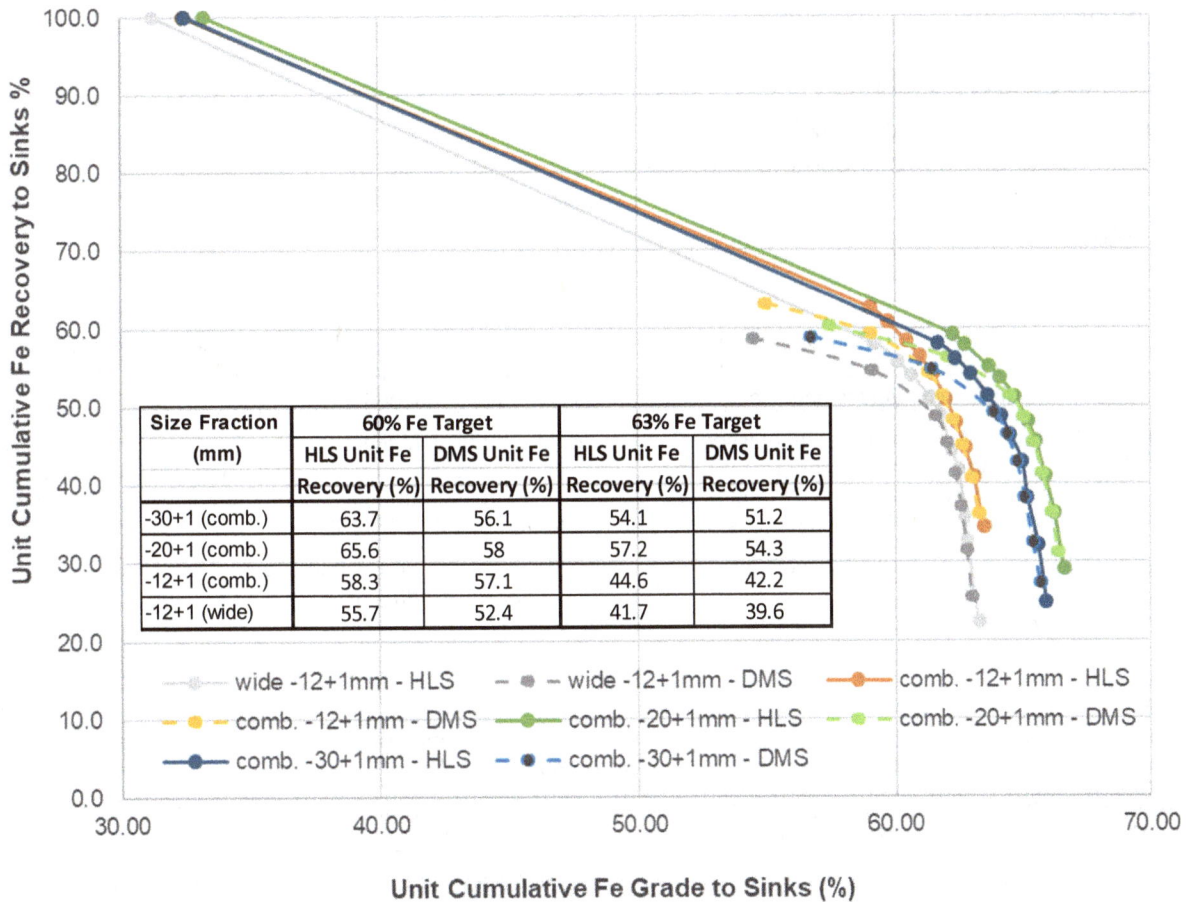

Size Fraction (mm)	60% Fe Target		63% Fe Target	
	HLS Unit Fe Recovery (%)	DMS Unit Fe Recovery (%)	HLS Unit Fe Recovery (%)	DMS Unit Fe Recovery (%)
-30+1 (comb.)	63.7	56.1	54.1	51.2
-20+1 (comb.)	65.6	58	57.2	54.3
-12+1 (comb.)	58.3	57.1	44.6	42.2
-12+1 (wide)	55.7	52.4	41.7	39.6

FIG 8 – Predicted iron grade-recovery curve for DMS in comparison to HLS at various top sizes.

CONCLUSIONS

- South African high-grade iron ore reserves are depleting, hence there is a need to exploit alternative resources in particular BIF which makes up greater than 65 per cent of current reserves in order to maximise resource utilisation and extend life-of-mine.

- Based on historical tests conducted globally, the premise has been that BIF material requires significant fine grinding in order to be upgraded. This is contrary to results obtained for South African BIF whereby coarse beneficiation is applicable to deliver an upgraded product meeting contaminant specifications.

- The C-grade material tested reported a headgrade of 32.9 per cent Fe, 17.2 per cent Si and 4.8 per cent Al with negligible phosphorus and manganese. Based on mineralogical analysis it was confirmed that the feed sample consists mainly of hematite (49.6 mass per cent) associated with illite and quartz as the major gangue minerals present.

- HLS test work revealed that optimum results can be obtained at a top size of -20+ mm achieving recoveries of 65.6 per cent at a grade of 60 per cent Fe and a drop to 57.2 per cent recovery at a slightly higher grade of 63 per cent Fe.

- DMS modelling produced similar results to that of laboratory HLS tests albeit at lower recoveries. This is expected on pilot applications due to inherent inefficiencies in operation.

- The promising results of this study indicate that coarse beneficiation of BIF material is possible. These ore types could be blended with superior grade products to improve overall yields and recoveries or sold directly at a 63 per cent Fe product grade whilst sacrificing recovery.

- For all the top sizes tested the fines concentrate contributes 16.7 per cent to 26.5 per cent to the overall mass yield and thus a simplified flow sheet comprising solely of Dense Media Separation could be used and the -1.18 mm fines stockpiled for later processing via gravity or magnetic separation upgrade.

ACKNOWLEDGEMENTS

The authors would like to acknowledge Mintek for funding the research project and for providing the necessary facilities and resources to successfully complete the study. In addition contribution from South Africa's primary iron ore producers is acknowledged in terms of sample availability for test work purposes.

REFERENCES

Anglo American, 2011. Section 2: Creating a growing and sustainable iron and steel value chain in South Africa, *The South African Iron and Steel Value Chain*. Available from: <https://www.angloamericankumba.com/~/media/Files/A/Anglo-American-Group/Kumba/investors/investor-presentation/iron-steel-1final.pdf>

Basson, I J, Thomas, S A J, Stoch, B, Anthonissen, C J, McCall, M-J, Britz, J, Macgregor, S, Viljoen, S, Nel, D, Vietze, M, Stander, C, Horn, J, Bezuidenhout, J, Sekoere, T, Gous, C and Boucher, H, 2018. The structural setting of mineralisation at Kolomela Mine, Northern Cape, South Africa, based on fully-constrained, implicit 3D modelling, *Ore Geology Reviews*, 95(April):306–324. doi:10.1016/j.oregeorev.2018.02.032.

Carney, M D and Mienie, P J, 2013. A geological comparison of the Sishen and Sishen South (Welgevonden) iron ore deposits, Northern Cape Province, South Africa, *Applied Earth Science*, 112:81–88. Available at: https://www.tandfonline.com/doi/abs/10.1179/0371745032501171.

Condie, K C, 2015. Earth as an Evolving Planetary System, *Earth as an Evolving Planetary System*. doi: 10.1016/C2010-0-65818-4.

Creamer Media, 2020. Iron-Ore 2020: A review of the iron-ore sector, Creamer Media. https://www.creamermedia.co.za/article/iron-ore-2020-a-review-of-the-iron-ore-sector-pdf-report-2020-06-26

Eriksson, P G, Hattingh, P J and Altermann, W, 1995. An overview of the geology of the Transvaal Sequence and Bushveld Complex, South Africa, *Mineralium Deposita*, 30(2):98–111. doi: 10.1007/BF00189339.

Gross, G A, 1965. Origin of precambrian iron formations, *Economic Geology*, pp 1063–1065. doi: 10.2113/gsecongeo.60.5.1063.

Rao, B V, 2004. Weibull partition surface representation for gravity concentrators, *Minerals Engineering*, 17(7–8):953–956. https://doi.org/10.1016/j.mineng.2004.03.001

Trendall, A F, 2002. The Significance of Iron-Formation in the Precambrian Stratigraphic Record, in *Precambrian Sedimentary Environments*, pp. 33–66. doi: 10.1002/9781444304312.ch3.

An innovative application of LJC Automatic Magnetic Flotation Separator in magnetite processing

B Xu[1], C Wang[2] and L Ding[3]

1. Engineer, LONGi Magnet Co. Ltd., Fushun Liaoning 113122, China. Email: info@ljmagnet.com
2. Senior Engineer, LONGi Magnet Co. Ltd., Fushun Liaoning 113122, China. Email: info@ljmagnet.com
3. Manager, LONGi Magnet Co. Ltd., Fushun Liaoning 113122, China. Email: lucyding@ljmagnet.com

ABSTRACT

In this paper, we introduce the term 'LJC' to refer to the LJC Automatic Magnetic Flotation Separator, which is an advanced technology developed by Longi Magnet, integrating various techniques such as hydro separation, magnetic flocculation, and magnetic columns. This technology combines gravity, hydro separation, and magnetic separation to achieve efficient and reliable separation processes. With its low maintenance requirements and high reliability, the LJC technology has proven to be effective in magnetite processing.

The LJC technology offers two primary benefits when compared to traditional magnetic drum separators. Firstly, it can enhance the quality of the targeted or improved concentrate at a coarser product size by minimising the carryover of gangue minerals into the concentrate. Secondly, it can significantly improve the concentrate quality (eg achieving >67 per cent Fe) while maintaining similar liberation size and mass yield. In China, this technology has been widely adopted in magnetite processing, replacing traditional magnetic drum separators and fines flotation in the final stage of the process.

Through collaboration with a magnetite producer in Western Australia, Longi Magnet successfully implemented the next-generation LJC Automatic Magnetic Flotation Separator to replace the original stage 3 magnetic drum separators. The application of the LJC technology resulted in an increase in product P_{80} from 32 μm to 43.1 μm, while maintaining a final concentrate grade of 65 per cent Fe. Additionally, the power consumption, measured in kWh per tonne of concentrate, was reduced by 10 per cent. By eliminating the bottleneck caused by the requirement of a coarser liberation size to achieve the desired concentrate grade, the plant's throughput was increased by significantly.

The LJC technology has demonstrated its ability to reduce the carbon footprint, improve concentrate grade and enhance mass recovery, offering an alternative solution for magnetite processing. The high-grade magnetite produced by the LJC technology is particularly suitable as a raw material for Directed Reduction Iron and green steel production. This paper introduces the principles of the LJC Automatic Magnetic Flotation Separator and provides details of its application in a magnetite operation in Western Australia.

INTRODUCTION OF LJC AUTOMATIC MAGNETIC FLOTATION SEPARATOR

The LJC Automatic Magnetic Flotation Separator (LJC) is an advanced technology developed by LONGi Magnet. It incorporates research on hydro separation, magnetic flocculation, magnetic elutriation, and other equipment to combine gravity, hydro separation, and magnetic separation processes.

The LJC separator consists of several components, including the feed box, electro-magnetic system, water distribution system, overflow cylinder launder, and underflow discharge system. The feed slurry is introduced from the top of the LJC separator, dispersing around the outside of the overflow cylinder. Upcurrent water is supplied to the bottom of the water distribution system (see Figure 1).

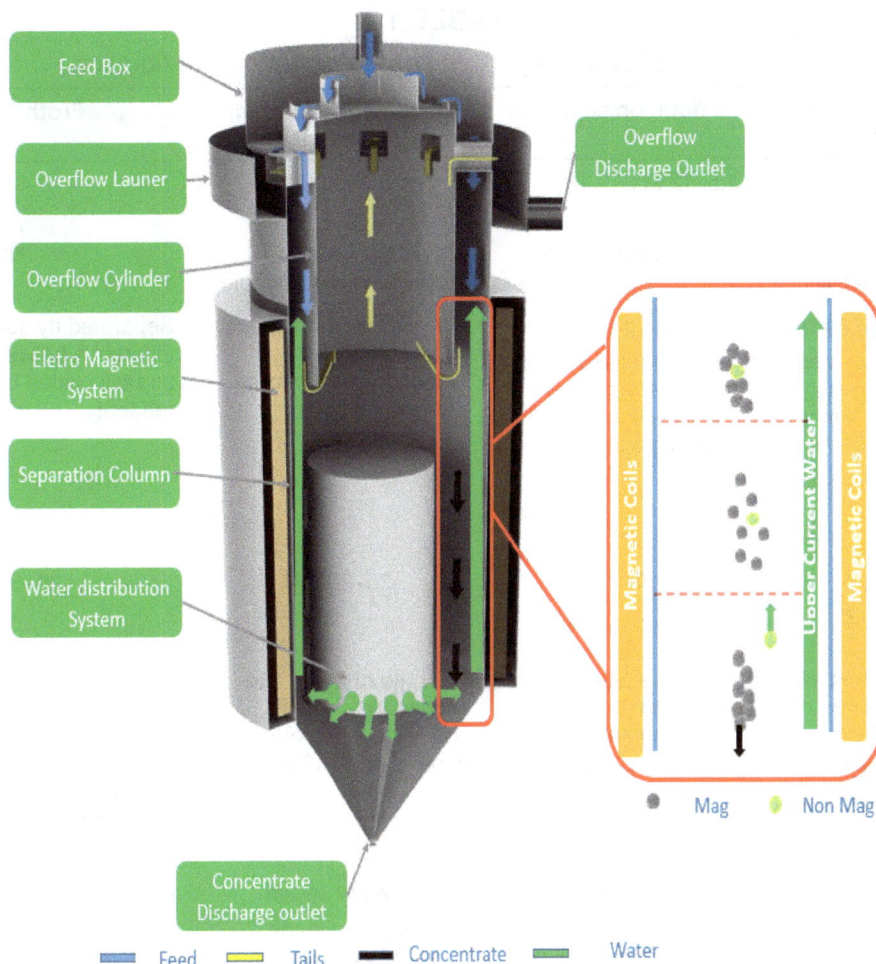

FIG 1 – Schematic of LJC automatic magnetic flotation separator.

Multiple low electro-magnetic fields are applied in the lower section of the separator in an alternating ON/OFF arrangement.

Inside the LJC separator, magnetic particles become magnetised in the electromagnetic fields and agglomerate to form a magnetic chain. The non-electromagnetic field breaks the magnetic chain, while flushing water removes entrapped gangue particles between the magnetite particles. The agglomeration effect enhances the density difference between magnetic and non-magnetic particles, allowing for the recovery of ultrafine magnetite particles and the rejection of coarse gangue with low magnetite content. This helps avoid misplacement of coarse gangue particles that may occur with traditional wet magnetic drum separators.

The innovative 'Feeding all around, centre overflowing' design of the feed and tails system, as well as the water distribution column in the middle, is a patented feature of LJC separators. This design effectively utilises the uneven distribution of magnetic fields within the separator. It facilitates the complete agglomeration, separation and re-agglomeration of magnetic particles for flushing, while reducing overall water consumption.

Each LJC separator is equipped with a control cabinet for local operation and parameter adjustment. Remote automatic control is also possible through Client PLC and SCADA system. The strength of the magnetic fields, flushing water flow rate, ore discharge volume, and concentrate discharge density can be automatically adjusted based on changes in the feed conditions. This makes the LJC separator a stable, automatic, and remotely controlled magnetite concentrator.

LJC separators have been widely adopted in magnetite processing in China as a replacement for traditional magnetic drum separators and fines flotation in the final stage. Table 1 provides a brief comparison among Wet Magnetic Drum Separators, LJC Automatic Magnetic Flotation Separator and Froth Flotation methods.

TABLE 1

Comparison of LJC, mag drum separators and froth flotation.

	LJC magnetic flotation separator	Wet magnetic drum magnetic separators	Froth flotation
Power consumption	1.5–12 kW	5.5–30 kW	2–55 kW
Mag field strength	0–300 Gs (electromagnetic field, adjustable)	800–4000 Gs (fixed magnets, nonadjustable)	N/A
Principle of separation	Mag susceptibility and gravity: • Mag particles magnetised by and agglomerated together forming a mag chain, travelling to underflow due to gravity. • Non-mag particles travelling towards to overflow by up current water	Mag and Non-mags are separated due to magnetic susceptibility	Separated by surface chemistry difference between concentrate and tails aided by chemical reagents
Processing capacity (tph/unit)	15–160 t/h	20–200 t/h	20–300 t/h
Application in magnetite separation	Recleaner or recleaner Stage of Magnetite Separation to improve concentrate grade and separation efficiency, mainly for particles $P_{60} < 75$ µm	All stage of magnetite separation: rougher, cleaner and recleaner particle size ≤12 mm	Mainly used for minimising the mag loss or improve the concentrate garden at final stage processing mainly after recleaner stage of mag dram separation, 45%–98% of -75 µm in feed
Advantage	• Low and adjustable mag field strength • Highly automated operation with multiple operation factors to optimise separation performance • Low/minimum gauges carryover to concentrate or mag loss to tails high separation efficiency • Low operation cost • Reliability and minimal maintenance	• Low water usage • Low tails volume, application to wider size ranges and all stage of mag separation • Low operation cost • Reliability and minimal maintenance	• Low/minimum gauges carryover to • Concentrate or mag loss to tails high separation efficiency
Disadvantage	• Slightly higher process water usage compared to mag drum separators • Only applied to final stage fine feed in magnetite separation	• Unadjustable mag field strength due to fixed magnets • Low separation efficiency • Impurities carryover or entrapment in concentrate	• High operation cost and environmental unfriendly due to chemical reagents required • Unstable operation with multiple factors to be considered • High fluctuation of performance with feed variation: flotation is feed mineralogy dependent not magnetite grade dependent process

APPLICATION OF LJC AT A MAGNETITE OPERATION IN WEST AUSTRALIA

Background and flow sheet

In one of the magnetite processing plants located in Western Australia, a comprehensive two-stage grinding and three-stage magnetic separation process is implemented. The initial objective of the plant was to produce a final concentrate with a 65 per cent Fe grade, targeting a particle size (P_{80}) of 32 µm. However, the plant's production capacity was significantly hindered by limitations within the grinding circuits, thereby impeding the attainment of the desired liberation size (P_{80} = 32 μm) necessary to achieve the targeted Fe grade (65 per cent).

To overcome this operational bottleneck and improve overall efficiency, an extensive collaborative research and testing initiative was undertaken between the magnetite producer and Longi Magnet. This initiative involved rigorous benchtop testing, followed by pilot plant trials and subsequently industrial plant trials. Subsequently, a strategic decision was made to replace the original stage 3 magnetic drum separators with the technologically advanced LJC Automatic Magnetic Flotation Separator (LJC). The primary objective behind this decision was to enhance the separation efficiency and alleviate the operational limitations that were impeding the plant's performance.

Figure 2 illustrates a simplified flow sheet of the project, highlighting the implementation of the LJC-10000 Automatic Magnetic Flotation Separators from Longi Magnet.

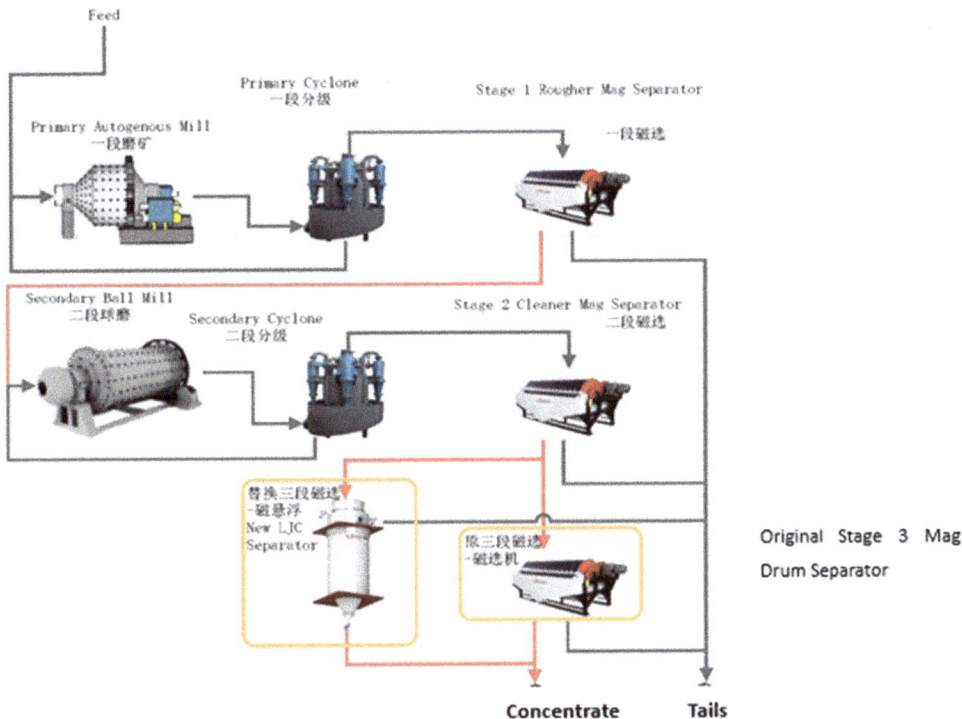

FIG 2 – Simplified flow sheet of LJC separator project at a magnetite process plant in WA.

Plant performance results analysis

To minimise disruptions to production, the project was executed in stages, with the gradual replacement of the stage 3 magnetic drum separators by the LJC separators. This approach provided an opportunity to directly compare the performance of the LJC separators with the original stage 3 magnetic drum separators.

In terms of Fe grade and recovery, the following results were observed under the same feed and upstream conditions.

Fe grade of concentrate

The Fe grade of the LJC Separator concentrate ranged from 1 per cent to 3.2 per cent, which was higher than the average Fe grade of 1.83 per cent obtained from the original three-stage magnetic separation concentrate. This comparison is illustrated in Figure 3.

Fe grade of tails

The Fe grade of the LJC Automatic Magnetic Flotation Separator tails was 2.1 per cent to 10.4 per cent lower than the average Fe grade of 6.16 per cent obtained from the original three-stage magnetic separation tails. This difference is depicted in Figure 4.

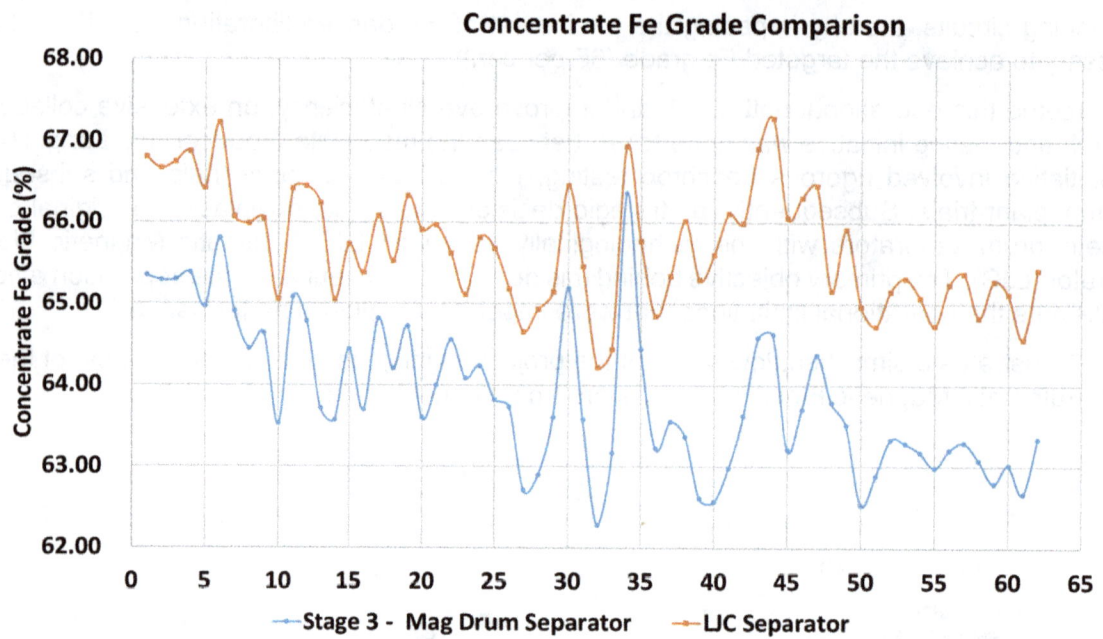

FIG 3 – Comparison chart of concentrate Fe grade.

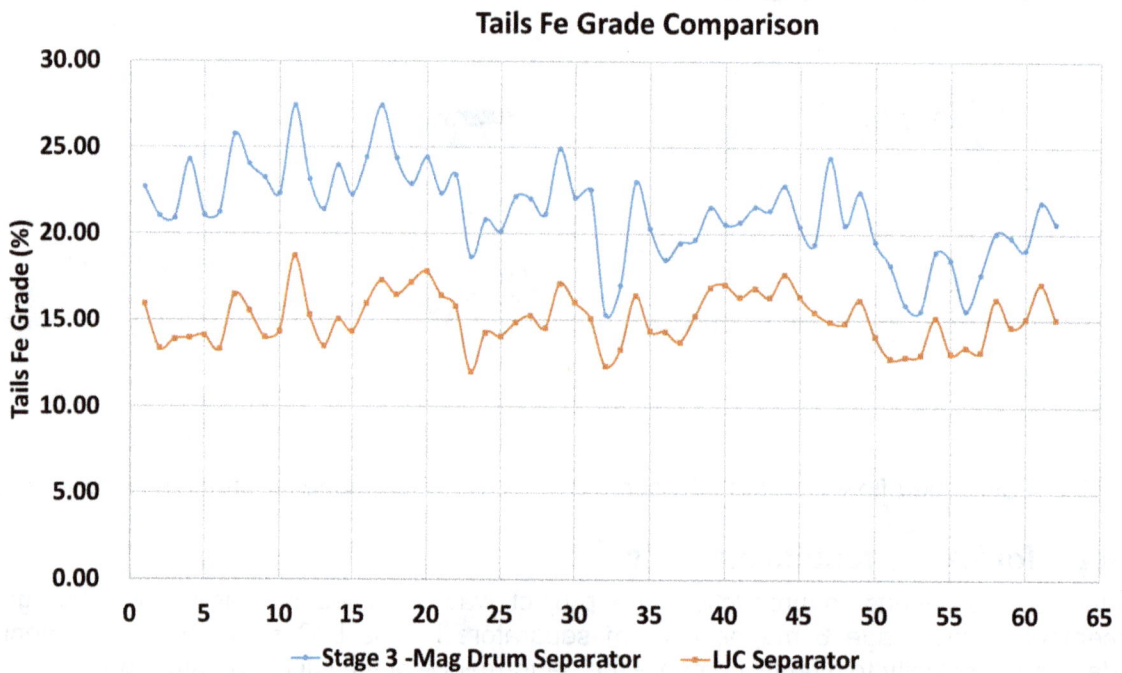

FIG 4 – Comparison chart of tailings Fe grade.

Fe elemental recovery

There was a negligible difference in the Fe elemental recovery of the entire circuit before and after the installation of the LJC Separators. This observation is presented in Figure 5.

These findings highlight the superior performance of the LJC separators in terms of achieving higher Fe grades in the concentrate and reducing Fe content in the tails, while maintaining comparable Fe elemental recovery levels throughout the processing circuit.

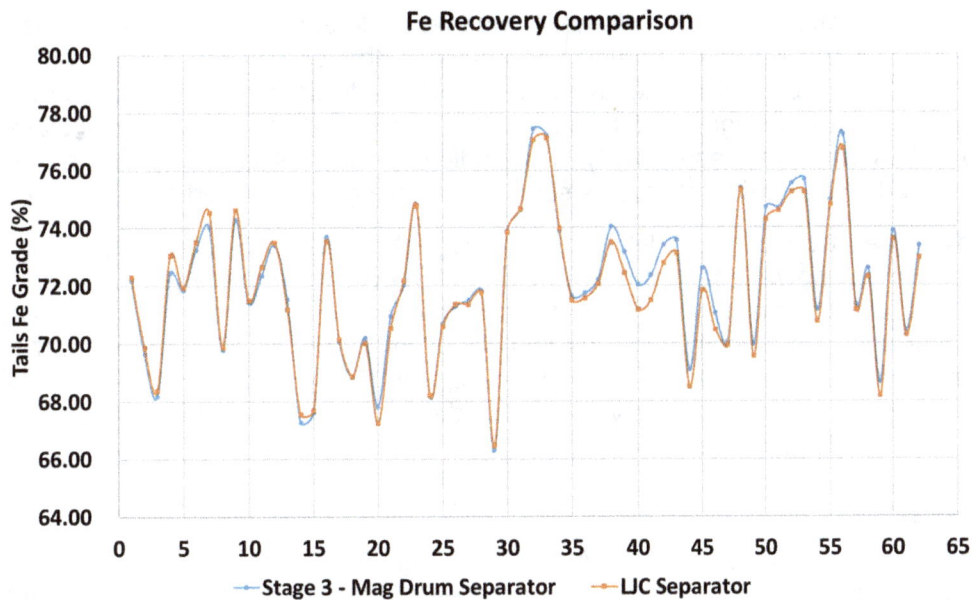

FIG 5 – Comparison chart of Fe recovery.

Processing capacity and feed particle size (F₈₀)

A comparative analysis was performed to assess the line processing capacity and feed particle size (F_{80}) between the original stage 3 magnetic drum separators and LJC Separators, as outlined in Figure 6.

Initially, a total of 16 wet magnetic drum separators were installed for the stage 3 magnetic separation of the selected line. Data collection took place throughout the project implementation, during which two to eight wet magnetic drum separators were replaced by the LJC separators. Further details can be found in Figure 6.

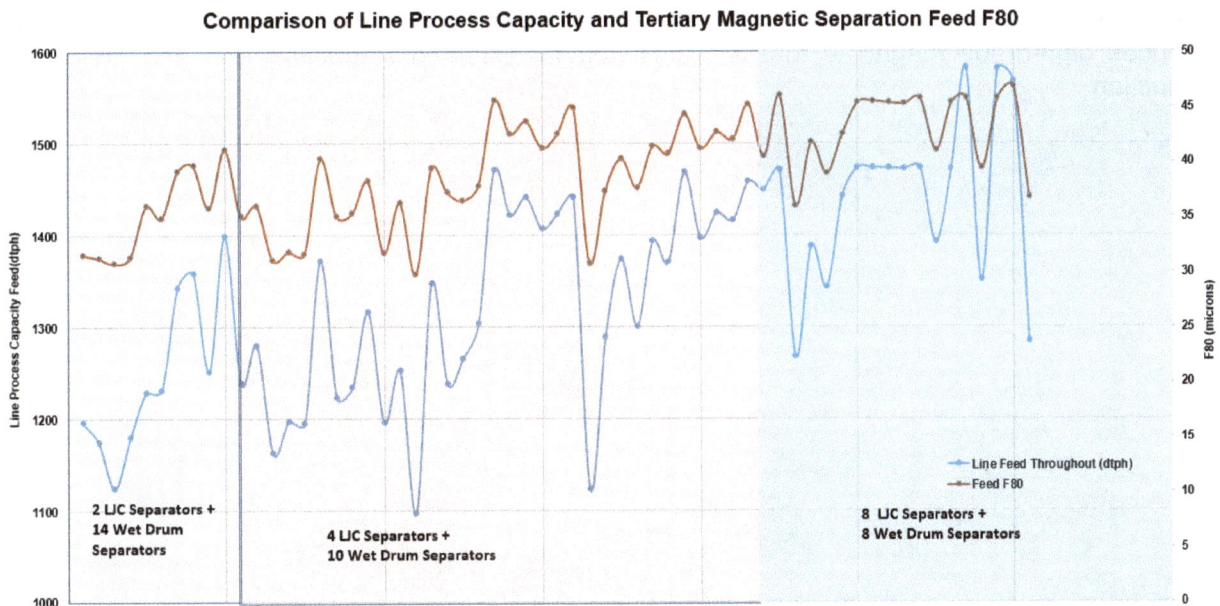

FIG 6 – Comparison of line process capacity and tertiary mag separation feed F_{80}.

- By implementing LJC Automatic Magnetic Flotation Separators, the magnetite processing plant has experienced a significant improvement in capacity while maintaining similar concentrate Fe grade (~65 per cent), Fe recovery, and Mass yield. The introduction of 2, 4, and 8 LJC separators has resulted in an increase in the feed throughput of the line. Specifically, the average feed throughput has risen from 1248 dt/h to 1323 dt/h (6 per cent) and 1442 dt/h (15 per cent) with the respective installation of 2, 4, and 8 LJC separators.

- The implementation of LJC Automatic Magnetic Flotation Separators has facilitated the processing of coarser feed particles, surpassing the capabilities of the original stage 3 magnetic drum separators. The average F_{80} of the tertiary magnetic separation feed has increased from an average of 35.15 µm to 38.47 µm and 43.23 µm with the installation of 2, 4, and 8 LJC separators, respectively. This indicates that the LJC separators can effectively handle a wider range of feed particle sizes, contributing to improved processing efficiency.

- Moreover, the introduction of LJC Automatic Magnetic Flotation Separators has led to a notable decrease in unit power consumption (measured in kWh/wmt) in the grinding separation area. This reduction is primarily attributed to the increased grinding particle size facilitated by the LJC separators. Specifically, with the installation of 2 and 8 LJC separators, the unit power consumption has decreased by 10.05 per cent, from 90.84 kWh/wmt to 81.80 kWh/wmt. This improvement in energy efficiency highlights the benefits of utilising LJC separators in the grinding separation process.

CONCLUSION

The successful implementation of LJC Automatic Magnetic Flotation Separators, replacing the stage 3 magnetic drum separators in the magnetite processing plant in Western Australia, showcased notable improvements in concentrate and tailings grades. Under equivalent ore feeding conditions and Fe recovery rates, the concentrate grade experienced an average increase of 1.83 per cent, while the average tailings grade saw a reduction of 6.16 per cent.

By maintaining a constant concentrate grade of 65 per cent, the adoption of LJC Automatic Magnetic Flotation separators facilitated an enlargement of the coarse grinding particle size from P_{80} = 32 µm to P_{80} = 43 µm. As a result, the processing capacity exhibited a noteworthy average increase of 15 per cent, accompanied by a 10 per cent decrease in unit power consumption when comparing eight to two of LJC separators installed. More improvement should be expected with all 16 units replaced.

The LJC Automatic Magnetic Flotation Separator, developed by Longi Magnet, is a proven and innovative technology that contributes to reducing the carbon footprint, enhancing concentrate grade, and improving mass recovery. It offers an alternative solution for magnetite processing and produces high-grade magnetite that is well-suited for Directed Reduction Iron and green steel production.

Mining and processing – tailings processing and storage

Slime waste to boost agricultural productivity

P Dixit, A K Mukherjee, A K Bhatnagar, P Patra and S Saha

1. Researcher, Tata Steel Ltd, Jamshedpur 831002, India. Email: prashant.dixit@tatasteel.com
2. Principal Scientist, Tata Steel Ltd, Jamshedpur 831002, India.
   Email: akmukherjee@tatasteel.com
3. GM OMQ, Tata Steel Ltd, Jamshedpur 831002, India. Email: atul.bhatnag@tatasteel.com
4. Professor, Bidhan Chandra Krishi Viswavidyalaya, Naida 741252, India.
   Email: drpatrapk@yahoo.co.in
5. Research Scholar, Bidhan Chandra Krishi Viswavidyalaya, Naida 741252, India.
   Email: sajalsaha.saha@gmail.com

ABSTRACT

TATA Steel's R&D Division at Jamshedpur has developed a slime beneficiation process to recover iron values from iron ore slime. This process recovers iron ore concentrate containing 2.2 per cent alumina from a feed containing 7–10 per cent alumina. The reject generated from beneficiation of iron ore slime is treated as waste as the iron content in the reject is below 45 per cent. Effective utilisation of this waste is required for making the iron ore slime beneficiation process a zero-waste technology. In this connection several studies were made towards utilisation of slime waste. One of these studies revealed that slime waste can be used as a soil conditioner. Slime waste contains particles of size less than 10 micron and when these particles are mixed with soil in definite proportions, the porosity and the permeability of the mixture goes down. As a result, the water holding capacity (WHC) improves. This concept was evaluated through a series of tests at greenhouse level to optimise the soil to slime ratio. Finally, a pilot scale field trial was carried out over a period of three months using paddy rice as the crop. Three separate plots with compositions of: soil + organic matter comprising Farmyard Manure (FYM), (ii) soil and slime waste (2:1) + organic matter (FYM), and (iii) soil and slime waste (2:1) + organic matter (FYM) + Nitrogen, Phosphorus and Potassium (NPK) fertiliser were used for the pilot trial. The results showed a 20 per cent increase in the productivity of paddy rice for plot (ii) as compared to plot (i). This was due to the beneficial effect of slime waste in the soil. The addition of inorganic fertiliser such as NPK in the ratio of (4:2:1) further enhanced the productivity of paddy rice.

INTRODUCTION

Iron ore is the major raw material for the iron and steel industry. Indian iron ore contains alumina (Al_2O_3) as the major impurity, which decreases the hot metal productivity of the blast furnace and adversely affects the cost of steel production. Therefore, the alumina content of the iron ore is lowered through beneficiation to an acceptable level. The beneficiation process generates rejects in the form of iron ore slimes. Presently, around 15–20 per cent of the feed treated in beneficiation plants is rejected as iron ore slimes in the tailing pond. However, the beneficiation of iron ore slimes was not commonplace in India due to its extremely fine particle size. Recently, Dixit *et al* (2015) and Thella, Mukherjee and Gurulaxmi (2012) developed a beneficiation process to recover iron ore concentrate from iron ore slimes. The reject generated from the slime beneficiation process contains iron values less than 45 per cent with an alumina content of 15 per cent or more. On average, 35 per cent of the iron ore slime ends up as rejects or waste at the end of the slime beneficiation process and further recovery of iron values from this waste is not possible due to its complex liberation characteristics. Therefore, the effective utilisation of this waste is required for making the iron ore slimes beneficiation process a zero-waste technology.

Numerous efforts have been undertaken by researchers worldwide to investigate the possibility of application of iron ore slimes as a soil modifier material. Qingzhong (2000) conducted pilot trials using magnetic iron ore tailings as a soil conditioner. It was demonstrated via a field plot trial and a field demonstration test that, when magnetic tailings are mixed with Farm Yard Manure (FYM) into the soil, the crop production significantly increases, ie the average yield of early rice increased by 12.63 per cent, the yield of midseason rice increased by 11.06 per cent and the yield of soybean increased by 15.5 per cent. The greenhouse experiment conducted by Roongtanakiat, Osotsapar and Yindiram (2008) at the Department of Soil Science, Faculty of Agriculture, Kasetsart University

Kamphaeng Saen Campus, Nakhon Pathomto, Thailand, to evaluate the effects of soil modification on growth, performance and accumulation of primary nutrients, as well as Fe, Zn, Mn and Cu in vetiver (perennial grass) plants is ongoing. Several invention utilise iron ore slimes as a filter media for purification of groundwater, suggest the impact of iron ore slimes on soil CEC (US patent 5543049, Hogen, 1994).

In the present investigation, efforts have been made to understand the effect of slime waste as a soil modifier for improving the soil fertility as well as crop yield.

MATERIALS AND METHODS

Material

Iron ore slime waste was generated in the Tata Steel pilot plant using the slime beneficiation process. Particle size measurement of the sample was performed using an online particle size analyser. The size distribution of the sample is shown in Figure 1, which shows that 80 per cent of the slime sample is below 5-micron size indicating the ultrafine nature of slime waste.

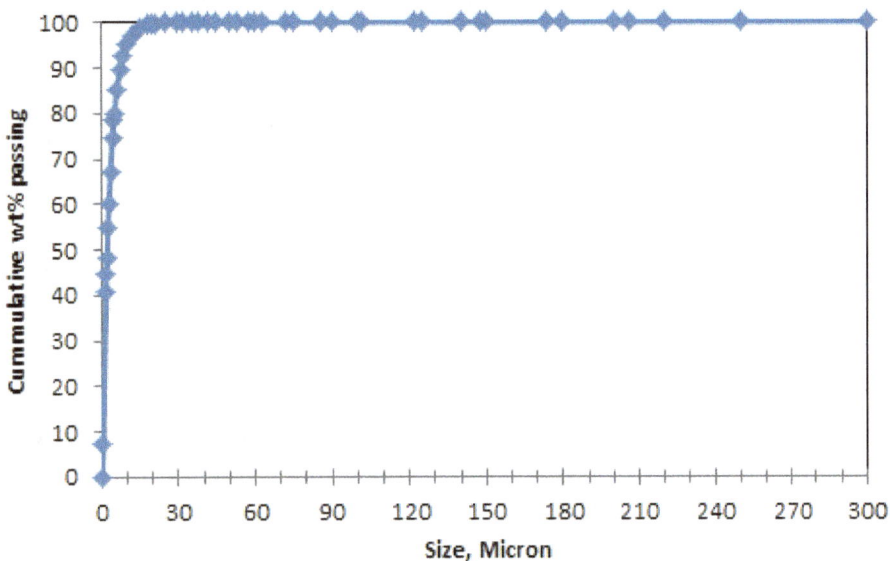

FIG 1 – Size distribution of slime waste.

Chemical analysis of the slime waste sample was carried out using an ICP analyser (Integra XL, IR Tech Pvt Ltd) and the results show that the sample contained 42.24 per cent Fe, 13.83 per cent Al_2O_3 and 9.54 per cent SiO_2.

Soil characterisation of the slime waste was carried out and the results from the physical and chemical characteristics are summarised in Table 1. The results show that the amount of total N, available N, exchangeable ammonium, available K and available P was 0.027 per cent, 156.8 mg/kg, 112 mg/kg, 15.4 mg/kg and 4.5 mg/kg respectively. With low organic carbon content of 0.08 per cent and cation exchange capacity, (CEC) of 4.5 cmol (p)/kg, the waste from iron ore slime beneficiation is fairly good for soil fertility and agricultural crops growth with tailored soil management practices.

TABLE 1

Results of physical and chemical soil characterisation of slime waste sample.

Characteristic	Parameter	Results
1	pH	7.44
2	EC (dS/m)	0.164
3	Max. Water Holding Capacity (%)	53.35
4	Exchangeable Ammonium (mg/kg)	112
5	Soluble Nitrate (mg/kg)	44.8
6	Available Nitrogen (mg/kg)	156.8
7	Total Nitrogen (%)	0.02688
8	Available Potassium (ppm)	15.4
9	Cation Exchange Capacity (Cmol/kg)	4.32
10	Organic Carbon (%)	0.08
11	Available Phosphorus (mg/kg)	4.5
12	Available iron (mg/kg)	158.98

Methods

Pilot scale demonstration of three treatments was carried out to understand the effect of slime waste on soil fertility as well as crop yield (see Figure 2). The optimised ratio of soil to slime waste (4:1) was selected based on a previous pot scale study.

FIG 2 – A pictorial view of the pilot scale demonstration of three soil treatments.

Treatments

The treatments evaluated were as follows:

Normal soil + FYM	(T1)
Normal soil+ iron ore slime rejects (in the ratio of 4:1) + FYM	(T2)
Normal soil+ iron ore slime rejects (in the ratio of 4:1) + FYM + NPK	(T3)

where NPK = nitrogen, phosphorus and potassium fertiliser.

Test factors and protocols:

- Crop: Rice

- Nitrogen, Phosphorus and Potassium (NPK) fertiliser in the ratio of 4:2:1

- Farmyard manure (FYM): 5 tonnes/hectare of field

- Duration: 90 days incubation

- Sampling days: 30, 60 and 90 days of incubation

- Analyses: pH, electrical conductivity (EC), organic carbon (OC%) and available N, P and K.

Results and discussion

Tests T1, T2 and T3 elucidate the application of slime waste as a soil modifier, for crop yield. Physical and chemical properties such as pH, EC, organic carbon content and the available nutrients nitrogen, phosphorus and potassium content present in crop soil for all three treatments were analysed after 30 days, 45 days and 90 days incubation periods.

Effect of slime waste on soil pH

The changes in soil pH for different treatments are presented in Figure 3. The pH value of the soil increased for 90 days after the transplant of seedling, for treatments T2 and T3. The highest pH value (pH = 8.27) was observed with a mixture of soil, slime waste and farmyard manure in the presence of NPK. Therefore, the addition of slime waste in the soil mixture may increase the soil pH, which is important for acidic soils. The increase in pH of rice growing soil due to the application of iron ore slimes can be ascribed to the reduction of Fe^{3+} to Fe^{2+} under the anaerobic environment of rice soil as suggested by Ponnamperuma (1972). The increased degree of reduction due to the application of higher rates of organic manure might have resulted in a greater increase in soil pH. However, this phenomenon needs to be substantiated by changes in redox potential (Eh) of the soil, which will measured in the next phase of the works.

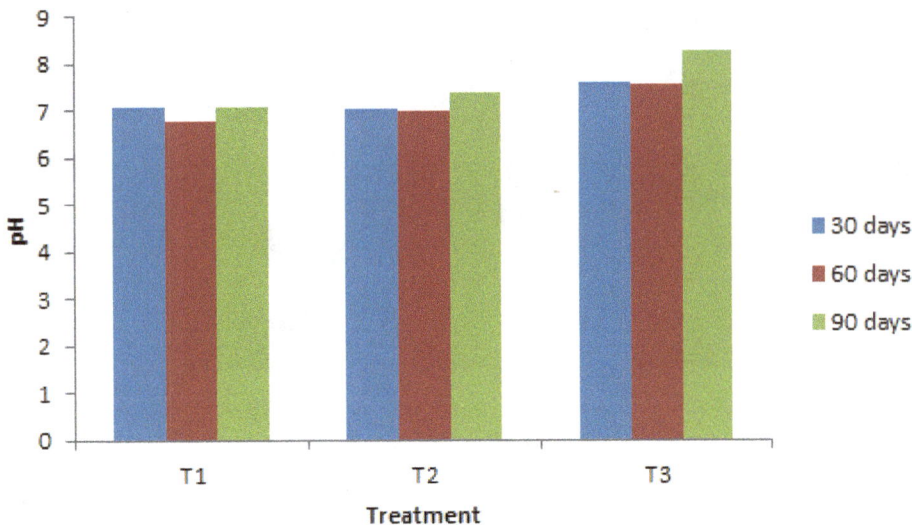

FIG 3 – Effect of different treatments on the pH of rice growing soil.

Effect of slime waste on soil EC

The changes in soil EC for different treatments are presented in Figure 4. It is observed that the EC of rice growing soil increased for 90 days after seedling transplant, for all treatments. The highest EC value of 332.1 micro-siemens per metre was observed for treatments with no addition of slime waste in the soil mixture. Further addition of slime waste in the same mixtures showed no significant reduction in EC; the same EC value of the 308.8 was measured. Therefore, it is possible the addition of slime waste to the soil mixture does not have much effect on the EC of soil.

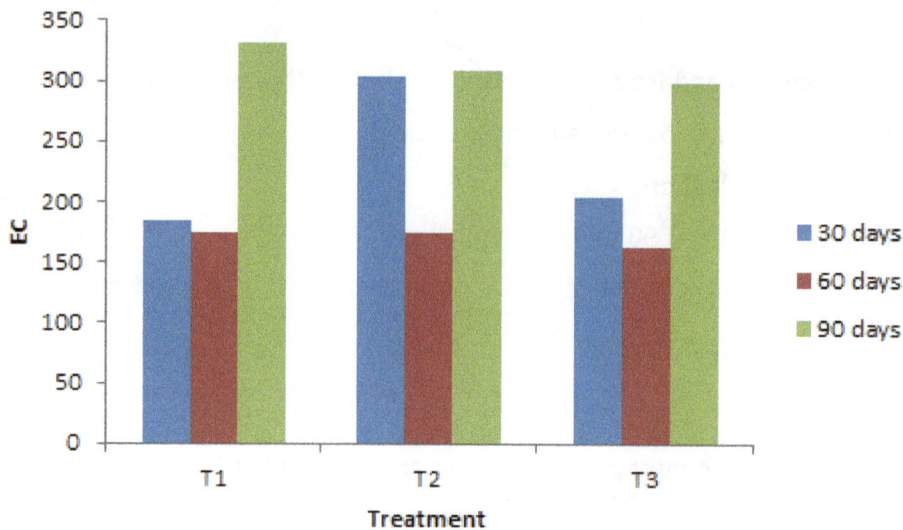

FIG 4 – Effect of different treatments on the electrical conductivity of rice growing soil.

Effect of slime waste on soil organic carbon

The changes in soil organic carbon for different treatments are presented in Figure 5. It is evident that the organic carbon content of rice growing soil increased for 90 days after seedling transplant at, for all treatments. The soil organic carbon content was found to higher in treatment T2 compared to treatment T1 due to the addition of slime waste to the soil. The increase in organic carbon of the rice growing soil due to the application of slime waste can be ascribed to the stabilisation of organic carbon with iron oxides and hydroxides generated from the slime waste.

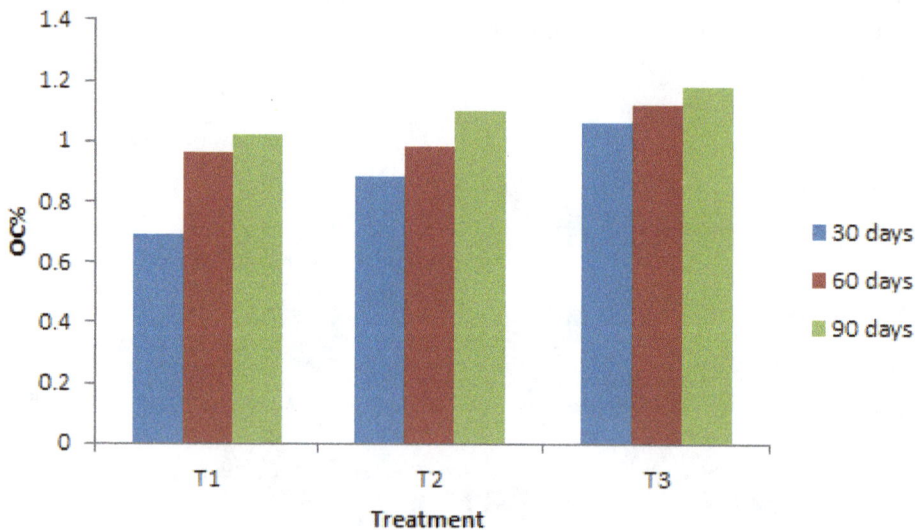

FIG 5 – Effect of different treatments on the organic carbon content of rice growing soil.

Effect of slime waste on available P in soil

The changes in available P for different treatments are presented graphically in Figure 6. It is apparent that the addition of slime waste and farmyard manure along with NPK at 90 days after transplantation contributes to highest value of available P (P value of 47.51 mg/kg was determined). Adding slime waste alone with NPK to the soil and farmyard manure mixture did not improve availability of P, compared to normal soil. Therefore, it may be concluded that NPK and the interaction of slime waste, farmyard manure with NPK resulted in greater stabilisation of organic carbon and occlusion of available phosphorus on soil reduction. Further work will fractionate phosphorus in the soil with and without addition of slimes, to provide information on the dynamic changes in different phosphorus fractions and determine mechanism for P retention.

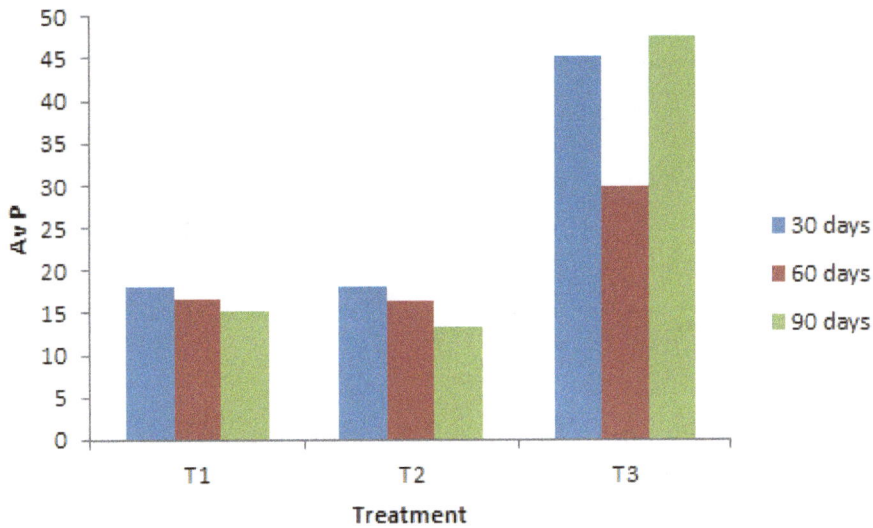

FIG 6 – Effect of different treatments on the available P content of rice growing soil.

Effect of slime waste on available K in soil

The changes in available K for different treatments are presented graphically in Figure 7. It is found that the addition of slime waste and farmyard manure along with NPK at 90 days after transplantation shows the highest value of available K (83.5 mg/kg). It is clear that it would be better to use NPK in the mixture of soil, slime waste and FYM to increase the available K content.

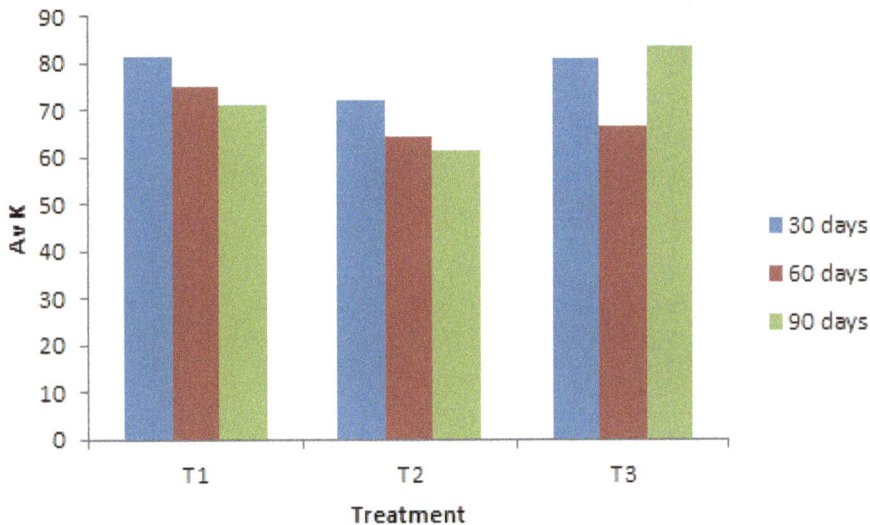

FIG 7 – Effect of different treatments on the available K content of rice growing soil.

Effect of slime waste on available N in soil

The changes in available N for different treatments are presented graphically in Figure 8. It is observed that the addition of slime waste and farmyard manure along with NPK at 90 days after transplantation shows the highest value of available N (55 mg/kg). It is clear that it would be better to use NPK in the mixture of soil, slime waste and FYM to increase the available K content.

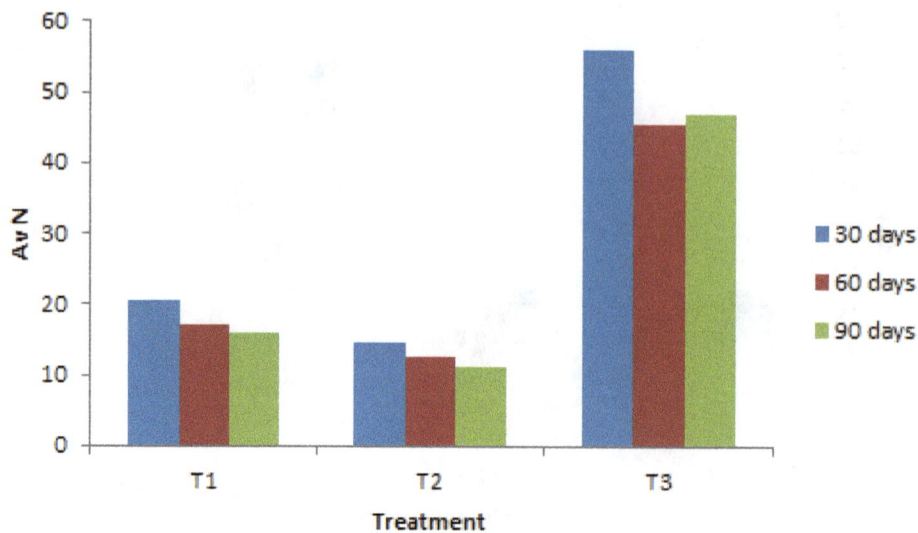

FIG 8 – Effect of different treatments on the available N of rice growing soil.

Productivity

Figure 9 shows the rice grain productivity data for treatments T1, T2 and T3. It is clear that the mixture of soil, slime waste and farmyard manure increase the rice grain yield by 0.45 t/hectare compared to the mixture of soil and farmyard manure. The mixture of soil, slime waste and farmyard manure in the presence of NPK fertiliser further improves the yield by 0.22 t/hectare compared to the mixture of soil, slime waste and farmyard manure. This confirms that slime can be used as a substance to improve the fertility of soil and crop productivity.

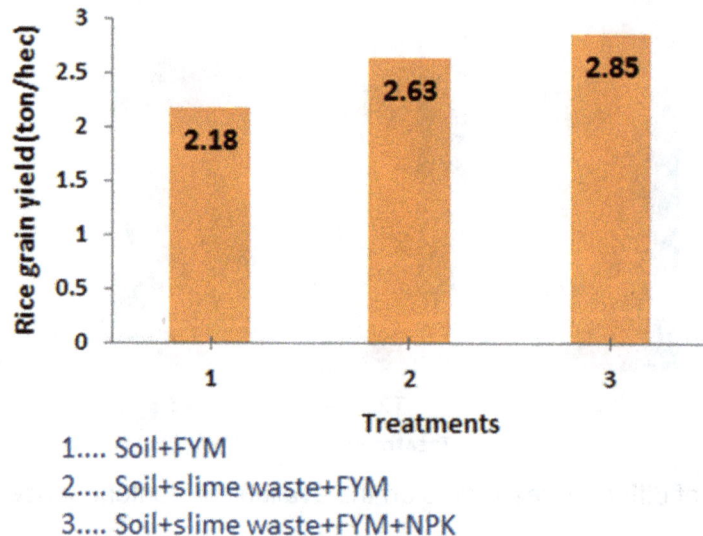

1.... Soil+FYM
2.... Soil+slime waste+FYM
3.... Soil+slime waste+FYM+NPK

FIG 9 – Productivity data obtained from different soil treatments.

CONCLUSIONS

About 35 per cent of iron ore slimes end up as rejects or waste at the end of slime beneficiation. This slime waste consists of 42.24 per cent Fe, 13.8 per cent Al_2O_3 and 9.54 per cent SiO_2. The particle size of slime waste is 80 per cent passing 5 microns. Pilot scale demonstration of three treatments was carried out using rice as the crop to understand the effect of slime waste on soil fertility as well as on crop productivity. Physical and chemical properties such as pH, EC, organic carbon content and available nutrients nitrogen, phosphorus and potassium content present in the cropped soil of all three treatments were analysed after incubation periods of 30 days, 45 days and 90 days. It was found that the mixture of soil, slime waste and farmyard manure has an advantage over the mixture of soil and farm yard manure in terms of increased pH and organic carbon content. Other properties such as available N, available P and available K of the mixture of soil, slime waste and farmyard manure were observed to be comparable to that for the mixture of soil and slime waste. The presence of NPK in the mixture of soil, slime waste and farmyard manure improved all properties in a significant manner. The productivity results showed a 20 per cent increase in the productivity of rice cultivated in the mixture of soil, slime waste and farmyard manure compared with that cultivated in a mixture of soil and slime waste. The addition of NPK fertiliser further enhanced the productivity of rice. This pilot scale trail demonstrates utility of slime waste as a soil modifier.

ACKNOWLEDGEMENTS

The authors are thankful to Tata Steel management for giving an opportunity to work on this project. The support and services provided by the staff of Bidhan Chandra Krishi Viswavidyalaya (BCKV) are also acknowledged.

REFERENCES

Dixit, P, Tiwari, R, Mukherjee, A K and Banerjee, P K, 2015. Application of response surface methodology for modeling and optimization of spiral separator for processing of iron ore slime, *Powder Technology*, 275:105–112.

Hogen, D R, 1994. US patent 5543049 – Microbial mediated water treatment. Available from: <https://patents.justia.com/patent/5543049>

Ponnamperuma, F N, 1972. The chemistry of submerged soils, *Advances in Agronomy*, 24:29–96.

Qingzhong, B A I, 2000. The status and strategy of tailings comprehensive utilization in China, in *Proceedings of Third Asia-Pacific Regional Training Workshop on Hazardous Waste Management in Mining Industry*, pp 151–159.

Roongtanakiat, N, Osotsapar, Y and Yindiram, C, 2008. Effects of soil amendment on growth and heavy metals content in vetiver grown on iron ore tailings, *Kasetsart Journal (Natural Science)*, 42:397–406.

Thella, J S, Mukherjee, A K and Gurulaxmi, S N, 2012. Processing of high alumina iron ore slimes using classification and flotation, *Powder Technology*, 217:418–426.

Reducing waste in the iron ore industry by beneficiating slimes using new flotation regime and two-stage processing routes

M Marques[1], K Silva[2], I Filippova[3], A Piçarra[4], N Lima[5] and L Filippov[6]

1. PhD student, Université de Lorraine, CNRS, GeoRessources, F54000 Nancy, France. Email: michelle.marques@univ-lorraine.fr
2. Research Engineer, Vale Mining Company, Iron Ore Beneficiation Development Team, 34006-270 Nova Lima, Minas Gerais, Brazil. Email: klaydison.silva@vale.com
3. Senior Scientist, Université de Lorraine, CNRS, GeoRessources, F54000 Nancy, France. Email: inna.filippova@univ-lorraine.fr
4. Study Engineer, Université de Lorraine, CNRS, GeoRessources, F54000 Nancy, France.
5. Research Engineer, Vale Mining Company, Iron Ore Beneficiation Development Team, 34006-270 Nova Lima, Minas Gerais, Brazil. Email: neymayer.lima@vale.com
6. Professor, Université de Lorraine, CNRS, GeoRessources, F54000 Nancy, France. Email: lev.filippov@univ-lorraine.fr

ABSTRACT

The iron industry produces a considerable amount of tailings per annum, due to its substantial volumes of production. The actual iron ore processing routes require the removal of slimes fractions. In addition to iron oxides and quartz, the iron ore slimes are generally composed of other minerals such as kaolinite and finely dispersed iron hydroxides which impact significantly the performance of reverse cationic flotation. In this study, we propose a two stage processing route that would allow further beneficiation of iron slimes which are not successfully beneficiated, avoiding their disposal as waste. The new amidoamine reagent was tested as a collector in a two-stage route to process two samples of iron ore slimes with d50 close to 10 µm and with high kaolinite content (13 to 20 per cent). The results showed that it was possible to obtain iron concentrate with 66.5 per cent iron and 48.4 per cent mass recovery for a feed containing 48.9 per cent Fe associated with hematite (39 per cent) and goethite (32 per cent). The collector efficiency was improved by mixing with frother for the slimes with higher goethite content of 55 per cent: the concentrate grade was increased from 58.8 per cent to 64.5 per cent with addition of the frother at the expense of reducing operation metallurgical recovery from 87.8 per cent to 75.6 per cent. The efficiency of the process was improved by applying an external energy impact to disperse the fine particles and clean the surface of the minerals.

The new amidoamine collector does not require a depressor such as corn starch and may provide the development of new concentration routes for complex ultrafine iron ores. The combination of the amidoamine-based flotation at first stage with high intensity magnetic separation allowed development of optimised two-stage processing routes to valorise the iron ore slimes rejected to the tailings in the actual flotation plant.

INTRODUCTION

One of the main challenges facing the Brazilian iron ore mining industry is related to the amount of slime generated before the reverse cationic flotation process. The development of a technology to concentrate iron ore slimes can improve the mass recovery at industrial plants, contributing to the mitigation of the environmental impacts caused by iron ore mining operations (Wolff, Costa and Dutra, 2011). The concentration of ultrafine tailings (also called slime) is considered one of the main problems of the iron mining industry in Brazil. Several studies have been developed aiming at finding an economically viable alternative to reduce the iron content in the slimes (Filippov, Severov and Filippova, 2014).

The main characteristics that make the concentration of Brazilian iron ore slime a complex problem are:

- The finer the particle size the higher the specific surface: this characteristic is responsible for the high iron content in the slimes' magnetic concentration tailings and also for the reduction in the probability of particle/bubble collision, as well as for some changes in the foam properties and pulp rheology, which influence the flotation process (Filippov, Severov and Filippova, 2014).

- Significant presence of clay minerals: this characteristic reduces the selectivity of the flotation process and consequently increases the contaminating minerals (mainly kaolinite and quartz) in the concentrate.

Usually, the mineralogy presented by the iron ore slimes is more complex than the conventional flotation feed. The desliming operation carried out before the conventional flotation process removes the aluminosilicates, generating slimes with a high kaolinite amount.

Some studies demonstrate the characteristics of kaolinite that make its concentration by cationic collectors, such as amines, extremely hard. According to Xu *et al* (2015), the kaolinite crystal has anisotropy, which makes its behaviour in relation to pH dependence on cationic flotation to be opposite to that observed for the oxides. Kaolinite is an aluminosilicate with two different surfaces parallel to the plane (001) in its structure, one tetrahedral plane being formed by SiO_4 tetrahedron and the second one formed by an octahedral plane AlO_6 (Liu *et al*, 2014). Due to this type of structure, the surface charge of each plane changes differently in relation to pH. The isoelectric point of the kaolinite alumina face is between six and eight while this value for the silica face is close to four (Gupta and Miller, 2010).

According to Yuehua *et al* (2004), the silica basal plane (001) exhibits high affinity for the cationic collector and it is responsible for the kaolinite flotation, while the kaolinite AlO_6 plane has weak interaction with the cationic reagent, remaining hydrophobic and lowering the flotation of the kaolinite with the cationic collectors.

In addition to the difficulties related to kaolinite flotation in iron ore slimes, the coating of particles of some minerals by ultrafine particles through electrostatic interaction (slime coating) reduces the selectivity of the concentration process. Slime coating could occur between kaolinite and hematite particles, since the basal plane (001) is negatively charged at pH greater than four (Liu *et al*, 2014), causing the attraction to the hematite particles with positive surface charge.

Amidoamine-based collectors are suitable candidates for developing efficient iron ore slime flotation processes. As reported in earlier studies, iron ore slimes, in addition to iron oxides and quartz, are generally composed of other minerals such as kaolinite and finely dispersed iron hydroxides (Wolff, Costa and Dutra, 2011; Lima *et al*, 2020). Matiolo *et al* (2020) carried small pilot column flotation tests with two different slimes samples from industrial plants located in the Iron Quadrangle (MG, Brazil), using amidoamine as collector and without any depressant. For the first sample tested by the author, it was possible to obtain iron concentrates with iron grades above 60 per cent with SiO_2 and Al_2O_3 contents below 5 per cent and 3 per cent, respectively, whereas for the second sample it was not possible to produce concentrates with iron grade higher than 55 per cent.

Silva *et al* (2021) carried out fundamental studies comparing the amidoamine collector N-[3-(Dimethylamino)propyl]dodecanamide) with etheramine and oleate collectors in the flotation of pure minerals such as hematite, quartz and kaolinite. The high selectivity of this reagent creates new perspectives for the development of new iron ore slimes concentration process routes. Filippov *et al* (2021) demonstrated the concentration of iron ore slimes at industrial pilot scale by using a 500 mm diameter flotation column with amidoamine collector which renders the use of depressant unnecessary.

This work proposes the adoption of the amidoamine N-[3-(Dimethylamino)propyl]dodecanamide to concentrate iron ore slimes with high kaolinite and iron hydroxides content in two flotation stages (reverse with amidoamine and direct with oleate). An improvement of the performance of the new collector for iron ore slimes flotation by combining it with a frother reagent was also investigated.

MATERIALS AND METHODS

Slimes samples

Lab-scale flotation tests were performed using two slimes samples from a Vale's industrial plant in the Iron Quadrangle, Brazil. The results of mineralogical analyses carried out by X-ray diffractometry using the powder method are presented in the Table 1.

TABLE 1

Mineralogical composition of the industrial plant slimes sample.

Mineral	% Estimated – Sample 1	% Estimated – Sample 2
Hematite	39	19
Goethite	32	55
Kaolinite	20	13
Quartz	9	11
Gibbsite	–	2

Table 2 shows the reagents used during the flotation tests.

TABLE 2

Reagents adopted in the laboratory flotation tests.

Stage	Reagent
Reverse Flotation Samples 1 and 2	Collector Amidoamine – (N-[3-(Dimethylamino)propyl]dodecanamide)
Direct Flotation Sample 1	Collector Sodium Oleate
Reverse Flotation Sample 2	Frother A65 – Polypropylene glycol

Exploratory tests were carried out and the best flotation test parameters are presented in Table 3.

TABLE 3

Parameters adopted in the laboratory flotation tests.

Stage	pH	Reagent	Dosage (g/t)	Flotation time (min)
Reverse Flotation Sample 1	8.5	Collector – Amidoamine	320	20
Direct Flotation Sample 1	7.0	Collector – Sodium Oleate	1425	10
Reverse Flotation Sample 2	10	Collector – Amidoamine	230–260	21–25
Reverse Flotation Sample 2	10	Frother – A65	54	25

For sample 1, the percentage of solids in the first flotation stage was 20 per cent and for sample 2, the percentage of solids was 25 per cent.

Two tests were carried out on sample 2: the first one using the collector amidoamine only and the second one using the combination of the amidoamine collector and the Frother A65 (with partial dosage in the beginning of the flotation and after 5 min and 10 min of flotation).

The typical particle size distributions of the slimes samples tested are reported in the Table 4, while Figure 1 illustrates the particle size distribution obtained by laser sizer (Malvern 3000) for sample 2.

TABLE 4

Particle size distribution of the slimes samples from the industrial plant.

ID	<75 μm	<45 μm	<25 μm	<10 μm
Sample 1	100	98.5	89.0	52.2
Sample 2	97.0	95.2	84.1	49.5

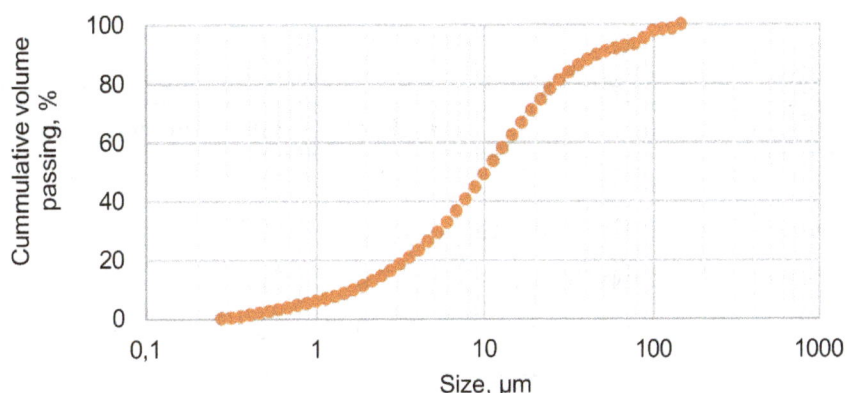

FIG 1 – Particle size distribution of sample 1.

RESULTS

The main concept for the two-stage flotation route tested on the actual ore samples (sample 1 and sample 2) was based on the preliminary study undertaken on pure minerals such as kaolinite, hematite and quartz (Silva *et al*, 2021) using conventional etheramine and oleate collectors as well as the novel amidoamine collector (Table 5).

TABLE 5

Recovery of hematite, quartz and kaolinite (single mineral flotation system) with different collectors at 300 g/t at pH 10: dosage of starch – 1000 g/t; flotation time – 6 min. Flotation was undertaken in a small impeller self-aspirated flotation cell (adapted from Silva *et al*, 2021).

Reagent-dosage (g/t)	Recovery to floated (%)		
	Hematite	Quartz	Kaolinite
Amidoamine	6	88.5	2
Etheramine	68	99.2	51
Oleate	95	-	1
Etheramine+starch	6	96.6	49

The flotation of hematite at pH 10 was not observed using amidoamine collector while etheramine promoted hematite flotation when starch was not used as depressor. A total flotation of hematite was achieved using the oleate collector for the same pH value. The authors demonstrated through FTIR spectroscopy studies (Silva *et al*, 2021) and DFT modelling (Silva *et al*, 2022; Filippov *et al*, 2022) that the most suitable explication concerning the mechanism responsible for selective flotation with the amidoamine studied (without starch) is the hindrance effect of this reagent on hematite surface.

Thus, the low adsorption capacity of the amidoamine on hematite surface does not allow a high density adsorption layer reducing their surface hydrophobicity and avoiding the flotation.

Hence, the two-stage flotation circuit in the reverse cationic flotation using amidoamine is used to remove mainly quartz to the froth product. The nonfloated product composed from hematite and kaolinite was then subjected to direct hematite flotation to obtain an iron concentrate while the kaolinite is not floated.

The following sections present the flow sheets and material balances for the two slime samples tested.

Slimes sample 1

Table 6 shows the best results obtained for the flotation of slimes sample 1 using a two-stage flotation circuit as shown in Figure 2.

TABLE 6

Mass balance obtained from the bench scale flotation test with the industrial plant slimes sample 1.

Product	Yield (%)	%Fe	Fe recovery,%
Feed	100.0	48.9	100.0
Reverse Flotation Concentrate	63.0	62.7	80.8
Reverse Flotation Reject	37.0	25.4	19.2
Direct Flotation Reject	14.6	50.1	15.0
Direct Flotation Concentrate	48.4	66.5	65.8

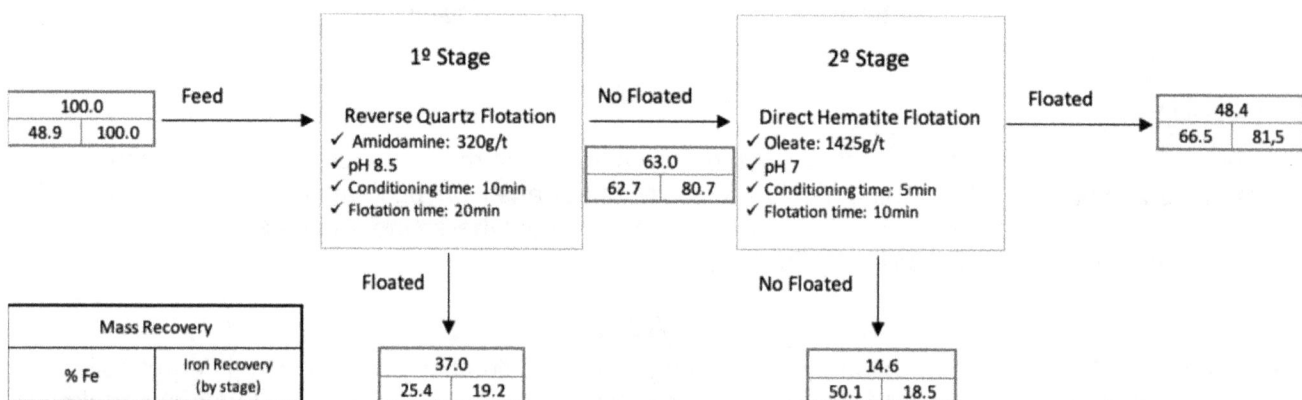

FIG 2 – Parameters and results of bench scale flotation tests using an industrial plant slime (sample 1).

The reverse flotation route using amidoamine as collector with no depressor produced an iron concentrate of 62.7 per cent. The iron content in the concentrate improved from 62.7 per cent to 66.5 per cent by direct flotation at the expense of a mass recovery reduction from 63.0 per cent to 48.4 per cent.

It is important to consider that the results presented by laboratory scale tests will likely be different in an industrial application due to the variability of slimes produced in an industrial circuit. This variability is due to the percentage of solids, size distribution and mineralogy as demonstrated by Filippov et al (2021).

Regarding an industrial application, a complete flow sheet with a scavenger stage (eg flotation or even magnetic concentration) needs to be developed. Depending on the results obtained in a scavenger circuit, the concentrate from this stage can be returned to the direct flotation feed or even used as the final product, as demonstrated in Figure 3.

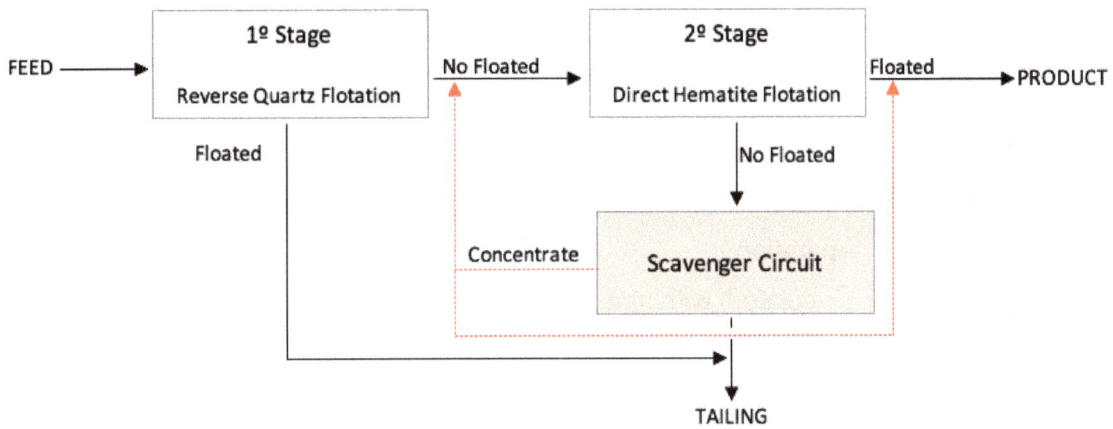

FIG 3 – Hypothetical flow sheet to concentrate iron ore slimes with high kaolinite content.

An issue presented by the tests with the slimes sample is the high consumption of oleate in the direct flotation (1425 g/t). A hypothesis for this high collector consumption is the high specific surface area for oleate adsorption due to the slimes size distribution, with approximately 50 per cent below 10 µm and the high iron content in the direct flotation feed.

The flow sheet in Figure 3 also needs to be compared with a route with only magnetic separation to evaluate the best iron ore slimes concentration process. Problems related to froth handling and carrying rates must also be considered. Further pilot and industrial tests need to be carried out to prove the applicability of this route.

Slimes sample 2

Figure 4 shows the best results obtained for the flotation of slimes sample 2, with and without frother. With the adoption of frother, the iron content in the product increased from 58.8 per cent to 64.5 per cent, but the iron content in the tailings also increased from 19.9 per cent to 29.4 per cent.

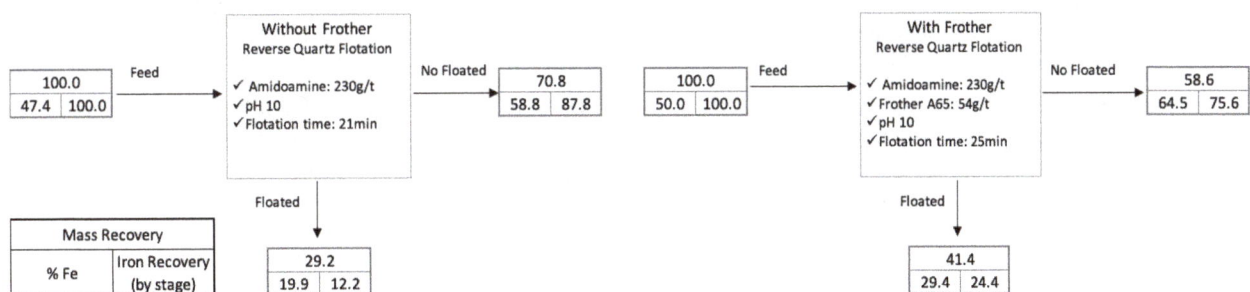

FIG 4 – Parameters and results of bench scale flotation tests using an industrial plan slime sample 2.

This result showed that it is possible to increase the efficiency of this amidoamine by mixing it with another type of reagent to obtain better product quality.

Further studies should be carried out to reduce the loss of iron in the tailings by testing other types of frothers.

Alternative two-stage circuit

Magnetic separation is a process that can be applied on an industrial scale to concentrate iron ore slimes. The efficiency of this process depends on characteristics such as size distribution and mineralogy. For example, the presence of goethite can hamper the process due to its lower magnetic susceptibility when compared to hematite. Considering that iron ore slimes are composed of ultrafine particles (almost 50 per cent below 10 microns), significant loss of iron to tailings can occur due to the 'competition' between magnetic force and hydraulic drag in the process, especially when the predominant iron oxide is goethite. Therefore, a combination of the flotation and magnetic separation can be an alternative to concentrate some iron ore slimes.

Flotation column pilot tests were undertaken according to the flotation regime obtained on samples 1 and 2 using amidoamine collector with no starch. Three different circuit configurations were evaluated:

- flotation only

- magnetic separation only

- flotation+magnetic separation.

The mass balance was calculated with the average value obtained from the optimised tests. Regarding the pulsating high intensity magnetic separator (VPHGMS), the best operating parameters were flow rate: 300 L/h, ring speed: three revolutions per minute, magnetic field: 1.3 T. The VPGHMS adopted in the pilot tests was the Longi LGS-EX, with a ring diameter of 500 mm (Figure 5). Flotation tests parameters were similar to that tested for sample 2.

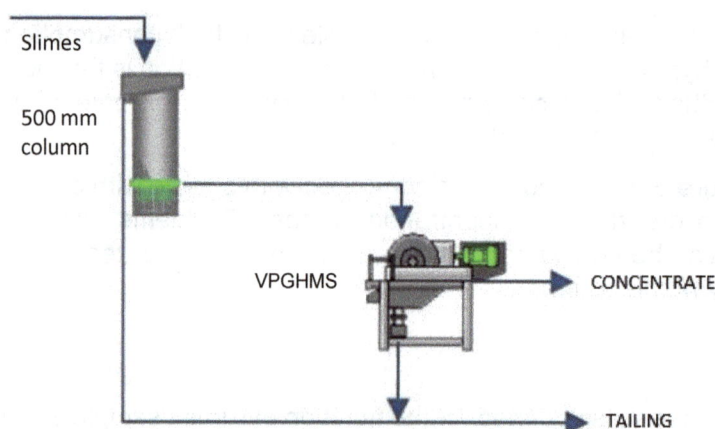

FIG 5 – Combined flotation-magnetic separation circuits for the slime concentration.

The exploratory studies of magnetic concentration and flotation showed that the synergy between these two distinct processes can be an adequate option to guarantee less iron loss to the tailings and higher iron content in the concentrate. Therefore, a suitable process route for the concentration of iron ore slime from the plant slimes product could be the rougher phase by column flotation with amidoamine (N-[3-(dimethylmino)propyl]dodecanamide) and the cleaner stage by magnetic concentration with VPGHMS. Table 7 presents the results of the pilot tests using the VPGHMS to concentrate the column flotation product.

TABLE 7

Pilot tests results considering two concentration stages (flotation as rougher and VPHGMS as cleaner).

Data	Flotation (%Fe)			VPHGMS (%Fe)			Global results				Yield (%)
	Feed	Product	Tailing	Feed	Product	Tailing	Final tailings		Final product		
							Fe,%	SiO$_2$,%	Fe,%	SiO$_2$,%	
Avg value	47.2	57.4	19.4	57.4	67.3	36.7	27.5	53.6	67.3	2.0	49.6

The combined circuit generated high quality iron concentrate of 67.3 per cent Fe with an average silica content close to 2.0 per cent (1.4 to 2.9 per cent), while the silica content in the iron concentrate was 4.4 per cent (2.4 to 7.7 per cent) for the circuit with the magnetic separation only.

In this study, different flotation routes to concentrate iron ore slimes were tested. These unconventional flow sheets were developed due to the specificities of the novel amidoamine collector which does not require the use of starch. Because of the selectivity of this reagent, loss of iron to the tailings is lower when flotation is compared with the magnetic concentration process, but the magnetic concentration generates products with lower SiO$_2$ content and with less variation in concentrate grade when compared with flotation. Thus, the combination of these two processes (flotation and magnetic concentration) tends to generate a route with lower iron losses to the tailings (compared to the route with magnetic concentration only) and better concentrate grades of the final product (when compared to a route where the flotation only is used). A similar approach was proposed in a recent work which presents combined concentration flow sheets using a pulsating high intensity magnetic separator (VPHGMS) and a final concentration stage using the gravity concentration using Reflux Classifier (Rodrigues *et al*, 2023). This work demonstrated the potential for achieving high Fe grade of more than 67 per cent, with high recovery and in turn a considerable process simplification.

CONCLUSION

The use of amidoamine-type reagents in combination with reagents for direct flotation of iron oxides showed, at lab-scale tests, an efficient separation of quartz and kaolinite gangue in the iron ore flotation. The results demonstrated the high selectivity of the amidoamine (N-[3-(Dimethylamino)propyl]dodecanamide) in the flotation of some types of iron ore slimes and reinforced the concept that flotation in the absence of iron oxides depressant may be possible, opening new alternatives for the concentration, by flotation, of slimes generated by the iron ore industry.

Considering the adopted collector dosages, the flotation of fine kaolinite with the amidoamine without iron depressants remained a challenge. The low floatability of kaolinite was partially tackled by the combination of reverse and direct flotations. This type of approach can be a solution for iron ore slimes flotation containing large amounts of kaolinite in the composition, however the feasibility of this process may be impaired by the high direct flotation collector consumption.

The addition of the A65 frother during the flotation of sample 2 has improved product quality, showing that this new type of collector has the potential to be improved when mixed with other types of reagents.

The use of a new amidoamine collector in combination with other reagents was found to be a potential route to process the iron ores slimes which are considered as tailings in actual industrial circuits, thereby reducing the volume of waste streams.

A two-stage combined flotation and high intensity magnetic separation circuit was tested at pilot scale. High quality iron concentrate was generated from a slime product with a d50 close to 10 μm.

ACKNOWLEDGEMENTS

This paper was supported by VALE S.A Mining Company in the context of developing new beneficiation routes for iron ore. The support from the Steval staff of GeoRessources, University of Lorraine is greatly acknowledged.

REFERENCES

Filippov, L, Severov, V and Filippova, I, 2014. An overview of the beneficiation of iron ores via reverse cationic flotation, *Int J Miner Process*, 127:62–69. https://doi.org/10.1016/j.minpro.2014.01.002.

Filippov, L O, Silva, K, Piçarra, A, Lima, N, Santos, I, Bicalho, L, Filippova, I V and Peres, A E C, 2021. Iron ore slimes flotation tests using column and amidoamine collector without depressant at industrial pilot scale, *Minerals*, 11(7):699.

Filippov, L O, Silva, L A, Pereira, A M, Bastos, L C, Correia, J C G, Silva, K, Piçarra, A and Foucaud, Y, 2022. Molecular models of hematite, goethite, kaolinite, and quartz: Surface terminations, ionic interactions, nano topography, and water coordination, *Colloids and Surface A*, 492(2016):88–99.

Gupta, V and Miller, J, 2010. Surface force measurements at the basal planes of ordered kaolinite particles, *J Colloid Interface Sci*, 344(2):362–371.

Lima, N, Silva, K, Souza, T and Filippov, L, 2020. The Characteristics of Iron Ore Slimes and Their Influence on the Flotation Process, *Minerals*, 10(8):675. https://doi.org/10.3390/min10080675

Liu, J, Sandaklie-Nikolova, L, Wang, X and Miller, J D, 2014. Surface force measurements at kaolinite edge surfaces using atomic force microscopy, *Journal of Colloid and Interface Science*, 420:35–40. https://doi.org/10.1016/j.jcis.2013.12.053

Matiolo, E, Couto, H, Lima, N and Silva, K, 2020. Improving recovery of iron using column flotation of iron ore slimes, *Min Eng*, 158:106608. https://doi.org/10.1016/j.mineng.2020.106608

Rodrigues, A, Delboni Jr, H, Silva, K, Zhou, J, Galvin, K P and Filippov, L O, 2023. Transforming iron ore processing – Simplifying the comminution and replacing reverse flotation with magnetic and gravity separation, *Min Eng*, 199:108112. doi.org/10.1016/j.mineng.2023.108112

Silva, K, Filippov, L O, Piçarra, A, Filippova, I V, Lima, N, Skliar, A, Marques, L and Filho, L S L, 2021. New perspectives in iron ore flotation: use of collector reagents without depressants in reverse cationic flotation of quartz, *Minerals Engineering*, 170(2021):107004.

Silva, K, Silva, L A, Pereira, A M, Bastos, L C, Correia, J C G, Piçarra, A, Bicalho, L, Lima, N, Filippova, I V and Filippov, L O, 2022. Comparison between etheramine and amidoamine (N-[3-(dimethylamino)propyl]dodecanamide) collectors: Adsorption mechanisms on quartz and hematite unveiled by molecular simulations, *Min Eng*, 180:107470. doi:10.1016/j.mineng.2022.107470

Wolff, A P, Costa, G M d and Dutra, F C, 2011. A comparative study of ultra-fine iron ore tailings from Brazil, *Mineral Processing and Extractive Metall Rev*, 32:47–59.

Xu, L, Hu, Y, Dong, F, Jiang, H, Wu, H, Zhen, W and Liu, R, 2015. Effects of particle size and chain length on flotation of quaternary ammonium salts onto kaolinite, *Mineral Petrol*, 109:309–316. doi:10.1007/s00710-014-0332-8

Yuehua, H, Wei, S, Haipu, L and Xu, Z, 2004. Role of macromolecules in kaolinite flotation, *Miner Eng*, 17:1017–1022. https://doi.org/10.1016/j.mineng.2004.04.012

Filtered tailings – an iron ore experience

L Vimercati[1] and R Williams[2]

1. Process Engineer, DiefenbachSrl, Bergamo 24020, Italy. Email: l.vimercati@diefenbachsrl.com
2. Business Development Manager, McLanahan Corp, Wangara WA 6023.
   Email: rwilliams@mclanahan.com.au

ABSTRACT

Transitioning to a green future requires the mining industry to adopt sustainable practices.

Filtered tailings are become a key expectation of the communities we operate in and are growing in a number or mining commodities including bauxite, nickel, gold and iron ore.

The filtration technologies required range from dewatering screens to vacuum and pressure filters. Particular challenges faced in iron ore applications include large scale which means several units of large size are often required. In addition, a high specific gravity that cause sedimentation issues and a high abrasion level make demands on achieving reliability level at above 90 per cent.

This paper describes the testing, selection criteria and specific features required to provide a rational option for equipment and plant design. An example of how this has been applied in an iron ore application in Brazil is used to demonstrate a successful outcome. This project installed four overhead beam filter presses each producing around 200 t/hr utilising a recessed chamber plate, feed pressures up to 15 bar, cake thickness of 5 mm.

The moisture lower than 20 per cent that was produced was considered to provide material with handling and geotechnical properties to guarantee a safe stacking.

Regular high pressure washing at 100 bar was considered imperative to maintain cloth life and process performance.

The correct management of the system (OPEX control) by PLC control ensures a balance between machine potentials and plant needs. With this equilibrium it is possible to limit mechanical wear preserving equipment lifetime.

INTRODUCTION

Transitioning to a green future requires the mining industry to adopt sustainable practices.

Filtered tailings are become a key expectation of the communities we operate in and are growing in a number or mining commodities including bauxite, nickel, gold and iron ore.

The filtration technologies required range from dewatering screens to vacuum and pressure filters. Particular challenges faced in iron ore applications include large scale which means several units of large size are often required. In addition, a high specific gravity that cause sedimentation issues and a high abrasion level make demands on achieving reliability level at above 90 per cent.

To ensure the proper process plant operations a correct equipment sizing is essential.

While seeming obvious the key equipment is sometimes considered a smaller item and can be subject to 'value engineering' optimisation.

A wrong filter press sizing can create a bottleneck on the plant line, causing problems that can affect both the upstream and downstream processes.

The accessorises required and often supplied with the filter have to be sized in order to be suitable for the filtration target.

A filter press is able to reach very high performance in terms of cake dryness however a number of variables need to be assessed and considered to reach optimal efficiency.

These include:

- chamber thickness

- feed pressure

- membrane or recessed chamber

- cloth wash pressure and frequency

- cake wash requirements

- cloth type.

It is important to carry out testing to select the most appropriate combination of all these process parameters.

Once the key process requirements are established machine availability, layout and power consumption must then be considered key points to be able to economically hit the target that is the tailings flow rate treatment to obtain a product that can be dry stackable.

There are a series of selection criteria than can have impact on both capital and operating costs.

The following should all be considered:

- Filter press designs commonly come in overhead beam or sidebeam configurations. A major advantage of the overhead beam style is the ability to hand a large number of plates (over 200) off the robust central support. This combined with a single plate opening device provides very high filtration area in a compact footprint reducing building costs. Sidebeam filters can offer a faster opening mechanism to allow reduction on cycle times but this can bring about additional OPEX through faster wear.

- Throughput considerations will affect the selection of both number of chambers and the plate size. Due to larger volumes, tailings filters typically utilise a 1500, 2000 or 250 mm sized plate and while most filter press manufacturers are designing for larger units these have not yet proven a cost-effective alternative.

- The requirement (or not) to provide cake drying via compressed air can offer an advantage over the use of membrane plates but the power consumption of large air compressors can be an additional OPEX burden.

- A key requirement is the selection of the feeding pump. This is a challenging duty requiring initially high-flow low-head but as the cycle advances the changes to a low-flow high-head one.

- Another main factor is necessity of high pressure cloth wash. The high pressure cloth wash installation affects the initial cost for the equipment but decrease the operative cost related to the cloth change. If a spare filter press is not available extra capacity has to be considered in order to make up for the time loss during wash cycle. This option reduces the cloth change frequency and its costs.

- Linked to the cloth washing above, cloths change is an important aspect also. There are different type of cloth installation systems, the target is to enable a fast installation type, also on overhead beam filter presses, in order to decrease maintenance time and any dead time caused by cloth failure.

This paper describes how these considerations were assessed for a specific case in an iron ore plant in Brazil. Brazil is a country with considerable focus on sustainable safe tailings management plants due to several serious incidents relating to the collapse of a tailings storage dam (Figures 1 and 2).

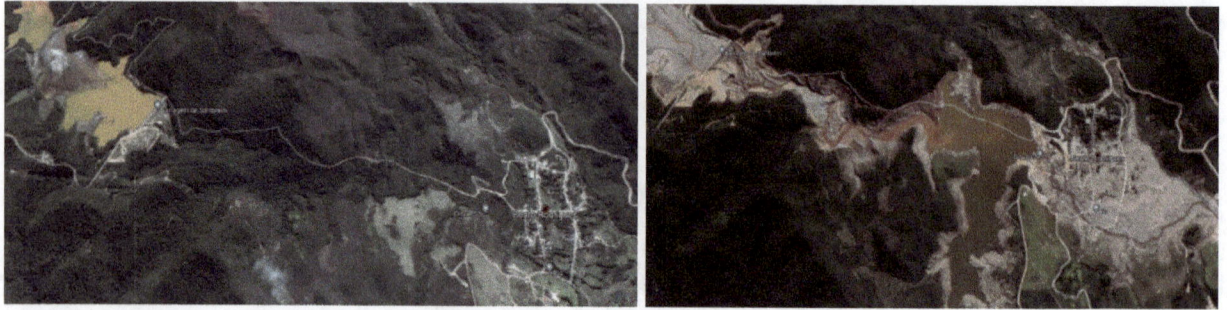

FIG 1 – Aerial view of the dam (before and after the collapse, 2015; Bento Rodriguez village).

FIG 2 – Aerial view of the dam (before and after the collapse, 2019; Córrego do Feijão village).

CASE STUDY

Iron ore tailings dewatering and stacking plant

The iron ore plant is located in Brazil (Minas Gerais) and the start-up was in 2021.

Inside the plant are installed four working filter presses in order to get customer's target, no standby units are currently planned (Figure 3).

FIG 3 – Filter press installation for iron ore tailings treatment.

The plant shown in Figure 4 includes four filter presses 2000 × 2000 mm with 206 plates installed (pos.1), suitable to treat almost 750 tons of dry solids per hour.

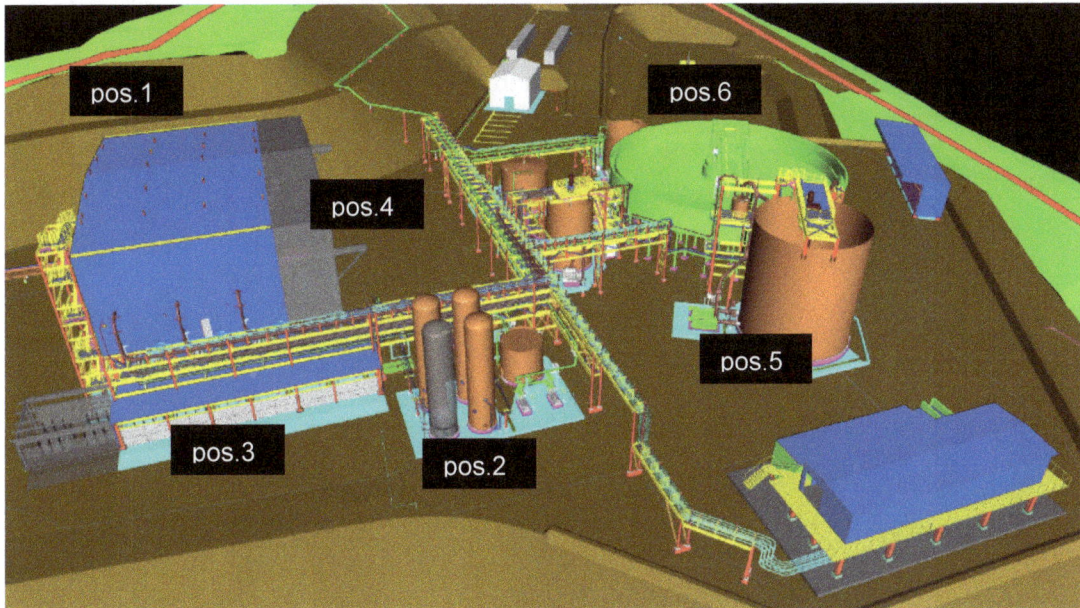

FIG 4 – Filtration tailing plant – 3D view.

In addition to the filters, the supply included:

- Air compressors and tanks for cake and core blowing:
  - Four 80 m^3 tanks, one each filter (pos.2)
  - Two air compressors each filter 19 m^3/min, 12 bar (pos.3)
- Cloth washing skid, one in common with four filters (pos.4)
- Drip trays (pos.1)
- Turbidity metres to check filtrate quality (pos.1)
- Inlet flowmeters (pos.1)
- Feeding pumps, flushing systems and slurry tank (pos.5)
- Slurry thickener (pos.6).

SIZING

Sizing was made starting from customer needs to treat 750 t/hr of solids and from information obtained from lab tests:

- Slurry solids per cent (w/w): 60 per cent
- Slurry solids per cent (v/v): 30 per cent
- Solid SG: 3.43 kg/dm^3
- Liquid SG: 1.00 kg/dm^3
- Slurry SG: 1.74 kg/dm^3
- Overall cycle time (hh.mm.ss): 00:21:05
- Final cake moisture: 14 w/w per cent
- Wet cake density: 2.34 kg/dm^3
- Dry cake density: 2.01 kg/dm^3 (after cake drying).

Laboratory and pilot test

Slurry test includes two different scales: lab and pilot.

Lab test

This first phase required a low quantity of product and can frame all main parameters as:

- Slurry
  - Specific weight
  - Solids concentration
  - pH
  - Temperature
  - PSD.
- Process
  - Chamber thickness
  - Chamber type
  - Filtration pressure
  - Cake washing water quantity and efficiency
  - Cake blowing air quantity and pressure
  - Squeezing coefficient and pressure
  - Cloth selection
  - Phase times
  - Surry conditioning.
- Cake and filtrate
  - Specific weight
  - Final moisture
  - Filtrate quality.

Pilot test

Results obtained from a lab skid (Figure 5) are perfectly scalable and data obtained can be used for industrial sizing.

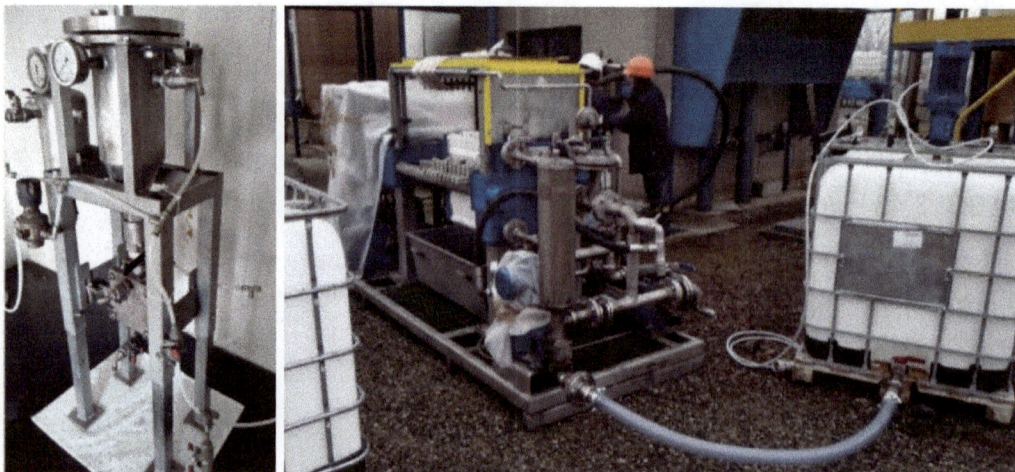

FIG 5 – Laboratory skid and pilot skid.

All data obtained by a lab test can be verified on pilot size.

Starting from information above and considering some sizing factors the necessary filter presses total volume is 134.48 m³.

- Cake volume for each filter: 33.62 m³
- Maximum cake volume available: 35.09 m³
- Total wet cake for each cycle: 314.7 t
- Total dry cake for each cycle: 270.6 t.

An important parameter for filling and filtration time is slurry flow rate.

Slurry flow rate

This slurry has two main physical characteristics which are in conflict with each other:

1. High abrasion level
2. High density.

Two conditions above mean that an equilibrium condition must be found so that the solids do not precipitate in the pipes and, at the same time, the wear of the parts in contact is reduced.

The identified value is between 2.65 and 3.00 m/s, with this setting we avoid solids precipitation and we reduce wear on pumps, piping, valves and cloths.

Maximum flow rate reached during filling and filtering phase is between 650–700 m³/h for each filter, considering an inlet diameter for each feeding side of DN200.

In this case the diameter is forced by filter plates that have a DN200 hole. Flow rate can be increased with a different plate design.

With this setting we can extend cloth lifetime and reduce related OPEX costs.

Slurry pump (Figure 6) design data:

- Maximum flow rate: 1000–1200 m³/h
- Maximum pressure: 13–15 bar
- Installed power: 370 kW

FIG 6 – Slurry feeding pump.

A graph with flow rate and pressure trend is shown in Figure 7.

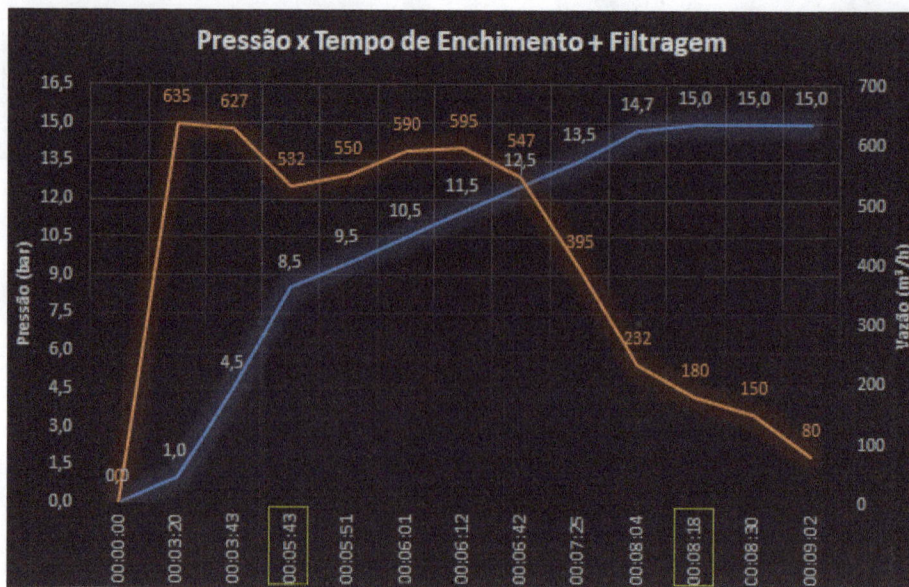

FIG 7 – Flow rate and pressure trend.

Filling and filtration times are closely related to slurry characteristics, an important parameter is the cake resistance and it can be obtained by a filtration test and it defines the relationship between necessary surface and time.

Cloths washing system

Another main point considered during sizing is the time necessary for a complete high pressure cloth washing cycle (Figure 8).

FIG 8 – Operating cloth washer.

Considering 1.5 to 2.0 [min] per plate, the overall washing cycle is about 6.87 hours each filter.

The HP washing cycle is not performed every cycle, usually once or twice per week and in some special cases it may need one cycle per day.

Considering the above, filter press volume is a bit higher to be able to treat the inlet flow rate in less time. This increased CAPEX cost is justified by the decreased OPEX cloth cost:

- cloth lifetime about 1500–2000 cycles

- cloth cost about €200–260 each

- about €53 600 per filter press for a set of cloths.

The high performance of the washer device and its skid can guarantee a good restoration of cloths filtering efficiency and an increasing of their lifetime.

Design data

The skid (Figure 9) has the following data:

- Pump flow rate: 400 L/min

- Number of nozzles installed on washer arm: 54

- Water pressure: 100 bar

- Water tank: 3000 L

- Installed power: 75 kW.

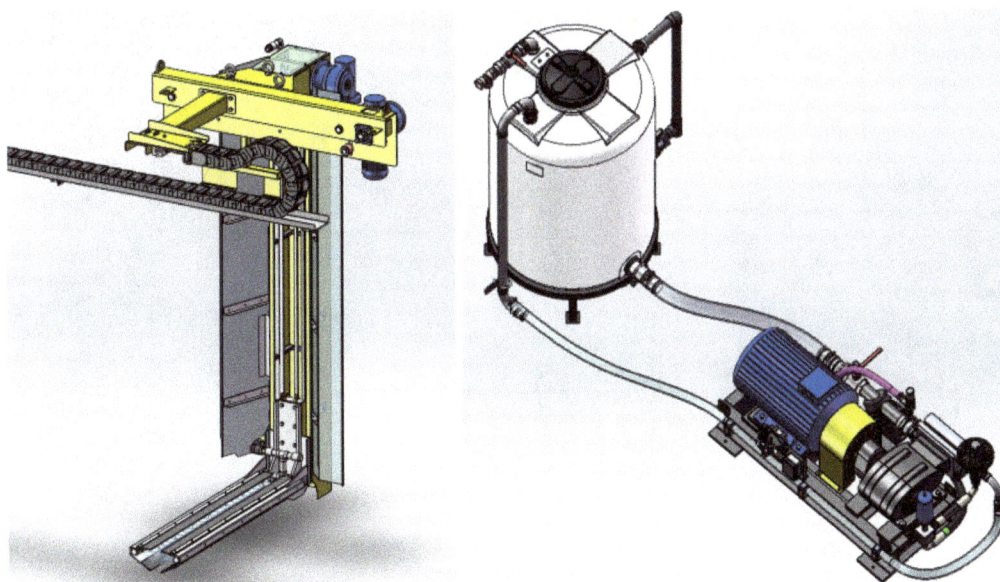

FIG 9 – Cloths washing device and skid.

Cake discharge system

The system installed on overhead beam is 'one by one' (Figure 10).

There are two different types, one with a trolley which moves inside the beam (7"/plate) and one with hooks that grab plates continuously (2.5–3"/plate).

The system installed in Brazil is continuous type with an average opening time of 2.7"/plate.

On new filters, following a detailed study we are now able to lower the opening time to 1.5"/plate on new filters. This improvement also provides a new cloth washer control logic.

FIG 10 – Cake discharge.

COMMISSIONING

During commissioning the optimisation of variable parameters was a key phase (Figure 11).

FIG 11 – Cake discharged during commissioning of different tailings plant.

The first one was the feeding pump, trying to save time during filling and flow rate. The initial set point was too high with a filling flow rate higher than 800 m³/h. The wear on the cloths was excessive causing cloth damage. A correct equilibrium was found with a flow rate between 600–700 m³/h. This enabled acceptable fill time and long cloth life.

The second phase was the cake blow, was observed that drying efficiency decreased plate by the plate moving away from the air entry point. The average value was satisfied but starting from this data the air inlet system has been improved on new tailings filter presses (Figure 12).

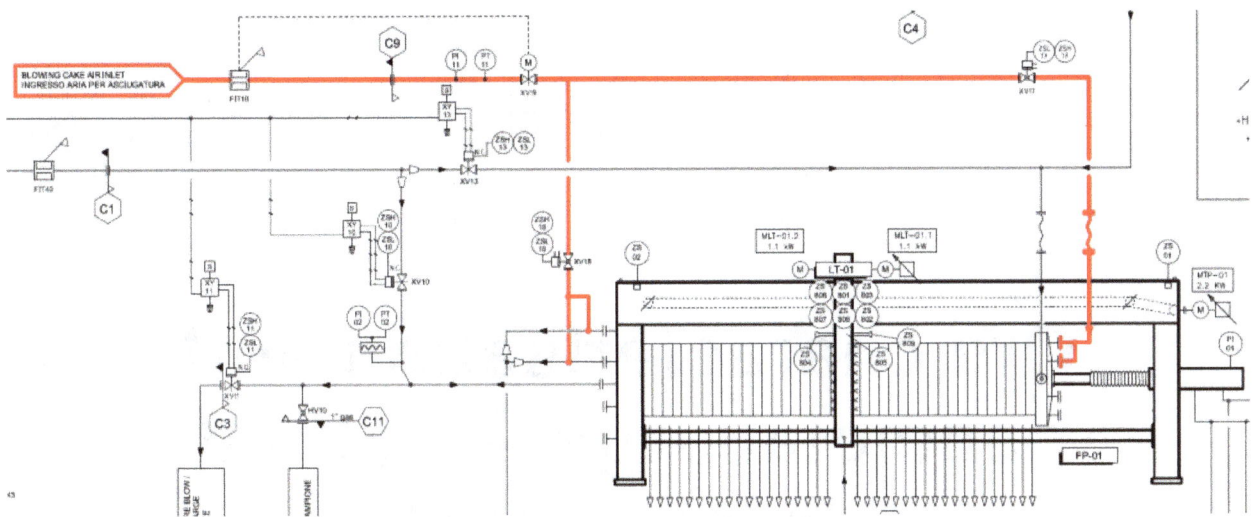

FIG 12 – P&ID of new cake blow line.

A second air inlet point has been added on the mobile head of the filter press, to standardise and ensure greater efficiency along the entre plates pack.

Also two proportional valves (4–20 mA) controlled have been installed in order to avoid the first big burst of air on the cloths.

Evening out the distribution of the air will also reduce the wear on the cloths caused by the high speed and abrasiveness of solid in the cake.

CONCLUSIONS

Dewatering by pressure filtration and dry stacking has become the default tailings management solution. The biggest challenge is to develop and improve filtering plants in order to minimise the costs optimising utilities consumptions without losing sight of the goal and ensuring maximum reliability.

Practical experience has demonstrated this to be viable for iron ore tailings. One of the key lessons learnt from this experience is to design the equipment taking into account the high abrasion anticipated. Neglecting this key aspect will lead to excessive operation costs.

Ore characterisation and geometallurgy – method development

Behaviour of iron ore granules – laboratory and commercial sinter plant granulation

O A Aladejebi,[1,7] S Mitra[2], T Singh[3], D J Pinson[4,7], S J Chew[5,7] and T Honeyands[6,7]

1. AAusIMM, Research Associate, Centre for Ironmaking Materials Research, School of Engineering, The University of Newcastle, Callaghan NSW 2308.
   Email: tosin.aladejebi@newcastle.edu.au
2. MAusIMM, Research Academic, Centre for Ironmaking Materials Research, School of Engineering, The University of Newcastle, Callaghan NSW 2308.
   Email: subhasish.mitra@newcastle.edu.au
3. AAusIMM, Research Associate, Centre for Ironmaking Materials Research, School of Engineering, The University of Newcastle, Callaghan NSW 2308.
   Email: tejbir.singh@newcastle.edu.au
4. Senior Technology and Development Engineer, Coke and Ironmaking Technology, BlueScope Steel, Port Kembla NSW 2505. Email: davidj.pinson@bluescopesteel.com
5. Principal Technology and Development Engineer, Coke and Ironmaking Technology, BlueScope Steel, Port Kembla NSW 2505. Email: sheng.chew@bluescopesteel.com
6. FAusIMM, Associate Professor, Centre for Ironmaking Materials Research, School of Engineering, The University of Newcastle, Callaghan NSW 2308.
   Email: tom.a.honeyands@newcastle.edu.au
7. ARC Research Hub for Australian Steel Innovation, University of Wollongong, Wollongong NSW 2522.

ABSTRACT

Currently, many commercial sinter plants are exploring the potential addition of high-grade iron ore concentrates to improve the sinter grade. A better understanding of the ore blend, granule behaviour and green bed properties is critical to making this change and maintaining productivity in the sinter plant. This study compared the performance of a laboratory scale granulation drum to a commercial scale unit. The blend used was an Asia-Pacific style sinter blend containing iron ore concentrates. The operating conditions of the two granulation drums (laboratory and commercial scale) were fixed at the same Froude Number (0.00337), space factor (15 per cent) and residence time (4.5 mins). The granules were characterised by measuring their mean size, size distribution and green bed voidage. It was found that the laboratory scale granules were representative of commercial sinter plant granules at similar moisture content. When the moisture was fixed at ~6.9 per cent, granulation time had no significant effect. A further fundamental investigation was undertaken to measure the water holding capacity and intra-particle porosity of the individual ores used in this study. The kinetics of water absorption were measured which showed that water absorbs quickly into the ore particles within the first 60 seconds, slowing down until final saturation was attained after six hours. A linear relationship between the water holding capacity and intra-particle porosity was noted for different ore particle sizes. The results of this study provide a baseline understanding of the current commercial sinter blend and encourage exploring the specific effect of increasing the proportion of ultrafine concentrates in sintering in a future study.

INTRODUCTION

A large proportion of the iron ore produced in the mine lacks the granulometric characteristics for direct blast furnace application. Typically, several types of iron ore fines and concentrates are blended, granulated, sintered and screened before being charged into the blast furnace. Granulation prior to sintering permits the formation of granules from different size ranges of the ore mixture to produce a desirable mineral microstructure (Formoso et al, 2003; Waters, Litster and Nicol, 1989). The inclusion of iron ore concentrates of high ferrous grade would improve the sinter grade. However, due to their typically small particle size (<150 μm), significant amounts of concentrates adversely affect the green bed permeability (Hsieh, 2017; Nkogatse and Garbers-Craig, 2022; Nyembwe, Cromarty and Garbers-Craig, 2016a, 2016b; Yang et al, 2019; Zhou et al, 2017).

The pattern of movement of the sinter mixture in the granulating drum affects the granule formation process, such that a steady tumbling motion is desirable to form granules of the desired size and strength (Ishikawa *et al*, 1982). Water is added to the raw materials mixture during the granulation process as the primary binder to facilitate the formation of the desired structure of adhering fines around nuclei particles. The amount of water required to attain the desired granule structure and optimum green bed permeability depends on the ore blend (Khosa and Manuel, 2007). The inherent properties of the ore blend influence the degree of granulation as well as the properties of the granules, ultimately determining the granule size, green bed voidage and green bed permeability (Ishikawa *et al*, 1982; Litster and Waters, 1988; Nyembwe, Cromarty and Garbers-Craig, 2017; Zhou *et al*, 2017).

Over the years, the granulation process has been improved by ironmakers to accommodate large proportions of concentrates in the blend (eg auxiliary binders, pre-granulation and segregated addition of selected fluxes and fuels). However, the fundamental principle of a steady tumbling motion in the granulation drum and the use of water to facilitate granule formation has not changed. Excessive water addition weakens the granules and affects the green bed voidage and permeability due to compaction. Further, during sintering, water migrates from the top layer of the bed to the bottom layer increasing the local moisture content of the granules in the bottom layer weakening the lower bed structure and decreasing bed permeability and sintering rate (Loo and Aboutanios, 2000; Singh *et al*, 2021, 2022). To better understand the granule formation and growth, the operating conditions in a commercial sinter plant granulating drum were replicated on a laboratory scale. Furthermore, some of the properties of the individual ore fines and granules produced were characterised.

MATERIALS AND METHODS

Raw materials and sinter mixtures

The materials used in this study were supplied by BlueScope Steel. The raw materials consisted of primary ore fines (three types), limestone, dolomite, return sinter fines and coke breeze. The particle size distributions of each iron ore were measured by sieving and are presented in Figure 1.

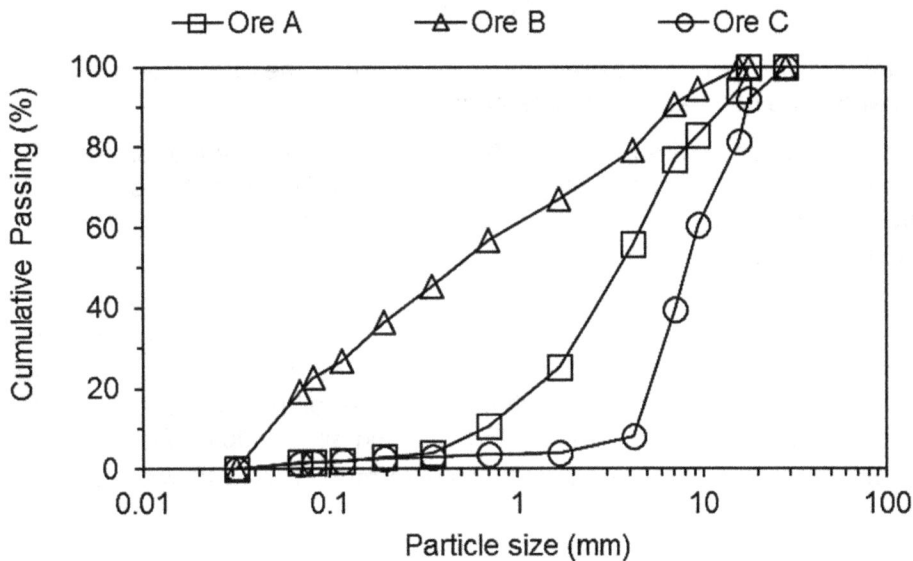

FIG 1 – Particle size distribution of the three iron ores.

The iron ore chemical composition, type, blend and Sauter mean diameter (SMD) of the three types of ore are presented in Table 1.

TABLE 1

Iron ore chemical composition (mass, per cent), type, blend (dry mass basis, per cent) and particle Sauter Mean Diameter (SMD).

Ore	Chemical composition							Ore type	Ore blend	SMD (mm)
	T_{Fe}	FeO	SiO_2	Al_2O_3	CaO	MgO	*LOI1000			
A	57.9	0.22	4.26	1.53	0.09	0.08	10.4	Pisolitic goethite	62.8	0.74
B	61.0	7.65	11.3	0.69	0.21	0.18	-0.32	Siliceous magnetite concentrate	33.8	0.11
C	61.2	0.32	3.0	1.61	0.44	0.02	6.17	Siliceous hematite	3.4	1.22

*LOI1000 is loss on ignition at 1000°C in air.

The iron ore blends were mixed with coke breeze to achieve a coke rate of 5.6 per cent (dry total feed), limestone and dolomite for a basicity of 1.65, a MgO content of 1.3 per cent and a return sinter fine proportion fixed at 20 per cent (dry total feed). Ore C is coarsest with a SMD of 1.22 mm, followed by Ore A with a SMD of 0.74 mm. Ore B, a siliceous magnetite concentrate, has a SMD of 0.11 mm (110 µm). The proportion of <150 µm particle size material in Ore B is about 36.4 per cent. During heating in air at 1000°C, Ores A and C have a high mass loss, while Ore B has a negative mass loss due to the oxidation of magnetite to hematite.

Apparatus and methods

Granulation

The laboratory granulation process was performed in a cylindrical drum of diameter 500 mm and length 310 mm. A green feed mixture weighing approximately 15.5 kg and representing a space factor of 15 per cent was granulated in the drum for 4.5 minutes. The Froude Number and the rotating speed of the granulation drum were 0.00337 and 15.4 rev/min, respectively. The Froude Number (Fr) as defined by Ishikawa *et al* (1982) is:

$$Fr = \frac{D \times N^2}{3600 \times g} \tag{1}$$

where D is the internal diameter of the granulation drum (m), N the rotational speed of the granulation drum (rev/min) and g the gravitational acceleration (9.81 m/s^2).

The space factor, residence time, Froude Number and the rotating speed of the laboratory granulation drum were chosen to match the operating conditions at full scale. A pre-determined amount of water was added to the mixture during granulation to promote the layering of fine particles onto the coarse particles.

After the granulation step was completed, a sub-sample of the granules was taken to measure the experimental moisture content, apparent density and size distribution. The moisture content was calculated from a 1 kg subsample dried at 105°C for 24 hours. The apparent granule density was determined by the volume displacement method in peanut oil using about 100 g of granules. The size distribution of the granules was measured by freezing about 500 g of granules in liquid nitrogen before sieving with the full $\sqrt{2}$ sieve series (16.0 to 0.25 mm). The SMD was calculated as:

$$SMD = \frac{1}{\sum_i \frac{x_i}{\overline{D_i}}} \tag{2}$$

where x_i is the mass per cent of granules in size interval i and $\overline{D_i}$ the mean size of the size interval i.

Green bed properties

Approximately 8 kg of granules were charged into a permeability rig of diameter 95 mm and length 580 mm. The bulk density of the granules was used to determine the green bed voidage of the packed bed as:

$$\varepsilon_{bed} = 1 - \frac{\rho_{bulk}}{\rho_{apparent}} \tag{3}$$

where ε_{bed} is the green bed voidage, ρ_{bulk} the bulk density of the granules in the cylindrical pot (kg/m³) and $\rho_{apparent}$ the apparent density of the granules measured by the oil displacement method (kg/m³).

The air flow rate at a bed pressure drop of 6 kPa across the 500 mm packed bed was recorded and used to calculate the Japanese Permeability Units (JPU) given by Equation 4.

$$JPU = \frac{Q}{A}\left(\frac{H}{\Delta P}\right)^{0.6} \tag{4}$$

where Q is the air flow rate across the packed bed (m³/min), A and H the bed cross-sectional area and bed height respectively (m², mm) and ΔP the pressure drop across the packed bed (mm H₂O).

Water holding capacity and intra-particle porosity of coarse iron ore

The ore samples were screened into four size ranges, -2.0 +1.0, -4.0 +2.0, -8.0 +4.0 and -11.2 +8.0 mm. The screened ore particles were washed thoroughly in water to remove the ultrafine particles that have the potential to interfere with the water holding capacity measurement. The washed particles were then dried at 105°C for 24 hours. Sub-samples of the dried particles were then used to measure the water holding capacity and apparent and skeletal densities as described below.

Skeletal density was determined using a nitrogen gas displacement pycnometer (Micromeritics, AccuPyc 1330). A total of 16 g of dried particles were loaded into a 10 cm³ container and sealed within the instrument compartment. Pulses of nitrogen gas were injected into the compartment containing the ore particles and then expanded into a secondary compartment to determine the sample solid phase volume and skeletal density.

Apparent density was evaluated using the volume displacement method. 50 g of the dried particles were initially soaked in oil for one hour to ensure all the open pores which were accessible had been filled. The oil-soaked particles were removed with excess oil gently wiped from the particle surface using paper towel before measuring apparent density by volume displacement method.

Intra-particle porosity can be expressed as:

$$\varepsilon_p = 1 - \frac{\rho_{apparent}}{\rho_{skeletal}} \tag{5}$$

where ε_p is the intra-particle porosity, $\rho_{apparent}$ the apparent density of the particles measured by the oil displacement method (kg/m³) and $\rho_{skeletal}$ the skeletal density of the particles measured using a nitrogen displacement pycnometer (kg/m³).

Water Holding Capacity was determined using the immersion test where ~50 g of the dried particles were soaked in water for set times. The soaked particles were kept at least 50 mm below the water surface to ensure full immersion. After the required soaking time t, excess water was gently wiped from the particle surfaces using paper towel and the wet weight w_{wet} recorded. The wet particles were then dried in the oven at 105°C for 24 hours and the dry weight w_{dried} recorded. The measurements were repeated three times with a standard deviation of <±0.2 per cent. The water holding capacity at time t, WHC_t was calculated as:

$$WHC_t = \frac{w_{wet} - w_{dried}}{w_{dried}} \times 100\% \tag{6}$$

By measuring the WHC over a series of immersion times, the kinetics of water absorption was determined for the -4.0 +2.0 mm size range. The WHC at 24 hours only was determined for the other size ranges.

RESULTS AND DISCUSSION

Granules and green bed properties

The granulation process in sintering involves the layering of fine particles onto the surface of coarse nuclei particles aided by moistening (Iveson et al, 2001; Kapur, Kapur and Fuerstenau, 1993; Litster and Waters, 1990). Figure 2 shows the granule size distribution for various moisture contents. Figure 3 shows the relationship between granules' SMD, green bed permeability (JPU) and green bed voidage as a function of moisture. Figures 2 and 3 also show the corresponding values for granules produced in a commercial scale granulating drum and sinter plant mixture granulated in the laboratory drum (both using normal operating raw materials including all steel plant recycle materials).

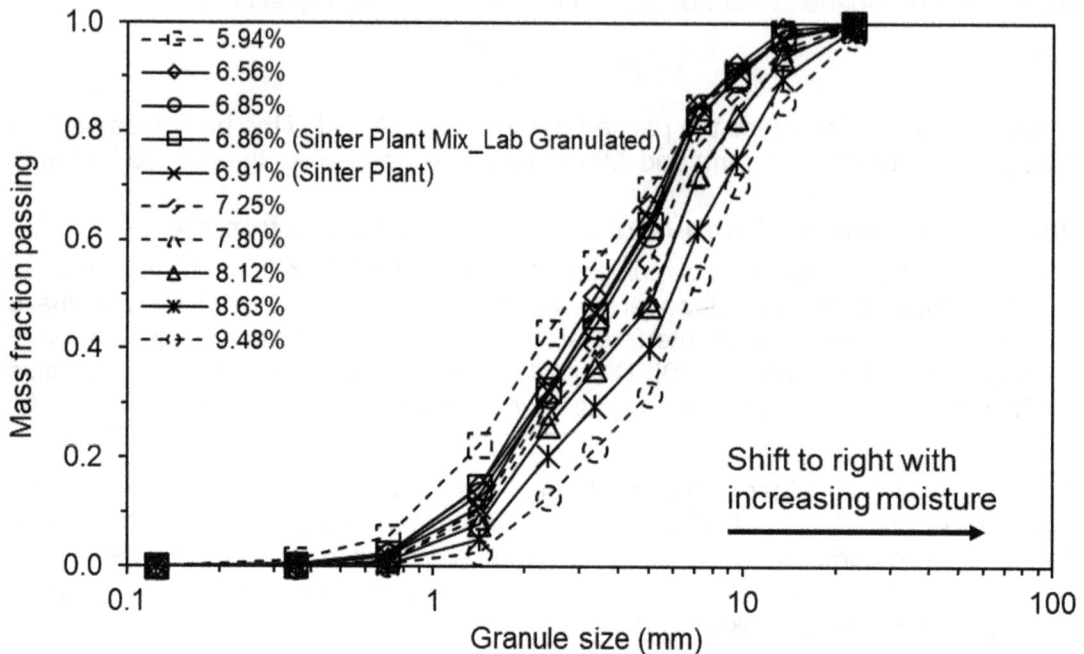

FIG 2 – Granule size distribution with moisture.

In Figure 2, the granule size distribution shifts to the right with increasing moisture indicating a coarser size and the granule SMD increases linearly with increasing moisture in Figure 3. The green bed permeability (JPU) increases with increasing moisture from 5.94 per cent to ~7.5 per cent and then decreases after ~7.5 per cent. However, the green bed voidage decreases with increasing moisture. The granule size distribution, SMD and the green bed voidage were quite representative of the commercial sinter plant at a similar moisture content of ~6.9 per cent. The permeability of the bed of commercial sinter plant granules was significantly higher than the bed of laboratory granules. This is unexpected given the similar voidage and SMD. It may be speculated that differences in the sample handling at the sinter plant compared to the laboratory influenced this JPU result.

Hinkley et al (1994) suggest that the green bed voidage is influenced by the granule size distribution and granule deformability as a function of moisture. That is, at a low moisture content during the granulation process, fines adhere to the surface of the nuclei particles, therefore, tightening the granule size distribution. This results in an increase in bed voidage. With a further increase in moisture, the voidage then passes through a maximum and decreases thereafter. This is due to a balance between the granule size distribution and granule deformability. At a higher moisture content, the adhering fines layer becomes thicker but weaker resulting in a further decrease in bed voidage. Figure 3 shows that at a moisture content between 5.94 per cent and ~7.5 per cent, green bed permeability increases from 16 to 20 corresponding to an increase in granule SMD (from 1.34 to ~2.2 mm) and a decrease in bed voidage (from 0.44 to 0.41). This suggests that the behaviour of the granules in this moisture range is controlled by a balance between narrower granule size distribution and particle deformability. At a moisture content above ~7.5 per cent, the green bed permeability decreases with a continuous increase in granule SMD and a decrease in bed voidage. This suggests that at a moisture content above ~7.5 per cent, particle deformability is the dominant factor.

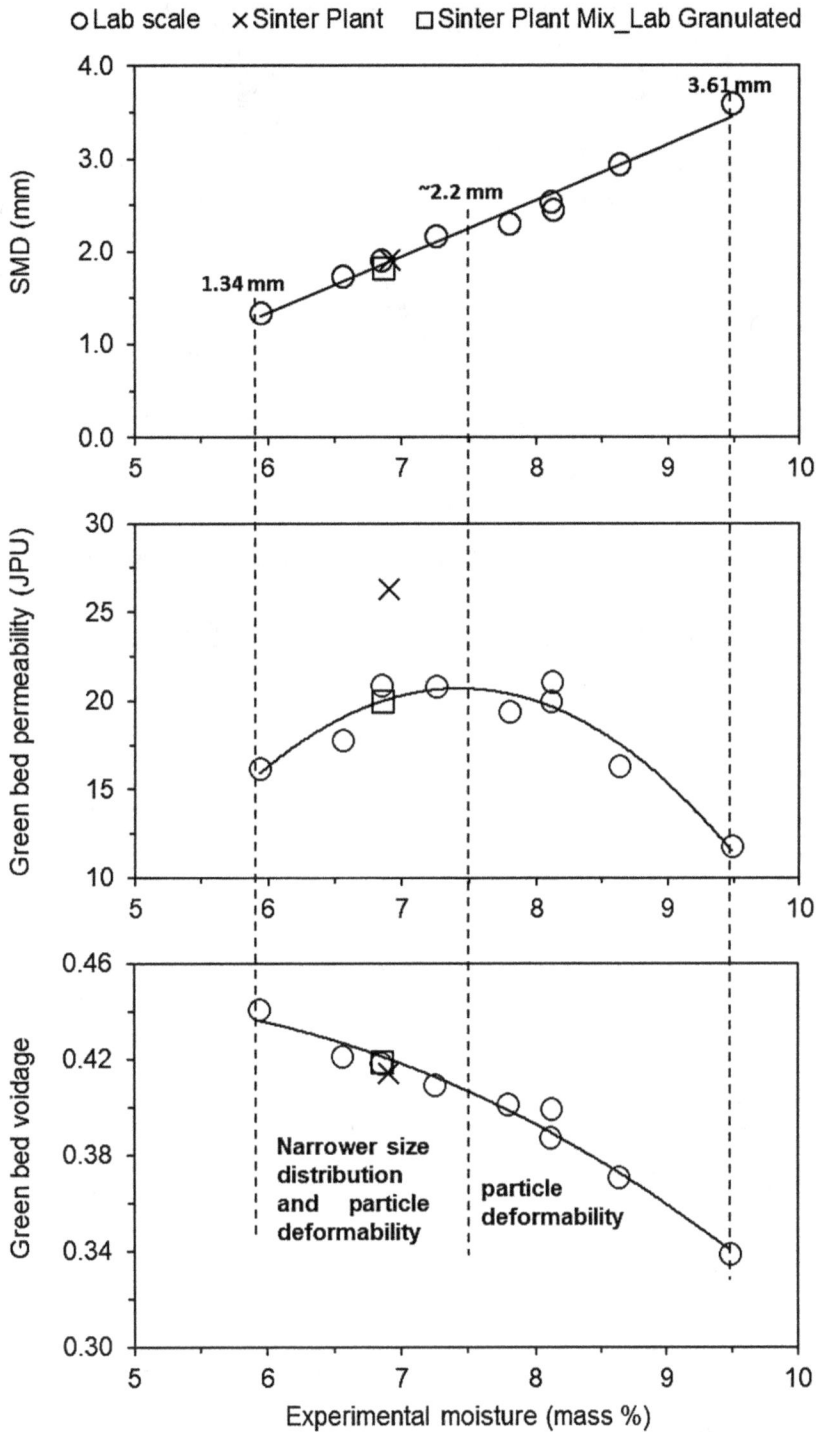

FIG 3 – Relationship between granule SMD, green bed permeability (JPU) and green bed voidage with moisture content.

Effect of granulation time

Figure 4 shows the granule size distribution with granulation time, while Figure 5 shows the temporal variation of granule SMD, green bed permeability (JPU) and green bed voidage, obtained at 6.9 per cent moisture content.

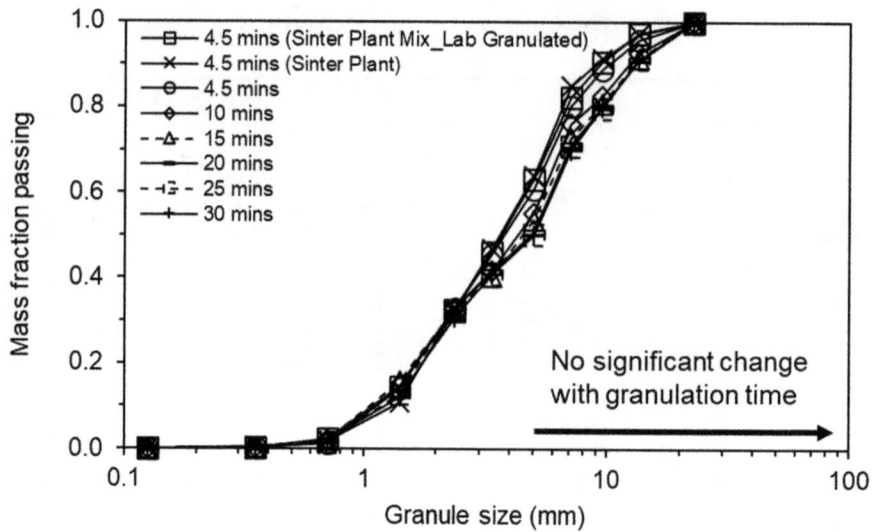

FIG 4 – Granule size distribution with time at a fixed moisture content of 6.9 per cent.

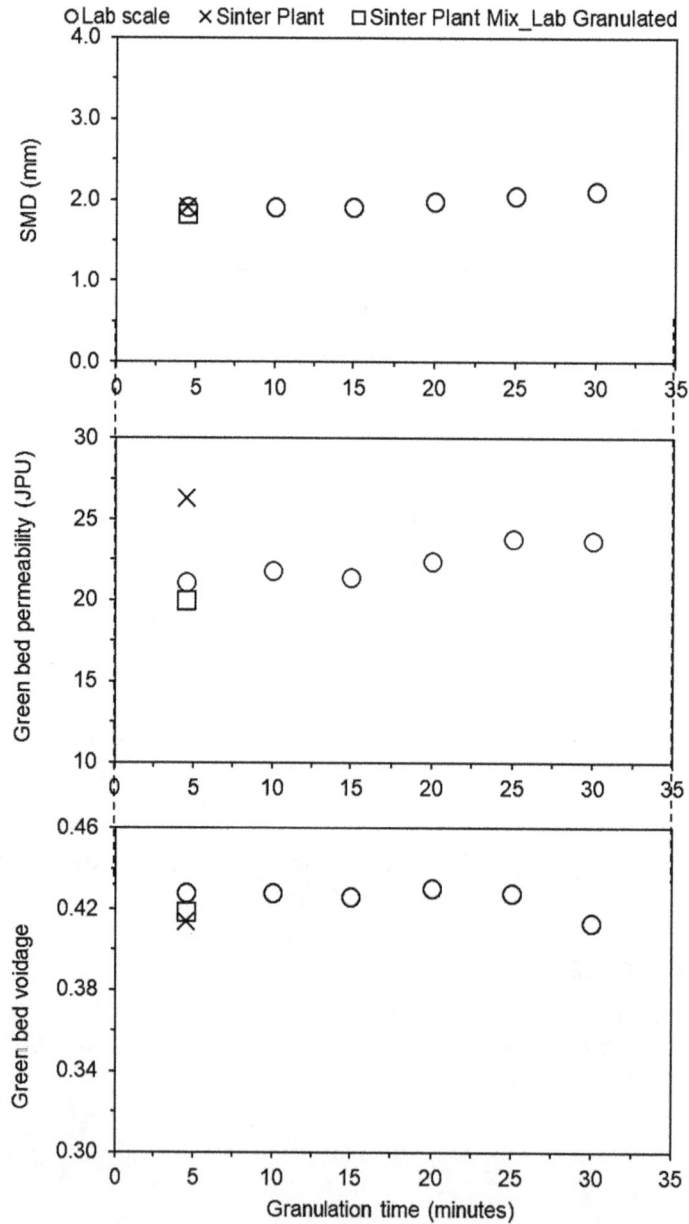

FIG 5 – Relationship between granules SMD, green bed permeability (JPU) and green bed voidage with time at a fixed moisture content of 6.9 per cent.

Figures 4 and 5 show that granulation time has no significant effect on the granule size distribution, granules' SMD, green bed permeability (JPU) and green bed voidage. This indicates that the Froude Number (0.0037) and space factor (15 per cent) alone set the operating conditions of the granulating drums (laboratory scale and commercial sinter plant) and ensure a steady tumbling motion within the cascading regime (Ishikawa et al, 1982). It can therefore be concluded that when granulation occurs within the cascading regime at a fixed moisture content, granulation times above 4.5 mins have no significant effect on the granules and green bed properties.

Water holding capacity and intra-particle porosity of coarse iron ore

The granulation process in commercial sinter plants occurs usually within a few minutes. Hence it is important to understand the rate at which water absorbs into the iron ore particles during granulation. Figure 6 shows the water holding capacity of coarse iron ore particles (-4.0 +2.0 mm size) with time.

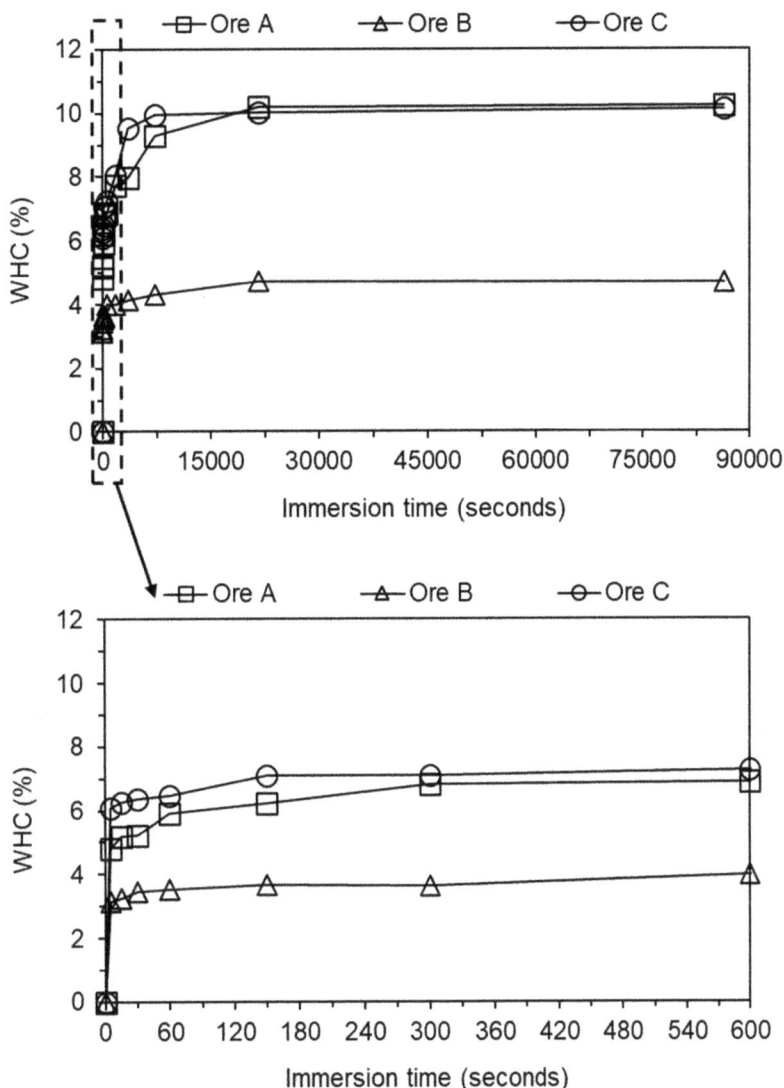

FIG 6 – Iron ore particles' water holding capacity as a function of time. Particle size was -4.0 + 2.0 mm.

Ores A and C have a similar high water holding capacity, while Ore B is significantly lower, demonstrating the strong dependence of WHC on ore type. Water absorbs quickly into the ore particles within the first 60 seconds, slowing down significantly until the final saturation was attained after six hours. In this study, the ore particles were washed and dried before the immersion testing to better understand the particles' water holding capacity behaviour with time. However, this is not the case for industrial ore fines used for granulation. In a typical commercial sinter plant, the ores are pre-wet during handling and storage to reduce dust. The initial moisture content of Ores A, B and C before granulation was 8.7 per cent, 1.5 per cent and 7.9 per cent, respectively. The water holding capacity of Ores A, B and C after 24 hours was 10.2 per cent, 4.7 per cent and 10.11 per cent, respectively. The initial moisture of the ores indicates that a smaller amount of water and less time were required to attain complete saturation of the particles during granulation. The growth of the granules through the layering of adhering fines onto the coarse nuclei particle is strongly influenced by the moisture available for granulation (Litster, Waters and Nicol, 1986; Litster and Waters, 1988; Zhou *et al*, 2017). This is further supported by the result of this study, where it was found that the granules' SMD and size distribution were not affected by increasing granulation time at a fixed moisture content as presented in Figures 4 and 5.

Figure 7 shows the water holding capacity after 24 hours and intra-particle porosity as a function of ore particle size. The ore particle size affects its water-holding capacity and intra-particle porosity.

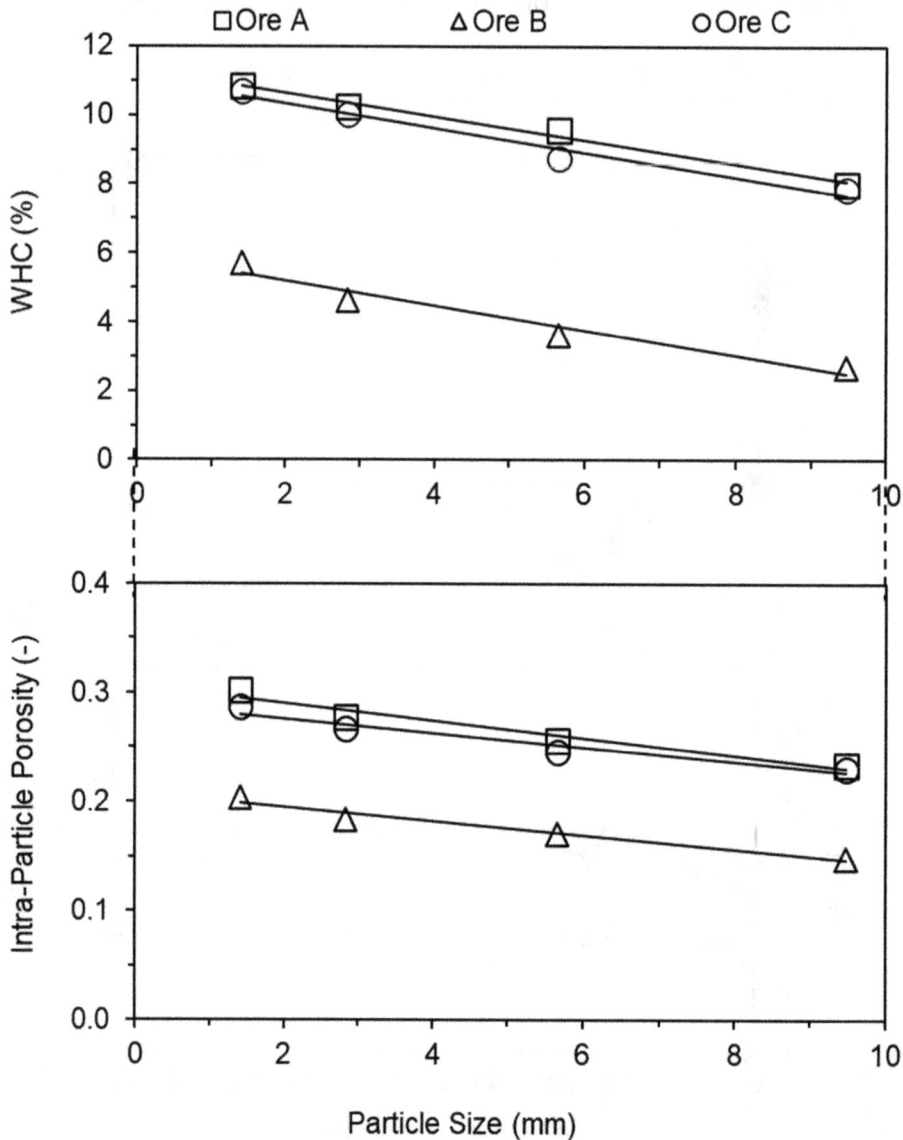

FIG 7 – Water holding capacity after 24 hours and intra-particle porosity as a function of ore particle size.

It can be seen that both the water holding capacity and intra-particle porosity of the ore particles decrease with an increase in particle size. Ores A and C have higher values and are similar when the particle size is the same, while Ore B has a lower value. The specific surface area of solid particles is defined as the total surface area per unit mass. Hence, connectivity between the particle size, WHC and pore network. It could be suggested that as the particle size increases the pore penetration depth remains constant due to the inability to displace already existing gas – even in a fully connected network. Figure 8 shows the relationship between the water holding capacity and intra-particle porosity of different ore types and particle sizes.

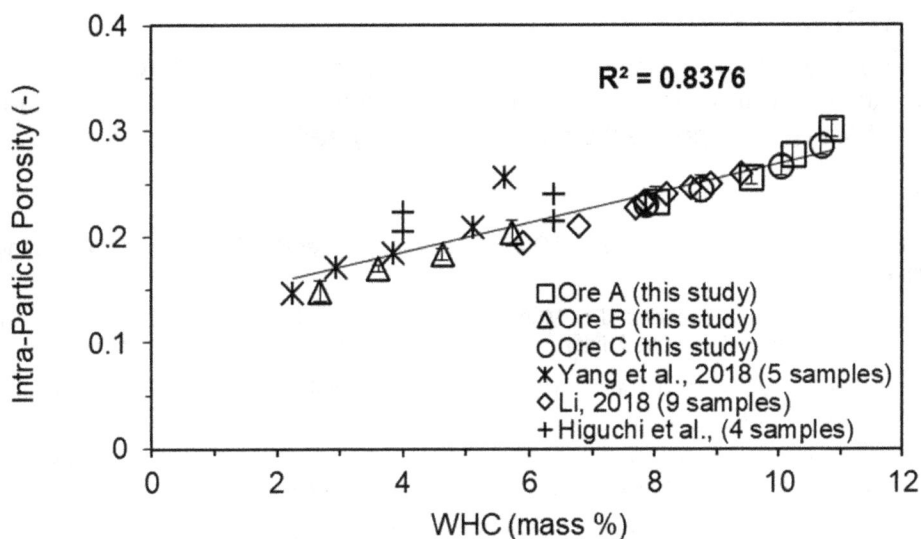

FIG 8 – Relationship between water holding capacity and intra-particle porosity of different ore types.

Figure 8 shows a linear relationship between the water holding capacity and intra-particle porosity of ores of different types (Higuchi, Lu and Kasai, 2017; Li, 2018; Yang *et al*, 2018). This result confirms that the interconnection between the intra-particle pore structure and water holding capacity influences the optimal water content during granulation which was earlier reported in previous studies (Ellis, Loo and Witchard, 2007; Formoso *et al*, 2003; Litster and Waters, 1988).

CONCLUSION

The behaviour of granules and green bed properties of a sinter blend containing iron ore concentrates during sintering was investigated and the following conclusions were drawn:

- The iron ore granules produced in a laboratory scale granulation drum provide a baseline to understand the current commercial sinter blend.

- Under similar operating conditions (Froude Number = 0.00337, space factor = 15 per cent and residence time = 4.5 mins), the laboratory scale granules were representative of commercial sinter plant granules produced at similar moisture content.

- In the moisture content range between 5.94 per cent and ~7.5 per cent, granules' behaviour was controlled by a balance between a narrower size distribution and particle deformability. Particle deformability was the dominant factor when the moisture content exceeded ~7.5 per cent.

- When the moisture content was fixed at ~6.9 per cent, granulation time had no significant effect on the granule size distribution and the green bed voidage.

- The water holding capacity and intra-particle porosity of ores of different sizes follow a linear correlation.

RECOMMENDATIONS

During this work some areas of further work were recognised and recommended for future study:

- Formulate a sinter blend containing a higher proportion of ore concentrates to improve the iron grade of the sinter for sustainable ironmaking. High-grade iron ore sinter lowers the reducing agent rate of blast furnaces, as well as the CO_2 emission levels in the steel industry.

- Evaluate granules and green bed properties of sinter blend containing higher proportions of ore concentrates to further explain the granule size distribution and granule deformability mechanisms.

ACKNOWLEDGEMENTS

The authors acknowledge funding from the Australian Research Council (ARC) through the Industrial Transformation Research Hubs Scheme under Project Number: IH200100005. The funding support and permissions from BlueScope Steel Ltd to publish are gratefully acknowledged.

REFERENCES

Ellis, B G, Loo, C E and Witchard, D, 2007. Effect of Ore Properties on Sinter Bed Permeability and Strength, *Ironmaking and Steelmaking*, 34:99–108.

Formoso, A, Moro, A, Pello, G F, Menéndez, J L, Muñiz, M and Cores, A, 2003. Influence of Nature and Particle Size Distribution on Granulation of Iron Ore Mixtures Used in a Sinter Strand, *Ironmaking and Steelmaking*, 30:447–460.

Higuchi, T, Lu, L and Kasai, E, 2017. Intra–Particle Water Migration Dynamics During Iron Ore Granulation Process, *ISIJ International*, 57:1384–1393.

Hinkley, J, Waters, A G, O'Dea, D and Litster, J D, 1994. Voidage of Ferrous Sinter Beds: New Measurement Technique and Dependence on Feed Characteristics, *International Journal of Mineral Processing*, 41:53–69.

Hsieh, L, 2017. Effect of Iron Ore Concentrate on Sintering Properties, *ISIJ International*, 57:1937–1946.

Ishikawa, Y, Kase, M, Sasaki, M, Satoh, K and Sasaki, S, 1982. Recent Progress in the Sintering Technology: High Reducibility and Improvement of Fuel Consumption, in *Proceeding of Ironmaking Conference, Annual Ironmaking Conference, Pittsburgh, USA*, pp 80–89.

Iveson, S M, Litster, J D, Hapgood, K and Ennis, B J, 2001. Nucleation, Growth and Breakage Phenomena in Agitated Wet Granulation Processes: A Review, *Powder Technology*, 117:3–39.

Kapur, P C, Kapur, P and Fuerstenau, D W, 1993. An Auto-Layering Model for the Granulation of Iron Ore Fines, *International Journal of Mineral Processing*, 39:239–250.

Khosa, J and Manuel, J, 2007. Predicting Granulating Behaviour of Iron Ores Based on Size Distribution and Composition, *ISIJ International*, 47:965–972.

Li, C, 2018. Numerical Analysis of the Packing Characteristics of Iron Ore Granules using Discrete Element Method, PhD Thesis, University of Newcastle Australia.

Litster, J D, Waters, A G and Nicol, S K, 1986. A Model for Predicting the Size Distribution of Product from a Granulating Drum, *Transactions of the Iron and Steel Institute of Japan*, 26:1036–1044.

Litster, J D and Waters, A G, 1988. Influence of the Material Properties of Iron Ore Sinter Feed on Granulation Effectiveness, *Powder Technology*, 55:141–151.

Litster, J D and Waters, A G, 1990. Kinetics of Iron Ore Sinter Feed Granulation, *Powder Technology*, 62:125–134.

Loo, C E and Aboutanios, J, 2000. Changes in Water Distribution when Sintering Porous Goethitic Iron Ores, *Mineral Processing and Extractive Metallurgy*, 109:23–35.

Nkogatse, T and Garbers-Craig, A, 2022. Evaluation of Iron Ore Concentrate and Micropellets as Potential Feed for Sinter Production, *Mineral Processing and Extractive Metallurgy Review*, 43:300–312.

Nyembwe, A M, Cromarty, R D and Garbers-Craig, A M, 2016a. Effect of Concentrate and Micropellet Additions on Iron Ore Sinter Bed Permeability, *Mineral Processing and Extractive Metallurgy*, 125:178–186.

Nyembwe, A M, Cromarty, R D and Garbers-Craig, A M, 2016b. Prediction of the Granule Size Distribution of Iron Ore Sinter Feeds that Contain Concentrate and Micropellets, *Powder Technology*, 295:7–15.

Nyembwe, A M, Cromarty, R D and Garbers-Craig, A M, 2017. Relationship between Iron Ore Granulation Mechanisms, Granule Shapes and Sinter Bed Permeability, *Mineral Processing and Extractive Metallurgy Review*, 38:388–402.

Singh, T, Li, H, Zhang, G, Mitra, S, Evans, G, O'Dea, D and Honeyands, T, 2021. Iron Ore Sintering in Milli-Pot: Comparison to Pilot Scale and Identification of Maximum Resistance to Air Flow, *ISIJ International*, 61:1469–1478.

Singh, T, Mitra, S, O'Dea, D, Knuefing, L and Honeyands, T, 2022. Quantification of Resistance and Pressure Drop at High Temperature for Various Suction Pressures During Iron Ore Sintering, *ISIJ International*, 62:1768–1776.

Waters, A G, Litster, J D and Nicol, S K, 1989. A Mathematical Model for the Prediction of Granule Size Distribution for Multicomponent Sinter Feed, *ISIJ International*, 29:274–283.

Yang, C, Zhu, D, Pan, J and Shi, Y, 2019. Some Basic Properties of Granules from Ore Blends Consisting of Ultrafine Magnetite and Hematite Ores, *International Journal of Minerals, Metallurgy and Materials*, 26:953962.

Yang, C, Zhu, D, Pan, J and Lu, L, 2018. Granulation Effectiveness of Iron Ore Sinter Feeds: Effect of Ore Properties, *ISIJ International*, 58:1427–1436.

Zhou, M, Zhou, H, O'Dea, D, Ellis, B G, Honeyands, T and Guo, X, 2017. Characterization of Granule Structure and Packed Bed Properties of Iron Ore Sinter Feeds that Contain Concentrate, *ISIJ International*, 57:1004–1011.

An approach to the calibration and validation of transfer chute DEM simulations handling cohesive iron ore fines

R Elliott[1], T Walkemeyer[2] and D Chang-Martin[3]

1. Principal Mechanical Engineer, Sedgman Onyx, Perth WA 6000.
   Email: richard.elliott@onyxprojects.com
2. Senior Mechanical Engineer – DEM Specialist, Sedgman Onyx, Perth WA 6000.
   Email: thomas.walkemeyer@onyxprojects.com
3. Senior Mechanical Engineer – DEM Specialist, Sedgman Onyx, Perth WA 6000.
   Email: darren.chang-martin@onyxprojects.com

ABSTRACT

Over the past decade, discrete element method (DEM) simulations have seen increasing adoption by the resources sector to inform the design of materials handling equipment such as transfer chutes, bins and hoppers. But, if 'all models are wrong, but some are useful' (Box, 1979) what approaches are available to check that a DEM simulation is useful? How to avoid garbage in equals garbage out.

This work considers what properties of bulk solid flow behaviour are important to transfer chute simulation. Cohesive bulk solids (such as sticky iron ore fines) can develop sufficient strength to form a stable cohesive arch, this phenomenon is not well predicted by measurements of angle of repose.

This work presents a two-part approach to the development of a useful DEM model that can be used to add value to the design and optimisation of bulk materials handling equipment.

Part 1 is the calibration of not just one, but a range of material models chosen to envelope the range of anticipated materials that will be handled. This work describes the DEM simulation of a batch of direct shear cell tests to develop material models that can be measured by the same metrics routinely used to characterise the flowability of cohesive iron ore fines.

Part 2 of this work describes systematic methods for validation of the transfer chute DEM simulations. The DEM model is only useful if it can adequately represent the typical problems observed within a transfer chute, such as blockages due to build-up and cohesive arching. Proper validation requires good observations. Validation examples of site observations are shown side-by-side with simulations to demonstrate the adequacy of the DEM model and calibration approach.

DEM simulations have the potential to add significant value by allowing engineers and designers to predict material flow-through equipment based on a range of material properties, allowing for stress testing designs and making iterative improvements.

INTRODUCTION

Avoidable delays and lost production regularly occur due to common materials handling problems often made worse by poor design choices. In the case of transfer chute design, unreliable operation of these assets can prove costly in the mining sector where large volumes of ore are handled at any given moment and uninterrupted production is of high importance.

Typical materials handling problems encountered by site personnel include:

- blockage/plugging of chutes or rate limitations
- non-central loading resulting in receiving conveyor belt drift
- material build-up/hangup causing belt run offs
- constant material spillage
- excessive chute wear in lined or unlined areas.

Another issue is the changing of feed to a facility as orebodies are exhausted and new ones take their place, often with characteristics the facility was not originally designed for. This is often seen in the iron ore industry as the move toward the mining of below water table ores results in a more cohesive feed than a facility was originally intended. This risks further lost production if interventions to modify or upgrade the facility are not made.

The use of DEM as a tool for the analysis and visualisation of bulk material flow in industrial applications has significantly increased over the past decade. When used correctly, DEM can be used to predict or diagnose these material handling problems and from these predictions, treatment measures can be designed and implemented. However, in the experience of the authors, many transfer chute DEM simulations still fail to capture or predict real-world flow problems that result in delays and lost production. Critically, the practice of validating these simulations to observed site behaviour seems to be lacking in industry, making it difficult for an independent reviewer to draw conclusions as to the adequacy and usefulness of the simulations for predicting real problems and demonstrating possible solutions.

This paper aims to describe the approach developed by the authors to reliably use DEM as a tool for the prediction of flow problems when handling cohesive iron ore fines material. The authors also present methods to validate critical aspects of the simulations to observed site behaviour. Only if the simulation can predict the problem can it be relied upon as a useful model to assist in the development of a solution.

QUANTIFYING FLOW BEHAVIOUR

An initial question that should be asked, what is the range of materials that will be handled and at what moisture contents? Secondly, what testing is required to quantitatively measure the flow behaviour of cohesive iron ore fines?

Cohesive bulk solids (such as iron-ore fines) can develop sufficient strength when consolidated to form a stable shape such as an arch or rat hole. This phenomenon is commonly associated with bins and hoppers, but consolidation stresses are also present within transfer chutes and here too stable build-up formations can occur. For the engineer responsible for the design of material handling equipment it is critical to be able to quantify the strength of a material and the conditions where it will or won't flow.

The design of transfer chutes has historically considered the flowing material stream as a lump mass with the principal resistances to flow arising from wall friction. The assumption being that the material will flow if the wall angle is sufficiently steep. This theory being adequately applied for many years for materials with low internal strength and good flowability, however the approach neglects resistance to flow arising from the internal strength of the material and fails to predict if a material will stop flowing due to its internal strength.

The foundation for the quantitative measurement of bulk solids strength and flow properties was laid by Jenike (1964) who developed the Jenike shear tester. This shear test method is still in wide use today with iron ore fines routinely tested in shear cells, measuring the stress at which the material yields (flows) over a range of consolidation pressures. The detail of the test methodology and theory is explained ASTM D6128-22 and D8081-17. Jenike also developed theory for solids flow and provided a methodology for the design of flow channels within hoppers and silos. These theories still underpin current methods in use today such as EN 1991-4-2006.

In general, the shear test is performed in three stages:

1. Pre-consolidation: a representative sample of material is placed in the shear cell and is consolidated by twisting the shear cell cover while applying a compressive load normal to the cover.

2. Pre-shear: a normal load is applied to the cover and the sample is sheared until a steady state shear stress has been reached.

3. Shear: the shear stress is removed and the normal load is reduced before the sample is sheared again until the shear force reaches a maximum value and then begins to decrease.

The shear stress during flow is plotted against the applied consolidation to determine the yield locus, as shown in Figure 1. The yield locus defines stress conditions required for material flow and the unconfined yield strength of the material.

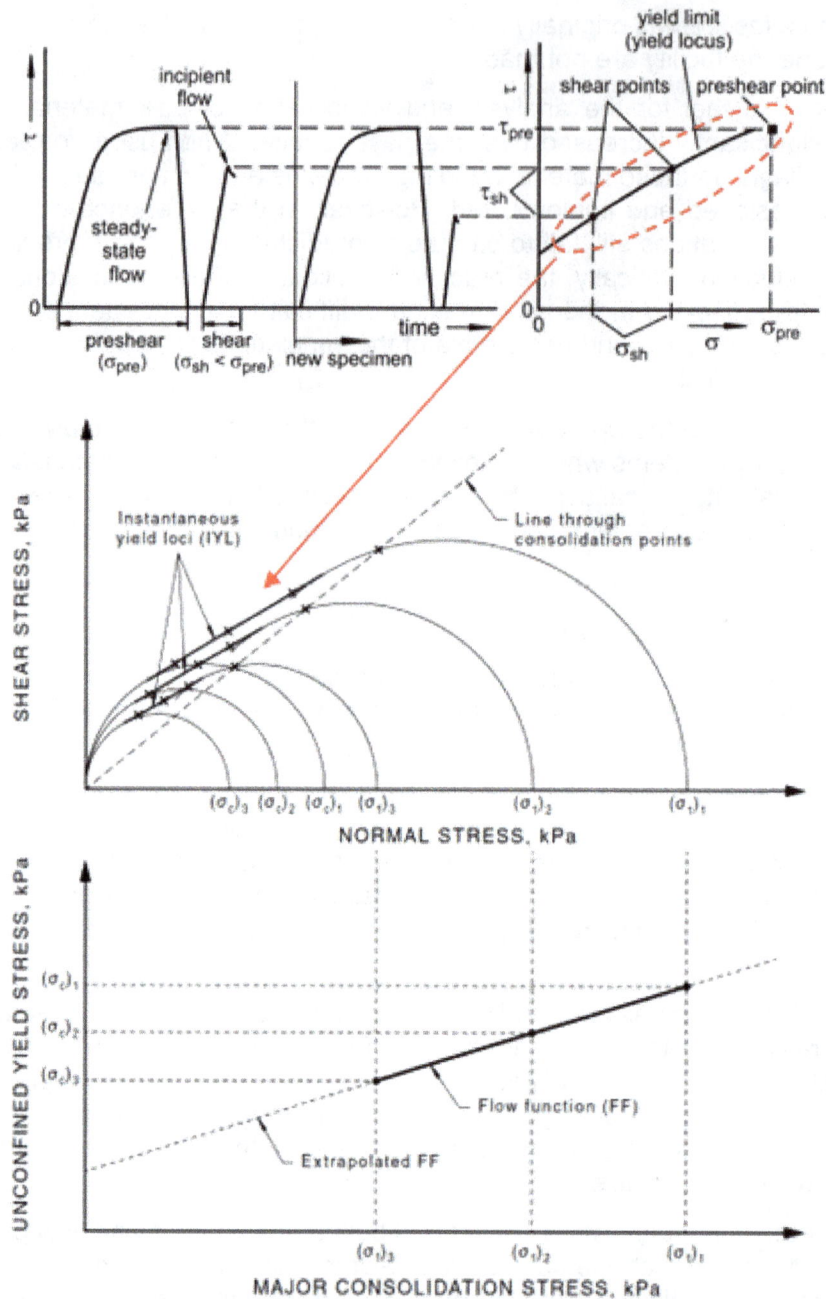

FIG 1 – Shear stress versus time, yield locus and flow function (AS 3880: Standards Australia, 2017; Schulze, 2014).

For cohesive materials, the yield locus will be dependent on the normal stress applied during pre-shear. Typically, three or more normal stresses are tested to produce a family of yield loci. Mohr circle theory is used to obtain the unconfined yield stress (σ_c) and major consolidation stress (σ_1). The flow function (FF) is then derived, as shown in Figure 1.

The outputs from the shear test allow the flowability of the bulk solid to be characterised by its unconfined yield strength, consolidation stress and kinematic/effective angles of internal friction. This also allows for a numerical characterisation of flowability to be calculated. The flowability (ff_c) is a ratio of the consolidation stress to unconfined yield strength and is calculated differently depending on source and testing organisation. The method provided by Schulze (2014) is presented as follows:

$$ff_c = \frac{\sigma_1}{\sigma_c}$$

Such a numerical characterisation also provides a link to the aforementioned quantitative statements of flow behaviour. Below is a quantitative classification of flow behaviour based on the flowability of a given sample from Schulze (2014):

$$ff_c < 1 \qquad \text{not flowing}$$

$$1 < ff_c < 2 \qquad \text{very cohesive}$$

$$2 < ff_c < 4 \qquad \text{cohesive}$$

$$4 < ff_c < 10 \qquad \text{easy-flowing}$$

$$ff_c > 10 \qquad \text{free-flowing}$$

These classifications are superimposed onto a flow function diagram as shown below. As can be seen from flow function A in Figure 2, the flowability of a given bulk solid is not a single value but is dependent on consolidation stress, typically increasing in flowability at greater consolidation stresses. It also follows that the shape of flow function and behaviour of a bulk solid strength as consolidation stress is changed, is characteristic of the bulk solid tested.

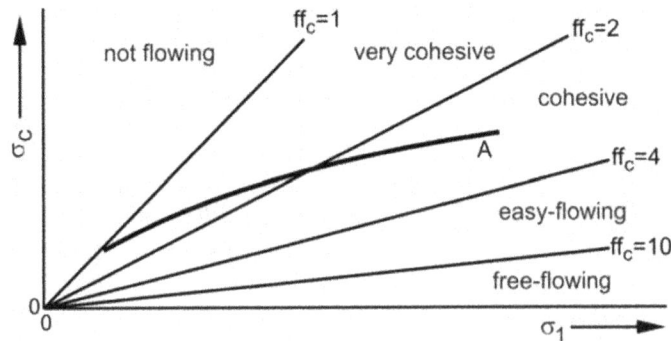

FIG 2 – Flow function and lines of constant flowability (Schulze, 2014).

Notes on angle of repose

The use of the angle of repose in the calibration of DEM material models remains popular in commercial applications and is regularly requested as part of a flow properties test report. Indeed, the initial demonstration of calibration to first time users by the tutorials of the major commercial DEM software codes have users simulate the angle of repose test. Despite its popularity, it is the view of the authors that calibration by simulation of angle of repose is not recommended for cohesive fines materials for the following reasons:

- The angle of repose of a static pile of material provides very little information on the strength and flowability of the material under different loads, as can be determined from the shear test described in the previous section.

- Particularly for cohesive bulk solids, the angle of repose is dependent on the test procedure and is therefore not an intrinsic property of the bulk solid. As there is no one standard procedure for the angle of repose test, it is difficult to know whether laboratory conditions have been replicated in the DEM simulation. Indeed, Schulze (2014) notes that the results of the angle of repose test is sensitive to particular boundary conditions, ie conditions at the free surface of the heap, geometry of test devices and preparation of the bulk solid. For instance, thorough stirring of bulk solids which easily fluidise leads to smaller angles of repose while greater compaction, eg overfilling the receptacle, would consolidate material and lead to higher angles of repose.

- Angles of repose may not be reasonably applied to cohesive bulk solids, which do not flow downwards continuously but avalanche from the receptacle, causing a varying angle of repose which is difficult to measure (Schulze, 2008).

Indeed, the limitations of the angle of repose test have been well known for some time. In one of the seminal works of the materials handling discipline, Jenike (1964) notes '*The angle of repose is not a measure of the flow-ability of solids. In fact, it is only useful in the determination of the contour of a pile, and its popularity among engineers and investigators is due not to its usefulness but to the ease with which it is measured*'. Such a statement is still true over half a century later when selecting a representative test for calibration of a DEM material model.

DEM VERSUS REAL WORLD

DEM allows for the flow behaviour of a material stream to be investigated at high levels of detail, with the simulations tracking interactions between individual particles and between particles and boundary walls with a variety of relevant physical contact models to suit the application. The total force on each particle is integrated over a time step to calculate the change in particle position and velocity and the simulation proceeds. The specific parameters input into the DEM numerical models are not typically directly derived from material test results. For instance, a user cannot directly enter physical values for the unconfined yield strength, bulk density, or even angle of repose, into the simulation. Rather, DEM requires calibrating the micro models to recreate the macro behaviour, iteratively setting the parameters that govern the contact model between individual particles in a process referred to as calibration.

Industrial application aims to replicate the real-world macro scale behaviour of the bulk solid, ie behaviour of the entire material stream or replication of an existing materials handling issue. As a DEM model simulates the behaviour of individual particles and stores information about each particle at every time step, the maximum number of particles and duration of simulation is limited by computational power, memory and storage. The number of particles is related to particle size, which places a ceiling on the minimum particle size achievable in a simulation. An implication of this is that the full particle size distribution (PSD) cannot be practically simulated for most industrial applications, eg the -4 mm the fines fraction as generally analysed in laboratory tests. It is current practice to reduce the number of particles by either scaling up the PSD uniformly by a given factor or scalping the PSD by neglecting particles below a certain cut size. As summarised by Coetzee (2018), in general, these practices do not greatly change results at lower scaling factors, while larger scaling factors tended to have a greater impact on the physical behaviour and resolution of the simulation.

Related to existing computational capacity is the use of real particle shapes in a simulation. Spherical particles remain the most popular shape used in simulations. This is due in part to their simplicity, but also as simulations are faster to perform than any other particle type due to any point on the surface of the particle being determined by knowing the diameter and coordinates of the centre (Hastie, 2010). A downside to this is consistently capturing the effects of non-spherical particles. While obvious, shape does play an important factor in simulating granular flows, with spherical particles shown to poorly predict non-spherical particles by various authors (Favier *et al*, 1999; Cleary, 1998; Li and Holt, 2005). However, current DEM codes address the issue by introducing rolling resistance to spherical particles to replicate the corner effects of non-spherical particles which has been shown to produce acceptable results and adequately replicate flow behaviour (Zhou *et al*, 1999; Pinson *et al*, 2004). Recent codes use polyhedrons to represent real particle shapes, potentially yielding a more realistic solution. However, the resulting penalty in simulation time can be as much as ten times longer than spherical particles (Potapov and Campbell, 1998).

However, as computational power improves over time, especially with the introduction of graphical processing units (GPUs) to solve simulations in recent years, the impact of these limitations will continue to be attenuated. As previously mentioned, the goal of industry DEM simulation is to recreate the macro behaviour of the bulk solid in sufficient granularity for use as a design tool rather than exhaustively recreating the micro effects. It is the view of the authors that this aim can be adequately fulfilled with the current state of computational power as will be demonstrated in following sections providing the model is calibrated and validated.

SIMULATION OF SHEAR CELL TESTING

The authors conduct simulations in the Rocky DEM software. For industrial scale DEM simulations of iron ore fines the authors typically use minimum spherical particle size of 16 mm diameter for most work based on the GPU processor performance and memory of a Nvidia Titan V GPU.

To test the suitability of the contact models and relevant parameters the authors simulate the Jenike shear cell test. The simulation follows the same shear cell methodology as outlined in ASTM D6128 and described in the previous section. This aim of this approach is to quantitatively describe the flowability of the simulated material in the same way as the real-world material and in particular to find adhesion parameters that work for the same particle size as will be used in the transfer chute simulations.

The shear test simulation reproduces the three stages of the actual test: pre-consolidation, pre-shear and test shear. To accommodate a 16 mm particle size required to commercially simulate transfers at the scale commonly used in the iron ore industry, the simulated shear test has been modelled at a scale factor of 6:1 such that the shear cell diameter is 570 mm rather than the typical size of 95 mm. This provides a ratio of shear cell size to particle size of 36. In real world shear cell testing the shear cell diameter should be at least 20 times the top size of the material tested, according to AS 3880 (Standards Australia, 2017). Scaling the shear cell increases the weight of material above the shear plane. the gravitational force in these shear cell simulations is reduced by a scale factor of 1:6 to avoid this increase.

Figure 3 shows the simulated shear force plots of an example material model with pre shear at a single consolidation pressure followed by the test shear at three reduced consolidation loads. These are then used to plot the yield locus for the simulated material.

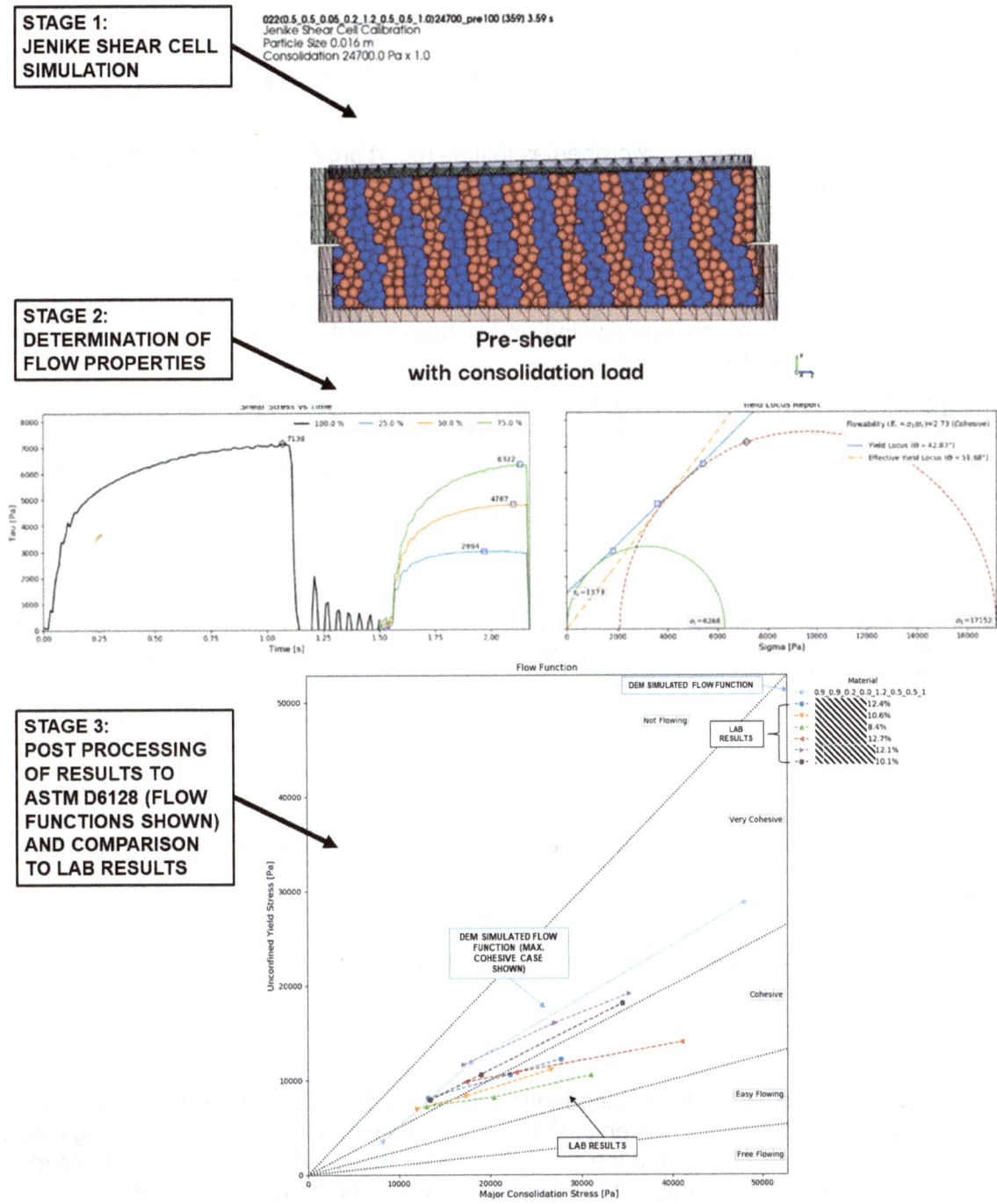

FIG 3 – Jenike shear cell test simulation in Rocky and an example of shear cell simulation results and determination of flow properties.

Normal and shear forces that have been measured from the shear test simulations then are processed according to the ASTM D6128 procedure to produce equivalent flow function and internal friction plots as shown in Figure 3 and compared with laboratory test shear results. The results have also been compared directly to the unprocessed laboratory shear test results where available.

As DEM material calibration is an iterative process, the simulated results are compared to the real-world flow properties testing report, with model input parameters adjusted accordingly until a match is found. The sensitivity of the simulated shear cell results to changes in the following parameters of the material model are tested through automated batches of shear cell simulations:

- rolling resistance

- static and dynamic friction coefficient

- adhesion (parameters depend on numerical model selected).

It is worth noting that a change to any of the DEM material model parameters above influences all simulated property results. The authors endeavour to find a balance of model parameters that provides a good fit between the simulated flow properties and the measured real-world flow properties.

As shown in Figure 3, a standard flow properties testing report provides insight into the range of flow behaviours of a given sample at varying moisture contents, creating a family of flow functions for a single sample. Additionally, depending on the number of samples provided to the testing laboratory, there would be multiple families generated per sample. Then creates a range of measured flow behaviours that can range from free flowing to very cohesive that a particular facility is expected to handle. By simulating the same shear tests used to generate the laboratory results, the envelope of expected flow behaviours at the free flowing and cohesive ends of the tested properties can be captured into the full-scale DEM simulations.

Following the delivery of numerous DEM projects over several years, the authors have collated a library of material model input parameters and corresponding shear cell results. This allows efficient calibration of future DEM material models either by directly matching the library entries to a given sample flow properties test report or using similar library entries as optimal starting points to begin the calibration process.

VALIDATION WITH SITE OBSERVATIONS

Verifiable validation of DEM simulations is challenging as it requires both sufficient access to the operating asset and gathered site data as well as iterative approach in adjusting the models to conform with reality. The most important step is to check and demonstrate that the simulation model was sufficient to capture the observed flow behaviour of the material and critically that the model shows flow problems where they exist.

As mentioned, to verify the model, it is necessary to gather information for the existing problems suitable for verification. It is here that challenges begin as details of the problem are typically lacking at the initial stages of the project. It is the experience of the authors that often, the only information given for a problem are limited to a qualitative statement:

- 'chute blocks'

- 'requires hosing every shift to prevent blockage and clean build-up'

- 'belt runs-off'

- 'spillage'.

Many materials handling problems are dependent on the ore type and moisture content resulting in intermittent issues that are hard to observe. Without detailed photos and videos depicting the actual problem it is not possible to verify that the simulation has replicated the behaviour and the model is accurate. This is especially the case with transfer design as guarding requirements often prevent personnel from accessing the problem areas in sufficient duration and position to diagnose problems.

Time lapse cameras are very useful in these scenarios as they can be mounted within guarded locations during shutdowns or maintenance windows and record data for sufficiently long to capture these intermittent issues. Figure 4 shows a comparison of the DEM simulated ore and time lapse camera footage. In the video, the time lapse camera recorded a transient surge of highly cohesive ore during which a blockage in the chute was observed, diagnosing the cause of the frequent blockages. It is worth noting that initial suspicions were that gradual build-up while handling the normal ore was the cause rather than a transient event. The close agreement between the time lapse footage and DEM model was achieved with proper calibration of the material model along with site operating data of the material feed and throughput.

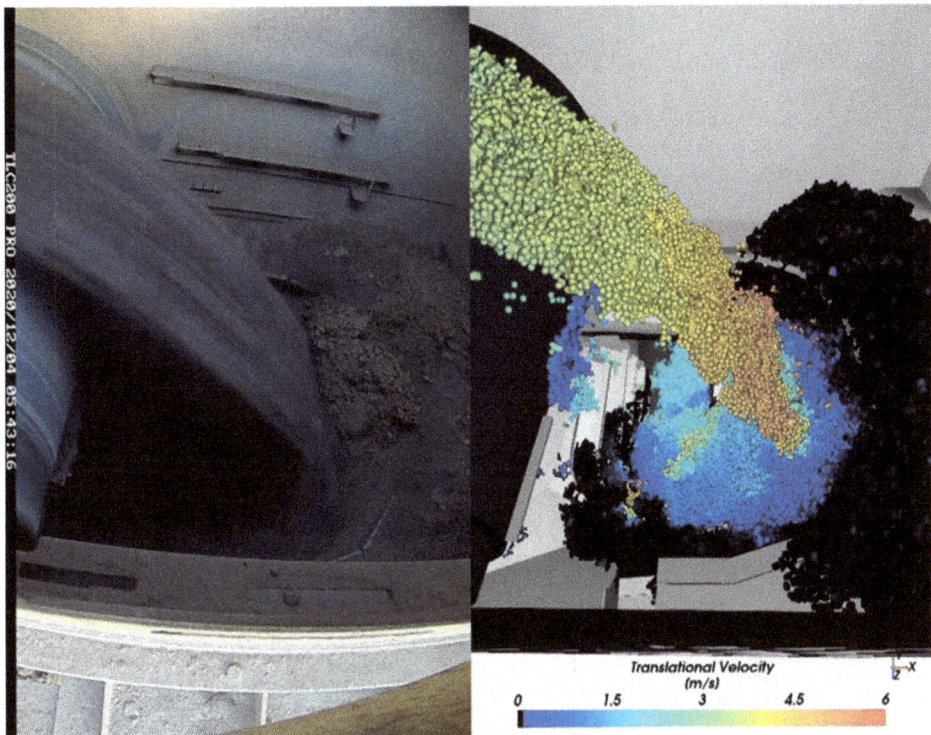

FIG 4 – Time lapse footage of transient surge of cohesive ore prior to a blockage event compared to DEM simulation replicating this behaviour.

Photogrammetry, flow and wear patterns

To significantly improve the precision by which the simulation results could be verified to site observations, photogrammetry seeks to recreate the camera pose and view angle by which the simulation model is seen to match available site photos or videos. This enables DEM simulated results to be directly overlaid onto real world site photos, allowing results and reality to be directly compared. A demonstration of the power of this technique is shown in Figure 5 which shows a close correlation between the DEM simulated high wear areas and their real-world counterparts. Another common area of difference between DEM simulations and reality is the trajectory off the discharge conveyor head pulley. Figure 6 demonstrates the use of the photogrammetry technique to validate the simulated ore trajectory, indicating that characteristics such as ore/belt adhesion, belt speed and bulk density are reflective of reality.

This allows end users to have greater confidence in the voracity of the simulation and also the efficacy of any future modifications to treat these material handling problems. In the current implementation, the photogrammetry technique is done manually, however, the authors are exploring various tools available on the market to assist or automate this process.

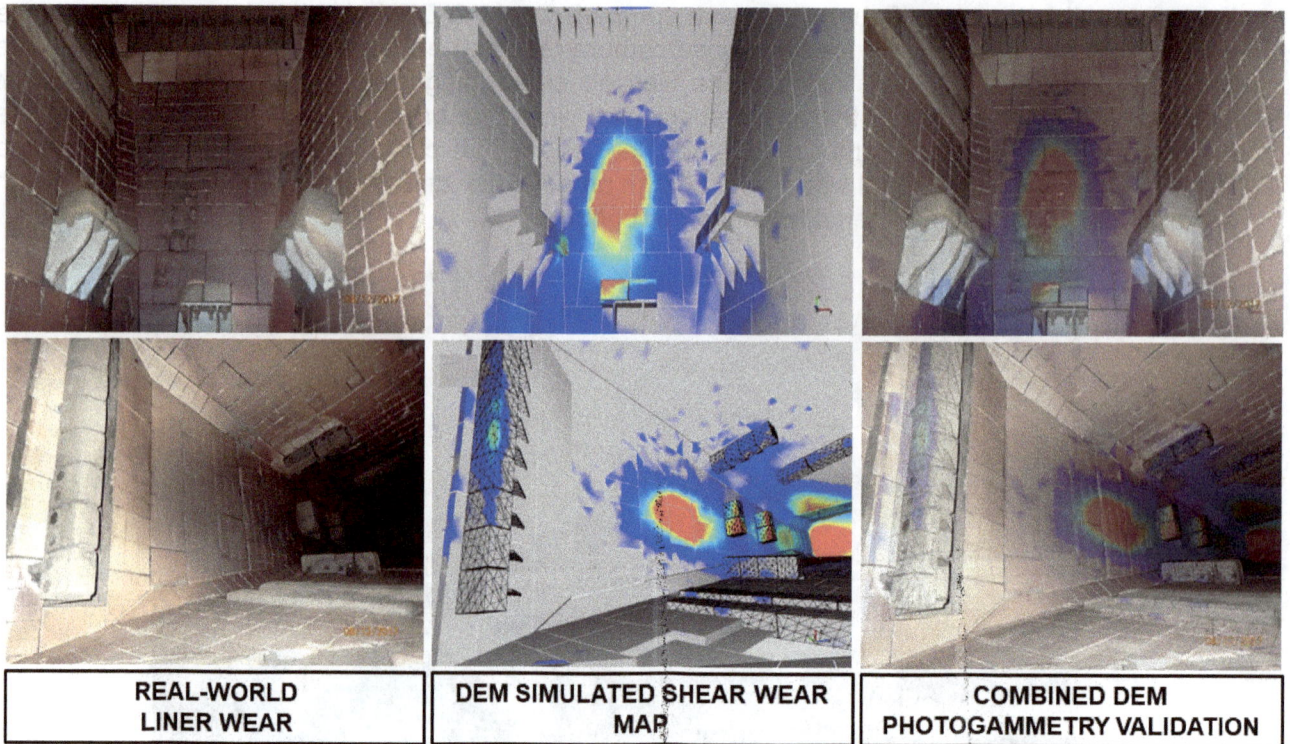

| REAL-WORLD LINER WEAR | DEM SIMULATED SHEAR WEAR MAP | COMBINED DEM PHOTOGAMMETRY VALIDATION |

FIG 5 – Photogrammetry technique demonstrating the validation of DEM wear maps to real world liner wear. As shown, the location of high wear areas corresponds to their real-world counterparts.

FIG 6 – Example of ore trajectory comparison using the Photogrammetry technique.

Build-up and blockage

While usage of time lapse cameras or videos of flow from safe access points provide an indication of the flow behaviour of the ore, the full picture of whether the simulations capture the real-world behaviour also require interrogation of the asset when it is not operating. This allows existing areas of build-up and exact locations of high wear areas to be ascertained and correlated to the DEM simulation. An example of the wear pattern correlation has been provided in the previous section in Figure 5. The capturing of existing areas of build-up is also important to diagnosing the source of blockages and design of treatment options. An example of this shown Figure 7 where the build up pattern within the chute after feed has stopped shows close agreement in the corresponding DEM simulation. Collection and validation to this data is important as build-up lower down within transfer chutes typically in corners, or at the rear of the boots are often subject to high consolidation pressures, which go hard over time and reduce the effective throat area of the chute, increasing the tendency for blockages to ensue. This may be indicative of wall/valley angles being too shallow and/or openings being too narrow.

FIG 7 – Material build-up pattern compared to DEM build-up. As shown, close agreement between the DEM simulation and real-life wear pattern was achieved with proper calibration of the DEM adhesive and cohesive properties.

By collecting and validating against this real world build-up or blockage data, both the root cause can be diagnosed and the DEM simulation can be confidently used to inform the design of treatment options. Figure 8 shows another real-world example of arching and subsequent blockage of an insert chute which is then replicated in DEM. The figure demonstrates both the importance of the collection of site data but also the validation itself. By recreating the observable phenomenon, in this case, the overflow of ore out of the front of the insert and at the same height over the insert cross beam, the underlying mechanism behind the blockage could be diagnosed with greater certainty (blockage occurs in an inaccessible area that would be difficult to observe).

Build-up and partial bridging at liner thickness change

Choking of flow initially at lower cross bar then at surge bar, resulting in blocked chute.

OVERFLOW AT FRONT OF INSERT

FIG 8 – Overflow from the front of an insert chute recreated in DEM. The mechanism for the overflow was also recreated in the simulation as shown, which resulted in replication of the overflow behaviour (ore flowing over the cross beam).

Ore presentation and belt drift

Conveyor belt drift/mistracking is a common material handling problem. While causes may stem fro maligned idlers, often the problem is due to off centre loading by the feed belt. To quantify this phenomenon in DEM, the authors have developed assessment methods for belt loading to report and demonstrate transient behaviour at both the loading zone and along the belt that could influence belt tracking behaviour.

Firstly, the shape and presentation of the material burden on the receiving conveyor is visually assessed for symmetry with the conveyor as an off centre burden profile is a cause of mistracking. This can be validated with site photos as described previously as well as LIDAR scanning of the burden itself. An example of this visual validation is presented in Figure 9.

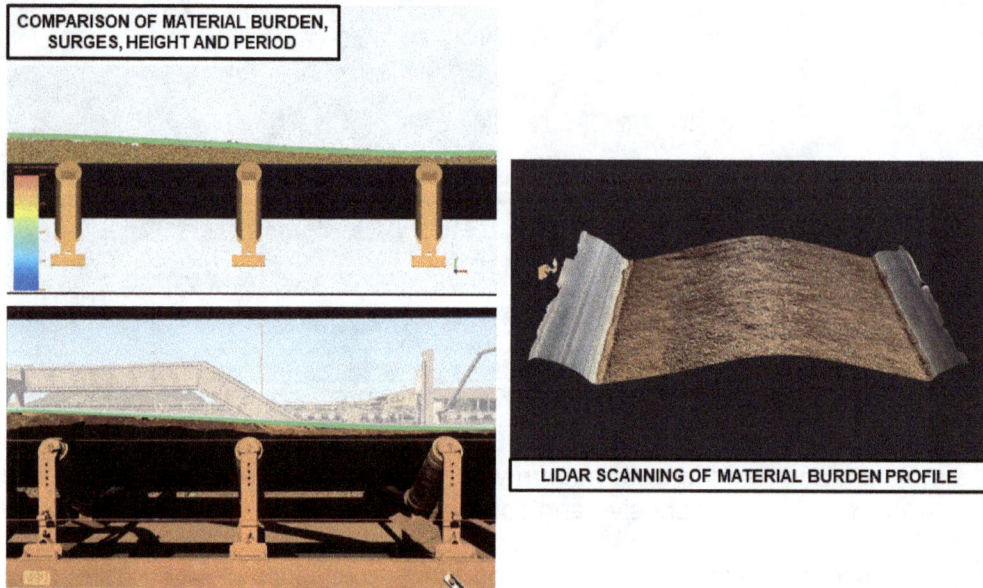

FIG 9 – On-belt validation and comparison (left), LIDAR scanning of burden profile (right).

Additional to burden profile offsets on the receiving conveyor are the lateral forces produced at the loading zone due to off centre presentation from the feed boot. These off-centre tracking forces can manifest as transient events which trip the receiving conveyor (eg on conveyor start-up) and may not be seen once the asset is operating for some time. In a DEM context, the method to quantify this even involves resolving the forces at each node into the following components.

- normal to belt surface
- along the belt (travel direction)
- across the belt surface (lateral).

These lateral forces on the belt surface are then summed across the width of the belt and then reported as a lateral force along the length of the belt. The authors are then able to capture the transient impact of these simulated forces graphically in the form of force plots as shown in Figure 9. Also shown in the Figure 10 is the validation of these simulated forces with time lapse footage showing the mistracking of the receiving conveyor during a loaded start-up. In this project, this enabled the root cause of the poor tracking (a transient start-up occurrence) of this particular conveyor to be diagnosed as previous attempts by others to recreate the mistracking phenomena proved unsuccessful due to a steady state assumption being taken to analyse the transfer.

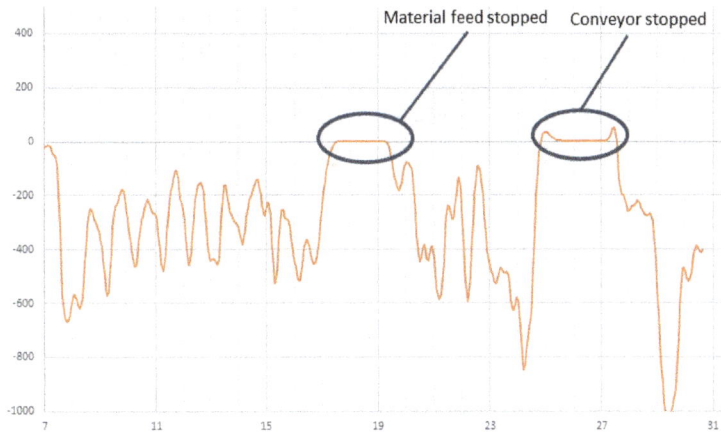

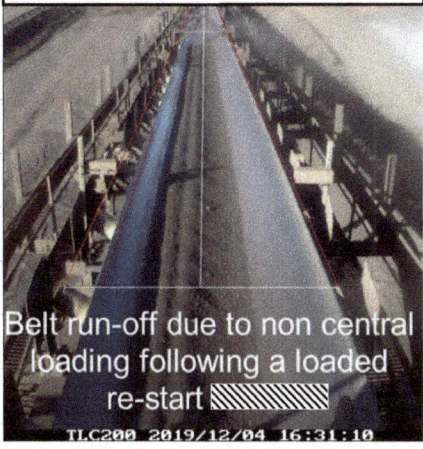

FIG 10 – Validation of DEM simulated transient lateral belt forces and time lapse footage showing belt mistracking during a loaded restart of the receiving conveyor.

Scraper observations

One final but often overlooked area of validation are the influence of belt scrapers on the performance of the transfer. Indeed, many commercial DEM transfer chute simulations neglect including scrapers into the models. This can lead to operational issues once designs have been constructed, especially in cases where scrapings build-up under the head pulley rather than being returned to the main flow or, in chutes containing ore sampling equipment, fall into the sample leg and contaminate the sample.

A DEM simulation is able to recreate the material spray and trajectory off the scrapers if the material model has been calibrated correctly. This means capturing the real-world cohesive behaviour of fines model along with the material to belt adhesion. An example of this form of validation is shown in Figure 11, where the material spray from the primary scraper has been replicated in the DEM simulation. Doing so not only provides another data point to check if the calibration matches reality but also ensure that any future designs/modifications successfully return the scrapings to the main flow.

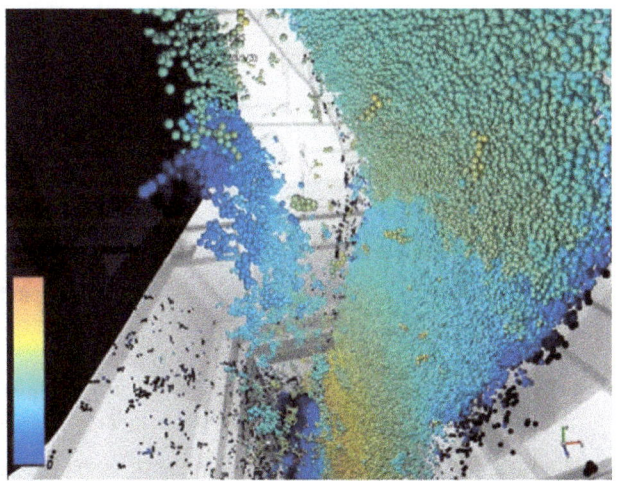

FIG 11 – Validation of fines material discharged off primary scraper.

CONCLUSIONS

DEM is a powerful tool in predicting the flow of granular materials and diagnosing materials handling problems. While DEM is not without its limitations, the impact of the limitations is being ameliorated as technology improves.

However, the efficacy of any given simulation hinges on the approach taken to calibrate the DEM model and validate its outputs. It is current practice for bulk solids to undergo flow properties testing by way of shear test, especially in the design of silos, bins and hoppers. This data is applicable for transfer chute design and provides quantitative insights into both the flowability of a given sample and the spectrum of behaviours that a facility will handle by varying the moisture content of the samples tested and testing this across a representative sample size of the orebody. The laboratory tests used to generate this data can be simulated in DEM for use in calibrating the material models to match the macro behaviours measured in the laboratory. The shear test data is preferred by the authors in the calibration of DEM material models over the commonly used angle of repose test that is popular in industry due to the repeatable, standardised nature of the shear test as well as its ability to comprehensively capture the flow behaviour of a bulk solid over varying consolidation pressures.

In tandem with proper calibration of the material model is validation of the simulations to real world site observations. These observations can be in the form of video, photos and site historian data. Several validation methods and areas to gather data have been provided. The use of photogrammetry leverages the visual nature of DEM simulations to correlate results with site videos and photographs. Tools such as time lapse cameras and Lidar scanners can be used to gather data that is not typically accessible or practical to record due to access restrictions on plant equipment. Suggested areas to validate a DEM simulation include: flow and wear patterns, areas of material build-up, receiving conveyor ore presentation and belt drift along with scraper discharge trajectories.

Once a baseline simulation of the existing problem is validated there can be confidence in the solutions developed using DEM as a tool for the iterative development and design optimisation of transfer chutes or any other materials handling equipment.

REFERENCES

ASTM International, 2017. ASTM D8081-17 – Standard Guide for Theory and Principles for Obtaining Reliable and Accurate Bulk Solids Flow Data Using a Direct Shear Cell, July 2017.

ASTM International, 2022. ASTM D6128-22 – Standard Test Method for Shear Testing of Bulk Solids Using the Jenike Shear Tester, January 2023.

Box, G E P, 1979. Robustness in the strategy of scientific model building, in *Robustness in Statistics* (eds: R L Launer and G N Wilkinson), pp 201–236 (Academic Press: New York).

Cleary, P W, 1998. Discrete Element Modelling of Industrial Granular Flow Applications, *Task, Quarterly – Scientific Bulletin*, 2(3):385–415.

Coetzee, C J, 2018. Particle upscaling: Calibration and validation of the discrete element method, *Power Technology*, 344:487–503.

Favier, J F, Abbaspom–Fard, M H, Kremmer, M and Raji, A O, 1999. Shape Representation of Asymmetrical, Non-Spherical Particles in Discrete Element Simulation Using Multi-Element Model Particles, in *Engineering Computations*, 16(4):467–480.

Hastie, D, 2010. Belt Conveyer Transfers: Quantifying and Modelling Mechanisms of Particle Flow, Doctor of Philosophy thesis, School of Mechanical, Materials and Mechatronic Engineering, University of Wollongong.

Jenike, A W, 1964. *Storage and flow of solids*, revised 1970 (University of Utah: Salt Lake City).

Li, L and Holt, R M, 2005. Approaching Real Grain Shape in the Simulation of Sandstone Using DEM, *Powders and Grains*, 2:1369–1373.

Pinson, D, Reed, J, Wright, B and Yu, A B, 2004. Application of Discrete Particle Simulation to Flow in a Transfer Chute, in *Proceedings of the Eighth International Conference on Bulk Materials Storage, Handling and Transportation*, pp 294–298 (Institution of Engineers, Australia).

Potapov, A V and Campbell, C S, 1998. A Fast Model for the Simulation of Non-Round Particles, *Granular Matter*, 1(1):9–14.

Schulze, D, 2008. *Powders and Bulk Solids: Behaviour, Characterization, Storage and Flow* (Springer: Berlin).

Schulze, D, 2014. Flow Properties of Powders and Bulk Solids [online]. Available from: https://dietmar-schulze.com/pdf/flowproperties.pdf [Accessed: 27 May 2023].

Standards Australia, 2017. AS 3880:2017 – Flow Properties of Coal, August 2017.

Zhou, Y C, Wright, B D, Yang, R Y, Xu, B H and Yu, A B, 1999. Rolling Friction in the Dynamic Simulation of Sandpile Formation, *Physica A*, 269:536–553.

Study on the dehydroxylation process of an ultrafine goethitic ore

J Mesquita[1], M Camarda[2], A Pirson[3] and F Vasconcelos[4]

1. Mining and Mineral Processing Senior Research Engineer, ArcelorMittal Maizières Research, Maizières Lès Metz 57208, France. Email: josue.mesquita@arcelormittal.com
2. Mineral Processing Technician, ArcelorMittal Maizières Research, Maizières Lès Metz 57208, France. Email: matthias.camarda@arcelormittal.com
3. Head of Characterization Team, ArcelorMittal Maizières Research, Maizières Lès Metz 57208, France. Email: arnaud.pirson@arcelormittal.com
4. Portfolio Leader – Mining, ArcelorMittal Maizières Research, Maizières Lès Metz 57208, France. Email: filipe.vasconcelos@arcelormittal.com

ABSTRACT

ArcelorMittal Prijedor is a goethitic iron ore mine located in the north-west of Bosnia and Herzegovina, which mainly supplies ArcelorMittal's European subsidiaries. Annual production of the asset stands at the level of 1.5–2.1 million tons of iron ore product which is sent to ArcelorMittal Zenica steelmaking plant. Two types of products with approximately 50 per cent Fe are generated in Prijedor through a process composed by crushing, screening, and magnetic separation stages. Within this process one of the tailings produced is called 'ultrafines'. The 'ultrafines' product corresponds to the desliming hydrocyclones overflow in Prijedor plant, which is classified in a size fraction below 25 µm and corresponds to about 5 per cent of feed. This material is richer in iron than the coarser size fractions of the ore but do not present a good performance in concentration processes. In this way, several studies have been carried out during the last years to find solutions to use the 'ultrafines' material without success. In this context, the possibility of dehydroxylate the goethite contained in the material transforming the mineral into hematite and thus increasing the iron grade of the final product was investigated. During the research, several dehydroxylation tests with different temperatures and time of exposure were performed. Using dihydroxylation process at temperatures higher than 800°C it was possible to produce a material with 43 per cent Fe from the original 39 per cent Fe 'ultrafines' sample. In addition, the dehydroxylation tests confirmed that goethite becomes completely transformed into hematite at 400°C (with an exposure time of 60 mins) and material is completely sintered at 1200°C. A simulation study based on dehydroxylation tests results showed that it would be possible to use a 'ultrafines' product with Fe grade higher than 45 per cent to produce a material with 50 per cent Fe through a complete dehydroxylation process.

INTRODUCTION

ArcelorMittal Prijedor, situated in the north-west of Bosnia and Herzegovina (Figure 1), is a goethitic iron ore mine primarily catering to ArcelorMittal's European subsidiaries. This mine operates as a 'joint venture' between ArcelorMittal Holdings and RZR Ljubija a.d. Prijedor, with the first holding a majority stake of 51 per cent. Annually, the mine produces approximately 1.5–2.1 million tons of iron ore with 50 per cent Fe, which is subsequently transported to ArcelorMittal Zenica, a steelmaking plant located within Bosnia and Herzegovina (ArcelorMittal, 2023).

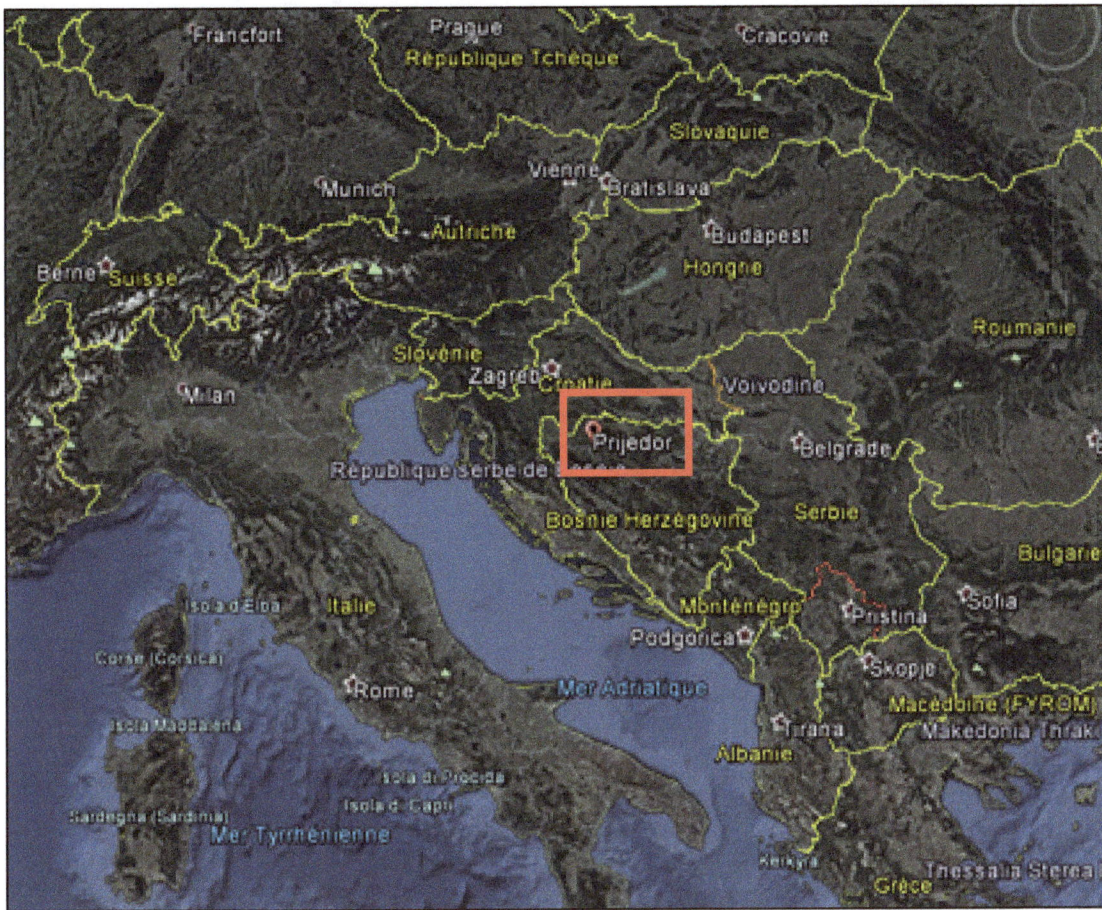

FIG 1 – ArcelorMittal Prijedor location.

According to Barbosa (2012), the Prijedor plant produces the two products presented in Table 1 using a process that involves crushing, screening, and magnetic separation stages (Figure 2). During this process, a by-product called 'ultrafines' is generated.

TABLE 1

ArcelorMittal Prijedor products characteristics.

| Product | Size range | Grade (%) | |
		Fe	SiO$_2$
APR	<40 >2 mm	52.00	9.00
BPR	<8 mm	50.00	10.50

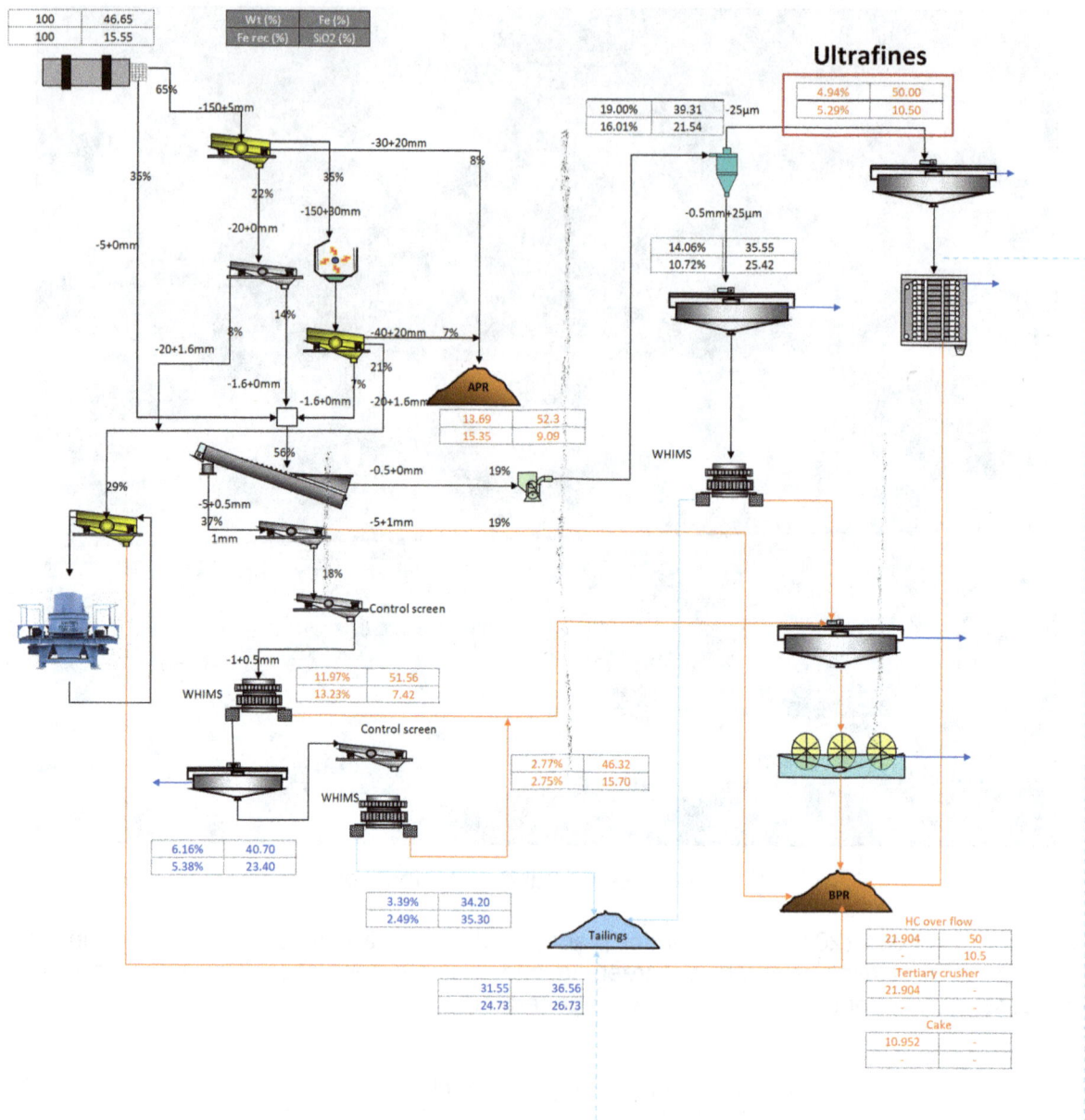

Ultrafines

Legend:
Wt (%)	Fe (%)
Fe rec (%)	SiO2 (%)

Feed:
| 100 | 46.65 |
| 100 | 15.55 |

Ultrafines:
| 4.94% | 50.00 |
| 5.29% | 10.50 |

−25μm:
| 19.00% | 39.31 |
| 16.01% | 21.54 |

−0.5mm+25μm:
| 14.06% | 35.55 |
| 10.72% | 25.42 |

APR:
| 13.69 | 52.3 |
| 15.35 | 9.09 |

WHIMS −1+0.5mm:
| 11.97% | 51.56 |
| 13.23% | 7.42 |

| 2.77% | 46.32 |
| 2.75% | 15.70 |

| 6.16% | 40.70 |
| 5.38% | 23.40 |

| 3.39% | 34.20 |
| 2.49% | 35.30 |

Tailings:
| 31.55 | 36.56 |
| 24.73 | 26.73 |

HC over flow:
| 21.904 | 50 |
| | 10.5 |

Tertiary crusher:
| 21.904 | |

Cake:
| 10.952 | |

Size fraction labels: −150+5mm, 65%, −30+20mm, 8%, 35%, 22%, −150+30mm, −20+0mm, 35%, −5+0mm, 14%, −40+20mm, 7%, 8%, 21%, −20+1.6mm, −1.6+0mm, 7%, −1.6+0mm, −20+1.6mm, 29%, 56%, −0.5+0mm, 19%, −5+0.5mm, 37%, −5+1mm, 19%, 1mm, 18%, Control screen, −1+0.5mm, WHIMS, Control screen, WHIMS, BPR, Tailings, WHIMS

FIG 2 – ArcelorMittal Prijedor process flow sheet.

The ultrafines product is the habitual name of the desliming hydrocyclones overflow in the concentrator, which is classified in 25 µm and corresponds to about 5 per cent of the feed mass. In general, this material is richer than the natural coarser size fractions of the ore but do not responds favourably to concentration processes. Besides that, when a rich ROM is fed into the plant, the ultrafines product presents grades near to 50 per cent and, after a filtration stage, it is sent directly to the BPR product pile (Barbosa *et al*, 2018). By the other side, most part of the time the product does not reach the quality requirements and it is diverted to the tailings dam after a thickening stage.

Over the past few years, numerous studies have been conducted to find viable solutions for the comprehensive utilisation of the ultrafine material, but unfortunately, no successful outcomes have been achieved (Barbosa, 2012). Therefore, the present work focuses on evaluating the feasibility of dehydroxylating the goethite present in the material, thereby transforming the mineral into hematite and consequently enhancing the iron grade of the final product to levels near to 50 per cent Fe.

DEHYDROXYLATION OF GOETHITE

For a long time (and at a detailed level) different areas of knowledge in which iron oxides are important raw materials have carried out fundamental and applied studies on iron-bearing minerals and their calcination/dehydroxylation processes. In this context, Cornell and Schwertmann (2003)

presented the main types of transformations that can occur between iron oxides, as well as their preferred environments (Table 2).

TABLE 2

Transformations between oxides (from Cornell and Schwertmann, 2003).

Precursor oxide	Product	Type of transformation	Environment of transformation
Goethite	Hematite	Thermal or mechanical dehydroxylation	Gas/vacuum
		Hydrothermal dehydroxylation	Solution
	Maghemite	Thermal dehydroxylation	Air + organic matter
Hematite	Magnetite	Reduction	Reducing gas
		Reduction-dissolution precipitation	Alkaline solution with N_2H_4
Magnetite	Maghemite/ hematite	Oxidation	Air
Maghemite	Hematite	Thermal conversion	Air

Second to Leonel (2011), the transformations that proceed without chemical changes are called isochemical. Transformations involving chemical modifications are dehydration (H_2O loss), dehydroxylation (OH loss) and oxidation/reduction (electron exchange).

In this context, the utilisation of abundant goethite ores is potentially important to many countries in the world, especially Australia. These ores contain many detrimental impurities and are difficult to upgrade to make suitable concentrates for the blast furnace (Jang *et al*, 2014). Thus, two major aspects were responsible for most of the studies on goethite dehydroxylation:

1. High value-added and relatively low-cost industrial application.

2. Need to use iron ores with higher LOI values.

In this sense, polymorphic oxides of FeOOH can be dehydrated to their corresponding Fe_2O_3 from the influence of temperature or mechanical action as presented here:

$$2FeOOH \rightarrow Fe_2O_3 + 2H_2O$$

The direct product of goethite dehydroxylation in pure phase is always hematite, whereas with lepidocrocite and maghemite an intermediate phase may occur. Dehydroxylation of different forms of FeOOH occurs over a wide temperature range (140–500°C), depending on the nature of the compound, its crystallinity, degree of isomorphic substitution and chemical impurities (Leonel, 2011).

The conversion of goethite to hematite starts with the release of the -OH groups, which creates a progressive alteration of the crystalline structures and it is facilitated by network of anions shared by the two minerals. This network remains intact, while the water is expelled, and the cations are rearranged. Three unit's cells of goethite form a unit cell of hematite (Jang *et al*, 2014).

According to Leonel (2011), Hematite obtained at low temperatures maintains the acicular morphology of the precursor crystals of goethite, but at temperatures higher than 600°C a sintering process leads to the formation of irregular crystals of hematite, as presented in Figure 3.

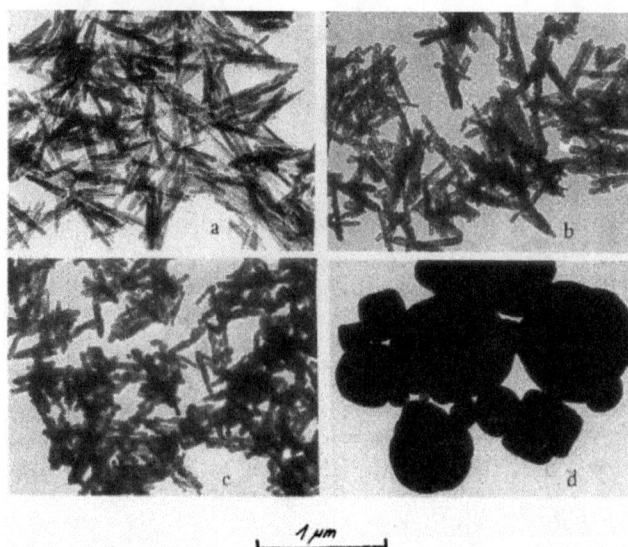

FIG 3 – Micrographics of goethite transformation at different temperatures (a) initial original sample; (b) 520°C; (c) 680°C and (d) 850°C (Balek and Šubrt, 1995).

A common feature of the dehydroxylation of all iron oxyhydroxides is the development of micro porosity due to the expulsion of water. This micro porosity development is followed, at high temperatures, by the coalescence of the micropores to mesopores. The formation of the pores is accompanied by the increase of the specific surface area, and for temperatures above 600°C the product sinters and the specific surface area is significantly reduced. During the dehydroxylation process, substitution of hydroxy bonds by oxy-bonds and sharing of faces between the octahedra occurs, leading to an increase in the density of the structure (Cornell and Schwertmann, 2003).

MATERIALS AND METHODS

Ultrafines sample

The 'Ultrafines' sample used in this study was collected in Prijedor plant in the end of 2016 with the objective of performing the dehydroxylation test work. The sample presented a yellowish colour with a considerable number of ultrafine particles – D_{80} of 20 µm and D_{50} of 5 µm (Figure 4).

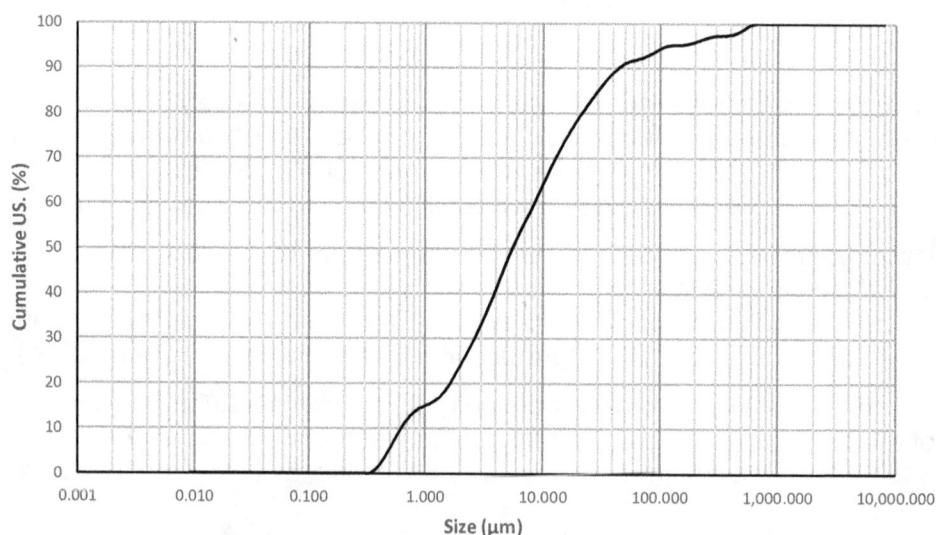

FIG 4 – Ultrafines sample size distribution.

The chemical analysis of the sample is presented in Table 3 and its mineralogical composition is presented in Table 4.

TABLE 3

Ultra-fines sample chemical analysis.

Fe (%)	FeO (%)	SiO$_2$ (%)	Al$_2$O$_3$ (%)	CaO (%)	MgO (%)	TiO$_2$ (%)	P (%)	Mn (%)	S (%)	Na$_2$O (%)	K$_2$O (%)	LOI (%)	Ba (ppm)	Sr (ppm)
39.21	<0.01	22.20	6.50	0.14	0.32	0.23	0.088	1.84	0.01	0.19	0.99	9.75	<10	60.00

TABLE 4

Ultrafines sample mineral composition.

Mineral/Element	Code	Formula	Proportion (%)
Goethite	GOE	FeO(OH)	64.4
Hematite	HEM	Fe$_2$O$_3$	0.0
Quartz	QTZ	SiO$_2$	13.2
Illite	ILL	(K,H$_3$O)(Al,Mg,Fe)$_2$(Si,Al)$_4$O$_{10}$[(OH)$_2$,(H$_2$O)]	12.4
Kaolinite	KAO	Al$_2$Si$_2$O$_5$(OH)$_4$	2.1
Manganese	Mn	Mn	2.9
Miscellaneous	Misc.	-	1.0

Material has 39.21 per cent Fe, 22.20 per cent SiO$_2$, no magnetite in this composition and a high content of LOI (9.75 per cent) due to the presence of hydrated minerals as goethite, Illite and kaolinite. Before conducting the calcination tests, the sample was subjected to drying in an oven at a fixed temperature of 105°C.

Tests methodology

Tests were performed aiming at evaluating the behaviour of ultrafines material under different temperatures (200°C to 1200°C) and exposure times (15 to 240 minutes). During the tests 40 g dry samples were put in a crucible and heated inside a laboratory scale furnace. Later, the dehydroxylation products were sent to chemical analysis and tested through XRD and laser particle size analyser.

Chemical analysis

Chemical analyses were performed using X-ray fluorescence (XRF) on fused beads for the major elements. Loss-on-ignition (LOI) was measured at 1000°C and the Fe^{2+} content was determined using potassium dichromate titration. All assays were performed at ALS laboratory in Ireland.

XRD

XRD analyses aimed at determining the mineralogy of dehydroxylated samples being performed using an equipment D2 Phaser distributed by Bruker.

Particle size distribution

The particle size distributions of the dehydroxylated samples were obtained using a laser analyser model Hydro 2000, distributed by Malvern Panalytical. An adjustment factor of 0.8 was applied to Malvern sizes based on the work of Xuan (2011) to convert them to screen meshes.

Dehydroxylation tests

Dehydroxylation tests were performed using a high temperature laboratory furnace model RHF 14/8, supplied by Carbolite Gero. The ramp rate, temperature and exposure time were controlled during the tests. A constant ramp rate of 40°C/min was adopted. The cooling stage was performed naturally with samples inside the furnace.

RESULTS

The results of dehydroxylation tests are presented in Table 5.

TABLE 5

Dehydroxylation tests results.

Test	Temp (°C)	Exposure time (min)	Loss of mass (%)	Grade (%)						
				Fe	FeO	SiO_2	Al_2O_3	P	S	LOI
1	-	-	-	39.21	<0.01	22.20	6.50	0.088	0.010	9.75
2	200	15	0.66	39.40	<0.01	22.30	6.58	0.091	0.010	9.58
3	250	15	1.36	39.27	<0.01	22.10	6.57	0.090	0.010	9.36
4	300	15	2.26	39.64	<0.01	22.30	6.64	0.091	0.011	8.54
5	350	15	3.15	39.93	<0.01	22.50	6.66	0.090	0.011	7.06
6	400	15	4.39	41.05	<0.01	23.40	6.87	0.094	0.010	5.62
7	400	60	6.95	41.99	<0.01	24.20	6.97	0.094	0.016	3.30
8	400	120	7.77	41.90	<0.01	23.80	6.97	0.094	0.014	3.18
9	500	15	6.98	41.90	<0.01	23.90	7.03	0.094	0.011	3.15
10	600	15	8.97	42.24	<0.01	23.80	7.05	0.096	0.010	1.91
11	600	60	8.45	42.68	<0.01	24.10	7.06	0.096	0.014	1.38
12	600	120	9.32	42.85	<0.01	24.20	7.09	0.096	0.014	1.73
13	700	15	9.48	42.78	<0.01	24.40	7.11	0.096	0.012	1.38
14	800	15	10.21	43.13	<0.01	24.40	7.21	0.099	0.012	0.51
15	800	30	10.37	43.20	<0.01	24.20	7.14	0.097	0.012	0.60
16	800	60	10.78	43.09	<0.01	24.20	7.20	0.099	0.011	0.35
17	800	120	10.49	43.18	<0.01	25.10	7.41	0.099	0.008	0.37
18	800	240	9.23	43.19	<0.01	24.20	7.26	0.102	0.007	0.41
19	1000	15	10.52	43.48	<0.01	24.70	7.24	0.100	0.003	0.19
20	1200	15	10.44	43.31	<0.01	24.50	7.29	0.099	<0.001	0.07

Results show that after the complete dehydroxylation of goethite is possible to achieve a product with approximately 43 per cent Fe from an initial sample of 39 per cent Fe (increase of four percentage points). This is due to the increase of the proportion in mass of the other elements with the removal of water (as the mass of water decreases the proportion of all other elements increases, also increasing their grades). In addition, it is worth to highlight that the LOI percentage is an indication of the amount of hydroxide removed from the sample. According to Jang *et al* (2014), the main cause of LOI losses is dehydration (loss of water adsorbed on the outer and inner surfaces at low temperature), and dehydroxylation (loss of water linked to the lattice trough calcination). In this context, the main process drivers are temperature and time, both analysed in detail in the topics below.

Influence of temperature

Figure 5 presents the changes of grades in relation to temperature during the dehydroxylation tests with 15 mins exposure time. Four different regions are observed in the graph and validated when observing the mineralogy of samples (Figure 6). A detailed description of each one is done next.

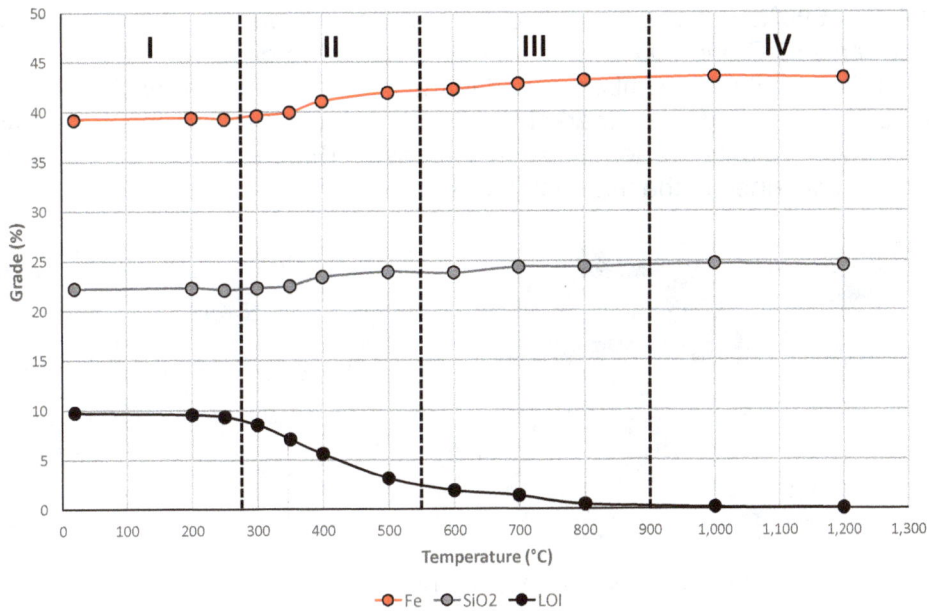

FIG 5 – Elements grades variation in relation to temperature.

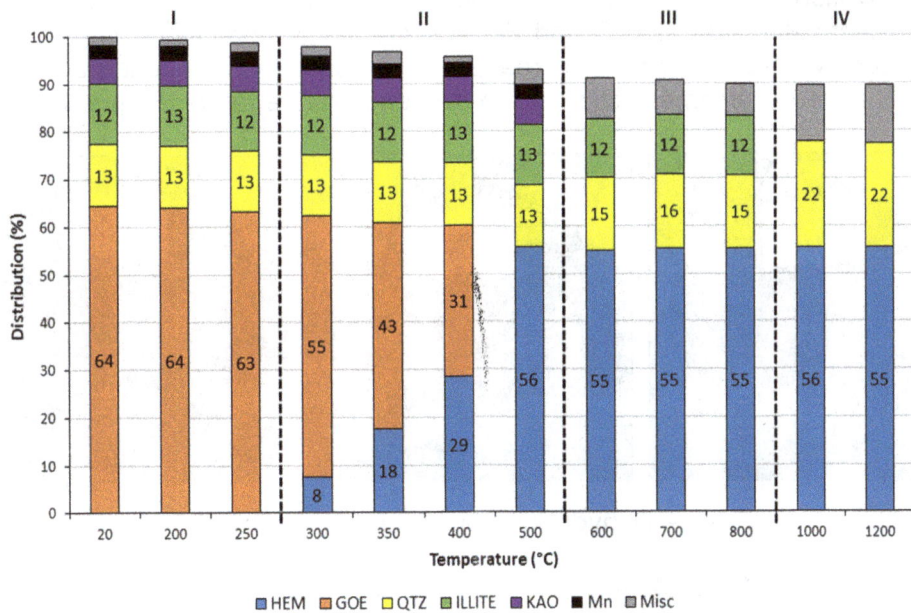

FIG 6 – Minerals distribution in relation to temperature.

Region I (below 300°C): No changes are observed in the material. In this temperature range, both grades and mineralogy remain at the same level of the original sample without any visual changes (Figure 7). A slight loss of mass is noticed, most probably due to a dehydration process.

FIG 7 – Dehydroxylation process – Region I.

Region II (300°C – 600°C): In this region goethite starts to gradually change into hematite from some point between 250°C and 300°C, with highest peak being observed between 400 and 500°C (which is in accordance with the numbers found in the literature (Jang *et al*, 2014). Fe/SiO$_2$ grades increasing and LOI decreasing are remarkable. Visual inspection shows the material ranging from yellowish to brownish colours as temperature increases (Figure 8). No changes are observed regarding the clays contained in the material (kaolinite, Mn and Miscellaneous).

FIG 8 – Dehydroxylation process – Region II.

Region III (600°C – 1000°C): In this region goethite seems to be completely converted to hematite and the decomposition of kaolinite occurs in some point between 500°C and 600°C. LOI grade achieves the lowest value after 800°C and Fe/SiO$_2$ grades are practically constant with a slight increase being observed as temperature increases (a Fe grade of 43 per cent is achieved). A higher agglomeration of material becomes noticeable in this temperature range (Figure 9).

FIG 9 – Dehydroxylation process – Region III.

Region IV (higher than 1000°C): In some point between 800°C and 1000°C illite disappears probably becoming muscovite (accounted as 'Miscellaneous'). After 800°C both Fe/SiO$_2$ grades and LOI remain constant and only physical transformations are observed (Figure 10). At 1200°C material is completely sintered and no additional increase in Fe grade is achievable through temperature change.

FIG 10 – Dehydroxylation process – Region IV.

In addition, the increase of the agglomeration level of particles was observed as temperature gets higher. In this context, Malvern was configured to perform three measurements and during all analysis the material was exposed to ultrasound. When observing the three measurements done is possible to see that the ultrasound waves affect the agglomeration of the samples over time causing its fall. After the second measurement almost all 'particles' higher than 100 μm disappear, indicating that ultrasound performed a de-agglomeration work. This behaviour was observed in all samples analysed.

In both cases the D_{20} did not change with the temperature while the D_{50} presents a small increase with temperature increasing. The biggest differences were observed in D_{80} values. When comparing the P_{80}'s of the different samples, the agglomeration process is evident (Figure 11). Despite this, the agglomeration seems not be critical for the process being easily reverted with the ultrasound.

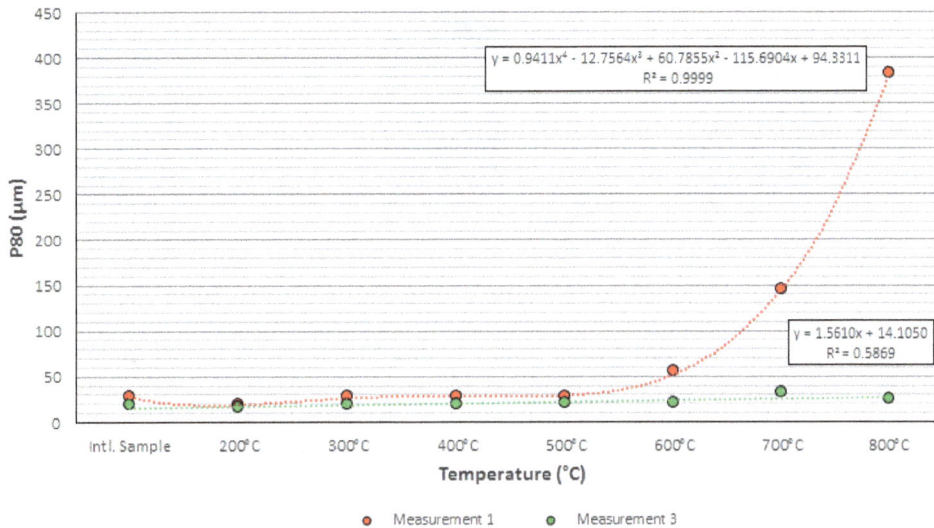

The chart shows P80 (μm) versus Temperature (°C) with two fitted curves:

$$y = 0.9411x^4 - 12.7564x^3 + 60.7855x^2 - 115.6904x + 94.3311$$
$$R^2 = 0.9999$$

$$y = 1.5610x + 14.1050$$
$$R^2 = 0.5869$$

FIG 11 – Comparison between P_{80} values in Malvern measurements.

Dehydroxylation kinectics

To evaluate the kinetics of the dehydroxylation process, different exposure times were evaluated considering three distinct temperatures (400°C, 600°C and 800°C).

Table 6 presents the results of kinetics study at 400°C.

TABLE 6

Calcination tests results in relation to time at 400°C.

Temp (°C)	Time (min)	Loss of mass (%)	Grade (%)					
			Fe	SiO$_2$	Al$_2$O$_3$	P	Mn	LOI
400	15	4.39	41.05	23.40	6.87	0.094	1.94	5.62
400	60	6.95	41.99	24.20	6.97	0.094	0.02	3.30
400	120	7.77	41.90	23.80	6.97	0.094	0.01	3.18

The table above shows that time has a considerable influence in the dehydroxylation process at 400°C. After 60 minutes system goes into balance and no additional changes are observed. When analysing the mineralogy of the products it is possible to conclude that this is due to the complete transformation of goethite into hematite (Figure 12). Besides that, practically all the other minerals remain in the system without any modifications.

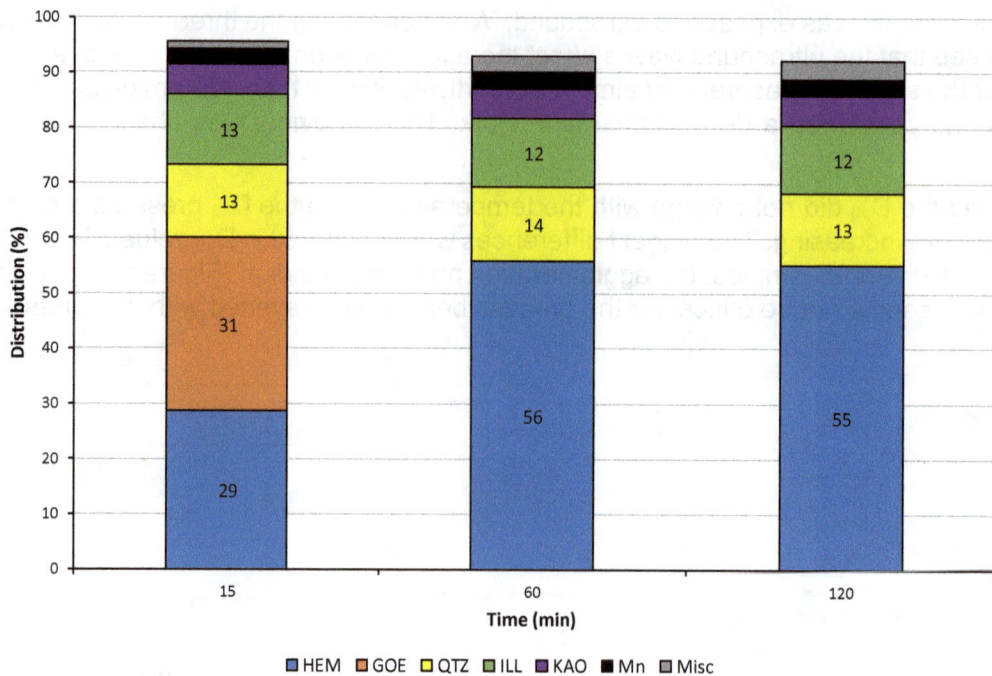

FIG 12 – Minerals distribution in relation to time at 400°C.

From 15 mins to 60 mins an increase of agglomeration level is observed (Figure 13). After 60 mins the agglomeration level remained the same. This fact indicates that agglomeration is directly related to the transformation of mineral phases.

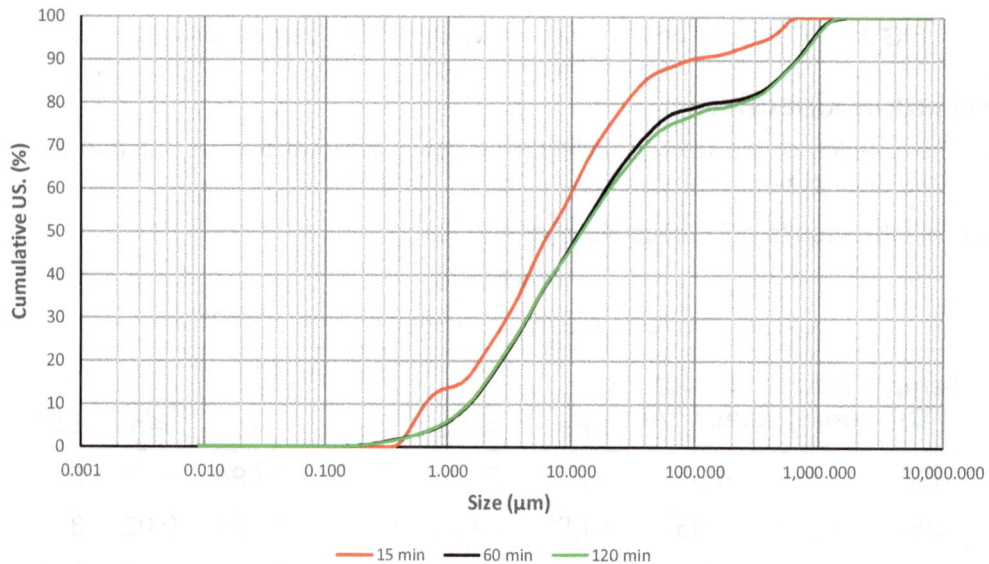

FIG 13 – Agglomeration in function of time at 400°C.

In relation to kinetics at 600°C and 800°C, no significant changes are observed with the variation of time, most probably due to the fact that all goethite is already dehydroxylated in the times evaluated.

A complete phases diagram summarising the results found in the study is presented in Figure 14.

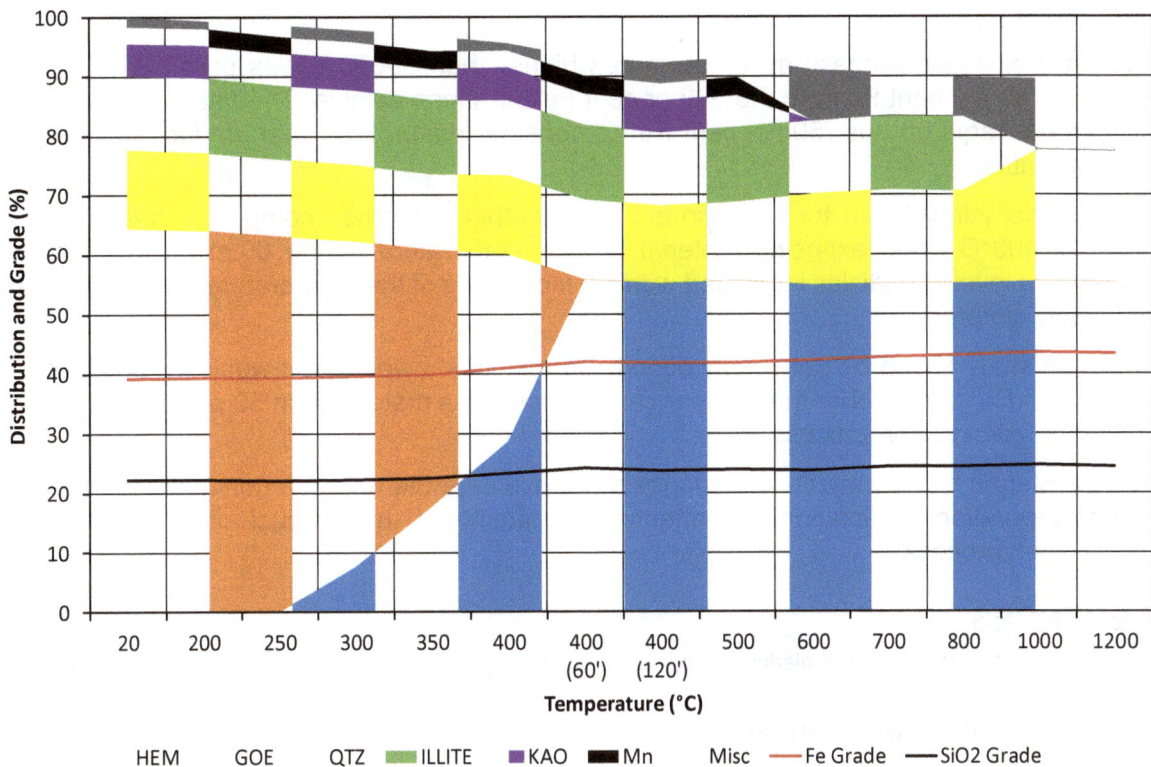

FIG 14 – Ultrafines dehydroxylated summary.

Considering the results achieved during this work, a simulation was performed to define the 'cut-off' grade of ultrafines product to have a dehydroxylated product with 50 per cent Fe (Figure 15).

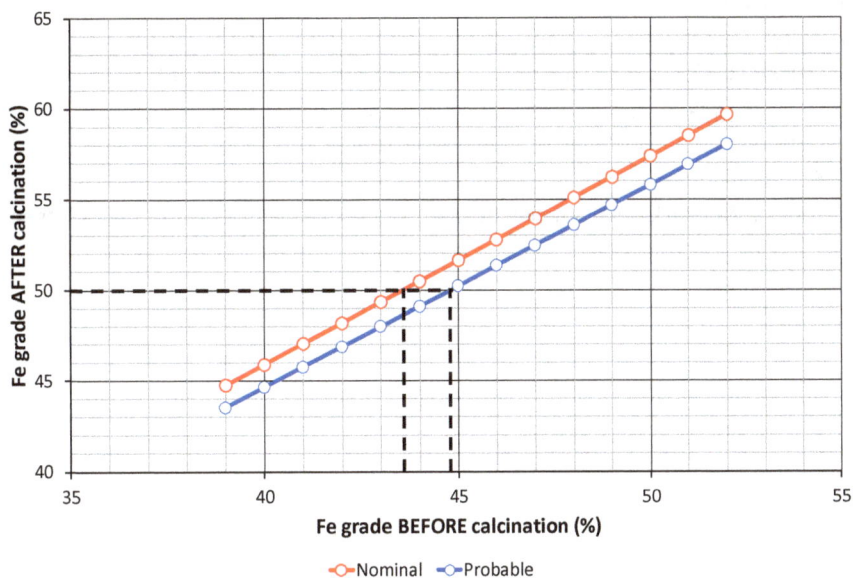

FIG 15 – Quality simulation of dehydroxylated products.

Nominal curve corresponds to the theoretical curve based on minerals chemical formula and probable curve corresponds to the curve built based on dehydroxylation tests results.

In theory, to achieve a product with 50 per cent Fe it would be necessary to feed the dehydroxylation process with an Ultrafines sample of 43.5 per cent Fe. Probable curve shows that a feed of approximately 45 per cent would be necessary to achieve the desired quality.

This exercise shows that it would be possible to use the Ultrafines portion with Fe grade higher than 45 per cent to produce a material with 50 per cent Fe through a dehydroxylation process. To verify if the project is feasible, an economic analysis must be done.

CONCLUSIONS

Using dehydroxylation process at temperatures higher than 800°C it was possible to produce a material with 43 per cent Fe from the 39 per cent Fe Ultrafines sample. The final grade achieved is lower than the required one (50 per cent Fe) to consider material as final product in ArcelorMittal Prijedor operation.

In addition, dehydroxylation tests confirmed that goethite becomes completely transformed into hematite at 400°C when exposing material to this temperature during 60 mins. In addition, the agglomeration level of particles increased with the increasing of the temperature. At 1200°C material is completely sintered.

A simulation study based on dehydroxylation tests results showed that it would be possible to use Ultrafines with Fe grade higher than 45 per cent to produce a material with 50 per cent Fe through a complete dehydroxylation process.

In the next stages of the project, it is suggested to evaluate the use of a concentration method as magnetic separation or flotation to enhance the quality of the product obtained through the dihydroxylation process.

REFERENCES

ArcelorMittal, 2023. ArcelorMittal Prijedor d.o.o. Prijedor [online]. Available from: http://prijedor.arcelormittal.com/ (Accessed: 16 May 2023).

Barbosa, M G, 2012. Buvac mine plant optimization report, ArcelorMittal Maizières Research Centre technical report.

Barbosa, M G, Lima, J, Souza, J M, Pirson, A, Guimarães, F V and Araujo, A C, 2018. Barite bearing iron ore: a literature review and mineral processing assessment, Poster presented to International Mineral Processing Congress (IMPC 2018), Moscow.

Balek, V and Šubrt, J, 1995. Thermal behaviour of iron (III) oxide hydroxides, *Pure & Appl Chem*, 67(11):1839–1842.

Cornell, R M and Schwertmann, U, 2003. *The iron oxides – structure, properties, reactions, occurrences and uses*, 2nd ed, pp 365–407 (Wiley-VCH GmbH & Co. KGaA).

Leonel, C M L, 2011. Estudo do processo de calcinação como operação unitária adicional na pelotização de minérios de ferro com altos valores de PPC, MSc. Thesis, University of Minas Gerais, Brazil.

Jang, K, Nunna, V R M, Hapugoda, S, Nguyen, A V and Bruckard, W J, 2014. Chemical and mineral transformation of a low grade goethite ore by dehydroxilation, reduction roasting and magnetic separation, *Minerals Engineering*, 60(2014):14–22.

Xuan, W, 2011. Calibration of Malvern laser diffraction by sieve analysis, ArcelorMittal Maizières Research Centre technical report.

Update of the geometallurgical model of a spirals circuit

J Mesquita[1], E Kleiderer[2], L Senez[3], R Belissont[4], Y Basselin[5], C Veloso[6] and F Vasconcelos[7]

1. Mining and Mineral Processing Senior Research Engineer, ArcelorMittal Maizières Research, Maizières Lès Metz 57208, France. Email: josue.mesquita@arcelormittal.com
2. Mineral Processing Technician, ArcelorMittal Maizières Research, Maizières Lès Metz 57208, France. Email: elena.kleiderer@arcelormittal.com
3. Characterization Research Engineer, ArcelorMittal Maizières Research, Maizières Lès Metz 57208, France. Email: louis.senez@arcelormittal.com
4. Senior Characterization Research Engineer, ArcelorMittal Maizières Research, Maizières Lès Metz 57208, France. Email: remi.belissont@arcelormittal.com
5. Characterization Technician, ArcelorMittal Maizières Research, Maizières Lès Metz 57208, France. Email: yves.basselin@arcelormittal.com
6. Master Planning Specialist, ArcelorMittal Sourcing, Luxembourg L-1160, Luxembourg. Email: carlos.veloso@arcelormittal.com
7. Portfolio Leader – Mining, ArcelorMittal Maizières Research, Maizières Lès Metz 57208, France. Email: filipe.vasconcelos@arcelormittal.com

ABSTRACT

Mont-Wright mining complex is an important asset of ArcelorMittal located in the province of Quebec, Canada. It is composed by Mont-Wright and Fire Lake mines, operating 365 days per annum, and producing 25 Mtpa of iron ore concentrate with 66 per cent Fe. Mont-Wright concentration plant is composed by seven lines with a flow sheet encompassing AG/SAG milling, primary and secondary classification stages, spiral concentrators (rougher, cleaner and recleaner stages) and dewatering unit operations. In this context, an equation called 'Wrec' was developed in 1971 with the objective of estimating the mass recovery of the different blocks in the beneficiation plant and support the mining planning works of Mont-Wright complex. This equation was based on the correlation between the results of heavy liquid tests performed with drill core samples and spiral concentrators pilot scale tests, returning an estimative of concentrator yield. However, since its first development, no updates have been done in the equation, which lead to some deviations from expected recoveries in the last years. This is probably due to the changes in the process and in the ore fed into the concentrator during the period. In this way, the study presented here corresponds to the execution of heavy liquid/spirals pilot scale tests with 26 composite samples collected at Mont-Wright deposit aiming at revising and updating 'Wrec' equation. Based on test work results, an updated formula was established to estimate the plant recovery and a comparison performed using industrial data showed that the old formula overestimated the real yield by 5.40 per cent, on average. At the end of the study the new equation was incorporated into the mining planning process.

INTRODUCTION

Mont-Wright mining complex is an important asset of ArcelorMittal located in the province of Quebec, Canada. It is composed by Mont-Wright and Fire Lake mines and Mont-Wright concentrator, which produces more than 25 million metric tons of iron ore concentrate every year (ArcelorMittal, 2023).

According to ArcelorMittal (2023), Mont-Wright mine is one of the most extensive open pit mines in North America with a surface area of 24 km[2], including mining equipment, crushing station, concentration plant, maintenance workshops and a train loading system. The production in the area started in 1974 and its remaining mineral resources are estimated at 5.1 billion metric tons (the current mining plan considers the continuity of operations in Mont-Wright complex up to 2053).

Fire Lake mine is an open pit mine in operation since 2006 located 55 km south of Mont-Wright mine. With an iron content higher than Mont-Wright, the contribution of this ore has an important role to play in increasing ArcelorMittal Canada production. Without a crusher or concentrator on-site, mining activities are like those carried out in Mont-Wright mine, but with smaller equipment. After the extraction, all the run-of-mine (ROM) from Fire Lake is shipped to Mont-Wright by train to be concentrated (ArcelorMittal, 2023).

Mont-Wright plant has a nominal throughput capacity of 10 000 tph and receives the ore from a banded iron formation with average Fe grade ranging from 25 to 30 per cent Fe. This material is fragmented and enriched to Fe grades higher than 66 per cent (Mesquita *et al*, 2019). The concentration plant is composed of seven lines with a flow sheet encompassing AG/SAG milling, primary and secondary classification stages, spiral concentrators (rougher, cleaner and recleaner stages) and dewatering operation, as presented in Figure 1. Once the ore has been processed at Mont-Wright, the concentrate is carried by train to Port Cartier pelletising plant, where one third of the final product is transformed on-site into pellets and the remaining production is sold on external market (ArcelorMittal, 2023).

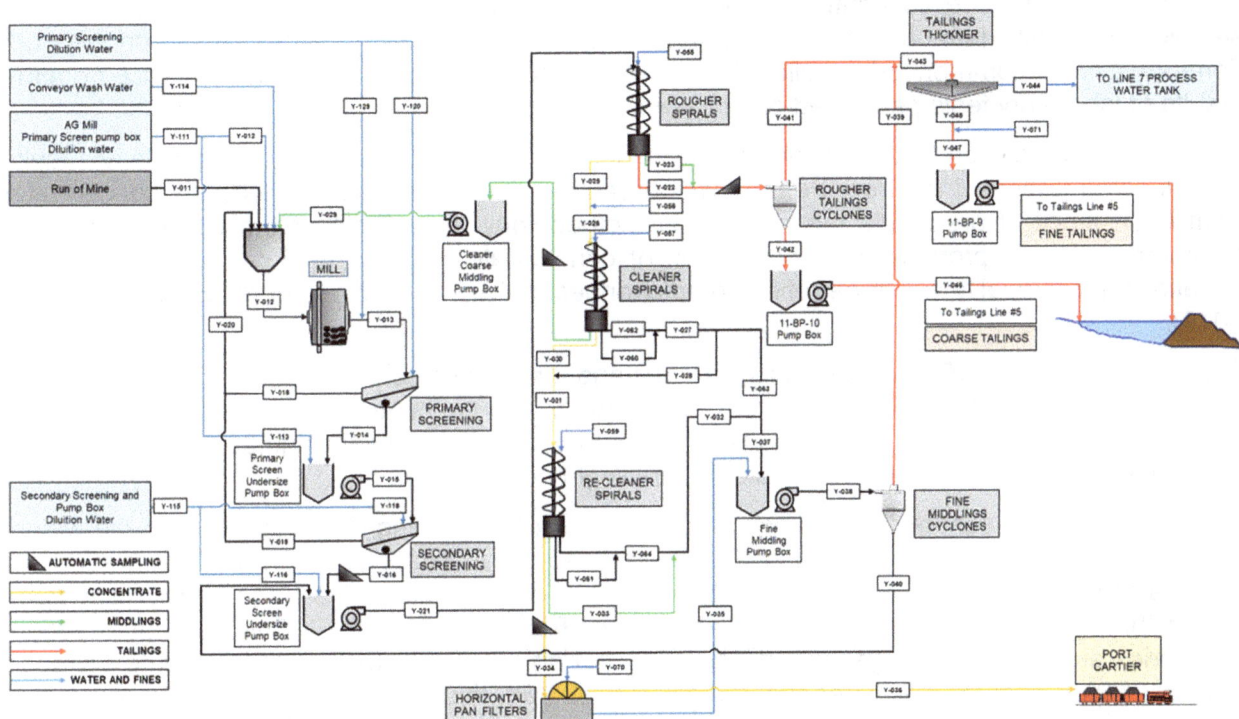

FIG 1 – Mont-Wright concentrator flow sheet.

The assessment of concentrator performance with various ores is crucial when considering the beneficiation process. It enables the development of a precise mining plan and ensures a stable production in the plant. By minimising performance losses caused by fluctuations in the feed or process, this approach guarantees the maximisation of economic gains (Ferreira *et al*, 2022).

Within this context, the utilisation of a prediction technique becomes indispensable for the design of a mining plan that guarantees the systematic extraction of ore throughout the life-of-mine, while ensuring consistent quality of the concentrator feed. Moreover, the prediction technique should provide a reasonably accurate estimate of plant performance and enable the use of simpler laboratory test procedures for analysing samples obtained from drilling campaigns (Silveira *et al*, 2022). By employing such a prediction technique, mining operations can achieve better control over ore extraction, maintain desired concentrator feed quality, and streamline the evaluation process through simplified laboratory tests.

With this purpose, a geometallurgical study was conducted in 1971 by QCM (Quebec Cartier Mining Company) and ARL (Applied research laboratory) aiming at predicting the weight recovery of Mont-Wright plant (called by them Wrec) from heavy liquid tests performed with drill core samples. Based on this study, a prediction formula was developed and has been used for the last 50 years in Mont-Wright complex (Bennet, 1971).

During this period the pits were deepened, the average characteristics of the ore has changed, Fire Lake mine production was incorporated into Mont-Wight concentrator feed, a seventh line was included in the plant and the models of spiral concentrators were changed. Despite all these factors, no updates were done in the equation developed by ARL so far.

Due to that, in the last couple of years some oscillations have been observed when comparing predicted values and industrial ones. For this reason, ArcelorMittal Mines Canada (AMMC) and ArcelorMittal Mineral Processing R&D Team (MMP) developed a study together to update the original equation considering the current situation of the mine/plant. The present paper presents the results obtained in this study.

WREC STUDY ORIGIN

In the early Mont-Wright development stages, between mid-1960s up to 1971, US Steel Applied Research Laboratory (ARL) has assisted Quebec Cartier Mining Company (QCM, now AMMC) in mine planning and ore concentration by devising methods for drill core assaying, testing bulk samples with a pilot plant and designing a flow sheet for concentrating the ore. Based on the work done, ARL has recommended to build a technique to predict the concentrator performance from the results of laboratory tests performed on drill cores (Bennet, 1971).

The designed prediction method involved a computer handling of laboratory heavy liquid tests results on drill cores to predict the Iron (Fe) and weight recoveries (Wrec) obtainable by the concentrator from each block of ore inside the mine. This methodology assumed that the three-stage spiral concentrator would produce a uniform grade of concentrate equal to 66.3 per cent Fe and that variations in ore quality would be reflected as changes in the iron and weight recoveries (Lacoste, 2019).

According to Bennet (1971), the calculated Wrec and the calculated iron recovery (Irec) were used to design a flow sheet and to specify crushing, grinding, and concentrating pieces of equipment for an industrial plant intending to produce 16 Mt a year of concentrate containing 66.3 percent Fe and 5.0 percent SiO_2 in the six original lines of Mont-Wright concentrator.

The Wrec equation continues to be used by AMMC Mining Planning Team to estimate the weight recoveries of the different blocks inside Mont-Wright deposit. In this context, approximately 1000 drill core samples are sent to COREM laboratory (located in Quebec City, Canada) every year to be tested through the Wrec procedure developed by ARL (COREM, 2016). The calculated weight recoveries are added to the geological database and an estimation technique is used to transport the information from the drill cores to the block model. With this information, AMMC Team perform the mining planning operation considering the production constrains adopted for Mont-Wright mine.

WREC FORMULA

In this section, a brief explanation about Wrec formula (Equation 1) is done. The explanation of the formula, as well as the details of its development are as follows.

$$Wrec = \frac{h}{66.3} * (1.33K - 27.33 + 0.4P)$$ (1)

Where:

K   Iron recovery in <850 μm >106 μm size fraction of heavy liquid concentrate (%)

P   Iron Recovery in <106 μm size fraction of ground drill core (%)

H   Fe grade of the drill core (%)

To start, it is important to highlight that the equation above is a direct rearrangement of the classical equation used to calculate the recovery of elements in the mineral processing area (Equation 2).

$$Irec = \frac{C}{H} * \frac{c}{h} * 100$$ (2)

Where:

Irec   Element recovery in concentrate (in this case Fe -%)

C   Mass of concentrate (% or mass unit)

c   Element (in this case Fe) grade in concentrate (%)

H   Mass of feed (% or mass unit)

h   Element (in this case Fe) grade in feed (%)

In the same way and for practical terms, it is possible to define the classical weight recovery formula as Equation 3.

$$Wrec = \frac{C}{H} * 100 \qquad (3)$$

Where:

Wrec   Concentrate weight recovery (%)

C   Mass of concentrate (% or mass unit)

H   Mass of feed (% or mass unit)

Substituting Equation 3 into Equation 2 it is possible to define Equation 4.

$$Irec = Wrec * \frac{c}{h} \qquad (4)$$

Where:

Irec   Element recovery in concentrate (in this case Fe -%)

Wrec   Concentrate weight recovery (%)

c   Element (in this case Fe) grade in concentrate (%)

h   Element (in this case Fe) grade in feed (%)

Rearranging Equation 4 it is possible to arrive at Equation 5, which has the same structure of the proposed Wrec Equation 1:

$$Wrec = \frac{h}{c} * Irec \qquad (5)$$

Where:

Irec   Element recovery in concentrate (in this case Fe -%)

Wrec   Concentrate weight recovery (%)

c   Element (in this case Fe) grade in concentrate (%)

h   Element (in this case Fe) grade in feed (%)

As the original study defined that the concentrate grade of Mont-Wright plant will be fixed at 66.3 per cent, Equation 5 becomes Equation 6.

$$Wrec = \frac{h}{66.3} * Irec \qquad (6)$$

Where:

Irec   Element recovery in concentrate (in this case Fe -%)

Wrec   Concentrate weight recovery (%)

h   Element (in this case Fe) grade in feed (%)

For metallurgical purposes, the *Irec* of a concentration process is usually calculated rather than the *Wrec*. Thus, a correlation was established between Fe recoveries obtained in pilot plant tests and the ones from bench scale laboratory test performed with the same samples (Lacoste, 2019).

In this context, the metallurgical tests showed no correlation between the feed characteristics and the iron recovery in the -106 µm fraction of the pilot plant, as usual for spiral concentrators (Mesquita *et al*, 2019). Based on the variability of the test and to improve estimates of the Fe recovery in the circuit, ARL decided to divide the total recovery in two different size fractions (>106 µm and <106 µm) as presented in Equation 7 and described by Bennet (1971).

$$Irec = (> 106\ \mu m\ Irec) + (< 106\ \mu m\ Irec) \qquad (7)$$

Where:

Irec   Element recovery in concentrate (in this case Fe -%)

>106 µm Irec   Recovery of Fe in the fraction >106 µm of the concentrate (%)

<106 µm Irec   Recovery of Fe in the fraction <106 µm of the concentrate (%)

From the data generated during ALR study, it was defined that 40 per cent of the Fe originally contained in the <106 µm size fraction of Mont-Wright Feed was recoverable. The results of the tests performed by ARL showed that the recovery of the <106 µm fraction into the concentrate depends more on the operating parameters of the spiral plant than on ore characteristics (Bennet, 1971).

In this way, the '<106 µm Irec' was defined as Equation 8. It is important to note that in the procedure developed by ALR, the <106 µm fraction of the ground drill cores are removed previously to the heavy liquid test and the P value corresponds to the iron recovery of this size fraction (COREM, 2016). In addition, the premise that the P value of the ground drill core will represent well the Fe distribution of the plant feed in <106 µm fraction was assumed.

$$< 106 \ \mu m \ Irec = 0.4P \tag{8}$$

Where:

<106 µm Irec   Recovery of Fe in the fraction <106 µm of the concentrate (%)

P   Distribution of iron in <106 µm size fraction of ground drill core (%)

To evaluate the Iron recovery in the fraction >106 µm, heavy liquid tests were performed using the >106 µm fraction of the samples tested in the pilot plant ground up to 850 µm. A very good correlation between both Iron recoveries in >106 µm fraction was observed ($R^2 = 0.997$) as presented in Figure 2. Based on that, the '>106 µm Irec' was defined as Equation 9.

$$> 106 \ \mu m \ Irec = 1.33K - 27.33 \tag{9}$$

Where:

>106 µm Irec   Recovery of Fe in the fraction >106 µm of the concentrate (%)

K   Iron recovery in <850 µm >106 µm size fraction of heavy liquid concentrate (%)

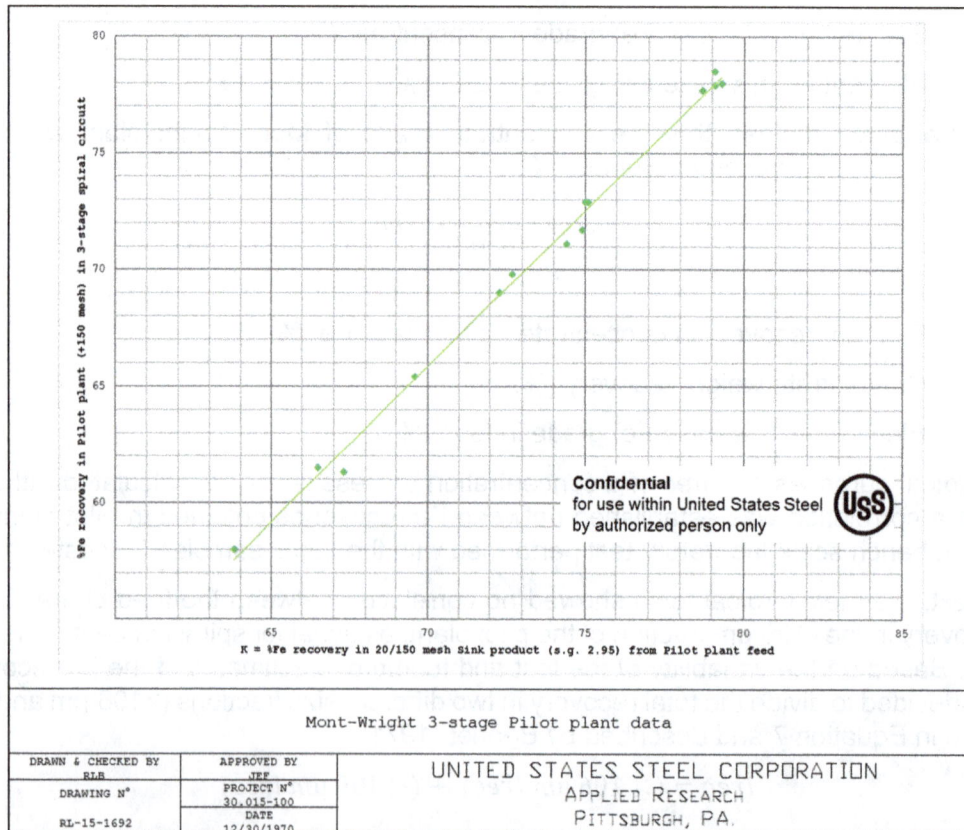

FIG 2 – Pilot scale recovery versus bench scale recovery (>106 µm size fraction).

Substituting Equations 8 and 9 into Equation 7, it is possible to define the equation used to estimate the 'Irec' (Equation 9) of the plant based on heavy liquid test results.

$$Irec = (1.33K - 27.33) + (0.4P) \qquad (10)$$

Where:

Irec   Element recovery in concentrate (in this case Fe -%)

K   iron recovery in <850 μm >106 μm size fraction of heavy liquid concentrate (%)

P   Distribution of iron in <106 μm size fraction of ground drill core (%)

Finally, substituting Equation 10 into Equation 6 it is possible to arrive in the final Wrec Formula, Equation 1.

MATERIALS AND METHODS

The research was divided in three different parts. The first axis corresponded to the execution of a descriptive statistics analysis with the historical concentrator data, the second one was focused on the execution of heavy liquid tests with the samples selected to be tested in the project and the last part corresponded to the execution of pilot scale spiral tests with the same samples. The update process of the Wrec formula was done using the information assessed from all stages.

Samples

The samples used in this study correspond to composites built using drill cores collected in both Mont-Wright and Fire Lake deposits. The drill cores were blended generating a total of 26 composite samples which were tested in heavy liquid bench scale tests and in a pilot circuit of spirals like the one present in Mont-Wright concentrator. The clustering of the samples was done analysing the contrast of iron and silica grades among the drill cores and the origin of the samples was also taken into consideration to have a wide coverage of different mineralogy inside the deposits. This was fundamental to have a maximum contrast in terms of Fe grade to support the Wrec formula update work. In this way, samples with Fe grade ranging from 14.74 per cent to 46.85 per cent were built (Figure 3).

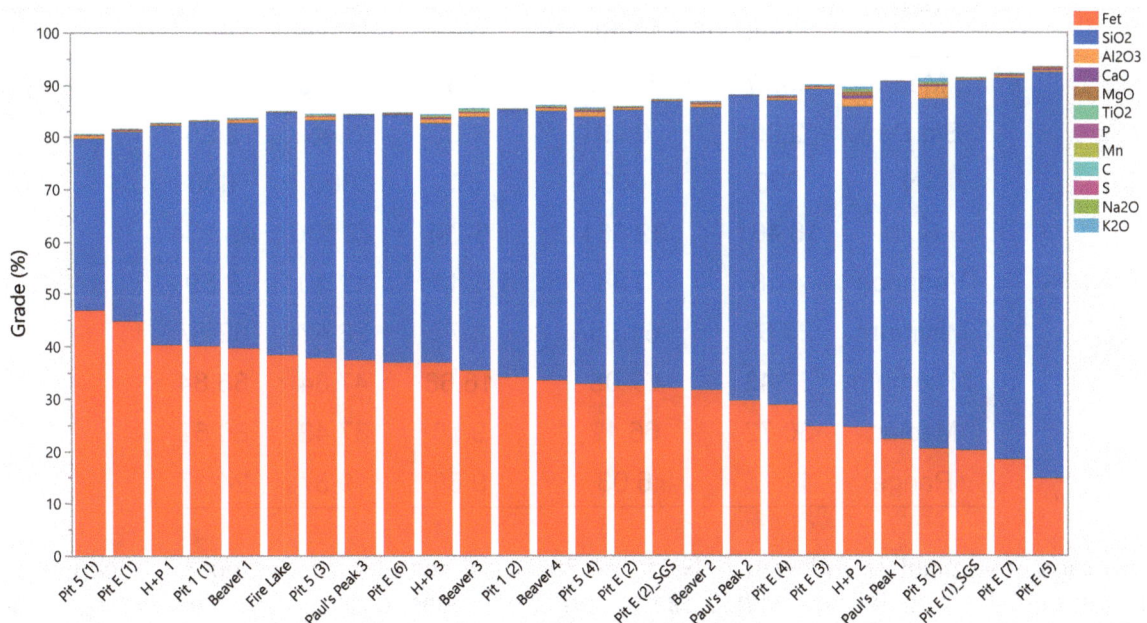

FIG 3 – Samples characteristics.

Descriptive statistics analysis

Aiming at understanding the details of Mont-Wright concentrator performance and getting important information for the Wrec formula update process, a complete descriptive statistic study was performed with industrial plant data.

The information used in the study corresponded to the monthly size-by-size results of Mont-Wright concentrator products (feed, concentrate and tailings) and encompassed the period between February 2012 and January 2022 (96 months). This information was used as a basis of comparison for the data generated in both heavy liquid and spirals pilot tests campaigns.

Spirals tests

Prior to spiral tests, the composite samples were ground to 2 mm achieving a product with a P_{80} near to 600 μm. The samples were tested in batch within a circuit composed by three stages of spirals (rougher, cleaner and recleaner) like Mont-Wright one. The parameters adopted in the tests were adjusted according to the operational parameters observed at Mont-Wright concentrator and the same models of equipment were adopted (HC33 for rougher stage and WW6E for both cleaner and recleaner stages – all machines supplied by Mineral Technologies).

Heavy liquid tests

In preparation for heavy liquid tests, the samples prepared for the spiral tests were ground to 850 μm and the fraction below 106 μm was removed through wet screening. Samples of approximately 50 g of the oversize product were analysed in the heavy liquid tests using a solution of LST (lithium heteropolytungstate) with density adjusted to 3 g/cm^3. The settling time of particles was fixed at 60 mins. The chemical analysis of screening undersize and both heavy liquid concentrate and tailings products were performed.

RESULTS

Mont-Wright concentrator historical data analysis

The results of the descriptive statistics analysis performed with the concentrator data are presented in Table 1.

TABLE 1

Mont-Wright concentrator descriptive statistics.

Item	Fe grade (%)			Recovery (%)	
	Feed	Concentrate	Tailings	Mass	Fe
Mean	30.77	66.10	9.52	37.55	80.56
Std Dev	2.05	0.29	1.05	3.48	2.95
N	600	600	600	600	600
Sum	18 461	39 661	5714	22 530	48 334
Variance	4.18	0.08	1.11	12.10	8.69
Minimum	25.68	61.05	6.77	28.47	62.67
Maximum	37.43	67.08	16.63	47.84	86.84
Median	30.72	66.12	9.50	37.42	80.62
Range	11.75	6.03	9.86	19.37	24.17

On average, the plant is fed with a ROM of 30.77 per cent Fe producing a concentrate with 66.10 per cent Fe and tailings with 9.52 per cent Fe. Those qualities lead to average yield of 37.55 per cent and Fe recovery of 80.56 per cent.

All the parameters analysed presented histograms with normal distribution as presented in Figure 4. In addition, higher variations were observed for Fe grades in both feed and tailings products (standard deviations of 2.05 and 1.05, respectively) while the concentrate grade is very stable over the time (standard deviation of 0.29).

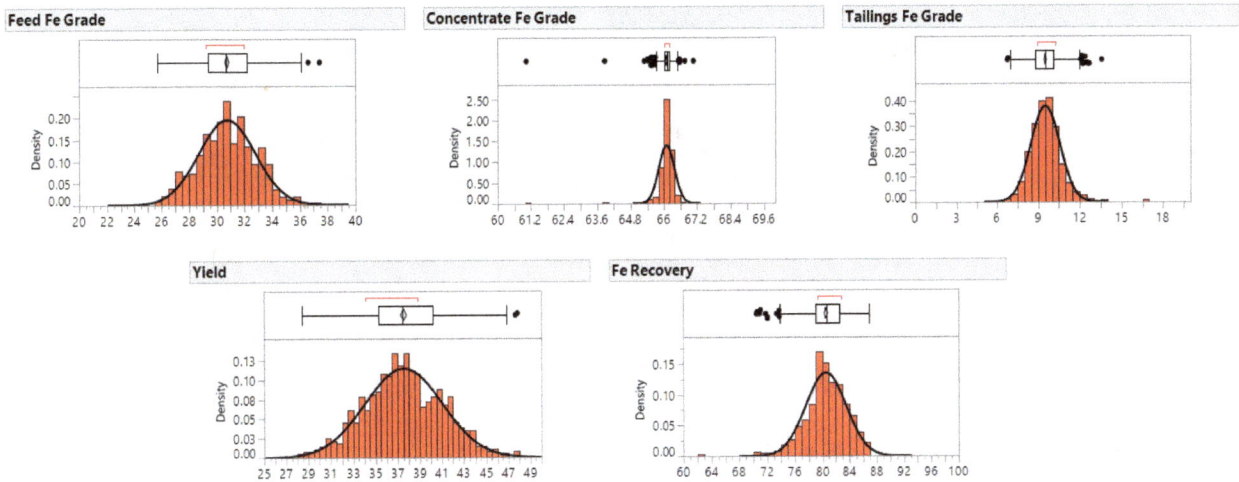

FIG 4 – Descriptive statistics histograms.

Regarding the yield and the Fe recovery, direct correlations are observed with the Fe grades of feed product (Figure 5). As expected, richer feed led to higher recoveries.

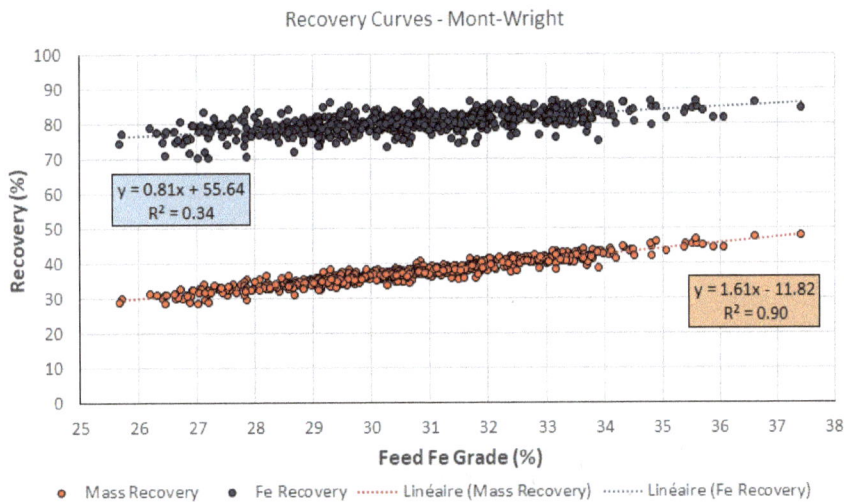

FIG 5 – Mont-Wright recovery curves.

Concentrate grade

In the original formula, ALS engineers considered that the quality target for Mont-Wright concentrate would be 66.30 per cent Fe. The assumption was done before the beginning of plant operation being based on bench/pilot scale tests results. In this context, the descriptive statistics analysis performed on plant data showed that the average quality produced in the concentrator today is 66.10 per cent Fe, as previously presented in Table 1. In this way, the concentrate quality of the original formula was changed based on the analysis above.

Fe recovery in <106 µm size fraction

In the same way as observed for the concentrate quality, the estimate of Fe recovery in the fraction below 106 µm used in the original formula (40 per cent) was based on laboratory tests. The descriptive statistics performed with plant data showed a slightly higher value, as presented in Figure 6. As such, the original value in the formula was changed from 0.40 to 0.46.

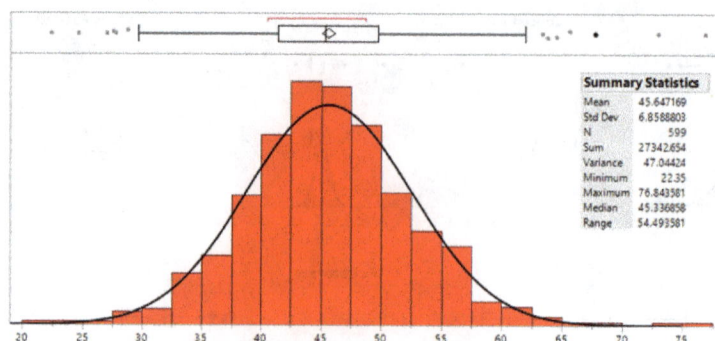

FIG 6 – <106 μm Fe recovery descriptive statistics.

Fe recovery in >106 μm size fraction

In the original Wrec formula, estimates of the Fe recovery in the fraction above 106 μm are based on the results of heavy liquid tests. The same procedure was followed to create a new regression curve based on the results of heavy liquid and spirals tests performed with the 26 composite samples. The updated regression model is presented in Figure 7, alongside its predecessor. The new regression presented a R^2 of 0.85 and eight outlier points were removed from the original data set (30 per cent of the information).

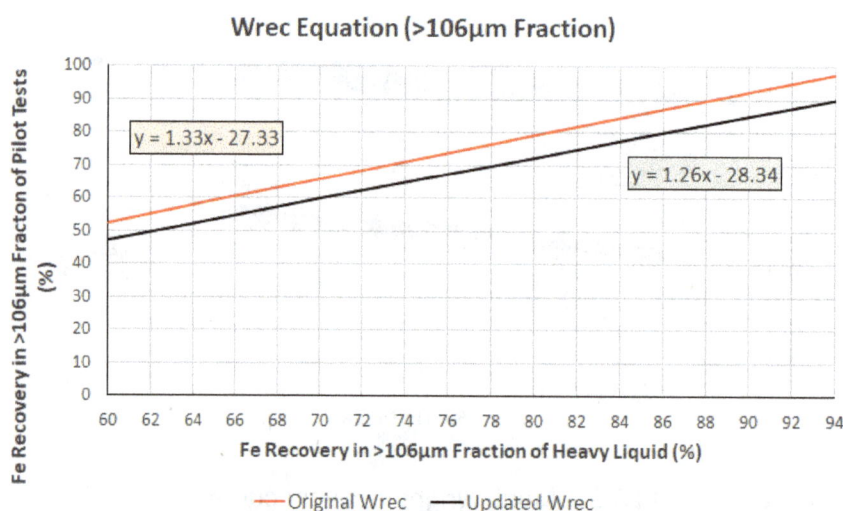

FIG 7 – Fe recovery curves comparison (>106 μm size fraction).

From the graph it is possible to see that the inclination of the new curve is very similar to the old one, but it is vertically offset, which indicates some change from the original study. In this way, the coefficients of the Wrec formula were updated based on the results achieved.

Updated Wrec formula

After modifications proposed in the previous sections, the updated Wrec formula becomes Equation 11.

$$Wrec = \frac{h}{66.10} * (1.26K - 28.34 + 0.46P)$$ (11)

Where:

K   Iron recovery in <850 μm >106 μm size fraction of heavy liquid concentrate (%)

P   Distribution of iron in <106 μm size fraction of ground drill core (%)

h   Fe grade of the drill core (%)

The updated formula is very similar to the original one, but the proposed changes result in slightly lower Wrec values than the previous equation. A comparison between both equations was done applying the different formulas for the database of 2019 Wroo campaign. The summary of the differences between the results (Old Wrec – Updated Wrec) is presented in Table 2. Data shows an average gap of 2.33 percentage points between the two formulas, which corresponds to a difference of 5.40 per cent.

TABLE 2

Wrec formulas results comparison.

Item	Difference (pp*)	Difference (%)
Mean	2.33	5.40
Std Dev	1.29	10.71
N	306	306
Sum	712	1,652
Variance	1.66	114.65
Minimum	-0.10	-125.70
Maximum	6.10	56.20
Median	2.40	6.60
Range	6.20	181.90

* pp = Percentage points.

Updated formula validation

In an attempt to assess the accuracy of Wrec formulas, a statistical analysis was done using the historical plant data and the Wrec values calculated using the two formulas applied to the database of 2019 Wrec campaign.

To execute this comparison, a filter was applied to the three groups of databased on the Fe grade of the samples. This filter was used to minimise the variability effect of drill cores being compared directly with the bulk samples feeding the plant, once both very low and very high-grades found in drill cores database could jeopardise the results of the analysis. In this way, only samples containing Fe grades within the range corresponding to the average plant feed grade (30.77 per cent Fe) plus/minus two standard deviations (4.10 per cent Fe) were included in the analysis. The comparison between the data is presented in Table 3.

TABLE 3

Wrec formula validation for a confidence interval of 95 per cent in Fe feed grade.

| Item | Fe feed grade (%) | | Yield (%) | | |
	Plant	Wrec	Plant	Old formula	Updated formula
Mean	30.82	31.06	37.51	40.68	38.01
Std Dev	1.87	2.05	3.20	3.33	3.09
N	695	83	578	83	83
Sum	21 418	2578	21 679	3376	3155
Variance	3.51	4.21	10.25	11.08	9.54
Minimum	26.70	26.70	28.75	33.05	31.22
Maximum	34.80	34.70	45.50	48.83	45.57
Median	30.50	32.30	36.18	37.84	35.47
Range	8.10	8.00	16.75	15.78	14.35

The data shows a similar Fe grade on feed in both cases and a good adherence of the updated formula to plant data.

CONCLUSIONS

The main objective of work was to revise and update the Wrec formula to ensure a good reconciliation between the predicted data and the concentrator performance. In this way, a previous statistical study was performed with the historical data from Mont-Wright concentrator to understand the overall operation of the plant.

The study showed that on average, the plant is fed with a ROM of 30.77 per cent Fe producing a concentrate with 66.10 per cent Fe and tailings with 9.52 per cent Fe. Those qualities lead to average yield of 37.55 per cent and Fe recovery of 80.56 per cent. In addition, data showed that the iron recovery decreases considerably below 106 μm inside the spirals circuit presenting an average value of 46 per cent.

In parallel, a campaign of bench and pilot scales tests were executed with 26 composite samples. Those samples were built using drill cores collected in both Mont-Wright and Fire Lake mines, grouped with the objective of representing the different regions of the deposits and encompassing a lager range of Fe grades. Based on test work results, an updated formula was established.

The old formula was compared to the updated one and an average decrease of 2.33 percentage points was observed in the Wrec values, which corresponds to a difference of 5.40 per cent. In addition, both groups of data were compared with plant figures and the updated formula presented a better adherence to the real values, indicating that the old formula was overestimating the mass recovery at Mont-Wright concentrator.

REFERENCES

ArcelorMittal, 2023. ArcelorMittal Mines et Infrastructure Canada (online). Available from: <https://mines-infrastructure-arcelormittal.com/> (Accessed: 24 May 2023).

Bennet, R L, 1971. Prediction of Mount Wright concentrator performance from tests on drill cores application to mine planning, USS Applied Research Laboratory technical report.

COREM, 2016. Analyse des échantillons AMMC, COREM technical report.

Ferreira, R G R, Frade, T M C, Machado, L C R and Guimarães, F R, 2022. Programa de geometalurgia do Minas Rio – Anglo American, paper presented to ABM Week 6° edition, São Paulo.

Lacoste, M, 2019. Metallurgical Testing and Results, ArcelorMittal Maizières Research Centre technical report.

Mesquita, J, Vasconcelos, F, Pirson, A, Correa de Araujo, A, Sylow, T and Gotelip, L, 2019. Tailings Valorization and Iron Recovery from Spiral Concentrators Circuit, in *Proceedings of the Iron Ore Conference 2019*, pp 752–761 (The Australasian Institute of Mining and Metallurgy: Melbourne).

Silveira, A L, Delano, E, Júnior, O and Cabral, T, 2022. Desenvolvimento da geometalurgia de curto prazo no Salobo, paper presented to ABM Week 6° edition, São Paulo.

Examining the relationship (if any) between elevated phosphorus and ore textural type in Pilbara iron ores

M I Pownceby[1], M J Peterson[2], J R Manuel[3] and N Karimian[4]

1. Geometallurgy Team Leader, CSIRO Mineral Resources, Clayton Vic 3168.
   Email: mark.pownceby@csiro.au
2. Senior Experimental Scientist, CSIRO Mineral Resources, Pullenvale Qld 4069.
   Email: michael.peterson@csiro.au
3. Senior Experimental Scientist, CSIRO Mineral Resources, Pullenvale Qld 4069.
   Email: james.manuel@csiro.au
4. Postdoctoral Fellow, CSIRO Mineral Resources, Clayton Vic 3169.
   Email: niloofar.karimian@csiro.au

ABSTRACT

The CSIRO ore classification scheme for Pilbara ores was used as a reference frame to target recognisable ore types/textures with potentially elevated phosphorus content for detailed analysis. Particles of each individual ore texture were separated from a Brockman high-P ore and each class was characterised using methods including XRF, XRD, EPMA and synchrotron-based XAS. Results were then correlated with optical microscopy and hand specimen identification to establish any relevant associated mineralogical phases and/or textures. This represents a more systematic approach to identification of high-P ore components based on particle and goethite type and texture. The information can then provide the basis for more targeted P-removal strategies factoring in the relationship between particle/ore texture and phosphorus deportment.

INTRODUCTION

A major part of Australia's future Pilbara iron ore resources is in high-P deposits. While phosphorus levels in existing Pilbara products are effectively managed by blending, this will become more difficult in the future as low-P are depleted. High-P levels in iron ore incur a price penalty because of a) the adverse embrittling effect of phosphorus on the quality of the end-product steel and b) the costs and emissions associated with dephosphorisation during steelmaking. Historically, market specifications for phosphorus in iron ore exported from Australia are typically 0.075 per cent P (Cheng *et al*, 1999) although this level has recently been raised to 0.09 per cent P with penalties for every 0.001 per cent increase in phosphorus above the acceptable limit. Australian ore producers and customers are therefore under rapidly increasing pressure to maximise ore grade to reduce energy usage and emissions associated with phosphorus impurities during processing. This has resulted in a more urgent need to develop mitigation strategies and therefore a renewed focus on the long-standing issue of high-P ore utilisation.

Insights obtained from previous analyses of the phosphorus content and its distribution in high-P iron ores, demonstrated that phosphorus can be present in ores in five principal forms (Ofoegbu, 2019):

1. As distinct crystalline phase(s) in the ore or in the associated gangue mineral(s).

2. As an amorphous phase(s) or adsorbed phase(s) in the gangue.

3. As distinct crystalline phase(s) in the iron-rich phase.

4. As amorphous phase(s) or adsorbed phase(s) in the iron-rich fraction of the ore.

5. As chemically substituted phosphorus within the iron-rich fraction.

or any combination of these. When present as a distinct, liberated crystalline or amorphous phase in the gangue mineral(s) (mechanisms 1 and 2), phosphorus removal from ores is more likely to be possible using physical beneficiation methods (depending on grain size). In contrast, the presence of phosphorus in the iron-rich fraction (mechanisms 3, 4 and 5) typically requires the application of other additional methods (eg chemical, or thermal) to achieve phosphorus removal or reduction. These extra processing steps come with associated economic and environmental costs but may be able to be incorporated in product strategy eg removal of Al, Si and P to produce a high-grade product for alternative low carbon processing.

It is well known that phosphorus in Pilbara hematite-goethite ores is associated with goethite (Graham, 1973; Dukino, England and Kneeshaw, 2000; Thorne *et al*, 2008; Manuel and Clout, 2017), usually also with Al and Si, but the mechanism of phosphorus incorporation is not fully understood (Pownceby *et al*, 2019). Furthermore, predictability of high-P goethite types and associated ore textures remains difficult due to the relatively low levels of phosphorus present and the lack of any clear direct indication (eg texture, colour, reflectivity) of its presence at elevated levels. Improved discrimination and identification of the occurrence and distribution of high-P ore components within deposits is essential for development of effective beneficiation strategies for high-P hematite-goethite ores.

In the current study, a systematic approach to identification of high-P ore components based on goethite particle type and texture was used to identify whether there was any consistent, recognisable ore type/texture that was correlated with elevated phosphorus. If any direct correlation can be determined, then the information could provide the basis for more targeted P-removal strategies factoring in the relationship between particle/ore texture and phosphorus deportment.

CHARACTERISATION METHODOLOGY

The sample examined in the study was a high-P Brockman type ore from the Pilbara region of Western Australia. Chemical analysis of the as-received ore indicated the ore comprised 62.4 per cent T_{Fe}, 3.0 wt per cent SiO_2 and 2.6 wt per cent Al_2O_3. The LOI was 4.3 per cent. The total phosphorus content of the sample was 0.28 wt per cent P.

The sample was hand-sorted into specific ore textural types based on the iron ore classification developed by Clout (2003) which uses textural groupings defined based on similarities in mineralogy, ore texture, porosity, mineral associations and hardness. The scheme is non-genetic and has been successfully applied to deposits in the Pilbara, Yilgarn and Gawler cratons. The main ore texture groups include dense martite/hematite, microplaty hematite, microplaty hematite-goethite, martite-goethite, goethite-martite and goethite-rich. Each group can be further subdivided into physically hard to softer subcategories (Table 1).

TABLE 1

The CSIRO ore group textural classification scheme, modified from Clout (2003).

Group	Dominant Mineralogy		Subgroup	Hardness	Porosity
	Matrix	**Infill**			
1	Dense Martite Hematite		a	Very hard	Very Low
		Silica	s		
2	Microplaty Hematite/ Martite		a	Hard	Medium
			b	Medium	Medium
			c	Soft/Friable	High
3	Martite	Goethite	a	Medium	Medium
			b	Soft	High
4	Goethite	Martite		Soft	High
5	Dense Hematite/goethite	Hydrohematite	a	Hard-very hard	Low
	Martite		b	Soft-Medium	Medium-High
6	Dense Martite	Goethite		Hard	Low
7	Dense Goethite	Martite		Hard	Medium
8	Microplaty Hematite	Goethite	a	Hard	Low
			b	Medium	Medium
			c	Soft	High
9	Ochreous Goethite			Low	Very High
10	Vitreous Goethite		a	Hard	Low
			b	Medium	Medium
11	Nanohematite	Martite/ Microplaty Hematite	a	Medium	Low
			b	Soft	Medium

CSIRO undertook hand sorting of -10+5 mm sized material to obtain representative samples of ore groups, according to Table 1. The particles were gently washed to remove particulate ultrafines then oven dried at 100°C. To identify the ore particle types, particles were individually examined under a stereomicroscope and their relative hardness was determined by breakage techniques involving either a geological hammer and steel bash plate or with the use of pliers. Particle streak was determined using a ceramic plate and relative magnetism was determined using a hand magnet. Following the hand sorting, 15 representative particle textural types, primarily from the goethite-rich groups, were selected for in-depth characterisation. These included particles from the goethite-rich ore groups Gp4, Gp7, Gp9 and Gp10. Some particles containing goethite as a minor matrix or infill component (Gp5A and 3B) and shale-rich particles were also selected for additional characterisation. In the following sections, samples are identified by their particle number (1–15) followed by the ore group number (Gp 1–11) eg sample 8–4 refers to Particle 8 containing predominantly Gp4 material.

X-Ray Fluorescence Spectroscopy (XRF)

For the XRF analyses, all samples were pulverised (see XRD section) before accurately weighing an amount of each of the samples into 95 per cent Pt/Au crucibles with approximately 5 g of 12:22 lithium tetraborate:metaborate flux previously dried at 550°C. Sample masses varied according to the size of the particle available for analysis and ranged from 0.03 g for sample 3–9 up to 0.3 g for samples 11 and 15 (shale samples). Each sample was pre-oxidised over an oxy-propane flame burner to bring the contents of the crucible to between 650°C to 700°C while oxygen was bled into the top of the crucible. The sample was held at this temperature for approximately five minutes before the mixture was fused into a homogeneous melt over an oxy-propane flame at a temperature of approximately 1050°C for approximately ten minutes. Air jets then cooled the mould and melt for approximately 300 seconds. The resulting glass discs were analysed on a Bruker S8 Tiger WDXRF system using a control program developed by Bruker and algorithms developed in-house by CSIRO.

X-Ray Diffraction (XRD) Phase Analysis

The samples were ground in ethanol in a McCrone micronising mill. The resulting slurries were oven dried at 60°C then thoroughly mixed in an agate mortar and pestle before being lightly back pressed into a stainless-steel sample holder for presentation to the X-ray beam. The XRD patterns of the powdered samples were collected with a PANalytical X'Pert Pro Multi-purpose Diffractometer using Fe filtered Co Ka radiation, automatic divergence slit, 2° anti-scatter slit and fast X'Celerator Si strip detector. The patterns were collected from 4 to 80° in steps of 0.017° 2θ with a counting time of 0.5 sec per step, for an overall counting time of approximately 35 min.

Phase identification was performed using PANalytical Highscore Plus© software (V4.8) which interfaces with the International Centre for Diffraction Data (ICDD) PDF4+ 2022 database. Semi-quantitative phase analysis was carried out *via* the Rietveld method using TOPAS V6 software.

Electron Probe Microanalysis (EPMA)

Polished blocks containing the representative, hand-sorted particles from the selected ore groups were analysed using automated EPMA mapping to generate element distribution maps that enabled the identification of any correlation between mineralogical and textural features.

The polished grain mounts were mapped at CSIRO using a JEOL Superprobe electron microprobe analyser (Model JXA 8500F) equipped with five wavelength dispersive (WD) spectrometers. Randomly selected areas on each sample, chosen to map as much of the particle as possible, were mapped on each individual particle. Operating conditions for the microprobe during mapping were an accelerating voltage of 20 kV, a beam current of 100 nA, a step size of 5 µm (in *x* and *y*), and a counting time of 10 ms per step.

Following mapping, the element distribution data were manipulated using the software package CHIMAGE (Wilson, MacRae and Torpy, 2008; Torpy *et al*, 2020) to provide an 'element distribution' map showing the distribution of all or selected elements within the mapped area.

X-ray Absorption Spectroscopy (XAS)

Fe K-edge extended X-ray absorption fine-structure (EXAFS) spectroscopy was employed at the Medium Energy X-ray Absorption Spectroscopy Beamline (MEX1) beamline at the Australian Synchrotron to examine the relative abundance of Fe oxide phases in selected mineral samples. The samples, along with Fe[(III)]-oxide reference standards, were prepared as pellets and sealed with Kapton tape after dilution with cellulose.

Transmission mode XAS spectra were collected at room temperature and in-line Fe-metal foil data was collected simultaneously for calibration purposes. Linear combination fitting (LCF) of k3-weighted EXAFS spectra in the 2-12 Å-1 range against Fe reference standards, including goethite, hematite, ferrihydrite and feroxyhyte, was used to quantify the relative abundance of Fe species in selected soil samples. The ATHENA program was used for standard background subtraction and edge-height normalisation (Ravel and Newville, 2005) and the F-test based on Hamilton's methodology (Calvin, 2013) was used to determine which Fe[(III)]- and/or Fe[(II)]-bearing phases contributed significantly to a fitted spectrum ($P<0.05$).

RESULTS AND DISCUSSION

Chemical analysis results from each of the particles are provided in Table 2. All data is reported as oxides and it is expected that errors in absolute wt per cent values may be high especially in samples where there were only low masses (<0.1 g) of material available for the analysis.

TABLE 2

Summary of chemical analysis results on the particles from the specific ore groups. Note that LOI was not determined due to insufficient sample size but can be deduced by difference (approximate only).

Sample No. – Ore Group	Specimen mass	XRF Sum (%)	TiO_2 (%)	Fe_2O_3 (%)	SiO_2 (%)	Al_2O_3 (%)	CaO (%)	K_2O (%)	MgO (%)	Mn_3O_4 (%)	Na_2O (%)	P_2O_5 (%)	SO_3 (%)
Goethite-dominant groups													
8–4	0.1828	91.7	0.26	82.8	4.39	2.81	0.02	0.01	0.04	0.17	0.70	**0.32**	0.04
9–4	0.2844	94.0	0.16	88.8	1.57	1.68	0.02	<0.01	0.07	0.24	0.21	**1.19**	0.04
7–7	0.0849	94.5	<0.03	90.8	0.86	0.63	0.05	<0.03	0.09	0.15	0.82	**0.98**	0.06
14–7	0.1019	77.2	0.09	73.3	1.43	1.13	0.03	<0.02	<0.02	0.06	0.34	**0.62**	0.06
4–7	0.2601	90.5	0.14	84.6	3.19	0.64	0.60	0.02	0.06	0.12	0.54	**0.21**	0.25
3–9	0.0332	89.4	0.44	67.6	5.33	7.80	0.46	0.11	1.01	0.31	4.16	**1.17**	0.49
1–10A	0.1584	95.3	0.21	87.4	2.09	2.50	0.17	0.04	0.24	0.77	1.15	**0.45**	0.17
2–10B	0.1348	88.6	0.18	72.8	7.29	5.64	0.19	0.03	0.46	0.27	0.46	**0.89**	0.20
Groups with goethite as a minor or infill component													
12–5A	0.0636	98.2	0.25	75.4	11.0	8.70	<0.04	<0.04	<0.04	0.21	1.16	**1.21**	0.10
5–5A	0.1169	92.0	<0.02	84.4	3.34	0.92	0.21	0.04	0.44	0.35	1.56	**0.40**	0.19
6–5A	0.0970	98.2	0.05	90.0	4.05	1.62	0.14	<0.03	0.35	0.10	0.78	**0.84**	0.12
10–3B	0.1424	100.6	0.04	95.4	1.47	0.75	0.03	0.03	0.22	0.04	1.94	**0.45**	0.08
13–3B	0.2034	99.7	<0.01	96.0	1.34	0.73	<0.01	0.03	0.08	0.03	1.08	**0.27**	0.03
Shale													
11-SH	0.3324	88.0	0.09	40.6	24.2	21.8	0.02	0.01	0.04	0.07	0.60	**0.44**	0.05
15-SH	0.3192	94.4	0.05	88.8	2.44	1.90	0.01	<0.01	0.02	0.07	0.15	**1.00**	0.02

The goethite-dominant group of particle types (Gp4, Gp7, Gp9 and Gp10) had P_2O_5 contents from XRF that varied between a minimum of 0.21 wt per cent (Particle 4 – Gp7) and a maximum of 1.19 wt per cent (Particle 9 – Gp4). Alumina contents of the goethite-rich ore types varied from 0.63 wt per cent (Particle 7 – Gp7) to 7.80 wt per cent (Particle 3 – Gp9). Silica contents also varied considerably, ranging from 0.86 wt per cent (Particle 7 – Gp7, up to 7.29 wt per cent (Particle 2 – Gp10).

X-ray diffraction results are provided in Table 3. They should be considered semi-quantitative given: (i) the small sample mass(es) available; and (ii) an internal standard was not used because the material after XRD was required for the XRF analyses. The lack of an internal reference standard meant that it was not possible to estimate the amorphous component and therefore the results of the phase identification and phase analysis are given in Table 3 as relative wt per cent of crystalline phases. For the samples where hydrohematite was identified, the proportions of hematite and hydrohematite should be considered semi-quantitative because there is nothing against which to calibrate the peak widths and shapes of the hematite phases.

TABLE 3

Results from the XRD analysis (relative wt per cent of crystalline phases).

Sample No. – Ore Group	Phase concentration (relative crystalline wt%)				
	Hematite	Hydrohematite	Goethite	Kaolin	Anatase
Goethite-dominant groups					
8–4	13		83	4	
9–4	39		61		
7–7	24		76		
14–7	49		51		
4–7	20		80		
3–9	5		91	3	<1
1–10A	67		33		
2–10B	5		85	10	<1
Groups with goethite as a minor or infill component					
12–5A	40	38		23	
5–5A	11		89		
6–5A	52	41	6		
10–3B	74		26		
13–3B	78		22		
Shale					
11-SH	4		43	52	<1
15-SH	41		54	4	

A plot of P_2O_5 and Al_2O_3 versus goethite content determined via XRD, shown in Figure 1, indicates an overall trend of both increasing P_2O_5 (Figure 1a) and increasing alumina (Figure 1b) with increasing goethite content in the particle, consistent with previous observations. In comparison, when plotting the same data for the groups with goethite as the matrix or infill phase, the same trend was not evident even though some particles within this data set (eg Particle 12 – Gp5) had very high P_2O_5 and Al_2O_3 contents (1.21 wt per cent and 8.70 wt per cent respectively). It is possible that in these particles, high alumina levels are likely to be due to the presence of kaolinite/gibbsite in addition to Al substituted in goethite (ie Al-silicates or oxides not directly associated with phosphorus).

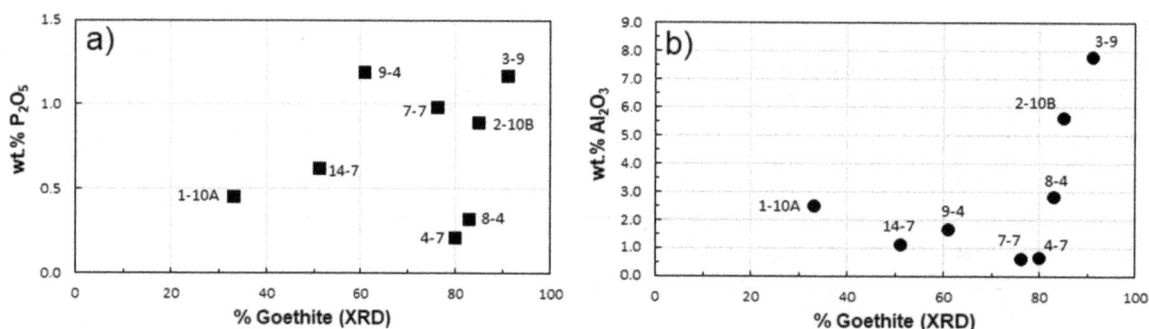

FIG 1 – Plots of wt per cent P_2O_5 (a); and wt per cent Al_2O_3 (b); versus goethite content determined by XRD in particles where the dominant ore type was goethite-rich. Each point is labelled according to the sample number and ore-type.

Results from the XAS analysis of the individual particle types are provided in Table 4. The linear combination analysis (LCF) of the Fe XAS data generally support the XRD results in terms of relative amounts of hematite and goethite in the respective particle types. It should be noted, however, that the absolute values differ significantly from the XRD results due to a combination of factors: (i) the Fe K-edge XAS results only measure the Fe oxide and oxyhydroxide phases in the particle – the technique is essentially blind to other phases not containing Fe which are therefore not accounted for; and (ii) the XAS method is more sensitive to the presence of other Fe oxide and oxyhydroxide phases that were not able to be determined via the XRD method. For example, the data in Table 4 indicate that other Fe oxide/oxyhydroxide phases such as ferrihydrite are present in nearly all samples, while feroxyhyte was detected in two of the particles.

TABLE 4

Fe speciation data expressed as a proportion of total Fe based on LCF fits of Fe K-edge EXAFS spectra. Data represent only the Fe oxide/hydroxide components of the particles – silicates are not measured by the technique. R^2 refers to the error in fit between the standard and calculated patterns.

Sample No. – Ore Group	Goethite %	Hematite %	Ferrihydrite %	Feroxyhyte %	R^2
Goethite-dominant groups					
8–4	48.9	34.2	16.9	0	0.004
9–4	29.5	70.5	0	0	0.003
7–7	67.8	14.5	17.7	0	0.004
14–7	40.7	44.1	15.2	0	0.003
4–7	72.4	12.6	15	0	0.003
3–9	60.4	8.3	31.3	0	0.009
1–10A	68.6	9.3	22.1	0	0.004
2–10B	70	8	22	0	0.004
Groups with goethite as a minor or infill component					
12–5A	22	78	0	0	0.002
5–5A	6.3	76.25	7.5	10	0.002
6–5A	69	21.7	9.3	0	0.002
10–3B	67	10.8	22.2	0	0.008
13–3B	21.5	78.5	0	0	0.002
Shale					
11-SH	0	78.7	6.3	15	0.002
15-SH	51.4	38.5	10.1	0	0.004

Iron hydroxides form a sequence of minerals differing in their thermodynamic stability; they include ferrihydrite (nominally $(Fe^{III})_2O_3 \cdot 0.5H_2O$ but other proposed formulas include $Fe_5OH_8 \cdot 4H_2O$ and $Fe_2O_3 \cdot 2FeO(OH) \cdot 2.6H_2O$), feroxyhyte $\delta FeOOH$, lepidocrocite $\gamma FeOOH$ and goethite $\alpha FeOOH$ (Vodyanitskii, 2010). All hydroxides are classified as thermodynamically unstable at near surface conditions, except for goethite. Feroxyhyte remains a rare mineral and with time, feroxyhyte spontaneously transforms into goethite by rapid oxidation of ferrous iron in the subsurface at the groundwater level (Carlson and Schwertmann, 1980). Ferrihydrite may be converted either into hematite or goethite although its transformation in natural systems can be blocked by chemical impurities adsorbed at its surface (for example silica, as most natural ferrihydrites are siliceous).

The phases ferrihydrite and feroxyhyte are difficult to determine using XRD due to their low crystallinity and nano-structured presentation and it may be speculated that previous reports of high amounts of amorphous, non-crystalline material in iron ores may be due in large part to the possible presence of these phases (or even lack of goethite crystallinity). According to Paige, Snodgrass and Nicholson (1997), the presence of phosphate can retard the transformation of ferrihydrite into crystalline products. These authors demonstrated that increasing phosphate from 0 to 1 mole per cent in solution results in an order of magnitude decrease in the rate of transformation to goethite. Higher levels of phosphate resulted in the ferrihydrite remaining amorphous.

The presence of feroxyhyte in at least two samples was unexpected as it is usually metastable at surface conditions and spontaneously reverts to goethite (see discussion above).

Plots of P_2O_5 and Al_2O_3 contents from XRF versus goethite determined by XAS showed no significant correlations (see for example Figure 2a for the P_2O_5 data plotted against goethite). We speculate this is due to the XRD data reporting all Fe oxy/oxyhydroxides as goethite and not recognising the presence of specific ferrihydrite and feroxyhyte phases. Both these phases are capable of adsorbing P_2O_5 and if the XAS data is plotted as (goethite+ferrihydrite ie total Fe oxy/oxyhydroxides) versus P_2O_5 (Figure 2b), a positive correlation becomes much clearer.

We note however that the data plots were all heavily impacted by the data for sample 9–4 which appeared to contain an anomalously low goethite content (29.7 per cent) in the Fe K-edge XAS data (*cf* 61 per cent in the XRD data). If this sample is excluded from the plots shown in Figure 2 then the XAS P_2O_5 versus goethite content plot indicates a slight trend of increasing P_2O_5 with increasing goethite content (*cf* Figure 2a) while when P_2O_5 is plotted against goethite+ferrihydrite, the trend for P_2O_5 versus total Fe oxide/oxyhydroxide is more pronounced (Figure 2b).

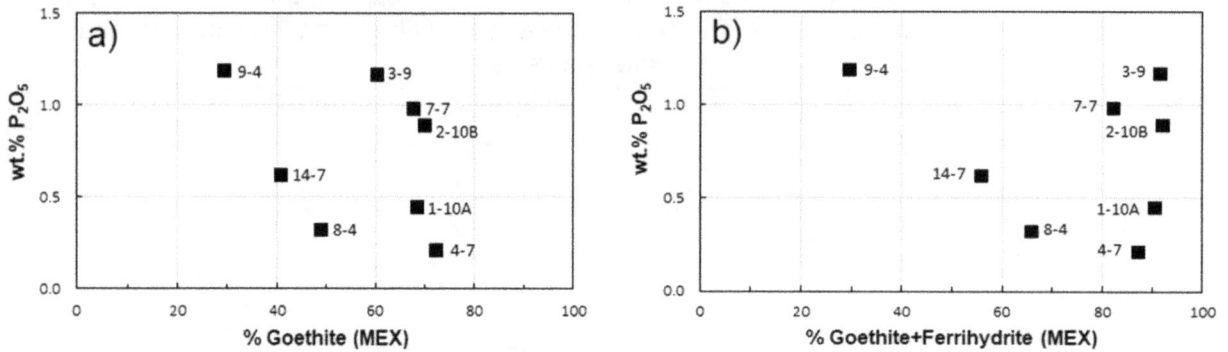

FIG 2 – Plots of wt per cent P_2O_5 versus goethite (a); and combined Goethite+Ferrihydrite (b) content in particles determined using XAS spectroscopy where the dominant ore type was goethite-rich. Each point is labelled according to the sample number and ore-type.

A question remains as to why sample 9–4 is so different in abundance between the XRD and XAS data. We speculate that this is due to mineralogical and chemical heterogeneity in the particle selected for analysis. This can be seen in the EPMA results described below and shown in the BSE image in Figure 3a. Particle 9–4 has obvious significant variation in the ratio of goethite:hematite, as well as variation in the distribution of phosphorus. The piece of the particle broken off and used for the Fe K-edge XAS analysis appears to have contained more hematite than the piece used for the XRD (and subsequent XRF) analysis. This sample heterogeneity was not evident in 2D (or on the exposed surfaces) when originally sorted. Whether or not the differences observed are due to sample heterogeneity is a problem that needs looking at more closely – there may in fact be something real going on in the difference between XRD and XAS results, particularly where the ratio H:G is reversed to G:H between the two methods.

The results from the EPMA mapping of the eight goethite-dominant particle types are provided in Figures 3–6. For each particle, the following information is provided: a backscattered electron (BSE) image, the Fe distribution map, the Al distribution map, the P distribution map and a combined Fe/P/Si distribution plot. Note that the concentrations of each individual element shown in Figures 3–6 are relative and have been thresholded to maximise contrast.

Comments on individual particle map results are provided in the following sections.

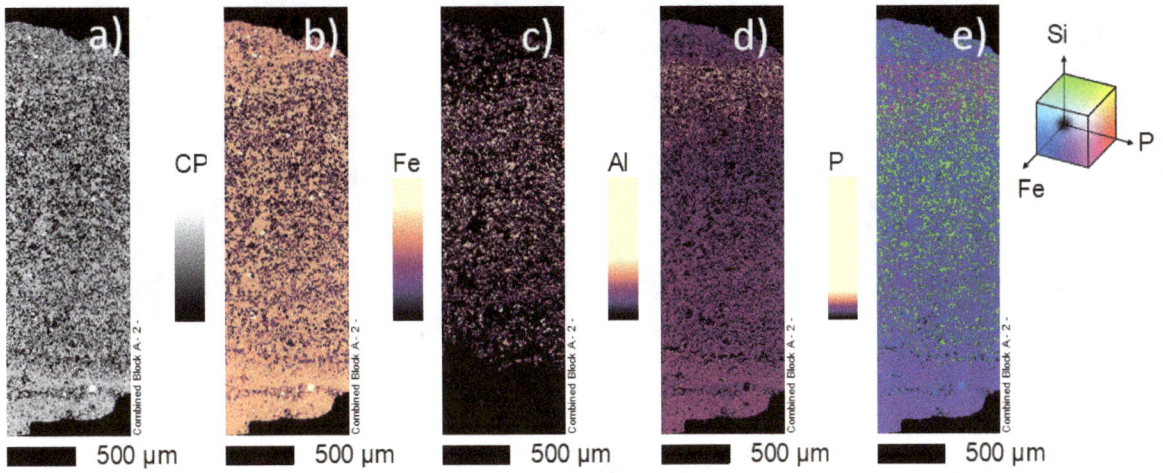

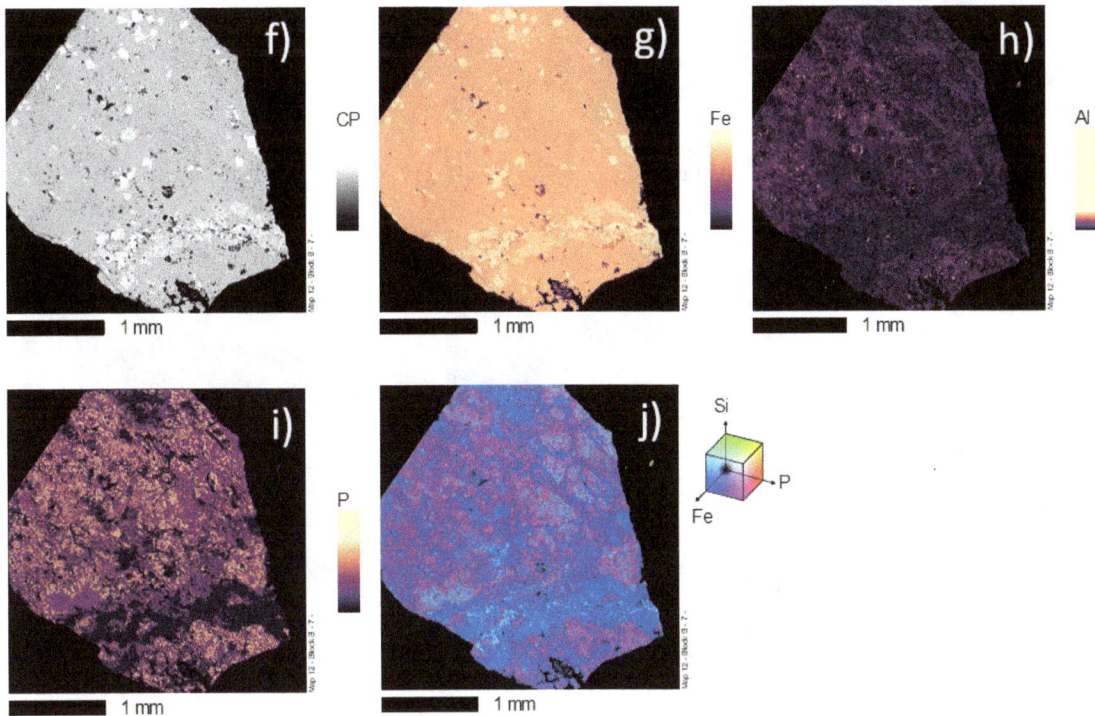

FIG 3 – EPMA maps showing the BSE image (labelled CP), Fe, Al and P element distribution maps and a three-colour Fe/Si/P distribution map. Images (a)–(e) are for Particle 2 (Gp10–85 per cent goethite by XRD, 0.89 per cent P_2O_5); and images (f)–(j) are for Particle 7 (Gp7–76 per cent goethite by XRD, 0.98 per cent P_2O_5).

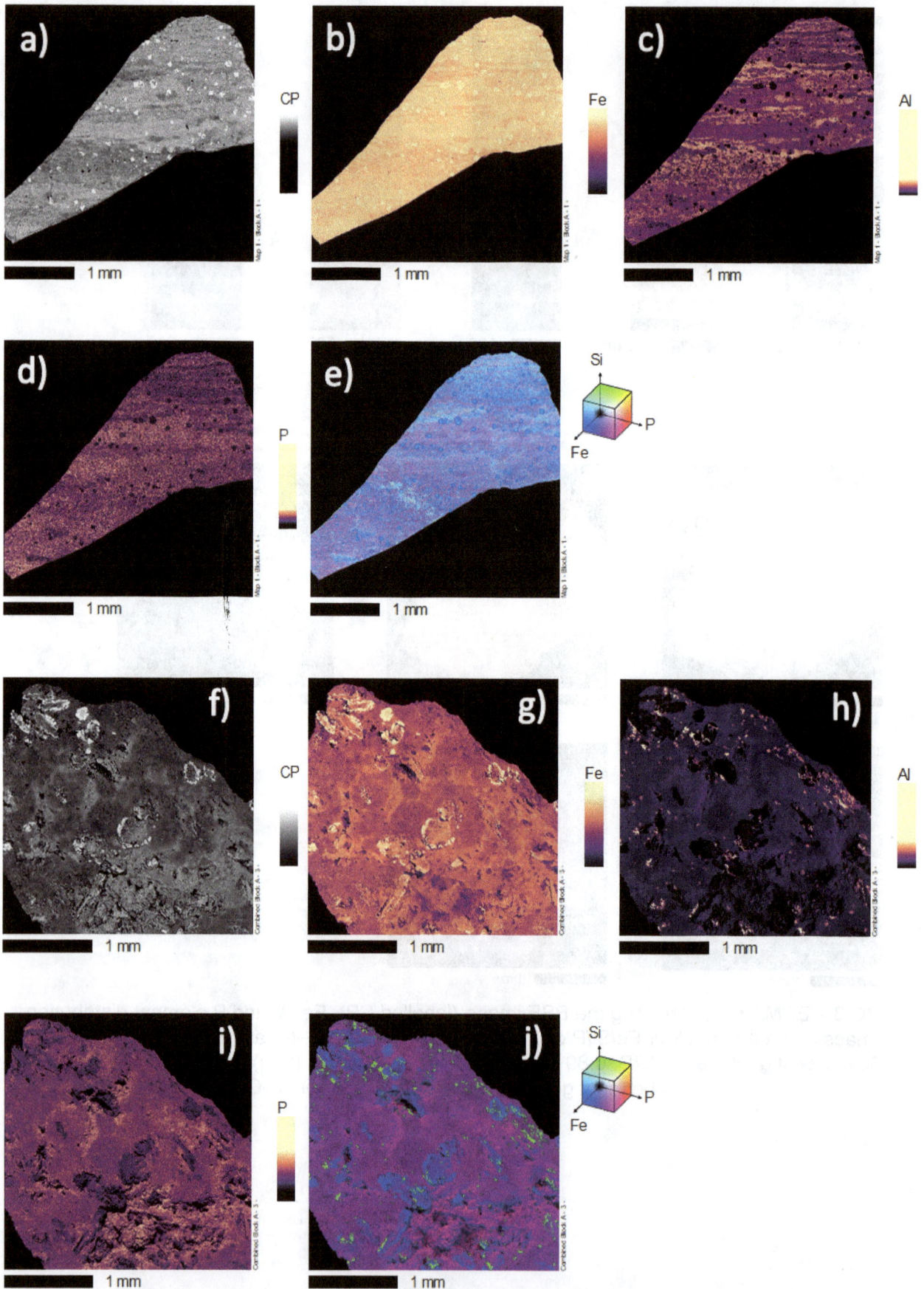

FIG 4 – EPMA maps showing the BSE image (labelled CP), Fe, Al and P element distribution maps and a three-colour Fe/Si/P distribution map. Images (a)–(e) are for Particle 1 (Gp10–33 per cent goethite by XRD, 0.45 per cent P_2O_5); and images (f)–(j) are for Particle 3 (Gp9–91 per cent goethite by XRD, 1.17 per cent P_2O_5).

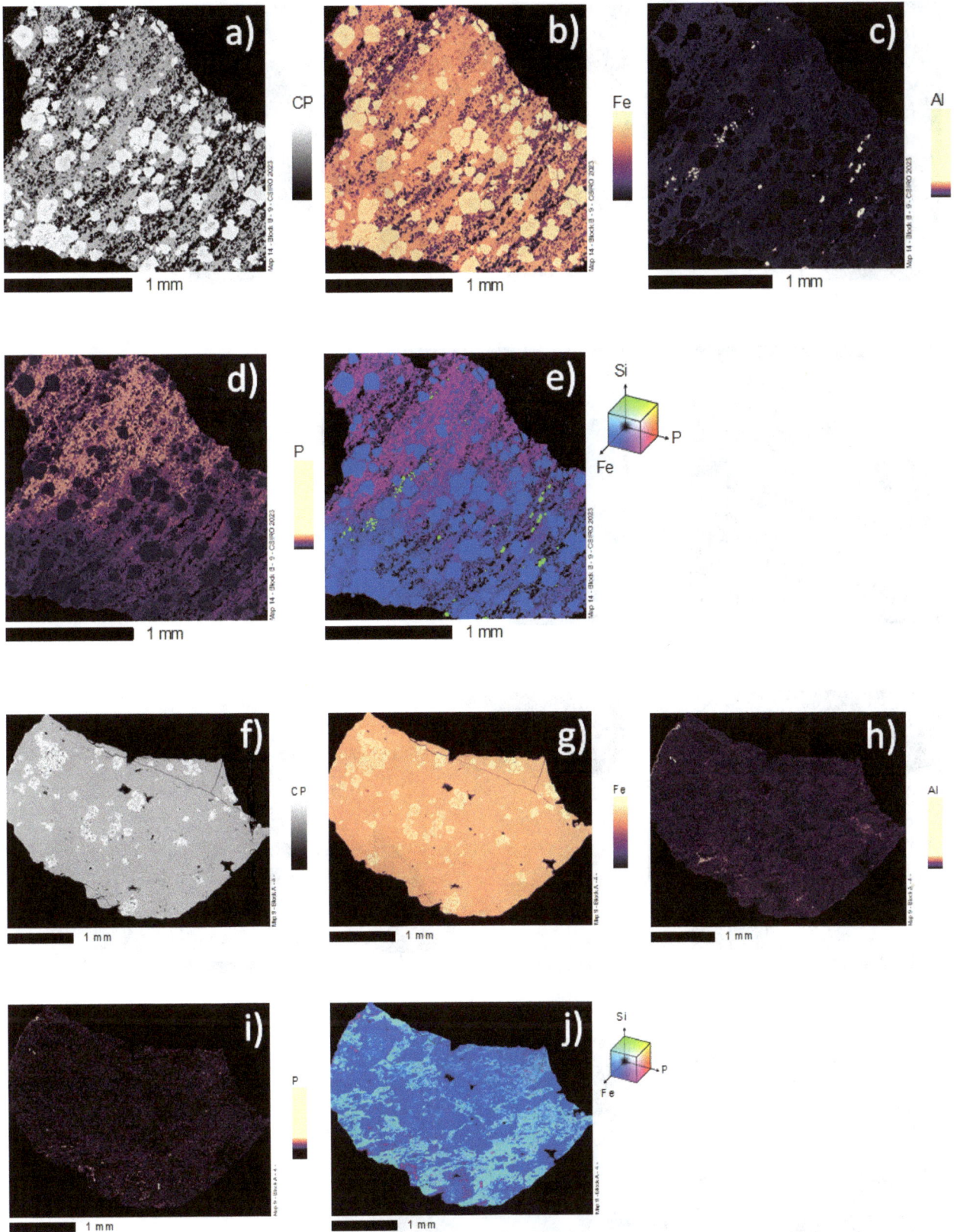

FIG 5 – EPMA maps showing the BSE image (labelled CP), Fe, Al and P element distribution maps and a three-colour Fe/Si/P distribution map. Images (a)–(e) are for Particle 9 (Gp4–61 per cent goethite by XRD, 1.19 per cent P_2O_5); and images (f)–(j) are for Particle 4 (Gp7–80 per cent goethite by XRD, 0.21 per cent P_2O_5).

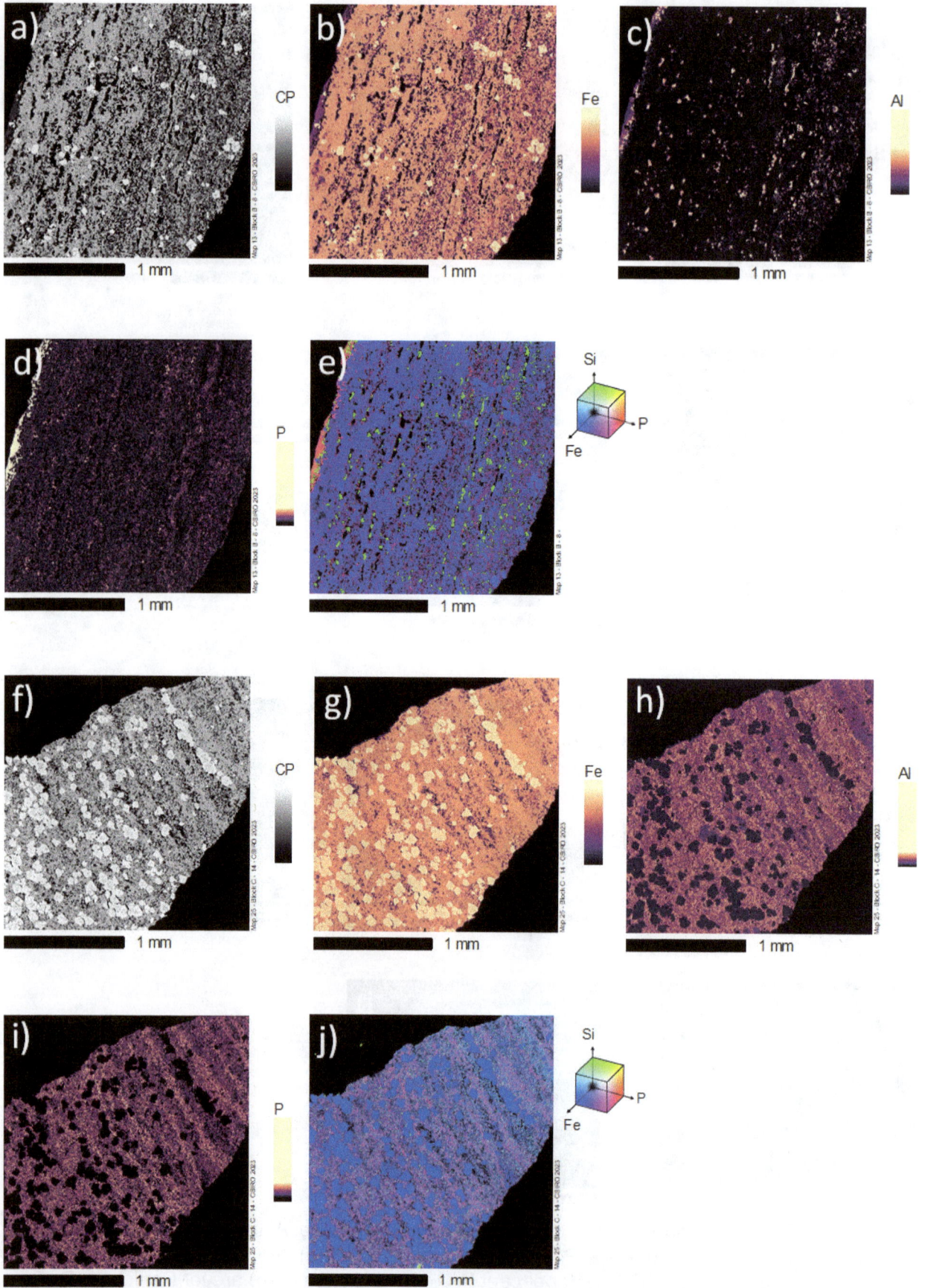

FIG 6 – EPMA maps showing the BSE image (labelled CP), Fe, Al and P element distribution maps and a three-colour Fe/Si/P distribution map. Images (a)–(e) are for Particle 8 (Gp4–83 per cent goethite by XRD, 0.89 per cent P_2O_5); and images (f)–(j) are for Particle 14 (Gp7–51 per cent goethite by XRD, 0.98 per cent P_2O_5).

Particle 2 (Group 10, 85 per cent goethite, 0.89 per cent P$_2$O$_5$)

Minor finely disseminated martite grains within a uniformly microporous goethite matrix of intermediate porosity (Figure 3a–3e). There is minor concentration of phosphorus in a band close to the top of the section, but otherwise there is a uniform distribution of phosphorus throughout the particle, at a relatively low level. Silicon is correlated with Al and is also distributed throughout the section, but as small, discrete inclusions within pore spaces (suggesting remnant particulate silica/aluminosilicate gangue).

Particle 7 (Group 7, 76 per cent goethite, 0.98 per cent P$_2$O$_5$)

The particle is characterised by disseminated martite grains within a dense goethite matrix, with a low level of microporosity and occasional larger, discrete pores (Figure 3f–3j). There appears to be a solution channel in the lower part of the section, that cross-cuts the NW-SE trending structure evident in the composite map, where there appears to be some precipitation of hydrohematite (nominally Fe$_2$O$_3$·nH$_2$O). The phosphorus map shows correspondence of a higher phosphorus level with the microporous areas within the section, with a lower level within denser areas and no phosphorus appeared associated with martite grains. The high-P outlines appear to reflect interlocking equigranular grains, possibly representing original (P-bearing) BIF gangue minerals, replaced by goethite. A few angular fragments with elevated Si level are evident within the goethite matrix and these are generally (but not exclusively) low in phosphorus.

Particle 1 (Group 10, 33 per cent goethite, 0.45 per cent P$_2$O$_5$)

Disseminated fine martite grains within a matrix of dense to finely microporous goethite with relatively well-defined boundaries between these sub-types and with remnant fine primary layering (horizontal) evident within the goethite (Figure 4a–4e). The phosphorus concentration broadly reflects the goethite porosity (ie phosphorus was at a higher level where porosity was lower) but appears to delineate detail of the original primary layering at the top of the particle. The Al map shows similar features and shows that the lower BSE reflectivity goethite phase (Figure 4a) has a higher Al level.

Particle 3 (Group 9, 91 per cent goethite, 1.17 per cent P$_2$O$_5$)

This particle appears to be comprised of a finely microporous (high porosity) goethite matrix, with some slightly denser areas (higher reflectivity) and apparent martite grain outlines, that have been almost entirely leached and replaced with goethite internally (Figure 4f–4j). Fine disseminated Al-silicate particles occur both within the matrix and within replaced martite grains. It appears that phosphorus is generally associated with goethite (as is Al, at apparently relatively low concentration), but both P and Al are largely absent from the interior of replaced martite grains. Silicon is correlated with Al and was localised in apparent remnant discrete gangue aluminosilicate grains within pore spaces.

Particle 9 (Group 4, 61 per cent goethite, 1.19 per cent P$_2$O$_5$)

The particle consists of disseminated martite grains with primary goethite defined layering and intermediate to high porosity (Figure 5a–5e). Phosphorus is clearly concentrated in the upper part of the section/image, apparently due either to leaching or ingress along porosity, although with no apparent clear change in texture/mineralogy at the (irregular) margin of this zone. There was some association with denser goethite layers. Little leaching of martite grains was apparent. Silicon correlated with Al and was localised in apparent remnant discrete gangue aluminosilicate grains within pore spaces.

Particle 4 (Group 7, 80 per cent goethite, 0.21 per cent P$_2$O$_5$)

The particle shows disseminated martite grains/clusters within a dense vitreous goethite matrix with isolated discrete pores (Figure 5f–5j). The phosphorus map shows a low concentration of phosphorus throughout most of the particle, with minor localised concentrations at micron scale; the Al distribution was similar. The combined three element map indicates the presence of two distinct dense goethite phases, with minimal variation in Fe content (ie high Si, minor Al, slightly lower Fe) and BSE brightness. A slightly higher level of phosphorus was associated with the Si-bearing goethite, mostly in the lower part of the section. Minor 'point' concentrations of phosphorus were not associated with Si and may represent remnant gangue grains, replaced by low-Si goethite (Figure 3j), except at the left-hand edge of the particle, where P, Al and Si were clearly associated. The low-Si goethite appears to be the later phase (surrounds pores, where partial infilling of void space has occurred). The two goethite phases may also represent different original BIF gangue phases (ie Si-bearing from silicates; low-Si possibly carbonate). This may suggest original P-bearing carbonate, or phosphorus retention by carbonate during early stages of leaching and during goethite replacement.

Particle 8 (Group 4, 83 per cent goethite, 0.89 per cent P_2O_5)

A moderately to highly microporous goethite matrix with disseminated martite grains, showing evidence of near-vertical primary layering (Figure 6a–6e). Phosphorus was at a slightly higher level in the more microporous layers towards the right of the section. A thin rim of P-rich particulate material was present on the left-hand edge of the particle, which appeared to be a very fine-grained mixture of ochreous goethite and kaolinite (Al-bearing). Si was correlated with Al and was locally concentrated within pores, apparently representing discrete remnant gangue particles (or kaolinite after Al-silicates).

Particle 14 (Group 7, 51 per cent goethite, 0.98 per cent P_2O_5)

In this particle, narrow, discontinuous bands of martite were interspersed between thin layers of finely microporous goethite, defining primary layering (Figure 6f–6j). The phosphorus and Al distribution was uniform, reflecting the internal porosity of the individual goethite layers.

Summary of EPMA mapping results

Phosphorus was consistently associated with goethite in both dense and porous textures, but in some cases at variable levels, even where little or no difference in goethite type/texture was evident, for example, Particle 7 (Figure 5f–5j). In this case, minimal difference was seen in the goethite reflectivity or texture (Figure 5f), but three distinct goethite compositions were evident, based on phosphorus content. In other cases, the phosphorus level was correlated with goethite porosity (typically higher where porosity was lower, eg Particles 1, 14), whereas the opposite was true in one case (Particle 8). There was a third group, where the phosphorus distribution varied independently from goethite type/porosity (notably Particle 7, but also Particles 2, 9). This variation confirms that the phosphorus level is not determined by goethite type/texture alone.

Some particles, for example Figure 5f–5j (Particle 7), showed a correlation between the Al and P distributions, in other cases with a separate association of Al and Si in discrete, disseminated, particulate gangue (eg Figure 3a–3e). The martite grains in the mapped particles had, in almost all cases, not undergone significant leaching and in all examples, phosphorus was not detected in martite, although it was present in goethite replacing leached martite (eg Figure 4j, Particle 3). Minor Si was often associated with martite.

In most mapped sections, the Si distribution did not correlate well with phosphorus distribution, although the relationship was not consistent throughout. In some cases (eg Particles 9, 3 and 2), Si was concentrated as disseminated, discrete 'particles', apparently representing remnant aluminosilicate gangue derived from the original BIF, often occupying pores within goethite.

Particles 9 and 3 both had high P contents (>1 per cent), but Particle 9 was also high in Al and Si, whereas Particle 3 was not. Conversely, particle 2 (Group 10b) was also goethite dominated (5 per cent hematite) and had relatively high ferrihydrite content (22.1 per cent), with high Al and Si and high P. The difference was, however, that Particle 2 had a higher kaolin content (10 per cent versus 3 per cent). Particle 12–5a, however (shale group), had high kaolin (23 per cent), but no goethite and no ferrihydrite so this appears to rule out a consistent hematite or goethite association with ferrihydrite.

Ferrihydrite was clearly associated with both dense brown goethite and microporous ochreous goethite, so there was no clear association with goethite type. Similarly, the highest values for ferrihydrite tended to occur in the samples with the highest goethite and/or Al/Si content, but there were also clear exceptions here. The only detected occurrences of feroxyhyte were in particles (eg Particles 5–5a and 11-sh) with low low/zero detected goethite content (XAS), although in these cases there was poor agreement between the hematite and goethite contents determined by XRD and those determined by XAS.

What was clear was that phosphorus occurs across different goethite types and textures. Phosphorus was always associated with goethite (often with Al and/or minor Si) and generally not where there was Si at higher level). These results show a tendency towards higher phosphorus content with higher goethite content, but not consistently so. This tends to support a 'genetic' origin, ie influenced by upgrading history.

In all particles, Al and Si were associated at 1:1, suggesting the presence of kaolinite and indicating the presence of remnant gangue. This demonstrates the need for closer examination of textures, partly to try and determine the occurrence of ferrihydrite/feroxyhyte. Based on these results, it seems likely that these phases are favoured by high P and/or Al and Si (ie suppressed H/G crystallisation), although there is not enough evidence to convincingly support this. In several cases the proportion of goethite detected by XRD is much lower than detected by synchrotron, so it is tempting to speculate that there may be a relationship between low totals of crystalline material detected by XRD (ie the presence of poorly crystalline goethite and possibly a high XRD amorphous content?) and high ferrihydrite/goethite from XAS. However, there are exceptions to this trend, suggesting that any relationship is complex. Nevertheless, we contend that the different results generated by the various techniques are clearly demonstrating something pertinent to phosphorus deportment in goethitic Fe ores and are thus significant when considering P-removal options.

CONCLUSIONS AND WHERE TO FROM HERE?

The approach of examining separate ore type particles provides a reference frame to systematically investigate the relationship between goethite and phosphorus and the combination of techniques used has provided important insights regarding the link between phosphorus distribution and mineralogy in Pilbara type ores. In the current study only a limited number of iron ore particle types were examined, however, they revealed important information regarding the deportment of impurities such as P, Al and Si. It is apparent there is a *general* relationship between goethite, Al and P, although this is not always consistent eg several of the non-goethite-dominant samples have low Al/Si, but high P, so P, Al and Si can vary independently. Al and Si are typically around 1:1 ratio in the samples (typical for Pilbara ores, *cf* kaolinite), but the EPMA maps show that Si is often not well correlated with P, which supports the general lack of correlation between silicate/Al-silicate gangue and phosphorus.

To resolve the question whether phosphorus is more likely to be associated with poorly crystalline goethite or ferrihydrite/feroxyhyte precursors there is also a need to correlate physical evidence for the presence of these phases and or poorly crystalline goethite with appearance under the microscope, as well as further investigate the correlations between high proportions of amorphous poorly crystalline phases with phosphorus occurrence. There is evidence in the literature (Hsu *et al*, 2020; Cornell and Schwertmann, 1996) for similar effects with other impurities (eg changes in goethite crystallite size and/or morphology with increasing alumina level). Some of the effects noted to potentially be associated with phosphorus may be enhanced or complicated by presence of Al in goethite in 'real' ore, so may also provide further insight into association of phosphorus and Al in Pilbara ores and whether this points to mechanism/s of incorporation of phosphorus in Fe-oxyhydroxides.

We recommend that the study be extended to sort and examine more samples within each ore type, so trends become more apparent. This would also provide more data for examining any trends with mineralogy and bulk chemistry.

With respect to beneficiation of high-P ores it is important to generate enough material for each ore group to enable detailed characterisation and test work. It is important to provide representative material that can be used to identify and characterise amorphous/poorly crystalline phases in ores, especially if these turn out to be associated with high-P, to target upgrading. It is likely that this type of phase will be fine and particulate (no large crystals), rather than dense in character and may be amenable to physical beneficiation. If this is the case, then the identification of 'indicator' textures/associations may be more useful as a practical guide to phosphorus distribution within high-P ores.

Assuming separation of the high-P ore components becomes possible we also recommend that targeted research should be applied to examine:

- Whether one or more ore groups are more susceptible to leaching of impurities eg if it could be demonstrated that that the goethite-rich microporous ore types are easily leached this material could be treated separately via say a heap leach process to remove P.

- The effects of heating on phosphorus deportment. Previous work in this area has been inconclusive as to the fate of phosphorus after heating and a larger amount of material for research would be useful.

ACKNOWLEDGEMENTS

We wish to acknowledge our CSIRO colleagues Shu Huang and Nathan Webster (XRD preparation and characterisation) and Cameron Davidson and Nick Wilson (EPMA preparation and mapping analysis). The assistance rendered by the staff of ANSTO at the Australian Synchrotron is also acknowledged, with special recognition given to Dr. Jessica Hamilton, the XAS beamline scientist who contributed significantly to the data collection process.

REFERENCES

Calvin, S, 2013. *XAFS for Everyone.* CRC press, doi:10.1201/b14843.

Carlson, L and Schwertmann, U, 1980. A natural occurrence of feroxyhyte (δ-FeOOH), *Clays Clay, Miner,* 28:272–280.

Cheng, C Y, Misra, V N, Clough, J and Muni, R, 1999. Dephosphorisation of Western Australian iron ore by Hydrometallurgical process, *Minerals Engineering.* 12:1083–1092.

Clout, J M F, 2003. Upgrading processes in BIF-derived iron ore deposits: implications for ore genesis and downstream mineral processing, *Applied Earth Science,* 112:89–95.

Cornell, R M and Schwertmann, U, 1996. *The Iron Oxides,* 571 p (Weinheim: VCH Verlagsgesellschaft).

Dukino, R D, England, B M and Kneeshaw, M, 2000. Phosphorus distribution in BIF-derived iron ores of Hamersley Province, Western Australia, *Transactions of the Institutions of Mining and Metallurgy: Section B,* 109(3):168–176.

Graham, J, 1973. Phosphorus in iron ore from the Hamersley Iron Formations, in *Proceedings of the AusIMM,* 246:41–42.

Hsu, L-C, Tzou, Y-M, Ho, M-S, Sivakumar, C, Cho, Y-L, Li, W-H, Chiang, P-N, Teah, H Y and Liu, Y-T, 2000. Preferential phosphate sorption and Al substitution on goethite, *Environmental Science: Nano,* 7:3497–3508.

Manuel, J R and Clout, J M F, 2017. Goethite classification, distribution and properties with reference to Australian iron deposits, in *Proceedings Iron Ore 2017;* pp 567–574 (The Australasian Institute of Mining and Metallurgy: Melbourne).

Ofoegbu, S U, 2019. Characterization studies on Agbaja iron ore: a high-phosphorus content ore, *SN Applied Sciences,* 1:204.

Paige, C R, Snodgrass, W J and Nicholson, R V, 1997. The effect of phosphate on the transformation of ferrihydrite into crystalline products in alkaline media, *Water, Air and Soil Pollution,* 97:397–412.

Pownceby, M I, Hapugoda, S, Manuel, J, Webster, N A S and MacRae, C M, 2019. Characterisation of phosphorus and other impurities in goethite-rich iron ores – possible P incorporation mechanisms, *Minerals Engineering,* 143(11):106022.

Ravel, B and Newville, M, 2005. Athena, Artemis, Hephaestus: Data analysis for X-ray absorption spectroscopy using IFEFFIT, *J Synchrotron Radiation,* 12:537–541.

Thorne, W, Hagemann, S, Webb, A and Clout, J, 2008. Banded iron formation-related iron ore deposits of the Hamersley Province, Western Australia, *SEG Reviews,* 15:197–221.

Torpy, A, Wilson, N, MacRae, C, Pownceby, M, Biswas, P, Rahman, M A and Zaman, M N, 2020. Deciphering the complex mineralogy of river sand deposits through clustering and quantification of hyperspectral X-ray maps, *Microscopy and Microanalysis,* 26(4):768–792.

Vodyanitskii, Yu N, 2010. Iron hydroxides in soils; a review of publications, *Eurasian Soil Science,* 43(11):1244–1254.

Wilson, N, MacRae, C and Torpy, A, 2008. Analysis of combined multi-signal hyperspectral datasets using a clustering algorithm and visualisation tools, *Microscopy and Microanalysis,* 14(2):764–765.

Ore characterisation and geometallurgy – new applications

Effect of calcium oxalate CaC$_2$O$_4$.H$_2$O on the Reduction and phosphorus partition of a high-P Brockman ore

S Hapugoda[1], L Lu[2] and H Han[3]

1. Senior Experimental Scientist, CSIRO, Brisbane Qld 4069. Email: sarath.hapugoda@csiro.au
2. Principal Research Scientist, CSIRO, Brisbane Qld 4069. Email: liming.lu@csiro.au
3. Research Officer, CSIRO, Brisbane Qld 4069. Email: hongliang.han@csiro.au

ABSTRACT

Charcoal produced from *Syzygium Zeylanicum* trees was widely used in ancient Sri Lankan iron smelting practices. Charcoal made from *Syzygium Zeylanicum* tree trunks, branches and leaves were analysed by optical mineralogy and SEM and found to contain unusually high concentrations of calcium oxalate (CaC$_2$O$_4$.H$_2$O – CaOx) crystals in plant cells. Partial and full reduction tests of a high-P Brockman ore were conducted at several temperatures with coke and coke doped with CaOx and CaCO$_3$. It was observed that CaOx promoted reduction and generated comparatively more metallic iron in reduced samples. Furthermore, CaOx encouraged phosphorous partition into slag phases than into the metallic iron, suggesting CaOx can a potential alternative fluxing material to calcium carbonate for treating high-P iron ores in ironmaking processes.

INTRODUCTION

Evidence for the earliest production of high carbon steel in South Asia is found in the Samanalawewa area of Sri Lanka where thousands of archaeological sites have been found (Seneviratne, 1985; Karunatilaka, 1991; Juleff, 1996, 1998). Wootz steel was produced in India and Sri Lanka in furnaces powered by monsoon winds and artefacts have been dated to 300 BC using radiocarbon dating techniques. In one of the most recent archaeological discoveries in Lanka, ironmaking furnaces along with samples of steel, dating to 200 BC, were discovered from a village in Hingurakgoda in north-west Sri Lanka. The discovery of these artefacts supported previous discoveries made in Anuradapura, Sigiriya, Ala Kola Weva, Kuratiyaya and Nikavatana (all locations in North-western Province, Sri Lanka) in support of ancient Syrian records that 'Sivhala' (Lanka), once was home to the world's best steelmaking technology. Recent studies have suggested that the ancient Lankan furnaces may have provided the source material for the legendary Damascus swords following export of steel to the Middle East. Coomaraswamy (1908) believed that iron slag heaps discovered in nearly every district of Ceylon (Sri Lanka) showed the vastness of the iron smelting industry in ancient times and may have extended to earlier times. For example, studies by Karunaratne and Adikari (1994) have shown that the earliest known age for iron smelting, based on C14 dating of iron slag found in a proto-historic context in an archaeological site from Aligala in Sigiriya, to be ca. 998–848 BC.

It has been reported that the *Syzygium Zeylanicum* (https://commons.wikimedia.org/wiki/Category:Syzygium_zeylanicum; https://indiabiodiversity.org/species/show/231269) tree species had been preferentially exploited for charcoal from the 3rd century BC onwards, ie Coomaraswamy (1908), De Silva (2011) and Solangaarachchi (1999, 2011). The local (Sinhalese) name for *Syzygium Zeylanicum* is 'Yakada Maran' which is translated as 'Iron Killer' plant. Although the use of charcoal from *Syzygium Zeylanicum* is described by many (ie Ashton *et al*, 1997), the actual reason for its use has not been investigated. Anoop and Bindu (2014) conducted a pharmacognostic and physico-chemical study on leaves of *Syzygium Zeylanicum* and reported the presence of calcium oxalate crystals in cell structures while Franceschi, Paul and Nakata (2005) described calcium oxalate formation in plants and that the Ca oxalate crystals formed specific shapes and sizes.

Hapugoda *et al* (2021) provided a detailed examination of charcoal produced from *Syzygium Zeylanicum* tree stem, branches, and leaves, and characterise slag iron from ancient ironmaking sites in Sri Lanka. Given that high quality steel was produced from Sri Lanka, it was hypothesised that the presence of calcium oxalate may be a key factor in iron production. From our observations we then conduct our own experimental studies into the use of calcium oxalate as a dopant during iron ore reduction. The effect of adding calcium oxalate on the partition of phosphorus, a key iron ore impurity element, was evaluated.

CHARACTERISATION OF CHARCOAL MADE FROM SYZYGIUM ZEYLANICUM

Charcoal from *Syzygium Zeylanicum* tree stems were prepared by burning dried pieces of each material above 100°C in a large steel barrel, covered with a lid. The charcoal produced from burning the tree stem and branches was typically observed to be solid pieces, Hapugoda *et al* (2021).

The charcoal (both from tree stems and branches) was set in epoxy resin, polished, and observed under the optical microscope, SEM (Scanning Electron Microscope) and EPMA (Electron Probe Micro Analyser). High magnification SEM backscattered electron (BSE) image (Figure 1a) and optical micrograph (Figure 1b) showed that calcium oxalate (CaOx) crystals developed excessively within cell structures parallel to the tree stems or branches and was present in almost every particle examined. The EPMA spot analysis results on individual calcium oxalate crystals showed a CaO content of between 45–57 wt per cent (average 47 wt per cent, with a std dev of 3.5 wt per cent from 20 spot analysis). Several SEM EDS spectra were obtained on Ca oxalate crystals – these confirmed that the crystals contained Calcium (Ca), Carbon (C) and Oxygen (O) only and were consistent with the material being CaC_2O_4.

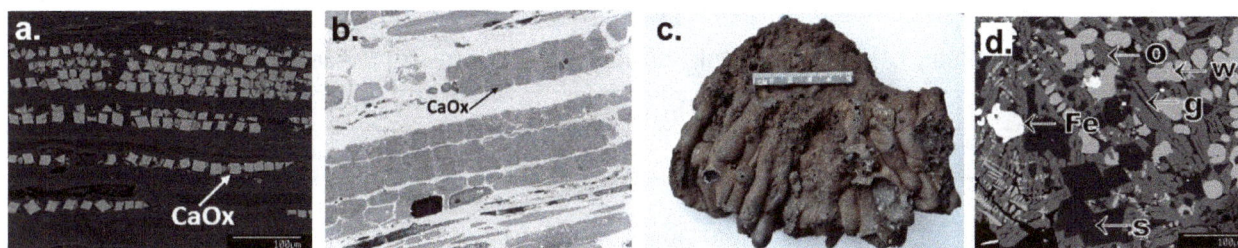

FIG 1 – (a) SEM backscattered electron image, (b) Optical micrograph showing high calcium oxalate (CaOx) crystal development in *Syzygium Zeylanicum*, (c) image of slag iron sample, (d) BSE image showing slag iron mineralogy.

CHARACTERISATION OF SLAG IRON SAMPLES – SLAG IRON CHEMISTRY AND MINERALOGY

Slag iron can be found in almost every district in Sri Lanka usually near famous archaeological sites, ancient temples, or ancient ruins. Several large iron slag samples were collected from three locations closer to archaeological sites at Madawala Ulpatha (Central Province GPS: X-186673.44, Y-264757.83, Lat: 7.61741, Long: 80.67118, Panwila (Central Province, GPS: X-194538.227 Y-246328.045, Lat: 7.36486, Long: 80.70917), Belihul Oya (Sabaragamuwa Province, Lat: 6.70494, Long: 80.78740). Samples were broken into several pieces and representative pieces from one such sample are shown in Figure 1c. These types of slag iron samples are common, well-preserved, and still fresh and very hard. In some cases, pores were filled with clay/soil with quartz (SiO_2) particles. Some samples were massive/hard with smaller pores, less soil attached and displaying a brownish yellow stain due to the presence of secondary goethite-Fe $(OH)_3$. Further, some samples displayed a coral-like shape with several tubular structures or rope-like structure, with larger pores within and with some soil attached to the sample surface. Majority of samples were dense and with few larger circular pores and had some goethite staining on the surface.

Table 1 contains average chemical analysis (XRF) data for several representative slag iron samples. Care was taken to remove as much soil and quartz particles attached to the samples before analysis. The chemical analysis data indicates that the samples had total Fe contents of between 45.0–51.0 wt per cent, SiO_2 between 18.7–21.6 wt per cent, Al_2O_3 between 7.3–8.5 wt per cent and CaO ranged from 0.75–2.45 wt per cent. As expected, the SiO_2 content was high in the slag and CaO was low indicating most likely low-basicity iron smelting conditions. The calcium (CaO) contents are low, and silica (SiO_2) contents are high in supergene goethite iron ores most likely used in iron smelting (Ranasinghe, 1986). Calcium for the smelting processes may have been added from widely available dolomitic crystalline limestone (marble–limestone) deposits in the island. The phosphorus content in the slag was significant and close to 0.4 per cent wt per cent (P_2O_5 = 1.8 wt per cent) and MgO (0.74–0.92 wt per cent) and TiO_2 (0.47–0.52 wt per cent) were also low. Potassium, K_2O (0.64–0.78 wt per cent) in the slag had most likely come from minor amounts of clay/soil attached to the samples. Manganese (Mn) is also significantly high in samples. Several slag iron samples were qualitatively analysed by XRD (X-ray diffraction analysis for the mineralogy and the major and minor minerals identified in the sample were: fayalite (Fe_2SiO_4), hercynite ($FeAl_2O_4$), wüstite (FeO), quartz (SiO_2), magnetite (Fe_3O_4), maghemite (Fe_2O_3), hematite (Fe_2O_3) and cohenite (Fe_3C). Several representative slag iron samples were mounted in 25 mm round polished blocks. Despite the slag samples being sourced from various locations, the textures and mineralogy of the slag samples were alike with typical textures characterised by a network of wüstite formed within a matrix of fayalite, spinel, and glass. The major and minor mineral phases identified in all the studied slag iron samples were: wüstite (FeO), olivine-fayalite ($FeSiO_4$), spinel (hercynite-$FeAl_2O_4$), metallic iron (Fe) and a glass phase. Under the optical microscope wüstite, hercynite, fayalite, metallic iron, glass, hematite, goethite, minor quartz (embedded) and some flux particles were recognised. Within two iron slag samples relict charcoal particles (most cell structures are replaced with goethite) were identified. In SEM BSE image shown (Figure 1d), predominantly wüstite in a fayalite/spinel (hercynite)/minor glass matrix can be seen. Most of the large spinel crystals displayed zoning and some spinel particles contained wüstite inclusions. The edges of some of the large spinel particles were recognised as being converted into wüstite. In the SEM backscattered electron image shown in Figure 1d, metallic Fe particles and wüstite can be seen in a fayalite/spinel (hercynite/glass) matrix.

TABLE 1

Slag Iron bulk chemistry (XRF) and Metal-Fe/Slag-glass chemistry (based on EPMA analysis).

Sample	Fe	SiO_2	Al_2O_3	TiO_2	Mn	CaO	P	S	MgO	K_2O
Slag iron (bulk)	45.0-55.0	18.0-22.0	7.0-9.0	0.3-0.5	0.7-1.5	1.0-3.0	0.4-0.6	0.02-0.04	0.6-0.8	0.6-0.8
phase	Fe	Si	Al	Ti	Mn	Ca	P	O	Mg	Total
Metal-Fe	100.63 (0.26)	0.01 (0.03)	0.02 (0.08)	0.01 (0.01)	0.25 (0.29)	0 (0)	0.01 (0.02)	0.27 (0.18)	0 (0.01)	101.21 (0.55)
Slag (glass)	19.23 (1.25)	16.21 (1.66)	9.07 (0.79)	0.49 (0.33)	1.41 (0.63)	8.19 (1.55)	1.7 (0.34)	39.24 (1.31)	0.02 (0.01)	95.56 (1.7)

The most significant observation from EPMA spot analysis was that phosphorus was mainly concentrated within the glass phase (Table 1, Figure 2) with an average content of 1.7 wt per cent phosphorus) and the phosphorus contents were significantly low in all the other phases (Table 1, Figure 2) specially in metallic iron. The EPMA data strongly suggest that in the Sri Lankan iron slag samples, phosphorus has been preferentially partitioned into the slag phase instead of reporting to the metal (Figure 2). This observation suggests that possibly, the presence of the calcium oxalate derived from the charcoal has a positive effect during ironmaking in that it leads to lower phosphorus levels in the hot metal. This observation was therefore explored further through preliminary reduction experiments whereby the high phosphorus iron ore was fluxed with calcium oxalate and calcium carbonate as comparison. The experimental procedure and results are detailed in the following section.

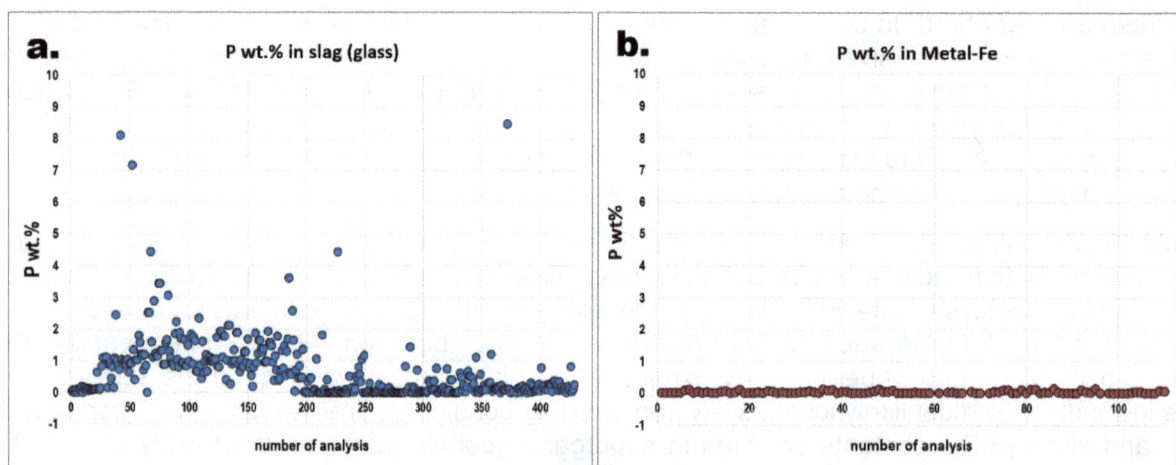

FIG 2 – EPMA analysis of Phosphorous in ancient slag iron samples, showing Phosphorous partition into slag phase (a) phosphorous in slag (glass); and (b) Phosphorous contents in metallic iron.

ORE CHEMISTRY AND MINERALOGY OF HIGH-P BROCKMAN ORE

Phosphorus is one of the most deleterious elements in iron ore as it follows iron during downstream reduction processes forming iron phosphides that make steel brittle. In this study a petrological examination (optical mineralogy ore characterisation) of a representative sample of West Australian Brockman-high phosphorous ores was conducted, which were described by Morris (1973), Graham (1973), Peixoto (1991), Dukino and England (1997), Dukino, England and Kneeshaw (2000), Clout, (2005), Thorne *et al* (2008) and MacRae *et al* (2010). The high phosphorus ore had iron (Fe) contents between 59.0 to 62.2 wt per cent and high phosphorous contents ranging from 0.132 to 0.165 wt per cent. Figure 3 displays the ore group abundance derived from optical point counting of two main varieties of Brockman ore (high phosphorus and low phosphorus). The Brockman High-Phosphorous ore samples are texturally variable, (Ostwald, 1981; Morris, 1985, 2002; Clout, (2005),Thorne *et al*, 2008) with most ore group categories returning notable point count totals (Figure 3).

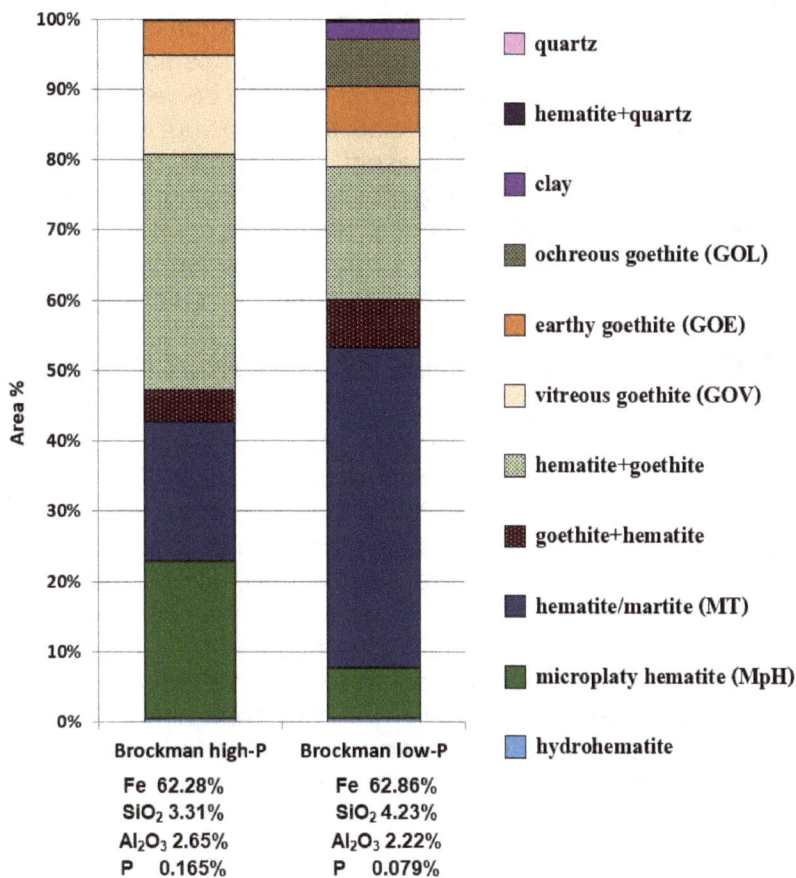

FIG 3 – Ore group abundance in Brockman high-phosphorus/low-phosphorus type iron ores.

Figure 3 displays the optical point counting data of ore groups within Brockman high-phosphorus and Brockman low-phosphorus type ores. The high-phosphorus ores tend to be dominated by goethite-rich ore groups. Hematite-goethite particle types were the most abundant and were largely present as moderately hard to microporous and friable ore groups. Hematite was more commonly evident as microporous martite, but also often observed as recrystallised hematite after martite, or as microplaty hematite. Haematitic ore groups comprised 40 per cent of the ore (Figure 3) with moderately hard particle types being the most abundant overall. Relict kenomagnetite was evident in some haematitic and hematite-goethite ore group particles. Goethite ore groups comprised 20 per cent of the ore. Hematite-goethite particle types comprised 38 per cent and Goethite-hematite particle types comprised 5 per cent of the ore. Examination of sized fractions (data not shown) indicated that the ratio of hard to friable particle types increased below 1 mm. Clay particles were minor overall, comprising ~1 per cent of the observed particles and were sometimes evident as ochreous or haematitic shale. Clay was also evident as partial pore infill in some microporous particle types or as thin coatings along particle rims. Some microporous goethite particles exhibited optical properties suggestive of Al and/or Si substitution for Fe in the goethite structure.

Three types of goethite have been identified in the ores (Manuel and Clout, 2017; Mohapatra *et al*, 2008). These were: vitreous goethite (GOV), ochreous goethite (GOL) and earthy goethite (GOE). GOV is macroscopically dark brown, lustrous, has a conchoidal fracture and is found as hard lump to coarse fines. Furthermore, GOV is hard, has low porosity, intermediate specific gravity, and an intermediate loss on ignition (LOI). Microscopical characteristics of GOV are zero to significant visible micro-/macro-porosity, intermediate reflectance and opaque and an amorphous-layered-crystalline morphology. GOL is macroscopically yellow with dull/chalky appearance and generally occur as fines or ultrafine due to its friable nature. It is microporous and has low specific gravity and a high loss on ignition (LOI). Microscopically GOL appears as a very finely microporous 'network', has low reflectance and with characteristic brown/yellow internal reflections. GOL also displays amorphous to uniform 'brushy/fibrous' textures and contains pseudomorphs after silicates/carbonates. GOE is an intermediate goethite type with a mixture of discrete GOV/GOL components and having intermediate porosity/physical properties. The studied Brockman iron ores were dense to moderately microporous and haematitic, with an increasing relative proportion of dense ore types with decreasing size fraction. Very hard dense hematite/martite (MT), dense GOV and moderately porous MT was dominant (increasingly so in the finer size fractions), but a significant amount of fine-textured, microplaty hematite (MpH) was also present. A significant amount of dense hematite/MT with goethite infill was present (~40 per cent), generally decreasing with decreasing size fraction. Goethite only ore types accounted for 20 per cent of the ores, except for a lower abundance in the +2 mm and +4 mm size fractions. Dense GOV steadily increased with decreasing size fraction, whereas microporous GOE and GOL was less abundant and constant in most size fractions.

LOCATION OF PHOSPHOROUS WITHIN HIGH-P BROCKMAN ORE

Based on EPMA data (Figure 4), the phosphorous was mainly located within goethite (GOV 0.01–1.6 wt per cent, avg 0.322 wt per cent, GOE 0.004–1.44 wt per cent, avg 0.214 wt per cent, GOL 0.0–0.92 wt per cent, avg 0.186 wt per cent) with hydro hematite (0.003–0.36 wt per cent, avg 0.098 wt per cent) and clays (kaolinite 0.003–0.261 wt per cent, avg 0.097 wt per cent, gibbsite 0.0–0.351 wt per cent, avg 0.068 wt per cent) containing only minor amounts of phosphorous. Rare apatite grains (P levels up to 18 wt per cent) were associated with microplaty hematite-MpH (0.0–0.083, avg 0.011) and complex REE minerals (P up to 12 wt per cent) within goethite were also found. Keno-magnetite (0.0–0.05, avg 0.02), Martite-MT (0.0–0.371 wt per cent, avg 0.049 wt per cent) and MpH ore types were almost free from phosphorous. Detailed EPMA analysis of various types of fully liberated and combined ore particles consisting of hematite/MT and goethite (GOV\GOE\GOL) indicated that phosphorous can be incorporated in any type of goethite. The highest levels of P in goethite (up to 1.6 wt per cent) were found in some GOV particles however the P content was, in general, variable in particle to particle within various goethite types in the same ore. For example, some GOV particles contained P up to 1.6 wt per cent while some GOV particles from the same ore contained less than 0.2 wt per cent P or were virtually phosphorous free. Phosphorous contents within various parts of an individual goethite particle were also variable. This may be because the high-P goethite particles were closer to a phosphorous releasing source (eg original apatite or REE minerals in the ore) whereas low-P goethite particles tend to be remote from any phosphorous releasing source. Various other forms of goethite were present in minor amounts in the Brockman ores (plain, cellular, platy, layered, wormicular, crystalline) and these were also analysed. There was no direct relationship with phosphorous content between these other goethite types. Rare apatite [$Ca_5(PO_4)_3(Cl/F/OH)$] grains were identified which have been completely replaced by goethite resulting in a high-P goethite. This type of rare high-P goethite particle contained up to 2.3 per cent wt per cent P.

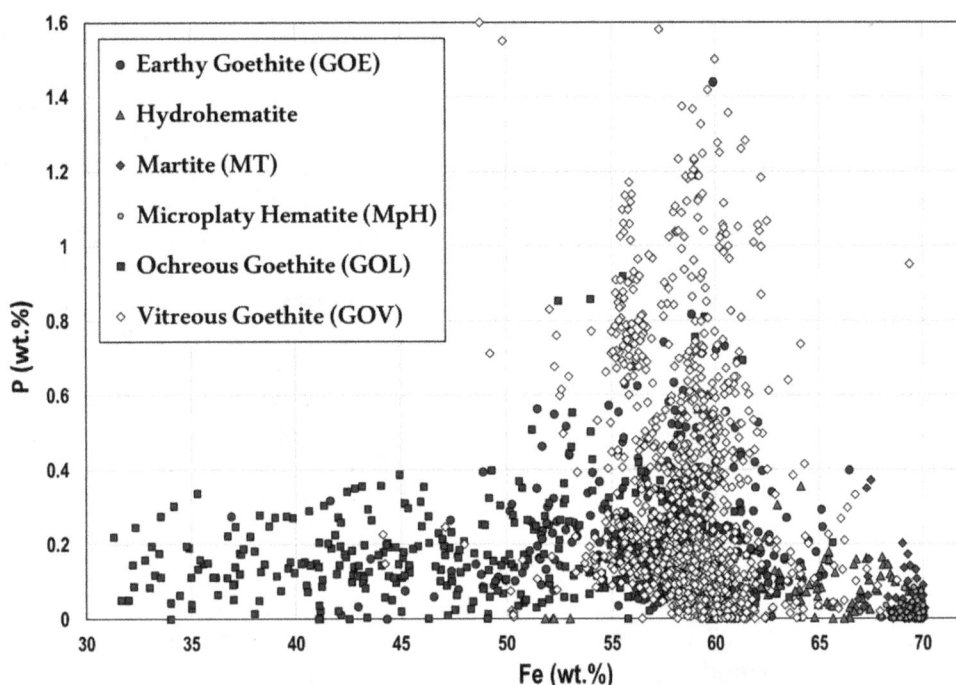

FIG 4 – Phosphorous (wt per cent) versus Fe (wt per cent) content (EPMA data) in various mineral phases from Brockman high-P ores.

PRELIMINARY REDUCIBILITY TESTS OF BROCKMAN HIGH-P IRON ORE WITH CALCIUM OXALATE

Preliminary reduction tests with high phosphorus (P = 0.135 wt per cent) ore sample, Dukino, England and Kneeshaw (2000), which was grinded to -1 mm size, doped with 10 per cent (wt) calcium oxalate monohydrate: $Ca(C_2O_4)$. H_2O were conducted using a microwave heating facility and a CO/CO_2 gas ratio of 40/60. Details regarding the microwave heating apparatus were previously presented in Nunna, Hapugoda and Pownceby (2017) and Nunna *et al* (2021). To examine the effect of the calcium oxalate on the partitioning of phosphorus in the reduced iron ore, we deliberately selected two temperatures as the complete decomposition of calcium oxalate is completed above 800°C according to the thermoanalytical results described by Földvári (2011). Results of heating experiments conducted in air and then in air with the addition of calcium oxalate indicated there was no FeO formed. However, when using the CO/CO_2 reduction atmosphere results, with or without calcium oxalate, showed a significant increase in the amount of FeO generated (up to 60 per cent increase) with the greatest amount occurring after doping with calcium oxalate. Like the FeO content increase observed in the chemical analysis data, the XRD determination of magnetite contents in reduced products confirm the increased effect of doping with calcium oxalate as magnetite content increased by up to 50–65 per cent when using the calcium oxalate with compared to the use of CO/CO_2 without calcium oxalate addition (no flux conditions). The results also indicated that small amounts of feldspar and SFCA (possibly the mineral khesinite) both formed only under the calcium oxalate doping conditions in air and in reducing gas atmospheres. It was observed that there were number of calcium phosphate crystals formed within the glass matrix in the calcium oxalate doped sample. To confirm the identity of (up to 17 wt per cent) high-phosphate (up to 17 wt per cent P) crystals, they were analysed by both SEM EDS and EPMA analysis and results indicated that they were in fact apatite crystals. Magnetite, hematite and SFCA were almost free from any phosphorus and, other than the individual apatite crystals, phosphorus was mainly located within the glass matrix which had significantly high phosphorus contents of up to 5.0 wt per cent phosphorus. There were comparatively very low, or close to zero phosphorus contents noted within SFCA particles.

PARTIAL AND FULL REDUCTION TESTS OF HIGH-P BROCKMAN ORE WITH CALCIUM OXALATE AND CALCIUM CARBONATE

To further test the effect of calcium oxalate on partitioning of phosphorus during ironmaking, complete (using coke) and partial reduction (without coke) tests to produce metallic iron and slag were conducted. Han *et al* (2014) explained a similar approach by completely reducing high phosphorus iron ores doped with $CaCO_3$ and separating high-P containing glass from metallic iron by magnetic separation. There is a possibility of applying the similar technique to remove the metal-Fe from high-P glass from this process (doped with calcium oxalate) or use the calcium oxalate treated high-P ores in ironmaking process directly. There is also a scope that calcium oxalate can be used in pellet making to increase the pellet strength. Su *et al* (2017), investigated the effect of high Na_2O addition for dephosphorisation efficiency of low-basicity converter slag and decrease the consumption of solid CaO. A reference sample of Brockman high-P ore with phosphorus content of 0.135 wt per cent was selected. Compressed pellets from grinded ore (down to -100 micron) were prepared by mixing 10 grams of -100 micron sized high-P ore, thoroughly mixed with either calcium oxalate or $CaCO_3$ (to examine if the effect was simply due to the presence of calcium) (Table 2). Pellet samples (test T1 and test T2) without coke were reduced in Nickel crucibles at 1300°C. Pellet samples: High phosphorus ore fluxed with either calcium carbonate or calcium oxalate and mixed with fine coke (tests T3 and T4) were placed in graphite crucibles and were reduced at 1250°C. Pellet samples: High-P ore fluxed with either calcium carbonate or calcium oxalate (tests T5, T6 and T7) were placed in graphite crucibles and were covered with equal amounts of coke and reduced. All the samples have the basicity of 2.0 except the T7 sample prepared with a higher basicity of 2.5. The difference between test T3/T4 and test T5/T6/T7 is therefore, T3/T4 pellets contained coke within pellets and the T5/T6/T7 pellets were covered with coke (outside). At first the all the test samples were slowly pre-heated in a horizontal tube furnace up to 900°C (100°C/min) and kept for 5 minutes and then the temperature was gradually increased (50°C/min) to relevant reduction temperature for 5 minutes to allow the ore to be reduced to metal-Fe and slag. Samples were then removed from the furnace and water quenched. After quenching the reduced samples were oven dried and mounted in epoxy resin and polished for optical microscopy, SEM and EPMA analysis.

TABLE 2

Test conditions, fluxes used and basicity values for the complete/partial reduction tests with a high-P ore doped with calcium carbonate ($CaCO_3$) or calcium oxalate ($CaC_2O_4.H_2O$).

Test No:	Ore	Flux	Basicity	Fuel	Cruicible	Temperature °C	Metal Fe
T1	Brockman High-P	$CaCO_3$	2.0	no-coke	Nickel	1300	absent
T2	Brockman High-P	$CaC_2O_4.H_2O$	2.0	no-coke	Nickel	1300	absent
T3	Brockman High-P	$CaCO_3$	2.0	coke	graphite	1250	present
T4	Brockman High-P	$CaC_2O_4.H_2O$	2.0	coke	graphite	1250	present
T5	Brockman High-P	$CaCO_3$	2.0	coke	graphite	1300	present
T6	Brockman High-P	$CaC_2O_4.H_2O$	2.0	coke	graphite	1300	present
T7	Brockman High-P	$CaC_2O_4.H_2O$	**2.5**	coke	graphite	1280-1300	present

RESULTS

Optical observations, EPMA and SEM analysis of reduced samples from test T1 and test T2

Optical observations indicated that metallic Fe was absent in both samples from test T1 ($CaCO_3$ fluxed) and test T2 (CaOx fluxed) reduced without using coke and wüstite/magnetite and glass were the main phases identified. It seems that wüstite content was slightly higher in CaOx doped sample with compared to the sample doped with $CaCO_3$. Figures 5 and 6 compares the phosphorous contents in glass phase and within iron oxides (wüstite/magnetite) of the high-P ore samples reduced at 1300°C with $CaCO_3$ and calcium oxalate respectively. It was noticed that in both samples high-phosphorous contents were reported in glass phases and wüstite/magnetite contained significantly low phosphorous levels and the phosphorous partition in glass matrix/Iron oxides for both samples look somewhat similar.

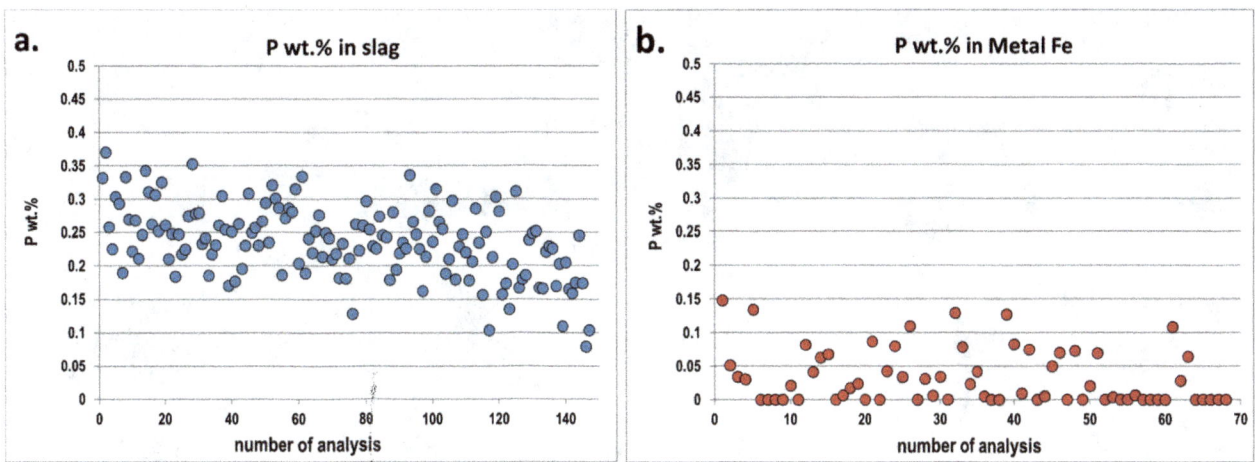

FIG 5 – Phosphorous contents in glass (a) and in wüstite/magnetite (b) in high-P ore sample reduced with $CaCO_3$ without using coke-test T1.

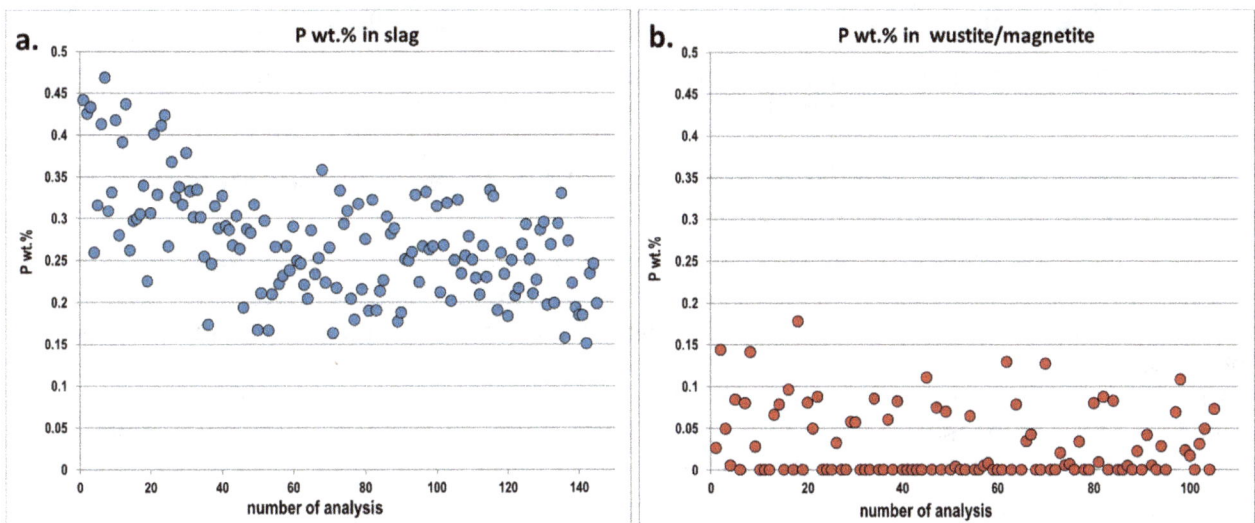

FIG 6 – Phosphorous contents in glass (a) and in wüstite/magnetite (b) in high-P ore sample reduced with CaOx without using coke-test T2.

Figure 7 displays optical micrographs from one of the reduced phosphorus-rich ore samples using coke. In reduced samples from tests T3, T4, T5, T6 and T7 abundant metallic Fe particles were recognised, and slag particles were comprised mainly of glass/silicates with fine wüstite grains.

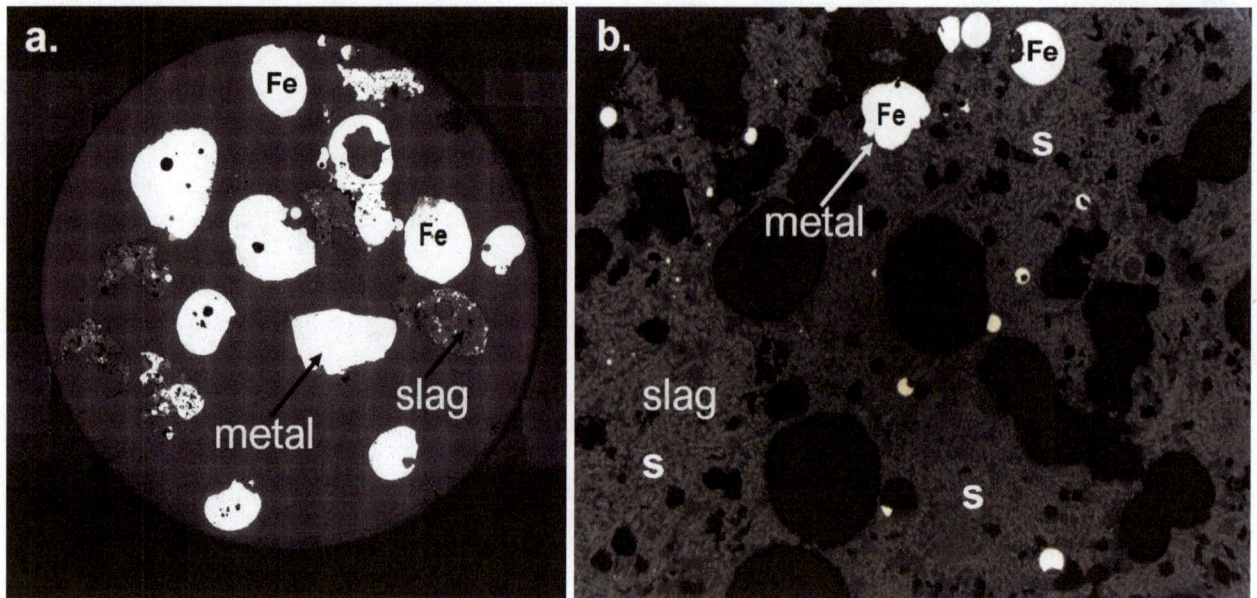

FIG 7 – Optical micrographs of the completely reduced (with coke) high phosphorus ore samples. Fe = metal-Fe, s = slag (glass) with wüstite.

Optical observations, EPMA and SEM analysis of reduced samples/with Coke from test T3 and test T4 (T = 1250°C)

For these two tests (T3 and T4) the high phosphorus ore, fluxes and coke were mixed prior to making the compressed pellets. Optical observations indicated that both samples have been reduced and abundant metallic iron particles were identified with slag (glass) particles. Metallic iron (Fe) particles and slag phases (glass and silicates) in test T3 ($CaCO_3$ fluxed) and test T4 (CaOx fluxed) reduced samples at 1250°C were analysed by EPMA and the measured phosphorus (P) contents in the metal-Fe and slag phases in respective samples are plotted in Figure 8a (calcium oxalate doped sample) and Figure 8b ($CaCO_3$ doped sample).

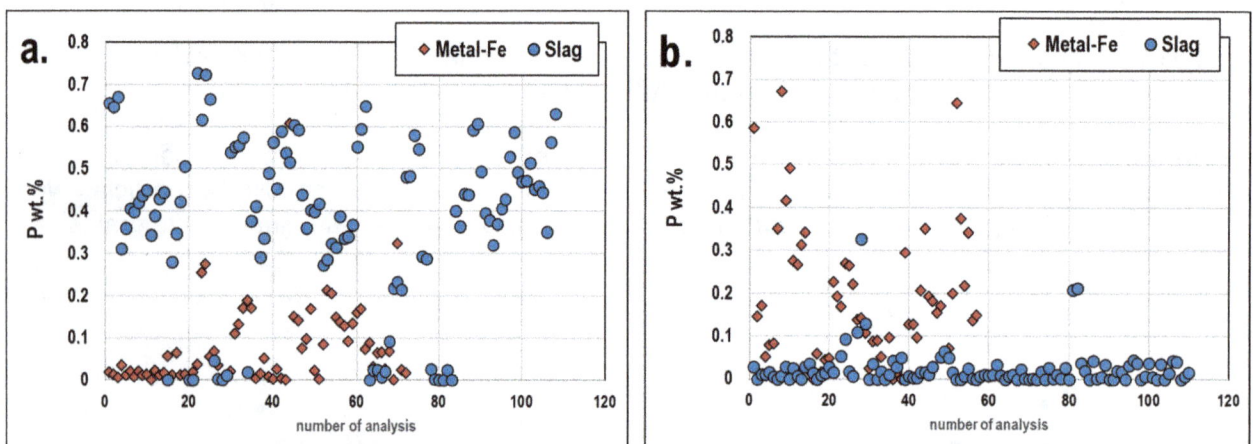

FIG 8 – Comparison of phosphorus (P wt per cent) contents within metallic Fe and glass in reduced samples, (a) phosphorous in metal-Fe and slag (glass) in CaOx doped/reduced sample, (b) phosphorous in metal-Fe and slag (glass) in $CaCO_3$ doped/reduced sample.

It was observed that phosphorus contents in the metal-Fe produced in the calcium oxalate doped sample (test T4) were significantly low whereas most of the slag (glass) analyses showed high contents of phosphorus within glass ranging up to 0.8 wt per cent. In comparison, the metal-Fe analysis results from the $CaCO_3$ doped sample (test T3) indicated higher levels of phosphorus incorporated in the metal-Fe too. Results also indicate that phosphorus in the slag (glass) phase from the calcium oxalate doped sample was somewhat evenly distributed while the glass in the $CaCO_3$ doped sample showed scattered values of phosphorus contents, ranging into some higher values and some very low values.

It was also noted that in both samples (calcium oxalate doped and $CaCO_3$ doped sample) a few very high phosphorus analyses within some of the larger metallic iron particles were reported by SEM EDS analysis. Additional EPMA spot analysis indicated these tiny regions contained phosphorus contents that ranged from 1.0–12.0 wt per cent of phosphorus. Further analyses confirmed that these were very small, localised areas within large metallic Fe particles and not spread throughout metal Fe particles. It was anticipated that this phosphorus was released to metallic Fe from relict rare earth particles or rare apatite found within the high phosphorus ore because none of the phosphorus EPMA analysis of hematite/goethite particles of high-phosphorus ore reported such high level of phosphorus, MacRae *et al* (2011) and the highest phosphorus levels in few goethite partials within high-phosphorus ore was reported from EPMA analysis was close to 1.0 wt per cent.

Optical observations, EPMA and SEM analysis of reduced samples/with Coke from test T5 and test T6 (T = 1300°C)

For these two tests (T5 and T6) the grinded high phosphorus ore and fluxes were thoroughly mixed prior to making the compressed pellets. The pellets were covered with coke inside the graphite crucible for the reduction. Both reduced samples were predominantly consisting of metallic-Fe, slag (glass), wüstite and minor magnetite EPMA and SEM EDS analysis was performed on a significant number of separate slag (glass) and metallic-Fe particles and the phosphorus contents (wt per cent) in slag and metallic-Fe is displayed in Figure 9 (test T5: $CaCO_3$ fluxed) and Figure 10 (test T6: CaOx fluxed). EPMA and SEM, EDS analysis of over large number of individual glass particles and individual metallic-Fe particles from both samples has shown excellent phosphorus partition between metallic iron and slag. Majority of glass particles analysed contained higher phosphorus contents with compared to the phosphorous content in the bulk ore, while majority of metallic-Fe had no phosphorus reported in them.

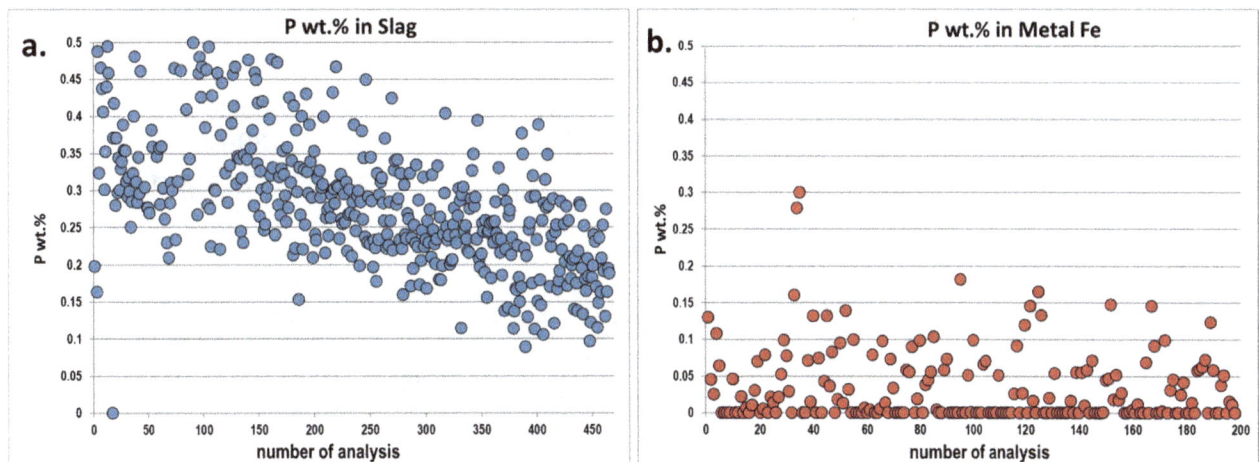

FIG 9 – Phosphorus in metal-Fe and slag (glass) in $CaCO_3$-fluxed, coke covered and completely reduced sample-test T5.

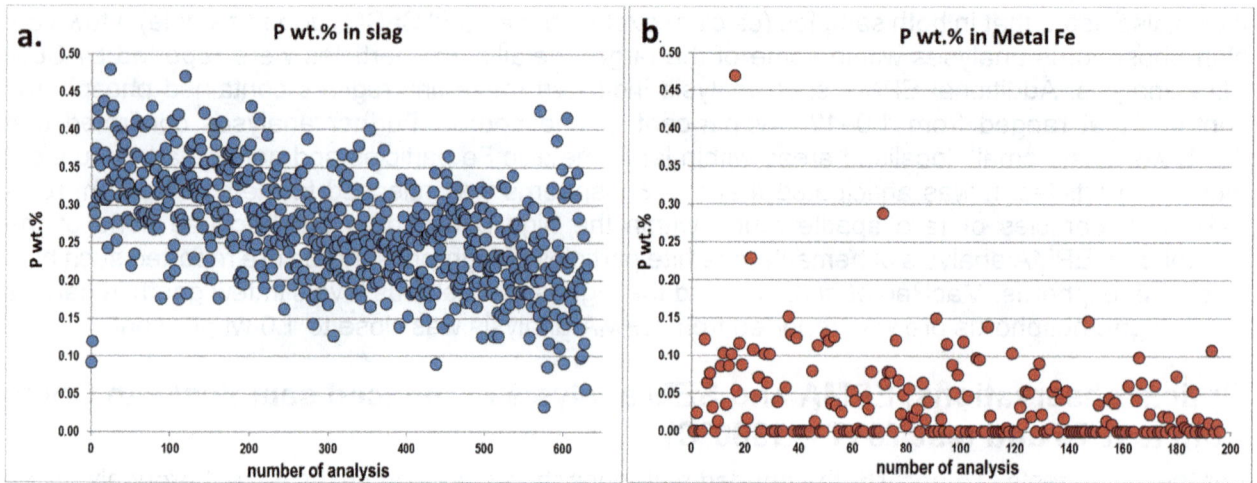

FIG 10 – Phosphorus in metal-Fe and slag (glass) in CaOx fluxed, coke covered and completely reduced sample-test T6.

Optical observations, EPMA and SEM analysis of reduced sample/with Coke from test T7 (T = 1300°C)

For this test (T7) the grinded high phosphorus ore and CaOx were mixed prior to making the compressed pellets, the coke was used around the pellets for the reduction. This sample has the highest basicity of 2.5 with compared to all other tests with basicity of 2.0. Table 3 display the average compositions for metallic-Fe, slag (glass), feldspar, magnetite and wüstite from EPMA/SEM EDS analysis. The Fe content in slag (glass) was significantly low and majority of silica was located within glass and feldspar while alumina was manly confined to feldspar.

TABLE 3

Average EPMA/SEM EDS analysis data of Fe-metal, slag (glass), feldspar, wüstite and magnetite in CaOx doped and sample-test T7 (basicity of 2.5) reduced at 1300°C.

phase	Phosphorus (P)	Iron (Fe)	Calcium (Ca)	Aluminium (Al)	Silicon (Si)	No. of Analysis
Metal-Fe	0.03	99.31	0.91	0.32	0.18	142
slag (glass)	1.57	3.24	20.25	0.22	10.87	129
Feldspar	0.03	2.21	32.5	17.55	8.22	33
Magnetite	0.03	70.1	3.1	0.68	0.14	17
Wüstite	0.02	75.7	0.65	0.32	0.05	16

EPMA and SEM, EDS analysis of over a significant number of individual glass particles and individual metallic-Fe particles in the sample has shown excellent phosphorus partition between metallic iron and slag (Figure 11a). Majority of glass particles analysed contained higher phosphorus contents with compared to the phosphorous content in the bulk ore, while majority of metallic-Fe had no phosphorus reported in them. Within some glass particles of the calcium oxalate doped sample large concentrates of calcium-aluminium feldspar was identified (Figure 11b). It appears that most of the metallic Fe particles are in contact with feldspar and EPMA analysis of feldspar indicated that feldspar did not contain any phosphorus in them.

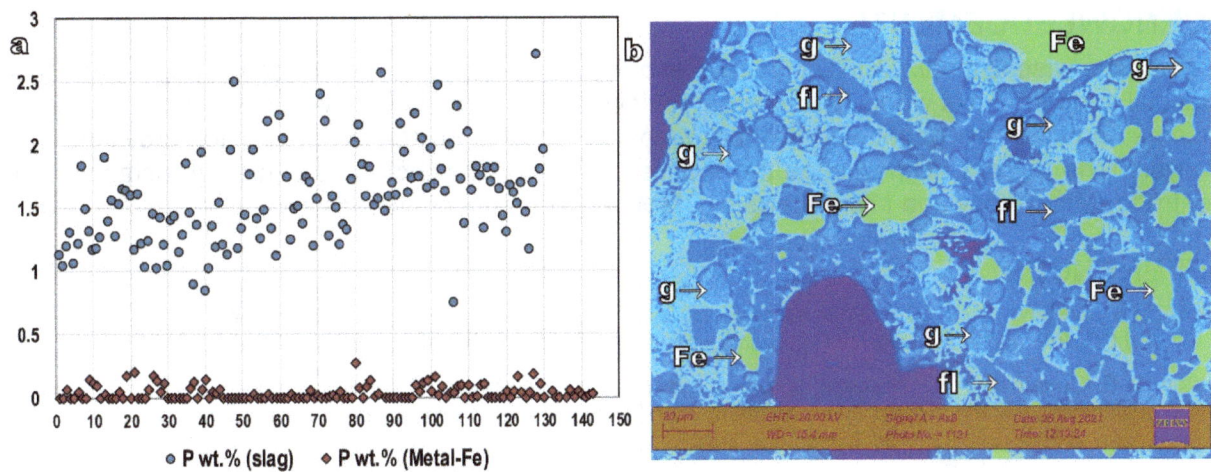

FIG 11 – Comparison of phosphorus (P wt per cent) contents (a) in slag and metal-Fe in reduced sample (test T7) and (b), metal-Fe (Fe), slag phase glass (g) and feldspar (fl) formed in CaOx fluxed/reduced sample.

CONCLUSIONS

Charcoal from *Syzygium Zeylanicum* plant has been used as a flux/reductant in ancient iron smelting processes in Sri Lanka. The iron produced after using charcoal or wood from such plant material is of exceptional quality with low impurities, in particular low phosphorus contents. Ethnographic evidence indicates that the pre-industrial smelters of Sri Lanka found it convenient to utilise charcoal as a fuel basis. Under the present study, analyses of slag samples from various locations showed that the phosphorus contents in metallic-Fe, wüstite, fayalite and spinel within slag was extremely low and phosphorus was mainly restricted into the slag (glass) phase. Further test work on charcoal produced from the *Syzygium Zeylanicum* plant indicated the presence of high levels of calcium oxalate phase in charcoal. Therefore, it was suspected that there is a significant effect on iron ore reducibility and phosphorus separation from high-phosphorus ores by using charcoal or wood which contained high percentage of calcium oxalate within.

The reduction tests were conducted using calcium oxalate and the results have shown that the calcium oxalate significantly enhanced the reducibility of a high-P iron ore by up to 50–65 per cent compared with no flux conditions. It has also shown that it could be a good alternative to the $CaCO_3$ for reducing high alumina ores as the raw ore used in this study had high alumina level of close to 4 wt per cent of Al_2O_3. At low temperatures (800°C and 900°C) under reducing conditions it was observed that the slag (glass) produced in calcium oxalate doped samples contained significantly high amounts of phosphorus and other Fe phases such as hematite, magnetite and SFCA contained less phosphorus compared to the phosphorus contents in slag (glass). In reduced samples (calcium oxalate doped and reduced at 800°C and 900°C), high-phosphorus glass was formed as small interstitial space filling grains between magnetite/hematite/SFCA particles and apatite/high phosphate crystals were also formed within glass. The attempt to separate glass from other was unsuccessful with grinding down the reduced sample to -100 microns and followed by the magnetic separation due to the fine size. Dissolving high phosphorus containing glass with acids and separate the liquid from reduced ore (high with phosphorus) to reduce the phosphorus may be a possible technique to investigate. Complete reduction tests of high-P ore doped with calcium oxalate confirmed that the phosphorus was mainly restricted in slag (glass) phases when the high-P ore was doped with calcium oxalate and completely reduced. When the ore, fluxes and coke were mixed, the Phosphorus separation/partition behaviour between calcium oxalate and calcium carbonate fluxed samples seems somewhat different. It was observed that mixing ore with fluxes only and using coke outside to reduce will provide better results for Phosphorous partition. Therefore, there is a good indication that calcium oxalate can be used to reduce high-P, high alumina ores at low temperatures and to retain phosphorus from mixing with metallic iron. Further investigations need to be conducted to check the effectivity of calcium oxalate in retaining phosphorus in slag by fine tuning the experimental conditions such as reduction temperature/gas atmosphere and basicity levels (balancing CaO/SiO_2) in the mixture.

ACKNOWLEDGEMENTS

The authors wish to thank CSIRO Mineral Resources for permission to publish the paper and for the financial support of this work. The external reviewers are thanked for their valuable comments and improvements to the paper. Dr Nathan Webster, CSIRO is acknowledged for providing XRD analysis. The authors acknowledge the facilities and the scientific and technical assistance of the Centre for Microscopy and Microanalysis (CMM) at University of Queensland.

REFERENCES

Anoop, M V and Bindu, A R, 2014. Pharmacognostic and Physico-Chemical Studies on Leaves of *Syzygium zeylanicum (L.) DC, International Journal of Pharmacognosy and Phytochemical Research,* 2014–2015; 6(4):685–689.

Ashton, M S, Gunathilake, S, De Zoysa, N, Dissanayake, M D, Gunathilake, N and Wijesundera, S A, 1997. *Syzygium zeylanicum,* yakada maran (S)/maranda (T), (DP 11:431), *A Field Guide to the Common Trees and Shrubs of Sri Lanka,* p 293, plant 33 (Wildlife Heritage Trust).

Clout, J M F, 2005. Iron formation-hosted iron ores in the Hamersley Province of Western Australia, in *Proceedings Iron Ore 2005 Conference,* pp 9–19 (The Australasian Institute of Mining and Metallurgy: Melbourne).

Coomaraswamy, A K, 1908. *Mediaeval Sinhalese Art: Being a Monograph on Mediaeval Sinhalese Arts and Crafts, Mainly as Surviving in the Eighteenth Century, with an Account of the Structure of Society and the Status of the Craftsmen,* 344 p, reprints: 1956, 2003 (Pantheon Books).

De Silva, M A T, 2011. *Evolution of Technological Innovations in Ancient Sri Lanka,* 444 p (Vijitha Yapa Publications).

Dukino, R D and England, B M, 1997. Phosphorus in Hamersley Range iron ores, in *Proceedings Ironmaking Resources and Reserves Estimation,* pp 197–202.

Dukino, R D, England, B M and Kneeshaw, M, 2000. Phosphorus distribution in BIF-derived iron ores of Hamersley Province, Western Australia, *Trans Inst. Min. Metall B,* B108:168–176.

Földvári, M, 2011. Handbook of thermogravimetric system of minerals and its use in geological practice, *Occasional Papers of the Geological Institute of Hungary,* 213:134.

Franceschi, V R, Paul, A and Nakata, P A, 2005. Calcium Oxalate in Plants: Formation and Function, *Annul. Rev. Plant Biol,* 56:41–71.

Graham, J, 1973. Phosphorus in iron ore from the Hamersley Iron Formations, in *Proceedings of the Australasian Institute of Mining and Metallurgy,* 246(June):41–42.

Han, H, Duan, D, Wang, X and Chen, S, 2014. Innovative Method for separating Phosphorus and Iron from high-Phosphorus Oolitic Hematite by Iron Nugget Process, *Metallurgical and Materials Transactions B, The Minerals and Material Society and ASM International,* 45B(Oct):1634.

Hapugoda, S, Nunna, V, Han, H, Pownceby, M I and Lu, L, 2021. Ancient to modern ironmaking – examining the effect on the behaviour of phosphorus and other impurities in iron ore by doping with calcium oxalate ($CaC_2O_4.H_2O$), in *Proceedings Iron Ore Conference 2021,* pp 13–31 (The Australasian Institute of Mining and Metallurgy: Melbourne).

Juleff, G, 1996. An ancient wind-powered Iron smelting Technology in Sri Lanka, *Nature,* 379(4):60–63.

Juleff, G, 1998. Early Iron and Steel in Sri Lanka: A study of the Samanalawewa Area (Materialien Zur Allgemeinen Und Vergleichenden Archaeologie), 422 p.

Karunaratne, P and Adikari, G, 1994. *Further studies in the settlement archaeology of the Sigiriya-Dambulla region* (eds: S Bandaranayake and M Morgren), pp 55–60, Postgraduate Institute of Archaeology, University of Kelaniya, Colombo, Sri Lanka.

Karunatilaka, P V B, 1991. Metals and metal use in ancient Sri Lanka, University of Peradeniya, *Lanka Journal of Humanities,* vol XVII and XVIII(1991–1992):104–118. Available from: <http://hdl.handle.net/123456789/2224>

MacRae, C M, Wilson, N C, Pownceby, M I and Miller, P R, 2010. Phosphorus and other impurities in Australian iron ores, *Microscopy and Microanalysis,* 16:896–897.

MacRae, C M, Wilson, N C, Pownceby, M I and Miller, P R, 2011. The Occurrence of Phosphorus and other Impurities in Australian Iron Ores, in *Proceedings Iron Ore Conference 2011* (The Australasian Institute of Mining and Metallurgy: Melbourne).

Manuel, J and Clout, J M F, 2017. Goethite classification, distribution and properties with reference to Australian iron deposits, in *Proceedings Iron Ore Conference 2017,* pp 567–574 (The Australasian Institute of Mining and Metallurgy: Melbourne).

Mohapatra, B K, Jena, S, Mahanta, K and Mishra, P, 2008. Goethite morphology and composition in banded iron formation, Orissa, India, *Resource Geology,* 58:325–332.

Morris, R C, 1973. A pilot study of phosphorus distribution in parts of the Brockman Iron Formation, Hamersley Group, Western Australia, *Western Australia Department of Mines Annual Report 1973,* pp 75–81.

Morris, R C, 1985. Genesis of iron ore in banded iron formation by supergene-metamorphic processes – a conceptual model, *Handbook of strata-bound and stratiform ore deposits* (ed: K H Wolf), pp 72–235 (Elsevier Science: The Netherlands).

Morris, R C, 2002. Genesis of high-grade haematitic orebodies of the Hamersley Province, Western Australia – A discussion, *Economic Geology*, 97:177–181.

Nunna, V, Hapugoda, S and Pownceby, M I, 2017. Study of microwave-assisted magnetising roasting and mineral transformation of low-grade goethite ores, in *Proceedings Iron Ore Conference 2017*, pp 575–582 (The Australasian Institute of Mining and Metallurgy: Melbourne).

Nunna, V, Hapugoda, S, Pownceby, M I and Sparrow, G J, 2021. Beneficiation of low-grade, goethite-rich iron ore using microwave-assisted magnetizing roasting, *Minerals Engineering, Minerals Engineering,* 166(2021):106826.

Ostwald, J, 1981. Mineralogy of Australian iron ores, *BHP Technical Bulletin,* 25:4–12.

Peixoto, G, 1991. Improvement of the reduction process in P content and other gangues in iron ore and its agglomerates, USA Patent Filed Nov. 14, Inter. Appl. No. PCT/BR9/00030.

Ranasinghe, S P, 1986. Iron Ore Deposits of Sri Lanka, Geological Society of Sri Lanka, Department of Geology, Univ. of Peradeniya, Sri Lanka, L J D Fernando Felicitation volume, pp 33–44.

Seneviratne, S, 1985. Iron Technology in Sri Lanka: a preliminary study of resource use and production techniques during the early iron age, *The Sri Lanka Journal of the Humanities,* 9:129–178.

Solangaarachchi, R, 1999. History of Metallurgy and Ancient Iron Smelting, *Vudurava,* 19(1):30–40.

Solangaarachchi, R, 2011. Ancient Iron Smelting Technology and the settlement pattern in the Kiri Oya basin in the dry zone of Sri Lanka, PhD thesis, University of Florida, 2011.

Su, C, Lv, N, Yang, J, Wu, L, Wang, H and Dong, Y, 2017. Effect of high Na_2O addition on distribution of phosphorus in low basicity converter slag, *J Iron Steel Res Int*, 26(2019):42–51. Available from: <https://doi.org/10.1007/s42243-018-0096-1>

Thorne, W, Hagemann, S, Webb, A and Clout, J, 2008. Banded iron formation-related iron ore deposits of the Hamersley Province, Western Australia, *SEG Reviews,* 15:197–221.

Process optimisation – a greener and cleaner iron ore value chain

Making a positive change – evolving the modern Integrated Remote Operations Centre (IROC) to optimise and support carbon-neutral value chains

J Bassan[1], G McCullough[2] and J Els[3]

1. Leader – Integrated Operations and Remotization, Hatch, Melbourne Vic 3000. Email: jarrod.bassan@hatch.com
2. Leader – Digital, Hatch, Brisbane Qld 4000. Email: george.mccullough@hatch.com
3. Principal – Iron and Steel, Hatch, Brisbane Qld 4000. Email: jeanne.els@hatch.com

ABSTRACT

Integrated Remote Operations Centres (IROCs) have a crucial role in optimising a mine's carbon emissions, especially as industry faces an imperative to achieve net-zero emissions on an accelerated timeline. To meet this time pressure, operations must adopt a comprehensive approach to energy-related carbon management and history has proven that modern technologies cannot be simply bolted into existing operations. An IROC is the most dependable solution to enable a swift transition from strategy to operational tactics that support carbon optimisation.

Achieving net-zero emissions requires addressing significant energy management challenges such as fleet electrification, integration of renewable energy sources and large-scale energy storage. These challenges demand a profound change in planning, managing and executing work at large operations. The key to overcoming these challenges is an integrated approach incorporating collaborative decision-making across the value chain and more sophisticated planning approaches.

Additionally, closer integration with downstream processing offers opportunities for carbon optimisation across the value chain. This includes the ability to supply tailored products that meet more varied feed-requirements of customers transitioning to less carbon-intensive processes and developing collaborative decision-making across the entire steelmaking value chain (by incorporating steelmakers, transport, coal/natural-gas/hydrogen supply, energy supply and iron-ore supply). However, this requires significant improvement in collaborative maturity and integrated decision-making capability.

Finally, the imperative to reduce carbon emission demands the introduction of a range of new technologies into existing operations. The new technologies will increase the operational complexity of the mines and IROCs will need to evolve capabilities to manage the increase complexity and support the effective operationalisation and integration of these new mining technologies.

This paper outlines the new capabilities required and the role of IROCs in effective integration across the demand value chain. The focus is on building on the mining industries' existing IROC experience and augmenting with processes, technologies and lessons from other industries.

INTRODUCTION

The mining industry in general is grappling with the urgent need to decarbonise operations. There is a particular spotlight on the iron and steel industry as it is directly responsible for 7 per cent of global greenhouse gas emissions. The steelmaking process contributes the bulk of these emissions with mining and beneficiating coming in a distant second at 0.8 per cent (International Energy Agency, 2020).

To decarbonise the iron and steel industry, there are three main goals:

1. Introduce non-traditional routes to produce steel at a significantly reduced carbon intensity than current methods, such as direct reduced iron (DRI) shaft furnaces into electric shaft furnaces (ESF).

2. Shift to steelmaking processes that are close to net zero, such as scrap melting or hydrogen DRI (green steel), as far as possible.

3. Reduce the carbon produced in mining, beneficiating and transporting the iron ore products.

In the Australian context, the steelmaking process is not in focus, so it is beyond the scope of this paper, except to consider the potential impacts on the demand for iron ore as steelmaking feedstock.

This paper focuses on the scope 1 and 2 emissions of the current iron ore mining operations (from mine to port) and beneficiation. We will also consider how new 'green' steelmaking methods will affect the future production and marketing of iron ore products.

The challenge of achieving net-zero iron ore production in Australia is magnified by the mega scale and complexity of the mining and processing operations. The result of this situation is there are many carbon emission vectors. Therefore, it is improbable that a single 'silver bullet' will be found to reduce or prevent all emissions at once.

Investment in decarbonisation is intensifying and as a result technological advancements in carbon emissions preventions are being accelerated but are decades from being commercialised and implemented. As a result, iron ore producers will need to recruit several solutions that together will reduce total carbon emissions towards 'net zero' status.

These solutions include electrification, hydrogen, autonomous operations, beneficiation and changes to marketing approaches.

Fortunately, the Australian iron ore industry has previously invested in capability to operate productively at ever increasing scale and complexity over the previous two decades. One of these key investments resulted in Australia leading the world in the establishment of Integrated Remote Operating Centres (IROCs).

IROCs are effectively the nerve centres of complex iron ore value chains. All operational decisions relating to mining, processing and logistics are made and directed from them.

The key objective of this paper is to explore how an IROC must evolve to have a critical role in supporting carbon neutral iron ore value chains. It will outline the forces shaping the iron ore industry in relation to decarbonisation, identify the foreseeable operational challenges associated with a low carbon iron ore value chain and identify what capabilities will be required in the future IROC to respond to these challenges.

AUSTRALIAN IRON ORE CONTEXT

Australia is the world's largest producer of iron ore, accounting for 36 per cent of total global production and has the world's largest economic Mineral Resources of iron ore (29 per cent of world total). Almost all of Australia's iron ore is produced in the Pilbara and 96 per cent is hematite (Fe_2O_3) ore, which typically blended and shipped, with typical iron content around 62 per cent, for use in blast furnaces for making pig iron as the first step in the steelmaking process (Summerfield, 2019).

The role of IROCs in Australian iron ore value chain today

Australian iron ore producers led the mining industry in the establishment of large scale IROCs and today most iron ore production in Australia is controlled from them.

The original drivers of IROCs in Australian were cost, safety and the need to coordinate logistics between mine and port (Schweikart, 2007). However, the increased visibility of operations became an enabler of increased productivity and production and enabled the operations centre to optimise the performance of the whole 'mines to ports' system. As autonomous and remotely operated vehicles were introduced, the IROCs evolved to support the introduction of technology and support standardisation across multiple sites (Parker, 2018).

There are some nuanced differences in the current Australian iron ore IROCs but in general they have the same role in the present day:

- Enhance safety by reducing the number of people on operational mine sites.

- Integrate value chain coordination to enable increased productivity and strive for production predictable outcomes.

- Manage complex operations and logistics thought the application and use of standard processes and technology, which also preserve corporate knowledge.

- Optimise for quality and volume through predictable product specification creation and order fulfilment at scale and matched to market demand.

- Support the deployment and integration of autonomous equipment into operations.

- Increasingly serve as talent attraction and retention instruments enabling people to work in highly appealing, modern environments in world-class cities like Perth.

As can be seen, the generalised role of an Australian iron ore IROC is coordinating all activities and operations to produce the most iron ore at a predictable quality and schedule. It is the nexus of operational decisions and critically important in the context of any operating model changes which may be required in the future.

Emissions in the iron ore value chain

A typical iron ore value chain for Pilbara hematite product is presented in Figure 1 with the energy inputs for the value chain. Diesel fuel and electricity (generated from diesel or LNG generators) are the principal energy sources for the iron ore value chain. Small quantities of electricity are contributed from renewable (solar) generation and some energy is contributed by the blasting process (usually via an emulsion of ammonium nitrate and fuel oil). This diagram presents the current state and the opportunities for decarbonisation.

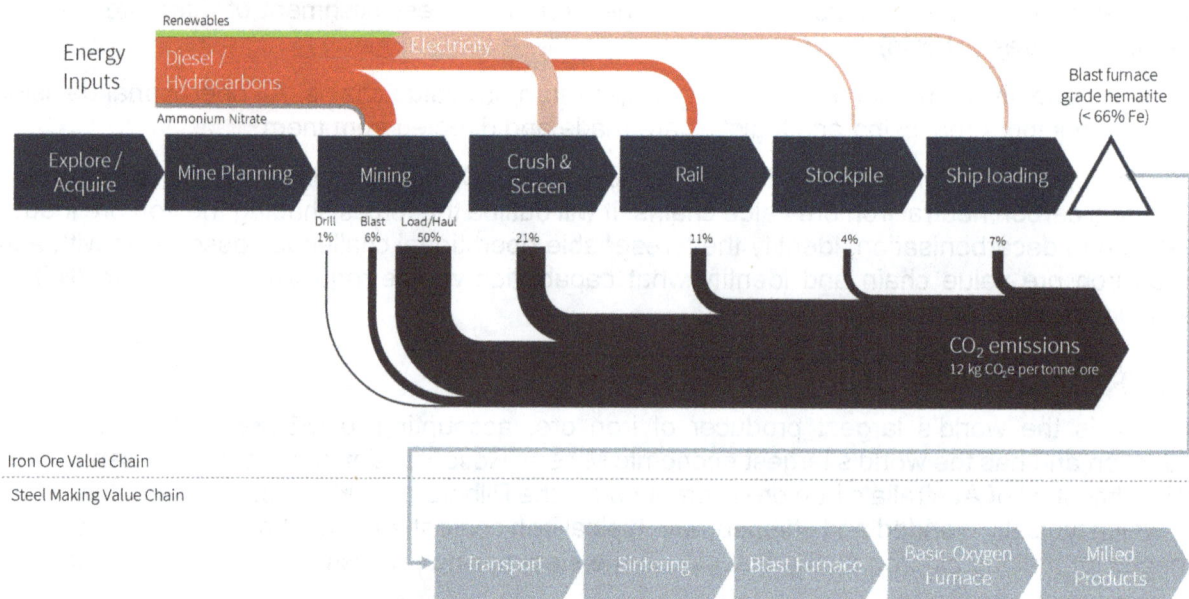

FIG 1 – Typical iron ore value chain today for Pilbara product.

Typical greenhouse gas (GHG) emissions for hematite iron ore mining, by process-step, are shown in Table 1. These were reported in a study by Norgate and Haque (2010) of four Western Australian iron ore operations, with emissions recorded as kg of CO_2-equivilant (CO_2e) per tonne of iron ore produced. Broadley consistent results were obtained in a separate study by Palamure (2016).

Producing a tonne of crude steel via the Blast Furnace (BF) and Basic Oxygen Furnace (BOF) (abbreviated as the BF-BOF) route emits around 2.2 t CO_2 (International Energy Agency, 2020).

TABLE 1

GHG emissions from Australian hematite/goethite mines, adapted from Norgate and Haque (2010).

Iron ore	CO$_2$e/t ore	
Process step	**kg**	**%**
Drilling	0.1	0.8
Blasting	0.7	5.9
Loading and hauling	6.0	50.5
Crushing and screening	2.5	21.0
Stacking and reclaiming	0.5	4.2
Rail transport	1.3	10.9
Port operations	0.8	6.7

The impact of global decarbonisation targets on Australian iron ore producers

The Australian government passed the Climate Change Act in 2022 which sets the target of reaching net zero emission by 2050. Metal mining accounts for 3 per cent of Australia's scope 1 emissions (the fourth biggest contributor after electricity (54 per cent) and coal mining (10 per cent) and metals) and decarbonisation is an imperative for the industry (Clean Energy Regulator, 2019).

Since 99 per cent of CO$_2$ emissions in steel are a result of the steelmaking process (only 0.8 per cent are from the mining of iron ore), there is considerable incentive (and pressure) on steel makers to decarbonise their process driven by policy, markets and investors (International Energy Agency, 2020). This will result in changing demands for iron ore products as steel makers seek feedstocks best suited to carbon optimisation in their processes.

Changing demand

Currently, 73 per cent of world steel production is via BF-BOF. Almost all ore from the Pilbara is used as feed stock to this process. Of the main steelmaking routes commercially used today (shown in Figure 2), BF-BOF is the highest carbon intensity (2.2 t CO$_2$/t crude steel) (Nichols and Basirat, 2022).

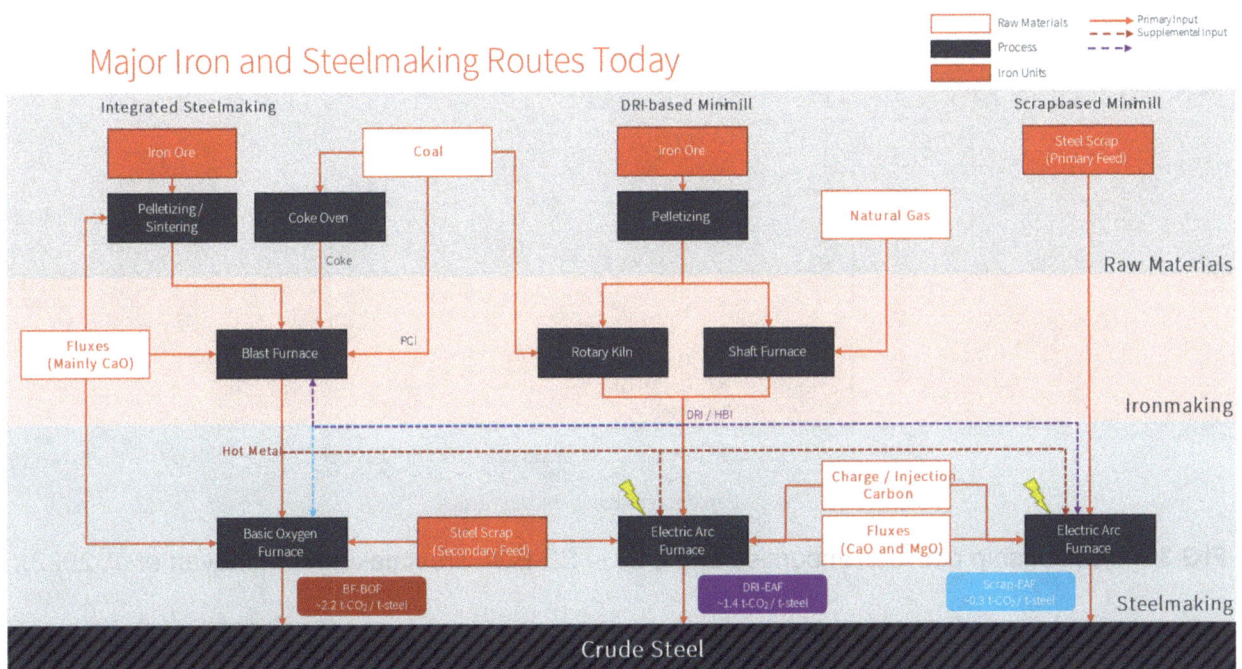

FIG 2 – Major steelmaking routes today.

Direct Reduced Iron and Electric Arc Furnace route (DRI-EAF) has much lower emissions the BF-BOF route, producing only 1.4 t CO_2/t crude steel and potentially zero emissions if powered by renewable electricity and green hydrogen. In 2020, DRI-EAF only accounted for 4.8 per cent of global steel production. The reasons are:

- Low availability of suitable iron ore feedstocks for DRI-EAF: DRI-EAF requires feedstock with high iron (at least 67 per cent Fe content) and low impurity (Linklater, 2021). Only a small percentage (3 per cent) of Australian iron ore meets this criterion. The goethite and hematite ores predominant in the Pilbara region are not amenable to economic beneficiation to >67 per cent Fe with current processes. The DRI-EAF process therefore requires feed from high quality pellets (produced from beneficiated magnetite ores) and scrap steel, both of which are in constrained supply.

- Availability and cost of green or blue H_2 at scale, which currently only accounts for 2 per cent of global hydrogen supply.

- Economic life of existing blast furnaces: existing blast furnaces have many decades of production life left – with the average age of being 13 years and typical expected service life of 40 years (International Energy Agency, 2020).

Therefore, the viable path to decarbonising steel involves a multi-decade transition period that focuses on optimising and retrofitting technologies to existing blast furnaces to reduce carbon emission (such as substitution of BOF with EAF, carbon capture and storage, heat recovery and waste stream recycling) (International Energy Agency, 2020).

Selecting iron ore with fewer impurities and higher Fe grade (and higher strength coke made from premium coking coal) will contribute to improve efficiency of existing blast furnaces (Ellis and Bao, 2020). This can be seen in the analysis by Madhavan *et al* (2022) (reproduced in Figure 3) which demonstrate the increased CO_2 emissions as the impurities of P and Si in the feed stock increases. There is also evidence that ocherous goethite ores can favourably improve the efficiency of the sintering process if the steelmaker is aware of their presence in the blend (Ware, Manuel and Raynlyn, 2013). It is expected that steel producers will become more selective and pay premiums for products that best suit their CO_2 optimisation strategy, although the demand patterns will change continually as individual steel plants will invariably select different technologies and implement them on different timelines. This is reflected in the predicted change in iron content over the coming decade, with a projected demand for higher grades increasing, as shown in Figure 4.

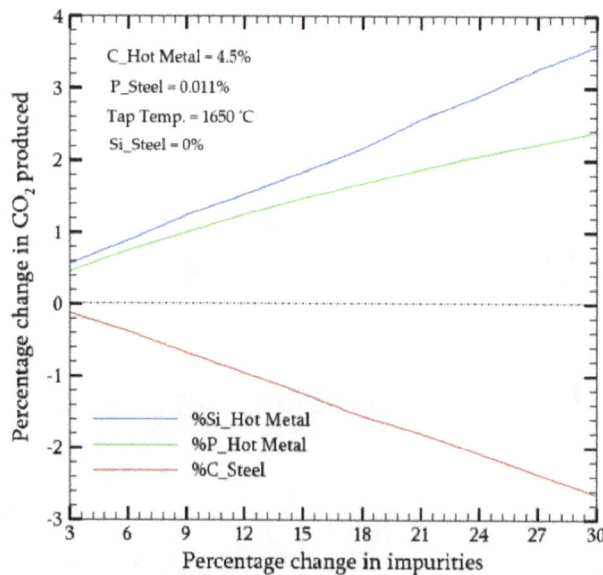

FIG 3 – Relationship between impurities and CO_2 in BF-BOF steelmaking (Madhavan *et al,* 2022).

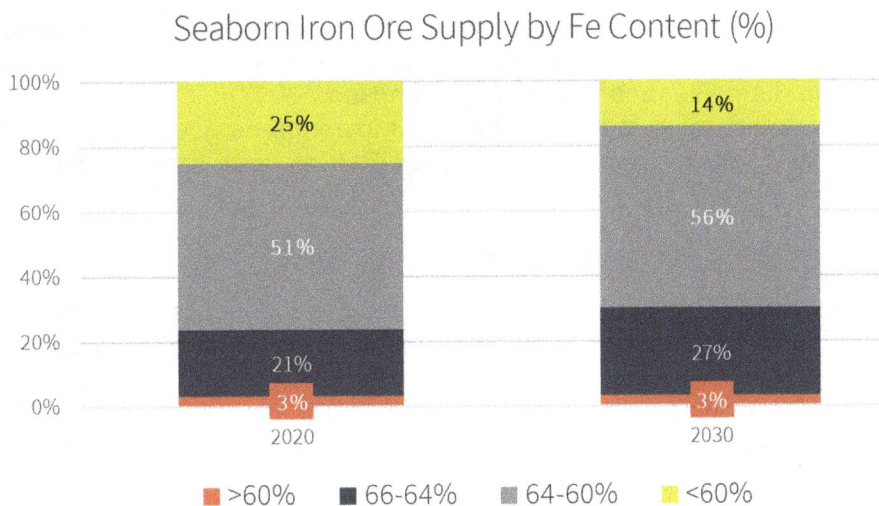

Seaborn Iron Ore Supply by Fe Content (%)

FIG 4 – Current and projected seaborn iron ore supply by Fe content, adapted from Nichols and Basirat (2022).

Consequently, there could be opportunities for some iron ore producers to take advantage of changing market conditions if they are adept at tailoring some of their products to quickly meet niche demand. Producers who are most capable of managing multiple products (including beneficiated products, as discussed later), blending to specific requirements and controlling grade throughout the mine to port value chain will be best equipped to take advantage of the market and maximise the economic potential of their assets or integrate new assets.

The implication of changing demand on IROCs

The role of IROCs will need to broaden to help mining companies adapt to changing demand for iron ore products. To fill this role, there are additional capabilities required by the future iron ore IROCs. These are:

- Collaborative decision-making process involving marketing, sales, mine planning, production scheduling and logistics scheduling process to enable faster response to market conditions.

- Ability to introduce new product blends and rapidly reconfigure all systems and processes without disrupting normal operations.

- Ability to maximise use of infrastructure and stockyard space to produce a flexible range of products.

- Better monitoring and control of grade throughout the value chain to support flexibility of offering more tailored products.

- Improved processes and tools for scheduling logistics that accommodate more complex scheduling requirements associated with tailored products.

Beneficiated magnetite products

Magnetite ores have much lower in-ground iron content (20–30 per cent Fe) than hematite ore but are amenable to being beneficiated into a DRI feedstock (Nichols and Basirat, 2022). Today, magnetite accounts for 4 per cent of Australian iron ore production (Summerfield, 2019), but is likely to increase in response to demand for DRI feed.

The mining process for magnetite faces the same challenges and opportunities for decarbonisation as hematite mines, but in addition, there is a requirement to operate a beneficiation processes.

Typically, the product is beneficiated as close to the mine as possible (to avoid the cost and logistical capacity required to transport large volumes of waste material in the unconcentrated ore). Thus, magnetite mines require construction of beneficiation plants and additional electrical energy to operate the plant.

As some magnetite deposits are found in proximity to hematite, it is likely that some operators will have to manage both products through the same logistics chain. An example is the Iron Bridge mine, which contains both hematite and magnetite (Simpson, 2017). It is also possible that new economically feasible processing routes are found for lower grade goethite rich ores (Nunna *et al*, 2021). Further discussion of the capability impacts on IROCs due to beneficiation is provided in the later section.

OPTIONS FOR A DECARBONISED IRON ORE VALUE CHAIN

Many options can be considered for decarbonisation, but the economics vary with numerous factors unique to each mine including remaining mine life, depth, pit design and availability of green energy sources. Consequently, different approaches will be applied at different mines and there could no longer be a 'standard' operating model for iron ore mines. As a result, the future IROC will have to manage a network of mines (supply sources) that operate a multitude of different mining methods and equipment and a mix of energy sources. Two possible decarbonised value chains are shown in Figures 5 and 6. Figure 5 shows the value chain for an electrified mining process, acknowledging that some electrical generation from spinning turbines will be required for the near future. Figure 6 shows the case for a hydrogen powered mining process that utilises 100 per cent green hydrogen and utilises hydrogen to provide the spinning reserve for the electrical grid.

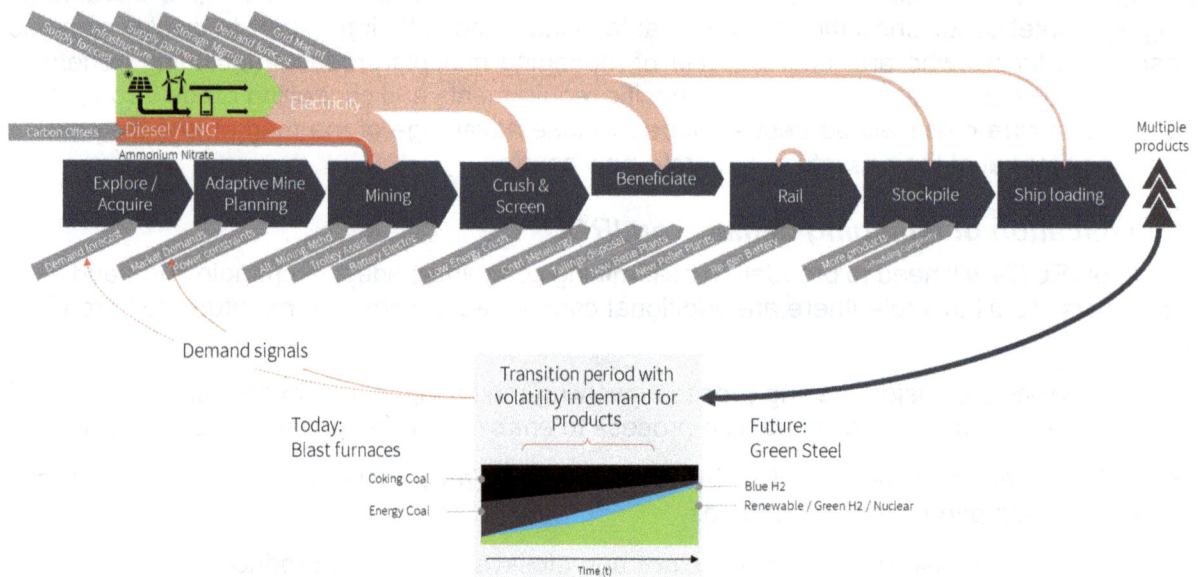

FIG 5 – Electrified iron ore value chain.

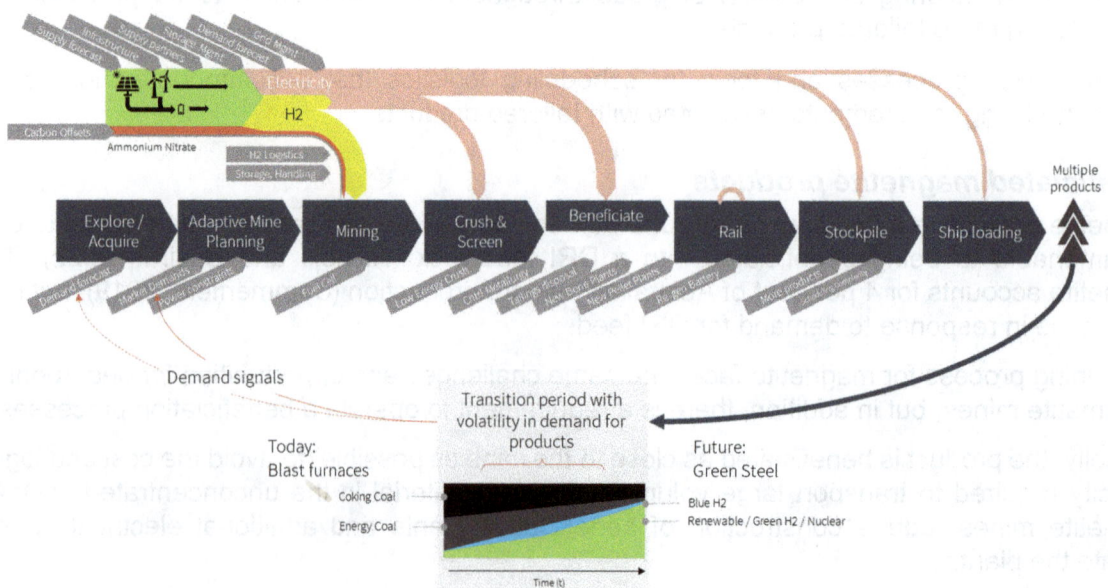

FIG 6 – Green hydrogen iron ore value chain.

The following sub-sections enumerate on the decarbonisation options under consideration for iron ore mining and identify the capabilities that will be required in an IROC to support effective decarbonisation strategies.

Mining

The mining process, including drilling, blasting, loading and hauling, is the most GHG intensive process in the iron ore value chain. The standard operational model today involves hauling between the loading unit (shovel or excavator) located at the bottom of the pit to a crusher located outside the pit on an uphill grade. Hauling is accomplished on large ultra-class trucks (up to 400 t) powered by diesel internal combustion engines (ICE).

Substitution of diesel as the energy source

There are several options for replacing diesel as the energy source for trucks (and other mobile equipment). Current large haulage trucks used in iron ore are diesel-electric, consisting of a diesel ICE turning an electric generator that provides power to electric motors propelling the truck. Mining companies and equipment suppliers are considering several options to reduce the impact from or eliminate the diesel, including:

- Cleaner diesel fuels or diesel alternatives (Dubov *et al*, 2019).

- Trolley assist (overhead electric cables supplying electricity to truck via pantograph) (Mazumdar, 2013).

- Battery electric vehicles (BEV) (Mazumdar, 2013).

- Hydrogen fuel-cell (Feng and Dong, 2020).

All options have advantages and disadvantages that influence their economic viability for different mining operations. Some technologies could be implemented in combination (for example, trolley-assist with battery, or trolley-assist with cleaner diesel are compatible combinations).

The options listed above retain the overall characteristics of truck-shovel mining method and are candidates for retrofitting existing mines because they are compatible with the mine design (although feasibility is required on a case-by-case basis considering haul road designs, grades, haul distances, ability to install infrastructure and other factors). They are also candidates for new mines, noting that there is more flexibility to optimise the mine-design and placement of necessary infrastructure in a new mine.

Much opportunity exists for further study to optimise operating practices for these technologies, especially BEVs. For example, Zhao *et al* (2021) have shown that up to 60 per cent cost saving is possible with an optimised algorithm for scheduling charging of BEV heavy-haul road transport vehicles. Initial implementations of BEVs and other new haulage technologies will have inherent inefficiency in their operating model, providing substantial room for continuous improvement.

The implication of diesel substitution on IROCs

To manage substituted energy sources, new capabilities will be required by the future iron ore IROCs. These are:

- Monitoring battery levels and charging/battery swap operations.

- Integrating battery charging decisions into dispatch optimisation (and autonomous haulage systems).

- Continually improving operating practices for new haulage technologies to drive efficiency, cost and throughput improvement.

Continuous reliable supply of electricity and/or H_2 is necessary and the role of the IROC in managing supply is discussed later in this paper.

Equipment that is more energy efficient

Many options exist to utilise equipment that requires less energy to perform a similar operation. Examples include high pressure grinding roller (HPGR) crushers which can substitute existing crushing technology and introduces minimal change to the operational complexity of the mine. Individual technology introductions may have minimal impact from an IROC perspective, but it is noted that there is a cumulative impact associated with many changes occurring. Each new technology will have an efficiency ramp-up curve and presents an opportunity for continuous improvement to improve efficiency of operation and throughput.

Alternative mining methods

Alternative mining methods to truck/shovel have been proposed for many years. One of the most promising options available (with a high technology readiness and applicability) is in-pit crushing and conveying (IPCC), which offers the opportunity to reduce truck haulage distances, electrify the extraction process and reduce overall energy consumption (Nehring *et al*, 2018).

IPCC achieves the goal of electrification but requires the introduction of a completely new operating procedures. Feasibility of IPCC is coupled tightly to the mine-design, therefore it is usually only an option for greenfield mines, where the opportunity exists to optimise the mine design and life-of-mine plan to the IPCC method.

Operation of site employing IPCC (or other alternative mining methods) could be expected to run from an IROC from 'first ore'. The operating characteristics of the new site would vary from existing truck and shovel mines and the IROC would serve as a useful mechanism to enable the most efficient implementation, continuous improvement and sharing of knowledge between sites (as additional IPCC sites are eventually bought online).

The implication of alternative mining methods on IROCs

To manage alternative mining methods, IROCs will need to develop additional capabilities:

- Coordinating maintenance and operation of fixed and semi-fixed assets located in the pit with HME operations.

- Managing occasional relocations of semi-fixed infrastructure in the pit and rapidly re-optimising operation after relocations.

- Continually improving operating practices to drive efficiency, cost and throughput improvement.

Beneficiation

Beneficiation is the process of removing impurities (gangue) and producing a concentrate with higher Fe percentage. Beneficiation is necessary for magnetite ores (which are found with much lower in-ground Fe grades of typically 20–30 per cent) and it can also be applied to improve lower grade hematite ores. Different beneficiation treatments are applicable to magnetite and hematite.

Agglomeration is the further process of turning magnetite concentrate into pellets through thermal treatment. The pellets can contain 65–70 per cent Fe and are suitable for use in DRI steelmaking. They contain low impurities and are sold as a premium product with higher price (Summerfield, 2019).

Several beneficiation plants operate or are under construction in Australia, but the majority (96 per cent) of Australian iron ore is hematite/goethite with sufficiently high Fe content to be blended, sold and shipped without being beneficiated (Summerfield, 2019).

As steel makers seek to decarbonise and existing reserves of the highest-grade iron ore are depleted, it can be expected that:

- Demand for ores with less gangue will increase.

- Demand for pellets as feed for DRI processes will increase.

- More beneficiation of low-grade ore will be required to maintain volume requirements.

As a result, it is reasonable anticipated that beneficiation plants will become more common in iron ore value chains. Beneficiation plants are more complex to operate then simple crushing/screening operations.

The implication of beneficiation on IROCs

Introducing beneficiation into the iron ore value chain will have significant implications for iron ore IROCs. They will need to introduce new capabilities:

- Managing and controlling complex metallurgical process and processing plants.

- Tightly coupling mining grade-control with plant operations. Efficiency and recovery of processing plant are dependent on consistent feed ore, requiring close collaboration between plant and mine to control grade and other characteristics.

- Energy management of the beneficiation plants (which introduce significant additional electrical load).

- Management of beneficiated products through the outbound logistics (particularly where the logistics chain is shared with hematite products).

- Management of tailings stream and tailings dams.

Process optimisation and improvement

Opportunity still exists to further optimise operations with little or no capital investment in plant and equipment. This has the effect of avoiding excess emissions while increasing output (ie reduce the kg of CO_2 per t of ore). It also has the desirable economic effect of reducing the cost per tonne of production. Any individual improvements will be incremental, but in aggregate a determined operational excellence program could amount to between 1–13 per cent percent business improvement (Constantini *et al*, 2022) and a potentially similar reduction or avoidance of CO_2.

Process optimisation will be driven by data analytics, machine learning and other solutions that enable operators, planners and supervisor to make better decisions and continually optimise the operations to reduce the energy consumption.

In a recent example, a mining operation conducted analysis of an already optimised concentration circuit. Using machine learning techniques and a digital twin, the opportunity to increase recovery by 0.8 per cent was identified, without impacting throughput. This is one of many examples of efficiency and productivity improvements that are typically accessible to mining operations when analytics is applied within a proper process improvement framework.

To achieve the outcome of optimisation and efficiency, it is necessary to:

- Have timely access to the good quality data so that it can analysed.

- Have the tools, analytical techniques and people available to interpret the data and draw the right conclusion.

- Take the correct action in a timely manner.

The implication of advanced process optimisation and improvement on IROCs

To enhance process optimisation through data, IROCs will need to include capabilities such as:

- Management of data and data quality so that good quality historical data is available for analysis and is available for use by data science techniques.

- Access to data analysis and data science skill sets and software tools for data science.

- Simulation and digital twin models to validate the conclusions of the data scientists and to simulate the impacts on operations and assist in developing new operating procedures.

- Ability to rapidly implement changes to operating procedures, technology components and work processes, including the retraining of personnel.

- Measurement and collection of data before and after implementing changes to validate value was delivered.

Energy supply

Scope 1 and 2 emissions in Australian iron ore products are from diesel and hydrocarbons (such as LNG). A small portion of additional emissions comes from the blasting process (usually from an emulsion of ammonium nitrate and fuel oil). Currently, only a small portion of electricity is generated from renewables although installation of additional capacity is planned.

The only effective pathways to decarbonisation require substitution of existing hydrocarbon energy sources (diesel and LNG) with non-GHG emitting sources. Interim transitional solutions can either substitute with a better alternative (eg green diesels which have lower carbon intensity) or reduce the energy required (more energy efficient methods), but these only offer marginal reductions and are transitional measures. Meaningful decarbonisation requires substitution of the energy sources with some combination of:

- Electricity from renewables (effectively wind or solar as the current viable options in the Pilbara).

- Green hydrogen, generated entirely from renewables.

- Nuclear (which is emerging as technically viable at small scale, but not currently an option in Australia).

Other renewable sources are not considered in this paper as they are less likely to be viable for the Australian iron ore industry.

Renewables from wind and solar

Wind and solar are currently the most practical renewables for the Australian iron ore industry. For example, the Pilbara region has good solar and reasonable wind characteristics. The region is considered as an extended case study in this section, although the conclusions apply wherever electricity will be sourced from wind and solar sources.

A locally interconnected grid (North West Interconnected System, NWIS) already exists in the Pilbara region (Figure 7). Many mining operations are already connected to the grid and some sites provide grid-connected generation capacity (both renewable and non-renewable). While the grid provides power for many domestic and commercial customers, the mining industry dominates the load on the network.

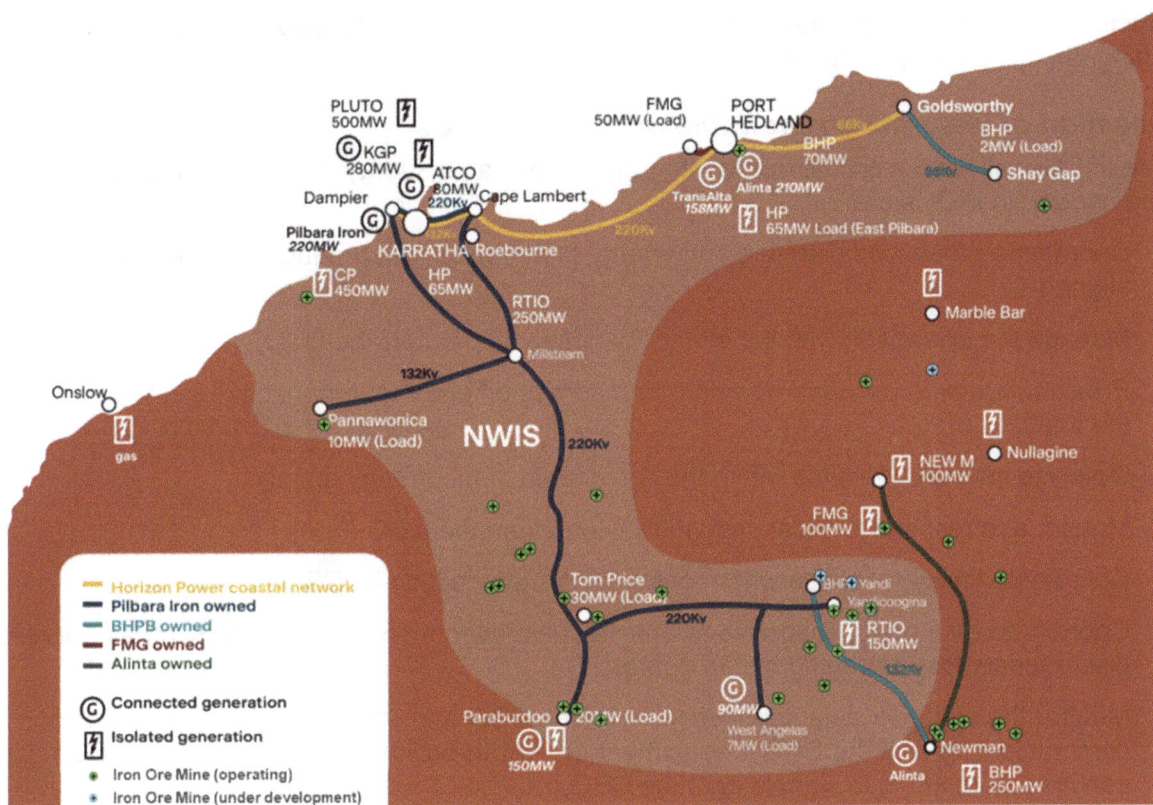

FIG 7 – NWIS grid in relation to the location of iron ore mines, adapted from Horizon Power (2021).

Load demand forecasts are remarkably crucial for energy suppliers and grid operators (Lanka *et al*, 2021). Managing a grid with less dispatchable synchronous generators (for example, diesel or LNG turbine generators) and more non-synchronous renewable generation sources (such as wind and solar) poses technical challenges, including:

- Forecasting supply and load demand on the grid.

- Managing battery storage capacity (provided by large capacity grid-forming batteries).

- Maintaining reserve margin (ie spare capacity) to absorb the next credible contingency.

- Managing demand and shedding load in scenarios where sufficient supply cannot be maintained.

The grid network operator (technically referred to as the distributed systems operator, DSO) will require closer operational interaction with mining companies to maintain quality, reliability and security of supply.

The mining company will need to work closely with the DSO to maintain the security of the grid and balance supply and demand as a greater portion of the supply comes from renewables. Some examples of strategies that could be employed are:

- Scheduling short maintenance shutdowns at night when solar is not available.

- Scheduling longer shutdowns during times of the year when less wind or solar predicted due to seasonal weather patterns.

- Integrating production schedules with renewable generation forecasts.

- Avoid running processes (other than the bottleneck) at full capacity during period of low generation.

- Adjusting recharging schedules for BEVs.

It will therefore be necessary to forecast the supply capacity of the system as an input to mine planning and scheduling. The forecasting windows typically used in energy markets can be mapped to similar planning windows used in mining. Since energy demand for a mine is a function of the production schedule, it is proposed that there will be increased collaboration between the mine planning and scheduling functions and the grid operator.

Implication of integrating renewable energy on IROCs

Future IROCs will have a critical role in integrating renewable energy sources. New capabilities that will be needed in IROCs are:

- Coordinate production planning and energy planning so that production and energy plans are aligned.

- Manage the interaction with the electricity network operator (the DSO).

- Provide regular supply and demand forecasts at different time horizons (week, day, hour) to the DSO.

- Optimise mining schedules within the constraints of the energy supply.

- Monitor and respond to broadcast notifications from the DSO.

- Coordinate timely response when requested to reduce load by DSO.

- Coordinate emergency response to unscheduled events on the network, such as outages or discretionary load shedding by the operator.

- Monitor the status of data connections with the DSO and ensure that devices are online and communicating with the DSO's equipment so that the company-managed generation assets can reliably respond to signals from the DSO.

Green hydrogen

Hydrogen offers the potential alternative energy source with characteristics closer to diesel, making it a viable option for powering trucks and other mobile equipment via fuel-cell technology (other options such as hydrogen internal combustion engines are not considered in this paper). It is also possible to use hydrogen to power turbine generators to produce electricity. Today, 99 per cent of H_2 produced is 'brown' hydrogen, which results in enormous quantities of CO_2 produced. Alternative process for 'blue' (capturing or reducing CO_2 emissions) or 'green' (no CO_2 emissions) exist but are more expensive and supply is constrained, accounting for only 1 per cent of world supply (International Energy Agency, 2019). All hydrogen (regardless of how it was produced), creates no GHG when converted into energy, such as in the fuel-cell or a turbine generator.

Hydrogen can be stored as a pressurised gas or cryogenic liquid, transported and used to refuel trucks with similar operating characteristics (in terms of refuelling time, power and load capacity) to current diesel-powered trucks. It is therefore being considered as potential option for powering mining truck fleets.

A key difference to diesel however is the relative cost of large-scale storage, with diesel being much cheaper then H_2 at the scales need to operate a mine. Nehring *et al* (2018) identified the main challenges for mine-scale adoption of H_2 being:

- Storage of hydrogen: storage of large volumes is excessive cost.

- Transmission and distribution via pipes are technically possible but may not be economically viable.

Two scenarios are considered for adapting to these challenges:

1. Small-scale on-site generation from renewable power.

2. Large off-site 'hydrogen hub' supplying multiple customers with limited on-site storage.

Option 1 requires the management of a gas plant, which introduces an additional level of operational complexity. In addition to direct production of H_2 from electrolysis of water (using renewable electricity), other options are possible, such as conversion of ammonium (which is easier to transport and store then H_2).

Option 2 requires management of logistics of transporting hydrogen, for example in pressurised or liquified canisters. This in turn introduces additional complexity in managing a reliable inbound logistics operation to supply hydrogen with small buffer capacity on-site.

Both options require some storage, transfer on-site and introduce operational complexity and management of additional risk controls related to explosive gas, as well as additional plant and equipment to be managed, operated and maintained.

It is likely that the green H_2 supply chain will have to be tightly coupled to mine production as large storage buffers are not economically viable due to the cost of H_2 storage.

Finally, it assumed that H_2 powered equipment will have different operating characteristics to the current diesel equipment. Re-optimisation of the operating processes will be required after the introduction of H_2 equipment.

Implication of green hydrogen on IROCs

The introduction of Green Hydrogen as an energy source will require the following capabilities in IROCs:

- Planning and forecasting the demand for H_2 in alignment with the production plan.

- Coordinating logistical supply of H_2 to the operations.

- Managing the operation related to the safe transfer and storage of H_2.

- Monitoring all controls implemented for critical risks associated with the storage and transfer of hydrogen.

- Developing and continually improving operating practices for H_2 powered equipment during its introduction to improve the 'ramp up' of a modern technology.

Nuclear

Small modular reactors (SMR) may become technically feasible in the short to medium term, but they currently are not an option in Australia (as they are not permitted under the Australian Radiation protection and Nuclear Safety Act 1998 and the Environment Protection and Biodiversity Conservation Act 1999) and there are no current frameworks in place allowing for the operation of SMRs in Australia (Cronshaw, 2020). There is currently renewed active debate on nuclear power in Australia and if they were to become permitted (and are technically and economically feasible), it is assumed that it would have the following impact:

- Any SMR would be connected to the grid and provide baseload with high-capacity factor.

- Supply and demand planning between the mining companies and grid operator would become less critical as grid would have a much higher capacity factor.

- An SMR on its own does not have 100 per cent supply uptime, with downtime required for maintenance and operational activities. While supply is significantly more reliable and predictable then wind/solar renewables, it will still require some supply/demand coordination and alterative generation capacity available.

- If operated by the mining company (as opposed to a generation company), the SMR introduces many operating procedures, processes and compliance challenges requiring a level of operational discipline not currently implemented at mining companies.

ANALYSIS OF IMPACT ON IROC

A model was developed to assess the level of additional operational complexity of the impacts of implementing the decarbonisation options discussed above and responding to changing market demand patterns. A total of 28 solutions were identified, grouped into eight pathways (see Table 2). The operational complexity, resulting from the introduction of each solution, was evaluated as a function of 16 weighted dimensions. The model also assessed the time of implementation, based on three factors and interdependencies, as shown in Figure 8.

TABLE 2

Decarbonisation solutions grouped by pathway.

Process	Solutions	Pathway
Explore/Acquire	Long-term demand forecast	Respond to Changing Demand
Adaptive Mine Planning	Market Demand	Respond to Changing Demand
Adaptive Mine Planning	Renewable Generation and storage	Renewables
Adaptive Mine Planning	Carbon offset pricing	Renewables
Mining	Trolley Assist Diesel Electric	Mining
Mining	Lower carbon fuels	Mining
Mining	Battery-electric (with renewables)	Mining
Mining	H_2 Fuel Cell	Mining
Mining	Alternative mining methods (eg IPCC)	Mining
Rail	Regenerative Battery	Other
Crush and Screen	Lower-energy crushing (eg HPGR)	Mining
Beneficiation	Operate new beneficiation plants	Beneficiation
Beneficiation	Operate new pellet plant	Beneficiation
Beneficiation	Manage tailings disposal	Beneficiation
Beneficiation	Metallurgical balancing and control	Beneficiation
Stockpile	Upgrade product tracking capability	Respond to Changing Demand
Stockpile	Upgrade scheduling and blend control	Respond to Changing Demand
Energy supply	Solar Infrastructure	Renewables
Energy supply	Wind Infrastructure	Renewables
Energy supply	Energy storage Infrastructure	Renewables
Energy supply	Green H_2 production	Hydrogen
Energy supply	Green H_2 supply logistics	Hydrogen
Energy supply	H_2 storage and handling	Hydrogen
Energy supply	Nuclear (small modular reactor)	Nuclear
Energy supply	Generation Forecasting	Renewables
Energy supply	Demand Forecasting	Renewables
Energy supply	Renewable grid management	Renewables
Energy supply	Storage management	Renewables
Process	Solutions	Pathway

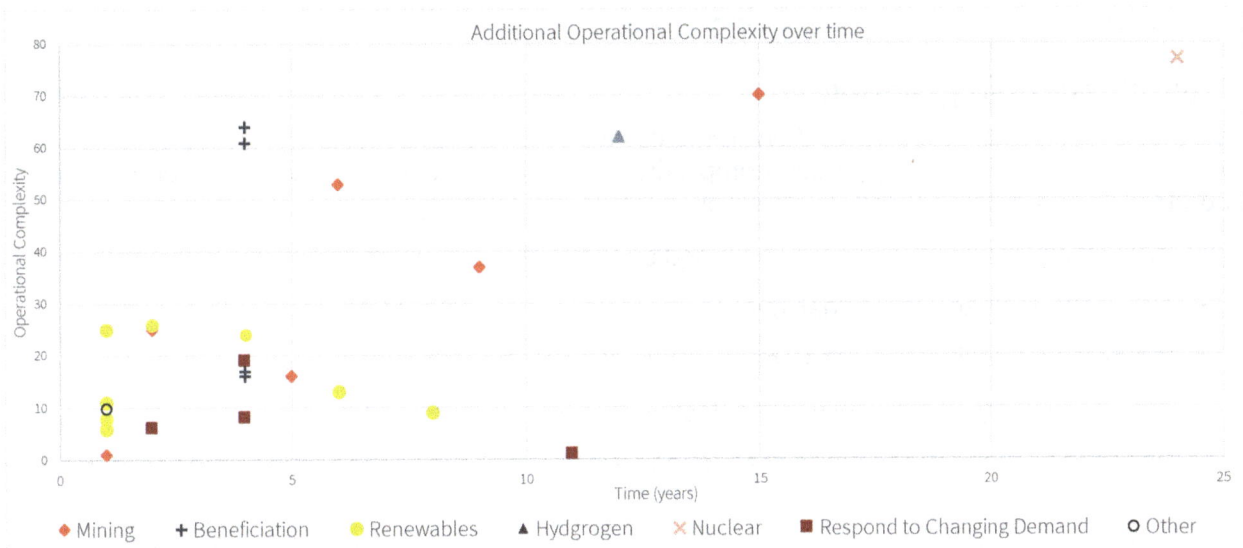

FIG 8 – Operational complexity of decarbonisation solutions and probable implementation timeline.

The model shows that, in aggregate, a significant increase in operational complexity will occur over the next 15 years (Figure 9). The biggest drivers of operational complexity are changes to mining methods (including electrification/hydrogen); introduction of more beneficiation processing; and management of renewables. These changes go beyond the original design criteria of today's IROCs and they will have to be reconfigured to adapt to the new demands.

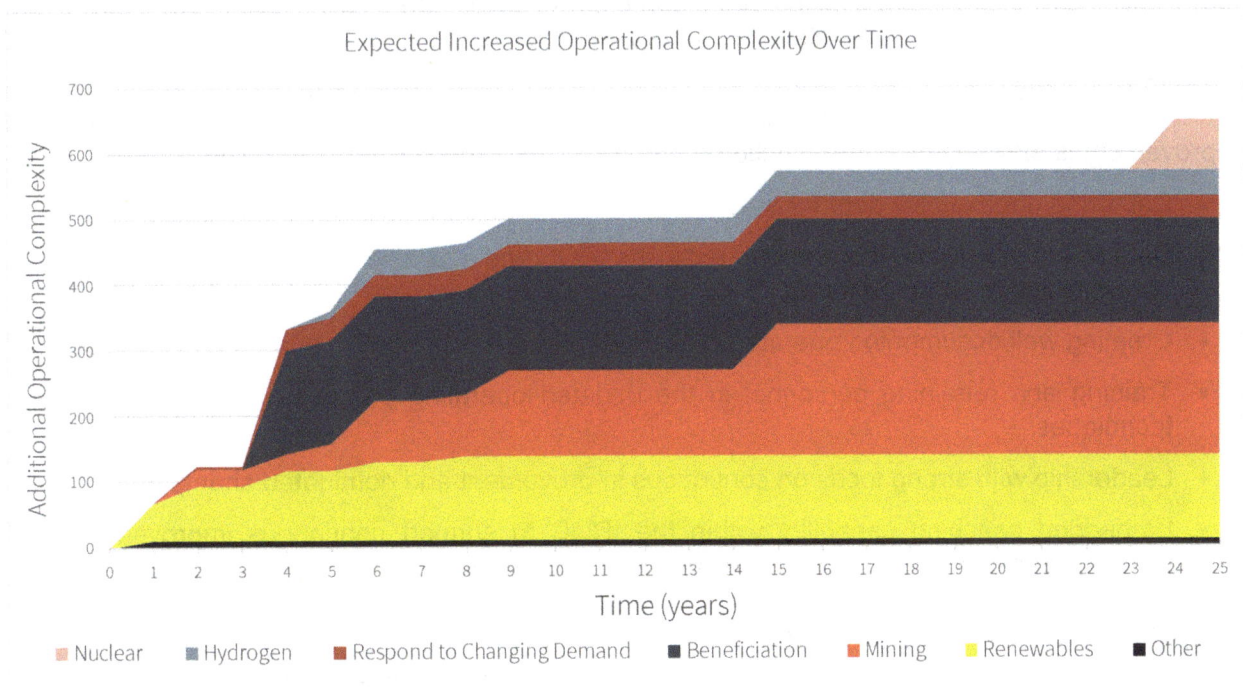

FIG 9 – Expected increase in operational complexity over time.

CHARACTERISTICS OF THE FUTURE IROC

In addition to the specific capabilities and considerations already identified in this paper applicable to the specific solution areas, there are four typical characteristics that will be essential for IROCs to support decarbonisation:

- support for technology adoption
- continuous improvement
- decision-making optimisation

- collaboration with internal and external partners.

Support technology adoption

As identified above, decarbonisation requires the rapid implementation of many innovative technologies. Each technology adds complexity to the mining operation with additional points of coordination and collaboration, with the main considerations being:

- integration into value-chain planning and scheduling
- coordination at an operational level
- the potential to disrupt mining operations or change the bottleneck
- the need to optimise the new technology
- the need to re-adjust the value chain to accommodate the operating characteristics of the new technologies.

The IROC will play a significant role in the above. The faster and more effectively the IROC can achieve operational integration of the innovative technologies, the less risk to production targets and better return on capital will be achieved. Therefore, it is proposed that IROCs should include capability that specialises in the introduction of new technologies. An effective way to achieve this is to incorporate it in the operating model of the IROC.

Continuous improvement

To enable process optimisation and efficiency improvements necessary of decarbonisation, the IROC must have strong capability in continuous improvement. This is also required to support the optimisation of new technologies after they are operationalised.

While most mining companies already have some operational excellence capability, in practice this may not be currently seen as a core capability of an IROC. It is proposed that continuous improvement is elevated to a core function of the IROC, with the key features being:

- Capturing good quality data for analysis.
- Data science skills and tools embedded in the IROC and available to be deployed to investigate opportunities for improvements.
- Creating well documented operated processes that are continually updated and improved.
- Training and retraining personnel in the updated operating process and improved control techniques.
- Leadership with strong focus on continuous improvement and nominated champions.
- Embedded coaching capability within the IROC to support continuous improvement and provide training in appropriate techniques and approaches.

Decision-making optimisation

Associated with continuous improvement is the requirement to improve the outcomes of decisions. This could include the ability to react faster to a condition, the ability to consider more information when deciding, or the ability to mathematically optimise for an outcome rather than relying on operator's assumptions, past experiences and beliefs.

Carbon neutral value chains add extra complexity to the decision process through the need to optimise for carbon; to consider new constraints (introduced by renewables, electrification etc); and to optimise product blends for profit. Human decision-makers cannot be expected to consistently achieve good outcomes given the number of constraints, the increasing complexity of the environment and time available to decide. To support carbon neutral, the IROC must incorporate:

- Definition of decision-making process and participants in decisions.
- Integration of analytics and decision support tools (such as optimisation engines, simulations, decision models derived from machine learning) into decision-making processes.

- Data-driven and evidence-based approach to measure the effectiveness of decisions.

- Continuous improvement of the decision models (linked into the overall continuous improvement mandate of the IROC.

- Data science capability to develop and continually improve analytics and decision support tools.

- Integration and information sharing between real-time operational systems, planning and scheduling systems and external partners in the value chain.

Collaboration

One of the features of the future decarbonised iron ore value chain will be reliance on external parties for energy, technology and tailored product demand. Consequently, collaboration and the ability to make optimised decisions that involve more external parties will become a necessary feature of the future IROC.

The linkage between value chain maturity and collaboration, in the context of IROCs, has been previously established (Farrelly *et al*, 2012), however today's iron ore value chain is only loosely coupled to suppliers and customers and they are not directly involved in decision-making processes. Until now, there has been little need to incorporate external collaboration.

The most mature examples of IROCs (outside of mining) are those characterised by well-defined and structured decision-making processes that extend to virtual partners outside the IROC. One leading example case study is an air traffic flow control centre that coordinates the activities of 66 air traffic control zones, military organisations from 25 countries and over 80 major airlines (EuroControl, 2019; Soyanov, 2022). Due to the vast number of partners involved, this operations centre is an exemplar of collaborative decision-making. They key features of a mature IROC with exceptional collaborative decision-making processes that are applicable to the mining context are:

- Processes for decision-making are well defined and documented. All parties agree and abide by the rules.

- Collaborative portal that supports decision-making. Partners can exchange structured information and make updates to their plans via a portal that is aligned to the collaborative decision-making (CDM) processes.

- Information is automatically checked when it is uploaded and shared to ensure plans conform to the planning rules.

- Deviations from the decision-making processes are followed up immediately (on the day).

CONCLUSIONS

The iron ore industry is commencing a period of profound change, driven by the need to decarbonise the iron ore value chain and the steelmaking process. The largest decarbonisation opportunity in the iron ore and steel value chain is in the steelmaking process and it will take several decades to create and commercialise the technologies to do this and there will be a lengthy transition while existing investments in steelmaking processes are optimised to reduce carbon. During the transition, there will an impact on the iron ore products that steel makers will require.

In the Australian context, there is an imperative to decarbonise the production of iron ore to meet policy targets. To realise this objective, several investments will be needed. These investments will also enable Australian iron ore producers to adapt to the change in market demand associated with downstream decarbonisation.

Integrated Remote Operating Centres (IROCs) are key components of the iron ore mining operating model. IROCs are effective in enabling efficient use of the logistics chain to coordinate production of standardised blends from mines employing a standard operating model. Modern IROCs have also supported the introduction of new autonomous and remotely operated equipment. Thus, IROCs are ideally placed to help Australian iron ore producers introduce decarbonising operational processes and technologies.

However, the current IROC model will need to incorporate additional capabilities to support the changing demands caused by decarbonisation. A defining feature of the carbon-neutral value chain will be its additional complexity above the current baseline in today's operations. The complexity will be driven by the introduction of innovative technologies (such as battery electric or hydrogen), additional processing of ore, the need to source renewable energy and changing requirements of customers (steelmakers). This paper has explored the impact of each of these technologies on IROCs and how they need to respond.

Four fundamental capabilities have been identified as required in decarbonising IROCs that will best enable adaptation to increased complexity: supporting the adoption of new technology; driving continuous improvement in operations; optimising decisions at all levels; and supporting collaboration up and down the value chain (including with internal and external suppliers).

Mining companies that can establish the capabilities identified in their IROC will be best placed to not just achieve their decarbonisation goals on time, but importantly, will be able to adapt fastest to changing conditions in the industry.

ACKNOWLEDGEMENTS

The authors would like to thank Michael Bobotis (Hatch) and Wai Kin Wong (Hatch) for their views and review input on climate change, impacts to the mining industry and sustainable energy grids. The authors acknowledge the use of data obtained from the publicly available Australian Mine Atlas, published by Geoscience Australia (Australian Government) in the analysis supporting this paper.

REFERENCES

Clean Energy Regulator, 2019. Australia's Scope 1 Emissions by Industry for NGER Reporters, The Australian Government. Available from: <https://www.cleanenergyregulator.gov.au/NGER/> [Accessed: 2 June 2023].

Constantini, X, Fookes, W, Neise, P, Pujol, F, Rubenstein, B and Sivecas, G, 2022. How Mining Companies Reach the Operational Excellence Gold Standard, McKinsey and Company.

Cronshaw, I, 2020. Australian Electricity Options: Nuclear, Parliament of Australia – Department of Parliamentary Services.

Dubov, D, Trukhmanov, D, Kuznetsov, I, Nokhrin, S and Sergel, A, 2019. Prospects for the Use of Liquefied Natural Gas as a Motor Fuel for Haul Trucks, *E3S Web Conference*, p 105.

Ellis, B and Bao, W, 2020, Nov 5. Pathways to Decarbonisation – Episode Two: Steelmaking Technology, BHP. Available from: <https://www.bhp.com/news/prospects/2020/11/pathways-to-decarbonisation-episode-two-steelmaking-technology> [Accessed: 2 June 2023].

EuroControl, 2019. Introducing the EuroControl Network Manager Operations Centre, EuroControl.

Farrelly, C T, Malherbe, G, Gonzalez, J, Bassan, J and Franklin, D C, 2012. The Network Centric Mine, in *Proceedings of the International Mine Management Conference*, pp 143–160 (The Australasian Institute of Mining and Metallurgy: Melbourne).

Feng, Y and Dong, Z, 2020. Integrated Design and Control Optimization of Fuel Cell Hybrid Mining Truck with Minimized Lifecycle Cost, *Applied Energy*, 270.

Horizon Power, 2021, June 29. Pilbara Network Facilities, Horizon Power. Available from: <https://nwis.com.au/media/4itnhu4z/pilbara-network.pdf> [Accessed: 2 June 2023].

International Energy Agency, 2019. *The Future of Hydrogen: Seizing Today's Opportunities* (IEA Publications: France).

International Energy Agency, 2020. *Iron and Steel Technology Roadmap* (IEA Publications: France).

Lanka, V V, Roy, M, Suman, S and Prajapati, S, 2021. Renewable Energy and Demand Forecasting in an Integrated Smart Grid, *Innovations in Energy Management and Renewable Resources*, pp 1–6.

Linklater, J, 2021, March. Adapting to Raw Materials Challenges: Part 1 – Operating Midrex Plants With Lower Grade Pallets and Lump Ores, Midrex. Available from: <https://www.midrex.com/tech-article/adapting-to-raw-materials-challenges-part-1-operating-midrex-plants-with-lower-grade-pellets-lump-ores/> [Accessed: 2 June 2023].

Madhavan, N, Brooks, G, Rhamdhani, M and Bordignon, A, 2022. Contribution of CO_2 Emissions from Basic Oxygen Steelmaking Process, *Metals*, 12(5).

Mazumdar, J, 2013. All Electric Operation of Ultraclass Mining Haul Trucks, IEEE Industry Applications Society Annual Meeting.

Nehring, M, Knights, P, Kizil, M and Hay, E, 2018. A Comparison of Strategic Mine Planning Approaches for In-pit Crushing and Conveying and Truck/Shovel Systems, *International Journal of Mining Science and Technology*, 28(2):205–214.

Nichols, S and Basirat, S, 2022, June. Iron Ore Quality a Potential Headwind to Green Steelmaking, Institute of Energy Economics and Financial Analysis. Available from: <https://ieefa.org/resources/iron-ore-quality-potential-headwind-green-steelmaking-technology-and-mining-options-are> [Accessed: 2 June 2023].

Norgate, T and Haque, N, 2010. Energy and greenhouse gas impacts of mining and mineral processing operations, *Journal of Cleaner Production*, 18(3):266–274.

Nunna, V, Hapugoda, S, Pownceby, M I and Sparrow, G J, 2021. Beneficiation of Low Grade, Goethite-rich Iron Ores by Microwave-assisted Magnetizing Roasting and Magnetic Separation, *Minerals Engineering*, 166.

Palamure, S, 2016. Energy Efficiency and Carbon Dioxide Emissions Across Different Scales of Iron Ore Mining Operations in Western Australia, thesis, Edith Cowan University.

Parker, K, 2018. The Operations Centre (OC), Rio Tinto Iron Ore Site Visit 2018. Western Australia: Rio Tinto.

Schweikart, V, 2007. Rio Tinto Iron Ore Remote Operations Centre, Autonomous Mining Systems Conference.

Simpson, C, 2017. The Iron Bridge Magnetite Deposits, in *Proceedings Iron Ore 2017 Conference,* pp 287–294 (The Australasian Institute of Mining and Metallurgy: Melbourne).

Soyanov, S, 2022, Nov 12. Operations Manager, EuroControl, interviewer: J Bassan.

Summerfield, D, 2019. *Australian Resource Reviews: Iron Ore 2019* (Geoscience Australia: Canberra).

Ware, N A, Manuel, J R and Raynlyn, T D, 2013. Fundamental Melting Behaviour of Hematite and Goethite Fine Ores in the Sintering Process, in *Proceedings Iron Ore 2013 Conference,* pp 485–486 (The Australasian Institute of Mining and Metallurgy: Melbourne).

Zhao, Z, Wu, G, Borboonsomsin, K and Kailis, K, 2021. Vehicle Dispatching and Scheduling Algorithms for Battery Electric Heavy-Duty Truck Fleets Considering En-route Opportunity Charging, *IEEE Conference on Technologies for Sustainability (SusTech),* pp 1–8 (IEEE: Irvine).

Moisture/particle interactions under oscillatory motion – model development

D Ilic[1], K Williams[2], V Gurung[3] and A Lavrinec[4]

1. Senior Research Associate, University of Newcastle, Callaghan NSW 2308.
   Email: dusan.ilic@newcastle.edu.au
2. Professor, University of Newcastle, Callaghan NSW 2308.
   Email: ken.williams@newcastle.edu.au
3. PhD Candidate, University of Newcastle, Callaghan NSW 2308.
   Email: virat.gurung@uon.edu.au
4. Research Associate, University of Newcastle, Callaghan NSW 2308.
   Email: aleksej.lavrinec@newcastle.edu.au

ABSTRACT

The moisture content of iron ore during handling can range from the dust extinction moisture at the point of arrival at the port to the transportable moisture limit at the point of loading ships for maritime transport. Moisture content and ore composition, specifically fines content, have a large influence on ore flowability. Typically, elevated moisture can result in bottlenecks, downtime and lost productivity due to build-up, conveyor carry-back and blockages, increasing maintenance costs. On the other hand, low moisture can result in increased dust emissions influencing the environmental footprint and product losses. Managing the use of water across the supply chain is therefore critical for service life of equipment, productivity, but also has environmental and social impacts, affecting the social license to operate.

Previously, experiments have shown that oscillations induced in iron ore during handling and transport can result in movement of moisture through the ore. This kind of moisture migration depends on the nature of these oscillations, such as the undulation of idlers during conveying, the inherent ore properties and the initial moisture content. Moisture has been observed to move either upwards and/or towards the bottom of the burden. These effects can be exploited to remove excess water from the ore influencing handleability and operation, however, the actual mechanisms of moisture migration are currently poorly understood.

The study presented herein involves the use of simulation to examine moisture and ore interactions. Simulations using discrete element method (DEM) and coupled with smoothed particle hydrodynamics (SPH) are implemented to investigate the behaviour of moisture in wet ore in the presence of oscillation. This is conducted at a fundamental level first using glass beads and water which is then compared to iron ore at high moisture content. The results of the modelling are discussed in view of experiments conducted.

INTRODUCTION

Moisture migration in wet bulk materials occurs at different stages of the supply chain including storage, processing and transport (Chen *et al*, 2020b). It can lead to a heterogeneous bulk profile or varying quality, loss of material functionality (Zafar *et al*, 2017) and in extreme cases severe problems like liquefaction (Ju *et al*, 2019). Therefore, increased understanding of moisture migration phenomena and its underpinning mechanisms in wet bulk solid materials is necessary to address and overcome possible handling problems in practice.

Moisture migration and the resulting heterogeneity in product quality has become a topic of interest in different industries including mining and minerals processing (Zafar *et al*, 2017; Vasic, Grbavcic and Radojevic, 2014; Chen, 2018). One example from the latter that is of significant relevance to this article is moisture migration during maritime transport. Around 3 billion tonnes (Bt) of granular material (eg iron ore, coal, grain, bauxite, phosphate) is transported annually by sea (Airey and Ghorbani, 2021). In the Australian context, mining and minerals account for almost 75 per cent of Australia's trade (Minerals Council of Australia, 2017) and over 10 per cent of GDP (DeCoff, 2022). By considering such dominance and economic significance of the resources industry, the prominence of understanding moisture migration mechanisms during maritime transport is evident.

In the recent past, several large oceanic bulk carriers were lost at sea, some with crew members, with one reported case of capsizing attributed to sudden loss of stability arising due to cargo liquefaction (Airey and Ghorbani, 2021; Ju *et al*, 2019). The phenomenon of liquefaction is related to an abrupt transformation from a solid, dry state to a liquid like state, due to the migration of moisture (Standard Club, 2021). At a more local scale, but essential to mineral and resource supply and trade is the example of problems causing moisture migration during transportation of bulk materials using belt-conveying systems. During conveying of wet ore, a significant amount of water can migrate due to the deflection of the belt between consecutive idler sets, thus changing the bulk material properties (Chen, 2018; Ilic, 2013). The properties of granular bulk solid materials are often determined according to their inherent moisture content and during storage, transport or handling processes, the moisture that is present within bulk solids is very much dynamic in its nature. The actual migration of moisture, that is movement and concentration in a specific region within a bulk material, occurs due to a range of factors including settlement, impact, changes in temperature, acceleration during handling and vibration and/or oscillatory motion exerted on the granular material.

The motivation for our project stems from previously observed moisture migration in wet, or rather, partially saturated bulk materials under oscillatory type motions, notably maritime transport and belt conveying. A partially saturated state occurs when the voids present between individual particles and grains are only partially filled or saturated with water (Chen, 2018). We can experimentally replicate this behaviour using an oscillating apparatus, whereby a partially saturated bulk solid sample is loaded in cells. The sample is initially placed inside a column of cells at a specific moisture content and then oscillated vertically. Following a period of oscillation (typically in the order minutes to hours), in this case corresponding to an acceleration of 6.0 m/s^2 and frequency of 2 Hz, with the addition of a specific amount of water above a starting moisture content (X), the moisture within the column migrates to the bottom as illustrated for an iron ore sample in Figure 1.

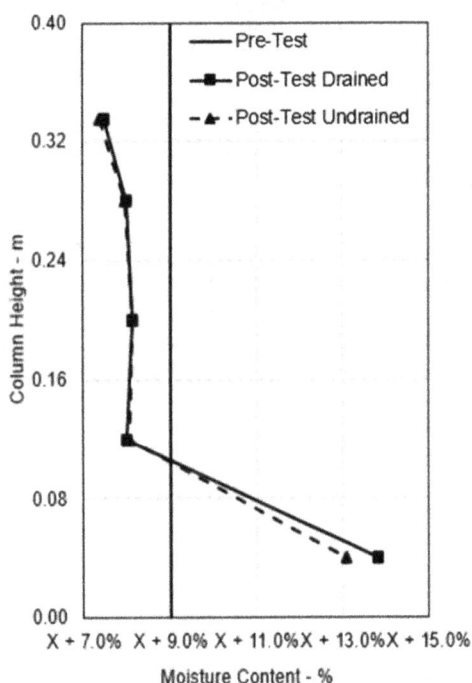

FIG 1 – Moisture migrating to the bottom of an iron ore, above 7.0 per cent added moisture.

The main objective with this work is to improve a theoretical model by using physical experimentation, modelling and simulation. The focus of this specific article is development of the simulation model.

The Discrete Element Method (DEM) is often used to model granular material systems. The method simulates movement of granular materials through calculations that trace individual particles (Tanaka, 2001). It is based on the integration of the equations of motion simultaneously for all particles by considering the contacting forces acting on the particles at a microscopic level and reflecting the motion of particles at a macroscopic level (Radjai and Dubois, 2011). Smoothed particle hydrodynamics (SPH) was initially developed to simulate non-axisymmetric phenomenon and solve astrophysical problems (Gingold and Monaghan, 1977; Lucy, 1977). In this meshless method, a continuous system is represented by a set of elements/particles (Wang *et al*, 2016) and unlike conventional grid-based methods numerical solutions are approximated using Lagrange formulation (Liu and Liu, 2010; Yang, Peng and Liu, 2014). With evolution of SPH, its application has almost naturally extended to the domain of fluid dynamics and solid mechanics, where, similarly to DEM, it is used to approximate equations of motion (Monaghan, 2005). To model transport phenomena, like DEM, individual elements with inherent physical properties representing the fluid are tracked (Gingold and Monaghan, 1977). This involves kernel interpolation to smooth and approximate particle/element interactions according to a defined radius, thus representing continuous fluid behaviour without a fixed grid or mesh (Liu and Liu, 2010; Wang *et al*, 2016; Ansys Inc., 2023).

A few simulation studies of moisture migration of partially saturated bulk solids exist in literature. For example, simulations investigated dislodging of particles under a hydraulic gradient (Tao and Tao, 2017) and preliminary studies were conducted at a small scale previously in our group investigating use of DEM-SPH coupling via a Python interface (Chen, 2018; Chen *et al*, 2020a). The latter included development of a liquid bridge model that was simulated and examined in detail. Our current research into moisture migration involves advancing the work, but with focus on exploring the development of a theoretical, continuum model. Consequently, using our experience of DEM application at full scale (eg Ilic, Lavrinec and Orozovic, 2020) we are commencing by modelling fluid-solid interactions at a fundamental scale. This includes exploring the applicability of the DEM-SPH approach to study moisture migration in bulk solids under dynamic conditions. In this paper, we present preliminary DEM-SPH modelling and experimental testing of idealised bulk solid/moisture interactions using glass beads and water during vertical oscillatory motion.

METHODS, MATERIALS AND APPARATUS

Preliminary experimental tests were conducted in a custom-built oscillation apparatus using glass beads. A single particle size fraction of glass beads at known moisture content was used. Modelling and simulation were conducted using a coupled DEM-SPH approach. The DEM involves calculation of forces between interacting solid particles, based on a contact model that describes an overlap in the tangential and normal directions. A rolling friction model was used to calculate motion resistance due to particle torque. The SPH method involves solving the Navier-Stokes equation to describe the flow of the fluid by interpolating elements with assigned physical properties and designated spatial position. These elements, representing the fluid phase, interact with neighbouring fluid elements, solid particles and boundary surfaces through forces that arise due to changes in pressure and viscosity (Ansys Inc., 2023). The coupling of the two methods is modelled by assuming solid-fluid interaction under no-slip conditions (Potapov, Hunt and Campbell, 2001).

Experimental testing and apparatus

Glass beads were initially air-dried for 24 hours to remove any inherent moisture that may be present. The moisture content of the glass beads sample was then increased (to 2.0 per cent by mass) by adding water to produce a partially saturated state. This was conducted by mixing a small mass of glass beads with water in a small bucket prior to loading into the cells. The small mass was approximated to fill each individual cell. The moisture level was slightly below the free drained saturation moisture content (which was previously determined to be 2.44 per cent by mass). The solids density of the glass spheres of 2504.8 kg/m³ was measured in a nitrogen displacement pycnometer.

Experiments were conducted in a hydraulic scissor-lift oscillation system (Chen, 2018), illustrated in Figure 2. The test involves vertically oscillating a previously loaded stack of acrylic cells (four cells, each 80 mm high, 140 mm diameter) with material. During oscillation, trays located underneath the columns of stacked cells, collect water as it passes through a perforated mesh. Peak amplitude and frequency were controlled via a computer interface to produce a sinusoidal motion.

FIG 2 – Oscillating apparatus.

Two different vertical motions ie 0.0396 m amplitude at 0.8 Hz frequency and 0.1187 m amplitude at 0.8 Hz, corresponding to peak accelerations of 0.5 and 1.5 m/s^2 respectively were investigated. We selected these parameters according to previous study in our group (Chen, 2018). The parameters represent heaving ship motion that can be experienced during maritime transport of iron ore. Following a test duration of 30 mins, the contents from each cell were removed onto trays and assessed for moisture content by oven drying at a temperature of 105°C.

Modelling and simulation

A simplified, small-scale CAD model was developed using Creo Parametric 9.0 and incorporated into a DEM environment using Rocky DEM software package (ESSS/Ansys Inc., 2023 R1). The geometry consisted of an 80 mm high, 20 mm diameter cell, situated immediately above a perforated mesh (staggered, 2 mm aperture size, 4 mm spacing). Initially, the cylinder was loaded with a mass of 23.7 g, 3.5 mm glass bead spherical particles, representing the solid phase. For the DEM representation of the phase, a Hertz-Mindlin contact model was used to calculate the forces acting in the horizontal and tangential directions and the Type C rolling friction model used to simulate torque resistance (Ajmal *et al,* 2020; Wensrich and Katterfeld, 2012).

Following this, the granular assembly was loaded with a mass of 0.48 g of 350 µm elements representing the water phase (2.0 per cent by mass), modelled using SPH and allowed to settle for 3.0 s. Each SPH element interacts with its neighbours based on a relative distance between them and a kernel function that is dependent on a smoothing length that is defined by the SPH element (Ansys Inc., 2023). The Wendland kernel function is used to compute localised fluid properties by integrating across the neighbouring SPH elements within the domain. Boundary forces on SPH elements are also treated as normal and tangential contacts, with normal force calculated according to a linear spring-dashpot model and the tangential force based on laminar viscous force arising out of relative tangential velocity (Ansys Inc., 2023).

After 3.0 s, vertical oscillation was induced into the system corresponding to that used in the experiments. The behaviour of the system was simulated for a period of 20.0 s of oscillation, during which, the movement of the solid and water phases was monitored and investigated. The analysis involved assessing six 10 mm high cylindrical regions within the mass modelled, from the bottom to the top of the burden. An illustration of the typical images observed during the simulation are shown in Figure 3 where the solid phase is coloured red and water phase is coloured blue. Once settled, the height of the solid particles was approximately 60 mm.

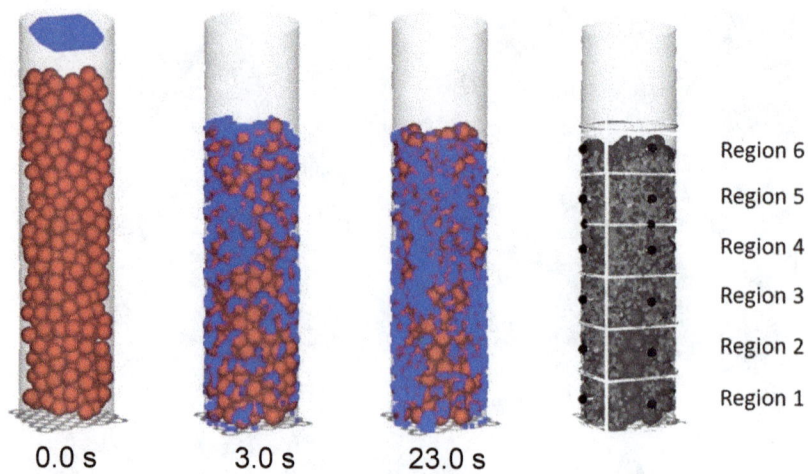

FIG 3 – Observations from initial DEM-SPH simulation and the analysis regions.

Following this, the simulation set-up was modified to distribute the water more evenly within the solid particle assembly compared to the previous simulation. This involved first generating incremental layers of 20 mm solid particles four times to a total height of approximately 80 mm. This produced a total mass of particles of 30.3 g. In between each increment, the water phase was generated, to a total of 0.64 g (or 2.0 per cent by mass) with both phases allowed to settle for 3.0 s. As the mass simulated in the revised set-up was larger compared to the initial case, the analysis involved assessing eight 10 mm high cylindrical regions within the mass modelled, from the bottom to the top of the burden. However, as only a very small number of particles/elements was present within Region 8, the results are presented with Region 7 and 8 combined. This revised set-up is illustrated in Figure 4.

Once settled, the height of the solid particles was approximately 70 mm. Upon reviewing the results, which are shown in the following section, the simulation with the higher acceleration of 1.5 m/s² was then extended for a further 40 s. During this extended simulation, the moisture content of different regions was monitored. The final orientation of the loaded cell from low to high bulk solid inside is presented in Figure 4. The centre of Region 1 at the bottom of the cell is located 5 mm above the perforated plate. The centre of each subsequent region above is equally spaced at 10 mm.

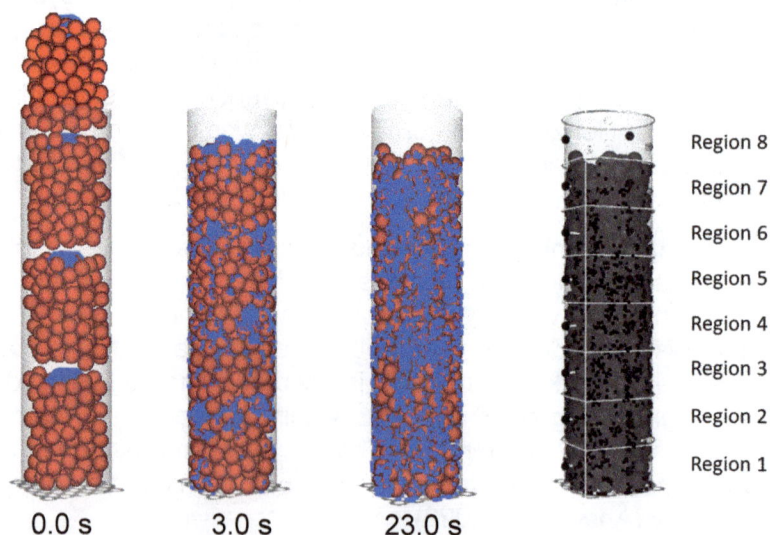

FIG 4 – Observations from the revised DEM-SPH simulation and the analysis regions.

To calculate the dynamics of the water phase in SPH, the water elements are modelled as weakly compressible and the mass is fixed. Pressure and velocity of the phase vary over time due to local volume changes according to controlled fluctuation in density (Ansys Inc., 2023). In this approach, an artificial equation of state calculates the pressure according to these density fluctuations. The

Morris viscosity formulation is used based on a laminar flow with low Reynolds number (Morris, Fox and Zhu, 1997). For even spatial distribution of SPH elements, a shifting correction is implemented according to Fick's law of diffusion (Ansys Inc., 2023). During DEM-SPH coupling, numerical solutions are obtained by using the same time step (whichever is the minimum). The simulation parameters of the study are summarised in Table 1.

TABLE 1

Summary of main simulation parameters.

Parameter	Value
Bulk solid fraction	0.6
Young's modulus (Pa)	1e6
Poisson's ratio	0.3
Coefficient of restitution	0.3
Particle density (kg/m^3)	2500
Particle diameter (mm)	3.5
Time step (s)	1.09e-6 to 2.76e-5
Coefficient of particle sliding friction	0.3
Coefficient of particle/wall sliding friction	0.3
Coefficient of rolling friction	0.1
Fluid density (kg/m^3)	1000
Fluid viscosity (Pa.s)	0.001
Fluid element size (μm)	350
Kernel minimum, distance factor	0.0001, 1.25
Shifting factor	2
Number of solid particles	540
Number of fluid elements	297031

RESULTS AND DISCUSSION

In this section, results from the experiments and the modelling study are presented. Relevant discussion is provided accordingly.

Experimental testing

The results from the experimental tests are shown in Figure 5. They show that after 30 minutes of oscillation, the bottom cell contains the highest moisture content. This indicates that for both accelerations the moisture progressively migrates towards the bottom, with some progressively lost to the tray below. It is also worth noting here that the mass of water collected in the tray below the cells was two times more for the lower acceleration of 0.5 m/s^2 (17.8 g) compared to the faster acceleration of 1.5 m/s^2 (8.9 g). This tends to explain the consistently lower moisture content measured at lower acceleration. Additionally, some additional error could have been introduced by pre-mixing the test sample prior to loading into the cells. This is currently being investigated, by for example, removing the test sample from the cells prior to conducting any vibration and assessing the moisture content.

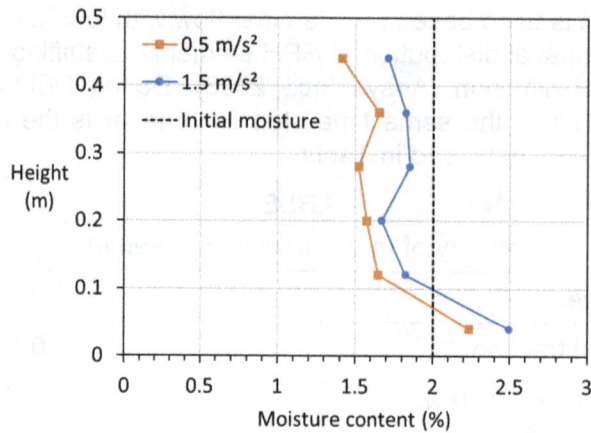

FIG 5 – Experimental testing of 3.5 to 4 mm glass beads.

Modelling and simulation

Typical results obtained from the initial DEM-SPH simulation are illustrated in Figure 6 and summarised in Table 2. Here the starting and final mass and moisture contents were calculated at a simulation time of 3.0 s and 23.0 s respectively. While a nominal amount of water equating to 2.0 per cent was applied to each case initially, by 3.0 s of settling time, some water had percolated through the sample, with each case decreasing to 1.72 to 1.73 per cent. In both cases, the results show the region at the top (Region 6) lost the most water. All the other zones gained moisture, apart from Region 2 (second from the bottom) which lost some moisture or remained relatively constant. These results illustrate the challenge in simulating a sample with a homogenous moisture profile. The results also show that over the 20.0 s of the simulated oscillation, a moisture loss of 0.21 per cent to 0.26 per cent was observed (more water loss for the lower acceleration).

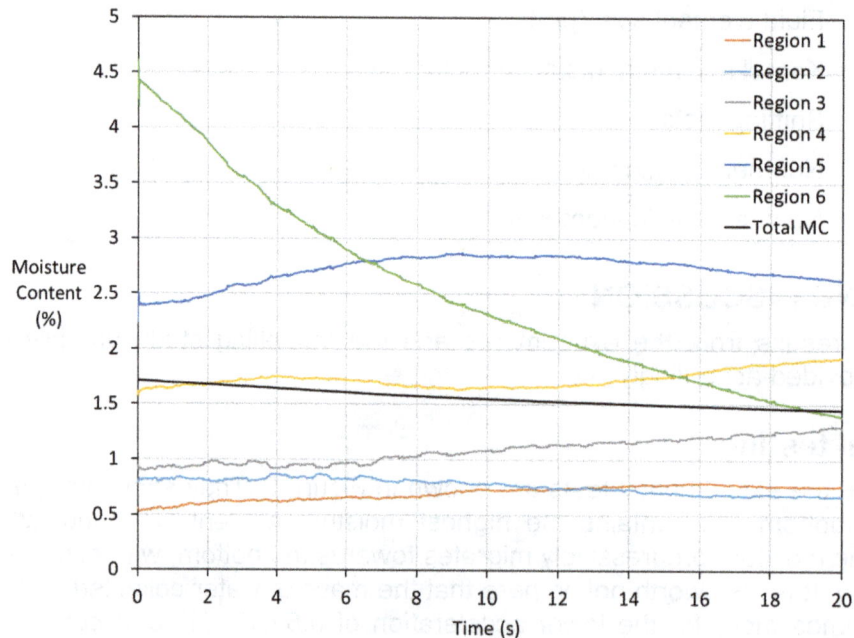

FIG 6 – Typical results obtained with initial DEM-SPH simulation during oscillation.

TABLE 2
Summary of results for the initial simulation.

Acceleration (m/s²)	0.5	1.5	0.5	1.5	0.5	1.5	0.5	1.5
Region	Starting mass (g)		Final mass (g)		Initial moisture content (%)		Final moisture content (%)	
1	3.82	3.82	3.87	3.98	0.56	0.61	0.77	0.71
2	4.10	4.10	4.15	4.21	0.81	0.68	0.68	0.70
3	4.10	4.21	4.21	4.32	0.91	0.98	1.29	1.32
4	3.98	3.98	4.10	4.21	1.58	1.43	1.93	2.14
5	4.04	4.04	4.10	4.15	2.24	2.54	2.63	2.76
6	3.70	3.59	3.31	2.86	4.26	4.28	1.40	1.41
Entire sample (g)	23.74				1.72	1.73	1.46	1.52

Subsequently, the revised DEM-SPH simulation for homogenous moisture distribution was investigated. The typical results are illustrated in Figure 7 and summarised in Table 3. Again, the mass and moisture content are assessed at 3.0 s and 23.0 s representing the start and end of the oscillation respectively. Following the first 3.0 s of settling time, the moisture content decreased from 2.06 to 2.03 per cent. For both accelerations, the results show Region 4 in the middle losing the most water. Interestingly, for both accelerations, the same trend of either losing or gaining moisture was observed and while in each case the top cell lost moisture, subsequent cells either lost or gained moisture in an alternating pattern. The results illustrate that a homogenous moisture profile cannot be attained using this set-up until oscillation is induced to promote movement. More moisture loss was again observed for the lower acceleration, which aligns with experimental observations. The results show a homogenous moisture profile being approached due to water movement following oscillation in the order of 20.0 s duration. Moisture was observed to be almost static following the initial 3.0 s of settling, prior to oscillation.

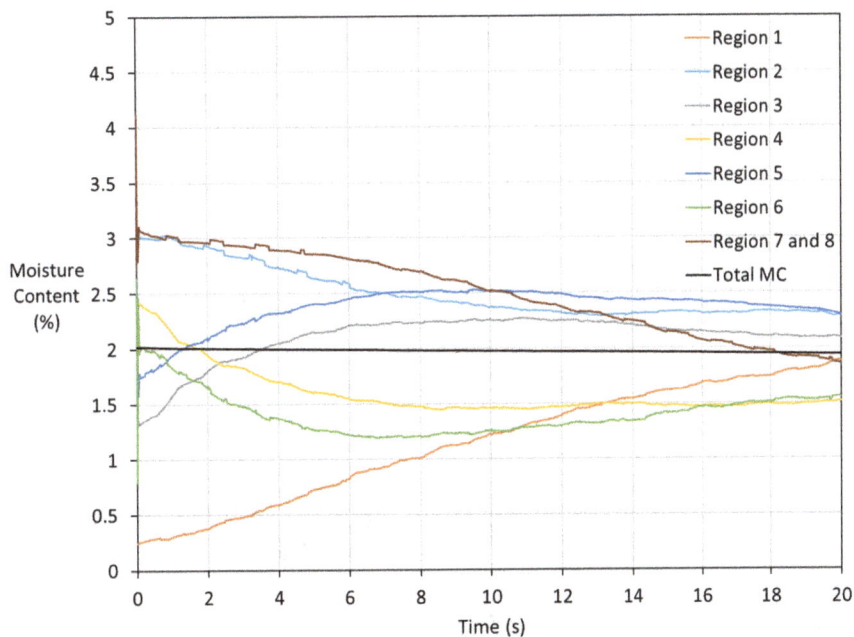

FIG 7 – Typical results obtained with revised DEM-SPH simulation during oscillation.

TABLE 3
Summary of results for the revised simulation.

Acceleration (m/s²)	0.5	1.5	0.5	1.5	0.5	1.5	0.5	1.5
Region	Starting mass (g)		Final mass (g)		Initial moisture content (%)		Final moisture content (%)	
1	3.93	3.87	3.98	3.98	0.27	0.25	2.14	1.88
2	4.43	4.21	4.43	4.55	3.06	3.18	2.12	2.29
3	4.15	4.38	4.27	4.43	1.08	0.81	2.30	2.08
4	4.38	4.15	4.32	4.27	2.76	3.03	1.50	1.52
5	4.10	4.27	4.04	4.49	1.26	0.99	2.23	2.29
6	3.98	3.93	4.10	3.93	2.62	2.94	1.39	1.55
7 and 8	5.33	5.50	5.16	4.66	2.80	2.66	1.75	1.86
Entire sample	30.31				2.03	2.02	1.92	1.94

Review of the above analysis, overall, the results indicate that oscillation is needed to move moisture vertically through the mass. In the absence of oscillation, the moisture of individual regions remained unchanged. The results also indicate that it is only following oscillation, that a homogenous moisture profile can be attained. Therefore, to assess additional movement of moisture throughout the solid particle assembly the simulation was extended for an additional 40.0 s. In the interest of time, this was simulated only for the case of higher acceleration. Using the data of the additional simulation time and combining with data presented previously in Table 3, the relevant moisture content at 3.0, 23.0, 43.0 and 63.0 s is plotted in Figure 8. The solid particle mass in each region analysed remained unchanged from that observed at 23.0 s.

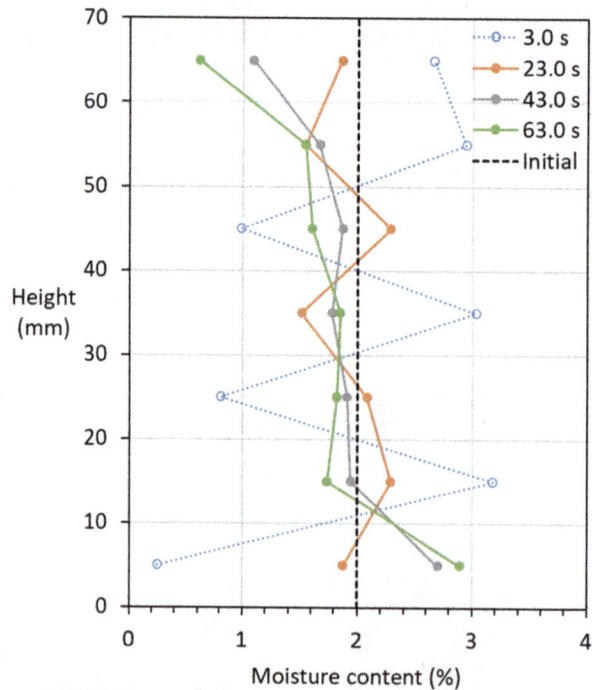

FIG 8 – Moisture content profile obtained from the revised and extended simulation (1.5 m/s²).

If we only consider the instance from a (fairly) homogenous moisture profile at 23.0 s, additional movement of moisture within the bulk particle assembly is observed after extended oscillation, to

43.0 s and 63.0 s. The bottom of the column, height of 5 mm corresponding to the centre of Region 1, shows the highest increase in moisture (0.82 per cent at 43.0 s and 1.01 per cent at 63.0 s). The combined region at the top, height of 65 mm corresponding to centre of Regions 7 and 8, showed the biggest decrease (0.77 per cent at 43.0 s and 1.24 per cent at 63.0 s). Overall, at the end of the simulation, a total of 0.36 per cent of moisture was lost or in other words moved downwards through the mass and passed through the perforated plate. This corresponds to a mass of 0.53 g. Qualitatively, the trends observed in this idealised, small-scale simulation generally align well with expectations generated from both the idealised glass beads experimental tests presented in Figure 5 and those of a wet iron ore illustrated in Figure 1.

While these preliminary results show insight into the movement of moisture within the bulk with much more control of initial conditions and detailed analysis compared to the experiments, it is obvious that further investigation is necessary. For example, the influence of surface tension has not yet been investigated. However, this also presents several challenges such as the physical scale of the simulation domain as well as the duration of the investigation. The extended simulation of 63.0 s took approximately one week to solve. On the other hand, moisture migration in practice generally occurs due to behaviour that takes minutes or hours. For this reason, a theoretical model is also being investigated that will translate the learnings from experimentation, fundamental interactions via simulation to full scale and industrial application.

CONCLUSIONS AND FURTHER WORK

Moisture migration can occur in bulk solid materials due to motion that can be induced during storage, handling and transportation. This can result in heterogenous moisture profiles, unexpected loss of strength/stability, increased problems due to handling and variable product quality. Preliminary experimentation under oscillatory motion has been undertaken using an idealised material, glass beads, in the presence of moisture. The experimental results illustrate moisture migrating towards the bottom of a column of material, which correlates to that observed with wet iron ore. A preliminary liquid/solid coupled DEM-SPH model has been developed with aim of assessing the interaction of moisture and particles, increasing the resolution of analysis and simulation of interactions, although at a much finer scale compared to physical experiments, both with respect to time and size. Simulation results showed that the method of generating the fluid phase is crucial in establishing a homogenous moisture profile and subsequently enabling a relevant assessment of moisture migration. This calls into question the homogeneity of the moisture profile that is present in the bulk material as it is physically loaded into the column for experimental oscillatory testing and will be investigated further. For example, once loaded into the cells/column, the material can be immediately removed for moisture content assessment to establish an indicative actual starting moisture profile.

The next steps in the development of the simulation model may include implementation of periodic boundaries, varying particle size distribution, increased head height of solids as well as a sensitivity analysis of the size of SPH elements modelled and/or influence of surface tension. The aim will be to calibrate the DEM-SPH parameters so that simulated behaviour is reflective of the behaviour observed in the experiments. Once this is established for the idealised material, it will be applied to iron ore samples of various properties, including particle size and moisture content. The modelling and simulation will be complemented with the development of an updated theoretical model.

ACKNOWLEDGEMENTS

The authors acknowledge the funding support from the Australian Research Council for the ARC Centre of Excellence for Enabling Eco-Efficient Beneficiation of Minerals, grant number CE200100009.

REFERENCES

Airey, D W and Ghorbani, J, 2021. Analysis of unsaturated soil columns with application to bulk cargo liquefaction in ships, *Computers and Geotechnics*, 140:104402.

Ajmal, M, Roessler, T, Richter, C and Katterfeld, A, 2020. Calibration of cohesive DEM parameters under rapid flow conditions and low consolidation stress, *Powder Technology*, 374:22–32.

Ansys, Inc., 2023. SPH Technical Manual.

Chen, J, 2018. Theoretical, experimental and numerical studies on dynamic moisture migration within bulk solids, PhD Thesis, University of Newcastle, Australia.

Chen, J, Orozovic, O, Williams, K, Meng, J and Li, C, 2020a. A coupled DEM-SPH model for moisture migration in unsaturated granular material under oscillation, *International Journal of Mechanical Sciences*, 169:105313.

Chen, J, Williams, K, Chen, W, Shen, J and Ye, F, 2020b. A review of moisture migration in bulk material, *Particulate Science and Technology*, 38(2):247–260.

DeCoff, S, 2022. Australia – Mining by the numbers 2021. Available from: <https://www.spglobal.com/marketintelligence/en/news-insights/research/australia-mining-by-the-numbers-2021> (accessed 24.5.2023).

Gingold, R A and Monaghan, J J, 1977. Smoothed particle hydrodynamics: theory and application to non-spherical stars, *Monthly Notices of the Royal Astronomical Society*, 181(3):375–389.

Ilic, D, 2013. Bulk solid interactions in belt conveying systems, PhD Thesis, University of Newcastle, Australia.

Ilic, D, Lavrinec, A and Orozovic, O, 2020. Simulation and analysis of blending in a conveyor transfer system, *Minerals Engineering*, 157:106576.

Ju, L, Vassalos, D, Wang, Q and Liu, Y, 2019. Solid bulk cargo instability during marine transport, *Ocean Engineering*, 186:106089.

Liu, M B and Liu, G R, 2010. Smoothed particle hydrodynamics (SPH): an overview and recent developments, *Archives of Computational Methods in Engineering*, 17(1):25–76.

Lucy, L B, 1977. A numerical approach to the testing of the fission hypothesis, *The Astronomical Journal*, 82:1013–1024.

Minerals Council of Australia, 2017. Submission to inquiry into national freight and supply chain priorities, Commonwealth of Australia, 26 p.

Monaghan, J J, 2005. Smoothed particle hydrodynamics, *Reports on Progress in Physics*, 68(8):1703.

Morris, J P, Fox, P J and Zhu, Y, 1997. Modelling low Reynolds number incompressible flows using SPH, *Journal of Computational Physics*, 136:214–226.

Potapov, A V, Hunt, M L and Campbell, C S, 2001. Liquid–solid flows using smoothed particle hydrodynamics and the discrete element method, *Powder Technology*, 116:204–213.

Radjai, F and Dubois, F, 2011. *Discrete-element modeling of granular materials* (Wiley-Iste).

Standard Club, 2021. Guidelines for the safe carriage of Nickel Ore. Available from: <https://www.standard-club.com/fileadmin/uploads/standardclub/Documents/Import/publications/loss-prevention-industry-expertise-handouts/3328315-guidelines-for-the-safe-carriage-of-nickel-ore_pdf.pdf> (Accessed 24 May 2023).

Tanaka, K, 2001. Numerical and experimental studies for the impact of projectiles on granular materials, *Handbook of Conveying and Handling of Particulate Solids*, pp. 263–270.

Tao, J and Tao, H, 2017. Factors Affecting Piping Erosion Resistance: Revisited with a Numerical Modeling Approach, *International Journal of Geomechanics*, 17(11).

Vasic, M, Grbavcic, Z and Radojevic, Z, 2014. Analysis of moisture transfer during the drying of clay tiles with particular reference to an estimation of the time-dependent effective diffusivity, *Drying Technology*, 32(7):829–840.

Wang, Z, Rong, C, Hong, W, Qiang, L, Xun, Z and Shu-Zhe, L, 2016. An overview of smoothed particle hydrodynamics for simulating multiphase flow, *Applied Mathematical Modelling*, 40(23–24):9625–9655.

Wensrich, C M and Katterfeld, A, 2012. Rolling friction as a technique for modelling particle shape in DEM, *Powder Technology*, 217:409–417.

Yang, X F, Peng, S L and Liu, M D, 2014. A new kernel function for SPH with applications to free surface flows, *Applied Mathematical Modelling*, 38(15–16):3822–3833.

Zafar, U, Vivacqua, V, Calvert, G, Ghadiri, M and Cleaver, J A S, 2017. A review of bulk powder caking, *Powder Technology* 313:389–401.

Geoscience 2033 – a vision for mineral exploration in ten years

L Karlson[1], C Itotoh[2] and K O'Halloran[3]

1. Principal Improvement, BHP WAIO Geoscience, Perth WA 6000.
   Email: lance.karlson@bhp.com
2. Manager Operations, BHP WAIO Geoscience, Perth WA 6000.
   Email: cornelius.itotoh@bhp.com
3. Superintendent Contracts, BHP WAIO Geoscience, Perth WA 6000.
   Email: kara.ohalloran@bhp.com

ABSTRACT

Global methods used to explore for minerals have progressed relatively slowly since the development of the first diamond core drill in 1863. Since then, key developments such as wireline geophysics (1920s), the tricone drill bit (1930s) and aeromagnetic surveys (1940s) all occurred in response to global events. In this project, BHP's WAIO Geoscience operations team hypothesise that a global movement towards a greener future as well as an increasing intolerance for safety events will drive another step change in exploration methods. We have leveraged our relationships with exploration contractor partners and industry leaders to develop a realistic vision for 2033 and beyond, where drill rigs, earthworks and geophysical units are operated and maintained remotely; emissions of dust and waste are eliminated; hazardous energy sources are controlled; procurement is based on sustainability rather than cost and carbon emissions are tracked with the same rigour as production metrics. We have combined these visions into a series of graphical illustrations with the aim of inspiring ourselves, our vendors and our peers to collaborate further on producing a step change in our industry.

INTRODUCTION

The aim of this study is to present a likely vision for changes in mineral exploration in the next ten years, which can in turn facilitate discussion on initiatives and roles that require additional resourcing and focus. While this is a purely speculative exercise, the authors have engaged with vendor representatives across the mineral exploration industry in Australia to discuss and assess the technological, social and environmental changes that are expected to soon impact our industry. This work has involved a detailed study of historical fatalities in the mineral exploration industry in Australia to best inform future safety initiatives, research into the relationships between mineral exploration technology changes and global events and a speculative look into technology opportunities well into the future.

GEOSCIENCE DEVELOPMENTS ARE DRIVEN BY GLOBAL EVENTS

Step changes in technological developments in mineral exploration have historically been associated with significant global events. This is illustrated in Figure 1, where the 'Technological Revolution' (1980–1910) which included developments such as electricity and industrial-scale manufacturing, undoubtedly led to the development of rotary drill rigs and aerial photography. This was followed by the Texas Oil Boom (1900–1940) which directly influenced the development of the tricone drill bit and the Schlumberger method (wireline resistivity). Breakthroughs in radioactivity research in the early 20th century also drove continuous developments in geophysical methods such as gamma ray logging, neutron and density logging. Also correlating with the 'Roaring Twenties' was the development of ground penetrating radar, seismic surveys and aerial magnetic surveys.

Rapid developments in mineral exploration techniques continued through to the 1960s, with combustion drill rigs, fracking, LiDAR and reverse circulation drilling all corresponding with the Arabian Oil Boom and the beginning of the Cold War. Since that time, the most significant developments have included GPS and the use of unmanned aerial vehicles (UAVs). Technologies have, in comparison, been developed slowly with minimal step changes. In fact, the drill rigs and geophysical technologies being used today largely mirror those used in the 1990s except for improved safety inclusions. While automation technologies in mineral exploration are being trialled or are currently available off the shelf, uptake has been arguably less progressive than the rapid changes of the early 20th century.

Technological developments in resource exploration 1880 - 2023

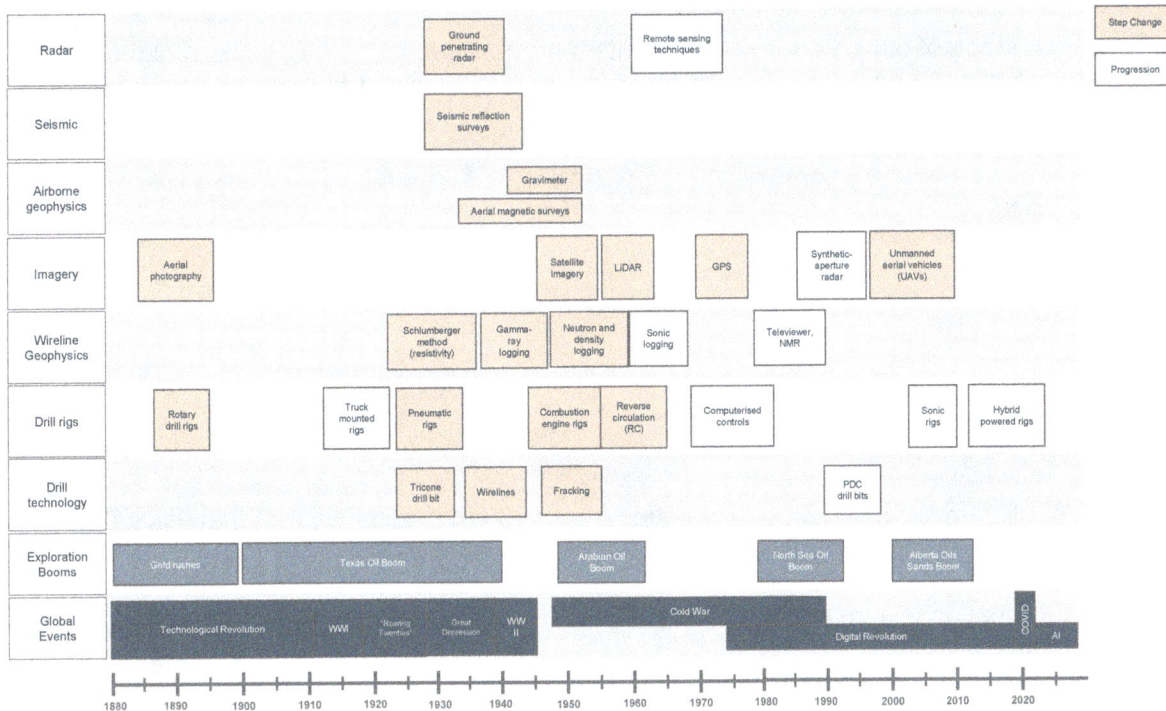

FIG 1 – Technological developments in resource exploration from 1880 to present (collated by authors).

The focus of this research is a vision for mineral exploration in the next ten years. Given recent trends, it can be argued that any future step changes will likely need to coincide with significant global events, social movements or external technological changes.

TAXONOMY OF MINERAL EXPLORATION FATALITIES

Understanding the taxonomy of historical fatalities in the mineral exploration industry allows us to determine what controls require development focus in the next ten years. A timeline of Australian mineral exploration fatalities shown in Figure 2 demonstrates that periods of increased drilling typically correlate with increased deaths. This correlation ends between 2018–2023 where zero fatalities have occurred despite metres increasing by 20 per cent. This absence of fatalities suggests that the mineral exploration industry in Australia has become significantly safer in the last five years, however it would be unwise to assume that further developments in safety controls are not required.

Australian Mineral Exploration Fatalities by Material Risk vs Metres Drilled*

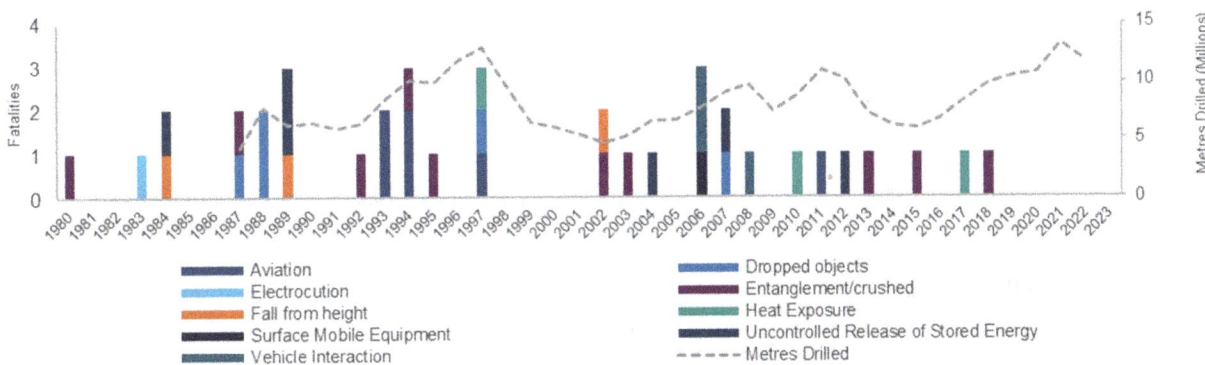

FIG 2 – Australian mineral exploration fatalities by material risk versus metres drilled (ABS, 2023). Fatalities data from Karlson (2023).

The above information is tabled in Figure 3 by material risks and damaging energy sources. This reveals that entanglement/crushing represents the highest proportion of Australian mineral exploration fatalities (10 deaths; 26 per cent). Nine of these ten events involved being caught in rotating drill strings, while the tenth fatality involved being crushed by a dozer belly plate during maintenance.

Uncontrolled release of stored energy represents the equal second (with aviation) highest cause of fatalities (six deaths; 16 per cent). Five of these six events involved high pressure air hoses, while the sixth event involved a downhole blowout.

Material Risk	Fatalities	% of Total
Entanglement/crushed	10	26%
Uncontrolled Release of Stored Energy	6	16%
Aviation	6	16%
Dropped objects	5	13%
Vehicle Interaction	3	8%
Heat Exposure	3	8%
Fall from height	3	8%
Surface Mobile Equipment	1	3%
Electrocution	1	3%
Grand Total	**38**	**100%**

Energy Source	Fatalities	% of Total
Machine	10	26%
Vehicular	9	24%
Gravitational	9	24%
Object	6	16%
Thermal	3	8%
Electrical	1	3%
Grand Total	**38**	**100%**

FIG 3 – Australian mineral exploration fatalities from 1980 to present by material risk category and energy source. From Karlson (2023).

The above information is summarised in the below infographic (Figure 4). Note that all fatalities are external to BHP.

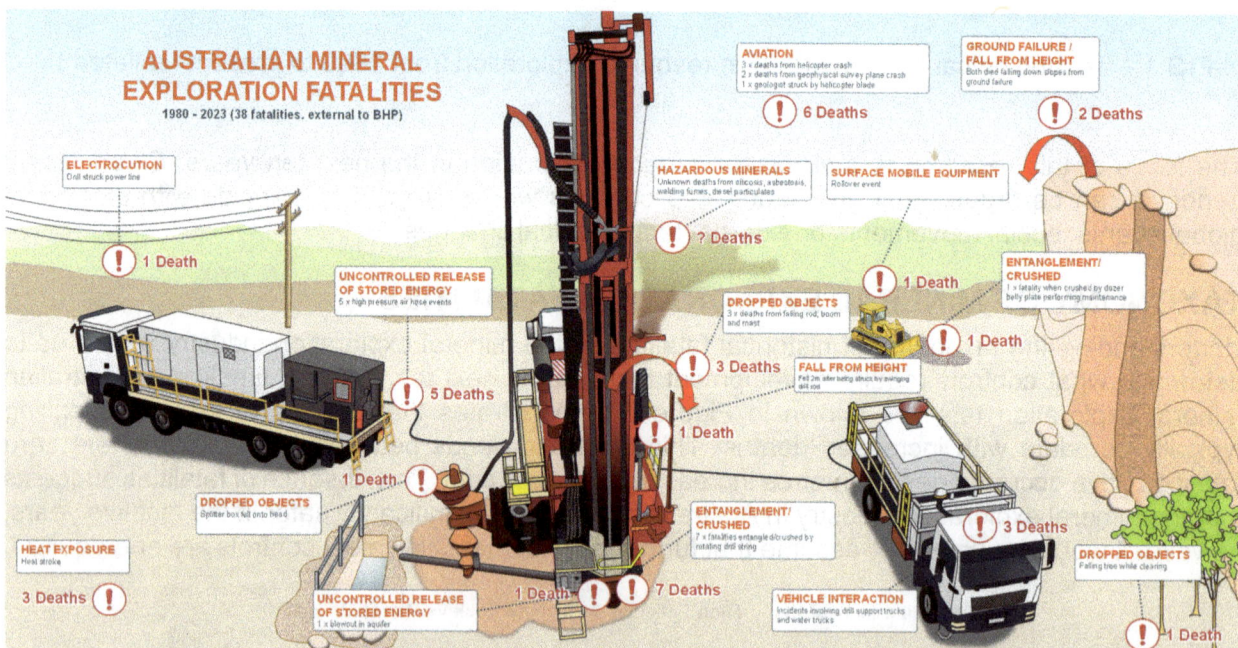

FIG 4 – Australian mineral exploration fatalities from 1980 to present infographic. Fatalities are external to BHP. From Karlson (2023).

CURRENT AND FUTURE STATE CATEGORIES

Discussions with representatives from drilling, earthworks and geophysics companies demonstrated that key changes in the mineral exploration industry are likely to be driven by the rapidly developing framework of environmental, social and corporate governance (ESG). While not all representatives were familiar with this framework, it was generally agreed that a movement towards a more sustainable Geoscience future was inevitable.

Other categories in which changes could be expected included safety and productivity, with a strong focus given to the potential effects of artificial intelligence (AI). All discussions and ideas collated have been summarised in Table 1, where current state is compared with expected future changes.

TABLE 1

Current issues and examples of future solutions in the mineral exploration industry.

Category	Current state	Future solution (examples)	Expected time frame for change
Safety	Sub-optimal drill methods	Optimisation of drill methods for geological setting (eg use of mud), informed by greater communication of geological models	5 years
Safety	Variable lifting and dropped objects controls	Casing and rod presenters that remove lifting and dropped objects risks	5 years
Safety	Manual exclusion zones	Radar enforced exclusion zones	5 years
Safety	Limited controls on operator fatigue and inattention	Engineering controls for operator fatigue and inattention	5 years
Safety	Manual emergency response triggers	Advanced emergency response solutions linked to people and equipment	5 years
Safety	Manual work on live equipment	Electronic management systems that control all hydraulics and high pressure. Elimination of work on all live equipment.	10+ years
Safety	Hands on operation of drill rigs, earthworks and geophysical units	Completely remote operation of all units, including during tasks such bit changes (drill rigs) and tool changes (geophysics units)	10+ years
Environmental, social, governance (ESG)	Emerging and/or underdeveloped environmental strategies and sustainable power use	Active incorporation of decarbonisation into procurement processes and capability uplift, including influencing key suppliers to set net zero targets	3 years
Environmental, social, governance (ESG)	No tracking of emissions	Monitoring of emissions as part of standard key performance indicators (KPIs)	3 years
Environmental, social, governance (ESG)	No standardised methods to track carbon usage	Standard methods and framework for tracking, recording and reporting carbon usage	3 years
Environmental, social, governance (ESG)	Dust controlled by water trucks and PPE with minimal dust suppressing technologies.	Dust suppression controls including wet drilling and rubber flaps at base of mast and water injection from drill head and bit.	3 years
Environmental, social, governance (ESG)	Diesel powered equipment	Hybrid power solutions/hydrogen power/battery power derived from solar	5+ years

Environmental, social, governance (ESG)	Size of drill pads and associated clearing of tracks	Rigs with smaller footprint facilitated by automation (eg no requirement for cabins) and reduced requirement for support trucks	5 years
Environmental, social, governance (ESG)	Water use (workshop, wash bay, drilling)	Reusable water solutions	5 years
Environmental, social, governance (ESG)	Diesel powered lighting plants	Solar powered transportable lighting	5 years
Productivity	Manual data interpretation, analysis and improvement	Integration of artificial intelligence (AI) in all administrative tasks and data analysis	2 years
Productivity	Geologist manually determines strata	Auto-interpretation of target strata via assay and geophysical tools built into drill string	3 years
Productivity	Manual and delayed data recording and processing	Real time data capture, eg geophysical data, plod data, measure while drill (MWD) data.	3 years
Productivity	Requirement for booster trucks to drill at greater depths/through harder lithologies	Greater air pressure capabilities of drill rigs	5–10 years

A consistent point made in collating these key expected changes was that the speed of transition will be reliant on the cost of changes. The drill rigs, geophysical units and earthworks machinery currently being built today do not include many of these technological changes and it is highly unlikely that these units will be retired before their expected lifespan to facilitate policy expectations. A more likely scenario is that change will come via retrofitting existing equipment with additional safety, environmental and productivity initiatives and that there will be no 'space age drill rig' by 2033.

LOOKING BEYOND 2033

The authors extended their discussions with vendor representatives to include their expectations of drill rigs beyond 2033 and what these rigs might look like. Feedback was used to produce a series of images generated with the assistance of artificial intelligence (AI) shown in Figures 5 and 6.

From 2030–2040 (Figure 5, images A and B), predicted features include automated drill rigs with no cabins required for operators; rigs powered by hybrid fuel types; XRF and geophysical logging integrated into the drill string and greater air pressure capabilities to negate the need for boosters.

From 2040–2050 (Figure 5, images C and D), predicted features include coil tube rigs mounted on automated trucks; hydrogen power sources; live geological interpretation and modelling with associated directional changes while drilling and significant increases in-depth and directional optionality.

FIG 5 – Speculative illustrations of drill rigs from 2030–2050. Concepts and designs created by the authors, with images generated with assistance from AI.

From 2050–2060 (Figure 6, images E and F), it is predicted that the traditional methods of sourcing downhole information by drill strings will be replaced with remote sensing and seismic methods that produce rock characteristics without the need for destructive sampling and analytical robots that can navigate through various terrains. This will ensure that no earthworks disturbance will be generated through the clearing of drill tracks and drill pads.

From 2060 onwards (Figure 6, images G and H), it is predicted that the remote sensing and analytical capabilities will continue to improve, with analysis able to be performed by drones that in turn produce models of subsurface features and generate mine plans for subsequent automatic extraction.

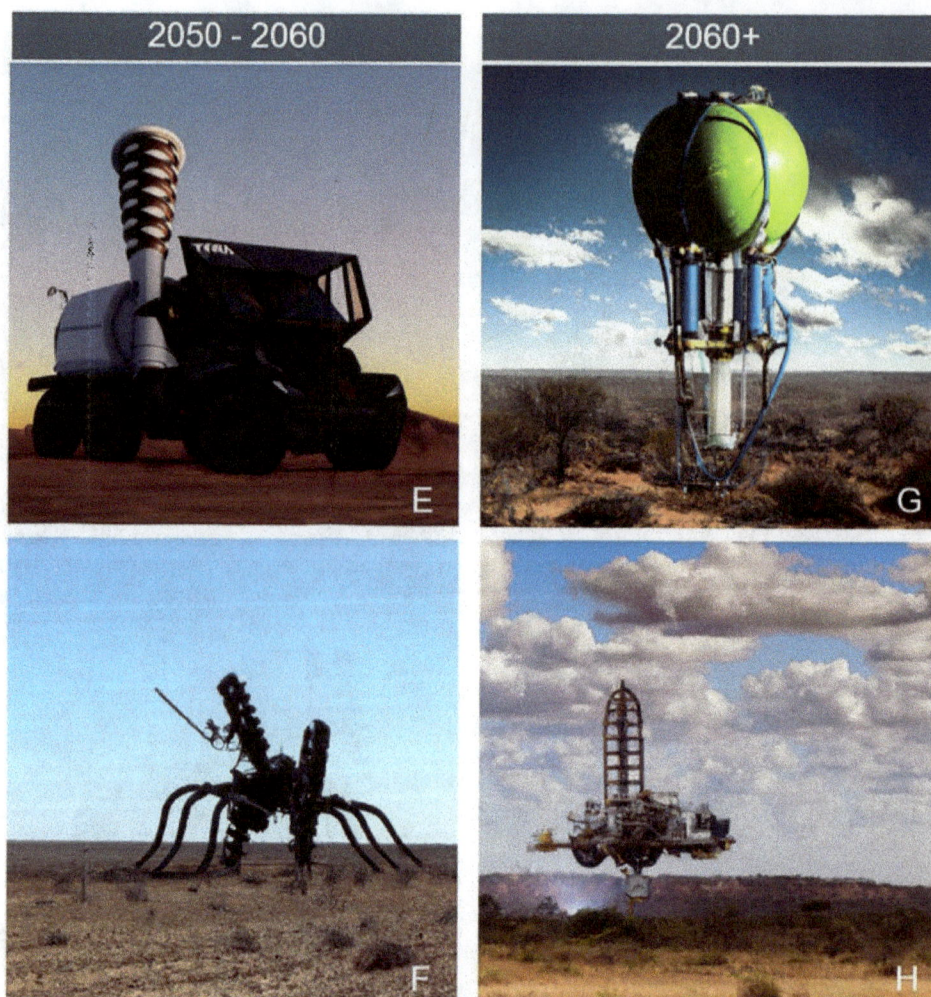

FIG 6 – Speculative illustrations of drill rigs from 2050 and beyond. Concepts and designs created by the authors, with images generated with assistance from AI.

CONCLUSIONS

The aim of this study was to present a vision for changes in mineral exploration in the next ten years, which can in turn facilitate discussion on mineral exploration companies' technological roadmaps. To ensure an appropriate context was established, the authors first produced a timeline of historical step changes in mineral exploration technology since 1880. It was concluded that key step changes correlated with significant global events and that changes since 1960 in the mineral exploration industry have been relatively minor. This suggests that we cannot expect rapid changes in the next ten years unless external developments such as artificial intelligence or major policy changes resulting from ESG initiatives cause major upheavals in the way exploration is conducted.

The study also included a detailed taxonomy of historical drilling fatalities in Australian mineral exploration since 1980. This showed that fatalities have remained below two per annum since 2007 and the industry has remained fatality-free since 2018. This is despite a 20 per cent increase in total mineral exploration metres drilled between 2018 and 2022. Entanglement/crushing is the leading cause of fatalities in the Australian mineral exploration industry (10 fatalities; 26 per cent), with fatalities occurring consistently from 1980 until 2018. Uncontrolled release of stored energy represents the equal second (with aviation) cause of fatalities in the Australian mineral exploration industry (six fatalities; 16 per cent).

With greater knowledge of historical trends in technology development as well as industry fatalities, the next stage of this study was to seek input from industry representatives on the types of changes they expect to see in the coming years. A recurring theme included impacts from ESG policies on the way mineral exploration work is conducted, recorded and reported. This was coupled with the expectation that operations will become increasingly automated, with operators removed from all sources of damaging energy.

Finally, industry partners were asked to provide input into their vision of what drill rigs may look like beyond 2030. This included expectations that mineral exploration techniques will be combined into singular units that perform a range of analytical tasks automatically. It was expected that these units would become less reliant on taking physical samples, performing analysis with reduced interaction with the environment and subsequently doing so via drones. These predictions should be taken as speculation only, however it is hoped that these visions prompt further discussion and allow companies to assign priorities to emerging areas of change.

ACKNOWLEDGEMENTS

The authors would like to acknowledge contributions by Clint Smith and Shane Goad from BHP, as well as inputs into current and future opportunities within the mineral exploration industry by contract partners of BHP WAIO Geoscience. These companies include Ausdrill, Carey Mining, Epiroc, Pentium, Piacentini and Son, Ranger Drilling, Ventia, Wallis Drilling and Welldrill.

REFERENCES

Australian Bureau of Statistics (ABS), 2023, 27 Feb. Mineral and Petroleum Exploration, Australia – Mineral exploration: metres drilled, seasonally adjusted and trend. Available from: <https://www.abs.gov.au/statistics/industry/mining/mineral-and-petroleum-exploration-australia/dec-2022#mineral-exploration>

Karlson, L, 2023. 2023 Geoscience Taxonomy Report, internal BHP publication.

HRC™ HPGR as an energy efficient alternative to horizontal mills in the iron ore processing circuit

E Nunes[1], B Raso[2], J Almeida[3], D Jacobson[4], R Mendes[5], G Cota[6], E Santos[7], T Machado[8] and E Pereira[9]

1. Application Manager – Crushing Systems, Metso, Sorocaba-SP, 18087–101, Brazil.
   Email: edis.nunes@mogroup.com
2. Application Intern – Crushing Systems, Metso, Nova Lima-MG, 34006–049, Brazil.
   Email: bruna.raso@mogrouppartners.com
3. Sr Process Technology Engineer, Metso, Sorocaba-SP, 18087–101, Brazil.
   Email: julio.almeida@mogroup.com
4. Process Engineering Director – Crushing Systems, Metso, Waukesha, WI-53186, USA.
   Email: dusty.jacobson@mogroup.com
5. Production Manager – JMN, Desterro de Entre Rios-MG, 35494–000, Brazil.
   Email: robert.mendes@jmendes.com.br
6. Technical Seller – Crushing, Metso, Nova Lima-MG, 34006–049, Brazil.
   Email: gabriel.cota@mogroup.com
7. Process Coordinator – JMN, Desterro de Entre Rios-MG, 35494–000, Brazil.
   Email: edson.santos@jmendes.com.br
8. Maintenance Supervisor – JMN, Desterro de Entre Rios-MG, 35494–000, Brazil.
   Email: tadeu.machado@jmendes.com.br
9. Laboratory Coordinator, Metso, Sorocaba-SP, 18087–101, Brazil.
   Email: edilson.pereira@mogroup.com

ABSTRACT

Grinding circuits are known to be the highest consumer of energy used in the ore comminution process, with high operational costs and sensitivity to ore conditions. The mining industry is being encouraged to look for more energy efficient processes while increasing the mass and metallurgical recovery; a specific example is projects with lower iron ore grades, which demands a higher level of liberation and resulting higher energy consumption. This has led to increased usage of high-pressure grinding roll (HPGR) technology as an alternative design within the comminution circuit. A growing use of the HRC™ HPGR technology has been used to generate concentrate through a more sustainable process. In the Brazilian state of Minas Gerais, the readaptation of iron ore beneficiation circuits has been initiated, replacing the traditional crushing-to-grinding flow sheet with circuits using crushing-to-HPGR flow sheets, with the main driver being reduced energy consumption. Utilising results from sampling campaigns and analysis of experimental data collected in an industrial plant, the work herein presents a case study from the JMN Mineração operating plant, where the HRC™ HPGR was used as an alternative to traditional grinding using horizontal mills. The results show that the HRC™ HPGR performance in this iron ore processing plant showed benefits in terms of reduced energy consumption as compared to the projected performance with a traditional grinding circuit.

INTRODUCTION

The global need for more cost-effective mineral processing techniques with less environmental impact has increased in recent years. Climate change has become a critical concern for the public and mining industry alike and as a result there is a focus on developing effective strategies to reduce greenhouse gas emissions and carbon footprints (Sullivan and Oliva, 2007). In addition, due to rising energy costs, the significance of effectively limiting energy consumption related to mineral processing has become paramount. Such factors have led to development of sustainable technologies aiming to achieve a balance between economic, environmental and social considerations, benefiting all stakeholders while minimising negative impacts on the environment and society (Vacchi et al, 2021; Weaver et al, 2017).

Mineral processing circuits consume around 5 per cent of the total electrical energy produced in the world, out of which 80 per cent is attributed to crushing and grinding (Fuerstenau and Abouzeid, 2007). The High Pressure Grinding Rolls (HPGR) technology has proven results for energy savings in this area. It is known that the most energy-efficient method for particle breakage involves compressing the particle bed between two plates (Schönert, 1979) and the HPGR uses this comminution technique with high-pressure interparticle compression crushing.

The HPGR advantages in the comminution process are well-documented. HPGR favours generation of micro-fissures in ores particles (Fernandes *et al*, 2014), which improves the processing circuit by reducing energy consumption in subsequent grinding stages, thereby decreasing costs and greenhouse gas emissions. While the application of HPGR preceding grinding is a well-studied circuit, the operation of a circuit with only crushers and HPGR is yet to be comprehensively validated in terms of operational benefits and suitability for future applications. Therefore, this study compares the HPGR key operating conditions based on full-scale plant data versus two simulated scenarios: one with a ball mill and the other with a rod mill. All three scenarios are evaluated using the same feed material and generating the same output. The primary aim is to evaluate the efficiency and potential advantages of the full HPGR comminution circuit.

JMN Mineração is currently running a plant with a HRC800™ HPGR and has been chosen as the selected study operation for this paper. Currently, the material that feeds the HPGR circuit is the jig concentrator tailings. Prior to the implementation of the HPGR circuit, this material was being discarded as waste; now, it is being processed to increase mass recovery – which has both an environmental and economic beneficial effect. This case study reviews the implementation and operation of the new HPGR circuit in order to quantify the benefits against a traditional grinding circuit approach.

Objectives

The primary objective of this study is to evaluate energy-savings in mineral processing by implementing a comminution circuit primarily based on crushing and HPGRs, instead of the conventional circuit with horizontal mills performing the final stages of comminution. With this target in mind, the work evaluates the HPGR performance in an operating plant comparing its energy consumption with the experimental values obtained with ball mill and rod mill simulations.

JMN Mineração Plant

JMN Mineração, a subsidiary of J. Mendes, has been operating since 2014 between the cities of Piracema and Desterro de Entre Rios, in Minas Gerais, Brazil. This location is within the Quadrilátero Ferrífero (QF) complex, a geological and mineral-rich region covering about 7000 km^2 in Central-Southern Brazil with extensive gold and iron ore reserves. The area's best-known mines are the Mina do Pico, Fazendão, Conceição, Brucutu, Alegria, Agua Limpa, Timbopeba, Corrego do Feijão, Capitão do Mato, Coelhos (JMN site) and others. Many geological and mining companies related to gold mining and iron ore mining have sites in the QF, such as Vale, Anglo American, Gerdau, Usiminas and Samarco.

The JMN Mineração mine uses open pit mining methods with primary mineral production being iron ore processed by crushing, grinding, spiral classification, jigs and magnetic separation. Its processing plant generates the following products: Lump Ore, Hematitinha, Sinter Feed and Concentrate. JMN has an ore capacity of 5 400 000 tons annually and a production capacity of 2 000 000 tons per annum. These resources are supplied throughout the Brazilian territory, mainly for the steel industry. In December 2021, the company completed a project involving HPGR and low-intensity magnetic separation implementation, which resulted in an increase in plant recovery (Brasil Mineral, nd). Figure 1 shows the JMN plant.

FIG 1 – JMN processing plant.

The first stage of the JMN's beneficiation plant is primary crushing. The product from the primary crushing circuit feeds a horizontal vibrating screen with two decks (44 mm top and 3 mm bottom screen decks). The first deck operates in closed circuit with the secondary crusher. Retained material from the second deck rfeeds two three-deck horizontal screens (25 mm, 16 mm, 3 mm decks from top to bottom). The undersize from all screens (-3 mm) goes to the spiral concentrators. Material between 44 mm and 25 mm is classified as Lump Ore and is partially sent to the tertiary crusher. The material between 25 mm and 16 mm is called Hematitinha and is either sold directly as product or may be diverted to tertiary crushing.

The size fraction between the 3 mm and 16 mm screens goes to two concentrator jigs. The jig concentrate material is contained in the Jig Concentrate hopper and the jig reject tailings below 3 mm goes to desliming hydrocyclones operating in closed circuit.

Prior to the HPGR circuit implementation, the plant operated without processing the jig rejects tailing material above 3 mm. After the HPGR installation, the plant operation has been restructured to reprocess this previous waste material in the new circuit. The Metso HRC™ 800 HPGR operates in closed circuit with a vibrating screen (3 mm separation) and the undersize goes to the spiral concentration stage. Figure 2 presents a representation of the JMN flow sheet.

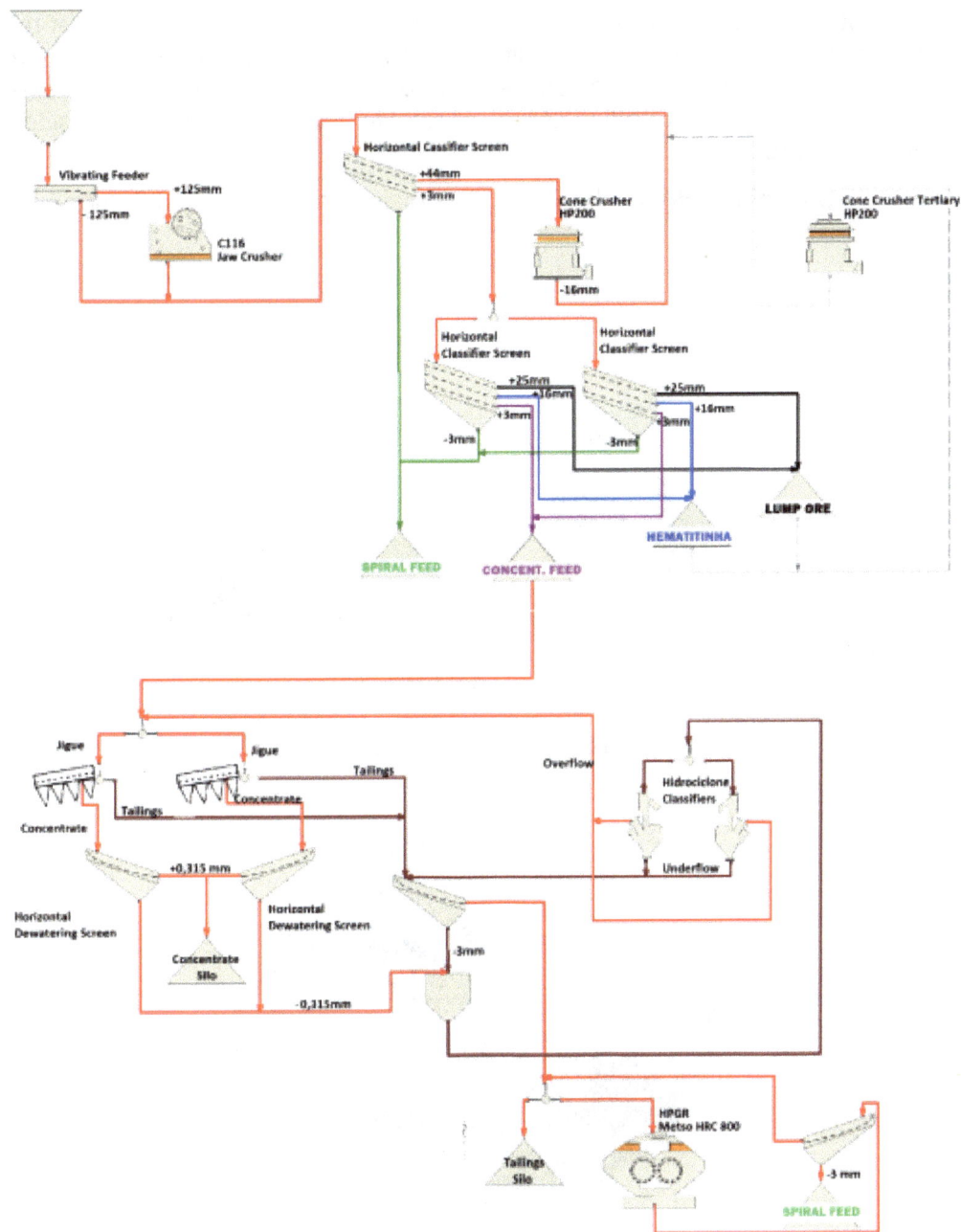

FIG 2 – JMN circuit.

HRC800™ HPGR overview

The HRC™ HPGR produced by Metso was selected as the preferred technology for this application for its advantages against traditional HPGR technology. The main structure of the crusher is the patented arch-frame design, which mechanically absorbs imbalance of loads in order to eliminate downtime caused by skewing (the condition where the axis of each roll does not stay parallel to each other due to uneven force distribution). The positioning of the HRC™ HPGR's hydraulic pistons sitting above the roll centres also deviates from the traditional concept. The cylinders only need to apply half the piston force on the crushing zone due to the mechanical advantage of the pivoting Arch-frame (Herman *et al*, 2015).

The structural stiffness allows a uniform operational opening across the entire roller width to be obtained even with a non-uniform feed, which permits the use of a flanged roll design (Oliveira, 2015). The flanges are designed to combat edge effects, a problem with traditional HPGRs where comminution is reduced at the edge of the roll (Herman *et al*, 2015).

The HRC™ 800 HPGR model utilised in this application has a roll diameter of 800 mm and a roll length of 500 mm. It has a nominal operating roll speed of 28.8 rev/min and a circumferential speed of 1.2 m/s. The speed range can be adjusted between 50 per cent and 110 per cent related to the nominal speed, providing operational flexibility. The hydraulic system can sustain a maximum pressure of 250 Bar. The specific force reaches a maximum of 4.5 N/mm² during operation. The equipment maximum feed size is rated at 32 mm, subject to application-specific considerations. The installed power is 2 × 110 kW.

Figure 3 shows a model of the HRC™ HPGR and its main components and Figure 4 illustrates the flanged rolls.

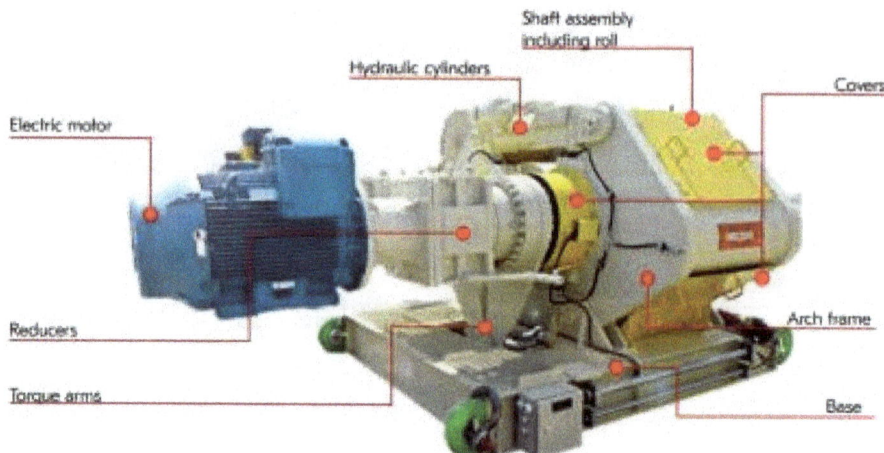

FIG 3 – HRC™ HPGR main components.

FIG 4 – HRC™ HPGR flanged rolls.

HRC™ HPGR operating indices

To assess the HPGR performance and efficiency in applications, Equations 1, 2 and 3 (Schönert, 1988) determine the main indices values (Almeida *et al*, 2023).

$$SP = \frac{F}{1000 \times D \times L} \quad (1)$$

The Specific Pressure is defined as the force applied to the rolls (kN), divided by the diameter (m) and the rolls width (m), where SP is the specific pressure (N/mm²), F is the specific force applied to the rolls, D is the roll diameter and L is the roll width.

$$SE = \frac{P}{Q} \quad (2)$$

The Specific energy is defined as the power (kW) divided by the throughput of solids (t/h), where SE is the specific energy (kWh/t), P is the power and Q is the throughput.

$$SC = \frac{Q}{D \times L \times V} \quad (3)$$

The Specific Capacity defined as the specific throughput divided by the diameter (m), the rolls width (m) and the roll peripheral speed (m/s), where SC is the specific capacity (ts/hm³) and V is the roll peripheral speed.

EXPERIMENTAL METHODOLOGY

Method

The methodology employed in this study began with the selection of five sampling locations in the JMN conveyor belt system to ensure representative data. Operating data from the day of sampling was obtained. During a plant crash stop, samples were collected and then subjected to particle size distribution analysis (PSD) and a Wi Bond test using a ball and rod mill.

After interpretation of the results, indices for a hypothetical ball mill circuit were calculated for comparison. Finally, the results were compared and discussed. The overall procedure adopted in this work is summarised in Figure 5.

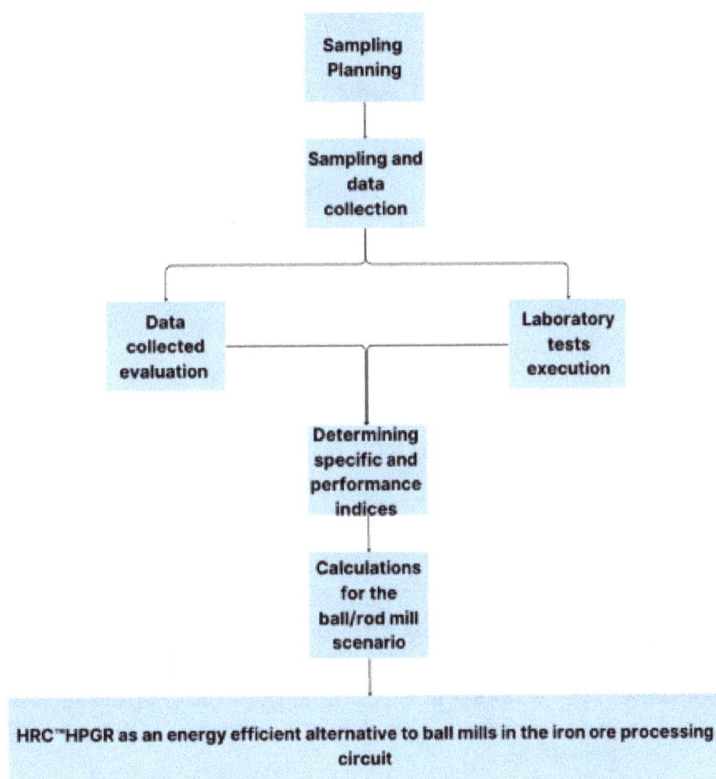

FIG 5 – Overall procedure adopted in this work.

Sampling campaign at the JMN Plant

The sampling campaign carried out at the JMN Beneficiation Plant included five sampling locations. The HRC™ HPGR roll pressure was set to 2.25 N/m³ and operated until a steady-state condition was achieved. During this period, data were obtained from the panel control system (CLP), including the power draw, roll speed and the operating gap between the rolls. The plant was shutdown and samples were collected around the points indicated in Table 1 and Figure 6.

TABLE 1

Sampling point descriptions.

Sampling point	Description
1	HRC800 new feed (+3 mm jig tailing)
2	HRC800 circuit product
3	Circulating load
4	HRC800 feed
5	HRC800 product

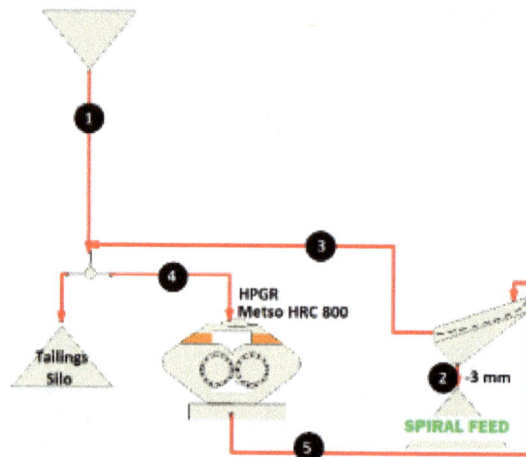

FIG 6 – Sampling point locations.

RESULTS AND DISCUSSION

In this section, the survey campaign and sample treatment outcomes are presented, along with the associated analyses. Following the sieving and WI laboratory work, a comparison between the industrial HPGR Circuit and a simulated scenario with horizontal mills was made. The key parameters compared were throughput, F_{80} (feed particle size 80 per cent passing), P_{80} (product particle size 80 per cent passing), required specific energy, actual specific energy consumption and the corresponding difference in required specific energy compared to the HPGR base case.

Sample campaign analysis

The sieve analysis was conducted in the laboratory located on the JMN premises. The HRC™ HPGR Feed F_{80} was measured at 10.22 mm and the circuit product (US Screening) P_{80} at 1.62 mm. The HRC™ HPGR feed contained a low amount of fines, with less than 10 per cent of the feed material below 2 mm. However the circulating load sample did contain a significant portion of minus 2 mm (25 per cent) material, which indicates some improvement in screening efficiency could be made. The circuit F_{80}/P_{80} reduction ratio was approximately 6:1, within the range considered for good performance. The material fraction below 2 mm increased from 9 per cent (feed) to 81 per cent (product). This data was used to construct a PSD curve as presented in the Figure 7.

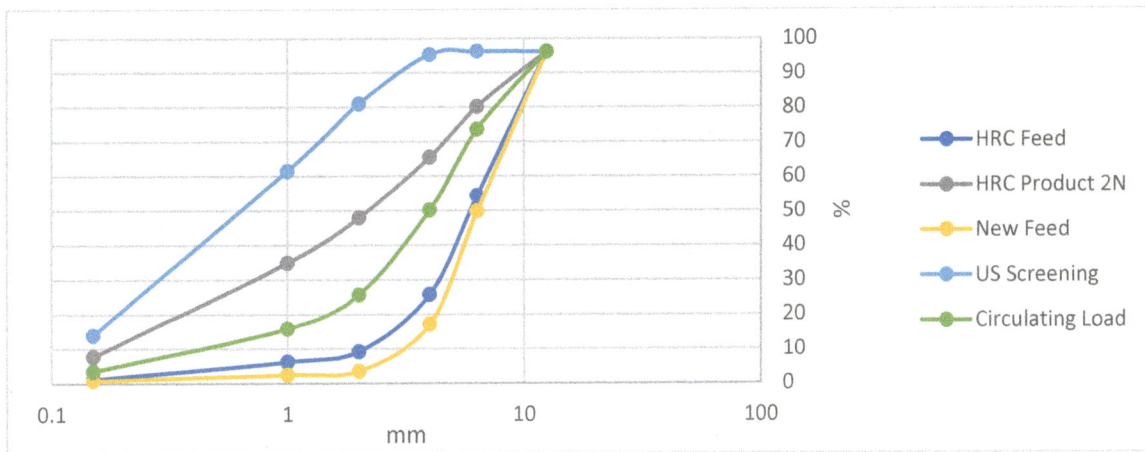

FIG 7 – PSD curve for the JMN HRC circuit.

The mass balance of the circuit was calculated to have 76 per cent recirculating load for a total HPGR feed of 143 t/h. This value was elevated slightly due to carryover of minus 2 mm in the circulating load. Overall, the operating conditions of the plant at the time of sampling are considered suitable for comparison to other circuit alternatives.

Wi Bond Test

The Work Index, as defined by Fred C. Bond, represents the comminution parameter that expresses the resistance of the rock or mineral to crushing and grinding processes. Through this test, the calculation of the required grinding power can be obtained and its value constitutes an ore characteristic and is used for designing industrial comminution plants (Todorovic *et al*, 2017).

Tests were performed for ball mills and rod mills and although the test is already standardised by the Brazilian MB-3253 Standard, there are still uncertainties about the test mesh choice and its interference in the final result. In one of their publications, Bond and Rowland state that the BWI should be measured on a test mesh according to the desired particle size of the mill product (Silva, 2022), which is different from that stated by Duque *et al* (2014); ie that the BWI should be measured on a standard test mesh of 150 μm (or close to this value).

For the Rod Mill Wi Test, 10 cycles were performed using a 1.18 mm mesh. The grindability value (grams/rotation) of the last three cycles was averaged to obtain a value of 10.871 g/t used by the software in the Wi determination. Also added to the calculations was the sieve measurement for passing 80 per cent of the feed and 80 per cent of the product (F_{80} = 9.75 mm and P_{80} = 0.85 mm). The Wi value obtained in the Wi Bond Test for a Rod Mill was 12.46 kWh/t.

For the Ball Mill Wi Test, it was decided to perform the test at 0.149 mm. Of the eight cycles performed, the averaged grindability value was 10.871 g/t. The F_{80} per cent was 2.5 mm and the P_{80} per cent was 0.12 mm. The Wi obtained in the Wi Bond Test for a Ball Mill was 14.04 kWh/t.

JMN HRC™ 800 HPGR specific energy and specific capacity

The HPGR circuit was established as the study base case. The machine computational data indicated that the HPGR was operating with a power of 43 kW per motor (86 kW total) at the fresh feed rate of 81 t/h. The rolls gap was operating at 18 mm and it was rotating at a speed of 16 rev/min. With these values, a total circuit specific energy of 1.06 kWh/t was calculated (Equation 2). This value aligns with anticipated projections, thereby confirming for this case the accuracy of the applied calculations and parameters. This validation is substantiated through the correlation between specific energy and specific pressure as delineated by Klymowsky *et al* (2002). Figure 8 displays the relationship stated by the author, with 'Minerio de ferro' most closely representing the ore processed at JMN.

For the specific capacity calculation, it was necessary to obtain the roll peripheral speed (0.67 m/s). Applying these values to Equation 3, a specific capacity of 302.30 t/hm³ was obtained.

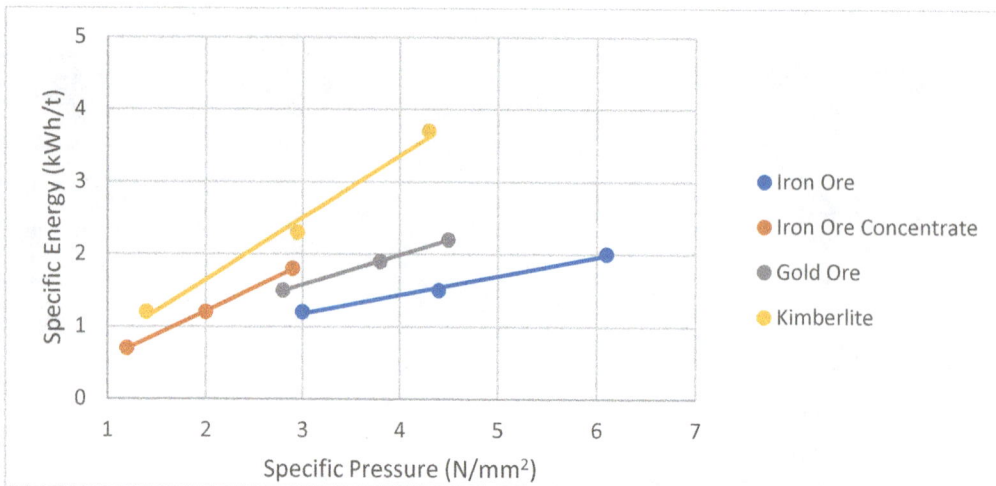

FIG 8 – Relation between specific energy and specific pressure.

Simulated Rod and Ball Mill application specific energy

To calculate the specific energy for the Ball Mill and Rod Mill simulation circuit, the Bond equation was used.

$$E = 10 \times BWi \times \left(\frac{1}{\sqrt{P80}} - \frac{1}{\sqrt{F80}} \right) \tag{4}$$

where E is the specific energy (kWh/t), BWi is the Bond Work Index (kWh/t), F_{80} is the size (µm) at which 80 per cent of the feed passes and P_{80} is the size (µm) at which 80 per cent of the product passes.

It is important to acknowledge the existence of certain factors that have the potential to add to the fundamental Bond equation. These factors, collectively referred to as Feed Equivalent Factors (EF factors), can introduce multiplicative adjustments, thereby influencing the calculated outcomes. While these EF factors offer a refining route to the estimation's accuracy, their incorporation warrants prudent consideration regarding their potential impact on power estimations for mills.

For this work, the specific energy (SE) values attributed to ball and rod mill calculations originate from the fundamental Bond equation adapted by integrating a feed equivalent factor, an adjustment introduced to accommodate the absence of fines within the feed material. If opting for all the comprehensive Bond equations factors, this would likely accentuate the disparity in power consumption discrepancies. The values for specific energy per ton for the two simulations are shown in Table 2, together with a comparison with the base case.

TABLE 2

Specific Energy for HPGR, Ball Mill and Rod Mill scenario.

Equipment	Throughput	F_{80} (µm)	P_{80} (µm)	Required Energy (kWh)	Specific Energy (kWh/t)	Required Specific Energy increase in relation to the base case (%)
HPGR Circuit – Base Case (survey)	81	10123	1629	86	1.06	-
Ball Mill Circuit – Simulated	81	10123	1629	181.5	2.24	211%
Rod Mill Circuit – Simulated	81	10123	1629	161	1.99	188%

DISCUSSION

The preceding sections presented a comparative energy analysis of utilisation of the HPGR circuit in operation against the calculated circuit energy consumption of a ball mill and a rod mill circuit. The data demonstrate that the HPGR circuit is characterised by a notable improvement in energy utilisation efficiency. These findings suggest that HPGR technology can provide a more streamlined energy utilisation process and minimise energy losses.

As Kapakyulu and Moys (2007) indicated, grinding mills are known for their inherent inefficiency, presenting challenges in control and incurring substantial steel and power consumption costs. Bouchard et al (2017) emphasized that less than 1 per cent of the electricity consumed by ball mills is attributed to comminution. By contrast, HPGR tend to apply more of their input energy directly to fracturing and grinding of the particles.

The findings of this study suggest that HPGR technology is a suitable alternative to grinding when processing material less than 16 mm in particle size down to product sizes near 1 mm P_{80} or less. Even when considering energy consumption differences in material handling, gains can be made by utilising compression comminution in place of attrition/grinding. Each project will require its own specific evaluation, however at a minimum HPGR technology has proven to have the ability to effectively process material down to the referenced P_{80} range. This can be used in justification studies to reprocess jig rejects to increase recovery, among other applications.

Energy consumption is a fundamental input into any project study, since this energy is predominantly obtained today from fossil fuel sources. This increases the amount of CO_2 released into the atmosphere. Along these lines, when the industrial operation is more energy-efficient and consumes less electricity per ton of processed material, the total amount of CO_2 emissions is reduced, which has a wider effect on the mining industry than simple operating cost analysis and has been stated as a focus of most mining companies.

CONCLUSION

The HPGR technology is a suitable alternative to grinding when processing material less than 16 mm down to product sizes near 1 mm P_{80} or less. The data demonstrate that the HPGR circuit is characterised by a notable improvement in energy utilisation efficiency (1.06 kWh/t versus 2.24 kWh/t and 1.99 kWh/t for a Ball Mill and a Rod Mill respectively).

The adoption of the proposed new circuit, ie HPGR in lieu of conventional grinding, stands as a milestone towards a more sustainable world as this circuit represents a tangible commitment to reducing energy consumption.

In a global landscape where awareness of carbon footprints and environmental responsibility is on the rise, this study showcases a compelling example of how innovation can be a pivotal ally in shaping a greener and more sustainable future. The convergence of efficiency and ecology within this new circuit not only eases strain on energy resources, but also extends an invitation to industries to embrace more conscientious practices.

REFERENCES

Almeida, J C F, Delboni, H, Bento, R, Reggio, A and Cremonese, E, 2023. Performance Analysis of HRCTM HPGR in Manufactured Sand Production, *Minerals*, 13(2):222. https://doi.org/10.3390/min13020222

Bouchard, J, LeBlanc, G, Levesque, M, Radziszewski, P and Georges-Filteau, D, 2017. Breaking down energy consumption in industrial grinding mills, in *Proceedings 49th Annual Canadian Mineral Processors Operators Conference*, pp 25–35.

Brasil Mineral, nd. About JMN Mineração (in Portuguese), Brasil Mineral. Available from: <https://www.brasilmineral.com.br/maiores/jmnmineracao> [Accessed: 14 Aug 2023].

Duque, T F M B, Schneider, C L, Mazzinghy, D B and Alves, V K, 2014. BWI em função da malha de teste, *Holos*, 3:112–121.

Fernandes, A L V, Oliveira, D M D, Machado, T F and Sobral, L G S, 2014. Influência das distintas formas de processamento mineral na biolixiviação de metais de base (Influence of different forms of mineral processing on the bioleaching of base metals), Environmental Technology Series, STA-74, CETEM/MCTI.

Fuerstenau, D W and Abouzeid, A Z M, 2007. Role of feed moisture in high-pressure roll mill comminution, *International Journal of Mineral Processing*, 82(4):203–210. 10.1016/j.minpro.2006.11.001

Herman, V S, Harbold, K A, Mular, M A and Biggs, L J, 2015. Building the world's largest HPGR–the HRC3000 at the Morenci Metcalf concentrator, in SAG 2015 Conference Proceedings.

Kapakyulu, E and Moys, M H, 2007. Modeling of energy loss to the environment from a grinding mill, Part II: Modeling the overall heat transfer coefficient, *Minerals Engineering*, 20(7):653–661.

Klymowsky, R, Patzelt, N, Knecht, J and Burchardt, E, 2002. Selection and sizing of high pressure grinding rolls, in *Proceedings of the Mineral Processing Plant Design, Practice and Control*, 1:636–668.

Oliveira, R N M D, 2015. Análise de desempenho do HRCTM HPGR em circuito piloto, Doctoral dissertation, Universidade de São Paulo.

Schönert, K, 1979. Aspects of the physics of breakage relevant to comminution, in Fourth Tewksbury Symposium.

Schönert, K, 1988. A first survey of grinding with high-compression roller mills, *International Journal of Mineral Processing*, 22(1–4):401–412.

Silva, G M B, 2022. Comparação de diferentes metodologias do ensaio de WI Bond para moinhos de bolas utilizando amostra com elevada quantidade de finos (Comparison of different methodologies of the WI Bond test for ball mills using a sample with a high amount of fines), thesis, School Of Mines, Federal University of Ouro Preto, Minas Gerais.

Sullivan, J and Oliva, M, 2007. Greenhouse gases an effective strategy for managing GHG emissions: follow this step-by-step guidance to fulfill the three critical elements of an effective GHG-management strategy [online], *Chemical Engineering*, 114(8):34. Available from: <https://link.gale.com/apps/doc/A168132097/AONE?u=brooklaw_main&sid=googleScholar&xid=8dab3980 [Accessed: 14 Aug 2023].

Todorovic, D, Trumic, M, Andric, L, Milosevic, V and Trumic, M, 2017. A quick method for Bond work index approximate value determination, *Physicochemical Problems of Mineral Processing*, 53(1):321–332.

Vacchi, M, Siligardi, C, Demaria, F, Cedillo-González, E I, González-Sánchez, R and Settembre-Blundo, D, 2021. Technological Sustainability or Sustainable Technology? A Multidimensional Vision of Sustainability in Manufacturing, *Sustainability*, 13(17):9942. doi:10.3390/su13179942

Weaver, P, Jansen, L, Van Grootveld, G, Van Spiegel, E and Vergragt, P, 2017. *Sustainable technological development* (Routledge).

Wet grinding, settling and filtering characteristics of Australian and Brazilian iron ores

J Pan[1], Q Zhou[2], D Q Zhu[3] and C C Yang[4]

1. Professor, School of Minerals Processing and Bioengineering, Central South University; Low Carbon and Hydrogen Metallurgy Reseach Centre (CSU-LCHMC), Central South University, Changsha 410083, China. Email: pjcsu@csu.edu.cn
2. Postgraduate student, School of Minerals Processing and Bioengineering, Central South University, Changsha 410083, China. Email: zqcsu@csu.edu.cn
3. Professor, School of Minerals Processing and Bioengineering, Central South University; Low Carbon and Hydrogen Metallurgy Reseach Centre (CSU-LCHMC), Central South University, Changsha 410083, China. Email: dqzhu@csu.edu.cn
4. Associate Professor, School of Minerals Processing and Bioengineering, Central South University; Low Carbon and Hydrogen Metallurgy Reseach Centre (CSU-LCHMC), Central South University, Changsha 410083, China. Email: smartyoung@csu.edu.cn

ABSTRACT

Oxidised pellets have become an indispensable high-quality charge for blast furnaces (BF) due to their uniform particle size, high iron grade, high mechanical strength, good metallurgical performance, as well as relatively low energy consumption and gaseous pollutant emissions in pellet production. Nevertheless, due to the rapid development of the global steel industry, high-quality pellet feeds are becoming more and more scarce. In order to broaden the source of pellet feeds and reduce the production cost of pellets, more and more steel mills are going to use coarse iron ore fines with relatively low iron grade and low impurities for preparation of desirable pellet feeds through a typical wet grinding-settling-filtering process. In this work, the grinding, settling and filtering behaviour of Brazilian and Australian iron ore fines are studied and compared, aiming at finding out the internal relationship between the mineralogical characteristics of different iron ore types and their grinding-settling-filtering performance. The results show that the grindability is heavily dependent on the hardness of iron ores, which is governed by the mineralogical characteristics of the raw materials. Usually the higher the hardness, the more grinding energy is required. Except for Brazilian specularite, Australian hematite-goethite type iron ore fines and Brazilian hematite ore fines exhibit good grindability with a Bond Work Index of about 10–15 kW•h/t. In addition, the settling and filtering performance of ground iron ores are largely affected by the density, mineralogical characteristics of iron ore fines (eg, quantity of clay minerals and goethite and porosity of ore particles), grinding fineness, etc. Generally, iron ore fines with higher density, lower quantity of soft minerals (clay and goethite) and larger particle size give superior settling and filtering performance. For iron ore fines with inferior grinding, settling and filtering performance, applying more energy-efficient grinding techniques, blending with ore types with good settling and filtering performance can reduce grinding energy consumption and improve settling and filtration rates to maximise the productivity of equipment and lower the operational cost.

INTRODUCTION

Oxidised pellets have become an essential charge material for blast furnace ironmaking due to their uniform particle size, high iron grade, excellent metallurgical properties and high mechanical strength. The raw iron ore concentrate used for pellet production undergoes mineral processing and grinding. However, Yie (2003) concluded that pellet production requires iron ore concentrate with suitable particle size distribution, specific surface area, low crystalline water mass fraction and reasonable chemical composition. The primary sources of iron ore for pellets are Australia, Brazil and Russia. Nonetheless, due to the rapid development of the global steel industry, high-quality and stable sources of iron ore are increasingly scarce. Workers (Hu and Gao, 2008; Sun, 2000) found that magnetite concentrate is the main raw material for producing oxidised pellets, but high-grade magnetite resources are becoming increasingly scarce and expensive. The rich ore fines used for sintering are generally hematite, which has a coarse particle size and abundant resources. To expand the availability of pellet feed and reduce production costs, steel mills are utilising coarse iron ore fines with high iron grade and few impurities to prepare ideal pellet feed through the conventional wet grinding-settlement-filtration process. This approach reduces the cost of ore preparation for pelletising and widens the source of iron ore concentrate for pelletising.

The primary objective of fine grinding of iron ore fines is to increase the specific surface area to meet the requirements of pellet feed and achieve uniform particle size distribution, while avoiding over-grinding. Mohamed et al (2003) showed that the optimum fineness of concentrates as pellet feed is within the range of 80–90 per cent smaller than 0.074 mm as well as 50–60 per cent smaller than 0.043 mm and the corresponding specific surface areas (SSAs) are 1500–2000 cm^2/g. Fu, Jiang and Zhu (1996) show that fine grinding of iron ore involves both wet and dry grinding processes, with the former requiring additional settling and filtering steps in production. Wet grinding is generally preferred over dry grinding because it is more energy-efficient (Peltoniemi et al, 2020). In addition, Gao et al (2022) point out that wet grinding also has advantages such as strong fluidity, high efficiency and no agglomeration. Wills and Napier-Munn (2006) conclude that the increased energy consumption of dry grinding is attributed to the extended grinding time required to achieve a product with a specific particle size distribution, with a majority of the energy being dissipated as heat. Studies have shown that the energy consumption differences between dry and wet grinding can be as high as 30–50 per cent. The settling and filtering performance after fine grinding and filtration of different iron ore types can vary considerably. Therefore, to enhance the settling and filtering performance of ore fines, it is crucial to understand the relationship between the material properties of different iron ores and their settling and filtering performance.

To obtain high-quality finely ground iron ore concentrate with improved settling and filtration performance, a more energy-efficient grinding technique can be used for iron ore fines with poor grinding, settling and filtering properties. By blending ores with those that possess excellent settling and filtering performance, a suitable mixture can be achieved that results in optimal grindability, settling and filtration properties. This approach meets the grinding and beneficiation process requirements while also reducing the energy consumption associated with grindability. The resulting high-quality finely ground iron ore concentrate is suitable for use as a raw material in pelletising.

There is a lack of research on the settling and filtration performance of wet-ground iron ore fines. This study investigates and compares the grinding, settling and filtration performance of iron ore fines from Brazil and Australia. The aim is to explore the intrinsic relationship between the process mineralogy and settling and filtration performance of different ore fines, optimise the fine grinding and settling and filtration process of Brazilian and Australian ores and improve their settling and filtration performance by rational blending. The goal is to optimise the process flow and improve resource utilisation.

EXPERIMENTAL

Raw materials

The experimental study utilised six types of iron ores, including three from Australia (designated as samples A1-A3) and three from Brazil (designated as samples B1-B3), as shown in Table 1. All six types of ores are typical high-grade, low-silicon ores, with total iron content exceeding 60 per cent and SiO$_2$ content below 5 per cent. They are characterised by high iron grade with minimal variation and low levels of harmful elements such as sulfur, phosphorus and non-ferrous metals. In particular, for the three Australian ore fines with high LOI, the iron grade of the pellets will be significantly increased after roasting. These ores only require grinding to achieve particle size and specific surface area requirements for pellet production, without the need for beneficiation. Therefore, from these perspectives, they are suitable raw materials for pellet production.

TABLE 1

Chemical composition of the six iron ore fines samples (%).

Sample	TFe	FeO	SiO$_2$	Al$_2$O$_3$	CaO	MgO	P	S	LOI
A1	60.64	0.66	3.33	2.23	0.06	0.10	0.085	0.026	6.46
A2	62.31	0.51	4.43	1.94	0.15	0.12	0.110	0.015	4.03
A3	60.18	0.59	4.87	2.22	0.21	0.09	0.083	0.036	7.35
B1	67.11	0.28	3.51	0.47	0.062	0.011	0.013	0.013	0.017
B2	62.15	0.92	4.80	1.57	0.2	0.13	0.077	0.011	3.75
B3	65.71	1.56	1.91	1.02	0.22	0.71	0.059	0.044	2.41

LOI – loss of ignition.

The particle size distribution of the iron ores varies among the different countries, as shown in Table 2 for the six types of iron ore fines. The particle size characteristics of five of the iron ore fines samples, excluding B1, are more suitable for sintering. However, if directly used as pellet feed, they do not meet the particle size requirements and require further fine grinding.

TABLE 2

Size distribution of the six types of iron ore fines (wt%).

Sample	Size distribution/mm					
	+8	-8+4	-4+1	-1+0.5	-0.5+0.25	-0.25
A1	11.4	27.21	28.35	3.99	8.19	20.87
A2	8.33	27.17	37.41	7.45	8.67	10.97
A3	10.28	19.48	34.94	11.27	11.47	12.56
B1	0	0	0	1.37	0.9	97.73
B2	8.72	13.02	31.77	12.40	13.03	21.06
B3	10.73	26.28	17.45	13.14	3.41	29.69

Methods

Grindability

In this paper, the standard Bond method was used to measure the Bond work index (Wi) by ball milling to estimate the grindability of iron ore fines.

The grindability index, or grindability, of minerals is an indicator of the difficulty of grinding ores and is a fundamental data requirement for industrial grinding machines. It is a characteristic constant of ores that can be determined experimentally. The most widely used method for measuring the grindability index is the 'Bond Work Index' proposed by F C Bond. Bond's 'crack hypothesis' suggests that the useful work consumed in grinding is proportional to the geometric mean of the volume and surface area of the product and gives the following famous Bond Work Index practical formula:

$$W = W_{ib} \left(\frac{10}{\sqrt{P_{80}}} - \frac{10}{\sqrt{F_{80}}} \right)$$

(1)

Where F_{80} and P_{80} are the widths of the square aperture sieve that allows 80 per cent of the feed and product to pass, respectively, in micrometers (μm); W is the energy required to crush a short ton (907.185 kg) of material of size F to size P; W_{ib} is the Bond Work Index, which represents the energy required to grind a theoretically infinite feed size to a product size of 80 per cent passing 100 μm (or 65 per cent passing 75 μm).

According to the standard Bond test, the Bond Work Index (W_{ib}) is obtained by dry grinding in a closed circuit ball mill until the circulating load reaches 250 per cent. The Bond ball mill work index (W_{ib}) (in kW•h/t) is calculated from the following equation:

$$W_{ib} = \frac{4.906}{P^{0.23} \cdot G_{bp}^{0.82} \cdot \left(\frac{1}{\sqrt{P_{80}}} - \frac{1}{\sqrt{F_{80}}} \right)}$$

(2)

where P is the particle size required for the test, in micrometers (μm); G_{bp} is the amount passing the specified test size per revolution, expressed in grams per revolution (g/r) (Ahmadi and Shahsavari, 2009; Free, McCarter and King, 2005; Gent et al, 2012; Magdalinovic et al, 2012; Ipek, Ucbas and Hosten, 2005).

Based on the recommended process parameters of grinding concentration, grinding media filling rate, and grinding time in the ball milling process according to the reference grinding process system, ball milling sample preparation for pelletising experiments was conducted. The wet ball mill had a size of Ø460 × 620 mm and a steel ball weight of 160 kg, with a filling rate of steel balls of 19.92 per cent. The grinding slurry concentration was 7 per cent and each grinding process used 40 kg of ore sample.

Settling characteristics

A certain amount of iron ore sample was weighed and the appropriate amount of water was added. Zhu et al (2018) point out that since particle size significantly affects the settling process, all samples used in the settling experiment should be ground to a similar particle size distribution, with 30 per cent of the particles in the slurry smaller than 0.074 mm. The prepared slurry was poured into a 1000 mL measuring cylinder and the upper part of the cylinder was blocked and shaken evenly. The measuring cylinder was then placed on a platform and the settling time of the iron ore was recorded using a stopwatch, while the height of the settling surface (ie solid-liquid interface height) was also recorded. Recordings were taken every minute for the first 20 minutes and every 10 minutes for the subsequent 40 minutes. The settling characteristic experiment is shown in Figure 1.

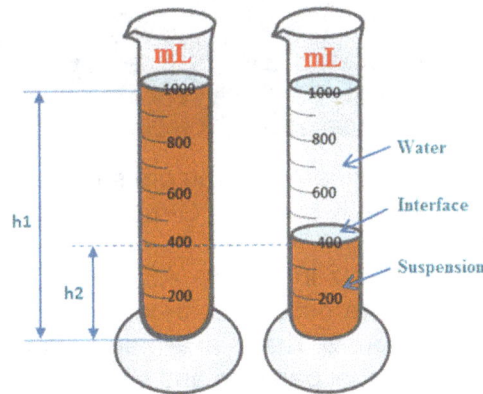

FIG 1 – Experimental diagram of method for determining settling characteristics.

Filtering characteristics

The bottom suspension obtained after settling of the slurry was further subjected to filtration experiments using the XTLZ-Φ260/Φ200 multi-purpose vacuum filter at the Wuhan Prospecting Machinery factory. Firstly, the iron ore fines was wet-ball milled, and then the corresponding size of finely ground concentrate was prepared at different concentrations. A 70 per cent concentration of the finely ground iron concentrate slurry was filtered under a vacuum of -0.08 MPa and the suction filtration time was set from 1 to 6 minutes. After that, the filtering was stopped and the moisture content of the filter cake was measured.

RESULTS AND DISCUSSION

Mineralogical characteristics

As shown in Figure 2, the major minerals in sample A1 are hematite and goethite, accounting for 36 per cent and 62 per cent of the sample, respectively. Sample A1 also contains a relatively high proportion of kaolinite (3 per cent) and quartz (3 per cent). The goethite and magnetite in A1 are interwoven with each other and there are many internal pores, which often contain magnetite, quartz and kaolinite particles.

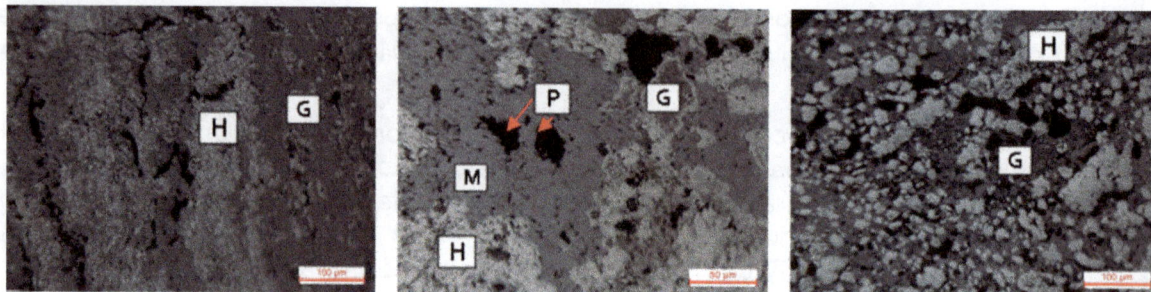

FIG 2 – Microstructure of the main minerals in iron ore sample A1 (H-hematite; G-goethite; M-magnetite; K-kaolinite; Q-quartz; P-pore).

As shown in Figure 3, the major minerals in sample A2 are hematite (68 per cent), goethite (29 per cent), quartz (1 per cent) and kaolinite (2 per cent). In sample A2, the goethite and magnetite are mainly interwoven with fine-grained hematite to form cluster-like aggregates. The hematite occurs as individual bean-shaped particles with large inter-particle pores. The quartz is distributed within the pores of the ore.

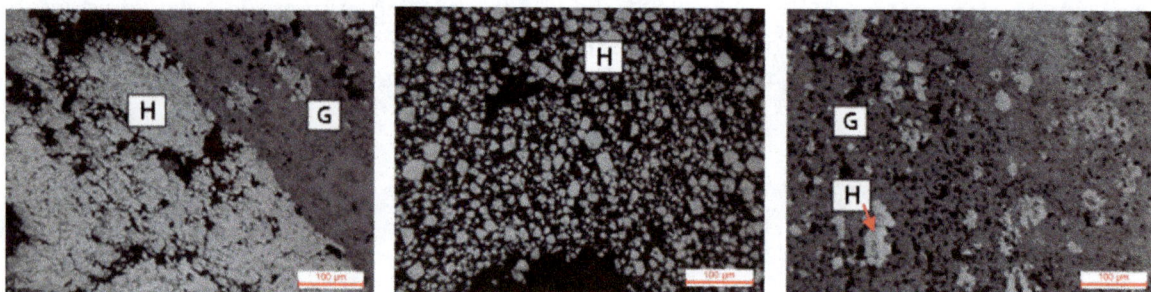

FIG 3 – Microstructure of the main minerals in iron ore sample A2 (H-hematite; G-goethite; M-magnetite; K-kaolinite; Q-quartz; P-pore).

As shown in Figure 4, the major minerals in sample A3 are hematite (46 per cent) and goethite (51 per cent). In sample A3, the goethite occurs as large continuous patches and is intercalated with hematite, magnetite and other ores. There are many internal pores within the grains of goethite. The hematite occurs as individual particles and some flattened particles embedded within the goethite grains. The magnetite mainly exists within the large grains of hematite and there are many pores.

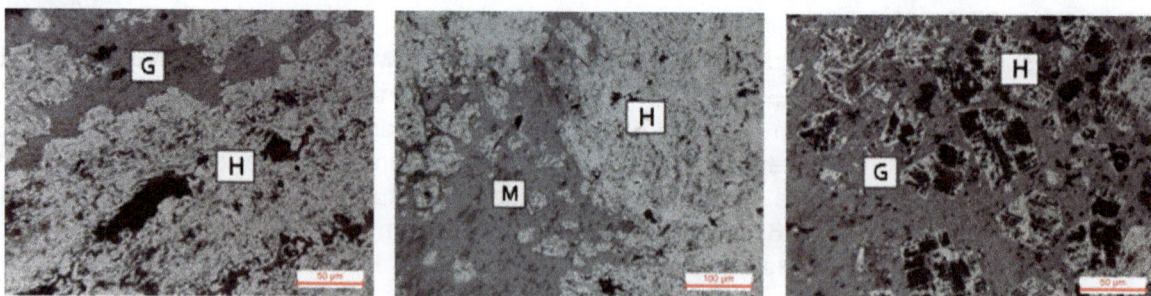

FIG 4 – Microstructure of the main minerals in iron ore sample A3 (H-hematite; G-goethite; M-magnetite; K-kaolinite; Q-quartz; P-pore).

As shown in Figure 5, the major minerals in sample B1 are hematite and specularite, with some quartz. The hematite occurs as coarse-grained or platy crystal aggregates with a dense structure and is mostly an independent mineral. Some hematite contains speckled specularite inclusions in a symbiotic and encased structure, while some specularite is interwoven with hematite in a reticular pattern. The gangue minerals are mainly quartz, occurring as independent minerals or filling intergrowths with hematite in a banded structure.

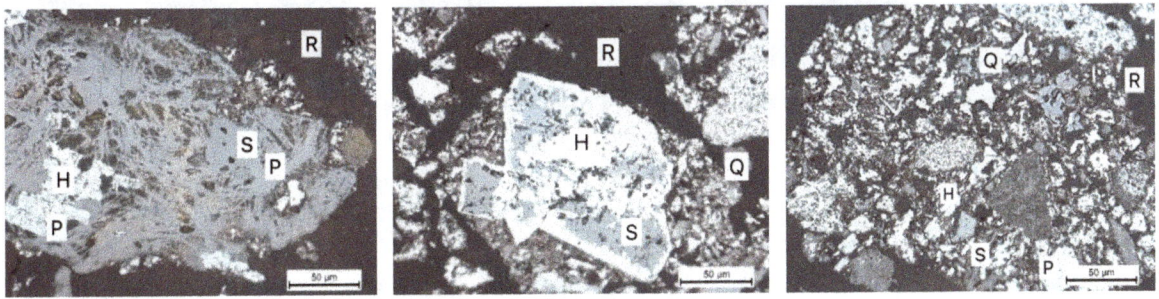

FIG 5 – Microstructure of the main minerals in iron ore sample B1 (H-hematite; S-specularite; R-resin; Q-quartz; P-pore).

As shown in Figure 6, the major mineral phase in sample B2 is hematite (specularite), with a content of up to 79 per cent, mainly occurring as granules and flakes. Sample B2 also contains a certain amount of goethite, typically closely symbiotic with hematite, with a content of around 13 per cent. The presence of goethite is the reason for the relatively high loss on ignition (LOI) of sample B2. Generally, the LOI of iron ore fines should be controlled within a certain range to avoid adverse effects on the pelletising and product mechanical strength during the firing process. In addition, B2 also contains small amounts of quartz and magnetite, mainly occurring as granules, with contents of about 4 per cent and 2 per cent, respectively.

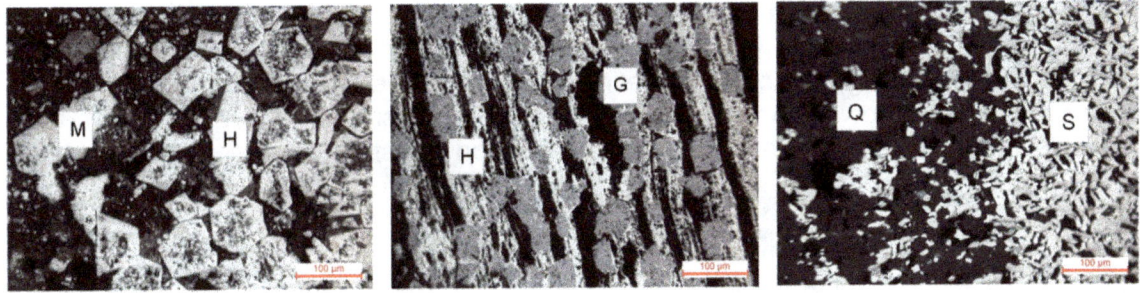

FIG 6 – Microstructure of the main minerals in iron ore sample B2 (H-hematite; S-specularite; M-magnetite; G-goethite; Q-quartz).

As shown in Figure 7, the major minerals in sample B3 are hematite and goethite, accounting for 90 per cent and 8 per cent, respectively, with small amounts of magnetite, kaolinite and quartz. The goethite occurs as large particles, typically surrounded by kaolinite and hematite, while the hematite particles vary in size, with larger particles typically occurring together with quartz and magnetite and smaller particles usually coexisting with kaolinite or embedded within the goethite. The quartz and kaolinite are typically distributed within the pores between the goethite and hematite particles.

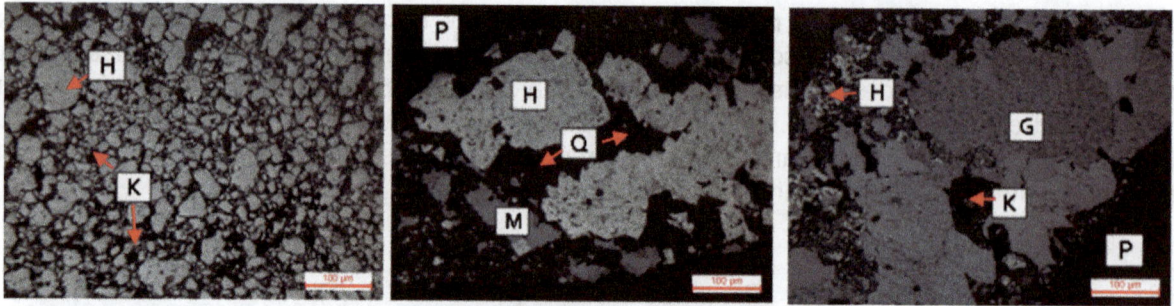

FIG 7 – Microstructure of the main minerals in iron ore sample B3 (H-hematite; G-goethite; M-magnetite; K-kaolinite; Q-quartz; P-pore).

Grindability

Long *et al* (2017) shows that coarse ores with inappropriate sizes require grinding prior to agglomeration. Grinding can increase the ore's specific surface area and surface activity, thus improving its agglomeration efficiency. The main influencing factors of the fine grinding process include ore properties, equipment performance, construction investment and production cost, especially ore properties. The Bond work index of imported iron ore concentrates varies greatly among different countries or regions, as shown in Figure 8. According to the Bond theory, the specific energy consumption of ore grinding is proportional to the Bond work index. Xing, Yang and Xi (2020) point out that strict control of the ore properties is necessary in a specific fine grinding process, otherwise it will directly lead to two serious consequences: either a significant decrease in mill capacity while ensuring the grinding product size, or a serious deviation of the grinding product size without adjusting the new feed rate to the mill. Therefore, the Bond work index is the most important factor influencing the selection of fine grinding process. De Bakker (2013) found that although fine grinding can liberate mineral particles that were previously considered to be unprocessable, it may result in high costs in terms of energy consumption and medium usage. For a specific type of iron concentrate purchased globally, its fine grinding performance should be studied in-depth. A reliable and practical process flow should be selected both to ensure the stability of grinding, filtering, and other operating factors and ensure that production costs such as material consumption, power consumption, and reagent consumption are within a reasonable range, achieving the optimal balance between construction investment and production cost.

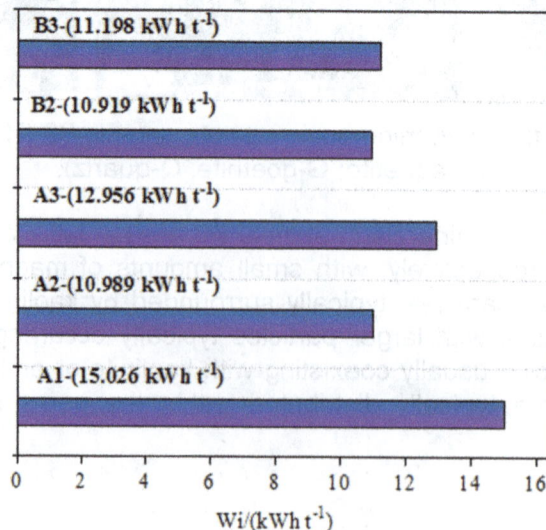

FIG 8 – Bond work index of the different iron ore fines samples.

According to the ranking of grindability based on the Bond work index (from high to low and corresponding energy consumption from low to high), B2>A2>B3>A3>A1, B2 exhibits good grindability and low grinding energy consumption, while A1 exhibits the poorest grindability and slightly higher grinding energy consumption. All five types of iron ore fines show good grindability, with Bond work indices of about 10–15 kW•h/t, which are considered to be materials that are relatively easy to grind.

Prior to entering the mill, the five types of ore fines were crushed to -3 mm using a jaw crusher and the particle size distributions of the crushed samples are shown in Table 3.

TABLE 3

Size distribution of the five iron ore fines after crushing to -3 mm/ per cent.

Sample	Particle size(mm) distribution/%							
	1–3	0.5–1	0.25–0.5	0.15–0.25	0.074–0.15	0.045–0.074	0.038–0.045	-0.038
A1	38.02	9.52	18.05	6.12	9.53	5.84	0.90	12.02
A2	32.07	8.55	9.36	3.14	9.70	9.57	1.26	26.36
A3	28.26	18.04	18.91	3.67	16.78	3.20	1.58	9.56
B2	25.40	13.11	12.40	3.24	9.79	9.26	8.25	18.55
B3	32.19	19.11	6.07	9.09	4.88	5.32	4.80	18.54

Wet ball milling experiments were conducted on the five types of iron ore samples obtained from the aforementioned crushing process, under the conditions of a fixed slurry concentration of 70 per cent and a dry ore mass of 2 kg. The relationship between the grinding fineness (content of -0.074 mm particles) and the grinding time was plotted, as shown in Figure 9.

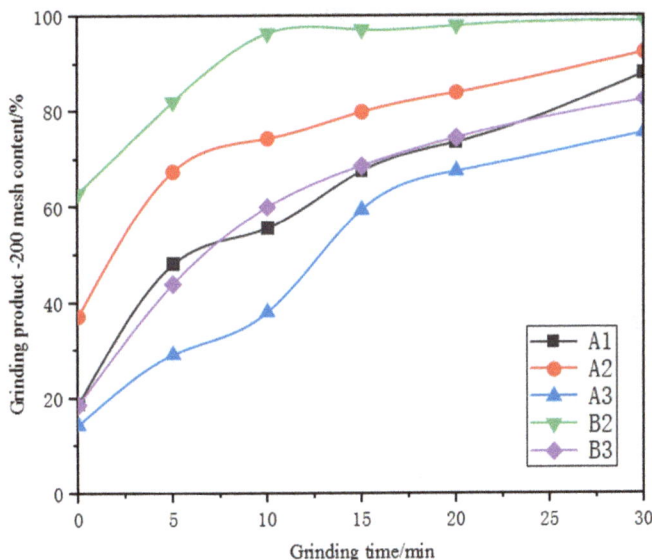

FIG 9 – Variation of grinding fineness (-0.074 mm content) with grinding time.

For the same grinding time, the B2 ore fines sample showed the best grinding effect, with the highest content of -200 mesh particles after grinding. After 5 minutes of grinding, the content of -200 mesh particles exceeded 70 per cent. Therefore, the grinding time for B2 should be controlled to be not too long during wet grinding to avoid over-grinding. The A1, A3 and B3 ore fines samples showed poor grinding performance, and after more than 15 minutes of grinding, the content of -200 mesh particles in the grinding product could only reach about 60 per cent. This is partly related to the slightly higher Bond work indexes and also to the low proportion of fine particles in the original crushed sample for A3.

Settling and filtration characteristics

Zhu *et al* (2018) have proposed that after wet ball milling, it is generally necessary to remove water from the ground concentrates to achieve a product containing 8–10 per cent moisture, which is suitable for the pelletising process. This dewatering is traditionally accomplished through a combination of settlement and filtration. Therefore, the settlement and filtration characteristics of the iron ore slurry are crucial to the pelletising process.

Settling characteristics

Su, Wang and Li (2017) point out that the process of solid particles suspended in a fluid separating from the fluid under the action of gravity is called settlement. Settlement testing of mineral slurries is one of the fundamental tests in mineral processing operations for tailings, concentrates, or intermediate materials, aimed at measuring the settling velocity of solid material groups in a slurry of a certain concentration, as a basis for the design of a beneficiation plant.

The settling performance of the five iron ore samples is shown in Figure 10. The static settling process of the slurry can be roughly divided into three stages: in the initial constant settling stage (within about 10–15 minutes), the particles in the grinding products of the five iron ore fines samples are less affected by the surrounding particles or the tube wall and the particle spacing is large. The fine particles undergo intense Brownian motion and collide and aggregate with each other, resulting in fast settling rates of 14.20, 13.40, 14.62, 15.00 and 16.33 cm/min for samples A1, A2, A3, B2 and B3 respectively. In the subsequent 5–10 minutes, the settling stage transitions to the interference settling stage, during which the settling rate of the ore particles decreases significantly. After a settling time of more than 20 minutes, the process transitions to the compression stage, during which Cai *et al* (2014) found that the settling rate is slowest due to the gradual increase in concentration and viscosity of the sediment layer. According to the Newtonian viscosity law, the drag force on the particles gradually increases, making the particles more susceptible to severe interference from the surrounding particles during settling. Overall, the static settling performance of the three Australian ore fines is poorer than that of the Brazilian ore fines and a settling time of more than 20 minutes is required.

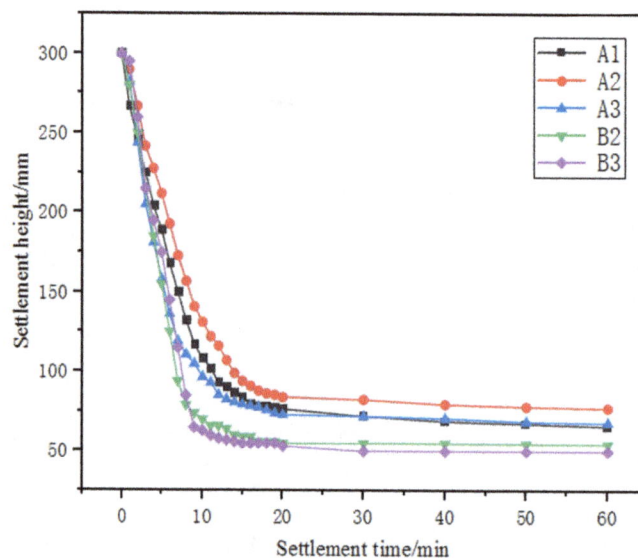

FIG 10 – Relationship between settlement height and settlement time of the five iron ore fines samples.

Filtering characteristics

Filtration (He *et al*, 2023) is an operation that separates solids and liquids by passing the liquid in a suspension through the channels of a porous medium under the action of an external force, while retaining the solid particles on the medium surface. The liquid that passes through the filtration medium under pressure is called the filtrate, while the solid particles trapped on the filtration medium surface form the filter cake or residue.

The influence of filtration time on the filtration performance of the five kinds of iron ore fines samples, with a grinding fineness of 70 per cent passing through 200 mesh, was investigated under the conditions of a slurry concentration of 70 per cent and a filtration vacuum of 0.08 MPa. As shown in the Figure 11, with the increase of filtration time, the residual moisture in the filter cake gradually decreases, but the rate of decrease and the filter cake moisture content vary among the five iron ore fines samples. Moreover, the change in residual moisture in the filter cake is not significant after the filtration time reaches 4 minutes. Even with a filtration time of up to 6 minutes, the filter cake moisture content of the five finely ground products remains above 10 per cent, indicating poor filtration performance and difficulty in filtration and dewatering. Due to moisture contents above 10 per cent, further drying is required before the samples can be directly submitted to the subsequent pelletising process.

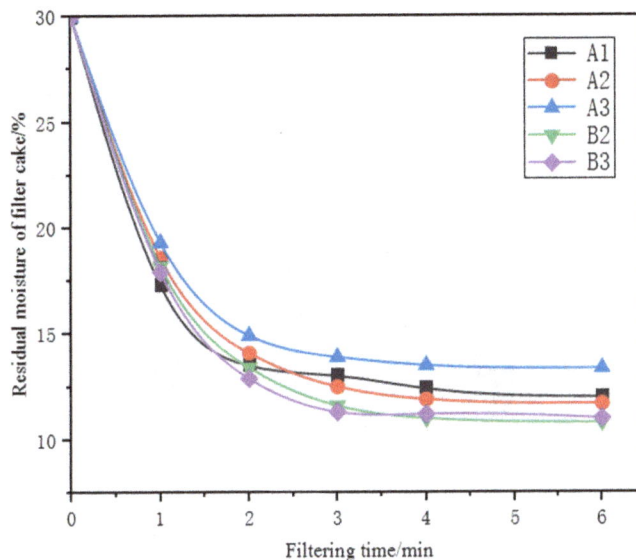

FIG 11 – Relationship between residual moisture of filter cake and filtering time of the five iron ore fines samples.

The relationship between grindability, settling and filtering performance and process mineralogy

The grindability, settlement and filtration performance of iron ore fines are greatly influenced by their density, mineralogical characteristics (such as the quantity of clay minerals and goethite and the porosity of the ore particles), and grinding fineness. Generally, iron ore fines with higher density, fewer soft minerals and larger particle size exhibit better settlement and filtration performance. However, it should be noted that no single factor has a decisive effect on the grinding, settlement and filtration performance, and these factors interact with each other. Therefore, other properties of the ore need to be considered as well. In addition, different grinding equipment and processes may have varying effects on the grindability of different iron ore fines, so selection and optimisation should be based on specific circumstances.

Analysis of the relationship between grindability and process mineralogy

The ore hardness is a reflection of the specific mineral composition of the ore itself and its physical and mechanical properties. Ores with a dense structure, tiny crystals and high hardness are more difficult to grind. Therefore, such ores require a longer grinding time in the grinding process to ensure that the required grinding fineness is achieved, but generally speaking, ore hardness mainly affects the processing capacity of the mill. On the other hand, ores with small hardness or well-developed decomposition are easy to crush and the processing capacity per unit volume of the mill is high.

Zhou and Shao (1997) have established the following relationship using linear regression:

$$P = 2.82 \times 10^7 \times H^{-2.02} \times D^{-1.22} \tag{3}$$

where the wear resistance P is inversely proportional to the Vickers hardness H and the hardness dispersion value D (The magnitude of the hardness dipersion value D reflects the degree of adaptability between the uniformity of ore hardness or the complexity of the mineral composition and processing technology.), indicating that the higher the hardness, the lower the P value and the more wear-resistant the rock. Conversely, if the rock is less wear-resistant, its hardness is lower.

As shown in Figures 8 and 12, the Vickers microhardness of iron ore fines is positively correlated with the Bond work index, and the grindability is largely dependent on the hardness of the iron ore. Iron ore fines sample A1 has the highest microhardness and the lowest grindability, so a longer grinding time is required to achieve the required grinding fineness. On the other hand, ore sample B1 has the lowest microhardness and the lowest Bond workindex among the five iron ore fines samples, resulting in lower grinding energy consumption.

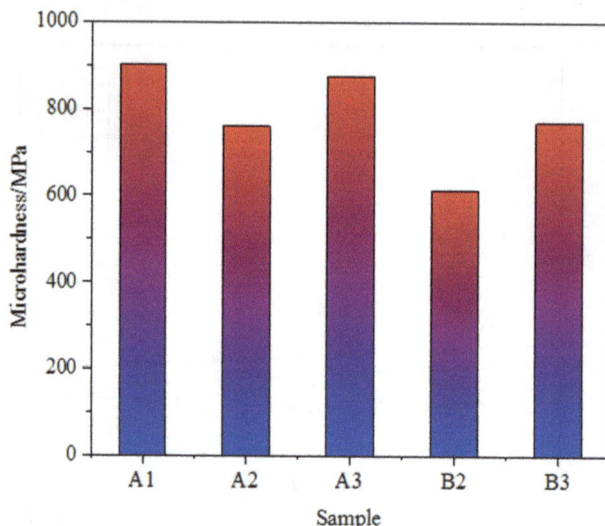

FIG 12 – Vickers microhardness of the iron ore fines samples.

In addition to microhardness, factors such as ore texture, pore size, goethite content and soft mineral content also affect the grinding effect. The higher the content of goethite in the ore, the more difficult it is to grind. This is due to the physical properties of goethite, which have high hardness and poor toughness, and can easily form small particles during the grinding process, leading to increased difficulty in grinding. At the same time, as the content of goethite increases, the frictional force between ore particles during grinding also increases, affecting the grinding effect. The high content of goethite in samples A1 and A3, both exceeding 50 per cent, may be one of the reasons for their poor grindability. The low content of goethite in sample B3 may be one of the reasons for its good grindability.

The mineral texture characteristics, such as mineral particle size, shape, association and spatial distribution, have a certain impact on grinding. If the minerals are interlocked, different types of minerals in the ore are interlocked and distributed, resulting in more interfacial boundaries between minerals and irregular shapes of mineral particles. This can lead to the formation of small mineral particles that are difficult to grind, resulting in poor grindability. If the minerals are enclosed, some minerals in the ore are wrapped by other minerals, and the hardness, toughness and fracture characteristics of these enclosed minerals may be different from those of the external minerals, making them difficult to grind. Sample A1 has interwoven goethite and magnetite, with many internal pores containing magnetite, quartz and kaolin particles. This may be one of the reasons for its high Bond work index. Sample A3 contains goethite intercalated with adjacent iron ores such as hematite and magnetite, with magnetite mainly existing inside the large hematite crystals.

Analysis of the relationship between settling performance and process mineralogy

The porosity of an ore refers to the ratio of the volume of pores in the ore to the total volume of the ore, usually expressed as a percentage. Porosity is an important physical property parameter of ores, reflecting the number and size of pores in the ore and has an important impact on ore processing, beneficiation and smelting processes. Generally, the greater the porosity of the ore, the smaller its density, and therefore the less resistance it experiences during settlement, resulting in slower settling speeds. As shown in Figure 10, Australian ore samples A1, A2 and A3 generally have higher porosity, which may be related to their slower settling speeds (the ore settling time exceeding 20 minutes).

A higher content of goethite in an ore typically has a negative impact on settlement and filtration performance. Minerals with a high content of goethite have smaller and elongated particles, resulting in slower settling speeds and potentially longer settling times. Therefore, it can be inferred that, under other identical conditions, the higher the content of goethite, the poorer the settlement ability. Australian ore samples A1, A2 and A3 generally have high hematite needle contents, all exceeding 40 per cent, which may be one of the reasons for their longer settling times.

Clay minerals, including kaolinite, typically have smaller particle sizes and are often plate-like or fibrous in shape. Therefore, ores with high kaolinite content generally have slower settling speeds.

Analysis of the relationship between filtering performance and process mineralogy

The porosity of an ore refers to the proportion of voids in the ore, and a higher porosity indicates more voids in the ore, which generally leads to better filtration performance. If the porosity of an ore is low, water molecules have difficulty penetrating into the interior of the ore, resulting in poor settling and filtration effects. Therefore, in general, a higher porosity leads to better filtration performance. However, the filtration performance of the three Australian ores is also affected by the content of goethite, kaolinite and particle size, despite their relatively large porosity.

Goethite is a fibrous or needle-like mineral that can affect the solid-liquid separation of the ore. When the goethite content in the ore is high, the shape of goethite may cause a staggered stacking phenomenon on the surface of the ore forming a denser structure, which leads to lower porosity in the ore. As a result, during settling and filtration, the water in the ore has difficulties draining smoothly, leading to poor filtration performance. Therefore, the low content of needle-like iron ore in samples B2 and B3 may be one of the reasons why they have a faster filtration rate and lower residual moisture content in the filter cake.

Kaolinite particles are generally small and tend to fill the voids, making it difficult for water to pass through and leading to poor filtration performance. The three Australian ores all contain varying amounts of kaolinite, which may partially explain why their filtration performance is poor. The kaolinite particles may fill the interstices between other particles, reducing the overall porosity of the ore and hindering the flow of water during filtration.

In summary, while the porosity of an ore is an important factor affecting its filtration performance, other factors such as the content of goethite and kaolinite, particle size and other factors should also be taken into consideration. By understanding and optimising these factors, it is possible to improve the filtration performance of ores and achieve better solid-liquid separation.

The effect of combined ore blending on settlement and filtration

Numerous production practices and experiments have shown that blending hard-to-grind and easy-to-filter materials with easy-to-grind and hard-to-filter materials for fine grinding can improve the filtration and settlement performance of the material, achieving the target for optimising filtration and settlement.

After wet ball milling, the moisture content of sample B3 exceeded 50 per cent and therefore required settlement and filtration treatment. Peng *et al* (2021) have shown that during the wet ball milling processing of sample B3, the particles underwent lattice deformation and surface defects, resulting in increased surface activity of the particles. The particle size improved after ball milling and the mass fraction of fine particles increased, the specific surface area increased and the hydrophilicity was improved. Additionally, during settlement, highly active kaolin particles produced electrostatic repulsive forces in the aqueous medium, increasing the slurry viscosity and hindering settlement. During filtration, irregularly shaped fine particles in sample B3 filled the pores of the filter cake, reducing its porosity, and the increased hydrophilicity of the finely ground kaolin particles made it difficult to remove moisture from the filter cake. Sample B3 after fine grinding had poor settlement and filtration performance, and it was difficult to obtain a filter cake with a moisture content of less than 10 per cent using existing filtration equipment. Using a large proportion of sample B3 alone was difficult, which affected production efficiency and smooth operation. In contrast, fine specular iron ore concentrate on its own was easy to settle and filter, and could achieve a filter cake moisture content of less than 10 per cent in 2 minutes of filtration time. Therefore, the settlement and filtration performance of finely-ground sample B3 was improved by blending it with coarser and regularly shaped sample B1 concentrate. Yi *et al* (2022) found that based on the differences in mineral surface properties, settlement and enhanced filtration technology based on carrier particles with regular particle size, fewer lattice defects and poor surface hydrophilicity could be developed. This is based on a combined ore blending settlement and filtration process as shown in Figure 13. And the comparison of settlement and filtration performance of single ores and blended ores can be represented in Figure 14. The addition of sample B1 concentrate to iron ore slurry was found to reduce both the settlement time and filter cake moisture content. Increasing the concentration of sample B1 resulted in a reduction of approximately 5 minutes in the settlement time. By finely grinding sample B3 and blending it with sample B1 concentrate, the filter cake moisture content could be controlled to below 10 per cent. By smart blending of materials and controlling the ratio of materials, finely ground imported fines suitable for pelletising could be produced. This solved the problem of fine particle pollution of water bodies and the difficulty in achieving the required moisture content of pelletising raw materials, which improved resource utilisation.

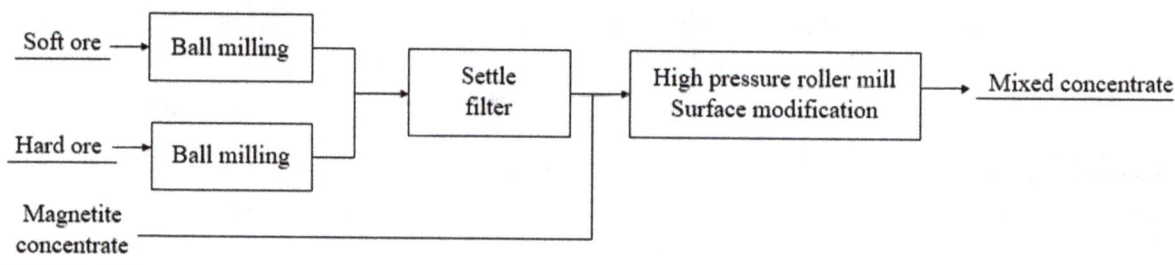

FIG 13 – Optimisation of the settlement and filtration process based on ore blending. Adapted from Yi *et al* (2022).

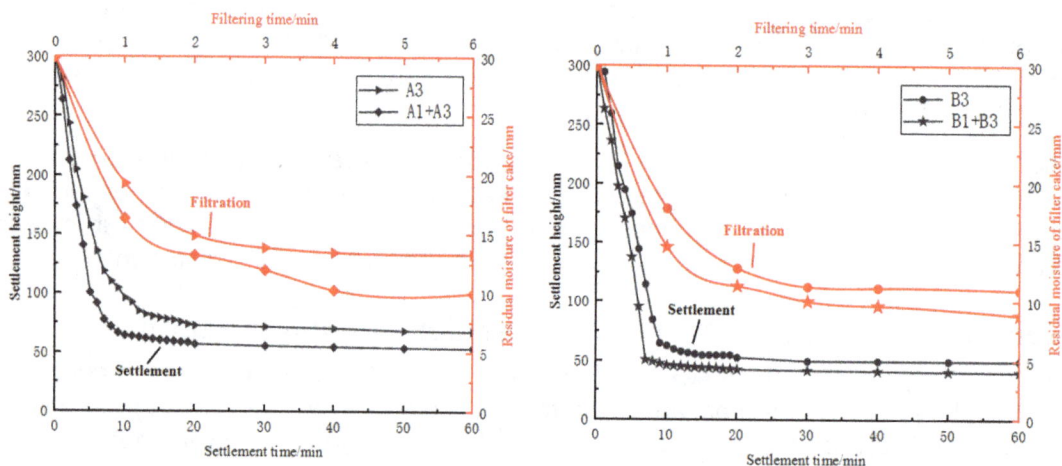

FIG 14 – Comparison of settlement and filtration performance of single ores and blended ores.

By blending sample A1 concentrate with Australian iron ore, the settlement time can be reduced from over 20 minutes to around 10 minutes, with a maximum settling rate of 23.55 cm/min during the first 10 minutes. For all cases, higher filtration pressure and longer filtration time resulted in lower residual moisture content of the filter cake. Filtration of samples A1 and A3 as single ores exhibited poor performance, with over 10 per cent residual moisture content even after 6 minutes of filtration. However, adding sample A1 improved the filtration performance of sample A3. Blending samples reduced the residual moisture content to around 10 per cent after 6 minutes of filtration. This suggests that blending of ores and concentrates can be a promising approach to improving the efficiency and quality of mineral processing.

CONCLUSIONS

1. By improving the technology of fine grinding, various imported ore resources can be well adapted and matched, providing more resource options for pellet feed blending. Moreover, imported fines ground by fine grinding have good pelletising performance and can be used in

pellet production with low harmful components, which is beneficial for blast furnace operations.

2. The Bond work index of the five iron ore fines samples investigated ranged from 10 to 15 kW·h/t, indicating good grindability and they can all be considered as materials that are relatively easy to grind. Among them, sample B1 had the lowest Bond work index and the best grindability, while sample A1 had the poorest grindability and slightly higher energy consumption during grinding.

3. The settling process for the slurry after grinding can be roughly divided into three stages: the constant settling stage, the interference settling stage and the compression stage. The static settling performance of the three Australian iron ore fines studied in this article was poorer than that of the Brazilian iron ore fines and the settling time required was more than 20 minutes. With increasing filtration time, the residual moisture in the filter cake gradually decreased, but the rate of decrease and the moisture content of the filter cake varied between the five types of iron ore fines studied. Even with a filtration time of up to 6 minutes, the filter cake moisture content of the five finely ground products remained above 10 per cent, indicating poor filtration performance and difficulty in filtration and dewatering.

4. There was an inherent relationship between the mineralogical characteristics of the ores and their grinding, settling and filtration performance. Grindability was largely determined by the hardness of the iron ore and in general, the harder the ore, the more grinding energy required. The settling and filtration performance of the iron ore fines samples was largely influenced by their density, mineralogical characteristics, grinding fineness, etc. Generally, iron ore containing fewer goethite minerals, fewer soft minerals (such as kaolinite) and coarse particle size has good settling and filtration performance.

5. The imported coarse iron ore fines have various characteristics and the differences in grindability and filtration performance are significant. By smart ore blending, it is possible to obtain ore fines with excellent settling and filtration performance, which can meet the requirements of the grinding and selection process, maximise the production rate of equipment, reduce operating costs and produce high-quality finely ground iron ore concentrate.

ACKNOWLEDGEMENTS

The authors would like to acknowledge the Analytical and Testing Center of Central South University which supplied the facilities for the measurements.

REFERENCES

Ahmadi, R and Shahsavari, S, 2009. Investigation of laboratory conditions effect on prediction accuracy of size distribution of industrial ball mill discharge by using a perfect mixing model, *Minerals Engineering*, 22:104–106.

Cai, F W, Zhou, X T, Xia, J P, Luo, Z Q, Man, L and Zhu, L, 2014. Study on Settling Character of Bauxite Tailings Slurry, *Bulletin of Chinese Ceramic Society*, 33(06):1544–1549.

De Bakker, J, 2013. Energy Use of Fine Grinding in Mineral Processing, *Metallurgical and Materials Transactions E,* 1(1):8–19.

Free, K S, McCarter, M K and King, R P, 2005. The use of a simplified comminution model to predict the power draw of grinding mills in industrial applications, *Minerals & Metallurgical Processing*, 22:96–100.

Fu, J Y, Jiang, T and Zhu, D Q, 1996. *Sintering and Pelletizing*, pp 313‑315 (Central South University of Technology Press: Changsha).

Gao, E X, Zhang, C, Li, Y P, Luo, J H, Gao, R Z and Rang, J C, 2022. Effects of Grinding Methods on the Flotation Kinetics of Sphalerite and Pyrite, *Metal Mine*, (09):100–106.

Gent, M, Menendez, M, Torano, J and Torno, S, 2012. Effect of grinding aids on the grinding energy consumed during grinding of calcite in a stirred ball mill, *Powder Technology*, 224:217–222.

He, J W, An, G, Wang, K, Kang, H J, Zhao, W C and Liu, C J, 2023. Shougang Jingtang imported fine grinding technology, *Hebei Metallurgy*, (02):33–36.

Hu, J and Gao, Z, 2008. Characteristics of Marra Mamba iron ore fines and the application technology in sintering process, *Research on Iron & Steel*, 36(5):25–28.

Ipek, H, Ucbas, Y and Hosten, C, 2005. Effect of particle size distribution on grinding kinetics in dry and wet ball milling operations, *Minerals Engineering*, 18:981–983.

Long, H M, Chun, T J, Wang, P, Meng, Q M, Di, Z X and Li, J X, 2017. Grinding kinetics of vanadium-titanium magnetite concentrate in a damp mill and its properties, *Metallurgical and Materials Transactions B*, 48(2):815–822.

Magdalinovic, N, Trumic, M, Trumic, G, Magdalinovic, S and Trumic, M, 2012. Grinding of hematite ore with a high specific surface area using a planetary ball mill, *International Journal of Mineral Processing*, 114–117:48–50.

Mohamed, O A, Shalabi, M E H, El-Hussiny, N A, Khedr, M H, Khedr, F and Mostafa, 2003. Effect of some additives on flowability and compressibility of lactose powders, *Powder Technol*, 130(2003):277–282.

Peltoniemi, M, Kallio, R, Tanhua, A, Luukkanen, S and Perämäki, P, 2020. Mineralogical and surface chemical characterization of flotation feed and products after wet and dry grinding, *Minerals Engineering*, 156:106500.

Peng, D S, Pan, J, Li, J, Tian, H Y, Yang, C C, Shi, X J and Sheng, W J, 2021. Experimental study on the preparation of fluxed pellets from fine grinding carajas powder, *Sintering and Pelletizing*, (04):50–57.

Su, G H, Wang, Z Q and Li, H Y, 2017. Application of modifier to improve sedimentation and filtration performance of superfine slurry of laterite nickel ore, *World Nonferrous Metals,* (16):247–248.

Sun, H, 2000. Proportioning of Laiyuan Rich Ores to Increase Sinter Strength, *Sintering and Pelletizing*, 25(5):43–45.

Wills, B A and Napier-Munn, T J, 2006. *Will's Mineral Processing Technology* (7th ed), pp 108–117, 267–352 (Elsevier Inc. Oxford: Great Britain).

Xing, W, Yang, H L and Xi, Z W, 2020. Study on Fine Grinding Technology of Imported Iron Concentrate, *Mining-Engineering*, 18(03):43–46.

Yi, L J, Jiang, L H, Li, J, Niu, C S, Chen, J H, Zhu, D Q, Yang, C C and Tian, H Y, 2022. Technology development and application of diversified low cost clean production of high quality pellet products, *Sintering and Pelletizing,* 47(01):95–103.

Yie, K W, 2003. The present state and the forecasting of our country pelletizing industry, *Sintering and Pelletizing 2003*, (01):1–4.

Zhou, C M and Shao, G Y, 1997. Weighted Vickers Hardness and Grindability of Granite, *Stone,* (01):18–20;30.

Zhu, D Q, Guo, Z Q, Pan, J and Wang, Z Y, 2018. Insights on pretreatment of Indian hematite fines in grate–kiln pelletizing process: the choice of grinding processes, *Journal of Iron and Steel Research International,* 25(5).

Immortality design of the allflux® upstream classifier in the iron ore industry – for a safe workplace and an economic, sustainable and highly efficient separation technology

M Steinberg[1] and M Amaranti[2]

1. MAusIMM, Managing Director, HAZEMAG allmineral Australia Pty. Ltd., Canning Vale WA 6155. Email: steinberg@allmineral.com
2. Project and Support Engineer, HAZEMAG allmineral Australia Pty. Ltd., Canning Vale WA 6155. Email: mario.amaranti@allmineral.com

ABSTRACT

Upstream Classifier are well-known in the iron ore industry for at least 15 years. At present, more than 60 units are in operation in Western Australia. The advantage of this technology is their capability to produce three products with one equipment – iron ore concentrate, iron ore middlings for further processing and slimes (tailings).

The principal of this beneficiation process – upstream classification and upstream separation – will always remain. However, the optimisation of the classification and separation efficiency of the equipment itself is a continuous process driven by enhanced automation and instrumentation, design changes and the use of upstream and downstream information and requirements.

Besides efficiency, the safety, maintenance and replacement conditions are very important subjects for the operation of an iron ore mine.

The aim of this paper is to present a new allflux® upstream classifier design, which has been developed together with several iron ore producer from Western Australia. The new design benefits from its sustainability approach by dividing the entire equipment into small reusable parts. As a result, maintenance is minor, replacement much faster, lifetime much longer and costs much lower.

Furthermore, it demonstrates new features to prevent hazardous working conditions such as confined space as well as new fixing devices to speed up the replacement of certain items and parts.

UPSTREAM CLASSIFIER

The allflux® process, as shown in Figure 1, makes sophisticated use of the laws of physics: the method is based on the fluidised bed principle. The allflux® separator separates lightweight material made of fine-grained substances; eg sand, iron ore fines. The allflux® treats the material in a two-step process, eliminating the need for pre-thickening. One single system combines high efficiency and high throughput.

The system is very simple: the allflux® generates an upward current in the coarse fraction chamber. Light and fine particles rise while coarse, heavy particles sink and are drawn off automatically from the bottom of the coarse chamber. In the fine fraction chamber, which is the second stage of the process, fine particles form an autogenously fluidised bed on which the light weight material to be separated floats and spills over a weir. The fine product is discharged automatically from the bottom of the fines chamber. The flow and control equipment facilitates the generation of three highly selectively graded products by one machine. Process control is automatic.

The first idea for the allflux® separator was born in 1988. The patent was registered in 1990 and the first industrial unit was installed in 1991. More than 150 units are in operation worldwide out of which more than 75 are supplied to the iron ore industry.

Since the introduction of the allflux® technology to the concrete sand industry more than 15 years ago, many more applications have been discovered. Fine coal recovery from ponds, iron ore and mineral sand concentration and high quality glass sand sizing are just a few examples of this unique technology.

FIG 1 – allflux®.

Advantages of the allflux®:

- high separating efficiency
- high capacity reaching 2000 m³/h
- consistent high product quality
- automatic operation and control of discharge systems and water feeds
- high solids content of discharged products
- low wear
- low energy consumption
- lesser area requirement compared to spirals
- lesser operating cost compared to spirals
- classification and concentration in a single unit.

Another benefit results from the relatively independent processing of coarse and fine material. It allows either fraction to be processed or stockpiled separately, or they may be blended to meet special applications. Even with variations in raw material composition, the ability to offer a consistent product is greatly enhanced.

Perhaps the most important advantage of the allflux® separator is an economic one. By combining high efficiency and high capacity with multiple processing stages, a reduction in specific production cost is realised. Overall plant size is minimised and hence investment and operation costs are reduced. The efficiency of the allflux® extends the boundaries for feasible fine particle processing, it may allow some previously uneconomic reserves to be mined.

The allflux® separator is a round, centre feed process vessel that is sized according to the hydraulic load. The process uses a unique combination of rising current and fluidised bed techniques and can be divided into three stages. The principle of an allflux® separator is schematically shown in Figure 2.

The allflux® separators are used in iron ore beneficiation to classify and simultaneously reduce the amount of silica, alumina and slimes. The classified products are prepared for optimised processing in the following upgrading process. While the allflux® technology has been in use for more than five years in South Africa, the Australian and Indian iron ore producers have discovered the value of this technology and are in the process of installing and commissioning.

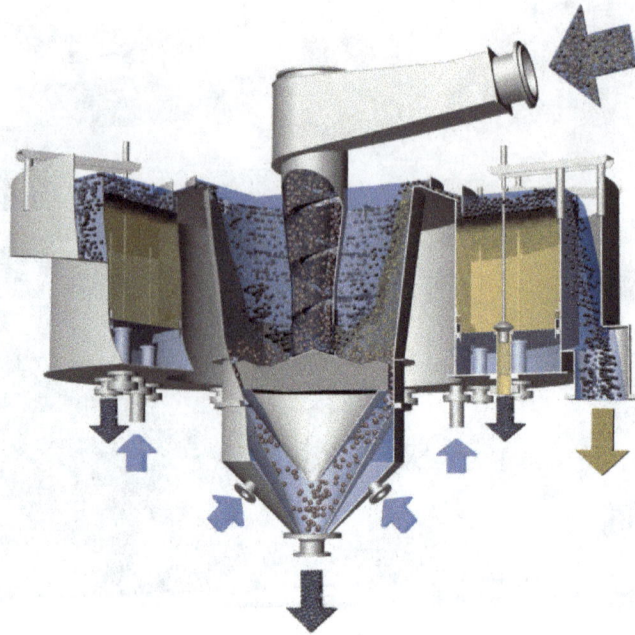

FIG 2 – allflux® schematic.

INTRODUCTION

More than 75 units of the allflux® upstream classifier are in operation in the iron ore industry. Iron ore producers operate within a dynamic market environment, characterised by continuous growth. As such, they prioritise the implementation of safety measures, the attainment of prolonged equipment lifespan and the reduction of capital and operating expenses whenever feasible.

Notably, the provision of refurbishment services holds significant potential to considerably enhance the operational lifespan of machines. By conducting yearly machine audits, it becomes possible to accurately determine the specific sections of the machine requiring refurbishment and subsequently plan and coordinate these activities with the maintenance shutdown schedules of the site.

As a result of close collaboration with the iron ore producer, leveraging their substantial experience in upstream classifier operation and identification of replacement requirements, significant modifications have been implemented to address the evolving needs of the iron ore producer. These key alterations encompass various aspects of the system, ranging from structural enhancements to improved operational features, thereby ensuring optimal performance and aligning with the specific demands of the iron ore production process.

The primary modifications can be summarised as follows:

- new rotable coarse section
- new locking method for side wall inspection hatches
- new style conidur plate hold down clamps
- new horizontal struts
- pulley arrangement to move the classifier into position.

NEW ROTABLE COARSE SECTION

The Coarse Section shown in Figures 3 and 4, composed of mild steel, incorporates a flanged connection for seamless assembly to the discharge cone. Positioned above the discharge cone, a displacement body is installed to ensure the uniform distribution of the feed slurry. The introduction of teeter water takes place below the displacement body within the Coarse Section.

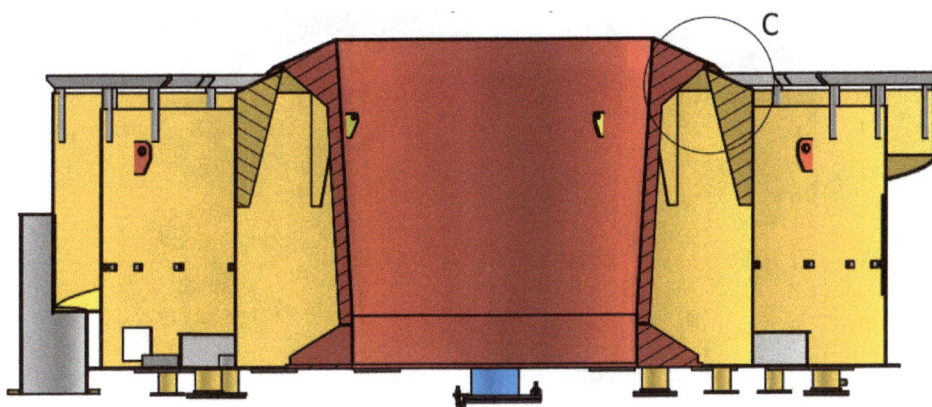

FIG 3 – Coarse Section marked in red.

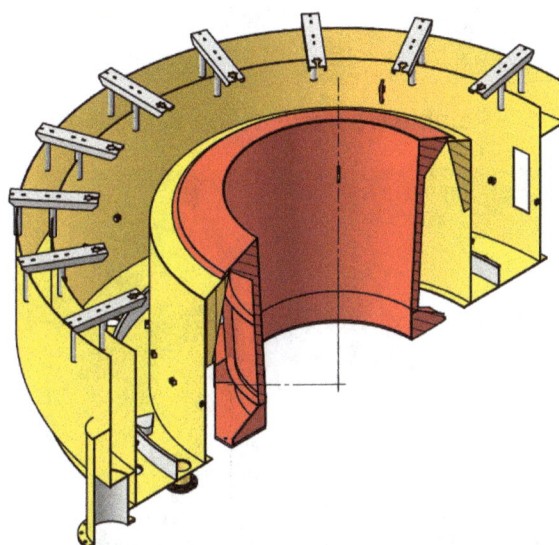

FIG 4 – Coarse Section marked in red.

The feed slurry enters the Coarse Section via a central feed pipe, conveyed from a launder and is directed onto the displacement body. Both the inlet section and its accompanying feed pipe are lined with a ceramic-rubber lining to enhance durability. The overflow from the Coarse Section flows directly into the middlings section.

For efficient discharge, a stainless-steel cone is employed in the Coarse Section. This cone is lined with a ceramic-rubber lining and polyurethane, providing exceptional resistance to abrasion. To facilitate maintenance and replacement procedures, the lower section of the cone is detachable and connected to a significant part of the cone via a flanged joint.

One notable advancement implemented in the refurbishment program for the upstream classifiers involved the complete rotatability of the Coarse Section. As part of this enhancement, the top section of the Coarse Section, resembling a 'mushroom' shape, underwent diameter reduction to enable the convenient extraction of the Coarse Section through the designated opening located above the classifiers at the site.

The newly implemented design allows for the removal of the Coarse Section as a single unit, as depicted in Figure 5 and 6. By simply removing the bolts that secure the Coarse Section to the Super Structure, the unit can be detached. This rotatable Coarse Section serves as a replacement for the worn-out counterpart, significantly minimising downtime. Consequently, maintenance and repairs can be conducted on the worn section without impeding the resumption of operations in the beneficiation plant.

FIG 5 – Rotable Coarse Section at the workshop.

FIG 6 – Rotable Coarse Section at site.

NEW LOCKING METHOD FOR SIDE WALL INSPECTION HATCHES

As part of the enhancements made, the size of the Side Wall Inspection Hatches (depicted in Figure 7) was increased, accompanied by the introduction of a new securing method. The objective was to simplify the removal of the hatches, thereby encouraging operators to take the necessary steps to clean the underside of the conidur plates.

FIG 7 – New inspection hatch.

To address the challenges posed by the previous method of securing the Side Wall Inspection Hatches, a new approach has been implemented using bolted-in struts. This new configuration replaces the previously used smaller bolts, which were both time-consuming and cumbersome to remove. The old design is shown in Figure 8.

FIG 8 – Old inspection hatch.

However, during trial removals of the new struts, it was discovered that they presented difficulty in being detached due to inadequate clearance between the strut insert and its seating point. To resolve this issue, a corrective measure was undertaken by grinding down the insert by approximately 2 mm (inclusive of approximately 1 mm of paint). This adjustment allows for easier removal of the two struts from their respective seats.

NEW STYLE CONIDUR PLATE HOLD DOWN CLAMPS

To replace the previous method of employing wooden wedges for securing the conidur plates, a new approach was implemented, utilising a bolt and block arrangement (depicted in Figure 9).

FIG 9 – New fix arrangement for conidur plates.

To substitute the previous wooden wedge system, a new hold-down system has been implemented. This system incorporates two slightly different sizes of clamps, with the shorter clamps featuring a larger foot arrangement specifically designed for use with the new horizontal struts. It should be noted that the shorter clamps can be utilised for securing the conidur plate, but not vice versa.

During the assembly process at the manufacturer's workshop, there were instances where these clamps were mistakenly applied interchangeably. While this does not adversely affect the functionality of the clamp, it is a matter that will be addressed in future installations to ensure the appropriate clamps are used in their designated positions.

As a precautionary measure aimed at mitigating the potential for wear, a plastic sleeve has been strategically positioned around the exposed thread as shown in Figure 10. This sleeve serves to create a protective barrier, minimising friction and safeguarding the integrity of the thread. By implementing this additional layer of defence, the risk of wear-related damage is significantly reduced.

FIG 10 – Plastic sleeve protection.

NEW HORIZONTAL STRUTS

In order to replace the previous stud-style horizontal 'hold down strut' (depicted in Figure 11), a novel strut design was employed, which eliminated the reliance on stud arrangements and instead utilised the newly mentioned clamping method for secure fastening. The introduction of this new strut design offers several notable advantages. Firstly, the removal process is significantly simplified, thereby facilitating maintenance procedures. Moreover, the replacement of the strut is notably easier compared to the previous studded strut, which proved to be problematic and time-consuming to replace. This enhanced design streamlines the overall maintenance workflow and contributes to operational efficiency.

FIG 11 – Old design studded strut.

As part of the innovative design, the newly introduced strut, shown in Figure 12, incorporates a distinctive feature in the form of a rubber seal located on the underside. This purposeful inclusion serves to effectively prevent any potential leakage, ensuring a reliable and robust sealing mechanism. By integrating this rubber seal into the strut, the possibility of unwanted fluid or material seepage is effectively eliminated, contributing to enhanced operational integrity and system performance.

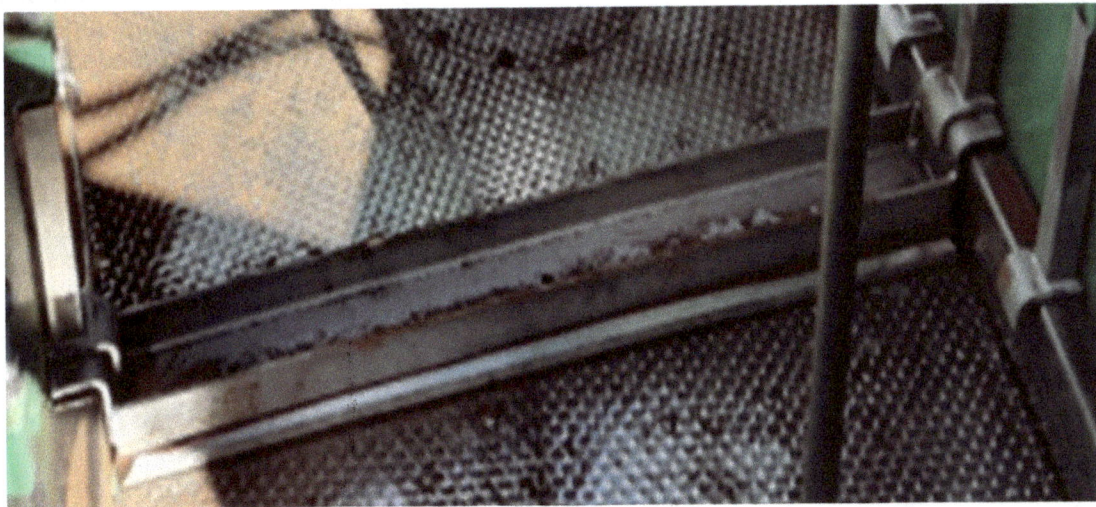

FIG 12 – New design strut.

PULLEY ARRANGEMENT TO MOVE THE CLASSIFIER INTO POSITION

To facilitate the precise movement of the Upstream Classifier to its designated location within the beneficiation plant, a well-designed skid arrangement (depicted in Figure 13) was employed. This skid arrangement featured a carefully orchestrated pulley system on each side of the classifier, manually operated by a winch mechanism, allowing for controlled pulling and positioning of the Upstream Classifier.

FIG 13 – Skid arrangement.

Through the strategic utilisation of a pulley arrangement, shown in Figure 14, positioned on both sides of the classifier, a meticulous operation was carried out to facilitate the safe and controlled movement of the classifier from its original location to its designated position within the beneficiation plant. This carefully orchestrated process involved leveraging the mechanical advantage provided by the pulley system to efficiently transfer and guide the classifier along the intended trajectory, ensuring precise placement and alignment within the de-sands building.

FIG 14 – Pulley arrangement.

CONCLUSIONS

The newly developed and implemented modifications of the upstream classifier are already in use at a big iron ore operation in Western Australia. In summary, the manufacturing and installation of the upstream classifier proved to be a highly successful endeavour. The methodology employed for the removal of the old classifier and installation of the new one operated smoothly, without encountering any notable obstacles.

A cost-saving approach was effectively implemented by leveraging the utilisation of existing hydraulic, hydraulic pack and electrical components. As a result, a new classifier, incorporating substantial improvements to facilitate future maintenance operations with greater ease and efficiency, was provided at a considerable cost savings compared to acquiring a new classifier with similar enhancements.

Moreover, the adoption of the new locking method for side wall inspection hatches enhances accessibility to the lower section of the unit. The larger size and increased frequency of these hatches streamline maintenance procedures, ensuring greater safety for operators. Furthermore, this improvement leads to a reduction in downtime and operational costs.

The introduction of the new style conidur plate hold down clamps and horizontal struts significantly simplifies the process of replacing conidur plates, making it faster, easier and safer. This notable enhancement translates into reduced downtime and decreased operational costs.

Additionally, the incorporation of a pulley arrangement to facilitate the movement and positioning of the classifier ensures a safer replacement process, effectively minimising downtime and ultimately contributing to cost savings in operations.

Collectively, these advancements result in enhanced efficiency, reduced downtime and diminished operational costs, underscoring the substantial benefits achieved through the implementation of these innovative modifications.

ACKNOWLEDGEMENTS

We would like to extend our heartfelt appreciation to all individuals and companies who have played a significant role in the successful completion of our project. Their contributions, support and collaboration have been invaluable throughout the journey.

We are immensely grateful to the team members who have dedicated their time, expertise and efforts to this endeavour. Their hard work, commitment and insightful contributions have been instrumental in achieving our goals.

We would also like to express our sincere gratitude to the various organisations and institutions that have provided us with resources, facilities and technical assistance. Their support and cooperation have been vital in conducting our research and experiments.

We would also like to extend our thanks to the reviewers and editors who have provided valuable feedback and suggestions, helping us improve the quality and clarity of our work.

Although we are unable to mention specific names, we are truly grateful for the collective efforts and contributions of all those involved. Their support has been essential in the successful completion of our project.

Red mud to green iron – proposed pathway for iron elution via bioleaching of bauxite residue

A Tolley[1]

1. Student, University of Queensland, Brisbane Qld 4169. Email: a.tolley@uqconnect.edu.au

ABSTRACT

Bauxite residue or 'red mud' is a by-product of the Bayer Process, the primary process of alumina refining. Due to its high alkalinity and variable composition, its storage and disposal poses high environmental risks to water sources, surrounding ecosystems and human health. The high alkalinity (pH 12–13) is a significant factor in its classification as hazardous waste; to mitigate this, Australian refineries have integrated neutralisation processes into bauxite refining.

Both sea water and carbon dioxide neutralisation processes are applicable. Multiple studies have demonstrated pH reduction from 12–13 to 7–10, depending on the method. The key advantages of CO_2 neutralisation are that it allows for carbon sequestration and is more effective when utilised as a pre-treatment for recovery of metals, while sea water neutralisation is more established and has been integrated into two Australian refineries.

This paper proposes the application of the bacteria *Exiguobacterium oxidotolerans* for bioleaching of iron from hematite in bauxite residue. The genus *Exiguobacterium* is characterised by its high levels of adaptability, halotolerance and alkaliphilic nature.

E. oxidotolerans is successful at eluting Fe^{2+} from hematite via reductive dissolution, which complexes with oxalic acid produced during metabolism. This aquatic bacterium presents a novel process pathway for iron elution due to direct interaction of *E. oxidotolerans* with hematite via the cell membrane and operates within the pH range of neutralised red mud. The proposed mechanism is extracellular electron transfer, a microbial metabolic process where electron transfer is facilitated between bacterial cells and extracellular material. Currently this is being investigated for promotion of seaweed growth in coastal environments as soluble iron aids and fortifies colonies.

This has potential downstream applications for cleaner iron and steel production. While theoretical, this paper explores the neutralisation, bioleaching and application of the Fe(II)- and Fe(III)-oxalate complexes formed.

INTRODUCTION

Bauxite residue or 'red mud' is a by-product of the Bayer Process, the primary method of refining alumina. Due to its high alkalinity and geospatially variable composition, its storage and disposal poses high environmental risks to water sources, surrounding ecosystems and human health. More than 70 million t are produced annually and many avenues have been explored to reduce the environmental impact, recover valuable metals and find alternative uses for the sheer volume of waste produced such as using it for building or refractory materials.

Iron primarily exists as hematite within red mud and is the component of interest due to its applications in steelmaking. Removal of iron can decrease the volume by up to 55 per cent, further rendering it a component of interest within red mud remediation. Additionally, iron-remediated red mud can then be used for refractory purposes, as iron decreases the melting point and thus makes it unsuitable for exposure to high temperatures.

This paper aims to highlight a novel bioleaching pathway to facilitate iron removal from red mud by drawing upon existing research and integrating it into a more sustainable iron value chain. *Exiguobacterium oxidotolerans* is a marine bacterium with a demonstrated capacity for eluting iron from hematite, specifically for depositing soluble iron in coastal environments. The following paper will illustrate how this bioleaching approach can be reconciled with red mud and the iron industry.

BAUXITE RESIDUE

Australian residue composition

During the Bayer Process, the high temperatures and pressures dissolve alumina-bearing minerals into highly alkaline solutions, while hematite remains insoluble alongside other oxide minerals (Archambo, 2021). The composition of red mud and subsequently the amount of hematite, varies with respect to the original orebody; this is one reason why finding a consistent use for red mud is difficult. Typically, hematite content is between 4–55 wt per cent (Archambo, 2021). Australia, as the second-largest producer of alumina globally, has a vested interest in remediating bauxite residue in some form.

Due to the variable composition, it should be noted that characterisation should be undertaken on the individual plant's scale for targeting key minerals. Snars and Giles (2009) investigated bauxite residue from 18 refineries, which included XRF analysis to determine the chemical compositions. The six Australian refineries, major components of the residue and pH are detailed in Table 1. The minor components (<1 per cent) were potassium oxide, sulfur trioxide, magnesia, phosphorous pentoxide and manganese(II) oxide.

TABLE 1

XRF analyses of red mud expressed as per cent oxide (Snars and Giles, 2009).

Plant	Location	pH	Fe_2O_3	Al_2O_3	SiO_2	Na_2O	TiO_2	CaO
Kwinana	WA	11.45	28.5	24	18.8	3.4	3.11	5.26
Pinjarra	WA	11.63	31.7	18.8	20.2	4.2	3.17	4.44
Wagerup	WA	11.99	29.6	17.3	30	3.2	2.65	3.64
Worsley	WA	12.56	56.9	15.6	3	2.2	4.46	2.39
Nabalco	NT	12.44	34.8	23.2	9.2	7.1	8.03	2.25
QAL	Qld	10.22	30.7	18.6	16	8.6	7.01	2.51

WA – Western Australia. NT – Northern Territory. Qld – Queensland.

The average pH of Australian bauxite residue is 11.72 and the hematite content is 35.37 per cent, with standard deviations 0.78 and 9.83 respectively. If Worsley is excluded due its significantly larger hematite content, the pH and hematite contents are 11.55 and 31.06 per cent with standard deviations of 0.74 and 2.15. While these figures are outdated, a more current analysis of Australian bauxite residue has not been completed.

Neutralisation

The high pH of red mud is the main reason for its classification as a hazardous waste (Gräfe, Power and Klauber, 2011). As such, significant research has been conducted into reducing the alkalinity and subsequently mitigate the residue's hazardous effects. One method integrated into Australian refineries is sea water neutralisation, which converts 'readily soluble, strongly caustic wastes into less soluble, weakly alkaline solids' (Rai *et al*, 2012). On average, sea water contains chlorine, sodium, sulfate, magnesium, calcium and potassium at a salinity of 3.5 per cent. Calcium and magnesium ions from sea water react with complex anions in red mud (hydroxyls, carbonates, aluminates) to precipitate and neutralise the residue (Archambo, 2021). The neutralising effect of calcium and magnesium ions decreases as the respective carbonates precipitate out around pH 8.5 (Rai *et al*, 2012).

Neutralisation is considered complete when the treated red mud is below pH 9.0. Discharged process sea water can be safely discharged back to the marine environment, as it is of comparable levels to world average sea water with regard to alkaline ions and concentrated toxic elements (Archambo, 2021).

This method of sea water neutralisation has been integrated into the Queensland Alumina Limited (QAL) and Yawun (Rio Tinto) refineries in Gladstone, Queensland (Australian Aluminium Council, 2020). Discharged sea water is at pH 7.8, while the process neutralises red mud to under pH 9 (Cristol and Greenhalgh, 2018).

Alongside sea water neutralisation, carbon dioxide has also been used to neutralise red mud with promising results. However, it is only effective for reducing a portion of the alkalinity, as the pH has demonstrated a tendency to 'rebound' due to the rate limiting step (dissolution of tri-calcium aluminate) pushing equilibrium towards hydroxide formation if all intermediaries fail to react. For this reason, CO_2 neutralisation is considered valuable as a pre-treatment for further processing due to the carbon sequestration and potential utilisation of intermediaries (Archambo, 2021).

BIOLEACHING

The remediation of red mud and extraction of iron can theoretically be coupled with the application of a novel bioleaching step. Aneksampant, Nakashima and Kawasaki (2020) studied the interaction between *Exiguobacterium oxidotolerans* and the hematitic surface, concluding that direct interaction between the two facilitates elution from insoluble Fe(III) to eluted Fe(II) under alkaline conditions. *E. oxidotolerans* contains many of the necessary characteristics for this bioleaching step and not been metallurgically applied.

Exiguobacterium

The genus *Exiguobacterium* exhibits high geographic, morphologic and physiological versatility, where the bacteria are Gram-positive, motile and aerobic or anaerobic depending on oxygen availability (Pandey, 2020). These bacteria are ultimately characterised by their versatility and adaptability. Despite at least 27 species, only eight genome sequences have been completed; however multiple stress tolerance related genes have been identified, confirming the environmental robustness of this genus. These stress response genes include oxidative stress, detoxification and carbon starvation, as well as genes relating to iron metabolism (Midha *et al*, 2016; Pandey, 2020). The species have been found to exist in a diverse range of niches throughout extreme environments, with strains isolated from deep sea hydrothermal vents, tidal flats, Himalayan glaciers, Siberian permafrost, soil and various processing effluents: potato, fish and beverage facility wastewater (Pandey, 2020).

This geographic diversity is coupled with characteristics that provide high utility for biotechnological applications. Several strains, including *E. oxidotolerans*, are halotolerant (up to 15 per cent NaCl), operate within a wide pH range of 5–11 and exhibit a high tolerance for presence of heavy metals and UV radiation. Adaptations include maintenance of membrane fluidity via production of branched-chain fatty acids, osmoregulation via solute uptake, cold shock protein expression and increased enzymatic catalytic affinity (Pandey, 2020). As extremophiles, they have also been extensively studied with regard to bioremediation and biodegradation of both organic and inorganic compounds (Pandey, 2020). Other applications include agro-ecological uses, particularly surrounding plant growth, yield and nutrient uptake.

Exiguobacterium oxidotolerans

Exiguobacterium oxidotolerans was isolated from the drain of a fish processing plant. As it's particularly halotolerant (up to 12 per cent NaCl), it was a species of interest for marine applications.

Soluble iron is essential for the coastal environment, as it fortifies seaweed colonies and promotes growth. Aneksampant, Nakashima and Kawasaki (2020) studied the interaction between *Exiguobacterium oxidotolerans* and hematite, for the purpose of imbuing coastal environments with soluble iron. The study concluded that direct interaction between the bacterial cell and the hematitic surface facilitates elution from insoluble Fe(III) to eluted Fe(II) through extracellular electron transfer (EET). EET's mechanism sees micro-organisms attach to solid extracellular materials, which directly transfers the electron through redox-active proteins (Kato, 2015). The c-type cytochrome is a pertinent redox-active protein, transferring the electron from inner to outer membrane via electron hopping through a series of cytochromes connecting external surfaces and respiratory chains (Kato, 2015).

It was proposed by Aneksampant, Nakashima and Kawasaki (2020) that hematite acts as a terminal electron acceptor. Through reductive dissolution and subsequent complexation, Fe(II)-oxalate and Fe(III)-oxalate are formed. Figure 1 outlines the proposed mechanism, with soluble complexes $Fe(C_2O_4)_2^{2-}$ and $Fe(C_2O_4)_3^{3-}$.

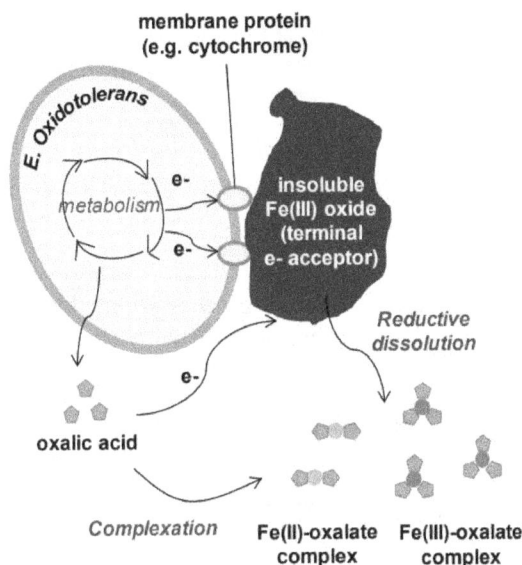

FIG 1 – Hypothesis of Fe(III) oxide direct elution (Aneksampant, Nakashima and Kawasaki, 2020).

Organic acids are produced as metabolites from *E. oxidotolerans*. Oxalic acid is notable as it is produced in much higher quantities and plays a key role in dissolving iron oxides, with it being studied extensively for acid leaching of iron (Yang *et al*, 2015, 2016; Archambo and Kawatra, 2022). Reductive elution of iron from hematite is induced by direct contact with oxalic acid. It was proposed that eluted Fe(II) could act as a nutrient to the bacteria, resulting in further production of oxalic acid, which would then be used as a chelator for Fe^{2+} (Aneksampant, Nakashima and Kawasaki, 2020).

It is suggested that *E. oxidotolerans* could find applications with red mud remediation. Other species within the *Exiguobacterium* genus exhibit maximum activity at pH 8.5 and over 60 per cent activity in pH values 5–11 (Rajaei *et al*, 2015). Notably, pH 8.5 is very close to the value of sea water-neutralised red mud (<pH 9.0), demonstrating applicability to the residue's environment. Additionally, further pH reduction was found during the application of *E. oxidotolerans* to hematite, likely due to the production of organic acids as bacterial metabolites (Aneksampant, Nakashima and Kawasaki, 2020). While this study was conducted between pH 6.8–8.0, Yumoto *et al* (2004) reported exhibition of growth between pH 7–10, further reinforcing the versatility of *E. oxidotolerans* and applicability to this process. The production of organic acids (notably oxalic acid) instigates a positive feedback loop, whereby an increase in acidity from microbial metabolism begets further production of organic acids and subsequently increases activity.

This could potentially be applied following carbon dioxide neutralisation, due to the wide pH range achieved during the neutralisation process. Archambo, Valluri and Kawatra (2020) reduced the pH of red mud from 12.5 to 7, with it rebounding to near 10 over a period of 7 days. This rebounding effect was observed across multiple studies, with varying success of maintaining a pH lower than 9.5 for any period exceeding 7 days (Rai *et al*, 2013; Archambo, Valluri and Kawatra, 2020). Rivera *et al* (2017) demonstrated a pH of 8.6 as a neutralisation step preceding acid leaching, as the compounds formed are water soluble. A similar pH of 8.45 was achieved using multiple CO_2 neutralisation cycles (Archambo, 2021), however the stability cannot be confirmed.

This rebound phase presents an interesting possibility: the pH rebound could be coupled with the bacterial pH decrease, for the purposes of secondary neutralisation as discussed by Archambo (2021). The production of organic acids lowered the pH during the experimental period (10 days) for *E. oxidotolerans*, with a corresponding increase in Fe^{2+} elution. However, this delay and subsequent rise could also be attributed to the lag and exponential phases of metabolic reactions, as Yumoto *et al* (2004) indicates a wider pH tolerance of 7–10 for microbial activity. As a secondary neutralisation process, combining the two would hypothetically see a pH of close to 9, where bauxite residue is considered neutralised.

To date, no study has been conducted between *E. oxidotolerans* and bauxite residue to confirm conditions are tolerable. Additionally, the presence of other compounds within red mud also present challenges with sustaining a bacterial population, especially those with anti-microbial or cytotoxic properties. While *E. oxidotolerans* has not been specifically studied with the major components of red mud in line with Table 1, other members of *Exiguobacterium* demonstrate at least some degree of tolerance to them.

- Aluminium oxide: *Exiguobacterium acetylicum* was used as a bioflocculant to remove aluminium salts and inorganic compounds, including aluminium oxide (Buthelezi, Olaniran and Pillay, 2009).

- Titanium oxide: an increase in biofilm formation was observed in exposed *Exiguobacterium acetylicum* cells (Mathur *et al*, 2017). While *E. acetylium* demonstrated sensitivty to titanium oxide nanoparticles, this was primarily associated with UVA, rather than visible and dark light (Mathur *et al*, 2015).

- Silica: *Exiguobacterium* have been found in silica-containing environments such as tidal flats and soil (Pandey, 2020).

- Disodium oxide: *Exiguobacterium* have been found in disodium oxide-containing environments such as tidal flats and soil (Pandey, 2020).

- Magnesium oxide: *Exiguobacterium* are typically halotolerant and have been found in deep sea hydrothermal vents and the ocean (Pandey, 2020).

- Calcium oxide: *Exiguobacterium* was found to be effective at demineralising shrimp that had been deproteinised with calcium oxide (Mathew *et al*, 2020).

APPLICATION

The conclusion of the bioleaching stage theoretically sees Fe(II)-oxalate and Fe(III)-oxalate complexes present, along with the remainder of the neutralised bauxite residue. Extraction of iron through oxalate complexation is a well-studied phenomenon, including within sediments and soils (McKeague and Day, 1966; Schwertmann, 1973; Phillips and Lovley, 1987; Laufer *et al*, 2020).

Iron is the core component of steel and 20 times more iron is used today than all other metals combined (Geoscience Australia, 2023). However, due to reliance on fossil fuels for achieving the high temperatures required during processing, the ironmaking industry is considered one of the main contributors to global warming (Santawaja *et al*, 2020). Iron- and steelmaking are furthermore classed as 'difficult-to-eliminate' emissions, highlighting a need for more sustainable production methods.

This extraction process can be directly integrated to a more sustainable iron production method. Santawaja *et al* (2020) developed an ironmaking process that exploits oxalic acid and iron oxalates, producing high purity iron at significantly lower temperatures. There are three key steps: (1) iron dissolution, (2) photochemical reduction, and (3) pyrolytic reduction. Equations 1–3 illustrate the dissolution and reductions:

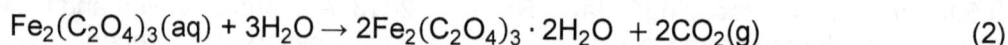

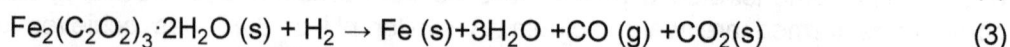

$$Fe_2O_3(s) + 3H_2C_2O_4 \,(aq) \rightarrow Fe_2(C_2O_4)_3(aq) + 3H_2O \qquad (1)$$

$$Fe_2(C_2O_4)_3(aq) + 3H_2O \rightarrow 2Fe_2(C_2O_4)_3 \cdot 2H_2O + 2CO_2(g) \qquad (2)$$

$$Fe_2(C_2O_2)_3 \cdot 2H_2O \,(s) + H_2 \rightarrow Fe\,(s) + 3H_2O + CO\,(g) + CO_2(s) \qquad (3)$$

This process makes use of the microbially-eluted iron-oxalate complexes and integrates it into the ironmaking supply chain. As this process is significantly more sustainable than conventional iron production and makes use of the products of the bioleaching step, it completes the process of iron from red mud and converting it into a more valuable form.

CONCLUSIONS

Recovering iron from red mud can reduce the initial volume by up to 55 per cent, rendering it the most promising avenue for remediation of red mud while also recovering a valuable resource.

Bioleaching of hematite from red mud presents a novel iron elution pathway via *E. oxidotolerans*. There is potential for application within both carbon dioxide (pH 7–10) and sea water (pH <9) neutralised red mud, simultaneously allowing for further neutralisation and carbon sequestration alongside iron extraction.

E. oxidotolerans exhibits characteristics that indicate it would be successful if applied to remediating red mud. The versatility of the genus, in conjunction with the wide operative pH range, production of iron-dissolving oxalic acid and no immediate incompatibility with major components of red mud, all support that *E. oxidotolerans* is a promising prospect for microbial remediation. However, further research is required to confirm hospitability of bauxite residue for bacterial elution.

The development of sustainable production methods is critical for the management of climate change. As the steelmaking industry is currently a significant contributor to that, sustainable extraction of iron from waste products offers an alternative source to traditional high-emission production. The final oxalate complexes produced from bioleaching directly substitute into new green iron production methods, further developing a more sustainable iron value chain.

REFERENCES

Aneksampant, A, Nakashima, K and Kawasaki, S, 2020. Microbial Leaching of Iron from Hematite: Direct or Indirect Elution, *Materials Transactions,* 61(2), pp 396–401.

Archambo, M, 2021. *New Horizons for Processing and Utilizing Red Mud* (Michigan Technological University).

Archambo, M and Kawatra, S, 2022. Extraction of Rare Earths from Red Mud Iron Nugget Slags with Oxalic Acid Precipitation, *Mineral Processing and Extractive Metallurgy Review,* 43(5):656–663.

Archambo, M, Valluri, S and Kawatra, S, 2020. Pretreatment of red mud with CO_2 for iron recovery, Annual SME Conference.

Australian Aluminium Council, 2020. *Bauxite Residue Storage using Seawater Neutralisation* [online]. Available from: <https://aluminium.org.au/wp-content/uploads/2020/11/201130-Sea-Water-Neutralisation.pdf>

Buthelezi, S, Olaniran, A and Pillay, B, 2009. Turbidity and microbial load removal from river water using bioflocculants from indigenous bacteria isolated from wastewater in South Africa, *African Journal of Biotechnology,* 8(14):3261–3266.

Cristol, B and Greenhalgh, R, 2018. *QAL Bauxite Residue Storage Using Sea Water Neutralisation,* 2nd International Bauxite Residue Valorization and Best Practices Conference.

Geoscience Australia, 2023. *Iron* [online]. Available from: <https://www.ga.gov.au/education/classroom-resources/minerals-energy/australian-mineral-facts/iron> [Accessed 4 June 2023].

Gräfe, M, Power, G and Klauber, C, 2011. Bauxite residue issues: III, Alkalinity and associated chemistry, *Hydrometallurgy,* 108(1–2):60–79.

Kato, S, 2015. Biotechnological Aspects of Microbial Extracellular Electron Transfer, *Microbes and Environments,* 30(2):133–139.

Laufer, K, Michaud, A, Røy, H and Jørgensen, B, 2020. Reactivity of Iron Minerals in the Seabed Toward Microbial Reduction – A Comparison of Different Extraction Techniques, *Geomicrobiology Journal,* 37(2):170–189.

Mathew, G, Mathew, D, Sukumaran, R, Sindhu, R, Huang, C, Binod, P, Sirohi, R, Kim, S and Pandey, A, 2020. Sustainable and eco-friendly strategies for shrimp shell valorization, *Environmental Pollution,* vol 267.

Mathur, A, Bhuvaneshwari, M, Babu, S, Chandrasekaran, N and Mukherjee, A, 2017. The effect of TiO_2 nanoparticles on sulfate-reducing bacteria and their consortium under anaerobic conditions, *Journal of Environmental Chemical Engineering,* 5(4):3741–3748.

Mathur, A, Raghavan, A, Chaudhury, P, Johnson, J, Roy, R, Kumari, J, Chaudhuri, G, Chandrasekaran, N, Suraishkumar, G and Mukherjee, A, 2015. Cytotoxicity of titania nanoparticles towards wastewater isolate *Exiguobacterium acetylicum* under UVA, visible light and dark conditions, *Journal of Environmental Chemical Engineering,* 3:1837–1846.

McKeague, J and Day, J, 1966. Dithionite- and oxalate-extractable Fe and Al as aids in differentiating various classes of soils, *Canadian Journal of Soil Science,* 46(1):13–22.

Midha, S, Bansal, K, Sharma, S, Kumar, N, Patil, P, Chaudhry, V and Patil, P, 2016. Genomic Resource of Rice Seed Associated Bacteria, *Frontiers in Microbiology,* vol 6.

Pandey, N, 2020. Chapter 10 – Exiguobacterium, in *Beneficial Microbes in Agro-Ecology*, pp 169–183 (Academic Press).

Phillips, E and Lovley, D, 1987. Determination of Fe(III) and Fe(II) in Oxalate Extracts of Sediment, *Soil Science Society of America Journal*, 51(4):938–941.

Rai, S, Wasewar, K, Mishra, R, Mahindran, P, Chadda, M, Mukhopadhyay, J and Changkoo, Y, 2013. Sequestration of carbon dioxide in red mud, *Desalination and water treatment*, 51:2185–2192.

Rai, S, Wasewar, K, Mukhopadhyay, J, Yoo, C and Uslu, H, 2012. Neutralization and Utilization of red mud for its better waste management, *Archives of Environmental Science*, 6:13–33.

Rajaei, S, Noghabi, K, Sadeghizadeh, M and Zahiri, H, 2015. Characterization of a pH and detergent-tolerant, cold-adapted type I pullulanase from Exiguobacterium sp. SH3, *Extremophiles*, 19:1145–1155.

Rivera, M, Ghania, O, Chenna, R, Koen, B and Van Gerven, T, 2017. Neutralization of bauxite residue by carbon dioxide prior to acidic leaching for metal recovery, *Minerals Engineering*, 112:92–102.

Santawaja, P, Kudo, S, Mori, A, Tahara, A, Asano, S and Hayashi, J, 2020. Sustainable Iron-Making Using Oxalic Acid: The Concept, A Brief Review of Key Reactions and An Experimental Demonstration of the Iron-Making Process, *ACS Sustainable Chemistry & Engineering*, 8(35):13292–13301.

Schwertmann, U, 1973. Use of oxalate for Fe extraction from soils, *Canadian Journal of Soil Science*, 53(2):244–246.

Snars, K and Giles, R, 2009. Evaluation of bauxite residues (red muds) of different origins for environmental applications, *Clay Science*, 46(1):13–20.

Yang, Y, Wang, X, Wang, M, Wang, H and Xian, P, 2015. Recovery of iron from red mud by selective leach with oxalic acid, *Hydrometallugy*, 157:239–245.

Yang, Y, Wang, X, Wang, M, Wang, H and Xian, P, 2016. Iron recovery from the leached solution of red mud through the application of oxalic acid, *International Journal of Mineral Processing*, 157:145–151.

Yumoto, I, Hishinuma-Narisawa, M, Hirota, K, Shingyo, T, Takebe, F, Nodasaka, Y, Matsuyama, H and Hara, I, 2004. Exiguobacterium oxidotolerans sp. nov., a novel alkaliphile exhibiting high catalase activity, *International Journal of Systematic and Evolutionary Microbiology*, vol 54.

Fundamental sintering characteristics of fine-grained hematite precipitate in a simulated sinter blend

N A Ware[1], G Reynolds[2] and L Lu[3]

1. Senior Experimental Scientist, CSIRO Mineral Resources, Pullenvale Qld 4069.
   Email: natalie.ware@csiro.au
2. Operations Manager, Queensland Pacific Metals, Brisbane Qld 4000.
   Email: greynolds@qpmetals.com.au
3. Senior Principal Scientist, CSIRO Mineral Resources, Pullenvale Qld 4069.
   Email: liming.lu@csiro.au

ABSTRACT

Queensland Pacific Metals (QPM) is progressing its Townsville Energy Chemicals Hub (TECH) project to produce nickel and cobalt sulfates for advanced battery materials. Apart from the nickel and cobalt sulfates, the process also generates a hematite by-product (~640 ktpa). The hematite by-product precipitated from the process, hereafter 'hematite precipitate', is high in Fe and particularly low in impurities. However, it is a fine <20 μm crystalline, hydrometallurgical precipitate and is produced as a filter cake containing a high moisture content of about 15 per cent. This is considerably different from natural hematite ore and is expected to change its performance during sintering. In the present study, laboratory scale sintering tests were undertaken to evaluate the potential impacts of substitution of the hematite precipitate for traditional magnetite concentrates and further for Australian hematite-goethite ores in a simulated industry base blend. Substitution of hematite precipitate for magnetite concentrates was found to slightly increase the sintering temperature required to achieve the strength of laboratory sinters. However, further substitution of hematite precipitate for Australian hematite-goethite ores slightly reduced the sintering temperature required to achieve the laboratory sinter strength. These results reflect the reduction in Al_2O_3 contents of the blends and are consistent with the trend observed in industry and pilot scale facilities. The reduction in sintering temperature generally correlates to a reduction in fuel rate in the larger scale sintering process. The pore structure of selected laboratory sinters was further examined and discussed in conjunction with sinter strength. The pore structure images highlighted the relative difference between the blends with the most obvious difference being seen at 1290°C which correlates well with the laboratory tumble index (TI). The laboratory sinter samples showed an increase in porosity and irregularity of pore shape, as hematite substitution increased demonstrating the correlation between a more porous structure and a lower TI.

INTRODUCTION

QPM is developing the TECH to process imported high-grade nickel laterite ore from New Caledonia. The ore will be processed through the facility to produce nickel and cobalt sulfates for nickel-cobalt-aluminium and nickel-cobalt-manganese Li-ion batteries. The TECH facility uses a patented recovery and recycling process called the DNi Process™ that uses a nitric acid leaching process for extracting nickel, cobalt and other valuable metals from the laterite ore. As the DNi Process™ recycles more than 98 per cent nitric acid and generates minimal waste products, the process is environmentally friendly with no requirement for tailing dams. Apart from the nickel and cobalt sulfates, the process also generates high purity alumina, hematite and other by-products. QPM is working actively to identify potential applications of these by-products to achieve its zero-waste ambition. QPM estimates to produce about 640 kt of hematite by product a year. The hematite by-product precipitated from the DNi Process™ is high in Fe and particularly low in impurities. However, it is highly crystallised, very fine, with an estimated top size of 20 μm and contains a high moisture content of about 15 per cent, which is considerably different from natural hematite ores. These properties are expected to affect its handleability and required agglomeration technologies for its application in ironmaking processes. Therefore, a comprehensive study was undertaken to evaluate the suitability of the hematite by-product for conventional and alternative agglomeration processes.

As part of this study, laboratory scale sintering tests were carried out to evaluate the impact of the substitution of the hematite precipitate for magnetite concentrates and further for Australian hematite-goethite ores in an industry simulated blend. Laboratory scale sintering tests carried out in a controlled environment provide fundamental information on the sinter matrix, which constitutes a large proportion of industry sinter structure. The test work was carried out using the -1 mm fraction of the sinter blend as this material is believed to represent the adhering fines which react with fluxes and melt to form the sinter matrix (Clout and Manuel, 2003). As the hematite by-product is very fine, it will contribute to the adhering fines and therefore, it is important to understand its behaviour in the formation of the sinter melt and matrix. This study was focused on the influence of adding hematite precipitate on the strength and porosity of laboratory sinters.

METHODOLOGY

A simulated industry blend typically used in Chinese steel mills (CSM) was selected as a base reference blend (Blend 1). It consisted of 25 per cent Brazilian hematite ores, 65 per cent Australian hematite and goethite ores and 10 per cent Chinese domestic magnetite concentrates. The hematite precipitate was first substituted for 5 per cent and 10 per cent magnetite concentrates in Blends 2 and 3 and then 10 per cent magnetite concentrates + 5 per cent Australian ores in Blend 4. The blends were each fluxed to a typical industry sinter silica level of 5 per cent and a sinter basicity of 1.8 using natural limestone and reagent grade SiO_2. Table 1 displays the chemistry of the hematite precipitate and Table 2 lists the composition of key species in the Laboratory sinter blends on a Loss on Ignition free (LOI_{free}) basis. The substitution of hematite precipitate for Chinese domestic magnetite concentrates in Blends 2 and 3 was found to reduce the Fe grade and increase the Al_2O_3 content of the sinter blends. In contrast, further substitution of hematite precipitate for Australian hematite and goethite ores has a negligible effect on the Fe grade and Al_2O_3 content of the sinter blend.

TABLE 1

Chemistry of the hematite precipitate.

Fe$_{TOTAL}$	SiO$_2$	Al$_2$O$_3$	P	S	CaO	TiO$_2$	Mn	MgO
58.435	0.225	1.945	0.005	0.211	0.010	0.020	1.280	0.400

TABLE 2

Constitution and chemical compositions of laboratory sinter blends (LOI_{free}).

	Hematite precipitate (%)	Magnetite concentrate (%)	Hematite goethite (%)	Fe$_{TOTAL}$	Al$_2$O$_3$	MgO
Blend 1	0	10	90	56.672	1.697	0.285
Blend 2	5	5	90	56.218	1.846	0.262
Blend 3	10	0	90	55.760	1.996	0.238
Blend 4	15	0	85	55.630	1.964	0.265

The component ores, concentrates, fluxes and hematite precipitate were dried in an oven at 105°C, weighed and mixed. Cylindrical tablets of about 4 g sinter mixture were compacted and sintered, in duplicate, in a horizontal tube furnace in a flow of N_2 + O_2 gas mixture that has an oxygen partial pressure of 5 × 10^{-3} atmospheres. Figure 1 shows the heating profile used for laboratory scale sintering. It was designed to simulate a typical sintering temperature profile where after an initial de-goethitisation step at 371°C, the tablets are then heated rapidly to the desired temperature and held there for three minutes before being quenched with the N_2 + O_2 mixture.

The fired tablets, or laboratory sinters, were then tumbled in pairs using a modified bond abrasion tester and hand screened to determine the percentage of +2 mm material, or compact tumble index (TI). Clout and Manuel (2003) established a correlation between the compact tumble index, pot and plant sinters suggesting this method can be used to predict the sinter quality of potential blends. A laboratory scale TI of 80 per cent +2 mm is considered to have achieved respectable strength and the temperature at which this is value is achieved is considered the optimum melt temperature.

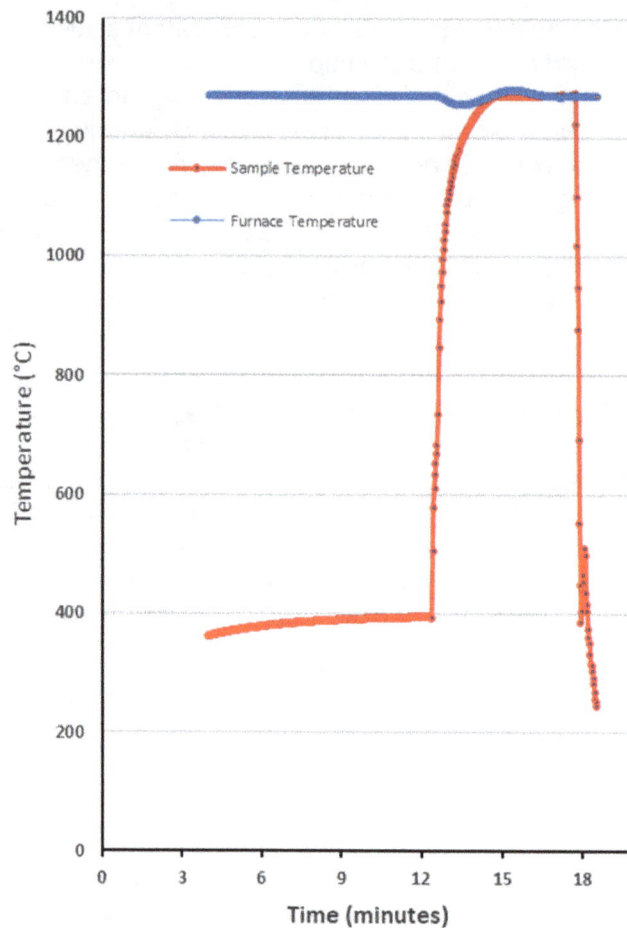

FIG 1 – Laboratory sintering heating profile, temperature (°C) versus time (minutes).

RESULTS AND DISCUSSION

Sinter matrix strength (TI)

Effect of substitution of hematite precipitate for magnetite concentrates on TI

The compacted tablets were fired at five temperatures with 10°C intervals from 1270–1310°C. The compact TI versus temperature curves are shown in Figure 2. The base blend (Blend 1) achieved a TI of 80 per cent (+2 mm) at 1290°C and maintained that strength over the temperature range to 1310°C. This is indicative of a typical result for a low temperature sinter blend.

Figure 2 also shows the effect of substitution of the hematite precipitate for different proportions of magnetite concentrates. At 5 per cent substitution level (Blend 2), the compact sinter also achieved a TI of 80 per cent at 1290°C and maintained this strength over the 1290–1310° temperature range. However, the blend (Blend 2) achieved a lower TI than the base blend at temperatures below 1290°C. At 10 per cent substitution (Blend 3), in which the hematite precipitate completely replaced the magnetite concentrates, the blend (Blend 3) required a slightly higher temperature of 1300°C to achieve TI of 80 per cent. Similarly, the blend had a considerably lower strength in the 1270–1290°C temperature range.

The results show that a slight increase in sintering temperature with substitution of hematite precipitate correlates with the slight increase in the Al_2O_3 level in each blend. This is consistent with industry experience (Lu, Holmes and Manuel, 2007) and pilot scale sintering studies (O'Dea and Ellis, 2015) that demonstrated increased alumina in the sinter blend correlated with an increased coke consumption (ie temperature) to maintain sinter quality.

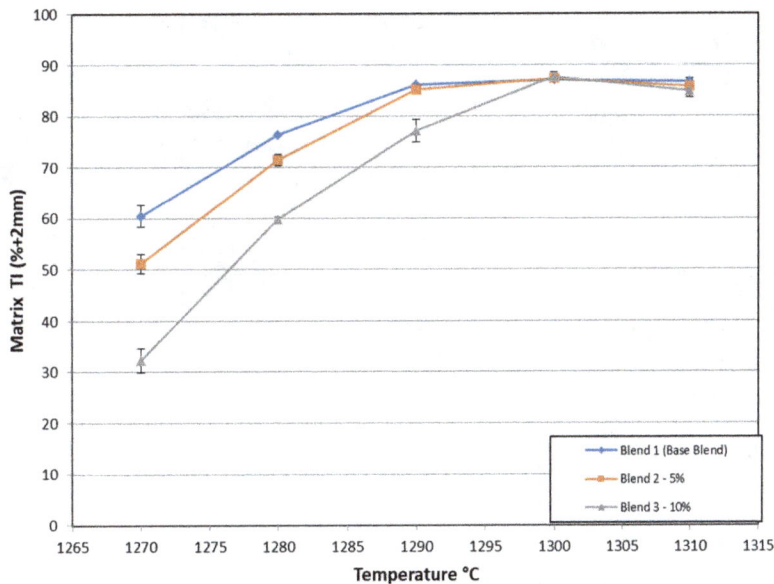

FIG 2 – Melt temperature curves for Blends 1–3 showing the effect of increasing hematite precipitate substitution for magnetite concentrate.

Effect of substitution of hematite precipitate for hematite-goethite ores on TI

Figure 3 compares the compact TI versus temperature curves for Blend 3 (10 per cent) and Blend 4 (15 per cent) which shows very little difference between the blends with only a subtle difference at 1290°C where Blend 4 achieved strength above 80 per cent + 2 mm while Blend 3 achieved a TI of only 77 per cent + 2 mm. A comparison of sinter images in Figure 4 for these two blends showed very little difference in terms of pore structure. The two blends had a slight difference in Al_2O_3 content with Blend 3 slightly higher at 1.996 per cent compared to 1.964 per cent for Blend 4. This may result in the slight reduction of tumble strength for Blend 3 at 1290°C and the slight increase in sintering temperature required to reach an acceptable TI.

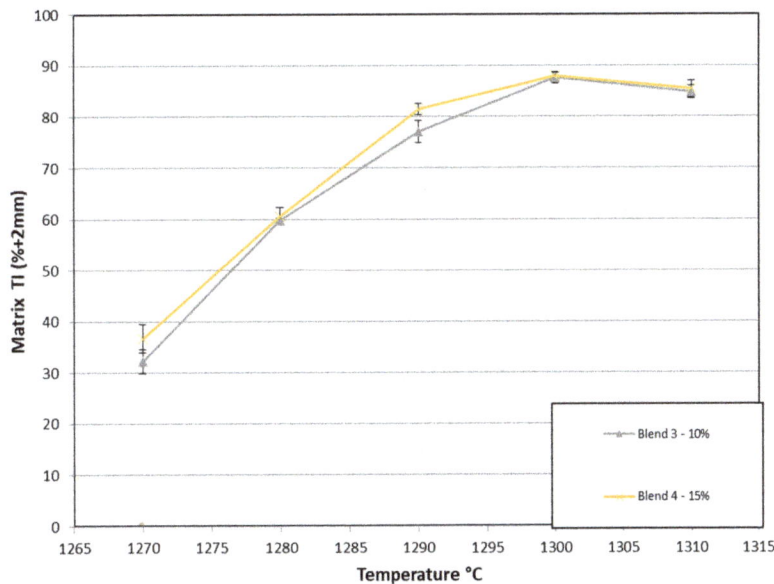

FIG 3 – Melt temperature curves for Blends 3 and 4 showing the effect of increasing hematite precipitate substitution for hematite-goethite ores.

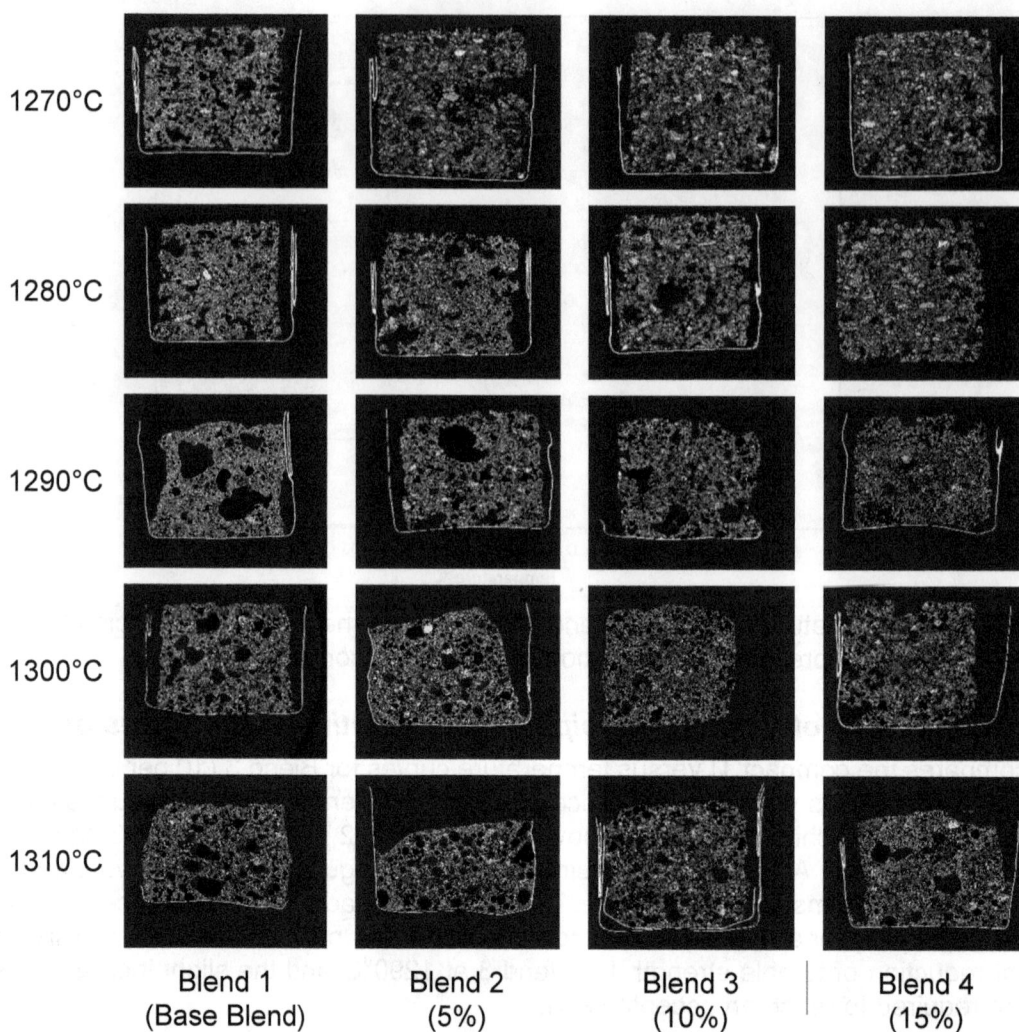

FIG 4 – Pore structure images of the entire tablet sample for each of the experimental sinter products (Sample width is approximately 10 mm).

Sinter porosity

Reflected light optical microscope-based cross-sectional images of the compacts fired at different temperatures are shown in Figure 4 providing a relative comparison of changes to the sinter matrix structure for each blend. Within the temperature series for each blend, the structure of each compact appears more melted, gradually losing shape and displaying more rounded pores, as the temperature is increased. These images are also useful in providing information on the relative difference between the blends and correlations can be made with changes in matrix strength for both set of results, magnetite substitution (Blends 1, 2 and 3) and hematite goethite substitution (Blends 3 and 4).

The most notable differences can be seen at 1290°C (middle row of Figure 4) where the base blend (Blend 1) and the 5 per cent hematite precipitate substitution blend (Blend 2) showed similar levels of melting consistent with their similar compact tumble index while the 10 per cent blend (Blend 3, no magnetite) showed less melting and increasing irregularity of pore shape, consistent with its comparatively lower TI.

The further substitution of hematite precipitate for Australian hematite-goethite ores appeared to have negligible impact on the pore appearance which is consistent with their similar TI values.

The sinter matrix comprised various bonding phases including SFCA (crystalline high temperature silico ferrite of calcium and aluminium) phases, secondary hematite and magnetite and glass. These bonding phases influence the strength and quality of the sinter (Pownceby *et al*, 2015; Webster *et al*, 2017). The fluidity of the melt is driven by chemistry and sintering temperature and a decrease in melt fluidity in turn leads to a more porous structure (Loo and Leung, 2003).

A further examination of the sinter mineralogy of the blends fired at 1290°C (Figure 5) using images at a higher magnification shows more obvious differences in porosity between Blends 1, 2 and 3, as hematite precipitate substitution for the magnetite concentrate increased. Figure 6 shows the similarity in structure between Blends 3 and 4 at 1290°C consistent with their similar matrix strength.

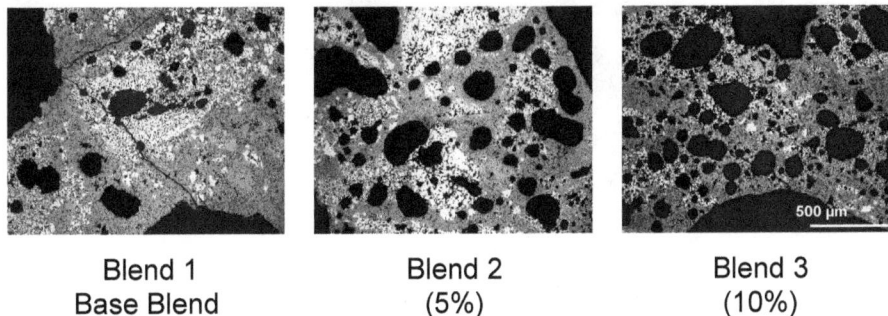

| Blend 1 | Blend 2 | Blend 3 |
| Base Blend | (5%) | (10%) |

FIG 5 – Reflected light optical microscope images showing structure in compact sinters fired at 1290°C and with different levels of hematite precipitate substitution of magnetite concentrates.

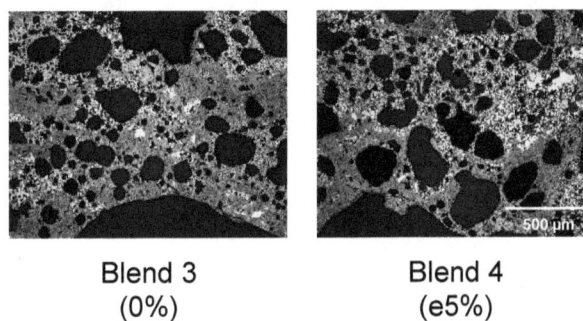

| Blend 3 | Blend 4 |
| (0%) | (e5%) |

FIG 6 – Reflected light optical microscope images showing structure in compact sinters fired at 1290°C and with different levels of hematite precipitate substitution of hematite-goethite ores.

CONCLUSION

The fine-grained QPM hematite precipitate, when substituting for magnetite concentrates in a simulated sinter blend, slightly increased the sintering temperature needed to achieve a laboratory TI of 80 per cent + 2 mm. When the hematite precipitate further substituted for an Australian hematite-goethite ore in the simulated blend, it slightly reduced the sintering temperature required to achieve the same laboratory sinter strength. The reduction in sintering temperature is suggested to correlate to a reduction in fuel consumption when scaled up to an industrial sintering process. An examination of pore structure showed the relative differences between blends and it was noted that at 1290°C, the compact sinters showed an increase in the number of pores and irregularity of pore shape as the hematite substitution was increased. The degree and irregularity of porosity evident in the compact sinters shows a good agreement with the TI, ie a more porous and irregular structure correlates with a lower laboratory TI.

ACKNOWLEDGEMENTS

The authors would like to thank Queensland Pacific Metals for the opportunity to carry out this work.

REFERENCES

Clout, J M F and Manuel, J R, 2003. Fundamental investigations of differences in bonding mechanisms in iron ore sinter formed from magnetite concentrates and hematite ores, *Powder Technology*, 130(1–3):393–399.

Loo, C E and Leung, W, 2003. Factors influencing the bonding phase structure of iron ore sinters, *ISIJ International*, 43(9):1393–1402.

Lu, L, Holmes, R J and Manuel, J R, 2007. Effects of alumina on sintering performance of hematite iron ores, *ISIJ International*, 47(3):349–358.

O'Dea, D and Ellis, B, 2015. New Insights into Alumina Types in Iron Ore and Their Effect on Sintering, in Proceedings 10th CSM Steel Congress and 6th Baosteel Biennial Academic Conference, China.

Pownceby, M I, Webster, N A S, Manuel, J and Ware, N, 2015. Iron Ore Geometallurgy – Examining the Influence of Ore Composition on Sinter Phase Mineralogy and Sinter Strength, in *Proceedings Iron Ore 2015*, pp 579–586 (The Australasian Institute of Mining and Metallurgy: Melbourne).

Webster, N A S, Pownceby, M I, Ware, N and Pattel, R, 2017. Predicting Iron Ore Sinter Strength through X-ray Diffraction Analysis, in *Proceedings Iron Ore 2017*, pp 331–334 (The Australasian Institute of Mining and Metallurgy: Melbourne).

Process optimisation – automation and machine learning

The critical role of mine planners in the evolution of closure at BHP

R Getty[1], J Heyes[2] and M Sanapala[3]

1. MAusIMM, Principal Global Closure, BHP, Perth WA 6000. Email: rebecca.getty@bhp.com
2. FAusIMM, Global Practice Lead Technical Capability, BHP, Perth WA 6000.
3. Global Practice Lead Technical Capability, BHP, Perth WA 6000.
   Email: murali.sanapala@bhp.com

ABSTRACT

Closure is an inevitable part of the life cycle of every mine. Successful closure outcomes critically depend upon early and ongoing integration of closure planning and mine planning. Early and holistic integration minimises residual risk and maximises opportunity to leave a cost-effective, positive legacy for the company, investors and future land users. Both disciplines are inherently interconnected and aim to maximise 'value' via trade-offs.

Historically, integration of closure into mine plans has been poorly implemented across the industry. There is still a perception that closure planning only starts as the end of a mine life approaches. This inevitably leads to tension when making business decisions, with progressive closure often being deferred due to short-term drivers and incentives.

To create a positive legacy after production has ceased, this tension can be managed by transparently and fully informing business decision-makers, such that the impacts of short-term decisions on long-term closure objectives are understood. BHP recently updated a business-wide standard 'Our Requirements for Closure and Legacy Management' designed to strengthen collaboration, education and integration of closure across the entire life cycle of the asset. Concurrently, the 'Life of Asset' definition was refined to include the postproduction phase, until closure objectives have been met.

This paper discusses the criticality of mine planning integration with closure and how updates to Our Requirements sharpen the focus from transactional to transformational planning. These concepts are applicable to any mining operation aiming to achieve successful closure across the full asset life cycle, from exploration to post-production.

INTRODUCTION

Mining is a temporary use of the land (Commonwealth of Australia, 2018; Keenan and Holcombe, 2021), with closure being an inevitable part of the life cycle of every mine. This continues after production stops and the land use transitions from mining to something else. Transition of the land to leave a positive legacy that is optimised, eg cost-effective, acceptable to future land users and maximises the value of the asset (full value including value for shareholders), is a primary goal of closure planning (ICMM, 2019; Hodge and Brehaut, 2022). Mine planners are critical internal stakeholders that enable this goal. Mine planners make tactical and strategic decisions that affect long-term closure outcomes, such as scheduling material movements to construct landforms and creating opportunities for progressive closure. Closure is complex and full of uncertainty over the long time frames of both the mine life and the post-mining period that closure designs must last for which can be over decades or centuries (Landform Design Institute, 2021).

Progressive closure is widely recognised by industry as good practice (ICMM, 2019). It reduces uncertainty, allows time for options to be explored, assumptions to be tested and costs optimised while revenue, equipment and people are available (Haymont, 2012; ICMM, 2019). Holistic integration of closure plans and mine plans from an early stage (before ground is even broken) and throughout the entire life cycle of the mine means decisions are transparent and informed so that opportunities can be maximised and residual risks can be minimised.

In 2006 the ICMM conducted a study in response to the growing literature on integrated closure planning and concluded:

> Integrated closure planning is a dynamic process which must commence in tandem with the other planning aspects of a mining process… It will be this approach to

sustainability planning that will give the greatest chance of longevity to an industry designed to exploit a non-renewable resource – proof that there can be long-term benefits to the people whose environments are affected by mining operations. (Fleurey and Parsons, 2006).

Whilst integrated planning is becoming more common (Slingerland and Wilson, 2015), especially when there is a key focus area for a mine such as integrated mine planning and tailings management (Badiozamani and Askari-Nasab, 2016) and integration of social dimensions (Worden, 2020), routine and upfront integration of mine and closure plans for land restoration is still maturing (Landform Design Institute, 2021). True integration is a transformational, rather than a transactional process. As quoted by Gerrard Whittle (2023), '*We should not change one part of the mining system without understanding the effect on the whole system.*'

BACKGROUND

Good practice closure planning has evolved reactively in response to poor closure outcomes. Closure planning that is integrated too late in a mine life cycle can be costly to remedy and may have irreversible impacts requiring long-term or in-perpetuity management (Sommerville and Ferguson, 2022). Environmental, social and economic impacts include degradation of land, water and biodiversity (by erosion and/or contamination), loss of cultural heritage or community identity, economic impoverishment, public health and safety issues or worse, impacts to human rights (Singh *et al*, 2022; Bainton and Holcombe, 2018). Failure to manage these residual risks can prevent tenure relinquishment, affect a company's reputation, restrict access to future resources and result in cost increases between 100–725 per cent (ICMM, 2019). Abandoned mine features where closure and transition has not occurred are currently estimated to be between 50 000 and 95 000 in Australia with about 18 000 features in Western Australia (Gutierrez, 2020; Werner *et al*, 2020; MINEDEX, 2023) where liabilities/costs are left to the government (and taxpayers), the most significant in Western Australia recently being the Ellendale diamond mine, which transferred reportedly $40 million in closure costs to the State in 2015 (ABC Rural, 2018).

Learnings from closure legacies indicate that early and ongoing integration of closure into design and operations is the most effective way to prevent poor closure outcomes (Sommerville and Ferguson, 2022; ICMM, 2019). Early planning and integration mean landforms can be designed and constructed to closure requirements from the start and fleet movements optimised to reduce expensive rework. High risk priorities (and materials) can be identified early and managed progressively, allowing adaption as needed and verification of performance assumptions and ongoing engagement with external stakeholders enables development of closure plans that are acceptable to future land users (ICMM, 2019). Optimised closure outcomes and objectives must be tailored to each site and are more efficient and cost-effective if considered as part of the day-to-day operations of a mine (ICMM, 2019; Shaw, Pedlar-Hobbs and Chubb, 2022).

The need for early and ongoing integration to effectively deliver positive closure outcomes was tendered more than 20 years ago, yet industry-wide experience of the impacts described above shows that integration of closure into mine plans has been poorly implemented (MMSD, 2002; ICMM, 2019; Getty and Morrison-Saunders, 2020). The primary reason is the tension between different time horizons. Closure planning has long-term objectives and production plans are inherently dynamic with a short-term focus. Common reasons why closure activities get deferred include the (incorrect) perception that closure planning only starts as the end of a mine life approaches, that there is time to start activities later, that the cost of closure is more appealing when scheduled later as costs are viewed with discount factors such as NPV and without production revenue who will pay for closure anyway (Dowd and Slight, 2006; Haymont, 2012)?

Informed and transparent decision-making is necessary to manage this tension so that the impacts of short-term business decisions on long-term objectives are understood. As said by Tyler and Heyes (2019):

> *Whilst the sustainability policies and principles of companies are set at the corporate level, it is the scientists and engineers who play a fundamentally important role in the sustainability of mining operations, our industry, and our place in the world.*

PROCESS AT BHP

The risk management hierarchy is well known and utilised within the mining industry. This same hierarchy can also be applied to the integration of closure plans and mine plans where the most effective elimination of closure risks starts at the design phase.

BHP's Western Australian Iron Ore (WAIO) Resource Engineering team has integrated mine planning and closure outcomes using the 'Finish and Fill' strategic approach. This approach is aimed to embed broader social and environmental values into the mine planning process and to support the BHP Climate Transition plan to reduce greenhouse gas (GHG) emissions within operations to net zero by 2050.

The approach challenges traditional open pit optimisation and subsequent mine designs, delivering optimised mine plans that extract the maximum resource value while focusing on completion of pit backfill, considering full life cycle costs rather than NPV optimal cash flow and social value. This led to an improved sequence of mining and progressive backfill that considered trade-offs and included closure and rehabilitation and future price sensitivities to ensure optimal resource recovery (and prevent sterilisation of resources). The project was run on several mine sites across WAIO with a significant overall reduction in closure costs, land disturbance and associated CO_2 emissions due to shorter hauls and reduced footprints. It also reduced the scale of pit lakes following closure which is highly valued by future land users (including Traditional Owners). The 'Finish and Fill' project improves on what was already proposed for South Flank, noting the need for *'operational discipline to ensure that these plans are followed, at the same time showing respect to the resources that we have been trusted to extract'* (Tyler and Heyes, 2019).

Incorporation of social value in the Capital Allocation Framework at leadership level and, for asset level plans, including NSW Energy Coal's pathway to closure is part of BHP's decision-making strategy. *'These decisions depend upon leaders and teams with the capability and authority to make considered choices – the same way they do on safety'* (BHP, 2022).

Alignment of social value with closure planning, in the same way as the industry addresses safety, is part of a broader culture of informed decision-making.

BHP recently updated (January 2023) a business-wide standard entitled 'Our Requirements for Closure and Legacy Management' to strengthen integration of closure across the entire life cycle of the asset. The process to update 'Our Requirements for Closure and Legacy Management' embodied the principals of integration and collaboration with extensive consultation across a broad range of internal stakeholders that included mine planning and the experiences of WAIO. Consultation was open and transparent to maximise improvement opportunities and early engagement highlighted that consistency of language was essential to support the business case for integrating closure. Early investigation highlighted some key gaps and 12 definitions for closure were added into BHPs Master Glossary (applicable to the entire business). This included 'entire life cycle of the asset' or 'Life of Asset', which was refined to include the post-production phase until closure objectives have been met.

Three closure culture goals have been defined to support 'Our Requirements for Closure and Legacy Management'. Closure planning must be risk-based, have integrated contributions and be early and progressive throughout the entire life cycle of the asset. Embedment of these culture goals will support fully informed and transparent decision-making, that includes social value to leave a positive legacy and empowers decision-makers to consider the long-term impacts when making business decisions:

> Whether it's capital allocation... the New South Wales Energy Coal pathway to closure... nature positive plans... or identifying a good idea on the ground – social value is part of all of these decisions (Caroline Cox, BHP, 2022).

The objective of mine planning is to maximise the value of resources in the ground, therefore it is necessary to strike a balance between entrepreneurial value creation, uncertainty and the constraints of technical complexity (such as geology, operational limits and safety) and external corporate commitments including social value (Smith, Faramarzi and Poblete, 2022).

CONCLUSIONS

Early and ongoing integration of closure plans and mine plans is critical to maximise the opportunity to leave a cost-effective, positive legacy for the company, investors and future land users. Mine planners are an important and influential stakeholder in this process. Recent experience at BHP has shown that significant, multiple benefits exist when closure and mine planning are integrated, thereby overcoming the perception that closure planning only starts as the end of a mine life approaches.

The evolution of a company-wide standard, ie 'Our Requirements for Closure and Legacy Management', has strengthened closure culture goals and supports proactive integration of planning and alignment with social value frameworks.

It utilises a common understanding of closure language and empowers decision-makers through being fully informed of long-term risks and opportunities.

Changing the focus from transactional to transformational planning is essential to enable creation of a positive legacy after production has ceased. These concepts are applicable to any mining operation aiming to achieve successful closure across the full asset life cycle, from exploration to post-production.

REFERENCES

ABC Rural, 2018. Diamond discovery in WA's remote Kimberley brings renewed hope to Australian diamond industry, *ABC Rural*. Available from <https://www.abc.net.au/news/rural/2018-01-18/diamond-discovery-kimberley-renewed-hope-industry/9331400> [Accessed: 28 July 2023].

Badiozamani, M M and Askari-Nasab, H, 2016. Integrated mine and tailings planning: a mixed integer linear programming model, *International Journal of Mining, Reclamation and Environment*, 30(4):319–346. doi:10.1080/17480930.2015.1092993

Bainton, N A and Holcombe, S, 2018. The Social Aspects of Mine Closure: A Global Literature Review, *Centre for Social Responsibility in Mining (CSRM)*, Sustainable Minerals Institute (SMI), The University of Queensland: Brisbane.

BHP, 2022. Social value investor presentation, BHP. Available from <https://www.bhp.com/news/media-centre/reports-presentations/2022/06/social-value-investor-presentation-28-june-2022> [Accessed: 28 July 2023].

Commonwealth of Australia, 2018. Resources 2030 Taskforce, Australian resources — providing prosperity for future generations, *Commonwealth of Australia*. Available from: <https://www.industry.gov.au/publications/resources-2030-taskforce-report> [Accessed: 28 July 2023].

Dowd, P and Slight, M, 2006. The Business Case for Effective Mine Closure, in *Mine Closure 2006: Proceedings of the First International Seminar on Mine Closure* (eds: A B Fourie and M Tibbett), pp 3–11 (Australian Centre for Geomechanics: Perth). https://doi.org/10.36487/ACG_repo/605_Dowd

Fleury, A and Parsons, A S, 2006. Integrated Mine Closure Planning, in *Proceedings of the First International Seminar on Mine Closure* (eds: A B Fourie and M Tibbett), pp 221–226 (Australian Centre for Geomechanics: Perth). https://doi.org/10.36487/ACG_repo/605_14

Getty, R and Morrison-Saunders, A, 2020. Evaluating the effectiveness of integrating the environmental impact assessment and mine closure planning processes, *Environmental Impact Assessment Review*, 82:106366, https://doi.org/10.1016/j.eiar.2020.106366

Gutierrez, M, 2020. Editorial for special issue: Sustainable use of abandoned mines, *Minerals*, 10(11):1015. https://doi.org/10.3390/min10111015

Haymont, R, 2012. Critical analysis and mine closure: why do things still go wrong in a swirl of feasibility, regulation and planning?, in *Proceedings of the Seventh International Conference on Mine Closure* (eds: A B Fourie and M Tibbett), pp 39–48 (Australian Centre for Geomechanics: Perth). https://doi.org/10.36487/ACG_rep/1208_05_Haymont

Hodge, R A and Brehaut, H, 2023. Towards a positive legacy: key questions to assess the adequacy of mine closure and post-closure, *Mineral Economics*, 36:181–186.

ICMM, 2019. *Integrated Mine Closure Good Practice Guide*, 2nd ed (ICMM, London).

Keenan, J and Holcombe, S, 2021. Mining as a temporary land use: A global stocktake of post-mining transitions and repurposing, *The Extractive Industries and Society*, 8(3):100924. https://doi.org/10.1016/j.exis.2021.100924.

Landform Design Institute, 2021. Mining With The End In Mind: Landform design for sustainable mining, *Landform Design Institute*. Available from: https://www.landformdesign.com/pdf/LDI-PositionPaper2021.pdf [Accessed: 28 July 2023].

MINEDEX, 2023. DMIRS Data and Software Centre. Available from: <https://dasc.dmirs.wa.gov.au/home?productAlias=MINEDEX> [Accessed: 28 July 2023]

Mining Minerals and Sustainable Development (MMSD), 2002. Breaking New Ground: Mining, Minerals and Sustainable Development. Available from: <https://www.iied.org/sites/default/files/pdfs/migrate/9084IIED.pdf>

Shaw, J, Pedlar-Hobbs, R and Chubb, D, 2022. The role of key performance indicators throughout the mine life in achieving closure objectives, in *Proceedings of the 15th International Conference on Mine Closure* (eds: A B Fourie, M Tibbett and G Boggs), pp 803–812 (Australian Centre for Geomechanics: Perth). https://doi.org/10.36487/ACG_repo/2215_58.

Singh, A, Bourgault, C, Kanse, L and Oldham, C, 2022. Developing the business case for responsible acid and metalliferous drainage (AMD) management (CRC TiME Limited: Perth).

Slingerland, N and Wilson, G W, 2015. End land use as a guide for integrated mine planning and closure design, in *Proceedings of the Tenth International Conference on Mine Closure* (eds: A B Fourie, M Tibbett, I Sawatsky and D van Zyl), 14 p (Australian Centre for Geomechanics: Perth). Available from: <https://www.researchgate.net/publication/304626811_End_land_use_as_a_guide_for_integrated_mine_planning_and_closure_design>

Smith, R, Faramarzi, F and Poblete, C, 2022. Strategic and tactical mine planning considering value chain performance for maximised profitability, in *Proceedings of IMPC Asia-Pacific 2022*, pp 1204–1225 (The Australasian Institute of Mining and Metallurgy: Melbourne). Available from: <https://www.researchgate.net/publication/363171774_Strategic_and_tactical_mine_planning_considering_value_chain_performance_for_maximised_profitability> [Accessed: 28 July 2023].

Sommerville, K and Ferguson, K, 2022. Let's reimagine our legacy of mining, in *Proceedings of the 15th International Conference on Mine Closure* (eds: A B Fourie, M Tibbett and G Boggs), pp 3–18 (Australian Centre for Geomechanics: Perth). https://doi.org/10.36487/ACG_repo/2215_0.01

Tyler, L and Heyes, J, 2019. Why should we 'think big' on closure?, in *Mine Closure 2019: Proceedings of the 13th International Conference on Mine Closure* (eds: A B Fourie and M Tibbett), pp 3–4 (Australian Centre for Geomechanics: Perth) https://doi.org/10.36487/ACG_rep/1915_01_Tyler

Werner, T T, Bach, P M, Yellishetty, M, Amirpoorsaeed, F, Walsh, S, Miller, A, Roach, M, Schnapp, A, Solly, P, Tan, Y, Lewis, C, Hudson, E, Heberling, K, Richards, T, Chia, H, Truong, M, Gupta, T and Wu, X, 2020. A Geospatial database for effective mine rehabilitation in Australia, *Minerals*, 10(9):745–767. https://doi.org/10.3390/min10090745

Whittle, G, 2023. Maximise your mine's potential: Integrated Strategic Planning for the mining industry, Whittle Consulting. Available from: <https://www.whittleconsulting.com.au/integrated-strategic-planning-for-the-mining-industry/> [Accessed: 28 July 2023].

Worden, S, 2020. Integrated mine closure planning: A rapid scan of innovative corporate practice, Centre for Social Responsibility in Mining, University of Queensland.

The 250 million ton per annum spatial digital twin

M Pomery[1]

1. Principal Global Geomatics, BHP, Perth WA 6000. Email: matthew.pomery@bhp.com

ABSTRACT

BHP's Western Australian Iron Ore Operations (WAIO) move approximately 250 million tons of ore per annum. The magnitude of the constant changes to the terrain models along with the diverse output and presentation requirements of numerous stakeholders presents significant challenges for the geospatial team.

This paper describes how the WAIO Survey team use a combination of technology and optimised workflows to capture, ingest, publish and access large data sets for diverse stakeholder requirements.

Machine learning for processing of data and the use of web based systems to track and prioritise task allocation of resources will be discussed in detail, in addition to the changes to the traditional roles of surveyors from the use of automation and mobile technology.

INTRODUCTION

In 2010 3D digital spatial data was well in use within mine planning and operations. Terrain data provided the base information for analysis and calculations, where an understanding of how a mine looked at a point in time was required.

Of the five BHP operating mines in the Pilbara, they all are operated by contracting companies with survey departments embedded in them. BHP has two surveyors overseeing contracts and governance of survey control, standards and procedures.

Creation of digital terrain surfaces is largely aligned to End Of Month (EOM) reporting for contract reconciliation. Entire sites are flown by a third party to validate productive movement and stockpile volume calculations. This is conducted on a quarterly basis. A weekly terrain file is required for planning purposes, requiring site survey teams to manage the update of data between aerial captures.

The data managed at each site was predominately vector based, strings and points derived from photogrammetry supplied by an aerial survey vendor. Each site survey team would update their respective data sets using Global Navigation Satellite System (GNSS) and total stations. Data was stored on-site servers, with archiving and larger project data sets maintained centrally in Perth. Data was sorted by project area only in central file shares.

Imagery is captured and published for use as a supplement to terrain data. This provides the situational context that terrain cannot, as terrain is managed as a 'bare earth' model meaning no buildings, vegetation or equipment is included. The frequency for updating imagery is aligned to the quarterly program.

2011–2015

Wider adoption of GNSS throughout the business requires improvement in spatial measurement. The establishment of Geodetic Datum of Australia 1994 (GDA94) based grids is implemented to replace the previous project and legacy site grids. This requires the reprojection of existing spatial data sets at each site affected, the creation of methods to reproject spatial data as required, along with governance to ensure users of data are aware of changes and implications. This effort is pivotal as it provides a solid base to work from in coming years.

Terrestrial laser scanners (TLS) became more commonly used throughout the surveying industry. At WAIO, mining grade scanners were introduced at all sites, with terrain data changing from strings to point clouds. This technology enabled the capture of terrain information at a much higher resolution, not previously achievable using GNSS or total station.

LiDAR technology for aerial surveys became attractive as a replacement to photogrammetry as capture and delivery times are reduced enabling faster reconciliations. LiDAR and imagery captures are staggered to allow for all sites to have all sites LiDAR captured in one day, then imagery another.

Contract staff become BHP employees following an organisational restructure. All site teams face some reduction in headcount as high precision GNSS Fleet Management Systems (FMS) are established. New hardware and software are introduced to standardise processes, differences in each site's approach to spatial data management are discovered. Site teams remain isolated from each other and work independent of each other.

Imagery data is compiled into a regional mosaic within the corporate Geographic Information System (GIS). This unlocks the ability to view available published imagery as a compilation, rather than a single image for a discrete area. This coincides with the introduction of web mapping applications, allowing internal users access to vector GIS data sets, where this was previously only achieved using desktop applications, server access permissions and some understanding of data. Web mapping is already being used globally through Google's effort and adoption of similar technology within the business gains support. Access to this information is exposed to a greater audience as a result.

Master Terrain Data (MTD) is created to support the vector and imagery data sets. This begins using the now monthly LIDAR captures at all sites. Wide area aerial mapping campaigns are undertaken to fill in the gaps around existing project data. Planning customers begin to use databased on complete coverage provided over exploration, mining, port and closure sites. This brings an improvement in how data is supplied to business planning cycles. With regional terrain and imagery now available, a more complete spatial digital twin is created.

GIS provides a rigorous and repeatable method to reproject data. Projection information for WAIO sites is managed through GIS.

2016–2018

Site survey teams are merged into one. Management is centralised to regional team hubs, reporting to a central manager. The focus remains on standardising and streamlining processes.

Much like the TLS previously, remotely piloted aircraft systems (RPAS) become the main survey capture tool as the industry adopted the technology. Site trials from the past few years prove RPAS allows ability to capture a larger area than TLS, with the added benefit of imagery and terrain products captured simultaneously to update the spatial digital twin. Data was still predominately managed within site repositories however. Site captured data is not widely shared or available for business planning departments not physically located on-site, rather via manual requests for information to supplement what is available.

With MTD updated monthly, a step change in efficiency came through a web application that allowed an area of interest to be selected and downloaded for use in other software applications. Additional scripts are required to be able to transform the output files into the different mine planning software formats, which is a significant improvement compared to manual compilation and creation of terrain files. Output projections and data formats are managed through selectable menus.

MTD begins to be used in internal web applications for terrain calculations and base layer context as a hillshade. As data is ingested into MTD, metadata could be standardised and stored with each capture. The business driver for this was to meet the Australasian Joint Ore Reserves (JORC) code requirements, with reference given to terrain information used to create ore models. This information is included in every extract. Sarbanes–Oxley Act (SOX) auditing also brings requirements to record our quality assurance (QA) steps undertaken to verify accuracy of data.

Site GNSS base stations are combined to become a continuously operating reference station (CORS) network for real time kinematic (RTK) corrections, as the demand for GNSS capability increases with Fleet Management Systems (FMS) uptake at sites.

Coupled to that uptake, the requirement to capture and publish mine site hazard locations creates the Mine Display data set. This is a vector-based representation for operators to be able to view hazard information on FMS screens in the cab. The development of these hazards layers has evolved at each site and data transferred manually from desktop applications to relevant FMS server for dissemination to equipment. This data is migrated to GIS to again help standardise, also to enable a level of automation for update to desktop applications for planning purposes and also to the mine controllers for upload to equipment. This is a significant improvement and sees orchestration of digital twin in three separate environments.

As road data is maintained in GIS, and MTD is available, automated measurement of road construction starts to be analysed. This is conducted at all sites, as data is maintained on a monthly basis for all road networks. Previous analysis may have occurred at a new intersection or an update to a section of road, this method is applied to all roads and sites, creating a data set for full coverage of operations. Measurement occurs at 1 m intervals along the road reporting road width, bund heights, grade and crossfall. This data is published to the web applications for information to mine services and production crews.

2019–2022

Access to GNSS traces from FMS were refined to enable hourly terrain files to increase the frequency that MTD could be updated. A secondary terrain surface was generated to meet the requirement for 'near-real time' data, where spatial data accuracy is not as important as currency. Capability built into the system, through the metadata attributed on ingestion, can filter different accuracy standards, for competing internal customer requirements.

Based on this near real time terrain, an in-house daily scheduling web application is created and can utilise hourly data updates for scheduling compliance analysis.

As Mine Display was able to shift site data management to GIS, remaining vector data sets still existing in planning desktop software applications, were able to be migrated using the same method. The standardisation of statutory mine plans was straightforward and the centralisation of the plan generation lead to streamlined publication. Data management of site survey teams is managed almost exclusively through a web application and data cascaded to relevant repositories for use in other applications as required by internal customers.

Improvements to the existing extracts web application brought the inclusion of the hourly FMS data. Proprietary mine planning software files are now part of the process resulting in the customer receiving the file with no further action. Imagery data was also made available using the same process.

Volume calculation becomes possible within the survey GIS web application. Base and footprint information managed in GIS, which lead to automated calculation and presentation of the results. Previously this was manually calculated using desktop mining software. This improvement results in six mine sites no longer working in isolation, rather a centralised effort where a single person performs reconciliation for all sites, in one calculation. SOX audits have driven improvement in the process to document all required QA checks and methodology used to calculate volumes. These have been built into MTD and recorded in each capture project for later demonstration in an audit.

The data captured by the aerial survey vendor changes to become simultaneous. Both LiDAR and imagery are captured at the same time, this removes the previous distortion of the spatial twin, as the terrain and imagery were at times a week apart in currency.

Of the terrain data captured within WAIO, the final piece of MTD was the incorporation of site captured data. Establishment of an automated ingestion routine includes any terrain data captured by RPAS or TLS. With this data available the migration of running terrain files, previously maintained by site survey teams, moves to scheduled extracts from MTD. This is possible by utilising GIS infrastructure that enables the ad hoc extracts. Site customers now have daily terrain files generated each morning that contains hourly FMS, site captured and aerial survey data, saved to relevant site file stores for consumption.

Multiple in-house web applications now consume MTD and imagery services as standard base layer information.

Site teams are typically made up of solo back-to-back roster surveyors, with a central Perth based geomatics team supporting GIS data and processes.

HOW DOES IT GROW

It is expected that physical data capture will become less of the surveyor's domain, instead providing the QA of what has been measured through other systems.

The spatial analysis of the data maintained is what can be built on in the future. Capturing all this information only truly becomes valuable once it can generate insight and inform decisions. There has been restriction of the development other internal groups improvement in the past, based on differing opinions on how information should be shared. This has resulted in a group simply bypassing what has been put in place completely, either sourcing the required data elsewhere, or duplicating another version. To limit this continuing, the survey department has taken the adoption of being a spatial provider, where it manages the capture and analysis, tailoring what is published based on customer requirements. This is a diversion from traditional mining operation rationale. 'Compliance to design' eg pit wall sign off, being able to quickly derive that analysis of whether a pit wall is over\under dug is something surveyors can own and improve how it is delivered. Currently this involves a lot of manual effort to produce.

Automated generation of road centrelines, based on MTD and adopting machine learning capability to identify key features on-site. The annotation of an image to determine road surfaces, water, vegetation means this can be applied to a RPAS capture, then have a terrain, image and vector data derived from the one process. Without manual effort to update a road centreline string means initiation of the road compliance script at a greater cadence. Currently this process runs monthly based on the required effort to maintain the road strings and occurs over a day or two. If the whole process is started by a new capture at site, road analysis could be updated within a matter of hours, based on a series of: MTD being updated, the roads being updated, then the analysis being run and all published to end users. The site survey team's focus will be on capturing and maintaining source data sets, while the centralised team derives end user products and support maintaining the system.

3D visualisation is a large part of current mine planning software. The limitations of this capability however are the size of the area you can display. Utilising MTD and imagery services, 3D viewing of a regional digital twin is currently possible, however becomes limited in functionality based on the required planning data sets not also being available. A step change of publishing these data sets to the GIS will see this become reality.

The datum used by WAIO and FMS, will need to continue to be refined as the demand for spatial measurement increases. The national datum is now a more recent epoch than the one current local projections are based on, so revision of these is required. The potential to reduce the amount in use in WAIO could see the adoption of a single projection. This would aid new FMS instances, along with ability to manage more area within the one. Using one local projection for software that cannot manage reprojection of data on-the-fly, would make regional planning a lot easier as well. Where current effort to reproject different data types and different sites, into a common projection would cease if only one was the standard.

Having data available through web applications means it can be, or is currently, mobile. As communication technology improves in the Pilbara, matching capability of what is available in Metro areas becomes achievable. The effort to continue publishing the business's spatial data in GIS is all leading towards this.

In recent years the increasing importance of environmental, social and governance (ESG) behaviours of companies and its associated impact on investing, have driven companies like BHP to take a more proactive approach to reducing carbon emissions. The importance of having spatial digital twins available for mining sites aids the modelling of potential changes to our concept of operations by decarbonising diesel use for example. This represents the largest technical challenge to BHP's decarbonisation pathway. Understanding the requirement for a fully electrified operation, quantifying future electricity demand and associated infrastructure requirements, starts by having a spatial twin to further model from.

Table 1 is a summary of what has been presented here.

TABLE 1

Summary of paper.

	Past	Present	Future
Data – terrain	Vector – points, lines, polygons	Point clouds, raster mosaics	Point clouds, raster mosaics
	GNSS, TLS, Photogrammetry	Past + RPAS, LiDAR, FMS	Present + remote sensing
	Proprietary software formats	Past + open source, GIS, web services	3D Web services
Data – imagery	Raster per capture	Raster mosaics, web services	Web services
Data – storage	Local file servers	Central file server	Cloud
Data access	Desktop	Desktop + web applications, mobile	Present + mobile acquisition and display
Team	5 × site teams 1 × Exploration 1 × Perth Capture focused Contract employment model	1 × team – Survey and Geomatics Capture and analytics BHP employees	1 × team Survey and Geomatics – Centralised processing/publishing and QA focused
Equipment	GNSS, total stations	Past + TLS, RPAS	Current + new industry standards
Datums	Multiple Local, National	Multiple Local (GDA94 based), National	Single local (GDA2020 based)

ACKNOWLEDGEMENTS

Kerry Turnock, Practice Lead Global Resource Knowledge, BHP, Perth

Christian Holland, Principal Program Management, BHP, Perth

Blair Chalmers, Manager Survey, BHP, Perth

New insights into the relationships between sinter mineralogy and physical strength using QXRD and optical image analysis

N A Ware[1], J M F Clout[2], J R Manuel[3], N A S Webster[4], E Donskoi[5] and M I Pownceby[6]

1. Senior Experimental Scientist, CSIRO Mineral Resources, Pullenvale Qld 4069.
   Email: natalie.ware@csiro.au
2. Principal, Clout Mining, Perth WA 6009. Email: john@cloutmining.com
3. Senior Experimental Scientist, CSIRO Mineral Resources, Pullenvale Qld 4069. Email: james.manuel@csiro.au
4. Group Leader, CSIRO Mineral Resources, Kensington WA 6151.
   Email: nathan.webster@csiro.au
5. Principal Research Scientist, CSIRO Mineral Resources, Pullenvale Qld 4069.
   Email: eugene.donskoi@csiro.au
6. Team Leader, CSIRO Mineral Resources, Clayton Vic 3168. Email: mark.pownceby@csiro.au

ABSTRACT

The relative influences of sinter mineralogy, sintering temperature, solid and pore macro- and microstructure on iron ore sinter physical strength are not yet fully understood. Whilst a range of analysis techniques have been used by researchers to evaluate sinter physical properties and sinter mineralogy, none have been able to directly link sinter mineralogy and physical strength under controlled conditions.

CSIRO has developed a unique laboratory method to study the fundamental melting properties of fine iron ores in a laboratory furnace under a controlled temperature and gas atmosphere environment designed to simulate the industrial sintering process. Compacted tablet samples for a 0.41 per cent and a 2.5 per cent Al_2O_3 series were fired in a tube furnace at various temperatures, cooled, then tumbled using a modified Bond abrasion test apparatus to determine a relative tumble index (TI) at each temperature. Quantitative XRD (QXRD) analysis was then carried out on the tumbled samples and optical image analysis (OIA) was conducted on polished sections of a separately fired compact in each case.

The amount of total SFCA was found to decrease with increasing firing temperature for both alumina series with the total amount of SFCA higher in the 2.5 per cent Al_2O_3 series compared to 0.41 per cent Al_2O_3 series samples. The increase in the amount of magnetite and hematite with increasing firing temperature was largely at the expense of a decrease in total SFCA. Overall, there was good agreement between the quantitative mineralogy determined from QXRD versus OIA for SFCA, hematite and magnetite. Qualitative analysis of high magnification images of the sinter correlated with both the OIA and QXRD results.

INTRODUCTION

Fine (<6.3 mm) iron ore cannot be directly charged to a blast furnace but must first be agglomerated in a high (1260–1300°C) temperature sintering process using coke as a fuel source and limestone or dolomite as flux. A typical ferrous burden to a blast furnace can comprise up to 75 per cent sinter and 10–15 per cent each of lump and pellets (Lu, Holmes and Manuel, 2007).

The ideal sinter product consists of 20 per cent unreacted hematite ore nuclei bonded together by 80 per cent of matrix consisting of hematite and the mineral silico-ferrite of calcium and aluminium (SFCA), as well as minor glass, larnite and magnetite. The physical and metallurgical quality and chemical composition of sinter impacts the blast furnace performance and sinter quality is directly influenced by both the physical and chemical properties of the ore blend as well as the conditions during the granulation and the sintering processes. The specific properties of the ore blend that can impact sintering quality include mineralogy, chemical composition, particle size distribution and particle porosity. In addition to the ore blend, sintering behaviour is also impacted by the type and the amount of fluxes used (eg limestone, dolomite), free moisture content required for granulation and fuel rate. The relationships between all these parameters are complex (Pownceby and Clout, 2002; Pownceby et al, 2015; Webster et al, 2012). The physical strength, microstructure and chemical composition of the matrix melt formed from the -1 mm adhering fines and fluxes is considered significant in predicting the performance of the sinter (Clout and Manuel, 2003).

The sinter matrix has been shown to form from reaction of Si-Ca-rich melt with hematite to form various phases including SFCA, secondary hematite, secondary magnetite, larnite and glass. The different types of SFCA phases which form are considered to be the key bonding phases that influence the sinter quality (Sasaki and Hida, 1982; Pownceby and Clout, 2002; Pownceby *et al*, 2015). Alumina content also plays a significant role in the formation of SFCA phases (Dawson, Ostwald and Hayes, 1985).

SFCA has generally been divided into two main groups including high Fe, low-SiO_2 SFCA-1 that has a platy crystal shape and a low Fe, high-Al SFCA which has a prismatic crystal morphology. Pownceby *et al* (2015) demonstrated that both SFCA and SFCA-1 form up to 1250–1270°C and above these temperatures only SFCA is favoured. Webster *et al* (2012) showed that as temperature increases SFCA melts to form magnetite and a liquid glass. Loo and Leung (2003) considered the main parameters controlling sinter melt characteristics to be viscosity, surface tension and density, showing that increased alumina results in less fluid melt and a more porous micro/macrostructure.

The majority of alumina in sinter studies has focused on gibbsite [$Al(OH)_3$] ores; gibbsite-type ores are mainly found in India and Africa, while in Australian ores the main alumina-bearing phase is kaolinite [$Al_2Si_2O_5(OH)_5$] (O'Dea and Ellis, 2015). In this work a systematic approach was taken to investigating and isolating the effect of increasing alumina on the sinter matrix melt and physical strength by doping a fine high-grade hematite ore with reagent grade kaolinite.

Blends were fired in a laboratory-controlled environment simulating the conditions in the initial stages of sintering as the flame front passes through the sinter bed. Matrix strength testing, optical image analysis and QXRD were carried out on the fired samples to determine tumble index (TI), porosity and SFCA, hematite and magnetite content to help further the understanding of the complex relationships during sintering.

METHODOLOGY

A high-grade hematite ore (Fe 68 per cent) was prepared using a batch stirred mill to achieve a size distribution of 80 per cent passing 80 µm to eliminate ore particle size in the matrix as a variable. One blend was prepared at natural ore alumina (0.41 per cent) content whilst the second blend was doped with reagent grade kaolinite. The two blends were fluxed to 1.8 basicity and 5.00 per cent SiO_2 using pulverised limestone and reagent grade silica. The chemistries of the raw materials are in Table 1 and blend target sinter blend compositions are shown in Table 2. There was no fuel (coke) added to the blends with heat required for reaction being provided externally by a resistance heating element tube furnace.

TABLE 1

Chemistry of materials used – as determined by X-ray fluorescence spectroscopy (XRF).

Raw materials	Fe	SiO_2	Al_2O_3	CaO	MgO	LOI_{900}
Hematite Ore	68.16	1.37	0.39	0.11	0.04	0.33
Limestone	0.28	1.96	0.44	53.61	0.44	42.71
Reagent-grade Kaolinite	0.43	47.05	37.48	0.03	0.24	12.79

TABLE 2

Blend Compositions of the pre-mixed synthetic sinter analogues studied.

Blend	Fe_T%	Al_2O_3%	SiO_2%	CaO	MgO%	Basicity
1	59.63	0.41	5.0	9.0	0.11	1.80
2	58.12	2.5	5.0	9.0	0.12	1.80

Laboratory scale sintering

The laboratory scale sintering method used in this work was previously reported in Clout and Manuel (2003). For each composition studied, two compacted cylindrical tablets of the limestone-fluxed hematite ores were fired in a laboratory furnace by holding at various temperatures for 3 min under a simulated typical sintering temperature profile (Figure 1) and at an oxygen partial pressure of 5×10^{-3} atmospheres (Hsieh and Whiteman, 1989). The two fired compacts were then tumbled together in a modified Bond Abrasion Tester for 8 min, screened and the compact Tumble Index (TI) recorded as the percentage retained above +2 mm. The tests were conducted in duplicate and the arithmetic mean calculated and plotted, with the standard deviation for all pairs being less than 4 per cent and most typically less than 2 per cent. Clout and Manuel (2003) showed a direct relationship between compact TI, pot grate and plant sinter TI with a compact TI of 80 per cent correlating to a pot grate TI of 65 per cent and a sinter plant TI of 74 per cent. The firing temperatures for each blend are selected to produce a melting curve (compact TI versus temperature) that show the compact strength of the blend both before and after it reaches 80 per cent TI. For each temperature, two additional compacts were fired: one was set in epoxy resin, sectioned longitudinally, polished for optical image analysis and a second used for quantitative X-ray diffraction (QXRD) analysis.

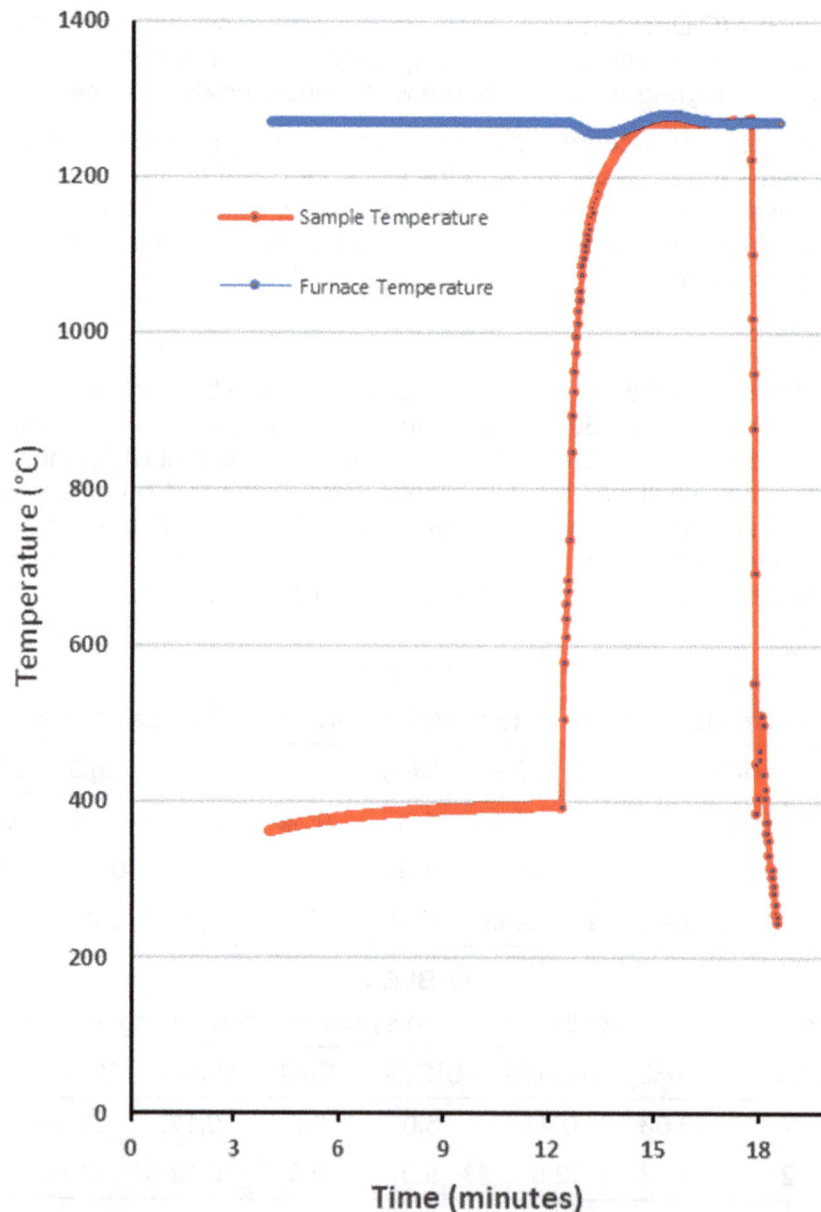

FIG 1 – Compact temperature heating profile used in the sintering tests: temperature (°C) versus time (minutes).

Optical image analysis (OIA) and quantitative X-ray diffraction analysis was carried out on sintered samples from Blend 1 (0.41 per cent Al_2O_3) and Blend 2 (2.5 per cent Al_2O_3). The QXRD was carried out using a Rietveld-based quantitative phase analysis method (Webster *et al*, 2017) and OIA was carried out on the entire tablet sample using CSIRO's Mineral/Recognition software (Donskoi *et al*, 2015). OIA results in area per cent results were converted to weight per cent using mineral specific gravities. Image analysis was carried out on the entire compact (Figures 2 and 3) as it was thought that analysing the maximum surface area would provide a more representative result for each sample. The compromise was that the image resolution was not high enough to distinguish between SFCA and SFCA-1 nor other types of mineral phases including quartz, glass and larnite. As a result, $SFCA_{TOTAL}$ is used to refer to combined SFCA and SFCA-1 identified phases.

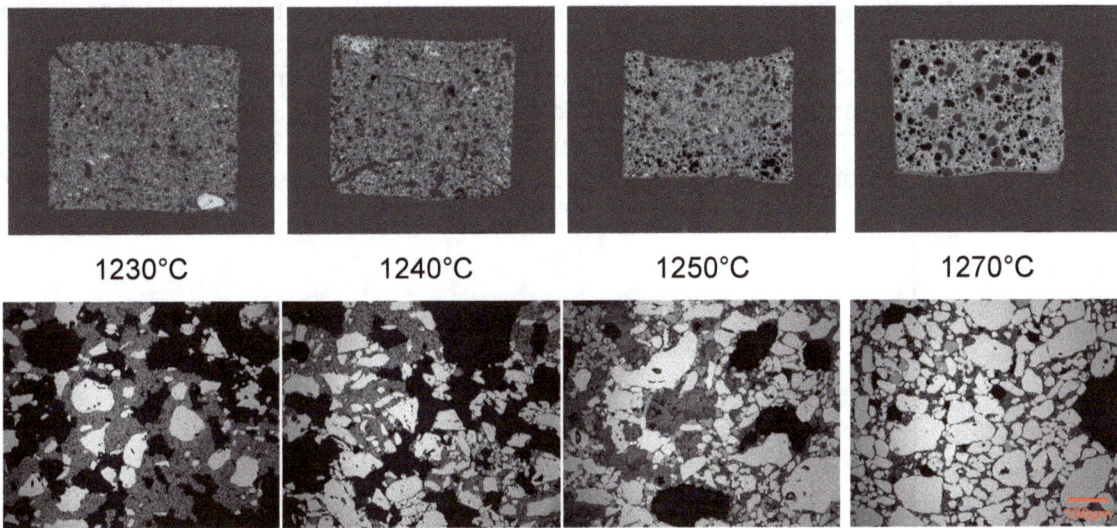

FIG 2 – Optical images showing the pore structures developed during sintering of Blend 1 – 0.41 per cent Al_2O_3 (natural alumina content) between 1230° and 1270°C.

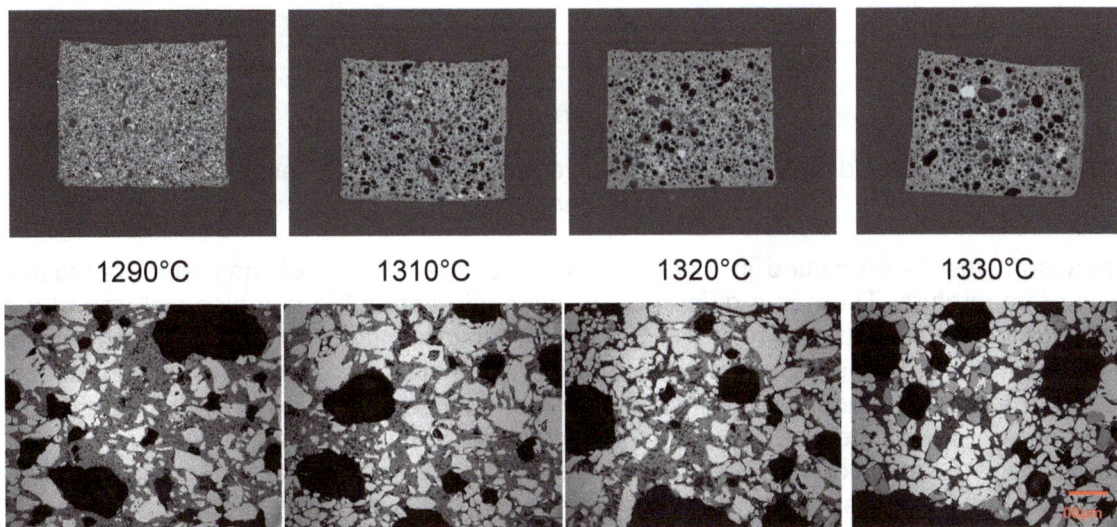

FIG 3 – Optical images showing the pore structures developed during sintering of Blend 2 – 2.5 per cent Al_2O_s (doped with addition alumina as kaolinite) between 1290°C and 1330°C.

RESULTS

Microstructure

Image analysis porosity measurements indicated no significant changes in total porosity however, a qualitative examination of the polished sections showed increased melt-solid and pore coalescence as temperature increases as well as a reduction in compact dimensions for the two blends (*cf* Figures 2 and 3). It is suggested that the fine nature of the hematite sample and fluxes may have also impacted the porosity results. Loo and Leung (2003) showed that the viscosity of the melt impacts bubble and melt coalescence with lower alumina levels decreasing the viscosity and leading to larger more spherical pores. Similarly, higher alumina levels decrease melt fluidity resulting in a greater number of irregular shaped pores. Melting behaviour seen in the current work is consistent with Loo and Leung (2003) to a point where melt surface tension is decreased sufficiently to enable the melt-solid and pore coalescence referred to above, at optimum temperatures.

Sinter mineralogy

The relative amount of $SFCA_{TOTAL}$ decreases as temperature increases for both the 0.41 per cent and 2.5 per cent Al_2O_3 blends (Figure 4). Furthermore, there is more $SFCA_{TOTAL}$ developed in the Blend 2 experiments with higher alumina contents consistent with the stabilising effect of alumina on SFCA formation. The increased alumina in SFCA also means that that SFCA is stable until higher temperatures before melting occurs. A key observation is that there is minimal difference between the OIA and XRD results for both blends indicating both techniques are suitable for quantifying the abundance of total SFCA. The OIA was carried out on the surface area of the sample while QXRD is carried out on the basis of mass which may account for the differences between OIA and QXRD.

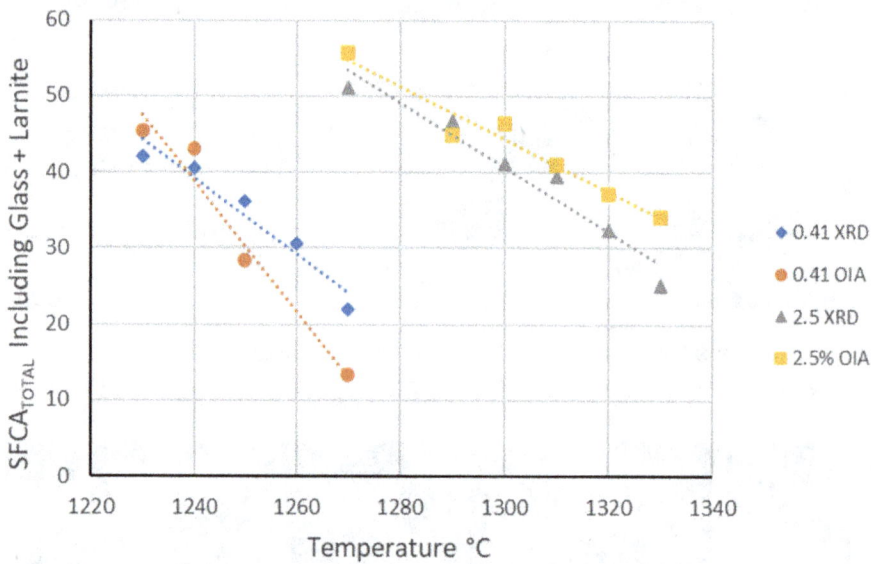

FIG 4 – $SFCA_{TOTAL}$ including glass and larnite versus temperature. Results from both OIA and QXRD analysis.

In comparison, results presented in Figure 5 show that the amount of magnetite increases with increasing temperature. There is a difference in magnetite content in samples sintered at the two different alumina levels with the higher 2.5 per cent kaolinite doped Al_2O_3 blend showing higher levels of magnetite at higher temperatures. Similarly, hematite also increases in abundance with increasing temperature although the effect is more pronounced for the lower Al_2O_3 blend (Figure 6). The decrease in abundance of $SFCA_{TOTAL}$ with increasing temperature appears therefore, for both blends, to be at the expense of increased abundance of hematite and magnetite with the 0.41 per cent Al_2O_3 blend showing a greater increase in hematite whereas the 2.5 per cent Al_2O_3 blend shows a greater increase in magnetite compared to hematite. Note optical examination of the samples reveals that the amount of secondary (rounded grains) hematite precipitated from melt increases with increasing temperature for both blends, reaching a maximum at the highest temperature (eg Figure 2 at 1270°C and Figure 3 at 1330°C) compared to the more angular primary ore hematite grains, which are more prevalent at the lower temperatures. The maximum temperature reached in both blends also coincides with the presence of coarse-grained prisms of SFCA in glass with larnite whereas at lower temperatures the SFCA is finer-grained and forms a matrix network between primary hematite grains.

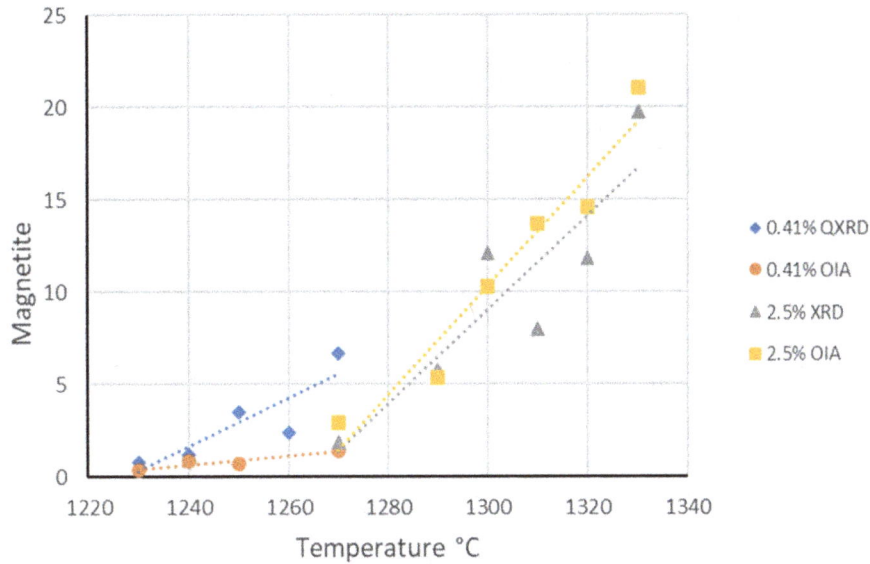

FIG 5 – Magnetite versus temperature. Results from both OIA and XRD analysis.

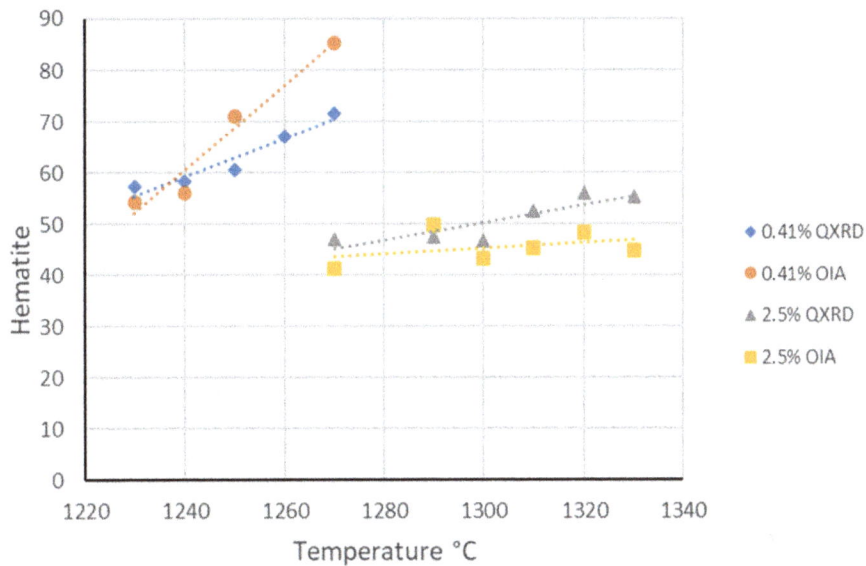

FIG 6 – Hematite versus temperature. Results from both OIA and QXRD analysis.

Sinter physical strength

Maximum sinter TI is reached at 1260°C and then declines slightly for the 0.41 per cent Al_2O_3 Blend 1 whereas for the 2.5 per cent Al_2O_3 Blend 2, TI continuously increases up to and including the maximum fired temperature of 1330°C (Figure 7). Whilst the drop in sinter TI for Blend 1 might be attributed in part to the decrease in $SFCA_{TOTAL}$ as bonding material due to the lower Al_2O_3 content forming a low alumina SFCA phase, for Blend 2, the continuous increase in TI is correlated negatively with $SFCA_{TOTAL}$ abundance (Figure 8).

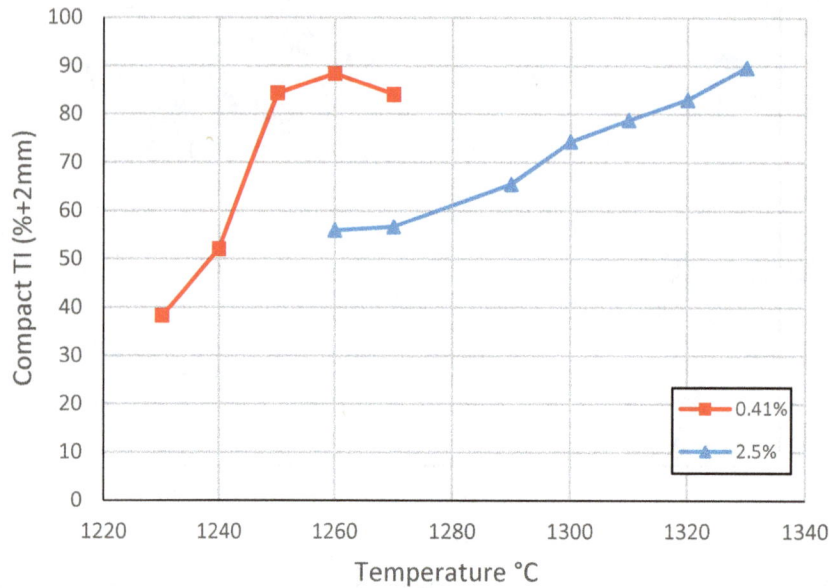

FIG 7 – Compact TI versus temperature.

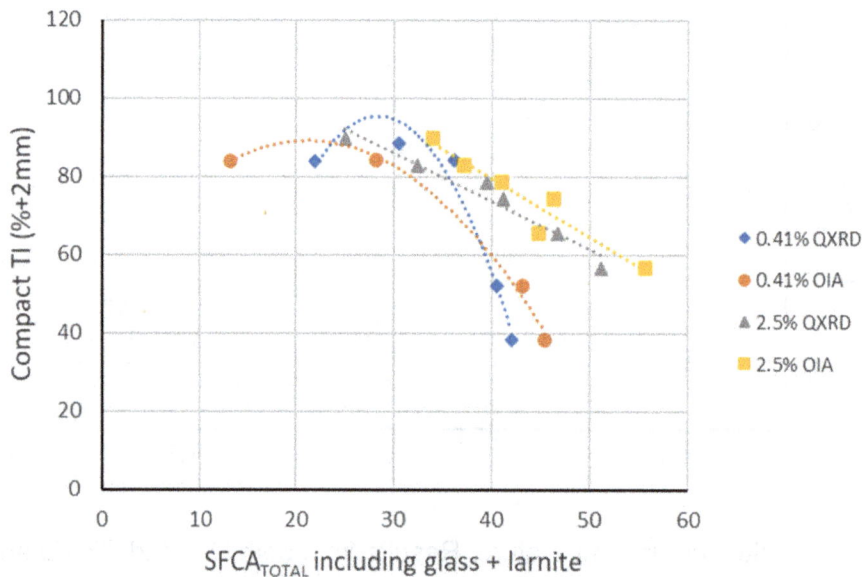

FIG 8 – Compact TI versus SFCA$_{TOTAL}$ including glass and larnite. Results from both OIA and QXRD methods.

DISCUSSION

Qualitative analysis of the high magnification optical images of the sintered products shown in Figures 2 and 3, correlated well with both the OIA and QXRD results (Figures 4–6). However, the type of SFCA was not quantified by the OIA method, although the highest temperature images indicate coarse-grained prismatic SFCA in glass is the dominant form of SFCA in Figure 2 at 1270°C and Figure 3 at 1330°C. Since the completion of this work, further and ongoing development of the Mineral/Recognition software, most notably automated identification of phases, has eliminated subjectivity in sample analysis and has allowed for distinction between SFCA-1 and SFCA. This will be reported in follow-up research.

Pownceby *et al* (2016) found that attainment of acceptable (optimum) matrix strength corresponds to the temperature transition between an SFCA-1 and an SFCA dominated matrix. Further to this, Honeyands *et al* (2017) reported that SFCA (prismatic) is mainly associated with magnetite while SFCA-1 (platy) is associated with primary and secondary hematite. The results in Figure 4 indicate that the relationship between sinter physical strength (TI) and SFCA abundance are consistent with these observations for the 0.41 per cent Al$_2$O$_3$ blend but in Figure 8 are negatively correlated for the

2.5 per cent Al_2O_3 blend. More research is required to resolve the impact of alumina and to explain the increase in TI with increasing temperature and decreasing in SFCA for the 2.5 per cent Al_2O_3 blend, including whether above 1300°C the impact of magnetite-magnetite diffusional bonding becomes more important for TI as recognised in hematite-magnetite concentrate sintering by Clout and Manuel (2003).

CONCLUSIONS

The results from the OIA and the QXRD analysis generally agree with each other with regards to the amount of SFCA, hematite and magnetite in the sinter samples. Both alumina series showed a decrease in SFCA while simultaneously increasing in magnetite content, with increasing temperature. The total amount of SFCA is higher in the 2.5 per cent Al_2O_3 sample than the natural (0.41 per cent) blend. A qualitative study of the higher magnification images of the sinter samples confirmed these results and showed a clear difference between blends with different alumina content.

There was close agreement between sinter TI and per cent melt phase for the low-alumina blend and an equivalent relationship was seen for the 2.5 per cent alumina blend (linear), indicating a difference in behaviour between the blends. The data reflected the relationship between TI and temperature in each case (a curve for low-Al, but linear increase for 2.5 per cent Al). In both cases, the data from XRD and OIA (total SFCA) showed good agreement.

Future work will concentrate on quantifying the types of SFCA phases produced in this study to help better understand the effect of SFCA phase type on the strength properties.

ACKNOWLEDGEMENTS

The authors would like to thank Mr Paul Nielsen for sample preparation and Ms Tirsha Raynlyn for assistance with the development of the image analysis methodology used in this study.

REFERENCES

Clout, J M F and Manuel, J R, 2003. Fundamental investigations of differences in bonding mechanisms in iron ore sinter formed from magnetite concentrates and hematite ores, *Powder Technology*, pp 393–399.

Dawson, P R, Ostwald, J and Hayes, K M, 1985. Influence of alumina on development of complex calcium ferrites in iron ore sinters, *Transactions – Institution of Mining and Metallurgy, Section C, Mineral Processing & Extractive Metallurgy*, 94(1):71–78.

Donskoi, E, Poliakov, A, Manuel, J R, Peterson, M and Hapugoda, S, 2015. Novel developments in optical image analysis for iron ore, sinter and coke characterisation, *Applied Earth Science*, 124(4):227–244.

Hsieh, L-H and Whiteman, J A, 1989. Sintering Conditions for SImulating the Formation of Mineral Phases in Industrial Iron Ore Sinter, *ISIJ International*, 29(1):24–32.

Honeyands, T, Manuel, J R, Matthews, L, O'Dea, D, Pinson, D J, Leedham, J, Monoghan, B J, Li, H, Chen, J, Hayes, P C, Donskoi, E and Pownceby, M I, 2017. Characterising the Mineralogy of iron ore sinters – state-of-the-art in Australia, in *Proceedings Iron Ore 2017*, pp 49–60 (The Australasian Institute of Mining and Metallurgy: Melbourne).

Loo, C E and Leung, W, 2003. Factors Influencing the Bonding Phase Structure of Iron Ore Sinters, *ISIJ International*, 43(9):1393–1402.

Lu, L, Holmes, R J and Manuel, J R, 2007. Effects of Alumina on Sintering Performance of hematite Iron Ores, *ISIJ International*, 47(3):349–358.

O'Dea, D and Ellis, B, 2015. New Insights into Alumina Types in Iron Ore and Their Effect on Sintering, in Proceedings 10th CSM Steel Congress and 6th Baosteel Biennial Academic Conference, China.

Pownceby, M I and Clout, J M F, 2002. The Importance of Fine Ore Chemical Composition and High Temperature Phase Relations – Applications to Iron Ore Sintering and Pelletising, in *Proceedings Iron Ore 2002*, pp 209–215 (The Australasian Institute of Mining and Metallurgy: Melbourne).

Pownceby, M I, Webster, N A S, Manuel, J and Ware, N, 2015. Iron Ore Geometallurgy – Examining the Influence of Ore Composition on Sinter Phase Mineralogy and Sinter Strength, in *Proceedings Iron Ore 2015*, pp 579–586 (The Australasian Institute of Mining and Metallurgy: Melbourne).

Pownceby, M I, Webster, N A S, Manuel, J R and Ware, N, 2016. The influence of ore composition sinter phase mineralogy and strength, *Transactions of the Institutions of Mining and Metallurgy: Section C*, 25:140–148.

Sasaki, M and Hida, Y, 1982. Consideration on the properties of sinter from the point of sintering reaction, *Tetsu-to-Haganè*, 68:563–571.

Webster, N A S, Pownceby, M I, Madsen, I C and Kimpton, J A, 2012. Silico Ferrite of Calcium and Aluminum (SFCA) Sinter bonding Phases: New Insights into Their Formation During Heating and Cooling, *Metallurgical and Materials Transactions B – Process Metallurgy and Materials Processing Science*, 43(6):1344–1357.

Webster, N A S, Pownceby, M I, Ware, N and Pattel, R, 2017. Predicting Iron Ore Sinter Strength through X-ray Diffraction Analysis, in *Proceedings Iron Ore 2017*, pp 331–334 (The Australasian Institute of Mining and Metallurgy: Melbourne).

Process optimisation – data science and predictive analytics

Effect of hydrogen on voidage and permeability of ferrous burden in softening and melting process

M M Hoque[1], S Mitra[2], N Barrett[3], D O'Dea[4] and T Honeyands[5]

1. Research Associate, Centre for Ironmaking Materials Research, The University of Newcastle, Callaghan NSW 2308. Email: mohammad.hoque@newcastle.edu.au
2. Research Academic, Centre for Ironmaking Materials Research, The University of Newcastle, Callaghan NSW 2308. Email: subhasish.mitra@newcastle.edu.au
3. PhD Student, Centre for Ironmaking Materials Research, The University of Newcastle, Callaghan NSW 2308. Email: nathan.barrett@uon.edu.au
4. Principal Technical Marketing, BHP Marketing Iron Ore, Brisbane Qld 4000. Email: damien.p.odea@bhp.com
5. Associate Professor, Centre for Ironmaking Materials Research, The University of Newcastle, Callaghan NSW 2308. Email: tom.honeyands@newcastle.edu.au

ABSTRACT

The steel industry is currently exploring hydrogen-enriched blast furnace operations as a means to reduce the greenhouse gas emissions. However, there is a need to investigate the gas permeability behaviour of mixed burdens comprising of lump and sinter as the current knowledge regarding their behaviour is limited, particularly with regards to hydrogen enrichment. In the present study, a series of interrupted softening and melting (S&M) under load tests were carried out in a hydrogen rich gas environment (~15 per cent) in the temperature range from 1000 to 1300°C by placing different types of ferrous ore layer including Newman Blend Lump (NBLL), sinter and sinter-NBLL mixed burden in between two coke layers. X-ray computed tomography (CT) was then used to scan the interrupted test samples. Scanned images were first analysed using an image processing workflow implemented in *ImageJ* and the material simulator GeoDict (v2022) to segment different phases such as coke, ferrous and void based on their grey scale values. A quantitative analysis was then performed on the segmented images to estimate the phase volume fraction as well as bed permeability. Bed void fraction thus obtained agreed well with the bed contraction data obtained from the S&M test. It was further noted that both ferrous layer void fraction and bed permeability decreased linearly with increasing temperature for all ore samples. The mixed burden voidage changed in a similar fashion to the sinter, demonstrating a high temperature interaction between the two materials.

INTRODUCTION

The iron and steelmaking industry contributes significantly to CO_2 emissions with approximately 7 per cent of the total global CO_2 emissions (Tan *et al*, 2019). To align with the global shift to 'low carbon economy', the sector is currently leaning to using hydrogen for reducing iron ore by partially replacing coke as a primary reducing agent (Ma *et al*, 2021; Chen and Zuo, 2021). The primary aim is to decarbonise the ironmaking process (Sun *et al*, 2020) which predominantly uses blast furnace route to produce about 75 per cent of hot metal worldwide (Liu, Zhang and Yang, 2015).

To assess the performance of ferrous burden in a typical blast furnace environment, S&M test is widely used in research (Liu *et al*, 2019). This test allows to assess ferrous burden behaviour at laboratory scale while providing valuable insights to the cohesive zone of blast furnace. This is achieved by placing a ferrous layer in between two coke layers in a reducing gas environment comprising simplified gas compositions and temperature profile that mimic a typical blast furnace operating condition. The gas compositions in these experiments typically include carbon monoxide (CO), carbon dioxide (CO_2) and nitrogen (N_2) (Liu *et al*, 2019), however, to understand the effect of hydrogen on reduction of ferrous burden, it is necessary to reassess and modify the test conditions for S&M experiments.

Until now, there have been only a few studies investigating the softening and melting behaviours of ferrous burden in the presence of hydrogen gas. For example, Qie *et al* (2018) explored the impact of H_2 addition on the softening and melting reduction behaviours of ferrous burden in a gas-injection blast furnace. They observed that the softening temperature of the burden decreased as the H_2 content in the reducing gas increased, while the melting and dripping temperatures of the burden increased with higher H_2 content in the reducing gas. Lan *et al* (2020) analysed the mechanism influencing the softening and melting behaviours of the ferrous burden in hydrogen-rich smelting. They found that an increased H_2 content in the reducing gas resulted in a downward movement and narrowing of the cohesive zone leading to improved permeability. Pan *et al* (2018) investigated the effect of different degrees of reduction on the cohesive zone and permeability of the mixed burden. They noted that as the reduction degree increased, the thickness of the cohesive zone decreased implying improved permeability of the burden with a corresponding downward shift in the location of the cohesive zone.

S&M test indirectly quantifies cohesive zone permeability in terms of ferrous burden voidage in the system which can be determined from X-ray computed tomography (CT). In recent time, several studies have reported ability of this method to study the complex three-dimensional (3D) microstructure of ferrous burden and quantify bed void fraction (Shatokha *et al*, 2009, 2010; Hjortsberg *et al*, 2013; Liu *et al*, 2018; Mitra *et al*, 2020). Previous studies (Liu *et al*, 2018; Mitra *et al*, 2020) from this research group have demonstrated the efficacy of CT scan method to characterise iron ore burdens. The present study specifically aims to utilise this approach to examine the effect of hydrogen enriched reduction on the bed void fraction in S&M test involving three different types of ferrous burden: lump, sinter and lump-sinter mixed burdens and quantify bed permeability of these burdens using a CT scan based computational fluid dynamics (CFD) modelling approach.

METHODOLOGY

Softening and melting under load test

In the present study, three different raw materials were used to carry out the S&M test, namely, sinter (SH1), Newman Blend Lump (NBLL) and mixed burden (MB1, composed of 79 per cent SH1 and 21 per cent NBLL). The chemical compositions and basicity of related raw materials are summarised in Table 1.

TABLE 1

Chemical compositions (wt per cent) and basicity of raw materials.

Materials	FeO	TFe	CaO	SiO$_2$	Al$_2$O$_3$	MgO	P	R$_2$	R$_4$
SH1	7.60	56.70	10.04	5.43	1.87	1.76	0.067	1.85	1.62
MB1	6.00	57.98	7.94	5.07	1.77	1.41	0.070	1.57	1.37
NBLL	-	62.80	0.05	3.70	1.40	0.10	0.080	0.01	0.03

NB: R$_2$: binary basicity, mass ratio of (CaO/SiO$_2$), R$_4$: quaternary basicity, mass ratio of (CaO+MgO)/(SiO$_2$+Al$_2$O$_3$).

The details of the experimental rig and testing conditions for the S&M under load test have been described in earlier studies (Liu *et al*, 2018; Loo, Matthews and O'Dea, 2011; Liu *et al*, 2019; Hoque *et al*, 2021). Briefly, iron ore samples (diameter: ~10.0–12.5 mm, height: 70 mm) were sandwiched between two layers of coke (diameter: ~10.0–12.5 mm) in a graphite crucible (60 mm inner diameter) and subjected to a programmed, time-dependent temperature, reducing gas (flow rate: ~14.0 L/min, 30 per cent CO + 15 per cent H_2 + 55 per cent N_2) and load (~1.0 kPa). The selection of the simplified hydrogen gas composition was based on the previous modelling prediction which indicated that hydrogen injected through the tuyeres mainly replaces the nitrogen component of the bosh gas (Barrett *et al*, 2022).

A series of interrupted tests were carried out at different temperatures varied from 1000°C to 1300°C wherein the heated samples were quenched quickly by turning off the furnace and passing N_2 gas through the bed. After cooling, the graphite crucibles containing the solidified bed of ferrous and coke particles were scanned using Synchrotron X-ray computed tomography (CT) scanning technology.

Synchrotron X-ray CT scanning

The X-ray CT imaging of the quenched samples was conducted at the Australian Synchrotron's Imaging and Medical Beamline (IMBL), utilising a 4.2 T superconducting multipole wiggler as the insertion device. The IMBL facility is equipped with a double crystal-Laue monochromator (Stevenson *et al*, 2017; Liu *et al*, 2018).

For this specific study, hutch 3B was utilised, with the sample positioned approximately 135 m from the X-ray source (Stevenson *et al*, 2017). The synchrotron operated at a ring current of 200 mA and an energy of 3 GeV. Due to the high absorbance of the S&M samples, a filtered white beam known as 'pink beam' was employed instead of monochromated X-rays. The use of 'pink beam' significantly increased the beam flux, thereby improving the signal-to-noise ratio of the measurements. A copper filter with a thickness of 12 mm was employed to produce a polychromatic beam with a weighted average energy of approximately 150–160 keV.

Projection images were acquired using a pco edge CMOS sensor with an optical lens and a gadolinium oxide scintillation screen (via a 45° mirror). The effective voxel size was 35.5 µm^3 and each scan consisted of 1800 projections collected over a 180° rotation with an exposure time of 0.12 s per acquisition. Given the large size of the samples, multiple scans were performed at different heights (Figure 1). The individual vertical slices for each angle were then combined into a single image before performing 3D CT reconstruction.

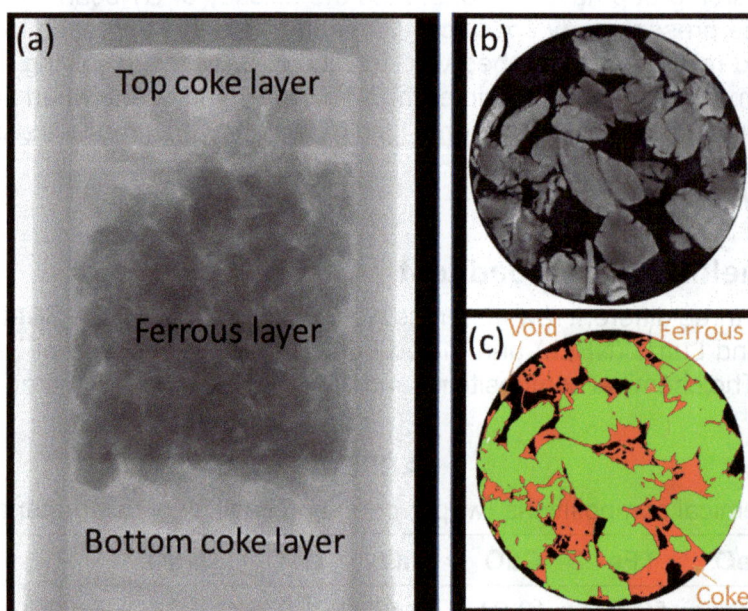

FIG 1 – Example of data processing: (a) original vertical synchrotron CT image (multiple scans); (b) reconstructed horizontal grey scale image in ferrous layer; (c) segmented and coloured horizontal image.

Image segmentation

After the reconstruction, each stack of 2D images (corresponding to each sample) were converted into a 3D grey scale volume that was used for further image processing. Each 3D volume consisted of a regular volumetric grid where each voxel had a unique grey scale that was a function of the attenuation of X-rays at that point. Prior to segmentation, the so-called ring effect due to beam hardening was removed from the images using a non-local mean (NLM) filter and appropriate thresholds for each phase were determined in *ImageJ*. Three different phases – coke, ferrous and void were then segmented using the material simulator GeoDict (v2022) (Figures 1b and 1c). After segmentation, a quantitative analysis was carried out over these images to estimate the volume fraction of coke, ferrous and void layers as a function of bed height and reconstruct the three-dimensional (3D) structure with these phase distributions. For more detailed description of the image processing, readers are referred to the work of Hoque *et al* (2021).

CFD modelling

To quantify the permeability of the ferrous burden, a computational fluid dynamics (CFD) model was developed using the material simulator GeoDict (v2022). The FlowDict module within GeoDict was employed to predict the bed permeability of each sample at various temperatures. A Left Identity Right (LIR) solver was used to generate non-uniform adaptive grid over the 3D structure resulting in low computational memory requirements. The gas flow field through the ferrous burden structure was simulated using the Stokes-Brinkman model with gas density and viscosity and flow rate as inputs. Periodic boundary conditions were applied along the principal flow direction while a no-slip boundary condition was applied along the tangential direction of the flow to represent the crucible wall.

RESULTS AND DISCUSSION

Figure 2 compares axial variations in the phase fractions of all three components – coke, ferrous and void with temperature for each burden sample at three different temperatures – 1000°C, 1100°C and 1300°C, respectively. Intersections of the coke (red dashed) and iron ore (blue dashed) lines indicate the individual phase boundary between top coke and ore layer and bottom coke and ore layer, respectively.

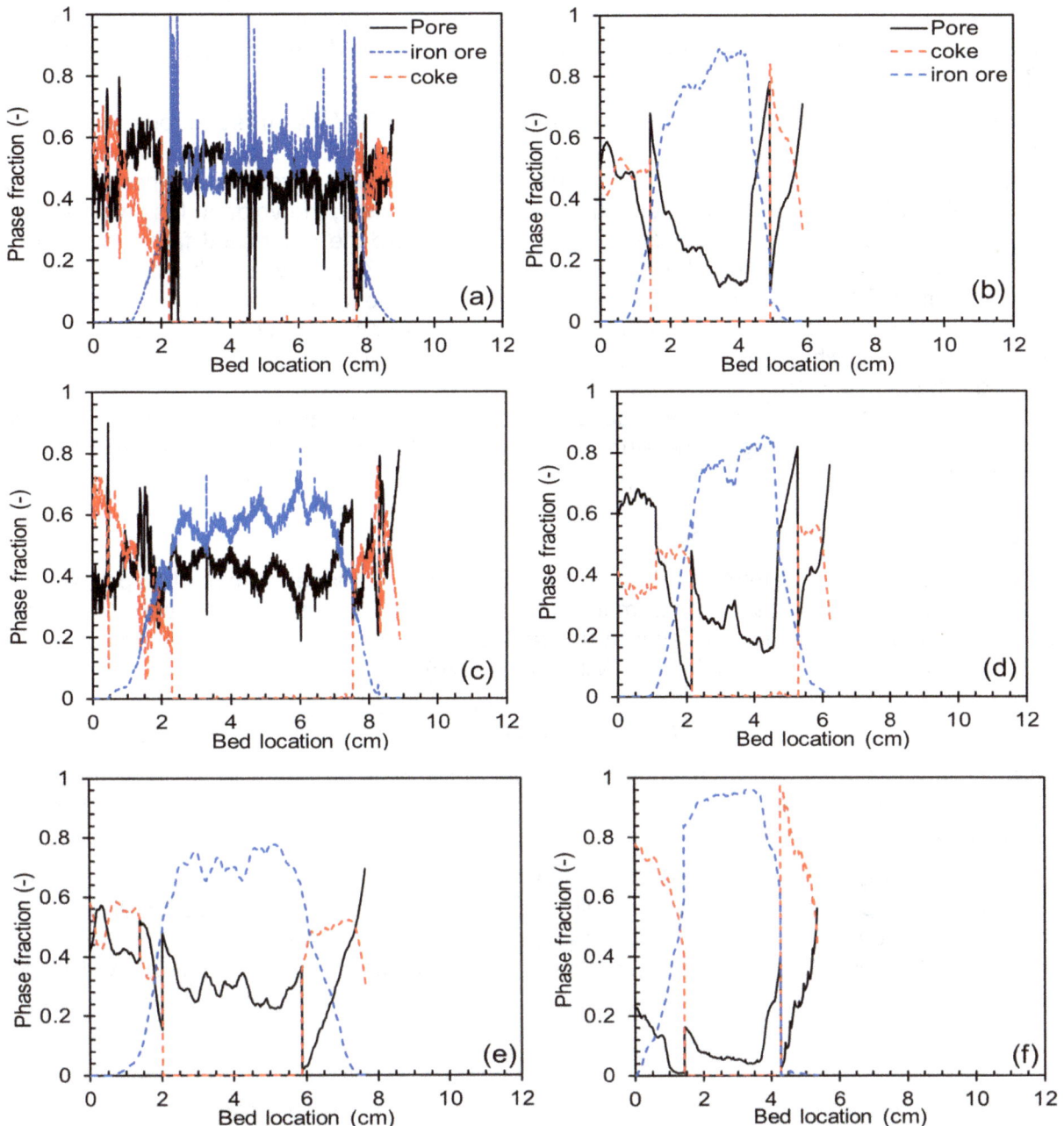

FIG 2 – Axial variations in phase fractions for (a) SH1 at 1000°C; (b) SH1 at 1300°C; (c) MB1 at 1000°C; (d) MB1 at 1300°C; (e) NBLL at 1000°C; and (f) NBLL at 1300°C in presence of 15 per cent H_2 gas.

With increasing temperature, it can be noticed that lump ore (NBLL) loses its voidage significantly even at a lower test temperature. For example, at 1000°C NBLL shows a decrease of ~75 per cent voidage (from ~0.4 to 0.3) while voidage in SH1 and MB1 remains almost unaffected at ~0.4 to 0.5.

This could be attributed to possible melt formation behaviour in the lump ore. An increase in the binary basicity value (R_2 in Table 1) in ore burden leads to the formation of compounds with higher liquidus temperatures suggesting that lump burden (R_2 = 0.01) will produce more melt compared sinter (R_2 = 1.85) at the same operating temperature, as discussed by Hoque *et al* (2021). However, in the CO-H_2 system the influence of higher reduction degree and enhanced carburisation of the metallic iron phase complicate this analysis. This is the subject of future work.

This behaviour is attributed to softening of burden due to formation of melt phases within the microstructure of the reduced ore particles which fills the surface pores of the particles as well as the internal voids spaces of the ferrous layer. Due to the series orientation of the ore and coke layers in the system in the direction of gravity, the upper coke layer remained relatively unaffected by the presence of melt; however, due to gravity effect and externally applied load, some presence of melt was detected in the bottom coke layer. This behaviour was more pronounced at the highest test temperature of 1300°C at which all ferrous burdens showed bed voidage (height averaged) less than <0.3.

Height-averaged void fraction values determined from tomography for all three ferrous burden types in presence of CO and H_2 enriched gas compositions are compared in Figure 3. For both CO and H_2 cases, a clear decreasing trend in the ferrous layer voidage with increasing temperature can be noted for all burdens. In case of CO reduction (no hydrogen case), the voidage loss was relatively less (~40 per cent) for sinter and mixed burden samples (voidage decreased from ~0.52 to 0.3) compared to the lump ore, where the loss was much higher (~ 75 per cent) with a decline from the initial value of ~0.2 to 0.05. This shows the benefit of mixing the lump with the sinter in the ferrous burden to maintain better voidage and permeability. In the 15 per cent hydrogen case, both sinter and mixed burden exhibited voidage loss ~53 per cent compared to the CO case, whereas the lump exhibited ~66 per cent loss. These results suggest that the decrease in voidage was less pronounced for lump when H_2 gas was added. The observation indicates that the lump permeability decreases less with increasing temperature in a H_2 gas addition environment compared to CO only. This was supported by the pressure drop values measured in the S&M tests for NBLL lump which showed that the maximum pressure drop was lower with 15 per cent H_2 gas compared to CO only gas, especially at 1300°C (see Table 2).

In order to validate the void fraction values obtained from the CT image segmentation process, they were compared with those estimated from the S&M test data. The combined inter-particle and intra-particle void fraction of the ferrous layer at different quench temperatures was calculated from the S&M test bed contraction data as follows (Mitra *et al*, 2020):

$$\varepsilon_{g,comb} = \left(\frac{H_1}{H_1+H_2}\right)\left(\frac{H_{Fe}-H_3-H_c}{H_{Fe}-H_c}\right) \tag{1}$$

where H_1 and H_2 are the equivalent height of inter-particle and intra-particle voidage in Fe layer, respectively; H_3 is the equivalent height of particle voidage in Fe layer; H_{Fe} is height of Fe layer and H_C is the height of Fe layer after contraction.

Based on the comparison in Figure 3, it is evident that the void fraction values obtained from the S&M test contraction align reasonably well with the CT-based void fraction values for all cases. The S&M test prediction errors, determined by the normal root mean square deviation (NRMSD), were less than 20 per cent, 18 per cent and 35 per cent for SH1, MB1 and NBLL for CO and H_2 gas, respectively.

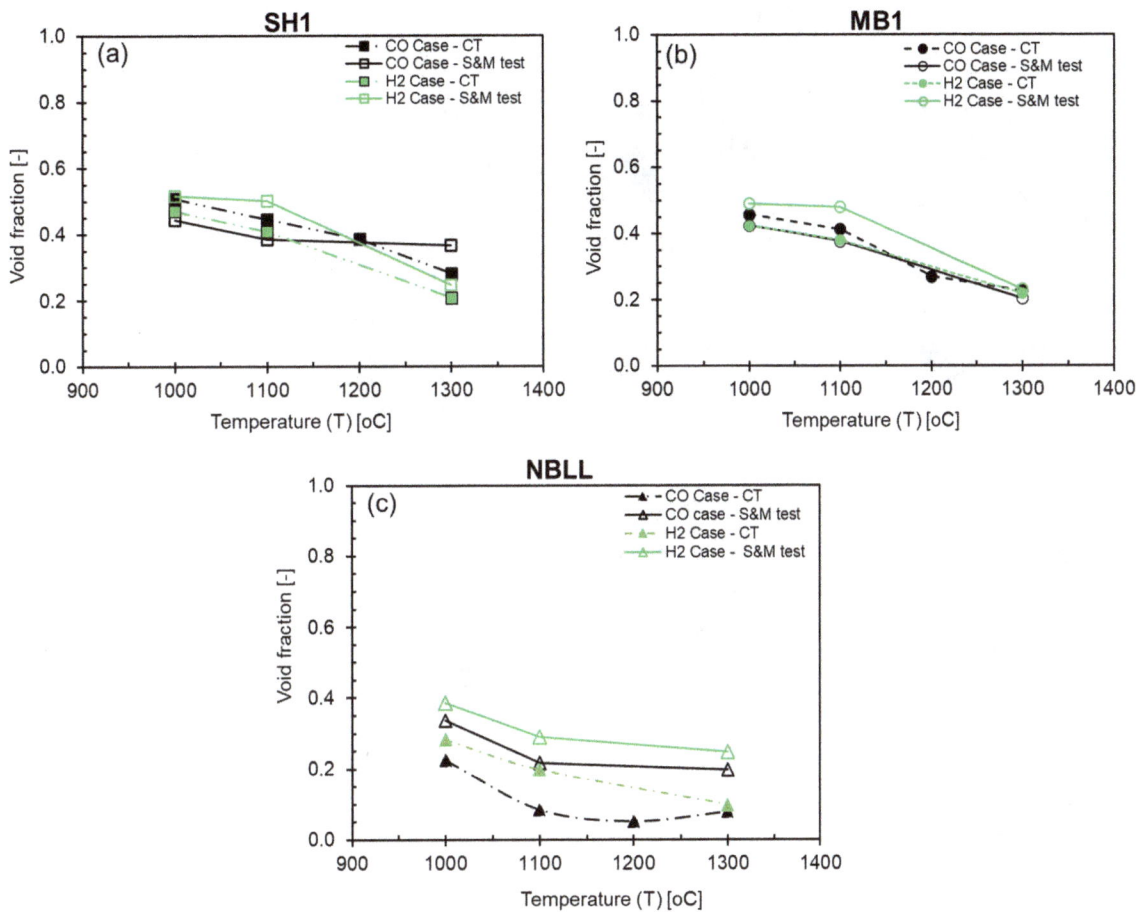

FIG 3 – Effect of temperature on the variation of ferrous layer voidage (bed area and height averaged value) values for SH1, MB1 and NBLL for H_2 and CO cases. The data of CO case were obtained from Hoque *et al* (2021).

TABLE 2

Pressure drop (kPa) of different quench ferrous burden samples at different temperatures in presence of CO and H_2 gas.

Temp (°C)	With CO gas			With 15% H_2 gas		
	SH1	MB1	NBLL	SH1	MB1	NBLL
1000	0.124	0.129	0.194	0.196	0.180	0.186
1100	0.102	0.154	0.367	0.172	0.190	0.261
1300	0.154	0.224	3.340	0.771	0.459	0.302

The reconstructed 3D structures with flow streamlines of the three ferrous burdens for three temperature cases – 1000, 1100 and 1300°C are shown in Figure 4. It can be noted that bed height decreased with increasing temperature with significant visible changes in the ferrous layer structure for lump compared to sinter and mixed burden where ferrous particle structures were reasonably intact even at higher operating temperatures. For the lump with H_2 case, the CT data indicates a greater loss of voidage at 1300°C than for the other burden types in both Figures 3 and 4. This is at odds with the pressure drop reported in Table 2 and the voidage calculated from the contraction using Equation 1. Further investigation is ongoing to understand this discrepancy.

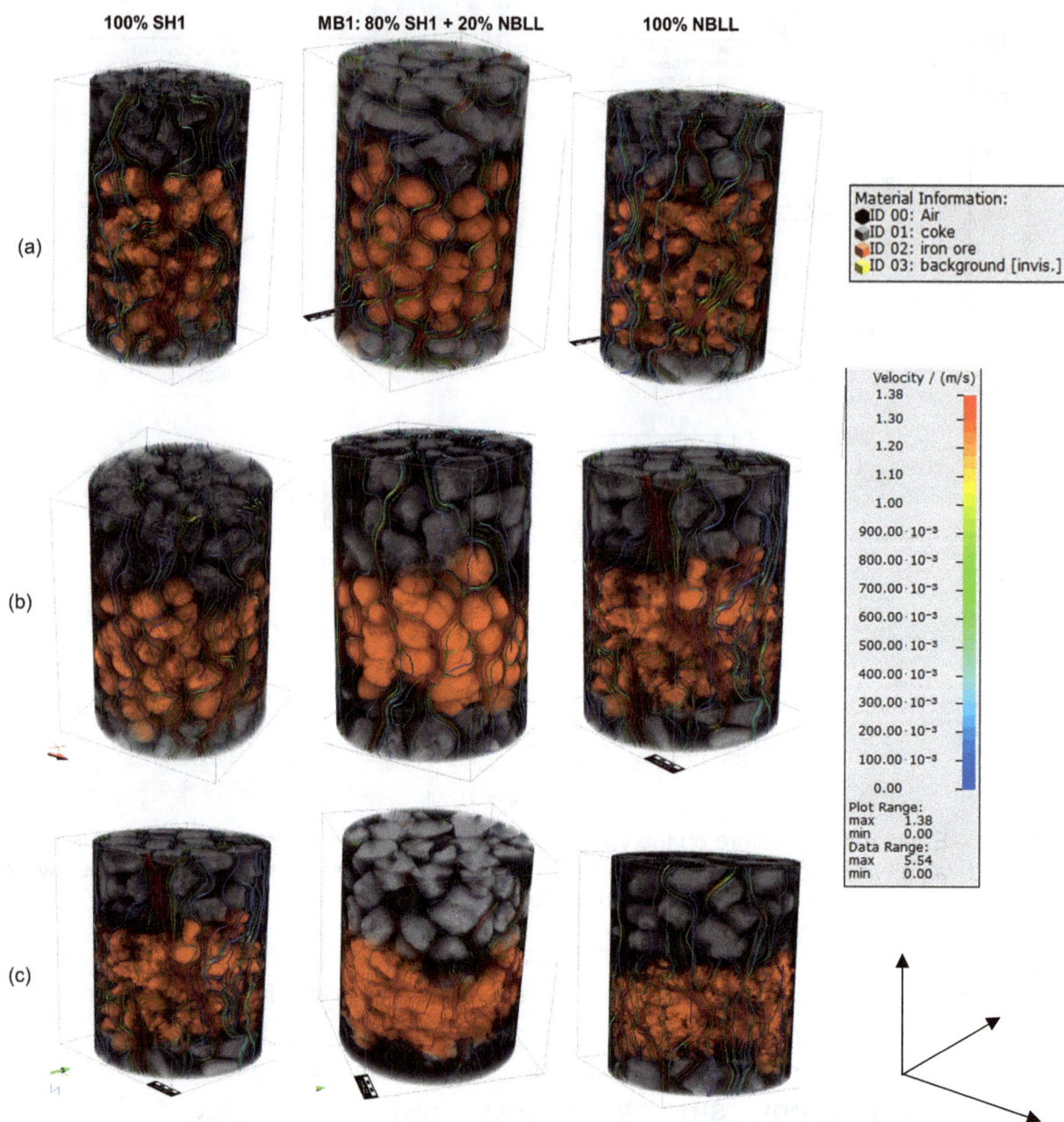

FIG 4 – 3D structure, void network and flow streamline of SH1, MB1 and NBLL samples in presence of H_2 gas (a) 1000°C; (b) 1100°C; and (c) 1300°C.

Figure 4 also presents the CFD model predicted gas streamlines around the coke and ore particles. At lower temperatures, gas flow evenly passes around the coke and ore particles where the structures remain intact. When the ferrous layer structures collapse due to liquid formation, a significant portion of the gas flow bypasses the middle ferrous layer (less permeable) and flows around the crucible wall. This behaviour is more pronounced in the lump case and in general for all cases at 1300°C. The CFD analysis suggests that the measured pressure drop in the S&M test at high levels of contraction should be interpreted with a degree of caution. A high pressure drop indicates a loss of ferrous layer voidage, however, the absolute value of the maximum pressure drop is also related to the state of voids near the crucible wall. Work is continuing to use the change in ferrous layer voidage with temperature information from the S&M test to better estimate the cohesive zone permeability in the blast furnace for different ferrous burdens (Mitra *et al*, 2020).

In the present study, the permeability of the ferrous burden at different temperature were estimated along the principal flow direction–z using the bed voidage and Darcy's law. Figure 5 shows the permeability of different ferrous burdens in the z-direction as a function of temperature. It can be observed that the bed permeability decreases linearly with bed temperature for all the cases. Although lump had lower permeability compared to other burdens, permeability loss was relatively high for SH1 (~33 per cent) and MB1 samples (~30 per cent) compared with the lump ore, where this loss was less (~25 per cent) as temperature increased.

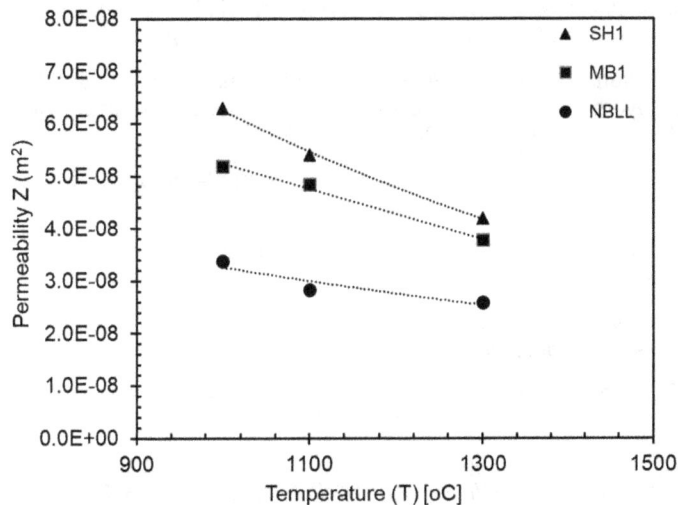

FIG 5 – Effect of temperature on the permeability of SH1, MB1 and NBLL samples in presence of 15 per cent H_2 gas.

It needs to be noted that it was not always possible to completely segregate the coke and void layer in the middle ferrous layer due to noise present in the CT samples analysis and very close grey scale value of these two phases even after careful segmentation. This can cause some variations in the predicted void fraction and bed permeability. This anomaly has been reported in our previous work (Mitra et al, 2020) in detail and is a subject of further work.

CONCLUSIONS

This study provides a novel CT scan-based methodology to obtain valuable insights to the morphological changes and permeability of different ferrous burdens from softening and melting tests performed in a hydrogen enriched reducing environment. The obtained bed void fraction data from CT measurements was benchmarked with the bed contraction data obtained from the S&M tests which showed reasonable agreement.

A consistent linear decrease in both the void fraction of the ferrous layer and bed permeability was noted with increasing temperature for all ore samples. The mixed burden results fell between the lump and sinter results, confirming that a high temperature interaction occurs. It was shown that the void fraction in the ferrous layer improves for lump compared to sinter and mixed burden when hydrogen gas is introduced.

ACKNOWLEDGEMENT

The authors would like to express their gratitude to BHP for providing funding for the Centre for Ironmaking Materials Research and granting permission to publish this paper. Additionally, the authors would like to acknowledge the Australian Synchrotron for their invaluable contribution by providing access to the CT tomography scanning facility.

REFERENCES

Barrett, N, Mitra, S, Doostmohammadi, H, O'Dea, D, Zulli, P, Chew, S and Honeyands, T, 2022. Assessment of blast furnace operational constraints in the presence of hydrogen injection, *ISIJ International*, 62(6):1168–1177.

Chen, Y and Zuo, H, 2021. Review of hydrogen-rich ironmaking technology in blast furnace, *Ironmaking and Steelmaking*, 48(6):749–768.

Hjortsberg, E, Forsberg, F, Gustafsson, G and Rutqvist, E, 2013. X-ray microtomography for characterisation of cracks in iron ore pellets after reduction, *Ironmaking and Steelmaking*, 40(6):399–406.

Hoque, M M, Doostmohammadi, H, Mitra, S, O'Dea, D, Liu, X and Honeyands, T, 2021. High temperature softening and melting interactions between Newman Blend Lump and Sinter, *ISIJ International*, 61(12):2944–2952.

Lan, C, Zhang, S, Liu, X, Lyu, Q and Jiang, M, 2020. Change and mechanism analysis of the softening-melting behavior of the iron-bearing burden in a hydrogen-rich blast furnace, *International Journal of Hydrogen Energy*, 45(28):14255–14265.

Liu, X, Honeyands, T, Evans, G, Zulli, P and O'Dea, D, 2019. A review of high-temperature experimental techniques used to investigate the cohesive zone of the ironmaking blast furnace, *Ironmaking and Steelmaking*, 46(10):953–967.

Liu, X, Honeyands, T, Mitra, S, Evans, G, Godel, B, Acres, R G and Acres Ellis, B, 2018. A Novel Measurement of Voidage in Coke and Ferrous Layers in Softening and Melting under Load Test Using Synchrotron X-ray and Neutron Computed Tomography, *ISIJ International*, 58(11):2150–2152.

Liu, Z J, Zhang, J L and Yang, T J, 2015. Low carbon operation of super-large blast furnaces in China, *ISIJ International*, 55(6):1146–1156.

Loo, C E, Matthews, L T and O'Dea, D P, 2011. Lump ore and sinter behaviour during softening and melting, *ISIJ International*, 51(6):930–938.

Ma, K, Deng, J, Wang, G, Zhou, Q and Xu, J, 2021. Utilization and impacts of hydrogen in the ironmaking processes: A review from lab-scale basics to industrial practices, *International Journal of Hydrogen Energy*, 46(52):26646–26664.

Mitra, S, Liu, X, Honeyands, T, Evans, G, O'Dea, D and Zulli, P, 2020. Pressure-drop modelling in the softening and melting test for ferrous burden, *ISIJ International*, 60(7):1416–1426.

Pan, Y Z, Zuo, H B, Wang, B X, Wang, J S, Wang, G, Liu, Y L and Xue, Q G, 2018. Effect of reduction degree on cohesive zone and permeability of mixed burden, *Ironmaking and Steelmaking*, 47(3):322–327.

Qie, Y, Lyu, Q, Liu, X, Li, J, Lan, C, Zhang, S and Yan, C, 2018. Effect of hydrogen addition on softening and melting reduction behaviors of ferrous burden in gas-injection blast furnace, *Metallurgical and Materials Transactions – B*, 49:2622–2632.

Shatokha, V, Korobeinikov, I, Maire, E and Adrien, J, 2009. Application of 3D X-ray tomography to investigation of structure of sinter mixture granules, *Ironmaking and Steelmaking*, 36(6):416–420.

Shatokha, V, Korobeinikov, I, Maire, E, Gremillard, L and Adrien, J, 2010. Iron ore sinter porosity characterisation with application of 3D X-ray tomography, *Ironmaking and Steelmaking*, 37(5):313–319.

Stevenson, A W, Crosbie, J C, Hall, C J, Häusermann, D, Livingstone, J and Lye, J E, 2017. Quantitative characterization of the X-ray beam at the Australian Synchrotron Imaging and Medical Beamline (IMBL), *Journal of Synchrotron Radiation*, 24(1):110–141.

Sun, G, Li, B, Yang, W, Guo, J and Guo, H, 2020. Analysis of energy consumption of the reduction of Fe2O3 by hydrogen and carbon monoxide mixtures, *Energies*, 13(8):1986.

Tan, X, Li, H, Guo, J, Gu, B and Zeng, Y, 2019. Energy-saving and emission-reduction technology selection and CO2 emission reduction potential of China's iron and steel industry under energy substitution policy, *Journal of Cleaner Production*, 222:823–834.

AUTHOR INDEX

AUTHOR INDEX